# Lecture Notes in Mechanical Engineering

**Lecture Notes in Mechanical Engineering (LNME)** publishes the latest developments in Mechanical Engineering—quickly, informally and with high quality. Original research reported in proceedings and post-proceedings represents the core of LNME. Volumes published in LNME embrace all aspects, subfields and new challenges of mechanical engineering. Topics in the series include:

- Engineering Design
- Machinery and Machine Elements
- Mechanical Structures and Stress Analysis
- Automotive Engineering
- Engine Technology
- Aerospace Technology and Astronautics
- Nanotechnology and Microengineering
- Control, Robotics, Mechatronics
- MEMS
- Theoretical and Applied Mechanics
- Dynamical Systems, Control
- Fluid Mechanics
- Engineering Thermodynamics, Heat and Mass Transfer
- Manufacturing
- Precision Engineering, Instrumentation, Measurement
- Materials Engineering
- Tribology and Surface Technology

To submit a proposal or request further information, please contact the Springer Editor of your location:

**China:** Dr. Mengchu Huang at mengchu.huang@springer.com
**India:** Priya Vyas at priya.vyas@springer.com
**Rest of Asia, Australia, New Zealand:** Swati Meherishi at swati.meherishi@springer.com
**All other countries:** Dr. Leontina Di Cecco at Leontina.dicecco@springer.com

To submit a proposal for a monograph, please check our Springer Tracts in Mechanical Engineering at http://www.springer.com/series/11693 or contact Leontina.dicecco@springer.com

**Indexed by SCOPUS. All books published in the series are submitted for consideration in Web of Science.**

More information about this series at http://www.springer.com/series/11236

K. M. Pandey · R. D. Misra ·
P. K. Patowari · U. S. Dixit
Editors

# Recent Advances in Mechanical Engineering

## Select Proceedings of ICRAME 2020

Springer

*Editors*
K. M. Pandey
National Institute of Technology Silchar
Silchar, India

R. D. Misra
National Institute of Technology Silchar
Silchar, India

P. K. Patowari
National Institute of Technology Silchar
Silchar, India

U. S. Dixit
Indian Institute of Technology Guwahati
Guwahati, India

ISSN 2195-4356 ISSN 2195-4364 (electronic)
Lecture Notes in Mechanical Engineering
ISBN 978-981-15-7710-9 ISBN 978-981-15-7711-6 (eBook)
https://doi.org/10.1007/978-981-15-7711-6



This Springer imprint is published by the registered company Springer Nature Singapore Pte Ltd.
The registered company address is: 152 Beach Road, #21-01/04 Gateway East, Singapore 189721, Singapore

K. M. Pandey · R. D. Misra ·
P. K. Patowari · U. S. Dixit
Editors

# Recent Advances in Mechanical Engineering

## Select Proceedings of ICRAME 2020

Springer

*Editors*
K. M. Pandey
National Institute of Technology Silchar
Silchar, India

R. D. Misra
National Institute of Technology Silchar
Silchar, India

P. K. Patowari
National Institute of Technology Silchar
Silchar, India

U. S. Dixit
Indian Institute of Technology Guwahati
Guwahati, India

ISSN 2195-4356 ISSN 2195-4364 (electronic)
Lecture Notes in Mechanical Engineering
ISBN 978-981-15-7710-9 ISBN 978-981-15-7711-6 (eBook)
https://doi.org/10.1007/978-981-15-7711-6



This Springer imprint is published by the registered company Springer Nature Singapore Pte Ltd.
The registered company address is: 152 Beach Road, #21-01/04 Gateway East, Singapore 189721, Singapore

# Preface

The First International Conference on Recent Advancement of Mechanical Engineering (ICRAME 2020) was held from 7 to 9 February 2020 at National Institute of Technology Silchar, Assam, India. The conference aimed to bring together experts from academic, scientific and industrial communities to address new challenges and present their latest research findings, ideas, developments and perspective of the future directions in the field of mechanical engineering. ICRAME 2020 invited researchers to participate in the conference. In this conference, ideas were discussed across the borders among the delegates. Participations of this conference were from all the neighbouring states of the Northeast India and also from other parts of India as well as abroad. There were different topics of interest considered in ICRAME 2020. These were related but not restricted to the following broad areas of mechanical engineering—thermal engineering, design engineering, manufacturing/production engineering and surface engineering. The recent developments in these areas were dealt with in this conference. The conference invited technical papers that addressed the state of the art in the mentioned areas of mechanical science and technology. The papers related to the theoretical modelling works, and analytical and numerical modelling including CFD, experimental investigations and also the state-of-the-art review papers in the relevant areas were considered in ICRAME 2020. The book proceedings publishes all the accepted and presented papers in the said conference. The following are the broad topics of the conference:

- Thermal Engineering: Bio-thermal, techniques in fluid flow, compressible flows, biofuels, advancement in renewable energy sources, solar thermal, renewable energy, off-grid renewable energy.
- Design Engineering: Computing in applied mechanics and product design, dynamics and control of structures/systems, fracture and failure mechanics, solid mechanics, differential/dynamical systems, modelling and simulation artificial intelligence: fuzzy logic, neural network, etc. Finite element analysis, advanced numerical techniques, advancements in tribology nanomechanics and MEMS, robotics.

- Manufacturing/Production Engineering and Surface Engineering: Casting, welding, etc. Intelligent and advanced manufacturing system, composites, conventional and non-conventional machining, ergonomics: human factors in seating comfort.

Silchar, India K. M. Pandey
Silchar, India R. D. Misra
Silchar, India P. K. Patowari
Guwahati, India U. S. Dixit

# Contents

**AI-Based ANN Modeling of Performance–Emission Profiles of CRDI Engine under Diesel-Karanja Strategies** . . . . . 1
P. Sandeep Varma, Subrata Bhowmik, Abhishek Paul, Pravin Ashok Madane, and Rajsekhar Panua

**ANFIS Prediction of Performance and Exhaust Emission Characteristics of CRDI Engine Fueled with Diesel–Butanol Strategies** . . . . . 11
Vivek Kumar Mishra, Subrata Bhowmik, Abhishek Paul, Ajay Yadav, and Rajsekhar Panua

**Development and Workspace Study of a 4-PP Planar Parallel XY Positioning Stage Using SMA Actuators** . . . . . 21
Rutupurna Choudhury, Deep Singh, Anuj Kumar, Yogesh Singh, and Chinmaya Kumar Sahoo

**Review of the Hybrid H-Savonius Rotor's Design and Performance** . . . . . 33
D. Sarkar, A. R. Sengupta, P. Bhadra, S. Alam, and B. Debnath

**Energy Storage Using Sensible Heat Storage Media: Thermal and Economic Considerations** . . . . . 41
Laxman Mishra, Abhijit Sinha, Prasanta Majumder, and Rajat Gupta

**Convective Flow of Nanofluid and Nanoencapsulated Phase Change Material Through Microchannel Heat Sink for Passive Cooling of Microelectronics** . . . . . 51
Manoj Kumar, Vikram Bisht, Sheshang Singh Chandel, Sumit Sinha-Ray, and Pradeep Kumar

**Fabrication of Treated and Untreated Coconut Fibre-Reinforced Epoxy-Based Composites of Different Fibre Content and Comparison of Their Tensile and Flexural Strengths** . . . . . 61
U. Deka, M. Bhuyan, C. Borah, S. Kakoti, and R. K. Dutta

**Study on Gate Location and Gate Number for Manufacturability of Polymer Gears** . . . . . . . . . . . . . . . . . . . . . . . . . . . . . . . . . . . 71
Bikram Singh Solanki, Harpreet Singh, and Tanuja Sheorey

**Fuzzy-PID Control of Hydro-motor Speed Used in Heavy Earth Moving Machinery** . . . . . . . . . . . . . . . . . . . . . . . . . . . . . . . . . . . 83
Shivdutt Sarkar, Mohit Bhola, Harsha Rowdur, and N. Kumar

**Phase Change in an Enclosure Under Different Combinations of Boundary Wall Condition: A Numerical Study** . . . . . . . . . . . . . . . . 95
Bhaskar Ranjan Tamuli, Sujit Nath, and Dipankar Bhanja

**Finite Element Modelling of Electromagnetic Crimping of Copper-Stainless Steel Tube-to-Tube Joint** . . . . . . . . . . . . . . . . . . . . 105
Deepak Kumar, Sachin D. Kore, and Arup Nandy

**Optimization of Protrusions for an Impinging Jet on a Curved Surface** . . . . . . . . . . . . . . . . . . . . . . . . . . . . . . . . . . . . . . . . . . 117
Alankrita Singh and B. V. S. S. S. Prasad

**Numerical Analysis on a Selection of Horn Material for the Design of Cylindrical Horn in Ultrasonic Machining** . . . . . . . . . . . . . . . . . . . 127
Guddakesh Kumar Chandan and Chinmaya Kumar Sahoo

**Emission Studies on a Diesel Engine Fueled with Mixed Biodiesel Produced from Non-edible Oils** . . . . . . . . . . . . . . . . . . . . . . . . . . . 137
Bhabani Prasanna Pattanaik, Chandrakanta Nayak, and Rahul Dev Misra

**Recent Advancement in Electromagnetic Forming Processes** . . . . . . . . . 147
Sagar Pawar, Sachin D. Kore, and Arup Nandy

**Review on Pool Boiling Heat Transfer Enhancement by Surface Fabrication Using Various Surface Coating Methods** . . . . . . . . . . . . . . 157
Sonali Priyadarshini Das, Raghavendra Singh, and Rahul Dev Misra

**Comparative Study of Extended and Unscented Kalman Filters for Estimating Motion States of an Autonomous Vehicle-Trailer System** . . . . . . . . . . . . . . . . . . . . . . . . . . . . . . . . . . . . . . . . . . . 165
Hussein F. M. Ali, Nader A. Mansour, and Youngshik Kim

**Numerical Study on Material Flow Behaviour in Friction Stir Welding of Low Carbon Steel** . . . . . . . . . . . . . . . . . . . . . . . . . . . . . . . . 175
Pardeep Pankaj, Abhishek Bhardwaj, Avinish Tiwari, Lakshmi Narayan Dhara, and Pankaj Biswas

**Comparing Wettability and Frictional Performance of Laser Micro-machined Discrete and Continuous Textures** . . . . . . . . . . . . . . . 185
V. Kashyap and P. Ramkumar

**Influence of Heat Input on Bead Geometry in Metal Inert Gas Welded Thick P91 Steel** . . . . . . 193
Lakshmi Narayan Dhara, Pardeep Pankaj, and Pankaj Biswas

**Opportunities and Challenges in Nanoparticles Formation by Electrical Discharge Machining** . . . . . . 203
Irshad Ahamad Khilji, Sunil Pathak, Siti Nadiah Binti Mohd Saffe, Shatarupa Biswas, and Yogesh Singh

**Preheating Path Selection Through Numerical Analysis of Laser-Aided Direct Metal Deposition** . . . . . . 211
Dhiraj Raj, Bipul Das, and Saikat Ranjan Maity

**Robust Synthesis of Path Generating Four-Bar Mechanism** . . . . . . 221
Sanjay B. Matekar and Ajay M. Fulambarkar

**Predicting the Surface Roughness in Single Point Incremental Forming** . . . . . . 233
Manish Oraon, Vinay Sharma, and Soumen Mandal

**Prediction of Droplet Size Distribution For Viscoelastic Liquid Sheet** . . . . . . 243
Saurabh Sharma, Debayan Dasgupta, Sujit Nath, and Dipankar Bhanja

**Linear Stability Analysis of Viscoelastic Liquid Sheet in Presence of Moving Gas Medium** . . . . . . 251
Debayan Dasgupta, Saurabh Sharma, Sujit Nath, and Dipankar Bhanja

**Development of Feature Extraction-Based Currency Recognition System Using Artificial Neural Network** . . . . . . 259
Deep Singh, Rahul Kumar, Rutupurna Choudhury, Ashutosh Padhan, and Yogesh Singh

**Design and Fabrication of Android Application-Based Grass Cutter Robotic System** . . . . . . 271
Rutupurna Choudhury, Deep Singh, Anuj Kumar, and Yogesh Singh

**Review on Various Coating Techniques to Improve Boiling Heat Transfer** . . . . . . 283
Amatya Bharadwaj and Rahul Dev Misra

**Effect of Particulate Type Reinforcements on Mechanical and Tribological Behavior of Aluminium Metal Matrix Composites: A Review** . . . . . . 295
V. S. S. Venkatesh and Ashish B. Deoghare

**A Review on Solar Drying Applications Using Latent Heat as Energy Storage Media** . . . . . . 305
Supreme Das, Agnimitra Biswas, and Biplab Das

**Finite Element Analysis of Stamping Process of Maraging Steel Built-Up Wing Panel of a Missile** . . . . . . . . . . . . . . . . . . . . . . . . . . . 319
P. Sridhar Reddy, B. V. R. Reddy, S. R. Maity, and K. M. Pandey

**Effect of Non-uniform Heating on Forced Convective Flow Through Asymmetric Wavy Channel** . . . . . . . . . . . . . . . . . . . . . . . . . . . . . 333
Sumit Kumar Mehta and Sukumar Pati

**Effect of Sintering Temperatures on Mechanical Properties of AA7075/B4C/Graphite Hybrid Composite Fabricated by Powder Metallurgy Techniques** . . . . . . . . . . . . . . . . . . . . . . . . . . . . . . . . . . 343
Guttikonda Manohar, Saikat Ranjan Maity, and Krishna Murari Pandey

**Natural Convection from Two Cylinders in an Enclosure with Sinusoidal Bottom Wall: A Numerical Study** . . . . . . . . . . . . . . . . 351
Dhrijit Kumar Deka, Gopal Chandra Pal, Sukumar Pati, and Pitambar R. Randive

**Identification of Parameters for Ultrasonic Machining (USM) on Drilling of Graphene Oxide/Pineapple Leaf Filler-Reinforced Epoxy Hybrid Composite Using TODIM Method** . . . . . . . . . . . . . . . . 361
Angkan Bania and Saikat Ranjan Maity

**Optimization of Planetary Gearbox Using NSGA-II** . . . . . . . . . . . . . . . . 367
Abhishek Parmar, P. Ramkumar, and K. Shankar

**Conjugate Heat Transfer Analysis for Flow Through Microduct Subjected to Non-uniform Heating** . . . . . . . . . . . . . . . . . . . . . . . . . . . 377
Abhijit Borah and Sukumar Pati

**Micro Electrical Discharge Milling of Titanium: Effects of Voltage and Tool Rotation Speed** . . . . . . . . . . . . . . . . . . . . . . . . . . . . . . . . . . . 387
Siddhartha Kar, Pallab Sarmah, and Promod Kumar Patowari

**Bio-plastic from Yam—An Ecofriendly Prospective** . . . . . . . . . . . . . . . . 397
Susheel Kumar, S. K. Pattanayak, and Krishna Murari Pandey

**Application of Box-Behnken Method for Multi-response Optimization of Turning Parameters for DAC-10 Hot Work Tool Steel** . . . . . . . . . . . 407
Sunil Kumar, Saikat Ranjan Maity, and Lokeswar Patnaik

**Cost Analysis of an Off-Grid Solar/Wind/Battery Based Renewable Energy System for Variable Load** . . . . . . . . . . . . . . . . . . . . . . . . . . . 417
Sujeet Singh, Krishna Murari Pandey, and K. K. Sharma

**Design and Development of a Mini Sugarcane Harvester** . . . . . . . . . . . . 427
M. A. Nagarjun, N. C. Mahendra Babu, and Manjunath M. Ullegaddi

**Simulation and Experimental Performance Studies of a Computerized Spark Ignition Engine** . . . . . . . . . . . . . . . . . . . . . . . . . . . . . . . . . . . . 441
Pritam Jyoti Saikia, Partha Pratim Dutta, and Paragmoni Kalita

**Machinability Study of Rubber using USM for Microdrilling** . . . . . . . . 455
Pallab Sarmah, Tapas Debnath, and Promod Kumar Patowari

**Design, Fabrication and Experimentation of Aqua Silencer for Diesel Genset** . . . . . . . . 463
Neelutpal Ghosh, Sameer S. Gajghate, Sagnik Pal, and Swapan Bhaumik

**Numerical Analysis OF PCM Within a Square Enclosure Having Different Wall Heating Conditions** . . . . . . . . 477
Pallab Bhattacharjee, Sujit Nath, and Dipankar Bhanja

**Numerical Modelling and Thermohydraulic Analysis of Circular Pipe Having Internal Vortex Generators** . . . . . . . . 487
Binay Bhushan Bora, Mohd Zeeshan, Sujit Nath, and Dipankar Bhanja

**Effect of Non-uniform Heating on Electroosmotic Flow Through Microchannel** . . . . . . . . 499
T. Sujith, Sumit Kumar Mehta, and Sukumar Pati

**Temperature-Dependent Random Frequency of Functionally Graded Spherical shells—A PCE Approach** . . . . . . . . 509
Vaishali and S. Dey

**Microwave Processing of Polymer Matrix Composites: Review of the Understanding and Future Opportunities** . . . . . . . . 517
Tejas Pramod Naik, Ram Singh Rana, Inderdeep Singh, and Apurbba Kumar Sharma

**A Comparative Study on the Hydrodynamic and Heat Transfer Behaviour of Conical Fluidized Bed with that of a Columnar Pressurized Circulating Fluidized Bed** . . . . . . . . 531
Hirakh Jyoti Das, Rituraj Saikia, and Pinakeswar Mahanta

**Study of the Effect of Welding Current on Heat Transfer and Melt Pool Geometry on Mild Steel Specimen Through Finite Element Analysis** . . . . . . . . 545
Mohd Aslam and Chinmaya Kumar Sahoo

**Study of Granular Food Material Drying in a Pilot-Scale Rotating Fluidized Bed with Static Geometry Dryer** . . . . . . . . 555
Pavitra Singh, Pankaj Kalita, Pinakeswar Mahanta, and Hirakh Jyoti Das

**Experimental Evaluation of Sand-Based Sensible Energy Storage System** . . . . . . . . 563
Prasant, Sujit Roy, Biplab Das, and Sumita Debbarma

**Purification of Biogas for Methane Enrichment Using Biomass Biochar and Biochar–Clay Composite** . . . . . . . . 573
Deep Bora, Lepakshi Barbora, and Pinakeswar Mahanta

**Experimental Investigation of Drying Characteristics of Tea in a Conical Bubbling Fluidized Bed Dryer** . . . . . . . . . . . . . . . . . . . . 583
Plabon Tamuly, Hirakh Jyoti Das, and Pinakeswar Mahanta

**Design and Development of a *XY* Positioning Stage Using Shape Memory Alloy Spring Actuator** . . . . . . . . . . . . . . . . . . . . . . . . . . . 593
Ashutosh Padhan and Yogesh Singh

**Selective Etching of Aluminium 6068 Using Photochemical Machining** . . . . . . . . . . . . . . . . . . . . . . . . . . . . . . . . . . . . . . . . . 601
Jaswant Kumar, Tapas Debnath, and Promod Kumar Patowari

**Microstructure and Mechanical Characterisation of Friction Stir Welded Inconel 718 Alloy** . . . . . . . . . . . . . . . . . . . . . . . . . . . . . . . 611
Sanjay Raj, Pardeep Pankaj, and Pankaj Biswas

**Effect of Dual-Height Plate Fins on Phase Change Material Cooling Technique: An Experimental Study** . . . . . . . . . . . . . . . . . . . . . . . . 619
Maibam Romio Singh, Asis Giri, and Pradip Lingfa

**Dynamic Analysis of Parametrically Excited Coupled Beam-Based Piezoelectric Energy Harvester** . . . . . . . . . . . . . . . . . . . . . . . . . . . 631
Ranit Roy, Anshul Garg, and Santosha Kumar Dwivedy

**An Overview of Wire Electrical Discharge Machining (WEDM)** . . . . . . 643
Shatarupa Biswas, Yogesh Singh, and Manidipto Mukherjee

**Comparative Analysis of Combustion Noise, Performance and Emission of LTC Diesel Engine with Multiple Injections** . . . . . . . . 653
Sanjoy Biswas and Achintya Mukhopadhyay

**Radial Basis Function-Based Probabilistic First-Ply Failure Analyses of Composite Spherical Shells** . . . . . . . . . . . . . . . . . . . . . . . . . . . . 667
Himanshu Prasad Raturi, Subrata Kushari, and Sudeep Dey

**Experimental Investigation of Dimensional Deviation of Square Punches Cut by Wire EDM of A286 Superalloy** . . . . . . . . . . . . . . . . . 677
Subhankar Saha, Saikat Ranjan Maity, and S. Dey

**Multi-optimization of μ-EDMed Arrayed Microrods Using Gray Relational Analysis** . . . . . . . . . . . . . . . . . . . . . . . . . . . . . . . . . . . . . 685
A. K. Singh, T. J. Singh, R. Pongen, and Promod Kumar Patowari

**Detection of Fault in a Bevel Gearbox Under Varying Speed Conditions** . . . . . . . . . . . . . . . . . . . . . . . . . . . . . . . . . . . . . . . . . . 697
Vikas Sharma, Anand Parey, Abhimanyu Pratap Singh, Atanu Paul, and Yogesh Singh

**CFD Analysis of the Performance of an H-Darrieus Wind Turbine Having Cavity Blades** . . . . . 711
Y. Kumar, A. R. Sengupta, Agnimitra Biswas, H. M. S. M. Mazarbhuiya, and Rajat Gupta

**The Effect of Negative Hardening Coefficients on Yield Surface Evolution** . . . . . 721
Praveen Kumar and Sivasambu Mahesh

**Influence of Camber Ratio and Thickness Ratio on the Airfoil Performance** . . . . . 729
Sujit Roy, Biplab Das, and Agnimitra Biswas

**Development of Modified Cyclic Plasticity Model to Simulate Cyclic Behaviour for SA333C–Mn Steel Under LCF Loading Conditions** . . . . 739
Vibhanshu Pandey, Partha Pratim Dey, Niloy Khutia, and Suneel K. Gupta

**High Speed Impact Studies of Kevlar Fabric with and without STF** . . . . . 749
M. Chinnapandi, Ajay Katiyar, Tandra Nandi, and R. Velmurugan

**Comparative Study of Nano and Micro Fillers in EPDM/Silicone Rubber for Outdoor Insulator Application** . . . . . 761
S. Bhavya, Unnam Mahesh, R. Velmurugan, and R. Sarathi

**Machinability Study for Slot Cutting on Glass Using Ultrasonic Machining Process** . . . . . 771
Anand Mohan Singh, Ranjan Majhi, and Promod Kumar Patowari

**Mechanical Design of a Modular Underwater Rov for Surveillance and Cleaning Purpose** . . . . . 779
Abhimanyu Pratap Singh, Atanu Paul, Yogesh Singh, and Koena Mukherjee

**Large Eddy Simulation of Turbulent Slot Jet Impingement on Heated Flat Plate** . . . . . 795
Ghulam Rabbani and Dushyant Singh

**Numerical Solution of Foreign-Gas Film Cooling in Supersonic Flow** . . . . . 807
Hitesh Sharma, Dushyant Singh, and Ashutosh Kumar Singh

**Deposition of Tungsten and Copper Particle on CFRP Composite** . . . . . 815
Rashed Mustafa Mazarbhuiya and Maneswar Rahang

**Performance Improvement of Turbine Blade Using Flow Control Techniques: A Review** . . . . . 823
Hussain Mahamed Sahed Mostafa Mazarbhuiya, Agnimitra Biswas, and Kaushal Kumar Sharma

**Parametric Analysis for Machining of Stainless Steel AISI (SS-430) Using Photo Chemical Machining** . . . . . . . . . . . . . . . . . . . . . . . . . . . 829
Gaijinliu Gangmei, Jaswant Kumar, Tapas Debnath, and Promod Kumar Patowari

**Fabrication and Characterization of Ramie Fiber Based Hybrid Composites** . . . . . . . . . . . . . . . . . . . . . . . . . . . . . . . . . . . . . . . . . . 839
Karanjit Kapila, Sutanu Samanta, and Sushen Kirtania

**Multiphase Numerical Modeling of PCM Integrated Solar Collector** . . . . . . . . . . . . . . . . . . . . . . . . . . . . . . . . . . . . . . . . . . . . 849
Bharat Singh Negi, Satyender Singh, and Sushant Negi

**Computer-Aided Analysis of Solidification Time and its Effect on Hardness for Aluminium Copper Alloy** . . . . . . . . . . . . . . . . . . . . . . 861
Sasmita Tripathy and Goutam Sutradhar

**Attribute of SiC Powder Additive Mixed EDM on Machining Performance and Surface Integrity Aspects of Inconel 625** . . . . . . . . . . 869
Ankan Das, Bhavani Tharra, V. V. N. Siva Rao Sammeta, and John Deb Barma

**Enhancement of Thermal Performance of Microchannels Using Different Channel Wall Geometries: A Review** . . . . . . . . . . . . . . . . . . . 881
Dipak Debbarma, K. M. Pandey, and Abhishek Paul

**Experimental Investigations of Beeswax Based Composite Phase Change Material** . . . . . . . . . . . . . . . . . . . . . . . . . . . . . . . . . . . . . . 891
Durgesh Kumar Mishra, Sumit Bhowmik, and Krishna Murari Pandey

**Investigation on Thermo-hydraulic Performance of Channel with Various Shapes of Rib Roughness: A Review** . . . . . . . . . . . . . . . . 901
M. K. Sahu, Kumari Ambe Verma, and K. M. Pandey

**Interfacial Instabilities in Rotating Hele-Shaw Cell: A Review** . . . . . . . . 911
Akhileshwar Singh, Krishna Murari Pandey, and Yogesh Singh

**Performance Analysis of a Scramjet Combustor with Cavity for Mach Numbers 3.0, 3.25 and 3.50 with Hydrogen as a Fuel** . . . . . . . . . . . . . . 919
Namrata Bordoloi, K. M. Pandey, and K. K. Sharma

**Study of Fuel Injection Systems in Scramjet Engine—A Review** . . . . . . 931
Kumari Ambe Verma, K. M. Pandey, and K. K. Sharma

**A Quantitative and Qualitative Review of Sustainable Manufacturing** . . . . . . . . . . . . . . . . . . . . . . . . . . . . . . . . . . . . . . . . . . . 941
Tejendra Singh and Jinesh Kumar Jain

**Impact Dynamics of a Viscoelastic Ferrofluid Droplet Under the Influence of Magnetic Field** . . . . . . . . . . . . . . . . . . . . . . . . . . . . . . . 961
Gaurav Kumar, Sudip Shyam, and Pranab Kumar Mondal

**Reacting Flow Solver for Martian Atmosphere Conditions** . . . . . . . . . . 969
P. Vicky Kumar, Anil Kumar Birru, and Vinayak Narayan Kulkarni

# About the Editors

**Prof. K. M. Pandey** obtained B.Tech in Mechanical Engineering from BHUIT, Varanasi now known as IIT BHU in 1980. Dr. Pandey also did M.Tech in Heat Power from the same institute in 1987. Prior to joining REC Silchar, Dr. Pandey served in BIT Mesra, Ranchi, as associate lecturer for a period of 3 years and 6 months. Dr. Pandey did his PhD in mechanical engineering in 1994 from IIT Kanpur. He has published and presented more than 325 papers in international and national journals and conferences. Dr. Pandey has also served in Colombo Plan Staff College Manila, Philippines, in the year 2002 as faculty consultant seconded by Government of India. Currently, he is working as professor in department of mechanical engineering of National Institute of Technology, Silchar, Assam, India. He has also served the department in the capacity of head for two terms of 3 years. He has also served as member of Board of Governors of BIT Silchar for two terms. His research interest areas are combustion, high speed flows, technical education, heat transfer, internal combustion engines, human resource management, gas dynamics and numerical simulations in CFD area from commercial softwares. Dr. Pandey has got more than 70 SCI-indexed journal papers and more than 150 research articles are indexed in Scopus. His h-index in Scopus is 17; and 23 in google scholar. He has guided 14 students for PhD and 89 students for M.Tech so far.

**Prof. R. D. Misra** received his B.E. in Mechanical Engineering from Jorhat Engineering College under Dibrugarh University in 1991, M.Tech. in Energy Studies from Indian Institute of Technology Delhi in 1996, and Ph. D. in Thermal Engineering from Indian Institute of Technology Roorkee in 2004. He has joined as faculty in Mechanical Engineering in the National Institute of Technology Silchar in 1992. He has served NIT Silchar as Lecturer from 1992 to 2003, as Assistant / Associate Professor from 2004 to 2009, as Professor from 2010 to 2018 and as Professor (HAG) from 2018 till date. He has published around 40 papers in referred international journals. Presently He is a professional member (life) of the Indian Society for Technical Education and The Institute of Engineers (Fellow).

**Prof. P. K. Patowari** is currently a professor at the Department of Mechanical Engineering, National Institute of Technology Silchar, Assam. He obtained his B.Tech degree (Mechanical) from NERIST, Itanagar, and M.E. (Prod.) from Jadavpur University, Kolkata, and Ph.D. from the Indian Institute of Technology, Kharagpur. His major areas of research interests include Advanced/Non-traditional Manufacturing Processes Micro-Manufacturing, Computer Aided Manufacturing (CAM), Computer Numerical Control (CNC), Micro Electro Mechanical Systems (MEMS). He has published more than 30 papers in referred international journals. Currently, he is the life member of professional bodies like Indian Society for Technical Education (ISTE) and Institute of Smart Structures and Systems (ISSS) and The Institute of Engineers (Fellow).

**Prof. U. S. Dixit** received his bachelor's degree in mechanical engineering from the erstwhile University of Roorkee (now Indian Institute of Technology (IIT) Roorkee) in 1987, his M.Tech degree in mechanical engineering from IIT Kanpur in 1993, and his PhD in mechanical engineering from IIT Kanpur in 1998. He has worked in two industries—HMT, Pinjore, and INDOMAG Steel Technology, New Delhi, where his main responsibility was designing various machines. Dr Dixit joined the Department of Mechanical Engineering, IIT Guwahati, in 1998, where he is currently a professor. He was also the Officiating Director of the Central Institute of Technology, Kokrajhar (February 2014−May 2015). Dr Dixit has been actively engaged in research in various areas of design and manufacturing over the last twenty-five years. He has authored/co-authored 115 journal papers, 109 conference papers, twenty-seven book chapters, and six books on mechanical engineering. He has also co-edited seven books related to manufacturing. He has guest-edited eleven special issues of journals. Presently he is an Associate Editor of the Journal of Institution of Engineers (India), Series C, and the Regional Editor (Asia) of the International Journal of Mechatronics and Manufacturing Systems. He has guided twelve doctoral and forty-four masters' students. Dr Dixit has investigated a number of sponsored projects and developed several courses. Presently, he is the Vice-President of the AIMTDR conference.

# AI-Based ANN Modeling of Performance–Emission Profiles of CRDI Engine under Diesel-Karanja Strategies

**P. Sandeep Varma, Subrata Bhowmik, Abhishek Paul, Pravin Ashok Madane, and Rajsekhar Panua**

**Abstract** The current investigation highlights the impact of Diesel–biodiesel blends on performance and exhaust emission profiles of a single-cylinder, common rail direct injection (CRDI) engine. Experiments were performed at constant engine speed (1500 rpm) and three engine loads (50, 75 and 100%) under high fuel injection pressure (900 bar) with volume proportions (10, 20 and 30%) of Karanja with Diesel. Utilizing CRDI engine experimental data, an artificial intelligence (AI)-affiliated artificial neural network (ANN) model has been created with the intention of forecasting brake thermal efficiency, oxides of nitrogen, unburned hydrocarbon and carbon monoxide emissions. From various tested ANN models, one hidden layer with three neurons along with logsig transfer function has been noticed to be optimum network for Diesel-Karanja paradigms under high fuel injection pressure. While developing the optimum model, standard Levenberg–Marquardt training algorithm has been employed. The optimum ANN model is capable to estimate the CRDI engine performance–emission profiles with an overall correlation coefficient value of 0.99742, wherein 0.99783, 0.99951 and 0.99969 for training, validation and testing datasets, respectively. Results made clear that the formulated AI-based ANN model is viable for predicting the existing CRDI engine performance and emission profiles of Diesel-Karanja blends operating under high fuel injection pressure.

P. S. Varma (✉) · P. A. Madane · R. Panua
Department of Mechanical Engineering, National Institute of Technology, Agartala, India
e-mail: psvarma.969@gmail.com

P. A. Madane
e-mail: madanepravin691@gmail.com

R. Panua
e-mail: rajsekhar_panua@yahoo.co.in

S. Bhowmik
Department of Mechanical Engineering, Indian Institute of Technology (Indian School of Mines), Dhanbad, India
e-mail: sbhowmik47@gmail.com

A. Paul
Department of Mechanical Engineering, National Institute of Technology, Silchar, India
e-mail: v1.abhishek@gmail.com



K. M. Pandey et al. (eds.), *Recent Advances in Mechanical Engineering*, Lecture Notes in Mechanical Engineering, https://doi.org/10.1007/978-981-15-7711-6_1

**Keywords** AI · ANN · CRDI engine · Performance–emission prediction

## 1 Introduction

The global energy demand in transportation sector has immensely relied on fossil fuels [1]. Fast reduction of petro-fuels and their deleterious effect on environment are creating huge concern over usage of fossil Diesel in conventional Diesel engines. However, Diesel engines are acknowledged for their operational reliability, higher thermal efficiency and lower carbon monoxide (CO) and unburned hydrocarbon (UBHC) emissions [2]. Despite these benefits, stringent emission mandates and limited fossil Diesel reserves may scotch the production and employability of conventional Diesel engine in coming years. Accordingly, most of the research is now intended toward finding an alternative renewable fuel source with acceptable performance and emission characteristics. Research studies [3–6] show that biodiesel has great potential to meet future global energy demand.

The properties of biodiesel primarily depend on the feedstock and processing technology, but usually it has ~11% (by mass) fuel-bound oxygen, high cetane number and low aromatic content when compared to fossil Diesel [5]. Biodiesel also exhibits high flash point temperature which is beneficial for safe storability and transportation [4]. However, the problematic issues of biodiesel are high viscosity and density which result in inferior spray characteristics [2]. Many researchers [3–6] have concluded that the usage of biodiesel in CI engines has resulted in approximately higher oxides of nitrogen ($NO_X$) emissions and decreased particular matter (PM), CO and UBHC emissions. In a study by Lee et al. [7] concurred that biodiesel blends have resulted in lower brake thermal efficiency (BTE)and torque but higher $NO_X$ emissions than pure Diesel operation.

In order to study the performance–emission profiles of a compression ignition (CI) engine over its entire operational range, arduous experimentation is required. To prevail over this problem, computational models are necessary. In this present study, artificial neural network (ANN) technique has been chosen because of its ability to learn, model curvilinear process and flexibility to changes in real time. Yusuf cay [8] developed an ANN model by using feed-forward back propagation method and demonstrated the viability of ANN model in forecasting the performances and exhaust fumes of gasoline engine. In another study Bhowmik et al. [9] devised an ANN model with high accuracy for estimating the indirect injection engine outcomes under ternary blends of Diesel, kerosene and ethanol. Paul et al. [10] discussed the effect of diesel–ethanol pilot fuel on performance–emission characteristics of compression ignition engine operating in dual-fuel mode with compressed natural gas as the main fuel. They created ANN model by using experimental data and stated that the model has good relationship between estimated and experimental values with an overall correlation coefficient ($R$) value of 0.99689. Bhowmik et al. [11] devised an ANN model for predicting output parameters of Diesel engine. The proposed model has given $R$ value which ranges from 0.999312 to 0.999852.

In this study, first experimentation was performed with Diesel-Karanja blends to evaluate performance and emission characteristics. Later, by utilizing this experimental data, an ANN model was formulated by considering the engine load and Karanja biodiesel share in the blend as input parameters and BTE, $NO_X$, UBHC and CO as output parameters.

## 2 Experimental Setup and Procedure

A single-cylinder, four-stroke, water-cooled, CRDI engine was used for the present investigation. The engine is synchronized to a crank angle sensor for measuring engine rpm. An eddy current dynamometer is synchronized to the CRDI engine for load measurement. By employing, NIRA©-based centralized data acquisition system, each and every instrument fitted to the engine is interfaced to the computer. Gaseous emissions from the CRDI engine are measured by using an AVL MDS 250 and an AVL 437 smoke meter. The engine experiments were performed at three different load conditions, namely 50, 75 and 100% for all the fuel blends (B10, B20 and B30), and high fuel injection pressure of 900 bar is employed for injecting fuels into combustion chamber. During the experimentation, speed of the CRDI engine is kept constant at 1500 rpm. Prior, to experimentation, the engine was first run on fossil Diesel fuel at the same operating points to acquire baseline data. To increase the authenticity, the engine experiments were conducted three times, and their mean value has been considered as the final output. Figure 1 encapsulates the schematic of CRDI engine setup.

## 3 ANN Modeling

ANN is a computational model based on biological processes, predominantly inspired by human brain. Its architecture involves three layers, for instance, input layer, hidden layer and output layer. The experimental data provided to ANN is divided into three sets, namely training dataset, validation dataset and testing dataset [11]. Training dataset is utilized to improve the generalization of the network in predicting the input–output relationship, validation data is used to lessen overfitting of network, and testing data is utilized to validate the generalization capability of model [12]. Complex nonlinear engineering problems can be simulated with the help of neural network tools.

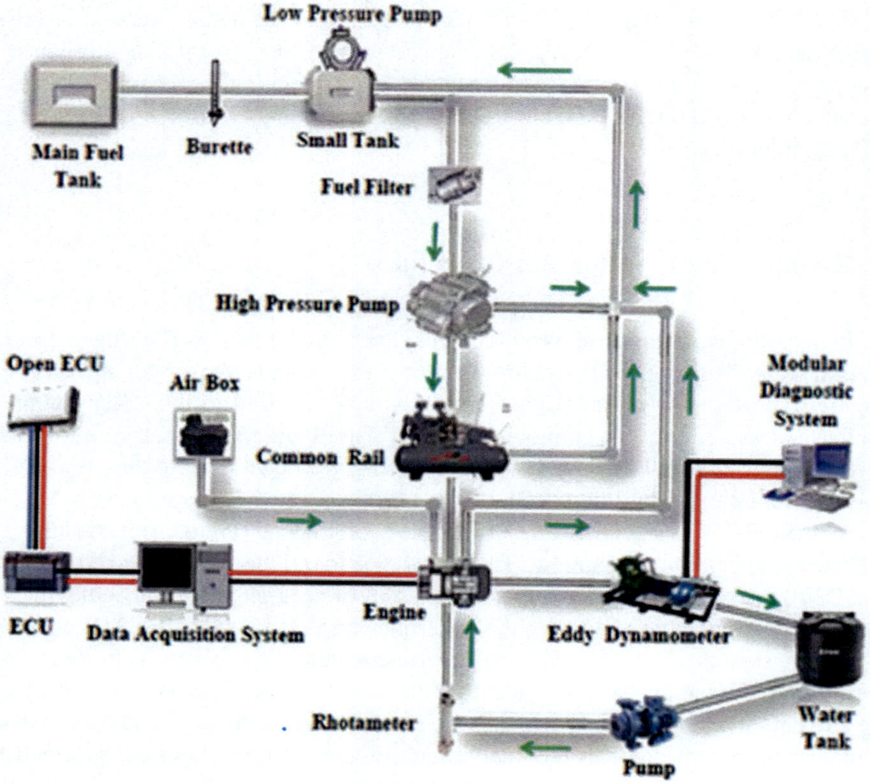


**Fig. 1** Schematic diagram of CRDI engine setup

### *3.1 Selection of Input and Output Parameters*

In this study, load and Karanja biodiesel share in the blend are considered as input parameters for predicting the output parameters, namely BTE, $NO_X$, UBHC and CO. While developing ANN model, 70% of the experimental data was defined for training, 15% of the experimental data was defined for validating, and the rest of the 15% data was utilized for testing the network.

### *3.2 Selection of Transfer Function*

The transfer function introduces curvilinear transformation into neural architectures so that the model is capable of having nonlinear match between input and output layers [14]. The performance of the neural network is greatly affected by selection of appropriate transfer function. Three basic transfer functions that are available in MATLAB© are logsig, tansig and purlin. Research studies [12–14] concluded that logsig transfer function is appropriate for predicting the output parameters of a Diesel engine.

### 3.3 *Selection of Training Algorithm*

In simple, training algorithm is the method followed for updating the connecting weights and bias in order to make improved generalization of input–output relationship. In this study, single hidden layer feed-forward neural network has been used by employing Levenberg–Marquardt back propagation training algorithm (trainlm). Many researchers [15–17] reported that trainlm is fast compared to other training algorithms and it has superior convergence.

## 4 Result and Discussion

Various network topologies were created by changing the number of neurons from two to twenty-five for each of the three basic transfer functions. All the constructed topologies were tested to measure their individual performance. The optimum model has been noticed to occur with logsig transfer function, and the topology comprises of three neurons in its hidden layer, two neurons in its input layer and four neurons in its output layer. Figure 2 shows the overall *R* value obtained for the optimum model. The overall *R* value of the optimum ANN model is 0.99742 wherein 0.99783, 0.99951 and 0.99969 for training, validation and testing datasets, respectively.

All the test fuels containing any proportion of biodiesel have shown decreased BTE compared to mineral Diesel operation. This is due to higher viscosity and existence of long chain of unsaturated fatty acid molecules in Diesel–Biodiesel blends compared to mineral Diesel. Among all the test fuels, the D90B10 fuel sample has shown maximum decrease in BTE, which when compared with 50% engine load mineral diesel operating condition it was found to be 15.9% lesser. The developed ANN model has estimated BTE with an *R* value of 0.998266. Figure 3 shows the comparison of ANN predicted BTE and experimental BTE. Hence, from the value of *R*, it can be inferred that ANN model can be employed for estimating the BTE of CRDI engine energized with Diesel–Biodiesel blends operating with high fuel injection pressure.$NO_X$ emissions of Diesel-Karanja blends have been observed to be lower than Diesel fuel. Among all chosen test fuels, the D70B30 fuel sample has shown maximum decrease in $NO_X$ emissions. It was found that this fuel sample has resulted in 32.3% lesser $NO_X$ emissions at 50% load condition when compared to pure Diesel. The fabricated ANN model has predicted $NO_X$ emissions with an *R* value of 0.9983; from this, we can conclude that the suggested ANN models has proved its viability in approximating the poisonous $NO_X$ emissions (as shown in Fig. 4). Biodiesel blends has resulted lesser UBHC emission than Diesel fuel. The minimum UBHC emissions among all the test fuels were noticed for D70B30 fuel sample.

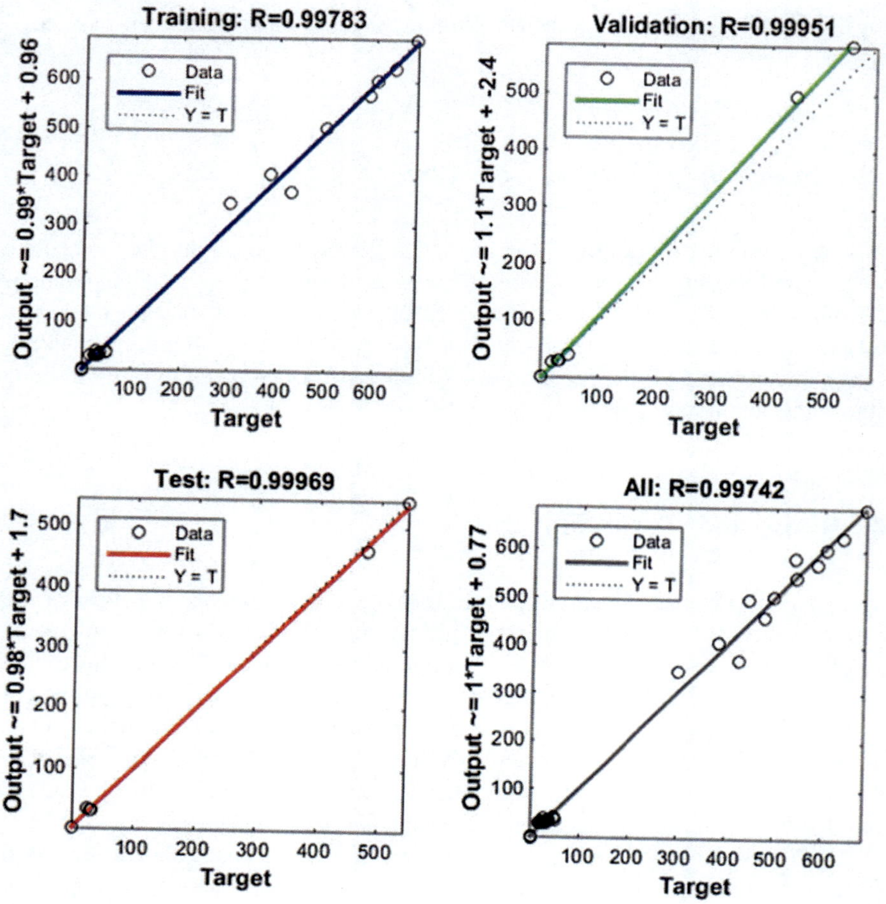


**Fig. 2** Overall correlation coefficient of developed model

When compared to mineral diesel operation, it was found that this fuel sample has shown 56.98% lesser UBHC emission at 100% engine load. The fabricated ANN model has predicted UBHC emission with an *R* value of 0.9880. The comparison of ANN model mapped UBHC and experimental UBHC is shown in Fig. 5. By comparing the *R* value of UBHC emission with that of BTE and $NO_X$ emissions, it can be deduced that the proposed model has lesser accuracy in estimating the UBHC emission compared to BTE and $NO_X$.

At most experimental conditions, CO emissions resulted from Diesel–biodiesel blends were higher compared to pure Diesel operation. This is due to higher viscosity and existence of long chain unsaturated fatty acid molecules in diesel–biodiesel blends compared to mineral Diesel. The developed optimum model has predicted CO emission with an *R* value of 0.9810. Figure 6 delineates the comparison of ANN model mapped CO and experimental CO. By comparing the *R* value of CO emission

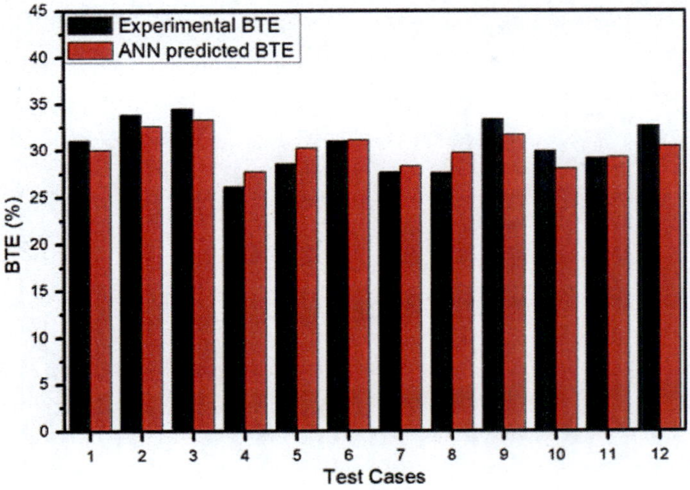


**Fig. 3** Comparison of experimentally measured BTE with ANN predicted BTE

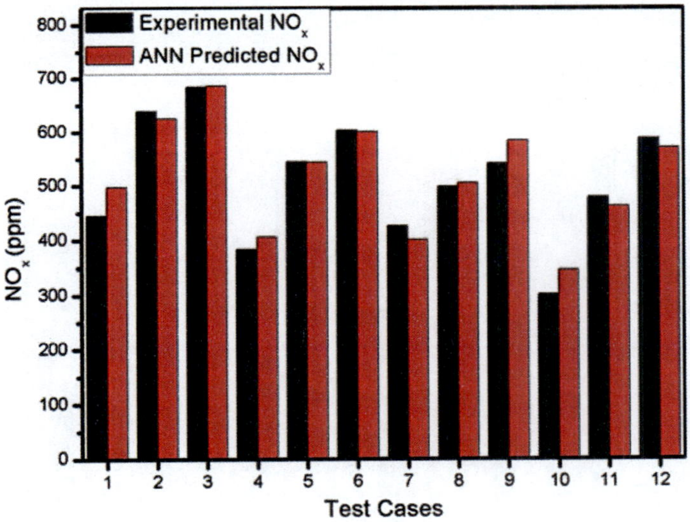


**Fig. 4** Comparison of experimentally measured $NO_X$ with ANN predicted $NO_X$

with that of BTE and $NO_X$emissions, it can also be deduced that the proposed model has lesser accuracy in estimating the CO emission compared to BTE and $NO_X$.

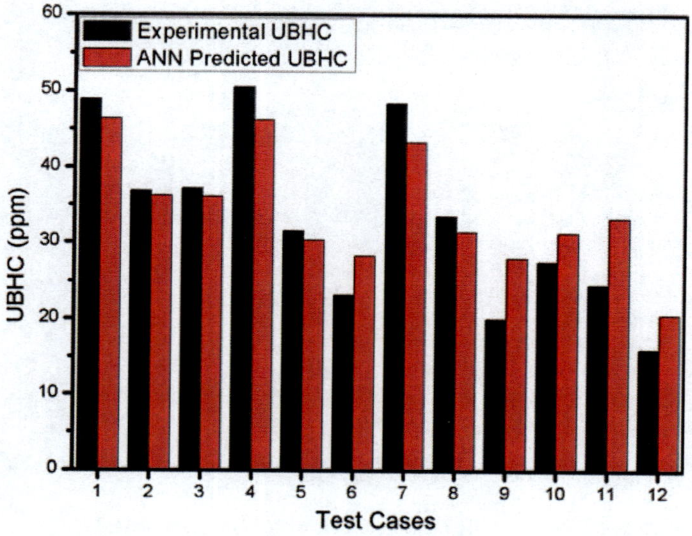


**Fig. 5** Comparison of experimentally measured UBHC with ANN predicted UBHC

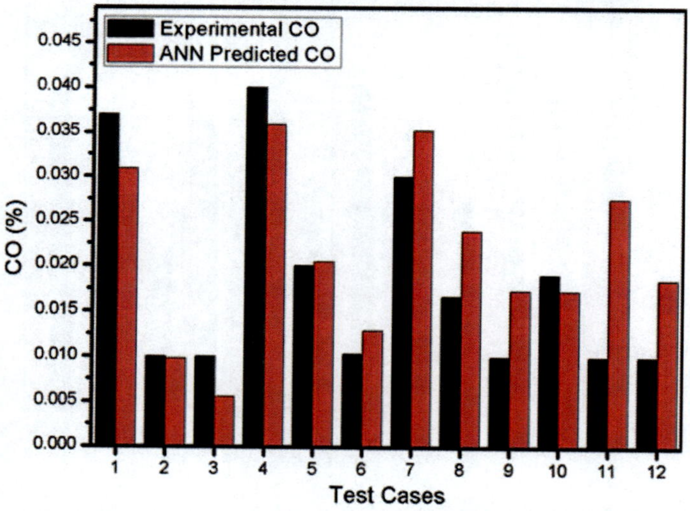


**Fig. 6** Comparison of experimentally measured CO with ANN predicted CO

## 5 Conclusion

The major findings from the experimental cum AI-based ANN model of CRDI engine fueled with various Karanja biodiesel share and engine loads under high fuel injection pressure are as follows:

- At every load condition, Diesel-Karanja blends have shown decreased BTE than Diesel fuel operation.
- $NO_X$ and UBHC emissions of Diesel-Karanja blends were lower than fossil Diesel at every load condition.
- At most experimental conditions, CO emissions resulted from Diesel-Karanja blends were higher than pure Diesel operation.
- The model developed with logsig transfer function and three neurons in its hidden layer has been noticed to be the optimum model for predicting performance–emission profiles of diesel–biodiesel blends under high fuel injection pressure.
- The optimum model has shown overall $R$ value of 0.99742 wherein 0.99783, 0.99951 and 0.99969 for training, validation and testing datasets, respectively.

ANN has proved its viability in predicting the performance and emission parameters of CRDI engine fueled with Diesel-Karanja biodiesel blends operating under high fuel injection pressure. By utilizing the developed AI-affiliated ANN model, the present investigation can be extended to map the output values at any distinct points of the input parameters under Diesel-Karanja strategies without conducting new experiments which will eliminate the experimental cost, time and effort.

**Acknowledgements** This research work was supported and funded by the Science and Engineering Research Board under the Department of Science and Technology, Government of India, under the Grant no. EEQ/2016/000058 for the project entitled "Effect of biodiesel-ethanol-Diethyl ether blends on performance, combustion and emissions of a CI engine under Hydrogen, CNG and LPG dual fuel mode."

## References

1. J.K. BaeChoongsik, Alternative fuels for internal combustion engines. Proc. Combust. Inst. **36**(3), 3389–3413 (2017)
2. A.K. Agarwal, A. Dhar, Experimental investigations of performance, emission and combustion characteristics of Karanja oil blends fuelled DICI engine. Renew. Energy **52**, 283–291 (2013)
3. MahendraP. PuneetVerma, Sharma, Review of process parameters for biodiesel production from different feedstocks. Renew. Sustain. Energy Rev. **62**, 1063–1071 (2016)
4. M. Suresh, C.P. Jawahar, Arun Richard, A review on biodiesel production, combustion, performance, and emission characteristics of non-edible oils in variable compression ratio diesel engine using biodiesel and its blends. Renew. Sustain. Energy Rev. **92**, 38–49 (2018)
5. G. Metin, C. Sayin, M. Canakci, The impact of fuel injection pressure on the exhaust emissions of a direct injection diesel engine fueled with biodiesel–diesel fuel blends. Fuel **95**, 486–494 (2012)
6. S.K. Kumar, V. Ramalingam, G. Ramalingam, Performance and exhaust emission characteristics of a CI engine fueled with Pongamiapinnata methyl ester (PPME) and its blends with diesel. Renew. Energy **33**(10), 2294–2302 (2008)
7. L. Sanghoon, C.S. Lee, S. Park, J.G. Gupta, R.K. Maurya, A.K. Agarwal, Spray characteristics, engine performance and emissions analysis for Karanja biodiesel and its blends. Energy **119**, 138–151 (2017)
8. C. Yusuf, Prediction of a gasoline engine performance with artificial neural network. Fuel **111**, 324–331 (2013)

9. S. Bhowmik, R. Panua, D. Debroy, Artificial neural network prediction of diesel engine performance and emission fueled with diesel–kerosene–ethanol blends: a fuzzy-based optimization. J. Energy Res. Technol. **139**(4), 042201 (2017)
10. A. Paul, S. Bhowmik, R. Panua, D. Debroy, Artificial neural network-based prediction of performances-exhaust emissions of diesohol piloted dual fuel diesel engine under varying compressed natural gas flowrates. J. Energy Res. Technol. **140**(11), 112201 (2018)
11. S. Bhowmik, A. Paul, R. Panua, S.K. Ghosh, D. Debroy, Performance-exhaust emission prediction of diesosenol fueled diesel engine: an ANN coupled MORSM based optimization, Energy **153**, 212–222 (2018)
12. U. Samet, M.B. Celik, Prediction of engine emissions and performance with artificial neural networks in a single cylinder diesel engine using diethyl ether. Eng. Sci. Technol. **21**(6), 1194–1201 (2018)
13. S. Bhowmik, R. Panua, S.K. Ghosh, D. Debroy, A. Paul, A comparative study of artificial intelligence based models to predict performance and emission characteristics of a single cylinder diesel engine fueled with diesosenol. J. Thermal Sci. Eng. Appl. **10**(4). 041004 (2018)
14. C. Yusuf, I. Korkmaz, A. Çiçek, F. Kara, Prediction of engine performance and exhaust emissions for gasoline and methanol using artificial neural network. Energy **50**, 177–186 (2013)
15. S. Bhowmik, A. Paul, R. Panua, S.K. Ghosh, D. Debroy, Artificial intelligence based gene expression programming (GEP) model prediction of Diesel engine performances and exhaust emissions under Diesosenol fuel strategies. Fuel **235**, 317–325 (2019)
16. S. Bhowmik, R. Panua, S.K. Ghosh, A. Paul, D. Debroy, Prediction of performance and exhaust emissions of diesel engine fuelled with adulterated diesel: an artificial neural network assisted fuzzy-based topology optimization. Energy Environ. **29**(8), 1413–1437 (2018)
17. M.A. Pravin, S. Bhowmik, R. Panua, Hybrid Taguchi–Fuzzy-based Performance exhaust emission trade-off study of variable compression ratio diesel engine fueled with undi-diesel blends, Environ. Progress Sustain. Energy, e13333 (2019)

# ANFIS Prediction of Performance and Exhaust Emission Characteristics of CRDI Engine Fueled with Diesel–Butanol Strategies

**Vivek Kumar Mishra, Subrata Bhowmik, Abhishek Paul, Ajay Yadav, and Rajsekhar Panua**

**Abstract** The present work investigates the ability of oxygenated Butanol on performance and exhaust emission characteristics of a single-cylinder, four-stroke, water-cooled, common rail direct injection (CRDI) engine. Experiments were performed at constant engine speed (1500 rpm) and six different load conditions, varying from 5 to 30 Nm. Based on CRDI engine experimental data, an artificial intelligence (AI)-affiliated adaptive neuro-fuzzy inference system (ANFIS) model has been formulated for predicting the output parameters, namely brake thermal efficiency (BTE), brake specific energy consumption (BSEC), oxides of nitrogen ($NO_X$), unburned hydrocarbon (UBHC) and carbon monoxide (CO) by considering the engine load and Butanol share in the blend as input parameters. With the increasing Butanol share in the Diesel–Butanol blend, the BTE and BSEC were significantly increased, and exhaust gas emissions, especially $NO_X$ and CO, were also reduced. The developed AI-based ANFIS model has the capacity of mapping the relationship between input–output parameters of the CRDI engine with good accuracies. In this study, the statistical performances obtained from ANFIS predicted model are (0.0000107–0.0000755) of mean square error, (0.000353–0.001533) of mean square relative error, (0.999722–0.999939) of correlation coefficient and (0.999444–0.999878) of absolute fraction of variance, which elevated the model capability to a higher stage under Diesel–Butanol strategies.

V. K. Mishra (✉) · A. Yadav · R. Panua
Department of Mechanical Engineering, National Institute of Technology, Agartala, India
e-mail: vivekmishra0212@gmail.com

A. Yadav
e-mail: ajaymips95@gmail.com

S. Bhowmik
Department of Mechanical Engineering, Indian Institute of Technology (Indian School of Mines), Dhanbad, India
e-mail: sbhowmik47@gmail.com

A. Paul
Department of Mechanical Engineering, National Institute of Technology, Silchar, India
e-mail: v1.abhishek@gmail.com


K. M. Pandey et al. (eds.), *Recent Advances in Mechanical Engineering*, Lecture Notes in Mechanical Engineering, https://doi.org/10.1007/978-981-15-7711-6_2

**Keywords** Artificial intelligence · ANFIS · CRDI · Performance–emission mapping

## 1 Introduction

In vehicles like buses, trucks and earth moving machineries, high torque is needed. The higher torque can be produced by the utilization of Diesel engines. Diesel engines have been broadly used to move heavy loads. However, the usage of conventional Diesel fuel in internal combustion engines is one of the major issues for air pollution, due to the high level of exhaust gas emissions, such as particulate matter (PM), oxides of nitrogen ($NO_X$). In recent days, to reduce exhaust gas emissions, from Diesel engines, researchers have shifted toward renewable and eco-friendly sources of alternative energy. Many alternative fuels are available like alcohols, biodiesel, etc. [1]. Among all alternative fuels, alcohols have shown better effects to decrease the emissions from conventional Diesel engines [1]. Alcohols are oxygenated fuel and contain a low amount of sulfur and carbon content as compared to conventional Diesel fuel. Alcohol fuels are restricted for their direct use in Diesel engines because of their poor cetane number [2, 3]. Among various alcohols, the autoignition temperature of Butanol (365 °C) is less than ethanol (479 °C) and methanol (434 °C) [4]. When the blend of Diesel–Butanol is used in the compression ignition (CI) engine, it a has high ability to ignite easily. Dogen [5] has concluded that with the increasing proportion of Butanol in Diesel, the performances of the CI engine are significantly improved along with $NO_X$ and smoke emissions. Nour et al. [6] investigated that the addition of Butanol share in Diesel, brake thermal efficiency (BTE) and brake specific energy consumption (BSEC) relatively improved alongside $NO_X$, unburned hydrocarbon (UBHC) and carbon monoxide (CO). Compared to other alcohol fuels, Butanol has a higher cetane number and less corrosion and oxygen content. Due to these properties, it is a more suitable additive in CI engine operation than other alcohols [5, 6]. By the process of fermentation of biomass, Butanol can be produced, especially from wastage of plants, corn and algae. Because of its inherent fuel properties and availability, Butanol nowadays is widely used in CI engine operation to partially replace the dependency on conventional Diesel and reduce higher exhaust emissions.

Using the artificial intelligence (AI)-based adaptive neuro-fuzzy inference system (ANFIS) model, the majority of the researchers have developed their model in order to predict the input–output relationship of CI engine [7–9]. Hosoz et al. [10] reported that ANFIS model is a combination of both neural network and fuzzy logic principles. Due to this combination, compared to other artificial intelligence models, this model is able to forecast very efficiently the engine output parameters in a short time. ANFIS also has a great ability to make the fundamental relationship between input and desired output parameters of any sector [11]. Bhowmik et al. [12] surveyed the indirect injection engine outputs for Diesel–Kerosene–Ethanol blends using the ANFIS approach. They reported that the developed model has an overall correlation coefficient (R) value greater than 0.98. The other parameters, such as mean square error (MSE), Theil U2, were also helpful to validate the developed the ANFIS model.

## 1.1 Motivation of Present Work

Global petro-fuel energy plays an important role in the economy growth of society. According to Krishna et al. [13], in most of the developing countries, the rate of energy consumption is expected to be increased by 84% by 2035. At present, due to a large number of automobile sectors, a huge amount of energy is consumed by heavy load vehicles in the form of Diesel fuel. With the increased Diesel consumption, the fossil Diesel reserves are reducing day by day and also the pollution level in the environment is increasing due to its higher exhaust gas emissions. To solve the above-mentioned issues, researchers [14, 15] have focused on an alternative source of energy which helps to partially reduce fossil Diesel consumption and decrease environment pollution. For this purpose, alcohol is considered one of the best sources of alternative fuel for Diesel engines because of its oxygenated property. In this present investigation, common rail direct injection (CRDI) engine has been used because its advanced features like control of injection pressure by use of electronic control unit (ECU) make it more efficient to control the poisonous emissions. Even though AI-affiliated models are not a new advancement in the area of internal combustion (IC) engine, however because of their high accuracy and less time consumption for forecasting the performance and exhaust gas emission parameters, it is being used by the various researchers. ANFIS model with Diesel–Butanol blends in the CRDI engine has not been explored.

## 1.2 Objectives of the Present Work

The present study proposed the following objectives:

- To investigate the output parameters of a CRDI engine under different loading conductions and Diesel–Butanol blends.
- To develop an ANFIS model for mapping the input–output relationship of the CRDI engine fueled with Diesel–Butanol strategies.
- To explore the viability of the ANFIS model to give the result of complex calculations in a short time with less cost.

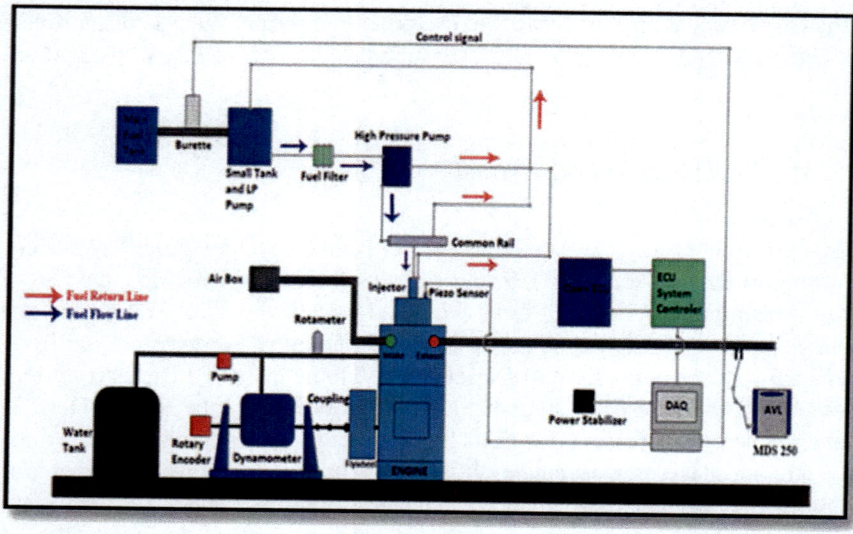


**Fig. 1** Schematic ofexisting CRDI engine setup

## 2 Experimental Setup and Methodology

The engine operation was performed on a Mahindra Jeeto 11 HP, single-cylinder, four-stroke, water-cooled CRDI engine. For maintaining the different loading conditions, the engine has been coupled with a dynamometer. To calculate the rotational speed of the engine, a crank angle encoder has been used with a one-degree step angle. Rotameter is provided for engine cooling water flow measurement. Figure 1 delineates the schematic of the CRDI engine setup. In this study, 5% and 10% (by volume) Butanol were included to mineral Diesel, and denoted as D95B5 and D90B10, respectively. The engine was operated at six different load conditions such as 5 Nm, 10 Nm, 15 Nm, 20 Nm, 25 Nm and 30 Nm at a constant engine speed of 1500 rpm.

## 3 Experimental Uncertainty Analysis

With the intention of scrutinizing the errors that arise while operating the CRDI engine, the root mean square methodology was used (as shown in Eq. (1)) [9]. Using Eq. (1), the total percentage of uncertainty (TSU) for the performance parameters is found to be 1.99. Table 1 shows the accuracy of the AVL MDS 250 exhaust gas analyzer. The study also added standard deviation for increasing the superiority of error analysis. Table 2 details the TSU and standard deviation for the exhaust emission parameters.

**Table 1** Accuracy of AVL MDS 250 gas analyzer

| Measurand | Range | Resolution | Accuracy |
|---|---|---|---|
| CO | 0–15% vol. | 0.01% vol. | <10.0% vol.: ±0.02% vol., ±3% o. M<br>≥10.0% vol.: ±5% o. M |
| $CO_2$ | 0–20% vol. | 0.01% vol. | <16.0% vol.: ±0.3% vol., ±3% o. M<br>≥16.0% vol.: ±5% o. M |
| HC | 0–30.000 ppm vol. | ≤2.000: 1 ppm vol. | <2000 ppm vol.: ±0.3% vol., ±3% o. M<br>≥5000 ppm vol.: ±5% o. M<br>≥10,000 ppm vol.: ±10% o. M |
| $O_2$ | 0–25% vol. | 0.01% vol. | ±0.02% vol<br>±1% o. M |
| NO | 0–5.000 ppm vol. | 1 ppm vol. | ±5 ppm vol<br>±1% o. M |

**Table 2** Average TSU and standard of exhaust gas emissions

| Parameter | Average TSU (%) | Average standard deviation |
|---|---|---|
| $NO_X$ | 1.247 | 0.7 |
| UBHC | 1.682 | 0.9 |
| CO | 0.845 | 0.3 |

$$\Delta U = \sqrt{\left(\frac{\partial U}{\partial x_1}\right)^2 \times \Delta x_1 + \left(\frac{\partial U}{\partial x_2}\right)^2 \times \Delta x_2 + \cdots + \left(\frac{\partial U}{\partial x_n}\right)^2 \times \Delta x_n} \quad (1)$$

## 4 ANFIS Modeling

ANFIS, in its essence, is a combined form of artificial neural network and fuzzy logic system. An ANFIS model is capable of mapping the complex relationships of any system in real life. Even though ANFIS was introduced three decades ago, due to its more accuracy and less time-consuming process, it is being more popular among all AI models in the field of internal combustion (IC) engines [9]. ANFIS model is made up of six layers. Each layer has multiple nodes (except the last layer), and the summation of all nodes consists of a network. Every layer is responsible for performing a particular function. In the ANFIS model, the fuzzy membership function is tuned with the assistance of the least-square and back-propagation gradient descent method [9]. In this present CRDI engine operation, two inputs such as engine load and Butanol share in the blend are considered to predict the output parameters, namely BTE, BSEC, $NO_X$, UBHC and CO. While developing the ANFIS model for Diesel–Butanol blends, 70% of the total data (30 cases) were preferred for training (21 cases) the model and rest of the data were preferred for validation (9 cases) of the model.

# 5 Result, Discussion and Validation of ANFIS Modeling

To develop the fundamental relationship between CRDI engine input and output parameter under Diesel–Butanol blends, ANFIS models have been formulated. The models have been validated in terms of mean square error (MSE < 0.01), overall correlation coefficient ($R > 0.98$) and mean absolute percentage error (MAPE < 5%). The comparison of ANFIS forecasted BTE and experimental measured BTE is demonstrated in Fig. 2a, b. The uncertainty for the $B$th is found to be 1.985. The MSE, MAPE, mean square relative error (MSRE), R and absolute fraction of variance ($R^2$) value are 0.000049, 1.79%, 0.00035, 0.999921 and 0.999841, respectively. In the same fashion, ANFIS predicted BSEC with respect to experimental measured BSEC is shown in Fig. 3a, b. The values of experimental uncertainty, MSE, MAPE, MSRE, $R$ and $R^2$ are 1.985, 0.000237, 2.92%, 0.000958, 0.999939 and 0.999878, respectively. ANFIS prediction for $NO_X$ emissions depicted in Fig. 4a, b with MSE value of 0.0000579, MAPE of 2%, MSRE value of 0.000636, $R$ value of 0.999903, $R^2$ value of

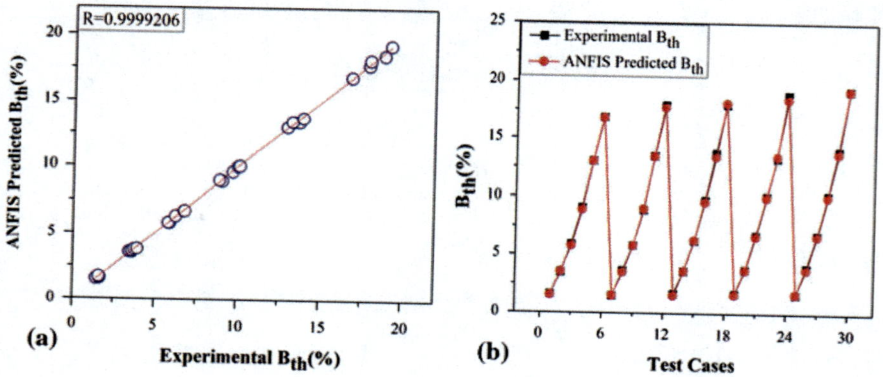


**Fig. 2 a, b** Comparison of ANFIS forecasted BTE and experimental BTE

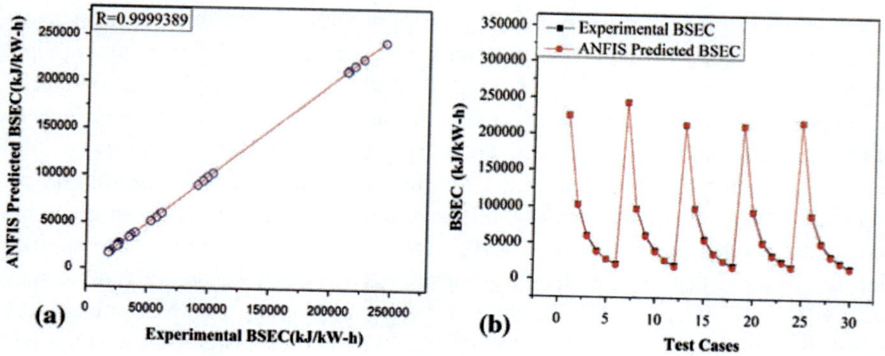


**Fig. 3 a, b** Comparison of ANFIS forecasted BSEC and experimental BSEC

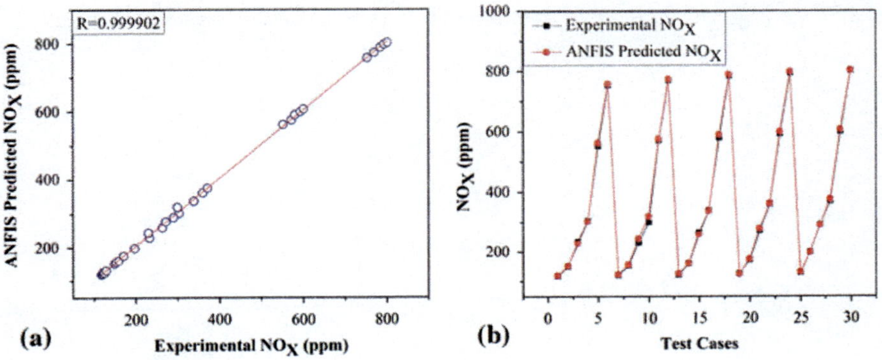


**Fig. 4** **a, b** Comparison of ANFIS forecasted $NO_X$ and experimental $NO_X$

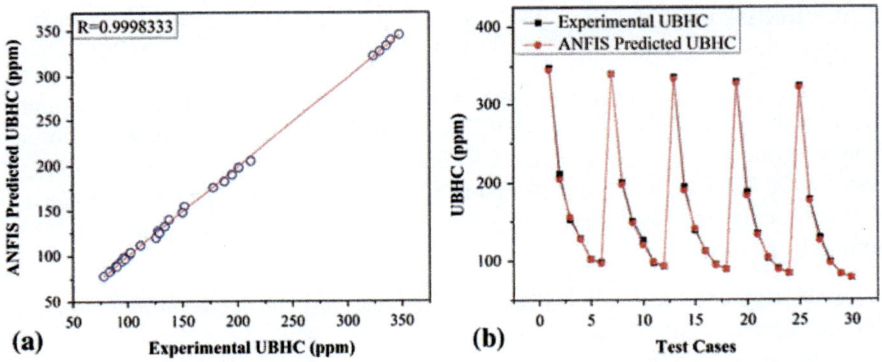


**Fig. 5** **a, b** Comparison of ANFIS forecasted UBHC and experimental UBHC

0.999806 and experimental uncertainty of 1.247. Figure 5a, b explains the difference between ANFIS forecasted UBHC and experimental UBHC with value MSE of 0.0000755, MAPE of 2.28%, MSRE of 0.000615, $R$ of 0.999833, $R^2$ of 0.999667 and experimental uncertainty of 1.682. ANFIS prediction of CO with experimental uncertainty of 0.845, $R$ value of 0.999722 with MSR value of 0.0000107, MSRE value of 0.001533 and $R^2$ value of 0.999444 is shown in Fig. 6a, b. All developed models have performed higher prediction quality with noteworthy errors. From the above, it is deduced that the ANFIS model has the capacity of mapping complex relationships of CRDI engine under Diesel–Butanol strategies with higher accuracies.

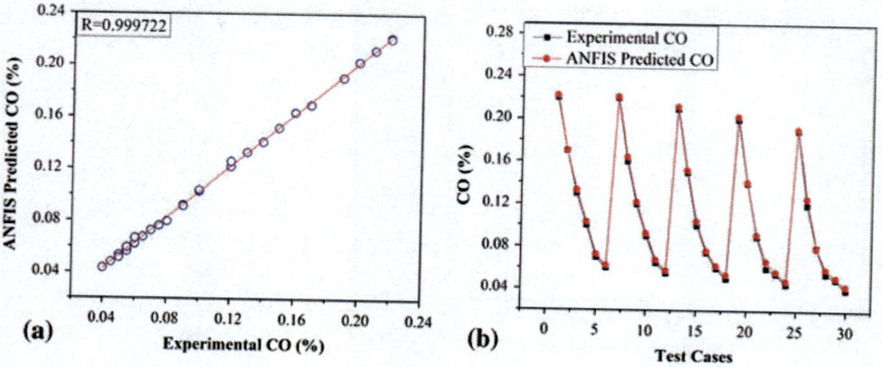


**Fig. 6** **a, b** Comparison of ANFIS forecasted CO and experimental CO

## 6 Conclusion

In this study, ANFIS models have been developed for forecasting the output parameters, namely BTE, BSEC, $NO_X$, UBHC and CO of CRDI engine fueled with Diesel–Butanol strategies. For this aim, the CRDI engine was run by varying engine load from 5 to 30 Nm with a step size of 5 Nm and at a constant speed of 1500 rpm. Based on experimental and developed ANFIS model, the following important outcomes are summarized below:

- Diesel–Butanol blends improved the exhaust gas emissions along with the improvement in BSEC and BTE.
- The developed ANFIS models are found to be optimal for predicting CRDI engine output parameters.
- Statistical analyses given by the ANFIS model have shown excellent correlations (*R*: 0.999722–0.999939) with praiseworthy errors.

**Acknowledgements** The present study was funded by the Science and Engineering Research Board (SERB) under the Department of Science and Technology (DST), Government of India, under Grant no. EEQ/2016/000058 for the project entitled "Effect of biodiesel-ethanol-Diethyl ether blends on performance, combustion and emissions of a CI engine under Hydrogen, CNG and LPG dual fuel mode."

## References

1. G. Li, S. Qian, H. Lee, Y. Hwang, R. Radermacher, Experimental investigation of energy and exergy performance of short term adsorption heat storage for residential application. Energy **65**, 675–691 (2014)

2. A. Paul, S. Bhowmik, R. Panua, D. Debroy, Artificial neural network-based prediction of performances-exhaust emissions of diesohol piloted dual fuel diesel engine under varying compressed natural gas flowrates. J. Energy Res. Technol. **140**(11), 112201 (2018)
3. S. Szwaja, J.D. Naber, Combustion of n-butanol in a spark-ignition IC engine. Fuel **89**(7), 1573–1582 (2010)
4. B. Thanapiyawanit, J-H. Lu, Cooling effect of methanol on an n-heptane HCCI engine using a dual fuel system. Int. J. Autom. Technol. **13**(7), 1013–1021 (2012)
5. D. Oğuzhan, The influence of n-butanol/diesel fuel blends utilization on a small diesel engine performance and emissions. Fuel **90**(7), 2467–2472 (2011)
6. N. Mohamed, A.M.A. Attia, S.A. Nada, Combustion, performance and emission analysis of diesel engine fuelled by higher alcohols (butanol, octanol and heptanol)/diesel blends. Energy Convers. Manage. **185**, 313–329 (2019)
7. P.A. Madane, S. Bhowmik, R. Panua, Hybrid Taguchi–Fuzzy-based performance-exhaust emission trade-off study of variable compression ratio diesel engine fueled with undi-diesel blends. Environ. Progr. Sustain. Energy, e13333 (2019)
8. S. Bhowmik, R. Panua, S.K. Ghosh, A. Paul, D. Debroy, Prediction of performance and exhaust emissions of diesel engine fuelled with adulterated diesel: an artificial neural network assisted fuzzy-based topology optimization. Energy Environ **29**(8), 1413–1437 (2018)
9. S. Bhowmik, R. Panua, S.K. Ghosh, D. Debroy, A. Paul, A comparative study of artificial intelligence based models to predict performance and emission characteristics of a single cylinder diesel engine fueled with diesosenol. J. Thermal Sci. Eng. Appl. **10**(4), 041004 (2018)
10. H. Murat, H.M. Ertunc, M. Karabektas, G. Ergen, ANFIS modelling of the performance and emissions of a diesel engine using diesel fuel and biodiesel blends. Appl. Therm. Eng. **60**(1–2), 24–32 (2013)
11. Ö İlker, M. Ciniviz, F. Candan, Estimating engine performance and emission values using ANFIS/ANFIS Kullanılarak motor PerformansveEmisyonDeğerleriTahmini. Int. J. Autom. Eng. Technol. **4**(1), 63–67 (2015)
12. S. Bhowmik, R. Panua, D. Debroy, A. Paul, Artificial neural network prediction of diesel engine performance and emission fueled with diesel–kerosene–ethanol blends: a fuzzy-based optimization. J. Energy Res. Technol. **139**(4), 042201 (2017)
13. J. Monisha, A. Harish, R. Sushma, T.P. Krishna Murthy, B. Mathew Blessy, S. Ananda, Biodiesel: A review. Int. J. Eng. Res. Appl. 902–912 (2013)
14. S. Bhowmik, A. Paul, R. Panua, S.K. Ghosh, D. Debroy, Performance-exhaust emission prediction of diesosenol fueled diesel engine: an ANN coupled MORSM based optimization. Energy **153**, 212–222 (2018)
15. S. Bhowmik, A. Paul, R. Panua, S.K. Ghosh, D. Debroy, Artificial intelligence based gene expression programming (GEP) model prediction of diesel engine performances and exhaust emissions under diesosenol fuel strategies. Fuel **235**, 317–325 (2019)

# Development and Workspace Study of a 4-PP Planar Parallel XY Positioning Stage Using SMA Actuators

**Rutupurna Choudhury, Deep Singh, Anuj Kumar, Yogesh Singh, and Chinmaya Kumar Sahoo**

**Abstract** The necessity to fabricate micron-sized objects at present is increasing rapidly. Planar parallel manipulators, an area of robotics, is also employed to develop motion stages for various applications. The present study proposes a 4-PP planar positioning motion stage. The end-effector of the proposed positioning stage undergoes motion along both the axis in a plane and restricts any angular motion. The planar parallel positioning device (manipulator) possesses four active input prismatic joints which are actuated by implementing Shape Memory Alloy (SMA) based smart material springs. The SMA spring-based actuators are very lightweight and provide higher work per unit mass in comparison to other actuators. The proposed manipulator possesses two degrees of freedom. This study presents the workspace analysis of the proposed manipulator actuated by smart materials. The study depicts the experimental workspace efficiency of 42.87% for the proposed motion stage in comparison to the feasible workspace region. The planar parallel motion stage has the ability to displace in microns within the workspace domain. This study also shows the adequacy of SMA spring-based actuators in the development of micro-positioning devices.

Note: P represents a prismatic joint.

**Keywords** Positioning stage · Planar parallel manipulator · Shape memory alloy

R. Choudhury · D. Singh · A. Kumar · Y. Singh (✉) · C. K. Sahoo
Mechanical Engineering, National Institute of Technology Silchar, Silchar, Assam 788010, India
e-mail: yogeshsingh15@gmail.com

R. Choudhury
e-mail: rutupurna.choudhury10@gmail.com

D. Singh
e-mail: deepsingh.0318@gmail.com

A. Kumar
e-mail: anujpal096@gmail.com

C. K. Sahoo
e-mail: chinmaya.cks@gmail.com



K. M. Pandey et al. (eds.), *Recent Advances in Mechanical Engineering*, Lecture Notes in Mechanical Engineering, https://doi.org/10.1007/978-981-15-7711-6_3

# 1 Introduction

Automation is one of the important parts of any industry. Nowadays, necessity of automation is vital in every field like industry, medical, transportation, etc. Robotics is one of the keys to automation. It plays a vital role in automating manual activities. Planar parallel manipulators (PPM) is an important content to robotics. An *XY* positioning stage is a PPM device used for movement and positioning of object. The *XY* positioning mechanism is driven by actuators for its motion and possesses wide applications including material handling and processing, micro-machining, fabrication, etc. *XY* motion stage is also used for precise positioning of objects or components. The literature consists of various planar positioning devices that actuate on conventional motors which are bulky and requires higher current as driving force. Presently, smart materials like shape memory alloy (SMA), piezoelectric material, etc., have gained remarkable recognition due to their great potential and self-transformation properties [1]. Fesperman et al. [2] developed an *XY* stage driven by linear motor which floats on a thin film of air. The object was displaced using three heterodyne displacement sensors. The object has the ability to displace 10 mm along both *X*- and *Y*-axes. The overall size of the position stage mechanism was 477 $m^2$. Shimizu et al. [3] developed millimetre-order *XY* micro-motion stage to position lightweight objects precisely. This positioning stage is driven by friction to recognize long range travel motion with the help of small stroke piezoelectric actuators. Lee et al. [4] designed a novel micro-*XY* motion stage mechanism with the application of thermal actuators. This motion was carried out by the thermal expansion force of the SU-8 thermal actuators. This actuator has the maximum displacement of 41 μm at 250 °C. This material is more suitable for an actuator that needs large displacement for use in micro-*XY* stages. Basic design issues like static, dynamic and tracking control are investigated by Qin et al. [5]. It addresses the fundamental problems of establishing a decoupled *XY* stage motion mechanism. The motion is carried out by two piezoelectric actuators. It was observed that the statically indeterminate structures are able to decouple the cross-axis coupling with increased stiffness and stability. Yao et al. [6] studied the design and the performance of parallel micro-positioning *XY* motion stage mechanism. This is also driven by piezoelectric actuators. The kinematic and dynamic study showed that the stage possesses larger workspace, good linearity and high bandwidth as compared to the serial kinematic designs. Wang et al. [7] studied the design of the flexure-based *XY* precision stage and obtained large task space under reduced driving force. Choi et al. [8] developed an *XY* stage motion mechanism by considering cross-strip flexure joints and over-constrained mechanism. The motion stage is composed of flexural joints, links and is driven by a linear motor. For large rotation cross-strip, flexural joints are used. Horizontal straightness, repeatability test and yaw are studied in this work. From the literature, it was observed that SMA actuators can be used as the best substitute for conventional actuators like servomotors, hydraulics and pneumatics [9] because of its unique behaviour and mechanical properties. So, this leads to the development of advanced and cost-effective actuators with a considerable reduction in complexity, weight and size. Chang-Jun et al. [10]

developed a micro-wheeled robotic device using SMA actuator. Similarly, Kim et al. [11] developed an earthworm-based micro-robot using SMA spring as actuator.

### *1.1 Contribution in the Present Paper*

This paper presents the SMA spring actuation-based 4-PP XY micro-motion stage mechanism and addresses the development of conceptual design and kinematics of the manipulator. In addition, feasible workspace analysis is also presented. The proposed motion stage consists of two legs configured prismatic-prismatic along *X*-direction and the other two legs configured prismatic-prismatic along *Y*-direction. The motion stage consists of four active prismatic joints at each leg actuated by nitinol (NiTi) SMA springs. The 4-PP *XY* micro-motion stage was fabricated, and the experiment was performed to determine the associated workspace and also the feasibility of smart materials for the development of the motion stage. The manipulator is designed in such a way that the angular orientation of the end-effector is restricted.

## 2 Shape Memory Alloy Actuator

Smart materials have the potential to recover large strains due to transformation in phase from martensite (lower temperature) to austenite (higher temperature) as in case of SMAs [12]. SMA or nitinol spring possesses a special behaviour called shape memory effect (SME) [9, 12]. This exceptional characteristic is the prime reason behind the wide application of the SMA in different fields like the surgical tool, aeronautical application and micro-actuators. According to the microscopic studies, SMA rearranges its crystal structure during phase transformation under the application of heat. This is the mechanism of SMA to recover its original shape. The major advantage of SMA is that it could restore its original memorized shape under thermal loading [9]. The SMA implemented as an actuator in the present study is nitinol tension spring of 0.75 mm wire diameter. It can elongate and contract maximum up to 140 mm and 29 mm, respectively. The material has Young's modulus and density of 120 MPa and 6.4 E-06 kg/mm$^3$, respectively.

## 3 Development of Smart Actuation-Based XY Micro-motion Stage

This paper presents the SMA actuation based XY micro-motion stage. The paper studies the workspace associated with the XY motion stage by implementing smart

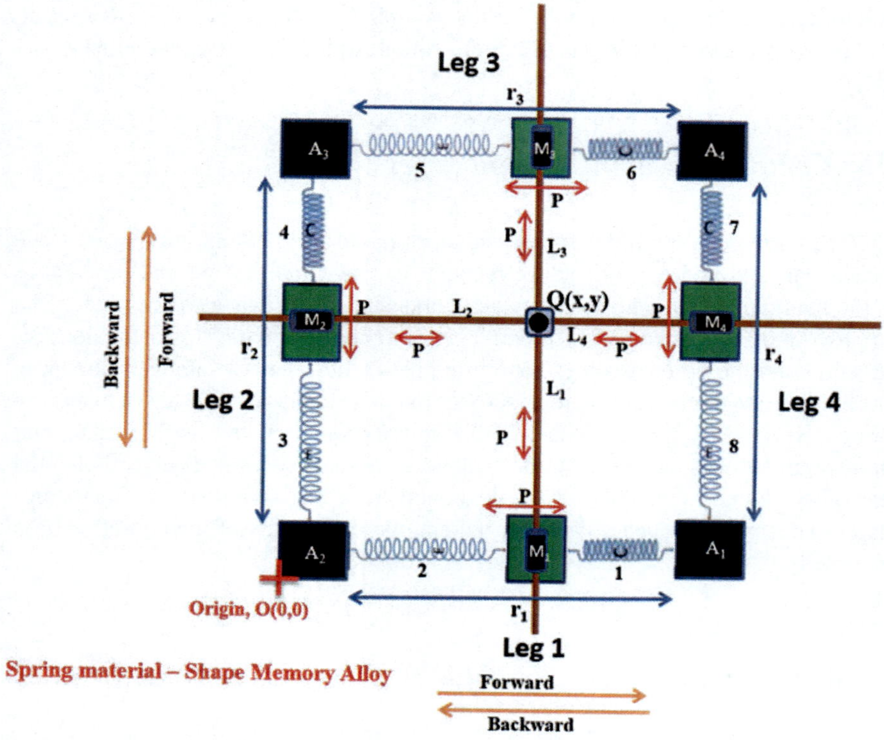


**Fig. 1** Schematic diagram of SMA spring actuation-based 4-PP *XY* motion stage mechanism

material, nitinol spring, as actuators for the active prismatic joints as depicted in Fig. 1.

Figure 1 illustrates the schematic diagram of shape memory alloy spring actuation-based *XY* robotic micro-motion stage. The *XY* motion stage consists of four legs of which each leg has prismatic-prismatic (PP) configuration. All the four legs of equal lengths together form a square configuration as depicted in Fig. 1. This device consists of eight numbers of nitinol SMA springs (1 – 8) as actuators for the active input prismatic joints ($M_1$, $M_2$, $M_3$ and $M_4$). Each leg consists of two springs at either side of the active prismatic joints which help to undergo bidirectional movement along its orientation axis. The contraction of the springs (1, 6) and (2, 5) results in forward and backward motion of the end-effector along *X*-axis, respectively. Similarly, the contraction of the springs (4, 7) and (3, 8) results in forward and backward motion of the end-effector along *Y*-axis, respectively. The blocks $A_1, A_2, A_3$ and $A_4$ are fixed to which one end of the SMA springs is connected forms the fixed base. The other ends of the SMA springs are connected to the active prismatic joints ($M_1$, $M_2$, $M_3$ and $M_4$). The links $L_1$ and $L_2$ form passive prismatic joints at $M_1, M_2, M_3$ and $M_4$ of each leg. The links $L_1$ and $L_2$ form a rigid joint at a point ($Q$) which is known as the end-effector of the motion platform. The $r_1$, $r_2$, $r_3$ and $r_4$ at each leg represent the stroke length of the active input translational joints which is directly proportional to the amount of contraction of either spring at the respective legs. The coordinates or the pose of the end-effector is represented as ($Q_x$, $Q_y$). The manipulator does not possess any revolute joints in order to avoid the rotation of the end-effector. Further, to avoid the end-effector rotation, the active prismatic joints ($r_1$ and $r_3$) and ($r_2$ and $r_4$) are actuated simultaneously and maintained the same value. '*C*' and '*E*' represented on the springs indicate its deformed nature as compressed and elongated, respectively.

### 3.1 *Kinematics of the Proposed* XY *Motion Stage*

The motion of the end-effector is dependent on the active input translational joints and vice versa. The position of the end-effector can be calculated based on the active translational joints ($r_1$, $r_2$, $r_3$ and $r_4$) based on the following equations:

$$x = r_1 = r_3,\ y = r_2 = r_4$$

## 4 Fabrication of 4-PP Planar Parallel Robotic Motion Stage

A four-legged 4-PP planar parallel robotic motion stage as depicted in Fig. 1 has been fabricated in-house to perform an experiment to determine the workspace. The experiment is performed to analyse the workspace associated with the manipulator. The fabricated model is depicted in Fig. 2.

The rigid manipulator's base represented by yellow area possesses fixed lengths of 220 mm and 220 mm along *X* and *Y*-axes, respectively. The fabricated manipulator is basically a wooden model comprised of NiTi SMA springs as actuators. The passive links used in the model is welding filler rod. The active input translational joints present at each leg move freely in the guide made of electrical wire casing. The available workspace for the end-effector motion is represented by the red box of 160 mm × 160 mm (25,600 mm$^2$) area.

## 5 Experimental Procedure

The fabricated prototype has been incorporated with nitinol (NiTi), an SMA, in order to actuate the active prismatic joints at each of the legs of the manipulator. The actuator, nitinol, is in the shape of spring with 19 helical windings of which maximum contraction length is ~29 mm under thermal loading. In order to actuate the SMA springs, an electrical current is supplied across it. Due to the resistance of SMA, its temperature rises, thereby leading to thermal loading. Due to a rise in

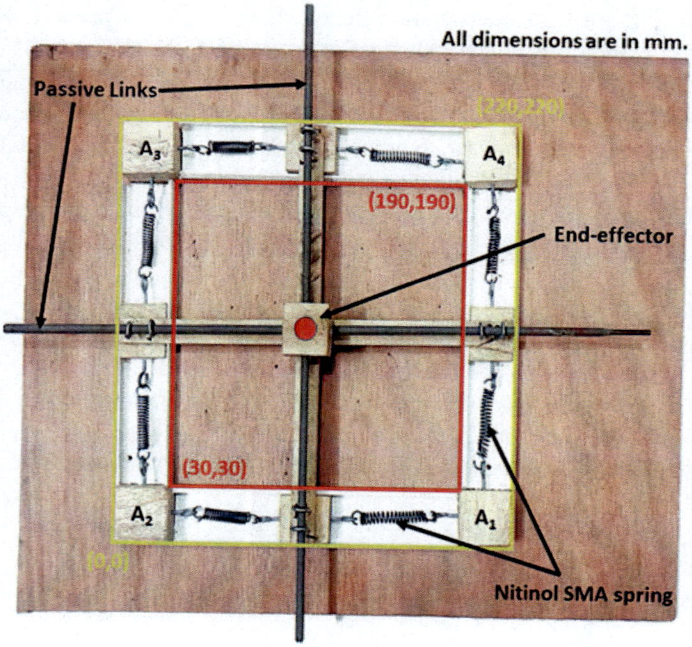


**Fig. 2** In-house fabricated model of 4-PP robotic XY motion stage

temperature, the SMA contracts and reaches its maximum contraction length under a certain range of current. The current supplied to the nitinol spring is DC which is supplied with the help of DC power supply (Model: Scientech 4180).

The experimental setup developed for the analysis of 4-PP SMA actuation-based micro-motion stage as depicted in Fig. 3 consists of various devices as specified in Table 1.

The electric current is supplied to only one out of two spring at each leg of the manipulators which leads to its contraction due to thermal load. This leads to the linear motion of the active prismatic joint at the legs of the manipulator. This motion is further transferred with the help of passive joints to the end-effector which reflects in the variation of its position. The new position so obtained is captured as an image in the digital camera (Nikon D5600). Further, AutoCAD is employed to interpret the position of the end-effector in *x* and *y* coordinates.

The four active translational or prismatic joints are termed as inputs to the motion stage while rest being passive. The passive joints are dependent on the motion of the active input translation joints. The linear motion at the active prismatic joints with the help of four passive prismatic joints leads to variation in pose (motion along *X*- and *Y*-axes) of the end-effector.

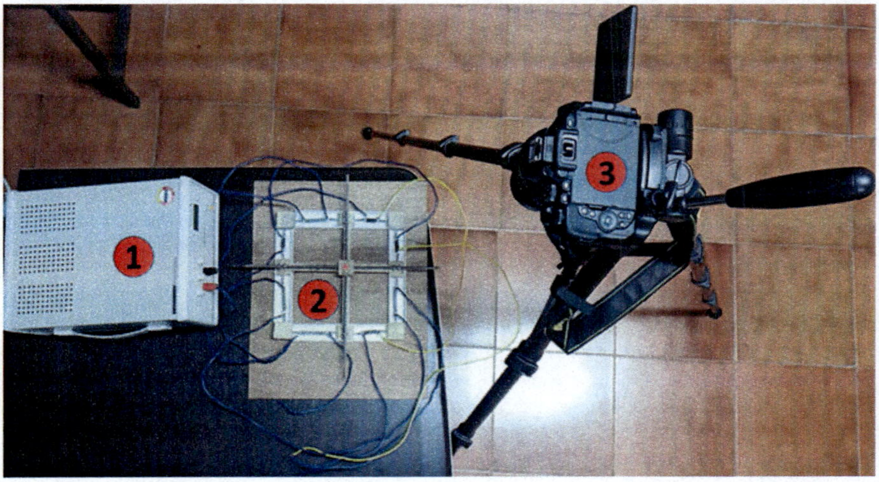


**Fig. 3** Experimental setup

**Table 1** List of devices for experimental setup

| S. No. | Devices | Specifications |
|---|---|---|
| 1 | DC power supply | Scientech 4180 (output: 0–30 V, 0–5 A) |
| 2 | Planar parallel manipulator | 4-prismatic-prismatic configuration |
| 3 | Digital camera with tripod | Nikon D5600 |

## 6 Results and Discussion

On application of thermal loading to the SMA springs, the SME occurs which results in a change in shape and size of the SMA to its original form. The spring contracts and its size reduces. The thermal loading is applied by supplying direct electrical current (3–5 A) to the spring which is also known as joule heating. The variation in size of the SMA spring varies the position of the active input translational joints which further affects the end-effector's pose with the help of passive prismatic joints. The forward movement of the active input translational joints in leg 1 and leg 3 together leads to variation in end-effector's pose toward positive $X$-axis, and its backward movement leads to the motion of the end-effector toward negative $X$-axis. Similarly, the forward movement of the active input translational joints in leg 2 and leg 4 together leads to variation in end-effector's pose along positive $Y$-axis, and its backward movement leads to the motion of the end-effector along negative $Y$-axis.

The pose of the end-effector is kinematically coupled to the active input translation joints ($r_1$, $r_2$, $r_3$ and $r_4$). Hence, the position of the end-effector is obtained based

**Table 2** Four sets of experimental data of the end-effector's pose of the 4-PP motion stage

| Direction of the actuation during heating of the SMA springs | Pose of the end-effector (mm) | | | | | | | |
|---|---|---|---|---|---|---|---|---|
| | Set-1 | | Set-2 | | Set-3 | | Set-4 | |
| | $Q_x$ | $Q_y$ | $Q_x$ | $Q_y$ | $Q_x$ | $Q_y$ | $Q_x$ | $Q_y$ |
| Initial position | 116.470 | 114.319 | 114.470 | 120.124 | 117.830 | 116.046 | 110.440 | 114.041 |
| $r_1, r_3$:$F$ | 121.390 | 116.781 | 121.510 | 122.217 | 124.120 | 115.661 | 108.900 | 115.085 |
| $r_1, r_3$:$B$ | 110.770 | 115.358 | 109.280 | 120.517 | 107.570 | 116.170 | 100.320 | 115.386 |
| $r_2, r_4$:$F$ | 114.610 | 124.716 | 113.590 | 126.167 | 115.490 | 122.773 | 101.880 | 120.193 |
| $r_2, r_4$:$B$ | 111.550 | 108.950 | 114.720 | 119.744 | 115.590 | 108.874 | 101.330 | 110.288 |
| $r_1, r_3$:$F$ and $r_2$, $r_4$:$F$ | 122.110 | 119.949 | 123.220 | 126.275 | 123.170 | 121.858 | 107.970 | 116.913 |
| $r_1, r_3$:$F$ and $r_2$, $r_4$:$B$ | 118.580 | 107.581 | 122.570 | 117.827 | 126.830 | 111.057 | 108.600 | 108.765 |
| $r_1, r_3$:$B$ and $r_2$, $r_4$:$F$ | 107.740 | 119.777 | 109.330 | 125.430 | 105.750 | 123.870 | 100.440 | 119.362 |
| $r_1, r_3$:$B$ and $r_2$, $r_4$:$B$ | 113.740 | 115.476 | 111.010 | 115.958 | 108.450 | 114.309 | 100.570 | 109.539 |

on various combinations of active prismatic joints and its direction of motion as specified in the first column of Table 2.

Initially, the image of the motion stage is captured and the pose of the end-effector was obtained. Then, the joints $r_1$ and $r_3$ were forward actuated which leads to the motion of the end-effector along positive $X$-axis as discussed previously. After full contraction of the SMA springs (1, 6), an image of the manipulator is captured and the pose of the end-effector was extracted. Furthermore, the backward motion of the same joints $r_1$ and $r_3$ were considered which led to the motion of the end-effector along negative $X$-axis. The pose of the end-effector was obtained for the maximum backward movement of the joints $r_1$ and $r_3$. Similarly, the pose of the end-effector was obtained for the forward and backward actuation of the active prismatic joints $r_2$ and $r_4$ which led to motion of the end-effector along $Y$-axis.

Till now, the motion of the end-effector was along a single axis. Now, the active prismatic joints $r_1$, $r_2$, $r_3$ and $r_4$ were actuated in the combination of forward and backward movements which led to the motion of the end-effector along both $X$ and $Y$-axes simultaneously. The image was captured, and the end-effector pose was extracted for each condition as specified in Table 2.

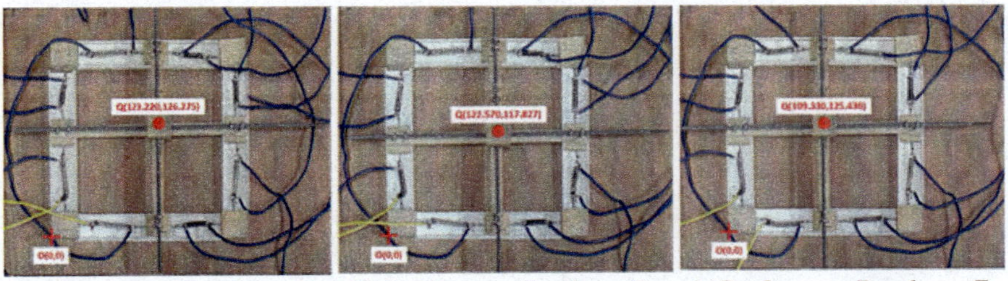

(a) Set-2 - $r_1$, $r_3$:F and $r_2$, $r_4$:F (b) Set-2 - $r_1$, $r_3$:F and $r_2$, $r_4$:B (c) Set-2 - $r_1$, $r_3$:B and $r_2$, $r_4$:F

**Fig. 4** End-effector pose of the proposed *XY* micro-motion stage

Four sets of experiments were conducted on the in-house fabricated 4-PP planar parallel micro-motion stage, and the obtained data are presented in Table 2 of which few experimental images are depicted in Fig. 4.

where *F* and *B* represent forward and backward actuation of the SMA springs, respectively.

The experiment showed gradual variation in the pose of the end-effector with gradual variation in the position of the active input translational joints. From the experimental data in Table 2, the minimum and maximum position of the end-effector along both *X* and *Y*-axes obtained are specified in Table 3. The range of deflection along both the axes denotes the stroke length of 26.51 mm and 18.694 mm along *X* and *Y*-axes, respectively. Hence, the total experimental workspace area is 495.578 $mm^2$ as depicted in Fig. 5.

Although the available workspace area is 160 mm × 160 mm, the presence of – (i) certain width of active prismatic joints, (ii) presence of hooks to which springs are connected and (iii) maximum possible contraction of spring is 29 mm – minimizes the workspace area. Hence, the workspace area in which the end-effector is feasible to undergo motion is 34 mm × 34 mm (i.e. 1156 $mm^2$) and is represented as feasible workspace in Fig. 5.

Figure 5 clearly depicts the variation in workspace regions of feasible workspace and experimental workspace which is an indication of the presence of certain inefficiency. The percentage of loss in the workspace is given as

$$\begin{aligned}\text{Percentage loss} &= (\text{Feasible Workspace} - \text{Experimental Workspace})\\ &\quad 100/\text{Feasible Workspace} = (1156 - 495.578)100/1156\\ &= 57.13\%\end{aligned}$$

**Table 3** Stroke length of the end-effector along the coordinate axes

| Direction | Deflection of the end-effector | | Stroke length (mm) |
|---|---|---|---|
| | Minimum (mm) | Maximum (mm) | |
| *X*-axis | 100.320 | 126.830 | 26.510 |
| *Y*-axis | 107.581 | 126.275 | 18.694 |

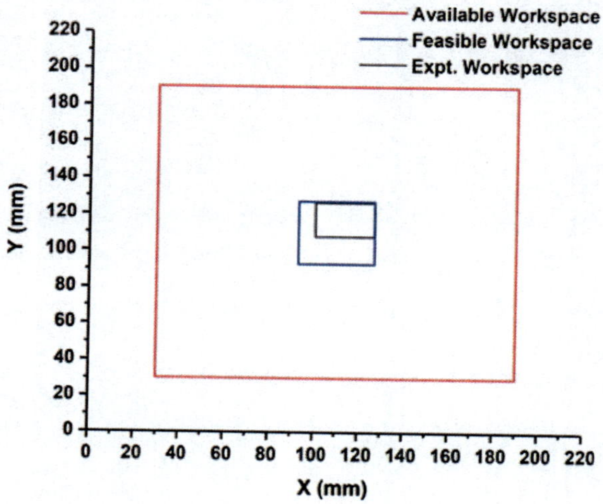


**Fig. 5** Workspace of 4-PP robotic micro-motion *XY* stage

Hence, the efficiency associated with the in-house fabricated model is 42.87%.

## 7 Conclusion

The major conclusions associated with this study have been outlined below:

1. A 4-PP planar parallel robotic micro-motion stage was fabricated.
2. The feasible and experimental workspace associated with the in-house fabricated model is 1156 $mm^2$ and 495.578 $mm^2$, respectively.
3. The experimental workspace of the in-house fabricated model showed an efficiency of 42.87%.

## References

1. K. Otsuka, C. Wayman, Shape memory material. Cambridge University Press, Cambridge. ISBN: 9780521663847 (1998)
2. F. Ronnie, O. Ozkan, H. Robert, R. Shalom, T. Tsu-Chin, P. James, L. Tiffany, B. John, C. Greg, Multi-scale alignment and positioning system–MAPS. Prec Eng **36**(4), 517–537 (2012)
3. Y. Shimizu, P. Yuxin, K. Junji, A. Toyohiro, I. So, G. Wei, L. Tien-Fu, Design and construction of the motion mechanism of an XY micro-stage for precision positioning. Sens. Actuators A **201**, 395–406 (2013)
4. J. Lee, D.W. Lee, Fabrication of a micro XY-stage using SU-8 thermal actuators. Microelectron. Eng. **86**(4), 1267–1270 (2009)
5. Y. Qin, B. Shirinzadeh, Y. Tian, D. Zhang, Design issues in a decoupled XY stage: static and dynamics modelling, hysteresis compensation, and tracking control. Sens. Actuators A **194**, 95–105 (2013)

6. Y. Qing, J. Dong, P.M. Ferreira, Design, analysis, fabrication and testing of a parallel-kinematic micropositioning XY stage. Int. J. Mach. Tools Manuf. **47**(6), 946–961 (2007)
7. P. Wang, Q. Xu, Design of a flexure-based constant-force XY precision positioning stage. Mech. Mach. Theor. **108**, 1–13 (2017)
8. Y.J. Choi, S.V. Sreenivasan, B.J. Choi, Kinematic design of large displacement precision XY positioning stage by using cross strip flexure joints and over-constrained mechanism. Mech. Mach. Theor. **43**(6), 724–737 (2008)
9. Y. Singh, S. Mohan, Development of a planar 3PRP parallel manipulator using shape memory alloy spring based actuators. In *Proceedings of the Advances in Robotics*, ACM, 10 (2010)
10. C. Jun, Qin, M.P. Sun, Y. Qin, A prototype micro-wheeled-robot using SMA actuator. Sens. Actuators A Phys. **113**(1), 94–99 (2004)
11. B. Kim, M.G. Lee, Y.P. Lee, Y. Kim, G. Lee, An earthworm-like micro robot using shape memory alloy actuator. Sens. Actuators A **125**(2), 429–437 (2006)
12. M.J. Jaronie, L. Martin, S. Aleksandar, A.G. Mark, A review of shape memory alloy research, applications and opportunities. Mater. Des. **56**, 1078–1113 (2014)

# Review of the Hybrid H-Savonius Rotor's Design and Performance

**D. Sarkar, A. R. Sengupta, P. Bhadra, S. Alam, and B. Debnath**

**Abstract** In recent times, the progress and activities on renewable energy sources are growing exponentially. Amongst the various kinds of renewable resources, wind energy is one of the most preferable due to its abundant availability, lesser cost, zero emission compared to other sources. Amongst the various kind of vertical axis wind turbines (VAWT), H-Darrieus rotor is more popular in the built environment for their simple constructions and higher power coefficient which also suffers from poor self-starting features. Again, the Savonius rotor is having the good self-starting ability but possess lesser power coefficient. To address all such limitations, existing investigations of hybrid H-Savonius rotor have been reported here in terms of the design, various parameters and aerodynamic performances which are used to improve their self-starting and efficiency. From this study, it is seen that the coaxial arrangement of the H-Savonius rotor is capable to exhibit higher efficiency and better self-starting characteristics than the staging assembly or the individual Savonius or H-rotor. Again, a newly designed hybrid H-Savonius rotor exhibits the maximum power coefficient of 0.414 at TSR2.5. Modification of the Savoniusblade and thicker H-rotor airfoil blade helps to increase the efficiency of the hybrid rotor. This present paper offers an overall idea on the research growth to improve the design and performance of the hybrid H-Darrieus rotor.

**Keywords** Hybrid H-Savonius rotor · Self-starting · Power coefficient · Torque coefficient · Computational fluid dynamics

## 1 Introduction

Energy is a key element for social and economic development and growth of a country. The world demands energy in the form of low cost. Since the beginning of the elevation, human society has been depending upon the conventional energy resources. The conventional fossils fuels are depleting so fast day by day, as it has limited resources.

D. Sarkar · A. R. Sengupta (✉) · P. Bhadra · S. Alam · B. Debnath
Department of Mechanical Engineering, JIS College of Engineering, Kalyani, West Bengal, India
e-mail: analsengupta88@gmail.com



K. M. Pandey et al. (eds.), *Recent Advances in Mechanical Engineering*, Lecture Notes in Mechanical Engineering, https://doi.org/10.1007/978-981-15-7711-6_4

Burning these fuels affects the environment critically by emitting greenhouse gases, which is the cause of global warming. The energy demand in the form of electricity is exponentially increasing very rapidly. Scientist and researchers are trying to meet the demands by using alternative renewable energy resources to power generation. This is why it motivates towards the development of wind energy in 1973 due to the high price of oil and limited fossils fuel resources [1]. Mankind has been using wind power for agriculture development, irrigation, to simplify mechanical work, navigation purposes, etc. In the past few decades, the researches and development activities in the field of renewable energy have been increased. Amongst various renewable energy sources, wind energy has got great attention as an alternative resource to coal, petroleum and nuclear energy which is more abundant, renewable, widely distributed, economically affordable, and most importantly, it has zero emission. To reduce the global energy-related carbon emission require a high growth rate of wind energy as a source of clean renewable energy.

### *1.1 Wind Turbine and Its Classification*

Amongst the renewable energy sources, wind energy is one of the options for green, well-economic energy generation. The wind turbine is a device which converts the wind energy for power generation. Depending upon the axis of rotation, wind turbines are generally classified into two categories: horizontal axis wind turbine (HAWT) and vertical axis wind turbine (VAWT). HAWTs are well known for their comparatively high efficiency over VAWTs, and it has been used for medium- to large-scale power generation. Again, VAWTs have advantages over the HAWTs such as lower installation and cost, compact design, easy to assemble, good self-starting ability, low cut-in speed, create less noise and omni-directional. VAWTs can be also classified into two types: Darrieus and Savonius turbines or rotor. H-Darrieus rotor is a variant of the curved-bladed Darrieus rotor, which has straight vertical blades [2]. In Fig. 1,

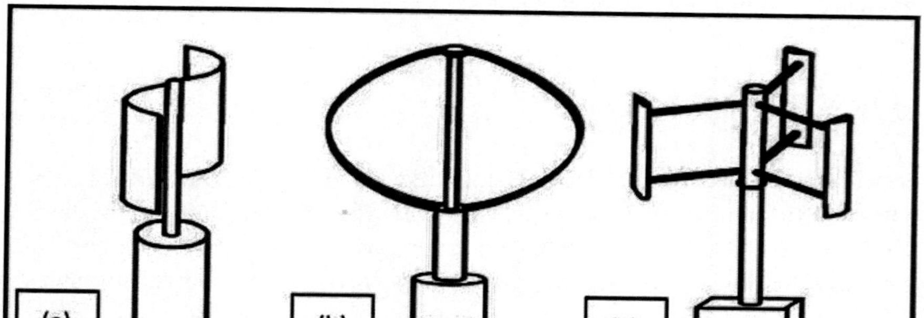


**Fig. 1** **a** Savonius rotor, **b** Darrieus rotor and **c** H-Darrieus rotor

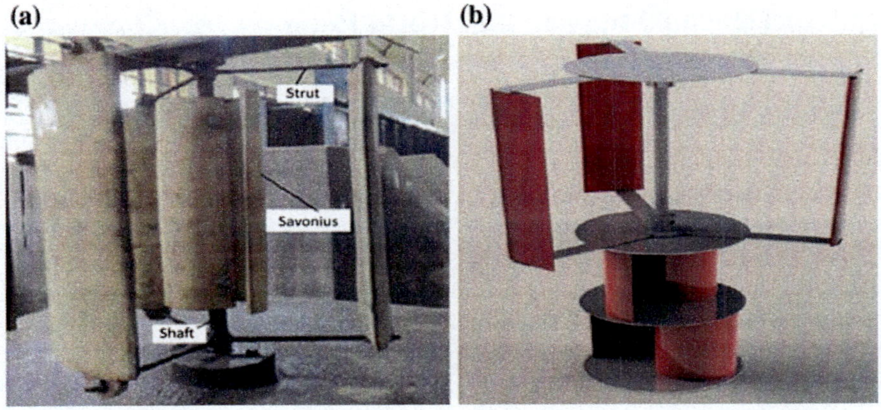


**Fig. 2** **a** Single-stage hybrid H-Savonius rotor and **b** multi-stage hybrid H-Savonius rotor

the classification of the wind turbines is reproduced from the available literature [3].

H-Darrieus rotors are lift force-based device. The energy is taken out from wind by the component of lift force working in the direction of rotation. Such turbines have the highest efficiency amongst the VAWTs, but it suffers from its poor starting torque. The Savonius rotors are drag forced-based device. The main advantages of such a rotor are its ability to self-staring in contrast to the other VAWTs. In Fig. 2, there are two diagram; one is single-stage hybrid H-Savonius rotor [4], and another is multi-stage hybrid H-Savonius rotor [5] which are presented in the classification of staging of hybrid rotor.

There is aspecial kind of hybrid rotor known as hybrid H-Savonius rotor, where both the lift and drag mechanisms are useful for the self-starting and performance improvement of such hybrid rotor [6]. The combined rotor is a combination of two different rotors (Darrieus and Savonius) which are mounted on the same shaft. Hybrid rotors are generally a combination of H-Darrieus and Savonius rotors. It is seen that a hybrid design of Darrieus with Savonius can make it fully self-starting, along with higher aerodynamic performances compared to any of the single rotor [7].

## 1.2 *Purpose of the Present Study*

Since the past few decades, the experimental and theoretical researches and applications on improving the design and overall performance of the VAWTs have been increased rapidly. Most of the works have been performed on either Darrieus or Savonius rotor, but only a few studies are there regarding the theoretical and experimental investigations of the hybrid H-Savonius rotor. In order to utilize the properties of both lift- and drag-type rotors, hybrid H-Savonius rotor came into the picture. The present review has covered most of the research works done on the design and performance improvement of hybrid H-Savonius rotor.

## 2 Experimental Study on the Hybrid Rotor

Compared to a Savonius rotor, one of the greatest advantages of the Darrieus rotor is its higher power coefficient (Cp). However, the Darrieus rotor suffers from its poor self-starting features. In order to use the advantages of both the rotors, the hybrid rotors came into existence to overcome the self-starting problem as well as to enhance their aerodynamic performances.

Bhuyan and Biswas [4] have experimentally investigated and found the results of three S818 unsymmetrical-bladed hybrid rotor having maximum Cp of 0.34 at tip speed ratio (TSR) 2.29 and Reynolds number of $1.92 \times 10^5$ for the optimum 0.15 overlap of the inner Savonius, followed by maximum Cp of 0.28 at TSR 2.42 and at same Reynolds number for the simple H-rotor. Mousavi et al. [8] studied experimentally the performance of a coaxial hybrid vertical axis wind turbine having DUW200 airfoils for the H-rotor. The performance has been compared with another hybrid turbine having a multi-staging arrangement and with simple H-rotor having same airfoils. The results showed that the coaxial system exhibits self-starting features and higher power coefficient compared to the others. Gawade and Patil [9] attempted an experimental study to measure the performance of individual Savonius and coaxial H-Savonius rotor. The maximum power coefficient for Savonius was obtained as 0.16, while for coaxial H-Savonius rotor, the maximum Cp of 0.39 was obtained. Here, the combined rotor showed the self-starting ability at wind speed of 3 m/s.

Abid et al. [10]concluded from their experimental study that a combination of NACA 0030 blade H-Savonius rotor showed the starting characteristics at low wind speed. Here, the three-bucket Savonius was mounted on top of the three-bladed H-rotor. The H-rotor blades with NACA 0030 airfoil has a higher thickness which resulted in an increase in the self-start capability of the rotor. Sahim et al. [11] studied on the hybrid H-Savonius rotor and showed how the radius ratio of the hybrid rotor affects the torque coefficient and power coefficient. It was observed that the higher radius ratio (RL = radius of Savonius/radius of Darrieus) resulted in lower power coefficient and higher torque coefficient. Siddiqui et al. [12] have experimentally analysed the performance of the five different arrangements of VAWT, viz. individual Savonius, Darrieus, two-stage Savonius in the middle of the H-rotor, two-stage Savonius on the top of the H-rotor and lastly Savonius below the H-rotor. It was found that the coaxial system showed the highest efficiency than the rest four. The highest power coefficient of about 0.41 was achieved at TSR of 1.6. Therefore, from the above experimental studies, it was found that coaxially placed H-Savonius rotor is the most efficient rotor system compared to the individual Savonius, individual Darrieus, H-Savonius rotor of multistage arrangements of such rotors.

## 3 Computational Fluid Dynamics (CFD) Analysis on the Hybrid Rotor

On this hybrid H-Savonius rotor, researchers have performed CFD simulations based on various design parameters and system arrangements to enhance their self-starting features and power performances. Some notable and important findings are briefly described below.

Hosseini and Goudarzi [13] have performed a CFD study on the design of an innovative hybrid H-Savonius rotor for obtaining an extended operative range and enhancing the self-starting capabilities. This hybrid rotor consisted of two buckets modified Savonius Bach-type rotor and three-bladed Darrieus rotor while the whole arrangement included setup of these two rotors in two stage. The result showed that the maximum value of Cp is 0.414 at TSR of 2.5. This design of the hybrid rotor being suitable for low and high wind speed also improved its self-starting ability and operating range. Gupta et al. [14] have studied the performance of a combination of three-bucket Savonius and three-bladed Darrieus rotor arrangement with the Savonius placed at the top over the Darrieus. It has been seen from this study that with the increase of overlap, the power coefficient started to decrease. The highest power coefficient obtained was 0.51 at TSR 0.61 without overlap, which is higher than the efficiency of the Savonius rotor at any overlap positions under the same test conditions. Letcher [15] has carried out an experiment in three different directions which are: CFD modelling, generator design and materials/manufacturing process. From the data, it was concluded that the power output of the combined setup is higher than the single Savonius and single Darrieus rotor. Some notable results of CFD analysis along with a comparison of various hybrid rotor systems have been discussed in Table 1.

In Fig. 3, Cpvs TSR graph of various designed hybrid rotors have been plotted. From this figure, it has been noticed that two-stage two buckets modified Savonius Bach-type rotor-combined three-bladed Darrieus rotor setup has the maximum Cp value of 0.414 at TSR 2.5.

Sharma et al. [16] also investigated on the hybrid three-bladed Darrieus–Savonius rotor, where Darrieus mounted on the top of the Savonius for overlap variation from

**Table 1** Notable results of CFD analysis along with a comparison of various hybrid rotor systems

| Turbine | Optimum TSR | Maximum Cp | Minimum TSR | Maximum TSR |
|---|---|---|---|---|
| Bhyuan et al. (with Savonius overlap 0) [6] | 2.29 | 0.19 | – | ~3 |
| Bhyuan et al. (with Savonius overlap 0.15) [6] | 2.29 | 0.34 | – | ~3 |
| Hosseini et al. [15] (hybrid rotor) | 2.5 | 0.414 | Self-starting | 4.5 |
| Sun et al. [19] | 1.8 | 0.33 | 1.25 | 2.25 |
| Liu et al. [20] | 2.24 | 0.41 | 0.4 | 2.58 |

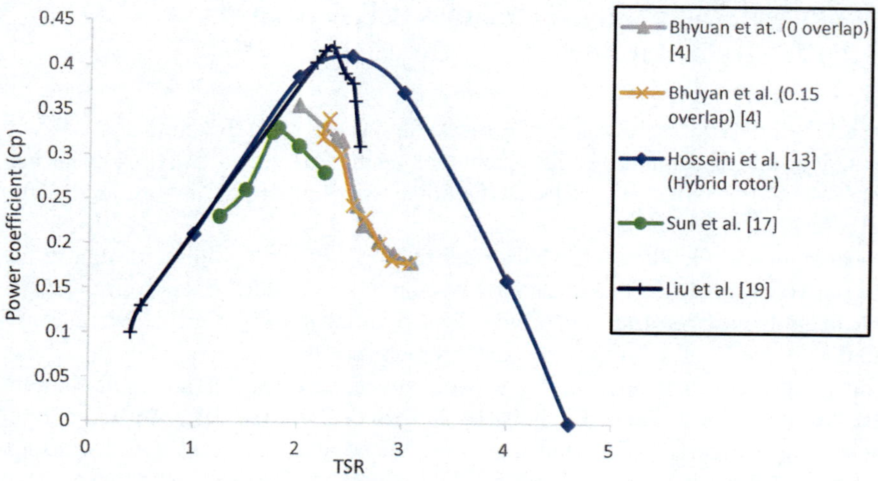


**Fig. 3** Cp versus TSR curve for various designs

10.8% to 25.8%. The maximum Cp of 0.53 was obtained at 0.604 TSR for an optimum overlap position of 16.8%. It was observed that Cp increases with the increase of overlap. However, there is an optimum value of overlap for which Cp is maximum, and beyond this, Cp started to decrease. The similar results are observed for Ct as well. Sun et al. [17] performed numerical analysis on the performance of the hybrid H-Savonius rotor with four different hybrid models; the numerical result indicated that the power coefficient (Cp) of the hybrid H-Savonius rotor is dropped down when the distance between its centre axis and the Savonius blades, whereas the starting torque can be significantly improved. The comparison of power coefficient amongst all the simulated turbines, the optimum TSR of four hybrid turbines that having maximum power coefficient is around 1.75. Amongst all four turbines, the Cp is highest of their hybrid turbine 1 (with zero distance in between the rotation axis and Savonius blade), the value of the highest obtained Cp is around 0.33 at TSR 1.80. Roshan et al. [18] investigated on the effects of non-dimensional parameters like initial overlap ratio ($\varepsilon$), arc angle ($\emptyset$) and curvature ($\alpha$) of the Savonius blades on the performance of the 18 hybrid H-Savonius models at 7 different TSRs. From the result of CFD simulation, it has been observed that model 12 having the maximum power coefficient (Cp) of 0.195 at TSR of 3 amongst all 18 models. Liu et al. [19]investigated that the larger modified Savonius (MS) rotor has a better self-starting capability. However, the power coefficient of hybrid Darrieus-modified Savonius (HDMS) VAWTs decreases when the size of MS rotor increased. An appropriate size of an MS rotor can maintain the power efficiency of HDMS compared to the Darrieus. The maximum power coefficient of the HDMS-1Xi VAWT is 0.41 which is 13% higher than the MS VAWT. Abdolahifar and Karimian [20] were done CFD analysis on the hybrid H-Savonius VAWT with the suitable wall and proper Savonius blade. In the comparison of the normal straight blade Darrieus rotor to the hybrid H-Savonius rotor, the hybridone produced 2.3% more average moment along with 40% less moment fluctuation at TSR 0.9.

## 4 Conclusions

To improve the overall efficiency and self-starting characteristics, hybrid H-Savonius rotor is being analysed by researchers which have shown the better result on low cut-in speed, wide operational range and especially in self-starting capability. From the present literature review, some important findings are listed below.

- Experimental study reveals that the coaxial arrangement of the H-Savonius rotor can exhibit higher efficiency and better self-starting features than the staging assembly or the individual Savonius or H-rotor.
- From the CFD result, the highest of power coefficient of 0.53 is obtained for an overlap position 16.8% of Savonius blade at TSR 0.604. The use of the modified Savonius rotor generated higher torque coefficient which helps to achieve self-starting characteristics of hybrid H-Savonius rotor.
- From the numerical analysis of an innovative new design of hybrid H-Savonius rotor, the maximum power coefficient of 0.414 at TSR 2.5 is achieved.
- The HDMS produced maximum Cp of 0.41 which is very closer to the new innovative-designed hybrid H-Savonius rotor, whereas the HDMS hybrid rotor produced 0.4% a lesser amount of power coefficient.

The present study gathered most of the investigations regarding the performances and designs of the hybrid H-Savonius rotor. It can act as a platform for future research in the field of Hybrid H-Savonius rotor.

## References

1. S. Roy, U.K. Saha, Review on the numerical investigations into the design and development of Savonius wind rotors. Renew. Sustain. Energy Rev. **24**, 73–83 (2013)
2. S. Eriksson, H. Bernhoff, M. Leijon, Evaluation of different turbine concepts for wind power. Renew. Sustain. Energy Rev. **12**(5), 1419–1434 (2008)
3. https://www.researchgate.net/profile/Eqwan_Roslan/publication/319242300/figure/fig2/AS:530482716319744@1503488343744/Types-of-VAWT-Savonious-Darrieus-and-Hrotor-types10.png (Accessed on 26/10/2019)
4. S. Bhuyan, A. Biswas, Investigations on self-starting and performance characteristics of simple H and hybrid H-Savonius vertical axis wind rotors. Int. J. Energy Convers. Manage. **87**, 859–867 (2014)
5. P. Rathod, K. Khatik, K. Shah, H. Desai, J. Shah, A review on combined vertical axis wind turbine. Int. J. Innov. Res. Sci. Eng. Technol. **5**(4) (2016)
6. N.H. Mahmoud, A.A El-Haroun, E. Wahba, M.H. Nasef, An experimental study on improvement of Savonius rotor performance. Alexandria Eng. J. **51**, 19–25. ISO 3297 (2007)
7. A. Kumar, A. Nikhade, Hybrid kinetic turbine rotor. Int. J. Eng. Sci. Adv. Technol. **4**(6), 453–463. ISSN: 2250-3376

8. S.M. Rassoulinejad-Mousavi, M. Jamil, M. Layeghi, Experimental study of a combined three bucket H-rotor with Savonius Wind Turbine. World Appl. Sci. J. **28**(2), 205–211 (2013)
9. S.G. Gawade, D.S. Patil, Comparitive study of a single stage Savonius with a combined Savonius-three bladed Darrieus. Int. J. Technol. Res. Eng. **2**(6). ISSN (Online): 2347–4718 (2015)
10. M. Abid, K.S. Karimov, H.A. Wajid, F. Farooq, H. Ahmed, O.H. Khan, Design, development and testing of a combined Savonius and Darrieus vertical axis wind turbine. Iranica J. Energy Environ. **6**(1), 1–4 (2015)
11. K. Sahim, D. Santoso, D. Puspitasari, Investigations on the effect of radius rotor in combined Darrieus-Savonius Wind Turbine. Int. J. Rotating Mach. 1–7 (2018)
12. A. Siddiquia, A.H. Memonb, S.N. Miana, R.K. Hatoona, M. Kamran, H. Shaikh, Experimental investigations of hybrid vertical axis wind turbine, in *4th International Conference on Energy, Environment and Sustainable Development 2016 (EESD 2016)*. Mehran University, Jamshoro, Sindh, Pakistan, Jan 2016
13. A. Hosseini, N. Goudarzi, Design and CFD study of a hybrid vertical-axis wind turbine by employing a combined Bach-type and H-Darrieus rotor systems. Energy Convers. Manage. **189**, 49–59 (2019)
14. R. Gupta, A. Biswas, K.K. Sharma, Comparative study of a three-bucket Savonius rotor with a combined three-bucket Savonius-three-bladed Darrieus rotor. Renew. Energy **33**(9), 1974–1981 (2008)
15. T. Letcher, The Ohio State University, Columbus, OH. Small Scale Wind Turbines Optimized for Low Wind Speeds. [Online] Available from: https://pdfs.semanticscholar.org/b7ff/253ec0da4f4d18e7a7cd3bdc5d77c19919ad.pdf?_ga=2.31398079.121060044.1577950283-1952833286.1573065156 (Accessed on 26/10/2019)
16. K.K. Sharma, A. Biswas, R. Gupta, Performance Measurement of a three-bladed combined Darrieus-Savonius rotor. Int. J. Renew. Energy Res. **3**(4), 13 (2013)
17. X. Sun, Y. Chen, Y. Cao, G. Wu, Z. Zheng, D. Huang, Research on the aerodynamic characteristics of a lift drag hybrid vertical axis wind turbine. Adv. Mech. Eng. **8**(1), 1–11 (2016)
18. A. Roshan, A. Sagharichi, M.J. Maghrebi, Nondimensional parameters' effects on hybrid Darrieus–Savonius wind turbine performance. J. Energy Resour. Technol. **142**(1): 011202, 12 (2019)
19. K. Liu, M. Yu, W. Zhu, Enhancing wind energy harvesting performance of vertical axis wind turbines with a new hybrid design: a fluid-structure interaction study. Renew. Energy**140**, 912–927 (2019)
20. A. Abdolahifar, S. Karimian, Aerodynamic performance improvement of hybrid Darrieus-Savonius vertical axis wind turbine. Amirkabir J. Sci. Technol. (2019)

# Energy Storage Using Sensible Heat Storage Media: Thermal and Economic Considerations

**Laxman Mishra, Abhijit Sinha, Prasanta Majumder, and Rajat Gupta**

**Abstract** Storage of energy is an important technology to bridge the time and space gap between the source/supply and sink/utilization of energy. Thermal energy storage has emerged as a means to capture heat from both low- and high-temperature sources. Storage of waste heat and solar thermal energy is easier and cheaper with the application of sensible heat storage materials. However, the knowledge of thermal and physical properties of sensible heat storage materials is important for economical and effective heat storage. Therefore, this paper presents the thermal and economic aspects of liquid and solid-state sensible heat storage materials. Thermal aspects are important for designing of the energy storage systems, while economic considerations are important in material selection and payback calculations. From the thermo-economic studies, it is found that water and rocks have great potential as liquid and solid sensible heat storage materials, respectively, primarily due to their low cost. Water also has impressive thermal properties which makes its storage density higher as compared to other liquids. Also, cast iron and steel present good potential as heat storage materials due to their high thermal capacity.

**Keywords** Cost · Sensible energy storage · Thermal capacity · Thermal conductivity

## 1 Introduction

With rapid development and improvement in standard of living, demand for energy has been increasing at an accelerated rate. Increased use of fossil-based energy sources has led to environmental concerns in the international society [1]. Also, the demand for energy varies depending on time of the day and month of the year. For example, demand of energy is higher during evenings as compared to mid-day. Thermal energy is in higher demand during winter months than those of summer.

L. Mishra · A. Sinha (✉) · P. Majumder · R. Gupta
Department of Mechanical Engineering, National Institute of Technology Mizoram, Aizawl, Mizoram 796012, India
e-mail: abhinit05@gmail.com



K. M. Pandey et al. (eds.), *Recent Advances in Mechanical Engineering*, Lecture Notes in Mechanical Engineering, https://doi.org/10.1007/978-981-15-7711-6_5

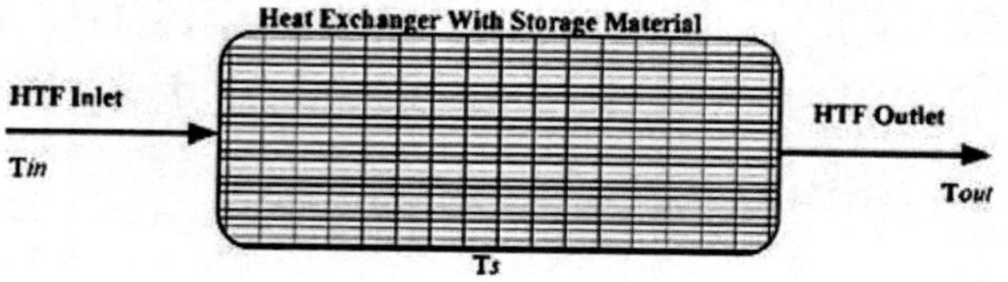


**Fig. 1** Sensible heat storage system

Similarly, sources of energy like solar and wind are intermittent in nature. The existence of discrepancy in demand and supply makes it important to store energy when it is in surplus so that it may be utilized during instances of high demand [2]. Storage of energy is also important as it can be collected from place of availability, stored, and supplied where and when it is required [3].

Various means and media have been used for storage of different forms of energy. Thermal energy is one of the important forms of energy essential for both household and industrial applications. It can be stored in three forms, viz. sensible heat, latent heat, and thermo-chemical energy [4]. Schematic of sensible heat storage system is shown in Fig. 1. The system is said to be charging when the material absorbs heat from the heat transfer fluid (HTF). This happens when the inlet temperature ($T_{\text{in}}$) of HTF is greater than storage temperature ($T_{\text{s}}$) of the storage material. Discharging occurs when $T_{\text{in}}$ is lesser than $T_{\text{s}}$. In this condition, the HTF gains heat from the storage material and its outlet temperature ($T_{\text{out}}$) is greater than $T_{\text{in}}$.

Each method of energy storage holds some basic advantage over others and is also associated with some drawbacks. Storing energy as sensible heat or latent heat is simple and relatively cheaper [5]; however, it cannot be stored for longer periods in these forms [1]. It has to be used within certain period of time after storage since it is lost to the ambient once the source of energy supply has been removed. Hence, they are mostly used for heating applications during intermittent cloudy hours or during the night hours of the same day. However, thermo-chemical storage overcomes this drawback. In such systems, thermal energy is stored in the form of chemical reactions, usually by removing the water of hydration from hydrated salts. The stored energy can be stored for months and recovered by mixing the separated chemical species back to get the original product along with release of stored heat. In sensible and latent mode of heat storage, there is loss of heat during storage. However, in thermo-chemical storage, since the salt and water are stored separately there is no loss of heat [6]. Also, the storage capacity or storage density of thermo-chemical storage is much higher than sensible and latent heat storage. However, this technology is not well developed yet which limits its practical application.

Effective methods of heat storage are also important for efficient use of solar energy which is a free source of energy [7, 8]. Application of sensible heat storage media enhances the absorption of solar flux and thus improves the daily output of a solar energy-based system [9]. Storage of thermal energy is necessary for an uninterrupted supply of power from solar thermal plants [10]. The heat storage materials are also useful for recovery of waste heat from thermal systems. Hence, this paper presents the thermal and economic aspects of sensible heat storage materials. The thermal properties of these materials are important for design of the energy storage system. While the economic aspects are important considerations that affect the choice of materials for energy storage.

## 2 Methodology

The thermal capacity has been obtained as the product of the density of the material and its specific heat capacity as shown in Eq. (1).

$$\text{Thermal Capacity} = \text{Specific Heat Capacity} \times \text{Density} \tag{1}$$

The thermal conductivities of the materials have been obtained from the literature. It is also important for faster and uniform distribution of heat within the bulk of the material which enhances its overall performance as energy storage material. The cost price of the materials has been obtained from online store [11]. The minimum price for a reasonable purity of the materials has been reported. The price is expressed as US dollars per ton of material.

## 3 Results and Discussion

Thermal and physical properties of materials are of important consideration while selecting a material for sensible heat storage. The amount of heat stored depends on thermo-physical properties of the material, viz. heat capacity and density. The rate of storage and retrieval depends on the thermal conductivity of the materials. Sensible heat storage materials have been divided into liquid materials and solids for the sake of convenience.

### *3.1 Liquid Sensible Heat Storage Materials*

Liquids like water, thermal oil, etc., have been widely used as thermal storage materials. A list of common liquid sensible heat storage materials and their thermo-physical properties are shown in Table 1. Water is abundantly available and is free natural resource. It also has high heat capacity which makes it popular among several applications.

Figure 2 shows the thermal capacity of some common liquid materials and liquid mixtures. Water has the highest thermal capacity of 4186 kJ/m$^3$K followed by water–ethylene glycol mixture in 50–50 fraction. Although engine oil has relatively lower thermal capacity of 1669 kJ/m$^3$K, it has been used by several researchers due to its stability in the temperature interval of its application and relatively lower price.

**Table 1** Thermo-physical properties of some common liquid sensible heat storage materials

| Material | Density (kg/m$^3$) | $C_p$ | References |
|---|---|---|---|
| Water | 1000 | 4.186 | [12] |
| Water–ethylene glycol (50:50) | 1050 | 3.47 | [13, 14] |
| Draw salt | 1733 | 1.55 | [13, 14] |
| Ethylene glycol | 1116 | 2.382 | [13, 14] |
| Molten salts | 500–2600 | 1.5 | [12] |
| Isobutanol | 808 | 3 | [13, 14] |
| Lithium | 510 | 4.19 | [13, 14] |
| Mineral oil | 800 | 2.6 | [12] |
| Propanol | 800 | 2.5 | [13, 14] |
| Synthetic oil | 900 | 2.1–2.3 | [12] |
| Butanol | 809 | 2.4 | [13, 14] |
| Dowtherms | 867 | 2.2 | [13, 14] |
| Ethanol | 790 | 2.4 | [13, 14] |
| Isopentanol | 831 | 2.2 | [13, 14] |
| Octane | 704 | 2.4 | [13, 14] |
| Engine oil | 888 | 1.88 | [13, 14] |
| Therminol 66 | 750 | 2.1 | [13, 14] |
| Sodium | 960 | 1.3 | [12] |

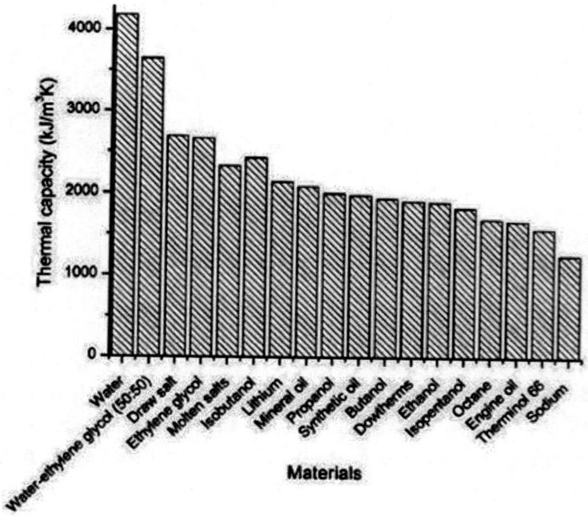


**Fig. 2** Thermal capacities of some common liquid sensible heat storage materials

## 3.2 Solid Sensible Heat Storage Materials

Several solids have been adopted for sensible heat storage. One major advantage of using solids is that they do not need any specialized mechanism for heat exchange between the material and charging/discharging medium (heat transfer fluid). Liquids, on the other hand, need better sealed heat exchangers to avoid leakage. Also, most solids are stable and do not undergo any major changes under repeated cycles of charging and discharging. Use of solids as sensible heat storage materials has several advantages including lower cost, wider range of temperature application, absence of leakage and corrosion, no requirement for freeze protection, and ability to use locally available materials for energy storage [15]. Solids are also easy to handle, and there is no loss of material with time (other than corrosion which is usually slow). Figure 3 shows thermal conductivity and thermal capacity of some solid sensible heat storage materials.

Table 2 shows the thermal capacity and thermo-physical properties of some solid sensible heat storage materials. These solid-state sensible heat storage materials have a vast range of thermal capacities ranging from cast iron (4980 kJ/m$^3$K) to sandstone (2000 kJ/m$^3$K). Materials below 2000 kJ/m$^3$K are usually not preferred as heat storage materials. They are used in some cases due to their lower price or some other specific requirements. Compounds of iron, viz. cast iron, steel, and magnetite, have highest thermal capacities. These are followed by gravely soil and wet earth

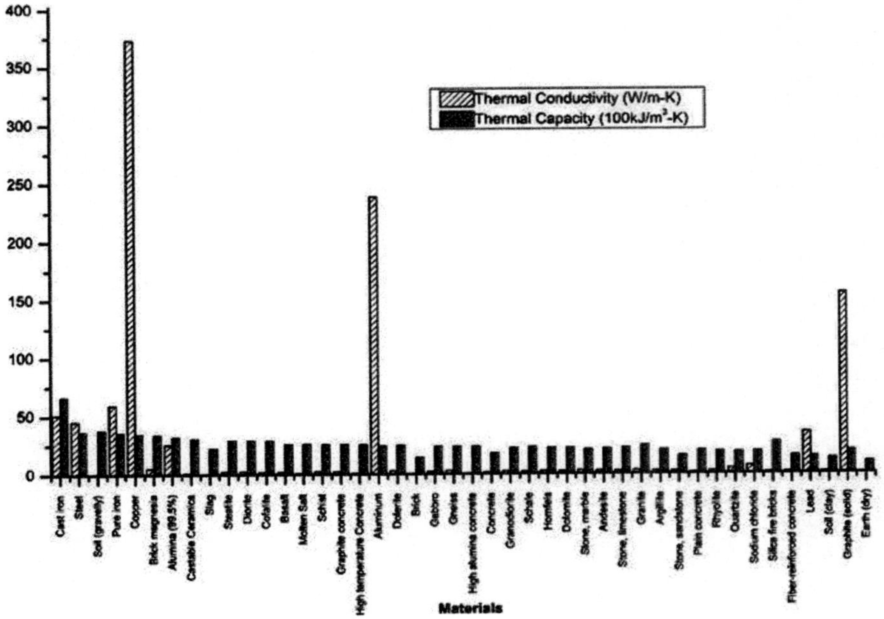


**Fig. 3** Comparison of thermal conductivity and thermal capacity of solid sensible heat storage materials

**Table 2** Thermal capacity and thermo-physical properties of solid sensible heat storage materials

| Material | Density ($kg/m^3$) | Thermal conductivity (W/m K) | Specific heat (kJ/kg K ) | Thermal capacity (× 100 $kJ/m^3$ K) | References |
|---|---|---|---|---|---|
| Cast iron | 7900 | 51.15 | 0.837 | 66.12 | [10, 12–14, 16, 17] |
| Steel | 7840 | 45 | 0.465 | 36.46 | [12–14, 17] |
| Soil (gravelly) | 2040 | 0.5 | 1.84 | 37.54 | [12–14, 17] |
| Pure iron | 7897 | 59.3 | 0.452 | 35.69 | [13, 14, 17] |
| Copper | 8954 | 373 | 0.383 | 34.29 | [12–14, 17] |
| Brick magnesia | 3000 | 5 | 1.13 | 33.90 | [12–14, 17] |
| Alumina (99.5%) | 3960 | 25.5 | 0.8 | 31.68 | [10, 12] |
| Castable Ceramics | 3500 | 1.35 | 0.866 | 30.31 | [12, 18–21] |
| Slag | 2700 | 0.6 | 0.836 | 22.57 | [10, 12] |
| Steatite | 2850 | 2.5 | 1.024 | 29.18 | [12] |
| Diorite | 2900 | 2.5 | 1 | 29.00 | [12] |
| Cofalite | 3120 | 1.75 | 0.917 | 28.61 | [10, 12, 18, 19] |
| Basalt | 2650 | 1.75 | 0.965 | 25.57 | [12] |
| Molten Salt | 1750 | 1.075 | 1.5 | 26.25 | [18, 19] |
| Schist | 2700 | 2.55 | 0.945 | 25.52 | [12] |
| Graphite concrete | 2680 | 2.43 | 0.95 | 25.46 | [2, 21] |
| High temperature concrete | 2750 | 1 | 0.916 | 25.19 | [12, 14, 19, 20] |
| Aluminum | 2707 | 238.4 | 0.896 | 24.25 | [12–14, 17] |
| Dolerite | 2800 | 2.6 | 0.885 | 24.78 | [12] |
| Brick | 1698 | 0.5 | 0.84 | 14.26 | [12–14, 17] |
| Gabbro | 2950 | 2.05 | 0.8 | 23.60 | [12] |
| Gneiss | 2700 | 2.9 | 0.8745 | 23.61 | [12] |
| High alumina concrete | 2400 | 0.2 | 0.98 | 23.52 | [10, 12] |
| Concrete | 2000 | 1.279 | 0.88 | 17.60 | [12–14, 17] |
| Granodiorite | 2700 | 2.35 | 0.835 | 22.55 | [12] |
| Schale | 2800 | 1.6 | 0.82 | 22.96 | [12] |
| Hornfels | 2700 | 2.25 | 0.82 | 22.14 | [12] |
| Dolomite | 2800 | 2.1 | 0.802 | 22.46 | [12] |
| Stone, marble | 2600 | 2.75 | 0.8 | 20.80 | [12–14, 17] |
| Andesite | 2650 | 2.55 | 0.815 | 21.60 | [12] |
| Stone, limestone | 2500 | 2.2 | 0.9 | 22.50 | [12–14, 17] |
| Granite | 2750 | 2.9 | 0.892 | 24.53 | [12, 13, 16, 17] |

(continued)

**Table 2** (continued)

| Material | Density ($kg/m^3$) | Thermal conductivity (W/m K) | Specific heat (kJ/kg K ) | Thermal capacity (× 100 $kJ/m^3$ K) | References |
|---|---|---|---|---|---|
| Argillite | 2450 | 2.2 | 0.838 | 20.53 | [12] |
| Stone, sandstone | 2200 | 1.8 | 0.71 | 15.62 | [12–14, 17] |
| Plain concrete | 2451 | 1.02 | 0.81 | 19.85 | [2, 14] |
| Rhyolite | 2450 | 1.95 | 0.785 | 19.23 | [12] |
| Quartzite | 2550 | 4.3 | 0.7265 | 18.53 | [12] |
| Sodium chloride | 2200 | 6.75 | 0.855 | 18.81 | [12] |
| Silica fire bricks | 1800 | 1 | 1.5 | 27.00 | [12] |
| Fiber-reinforced concrete | 2440 | 1.16 | 0.63 | 15.37 | [2, 14] |
| Lead | 11,340 | 35.25 | 0.131 | 14.86 | [12–14, 17] |
| Soil (clay) | 1450 | 1.3 | 0.88 | 12.76 | [12–14, 17] |
| Graphite (solid) | 2200 | 155 | 0.879 | 19.34 | [10, 12] |

with thermal capacities of 3772 $kJ/m^3K$ and 3680 $kJ/m^3K$, respectively. Both iron compounds and soil are very stable materials which can be used for several cycles of heating and cooling without causing major changes in their thermal properties. This contributes to their wide acceptance as sensible heat storage materials.

### *3.3 Thermal Conductivity*

Thermal conductivity of energy storage materials is an important factor while making selection of materials. This is because the rate of storage and retrieval of heat to and from the material depends largely on its thermal conductivity. A material with higher thermal conductivity is more effective in storing energy and is preferred over other materials with lower conductivities. Thermal capacity, on the other hand, determines how much of heat energy can be stored by a given material. From Fig. 3, it can be seen that copper has the highest thermal conductivity and also thermal capacity which is desired from a sensible heat storage material. While pure iron, steel, and cast iron have lower thermal conductivities and appreciably high thermal capacities. Aluminum and graphite have high thermal conductivities and good thermal capacities. Rocks and pebbles have average thermal capacity and low thermal conductivity. However, they are widely used due to their economic aspects, easy availability, and thermal stability. Copper, aluminum, and graphite are less popular in spite of their good thermal properties. This is because of the economic considerations which is discussed in the following section.

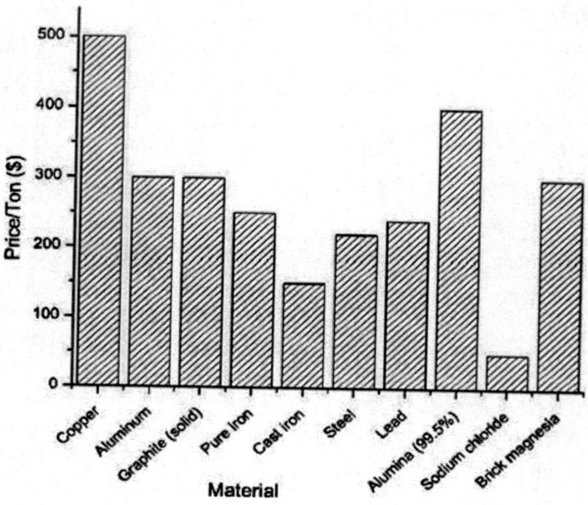


**Fig. 4** Cost of some common solid sensible heat storage materials

## 3.4 *Economic Aspects*

Cost of the energy storage material is one of the most important factors which often dominates over thermo-physical properties. Materials with lower cost are preferred over others since it leads to lower initial investment and return on investment is faster. Some materials may not be used for energy storage in spite of having high thermal capacity and thermal conductivity because of its high cost. Figure 4 shows the cost price of some common sensible heat storage materials in US dollars per ton. It can be observed from the figure that copper has the highest price which makes it less popular as energy storage materials even though it has high thermal conductivity and thermal capacity as seen from Fig. 3. On the other hand, iron derivatives like steel and cast iron have been in wide use due to their low cost and high thermal capacity.

# 4 Conclusion

The thermal capacities of some solid and liquid sensible heat storage materials have been presented. Thermal conductivities and cost of some common solid materials have also been discussed. Based on thermo-physical properties, thermal conductivity, and cost of materials, it is found that certain materials having average thermal properties are more popularly used than others with better thermal properties due to their low cost and availability. Among liquids, water is widely used due to its high thermal capacity of 41.9 $MJ/m^3K$ and low cost. Among solids, rocks and pebbles find application in several systems due to their low cost and easy availability. Iron derivatives like steel (36.4 $MJ/m^3K$) and cast iron (66.12 $MJ/m^3K$) have good thermal capacities, average thermal conductivities of 45 W/m K and 51.15 W/m K, respectively, and relatively low price. Hence, they are more popular than copper or aluminum which have excellent thermal properties but are costly.

## References


1. M. Eslamiand M.A. Bahrami, Sensible and latent thermal energy storage with construct alfins. Int. J. Hydro. Energy **42**(28), 17681–17691 (2017)
2. L. Mishra, A. Sinha, R. Gupta (2020) *Thermo-Economic Study of Phase Change Materials (PCMs) for Thermal Energy Storage*, ed by B. Biswal, B. Sarkar, P. Mahanta, Advances in Mechanical Engineering Lecture Notes in Mechanical Engineering (Springer, Singapore, 2020), pp. 1217–1226
3. J. Schroder, Thermal energy storage and control. J. Eng. Ind. **74**, 893–896 (1975)
4. L. Mishra, A. Sinha, R. Gupta, Recent developments in latent heat energy storage systems using phase change materials (PCMs)—a review, in *Springer Transactions in Civil and Environmental Engineering*. ed. by H. Drück, R. Pillai, M. Tharian, A. Majeed (Springer, Singapore, 2019), pp. 25–37
5. L. El-kaddadi, M. Asbik, N. Zari, B. Zeghmati, Experimental study of the sensible heat storage in the water/$TiO_2$ nanofluid enclosed in an annular space. Appl. Therm. Eng. **122**, 673–684 (2017)
6. F. Trausel, A.J. De Jong, R. Cuypers, A review on the properties of salt hydrates for thermochemical storage. Energy Procedia **48**, 447–452 (2014)
7. A. Abhat, Low temperature latent heat thermal energy storage: Heat storage materials. Sol. Energy **30**(4), 313–332 (1983)
8. D.V.N. Lakshmi, A. Layek, P.M. Kumar, Performance analysis of trapezoidal corrugated solar air heater with sensible heat storage material. Energy Procedia **109**, 463–470 (2017)
9. P. Patel, R. Kumar, Comparative performance evaluation of modified passive solar still using sensible heat storage material and increased frontal height. Procedia Technol. **23**, 431–438 (2016)
10. S. Khare, M.D. Amico, C. Knight, S. Mcgarry, Solar energy materials and solar cells selection of materials for high temperature sensible energy storage. Sol. Energy Mater. Sol. Cells **115**, 114–122 (2013)
11. Alibaba. [Online], Available from: https://www.alibaba.com, last accessed 2018/01/03
12. R. Velraj, Sensible heat storage for solar heating and cooling systems, eds. by R.Z. Wang, T.S. Ge, *Advances in Solar Heating and Cooling* (Woodhead Publishing, UK, 2016), pp. 399–428
13. D.L. Perry, *Handbook of Inorganic Compounds* (CRC Press, Taylor & Francis Group, Boca Raton, FL, 2011), pp. 33487–2742
14. R. Tiskatine, R. Oaddi, R. A. El. Cadi, A. Bazgaou, L. Bouirden, A. Aharoune, A. Ihlal, Solar energy materials and solar cells suitability and characteristics of rocks for sensible heat storage in CSP plants. Solar Energy Mater. Solar Cells **169**, 245–257 (2017)
15. C. Odenthal, W.D. Steinmann, M. Eck, The cellflux concept as an alternative solution for sensible heat storage. Energy Procedia **69**, 957–967 (2015)
16. L. Nkhonjera, T. Bello-Ochende, C.K. King, A review of thermal energy storage designs, heat storage materials and cooking performance of solar cookers with heat storage. Renew. Sustain. Energy Rev. **75**, 157–167 (2016)
17. A. Dinker, M. Agarwal, G.D. Agarwal, Heat storage materials, geometry and applications: a review. J. Energy Inst. **90**(1), 1–11 (2015)
18. X. Py, N. Calvet, R. Olives, A. Meffre, P. Echegut, C. Bessada, E. Veron, S. Ory, Recycled material for sensible heat based thermal energy storage to be used in concentrated solar thermal power plants. ASME J. Solar Energy Eng. **133**(3), 031008–031008-8 (2011)
19. M.E. Navarro, M. Martınez, A. Gil, A.I. Ferna, L.F. Cabeza, R. Olives, X. Py, Solar energy materials & solar cells selection and characterization of recycled materials for sensible thermal energy storage. Sol. Energy Mater. Sol. Cells **107**, 131–135 (2012)

20. D. Laing, W.D. Steinmann, R. Tamme, C. Richter, Solid media thermal storage for parabolic trough power plants. Sol. Energy **80**(10), 1283–1289 (2006)
21. C. Ferone, F. Colangelo, D. Frattini, G. Roviello, R. Cioffi, R. di Maggio, Finite element method modeling of sensible heat thermal energy storage with innovative concretes and comparative analysis with literature benchmarks. Energies **7**(8), 5291–5316 (2014)

# Convective Flow of Nanofluid and Nanoencapsulated Phase Change Material Through Microchannel Heat Sink for Passive Cooling of Microelectronics

**Manoj Kumar, Vikram Bisht, Sheshang Singh Chandel, Sumit Sinha-Ray, and Pradeep Kumar**

**Abstract** In this work, a microchannel heat sink (MCHS) of dimension 40 mm × 40 mm × 14 mm with a circular copper tube (ID 0.8 mm) is designed, mimicking microelectronic component. The MCHS is subjected to a constant temperature condition which is mitigated via convective flow of single-phase fluid and different nanofluids. The investigation focused upon the heat transfer characteristics of all the coolants and to elucidate the effect of non-spherical nanoparticles in nanofluid. Here, the fluid flow and heat transfer parameters of the MCHS are focused experimentally at a constant Reynolds number of 1170 (velocity 1.19 m/s). Four coolants were used to study the heat transfer characteristics of the MCHS which was subjected to constant surface temperature of 125 °C, namely de-ionized (DI) water (hereafter abbreviated as water), Water + 0.0005% (v/v) CuO, Water + 0.0005% (v/v) multi-wall carbon nanotube (MWCNT) and DI water + 0.0005% (v/v) wax-intercalated MWCNT (WICNT). The wax is the phase change material (PCM) in this context which was encapsulated within MWCNT via self-sustained diffusion. Carbon nanotubes had an aspect ratio of 200. During the experiment, the temperature of particular depth within the MCHS from the surface was monitored. The maximum allowable limit of that particular point, aka the part of the integrated circuit (IC), was fixed between 70 and 50 °C, while the lower one is considered safe, whereas the upper one is critical for IC. It was seen that over a long period of unsteady-state investigation, the effect of WICNT was more pronounced in reducing the time required to lower the temperature to the safe operating limit. The cooling time of MCHS is reduced by 11.54% and 20.77% using Water + 0.0005% (v/v) multi-wall carbon nanotube and wax-intercalated MWCNT, respectively, over water which explains that WICNT

Manoj Kumar, Vikram Bisht and Sheshang Singh Chandel—Contribution equally.

M. Kumar · V. Bisht · S. S. Chandel · S. Sinha-Ray (✉) · P. Kumar
School of Engineering, Indian Institute of Technology Mandi, Mandi, HP 175075, India
e-mail: sumitsinha@iitmandi.ac.in

S. Sinha-Ray
Department of Mechanical and Industrial Engineering, University of Illinois at Chicago, Chicago, Illinois 60607-7022, USA



K. M. Pandey et al. (eds.), *Recent Advances in Mechanical Engineering*, Lecture Notes in Mechanical Engineering, https://doi.org/10.1007/978-981-15-7711-6_6

enhances the heat transfer which is beneficial to increase the total operation time or computation time at the same memory usage rate.

**Keywords** Multi-wall carbon nanotube · Phase change materials · Microchannel convective cooling · Nanofluid · Nanoencapsulation

## 1 Introduction

Modern personalized device assistants (PDAs) require high power and fast processing to deal with sophisticated calculations. With the advancement in this field, these systems have become portable with rapid miniaturization, compact design, yet billions of integrated circuits (IC) in one motherboard, per se. However, the reduction in microelectronics size peril of heat dissipation arises. This causes further degradation of electronics such as inducing thermal stresses which results in cracking of soldered joints, thermal runaway and electric overstress resulting in more heat accumulation [1]. Therefore, in order to increase heat dissipation, it requires an efficient cooling system. There are two broad cooling techniques to increase the heat transfer coefficient, i.e., active cooling and passive cooling. One of the most used forms of active cooling is using forced convection of air [2]. Passive cooling includes a heat pipe, heat spreader, installing fins on the heat sink (passive cooling), etc. which indirectly either need power from board or are not effective enough. Going through the basic relation of heat transfer coefficient $h = (Nu \cdot \kappa/D)$ [2], which suggests that to increase h, either D (characteristic length) needs to be decreased or $\kappa$ (thermal conductivity) of fluid is increased. By using the approach of decreasing characteristics length, the microchannel heat sink was first used by Tuckerman and Pease [2] which gained attention because of its capability to remove large amounts of heat from a small area. The detailed study of micro heat pipes, their models, classification and working fluids can be found in Ref. [3]. Further, in order to enhance performance different configurations of microchannels [4] rectangular, triangular, rectangular curved, triangular curved and trapezoidal are taken. Another approach is by increasing the thermal conductivity of the fluid. Frequently and easily available fluid is water. Therefore, to make water more conducting in order to enhance microchannel efficiency is enhancing water properties by selecting suitable additives and/or synthesizing nanofluids. After the invention of nanofluids [5], they seem to be promising for cooling applications [6, 7]. Using these ideas, various works on nanofluids for microchannel have been undertaken. Using copper and diamond nanoparticles in water resulted in a 10% enhancement in heat dissipation as observed by Jang et al. [8]. With the addition of CuO nanoparticles, the nanofluid absorbs more heat at a low flow rate as discussed by Chein et al. [9]. Al2O3 nanoparticles [10, 11] produced heat transfer enhancement of 21.6%. Recent growth in industrial applications is also seen where several industries, inclined to automotive or heating industries, are partnering with well-known research laboratories like.

Argonne National Laboratory (ANL) for industrial grade cooling applications [12] or for quenching applications [13]. The hydrodynamic property of nanofluid depends on various factors like volume fraction, particle size which has been shown by [14]. An unconventional nanomaterial for nanofluid is carbon nanotubes (CNTs) which is used for nanofluid preparation demonstrating a 2.9 °C additional cooling than base fluid water [15]. CNTs are expected to be the next-generation material for interface materials electronics in order to overcome contact thermal resistance because of their high thermal conductivity [16].

PCMs are lucrative options to extract a large amount of heat and maintaining temperature for specific applications, like in solar heating or building heating applications. Details of several PCMs and their thermal properties are mentioned in Ref. [17]. Apart from solar heating, PCMs are extremely useful for electronic cooling applications, like in a heat sink with or without fins or porous matrix filled with PCM [18–20]. The thermal properties of PCMs affect the heat dissipation rate. Low thermal conductivity limits their usage despite having large latent heat, and hence, various works have been undertaken to study the enhancement of thermal conductivity of PCMs by embedding nanoparticles within nanoconfinements like CNTs and carbon nanofillers [21], TiO2 nanoparticles dispersed in n-octadecane, etc. A new microencapsulated PCMs based on Al2O3 [22] nanoparticles were synthesized through emulsion polymerization, and it was found that microencapsulation enhanced thermal conductivity.

Here in this work the effective cooling of microchannel heat sink via various coolants, namely DI water, Water + 0.0005% (v/v) CuO, Water + 0.0005% (v/v) multi-wall carbon nanotube (MWCNT) and Water + 0.0005% (v/v) wax- intercalated MWCNT(WICNT), is studied. The wax intercalation is achieved via self-sustained diffusion which has followed the procedure from ref. 13. The work will demonstrate the use of nanofluid for enhancement of the life cycle of a microelectronic component by reducing the time required to bring a superheated integrated circuit from a critical temperature of 70 °C to lower the safe temperature of 50 °C. This is to further add that, over the course of this article, the additional positive effects of intercalating wax in CNTs will be discussed which will further improve the cooling effects of nanofluids flowing through microchannels over a microelectronic component. Following this route, within a given time frame, a microelectronics is expected to perform more computational cycles even under higher memory consumption, before falling under critical operating condition.

## 2 Experiments

In this study, nanofluids have been used to increase the thermal performance of microchannel with the aim to use the potentials of high thermal conductivity of CNTs and storage property of paraffin wax. Magnetic stirrer set at 125 °C used as an equivalent chip surface/Heater. Copper block of dimensions 40 mm × 40 mm × 14 mm was fabricated as the heat sink with two circular copper tubes of ID 0.8 mm

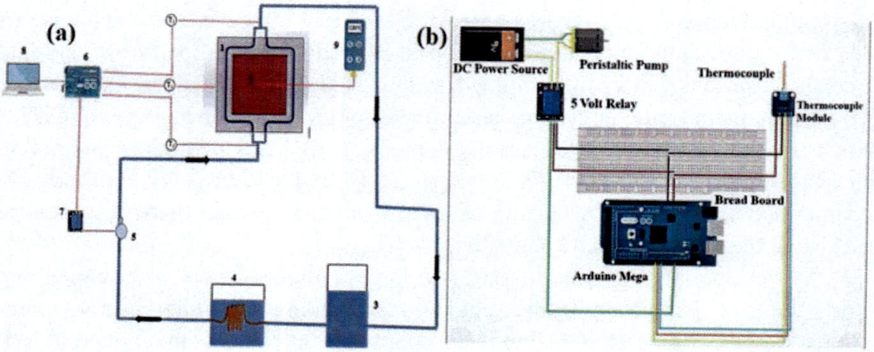


**Fig. 1** **a** Schematic diagram of the microchannel heat sink and its components and **b** electrical circuit of the driving mechanism. The details of individual components are mentioned in Table 1.

x length 120 mm engraved in it as a microchannel passive cooling arrangement. The entire working scheme is described in Fig. 1. Arduino Mega (6) is used to measure the temperature of the microchannel. As thermocouple T3 reads 70 °C (upper critical temperature), relay (7) is used to switch on the pump (5) and working fluid is forced to flow in microchannel until temperature reaches 50 °C (lower safe temperature). The temperature at inlet and at the outlet is measured by thermocouples T2 and T1, respectively. Outlet working fluid is recycled to a reservoir (3), and then before re-entering, it passes through the cooling coil (4) in order to keep the same inlet temperature for all cycles. Four different working coolants are taken: water, copper oxide (CuO) nanofluid in water, CNT suspension (0.0005 vol%) in water, paraffin wax (as obtained from Sigma-Aldrich)-intercalated CNT suspension (0.0005 vol%) in water. CNT suspension was prepared by mixing surfactant sodium dodecyl sulfate (SDS) solution (five times weight of CNT) in water; however for WICNT suspension, the SDS concentration was increased to 13 times to tackle the hydrophobicity of wax, which may stick to exterior of CNTs. Wax-intercalated CNT is prepared by a self-sustained mechanism [23] which is as follows 1.5 wt% wax solution made in benzene was sonicated for 30 min. After sonication, 30 mg of CNT was added to the solution and further sonicated for 3 min and the solution was left open in fume hood. As benzene evaporates, wax diffuses into CNT and the driving force for this process keeps increasing as the wax concentration keeps on increasing in evaporating solution. Thereby, wax-intercalated CNT got prepared and further characterization was done using differential scanning calorimetry for melting point and latent heat of sample calculation, as shown in Fig. 2.

## 3 Results and Discussion

The DSC data of wax-intercalated CNTs is shown in Fig. 2, where it can be seen that the melting peak was observed at 59.45 °C, with melting starting at 57 °C and

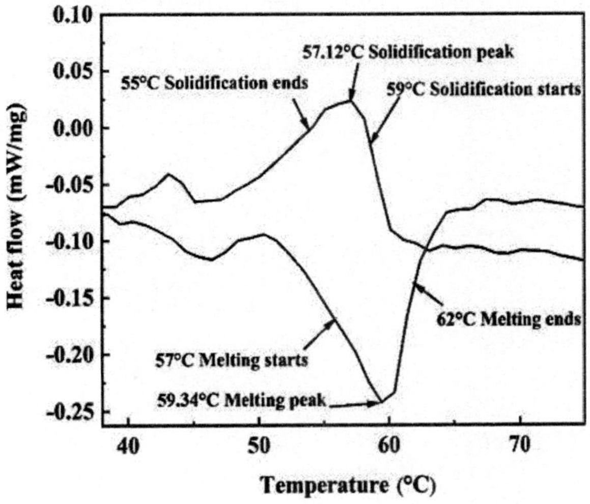


**Fig. 2** DSC thermogram of wax intercalated CNTs

**Table 1** Components as numbered in Fig. 1a

| | | |
|---|---|---|
| 1. Microchannel | 2. Hot surface | 3. Reservoir |
| 4. Cooling area | 5. Pump | 6. Arduino Mega |
| 7. Relay | 8. PC | 9.Temperature monitor |

ending at 62 °C. The calculated latent heat was 135.49 J/g, which was near similar to 141 J/g as mentioned in ref. 13. This indicates to a filling of 90% within the CNT cores.

As mentioned in Sect. 2, the experiment was conducted between two critical temperatures 70 °C and 50 °C, which represented an upper critical temperature for microelectronics and safe temperature, respectively. The intention of this experiment was to understand whether the effect of CNT and wax-intercalated CNT can be prominent in reducing the time required for cooling than water, which was otherwise subjected to the same conditions. The effect of CNT should be more pronounced on the thermal conductivity of water following Hamilton–Crosser model [24], which describes the enhancement of thermal conductivity ($\kappa$) of any base fluid due to the addition of non-spherical particles.

$$\kappa_{nf} = \frac{\left[\kappa_p + (n-1)\kappa_f - (n-1)\phi\left(\kappa_f - \kappa_p\right)\right]}{\left[\kappa_p + (n-1)\kappa_f + \phi\left(\kappa_f - \kappa_p\right)\right]} \tag{1}$$

where $\kappa$ is thermal conductivity, $\phi$ is a particle concentration of nanofluid in % (here, 0.0005), n is an experimental parameter.

$n$ is an experimental parameter, called as 'shape factor', where $n = 22.97$ for CNT and defined as $n = 3/\psi$ where $\psi$ is known as sphericity. It is defined as surface area of a sphere with the same volume as the particle to the surface area of the particle.

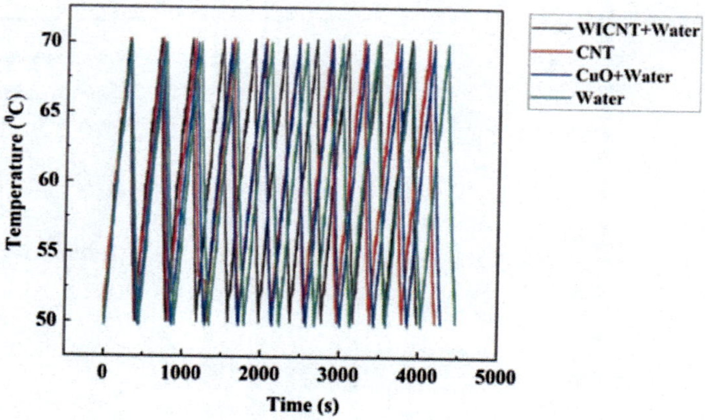


**Fig. 3** Heating–cooling cycles for a different coolant where the upper temperature is 70 °C and lower bound is 50 °C. Individual coolant cycle legends are embedded in the picture

Subscripts $nf, f, p$ signify nanofluid, base fluid and particle, respectively. The thermal conductivity of CNTs was taken as ~300 W/m K [25].

Following Eq. 1, the thermal conductivity of CNT-nanofluid was found to be 0.6048 W/m K, mere 1.08% greater than base fluid-water (0.6 W/m K). However, as per Ref. [13], the uniaxial flow of CNTs in microchannels actually allows them to align in the direction of flow and hence an apparent increase the volume percentage at any local cross section could be experienced, which may result in higher temperature reduction. Although the magnitude of thermal conductivity increment was not lucrative by mere calculation, but from the arguments of Ref. [13], an indicative efficient heat conduction could be predicted. During the experiments, the base plate was kept at 125 °C.

The heating–cooling cycle was run ten times, where the entire process started to come to a steady state on from seventh cycle, which can be seen in Fig. 3. The performance of all the coolants is shown in a single graph, where the difference in cooling time starts to become more prominent. Figure 4 explains the heating cooling curve for different fluid for the tenth cycle, which is shown here for better visualization and for sake of brevity. From the graph, it can be concluded that the heating time is relatively close for all fluids, but cooling time was observed to be significantly different. Due to the possible alignment and possible percolative network of CNTs in CNT + water flow, it takes less time for cooling compared to water. For WICNT + Water, additional cooling is obtained due to wax that absorbs a significant amount of energy during the melting process. This is near analogous to the results mentioned in Ref. [13], where the effect of cooling was studied in terms of reduction in temperature at steady state, but it was evident from their result that at the transient condition, the CNT nanofluid was able to remove heat quicker than base fluid (DI water). Here, in this study, a similar effect was observed over a fixed temperature difference in the transient condition. Given the flow rates were near similar. To visualize better, within a domain of 70–60 °C, all the coolants' performances are plotted in Fig. 4b,

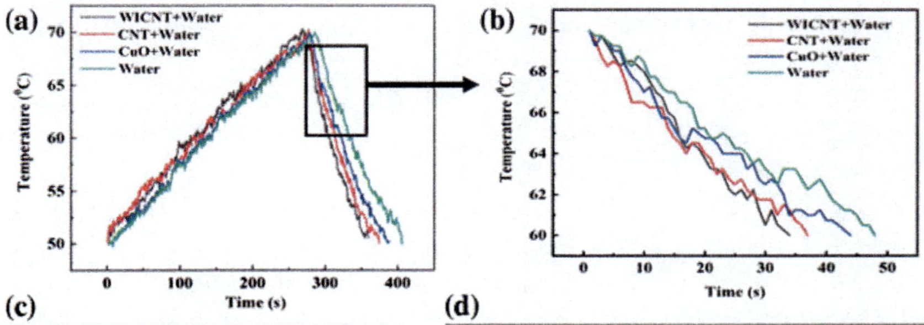


(c)

| Coolant | Avg. Cooling Time (70°C-60°C) (s) |
|---|---|
| Water | 49 |
| CuO+Water | 42 |
| CNT+Water | 36 |
| WICNT+Water | 33 |

(d)

| Coolant | Slope (°C/s) | Cooling rate enhancement (%) | Adjusted $R^2$ |
|---|---|---|---|
| Water | -0.19 | - | 0.97 |
| CuO+Water | -0.2 | 5.26 | 0.972 |
| CNT+Water | -0.235 | 23.68 | 0.97 |
| WICNT+Water | -0.32 | 68.4 | 0.967 |

**Fig. 4** Comparison of the heating–cooling curve for different fluid at **a** the tenth cycle, **b** cooling variation for 70–60 °C, **c** average cooling time to reach from 70 to 60 °C, **d** curve fitting for different coolant

where it can be seen that CNT or WICNT suspension performed much better than other coolants. Hence, this is an indication of the applicability of CNT suspension as an effective coolant for the given situation. However, the additional effect of wax melting was clear for WICNT suspension. Figure 4c shows the average cooling time to achieve 60 °C for different coolants used in microchannel cooling. WICNT + Water takes 16 s less time to achieve a safe working temperature than DI water and 3 s lesser than CNT + water; definitely, melting of wax was evident in this working domain, which correlates to data obtained from DSC thermogram.

The effect of wax melting on cooling behavior, in the temperature range of 68–60 °C, is shown in Fig. 4d. The cooling rate of different coolant is shown as a zoomed-in curve from the data elucidated in Fig. 4b. It is a safe assumption of linear decrement in temperature over time, and hence, a linear fitting is a well-suited elaboration of data to provide the temperature reduction gradient. The adjusted-$R^2$ value of data fitting close to ~0.97 for all the cases is a clear statement of that [c.f. Figure 4d]. Figure 4d concludes the cooling effect of MCHS became more significant in case of WICNT + Water suspension compared to CNT + Water suspension, where the cooling rate was seen to be 68.4% more than base fluid (DI water) in the former case and 36.17% additionally enhanced than pure CNT + Water suspension. The effect is an ensemble effect of fresh PCM being pumped into the system which is subjected to a temperature greater than its melting point, as suggested by Fig. 2. The readers may note that the particle characteristic reaction time ($\tau$) is much smaller than the particle retention time ($t$), the latter being simply the process time scale, the former can be deemed to be insignificant than the latter, and hence, PCM melting will be in effect.

The time required for heat transfer across wax in CNT, aka characteristic reaction time, which is a function of thermal diffusivity ($\alpha$) and characteristic length ($L_c$, in this case, the inner diameter of CNT) is given by [26]

$$\text{Characteristic reaction time } \tau = \frac{L_c^2}{\alpha} \quad (2)$$

The melting time of confined wax is calculated using the characteristic reaction time. In the present study, value of thermal diffusivity of wax was taken as ~$10^{-7}$ $m^2/s$ following Ref. [26] and characteristic length is the inner diameter of CNT (35 nm) which gives the value of 12 ns. It is also safe to mention, following Ref. [13], that CNTs agglomerate during the flow in a microchannel and hence the local volume fraction increases drastically which will lead to near metallic thermal diffusivity across the channel and hence, CNTs near the wall will be subjected to wall temperature almost instantly. On the other hand, the particle retention time as shown in Eq. 3, yields a value of 76 ms.

$$\text{Particle retention Time } t = \frac{L}{V}. \quad (3)$$

where $L$ is the length of the microchannel (90 mm) and $V$ is the velocity at the inlet of the microchannel (1.19 m/s). This proves that the processing time was immaterial in regards to wax melting and hence over a period of 33 s, which is mentioned in Fig. 4c, provides ample time for the wax to interact with the hot fluid surrounding CNTs and melt to absorb more heat. The mere fact that near 55–52 °C, where the wax is almost solidified (c.f. Fig. 2), the cooling curve of WICNT + water approaches toward CNT + water.

Table 2 explains the cycle performance of MCHS for different coolants. Results are analyzed in terms of cooling enhancement (70–50 °C) by reducing the cooling time, which shows that once the process is stabilized, hereinafter 6th-7th cycle, WICNT + water, allows ICs to run for additional time under the same duty cycle, or on the same note the microelectronics can work at a higher capacity than any other coolant mentioned herein under a stipulated time.

**Table 2** Cooling time reduction (70–50 °C) of different coolants as compared to base fluid water

| Coolant | Cooling time reduction (%) |
|---|---|
| Water | 0 |
| CuO + Water (0.0005% v/v) | 6.92 |
| CNT + Water (0.0005% v/v) | 11.54 |
| WICNT + Water (0.0005% v/v) | 20.77 |

## 4 Conclusion

Experiments were carried out to observe the heating cooling curve of microchannel heat sink using different nanoparticles like CuO, CNT and WICNT with water as base fluid. The automated system is used to increase the effective working of the cooling techniques during the working condition of microelectronic. The study of the heating and cooling behaviors is experimentally monitored and the enhancement of cooling, aka cooling time reduction, is shown in Sect. 3. The data explains the total working efficiency in terms of the number of cycle gain which directly elucidates the cooling enhancement of coolant in the microchannel heat sink. The present study shows the cooling performance enhancement in terms of cooling time, and the WICNT + Water gives the maximum cooling time reduction of 20.77% as compared to water. The cooling rate of WICNT + Water is 68.4% more than base fluid (water) and 36.17% more than CNT + Water during critical temperature. This result leads to a potential increase in the number of cycles or increases the operating performance of the chip. It can also be concluded from the above calculation that different types of wax, with a variety of melting temperatures, can be used in the nanoencapsulated systems to expand the domain of working temperature to extend for a wide range of electronic cooling, depending upon the critical functioning of microelectronics. The study hence proves to open a plethora of options to exercise the passive cooling of microelectronics, where the integral effect of efficient phase change couple with forced convection can be utilized.

**Contribution of Authors** M.K. and S.S.C. performed the heat transfer experiments and analyzed data. V.B. conducted the experiments with CNTs. P.K. and S.S-R. conceived the idea. M.K., V.B. and S.S.C. equally contributed in writing the manuscript.

**Acknowledgements** M.K. and V.B. acknowledge the scholarship from Ministry of Human Resource and Development, Govt. of India and S.S.C. acknowledges the funding support from DST-SERB (IITM/SERB/SSR/215).

**Conflict of Interest** Authors declare no conflict of interest.

## References

1. Y. Shabany, *Heat Transfer: Thermal Management of Electronics* (CRC press, 2009)
2. D.B. Tuckerman, R.F.W. Pease, High-performance heat sinking for VLSI. IEEE Electron. Device Lett. **2**(5), 126–129 (1981)
3. M. Groll, M. Schneider, V. Sartre, M.C. Zaghdoudi, M. Lallemand, Thermal control of electronic equipment by heat pipes.Revue géńérale de thermique **37**(5): 323–352 (1998)
4. C.B. Sobhan, R.L. Rag, G.P. Peterson, A review and comparative study of the investigations on micro heat pipes. Int. J. Energy Res. **31**(6–7), 664–688 (2007)
5. S.U.S. Choi, A.E. Jeffrey, *Enhancing Thermal Conductivity of Fluids with Nanoparticles*. No. ANL/MSD/CP-84938; CONF-951135-29. Argonne National Lab., IL (United States), 1995.
6. X. Wang, X. Xu, S.U.S. Choi, Thermal conductivity of nanoparticle-fluid mixture. J. Thermophys. Heat Transfer **13**(4), 474–480 (1999)

7. S. Lee, S.S. Choi, S.A. Li, J.A. Eastman, Measuring thermal conductivity of fluids containing oxide nanoparticles. J. Heat Transfer **121**(2), 280–289 (1999)
8. S.P. Jang, S.U.S. Choi, Cooling performance of a microchannel heat sink with nanofluids. Appl. Therm. Eng. **26**(17–18), 2457–2463 (2006)
9. R. Chein, J. Chuang, Experimental microchannel heat sink performance studies using nanofluids. Int. J. Therm. Sci. **46**(1), 57–66 (2007)
10. J.-Y. Jung, H.-S. Oh, H.-Y. Kwak, Forced convective heat transfer of nanofluids in microchannels. Int. J. Heat Mass Transf. **52**(1–2), 466–472 (2009)
11. T.C. Hung, W.M. Yan, X.D. Wang, C.Y. Chang, Heat transfer enhancement in microchannel heat sinks using nanofluids. Int. J. Heat Mass Transf. **55**(9–10), 2559–2570 (2012)
12. https://www1.eere.energy.gov/manufacturing/industries_technologies/nanomanufacturing/pdfs/ooling.pdfnanofluids_industrial_c
13. https://www.uk-cpi.com/case-studies/cooling-of-nanoquenching-fluids-for-the-aerospace-industry
14. P.K. Singh, P.V. Harikrishna, T. Sundararajan, S.K. Das, Experimental and numerical investigation into the hydrodynamics of nanofluids in microchannels. Exp. Thermal Fluid Sci. **42**, 174–186 (2012)
15. S. Sinha-Ray, S. Sinha-Ray, H. Sriram, A.L. Yarin, Flow of suspensions of carbon nanotubes carrying phase change materials through microchannels and heat transfer enhancement. Lab Chip **14**(3), 494–508 (2014)
16. P.K. Schelling, S. Li, E.G. Kenneth, Managing heat for electronics. Mater. Today **8**(6), 30–35 (2005)
17. Jaconelli, L. and Anton, P. "Numerical Verification of Mobile Thermal Energy Storage Performance." (2017).
18. F.L. Tan, C.P. Tso, Cooling of mobile electronic devices using phase change materials. Appl. Therm. Eng. **24**(2–3), 159–169 (2004)
19. R. Srikanth, P. Nemani, C. Balaji, Multi-objective geometric optimization of a PCM based matrix type composite heat sink. Appl. Energy **156**, 703–714 (2015)
20. Y. Cui, C. Liu, S. Hu, X. Yu, The experimental exploration of carbon nanofiber and carbon nanotube additives on thermal behavior of phase change materials. Sol. Energy Mater. Sol. Cells **95**(4), 1208–1212 (2011)
21. S. Motahar, N. Nikkam, A.A. Alemrajabi, R. Khodabandeh, M.S. Toprak, M. Muhammed, Experimental investigation on thermal and rheological properties of n-octadecane with dispersed TiO2 nanoparticles. Int. Commun. Heat Mass Transfer **59**, 68–74 (2014)
22. X. Jiang, R. Luo, F. Peng, Y. Fang, T. Akiyama, S. Wang, Synthesis, characterization and thermal properties of paraffin microcapsules modified with nano-Al2O3. Appl. Energy **137**, 731–737 (2015)
23. S. Sinha-Ray, R.P. Sahu, A.L. Yarin, Nano-encapsulated smart tunable phase change materials. Soft Matter **7**(19), 8823–8827 (2011)
24. R.L. Hamilton, O.K. Crosser, Thermal conductivity of heterogeneous two-component systems. Ind. Eng. Chem. Fundam. **1**(3), 187–191 (1962)
25. T.Y. Choi, D. Poulikakos, J. Tharian, U. Sennhauser, Measurement of the thermal conductivity of individual carbon nanotubes by the four-point three-ω method. Nano Lett. **6**(8), 1589–1593 (2006)
26. E. George, R. Shah, D. Forester, *Fuels and Lubricants Handbook: Technology, Properties, Performance, and Testing* (ASTM International, USA, 2003), p. 224

# Fabrication of Treated and Untreated Coconut Fibre-Reinforced Epoxy-Based Composites of Different Fibre Content and Comparison of Their Tensile and Flexural Strengths

**U. Deka, M. Bhuyan, C. Borah, S. Kakoti, and R. K. Dutta**

**Abstract** The use of natural fibres has found impetus in recent times in place of conventional composites. Coconut is widely grown in tropical and subtropical regions. In the present work, an attempt has been made to study the use of coconut fibre as reinforcements in epoxy-based composites. The tensile and flexural properties of the coconut fibre-reinforced composite are studied to investigate its external load-carrying capacity. The quantity of fibre content was varied from 1% by weight, 3% by weight and 5% by weight of coconut fibre. These were employed as reinforcements in epoxy resin for both alkali-treated and untreated specimens. Alkali treatment enhanced the mechanical properties of the composites by providing better adhesive properties between matrix and fibre. The results revealed that composite having 3% by weight of treated coconut fibre showed the best tensile and flexural strength out of all the samples. In the light of these findings, it can be concluded that coconut fibre-reinforced composites may be used in various domestic, construction and industrial purposes. However, there is a dearth of research on how to further enhance their mechanical properties which might help in several applications by also reducing the quantity of natural waste produced by coconut fibres.

**Keywords** Alkali treatment · Coconut fibre · Fibre pull out · Flexural strength · Tensile strength

U. Deka (✉) · M. Bhuyan · C. Borah · S. Kakoti · R. K. Dutta
Mechanical Engineering, Assam Engineering College, Guwahati, India
e-mail: uddiptaadeka1996@gmail.com

M. Bhuyan
e-mail: monamibhuyan1994@gmail.com

C. Borah
e-mail: chayanikaborah013@gmail.com

S. Kakoti
e-mail: sangitakakoti25@gmail.com



K. M. Pandey et al. (eds.), *Recent Advances in Mechanical Engineering*, Lecture Notes in Mechanical Engineering, https://doi.org/10.1007/978-981-15-7711-6_7

## 1 Introduction

A budding environmental consciousness for the need for sustainable development has burgeoned during the recent years. This, in turn, has promoted interest in using natural fibres as reinforcements in polymer composites instead of synthetic fibres. The composites of natural fibres impart stiffness and strength to the product. Moreover, they have numerous advantages like low density, low cost, low energy content, easy availability and recyclability [1]. Thus, natural fibres can be used to replace conventional materials like plastic and wood in different structural and non-structural applications.

Various investigations have been carried out to study the variation of properties of natural fibres as reinforcements in the polymer. The chemical treatment of the fibres showed to have enhanced the mechanical properties of the composites [2, 3]. Yan et al. studied how the mechanical and microstructural properties are influenced by alkali treatment of the fibre-reinforced cement composites [4]. Kumar et al. [5] prepared composites by taking different weight ratios of coir fibres with epoxy resin. They then performed tensile, impact and hardness tests on them. Results showed that the addition of fibres improved the strength of the composite. Comparison studies have been made on the effects of different types of chemical treatments of the reinforced fibres [6]. Biswas et al. [7] found that the length of the fibres also significantly influences the mechanical properties of the composites such as tensile strength, flexural strength, impact strength and micro-hardness of the composites. Sharifah et al. studied how the flexural strength depends on the location of high resin density areas and orientation of fibres [8]. Furthermore, Sumesh et al. [9] studied how temperature and compression pressure also affected the properties of coconut fibre-reinforced epoxy composites. The dynamic mechanical properties of these composites have also been studied [10]. Sarki et al. studied coconut shell-filled composites which were prepared from epoxy polymer matrix containing up to 30 wt% coconut shell fillers. The effects of coconut shell particle content on the mechanical properties of the composites were investigated [11]. Ramaraj et al. studied the poly(vinyl alcohol) (PVA) composites with 10, 20, 33 and 55 wt% of coconut shell powder, and it was observed that the introduction of coconut shell powder varies the tensile strength and affects percentage of elongation, tear and burst strengths, moisture content, density and swelling capacity [12]. Yuhazri et al. developed a unique bulletproof vest made of coconut fibre, which provides all the protection that can be found in a regular vest. It is not only economical but also lighter. As opposed to a traditional bulletproof vest which weighs about 9 kg, this particular vest weighs only 2 kg and is also less in cost [13]. Yuhazri et al. utilized coconut fibres in the manufacturing of motor cycle helmet. They used epoxy resins from thermoset polymer as the matrix materials and coconut fibres as the reinforcement [14]. Although an extensive study has been done on the different properties of coconut fibre-reinforced epoxy composites, there is a bereft of research on how the smallest variations can change the mechanical properties of the composites as well as devising methods for making these composites more robust for practical applications. Factors such as curing time also need to be studied

**Fig. 1** Coconut fibres

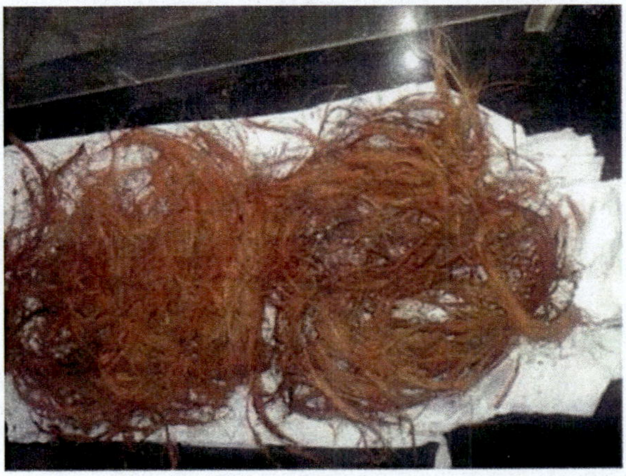

and expounded. In the present research work, we have carried out the mechanical and flexural testing of treated coconut fibre and untreated coconut fibre-reinforced epoxy composite of different compositions.

## 2 Materials and Method

### *2.1 Coconut Fibres*

Coconut fibres were extracted manually from the husk of *Cocos nucifera.* The density of the fibres was found to be 1.2 g/cm$^3$. Figure 1 shows the extracted coconut fibres.

### *2.2 Resin*

In the present investigation, Araldite XIN-100 IN epoxy and Hardener XIN-900 IN, both supplied by Amazon India, are used.

In the present investigation, the volume percentage of epoxy used with hardener is 10:1 [5].

### *2.3 Chemical Treatment of Coconut Fibres*

In the present study, the fibres were treated with a NaOH solution. After extraction of the fibres, they were cut to a length of 3 mm. Then, they were soaked in water for 60 min and were heated at 50 °C until a constant weight was obtained. The process of

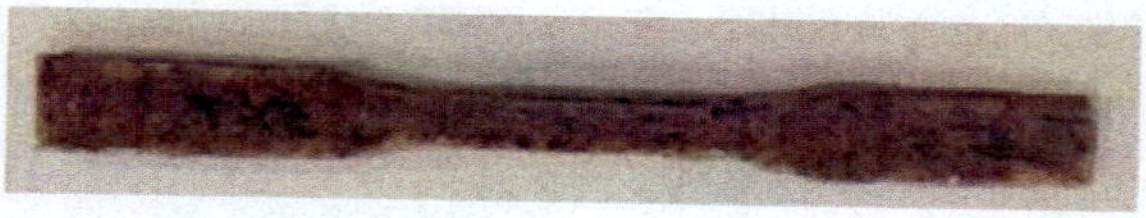

**Fig. 2** Tensile test sample

**Fig. 3** Flexural test sample

heating was done to ensure that there is no moisture left in the fibres. Some portion of the total fibres was kept separately as untreated fibres. The other portion was soaked in a 0.25 M NaOH solution for 60 min. After the alkaline treatment, the fibres were rinsed with water to extract out the remaining alkali content and again dried at 65 °C in a hot air oven for 7 h to remove moisture [4]. The treated and untreated fibres were ready to be used for making composites.

## 2.4 Composite Preparation

The composites were cast in the mould by hand lay-up process. Cardboard moulds were prepared according to the ASTM D638 Type I for the tensile test and ASTM D790 for the flexural test (Figs. 2 and 3). The inside surface of the moulds was coated with silicon spray for easy removal of the composites from the moulds. The epoxy resin is mixed with a hardener in the volume ratio of 10:1 with a mechanical whisk for 5–7 min. The prepared mixture is then poured into the cardboard moulds. The coconut fibres were randomly placed over the resin. After 20 min, another thin layer of resin is poured over the fibres. Then, the composites were left to cure for 3 days so that the fibres were impregnated with epoxy. After the removal of the composites from their moulds, they were heated at 65 °C for 7 h to eliminate any moisture left in the fibres [4]. The weight percentage of fibres was varied at 1, 3 and 5% to the weight of the resin.

## 2.5 Testing of Composite Material

### 2.5.1 Tensile Testing

For finding the tensile properties of coconut fibre-reinforced epoxy-based composite material, an INSTRON-8801 Dynamic Universal Testing Machine was used. Seven different samples (three untreated and four treated) with varying weight percentages were tested using this machine shown in Fig. 4.

**Fig. 4** Tensile testing of composite in INSTRON-8801

### 2.5.2 Flexural Testing

For the flexural test, a jig was attached to the INSTRON-8801 Dynamic Universal Testing Machine. The lower part of the jig was mounted and set according to the span length before it was attached to the INSTRON machine shown in Fig. 5.

## 3 Results and Discussion

The results obtained from the tensile and flexural tests are analysed to make a comprehensive study on the effect of both treated and untreated fibres in epoxy composites for 1, 3 and 5% by weight composites.

The various results are given in Tables 1 and 2.

From Fig. 6, it is seen that the composite sample having 3% by weight of treated coconut fibres as reinforcement material has the maximum strength of 15.06 MPa. Figure 7 shows that flexural strength is maximum for the composite having 3% by weight of treated coconut fibres as a reinforcement which was found to be 50.97 MPa. Both tensile and flexural strength showed improvement on addition of fibres to pure epoxy.

The mechanical properties of reinforced composites may be attributed to factors like adhesive forces between matrix and fibres which ultimately account for the effectiveness of load transfer, orientation, and length of fibres, aspect ratio and inherent properties of both fibres and matrix [15]. The increase in weight percentage of fibre

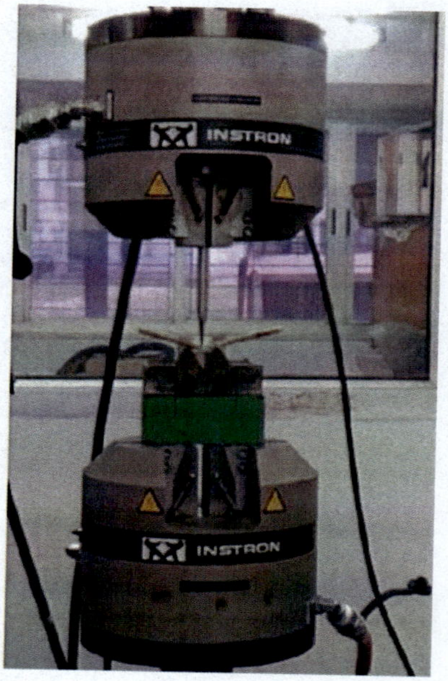


**Fig. 5** Flexural testing of composite

**Table 1** Tensile strength of composites with different weight % of the fibre

| Sample | Weight % of the fibre (%) | Tensile load, *F* (N) | Extension (mm) | Tensile strength (MPa) | Tensile modulus, *E* (MPa) |
|---|---|---|---|---|---|
| Pure epoxy | 0 | 348.70 | 2.55 | 7.74 | 310.78 |
| Coconut fibres treated with NaOH solution | 1 | 534.71 | 5.26 | 11.88 | 350.53 |
| | 3 | 677.67 | 1.92 | 15.06 | 733.33 |
| | 5 | 512.40 | 2.79 | 11.39 | 447.52 |
| Untreated coconut fibres | 1 | 490.48 | 2.25 | 10.90 | 421.34 |
| | 3 | 535.20 | 1.80 | 11.90 | 768.11 |
| | 5 | 518.90 | 2.70 | 11.53 | 400.00 |

**Table 2** Flexural strength of composites with different weight % of the fibre

| Sample | Weight % of the fibre (%) | Flexural load, *F* (N) | Extension (mm) | Flexural strength (MPa) | Flexural modulus, *E* (MPa) |
|---|---|---|---|---|---|
| Pure epoxy | 0 | 35.10 | 5.76 | 12.19 | 144.38 |
| Coconut fibres treated with NaOH solution | 1 | 115.43 | 5.25 | 40.08 | 613.33 |
| | 3 | 146.80 | 5.07 | 50.97 | 847.24 |
| | 5 | 73.18 | 4.88 | 25.40 | 341.26 |
| Untreated coconut fibres | 1 | 79.61 | 4.67 | 27.64 | 205.52 |
| | 3 | 83.12 | 6.00 | 28.86 | 286.22 |
| | 5 | 92.84 | 5.63 | 32.24 | 511.50 |

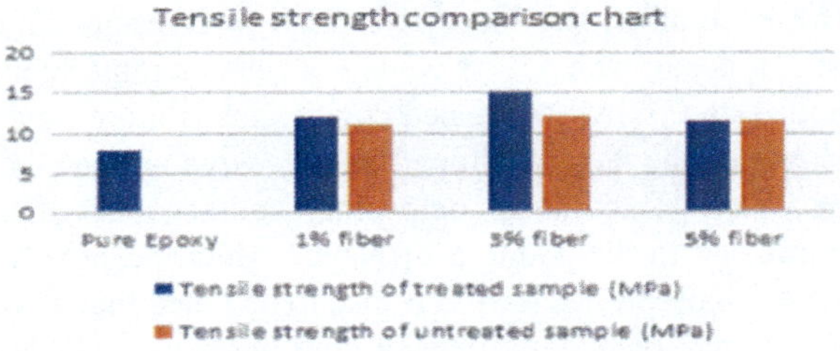


**Fig. 6** Comparison of tensile strength

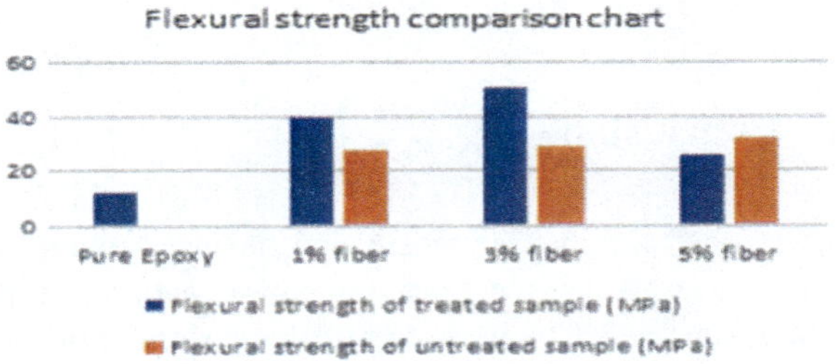


**Fig. 7** Comparison of flexural strength

in epoxy resin tends to increase the tensile strength. This is mainly due to increased stress transfer from the polymer matrix to the fibres [16, 17]. But after 3% by weight of fibre reinforcement, its strength decreases for 5% by weight of fibre reinforcement. This might be because as fibre wt% increases, the binding force and interface between fibres to resin starts to decrease. This may be due to improper adhesion and agglomeration of fibres. Moreover, both elongation and tensile strength increase for composites reinforced with 1% by weight-treated coconut fibres when compared with pure epoxy, which can imply that it is possible to have such reinforced composites with both higher strength and flexibility. The trend seen in flexural strength is caused due to bending and breaking of fibres [18].The flexural strength of untreated composites increases from 1 to 3 to 5% wt/wt composites. This implies that increased fibre content resists the bending of the composite. This can be counterintuitive to research where it has been found that increasing reinforcement can result in poor dispersion of fibres and the formation of voids resulting in lower flexural strength [19]. Alkali treatment of fibres helps to reduce the percentage of lignin and hemicelluloses and increase that of cellulose, thus allowing better stress transfer to the reinforcement by removing the hydrophilic nature of coconut fibres [5, 20]. Alkali treatment also makes the breaks the fibres into further smaller fibres, thus increasing the surface roughness and aspect ratio of the fibres. This improves fibre wet ability and is found to improve the mechanical properties of the fibres [21]. Tensile strength of treated fibre-reinforced composite is found to be more than their untreated counterpart for 1% by weight and more significantly so for 3% by weight fibre-reinforced specimen. The untreated samples have a larger number of voids and weaker bonding than the treated samples [17] for both 1% by weight and 3% by weight samples. Higher strength of 1 and 3% wt/wt composites implies complete matrix fusion which facilitates fibre permeation and formation of strong fibre/matrix interfacial bonding [5].

However, alkali treatment does not have a prominent effect on the tensile strength of composite with 5% by weight of fibre as reinforcement. The flexural strength of both 1 and 3% wt/wt treated composites is more than their untreated counterparts. However, the flexural strength of 5% wt/wt untreated composite is greater than that of 5% wt/wt untreated composite. One reason for this might be damage caused to fibres due to the addition of NaOH which negatively affected the fibre and resin adhesive properties.[22]. It is also to be noted that flexural strength also depends on the orientation and presence of fibre rich areas, which might have been in play here as well.

## 4 Conclusion

It was observed that the tensile and flexural properties are influenced by the fibre content as well as by chemical (alkaline) treatment of the composite. The composite sample having 3% by weight of treated coconut fibre as reinforced material showed the highest tensile strength (15.06 MPa) as well as the highest flexural strength (50.97 MPa). Alkali treatment improved the properties in general, but it did not show a favourable effect on the properties of a composite having 5% by weight of coconut fibre as reinforced material. This implies that chemical treatment might not always enhance the properties of the composites. The elastic modulus was noted to be following a trend (rate of change) different from that of the maximum tensile strength. Further investigation is necessary to account for this observation. The influential design factor is only limited to composition in this particular experiment. However, more robust design modules are necessary to study the effect of chemical treatment, curing time, the temperature on the properties of composites with different fibre contents. The future scope of the use of coconut fibres as reinforcements in composites is very wide and can pave the way for the manufacturing of robust, economical, environmentally friendly and durable alternatives to synthetic fibre based composites. Moreover, these reinforced composites pave the way to make use of coconut fibres which are usually discarded as waste products as useful raw materials for robust materials with excellent properties and strength.

## References

1. P. Wambua, J. Ivens, I. Verpoest, Natural fibres: can they replace glass in fibre reinforced plastics? Compos. Sci. Technol. **63**(9), 1259–1264 (2003)
2. A.K. Bledzki, J. Gassan, Composites reinforced with cellulose based fibres. Prog Polym Sci **24**, 221–274 (1999)
3. B.F. Yousif, K.C. Ming, Flexural properties of treated and untreated kenaf/epoxy composites. Mater. Des. **40**, 378–385 (2012)
4. L. Yan, N. Chouw, L. Huang, B. Kasal, Effect of alkali treatment on microstructure and mechanical properties of coir fibres, coir fibre reinforced-polymer composites and

reinforced-cementitious composites. Constr. Build. Mater. **112**, 168–182 (2016)
5. S.M.S. Kumar, D. Duraibabu, K. Subramanian, Studies on mechanical, thermal and dynamic mechanical properties of untreated (raw) and treated coconut sheath fiber reinforced epoxy composites. Mater. Des. **59**, 63–69 (2014)
6. A.H. Abdullah, F.F.A. Mutalib, M.F. Mat, Tensile and fracture toughness properties of coconut spathe fibre reinforced epoxy composites: effect of chemical treatments. Adv. Mater. Res. **1133**, 603–607 (2016)
7. S. Biswas, S. Kindo, A. Patnaik, Effect of fiber length on mechanical behavior of coir fiber reinforced epoxy composites. Fibres Polym. **12**(1), 73–78 (2011)
8. H. Aziz Sharifah, Martin P. Ansell, The effect of alkalization and fibre alignment on the mechanical and thermal properties of kenaf and hemp bast fibre composites: Part 1 – polyester resin matrix. Compos. Sci. Technol. **64**, 1219–1230 (2004)
9. K.R. Sumesh, K. Kanthavel, The influence of reinforcement, alkali treatment, compression pressure and temperature in fabrication of sisal/coir/epoxy composites: GRA and ANN prediction. Polym. Bullet. (2019)
10. N. Saba, M. Jawaid, O.Y. Alothman, M.T. Paridah, A review on dynamic mechanical properties of natural fibre reinforced polymer composites. Constr. Build. Mater. **106**, 149–159 (2016)
11. J. Sarki, S.B. Hassan, V.S. Aigbodion, J.E. Oghenevweta, Potential of using coconut shell particle fillers in eco-composite materials. J. Alloy. Compd. **509**, 2381–2385 (2011)
12. B. Ramaraj, P. Poomalai, Ecofriendly Poly (vinyl alcohol) and coconut shell powder composite films: physico-mechanical thermal properties, and swelling studies. J. Appl. Polym. Sci. **102**, 3862–3867 (2006)
13. M.Y. Yuhazri, M.D.M Palil, High impact hybrid composite material for ballistic armor. J. Adv. Manuf. Technol. **2**(1), 1–10 (2008)
14. M.Y. Yuhazri, M.M.P. Dan, Helmet shell using coconut fibre (Deco-Helmet). J. Adv. Manuf. Technol. **1**(1), 23–30 (2007)
15. M. Baiardo, E. Zini, M. Scandola, Flax fibre–polyester composites. Compos. A Appl. Sci. Manuf. **35**(6), 703–710 (2004)
16. S. Fu, B. Lauke, Y. Maya, Stress transfer in short fibre-reinforced polymers. Science and engineering of short fibre-reinforced. Polym. Compos. 67–89 (2019)
17. B.F. Abu-Sharkh, R. Kahraman, S.H. Abbasi, I.A. Hussein, Effect of epolene E-43 as a compatibilizer on the mechanical properties of palm fiber-poly(propylene) composites. J. Appl. Polym. Sci. **92**(4), 2581–2592 (2004)
18. S. Vijayakumar, T. Nilavarasan, R. Usharani, L. Karunamoorthy, Mechanical and microstructure characterization of coconut spathe fibres and Kenaf bast fibres reinforced epoxy polymer matrix composites. Procedia Mater. Sci. **5**, 2330–2337 (2014)
19. N. Venkateshwaran, A. ElayaPerumal, A. Alavudeen, M. Thiruchitrambalam, Mechanical and water absorption behaviour of banana/sisal reinforced hybrid composites. Mater. Des. **32**(7), 4017–4021 (2011)
20. M.J. John, R.D. Anandjiwala, Recent developments in chemical modification and characterization of natural fiber-reinforced composites. Polym. Compos. **29**(2), 187–207 (2008)
21. G. Goud, R.N. Rao, Effect of fibre content and alkali treatment on mechanical properties of Roystonea regia-reinforced epoxy partially biodegradable composites. Bull. Mater. Sci. **34**(7), 1575–1581 (2011)
22. A. Oushabi, S. Sair, F. Oudrhiri Hassani, Y. Abboud, O. Tanane, A. El Bouari, The effect of alkali treatment on mechanical, morphological and thermal properties of date palm fibres (DPFs): Study of the interface of DPF–Polyurethane composite. South Afr. J. Chem. Eng. **23**, 116–123 (2017)

# Study on Gate Location and Gate Number for Manufacturability of Polymer Gears

**Bikram Singh Solanki, Harpreet Singh, and Tanuja Sheorey**

**Abstract** In recent years, the utility of polymer gear increased tremendously due to their substitutability over the conventional metallic gear for light load applications. The gear is subjected to complex working conditions during the meshing; therefore, manufacturing technique may play a crucial role to produce the gear with high material characteristics and dimensional accuracy. The injection molding process is one of the most useful techniques for mass manufacturing of the plastic parts and gear as well because this process offers lower production cost and superior profile accuracy. However, the injection molding product may contain some common defects, namely weld line, shrinkage, residual stress, deflection, etc. This study emphasized to find out the best gate location and gate number for the spur gear made of KEPITAL®F20-03 material. A set of designs and locations of the gate have been studied using molding software with the aim to minimize molding defects. The result shows that the polymer gear with four-gate central injection system (CIS) possessed better weld line location, the maximum value of minimum weld line meeting angle of 13.292°, the minimum value of maximum von Mises residual stress of 118.352 MPa and maximum weight of 11.57 g. However, the polymer gear with two-gate CIS possessed minimum value of maximum volumetric shrinkage of 14.454% and deflection of 0.26 mm.

**Keywords** Gear · Weld line · Shrinkage · Residual stress · Deflection

B. S. Solanki (✉) · H. Singh · T. Sheorey
Department of Mechanical Engineering, IIITDM, Jabalpur, India
e-mail: 1713601@iiitdmj.ac.in

H. Singh
e-mail: hps.dme@iiitdmj.ac.in

T. Sheorey
e-mail: tanush@iiitdmj.ac.in



K. M. Pandey et al. (eds.), *Recent Advances in Mechanical Engineering*, Lecture Notes in Mechanical Engineering, https://doi.org/10.1007/978-981-15-7711-6_8

## 1 Introduction

The metallic gears are being used since a century ago to transfer the power from driving shaft to driven shaft. In recent decades, these metallic gears are replaced by polymer gears for light load applications. Since, polymer gears offer low weight, low cost, self-lubrication, noiseless operation, high corrosion resistance, high accuracy, high efficiency and design freedom [1]. Therefore, replacing the metallic gear from polymer gear can reduce 70% mass, 80% inertia and up to 9% fuel consumption [2]. Due to these advantages, polymer gears are used in automotive applications, household machinery, food and textile industries [3]. The polymer gears are manufactured using an abrasive water jet, milling, shaping, hobbing and injection molding machine. Out of these machines, the injection molding is mostly used to manufacture the polymer gear. Because this machine converts plastic pellets and plastic powder into finished gear with a superior surface feature, low production cost, high-dimensional and profile accuracy. However, the injection-molded gear may contain defects like weld lines, air traps, shrinkage, residual stresses, deflection, etc. The weld lines are formed when two different fluid streams flowing in the opposite direction and meet together due to geometrical constraints in the mold. The meeting flow fronts might have different temperature and different meeting angle. Due to this, weld lines may contain anisotropic material properties, which can be a cause of failure. Therefore, meeting flow fronts should have a minimum temperature difference and maximum meeting angle. Imihezri et al. [4] have increased the gate number from one to three, the number and length of weld lines increased in V- and X-type rib clutch pedal made of polyamide. Similarly, Zhaoet al. [5] have reported the minimum number of weld lines formed in the inner decoration part of the automobile with a single gate. Therefore, the formation of weld lines depends upon gate number, and it should be optimized. In polymer gear, the weld lines are hazardous if these appear at flank, face and a tip portion of gear teeth. Since these portions of teeth are subjected to severe mechanical and thermal loading during the meshing of gear. The volumetric shrinkage is a volumetric contraction in the cavity volume of injection molded part subjected to the cooling process. In the cooling stage, the specific volume of polymer reduces and causes contraction in part dimensions. If the reduction in a component is greater than the acceptable tolerance limit of part, it causes a huge loss to the manufacturer. Mehat et al. [6] have compared gear shrinkage in PA66-GF and recycled PA66-GF. They found gear shrinkage depends 56% on blending ratio, 24.1% on mold temperature and 10.6% on cooling time. Yoon et al. [7] have reported a 9% improvement on shrinkage of P-type gear than S-type gear at pressure released time of 4 s. Similarly, Wang et al. [8] have minimized POM micro-gear shrinkage to 0.06% by optimizing process parameters. In shrinkage of micro-gear, packing time was the most influenceable parameter with a 32.47% contribution. Taifor et al. [9] have analyzed that volumetric shrinkage increased as long as melt temperature increased; however, it decreased as the mold temperature and packing time increased. The residual stresses are the process-induced stresses frozen in the molded parts. The residual stresses influence the polymer part in a similar way to externally applied mechanical and thermal stresses. If, the magnitude of the induced residual stresses becomes higher than the material strength causes part either to change shape or crack generation. These stresses are flow-induced residual stresses and thermal induced residual stress. However, the magnitudes of flow-induced residual stresses are usually less than the thermal-induced stresses. Shruik et al. [10] have found that in cooling stage, the skin of polymer parts cool instantly and solidify while the core remains in a molten state. As cooling continues, the core heat transfer through a solidified outer surface of parts which act as a barrier in heat transfer. Due to non-uniform temperature distribution between the core and skin of parts, compressive stresses at the skin and tensile stresses at the core region were reported. Similarly, Siegmann et al. [11] have noticed that the compressive stresses at surface and tensile stresses at the core of part. Moreover, Isaza et al. [12] have found that mold and melt temperature as the most significant parameter to control the residual stresses. Huang et al. [13] have observed a thinner portion of SD card cools easily because less heat accumulates in this region but prone to high residual stress, which causes the polymer part to deflect more.

## 2 Numerical Study

### 2.1 *Polymer Gear*

For the study, an involute profile spur gear was used as model. It was designed with Solidworks 2014 (Fig. 1). The gear dimensions are given in Table 1. The keyway of dimensions 10 mm x 3 mm x 2 mm was provided to fix the gear with a shaft of 12 mm diameter. The numerical analysis was performed using CIS and surface

**Fig. 1** Gear model

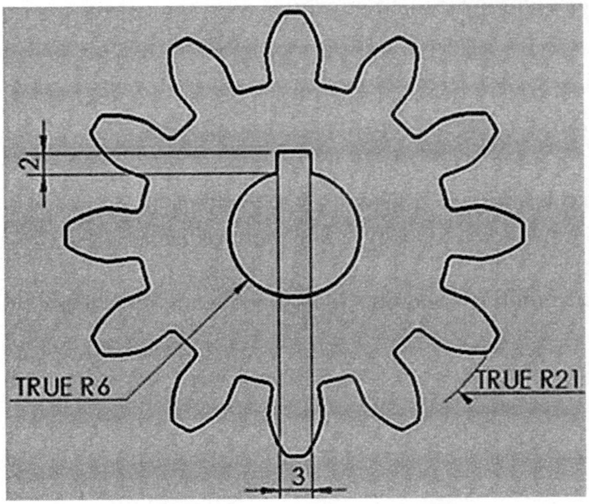

**Table 1** Gear dimensions

| Dimensions | Values |
|---|---|
| Module (mm) | 3 |
| Number of teeth | 12 |
| Face width of gear tooth (mm) | 10 |
| Pressure angle (°) | 20 |
| Tip diameter (mm) | 42 |

**Fig. 2** Gear mesh

injection system (SIS). In both types of injection system two, three and four gates were employed alternatively.

## 2.2 *Polymer Material*

In this study, KEPITAL F20-03, Acetal (POM) copolymer material was used. This is a medium viscosity grade thermoplastic material manufactured by Korea Engineering Plastics Co. Ltd. for general-purpose injection molding applications [14]. The molding condition and properties of KEPITAL F20-03 are shown in Table 2.

**Table 2** POM properties [14]

| Property name | Properties |
|---|---|
| Material structure | Crystalline |
| Melt temperature (°C) | 165 |
| Melt density (g/cc$^3$) | 1.41 |
| Melt mass flow rate (g/10 min) | 9 |

### *2.3 Injection Moulding Process Parameters*

Injection molding process parameters were fixed on the basis of material datasheet of KEPITAL F20-03 [14] and the study carried out by Ramakrishna et al. [15]. The numerical analysis was performed using gate diameter 1 mm, injection time 1.75 s, injection pressure 100 MPa, melt temperature 180 °C, packing pressure 100 MPa, packing time 25 s and cooling time 20 s, air temperature 25 °C, injection volume 8. 58267 cc$^3$ and ejection temperature 114.8 °C.

### *2.4 Tools and Outcomes*

Gate location and gate number effect evaluated to minimize molding defect (weld lines, volumetric shrinkage, residual stress and deflection) on POM gear. Mesh type used was "mixed 3D mesh" (Fig. 2). Gear model consists of 1,949,807 solid elements. The best gate location and best gate number for gear were determined numerically using the "fill-pack-warp" analysis sequence.

## 3 Results and Discussion

### *3.1 Weld Line*

In gate location and gate number analysis, 272, 274 and 364 weld lines were formed in gear with three-gate SIS, two-gate SIS (Fig. 3a) and four-gate CIS (Fig. 3b), respectively. Therefore, gear with three-gate SIS contains a minimum number of weld lines. However, in all types of SIS, long weld lines were formed on the face and flank of four teeth which would produce anisotropic material property at those teeth. This may encourage early fatigue and wear failure of gear teeth. The maximum value of minimum flow front meeting angle of 13.292° was found in gear with four-gate CIS followed by 9.639° and 7.085° with two-gate CIS and two-gate SIS, respectively. Moreover, gear with two-gate CIS contains weld lines with the maximum flow front meeting angle 134.99° resulting in better mechanical strength but concentrated weld lines at two teeth lead to failure of the gear part. Therefore, in injection system

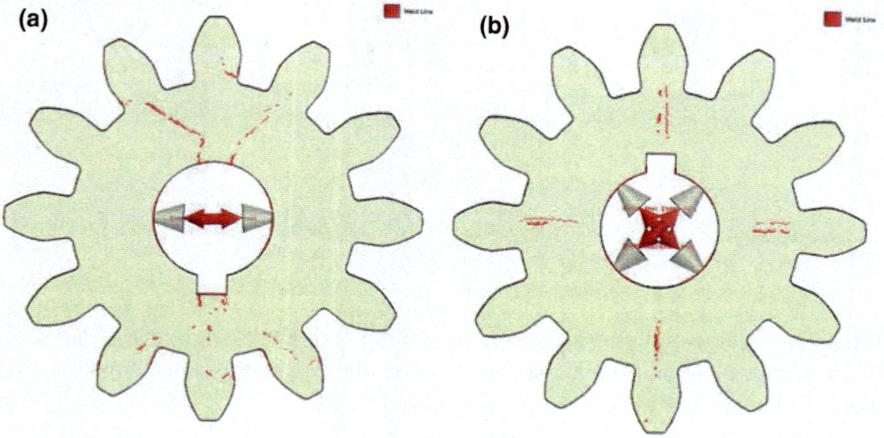


**Fig. 3** Weld line distribution on gear with (a) two-gates CIS and (b) four-gates CIS

analysis, the gear with four-gate CIS contains 374 weld lines, maximum of minimum flow front meeting angle and five teeth which were affected by small weld lines. These can be further minimized by optimizing the process parameters.

## 3.2 *Volumetric Shrinkage*

In this study, the minimum value of maximum volumetric shrinkage of 14.454% and the minimum value of average volumetric shrinkage of 5.59% were found at the end of packing stage in gear with two-gate CIS (Fig. 4). Similarly, the minimum value of maximum volumetric shrinkage of 17.358% was also obtained at the end of cooling stage in gear with two-gate CIS. A larger value of maximum volumetric shrinkage may indicate the chances of sink mark and voids inside the cavity. However, no voids were found inside the gear cavity. In a 35.618% portion of gear, small amount of positive sink mark was observed with the maximum value of 0.059 at a thicker section; moreover, in remaining portion, negative sink marks were observed with a value of −0.0008 at teeth and surface of the gear. Positive sink mark indicates that more packing pressure is required at the thicker section, whereas negative sink mark indicates over packing of teeth and surface of gear.

The minimum value of volumetric shrinkage was observed in gear with two-gate CIS because this injection system provided lowest average temperature of 85.10°C and highest average pressure of 35.087 MPa at the end of packing stage. Further, it produced the highest average polymer density (Fig. 5) of 1.348 g/cc at the end of packing stage and 1.391 g/cc at the end of cooling stage. The maximum volumetric shrinkage was also found at thicker section of gear with all types of injection system. This is due to the accumulation of heat at that section. As the polymer possesses lower thermal conductivity, therefore, surface of polymer cools easily while the core

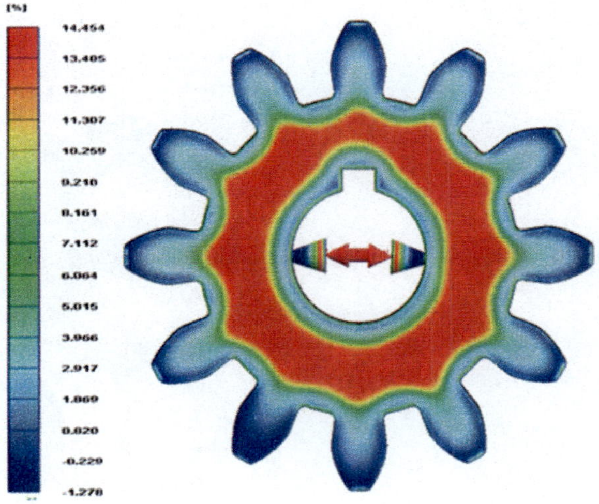


**Fig. 4** Volumetric shrinkage

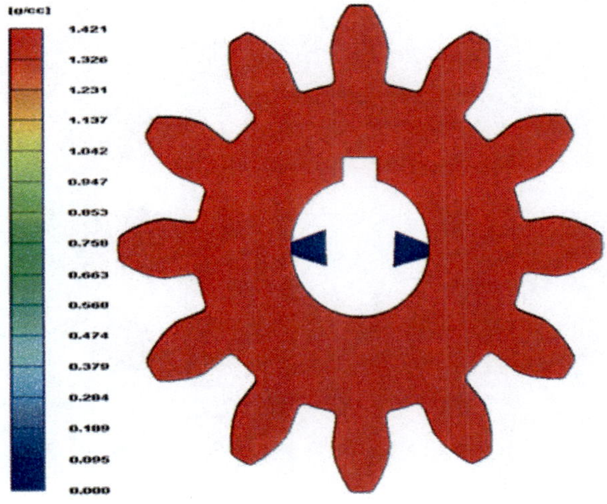


**Fig. 5** Density

remains hot and provides enough time to achieve a highly oriented structure at the core [10]. The effect of gates location and gates number on the maximum volumetric shrinakge of gear are shown in Fig. 8.

## 3.3 Residual Stress

In this study, the minimum value of maximum thermal induced von Mises stress and minimum value of average thermal induced von Mises stress were found as 14.989 MPa and 37.284 MPa in two-gate CIS (Fig. 6). Similarly, the magnitude

**Fig. 6** Residual stress

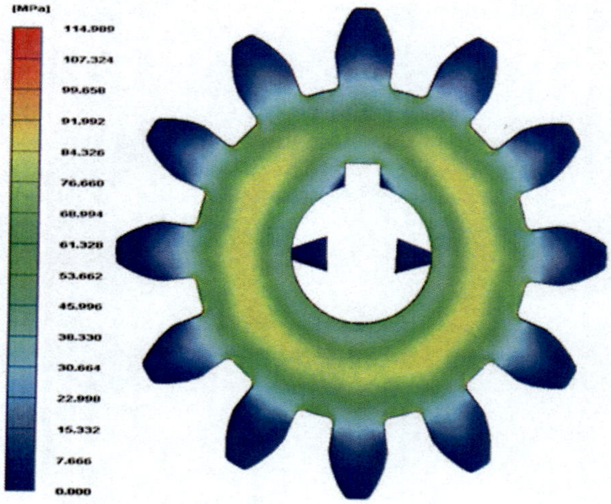


**Fig. 7** Deflection

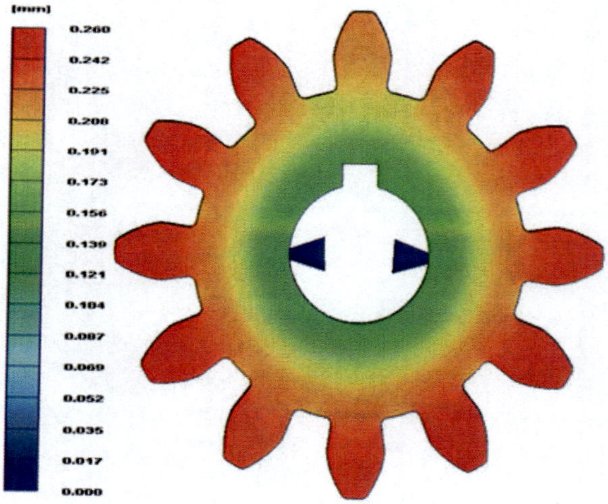


of 118.352 MPa and 38.03 MPa were found in four-gate CIS, respectively. The minimum von Mises stress was found in two-gate CIS because this injection system provides a minimum average temperature of 85.108°C and minimum average volumetric shrinkage of 5.593% at the end of packing stage. The effect of gates location and gates number on residual stress are shown in Fig. 9.

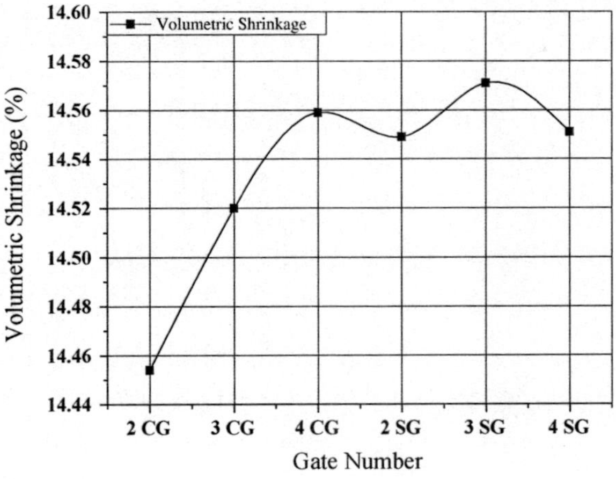


**Fig. 8** Volumetric shrinkage

## *3.4 Deflection*

In this study, the minimum value of maximum deflection of 0.26 mm found at tip of gear teeth in two-gate CIS (Fig. 7) followed by 0.272 mm in three-gate CIS. However, the minimum value of average deflection in two-gate and four-gate CIS was found as 0.204 mm and 0.212 mm, respectively. The minimum deflection was observed at a thicker section of gear, while the maximum deflection was observed in teeth of gear with all the six analyzed injection system. The maximum deflection was found at the tip of gear teeth because this zone has a larger surface area to transfer heat in comparison to a thicker section, where large heat accumulates. Therefore, gear teeth solidify easily but the thicker section of gear remains hot, which causes tensile stress at a thicker section and compressive stress at the tip of teeth. As gear teeth have low rigidity than thicker sections, these stresses produce more deflection at teeth [10]. The effect of gates location and gates number on deflection of gear are shown in Fig. 10.

# 4 Conclusions

The gate location and gate number play a significant role in the manufacturing of polymer gear. In this study, the numerical analysis was performed to observe the effect of gate location and gate number on manufacture ability of high-quality polymer gear. The following conclusions can be drawn from the present study:

- The minimum number of weld lines found in gear with three-gate SIS; however, the maximum value of minimum flow front meeting angle observed in gear with four-gate CIS.

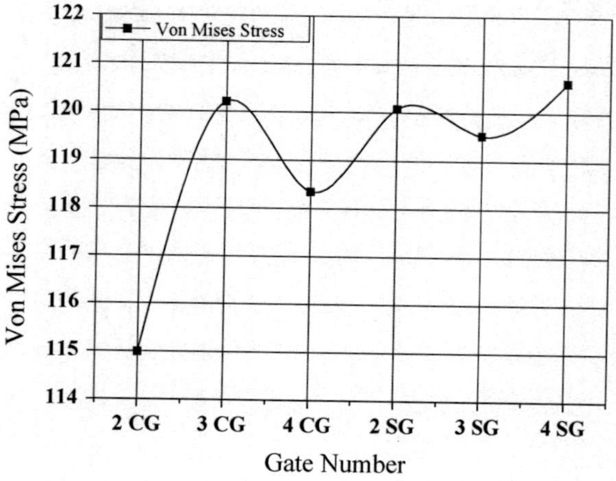


**Fig. 9** Residual stress

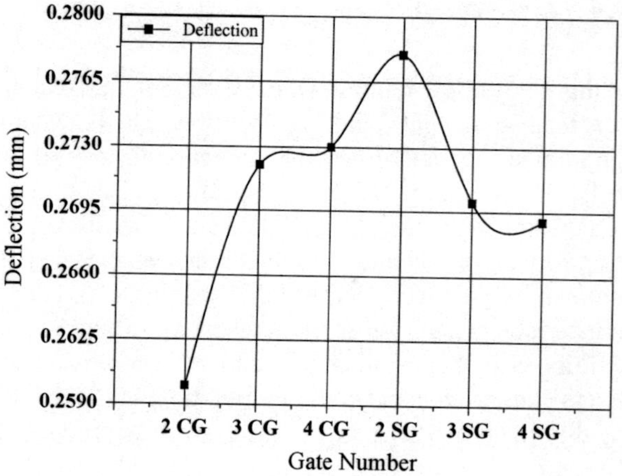


**Fig. 10** Deflection

- The minimum volumetric shrinkage and residual stress found in a thicker section of gear with two-gate CIS; moreover, the minimum deflection of 0.26 mm also found at teeth of same gear.

The gear with two-gate CIS contains minimum volumetric shrinkage, residual stress and deflection, but many weld lines were formed at gear teeth, which would degrade the strength. However, the gear with four-gate CIS contains minimum weld line at gear teeth; hence, this can be recommended for future work.

## References

1. A. Ozdemir, O. Uluer, A. Guldas, Flow Front advancement of molten thermoplastic materials during filling stage of a mold cavity. Polym. Testing **23**(8), 957–966 (2004)
2. L. Snyder, At the "PEEK" of the polymer food chain. Gear Technol. (2010), [Online] Available from: https://www.geartechnology.com/issues/0610x/victrex.pdf. Accessed 12 Oct 2019
3. H. Imrek, Performance improvement method for Nylon 6 spur gears. Tribol. Int. **42**(3), 503–510 (2009)
4. S.S.S Imihezri, S. MSapuan, S. Sulaiman, M.M. Hamdan, E.S. Zainuddina, M.R. Osmana, M.Z.A. Rahman, Mould flow and component design analysis of polymeric based composite automotive clutch pedals. Mater. Des. **171**, 358–365 (2006)
5. L. Zhao, C. Binghui, Y. Min, Z. Shangbing, Application of mold flow software in design of injection mold, in *International Conference on Mechanic Automation and Control Engineering* (Wuhan, China, 26–28 June 2010)
6. N.M. Mehat, N.S. Zakaria, S. Kamaruddin, Optimization of recycled glass fibre-reinforced plastics gear via integration of the Taguchi method and grey relational analysis. Mater. Sci. Eng. 1–10 (2017)
7. J.D. Yoon, S.W. Cha, T.H. Chong, Y.W. Ha, Study on the accuracy of injection molded plastic gearwith the assistance of supercritical fluid and a pressurized mold. Polym.-Plast. Technol. Eng. **46**(9), 815–820 (2007)
8. M.W. Wang, F. Arifin, T.T. Huynh, Optimization of molding parameters for a micro gear with Taguchi method, second forum in research science and technology. J. Phys. 1–6 (2019)
9. M. Taifor, E. Öztürk, N. Yildirim, Simulation and optimization of a plastic injection process for the manufacture of plastic gears with minimized defects in the product, in *International Conference on Mathematics Engineering Natural and Medical Science*s (Gaziantep, Turkey, 22–25 Nov 2018), pp. 1–13
10. L.C.E. Struik, Orientation effects and cooling stresses in amorphous polymers. Polym. Eng. Sci. **18**(10), 799–811 (1978)
11. A. Siegmann, A. Buchman, S. Kenig, Residual stresses in polymers I: the effect of thermal history. Polym. Eng. Sci. **22**(1), 40–47 (1982)
12. C.A.V. Isaza, J.D. Sierra, J. Posada, J.F. Botero-Cadavid, Analysis and modeling of simulated residual stress of mold injected plastic parts by using robust correlations. Revista Materia **22**(4), 1–13 (2017)
13. S.M. Huang, R.Y. Jou, J.C. Lin, K.S. Lee, Study of mold design on SD card with 3D mold flow analysis. Key Eng. Mater. **364–366**, 629–633 (2008)
14. Material Data Center, Datasheet of KEPITAL® F20–03-POM-Korea Engineering Plastics Co. Ltd. [Online] Available from: https://www.materialdatacenter.com/ms/en/tradenames/Kepital/Korea+Engineering+Plastics+Co%252E+Ltd/KEPITAL%C2%AE+F20-03/990ff15c/1408. Accessed 29 Dec 2019.
15. R. Ramakrishna, K. Mao, Minimization of shrinkage in injection molding process of acetal polymer gear using taguchi doe optimization and anova method. Int. J. Mech. Ind. Technol. **4**(2), 72–79 (2017)

# Fuzzy-PID Control of Hydro-motor Speed Used in Heavy Earth Moving Machinery

**Shivdutt Sarkar, Mohit Bhola, Harsha Rowdur, and N. Kumar**

**Abstract** Hydrostatic transmission (HST) system is widely used in the Heavy Earth Moving Machineries (HEMM). The hydro-motor used in the HST system provides the rotary motion for their traction, swing or other functions used in HEMM catering to different load profiles. This causes continuous variation in speed of the hydro-motor which reduces the efficiency of the HST system. Hence, controlling the speed of the hydro-motor is very much indeed. Constant speed of the hydro-motor can be achieved by controlling the input flow of it. Input flow can be controlled by controlling the displacement of the variable displacement hydraulic pump or by varying the flow through a proportional directional control valve (PDCV). The main objective of this research work is to design the PID controller which regulates the input flow to the hydro-motor so that the speed of the hydro-motor remains constant irrespective of the varying external load. To increase the PID controller performance, fuzzy control algorithm has been employed using MATLAB/Simulink environment. The main function of the fuzzy controller is to online tune the PID parameters which increase the performance of the drive. Comparison has been drawn based on the results obtained using Fuzzy-PID controller to the simple PID controller. The results show that Fuzzy tuned PID controller provides minimum settling time and less overshoot when compared to simple PID controller.

## 1 Introduction

In HST fluid power is used as a medium to transmit the mechanical energy at the input shaft of the pump to the output shaft of the hydraulic motor. The main reason

S. Sarkar (✉) · M. Bhola · H. Rowdur · N. Kumar
Department of Mining Machinery Engineering, Indian Institute of Technology (Indian School of Mines), Dhanbad, Jharkhand 826004, India
e-mail: iitdhanbadshivdutt@gmail.com

M. Bhola
e-mail: mohitbhola.90@gmail.com



K. M. Pandey et al. (eds.), *Recent Advances in Mechanical Engineering*, Lecture Notes in Mechanical Engineering, https://doi.org/10.1007/978-981-15-7711-6_9

for the deployment of HST system is due to its high power to weight ratio, stepless speed variations, and stable control. Essentially, an HST system can be of two types: one is closed circuit and another one is open circuit. The hydro-motor (rotary actuator) used in the HEMM provides the rotary motion for their traction, swing or other operation used in HEMM catering to the different load profiles during their operation. The speed of the hydro-motor is achieved by controlling the input flow to it. Input flow to the hydro-motor can be controlled in 3 ways:

- By controlling the displacement of the Hydraulic Pump (VDP).
- By controlling the speed of the pump with the help of variable frequency drive (VFD)
- By varying the flow through a proportional directional control valve (PDCV).

In fact, there are many types of controllers that exist in practical to get the desired output. These controllers calculate the error and nullifies it to provide the desired output. In this paper, an open circuit HST system is considered which comprises a hydrostatic pump, directional control valves, pressure control valves, and the hydro-motor. The speed of the hydro-motor is controlled irrespective of the load using PID control. The parameters of the PID need to be tuned "on-line" depending on the operation situation. Perfect tuning of the PID coefficients can make the system stable and efficient. For this, fuzzy logic controller (FLC) serves the purpose. In this regard, some of the recent works carried out on hydro-motor speed control by the eminent researchers have been discussed.

Yang et al. [1] made a comparative study on PID control and fuzzy control of the hydro-motor speed of an HST type wind turbine. The hydro-motor speed is kept almost constant by controlling the input flow at the two hydro-motors with the help of fuzzy controller. It is concluded that better dynamic response can be achieved using FLC compared to PID controller. Sailan et al. [2] presented modeling, design and implementation of steering system by using fuzzy-PID controller on an automated all-terrain vehicle (ATV). Steering wheel angle and brake pedal are the input parameters used to control the spool movement of the built-in proportional valve. Depending on the movement of the spool, the flow entering into the hydro-motor will increase or decrease which results in the change in speed of the hydro-motor.

Similar sort of work was carried out by Tran et al. [3] used fuzzy self-tuning PID controller for investigating the speed stability of the shaft. Another research carried out by Xu et al. [4] studied the control system of the coordinate controller of the fin stabilizer used in marine applications by using two electro-hydraulic proportional pumps. The work carried out by Varshneyet al. [5] has found that there is reduction in fluctuation of brushless direct current (BLDC) motor speed up to 50% and 80% during suddenly and gradual removal of load, respectively. This gets possible with the incorporation of the fuzzy-PID controller over traditional PID control algorithm.

Barai et al. [6] carried out his research on hydraulically actuated hexapod robot and its motion is effectively controlled using fuzzy logic tuned PID controller. It has been found that the fuzzy logic controller is also suitable for nonlinear plants like electro-hydraulic servo systems which have many non-linearities associated with it. Sinthipsomboon et al. [7] proposed a combination of fuzzy -PID or simple FLC for the hydro-motor speed control in servo electro-hydraulic systems (SEHS) depending on the deviation in the resulted error. In case of potential boom energy recovery of the hybrid excavator using a fuzzy-PID control system was performed by Dai et al. [8]. Fuzzy tuned PID control is used to work the generator in its most efficient zone. Results obtained with the fuzzy-PID control are more accurate when compared to results obtained through only PID controller.

Based on the above literature works, FLC is used to tune the PID parameters for the robust development of the controller according to the motor speed error and its derivative. Fuzzy-PID controller provides better performance and greater adaptability for the control system. In this respect, the current research work is focussed on controlling the speed of the hydro-motor using PID whose parameters are tuned by FLC.

## 2 Proposed System

As per the test setup figure (a), The primary objective of the current work is to control the hydro-motor speed by using Proportional Directional Control Valve (PDCV). Referring to Fig. 1, The electric motor drives a fixed displacement pump. The hydraulic fluid from the main pump flows to hydro-motor through a PDCV. To control speed of hydro-motor, solenoid controls the displacement of the DCV spool

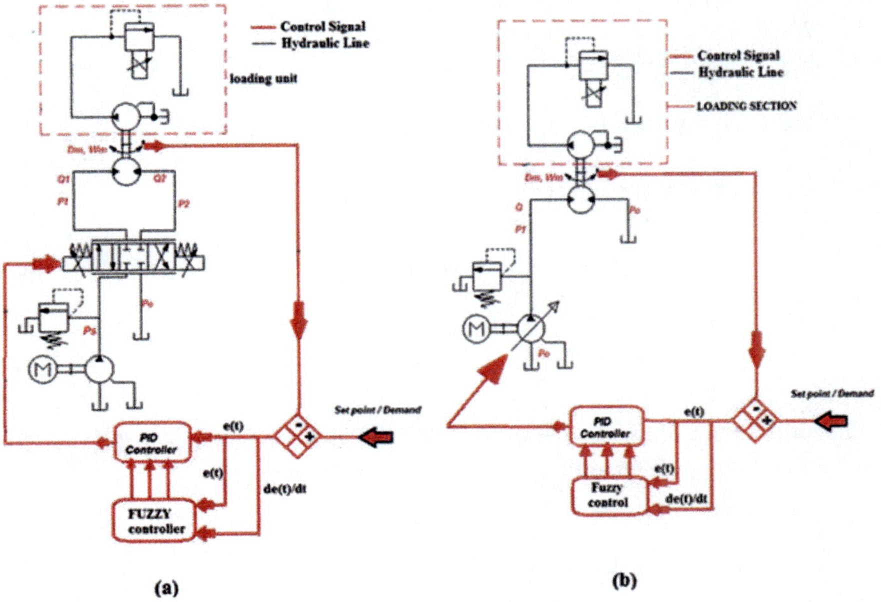


**Fig. 1** Schematic diagram of **a** proportional DCV control HST system and **b** pump control HST system

along its axis. Thus, by varying the orifice area, the amount of fluid flowing through hydro-motor is controlled. The stroke of spool is dictated by a PID controller working in a closed loop with angular velocity as feedback signal. And for linear and better control of the PID parameters, fuzzy logic is used.

Figure (b) proposes another possible way of controlling the hydromotor speed by controlling the flow discharged by the pump. In this case, the main pump used is the VDP. A VDP is driven by an electric motor, fluid from pump flows to hydro-motor via a flexible hydraulic hose. A pressure relief valve is incorporated to prevent the damage of system in case if pressure rises above its set point. A rotating load is attached to the hydraulic motor shaft. Electric motor is rotating at a fixed angular speed. The main flow from the VDP is controlled by regulating the swash plate angle. By varying the swash plate angle the volumetric displacement of pump is varied. By varying the volumetric displacement, amount of fluid flowing to the hydraulic motor is varied according to the requirement. For continuously varying load the swash plate angle needs to be changed accordingly. For this purpose, a Fuzzy-PID controller with angular velocity feedback is provided. The Fuzzy PID controller is tuned to give the best results in terms of angular velocity by continuously changing swash plate angle accordingly.

## 3 Mathematical Modelling of the Proposed System

The angular velocity of motor shaft can be calculated using this equation according to the specifications and working conditions.
$\omega_m = \int \frac{(P_1 - P_2)D_m - M_{\text{load}} - M_{\text{loss}}}{J}$ (1).

**Equations Used for Speed Control of Hydro-motor Through Proportional DCV**

Inlet and outlet flow across the hydro-motor is defined by

$$Q_1 = \frac{v}{\beta}\frac{\mathrm{d}P_1}{\mathrm{d}t} + D_m\omega_n + \frac{P_1 - P_0}{R_{\text{leakage}}} \tag{2}$$

$$Q_2 = D_m\omega_n - \frac{(P_2 - P_o)}{R_{\text{leakage}}} - \frac{v}{\beta}\frac{\mathrm{d}P_2}{\mathrm{d}t} \tag{3}$$

Equations of flow through proportional DCV are derived from the application of flow continuity through the orifices of valve and can be given as

$$Q_1 = C_\mathrm{d} w x_v(t)\sqrt{\frac{2}{\rho}(P_s - P_1)} \tag{4}$$

$$Q_2 = C_\mathrm{d} w x_v(t)\sqrt{\frac{2}{\rho}(P_2 - P_0)} \tag{5}$$

Here, $C_d$ is discharge coefficient, $x_v(t)$ is spool displacement,
Similarly, pressure across the hydro-motor defined below

$$P_2 = \frac{\beta}{v_2} \int D_m \omega_m - \frac{(P_2 - P_o)}{R_{\text{leakage}}} - Q_2 \tag{6}$$

$$P_1 = \frac{\beta}{v_1} \int Q_1 - D_m \omega_m - \frac{(P_1 - P_o)}{R_{\text{leakage}}} \tag{7}$$

**Equations Used in Speed Control of Hydro-motor Using Variable Displacement Pump**

Inlet and outlet flow of the hydro-motor is defined by considering port valve resistance at the inlet port of hydro-motor.

$$D_P \omega_P = Q_{\text{out}} + \frac{P_1 - P_0}{R_{\text{leakag}}} + \frac{V}{\beta}\frac{\mathrm{d}p_1}{\mathrm{d}t} \tag{8}$$

$$Q_{\text{out}} = \frac{P_1 - P_2}{R_{\text{VI}}} \tag{9}$$

Similarly, pressure across the hydro-motor defined with considering external leakage and port valve resistance

$$P_2 = \frac{\beta}{V} \int \left( \frac{P_1 - P_2}{R_{\text{VI}}} - \frac{P_2 - P_0}{R_{\text{leakage}}} - D_m \omega_m \right) \tag{10}$$

$$P_1 = \frac{\beta}{V} \int \left( D_P \omega_P - \frac{P_1 - P_2}{R_{\text{VI}}} - \frac{P_1 - P_0}{R_{\text{leakage}}} \right) \tag{11}$$

## 4 PID Controller

The structure of the (Proportional Integral and Derivative) PID control algorithm used for the angular velocity control of HST system is given below.

$$u(t) = k_p.e(t) + k_i \int_0^t e(t)\mathrm{d}t + k_d \frac{\mathrm{d}e(t)}{\mathrm{d}t} \tag{12}$$

The initial tuning of the PID control parameters was made based on the **Ziegler–Nichols** tuning method and the gain coefficients are $K_p = 0.04$, $K_i = 0.75$ and $K_d = 0.001$. Likewise, in Fig. 1b for pump control HST system values of PID parameters are $K_P = 5$, $K_I = 2$ and $K_D = 0.5$.

## 5 Design of the Fuzzy-PID Controller

The basic schematic structure of the fuzzy-PID controller and rule base is shown in Figs. 2 and 3 respectively. The inputs for the fuzzy block are the speed error '$e(t)$' and the time differential of the speed error 'd$e(t)$/d$t$'. These parameters generate the tuned PID controller coefficients.

### Membership Function of Input and Output of Fuzzy-PID Control in MATLAB/SIMULINK

### Fuzzy Rules to Calculate the Coefficients of the Parameters of the PID Controller

Figure 4 provides the logic for the calculation of the coefficients of the PID controller. The range of PID parameters based on the speed error and its rate is defined.

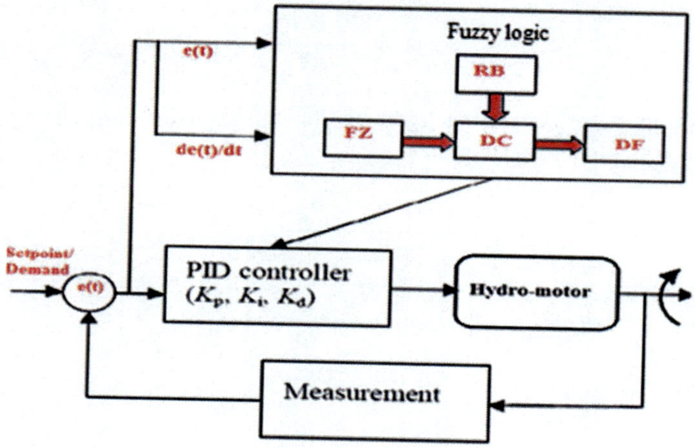


**Fig. 2** Self-tuning parameter fuzzy-PID control structure

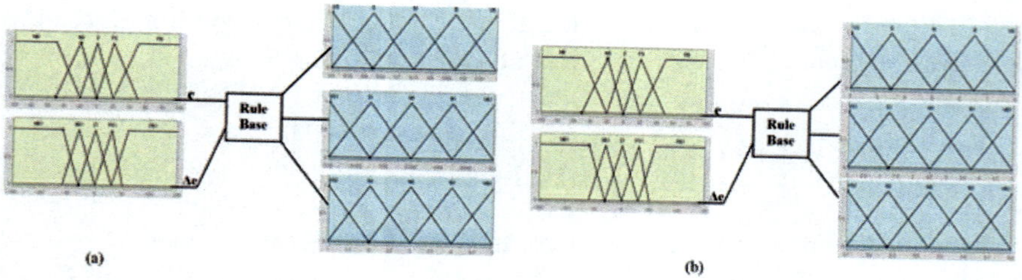


**Fig. 3** Membership function for **a** proportional DCV control system and **b** variable displacement pump control

*Fuzzy rules for Proportional gain ($K_P$)*

| e \ Δe | NB1 | NS1 | Z1 | PS1 | PB1 |
|---|---|---|---|---|---|
| NB | VB | VB | B | B | M |
| NS | V | B | B | M | S |
| Z | B | M | M | M | S |
| PS | B | M | S | S | VS |
| PB | M | S | S | VS | VS |

Fuzzy rules for integral gain ($K_I$)

| e \ Δe | NB1 | NS1 | Z1 | PS1 | PB1 |
|---|---|---|---|---|---|
| NB | VS1 | VS1 | S1 | M1 | M1 |
| NS | VS1 | S1 | M1 | M1 | M1 |
| Z | S1 | S1 | M1 | B1 | B1 |
| PS | S1 | M1 | M1 | B1 | VB1 |
| PB | M1 | M1 | B1 | VB1 | VB1 |

Fuzzy rules for derivative gain ($K_D$)

| e \ Δe | NB1 | NS1 | Z1 | PS1 | PB1 |
|---|---|---|---|---|---|
| NB | M2 | S2 | VS2 | M2 | M2 |
| NS | M2 | S2 | S2 | S2 | M2 |
| Z | M2 | S2 | S2 | S2 | M2 |
| PS | B2 | M2 | M2 | M2 | B2 |
| PB | VB2 | B2 | M2 | B2 | VB2 |

**Fig. 4** Fuzzy rules calculated for online tuning of PID parameters

# 6 Simulation Work and Results

The simulation work has been carried out on MATLAB/Simulink is discussed below.

## *6.1 MATLAB/SIMULINK Model of the Fuzzy-PID Control Test Setup for Proportional (DCV) Control*

Figure 5 shows the hydraulic circuit developed for the HST drive operated by PDCV developed in MATLAB Simulink environment. Hydro-motor speed error and its rate are the input parameters fed to fuzzy system in order to optimize the PID parameters for better response and control of the hydro-motor speed.

Figure 6 shows the actual speed response with demand speed using conventional PID and FuzzyPID controller tuned by applying Ziegler–Nichols method. It has

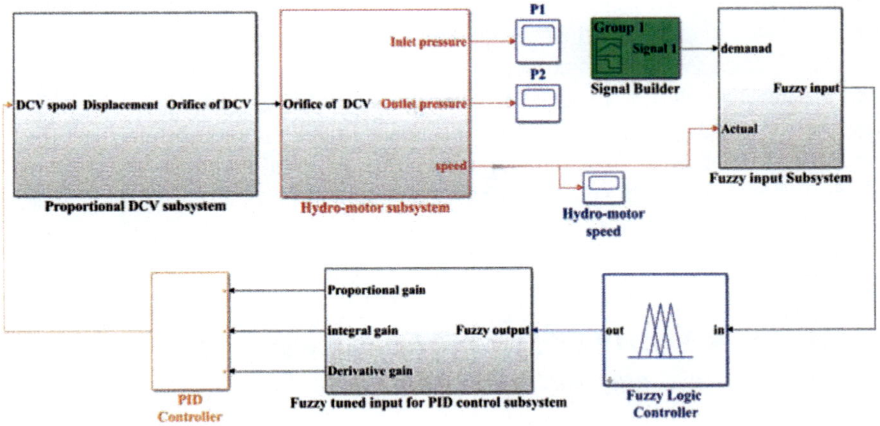


**Fig. 5** Proposed MATLAB/SIMULINK model of proportional DCV control system

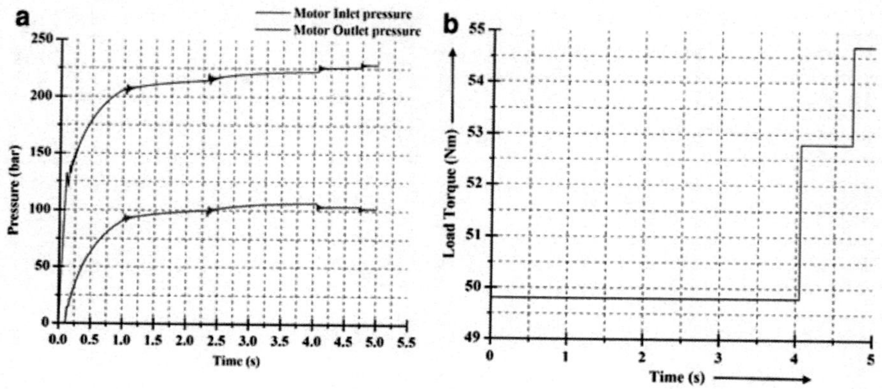


**Fig. 6** **a** Hydro-motor load-torque versus time, **b** pressure difference across hydro-motor versus time

been observed that there is a significant amount of oscillations during step demand of the speed. After 2.27 s, the load on the hydro-motor varies results in fluctuation of the speed at the time 4.1 and 4.6 s, respectively. So, to minimize the settling time, overshoot and making the system linearly more stable conventional PID is tuned by fuzzy logic.

The comparison has been made between the conventional PID controller and fuzzy-PID controller. It has been observed that there is a significant reduction in settling time and overshoot in the speed response of the hydro-motor using fuzzy-PID controller.

**Load and Pressure Varying Graph with Time**

The pressure difference across the hydro-motor at different load and speed are shown in Fig. 6. As the graph shows when the load varies, inlet pressure $P_1$ increase at the same time outlet pressure $P_2$ decrease. Because when $P_1$ increase then inlet flow will decrease, hence to retain the flow across the motor the pressure $P_2$ decreases.

Figure 7 represents the comparison in response to the hydromotor speed variation when system is employed under conventional PID and Fuzzy-PID controller. The blue line represents system response with conventional PID control and red line with Fuzzy-PID control. From the graph, it can be observed that the peak rise time and settling time of the PID control is high compared to fuzzy-PID control. Hence, fuzzy-PID control is more efficient and better in response compared to PID controller.

## 6.2 *MATLAB/SIMILINK System Model of Fuzzy-PID Control Test Setup for Pump Control*

Figure 8 shows the hydraulic circuit developed for the HST drive operated by controlling the displacement of the VDP developed in MATLAB Simulink environment.

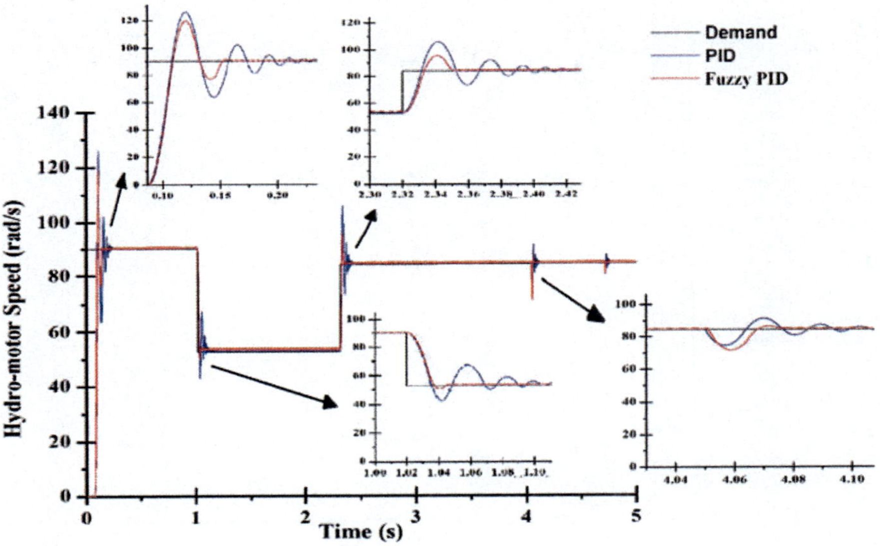


**Fig. 7** System step response for the Fuzzy-PID control

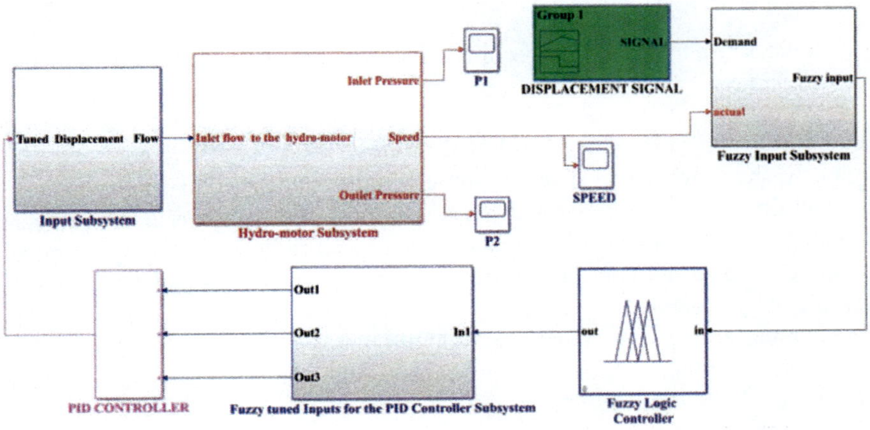


**Fig. 8** Proposed MATLAB/SIMULINK model of variable displacement pump control system

Hydro-motor speed error and its rate are the input parameters fed to fuzzy system in order to optimize the PID parameters for better response and control of the hydro-motor speed. Flow coming out the VDP is regulated based on the demand of the hydro-motor speed.

From Fig. 9 it can be observed that response of the system following a first-order system and fuzzy-PID controller is having better response compared to PID control

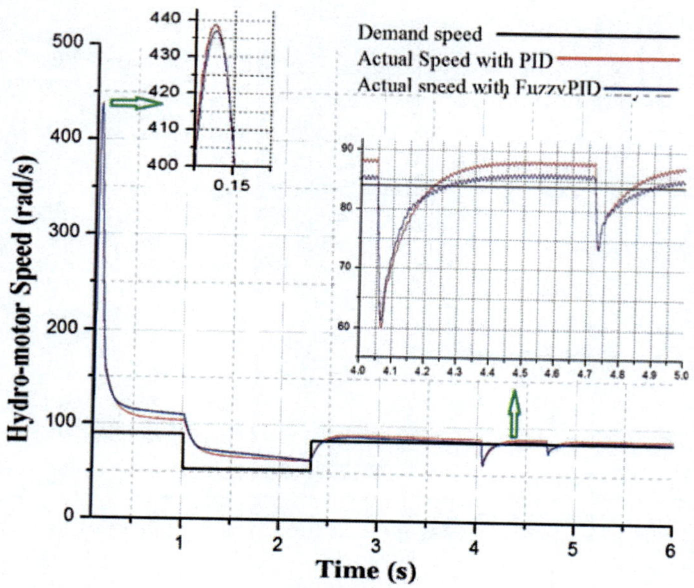


**Fig. 9** System step response for the Fuzzy-PID control

system. Peak rise time and time required to achieve SteadyState has been shown in zoomed portions.

**Load and Pressure Varying Graph with Time**
Figure 10 shows the actual speed response with demand speed using conventional PID and Fuzzy-PID controller and it has been observed that there is a significant reduction in settling time and overshoot in the speed response of the hydro-motor using fuzzy-PID controller. The pressure loss between inlet pressure ($P_1$) and inside pressure of hydro-motor ($P_2$) is due to port valve resistant, nearly about 2 bar. Both test setup performed at the same load of hydro-motor used in proposed model and result has been found that the response time of VDP control system is more than Proportional DCV control system.

## 7 Conclusion and Future Work

This manuscript mainly focuses on the controlling the hydro motor speed irrespective of change in external load torque and speed demand. This is accomplished with the help of PID controller and fuzzy-PID control. The comparative analysis has been drawn based on the simulation results obtained from MATLAB Simulink environment. From the above performance of Proportional(DCV) control system and variable displacement pump control (VDPC) system. It has been concluded that the speed response of hydro-motor is precisely controlled by fuzzy-PID controller when

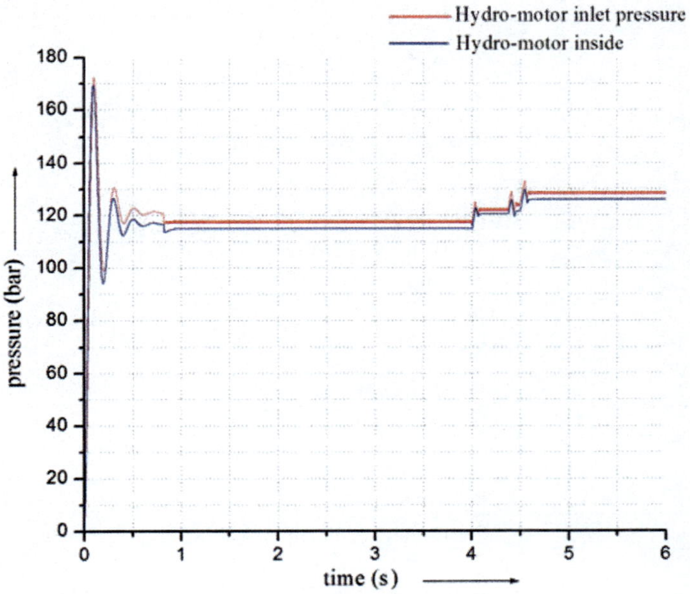


**Fig. 10** Comparison between actual and demand pressure

compared with PID controller. The response of fuzzy-PID controller is much better than PID controller in sense of various parameters like settling time, overshoot and achieving the steady-state. Hence, it is much necessary for the online tuning process of the PID parameters using fuzzy logic control, for smooth control of hydro-motor for varying speed and load profiles. Due to irrespective change in external load torque and speed demand of an HST system its stability, efficiency, accuracy & production getting affected so this paper proposed a novel control system to control the all these following parameters like Delay time, Rise time, Peak time, Time constant, Settling time and Maximum overshoot as well that could optimize HST system at any circumstances.

| Appendix | Factors |
|---|---|
| 4.45e-6 $m^3$/rad | Displacement of Hydro-motor ($D_m$) |
| 350e5 Pa | Constant pressure supply by pump ($P_s$) |
| 0 Pa | Atmospheric Pressure (pressure of tank) ($P_o$) |
| 1e9 Pa | The bulk modulus of oil ($\beta$) |
| 0.001 $kgm^2$ | Inertia load at Hydro-motor output shaft ($J$) |
| 0.01 m | Valve gradient ($W$) |
| 0.8e-3 m | The maximum opening width of Valve ($x_v(t)$) |
| 1000 | Filter coefficient ($N$) |
| 1e12 n-s/$m^5$ | External leakage ($R_{leakage}$) |

(continued)

(continued)

| Appendix | Factors |
|---|---|
| 1e11 n/m$^5$ | Bulk stiffness ($\frac{\beta}{v}$) |
| 0.019 kgm$^2$ | Inertia due to viscusfriction ($M_{\text{loss}}$) |
| 16e-10 n-s/m$^5$ | Port valve resistance |
| VS | Very small |
| S | Small |
| M | Medium |
| B | Big |
| VB | Very big |
| NS | Negative small |
| NB | Negative big |
| Z | Zero |
| PS | Positive small |
| PB | Positive big |

## References


1. B. Xu, L. Sha, J. Zhang, R. Ding,.The coordinate controller of the two electro-hydraulic proportional pumps controlled system for a heavy load and large inertia fin stabilizer, in *JFPS International Symposium on Fluid Power* (2014)
2. H. Yang (2016), Comparative Study on PID control and fuzzy control of motor speed of hydraulic transmission type wind turbine, in *IEEE International Conference on Power and Renewable Energy*
3. D.G. Varshney, B. Dwivedi, Speed response of brushless DC motor using fuzzy-PID controller under varying load condition. J. Electro. Syst. Inf. Technol. **4**, 310–321 (2017)
4. K. Sinthipsomboon, I. Hunsacharoonroj, J. Khedari, A hybrid of fuzzy and fuzzy self-tuning PID controller for servo electro-hydraulic system. Intech open. https://doi.org/10.5772/48614(2012)
5. R.K. Barai, K. Nonami, optimal two-degree-of-freedom fuzzy control for locomotion control of a hydraulically actuated hexapod robot. Inf. Sci. (2007)
6. X. Dai, C. Zhang, S. Li, Fuzzy PIO control for boom energyrecovery on hybrid hydraulic excavator, in *International Conference on Computer Science and Automation Engineering* (2011)
7. K. Sailan, K.D. Kuhnert, H. Karelia, Modeling, design, and implement of steering fuzzy-PID control system for DORIS robot. Int. J. Comput. Commun. Eng. (2014)
8. T.A. Tran, Y. Yuan, Marine engine rotational speed control automatic system based on Fuzzy-PID logic controller, in *International Conference on Transportation Information and Safety* (2017)

# Phase Change in an Enclosure Under Different Combinations of Boundary Wall Condition: A Numerical Study

**Bhaskar Ranjan Tamuli, Sujit Nath, and Dipankar Bhanja**

**Abstract** The research presented here studies heat transfer process in phase change material (PCM) in an enclosure under three different combinations of boundary heated surface. Isothermal conditions were imposed at the boundary, with adjacent sides and opposite sides heated. Numerical techniques were implemented to solve the governing equation employing "effective heat capacity" method for modelling of the phase change. The fastest melting was observed for adjacent heating walls, approximately 28.08 and 13.79% faster than other two. It was observed that natural convection is the main reason driving the pace of melting and has different impacts under different conditions. Natural convection is not significant while providing heat through vertical walls which also corresponds to slowest response time. The natural convection is also responsible for the wave-like shape attained by the solid–liquid interface. The pattern of melting signifies a symmetric melting pattern for the vertically heated sides.

**Keywords** Phase change material · Effective heat capacity · Melt fraction

## Nomenclature

$c_p$ Specific heat of PCM (kJ/kg K)
$g$ Gravitational acceleration (m/s$^2$)
$L_F$ Latent heat of fusion (kJ/kg)
$k$ Thermal conductivity of PCM (W/m K)

B. R. Tamuli (✉) · S. Nath · D. Bhanja
Department of Mechanical Engineering, National Institute of Technology Silchar, Silchar, India
e-mail: tamulibhaskarranjan@gmail.com

S. Nath
e-mail: sujitnath2008@gmail.com

D. Bhanja
e-mail: dipankar.bhanja@gmail.com


K. M. Pandey et al. (eds.), *Recent Advances in Mechanical Engineering*, Lecture Notes in Mechanical Engineering, https://doi.org/10.1007/978-981-15-7711-6_10

$p$ Pressure (Pa)
$T$ Temperature (K)
$T_m$ Melting temperature (K)
$t$ Time (s)
$\upsilon$ Velocity (m/s)
$\beta$ Coefficient of thermal expansion (1/K)
$\theta$ Melt fraction
$\mu$ Viscosity of liquid PCM (Pa s)
$\rho$ Density of fluid (kg/m$^3$)

## 1 Introduction

Flexibility in energy usage is one of the desired qualities for a better energy security. Unlike other forms of energy, heat energy dissipates continuously and is difficult to store. This problem is countered by energy storage systems which essentially stores energy in the form of sensible or latent heat. Latent heat energy storage (LHES) is more suitable than sensible energy storage due to its higher volumetric energy density [1]. Phase change material (PCM) has been quite commonly used for the purpose of energy storage and more research is now being focused in it. Apart from energy storage, it has found different applications in the fields of electronic cooling, solar PVT, building integrated cooling system, space heating, etc. [2, 3].

Heat transfer in enclosures has been widely studied as it resembles many engineering problems e.g. solar energy collection, geothermal applications, reactor cooling, insulation systems, etc. [4]. Natural convection plays a crucial role in the convection inside these enclosures. Enclosures filled with PCM are also studied by different researchers over the years. Experimental investigation on a rectangular enclosure with lauric acid was investigated by Kamkari et al. [5] on the angular orientation and reported that decrease in angular placement with horizontal enhances the chaotic flow resulting in the formation of Benard convection cells. Enclosure with PCM was also studied widely and one such numerical study is performed by Fadl and Eames [6] with a different heat flux applied to one boundary. Most of the PCM suffers the disadvantage of low thermal conductivity. Different techniques have been employed to tackle this problem such as inserting of metal fins, nano-particles and wire mesh. The phase change process can be subdivided into four stages: conduction, transition, strong convection and weak convection. Experimental study by Shokouhmand and Kamkari [7] reported that the convection intensity decreases with the evolution of melt front. Though conduction is equally dominant in phase change phenomena, it is the buoyancy-driven natural convection which is responsible for interface shape and the melting rate. The melting rate seems to be affected by horizontal fins with different lengths attached to a PCM enclosure. Compared with equal length fins, decreasing fin length ratio increases the melting rate and the melting time is reduced significantly. The melt front propagation is subjected to the boundary condition applied and it greatly affects the heat transfer and flow physics. For latent heat storage (LHS) system incorporating PCM, the natural convection plays an important role in charging rather than discharging [8]. Hence, understanding the details of natural convection and its variation with different applied boundary conditions is of extreme importance. Impact of natural convection in an enclosure with PCM is related to metal structure through which heat is provided and its orientation. Location of the applied heated surface has been varied in a study by Dadvand et al. [9] at three different locations and their result signifies that placing of heated surface at top promotes high melting rate.

The above-mentioned literatures provide the evidence that natural convection plays an important role in the melting process and it is applicable only after the liquid phase is attained. This work presents a comparison of three different orientations of boundary conditions in a view to study the natural convection induced in the process. Observation of the melting pattern and the set-up of natural convection under the influence of boundary conditions is the prime focus in this study. This is aimed to be achieved by numerical techniques employing suitable models.

## 2 Physical Model

Figure 1 shows a schematic diagram of the different cases studied in the presented paper. A square enclosure with the dimensions $H = 0.05$ m is filled with a phase change material. Isothermal boundary condition is applied at the boundary of the enclosure. Three cases are studied here, where the isothermal conditions are applied in three different ways. In Case 1, two adjacent sides are kept at constant temperature while the other two are maintained at adiabatic condition. In Cases 2 and 3 being similar, isothermal boundary condition is implemented at two opposite sides. In Case 2, two vertical sides are maintained in isothermal condition whereas same situation has been made for two horizontal sides in Case 3. Natural convection set up differently according to spatial position. Therefore, above-mentioned cases have

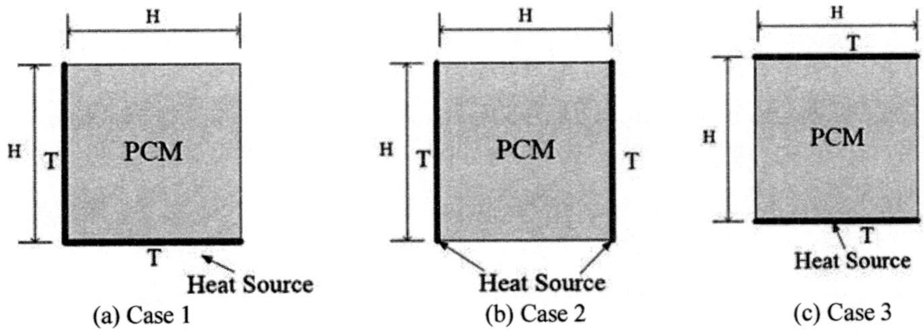


**Fig. 1** Schematic representation of the enclosure studied

**Table 1** Thermo-physical properties of lauric acid [10]

| | |
|---|---|
| Specific heat capacity (kJ/kg K) | 2.18 (solid)—2.39 (liquid) |
| Density (kg/m$^3$) | 940 (solid)—885 (liquid) |
| Thermal conductivity (W/m K) | 0.16 (solid)—0.14 (liquid) |
| Viscosity (Pa s) | 0.004 |
| Latent heat of fusion (kJ/kg) | 187 |
| Thermal expansion coefficient (1/K) | 0.0008 |
| Melting temperature range (°C) | 43.5–48.2 |

been selected with a view to understand the effect of heat input through different locations in the enclosure. Lauric acid is selected as the phase change material here. The thermo-physical properties are given in Table 1.

### *2.1 Governing Equation*

2-D transient numerical study was performed for the investigation of the PCM in the enclosure. The PCM domain is assumed to be melt over a range of temperature. The governing equations are shown below

Continuity equation

$$\nabla.\vec{v} = 0 \tag{1}$$

Momentum equation

$$\frac{\partial \vec{v}}{\partial t} + (\vec{v}.\nabla)\vec{v} = \frac{1}{\rho}\left(-\nabla p + \mu\nabla^2\vec{v} + \vec{F} + \vec{S}\right) \tag{2}$$

Energy equation

$$\rho c_p \frac{DT}{Dt} = k\nabla^2 T \tag{3}$$

The buoyancy effect due to natural convection is combined in the body force term F in momentum equation and Boussinesq approximation is assumed.

$$\vec{F} = \rho\vec{g}\beta(T - T_m) \tag{4}$$

The PCM is treated as liquid in the developed model; hence to nullify the effect of velocity in the solid PCM, Darcy's source term is used which is represented by $\vec{S}$.

$$\vec{S} = \frac{(1-\theta)^{2}}{\left(\theta^{3}+e\right)} A_{\text{mush}} \vec{g} \tag{5}$$

$A_{\text{mush}}$ is mushy zone constant and it can have any arbitrary value ranging from $10^3$ to $10^7$. Based on the literature, the value is taken as $10^6$. This parameter is responsible for velocity transition in mushy region.

Theta ($\theta$) is the melt fraction and expressed as:

$$\theta = \frac{T - T_s}{T_l - T_s} \tag{6}$$

The model is based on effective heat capacity method, which combined both specific heat and latent heat of a material into one term called effective heat capacity.

$$c_p = \begin{cases} c_{p_s} & \text{if } \; T < T_s \\ \frac{c_{p_s}+c_{p_l}}{2} + \frac{L_F}{T_l - T_s} & \text{if } \; T_s \le T \le T_l \\ c_{p_l} & \text{if } \; T > T_l \end{cases} \tag{7}$$

The above equations are solved using COMSOL Multiphysics [10]. The special treatment required to handle the extra source term and the specific heat is allowed by the phase change module of the software.

## 2.2 *Initial and Boundary Condition*

Initially, ($t = 0$) the whole domain is maintained at initial temperature, $T_o = 298$ K. At any time, $t > 0$, isothermal boundary condition ($T$) is applied at specified boundaries as mentioned in Fig. 1. All the other sides of the enclosure are remained insulated. The simulation is carried out at dimensionless Stefan number 1.20 corresponding to temperature ($T$) 393 K which ensures smooth handling by the computational software to observe results. Stefan number (*Ste*) is defined as the ratio of sensible heat to the latent heat and expressed as:

$$\text{Ste} = \frac{c_{p_s}(T_m - T_o) + c_{p_l}(T - T_m)}{L_F}$$

## 2.3 *Model Validation*

The developed model is validated by comparing against the work of Abdi et al. [10]. The physical model and the geometrical arrangements considered are same in

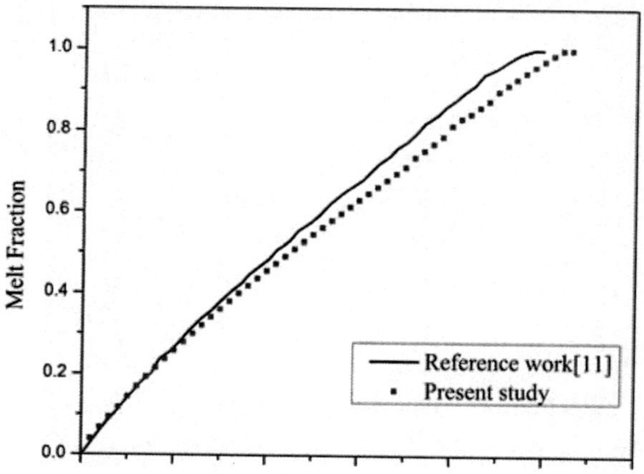


**Fig. 2** Comparison of developed model with Abdi et al. [11] for bottom heated cavity

their published work. A rectangular enclosure of length 0.12 m and width 0.5 m surrounded by a layer of plexiglass of thickness 0.025 m is the studied domain. Isothermal condition of Ste = 0.55 is maintained at the bottom wall of the enclosure. Figure 2 shows the melt fraction curve for the reference work and the present work. The result depicts a well agreement of current numerical result with published work.

## 3 Results and Discussion

The present study intends to compare and evaluate the melting phenomenon due to application of heat at different boundaries of the enclosure. In addition, how this variation of isothermal boundaries induces the natural convection is also under the scope of the investigation. Before proceeding with the intended analysis, grid independency test was performed. Four models with 10,000, 15,000, 20,000 and 22,500 grid points are studied. The grid independency test showed that the relative change in melting time for each model with bottom heated temperature ($T = 343$ K) was 1.5, 0.06 and 0.03%. Hence, little changes in the result were observed when model is upgraded from 15,000 grid points onwards. Keeping both the accuracy and computational cost in mind, the model with 15,000 grids is chosen as the point of grid independency. The convergence criteria are taken as $10^{-6}$.

Figure 3 depicts the liquid fraction or the melt fraction growth over time due to the application of heat. As heat is supplied through the boundaries of the enclosure to the PCM inside, which is kept at an initial temperature of 298 K, the temperature of the PCM starts increasing. The temperature variation inside the enclosure is not uniform, resulting in some portion of the PCM being melted while some part still in solid phase. Henceforth, there is coexistence of two phases (solid and liquid) which can be expressed by the parameter melt fraction or the liquid fraction. Melt fraction study for the present work reveals that adjacent heated boundary sides are faster than other two cases. Case 2 and Case 3 show 28.08 and 13.79% slower response than Case

**Fig. 3** Melt fraction graph

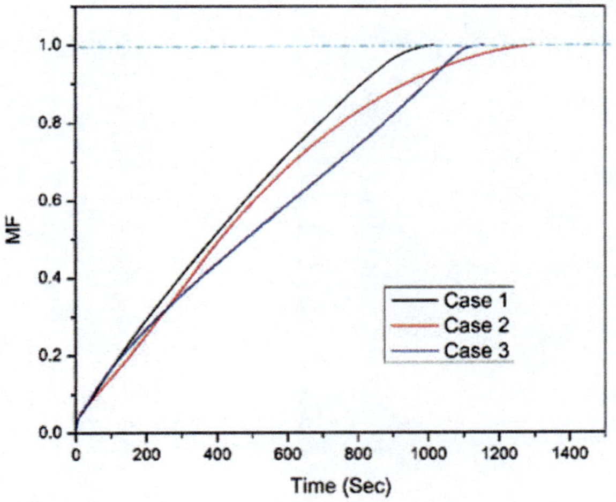


1. Natural convection can be attributed for this difference in melting time observed in these three cases. Interestingly, the melt fraction curve for Cases 2 and 3 intersects with each other which points out that for most of the time period faster melting rate was observed for two parallel vertical heated sides than horizontal sides. However nearing the completion of the melting process, the convection in horizontally heated case becomes dominant resulting in faster completion of the melting process. This phenomenon can be explained with contour plots.

Figure 4 shows the visualization of the same parameter as mentioned earlier, but it will give a more comprehensive idea about it. At the very early phase of the melting process, conduction mode is more dominated as seen by symmetric propagation of melt fraction. At 300 s, formation of Benard convection cell in the horizontal heated sides can be observed in Cases 1 and 3 by the wave-like interface of melt fraction. At this instance, natural convection sets up in the PCM which accelerated the heat transfer process. In Case 3, the top heated side does not contribute to natural convection and only conduction mode prevails from the top side which leads to relatively slower melting process than Case 2. After some time, the effect of natural convection diminishes in vertically heated case leading to weak convection phase which shows that the total melting time required becomes highest among all the cases. At 1000 s, almost all the PCM is converted to liquid part, whereas a significant portion of solid can be seen in Cases 2 and 3.

Figure 5 describes the temperature contour of the enclosure at different time instances. This gradual depiction of contours also gives an idea of evolution of melting pattern in different cases. In Cases 1 and 3, at the bottom horizontal side, Benard convection cells are found forming which will eventually give rise to natural convection. Due to the close proximity of the heated sides, the convective current mixing is possible in Case 1 which leads to lowest melting time. In Case 3, there is no convection induced due to top heated wall and the heat transfer occurs by conduction. For bottom plate, natural convection appears which makes the melting

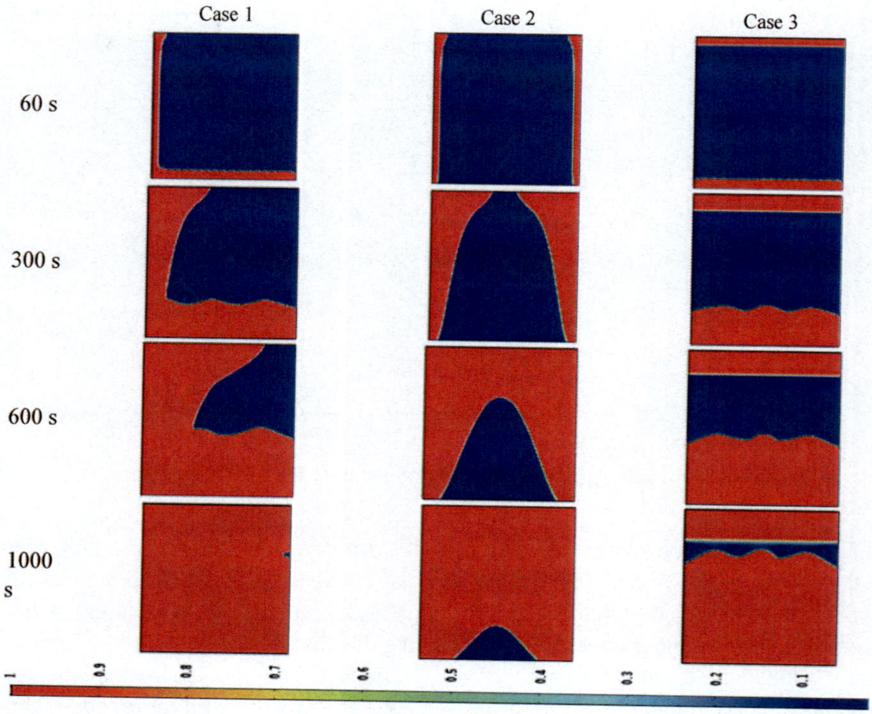


**Fig. 4** Melt fraction contour

process quicker in upward direction. However, evolution of melting pattern in Case 2 is symmetrical about vertical axis. The melt front produced by the heated surface at both right and left walls, progressed towards each other and finally merged as bell curve-shaped melting interface is formed. The contour plot 5 also infers about the variation of the temperature inside the enclosure. The higher temperature region is generally shifted towards the heated boundaries. But in Case 2, higher temperature regions are identified at the top of the enclosure rather than left and right walls and with time it gradually propagates to bottom of the enclosure.

## 4 Conclusion

The study presented here investigates PCM in an enclosure numerically under different combinations of heated surface. Isothermal conditions are maintained at two adjacent and parallel sides of the enclosure at $Ste = 1.2$. During phase change process, both conduction and convection phenomena come to the picture, but which mode is dominant over the other ultimately decides the melting rate. The results show that when heat is supplied through the adjacent sides, fastest melting rate is achieved.

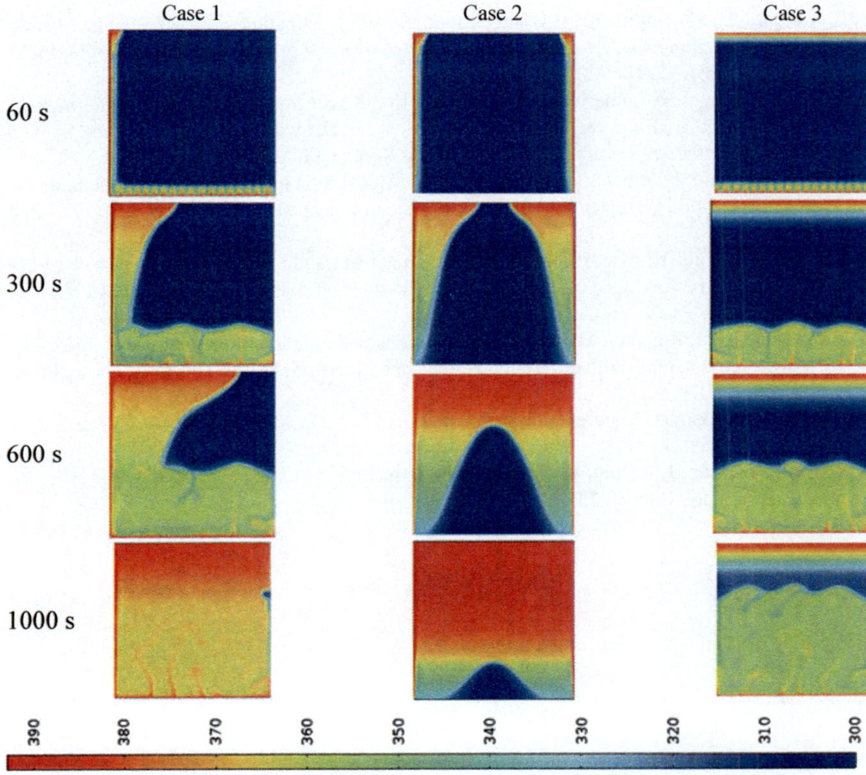


**Fig. 5** Temperature contour

The combination of natural convection induced by both the plates contributes to it. Parallel heated sides, i.e. both vertically heated and horizontally heated cases, show 28.08 and 13.79% slower response than adjacent heated sides. The melting patterns deform when natural convection becomes dominant, otherwise it follows a steady symmetrical path.

## References

1. S. Tiari, S. Qiu, Mahdavi, M, Numerical study of finned heat pipe-assisted thermal energy storage system with high temperature phase change material. Energy Convers. Manage. **89**, 833–842 (2015)
2. R. Kalbasi, M. Afrand, J. Alsarraf, M.D. Tran, Studies on optimum fi ns number in PCM-based heat sinks. Energy **171**, 1088–1099 (2019)
3. M.M. Alkilani, K. Sopian, M.A. Alghoul, M. Sohif, M.H. Ruslan, Review of solar air collectors with thermal storage units. Renew. Sustain. Energy Rev. **15**, 1476–1490 (2011)
4. L.M. Al-balushi, M.J. Uddin, M.M. Rahman, Natural convective heat transfer in a square enclosure utilizing magnetic nanoparticles. Propuls. Power Res. **8**, 194–209 (2019)

5. B. Kamkari, H. Shokouhmand, F. Bruno, Experimental investigation of the effect of inclination angle on convection-driven melting of phase change material in a rectangular enclosure. Int. J. Heat Mass Transf. **72**, 186–200 (2014)
6. M. Fadl, P. Eames, A numerical investigation into the heat transfer and melting the melting process of lauric acid in a rectangular enclosure with three values of wall heat flux, in *10th Internation Conference on Applied Energy* (Hong Kong, China, 22–25 Aug 2018)
7. H. Shokouhmand, B. Kamkari, Experimental investigation on melting heat transfer characteristics of lauric acid in a rectangular thermal storage unit. Exp. Therm. Fluid Sci. **50**, 201–212 (2013)
8. Y.B. Tao, Y.K. Liu, Y.L. He, Effects of PCM arrangement and natural convection on charging and discharging performance of shell-and-tube LHS unit. Int. J. Heat Mass Transf. **115**, 99–107 (2017)
9. A. Dadvand, N.H. Boukani, M. Dawoodian, Numerical simulation of the melting of a NePCM due to a heated thin plate with different positions in a square enclosure. Therm. Sci. Eng. Prog. **7**, 248–266 (2018)
10. COMSOL multiphysics reference manual, version 5.3, COMSOL Inc. https://www.comsol.com
11. A. Abdi, V. Martin, J.N. Chiu, W, Numerical investigation of melting in a cavity with vertically oriented fins. Appl. Energy. **235**, 1027–1040 (2019)

# Finite Element Modelling of Electromagnetic Crimping of Copper-Stainless Steel Tube-to-Tube Joint

**Deepak Kumar, Sachin D. Kore, and Arup Nandy**

**Abstract** Electromagnetic crimping is a high-speed joining by forming method that deforms electrically conducting materials using an electromagnetic field. This paper explores the finite element modelling of electromagnetic crimping of Cu-SS tube-to-tube joint with the use of LS-DYNA™ software which utilizes finite element method combined with the boundary element method. Simulations are performed at five different discharge energy values and stand-off distance, as well as the overlapped length of the tube is kept constant. Effect of discharge energies on the magnetic field, radial displacement and impact velocity and effect of plastic strain have been studied. The maximum magnetic field of 17 T is obtained at 6.2 kJ of discharge energy. The developed model can be used as a primary study to investigate interference-fit tube-to-tube joining by electromagnetic forming.

**Keywords** Electromagnetic crimping · Tube-to-tube joint · Interference-fit · Cu-SS · Finite element modelling · LS-DYNA

## List of Symbols

| | |
|---|---|
| $E$ | Discharge energy |
| $R$ | Resistance |
| $L$ | Inductance |
| $C$ | Capacitance |
| $I$ | Current |

D. Kumar (✉) · A. Nandy
Department of Mechanical Engineering, Indian Institute of Technology Guwahati, Guwahati, India
e-mail: kumar176103003@iitg.ac.in

A. Nandy
e-mail: arupn@iitg.ac.in

S. D. Kore
School of Mechanical Sciences, Indian Institute of Technology Goa, Goa, India
e-mail: sachin@iitgoa.ac.in



K. M. Pandey et al. (eds.), *Recent Advances in Mechanical Engineering*, Lecture Notes in Mechanical Engineering, https://doi.org/10.1007/978-981-15-7711-6_11

| | |
|---|---|
| $B$ | Magnetic flux density |
| $E$ | Electric field intensity |
| $H$ | Magnetic field intensity |
| $\sigma$ | Electrical conductivity |
| $A$& $B$, $C$, $n$ | Material constants |
| $T$ | Absolute temperature |
| $\delta$ | Skin depth |
| $C_1$ | Structural damping matrix |
| $\ddot{u}$ | Nodal acceleration vector |
| $u$ | Nodal displacement vector |
| $\delta$ | Damping coefficient |
| $V$ | Discharge voltage |
| $J$ | Current density |
| $p(t)$ | Magnetic pressure |
| $T_m$ | Melting temperature |
| $J_s$ | Source current density |
| $\rho$ | Total charge density |
| $t$ | Time |
| $\mu_m$ | Magnetic permeability |
| $F$ | Lorentz force |
| $\varepsilon$ | Electrical permittivity |
| $\sigma_y$ | Equivalent plastic stress |
| $T_r$ | Room temperature |
| $M$ | Structural mass matrix |
| $K$ | Structural stiffness matrix |
| $\dot{u}$ | Nodal velocity vector |
| $\vec{F}$ | Applied load vector |
| $\omega$ | Damped angular frequency |
| $V_0$ | Initial discharge voltage |

## 1 Introduction

Electromagnetic forming is a high-speed forming method, which utilizes the electromagnetic field to deform an electrically conducting material like copper, aluminium, etc. Compared to conventional forming, high-strain rate forming allows materials like aluminium to improve formability [1]. Electromagnetic forming can have one of the most promising applications in joining and forming of tubular parts [2–4]. Two variations of joining by electromagnetic forming have been proposed as electromagnetic crimping and electromagnetic welding [5]. Electromagnetic crimping is a joining by an electromagnetic forming technique where parts are joined without any metallic bond formation. This method is preferred for asymmetrical objects. Based on the process parameters as well as the shape and material of joining parts, the mechanism of joining can be form-fit, interference-fit or a combination of both. Form-fit joining is based on mechanical interlocking between two parts, while interference-fit is based on interference stress between the parts. Strength of such joints depends on the interference stress between the workpiece, coefficient of friction and area of contact between the workpiece [6]. Hyping Yu et al. have developed a sequential coupling model using Ansys explicit dynamics for electromagnetic forming [7]. In this paper, a finite element model has been developed using electromagnetic module of LS-DYNA™ software to simulate electromagnetic crimping of copper (outer) tube onto SS 304 (inner) tube creating an interference-fit tube-to-tube joint. Variation of the magnetic field, radial displacement, impact velocity and effective plastic has been studied at five different discharge energies. Discharge energy levels are selected based on experimental investigation. Pull-out strength of the Cu-SS tubular joint gets stagnant after 6.2 kJ of discharge energy. Therefore, analysis is performed for 3.9 kJ (lower) to 6.2 kJ (higher) discharge energy levels.

## *1.1 The Theoretical Background of Electromagnetic Crimping*

Electromagnetic crimping can be analysed in terms of an RLC circuit as shown in Fig. 1. Capacitor bank, which acts as a power source, discharges a high-amplitude, damped sinusoidal current through the solenoid coil. Current flowing in the coil generates a magnetic field as per Maxwell's law. A single-step field shaper is inserted between the solenoid coil and copper tube. As per Faraday's law of eddy current induction, the magnetic field induces a secondary current on the surface of the field

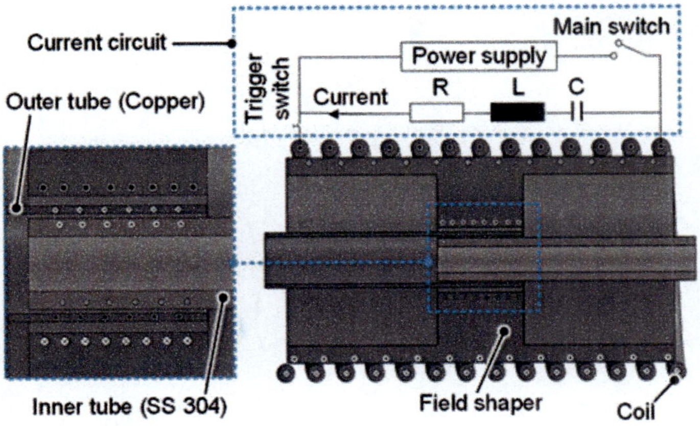


**Fig. 1** Schematic diagram of the electromagnetic crimping set-up

shaper which flows inwards towards the axis due to slit of the field shaper generating its own magnetic field. This induces a tertiary current on the surface of the copper tube leading to another magnetic field generation in the opposite direction. Interaction of opposite magnetic fields causes high Lorentz force acting on the outer tube leading to plastic and elastic deformation of the copper and SS 304 tube, respectively.

Physics of the electromagnetic crimping process can be explained by Maxwell's equation which is the governing equation of electromagnetic crimping as shown from Eqs. 1 to 6.

$$\vec{\nabla} \times \vec{E} = \frac{-\mathrm{d}\vec{B}}{\mathrm{d}t} \tag{1}$$

$$\vec{\nabla} \times \vec{H} = \vec{J} + \varepsilon \frac{\mathrm{d}\vec{E}}{\mathrm{d}t} \tag{2}$$

$$\vec{\nabla} \cdot \vec{B} = 0 \tag{3}$$

$$\vec{\nabla} \cdot \vec{E} = \frac{\rho}{\varepsilon} \tag{4}$$

$$\vec{J} = \sigma \vec{E} + \vec{J}_S \tag{5}$$

$$\vec{B} = \mu \vec{H} \tag{6}$$

Energy stored in the electromagnetic forming system can be represented as

$$E_1 = \frac{1}{2} C V^2 \tag{7}$$

$$\vec{B} = \vec{\nabla} \cdot \vec{A} \tag{8}$$

$$\vec{J} = \vec{\nabla} \times \frac{\vec{B}}{\mu_m} \tag{9}$$

$$\vec{J} = \vec{\nabla} \times \frac{1}{\mu_m} \left( \vec{\nabla} \times \vec{A} \right) \tag{10}$$

$$\vec{F} = \vec{J} \times \vec{B} = \left( \vec{\nabla} \times \frac{\vec{B}}{\mu_m} \right) \times \vec{B} \tag{11}$$

$$p(r, t) = \int_{ro}^{ri} F(r, t)\mathrm{d}r = \frac{1}{2} \mu \left( H_{\mathrm{gap}}^2(t) - H_{\mathrm{pen}}^2(t) \right) \tag{12}$$

$$p(t) = \frac{1}{2}\mu H_{\text{gap}}^2(t) \tag{13}$$

The Lorentz force acting on the outer tube is dependent on the skin effect. Discharge frequency is a function of circuit inductance and capacitance of the electromagnetic forming system. Skin depth is a function of electrical conductivity, frequency and magnetic permeability of the workpiece. Skin depth can be described as

$$\delta = \sqrt{\frac{1}{\pi \mu_0 f \sigma}} \tag{14}$$

## 2 Finite Element Modelling

A finite element model has been developed using LS-DYNA™ electromagnetic module, which uses a combination of finite element method and boundary element method to solve the electromagnetic forming problem. BEM solves electromagnetic problems for regions, which does not have eddy current and therefore eliminates the use of meshing for air, which further reduces the complexity of the model. The simulation set-up is shown in Fig. 2. Dimensions of the coil and workpiece are shown in Tables 1 and 2, respectively.

Schematic diagram of the flow chart of the finite element model is shown in Fig. 3.

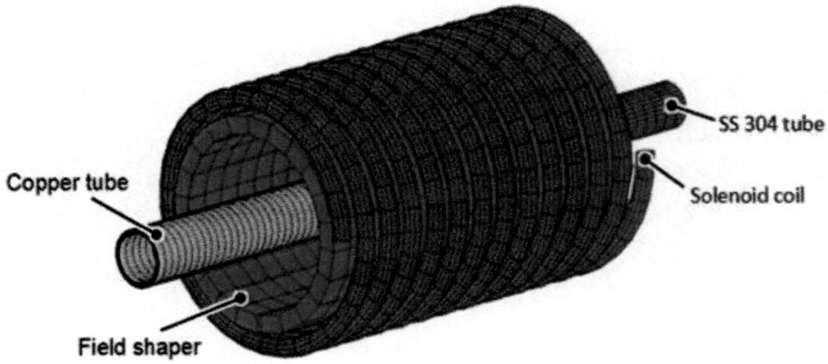


**Fig. 2** Simulation set-up

**Table 1** Dimensions of the solenoid coil

| Coil material: copper | |
|---|---|
| No. of turns: 13 | Outer diameter: 57 mm |
| Coil length: 91 mm | Inner diameter: 47 mm |
| Outer diameter: 55 mm | Diameter of wire: 4 mm |

**Table 2** Dimensions of the outer and inner tube

| Parts | Materials | Outer diameter (mm) | Inner diameter (mm) | Thickness (mm) |
|---|---|---|---|---|
| Outer tube | Copper | 12.7 | 10.9 | 0.9 |
| Inner tube | SS 304 | 10.0 | 6.0 | 2.0 |

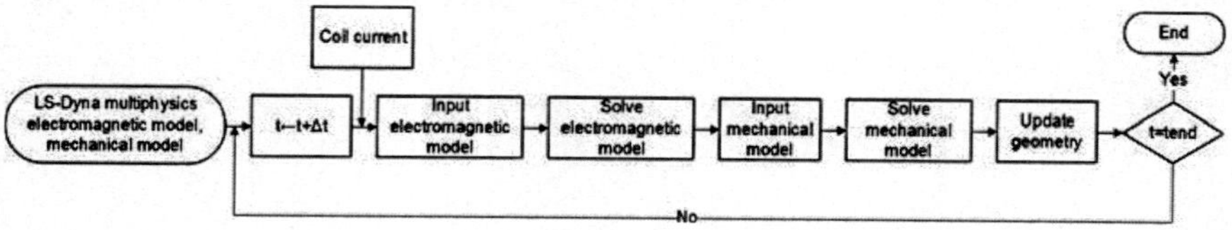


**Fig. 3** Flow chart showing the simulation steps

LS-DYNA solves the model in two steps. First, it solves the electromagnetic model to compute the Lorentz force from equation acting on each node of the outer tube and use it as an input to the mechanical solver, which computes the deformation. The mechanical solver uses transient dynamic equilibrium equation to compute the deformation at each time step [8].

$$\boldsymbol{M}\ddot{\boldsymbol{u}} + \boldsymbol{C}_1\dot{\boldsymbol{u}} + \boldsymbol{K}\boldsymbol{u} = \vec{\boldsymbol{F}} \tag{15}$$

which further updates the geometry back into the electromagnetic model, and the cycle continues with each time increment until the simulation reaches on end time. All the parts are modelled with brick element. The number of elements of each part is shown in the figure. Schematic of the simulation set-up is shown in the figure. The field shaper, and outer and inner tubes are modelled with eight-node brick element, whereas solenoid coil is modelled with four-node rigid brick element. Electromagnetic crimping process is a high-strain rate process, and a simplified Johnson–Cook material model, which is a widely used material model for high temperature and high-strain rate applications, is used. Here, the effect of temperature is neglected. The simplified Johnson–Cook material model is given in Eq. 16. Johnson–Cook parameters for copper and SS 304 are given in Table 3.

$$\sigma_y = (A + B\varepsilon^n)\left(1 + C\ln\left(\frac{\dot{\varepsilon}}{\dot{\varepsilon}_o}\right)\right) \tag{16}$$

**Table 3** Johnson–Cook material constants for copper and SS 304

| Material | $A$ (MPa) | $B$ (MPa) | $n$ | $C$ | $T_m$ (K) | $m$ |
|---|---|---|---|---|---|---|
| Copper | 90 | 292 | 0.31 | 0.025 | 1331 | 1.09 |
| SS 304 | 350 | 275 | 0.36 | 0.022 | 1723 | 1 |

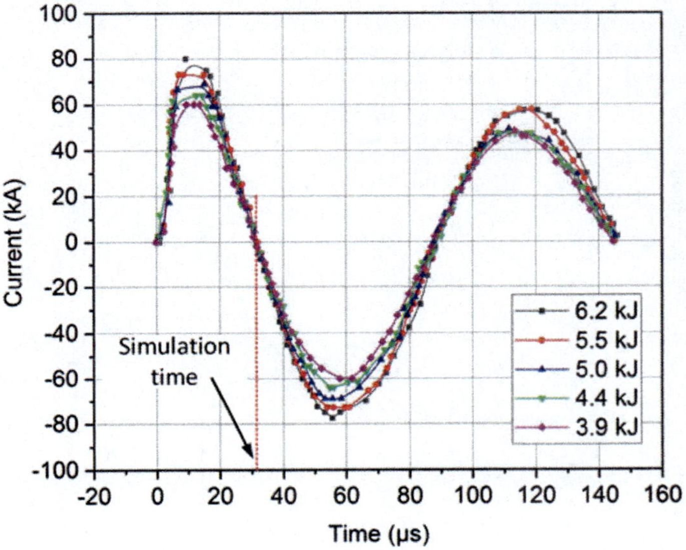


**Fig. 4** Discharge current waveform at different discharge energies

Discharge current flowing in the solenoid coil is used as a loading curve in the electromagnetic model of LS-DYNA™ to solve the electromagnetic model, which is measured experimentally with the help of Rogowski coil and an oscilloscope as shown in Fig. 4. Current can be explained by a damped sinusoidal function as

$$I = \frac{V_0}{\omega L} e^{-\delta t} \sin \omega t \tag{17}$$

It has been observed that major deformation happens during electromagnetic crimping and during the first half period of the first cycle of discharge current waveform. Therefore, to reduce the computation time, simulation is performed up to the first pulse only. Discharge current waveform is shown in the figure. To make the results independent of the time step and mesh size, mesh convergence study has been performed at four different mesh sizes as shown in Fig. 5.

Results are measured at four different nodes, and convergence is achieved at all the nodes. Final selected mesh size is 0.5 mm. The total number of elements and nodes in the final simulation model is 47,948 and 56,363, respectively.

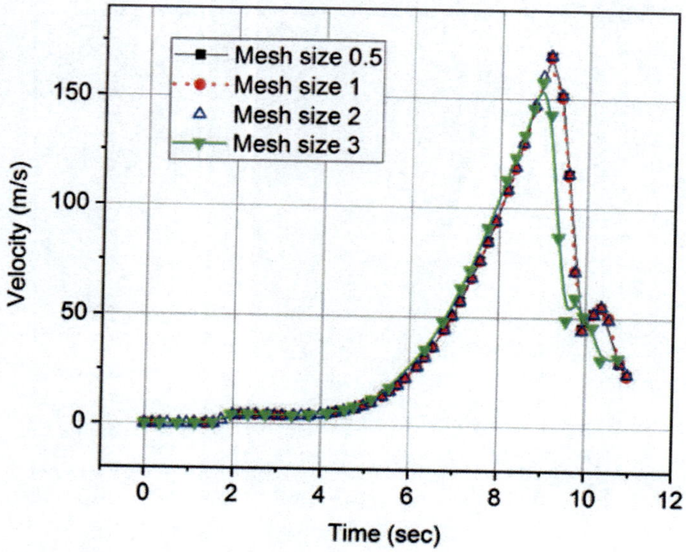


**Fig. 5** Variation of impact velocity with time at 3.9 kJ of discharge energy for 4 different mesh sizes

# 3 Result and Discussion

## 3.1 *Magnetic Field*

The magnetic field is dependent on many factors, such as the number of turns per unit length of the solenoid coil, discharge current flowing in the solenoid coil and the steps of the field shaper which intensifies the magnetic field strength. Discharge current is dependent on discharge energy/voltage as per Eq. 17, and higher current leads to higher magnetic field strength. Therefore, the magnetic field varies with time as per the variation of the current pulse with time as shown in Fig. 4. From the results, it is observed that the magnetic field strength increases with an increase in discharge energy. The maximum magnetic field of 17 T is observed at 6.2 kJ of discharge energy as shown in Fig. 6a.

## 3.2 *Radial Displacement*

Radial deformation of the outer tube increases with an increase in discharge energy. Maximum radial deformation is observed at 6.2 kJ of discharge energy as shown in Fig. 6b.

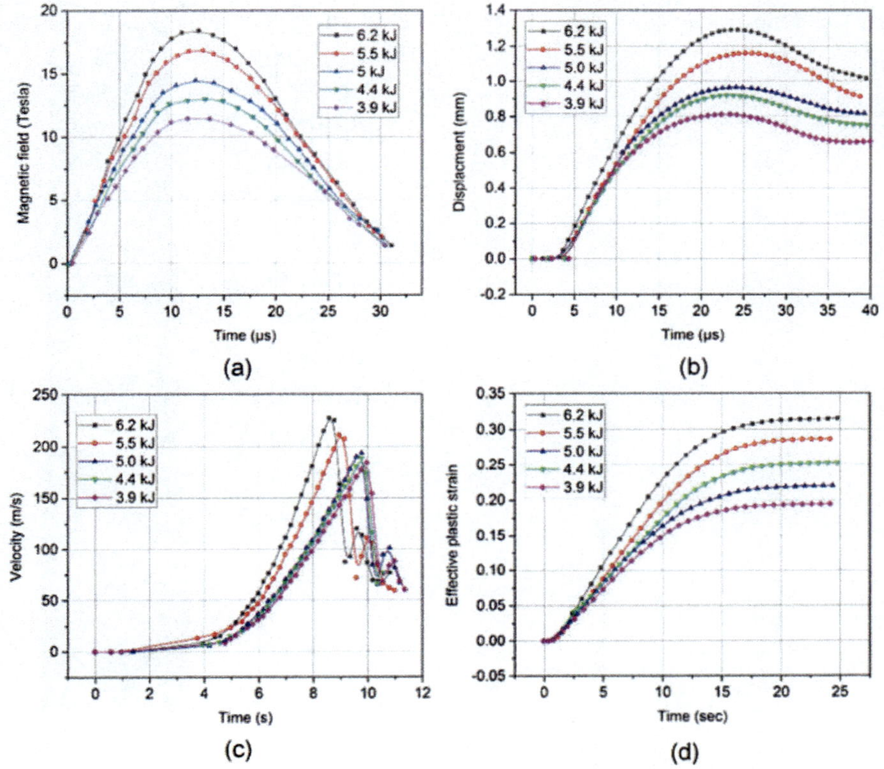


**Fig. 6** Variation of **a** magnetic field, **b** displacement, **c** velocity and **d** effective plastic strain in the outer tube with time in the crimping zone

## 3.3 *Impact Velocity*

The velocity of the impact of the outer tube onto the inner tube is averaged in the crimping zone, and the maximum impact velocity of this is observed at 6.2 kJ of discharge energy. Impact velocity increases with an increase in discharge energy as shown in Fig. 6c.

## 3.4 *Effective Plastic Strain*

Increase in discharge energy leads to an increase in impact velocity, which causes an increase in the effective plastic strain of the outer tube as shown in the figure. The maximum effective plastic strain of 0.31 is observed at 6.2 kJ of discharge energy. The fringe pattern showing the effective plastic strain of the electromagnetically

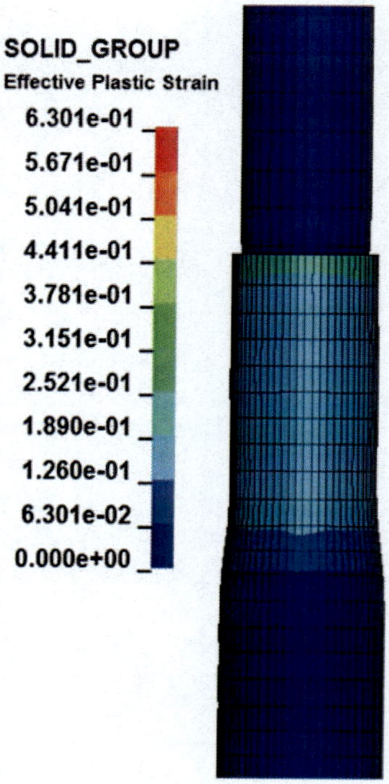


**Fig. 7** Effective plastic strain in the electromagnetically crimped joint at 6.2 kJ of discharge energy

crimped sample at 6.2 kJ is shown in Fig. 6d, and the fringe pattern of the crimped joint is shown in Fig. 7.

## 4 Conclusion

A finite element model has been developed for electromagnetic crimping of Cu-SS tube-to-tube joint using LS-DYNA™ software. Simulations are performed at five discharge energy values, and the copper tube is electromagnetically crimped onto SS 304 tube. FEM–BEM combination is very suitable because it reduces the computation time. Increase of discharge energy leads to an increase in deformation of the outer tube, which increases the interference pressure leading to higher joining strength.

## References

1. V. Psyk, D. Risch, B. Kinsey, A. Tekkaya, M. Kleiner, Electromagnetic forming—a review. J. Mater. Process. Technol. **211**, 787–829 (2011)
2. G. Areda, S.D. Kore, Numerical study and experimental investigation on electromagnetic crimping for tube-to-rod configuration. Int. J. Precis. Eng. Manuf. **20**, 181–191 (2019)
3. A. Rajak, S.D. Kore, Experimental investigation of aluminium–copper wire crimping with electromagnetic process: Its advantages over conventional process. J. Manuf. Processes **26**, 57–66 (2017)
4. S. Pawar, S.D. Kore, Electromagnetic forming and perforation of aluminium tubes. J. Mech. Sci. Technol. **33**, 5999–6007 (2019)
5. C. Weddeling, O. Demir, P. Haupt, A. Tekkaya, Analytical methodology for the process design of electromagnetic crimping. J. Mater. Process. Technol. **222**, 163–180 (2015)
6. M. Kleiner, M. Marré, C. Beerwald, W. Homberg, D. Löhe, P. Barreiro, V. Schulze, Investigation of force-fit joints produced by electromagnetic tube compression. Annal. German Acad. Soc. Prod. Eng. WGP **XIII**, 227–230 (2006)
7. H. Yu, C. Li, Z. Zhao, Z. Li, Effect of field shaper on magnetic pressure in electromagnetic forming. J. Mater. Process. Technol. **168**, 245–249 (2005)
8. T. Itoch, Damped vibration mode superposition method for dynamic response analysis. Earthquake Eng. Struct. Dynam. **2**(1), 47–57 (1973)

# Optimization of Protrusions for an Impinging Jet on a Curved Surface

**Alankrita Singh and B. V. S. S. S. Prasad**

**Abstract** Numerical simulations using ANSYS Fluent 17.2 are carried out to explore the heat transfer characteristics of jet impingement on a protruded leading edge of a gas turbine blade. Many researchers have found that inclusion of protrusion does not improve Nusselt number though this contributes to augmentation in heat transfer due to an increase of surface area. In certain cases, the protrusion has also been found to degrade the heat transfer performance. Therefore, this work focuses on the effect of single protrusion to determine the effect of position of the protrusion on a concave surface and finally obtaining an optimized position of protrusions. The position of the single protrusion is varied at several angles with respect to the centre of the plate to explore the corresponding heat transfer characteristics. Thereafter, multiple protrusions are considered at desired positions to determine optimum effective location of protrusions for maximum heat transfer. The optimal position is determined using Genetic Algorithm.

**Keywords** Impingement cooling · Genetic Algorithm · concave surface · Protrusion · Numerical simulations

## 1 Introduction

Jet impingement is an effective cooling technique to keep thermo-mechanical stresses of turbine blades within safe limits, while operating at higher turbine inlet temperatures (TIT). Impinging jets involve complex flow physics and also inherit complex heat and mass transfer phenomena. Because of their flexibility, they are used in various industries like metal annealing, paper and textile drying, electronics cooling and turbine blade cooling.

A. Singh (✉) · B. V. S. S. S. Prasad
Department of Mechanical Engineering, Indian Institute of Technology Madras, Chennai, India
e-mail: alankrita_92@rediffmail.com

B. V. S. S. S. Prasad
e-mail: prasad@iitm.ac.in



K. M. Pandey et al. (eds.), *Recent Advances in Mechanical Engineering*, Lecture Notes in Mechanical Engineering, https://doi.org/10.1007/978-981-15-7711-6_12

Compared to the studies of jet impingement on smooth flat and concave surfaces, very few work can be found on protruded surfaces. Similar to pins and ribs, protrusions are provided to generate high turbulence around them. Xie et al. [1] numerically, investigated three-dimensional convective heat transfer in rectangular two-pass channels which consisted of arrays of circular pins, hemispherical dimples and protrusions. The pins, dimples and protrusions enhanced heat transfer with respect to smooth tip channel at a moderate subsidiary pressure loss. They also concluded that protrusions exhibit moderately higher heat transfer compared to dimples at higher Reynolds number. Zhang et al. [2] carried out PIV and numerical simulations for single jet impingement on a spherical protrusion over a flat surface. They observed an improvement in the Nusselt number with the increase of protrusions area. An elevation in heat transfer with the use of dimples and protrusions was also observed by Shen et al. [3], in a rectangular passage with ejection slots. Dobbertean and Rahman [4] conducted a numerical study on heat transfer from an impinging jet on a rectangular and triangular ribs over flat plate for different materials. For all the cases, they found an increase in the Nusselt number values with an increase of the depth of indentation of the protrusion. However, for the case of rectangular step, there was a limit on the depth of indentation because of slower recirculation. Jing et al. [5] observed an enhancement in cooling by various arrangements of protrusions in non-flat channels. Taslim et al. [6] conducted experimental studies on the inclusion of a protrusion on a concave surface. They found that the major contribution of heat transfer augmentation was due to increased surface area for heat transfer.

Huang et al. [7] investigated the effect of a dimple on the thermal performance. They concluded from their study that an optimized arrangement of dimples with the right size is necessary to achieve better heat transfer. In view of the foregoing discussion, the present study is devoted to study the effect of location of a single protrusion on a concave surface to determine the desired position for inclusion of protrusion for the purpose of heat transfer augmentation.

## 2 Methodology

### 2.1 *Physical Model*

In spite of the fact that the leading edge of gas turbine is not of circular shape, the use of circular profile is regarded to be the best model for the objective of specifying the geometric parameters. With this view, the present modelling is done on semi-circular concave surface. The present work focuses on evaluating and comparing numerical simulations of the heat transfer characteristics of jet impingement cooling from smooth and protruded semi-circular concave plate at various Reynolds number and curvature ratio. For the purpose of validation, some of the dimensions of geometry are based on the numerical study of Yang et al. [8]. The geometrical model used in simulation for present study is shown in Fig. 1. Two-dimensional uniformly

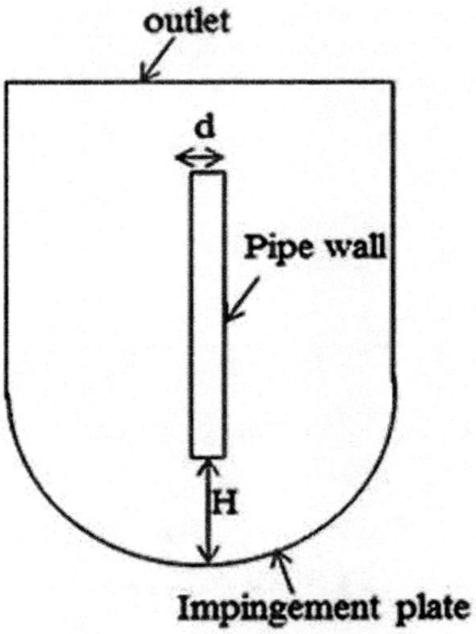


**Fig. 1** Computational model used in the present study

heated ($T$ = 335 K) smooth semi-circular concave plate ($D$ = 160 mm) is cooled by perpendicular jet impingement by a circular nozzle ($d$ = 2 mm) at the centre of plate. The velocity of air jet coming out from nozzle is specified such that Reynolds number falls in the required range (5000 < Re < 50,000) at an atmospheric temperature of 300 K. The same computation is repeated with slight modifications in the concave plate geometry. The smooth semi-circular concave plate previously assumed is modified by adding a circular protrusion (diameter of protrusions: 2–5 mm).

## 2.2 *Computational Details*

Because of symmetry, a two-dimensional computational model is chosen for the present investigation. The governing equations are the Reynolds averaged continuity, momentum and energy equation. The turbulence model is selected after comparing available reports on jet impingement with present numerical simulation as discussed in validation in the later section. The SST k-ω model is chosen to investigate jet impingement cooling as studies in the past have proved that for jet impingement study on concave surfaces, the best compromise between accuracy and solution speed can be obtained with this model.

The fluid flow is assumed to be steady and incompressible with constant fluid properties. The flow is fully developed with a turbulence intensity of 5% and turbulent viscosity of 10. Viscous dissipation and buoyancy effects are neglected.

Finite volume-based solver Fluent 16.1 is used to carry out two-dimensional steady-state simulation. All equations are solved by pressure-based solver by using second-order upwind scheme for discretization. Semi-implicit method for pressure

**Table 1** Boundary conditions employed in the present study

| Location | Variables | Values |
|---|---|---|
| Inlet pipe | Temperature Reynolds number | 300 K 5000–50,000 |
| Impingement plate | Constant temperature | 335 K |
| Pipe wall | Heat flow | Insulation |
| Outlet | pressure | Pressure outlet |

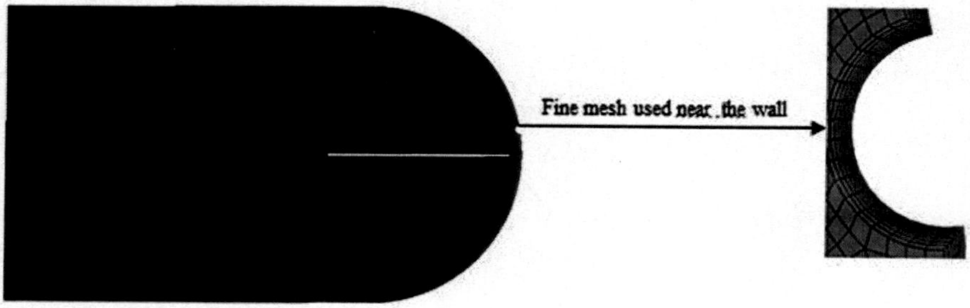


**Fig. 2** Mesh adopted in present study

linked equation (SIMPLE) algorithm is used. The convergence of solution is considered when sum of the scaled residuals falls below $10^{-4}$ for momentum, continuity and turbulence equation and $10^{-7}$ for energy equation.

The two-dimensional model used for present computational study is shown in Fig. 1 and the boundary conditions are described in Table 1. Meshing is done in Ansys Workbench 16.1. Finer meshes are used near the jet pipe surface and the concave surface as shown in Fig. 2. For all cases considered, the maximum wall $Y+$ was maintained to be less than 1.

## 2.3 Validation of Results

The present numerical model is validated with the experimental results of Yang et al. [9] for $H/d = 3$ and several values of Reynolds number and the results are shown in Fig. 3. From the figure, it is seen that there is a good agreement between present study and experimental results of Yang et al. [9]. The slight deviation (4.7%) may be because of round off and experimental errors. Different sizes of mesh were chosen to carry out a grid independence study. Simulations on grid with 92,000 and 0.17 million nodes showed grid independence for single and multiple protrusions over a concave surface, respectively.

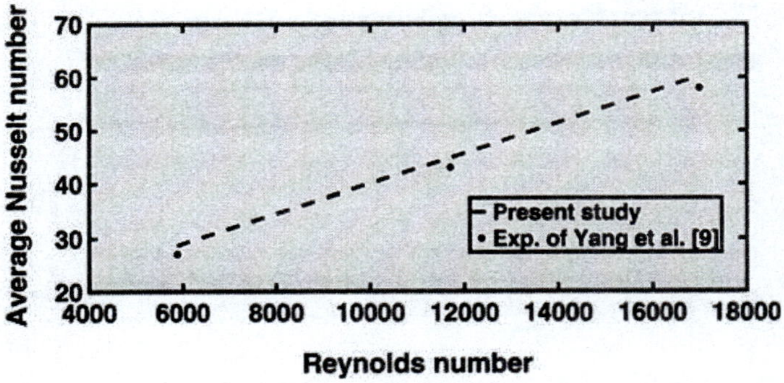


**Fig. 3** Numerical validation of present study

## *2.4 Background of Present Study*

Smooth ($d = 0$) and various sizes of multiple ($N = 21$) semi-circular protrusions were considered over a concave surface for the jet impingement study. Figure 4 represents behaviour of heat transfer with variation in dimension of protrusion. It is observed that smooth plate corresponding to $d = 0$ mm gives the highest heat transfer and all the sizes of protrusions fail to enhance heat transfer compared to smooth plate. It was expected that protrusions would enhance heat transfer because of increase in surface area and turbulence but the reverse happened. The local Nusselt number plot in Fig. 5 shows that protrusions are helpful in the augmentation of local Nusselt number. Keeping this in mind, the behaviour of single protrusion at various positions on the concave plate is studied to determine the desired location of inclusion of protrusion for heat transfer augmentation.

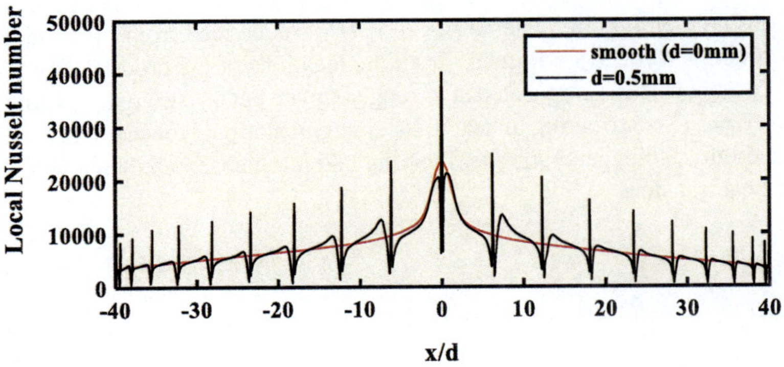


**Fig. 4** Variation of local Nusselt number over a concave surface

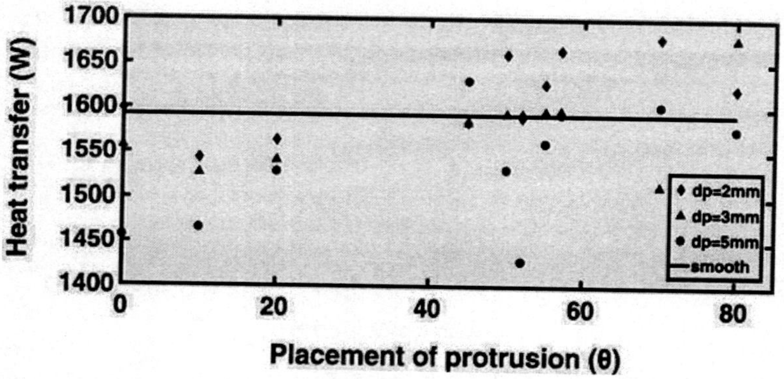


**Fig. 5** Variation of heat transfer with location of protrusion

# 3 Results and Discussion

A comparison of actual heat transfer results with the amount of heat being released after the impingement effect may throw more light on the physics of the problem. Nusselt number being non-dimensional number can be misleading as it does not incorporate the increase in surface area.

## 3.1 *Effect of Heat Transfer with Location of Single Protrusion*

Figure 5 shows the variation of heat transfer with the location of single protrusion over concave surface for different diameters of protrusion. For most of the cases, protrusions are helpful only when $50° < \theta > 70°$. It is noticed from Fig. 6 that two counter rotating vortices are formed for all the locations of the protrusion. For $\theta = 57°$ and $10°$, the vortices are located at the edge of plate but the strength of vortices is lower for $\theta = 57°$ contributing to heat transfer augmentation. However, for $\theta = 70°$, the vortices are stable, generating less mixing and turbulence and thereby resulting in lower heat transfer.

## 3.2 *Curvature Ratio on Heat Transfer*

Figure 7 represents the variation of heat transfer with two different dimensions of plate. It is seen that for the same curvature ratio the geometrical dimension of plate and the jet influence the impingement heat transfer. A lower jet diameter results in lower heat transfer because of lower jet momentum and energy. The curvature of the

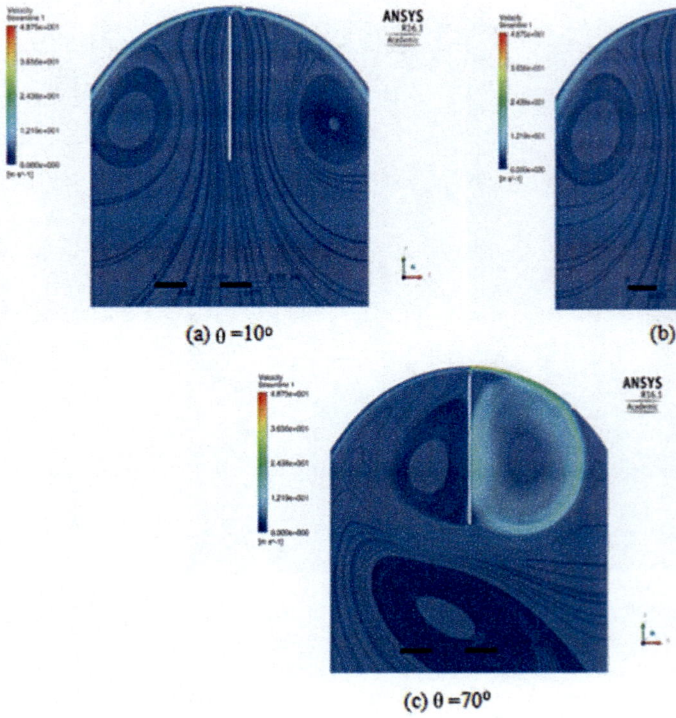


**Fig. 6** Velocity streamline for $H/d = 3$, Re = 6461, $d_p = 3$ mm at various location on a plate of $D = 260$ mm

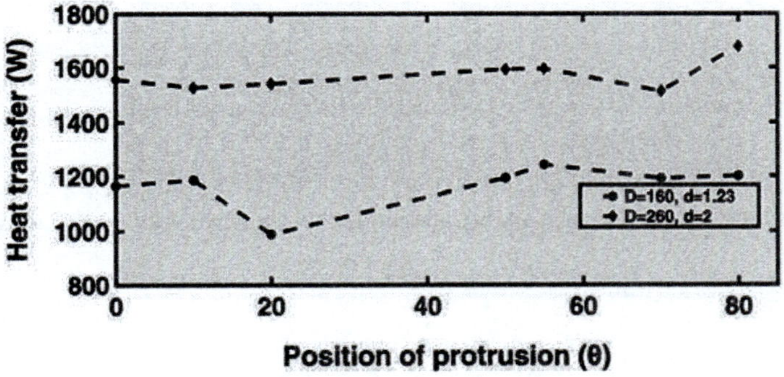


**Fig. 7** Velocity streamline for $H/d = 3$, Re = 6461, $d_p = 3$ mm at various location on a plates with $D/d = 130$

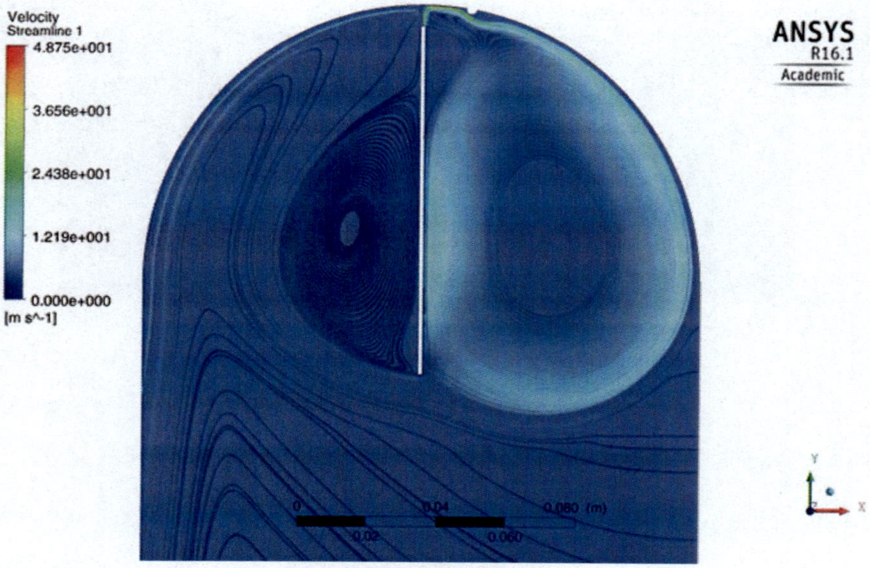


**Fig. 8** Velocity streamlines for $D = 160$ mm, $d = 1.23$ mm, $\theta = 10°$

plate also affects the heat transfer. A larger curvature ($D = 260$ mm) shows symmetric streamlines with single protrusion (Fig. 6b), whereas for a smaller curvature ($D = 160$ mm), the effect of protrusion is showing more asymmetry in streamlines as seen in Fig. 8.

### 3.3 *Optimization of Multiple Protrusions*

The optimization is carried out using genetic algorithm for determining the effective angle and diameter of the protrusion at which the heat transfer is maximum. The objective function created by nonlinear regression is:

$$\text{maximize}: \ y = -1.822 - 9.751 \times 10^{-4}x_1 + 0.156x_2 - 1.987 \times 10^{-4}x_1x_2 + 4.201 \times 10^{-6}x_1^2 - 1.713 \times 10^{-2}x_2^2$$

where $x_1$ and $x_2$ are angle and diameter of protrusion, respectively. Through the optimization procedure, the optimum set of conditions turns out to be of 1660.6 W for $\theta = 78.1°$ and $d = 2.01$ mm.

Based on the optimum effective location of protrusions, multiple protrusions are placed at desired location as given in Table 2. For particular case, it is seen that multiple protrusions at desired location always give higher heat transfer compared

**Table 2** Position of protrusions and its corresponding heat transfer values for 2 mm diameter of protrusion

| S. No. | $\theta$ | Heat transfer (W) |
|---|---|---|
| 1 | smooth | 1591.8 |
| 2 | ±(45°, 50°,55°) | 1756 |
| 3 | ±(50°, 53°, 56°) | 1660.8 |
| 4 | ±(60°, 65°, 70°, 75°) | 1670 |
| 5 | ±(68°, 73°, 78°) | 1666.5 |
| 6 | ±(65°, 69°, 73°, 77°) | 1666.6 |

to the smooth surface. However, more detailed studies are required to generalize this conclusion and full solution space has to be searched.

## 4 Conclusions

Based on detailed numerical solutions of heat transfer from a circular jet impinging on a concave surface, it is seen that random location of protrusions throughout a concave surface may not necessarily increase the impingement heat transfer. Analysis of heat transfer by considering only single protrusion at various locations of the plate must be done to determine the desired position of allocation of protrusions to achieve the highest heat transfer. The curvature ratio of a protruded plate also plays a vital role in the impingement heat transfer.

## References

1. G. Xie, B. Sunden, W. Zhang, Comparisons of pin/dimples/ protrusions cooling concepts for a turbine blade tip wall at high Reynolds number. J. Heat Transf. **113**, 1–9 (2011)
2. D. Zhang, H. Qu, J. Lau, J. Chen, J. Xie, Flow and heat transfer characteristics of single jet impinging on protrusioned surface. Int. J. Heat Mass Transf. **58**, 18–28 (2013)
3. Z. Shen, Y. Xie, D. Zhang, Experimental and numerical study on heat transfer in trailing edge cooling passages with dimples/protrusions under the effect of side wall slot ejection. Int. J. Heat Mass Transf. **92**, 1218–1235 (2016)
4. M.M. Dobbertean, M.M. Rahman, Numerical analysis of steady state heat transfer for jet impingement on patterned surfaces. Appl. Therm. Eng. **103**, 481–490 (2016)
5. Q. Jing, D. Zhang, Y. Xie, Numerical investigations of impingement cooling performance on flat and non-flat targets with dimple/protrusion and triangular rib. Int. J. Heat Mass Transf. **126**(A), 169–190 (2018)
6. M.E. Taslim, L. Setayeshgar, S.D. Spring, An experimental evaluation of advanced leading edge impingement cooling concepts. J. Turbomach. **123**, 147–153 (2001)
7. X.M. Huang, W. Yang, T.Z. Ming, W.Q. Shen, X.F. Yu, Heat transfer enhancement on a microchannel heat sink with impinging jets and dimples. Int. J. Heat Mass Transf. **112**, 113–124 (2017)
8. Y.-T. Yang, T.-C. Wei, Y.-H. Wang, Numerical study of turbulent slot jet impingement cooling on a semi-circular concave surface. Int. J. Heat Mass Transf. **54**, 482–489 (2011)

9. G. Yang, M. Choi, J.S. Lee, An experimental study of slot jet impingement cooling on concave surface: effects of nozzle configuration and curvature. Int. J. Heat Mass Transf. **42**, 2199–2209 (1999)

# Numerical Analysis on a Selection of Horn Material for the Design of Cylindrical Horn in Ultrasonic Machining

**Guddakesh Kumar Chandan and Chinmaya Kumar Sahoo**

**Abstract** Ultrasonic machining is one of the advanced machining processes, utilized for machining hard and brittle materials viz. ceramics, glass, titanium–aluminium composites, etc. for drilling operation with high precision. Horn is one of the important components in the ultrasonic machining process, which transfers longitudinal vibration from the transducer to the tool end. The present investigation considers the design of a simple three-dimensional cylindrical horn using different materials (aluminium, titanium, steel, stainless steel and mild steel) in dynamic conditions. COMSOL multiphysics, a finite element software, is used to investigate the effect of the materials on the horn performance. The results are presented in terms of amplitude, mode shape, and von Mises stress. The analysis results showed that aluminium is one of the suitable materials for horn design followed by titanium, steel, stainless steel, and mild steel. The aluminium horn showed a high amplitude of vibration at the horn end (20.12 μm) at the frequency of 18,445 due to the low damping coefficient. Other materials titanium (12.266 μm), steel (7.134 μm), mild steel (7.036 μm) and stainless steel (6.145 μm) have also shown reasonable amplitude at appropriate applied natural frequency.

**Keywords** Ultrasonic horn · COMSOL multiphysics · Modal analysis

## 1 Introduction

The recent development of hard, tough, and brittle materials for aerospace automotive, textile, medical, etc. industries creates challenges for the manufacturing industry. The hard and heat resistant material and brittle materials show difficulty in machining using a conventional machining process. Advanced machining processes are the

G. K. Chandan (✉) · C. K. Sahoo
Department of Mechanical Engineering, National Institute of Technology Silchar, Silchar, Assam 788010, India
e-mail: gkchandancad@gmail.com

C. K. Sahoo
e-mail: chinmaya.cks@gmail.com


K. M. Pandey et al. (eds.), *Recent Advances in Mechanical Engineering*, Lecture Notes in Mechanical Engineering, https://doi.org/10.1007/978-981-15-7711-6_14

most suitable for machining the hard and brittle material. Ultrasonic machining (USM) is one of the preferred methods for machining hard and brittle material due to its high accuracy and surface finish. The USM process depends on the parameters like frequency, amplitude, the flow of abrasives, temperature and design of horn. In USM, the horn is one of the major components used for transmission and amplification of the ultrasonic vibration from the transducer to the cutting tool. The machining process is largely dictated by the design and material of the horn. Amin et al. [1] studied four types of horn geometry (cylindrical, conical, stepped, exponential) for horn materials (Steel AISI-4063) and suggested that a new horn design considering the upper end with hole and tooltip at the free end. The suggested horn profile has an advantage in the case of material removal rate and safer working stress over other horn profiles mentioned earlier [1]. Nad [2] studied dynamic characteristics of the horn (amplification factors and natural frequencies) in the resonant state for different geometrical shapes viz. cylindrical, tapered, exponential, stepped. The study revealed that the efficiency and performance of ultrasonic machining systems depend on the design of the horn [2]. Rani et al. [3, 4] investigated the effect of horn geometry (i.e. cylindrical, Gaussian, catenoid, stepped, bezier) on the generation of von Mises stresses and amplitude during ultrasonic welding. Bezier profile horn has shown a high amplitude within the permissible stress limit. They further designed and experimentally tested horn for ultrasonic welding; the results showed that the titanium horn achieved high amplitude with a low rise in temperature compared to aluminium, mild steel, and stainless steel horn [3, 4]. Roy et al. [5] concluded that circular hollow horns made with titanium showed better magnification factors and lower stress compared to conical and exponential horns. Lin [6] studied ultrasonic sinusoidal horns made up of hard aluminium. He further studied movement of sinusoidal horn in the torsional and longitudinal direction. The result reveals about the sinusoidal horn had a large mechanical amplitude [6]. Lin et al. [7] purposed an adjustable longitudinal step-type ultrasonic horn using aluminium alloy. The author used a non-traditional way to improve the steeped Al alloy horn performance by varied the electric impedance and location of piezoelectric material [7]. Rosca et al. analysis [8] designed and characterized by an axisymmetric ultrasonic horn using steel. It had the specific working frequency, nodal point position, and amplification factor. The designed ultrasonic horn showed a frequency of around 19,900 Hz with an amplification factor at 5. Results revealed that the purposed horn has higher accuracy in terms of placing of nodal points which altogether help in the better analysis [8]. Wang et al. [9, 10] developed a bezier profile horn using stainless steel for the cutting process. The design showed a reduction of penetration force during cutting material with high displacement [9, 10]. Nguyen et al. [11] studied different horn geometry (i.e. stepped, catenoidal, bezier, ounbs, nurbs) made up of stainless steel for ultrasonic welding. The joints welded by the bezier horn had maximum bonding strength while comparing with the catenoidal and stepped horn [11]. Seah et al. [12] designed conical and stepped horns using the empirical equation. The experimental results revealed that stepped horns suffer from stress concentration [12]. Xiao et al. [13] designed and manufactured an ultra-long horn using titanium alloy for machining deep hole parts with 1/2 wavelengths. For design, the four-terminal network method is used and calculates the transition of different geometric horn studied cases viz. conical and sinusoidal. Their results showed that resonant frequency, width, stiffness for the conical case are larger. In the case of a sinusoidal horn, the vibration frequency is stable and has higher magnification [13]. Fu et al. [14] designed a barbell horn operated at full wavelength longitudinal vibration. Full wavelength horn is helpful to improve in the transfer of the acoustic energy of the ultrasonic oscillatory system operated in the longitudinal vibrational mode and obtained the resonance frequency equations [14]. Xu et al. [15] designed a cup-shaped horn with wavelength with the consideration of natural frequency equations. It revealed that the cup-shaped horn has a distinctive equivalent circuit with good vibrational performance. The horn showed a high amplitude of the operating mode with uniformity of amplitude at the working surface [15]. Kumar et al. [16] studied about free vibration response of laminated composite and sandwich shell model based on higher-order zigzag theory (HOZT). The proposed FEM satisfies the inter-laminar shear stress, hence most suitable to model sandwich shells along with composite shells [16]. Anish et al. [17] investigated the influence of openings and additional mass on free vibration analysis of laminated composite sandwich skew plates using improved higher-order shear deformation theory (IHSDT). FE model based on IHSDT has been coded in FORTRAN [17].

Numerous studies have been done on ultrasonic horn design and material. Research groups have studied the effect of ultrasonic vibration on limited materials. An attempt has been made to analyze the characteristics of different horn materials at similar conditions using the finite element method. The numerical analysis provides a clear view of the effect of horn material on longitudinal displacement, von Mises stress, and frequency. For the current analysis, a simple cylindrical horn has been selected. The cylindrical horn can be easily manufactured and it is easy to modify the existing design of cylindrical horn, which makes it a preferred choice in industrial applications.

## 2 Numerical Analysis

A cylindrical horn of diameter 17.5 mm and length 130 mm was developed as shown in Fig. 1. The structural and harmonic response of the cylindrical horn model was analyzed using a commercial finite element analysis software (COMSOL multiphysics). To study the effect of material on the ultrasonic horn, five different materials viz. aluminium, titanium, mild steel, stainless steel, and steel were considered. The different properties of the material are tabulated in Table 1. The relationship between Young's modulus ($E$) and density ($\delta$), speed the sound (Cc) can be represented as Cc $= (E/\delta)^{1/2}$.

Following assumption are considered for the numerical analysis of cylindrical horn.

i. The material is assumed to be linearly elastic in nature.

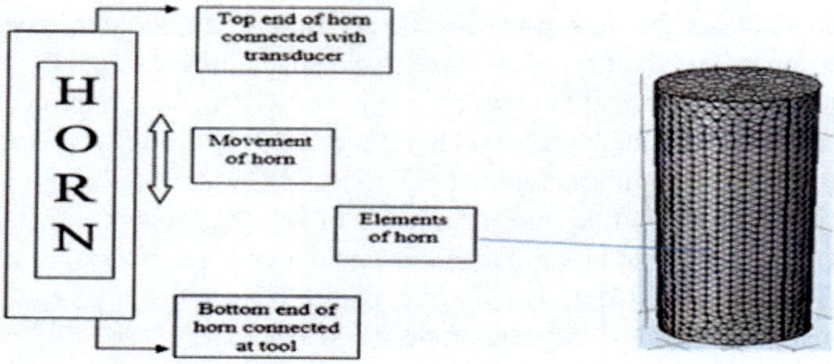


**Fig. 1** Ultrasonic horn and its components and cylindrical model (meshed) used for numerical analysis

**Table 1** Properties of the different horn materials

| S. No. | Material | Density (kg/m$^3$) | Young's modulus | Poisson's ratio | Speed of sound (m/s) |
|---|---|---|---|---|---|
| 1 | Aluminium | 2740 | 74.5 | 0.33 | 5215 |
| 2 | Titanium | 4430 | 114 | 0.34 | 5073 |
| 3 | Steel (AISI-4063) | 7800 | 210 | 0.33 | 5064 |
| 4 | Mild steel | 7870 | 205 | 0.29 | 5104 |
| 5 | Stainless steel | 7900 | 200 | 0.3 | 5032 |

ii. Displacement amplitude is equal to the wavelength of the horn.
iii. Horn motion is irrotational.
iv. The walls of the horn are perfectly rigid and smooth.
v. The effect of forces like viscosity, gravity, external force, and friction has not considered.
vi. The horn was subjected to a frequency range of around 18–20 kHz.

## 2.1 Governing Equation

The governing equations are used that have been numerically solved can be expressed as:

Equation of motion for free vibration,

$$\mathrm{M}\ddot{u} + \mathrm{B}\mathring{u} + K\mathrm{u} = 0 \tag{1}$$

$M, B, K, u, \mathring{u}, \ddot{u}$ denotes inertial force, damping force, stiffness, displacement, velocity, and acceleration, respectively.

For natural frequency in the object,

$$-\rho\omega^2 u = \nabla.S \tag{2}$$

$\rho$, $\omega$, $S$, $u$ indicates the density of the material, angular frequency, stress, displacement, respectively.

For the travel of frequency in the wall of the following equation used,

$$\nabla.\left(-\frac{1}{\rho_c}(\nabla P_t - q_d)\right) - \frac{K_{\text{eq}}^2 P_t}{\rho_c} = Q_m \tag{3}$$

$\rho_c$, $P_t$, $q_d$, $K_{\text{eq}}$, $Q_m$ shows the density of the material, total pressure, dipole domain source, wavenumber used in the equation, monopole domain source, respectively [18, 19].

## 2.2 Boundary Condition

The boundary conditions used to solve Eqs. (1)–(3) are summarized as follows:

i. Horn is assumed to be rigidly fixed with transducer end.
ii. Boundary normal component of velocity and acceleration is considered as zero, i.e.

$$-n.\left(-\frac{1}{\rho_c}(\nabla P_t - q_d)\right) = 0$$

iii. The damping of the materials considered zero.
iv. A frequency amplitude is applied at the transducer–horn interface.

## 2.3 Numerical Solution Methodology and Mesh Quality Evaluation

A commercial finite element analysis software (COMSOL multiphysics) was used to solve the governing Eqs. (1)–(3) along with the boundary conditions of cylindrical horn made of different materials. Prior to detailed modal analysis, a mesh sensitivity evaluation is conducted for the present study. The free tetrahedral mesh is generated for the entire present geometry. Mesh size varied between 5505–43,490, for analysis to check the effect of meshing and find the correct size of the mesh for the analysis. The effect of mesh size and displacement of the horn (amplitude of vibration) subjected to ultrasonic frequency has been shown in Table 3. The corresponding results are presented in the form of a graph in Fig. 2. The present study is divided into two studies. One is to find natural frequency of cylindrical horn. The computational time taken for first study is 478 s. Another one is to determining amplitude and von Misses stress for cylindrical horn. The computational time taken for another study is

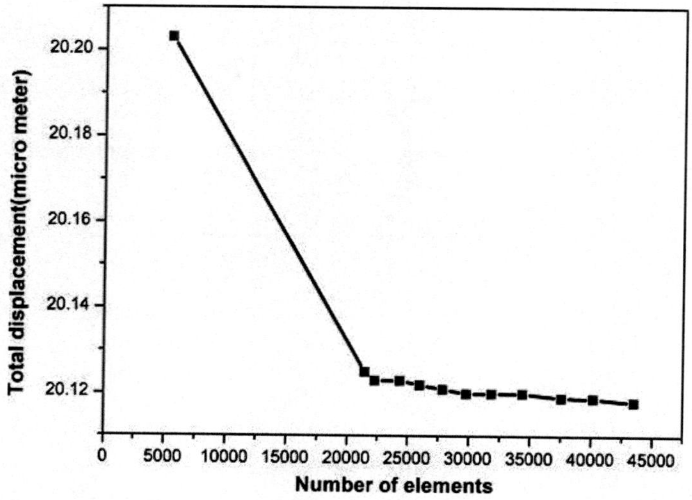


**Fig. 2** Variation of amplitude with the number of elements

144 s. There was 12 different mesh used for the selection of the size of mesh element over 5 different materials. There is a sharp drop in result for the mesh elements 5505–21,428 due to increase in mesh elements. The results also showed minimization of error by increasing the mesh element size (in between 40,158 and 43,490), the error in the 0.004% error with the increase in the number of mesh elements from 40,158 to 43,490. However, the increase in the number of elements enhances the computation time. A trade-off is finalized between computational accuracy and computational time. As a result, the number of domain elements, boundary elements, edge elements are selected as 40,158, 2980, 204, respectively. Furthermore, the maximum element size i.e. 3.7 mm is considered for the entire domain of cylindrical horn. Shear locking is avoided in present work by using finer element meshes using mixed interpolated tensorial components (MITC) schemes [20].

## *2.4 Model Validation*

The correctness of the numerical study has been validated by comparing the results of the present study (Al horn design) with earlier published research papers under similar conditions [3]. Figure 3 shows the amplitude and von Misses stress obtained for aluminium horn compared with earlier studied. A negligible difference in results can be obtained.

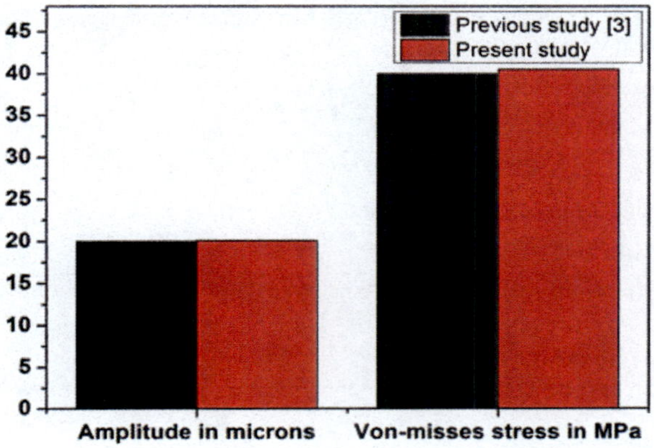


**Fig. 3** Comparison of displacement amplitude and von Mises stress with a previous study [1]

## 3 Result and Discussion

In this study, the effect of the different material configuration of the cylindrical horn is numerically analyzed. Five different materials are carried out for investigation viz. aluminium, titanium, stainless steel, mild steel, and steel. The natural frequency of the materials varied around 18–20 kHz. The results are in the form of dynamic conditions for horns in the modal analysis are presented in the subsequent sections as shown in Table 2.

### *3.1 Modal Analysis*

The natural frequency of the horn is deduced from the modal analysis where different mode shapes are generated by a mechanical system at a specific frequency due to the specific pattern of vibration which is shown in Table 3 and Fig. 4. Table 3 clearly shows that the natural frequency of the horn completely depends on the material properties which varies between 18,445 and 19,354 Hz for aluminium, steel (AISI-4063), mild steel, and stainless steel. It would be observed that the horn made of aluminium

**Table 2** Dynamic analysis results like natural frequency, amplitude, and von Mises stress of different horns materials

| S. No. | Material | Natural frequency (Hz) | Amplitude (μm) | von Mises stress (MPa) |
|---|---|---|---|---|
| 1 | Aluminium | 18,445 | 20.12 | 40.092 |
| 2 | Titanium | 19,512 | 12.266 | 40.478 |
| 3 | Steel (AISI-4063) | 18,354 | 7.134 | 40.091 |
| 4 | Mild steel | 18,332 | 7.036 | 41.005 |
| 5 | Stainless steel | 19,354 | 6.145 | 40.905 |

**Table 3** Different mode shapes and respective frequencies (Hz) for aluminium, titanium, steel, mild steel, stainless steel

| Material | Mode shape (1) | Mode shape (2) | Mode shape (3) | Mode shape (4) |
|---|---|---|---|---|
| Aluminium | 17,706 | 18,445 | 20,058 | 28,893 |
| Titanium | 14,452 | 19,512 | 22,690 | 23,606 |
| Steel (AISI-4063) | 17,619 | 18,354 | 19,477 | 28,751 |
| Mild steel | 17,402 | 18,332 | 19,631 | 28,444 |
| Stainless steel | 18,003 | 19,354 | 28,001 | 28,972 |

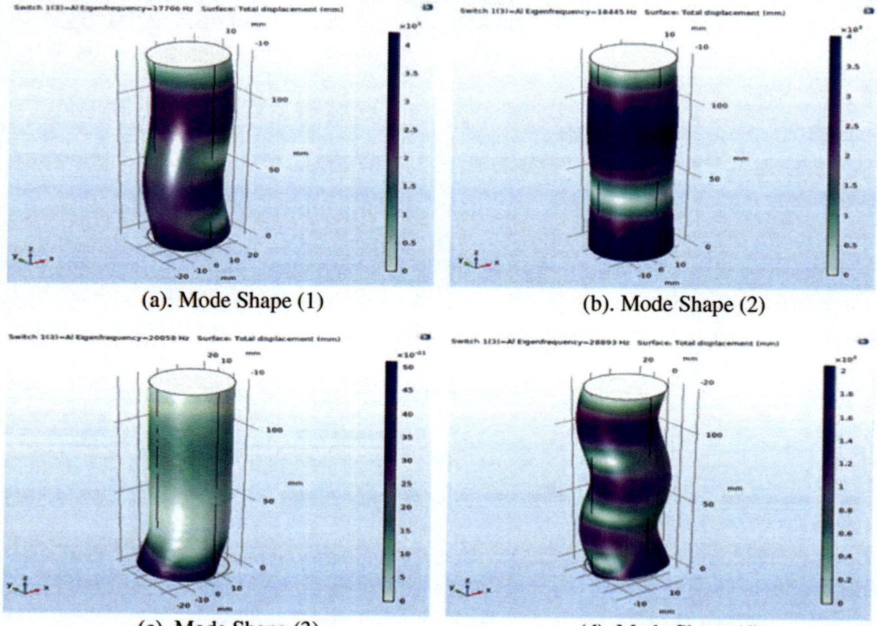

(a). Mode Shape (1) (b). Mode Shape (2)

(c). Mode Shape (3) (d). Mode Shape (4)

**Fig. 4** Different mode shapes of cylindrical shaped aluminium horn at different eigen frequency

showed the high amplitude of vibration, i.e. maximum displacement of 20.12 μm at the end of the horn. Similarly, titanium showed amplitude about 12.266 μm, mild steel, and stainless steel shows the amplitude of 7.036 and 6.145 μm, respectively, at the end of the horn. The von Mises stress developed in these materials is around 40 MPa. The aluminium showed better behaviour of ultrasonic wave transfer and high amplitude at the end of the horn. However, the selection of the material for horns is also on mode shape. Mode shape for the Al horn showed (Fig. 4a–d) that close to natural frequency, i.e. at 18,445 Hz (mode shape 2). The numerical simulation showed that the improper axial movement is observed for the aluminium horn especially for mode 1, 3 and 4 (refer Fig. 4). Similar results are also observed for material made of titanium, stainless steel, and mild steel. Effect of mode shape for Al has already reported by earlier published articles.

### 3.2 *Harmonic Analysis*

From the base end of the transducer, frequency is applied at the top of the horn. The analysis indicates (from the mode shape) that there is a flow of frequency from the top end to the bottom end of the horn uniformly at its natural frequencies. The amplitude of vibration gradually increases along the axis of the horn approaches to a minimum and again increases reach to the maximum at the lower part of the horn due to that there is a formation of wave shape in the horn. It is due to the wavelength of the horn which travels between horns [2]. It is also shown that the amplitude of vibration around 20.12, 12.266, 7.134, 7.036, 6.145 μm for aluminium, titanium, steel, mild steel, and stainless steel, respectively, at the bottom portion of the horn. The variation in the amplification factor is due to the relationship between the density of the material and Young's modulus of the materials [4]. The von Mises stress depends upon the geometrical shape of the horn, and in this study, a simple cylindrical horn is considered. von Mises stress depends upon yield stress of the material, and since in this horn there is the load applied axially, there is the very minute difference between von Mises stress upon different materials. Generally, von Mises stress is considered when complex loading applies throughout the system.

## 4 Conclusion

The present work showed that the simple cylindrical horn made of aluminium, titanium, stainless steel, and mild steel can be successfully utilized for ultrasonic frequency transmission. The increase in the amplitude of the vibration at the end portion of the cylindrical horn provides the opportunity to use it for ultrasonic machining, welding, etc. The result of the numerical analysis can be considered as follows:

I. Aluminium is observed to be one of the best material among mild steel, stainless steel, titanium due to high amplitude at the end of the horn.
II. Mode-shaped analysis confirms that for aluminium mode2 (frequency: 18,845) found to be suitable. In other modes, the proper longitudinal transmission of the vibration has not occurred. The improper flow of the frequency makes the horn unsuitable for utilization.
III. The amplitude at the end of the horn is observed to be 20.12, 12.266, 7.134, 7.036, and 6.145 for aluminium, titanium, steel, mild steel, and stainless steel, respectively. The aluminium horn showed is the high amplitude of vibration

due to its low damping coefficient and followed by titanium, steel, mild steel, and stainless steel.

## References


1. S.G. Amin, M.H.M. Ahmed, H.A. Youssef, Computer-aided design of acoustic horns for ultrasonic machining using finite-element analysis. J. Mater. Process. Technol. **55**, 254–260 (1995)
2. M. Nad, Ultrasonic horn design for ultrasonic machining technologies. Appl. Comput. Mech. **4**, 79–88 (2010)
3. M. Roopa Rani, R. Rudramoorthy, Computational modeling and experimental studies of the dynamic performance of ultrasonic horn profiles used in plastic welding. Ultrasonics **53**, 763–772 (2013)
4. M. Roopa Rani, K. Prakasan, R. Rudramoorthy, Studies of thermo-elastic heating of horns used in ultrasonic plastic welding. Ultrasonics **55**, 123–132 (2015)
5. S. Roy, Design of a circular hollow ultrasonic horn for USM using finite element analysis. Int. J. Adv. Manuf. Technol. **93**, 319–328 (2017)
6. S. Lin, Study of the longitudinal torsional composite mode exponential ultrasonic horns. Ultrasonics **34**, 757–762 (1996)
7. S. Lin, H. Guo, Xu. Jie, Actively adjustable step-type ultrasonic horns in longitudinal vibration. J. Sound Vib. **419**, 367–379 (2018)
8. I.-C. Rosca, M.-I. Pop, N. Cretu, Experimental and numerical study on an ultrasonic horn with shape designed with an optimization algorithm. Appl. Acoust. **95**, 60–69 (2015)
9. D.-A. Wang, H.-D. Nguyen, A planer Bezier profiled horn for reducing penetration force in ultrasonic cutting. Ultrasonics **54**, 375–384 (2014)
10. D.-A. Wang, W.-Y. Chuang, K. Hsu, H.-T. Pham, Design of a Bézier-profile horn for high displacement amplification. Ultrasonics **51**(2), 148–156 (2011)
11. H.-T. Nguyen, H.-D. Nguyen, J.-Y. Uan, D.-A. Wang, A nonrational B-spline profiled horn with high displacement amplification for ultrasonic welding. Ultrasonics **54**(8), 2063–2071 (2014)
12. K.H.W. Seah, Y.S. Wong, L.C. Lee, Design of tool holders for ultrasonic machining using FEM. J. Mater. Process. Technol. **37**, 801–816 (1993)
13. C. Xiao, B. Han, Research and design of ultra-long ultrasonic horn. J. Inst. Eng. India Ser. C, 1–8 (2018)
14. Fu. Zhiqiang, X. Xian, S. Lin, C. Wang, Hu. Wenxu, G. Li, Investigations of the barbell ultrasonic transducer operated in the full-wave vibrational mode. Ultrasonics **52**, 578–586 (2012)
15. Xu. Long, Investigation of a cup-shaped ultrasonic transducer operated in the full-wave vibrational mode. Ultrasonics **59**, 108–118 (2015)
16. A. Kumar, A. Chakrabarti, P. Bhargava, Vibration of laminated composites and sandwich shells based on higher order zigzag theory. Eng. Struct. **56**, 880–888 (2013)
17. Anish, Ajay Kumar, and Anupam Chakrabarti, Influence of openings and additional mass on vibration of laminated sandwich rhombic plates using IHSDT. J. Thermoplastic Compos. Mater. **33**(1), 3–34 (2018)
18. COMSOL, Structural mechanics module user's version 5.4, 1998–2018
19. COMSOL, Acoustics module user's guide version 5.4, 1998–2018
20. K.-J. Bathe, E.N. Dvorkin, A four-node plate bending element based on Mindlin/Reissner plate theory and a mixed interpolation. Int. J. Numer. Meth. Eng. **21**(2), 367–383 (1985)

# Emission Studies on a Diesel Engine Fueled with Mixed Biodiesel Produced from Non-edible Oils

**Bhabani Prasanna Pattanaik, Chandrakanta Nayak, and Rahul Dev Misra**

**Abstract** Over exploitation of fossil-based fuels for energy has given rise to serious issues like environmental degradation and global warming. Replacement of petroleum-based fuels with biofuels, such as vegetable oils, biodiesels, etc. has drawn the attention of researchers in order to address the a fore-mentioned issues in recent years. The present experimental investigation aims to produce mixed biodiesel (MXBD) from a mixture of three non-edible oils, viz. jatropha, karanja, and mahua and its application in a compression ignition engine to study the engine performance and exhaust emission behaviors. Results showed that MXBD exhibited superior engine performance and emissions over diesel, jatropha biodiesel (JBD), karanja biodiesel (KBD), and mahua biodiesel (MBD). Brake thermal efficiency with MXBD was greater compared to diesel and other biodiesels, whereas brake specific energy consumption with MXBD was lesser than other biodiesels and a little more than diesel at all engine loads. CO, HC, and smoke emissions were lowest with MXBD compared to all other considered fuels at entire engine loads. NOx emissions with MXBD were found to be marginally higher than those with other test fuels. Thus, the present work establishes MXBD as a better alternate fuel option for diesel engines over JBD, KBD, and MBD in terms of both engine performance and exhaust emissions.

**Keywords** Diesel engine · Engine performance · Exhaust emissions · Mixed biodiesel · NOx emissions

---

B. P. Pattanaik (✉)
Department of Mechanical Engineering, KMBB College of Engineering and Technology, Bhubaneswar, Odisha 752056, India
e-mail: bpprdmnits@gmail.com

C. Nayak
Department of Mechanical Engineering, Gandhi Institute for Technological Advancement, Bhubaneswar, Odisha 752054, India
e-mail: cknayak1977@gmail.com

R. D. Misra
Department of Mechanical Engineering, National Institute of Technology Silchar, Silchar, Assam 788010, India



K. M. Pandey et al. (eds.), *Recent Advances in Mechanical Engineering*, Lecture Notes in Mechanical Engineering, https://doi.org/10.1007/978-981-15-7711-6_15

# 1 Introduction

The worldwide concern for depletion of fossil fuels, environmental pollution, increasing fuel prices, and climate change has led to intensive search for alternative fuels. At present, attention is being focused on development of renewable and clean burning fuels. Use of biofuel in engines has been an emerging technology in the past few years owing to its inherent characteristics, such as biodegradability, non-toxicity, and lower exhaust emissions [1]. In this context, biodiesel ormethyl/ethyl ester derived from straight vegetable oils (SVO) and animal fat via trances terification is regarded as a bio-degradable, non-toxic, and environment-friendly alternative fuel for diesel engines. Biodiesel is low in sulfur and aromatics content and is comprised of oxygen in its molecular structure [2]. Combustion of biodiesel leads to lesser carbon monoxide (CO), unburned hydrocarbons (HC), sulfur dioxide (SO2) and particulate matter (PM) emissions in contrast to diesel [2–4]. Further, its improved cetane number leads to better combustion in compression ignition (CI) engines. However, biodiesel shave poor low-temperature properties, such as relatively higher cloud and pour points, and their combustion might release higher nitrogen oxides (NOx) emissions [5, 6]. Thus, biodiesels can be used as blends in CI engines minus any engine alterations. However, properties of the blends must be compatible with the international biodiesel standards, i.e., EN14214+A1 and ASTM 6751-12 [7]. Else, it may cause serious damage to the engine components, such as fuel injector and filter choking and plugging, corrosion of fuel flow components, increased carbon deposits, etc [8]. Thus, biodiesel production from non-edible sources may be an effective solution to address the concerns related to the uses of petroleum diesel fuel. Many researchers have published substantial number of research articles on successful use of biodiesel produced from non-edible sources as fuel in CI engines [9–11]. However, the limited availability of non-edible vegetable oil sources for making of biodiesel in various regions across the globe raises questions over its commercial production.

Arunprasad and Balusamy [12] conducted the performance and emission study on a CI engine run with mixed biodiesel produced from four non-edible SVO sources, namely *Jatropha Curcas*, *Pongamia Pinnata*, *Thevetia Peruviana*, and *Azadirachta Indica*. As reported by them, the mixed biodiesel exhibited comparable engine performance along with lesser HC, CO, and smoke emissions besides little higher NOx emission in contrast to diesel fuel. Khan et al. [13] used mixed biodiesel produced from castor and karanja oils as blended fuel in a CI engine and reported decline in CO, CO2, HC, NOx, and smoke emissions with rise in percentage of biodiesel in the fuel blends. Mishra and Nayak [14] investigated on the application of post-mixed biodiesel blends, produced from karan ja and mahua SVO sources, in a turbo charged direct injection (DI) diesel engine and revealed that the post-mixed biodiesel blends displayed significantly lower HC, CO, and smoke emissions along with comparable engine performance at the cost of marginally higher NOx and CO2 emissions. Likewise, lesser HC, CO, and smoke emissions accompanied by marginally increased NOx emissions were reported by Öztürk [7], who studied the use of mixed biodiesel, derived from canola oil and hazelnut soap stock sources, as a blended fuel in a CI engine. In the present work, efforts were made to study and establish the use of mixed biodiesel fuel, derived from various non-edible SVO sources, as a prospective alternative fuel for CI engine applications. Further, the present work features a few novelties that are highlighted in the subsequent sentences. Production of biodiesel from purely non-edible sources nullifies the possibilities of food versus fuel controversy. Establishment of non-edible mixed vegetable oil as a potential source for biodiesel production successfully addresses the feedstock availability issues. Lastly, the economy of biodiesel production process is improved as low-cost non-edible sources can be used for preparation of mixed vegetable oil as a source for biodiesel production.

## 2 Materials and Methods

Three different non-edible SVOs, namely Jatropha, Karanja, and Mahua were selected for production of biodiesel. All the three selected SVOs were initially filtered and then thoroughly mixed in equal volume proportions to prepare a mixture of non-edible SVOs named as mixed SVO. A specified quantity of mixed SVO was fed into the biodiesel reactor along with measured quantities of methanol (CH3OH) and potassium hydroxide (KOH). Transesterification reaction of the mixed SVO was then carried out at 60 °C for a period of 1 h. After the reaction time is over, the reaction was stopped by switching off the heater and the stirring motor and the reaction product was allowed to settle down inside the reactor for a period of four hrs. After the settling time ended, the methyl ester or biodiesel produced was collected, water washed, and dried to obtain pure biodiesel. The biodiesels produced from mixed SVO, Jatropha, Karanja, and Mahua oils are referred as MXBD, JBD, KBD, and MBD, respectively. Fuel characterization for all the developed biodiesels and diesel was carried out following standard ASTM methods and the results are displayed in Table 1.

**Table 1** Fuel properties

| S. No. | Properties | Diesel | MXBD | JBD | KBD | MBD | ASTM standard |
|---|---|---|---|---|---|---|---|
| 1 | Density at 20 °C, kg/m$^3$ | 835 | 848 | 852 | 865 | 878 | D4052 |
| 2 | Kinematic viscosity at 40 °C, cSt | 2.67 | 4.36 | 4.45 | 4.73 | 5.02 | D445 |
| 3 | Flash point, °C | 71 | 145 | 152 | 158 | 172 | D93 |
| 4 | Calorific value, MJ/kg | 44.58 | 41.8 | 41.6 | 40.75 | 39.35 | D240 |
| 5 | Cloud point, °C | 6.6 | 10.8 | 11.2 | 11.8 | 12.5 | D2500 |
| 6 | Pour point, °C | 3.2 | 4.8 | 4.9 | 5.4 | 6.2 | D97 |
| 7 | Cetane index | 49 | 54.5 | 54 | 51 | 50.5 | D4737 |
| 8 | Ash content, % | 0.001 | 0.004 | 0.005 | 0.007 | 0.007 | D976 |

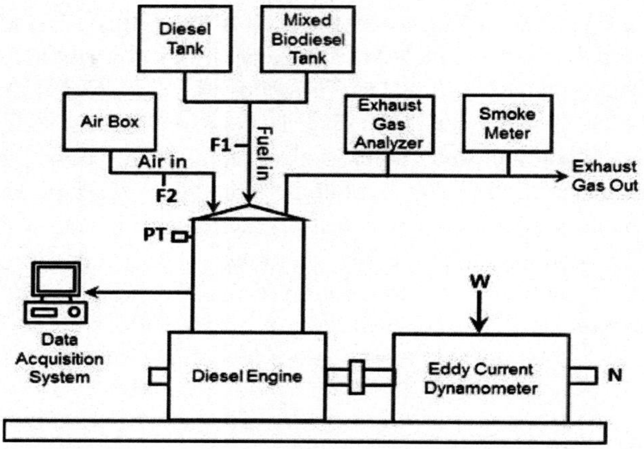


F1: Fuel flow measuring sensor, F2: Airflow measuring sensor, PT: Pressure sensor, W: Load, N: rpm pick-up and crank angle encoder

**Fig. 1** Schematic diagram of the test engine setup

A computerized four-stroke single-cylinder water-cooled DI diesel engine was used for engine experiments using the prepared test fuels. The test engine was attached to an eddy-current dynamometer for applying loads on the engine. The exhaust line of the test engine was appropriately reformed and attached to the AVL Digas 444 exhaust emission analyzer and AVL 437 smoke meter for estimation of various emission parameters. The fuel supply line was attached with two distinct fuel tanks for the supply of diesel and biodiesel fuels. The fuel flow and airflow rates were measured through an optical sensor and pressure transducer, respectively. Thermocouples were used to know the inlet air, exhaust gas, and cooling water temperatures. A schematic diagram displaying all the main components of the test setup is displayed in Fig. 1. Further, the systematic technical specifications of the test engine are described in Table 2.

## 3 Results and Discussion

### *3.1 CO Emissions*

It is observed that the CO emissions for the entire test fuels and diesel gradually decrease from zero load up to 85% load and then tend to increase at a sharp rate up to full load which may be ascribed to the fact that as the load increases, better turbulence is achieved which inturn results in better mixing and availability of more oxygen leading to better combustion and reduced CO emissions . However, increase in load beyond 85% leads to increased fuel injection, formation of rich mixture, and availability of less oxygen. These factors cause incomplete combustion and higher

**Table 2** Technical specifications of the test engine

| S. No. | Particulars | Specifications |
|---|---|---|
| 1 | Make and model | Kirloskar AV-1 |
| 2 | Engine type | Computerized single-cylinder 4-stroke DI diesel engine |
| 3 | Max. power | 5.2 kW |
| 4 | Max. speed | 1500 rpm |
| 5 | Compression ratio | 17.5 |
| 6 | Bore × stroke | 110 mm × 87.5 mm |
| 7 | Swept volume | 661 cc |
| 8 | Injection pressure | 220 bar |
| 9 | Injection nozzle opening | 23°bTDC |
| 10 | Dynamometer | Eddy-current type |

CO emissions up to full load. The lowest CO emissions for all the fuels are observed at 85% load. All the biodiesels showed lower CO emissions in comparison with diesel at all loads, which is credited to the existence of oxygen in biodiesels leading to better combustion and lower CO emissions. MXBD showed lowest CO emissions at all engine loads owing to its lower viscosity, higher calorific value, higher cetane index, and superior combustion. The CO emissions with MXBD are observed to be 23.8%, 7.2%, 14.3%, and 16.7% lower than that with diesel, JBD, KBD, and MBD, respectively, at 85% load (Fig. 2).

## 3.2 HC Emissions

Results showed that the HC emissions decrease with rise in load until 85% and then follow an increasing trend until full load. With increase in engine load, increased turbulence is achieved which is responsible for better mixing of the fuel and better combustion leading to lower HC emissions. Higher HC emissions at lower loads is because of poor combustion owing to poor mixing and lower cylinder temperatures. On the other hand, beyond 85% load incomplete combustion takes place due to formation of rich mixture as a result of increased fuel injection and availability of less oxygen. All the biodiesels exhibited lower HC emissions in comparison with diesel at all loads. This is credited to the existence of oxygen in biodiesels and their higher cetane index, which is accountable for improved combustion and reduced HC emissions [15]. Further, among all the biodiesel fuels, MXBD showed lowest HC emissions at all loads. This is attributable to the higher calorific value, higher cetane index, and lesser viscosity of MXBD over other biodiesels leading to better combustion and reduced HC emissions. At 85% load, the HC emissions with MXBD are 30.8%, 8.1%, 16.3%, and 18.5% lower than that with diesel, JBD, KBD, and MBD, respectively (Fig. 3).

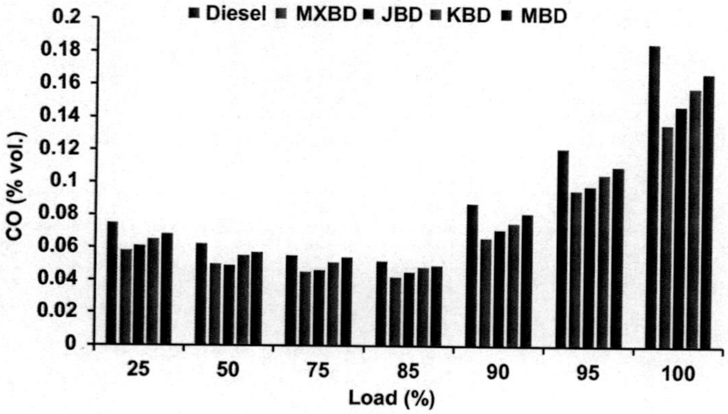


**Fig. 2** Variation of CO emissions with load

### *3.3 NOx Emissions*

It is observed that NOx emissions for all the fuels increase with rise in load. With rise in load, the inside cylinder temperature also rises leading to increased NOx emissions. The highest NOx emissions were recorded at full load conditions. In addition, all the biodiesels exhibited higher NOx emissions in comparison to diesel at all engine loads. The probable cause for increased NOx emissions with biodiesels, compared to that with diesel, could be the reduced ignition delay and early commencement of combustion owing to their large value of cetane index and presence of oxygen. The same increases the premixed combustion period resulting in an extended residence time of higher temperature within the engine cylinder [16]. The NOx emissions with MXBD are observed to be marginally higher than other biodiesels at all loads. The probable cause for the same may be the higher calorific value and greater cetane index of MXBD, over the other biodiesels, resulting in shortest ignition delay and earliest start of combustion leading to highest in-cylinder temperatures and higher NOx emissions. At 85% load, the NOx emissions with MXBD are found to be 7.56%, 0.6%, 1.13%, and 1.3% higher than those with diesel, JBD, KBD, and MBD, respectively. Similarly, the NOx emissions with MXBD are observed to be 8.87%, 0.86%, 1.6%, and 1.93% higher than those with diesel, JBD, KBD, and MBD respectively, at full load conditions (Fig. 4).

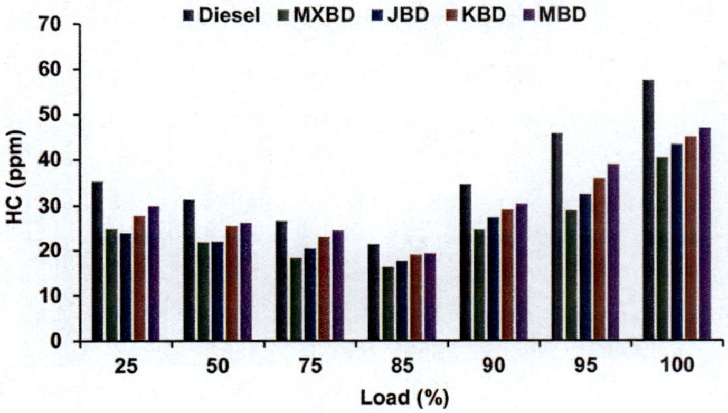


**Fig. 3** Variation of HC emissions with load

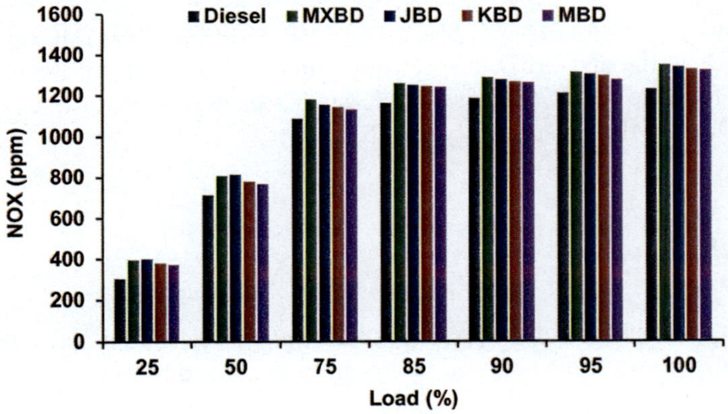


**Fig. 4** Variation of NOx emission with load

## *3.4 Smoke Emissions*

Smoke or soot is mainly composed of carbon particles.It is noticed that smoke opacity attains higher value with upsurge in engine loading for all the considered test fuels. With rise in load, the rate of fuel injection increases, which results in inadequate mixing and incomplete combustion, thereby leading to greater smoke or soot formation [7]. Lower smoke opacity with biodiesels shows improved combustion of biodiesels because of their higher cetane index and oxygen content [14]. Another reason to the lower smoke opacity with biodiesels is their lesser carbon to hydrogen proportion, which leads to reduced soot creation. The observed smoke opacity trend is in agreement with the existing literature [12, 14].The smoke opacity with MXBD is found to be lowest among all the biodiesels, which is credited to its higher cetane index, greater calorific value, and low viscosity resulting in better combustion. The

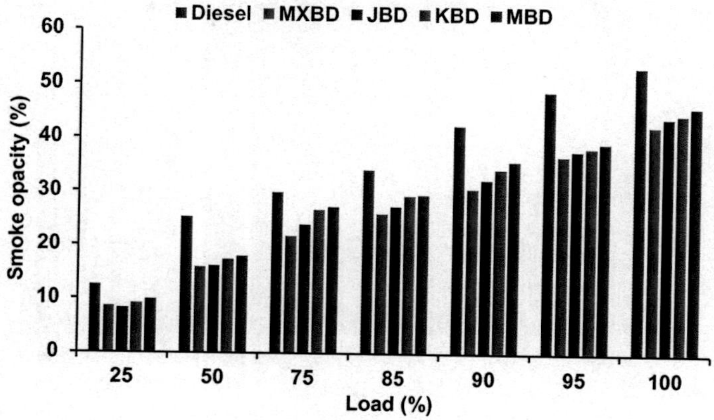


**Fig. 5** Variation of smoke opacity with load

smoke opacity with MXBD is 31.8%, 5.4%, 13%, and 13.8% lower than those with diesel, JBD, KBD, and MBD, respectively. Similarly, at full load, the same with MXBD is found to be 25.7%, 3.6%, 5.2%, and 8.3% lower in comparison to diesel, JBD, KBD, and MBD, respectively (Fig. 5).

## 4 Conclusions

The primary conclusions of the present experimental investigation drawn through a painstaking analysis of the obtained results are summarized below .

- All the considered biodiesels exhibit lower exhaust emissions as HC, CO, and smoke opacity in contrast to diesel at all engine loads.
- The NOx emissions for biodiesels are little higher at all loads in contrast to diesel.
- MXBD exhibited excellent fuel properties among all the prepared biodiesels.
- The HC, CO, and smoke emissions with MXBD are significantly lower in contrast with diesel, JBD, KBD, and MBD. However, the NOx emissions with MXBD are observed to be marginally higher than that with other considered biodiesels.

## References

1. B.P. Pattanaik, R.D. Misra, Effect of reaction pathway and operating parameters on the deoxygenation of vegetable oils to produce diesel range hydrocarbon fuels: a review. Renew. Sustain. Energy Rev. **73**, 545 (2017)
2. B.P. Pattanaik, C. Nayak, B.K. Nanda, Investigation on utilization of biogas and Karanja oil biodiesel in dual fuel mode in a single cylinder DI diesel engine. Int. J. Energy Environ. **4**, 279 (2013)

3. N. Joy, Y. Devarajan, B. Nagappan, A. Anderson, Exhaust emission study on neat biodiesel and alcohol blends fueled diesel engine. Energy Sour Part A Recovery Util. Environ. Effects **40**, 115 (2018)
4. N. Damanik, H.C. Ong, C.W. Tong, T.M. Mahlia, A.S. Silitonga, A review on the engine performance and exhaust emission characteristics of diesel engines fueled with biodiesel blends. Environ. Sci. Pollut. Res. **25**, 15307 (2018)
5. C. Attaphong, L. Do, D. Sabatini, Vegetable oil-based microemulsions using carboxylate-based extended surfactants and their potential as an alternative renewable biofuel. Fuel **94**, 606 (2012)
6. E.A. Espinosa, R.P. Rodríguez, R. Sierens, S. Verhelst, Emulsification of waste cooking oils and fatty acid distillates as diesel engine fuels: an attractive alternative. Int. J. Sustain. Energy Plann. Manage. **9**, 3 (2016)
7. E. Öztürk, Performance, emissions, combustion and injection characteristics of a diesel engine fuelled with canola oil–hazelnut soapstock biodiesel mixture. Fuel Process. Technol. **129**, 183 (2015)
8. M. Canakci, H. Sanli, Biodiesel production from various feedstocks and their effects on the fuel properties. J. Ind. Microbiol. Biotechnol. **35**, 431 (2008)
9. J. Hwang, D. Qi, Y. Jung, C. Bae, Effect of injection parameters on the combustion and emission characteristics in a common-rail direct injection diesel engine fueled with waste cooking oil biodiesel. Renew. Energy **63**, 9 (2014)
10. A.K. Agarwal, A. Dhar, J.G. Gupta, W.I. Kim, K. Choi, C.S. Lee, S. Park S, Effect of fuel injection pressure and injection timing of Karanja biodiesel blends on fuel spray, engine performance, emissions and combustion characteristics. Energy Convers. Manage. **91**, 302 (2015)
11. A.R. Dey, R.D. Misra, Effect of infiltration of bio-lubricant on the performance of a compression ignition engine fuelled with biodiesel blends. Clean Technol. Environ. Policy 19, 553 (2017)
12. S. Arunprasad, T. Balusamy, Experimental investigation on the performance and emission characteristicsofadieselenginebyvaryingtheinjectionpressureandinjectiontimingusingmixed biodiesel. Int. J. Green Energy **15**, 376 (2018)
13. A. Mangad, M.M. Roy, W. Wang, Improvementincoldflowpropertiesofbiodieselanditseffects on diesel engine performance and emissions. Int. J. Energy Environ. **9**, 233 (2018)
14. P.C. Mishra, S.K. Nayak, Pre-and post-mixed hybrid biodiesel blends as alternative energy fuels- an experimental case study on turbo-charged direct injection diesel engine. Energy **160**, 910 (2018)
15. A.K. Agarwal, D.K. Srivastava, A. Dhar, R.K. Maurya, P.C. Shukla, A.P. Singh, Effect of fuel injection timing and pressure on combustion, emissions and performance characteristics of a single cylinder diesel engine. Fuel **111**, 374 (2013)
16. C.C. Enweremadu, H.L. Rutto, Combustion, emission and engine performance characteristics of used cooking oil biodiesel—a review. Renew. Sustain. Energy Rev. **14**(9), 2863 (2010)
17. B.S. Chauhan, N. Kumar, H.M. Cho, A study on the performance and emission of a diesel engine fuelled with Jatropha biodiesel oil and its blends. Energy **37**, 616 (2012)
18. L. Zhu, C.S. Cheung, W.G. Zhang, Z. Huang, Combustion, performance and emission characteristics of a DI diesel engine fueled with ethanol–biodiesel blends. Fuel **90**, 1743 (2011)
19. B.P. Pattanaik, R.D. Misra, Experimental studies on production of deoxygenated vegetable oils and their performance evaluation in a compression ignition engine. Biomass Convers. Biorefinery **8**, 899 (2018)
20. C. Sayin, M. Gumus, Impactofcompressionratioandinjectionparametersontheperformanceand emissions of a DI diesel engine fueled with biodiesel-blended diesel fuel. Appl. Therm. Eng. **31**, 3182 (2011)
21. P.K. Sahoo, L.M. Das, M.K.G. Babu, P. Arora, V.P. Singh, N.R. Kumar, T.S. Varyani, Comparative evaluation of performance and emission characteristics of jatropha, karanja and polanga based biodiesel as fuel in a tractor engine. Fuel **88**, 1698 (2009)

# Recent Advancement in Electromagnetic Forming Processes

**Sagar Pawar, Sachin D. Kore, and Arup Nandy**

**Abstract** Electromagnetic forming (EMF) is a contactless forming process where electromagnetic forces are applied to achieve the required deformation. It is also called as a pulsed magnetic forming technology where the main driving force is Lorentz force. Generally, EMF is used to deform highly conductive materials where tubes or sheets are expanded or compressed and magnetic forces are used to shape or join or cut the materials. It is a high-speed, high-energy rate forming process. As it involves a high strain rate, therefore, with the use of it the forming limits of several materials can be extended. In this paper, the EMF processes are reviewed by considering the recent advancements in different applications of electromagnetic forming, which includes perforation of tubes, crimping of tubes for different applications, joining by forming of tubes for torsional applications, etc. The working principle of different electromagnetic manufacturing techniques is explained. Based on this survey, it is identified that the recent developments of the electromagnetic forming research show a wide range of industrial applications. It can be commercialized to have sustainable manufacturing technology.

**Keywords** Crimping · Electromagnetic forming · Joining by forming · High-speed forming · Perforation · Simulations

---

S. Pawar (✉) · A. Nandy
Department of Mechanical Engineering, Indian Institute of Technology Guwahati, Guwahati, Assam, India
e-mail: s.pawar@iitg.ac.in

A. Nandy
e-mail: arupn@iitg.ac.in

S. D. Kore
School of Mechanical Sciences, Indian Institute of Technology Goa, Ponda, Goa, India
e-mail: sachin@iitgoa.ac.in



K. M. Pandey et al. (eds.), *Recent Advances in Mechanical Engineering*, Lecture Notes in Mechanical Engineering, https://doi.org/10.1007/978-981-15-7711-6_16

# 1 Introduction

In the metal forming process, external forces are applied on the metal in such way that the stresses induced are greater than the yield stress and less than the ultimate stress, so that the metal is experiencing plastic and permanent deformation to change the shape. The external forces applied for the forming process is called deforming force. Forming processes are classified into two types, conventional forming process and high-energy rate forming processes. In high-energy rate forming processes, a large amount of energy is applied for a very short interval of time. Many metals tend to deform more rapidly under extra fast application of a load, which results in the usefulness of these processes to deform large size parts out of most metals, including those who are difficult to deform. The parts are deformed at a fast rate, so these processes are called a high-speed rate or high velocity forming processes. The electromagnetic forming process (EMF) is also a high-energy rate forming process. In the EMF process, high Lorentz forces are produced due to the interaction between two magnetic fields. It is a high-energy system that can discharge its energy within a short duration of time, and the conductive workpiece is driven into the die or free-formed by magnetic pressure [1, 2]. The working principle is well explained by Maxwell [3]. In early 1924, Kapitza is the first one who used magnetic forces to deform the solid conductor [4]. After that Harvey and Brower explained the detailed working principle as well as explained some application of process [5, 6]. Psyk et al. [7] in the year 2011 did the extensive review of basic research on the principle of EMF as well as the study related to the involvement of different process parameters. It is observed that various applications of the EMF can be possible by various arrangements and shapes of coil and workpiece. Very high velocities like about 250 m/s and high strain rates in the range of $10^4\ s^{-1}$ are observed. Due to the high strain rates, the mechanical properties of the material can be improved compared to the conventional or low strain rate forming processes.

In electromagnetic compression of the tube, the coil is placed around the tube, and for electromagnetic expansion, the coil is placed inside the tube. These two processes, as well as the electromagnetic expansion of sheet, are shown in Fig. 1. In other applications, based on different requirements, various kinds of coil geometries can be used. In the following review, the recent advancements and their working principles are discussed. The focus of this review is set on the advancement in the following processes,

- Electromagnetic forming.
- Electromagnetic joining by forming (crimping).
- Electromagnetic forming and perforation of tubes.
- Electromagnetic forming and perforation of sheets.

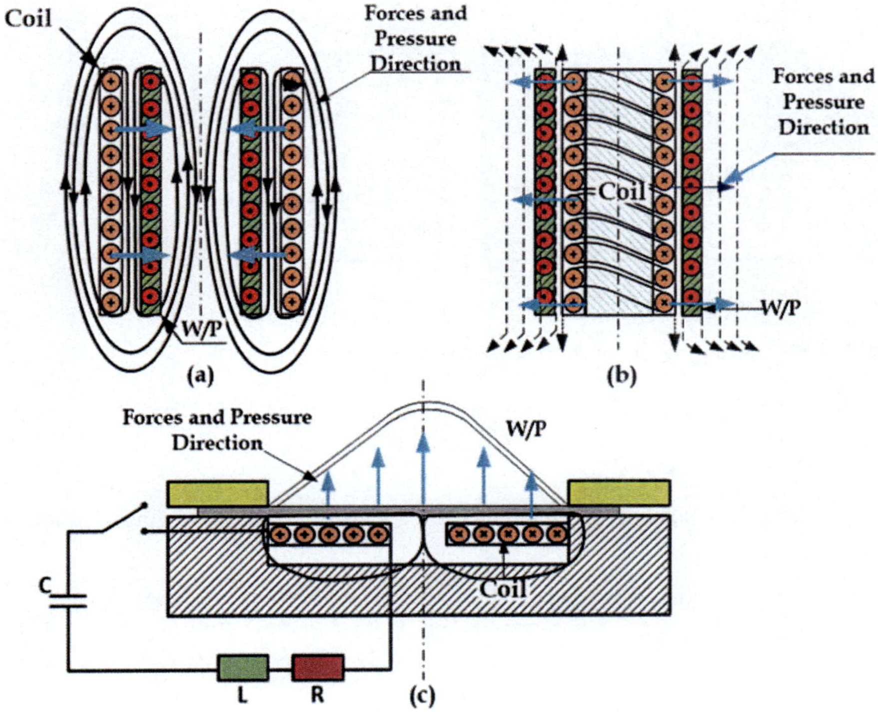


**Fig. 1** **a** EM compression of the tube, **b** EM expansion of tube and **c** EM expansion of the sheet

## 2 Working Principle of the Electromagnetic Forming Process

The electromagnetic forming is based on the resonant circuit configuration. The high magnetic pressure which is required to deform the electric conductive material is obtained with the help of a magnetic pulse generator [8]. The schematic representation of electromagnetic forming is shown in Fig. 2 where the electromagnetic forming machine is represented by resistor$R_i$, inductance $L_{i,}$ and capacitance $C$ circuit. Here, the coil is represented as a tool.

EMF system consists of a bank of capacitors, a workpiece, and a coil. The energy is stored in the capacitor bank and it is discharged through the coil which produces an alternative damped electric current in the coil. The magnetic field produced by the coil induces the eddy current in the workpiece. The produced eddy current flows through the workpiece and establishes another magnetic field around it. The secondary magnetic field is opposite in nature which opposes the primary magnetic field and causes deforms the workpiece. When charging voltage $V$ is supplied to the capacitor bank of capacitance $C$, the stored energy ($E$) in the capacitor bank is given by,

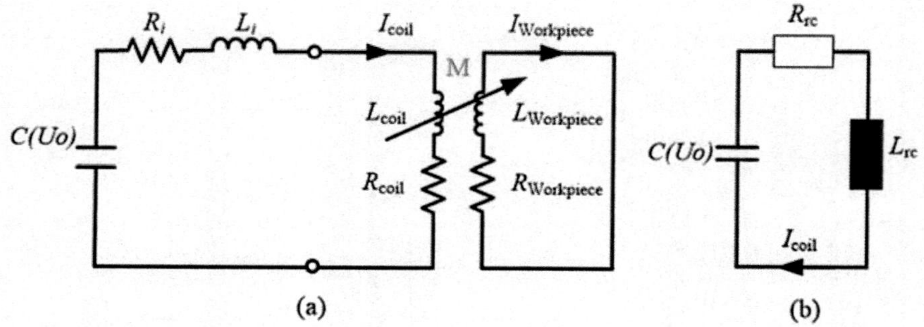


**Fig. 2** **a** EMF circuit diagram and **b** reduced version of EMF circuit diagram

$$E_c(t) = \frac{1}{2}CV^2 \tag{1}$$

The current flowing through the circuit can be described by the following equations [18],

$$L_c\frac{\partial I_c(t)}{\partial t} + M\frac{\partial I_c(t)}{\partial t} + R_c I_c + \frac{1}{C}\int I_c(t)\mathrm{d}t = 0 \tag{2}$$

$$\frac{\partial(L_w I_w)}{\partial t} + \frac{\partial(M I_c)}{\partial t} + R_w I_w = 0 \tag{3}$$

By solving the above equations with the initial conditions $I(0) = 0$ and $L\frac{dI}{dt}(0) = V$ gives following equation of current as a function of time,

$$I(t) = I_m e^{-\beta t}\sin(\omega t) \tag{4}$$

$$\beta = \frac{R}{2L}, \omega = 2\pi f = \sqrt{\frac{1}{LC} - \beta^2} \text{ and } I_m = \frac{V}{\omega L} \tag{5}$$

With $I_m$ is the maximum intensity of the current, $\omega$ is the angular frequency and $\beta$ is the damping coefficient. The electromagnetic phenomenon involved is taken care by the following Maxwell's equations,

$$\nabla \times H = J \tag{6}$$

$$\nabla \times E = -\frac{\partial}{\partial t}B \tag{7}$$

$$\nabla \cdot B = 0 \tag{8}$$

$$J = \gamma E \tag{9}$$

$$B = \mu H \tag{10}$$

where $J$ is the current density, $B$ is the magnetic flux density, $E$ is the electric field, $H$ is the magnetic field, $\mu$ is the permeability and $\gamma$ is the electric conductivity of the material.

The Lorentz forces ($F$) can be calculated by the Eq. (11),

$$F = J \times B \tag{11}$$

## 3 Joining by Electromagnetic Forming

The electromagnetic crimping is a type of joining by the electromagnetic forming process, which is used to join two metal tubes. In this process for joining, no other material is used. Different types of fit joints can be possible by this process. The basic physics is the same as the electromagnetic forming process. Here, for the joining, the coil is placed over the workpiece. The schematic representation of the EM crimping process is shown in Fig. 3a. To increase the joint strength different types of groves have been created on the rod surface. Wedeling et al. have reported the variation in dimensions of the joint on the pull-out strength [9] and concluded that triangular grooves are weakest among the other.

To achieve more joint strength, many researchers have studied the effect of the number of grooves on the same rod as well as by variation in other process parameters [10–12]. In EM crimping, field shaper is very important. It is used to concentrate the magnetic field into the required region to achieve an efficient joint with more strength. The schematic representation of EM crimping with field shaper is shown in Fig. 3b.

Recently, some researchers have studied electromagnetic crimping processes with different shapes of field shapers [13–15]. Rajak et al. have compared conventional crimping with the EM crimping process and showed the advantages of EM crimping over the conventional crimping process [16]. The different types of field shapes are shown in Fig. 4 and it is observed that among all, step type field shaper is best.

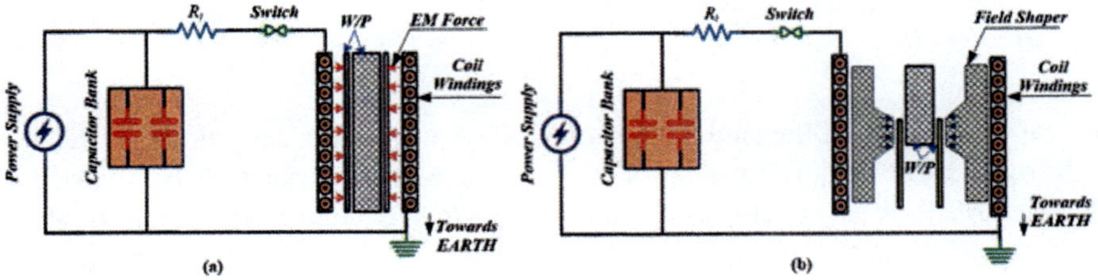


**Fig. 3** **a** Schematic representation of EM crimping and **b** EM cramping with field shaper

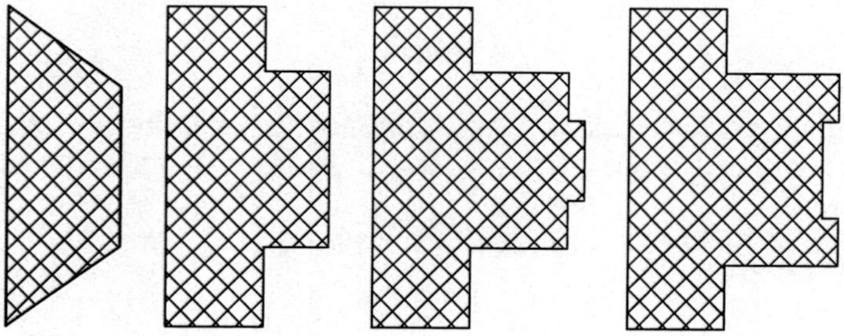


**Fig. 4** Different field shapers: **a** conical, **b** step, **c** convex and **d** concave

Researchers have well documented the use of electromagnetic crimping in joining dissimilar material used in various industrial applications.

## 4 Electromagnetic Shearing

Electromagnetic shearing is a recently developed non-conventional manufacturing process that can fulfill the increasing demands of the flexible manufacturing market. This process is carried out at a very high velocity of about 200 m/s and hence called a high-speed shearing process. In the automobile industries, it is very difficult to shear the lightweight material. In conventional shearing, setting the optimum clearance is very difficult, and it will lead to the formation of burrs on the edges of the sheared panels. While in the case of electromagnetic cutting, the setting of optimal cutting clearance need not be required. One side tooling provides high flexibility as well as saves the production cost. The advantages of EM shearing over conventional are explained by Golowin et al. [17]. Electromagnetic shearing has the number of benefits over the other methods like it can shear very thin materials with almost zero clearance angles and can cut the workpiece into any desired shape. Some of the recent developed electromagnetic shearing techniques are explained below.

### *4.1 Electromagnetic Forming and Perforation (EMFP) of Sheets*

The feasibility study of the simultaneous EMFP of the sheet is carried out by Patel and Kore [18]. The detailed analysis of the variation of process parameters is carried out. The EMFP of sheet setup is the design and manufactured. Since it is single side shearing method, the requirement of matching the tooling set is eliminated, which will reduce the production time. This feasibility test has been carried out to manufacture deformed and perforated part used in washing machine. The schematic

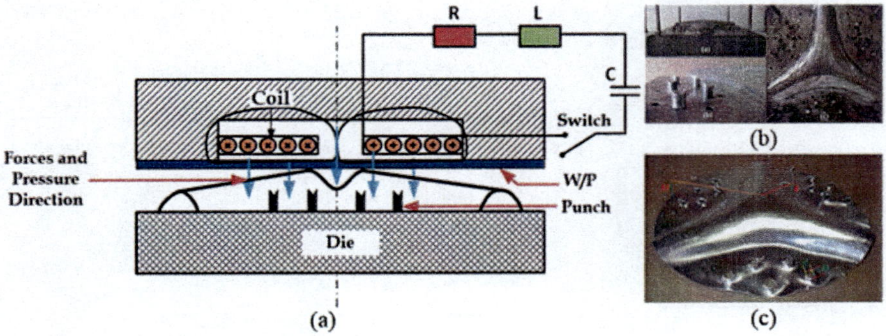


**Fig. 5** **a** Electromagnetic forming and perforation of the sheet, **b** die views [18] and **c** deform sample [18]

representation of the electromagnetic forming and perforation of the sheet process are shown in Fig. 5a. The die used and sheet after the experiment are shown in Fig. 5b, c.

### *4.2 Electromagnetic Forming and Perforation of Tubes*

The EMFP of tubes is a simultaneous process, and it is an application of the electromagnetic forming process. It has large applications where deformed perforated tubes are used, such as in muffler tubes, oil and gas industries, as well as in filters. Pawar et al. [19] have developed a setup for electromagnetic forming and perforation process. The detailed study of the shape of punch on the quality of perforation has been studied [20]. The schematic representation of electromagnetic forming and perforation of tubes is shown in Fig. 6a. The tubes after experiments are shown in Fig. 6b.

## 5 Conclusion

This review shows the recent advancements in electromagnetic forming techniques. It is observed that the electromagnetic forming process can be used for the manufacturing of different types of products. The advancement in joining by electromagnetic forming techniques as well as their advantage over conventional manufacturing process is valuable, and in the future, it may replace the convention way of joining. The feasibility study of simultaneous forming and perforation of sheets and tubes motivates to develop new EM forming setup and manufacture more complex and near-net-shape products with the help of EM forming techniques. This combination

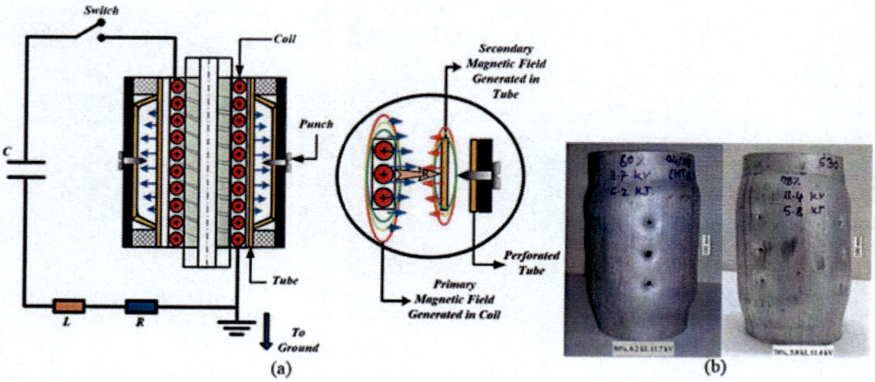


**Fig. 6** **a** Schematic representation of electromagnetic forming and perforation of the tube and **b** perforated tubes [19]

may eliminate the use of different processes involved in the manufacturing of single products and can lead to cost-effectiveness and higher production rates.

## References

1. A.G. Mamalis, D.E. Manolakos, A.G. Kladas, A.K. Koumoutsos, Electromagnetic forming and powder processing: trends and developments. Appl. Mech. Rev. **57**(4), 299–324 (2004)
2. J.P. Correia, M.A. Siddiqui, S. Ahzi, S. Belouettar, R. Davies, A simple model to simulate electromagnetic sheet free bulging process. Int. J. Mech. Sci. **50**(10–11), 1466–1475 (2008)
3. Maxwell, A treatise of electricity and magnetism, 1873, Macmillan and Co.
4. P.L. Kapitza, A method for producing strong magnetic fields. Proc. Roy. Soc. Ser. A **105**, 691–710 (1924)
5. G.W. Harvey, D.F. Brower, Inventors; general dynamics corp, assignee. Metal forming device and method. United States patent US 2,1961,976,907
6. D.F. Brower, *Metals Handbook*, vol. 4 (ASM, 1969). Metals Park
7. V. Psyk, D. Risch, B.L. Kinsey, A.E. Tekkaya, M. Kleiner, Electromagnetic forming—a review. J. Mater. Process. Technol. **211**(5), 787–829 (2011)
8. D. Ertelt, Berechnung von Last- und Entladungsgrößen bei der magnetischen Hochleistungs-Impuls-Umformung. Wiss. Z. Techn. Hochschule Magdeburg **26**(4), 57–61 (1982)
9. C. Weddeling, *Electromagnetic Form-Fit Joining*. PhD Thesis. Department Of Mechanical Engineering, TU Dortmund University, Germany (2014)
10. W. Christian, K.O. Demir, P. Haupt, E.A. Tekkaya., Analytical methodology for the process design of electromagnetic crimping. J. Mater. Process. Technol. **222**, 163–180 (2015)
11. W. Christian, V. Walter, P. Haupt, E.A. Tekkaya, V. Schulze, K.A. Weidenmann, Joining zone design for electromagnetically crimped connections. J. Mater. Process. Technol. **225**, 240–261 (2015)
12. K. Faes, O. Zaitov, E. De Waele, Electromagnetic pulse crimping of axial form fit joints, in *5th International Conference on High Speed Forming*, pp. 229–242 (2012)
13. A.K. Rajak, R. Kumar, H. Basumatary, S.D. Kore, Numerical and experimental study on effect of different types of field-shaper on electromagnetic terminal-wire crimping process. Int. J. Precis. Eng. Manuf. **19**(3), 453–459 (2018)

14. R. Kumar, S.D. Kore, Effects of surface profiles on the joint formation during magnetic pulse crimping in tube-to-rod configuration. Int. J. Precis. Eng. Manuf. **18**(8), 1181–1188 (2017)
15. G.T. Areda, S.D. Kore, Numerical study and experimental investigation on electromagnetic crimping for tube-to-rod configuration. Int. J. Precis. Eng. Manuf. **20**(2), 181–191 (2019)
16. A.K. Rajak, S.D. Kore, Experimental investigation of aluminium–copper wire crimping with electromagnetic process: Its advantages over conventional process. J. Manuf. Processes **26**, 57–66 (2017)
17. S.M. Golowin, *Path Actuators for Magnetic Pulse Assisted Forming and Punch-Less Electromagnetic Shearing Doctoral Dissertation* (The Ohio State University, 2008)
18. C. Patel, *Application of Electromagnetics in Forming of Tubes and Sheets.* Doctoral dissertation (Indian Institute of Technology Guwahati, Guwahati, India, 2017)
19. S. Pawar, S.D. Kore, A. Nandy, *Magnetic Pulse Forming and Punching of al Tubes—A Novel Technique for Forming and Perforation of Tubes* (Advances in Forming, Machining and Automation Springer, Singapore, 2019), pp. 67–77
20. S. Pawar, S.D. Kore, Electromagnetic forming and perforation of Al tubes. J. Mech. Sci. Technol. **33**(12), 5999–6007 (2019)

# Review on Pool Boiling Heat Transfer Enhancement by Surface Fabrication Using Various Surface Coating Methods

**Sonali Priyadarshini Das, Raghavendra Singh, and Rahul Dev Misra**

**Abstract** This review paper includes different coating methods with different nanocoating material for enhancing the surface properties. Due to surface properties (wettability, surface contact angle, roughness, porosity, etc.) and thickness of nanocoating, heat transfer rate increases. Nucleate boiling heat transfer and critical heat flux are major factors which decides the heat transfer rate. So if these factors are controlled, then heat transfer rate automatically controlled. Future scope in this field is also presented in this paper. Coating methods, by which non-metal material coated on metal are also listed below.

**Keywords** Nucleate boiling heat transfer (NBHF) · Critical heat flux (CHF) · Heat transfer

## 1 Introduction

Many methods are there for removing thermal energy in large quantity and among those method boiling is the most efficient one helps in the maintenance of very low wall temperature as compared to other methods. Boiling heat transfer finds its applications in many engineering fields like cooling of turbine blades, supercomputers, etc.

The boiling curve, as first stated by Nukiyama [1], demonstrates the important features of the boiling heat transfer as shown in Fig. 1.

S. P. Das (✉) · R. Singh · R. D. Misra
Department of Mechanical Engineering, National Institute of Technology Silchar, Silchar, Assam 788010, India
e-mail: dassonalipriyadarshini93@gmail.com

R. Singh
e-mail: raghavendrasingh301@gmail.com

R. D. Misra
e-mail: rdmisra@mech.nits.ac.in


K. M. Pandey et al. (eds.), *Recent Advances in Mechanical Engineering*, Lecture Notes in Mechanical Engineering, https://doi.org/10.1007/978-981-15-7711-6_17

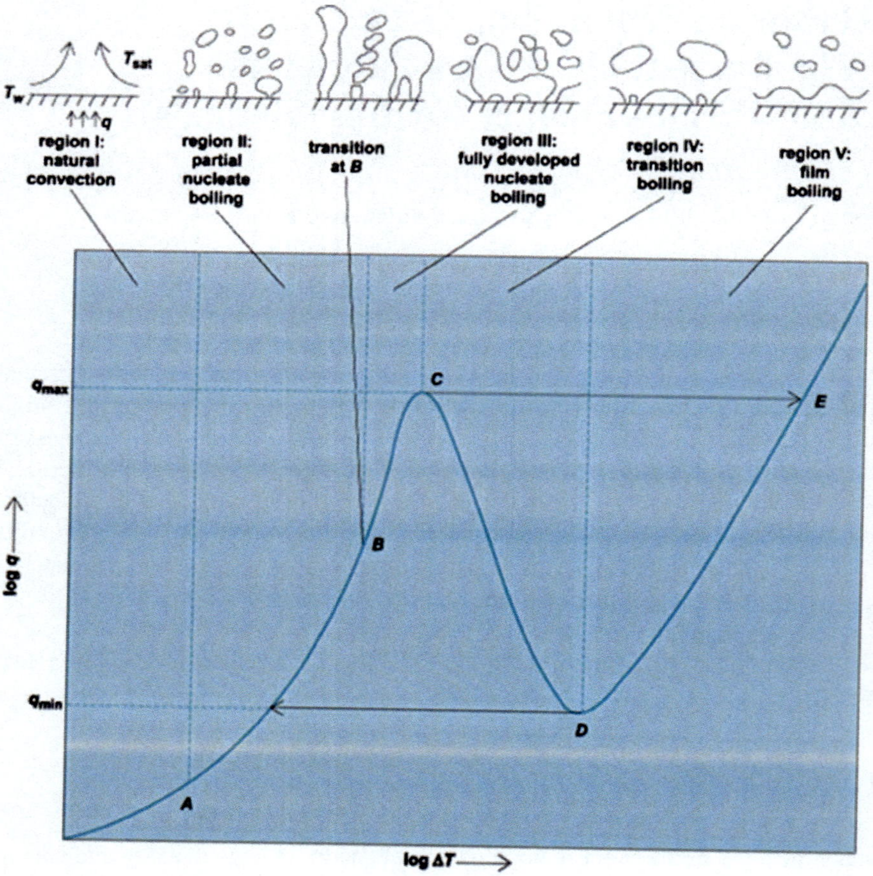


**Fig. 1** Pool boiling curve [1]

From the boiling curve, it is observed that the nucleate boiling heat transfer (NBTH) region to be the most important region where maximum heat is being transferred after which the critical heat flux (CHF) is reached and burn out occurs due to excessive surface temperature. So our main focus is to increase both the parameters. This could be achieved by surface fabrications.

Many surface fabrication processes are as such:

- Mechanical machining of the surface
- Coating of the surface
- Chemical processes
- Micro/nanoelectro mechanical system (MEMS/NEMS) method.

Webb [2] has suggested in their experiment that the boiling performance can be increased or enhanced by mechanical working or by chemical etching, by coating of nanoparticles of metal, non-metal or metal oxide or by attaching nucleation site promoters. He also found that the shape of the coating grit barely affects the boiling parameters, but the coating thickness has a great influence on the parameters. Webb [3] again did nanocoating this time using spherical structure and found that the geometry of nanoparticle has no effect on the boiling performance.

Milton [4] researched that porous metallic coating has been an enhancement technique. Various geometric parameters have been considered to produce higher enhancement such as particle size, particle shape, coating thickness and porosity.

Chang and You [5], as summarized the work of You et al. [6] and Chang and You [7] on micro-porous coating, and did used five different coating over copper base that were aluminum, diamond, silicon and copper nanoparticle. Saturated FC-72 was used as the boiling liquid at atmospheric pressure. This lead to 90% reduction in wall super heat, 30% increment in CHF and 100% enhancement of NBTH.

This paper constitutes the work of various researchers using different coating methods and coating materials for enhancing the NBTH and CHF of pool boiling heat transfer.

## 2 Nucleate Boiling Heat Transfer (NBTH)

In NBTH region, a large amount of heat is being transferred as a virtue of vigorous bubble activities, i.e., bubbles generations, growth and detachment from the heating surface.

The heat flux partitioning model expresses boiling heat flux as the sum of the following three parameters.

1. Evaporation heat flux ($q_e''$)
2. Quenching heat flux ($q_q''$)
3. Convective heat flux ($q_c''$).

The nucleate boiling heat flux can be expressed as

$$q''_{\mathrm{NB}} = q''_e + q''_q + q''_c \quad (1)$$

Many parameters affect the heat flux like the nucleation (active) site density, frequency and diameter of the departing bubbles, etc.

Parameters like surface roughness, wettability, cavity size, active nucleation site density, bubble departure diameter [8] and frequency highly affect NBTH characteristics [9–13]. For enhancing and optimizing NHBT, we need a good knowledge of surface characteristics effects on NBTH.

Many studies have claimed surface roughness to be major factor affecting NBTH. Westwart [14] has reported that the surface roughness affects the boiling curve, hence, many theories emerged to validate that the nucleation site density can be changed by surface roughness.

Overall the recent studies have shown that roughness, wettability, etc., surface characteristics affect the surface superheat, NBTH and CHF.

## 3 Critical Heat Flux (CHF)

The upper limit of NBTH is the CHF and its very important to predict and enhance CHF for the efficiency and safety of the boiling system.

Firstly, You et al. [15] have enhanced the CHF by using nanofluid in the boiling medium. Then, Zuber [16] has explained the CHF phenomenon in his hydrodynamic theory. Liter and Kaviany [9] have enhanced CHF by using modulated porous layer coating which increased CHF by three times. Kandlikar [11] has developed upper surface with horizontal and vertical orientation. The experiment was done for water, refrigerant and cryogenic liquid. It showed that how varied contact angle could affect the boiling. Similarly, many researchers have investigated that capillary wicking, surface grain size and thermal conductivity of the surface affect CHF. Eastman [17] has used copper nanoparticle in ethylene glycol and found the thermal conductivity to increase by 40%.

None of the research till date are able to accurately enhance CHF and NBTH, hence, there is a lot of scope in this field.

## 4 Surface Fabrication by Coating Methods

The surface can be coated by various methods like vapor deposition technique where a small sample of the coating material is taken in the vacuum chamber and is evaporated such that the vapor sticks to the bare surface. Similarly, we could use the basic electroplating method or chemical etching. We could also use plasma spray method where the nanoparticle is well mixed with the epoxy is dispersed over the bare surface and is cured. We could also use the nanofluid for coating.

You et al. [15] have coated diamond nanoparticle over copper surface and the boiling liquid was FC-72. Five different sizes of diamond was taken, i.e., 2, 10, 20, 45 and 70 $\mu$m. It was observed that when the current was below 25 W/cm$^2$ the coating with larger grit size showed enhanced boiling phenomena due to increased nucleation site density (active), but when the current density was increased to 25 W/cm$^2$ the performance started degrading as increased thermal resistance by the coating material. Kim et al. [18] have also used diamond as their coating material and also found the same trends in the boiling curve with 12 W/cm$^2$ current density supply. They also found that CHF got delayed due to increased heat transfer which led to hydrodynamic stability (Table 1).

Many researchers have used metal or metal oxide coating for enhancement of pool boiling heat transfer. Tehver et al. [30] have used Al, bronze, Cu, corundum as coating material over Al and Cu heating rod by help of plasma sprayed coating method while the boiling was done in Freon. Liter and Kaviary and Hwang and Kaviary [31] have used copper nanoparticles to coat over copper plate by dry diffusion and sintering in nitrogen and hydrogen atmosphere with pentane as boiling liquid. Li et al. [32] have

**Table 1** Lists the work of many researcher using nanofluid for coating

| Ref. | Base surface | Nanomaterial | Base boiling fluid |
|---|---|---|---|
| [19] | Nichrome | $SiO_2$ | Water |
| [18, 20–23] | Nichrome | $Al_2O_3$, $TiO_2$ | Water |
| [12, 20] | Stainless steel | $Al_2O_3$, $ZrO_2$, $SiO_2$ | Water |
| [24] | Copper | $ZrO_2$ | Water |
| [25] | Nickel | $SiO_2$ | Water |
| [26, 27] | Brass | $Al_2O_3$ | Water |
| [28] | Copper | ZnO | Ethylene glycol |
| [29] | Copper | $Al_2O_3$ | Water |

used copper nanoparticle of size 250 μm for coating over copper in argon atmosphere with water as the boiling fluid.

Apart from metal oxides, researchers have also used $TiO_2$, grapheme, etc., for coating. Takata et al. [33] have used $TiO_2$ as coating material over copper plate. The method he used was dipping and sputtering. A mixture of $TiO_2$ and $SiO_2$ (20 wt%) was used and baking was done at a temperature of 150 °C. He found that the CHF and NBTH coefficients have increased substantially as compared to the bare copper (Tables 2 and 3).

Ahn [44] used multiwalled carbon nanotubes (MWCNT) for coating over the silicon wafer by chemical vapor deposition method. They found the CHF to be increased by 28%.

**Table 2** Surface coating using metal or metal oxide

| Ref. | Base surface material | Coating material | Boiling fluid |
|---|---|---|---|
| [30] | Al, Cu | Al, bronze, Cu, corundum | Freon |
| [32] | Cu | Cu | Water |
| [34] | Al | $Al_2O_3$ | FC-72 |
| [35] | Al, Cu | Cu, Mo, Al, Zn | R134a, R407c |
| [36] | Cu | Cu | Pentane |
| [37] | Cu | Cu | Water |
| [38] | Al, Cu | ZnO | Water |
| [39] | Pt | $Al_2O_3$ | Water |

**Table 3** Surface coating using $TiO_2$, ghaphene, etc.

| Ref. | Base surface material | Coating material | Boiling fluid |
|---|---|---|---|
| [33] | Cu | $TiO_2$ | Water |
| [40] | Cu | Graphite fiber reinforced copper | Freon 113 |
| [41] | Al, Cu | Graphite fiber reinforced Al/Cu | Pentane |
| [42, 43] | Cu | Graphite | HFE-7100, FC-72 |
| [33, 44] | Si wafer | CNTs | PF-5060 |
| [45] | Cu | CNTs, Nafion | Water |
| [46, 47] | Si wafer | Teflon | Water |

## 5 Conclusion

From this discussion, we get to know that the NBHT can be enhanced by the following points.

- Increasing nucleation (active) site density
- Increasing wettability
- Bubble formation, growth and development.

Similarly, CHF can be increased by the following points.

- High wettability
- Increased surface roughness
- Microstructure.

Overall it can be concluded that if nano/microcoating is done we get more surface for bubble formation, nucleation growth and dispatch, hence, increasing the boiling performance. And the coating thickness should be of adequate thickness so that there is no thermal resistance and the burnout is delayed.

Still many obstacles like durability of the coating material, corrosion, thermal stress, mechanical and chemical wear out fouling, etc., that could ensure the longevity of the boiling experiment setup. Therefore, enhanced fabrication technique is required which could ensure the long term application of the boiling heat transfer for cooling purpose.

## References

1. N. Shiro, The maximum and minimum values of the heat q transmitted from metal to boiling water under atmospheric pressure. Int. J. Heat Mass Transf. **27**(7), 959–970 (1984)

2. C. Pais, R.L. Webb, Literature survey of pool boiling on enhanced surfaces. ASHRAE Trans. **16**(pt 1), 79–89 (1991)
3. L.H. Chien, R.L. Webb, A nucleate boiling model for structured enhanced surfaces. Int. J. Heat Mass Transf. **41**(14), 2183–2195 (1998)
4. Milton, Heat exchange system. Foreign Aff. **91**(5), 1689–1699 (1968)
5. J.Y. Chang, S.M. You, Boiling heat transfer phenomena from micro-porous and porous surfaces in saturated FC-72. Int. J. Heat Mass Transf. **40**(18), 4437–4447 (1997)
6. T.W.S.S.M. You, Enhance Boiling Heat Transfer with application to cooling of electonic equipment.Pdf. (1992)
7. J.Y. Chang, S.M. You, Enhanced boiling heat transfer from micro-porous surfaces: Effects of a coating composition and method. Int. J. Heat Mass Transf. **40**(18), 4449–4460 (1997)
8. R.W. Bowring, *Physical Model, Based on Bubble Detachment, and Calculation of Steam Yoidage in the Sub cooled Region of a Heated Channel*, pp. 135–142 (1962)
9. S.G. Liter, M. Kaviany, Pool-boiling CHF enhancement by modulated porous-layer coating: Theory and experiment. Int. J. Heat Mass Transf. **44**(22), 4287–4311 (2001)
10. Y. Haramura, Y. Katto, A new hydrodynamic model of critical heat flux, applicable widely to both pool and forced convection boiling on submerged bodies in saturated liquids. Int. J. Heat Mass Transf. **26**(3), 389–399 (1983)
11. S.G. Kandlikar, A theoretical model to predict pool boiling CHF incorporating effects of contact angle and orientation. J. Heat Transf. **123**(6), 1071–1079 (2001)
12. S.J. Kim, I.C. Bang, J. Buongiorno, L.W. Hu, Surface wettability change during pool boiling of nanofluids and its effect on critical heat flux. Int. J. Heat Mass Transf. **50**(19–20), 4105–4116 (2007)
13. S.D. Park et al., Effects of nanofluids containing graphene/graphene-oxide nanosheets on critical heat flux. Appl. Phys. Lett. **97**(2) (2010)
14. J.W. Westwater, Boiling Heat Transf. **101**(5), 370–377 (1959)
15. S.M. You, J.H. Kim, K.H. Kim, Effect of nanoparticles on critical heat flux of water in pool boiling heat transfer. Appl. Phys. Lett. **83**(16), 3374–3376 (2003)
16. N. Zuber, Hydrodynamic Aspects Of Boiling Heat Transfer (Thesis) (1959)
17. J.A. Eastman, S.U.S. Choi, S. Li, W. Yu, L.J. Thompson, Anomalously increased effective thermal conductivities of ethylene glycol-based nanofluids containing copper nanoparticles. Appl. Phys. Lett. **78**(6), 718–720 (2001)
18. H. Kim, J. Kim, M.H. Kim, Effect of nanoparticles on CHF enhancement in pool boiling of nano-fluids. Int. J. Heat Mass Transf. **49**(25–26), 5070–5074 (2006)
19. P. Vassallo, R. Kumar, S. D'Amico, Pool boiling heat transfer experiments in silica-water nano-fluids. Int. J. Heat Mass Transf. **47**(2), 407–411 (2004)
20. S.J. Kim, I.C. Bang, J. Buongiorno, L.W. Hu, Effects of nanoparticle deposition on surface wettability influencing boiling heat transfer in nanofluids. Appl. Phys. Lett. **89**(15) (2006)
21. I.C. Bang, S. Heung Chang, Boiling heat transfer performance and phenomena of $Al_2O_3$-water nano-fluids from a plain surface in a pool. Int. J. Heat Mass Transf. **48**(12), 2407–2419 (2005)
22. H.D. Kim, J. Kim, M.H. Kim, Experimental studies on CHF characteristics of nano-fluids at pool boiling. Int. J. Multiph. Flow **33**(7), 691–706 (2007)
23. H. Kim, M. Kim, Experimental study of the characteristics and mechanism of pool boiling CHF enhancement using nanofluids. Heat Mass Transf. Und Stoffuebertragung **45**(7), 991–998 (2009)
24. M. Chopkar, A.K. Das, I. Manna, P.K. Das, Pool boiling heat transfer characteristics of $ZrO_2$-water nanofluids from a flat surface in a pool. Heat Mass Transf. Und Stoffuebertragung **44**(8), 999–1004 (2008)
25. E. Forrest, E. Williamson, J. Buongiorno, L.W. Hu, M. Rubner, R. Cohen, Augmentation of nucleate boiling heat transfer and critical heat flux using nanoparticle thin-film coatings. Int. J. Heat Mass Transf. **53**(1–3), 58–67 (2010)
26. D. Wen, M. Corr, X. Hu, G. Lin, Boiling heat transfer of nanofluids: The effect of heating surface modification. Int. J. Therm. Sci. **50**(4), 480–485 (2011)
27. D. Wen, Influence of nanoparticles on boiling heat transfer. Appl. Therm. Eng. **41**, 2–9 (2012)

28. M. Kole, T.K. Dey, Investigations on the pool boiling heat transfer and critical heat flux of ZnO-ethylene glycol nanofluids. Appl. Therm. Eng. **37**, 112–119 (2012)
29. J.Y. Jung, E.S. Kim, Y.T. Kang, Stabilizer effect on CHF and boiling heat transfer coefficient of alumina/water nanofluids. Int. J. Heat Mass Transf. **55**(7–8), 1941–1946 (2012)
30. J. Tehver, H. Sui, V. Temkina, Heat transfer and hysteresis phenomena in boiling on porous plasma-sprayed surface. Exp. Therm. Fluid Sci. **5**(6), 714–727 (1992)
31. G.S. Hwang, M. Kaviany, Critical heat flux in thin, uniform particle coatings. Int. J. Heat Mass Transf. **49**(5–6), 844–849 (2006)
32. C.H. Li et al., Comparison study of liquid replenishing impacts on critical heat flux and heat transfer coefficient of nucleate pool boiling on multiscale modulated porous structures. Int. J. Heat Mass Transf. **54**(15–16), 3146–3155 (2011)
33. Y. Takata et al., Effect of surface wettability on boiling and evaporation. Energy **30**(2–4), 209–220 (2005)
34. S. Vemuri, K.J. Kim, Pool boiling of saturated FC-72 on nano-porous surface. Int. Commun. Heat Mass Transf. **32**(1–2), 27–31 (2005)
35. S.S. Hsieh, C.J. Weng, Nucleate pool boiling from coated surfaces in saturated R-134a and R-407c. Int. J. Heat Mass Transf. **40**(3), 519–532 (1997)
36. J.H. Kim, K.N. Rainey, S.M. You, J.Y. Pak, Mechanism of nucleate boiling heat transfer enhancement from microporous surfaces in saturated FC-72. J. Heat Transfer **124**(3), 500–506 (2002)
37. C. Li, Z. Wang, P.I. Wang, Y. Peles, N. Koratkar, G.P. Peterson, Nanostructured copper interfaces for enhanced boiling. Small **4**(8), 1084–1088 (2008)
38. T.J. Hendricks, S. Krishnan, C. Choi, C. Chang, B. Paul, Enhancement of pool-boiling heat transfer using nanostructured surfaces on aluminum and copper. Int. J. Heat Mass Transf. **53**(15–16), 3357–3365 (2010)
39. B. Feng, K. Weaver, G.P. Peterson, Enhancement of critical heat flux in pool boiling using atomic layer deposition of alumina. Appl. Phys. Lett. **100**(5), 98–101 (2012)
40. Y. Wen-Jei, H. Takizawa, D.L. Vrable, Augmented boiling on copper-graphite composite surface. Int. J. Heat Mass Transf. **34**(11), 2751–2758 (1991)
41. H.S. Liang, W.J. Yang, Nucleate pool boiling heat transfer in a highly wetting liquid on micro-graphite-fiber composite surfaces. Int. J. Heat Mass Transf. **41**(13), 1993–2001 (1998)
42. M.S. El-genk, J.L. Parker, *Pool Boiling in Saturated and Subcooled-7100 Dielectric Fluid from a Porous Graphite Surface*, pp. 655–662 (2004)
43. J.L. Parker, M.S. El-Genk, Effect of surface orientation on nucleate boiling of FC-72 on porous graphite. J. Heat Transf. **128**(11), 1159–1175 (2006)
44. H.S. Ahn, N. Sinha, M. Zhang, D. Banerjee, S. Fang, R.H. Baughman, Pool boiling experiments on multiwalled carbon nanotube (MWCNT) forests. J. Heat Transf. **128**(12), 1335–1342 (2006)
45. X. Dai et al., Enhanced nucleate boiling on horizontal hydrophobic-hydrophilic carbon nanotube coatings. Appl. Phys. Lett. **102**(16) (2013)
46. A.R. Betz, J. Xu, H. Qiu, D. Attinger, Do surfaces with mixed hydrophilic and hydrophobic areas enhance pool boiling? Appl. Phys. Lett. **97**(14), 1–4 (2010)
47. A.R. Betz, J.R. Jenkins, C.J. Kim, D. Attinger, Significant boiling enhancement with surfaces combining superhydrophilic and superhydrophobic patterns, in *Proceedings of IEEE International Conference on Micro Electro Mechanical Systems*, pp. 1193–1196 (2011)

# Comparative Study of Extended and Unscented Kalman Filters for Estimating Motion States of an Autonomous Vehicle-Trailer System

**Hussein F. M. Ali, Nader A. Mansour, and Youngshik Kim**

**Abstract** Kalman filters are used for motion state estimation of an autonomous vehicle-trailer system, which can be utilized directly to motion control and autonomous navigation. The autonomous vehicle-trailer system consists of an autonomous vehicle and a passive trailer coupled to the vehicle by a trailer hitch. The vehicle-trailer system is equipped with the global positioning system (GPS), encoder-based odometry, and hitch angle sensors. A Simulink model is first developed for the system kinematics. The vehicle states are then estimated using extended Kalman filter (EKF) and unscented Kalman filter (UKF). Simulation results are compared and discussed based on the root mean square error (RMSE) and the simulation time. The results indicate that both EKF and UKF algorithms have very close RMSE for the position $x$ and $y$, whereas the processing time is increased by 17.7% for the UKF.

**Keywords** Kalman filter · Motion state estimation · Localization · Sensor fusion · Vehicle-trailer

## 1 Introduction

Recently, significant research work has been conducted and huge capital has been invested in the field of autonomous vehicles (AV) or also known as self-driving vehicles. Autonomous vehicles were typically endorsed by customers for many reasons. They can be safer by reducing non-deliberate mistakes and risky behavior of the

H. F. M. Ali · N. A. Mansour
Mechanical Engineering, Benha University, Benha, Egypt
e-mail: hussein.ali@ejust.edu.eg

N. A. Mansour
e-mail: nader.mansour@bhit.bu.edu.eg

H. F. M. Ali · N. A. Mansour · Y. Kim (✉)
Mechanical Engineering, Hanbat National University, Daejeon, Korea
e-mail: youngshik@hanbat.ac.kr



K. M. Pandey et al. (eds.), *Recent Advances in Mechanical Engineering*, Lecture Notes in Mechanical Engineering, https://doi.org/10.1007/978-981-15-7711-6_18

**Fig. 1** Autonomous vehicle-trailer system

drivers. They can reduce traffic congestion, pollution, and fuel consumption. Additionally, they can increase the benefit of the time wasted through driving. In the same context, autonomous vehicle-trailer systems may be more important to transport goods for long driving periods among excessively remote cities. The vehicle-trailer system consists of an autonomous vehicle and a passive trailer which are coupled by a trailer hitch as shown in Fig. 1 [1].

The term "autonomous" implies that the vehicle is computer-controlled and driven depending on a variety of sensors that help the vehicle perceive its surroundings. The accuracy of these sensor data is necessary for the vehicle to make proper decisions instantaneously. However, it is usually very hard to obtain accurate data from the sensors due to the existence of noise and uncertainty. For that reason, measurements from different sensors are used together through estimation algorithms to obtain the best estimate of motion states such as positions and velocities required to solve the navigation and control problems.

Based on the dynamics model of the system, the state estimation approach can be categorized into linear and nonlinear model. Dynamic model-based approaches depend on differential and difference equations along with the measured data to generate estimates of the unmeasured variables [2]. State observers and Kalman filter approaches are presented in [3, 4] to estimate variables of linear systems. However, most of the systems encountered in engineering domains are nonlinear. For the nonlinear vehicle systems, different estimation algorithms have been implemented. Extended and unscented Kalman filters are used to estimate vehicle velocity as reported in [5, 6]. Nonlinear observer (NLO) and reduced nonlinear observer (RNLO) estimation approaches are also applied in [7] to estimate vehicle states.

Additionally, dual extended Kalman filter (DEKF) and nonlinear unknown input observer (NUIO) strategies are proposed in [8, 9]. Furthermore, artificial neural network (ANN) is used to estimate the sideslip angle as discussed in [10].

In this paper, we provide simulation-based estimation for the motion states of the vehicle-trailer system using both extended and unscented Kalman filters. Random noise has been added to the input and the measured signals to simulate a real noisy environment. We also provide comparison study on the two algorithms in terms of processing time, effect of noise, and quality of the estimated states. These estimated states can be used later for navigation and control purposes. This paper is organized such that the vehicle-trailer model is presented in Sect. 2. In Sect. 3, we discuss Kalman filtering estimation techniques including extended and unscented Kalman filter. Simulation results are presented in Sect. 4, and finally conclusions and future plans are provided in Sect. 5.

## 2 Modeling of Vehicle-Trailer

This section presents the continuous-time kinematic model for the vehicle-trailer system, which we used to formulate the discrete-time linear state-space model for the EKF and UKF in the next section. Based on the system as illustrated in Fig. 2, we can derive the general vehicle-trailer system mathematical model considering steering kinematics. The state variables are the Cartesian coordinates, $(x, y)$ at the point $C_1$ (the center of the vehicle rear axle), the heading angle,$\theta$, and the hitch angle, $\psi$. The state equations of the vehicle-trailer system:

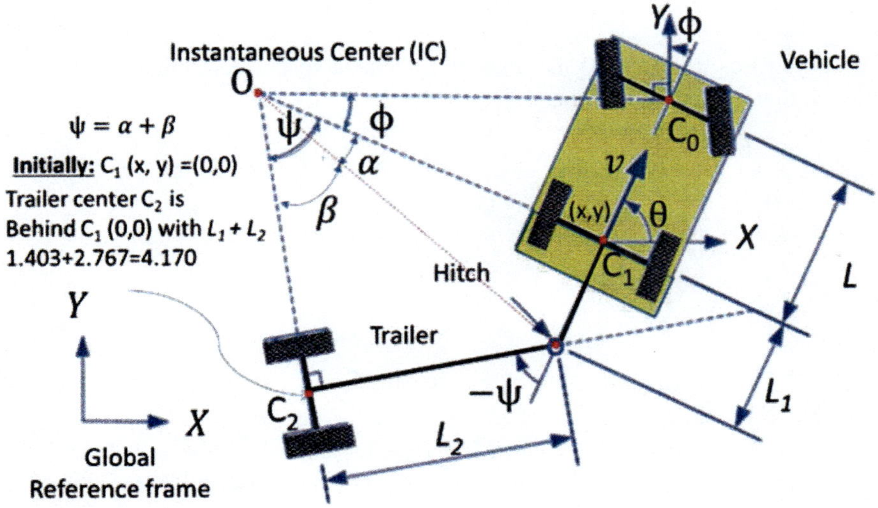


**Fig. 2** Kinematics of vehicle-trailer system

$$\dot{x} = v\cos\theta, \quad \dot{y} = v\sin\theta$$
$$\dot{\theta} = \omega, \quad \dot{\psi} = \frac{v}{L_2}\left[\frac{\tan\phi}{L}(L_2 + L_1\cos\psi) - \sin\psi\right] \tag{1}$$

where

- $v$ the vehicle linear velocity at the rear axle center $C_1$, (control input 1 in this model)
- $\omega$ the vehicle angular velocity ($\dot{\theta}$) at $C_1$, (control input 2 in this model)
- $\phi$ the steering angle
- $L$ the distance between the front and rear axles
- $L_1$ the hitch length
- $L_2$ the trailer length.

In Fig. 2, using curvature definition, trigonometric is identified for the right-angled triangle $OC_0C_1$, the curvature can be expressed at the point $C_1$ by:

$$\kappa = \frac{1}{\overline{OC_1}} = \frac{\dot{\theta}}{v} = \frac{\tan\phi}{L} \tag{2}$$

Consequently, the state model can be formulated for the vehicle-trailer system by applying (2) to (1) in vector–matrix form by,

$$\begin{bmatrix} \dot{x} \\ \dot{y} \\ \dot{\theta} \\ \dot{\psi} \end{bmatrix} = \begin{bmatrix} \cos\theta & 0 \\ \sin\theta & 0 \\ 0 & 1 \\ -\frac{\sin\psi}{L_2} & \frac{L_2+L_1\cos\psi}{L_2} \end{bmatrix} \begin{bmatrix} v \\ \omega \end{bmatrix} \tag{3}$$

Based on Eqs. (2) and (3) the discrete-time state-space representation for the vehicle-trailer system is formulated:

$$X^k = f\left(X^{k-1}, u^k\right) = A(k)X^{k-1} + B(k)u^k \tag{4}$$

$$X^k = \begin{bmatrix} x^k \\ y^k \\ \theta^k \\ \dot{x}^k \\ \dot{y}^k \\ \dot{\theta}^k \\ \psi^k \\ \dot{\psi}^k \end{bmatrix} = \begin{bmatrix} x^{k-1} + v^k\cos\theta^{k-1}\Delta t \\ y^{k-1} + v^k\sin\theta^{k-1}\Delta t \\ \theta^{k-1} + \omega^k\Delta t \\ v^k\cos\theta^{k-1}\Delta t \\ v^k\sin\theta^{k-1}\Delta t \\ \omega^k \\ \psi^{k-1} + \left(-v^k\sin\psi^{k-1}/L_2 + \omega^k\left(L_2 + L_1\cos\psi^{k-1}\right)/L_2\right)\Delta t \\ -v^k\sin\psi^{k-1}/L_2 + \omega^k\left(L_2 + L_1\cos\psi^{k-1}\right)/L_2 \end{bmatrix} \tag{5}$$

## 3 Kalman Filter

This section presents the basic concept of the linear Kalman filter which is a stochastic model based on the state observer of deterministic system Eq. (6). This method is used to model input noise and also the measurement noise Eq. (7), where $u$ contains noise and also the measurement state contain noise Eq. (8). The Kalman filter algorithm is based on two steps. The first step is the prediction based on the states model (including input noise) and the second step is the update based on the measurement (including measurements noise) as illustrated in Fig. 3.

$$\text{State observer:} \quad \hat{x}_{k+1} = A\hat{x}_k + Bu_k + K\left(y_k - C\hat{x}_k\right) \text{ Deterministic system} \tag{6}$$

$$\text{Kalman Filter:} \quad \hat{x}_k = A\hat{x}_{k-1} + Bu_k + K_k\left(y_k - C\left(A\hat{x}_{k-1} + Bu_k\right)\right) \quad \text{Stochastic system} \tag{7}$$

The measurement noise covariance $R$ and the control input noise covariance $R_v$ are determined using diagonal matrices,

$$R = \text{diag}\left[\sigma_x^2,\ \sigma_y^2,\ \sigma_\theta^2,\ \sigma_\psi^2\right], \quad R_v = \text{diag}\left[\sigma_v^2,\ \sigma_\omega^2\right] \tag{8}$$

Extended Kalman filters (EKF) is for nonlinear systems which can be linearized. EKF drawbacks are:

- It is difficult to calculate the Jacobians (if they need to be found analytically)

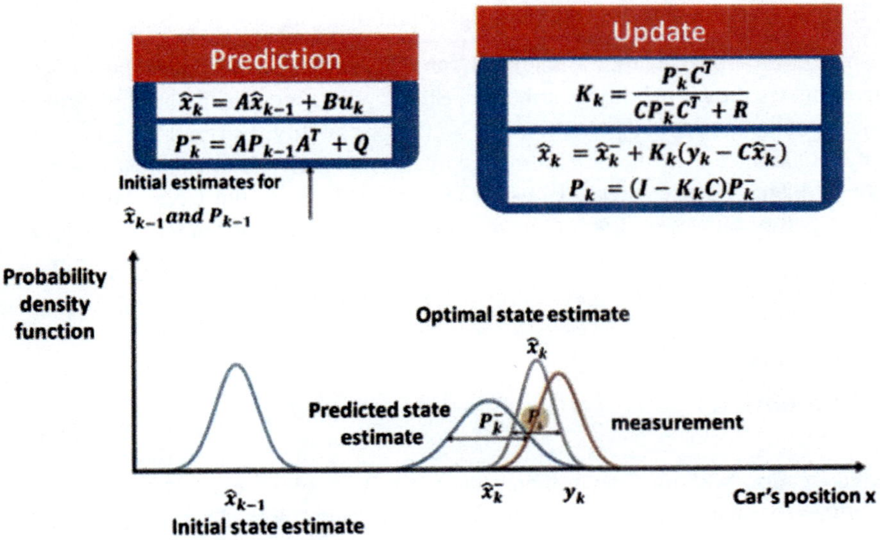


**Fig. 3** Linear Kalman filter two steps estimation: prediction then update [11]

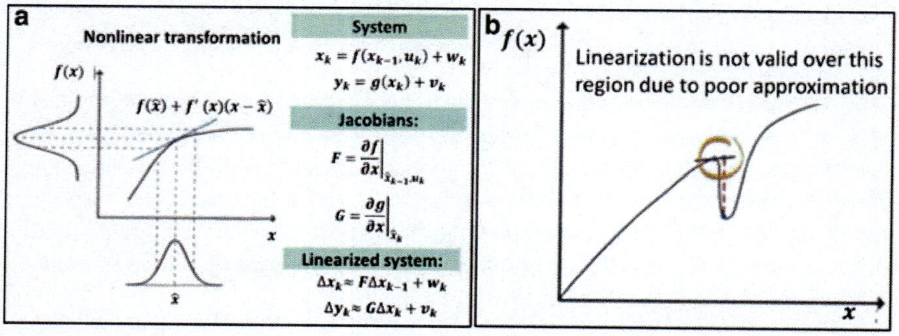


**Figure 4** **a** Extended Kalman filter main concept and **b** limitation [11]

**Table 1** Comparison between different type of Kalman filters

| State estimator | Model | Assumed distribution | Computational cost |
|---|---|---|---|
| KF | Linear | Gaussian | Low |
| EKF | Nonlinear (linearizable) | Gaussian | Low (if Jacobians are computed analytically) Medium (if Jacobians are computed numerically) |
| UKF | Nonlinear | Gaussian | Medium |
| PF | Nonlinear | Non-Gaussian | High |

- There is a high computational cost (if computer numerically)
- EKF only work on a system that have differentiable model
- EKF is not optimal if the system is highly nonlinear, as shown in Fig. 4.

Instead of approximating a nonlinear function, unscented Kalman filter (UKF) approximates the probability distribution. Sigma points are selected such that their mean $\mu$ and covariance $P$ are the same as the probability distribution. The mean $\mu$ and covariance $P$ of the transformed sigma points are used to calculate the new state estimate. Particle filter (PF) uses the same concept as the UKF, which uses sample points. However, it is for any arbitrary distribution (not limited to Gaussian assumption) and it needs larger number of points [11]. Table 1 compares Kalman filters based on model type, distribution and the computation cost.

## 4 Simulation

In this section we simulate motion state estimation of a vehicle-trailer system applying EKF and UKF. The discrete kinematic model, Eq. (5), for the vehicle-trailer system is implemented using MATLAB/Simulink as shown in Fig. 5. We then consider input noises in vehicle's linear and angular velocities, $v$ and $\omega$, and

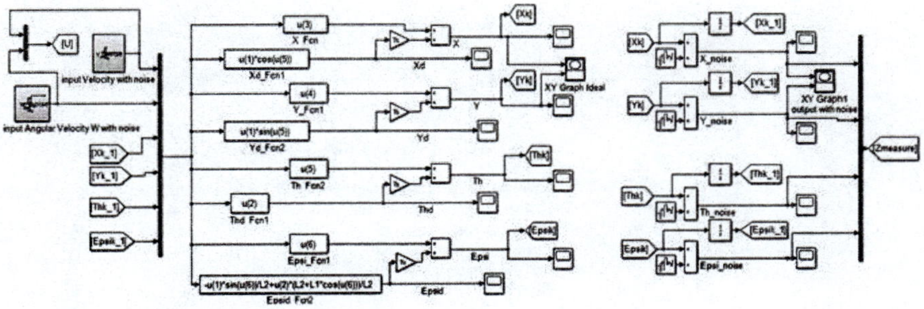


**Fig. 5** Vehicle-trailer nonlinear model using Simulink

measurement noises in the GPS position, $(x, y)$, the heading angle, $\theta$, and the hitch angle, $\psi$.

For relatively higher noises in the GPS position $(x, y)$ where $\left[\sigma_x, \sigma_y, \sigma_\theta, \sigma_\psi\right] = [0.5, 0.5, 0.1, 0.0149]$ and $[\sigma_v, \sigma_\omega] = [0.034, 0.034]$, we observe more noises in the estimated vehicle position rather than the trailer, as shown in Fig. 6.

While for less noises in the GPS postion where $\left[\sigma_x, \sigma_y, \sigma_\theta, \sigma_\psi\right] = [0.0124, 0.0037, 0.1, 0.0149]$ and $[\sigma_v, \sigma_\omega] = [0.034, 0.034]$, higher noises in inputs $(v, \omega)$ and states $(\theta, \psi)$ rather than $(x, y)$ cause more noises in the estimated trailer position compared to the vehicle, as shown in Fig. 7.

These simulations are repeated 10 times and the average elapsed time is 5.49 s for EKF and is 6.46 s for UKF. Moreover, to study the effect of input values on the performance of EKF and UKF, the same simulations are repeated with different inputs in terms of amplitude and frequency. Hence, multiples of (1, 2, 3, 5, and 10) of

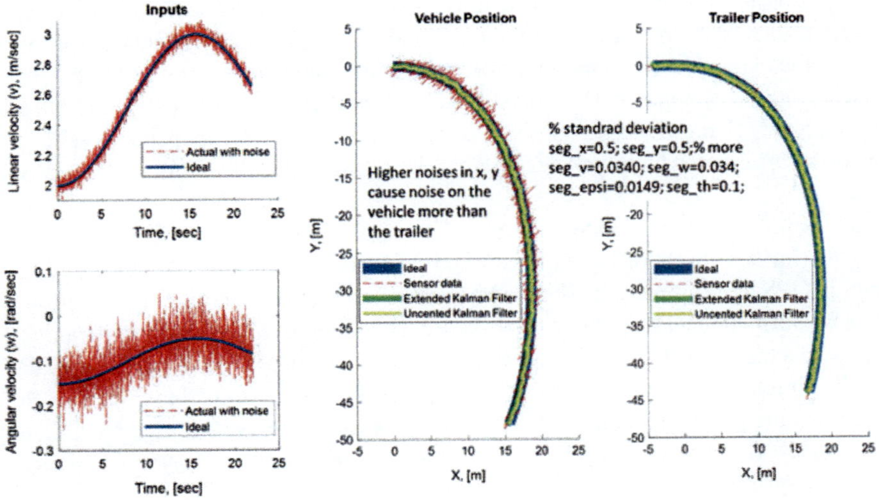


**Fig. 6** Simulation set 1: higher GPS noise

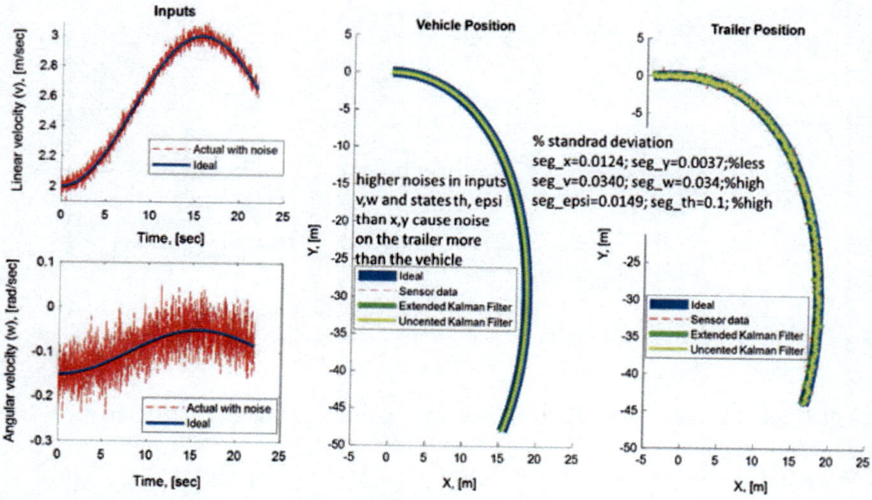


**Fig. 7** Simulation set 2: lower GPS noise

the amplitude and frequency are applied. The simulation results show slight change (i.e., $10^{-4}$ m) in the performance of EKF and UKF. The RMSE for x position is 0.0971 m and for *y* position is 0.0933 m.

## 5 Conclusion and Future Work

Kalman filtering-based motion state estimation methods are used for an autonomous vehicle-trailer system. The autonomous vehicle-trailer system consists of an autonomous vehicle and a passive trailer which are coupled by a trailer hitch. The vehicle-trailer system is equipped with the GPS, encoder-based odometry, and hitch angle sensors. Using GPS and odometry sensors, we obtain two independent vehicle motion information which includes orientation, position (localization), and velocity data.

After implementing the kinematic Simulink model for the system, we investigated the effect of two levels of inputs noises and measurements noises. We noticed that the high GPS noise impact the vehicle localization more than the trailer localization. Also, we verified that the Kalman filters succeeded to remove the noise significantly.

Moreover, EKF and UKF results are compared and discussed based on the RMSE and the simulation time. The results indicate that both algorithms have very close RMSE for the position x and y, whereas the processing time is increased by 17.7% for the UKF. So, we can apply both of them while UKF is more suitable for highly nonlinear systems.

For future work, we will investigate the performance of the EKF and UKF while considering tire friction effects by implementing a tire model such as Fiala model.

**Acknowledgements** This research is supported by a project grant (UD180045RD) from the Agency for Defense Development, Korea (ADD).

Also, the first author is funded by the Korea Research Fellowship (KRF) program by the National Research Foundation (NRF) with KRF Grant (2019H1D3A1A01071124).

## References

1. Y. Kim, Motion state estimation for an autonomous vehicle-trailer system using Kalman filter-based multisensor data fusion. Int. J. Asia Life Sci. **11**, 81–92 (2015)
2. S.C. Patwardhan, J. Prakash, S.L. Shah, Soft sensing and state estimation: review and recent trends. IFAC Proc. **40**(19), 65–72 (2007)
3. D. Luenberger, Observers for multivariable systems. IEEE Trans. Autom. Control **11**(2), 190–197 (1966)
4. R.E. Kalman, A new approach to linear filtering and prediction problems. J. Basic Eng. **82**(D), 34–45 (1960)
5. A. Katriniok, D. Abel, Adaptive EKF-based vehicle state estimation with online assessment of local observability. IEEE Trans. Control Syst. Technol. **24**(4), 1368–1381 (2016)
6. J. Stephant, A. Charara, D. Meizel, Virtual sensor: Application to vehicle sideslip angle and transversal forces. IEEE Trans. Ind. Electron. **51**(2), 278–289 (2004)
7. H.Y. Guo, H. Chen, D.P. Cao, W.W. Jin, Design of a reduced order non-linear observer for vehicle velocities estimation. IET Control Theor. Appl. **7**(17), 2056–2068 (2013)
8. T.A. Wenzel, K.J. Burnham, M.V. Blundell, R.A. Williams, Dual extended Kalman filter for vehicle state and parameter estimation. Vehicle Syst. Dyn. **44**(2), 153–171 (2006)
9. L. Imsland, T.A. Johansen, H.F. Grip, T.I. Fossen, On non-linear unknown input observers-applied to lateral vehicle velocity estimation on banked roads. Int. J. Control **80**(11), 1741–1750 (2007)
10. D. Piyabongkarn, R. Rajamani, J.A. Grogg, J.Y. Lew, Development and experimental evaluation of a slip angle estimator for vehicle stability control. IEEE Trans. Control Syst. Technol. **17**(1), 78–88 (2009)
11. The Math Works, Inc. Math Works, 2019. [Online] Available from: https://kr.mathworks.com/videos/series/understanding-kalman-filters.html. Accessed 30 Nov 2019

# Numerical Study on Material Flow Behaviour in Friction Stir Welding of Low Carbon Steel

**Pardeep Pankaj, Abhishek Bhardwaj, Avinish Tiwari, Lakshmi Narayan Dhara, and Pankaj Biswas**

**Abstract** In friction stir welding (FSW) process, material flow is the most important aspect which affect the mechanical properties and microstructure of the welded joints. The good plasticized material flow reduces the formation of defects in the welded joint. In the present study, a three-dimensional volume of fluid (VOF) model based on ANSYS 14.5 FE software package was developed to predict the effect of traverse speeds (i.e. 90, 132 and 180 mm/min) on material flow behaviour during FSW of low carbon steel. Stain and temperature-dependent material properties were incorporated in developed material flow model. It is observed that the tool traverse speed strongly influenced the mixing of plasticized material in FSW of low carbon steel. The velocity of material flow was reduced as the distance increases away from the rotating axis of the probe or weld zone. The velocity vector of plasticized material was different at different planes throughout the welded joint. The material in plane nearby the top surface exhibited the maximum velocity than the plane close to the bottom surface. Experiment was also carried out using tungsten carbide tool to validate the material flow model. The transient thermal profile obtained from FE analysis and experiment was agreed properly well for peak temperature with a maximum percentage error of 6.72%.

---

P. Pankaj (✉) · A. Bhardwaj · A. Tiwari · L. N. Dhara · P. Biswas
Department of Mechanical Engineering, Indian Institute of Technology Guwahati, Guwahati, Assam 781039, India
e-mail: pankajpardeep22@gmail.com

A. Bhardwaj
e-mail: abhishektrivedi120@gmail.com

A. Tiwari
e-mail: avit252@gmail.com

L. N. Dhara
e-mail: nndhara@gmail.com

P. Biswas
e-mail: pankaj.panu012@gmail.com



K. M. Pandey et al. (eds.), *Recent Advances in Mechanical Engineering*, Lecture Notes in Mechanical Engineering, https://doi.org/10.1007/978-981-15-7711-6_19

**Keywords** Material flow behaviour · Volume of fluid · Sticking and sliding · Friction stir welding · Low carbon steel

## 1 Introduction

Several manufacturing techniques, i.e. welding, brazing, soldering, clinching, adhesive bonding, injected metal assembly and discrete fastening are able to join the similar and dissimilar material [1]. In FSW, the thermal analysis and material flow analysis are the important characteristics during welding of low carbon steels which need to be numerically investigated and validated. FSW is a solid-state welding technique to joins the workpieces without melting and using filler metals. The stir zone exhibits the highest temperature which is generally 80% of the liquidius temperature of the materials being joined [2]. Friction stir (FS) welded joints exhibit the excellent mechanical properties due to the absence of defects (i.e. porosity, slag inclusions and spatter) and lower temperature generation compared to the fusion welding processes. During FSW process, the rotating tool creates the joint action of frictional heat, plastic deformation and stirring phenomena. The heat generation between the workpiece and tool is the major element which acts as a driving force for friction stir welding technique. The maximum temperature generated by the heat flux must be high enough such that it softens the material for stirring action by pin but it should be low enough so that material does not liquefy. This generated heat is conducted to both workpiece and tool. The weld quality, shape and size of weld, residual stress and deformation in the FS welded specimen, tool life and its efficiency for the welding process is depicted by the extent of the heat conducted into the workpiece material [3]. Insufficient heat could lead to wear of the tool pin resulted in insufficient plastic deformation. Therefore, it is most important to get the understanding about the material flow aspect for improving the weld quality for FSW process. Buffa et al. [4] developed the FE model using DEFORM-3DTM software to predict the thermal history during FSW of AA7075 aluminium alloy. The FE model was assumed the FSW tool as rigid body and workpiece as a rigid viscoplastic. Das et al. [5] assumed the sticking and sliding phenomena between the workpiece and FSW tool in developed FE thermal model for FSW of AA6061 plates. Their established heat source model was fairly agreed the experimental results with percentage error of 5% for maximum temperature. Kadian et al. [6] investigated the flow behaviour in friction stir welded AA6061 by using a threaded tool. The higher relative velocity at tool/workpiece interface generates the more heat in the advancing side than the retreating side. S.D. Ji et al. [7] developed the FE model based on ANSYS Fluent software to investigate the effect of pin and shoulder geometry on plasticized material movement during friction stir welding. It was found that the material flow velocity decreases as distance increase away from welded region. Pankaj et al. [8] performed the FSW to join the DH36 steel and mild steel at welding speed of 2.2 mm/s and rotational speed of 600 rpm. They reported the maximum hardness value in TMAZ of DH36 steel. Tiwari et al.

[9] used the FSW technique to join the low carbon steel with tool made up of tungsten carbide. The tensile strength of the FS welded specimens was increased by decreasing the rotational speed and increasing the traverse speed due to grain refinement in macrostructure. Kadian et al. [10] studied the material flow behaviour using volume of fluid (VOF) approach during FSW of aluminium and copper alloy. They reported the plasticized material flow strongly influenced by welding parameters, i.e. rotational speed and traverse speed.

From the literature survey, it is observed that numerous models were developed for FSW process to predict the material flow behaviour, most of these were not concerned with the FSW of aluminium alloys. In this study, 3D material flow model based on volume of fluid (VOF) approach was established to investigate the material flow during FSW of low carbon steel. The experimental work was also performed to validate the numerical model.

## 2 Finite Volume Model

The finite volume model was established to investigate the flow of plasticized material during FSW of low carbon steel. The sheets having length of 100 mm, width of 75 mm and thickness of 4 mm were welded in square butt joint configuration. In this study, FSW tool with cylindrical pin diameter of 9 mm and shoulder diameter of 25 mm was used. Except the bottom surface, the heat transfer coefficient of 25 W/m$^2$ K was applied on the remaining surfaces. However, the higher heat transfer coefficient (i.e. 2000 W/m$^2$ K) was applied at the bottom surface to achieve the realistic results. The schematic illustration of the developed model for material flow analysis is shown in Fig. 1a. This diagram shows the coordinate system with all solid and

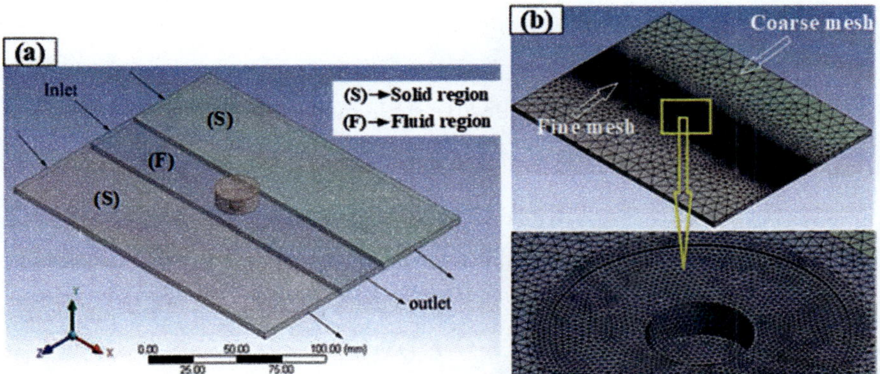


**Fig. 1** **a** Schematic illustration for material flow analysis, **b** mesh view of model

fluid regions in the worksheet for material flow analysis. A mesh sensitivity analysis was implemented to predict the optimum meshing parameters such as element size and number of elements. The tetrahedral and brick elements were used to mesh the tool wall in the worksheets. A fine meshing was applied near the tool pin zone in order to record the detailed flow behaviour of the pin as shown in Fig. 1b. Coarse mesh was applied away from this region to reduce the computational time.

Based on the following heat generation equation, the heat input was applied as a heat flux in developed material flow model.

$$Q = (1 - \delta)\mathrm{Q}_{\mathrm{sliding}} + \delta \mathrm{Q}_{\mathrm{sticking}} \quad (1)$$

$$\begin{aligned} Q = {} & (1 - \delta)\left\{\frac{2}{3}\pi\mu^{*}\omega P\left(R^{3} - r_{p}^{3}\right) + \frac{2}{3}\pi\mu^{*}\omega P r_{p}^{3} + 2\mu^{*}\pi\tau^{*}\omega r_{p}^{2} l\right\} \\ & + \mu^{*}\pi S_{ys}^{*} R h_{1} v + \mu^{*} S_{ys}^{*} \pi r_{p} l v \\ & + \delta\left\{\frac{2}{3}\pi\omega\tau^{*}\left(R^{3} - r_{p}^{3}\right) + \frac{2}{3}\pi\omega\tau^{*} r_{p}^{3} + 2\pi\tau^{*}\omega r_{p}^{2} l\right\} \end{aligned} \quad (2)$$

where $\tau^*$ is the shear strength of the materials at about 80% of its melting point temperature, $S_{ys}^*$ is the yield strength of the materials at about 80% of its melting point temperature,$\omega$ is the angular speed of tool, $r_p$ is the tool pin radius, $R$ is the cylindrical shoulder radius and $l$ is the pin height. In present FE model, $\delta = 0.58$ was taken for considering both sticking and sliding conditions [11]. The constant value of coefficient of friction ($\mu^*$) was taken as 0.3 and the value of plunging force was considered as 25 kN. Temperature dependent material properties of mild steel were taken from published literatures [12, 13]. The viscosity of the material in the term of flow stress and strain rate is given by:

$$\mu = \frac{\sigma_R}{3\varepsilon} \quad (3)$$

where $\sigma_R$ is the flow stress, $\varepsilon$ is the effective strain rate and $\mu$ denotes the viscosity of the material. The following Zener–Holloman equations is used to determine the flow stress.

$$\sigma_R = \frac{1}{\alpha} ln\left[\left(\frac{Z}{A}\right)^{1/n} + \left(1 + \left(\frac{Z}{A}\right)^{\frac{2}{n}}\right)^{\frac{1}{2}}\right] \quad (4)$$

$$Z = \varepsilon \exp\left(\frac{Q}{RT}\right) \quad (5)$$

where $Z$ is the Zener–Holloman parameter, $R$ is the universal gas constant, $T$ is the temperature and $\alpha$, $n$, $A$ are the material properties were taken from published paper

[14]. The momentum, mass and energy are generally monitored by the FSW process. Considering the material mass constant, the equation of mass conversation is written as follow.

$$\frac{\partial u}{\partial x} + \frac{\partial v}{\partial y} + \frac{\partial w}{\partial z} = 0 \tag{6}$$

$$\rho u \frac{\partial u}{\partial x} + \rho v \frac{\partial u}{\partial y} + \rho w \frac{\partial u}{\partial z} = F_x - \frac{\partial p}{\partial x} + \mu \left( \frac{\partial^2 u}{\partial x^2} + \frac{\partial^2 u}{\partial y^2} + \frac{\partial^2 u}{\partial z^2} \right) \tag{7}$$

$$\rho u \frac{\partial v}{\partial x} + \rho v \frac{\partial v}{\partial y} + \rho w \frac{\partial v}{\partial z} = F_y - \frac{\partial p}{\partial y} + \mu \left( \frac{\partial^2 v}{\partial x^2} + \frac{\partial^2 v}{\partial y^2} + \frac{\partial^2 v}{\partial z^2} \right) \tag{8}$$

$$\rho u \frac{\partial v}{\partial x} + \rho v \frac{\partial v}{\partial y} + \rho w \frac{\partial v}{\partial z} = F_y - \frac{\partial p}{\partial y} + \mu \left( \frac{\partial^2 v}{\partial x^2} + \frac{\partial^2 v}{\partial y^2} + \frac{\partial^2 v}{\partial z^2} \right) \tag{9}$$

where the velocity components in the $x$, $y$ and $z$ directions are represented by $u$, $v$, and $w$, respectively. The energy conservation equation is represented as follow. $F_x$, $F_y$ and $F_z$ denotes the forces in $x$, $y$ and $z$ direction, respectively.

$$\rho C_p \frac{\partial T}{\partial x} + \rho C_p \frac{\partial T}{\partial y} + \rho C_p \frac{\partial T}{\partial z} = k \left( \frac{\partial^2 T}{\partial x^2} + \frac{\partial^2 T}{\partial y^2} + \frac{\partial^2 T}{\partial z^2} \right) \tag{10}$$

where $T$ denotes the temperature of the fluid. $\rho$ and $C_p$ represent the density and specific heat capacity of the material respectively.

## 3 Results and Discussion

In the present study, a three-dimensional material flow model based on ANSYS Fluent 14.5 FE software package developed to predict the effect of traverse speeds (i.e. 90, 132 and 180 mm/min with constant rotational speed of 600 rpm) during FSW of low carbon steel. Figure 2 shows the temperature contour of the FS welded low carbon steel on the top surface at 600 rpm and 180 mm/min.

Figure 2 shows that the maximum temperature was reached about 1091 K, which is lesser than the melting point temperature of low carbon steel. Hence, it is confirmed that the FSW is a solid-state welding process. As going far from the weld centre line, the temperature was decreased as shown in Fig. 2. To obtain the sound quality weld, it is very important that the mixing or flow behaviour of the plasticized material should be sufficient during welding process. The better plasticized material reduces the formation of defects/voids in the FS welded joint. Figure 3 represents the material flow behaviour with velocity vector at varying traverse speed.

**Fig. 2** Thermal contour during FSW of low carbon steel

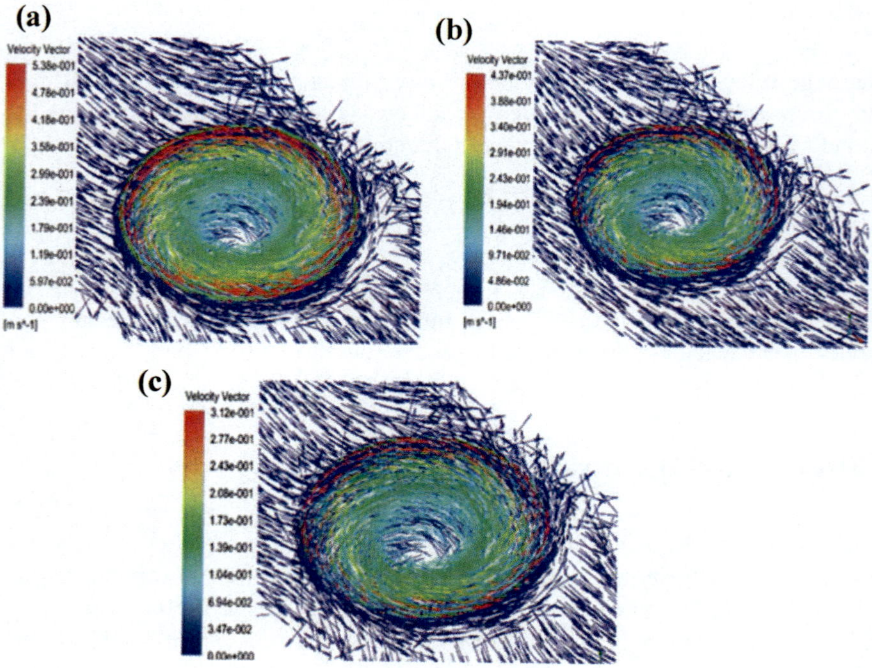


**Fig. 3** Material flow behaviour during FSW at **a** 90 mm/min, **b** 132 mm/min, **c** 180 mm/min

It is observed that material exhibits the higher value of velocity vector in the shoulder edge as compared to the other shoulder surface and the material rotational direction is similar as the FSW tool as shown in Fig. 3. The material flow is also influenced by the tool pin. Towards the pin region, the velocity vector value keeps on decreasing because the shoulder is responsible for maximum amount of deformation in weld nugget. The material flow pattern is almost equivalent but the magnitude is different at varying traverse speed. It is also found that the value of velocity vector

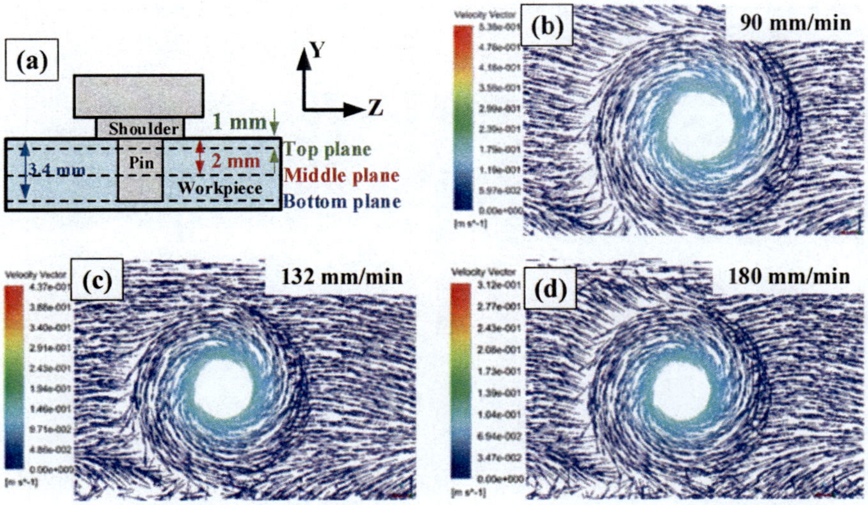


**Fig. 4** **a** Different planes from top surface, material flow behaviour during FSW in middle plane at, **b** 90 mm/min, **c** 132 mm/min, **d** 180 mm/min

increased by decreasing the traverse speed due to large amount of deformation at lower traverse speed. Particles at the outer pin surface exhibited the higher velocity as compared to the pin centre. The strength of the welded joints strongly influenced by the temperature distribution. Increase in temperature reduce the strength of the material and increase the material flow velocity. The uneven temperature distribution during FSW process resulted in gradient distribution to plasticized material flow. In order to get the better understanding of material flow, some planes, i.e. top plane, middle plane and bottom planes were considered at 1 mm, 2 mm and 3.4 mm away from the top surface, respectively, as shown in Fig. 4a. The material flow behaviour in the middle plane at varying traverse speed is shown in Fig. 4b–d. Figure 5 shows the velocity vectors against the top and bottom at 1 mm and 3.4 mm away from the top surface, respectively.

The flow of plasticized material was different at different planes throughout the welded joint as shown in Figs. 4 and 5. The material at top plane nearby the shoulder/workpiece interface exhibited the maximum velocity than the middle plane and bottom plane which was close to the bottom surface as shown in Fig. 5. The motion of shoulder mainly affected the top plane and bottom plane was strongly influenced by the stirring of the pin. The rotating FSW tool revolved the plasticized material and deposited this material in a circulated manner behind the pin. The transient temperature data was measured at 600 rpm and 90 mm/min using K-type thermocouples connected with the data acquisition system as shown in Fig. 6a. From experiment, the transient thermal profile obtained at 18 mm far from weld centre line was compared with numerical results as shown in Fig. 6b.

The transient thermal profile obtained from FE analysis was compared fairly well with experiment for peak temperature with maximum percentage error of 6.72% as

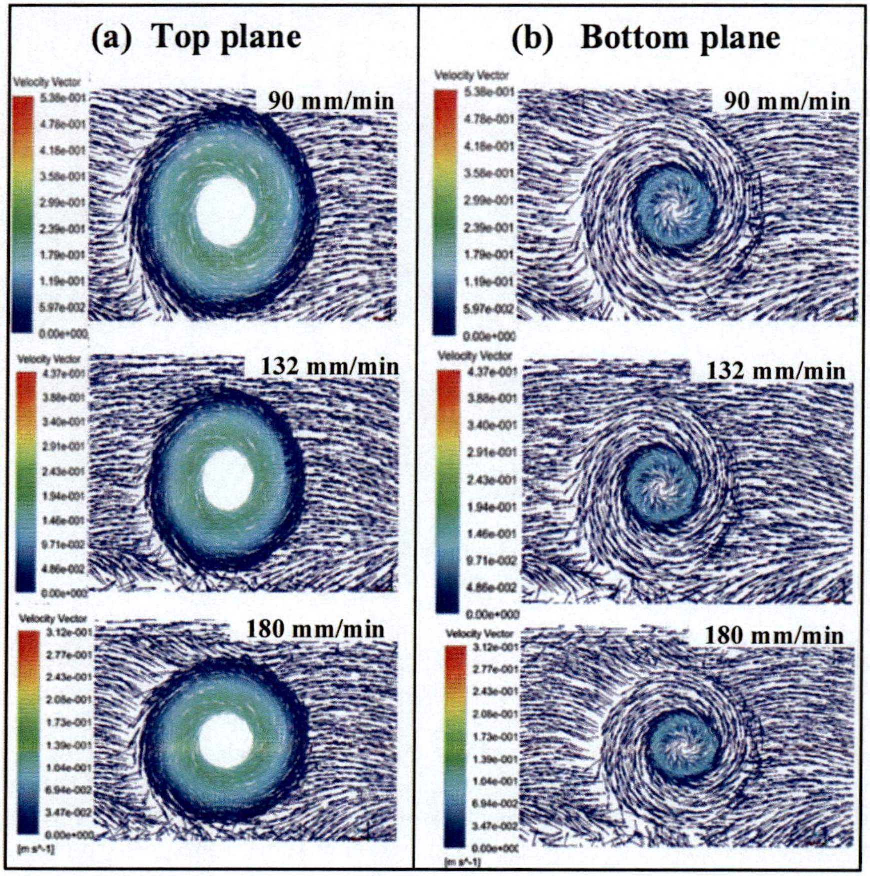


**Fig. 5** Material flow behaviour during FSW at **a** top plane, **b** bottom plane

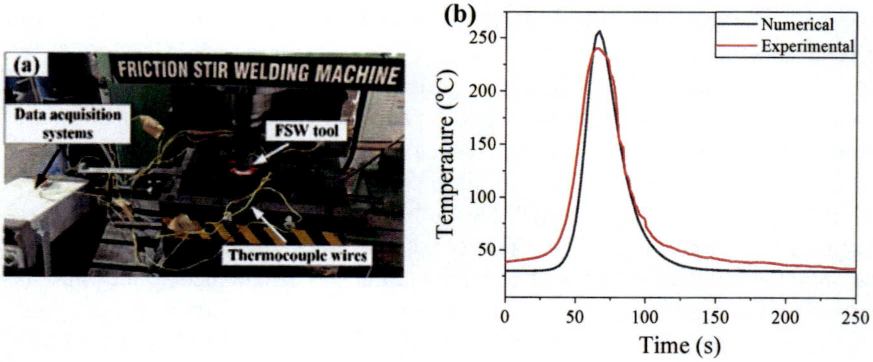


**Fig. 6** **a** FSW machine and temperature measuring setup, **b** comparison of experimental and numerical thermal profiles

shown in Fig. 6. Therefore, the good comparison showed that the established 3D material flow model can be able to investigate the material flow behaviour in the FSW of low carbon steels.

## 4 Conclusions

In present study, 3D volume of fluid (VOF) model was established to predict the material flow behaviour during FSW of mild steel. Experiment was also implemented to confirm the established material flow model. The numerical work can be drawn the following conclusions.

- The developed material flow model was validated with experimentally obtained thermal profile for peak temperature with a maximum percentage error of 6.72%.
- The tool traverse speed significantly affected the magnitude of the material flow velocity. The higher velocity vector was obtained at the low traverse speed.
- Increasing distance away from the rotating axis of the probe or weld zone reduces the material flow velocity.
- The material at the plane nearby the top surface exhibited the maximum velocity than the plane which was close to the bottom surface. The motion of shoulder mainly influenced the top plane and bottom plane was strongly influenced by the stirring of the tool.

## References


1. C.A. Hernandez, V.H. Ferrer, J.E. Mancilla, L.C. Martinez, Three-dimensional numerical modeling of the friction stir welding of dissimilar steels. Int. J. Adv. Manuf. Technol. **93**(5–8), 1567–1581 (2017)
2. P. Colegrove, 3 Dimensional Flow and Thermal Modelling of the Friction Stir Welding Process (Doctoral dissertation) (2001)
3. Y.J. Chao, X. Qi, W. Tang, Heat transfer in friction stir welding—experimental and numerical studies. J. Manuf. Sci. Eng. **125**(1), 138–145 (2003)
4. G. Buffa, J. Hua, R. Shivpuri, L. Fratini, Design of the friction stir welding tool using the continuum based FEM model. Mater. Sci. Eng., A **419**(1–2), 381–388 (2006)
5. N.K. Das, A.K. Kadian, A. Tiwari, P. Pankaj, P. Biswas, *Transient Thermal Analysis on Friction Stir Welding of AA6061. In Manufacturing Engineering* (Springer, Singapore, 2019), pp. 67–82
6. A.K. Kadian, P. Biswas, A comparative study of material flow behavior in friction stir welding using laminar and turbulent models. J. Mater. Eng. Perform. **24**(10), 4119–4127 (2015)
7. S.D. Ji, Q.Y. Shi, L.G. Zhang, A.L. Zou, S.S. Gao, L.V. Zan, Numerical simulation of material flow behavior of friction stir welding influenced by rotational tool geometry. Comput. Mater. Sci. **63**, 218–226 (2012)
8. P. Pankaj, A. Tiwari, S. Suman, A. Kumar, R. Bhattacharjee, S. Majumder, P. Biswas, *Dissimilar Friction Stir Welding of DH36 Shipbuilding Steel and Mild Steel. In Advances in Additive Manufacturing and Joining* (Springer, Singapore, 2020), pp. 397–408
9. A. Tiwari, P. Singh, P. Pankaj, P. Biswas, S.D. Kore, FSW of low carbon steel using tungsten carbide (WC-10wt.% Co) based tool material. J. Mech. Sci. Technol. **33**(10), 4931–4938 (2019)

10. A.K. Kadian, P. Biswas, The study of material flow behaviour in dissimilar material FSW of AA6061 and Cu-B370 alloys plates. J. Manuf. Processes **34**, 96–105 (2008)
11. A.K. Kadian, P. Biswas, Effect of tool pin profile on the material flow characteristics of AA6061. J. Manuf. Processes **26**, 382–392 (2017)
12. P. Pankaj, A. Tiwari, R. Bhadra, P. Biswas, Experimental investigation on $CO_2$ laser butt welding of AISI 304 stainless steel and mild steel thin sheets. Opt. Laser Technol. **119**, 105633 (2019)
13. P. Biswas, N.R. Mandal, O.P. Sha, M.M. Mahapatra, Thermo-mechanical and experimental analysis of double pass line heating. J. Mar. Sci. Appl. **10**(2), 190–198 (2011)
14. R.G.G.R. Nandan, G.G. Roy, T.J. Lienert, T. Debroy, Three-dimensional heat and material flow during friction stir welding of mild steel. Actamaterialia **55**(3), 883–895 (2007)

# Comparing Wettability and Frictional Performance of Laser Micro-machined Discrete and Continuous Textures

**V. Kashyap and P. Ramkumar**

**Abstract** TiAl4V is used widely in aerospace and biomedical application due to its high specific strength and good bio-compatibility. Its poor tribological performance restricts usage for hip implant articulation. Various surface characteristics such as surface roughness and wettability affect the tribological behaviour of Ti6Al4V sliding. Surface texturing is the recent technique to modify the surface features and improve the wettability. This study aims to compare the wettability and coefficient of friction (CoF) of discrete and continuous texture under bio-lubricated condition. Dimple and crosshatch textures are fabricated using laser surface texturing (LST) technique. The geometrical parameters such as depth, pitch and area density have been kept the same. Wettability associated with both the textures are analysed by measuring surface contact angles using goniometer. Further, friction behaviour is evaluated for all the textured and non-textured surfaces under biological environment using reciprocating pin-on-disc tribometer. Results show a significant reduction in contact angle for crosshatch texture compared to dimple and non-textured surface. Also, both the texture reduced the friction by 24% compared to non-textured surface.

**Keywords** Contact angle · Crosshatch · Dimple · Friction · Laser surface texturing (LST) · Ti6Al4V · Wettability

## 1 Introduction

Ti6Al4V grade-5 material is widely used in various aerospace and orthopaedic applications due to its remarkable specific strength, chemical inertness, bio-compatibility and corrosion resistance [1, 2]. However, currently the poor tribological property of this material is being investigated intensively using various surface modifications

V. Kashyap · P. Ramkumar (✉)
Machine Design Section, Department of Mechanical Engineering, Indian Institute of Technology Madras, Chennai, India
e-mail: ramkumar@iitm.ac.in

V. Kashyap
e-mail: vivekvedkashyap@gmail.com



K. M. Pandey et al. (eds.), *Recent Advances in Mechanical Engineering*, Lecture Notes in Mechanical Engineering, https://doi.org/10.1007/978-981-15-7711-6_20

such as nitriding, coating and texturing [3–6]. These modifications greatly affect the various surface characteristics such as surface finish, wettability and tribological performance under different contact interfaces [7–9]. These surface characteristics play an important role determining the lifespan of any component under service. In particular, wettability is one of the key factors in improving lubrication and fluidic behaviour of a tribo-contact [1, 9]. It is the balanced intermolecular interaction of adhesive (liquid to solid) and cohesive (liquid to liquid) forces to finally maintain the liquid contact over a solid surface. The same is measured in terms of surface contact angle which is the angle between solid plane and tangent of liquid vapour interface.

To improve the wettability and other tribological properties of an articulating interface, surface texturing with different shapes is being investigated over past few years [10, 11]. Discrete textures such as circular, triangular and elliptical shapes are already studied for their positive influence over tribological performance [12]. Also, effect of geometrical parameters is being studied to obtain better wettability and maximum frictional benefits under engine application [13, 14], mechanical seals [15, 16] and biomedical application [17]. However, recent studies for continuous/connected textures have also shown tremendous tribological benefits [18]. But the effect of various textured surface characteristics such as feature geometry, area density and distribution is not having been investigated to the fullest. Moreover, a comparative study is needed to determine the difference in the wettability and tribological behaviour in discrete and continuous textures to finally select a desired functional surface for any specific application. Hence, discrete dimple texture and continuous crosshatch texture of similar geometry are fabricated using laser surface texturing technique. To improve the usefulness of Ti6Al4V under biological condition, the wettability and frictional performance of these textures are evaluated and compared with non-textured surface using goniometer and reciprocating pin-on-disc tribometer, respectively.

## 2 Experimental Methodology

### *2.1 Sample Preparation and Texture Fabrication*

Ti6Al4V square samples with a size 20 × 20 × 5 mm$^3$ is EDM cut and mirror polished to a surface roughness $R_a$ of 0.05 μm measured by Bruker non-contact optical profiler. Later, Nd:YAG nanosecond solid beam laser of 1064 nm wavelength at 3.5 kHz frequency was used to fabricate the textures over polished samples. The texture pitch and depth for both the textures were 160 μm and 5 μm, respectively. Whereas width of 20 μm and diameter of 85 μm were opted to maintain an equal texture area under single pitch square. Therefore, a constant texture area density of 25% is achieved for both the textures. Both the fabricated textures are shown in Fig. 1. All the geometrical parameters and surface roughness factors are provided in Table 1.

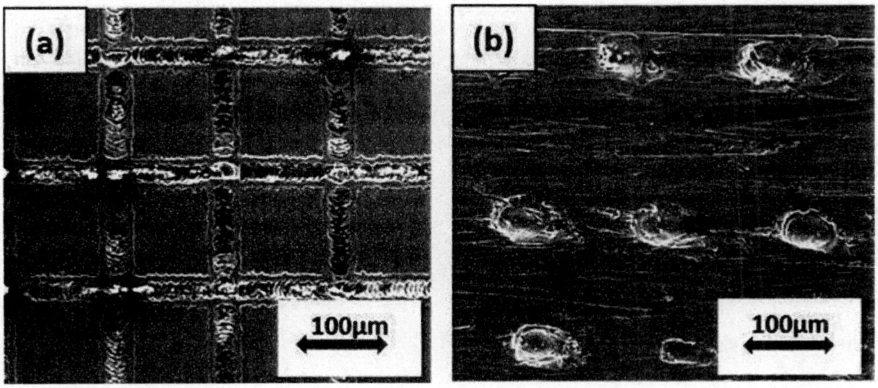


**Fig. 1** SEM image of **a** crosshatch and **b** dimple texture

**Table 1** Geometrical parameters of the textures

| Texture | Dia./width (μm) | Depth (μm) | Pitch (μm) | Aspect ratio | Roughness factor[a] |
|---|---|---|---|---|---|
| Dimple | 85 | 5 | 160 | 0.06 | 1.052 |
| Crosshatch | 20 | 5 | 160 | 0.25 | 1.109 |

[a]Roughness factor ($r$) for dimple and crosshatch texture are given by Eqs. (1) and (2), respectively

$$r = \frac{p^2 + \pi dh}{p^2} \tag{1}$$

$$r = \frac{p^2 + 4(p - w)h}{p^2} \tag{2}$$

where

- $p$ pitch of the textures.
- $d$ diameter of dimple texture.
- $w$ width of crosshatch texture.
- $h$ height of the texture.

## 2.2 *Contact Angle Measurement*

Ideally, the wettability of a surface is represented by a static contact angle or Young's contact angle $\theta_Y$. However, it really difficult to measure a single static angle value as the angle is greatly affected by surface roughness surface, time-dependent oxidation, contamination and continuous evaporation of the liquid droplet used during the measurement [19–21]. Hence, instantaneous contact angle ($\theta i$) is measured by goniometer using drop shape analysis (DSA) technique where the instantaneous

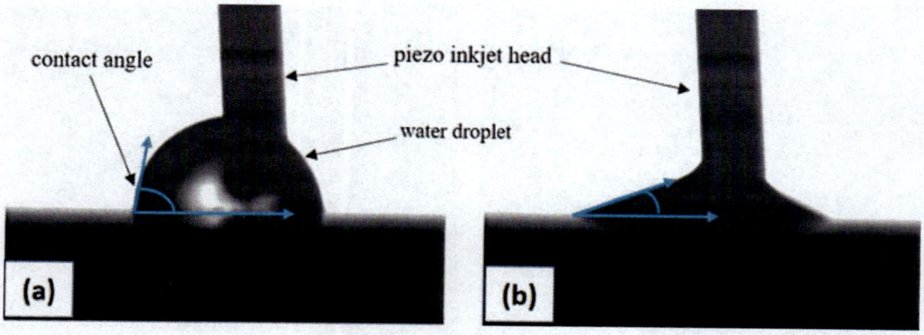


**Figure 2** **a** Advancing angle, **b** receding angle

angle is given as Eq. (3).

$$\theta_i = 2 \times \tan^{-1}\left(\frac{2h_i}{d_i}\right) \tag{3}$$

Contact angles of all the textured and non-textured surface were measured by sessile drop technique using contact angle goniometer. The accuracy of the contact angle metre is 0.1°, and all the measurements were taken at room temperature (27 °C). Ultrasonically cleaned samples were mounted over the micro-stage, and the pure water droplet is formed gradually using piezo inkjet head. The volume of pure water sessile droplet is gradually increased at a rate of 1.2*n* L/s for 20 s, and the dynamic contact angles were measured by capturing the droplet images at every 1/5th of sec. The maximum contact angle during liquid front advancing is recorded as the angle of advancing (as shown in Fig. 2a). Similarly, the minimum contact angle while retracting the droplet volume is measured as angle of receding (as shown in Fig. 2b). These two angles are the limiting values of the equilibrium contact angle, and the difference between these angles is known as contact angle hysteresis which is further used to obtain the final equilibrium contact angle. The results from the wettability test would be a determining factor for the lubricant film retention at any functional interface and hence affecting the tribological performance of the articulating components.

## 2.3 Tribology Test for Textured and Non-textured Surfaces

Ti6Al4V material has been greatly studied in orthopaedic application due to its good bio-compatibility but its usage is limited due to poor tribological performance. Due to unsatisfactory performance of a metal-on-polymer (MoP)-type hip implant, hard-on-hard (MoM and MoC) type of contact tribo-pairs are being investigated along with the implementation of different textured surfaces. Hereby, the frictional performance of discrete dimple texture was compared with the continuous crosshatch texture using reciprocating pin-on-disc tribometer (Rtec Instruments, USA). The alumina pin of 10 mm length and 6 mm diameter was made slid against all the textured and non-textured surfaces under bio-lubricated condition. Fetal bovine serum (FBS) with 25% concentration and viscosity which is 1.36 cP was heated at 40 °C for 30 min to replicate the synovial joint fluid condition. Stroke length of 10 mm and frequency of 1 Hz were selected to achieve average walking speed. A load of 10 N was applied, equivalent to 100 MPa Hertzian pressure, according to the reported contact pressure at the hip joint interface [5]. Generally, a maximum contact pressure of 10 MPa occurs at the joint [21] but even higher contact pressures can be observed depending upon the material combination, area of contact and applied load [22]. The test duration was 30 min including 5 min of running-in period; all the experiments were repeated thrice for better repeatability, and friction force was measured for all the textured and non-textured surfaces.

## 3 Results and Discussion

### 3.1 Contact Angle and Surface Wettability

The observed advancing and receding contact angles for discrete, continuous and non-textured samples are presented in Fig. 3. The advancing angle for non-textured sample is lesser than 90° suggesting that the surface is hydrophilic [23]. Advancing and receding angles for dimple and crosshatch texture were comparatively lesser than

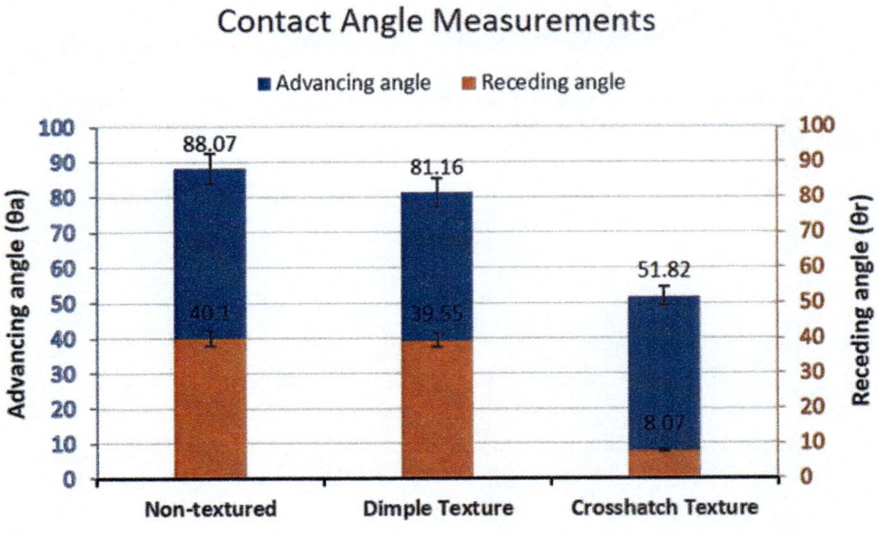


**Fig. 3** Advancing and receding contact angles for textured and non-textured surface

that of non-textured surface. This indicates that both of the textures have a reasonably good wettability and affinity for liquid film formation over the surface. The variation in advancing and receding angles for both textured surfaces correctly follows Wenzel criteria and therefore making non-textured hydrophilic surface even more hydrophilic. Also, the improved wettability of crosshatch texture can be associated to the increased roughness factor value (as provided in Table 1) when compared to dimple texture of similar geometry. Discrete textures may also tend to trap air within the texture and end up increasing the contact angle [24], whereas crosshatch textures are fully connected and hence are free from the air entrapment. Overall, continuous texture has shown better surface wettability compared to discrete and non-textured surface and can significantly improve the lubricating conditions at the hip implant interface.

### 3.2 Friction Response Under Biological Environment

Coefficient of friction (CoF) plays an important role in estimating the overall performance on a hip implant interface. Figure 4 shows the friction coefficient response for all the textured and non-textured surfaces. The non-textured sample showed the highest friction. Whereas, both textures showed relatively lower friction and reduced

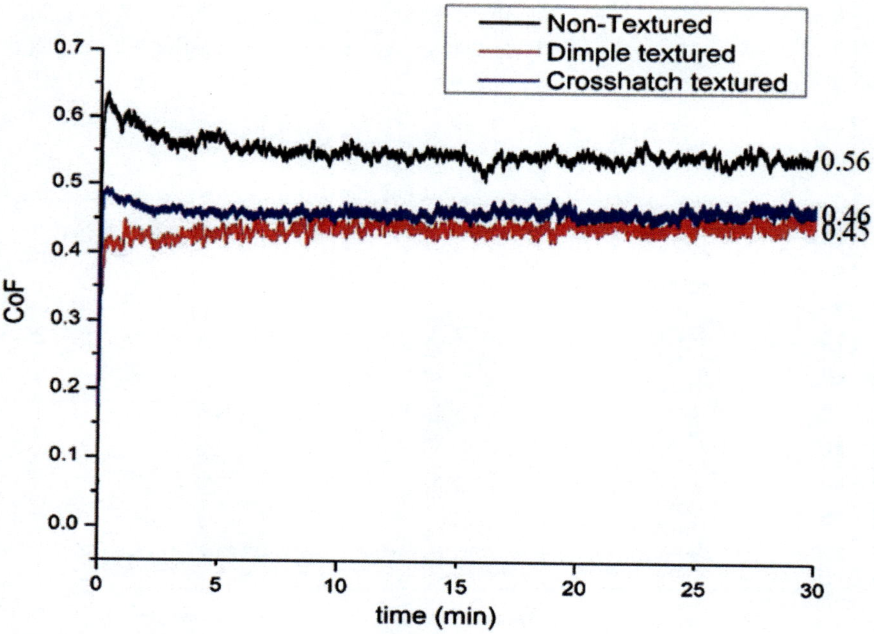


**Fig. 4** Coefficient of friction for all textured and non-textured surfaces for 10 N load

by almost 24%. Under the opted load and speed, the fluid film thickness of 10 nm is expected [25], which is lesser than the roughness value at the interface. Thus, the articulation occurs under boundary lubrication regime. Here, the friction reduction with both textures can be directly related to the reduced real area of contact and additional hydrodynamic pressure generation [18, 26, 27]. For both the textures, not only the reduced real area, but also the ability to capture the wear debris would have helped in reducing friction for the textured surfaces. In overall, both the textures are equally capable of reducing friction but the improved wettability in crosshatch texture may retain a thin lubricant layer at the interface. Hence, the continuous textures may benefit for improved durability by providing additional lubricity at the hip implant interface.

## 4 Conclusion

Dimple and crosshatch textures of similar geometry were fabricated using laser surface texture technology. Wettability and frictional characteristics were evaluated and compared with non-textured sample. Following conclusions can be drawn:

- Surface texturing greatly affects the wettability of the surface and follows Wenzel criteria. Moreover, the increase in roughness factor increases the wettability even for the textures with different shapes.
- Crosshatch texture shows significantly lower contact angles (both advancing and receding) compared to dimple and non-textured surface. Therefore, continuous texture significantly improves the wettability compared to discrete one.
- The frictional associated with both the textures were comparable and almost 24% frictional reduction was observed compared to non-textured surface for 10N load. In overall, crosshatch textures can be preferred over dimple textures as improved wettability would show better lubricant film formation at the hip implant interface.

## References

1. K. Balani, V. Verma, A. Agarwal, R. Narayan, *Biosurfaces A Materials Science and Engineering Perspective* (John Wiley & Sons Inc., Hoboken, New Jersey, 2015)
2. C. Leyens, M. Peters, *Titanium and Titanium Alloys Fundamentals and Applications* (Wiley-VCH Verlag GmbH & Co. KGaA, Weinheim, 2003)
3. X. Dangsheng, G. Zhan, J. Zhongmin, Friction and wear properties of UHMWPE against ion implanted titanium alloy. Surf. Coat. Technol. **201**, 6847–6850 (2007)
4. L. Lukasz, W. Justyna, M. Jacek et al., Retrieval analysis of titanium nitride (TiN) coated prosthetic femoral heads articulating with polyethylene. J. Mech. Behav. Biomed. Mater. **55**, 127–139 (2016)
5. D. Choudhury, H. Ay Ching, A.B. Mamat et al., Fabrication and characterization of DLC coated micro dimples on hip prosthesis heads. J. Biomed. Mater. Res. Part B Appl. Biomater. **103**, 1002–1012 (2014)

6. Y. Dong, P. Svoboda, M. Vrbka et al., Towards near permanent CoCrMo prosthesis surface by combining micro-texturing and low temperature plasma carburizing. J. Mech. Behav. Biomed. Mater. **55**, 215–227 (2015)
7. X. Liu, P.K. Chu, C. Ding, Surface modification of titanium, titanium alloys, and related materials for biomedical applications. Mater. Sci. Eng. **47**, 49–121 (2004)
8. T. Ibatan, M.S. Uddin, M.A.K. Chowdhury, Recent development on surface texturing in enhancing tribological performance of bearing sliders. Surf. Coat. Technol. **272**, 102–120 (2015)
9. S.R. Paital, Z. Cao, W. He, N.B. Dahotre, Wetting effects on in vitro bioactivity and in vitro biocompatability of laser micro-textured Ca-P coating. Biofabrication **2**, 025001, 14. pp (2010)
10. Y. Lu, P. Guo, P. Pei, K.F. Ehmann, Experimental studies of wettability control on cylindrical surfaces by elliptical vibration texturing. Int. J. Adv. Manuf. Technol. **76**, 1807–1817 (2015)
11. L. Qin, P. Lin, Y. Zhang, G. Dong, Q. Zeng, Influence of surface wettability on the tribological properties of laser textured Co-Cr-Mo alloy in aqueous bovine serum albumin solution. Appl. Surf. Sci. **268**, 79–86 (2013)
12. X.L. Wang, S.M. Hsu, An integrated surface technology for friction control: a new paradigm effects of geometric shapes on friction, in *The 4th China International Symposium on Tribology* (Xi'an, 2004), pp. 12–20
13. P. Mishra, P. Ramkumar, Effect of additives on a surface textured piston ring–cylinder liner system. Tribol. Mater. Surf. Interf. **13**(2), 67–75 (2019)
14. I. Etsion, E. Sher, Improving fuel efficiency with laser surface textured piston rings. Tribol. Int. **42**, 542–547 (2009)
15. I. Etsion, Y. Kligerman, G. Halperin, Analytical and experimental investigation of laser-textured mechanical seal faces. Tribol. Trans. **42**, 511–516 (1999)
16. Y. Kligerman, I. Etsion, Analysis of the hydrodynamic effects in a surface textured circumferential gas seal. Tribol. Trans. **44**, 472–478 (2001)
17. D. Choudhury, T. Roy, I. Krupka, et al. Tribological investigation of ultra-high molecular weight polyethylene against advanced ceramic surfaces in total hip joint replacement. J. Eng. Tribol. **229**(4), 410–419 (2015)
18. V. Kashyap, P. Ramkumar, Feasibility study of micro-groove crosshatch surface texturing on Ti6Al4V for improved biotribological performance in metal-on-polymer hip implant tribology. Mater. Surf. Interf. **13**(3), 150–160 (2019)
19. M. Ramiasa, J. Ralston, R. Fetzer, R. Sedev, The influence of topography on dynamic wetting. Adv. Colloid Interf. Sci. **206**, 275–293 (2014)
20. Y. Yuan, T.R. Lee, Contact angle and wetting properties, in *Surface Science Techniqued*. ed. by G. Bracco, B. Holst (Springer, Berlin Heidelberg, 2013), pp. 3–34
21. N.Y.P. Afoke, P.D. Byers, W.C. Hutton, Contact pressure in the human hip joint. J. Bone Joint Surg. **69**, 536–541 (1987)
22. T. Roy, D. Choudhury, S. Ghosh, A.B. Mamat, B. Pingguan-Murphy, Improved friction and wear performance of micro dimpled ceramic-on-ceramic interface for hip joint arthroplasty. Ceram. Int. **41**, 681–690 (2015)
23. J. Park, U. Pasaogullari, L. Bonville, Wettability measurements of irregular shapes with Wilhelmy plat method. Appl. Surf. Sci. **427**, 273–280 (2018)
24. T.J. LichaoGao, C.A. McCarthy, Hysteresis explained. Langmuir **22**, 6234–6237 (2006)
25. A. Mavraki, P.M. Cann, Lubricating film thickness measurements with bovine serum. Tribol. Int. **44**, 550–556 (2011)
26. S. Mezghani, I. Demirci, H. Zahouani, M. El Mansori, The effect of groove texture patterns on piston-ring pack friction. Precision Eng. **36**, 210–217 (2012)
27. D.W. Bechert, M. Bruse, W. Hage, R. Meyer, Fluid mechanics of biological surfaces and their technological application. Naturwissenschaften **87**, 157–171 (2000)

# Influence of Heat Input on Bead Geometry in Metal Inert Gas Welded Thick P91 Steel

**Lakshmi Narayan Dhara, Pardeep Pankaj, and Pankaj Biswas**

**Abstract** Heat input during welding is the controlling factor for weld bead geometry (bead width, reinforcement height, and depth of penetration). Heat input plays an important role to evaluate the quality of the joints. Present experimental investigation aims to determine the effect of heat input on weld bead geometry. Gas metal arc welding of creep strength enhanced ferritic (CSEF) steel P91 with filler wire ER90SB9 using pure argon gas has been done. It has been found out that with increasing heat input, depth of penetration increases. However, variation of bead width and reinforcement with heat input is not so much clear. Based on bead-on-plate experiment, heat input of 1.337 kJ/mm was recommended for joining of 6 mm thick P91 steel plate. For weldments characterization, optical macroscope, optical microscope, Vickers microhardness tester were used.

**Keywords** Welding · Bead-on-plate · Weld bead geometry · P91 steel

## 1 Introduction

The creep strength enhanced ferritic (CSEF) steels are special kinds of high alloy steels that contain 9–12% Cr, small amounts of Mo, V, Nb, W, Co, B, N, and Ni [1]. It possesses more creep resistance property than austenitic stainless steel [2]. In this modern era, CSEF steels are used in various nuclear and thermal plants at large scale due to its low thermal coefficient, good thermal conductivity, adequate creep rupture strength at elevated temperature as well as good corrosion resistance

L. N. Dhara (✉) · P. Pankaj · P. Biswas
Mechanical Engineering, IIT, Guwahati, Guwahati, India
e-mail: nndhara@gmail.com

P. Pankaj
e-mail: p.pardeep@iitg.ac.in

P. Biswas
e-mail: pankaj.biswas@iitg.ac.in



K. M. Pandey et al. (eds.), *Recent Advances in Mechanical Engineering*, Lecture Notes in Mechanical Engineering, https://doi.org/10.1007/978-981-15-7711-6_21

and stress corrosion cracking [3–5]. P91 is one grade of CSEF steel which was first developed at Oak Ridge National Laboratory (ORNL) [6].

Gas metal arc welding or metal inert gas welding (GMAW or MIG) is extensively used in many industries due to its high reliability, ease of use, high deposition rate, the absence of flux, all position capability, and suitability for both ferrous and nonferrous metals and alloys [7]. GMAW is used quite prevalently to weld a thick P91 steel plate. Welding demands a high depth of penetration with optimum weld bead width for joining thick plate. Heat input is responsible for producing such a required weld bead profile [8]. Therefore, it is necessary to study the influence of heat input on the weld bead geometry.

Shen et al. [9] conducted a series of measurements on submerged arc welded plates of ASTM A709 Grade 50 steel to know the variation in heat input achieved using single and double wires with bead reinforcement, bead width, penetration depth, contact angle, heat-affected zone (HAZ) size, deposition area, penetration area, and total molten area. They found that the bead reinforcement, bead width, penetration depth, HAZ size, deposition area, and penetration area increased with increasing heat input, but the bead contact angle decreased with it [9]. Mondal et al. [10] investigated the influence of heat input on weld bead geometry using duplex and

stainless steel wire electrode E2209 T01 on low alloy steel specimens. They observed that parameters of welding geometry are on the whole linearly related to heat input [10]. Saha et al. examined the role of heat input on bead width, reinforcement and depth of penetration of fiber laser-welded nitinol shape memory alloy. They reported that welding depth is changed from full penetration to partial penetration with a decrease in heat input [11]. Saha et al. studied the effect of heat input on austenitic steel weld bead on low carbon steel. They concluded that weld bead width increases linearly with an increase in heat input, whereas reinforcement height and depth of penetration do not increase with the increase in heat input [12].

Various researchers have published their work on the relationship between heat input and weld bead geometry. However, there is a little bit of work that has done to establish a relationship between heat input and weld bead geometry in GMAW of P91 steel with 100% argon as a shielding medium.

## 2 Experimental Details

In the present work, the ER90SB9 weld bead was produced on P91 steel. Process parameters such as welding current and torch travel speed were selected in three levels (i.e., 120 A, 140 A, 160 A, and 155 mm/min, 175 mm/min, 200 mm/min) while voltage was kept constant at 27 V throughout the experiment. In the second part of the experiment, welding was performed using the best heat input at optimum process parameters. Post-weld heat treatment (PWHT) of the welded specimen was carried out at 746 °C for 4 h. For weldments characterization, optical macroscope, optical microscope, Vickers microhardness tester were used. To examine the shape of the weld bead, the polishing of samples were carried out by different grade emery papers

**Table 1** Chemical composition of P91

| | | | | | | |
|---|---|---|---|---|---|---|
| %C | %Si | %Mn | %P | %S | %Zr | %Al |
| 0.109 | 0.462 | 0.564 | 0.0185 | 0.003 | 0.0025 | 0.038 |
| %Ni | %Co | %Cu | %Nb | %P | %V | %Sn |
| 0.26 | 0.0059 | 0.263 | 0.0096 | 0.085 | 0.22 | 0.0137 |
| %Pb | %As | %Mo | %Cr | %Fe | | |
| 0.0126 | 0.0662 | 0.953 | 9.11 | 87.81 | | |

followed by cloth polishing with diamond paste as an abrasive particle. Samples were etched by Vilella's reagent (1 gm picric acid, 100 ml ethanol, and 5 ml hydrochloric acid) for 50 s.

## *2.1 Base Material*

P91 steel of 75 mm × 150 mm × 6 mm was used as a base material. The chemical composition is shown in Table 1.

## *2.2 Electrode Material*

ER90SB9 was used as electrode/filler material for this present experiment. The chemical composition of ER90SB9 electrode is shown in Table 2.

**Table 2** Chemical composition of filler material

| | | | | | | |
|---|---|---|---|---|---|---|
| %C | %Si | %Mn | %P | %S | %Cr | %Mo |
| 0.090 | 0.270 | 0.540 | 0.0289 | 0.0076 | 8.80 | 2.0906 |
| %Ni | %Al | %Co | %Cu | %Nb | %Ti | %V |
| 0.470 | 0.0105 | 0.0741 | 0.3417 | 0.0023 | 0.0316 | 0.0475 |
| %W | %Cr | %Ce | %B | %Fe | | |
| 0.0261 | 8.80 | 0.0103 | 0.0010 | Rest | | |

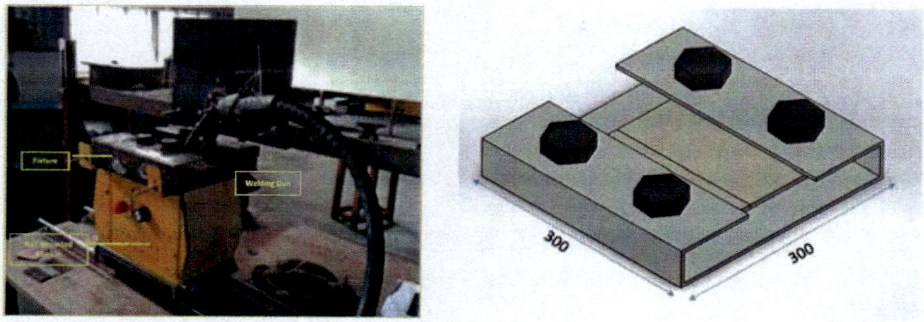


**Fig. 1** Experiment setup and work holding fixture

## 2.3 *Welding Technique*

Gas metal arc welding using 100% argon as shielding gas was used to produce weld bead. Gas flow rate was constant 14 lit/min. Migatronic Pvt Ltd, Denmark, made GMAW machine (Model-SIGMA GALAXY 501c) was used for experiment (Fig. 1).

## 2.4 *Process Parameter*

Heat input depends on process parameters like welding voltage, welding current, and torch travel speed as shown in Equation 1. The heat input and other process parameters that are used for the present investigation are shown in Table 3.

$$Q = \frac{V \times I \times 60}{S \times 1000} \times \eta \quad (1)$$

where,

**Table 3** Process parameter of the welding process

| Voltage, V (V) | Current, I (A) | Travel speed, S (mm/min) | Heat input (kJ/mm) |
|---|---|---|---|
| 27 | 120 | 200 | 0.777 |
| 27 | 120 | 175 | 0.888 |
| 27 | 120 | 155 | 1.003 |
| 27 | 140 | 175 | 1.036 |
| 27 | 160 | 200 | 1.036 |
| 27 | 140 | 155 | 1.170 |
| 27 | 160 | 175 | 1.184 |
| 27 | 160 | 155 | 1.337 |

$Q$ Heat Input (kJ/mm) V = Voltage (V).
$I$ Current (A).
$S$ Travel Speed (mm/min).
$\eta$ Efficiency (In this experiment, it is taken as 0.8).

## 3 Results and Discussions

Results obtained from visual observation are shown in Table 4. Observation results of weld bead geometry are represented in Table 5. Low-to-medium spatter was observed during the bead-on-plate experiment. The continuous deposition was made on the base plate. One blowhole defect was seen in a weld sample.

The values of the weld bead parameters, variation of weld bead parameters with heat input, and macrograph of weld bead geometry are shown in Table 5, Figs. 2, and Fig. 3, respectively. Figure 2 shows that depth of penetration increases with increasing heat input. The maximum width of depth of penetration is obtained data maximum heat input of 1.337 kJ/mm. With an increase in heat input, the volume of molten electrode and work region increases causing the possible increase in depth of penetration. However, the trend of weld bead width and height of reinforcement with heat input are not so much clear in this experiment domain. This is maybe due to 100% argon shield. Ebrahimnia et al. [13] reported that pure argon is not suitable for welding steel since it cannot provide the desired arc stability and desired weld bead characteristics [13]. Maximum weld bead width and reinforcement are obtained at a heat input of 1.003 kJ/mm. High depth of penetration with reasonable good weld bead width is desired in welding. It is seen that at welding heat input

**Table 4** Visual inspection of bead-on-plate experiments done by GMAW with a shielding gas of 100% argon.

| Sl. No. | Sample No | Voltage (V) (V) | Current (I) (A) | Travel Speed (S) (mm/min) | Heat input (kJ/mm) | Blow hole | Continuity in deposition | Spatter |
|---|---|---|---|---|---|---|---|---|
| 1 | A1 | 27 | 120 | 200 | 0.777 | Nil | Cont | Nil |
| 2 | A2 | 27 | 120 | 175 | 0.888 | Nil | Cont | Nil |
| 3 | A3 | 27 | 120 | 155 | 1.003 | Nil | Cont | Nil |
| 4 | A4 | 27 | 140 | 175 | 1.036 | Nil | Cont | Med |
| 5 | A5 | 27 | 160 | 200 | 1.036 | yes | Cont | Nil |
| 6 | A6 | 27 | 140 | 155 | 1.170 | Nil | Cont | Nil |
| 7 | A7 | 27 | 160 | 175 | 1.184 | Nil | Cont | Med |
| 8 | A8 | 27 | 160 | 155 | 1.337 | Nil | Cont | Low |

*Cont.* continuous, *Med.* medium

**Table 5** Values of weld bead geometry components from bead-on-plate experiments

| Sample No | Voltage (V) (v) | Current (I) (A) | Travel Speed (S) (mm/min) | Heat input (kJ/mm) | Weld bead width (W) (mm) | Height of reinforcement (R) (mm) | Depth of penetration (P) (mm) | PSF (W/P) | RFF (W/R) |
|---|---|---|---|---|---|---|---|---|---|
| A1 | 27 | 120 | 200 | 0.777 | 7.48 | 4.37 | 1.41 | 5.30 | 1.71 |
| A2 | 27 | 120 | 175 | 0.888 | 5.87 | 3.79 | 1.93 | 3.04 | 1.54 |
| A3 | 27 | 120 | 155 | 1.003 | 11.22 | 4.45 | 2.74 | 4.09 | 2.52 |
| A4 | 27 | 140 | 175 | 1.036 | 6.10 | 3.34 | 3.13 | 1.94 | 1.82 |
| A5 | 27 | 160 | 200 | 1.036 | 10.37 | 4.00 | 4.62 | 2.24 | 2.59 |
| A6 | 27 | 140 | 155 | 1.170 | 7.40 | 4.76 | 5.05 | 1.46 | 1.55 |
| A7 | 27 | 160 | 175 | 1.184 | 9.39 | 4.86 | 5.06 | 1.85 | 1.93 |
| A8 | 27 | 160 | 155 | 1.337 | 8.45 | 3.75 | 6.08 | 1.38 | 2.25 |

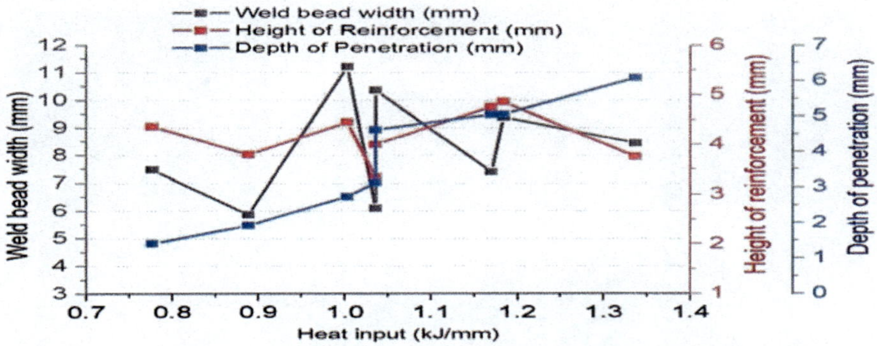


**Fig. 2** Plot of the variation of weld bead width, the height of reinforcement, and depth of penetration with heat input

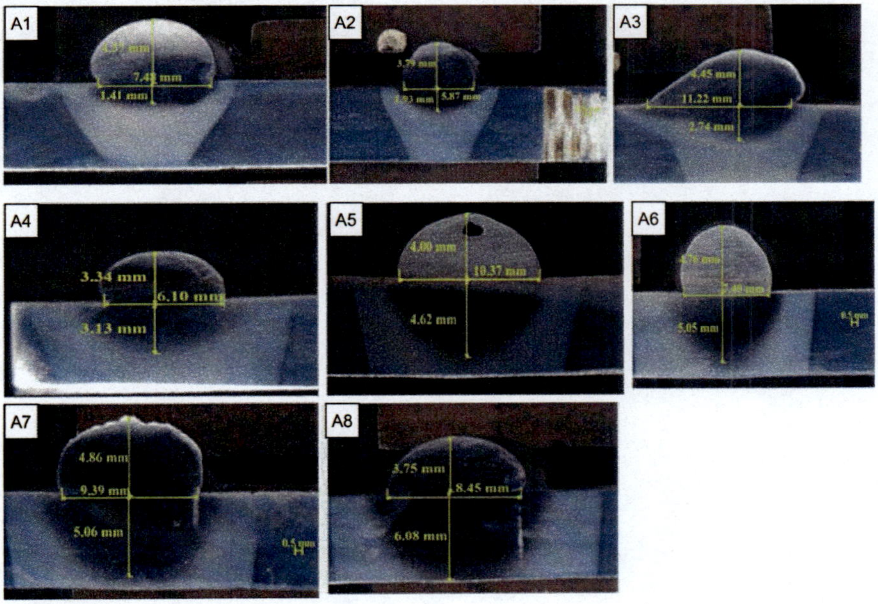


**Fig. 3** Macrograph of weld bead geometry

of 1.337 kJ/mm, good penetration and weld bead width at optimum reinforcement height are observed.

Figure 4 shows the macrograph and micrograph of the different zone of the welded sample with or without post-weld heat treatment (PWHT). Coarse grain is observed at fusion zone, whereas the base metal has finer grains. According to Hall–Petch relationship, finer grains will have greater yield strength [14]. So this implies that the yield strength of weld metal is lesser than the base metal.

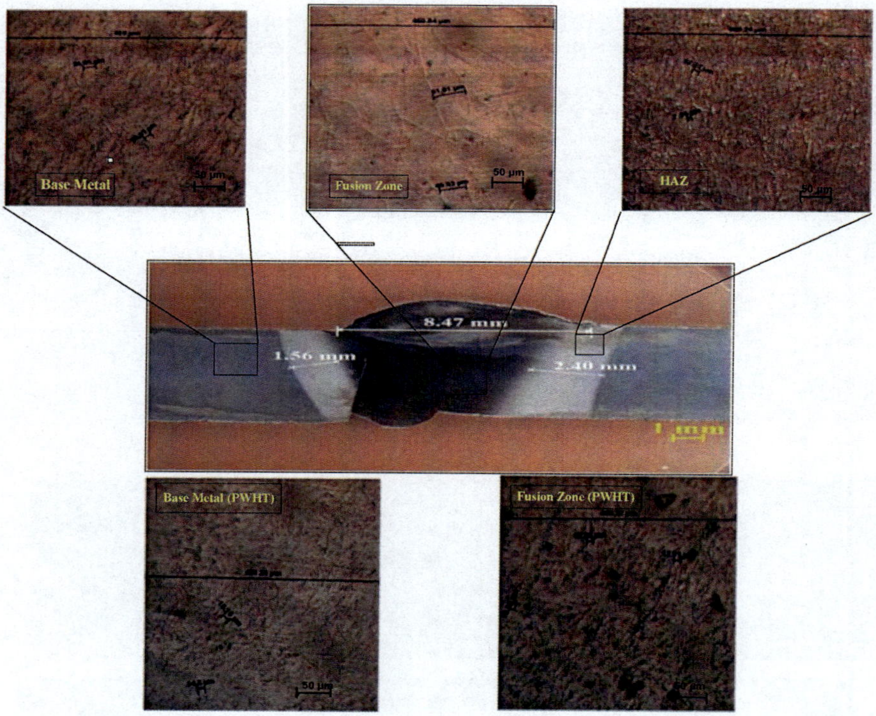


Fig. 4 Macrograph and micrograph of different zone of welded sample before and after PWHT

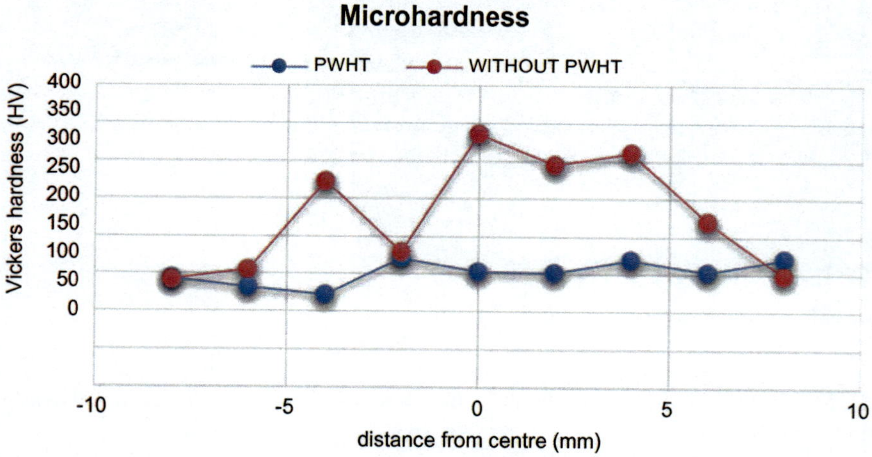


Fig. 5 Variation of microhardness of weld joint as-welded and PWHT condition

Variation of microhardness for the weld zone is shown in Fig. 5. The hardness was measured across the weld zone at 3 mm below the reinforcement outer surface. It is also seen clearly that the Vickers hardness is less in the case of post-weld heat treatment. Post-weld heat treatment relieves the material of internal stresses and softens the material. Hardness leads to brittleness, and by performing the heat treatment after welding, the hardness has reduced.

## 4 Conclusions

Following important inferences may be derived from this experimental investigation:

- The weld bead parameter is highly influenced by heat input. The depth of penetration obtained during GMAW of P91 steel gets increased on increasing heat input.
- PWHT resulted in a considerable decrease in the hardness of the weld zone.
- Finally, it can be concluded that a heat input of 1.337 kJ/mm may be recommended for welding of P91 steel to have an optimum depth of penetration and good tensile strength within the experimental domain.

## References


1. C. Pandey, M.M. Mahapatra, P. Kumar, N. Saini, Effect of normalization and tempering on microstructure and mechanical properties of V-groove and narrow-groove P91 pipe weldments. Mater. Sci. Eng. **A685** 39–49 (2017)
2. X. Li, M. T. Cabrillat, Y. Lejeail, Study of Modified 9Cr-lMo Welds (2006)
3. A.K. Singh, V. Dey, R.N. Rai, Study on the effect of high-temperature ceramic fiber insulating board to weld grade p-91 steel. J. Manuf. Process. **25**, 1–7 (2017)
4. B. Silwal, L. Li, A. Deceuster, B. Griffiths, Effect of post weld heat treatment on the toughness of heat-affected zone for Grade 91 steel Weld. Res. **92**, 80s–87s (2013)
5. C. Pandey, M.M. Mahapatra, P. Kumar, A Giri microstructure characterization and Charpy toughness of P91 weldment for as-welded, post-weld heat treatment and normalizing & tempering heat treatment. Met. Mater. Int. **23**, 900–914 (2016)
6. C. Pandey, M.M. Mahapatra, P. Kumar, N. Sahani, Some studies on P91 steel and heir weldments. J. Alloys Comp. **743**, 332–364 (2018)
7. M. Shoeb, M. Parvej, P. Kumari, Effect of MIG welding input process parameters on weld bead geometry on HSLA steel. IJEST **5**, 200–2012 (2013)
8. N. Murugan, R.S. Parmar, Effect of welding condition on microstructure and properties of Type 316L stainless steel submerged arc cladding. Welding Research Supplement, AWS, 210s–220s (1997)
9. S. Shen, I. Oguocha, S. Yannacopoulos, Effect of heat input on weld bead geometry of submerged arc welded ASTM A709 Grade 50 steel joints. J Mater Process Tech. **212**(1), 286–294 (2012)
10. A. Mondal, M.K. Saha, R. Hazra, S. Das, Influence of heat input on weld bead geometry using duplex stainless steel wire electrode on low alloy steel specimens, 2016, J. Cogent Eng. **3**, 1–14 (2016)

11. P. Saha, S. Datta, S.K. Patihar, Effects of heat input on weld-bead geometry, surface chemical composition, corrosion behavior and thermal properties of fiber laser-welded nitinol shape memory alloy. JEMP **16**, 2754–2763 (2019)
12. M.K. Saha, R. Hazra, A. Mondal, Effect of heat input on geometry of austenitic stainless steelweld bead on low carbon steel. J. Instit. Eng. (India) **100**, 607–615 (2019)
13. M. Ebrahimnia, M. Goodarzi, M. Nouri, M. Sheikhi, Study of the effect of shielding gas composition on the mechanical weld properties of steel ST 37–2 in gas metal arc welding. Mater. Des. **30**(9), 3891–3895 (2009)
14. V. Bata, E.V. Pereloma, An alternative physical explanation of the Hall-Petch relation. Acta Meterilia **52**, 657–665 (2004)

# Opportunities and Challenges in Nanoparticles Formation by Electrical Discharge Machining

**Irshad Ahamad Khilji, Sunil Pathak, Siti Nadiah Binti Mohd Saffe, Shatarupa Biswas, and Yogesh Singh**

**Abstract** Nanoparticles (NPs) have drawn immense attention due to the full range of new applications in various fields of industries such as electronics, optical, biomedical, pharmaceutical and cosmetics. NPs gained importance due to their exceptional properties like antibacterial activity, high resistance to oxidation, exceptional adhesive properties, better thermal conductivity and many more. Various interdisciplinary researches have been done in the field and still going on. The aim of this paper is to briefly describe the details of NPs processing methods, their benefits and limitations and the need of new process in the field. In this paper, electrical discharge machining (EDM) has been presented as possible new process for the synthesis of NPs. The challenges in the development of EDM as a NPs synthesis process have also been discussed in this paper.

**Keywords** Nanoparticles · EDM · Physical · Chemical · Mechanical

## 1 Introduction

### *1.1 Nanoparticles*

In the current trend, enormous attention has been paid in the field of nanotechnology where the matter is considered and deployed on the atomic or molecular level. Nanotechnology already has and will remain to have, a significant impact on our life as well as on the global economy [1]. The main reason for the fascination in nanometer-sized structures is their novel properties such as enhanced surface tension, larger surface area, adhesion, lubrication, stabilization control flocculation

I. A. Khilji · S. Pathak (✉) · S. N. B. M. Saffe
Faculty of Manufacturing and Mechatronics Engineering Technology, University Malaysia Pahang, Lebuhraya Tun Razak, Darul Makmur Pahang, 26300 Kuantan, Malaysia
e-mail: sunilpathak87@gmail.com

S. Biswas · Y. Singh
Mechanical Engineering Department, NIT, Silchar, India


K. M. Pandey et al. (eds.), *Recent Advances in Mechanical Engineering*, Lecture Notes in Mechanical Engineering, https://doi.org/10.1007/978-981-15-7711-6_22

of colloidal dispersion [2]. Surface effects and quantum mechanical effects stand up in nanostructures due to their large surface-to-volume ratio and their dimensions are of the same order as of the electron wavelength, respectively. Such effects are utilized to produce new, improved materials, as well as novel medical, optical and electronic devices. A microscopic particle having a size less than 100 nm in any one of its dimensions is known as nanoparticle (NP). It can be classified into different classes based on their properties, shapes or sizes that include fullerenes, ceramic NPs, metal NPs and polymeric NPs. Due to the wide divergence of their use, extensive researches have been conducted for their optimization and finding solutions to obtain NPs of novel materials. NPs are of tremendous methodical curiosity as they form a link between the bulk materials and their atomic or molecular structures [3]. The properties of materials change as their size approaches the nanoscale and as the percentage of atoms at the surface of a material becomes significant[4]. Nanoparticles also exhibit several unique features relative to the bulk material such as (a) quantum confinement in particles associated with semiconductors; (b) interface permittivity in incident lights for some metallic particles; and (c) super-paramagnetic in magnetic materials [5]. An example on unique properties of NPs: bulk copper (Cu) in the form of wire or ribbon, the bending starts at about the 50 nm scale with the undertaking of Cu atoms, while the same Cu NPs with less than 50 nm are superhard materials that do not possess similar flexibility and ductility compared to bulk copper. A brief detail on essential requirements from the NPs is presented in Fig. 1.

Depending upon the applications, the synthesis of NPs may differ. For instance, NPs prepared from organic polymers are often used for drug delivery, whereas semiconducting nanoparticles are used for optoelectronic and electronics applications. The nanoparticles are generated during combustion in car engines and are

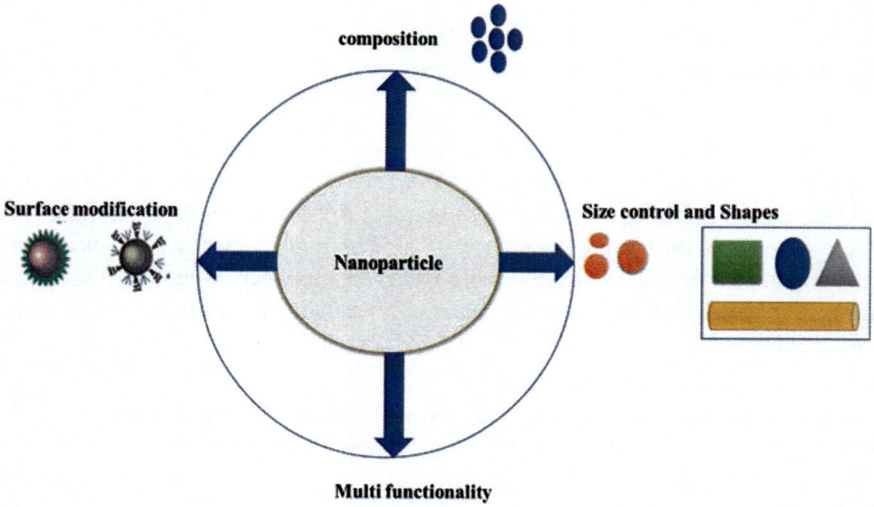


**Fig. 1** Essential requirements of NPs

**Table 1** Summary of applications for metal NPs

| Metal | Benefits of NPs |
|---|---|
| Silver [10] | Absorbs and scatters light, stable, antibacterial, disinfectant, used in coating |
| Gold [11] | Interactive with visible light, reactive, highly conductive, used in coating |
| Aluminium[12] | High reactivity, sensitive to moisture, heat and sunlight, large surface area, non-magnetic |
| Iron[13] | Reactive and unstable, sensitive to air (oxygen) and water |
| Cobalt[14] | Unstable, magnetic, toxic, absorbs microwaves, magnetic |
| Cadmium[15] | The semiconductor of electricity, insoluble non-magnetic, low reactive |
| Lead[16] | High toxicity, reactive, highly stable |
| Copper[17] | Ductile, very high thermal and electrical conductivity, highly flammable solids |
| Zinc[18] | Antibacterial, anti-corrosive, antifungal, UV filtering |
| Aluminium oxide[12] | Increased reactivity, sensitive to moisture, heat and sunlight, large surface area |
| Silicon dioxide[19] | Less toxic, able to functionalize many molecules |

organic *soot particles*. However, soot particles are also commercially produced to obtain a substance known as carbon black [6]. *Metal oxide* particles are found in paints, cosmetics and sunscreens, while metal nanoparticles have a broad range of applications [7]. In specific, silver is used for its antibacterial properties, gold and other transition metals such as palladium, platinum and rhodium possess inert and catalytic properties [8]. The nanoparticles used as seeds for *semiconductor nanowires* were mainly gold particles, while gold, titanium dioxide ($TiO_2$) and soot particles were used for nano-safety research [9]. A summary of applications for metal NPs is presented in Table 1.

## 1.2 Methods for Synthesis of NPs

Various methods can be utilized to engineer NPs of different materials. Fundamentally, there are three methods for generating NPs, namely (i) physical; (ii) chemical; and (iii) mechanical. The mentioned processes can be further divided into several processes depending upon the need, but the principle working remains the same. In chemical synthesis of NPs, solvents and hard chemicals in the form of liquid/gas were generally used to produce NPs. This process results in bulk generation, easy and non-laborious work environment. During the mechanical processing's grinding of a bulk material in the form of small particles the have been done, it is a less complicated process compared to others, uncontaminated particles can be produced.

Whereas the physical processes have the capability to produce controlled particle size and can produce pure NPs.

These processes were in use for NPs synthesis from decades, but still, they suffer from there inherent limitations. For example, (i) physical concept-based processes are costly, less productive (i.e. in terms of time to NPs generations), sophisticated process design and highly skilled labours are required for operation; (ii) chemical-based processes generate high chemical waste and toxic emissions which are hazardous for the environment and humans, difficulties in generating the pure NPs; (iii) mechanical and its based constituent processes such as milling and grinding also lack with non-similar size and shape of NPs, larger floor area and high energy consumption are also one of the major limitations for such processes. A brief detail on the advantages and limitations of NPs synthesis processes has been presented in Table 2, while a comparison of these processes based on valuable economical and structural prospects has been presented in Table 3.

To overcome these limitations of available NPs synthesis processes, a new and novel procedure for generating NPs by electrical discharge machining (a type of spark erosion process) has been presented in this paper. Opportunities and challenges in EDM for NP synthesis have been briefly discussed.

**Table 2** Comparative table for the technique for the synthesis of the nanoparticle

| Physical processes | Chemical processes | Mechanical processes |
|---|---|---|
| Physical vapour deposition, laser ablation, Sputter deposition, Electric arc depositions, ion implantation | Sol–Gel, Electrodeposition, colloidal methods, water–oil microemulsions method, hydrothermal synthesis etc | Attrition Ball Mill, Vibrating Ball Mil, High Energy Ball Mil, vibrator |
| Advantages: Fewer losses compare with other processes, chemical-free process | Advantages: Easy, less time-consuming, large quantities of material can be obtaining, variety of size and shape, self-assembly or patterning possible | Advantages: Pure NP, chemical-free process |
| Limitations: Losses, time-consuming, required high-class setup, control over the process is a typical task, and high skilled labour is required | Limitations: High cost, high chemical waste, contaminated NPs, hazardous for environment and human | Limitations: More space required, high energy required, the non-uniform shape of particles, phase change possibilities |

**Table 3** Comparison of the available NPs synthesis techniques

| Type of Processes | Cost | Time | Oxidation | Waste material |
|---|---|---|---|---|
| Chemical | High | High | High | High |
| Mechanical | High | High | Low | High |
| Physical | High | High | Low | Low |

# 2 Electrical Discharge Machining

Electrical discharge machining (EDM) is an electro-thermal non-traditional machining process, generate electrical spark with the help of electric energy, and thermal energy is used to remove the material for the workpiece material [21, 22]. Electrically conductive materials machined by EDM where the thermal energy of spark detaches the material through repeated sequences of electrical discharge between the small gap of an electrode and a workpiece. A voltage is usually applied between two electrodes (i.e. cathode and anode), and a temperature of around 10,000 °C has been generated between electrodes during the spark. EDM process removes undesirable material in the form of debris and produces the shape of the tool surface as a metal portion by the recurring electrical spark between the tool and the workpiece in the presence of dielectric liquid. These particles are flushed away from the narrow machining zone due to the continuous flow of the dielectric fluid (generally kerosene, transformer oil, distilled water, etc.) [21]. The study of debris or flushed particles may play an important role in accepting the mechanism of alloy formation in EDM process. Additionally, the reuse of debris in engineered components will solve a severe problem in mechanical industries and this will also add on more procedures in the generation of oxides free NPs. The details on the working mechanism of EDM process can be found in [23].

## *2.1 Opportunities and Challenges in NPs Synthesis by EDM*

A very few researchers have worked and reported the NPs production by EDM. However, they were not able to report the usefulness of the collected debris in the application, but they have successfully able to produce micrometer-sized particles by EDM.

Ultrashort pulse powered low energy EDM has been utilized by Roy et al. [24] for production of NPs. Their study was focused on observing the shape of the produced NPs; they reported a mixed type of results with some particles being spherical (single phase) and majority as non-spherical (random). Khanra et al. [23] have used EDM for mild steel as workpiece and ZrB2–Cu composite as an electrode with kerosene as the dielectric fluid. They reported multiple types of NPs (i) spherical shaped single phase of workpiece material while (ii) non-spherical (random) consists of a mixture of work material and tool material. Effects of EDM parameters and cooling procedure on particle synthesis and tool wear have been studied [25], and they reported that the pulse-on time and voltage plays an essential role in tool wear. Ayers et al.[26] studied the effects of EDM parameters on the composition and morphology of titanium carbide (TiC) metal powder by EDM process, and they reported TiC and WC NPs having single phase were successfully produced by EDM with kerosene as a dielectric fluid and graphite as a tool. Kumar et al. [27] studied the effect of mixing alumina particles in a dielectric in EDM of Inconel alloy and reported improvements in the formation of debris. In the most recent work, Katiyar et al. [28] have studied the influence of pulse parameters and applied voltage on the synthesis of Fe–Cu NPs in the EDM process. They reported that pulse-on time plays a vital role in deciding the shape of the particles (i.e. spherical/non-spherical), and they observed the minimum size of spherical NP generated is 20 nm with no oxidation and contamination.

Considering, very fewer literature and huge potential in the process, it is viable that the EDM can be more frequently used an NPs synthesis process for a wide range of hard-to-cut materials, namely tungsten carbide, titanium and its alloys, Inconel and its alloys and many other important engineering materials. However, with the viability, there are plenty of challenges. Few of the most vibrant challenges are mentioned below:

- Obtaining a set value of EDM parameters by their optimization for the synthesis of NPs with uniform shape and size. A detailed list of parameters along with the set responses for NP synthesis is presented in Fig. 2.
- Avoiding mixing of workpiece and tool materials during the synthesis.
- Development of proper methodology for collection and extraction of NPs from the dielectric chamber.
- Development of theoretical model and simulations-based studies for proper rectification of the occurred problem during experiments.
- Avoiding oxidization of the generated NPs, by maintaining a closed environment.
- Repeatability of the results and testing for more difficult to cut materials.

The mentioned challenges can be achieved with the help of in-depth research in the field by selecting the most affecting parameters of the EDM and also by developing a novel collector for the collection of NPs and their extraction.

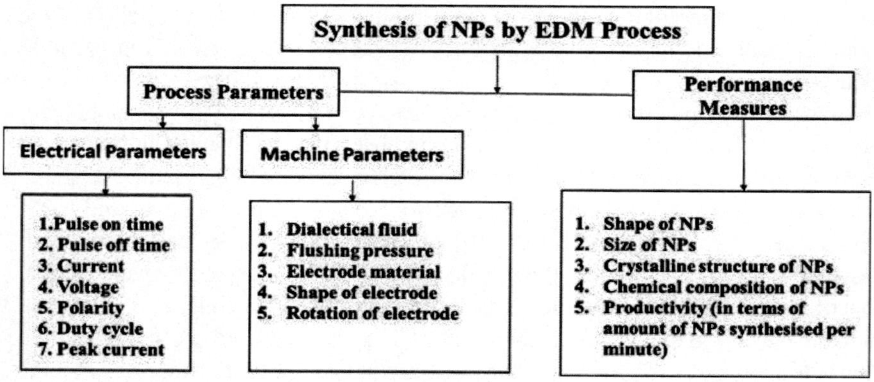


**Fig. 2** Important EDM parameters, along with the set responses for NP synthesis

## 3 Conclusions

In the present work, a brief introduction on the nanoparticles and their synthesis methods has been presented. From the literature and discussion, it can be concluded that EDM has enormous potential to be developed as an effective process for the synthesis of uniform shape NPs without any contamination. It was also observed that EDM has the capability of developing very small-sized spherical particles with a minimum of 20 nm. It was found from the literature that the pulse parameters and applied voltage play a crucial role in NPs shape and size. The development of the EDM process variant with effective utilization is suggested to improve the non-conventional machining process applicability in NP synthesis field. Still, there are some sustainability issues required to be addressed in future work particularly to the EDM process to investigate more sustainable dielectrics. Further research and developments are encouraged to improve quality, productivity and sustainability of EDM. The growth of EDM for reasonable production for the small-scale manufacturers as well across the world to reach triple bottom line sustainability is quite probable.

**Acknowledgements** Authors are thankful to Universiti Malaysia Pahang for funding the present work through RDU Grant (RDU1822023) and (RDU1903137).

## References

1. M.C. Roco, Broader Societal Issues of Nanotechnology. J. Nanoparticle Res. **4**(5), 181–189 (2003)
2. S. Liufu, H. Xiao, Y. Li, Investigation of PEG adsorption on the surface of zinc oxide nanoparticles. Powder Technol. **145**, 20–24 (2004)
3. C.A. Silvera Batista, R.G. Larson, N.A. Kotov, Nonadditivity of nanoparticle interactions. Science **6257**(350), 124–247 (2018)
4. A. Koçak, B. Karasu, General evaluations of nanoparticles. El-Cezeri Fen ve Mühendislik Derg **1**(5) 1 191–236 (2018)
5. J. Jeevanandam, A. Barhoum, Y.S. Chan, A. Dufresne, M.K. Danquah, Review on nanoparticles and nanostructured materials: history, sources, toxicity and regulations. Beilstein J. Nanotechnol. (9), 1050–1074 (2018)
6. C.M. Long, M.A. Nascarella, P.A. Valberg, Carbon black versus black carbon and other airborne materials containing elemental carbon: physical and chemical distinctions. Environ. Pollut. (181), 271–286 (2013)
7. X. Yu, T.J. Marks, A. Facchetti, Metal oxides for optoelectronic applications. Nat. Mater. **4**(15), 383–396 (2016)
8. J.L. Clement, P.S. Jarrett, Antibacterial silver. Met. Based. Drugs **6**(5) 467–82 (1994)
9. C.W. Schmidt, Nanotechnology-related environment, health, and safety research: examining the national strategy. Environ. Health Perspect. **4**(117), A158–61 (2009)
10. J.C. Hulteen, D.A. Treichel, M.T. Smith, M.L. Duval, T.R. Jensen, R.P. Van Duyne, Nanosphere Lithography : Size-Tunable Silver Nanoparticle and Surface Cluster Arrays. pp. 3854–3863 (1999)
11. B. Syed, N.M.N. Prasad, S. Satish, Endogenic mediated synthesis of gold nanoparticles bearing bactericidal activity. J. Microsc. Ultrastruct. **3**(4), 162–166 (2016)

12. P. Geetha, M.S. Latha, S.S. Pillai, B. Deepa, K. Santhosh Kumar, M. Koshy, Green synthesis and characterization of alginate nanoparticles and its role as a biosorbent for Cr(VI) ions. J. Mol. Struct. (1105), 54–60 (2016)
13. G.M. Sulaiman, AT. Tawfeeq, A.S. Naji, Biosynthesis, characterization of magnetic iron oxide nanoparticles and evaluations of the cytotoxicity and DNA damage of human breast carcinoma cell lines. Artif. Cells Nanomedicine Biotechnol. **6**(46), 1215–1229 (2018)
14. C.C.L. McCrory, S. Jung, J.C. Peters, T.F. Jaramillo, Benchmarking heterogeneous electrocatalysts for the oxygen evolution reaction. J. Am. Chem. Soc. **45**(135), pp. 16977–16987 (2013)
15. G. Rudko, A. Kovalchuk, V. Fediv, W.M. Chen, I.A. Buyanova, Enhancement of polymer endurance to UV light by incorporation of semiconductor nanoparticles. Nanoscale Res. Lett. **81**(10)
16. J. Osuntokun, P.A. Ajibade, Morphology and thermal studies of zinc sulfide and cadmium sulfide nanoparticles in polyvinyl alcohol matrix. Phys. B Condens. Matter **496**(496), pp. 106–112 (2016)
17. C.H. Ryu, S.J. Joo, H.S. Kim, Two-step flash light sintering of copper nanoparticle ink to remove substrate warping. Appl. Surf. Sci. **384**, 182–191 (2016)
18. K.I. Bogutska, Y.P. Sklyarov, Y.I. Prylutskyy, Zinc and zinc nanoparticles : biological role and application in biomedicine. (1), pp. 9–16 (2013)
19. ÜH. Kaynar, I. Şabikoğlu, S.Ç Kaynar, M. Eral, Modeling of thorium (IV) ions adsorption onto a novel adsorbent material silicon dioxide nano-balls using response surface methodology. Appl. Radiat. Isot. **115**, 280–288 (2016)
20. W.M.M. Mahmoud, T. Rastogi, K. Kümmerer, Application of titanium dioxide nanoparticles as a photocatalyst for the removal of micropollutants such as pharmaceuticals from water. Curr. Opin. Green Sustain. Chem. **6**, 1–10 (2017)
21. A.A. Abdu Aliyu, J. Mohd Rohani, A.M. Abdul Rani, H. Musa, Optimization of electrical discharge machining parameters of sisic through response surface methodology. J. Teknol. (79)1 (2019)
22. S.S. Mahapatra, A. Patnaik, Parametric optimization of wire electrical discharge machining (WEDM) process using Taguchi method. J. Brazilian Soc. Mech. Sci. Eng. **4**(28), 422–429 (2006)
23. M.R. Shabgard et al., J. Mater. Process. Technol. **3**(229), 1531–1538 (2006)
24. Vikas, A. K. Roy, K. Kumar, Effect and optimization of various machine process parameters on the surface roughness in EDM for an EN41 material using grey-Taguchi. Procedia Mater. Sci. (6), 383–390 (2014)
25. S. Abdulkareem, A. Ali Khan, M. Konneh, Cooling effect on electrode and process parameters in EDM. Mater. Manuf. Process. **6**(25), 462–466 (2010)
26. J.D. Ayers, K. Moore, Formation of metal carbide powder by spark machining of reactive metals. Metall. Mater. Trans. A. **6**(15), 1117–1127 (1984)
27. A. Kumar, A. Mandal, A.R. Dixit, A.K. Das, Performance evaluation of $Al_2O_3$ nano powder mixed dielectric for electric discharge machining of Inconel 825. Mater. Manuf. Process. **9**(33), 986–995 (2018)
28. J.K. Katiyar, A.K. Sharma, B. Pandey, Synthesis of iron-copper alloy using electrical discharge machining. Mater. Manuf. Process. **14**(33), 1531–1538 (2018)

# Preheating Path Selection Through Numerical Analysis of Laser-Aided Direct Metal Deposition

**Dhiraj Raj, Bipul Das, and Saikat Ranjan Maity**

**Abstract** The process of laser cladding offers superior metallurgical and mechanical properties of the product. This is possible through the additive nature of the process but brings challenges in controlling the surface requirements and defects. The generation of thermal stresses in the cladding material as well as the substrate materials results in delamination and formation of a substantial crack in the remelted zone. One of the strategies followed to control the defect formation and also to manage the microstructure in the cladding material, preheating is often applied. Moreover, preheating also facilitates the localized melting of the cladding material over the substrate for better melting and remelting. In the current investigation, a laser source is simulated in COMSOL and two paths for preheating a cavity on stainless steel materials are used. The power of the laser source is kept fixed at 70 W and the other parameter like laser traverse speed and beam spot diameter has been fixed at 1.5 mm/s and 1 mm respectively. The numerical solution of the system of equations resulted in time–temperature information. Later, the heat transfer gradient is computed for the paths from the time–temperature information acquired through numerical simulation and the best one is selected to obtain a uniform heat distribution. The study revealed that the path which followed a circular pattern results in a better heat transfer gradient for effective preheating of the substrate material.

**Keywords** Laser cladding · Stainless steel · Temperature profile · Preheating

## 1 Introduction

Technological developments in the area of materials and alloy having advanced properties have induced many problems and difficulties regarding the machining of these materials. The scientist and engineers have started using a laser as a heat source for machining and achieving high dimensional accuracy of these materials to meet the requirement of several industries [1]. In today's industrial world, a lot of

D. Raj (✉) · B. Das · S. R. Maity
Department of Mechanical Engineering, National Institute of Technology, Silchar, India
e-mail: dhirajraj12054@gmail.com



K. M. Pandey et al. (eds.), *Recent Advances in Mechanical Engineering*, Lecture Notes in Mechanical Engineering, https://doi.org/10.1007/978-981-15-7711-6_23

developments have been done in the laser field because of its frequent use for making hard and complex shape materials. Laser technology is being widely used after the development of the high-power laser in various fields, such as in medical field laser therapy, laser cladding in industry, military field, and in many other areas [2].

Laser cladding process is a process in which fusion occurs with the help of the laser beam. In this process, one material is used as a cladding material and the other one is used as substrate material. The laser beam is focused on the material with the help of a focusing system. When the laser strikes the material, fusion occurs and molten pool formation starts. The cladding material is either pre-placed on the substrate or fed simultaneously with the formation of a molten pool. The surface tension gradient drives the molten material flow. The substrate and clad material get mixed in the molten state and it is allowed to cool by quenching process which led to the formation of fine-grained microstructure. Laser cladding technology is highly preferable because of its advantages over other processes like its high energy density, small heat-affected zone is formed, metallurgical bonding between the substrate and the powder is excellent, high depositing efficiency [3]. It is also mainly used for improving the surface properties of metal materials by depositing a coating of other metal on it. The coatings deposited by the laser cladding process consist of finer and denser microstructure with a lower dilution rate [4]. It is the technique that controls the porosity problem, insufficient process temperature, and distortion of the workpiece to some extent in comparison with the other techniques. The laser cladding process along with the proper set up has been shown in Fig. 1.

Laser cladding is considered as one of the best technique of improving the surface quality because the coating deposited by this technique has higher hardness as compared to the substrate material. The clad process has a major effect on the quality and characteristics of the layer of the surface formed after the cladding process. The parameters that affect the clad process also affect the clad profile, dilution, finishing off the surface and other defects like crack, porosity, etc. Some of the input parameters are scanning speed, laser power, beam spot diameter, overlap ratio, the direction of clad, etc. [6].

## 2 Methodology

The AISI 410 stainless steel plate with dimension 40 mm × 40 mm × 5 mm is used as a base material. A cavity is made around the crack area on the plate having a maximum radius of 6.6 mm and depth of 2.5 mm as shown in Fig. 2. The chemical compositions of 410 stainless steel are presented in Table 1 [7]

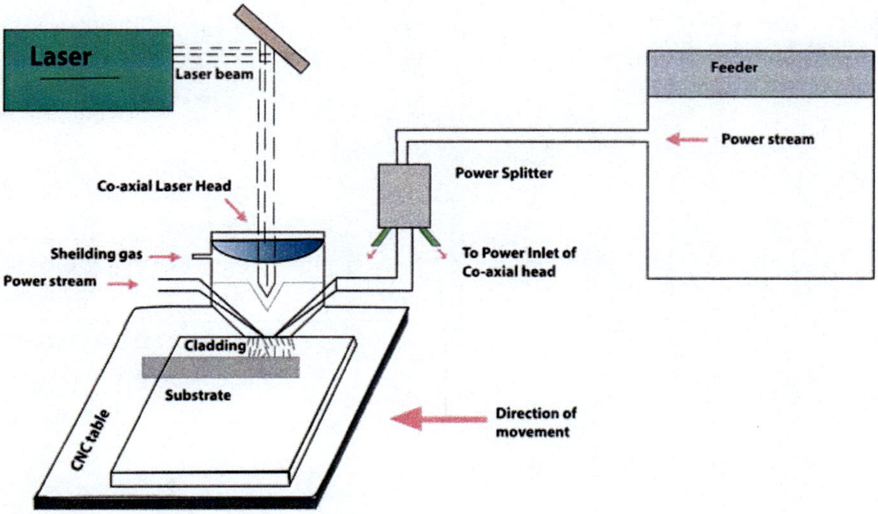


**Fig. 1** Schematic diagram of laser cladding process [5]

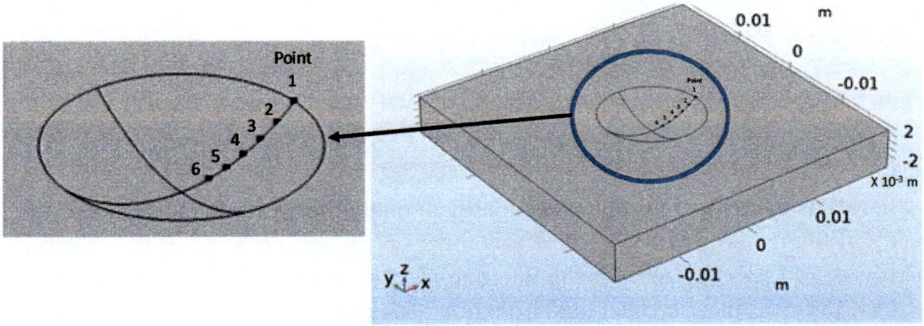


**Fig. 2** Geometry of the 410 stainless steel used for preheating

**Table 1** Chemical composition of AISI 410 stainless steel

| Material | C | Mn | Si | P | S | Cr | Ni |
|---|---|---|---|---|---|---|---|
| 410L stainless steel | 0.15 | 1 | 1 | 0.04 | 0.03 | 11.5–13.5 | 0.75 |

## 2.1 *Heat Transfer Equation and Boundary Conditions*

In this model, the basic Fourier heat conduction equation is used for heat transfer. All surfaces except the cavity area were assumed to be thermally insulated. Convection and radiation phenomenon is neglected in this model. The governing equation for heat conduction is given by Eq. (1). Equation (2) is the initial condition and Eqs. (3)

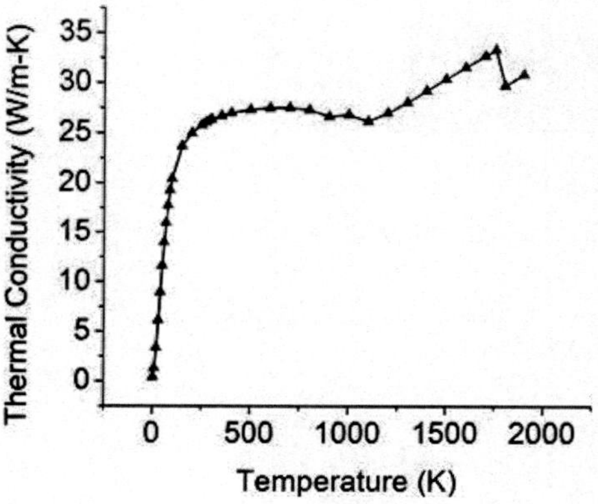


**Fig. 3** Thermal conductivity of 410 stainless steel [9]

and (4) are the boundary conditions.

$$\frac{\partial}{\partial x}\left(k\frac{\partial T}{\partial x}\right)+\frac{\partial}{\partial y}\left(k\frac{\partial T}{\partial y}\right)+\frac{\partial}{\partial x}\left(k\frac{\partial T}{\partial z}\right)+q=\rho c\frac{\partial T}{\partial t} \quad (1)$$

$$T(x,y,z)=T_0 \quad (2)$$

$$n.q=0 \quad (3)$$

$$n.q=q_0 \quad (4)$$

where $T$ is the temperature, $k$ is the thermal conductivity of the material, $\rho$ is the density of the material (7800 kg/m$^3$), $c$ is the specific heat capacity of the material which was 500 J/Kg-K, and $q$ is the internal heat. $T_0$ is the initial temperature, i.e., 293.15 K [8], $q_o$ is the total heat flux. Equation (3) represents the boundary condition of the surfaces which is thermally insulated. Equation (4) represents the boundary condition of the cavity area where heat flux was given. The thermal conductivity of the material varies with the change of temperature which is shown in Fig. 3.

## 2.2 Heat Source Model

Comsol software is used to build a finite model for this study. In this model, a moving laser beam is used as a point heat source which is having a Gaussian distribution intensity profile. In this model, point heat source is used instead of volumetric heat source because we are only preheating the surface of the material and the temperature will be below its melting point so melting pool will not be formed. In modellling of the whole process, an extra-fine physics-controlled mesh is used for computational efficiency and simulation accuracy as shown in Fig. 4. The 3D finite element entity possesses

44,679 tetrahedron elements and 5142 triangular elements. Minimum element quality is 0.1972. The Gaussian heat flux ($Q_{\text{in}}$) is given by Eq. (5)

$$Q_{\text{in}} = \frac{2nQ_0}{\Pi R_b^2} \exp(-(2R^2)/R_b^2) \quad (5)$$

where $Q_o$ is the total power, $n$ is the absorption co-efficient, i.e., 0.25. $R_b$ is the radius of focus of the beam, and $R$ is the radial co-ordinate of the focus of the beam which is given by Eq. (6). $x_1$ and $y_1$ are the coordinate of the center of the focus of the laser beam [10].

$$R = \sqrt{(x - x_1^2)} + \sqrt{(y - y_1^2)} \quad (6)$$

In this study, the cavity is preheated by using two different paths. In the first path, it is heated by moving the point laser heat source continuously on a circular path inside the cavity. In the second path, point laser is used to heat only nine different points inside the cavity. For a circular path, the radius of the path is calculated from the designed geometry of the cavity. The $x_1$ and $y_1$ are given by the parametric equation of circle as described in Eq. (7). For preheating the whole cavity, six paths are considered with different radius and different velocity.

$$x_1 = R * \cos(\omega * t),\ y_1 = R * \sin(\omega * t) \quad (7)$$

For point heating, the center of the focus of the beam, $x_1$ and $y_1$ are defined for all nine points and simulations are computed. The process parameters of laser point heat source used for preheating of the cavity are laser power $Q_o = 70$ W, scanning velocity $V = 1.5$ mm/s and beam spot diameter $D = 1$ mm.

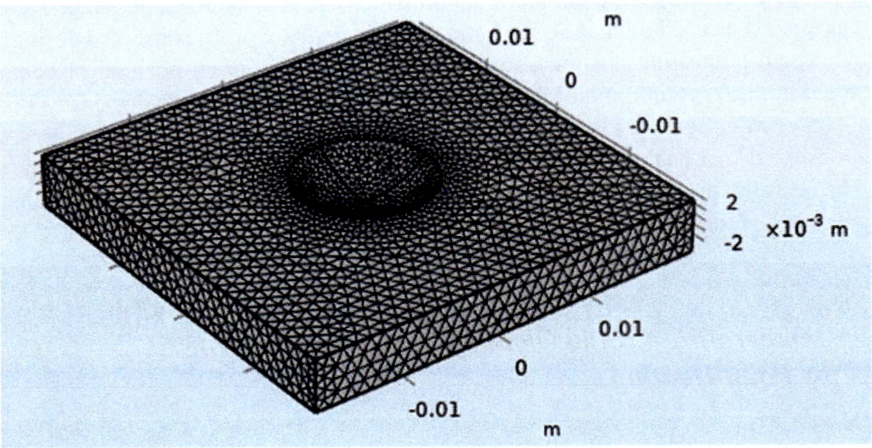


**Fig. 4** Finite element mesh model

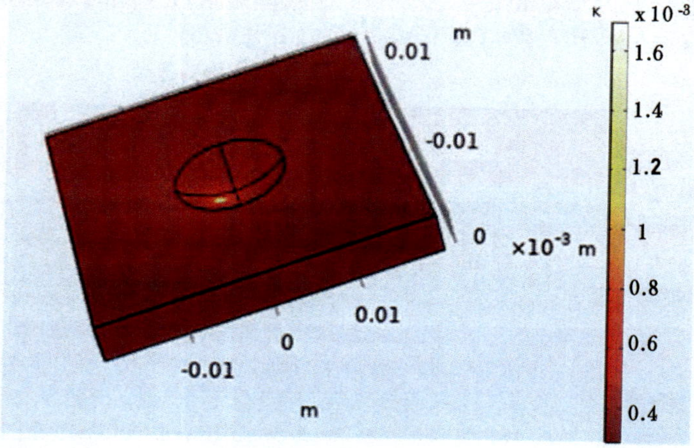


**Fig. 5** Temperature distribution model at $t = 103.5$ s

# 3 Results and Discussion

## 3.1 *Temperature Profile Obtained During Preheating by Circular Path*

The temperature distribution model obtained after the final simulation in Comsol is shown in Fig. 5. The maximum simulated temperature obtained is 1658 K which is lower than the starting range of melting temperature of 410 stainless steel. The temperature profile with respect to time is calculated at six points: point 1, point 2, point 3, point 4, point 5, and point 6 which are inside the cavity as shown in Fig. 2. The point lies on the circular path on which the heat source is moving. The distance between each point is about 1 mm. The temperature profile observed during circular path preheating is shown in Fig. 6. Point 1 is the initial position of heat source. Similarly, point 2, point 3, point 4, point 5, and point 6 are the positions of heat source after completion of first, second, third, fourth, and fifth circular turns. It is observed that peak temperature is obtained at the time when the heat source will be nearer to these points and when the distance of heat source from these points increases, then the peak temperature decreases.

## 3.2 *Temperature Profile Obtained During Preheating by Point Heating*

The temperature distribution model obtained after the final simulation in Comsol is shown in Fig. 7. The temperature profile during this simulation is calculated at the

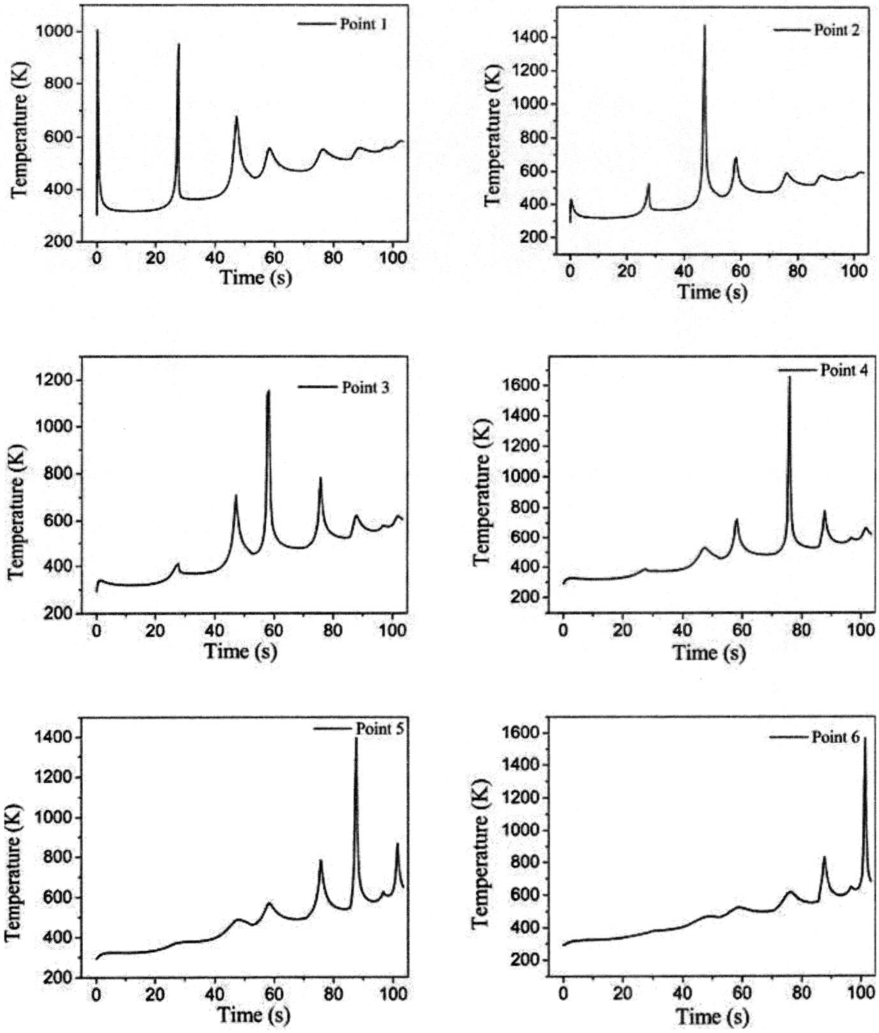


**Fig. 6** Temperature profile of preheating of cavity by circular path

same point on which it was calculated during the preheating by the circular path. In the preheating of cavity, nine points are taken randomly for point heating so that it almost covers the entire cavity volume and all the points are heated for 11.5 s. The temperature profile observed during point heating preheating is shown in Fig. 8 It is observed that the maximum temperature is obtained at point 1 as 1258.5 K. The reason behind the high temperature obtained at point 1 is that it is one of the points out of nine points on which point heating is performed. The peak temperature at all the points except point 1 is much lower as compared to that observed during preheating

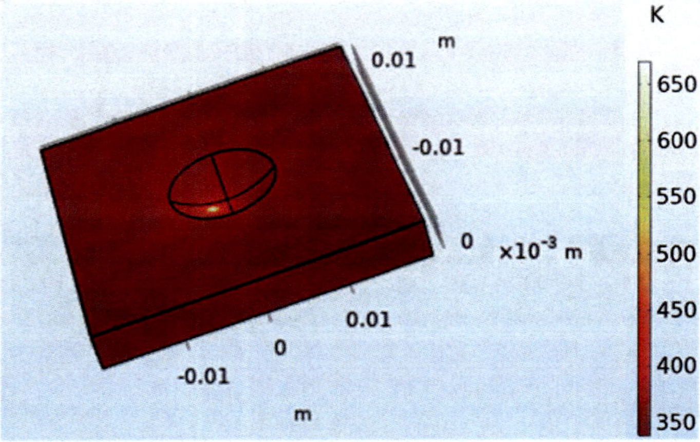


**Fig. 7** Temperature distribution model at $t = 103.5$ s

by the circular path. It is because during the point heating, the temperature rise is not high.

The numerical temperature gradient has been calculated for both the paths at all the six points on which temperature profile is obtained using MATLAB that is shown in Fig. 9. Thus, we can conclude from Fig. 9 that the gradient of the circular path is higher at all the six points as compared to that of point heating. The rise in temperature will be higher and heating effect will be more in the circular path as compared to point heating. So, we can conclude that the circular path heating scheme will be better for preheating the cavity.

## 4 Conclusion

In this research work, a numerical investigation is done for selecting a suitable path for preheating using a laser heat source. Two different paths are taken for preheating the cavity. The laser heat source is used as a point heat source and modeled using COMSOL software. The temperature profile was obtained for both the paths at six different points. The numerical temperature gradient is calculated for both the path at all the six points and both the gradients are compared. It is observed that the gradient of the circular path was higher as compared to the point heating which results in higher temperature rise and better heating effect of the cavity. So, we can conclude that among the two paths taken, the circular path heating scheme is better and more suitable for preheating the cavity.

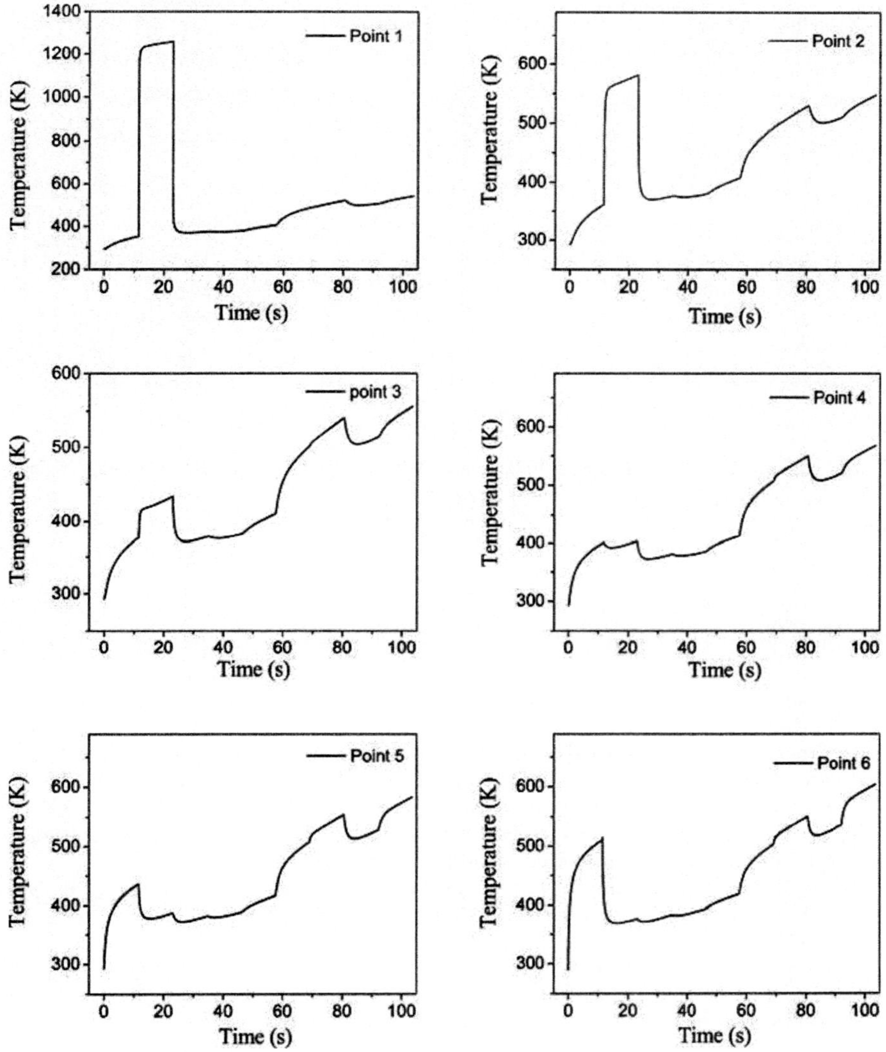


**Fig. 8** Temperature profile of preheating of cavity by point heating

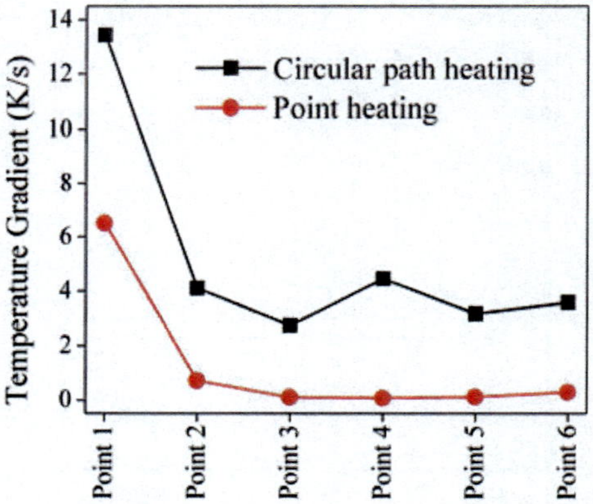


**Fig. 9** Comparison of numerical temperature gradient of the paths selected for preheating

**Acknowledgements** The authors would like to gratefully acknowledge the financial support from TEQIP-III, National Institute of Technology Silchar for carrying out the research work.

## References

1. V.K. Jain, *Advanced Machining Process*, 2nd edn. (Allied Publishers Private Limited, New Delhi, 2009), pp. 280–308
2. Z. Wang, J. Zhang, P. Zhang, H. Zhou, Ti. Zhou, Effect of the 75ferrosilicon on the laser cladding on gray cast iron. Opt. Laser Technol. **113**, 64–71 (2019)
3. Li, M., Zhao, W., Hou, K., Kou, D., Yuan, Z., Zhang, X., Xu, J., Hu, Q., Wang, D., Zeng, X.: A comparison of microstructure and mechanical properties of laser cladding and laser-induction hybrid cladding coatings on full-scale rail, Mater. Sci. Eng. **A748**, 1–15 (2019)
4. H. Tengfei, M. Xiao, Y. Zhang, Y. Shen, Effect of Cr content on microstructure and properties of Ni-Ti-xCr coatings by laser cladding. Optik **179**, 1042–1048 (2019)
5. Awasthi, R., Santosh K., Srivastava, D., Dey, G.K.: Solidification and microstructural aspects of laser-deposited Ni-Mo-Cr-Si alloy on stainless steel. Pramana **75**(6), 1259–1266 (2010)
6. Masoud , B., Reza S.R., Ali Ghasemi, ND.: YAG laser cladding of Co–Cr–Mo alloy on γ-TiAl substrate Opt. Laser Technol. **80**, 145–152 (2016)
7. George M.: Azo materials. Oct 23 2001 [Online] Available from: https://www.azom.com/article.aspx?ArticleID=970. Accessed 5 Nov 2019
8. Du, Y., You, X., Qiao, F., Guo, L., Liu., Z.: A model for predicting the temperature field during selective laser melting. Res. Phys. **12**, 52–60 (2019)
9. Ho, C.Y., Chu, T.K.: Electrical resistivity and thermal conductivity of nine selected AISI stainless steels. In: Thermophysical and Electronic Properties Information Analysis Center Lafayette. No. CINDAS-45 (1977)
10. Li. Chang, Y. Zhibin, J. Gao, J. Zhao, X. Han, Numerical simulation and experimental study of cladding Fe60 on an ASTM 1045 substrate by laser cladding. Surf. Coat. Technol. **357**, 965–977 (2019)

# Robust Synthesis of Path Generating Four-Bar Mechanism

Sanjay B. Matekar and Ajay M. Fulambarkar

**Abstract** Synthesis of path generating four-bar mechanism is performed with the objective to minimize the error between generated and specified path points. The deterministic synthesis of mechanism determines the link dimensions without considering the effect of uncertainties. The deterministically synthesized mechanism will not perform as desired due to the uncertainties. Tolerance design has been performed to improve robustness of such mechanisms by allocating suitable link tolerances and joint clearances. The robustness of such mechanisms has been analyzed by Taguchi method. Thus, the robust design of mechanism has been performed in stages. In this paper, a methodology is presented for optimum robust synthesis of path generating four-bar mechanism. Differential evolution has been used as evolutionary algorithm for the synthesis. The methodology simultaneously optimizes link lengths and corresponding tolerances, along with joint clearances at both ends of coupler link. The modified geometry of four-bar mechanism considering link tolerances and joint clearances has also been discussed. The methodology has been applied for synthesis of four-bar mechanism for tracing a straight line path. It has been observed that the synthesized mechanism has the minimum objective function value. It has also been observed that the variation in objective function value due to uncertainties is minimum, as the standard deviation of objective function value is minimum. Thus, the methodology has resulted in robust path generating four-bar mechanism.

**Keywords** Differential evolution · Optimization · Robust design · Synthesis of four-bar mechanism · Uncertainties in mechanism

---

S. B. Matekar (✉)
Mechanical Engineering Department, Sinhgad College of Engineering, Pune, India
e-mail: sanjay.matekar@pccoepune.org

Mechanical Engineering Department, Pimpri Chinchwad College of Engineering, Pune, India

A. M. Fulambarkar
Symbiosis Skills and Open University, Pune, India
e-mail: fajay@rediffmail.com



K. M. Pandey et al. (eds.), *Recent Advances in Mechanical Engineering*, Lecture Notes in Mechanical Engineering, https://doi.org/10.1007/978-981-15-7711-6_24

# 1 Introduction

Robust design of mechanism ensures that there will be minimum deviation from the intended performance. Optimum deterministic synthesis of path generating four-bar mechanism determines the link dimensions. The effect of uncertainties in the mechanism is not considered in this synthesis. Tolerance design has been performed to assign suitable tolerances and joint clearances to these mechanisms to make them robust. In this paper, the optimum robust synthesis of path generating four-bar mechanism is discussed.

The objective function for deterministic synthesis has been defined as root mean square (RMS) of error between the generated path points and the prescribed path points by Kunjur and Krishnamurthy [1] and Matekar and Gogate [2]. The deterministically synthesized mechanism would deviate from the prescribed task due to uncertainties. The uncertainties in mechanism are controllable (tolerance and joint clearance) and uncontrollable (environmental and operating conditions). The effect of controllable uncertainties on the displacement of slider in a slider crank mechanism has been investigated by Matekar and Fulambarkar [3].

Tolerance design has been applied to identify suitable values of controllable uncertainties under the influence of uncontrollable uncertainties for the deterministically synthesized mechanisms. The tolerances for path generating four-bar mechanism have been obtained using orthogonal arrays, by Chen and Huang [4], Huang and Zhang [5], Huang et al. [6]. Robustness of such mechanisms can be analyzed using Taguchi quality loss function defined by Phadke [7].

Researchers have proposed methodologies for robust design synthesis of mechanisms. A robust design optimization objective function has been defined considering tolerances of design variables by Lee and Park [8]. The methodology has been applied for structural design. A two-step approach has been proposed for robust design of mechanism by Caro et al.[9]. Here, in first step of synthesis link, dimensions have been obtained, and in second step, optimal tolerances have been designed.

From the literature review, it has been observed that robust design of four-bar mechanism has been performed in stages. First, the link dimensions have been determined and then the tolerances have been obtained. Thus, if the synthesis is performed to determine link dimensions and tolerances simultaneously will result in a robust mechanism. In this paper, robust synthesis of path generating four-bar mechanism has been discussed for determining link lengths with optimum tolerance and joint clearances. A four-bar mechanism has been synthesized for generating specified vertical straight line. Differential evolution has been used as the optimization algorithm.

# 2 Path Generating Four-bar Mechanism

Figure 1 shows the parameters of a path generating four-bar mechanism at ideal condition, i.e., without any joint clearance and link tolerance. The coupler point 'C'

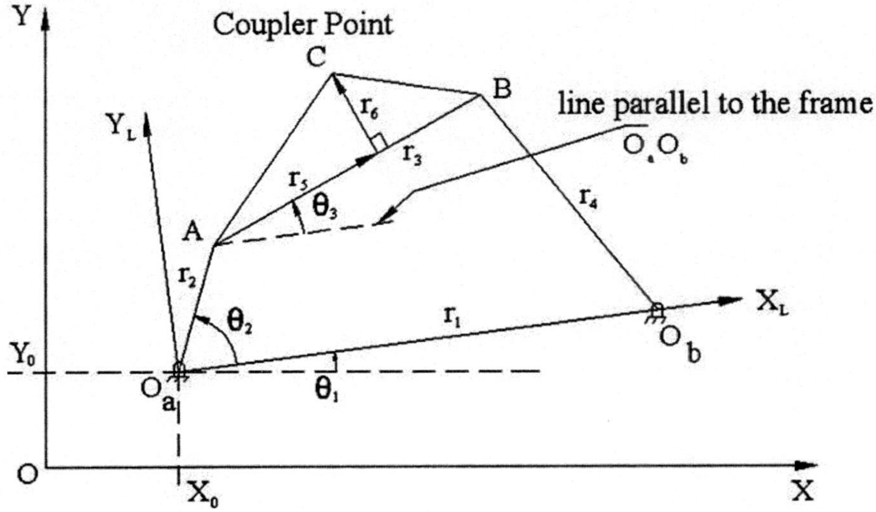


**Fig. 1** Path generating four-bar mechanism at ideal condition [3]

traces the specified path.$\theta_1$ is the angle made by fixed link with horizontal axis. $r_1$,$r_2$,$r_3$ and $r_4$ are the link lengths of the fixed, input, coupler and output links, respectively. $\theta_2$ is the angle made by input link with fixed link. $r_5$ and $r_6$ are the coordinates of the coupler point C in coupler link. $X_0$ and $Y_0$ are the coordinates of fixed pivot. $\theta_2^1$ is the input link angle at first desired path point. $\Delta\theta_2$ is the angular rotation of input link to traverse the desired path points.

## 2.1 Equivalent Four-bar Mechanism

For robust synthesis of mechanism, link tolerances and joint clearances at both ends of coupler link have been considered as control parameters. The geometry of the mechanism gets modified due to these control parameters. The modified path generating four-bar mechanism is shown in Fig. 2. Where all the link lengths are modified as ‘optimum link length $\pm$ link tolerance’. The joint clearances at both ends of coupler, $Rc_1$ and $Rc_2$, are considered. The joint clearance are considered as virtual links in the equivalent mechanism.

Figure 3 shows a joint between links ‘*i*’ and ‘*j*’, formed by inserting pin at end of link ‘*j*’ in the hole at end of link ‘*i*’. Joint clearance $R_c$ is generated, which is the difference between radius of the hole and the pin. The radial clearance circle is also shown in Fig. 3. The joint clearance has been considered as virtual link $R_{ci}$ in the mathematical model. It has been considered that, as the mechanism moves the virtual link will orient anywhere in the radial clearance circle. Thus, the angle of clearance link ‘$\alpha_i$’ with respect to horizontal axis, as shown in Fig. 4, is considered

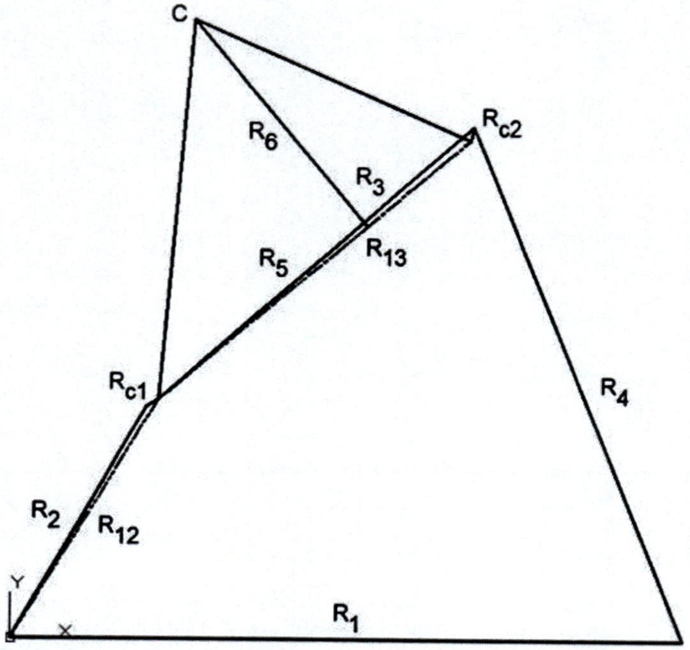


**Fig. 2** Modified path generating four-bar mechanism

**Fig. 3** Joint clearance

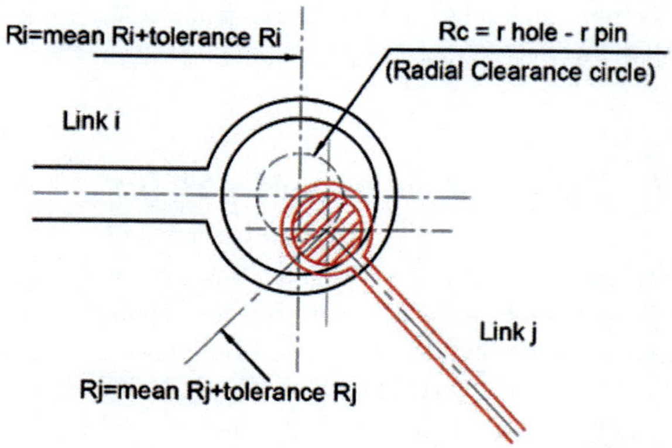


to be randomly oriented between 0° to 360° in the clearance circle.

Figure 4 shows the equivalent link where the mechanism links are connected by the virtual clearance link. Point '*P*' is the connecting point of two links. The coordinates of '*P*' are calculated using the modified link length and virtual clearance link with respective angles with horizontal $\varphi_i$ and $\alpha_i$, as in (1) and (2). The equivalent link length $R_{i1}$ is calculated using the point coordinates, as in (3). The link angle $\varphi_{i1}$ also gets modified as in (4).

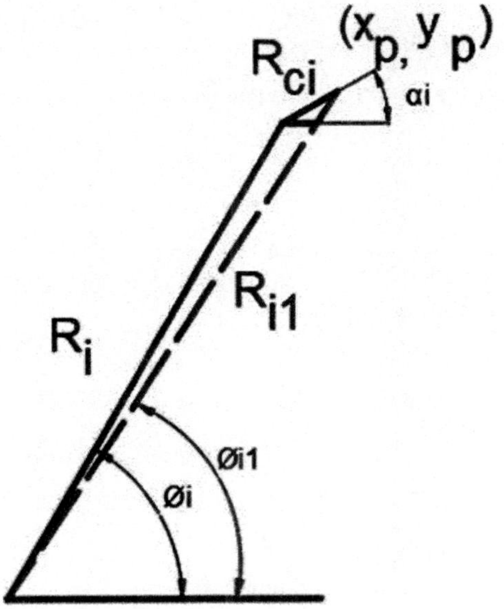


**Fig. 4** Fig. 4. Equivalent link

$$x_p = R_i * \cos\varphi_i + R_{ci} * \cos\alpha_i \tag{1}$$

$$y_p = R_i * \sin\varphi_i + R_{ci} * \sin\alpha_i \tag{2}$$

$$R_{i1} = \sqrt{x_p^2 + y_p^2} \tag{3}$$

$$\varphi_{i1} = a\tan\left(y_p/x_p\right) \tag{4}$$

# 3 Robust Synthesis of Path Generating Four-bar Mechanism

The robust synthesis of path generating four-bar mechanism ensures that the mechanism will trace the specified path with minimum deviation. The synthesis has been performed to identify optimum design parameter values along with the applicable optimum tolerances and joint clearances under the influence of uncertainties.

### 3.1 Parameters for Robust Synthesis

In this paper, two types of parameters, controllable and uncontrollable, have been considered. Controllable parameters are those which are controlled by designer. Following controllable parameters have been considered for the synthesis:

Link lengths, $r_1$ to $r_6$
Tolerance of each link, $tr_1$ to $tr_6$
Joint clearance at both ends of coupler link, $rc_1$ and $rc_2$
Position of fixed pivot ($X_0$, $Y_0$)
Orientation of fixed link ($\theta_1$)

Uncontrollable (noise) parameters cannot be control by designer but influence the position of coupler point. Following noise parameters have been considered for the synthesis:

1. Link deformation due to environmental conditions considered as 5 microns/mm of link length.
2. Orientation of radial clearance, varying randomly in clearance circle from 0° to 360°.

### 3.2 Objective Function for Robust Synthesis

Equation (5) shows the objective function for synthesis of path generating four-bar mechanism, which is same as of deterministic synthesis. It is defined as the minimization of root mean square of error ($E_{\mathrm{RMS}}$) between the specified path points,$C_s^i$ and the generated path points, $C_g^i$. Here, '$n$' is number of points specified on the path. The coupler point coordinates have been calculated using modified path generating four-bar mechanism.

$$E_{\mathrm{RMS}} = \sqrt{\frac{1}{n}\sum_{i=1}^{n}\left[\left(C_{xs}^i - C_{xg}^i\right)^2 + \left(C_{ys}^i - C_{yg}^i\right)^2\right]} \quad (5)$$

### 3.3 Procedure for Robust Synthesis

The robust synthesis has been performed using optimization algorithm. Following values have been decided before start of the optimization process.

1. The lower and upper limits of control parameters.
2. The population size for optimization, 'Out_mech'.
3. No of mechanism to be evaluated in the inner loop, 'In_mech'.

Following procedure has been followed for the optimum robust synthesis of mechanism. A MATLAB program has been written as per the procedure.

1. A set of control parameter values have been generated randomly, equal to 'Out_mech' between the specified lower and upper limits.
2. Each candidate in the set has been analyzed for Grashof's criterion using the link length parameter values.
3. The qualified candidates in step 2 have been analyzed for robustness in the inner loop.
4. From each qualified candidate, a set of equivalent mechanisms equal to 'In_mech' have been generated. The link length has been varied using normal distribution. Where, link length parameter value has been considered as mean, and corresponding link tolerance parameter value has been considered as normal distribution.
5. Both the noise parameter values have also been generated randomly, equal to 'In_mech', within the specified ranges as discussed in Sect. 3.1.
6. Each equivalent mechanism generated in step 4 is analyzed with each candidate from the noise parameter set. Thus, the analysis has been performed for 'In_mech' × 'In_mech' times. The link deformation has been applied to modified link length of the equivalent mechanism. The clearance link direction has been considered while calculating equivalent link.
7. The objective function value for each analysis from step 6 has been stored and the maximum value has been considered as the objective function value for that candidate. Thus, the worst performance has been considered as the objective function value for that candidate.
8. After analyzing each qualified candidate from step 2 in the inner loop, the candidate having minimum objective function value has been considered as the robust mechanism for that run.

Figure 5 shows the flowchart of the outer–inner loop procedure to synthesize the robust mechanism.

## 4 Robust Synthesis of Mechanism for Tracing Vertical Line

Figure 6 shows the path generation task, as to trace a vertical straight line. The line is specified from point (200, 200) to (200, 450). 15 equispaced points on the line have been considered for the synthesis. Table 1 lists the lower and upper limits of the control parameters (discussed in Sect. 3.1) used for the synthesis.

The optimization has been performed as per the procedure discussed in Sect. 3.3 using differential evolution as the optimization algorithm. For each run, the population size has been considered as 600 and the number of mechanism evaluated for the inner loop has been 20. The optimization process has been repeated for 50 times, resulting in 50 optimum mechanisms. The objective function values of all 50 mechanisms have been sorted in ascending order to determine the best optimum

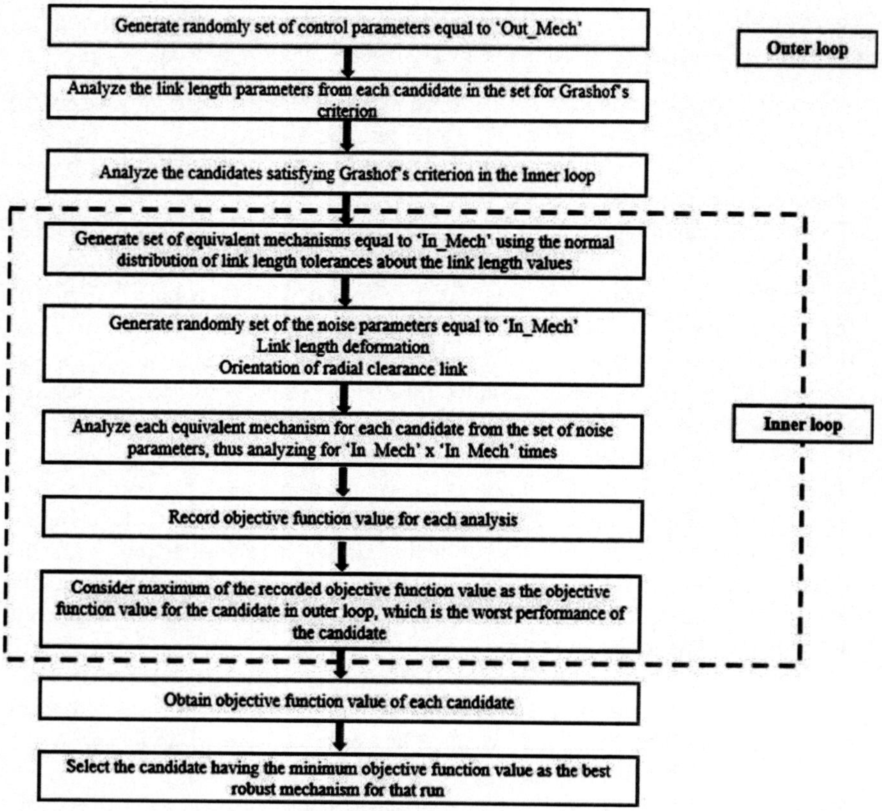


**Fig. 5** Outer–inner loop for robust synthesis

mechanism. Table 2 lists the parameter values of the best synthesized mechanism along with its objective function value.

Figure 7 shows the best synthesized optimum mechanism, where the coupler point is at the start point of line. The complete coupler curve traced by the coupler is shown. Figure 8 shows the points traced by the mechanism along with the specified points. It can be observed that the coupler point traces the required points very closely.

The variation in objective function value has been obtained by analyzing the best optimum parameter values. The link length tolerances have been varied in the designed ranges and equivalent mechanisms have been generated. These have been analyzing for variation in noise parameters. Total 400 objective function values have been obtained. Figure 9 shows the normal distribution of the 400 objective function values. It can be observed that the mean of objective function is 1.261 and the standard deviation is 0.052.

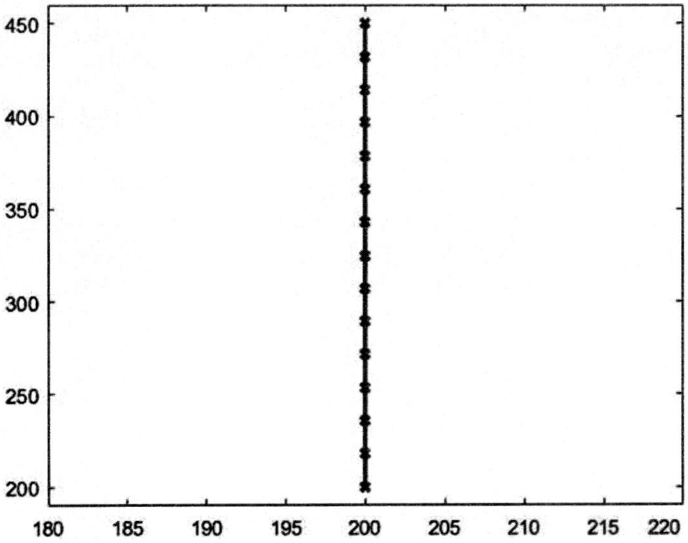


**Fig. 6** Specified path generation task

**Table 1** Limits of design parameters

| Design parameter | Lower limit | Upper limit |
|---|---|---|
| $r_1$ (mm) | 50 | 1200 |
| $r_2$ (mm) | 50 | 400 |
| $r_3$ (mm) | 50 | 800 |
| $r_4$ (mm) | 50 | 1000 |
| $r_5$ (mm) | 30 | 400 |
| $r_6$ (mm) | 30 | 400 |
| $tr_1$ (% of link length) | 0.001 | 0.01 |
| $tr_2$ (% of link length) | 0.001 | 0.01 |
| $\mathbf{tr_3}$ (% of link length) | 0.001 | 0.01 |
| $tr_4$ (% of link length) | 0.001 | 0.01 |
| $tr_5$(% of link length) | 0.001 | 0.01 |
| $tr_6$(% of link length) | 0.001 | 0.01 |
| $rc_1$(mm) | 0.1 | 1 |
| $\mathbf{rc_2}$(mm) | 0.1 | 1 |
| $X_0$ (mm) | −100 | 400 |
| $Y_0$ (mm) | −100 | 400 |
| $\theta_1$(rad.) | 0 | 6.28 |

**Table 2** Parameter values of best optimum mechanism

| Design variables | Optimum value |
|---|---|
| $r1$ (mm) | 1200 |
| $r2$ (mm) | 474 |
| $r3$ (mm) | 675 |
| $r4$ (mm) | 1000 |
| $r5$ (mm) | 165 |
| $r6$ (mm) | 275 |
| Link 1 Tolerance (%) | 0.003 |
| Link 2 Tolerance (%) | 0.008 |
| Link 3 Tolerance (%) | 0.002 |
| Link 4 Tolerance (%) | 0.007 |
| Link 5 Tolerance (%) | 0.008 |
| Link 6 Tolerance (%) | 0.001 |
| $X_0$ (mm) | 19.82 |
| $Y_0$ (mm) | 20 |
| $\theta_1$(rad.) | 0.1065 |
| Joint clearance (crank side) (mm) | 0.281 |
| Joint clearance (output side) (mm) | 0.105 |
| Objective function value | 1.403 |

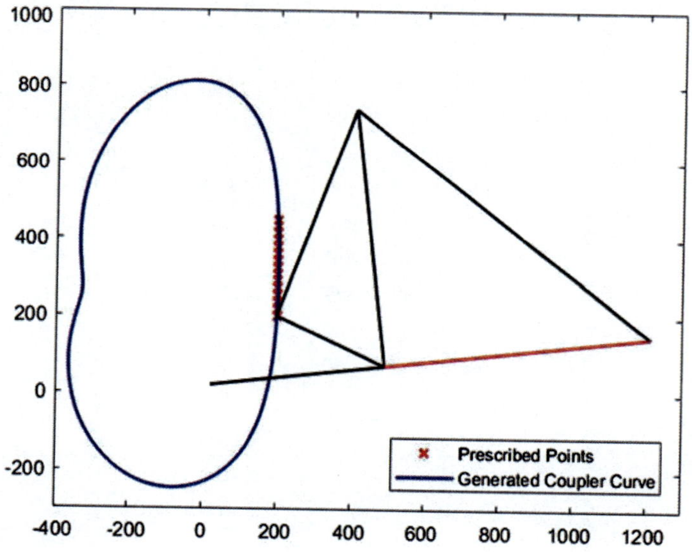


**Fig. 7** Best mechanism with coupler curve

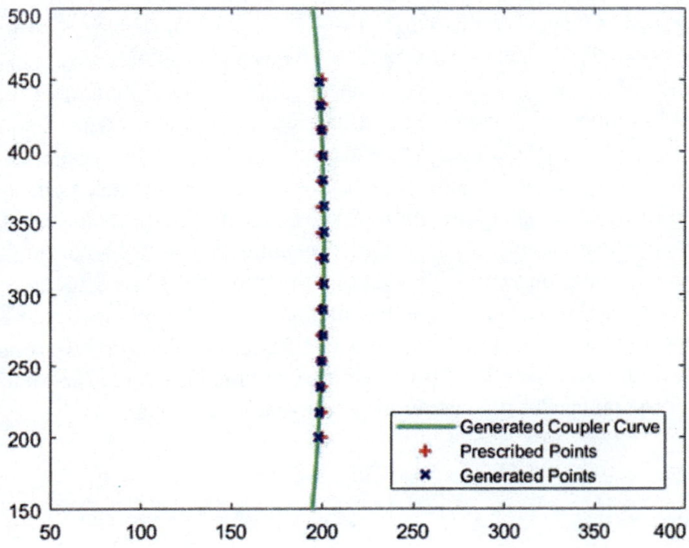


**Fig. 8** Path traced by best mechanism

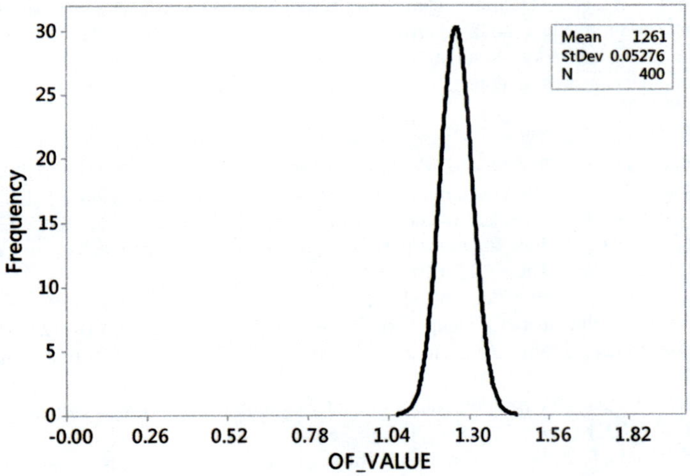


**Fig. 9** Normal distribution curve of objective function

# 5 Conclusion

It is necessary to define the link length tolerances and joint clearances for the mechanism for ease of assembly and operation. When the link tolerances and joint clearances are obtained in the design stage, it increased the robustness of the mechanism. It was observed from the literature review that the robust design of mechanism has been performed in stages. In this paper, robust synthesis of path generating mechanism using optimization has been discussed. The robust synthesis results in optimum link dimensions and corresponding tolerances along with the joint clearance values of joints at both ends of connecting rod has also been synthesized. The modified equations for link lengths and link angles have also been discussed. These have been used to determine the objective function value. A four-bar mechanism has been synthesized for vertical path generation task. The robust synthesis has resulted in an optimum mechanism having objective function value of 1.403. The normal distribution of objective function value indicated that the mean has been 1.261 and standard deviation has been 0.052. Thus, it can be concluded that the synthesized mechanism has been optimum as it has minimum objective function value and also robust as the standard deviation has been less. It can be also be concluded that the methodology is suitable for synthesis of path generating four-bar mechanism.

## References


1. S. Krishnamurty, A. Kunjur, Genetic algorithms in mechanism synthesis J. Appl. Mech. Robot. **4**(2), 20 (1997)
2. S. Matekar, G. Gogate, Optimum synthesis of path generating four-bar mechanisms using differential evolution and a modified error function. Mech. Mach. Theory **52**, 162 (2012)
3. S. Matekar, A. Fulambarkar, Simulation and experimental validation of slider displacement in slider crank mechanism with connecting rod tolerance and joint clearance. Int. Rev. Mech. Eng. **12**(12), 970 (2018)
4. F. Chen, H. Huang. Application of Taguchi method on the tolerance design of a four bar function generating mechanism. In: ASME 2005 International design engineering technical conferences and computers and information in engineering conference. Volume 7: 29th Mechanisms and Robotics Conference, Parts A and B Long Beach, 24–28. (California, USA, 2005). 727–733
5. X. Huang, Y. Zhang, Robust tolerance design for function generation mechanisms with joint clearance. Mech. Mach. Theory **45**(9), 1288 (2010)
6. X. Huang, Y. Zhang, Lu. Hao, Le. Tang, He. Li, Tolerance design for four bar function generating mechanisms with joint clearance using Taguchi method. Adv. Mater. Res. **118–120**, 518 (2010)
7. M.S. Phadke, Quality Engineering using Robust Design, 2nd edn. (Pearson Education, 2012), pp. 18–24
8. K.-H. Lee, G.-J. Park, Robust optimization considering tolerances of design variables. Comput. Struct. **79**, 80 (2001)
9. S. Caro, F. Bennis, P. Wenger, Tolerance synthesis of mechanisms: a robust design approach J. Mech. Des. **127**, 87 (2005)

# Predicting the Surface Roughness in Single Point Incremental Forming

**Manish Oraon, Vinay Sharma, and Soumen Mandal**

**Abstract** The demand for customized products slowly shifted the manufacturing industry from mass production to rapid prototyping or batch production that needed a new sheet forming technique. In recent years, the researchers pay attention to single point incremental forming (SPIF), an emerging sheet forming technique. It is a quite flexible sheet forming process which eliminates the dedicated die or punch. The surface quality of the finished product in SPIF is one of the domains because it is concerned with the customer's demand. The tool feed and variation in depth of deformation resembles in the finished product. For minimizing the waviness of the finished part, ANN-based modeling is done in the present study. A good agreement with experimental data and overall regression performance of ANN 98.99% and an acceptable error of −0.136 is found.

**Keywords** SPIF · Surface roughness · Artificial neural network · Input variables

## 1 Introduction

The sheet metal forming is the common forming process in which the metal sheet is shaped by using specialized dies and punch. The demand for the customized product increased the cost of production because a dedicated die is needed for processing. The making of customized die only for prototypes, models, or even for low batch production is a critical issue for taking any industry. To overcome this complex situation, in the early 90s, the incremental sheet metal forming technique has been introduced

M. Oraon (✉) · S. Mandal
Department of Production Engineering, BIT, Mesra Off Campus, Patna, India
e-mail: moraon@bitmesra.ac.in

S. Mandal
e-mail: soumenbhu@gmail.com

V. Sharma
Department of Production Engineering, BIT, Mesra Off Campus, Deoghar, India
e-mail: vinay@bitmesra.ac.in


K. M. Pandey et al. (eds.), *Recent Advances in Mechanical Engineering*, Lecture Notes in Mechanical Engineering, https://doi.org/10.1007/978-981-15-7711-6_25

in which the metal sheet is formed by using a simple tool [1]. There are numbers of experimental results, finite element (FE) modeling, and simulation-based models are proposed by researches for the optimization of ISMF [2–4] but yet, no adequate solution is adopted [5, 6]. In SPIF, the induced forces are increased accordingly with the metal sheet thickness, tool diameter, wall angle and profoundly increases with increasing the vertical step size [7]. It is found that within the explored limits, vertical step size has the least significant impact, and therefore, it can be increased without great penalty, whereas the deformation force is primarily dependent on the wall angle and tool diameter [8]. The ANN modeling is done for the investigation of vertical force component ($Fz$) during the SPIF of two metals, e.g., aluminum AA3003-O and Cu–Zn alloy. The feed-forward backpropagation (BPNN) algorithm is adopted for simulation where the sigmoid transfer function is taken at the hidden layer and linear transfer function at the output layer. The result shows that the predicted $Fz$ through ANN is found very close to the experimental data set with mean absolute error of −0.215 [9]. The titanium sheet is formed into a medical implant through SIPF and measured its interior surface roughness with considering the forming tool diameter, feed rate, and the coefficient of friction. It is observed that the roughness of the inner surface can be minimized with forming at moderate feed rate [10], whereas the surface profile of stainless steel DC01, stainless steel 304, and aluminum AA1050 is improved by doing the SPIF to the large diameter of the tool and at higher RPM [11]. In the succession, surface roughness is measured and found that the step depth increment is the prominent input variable [12]. Moreover, the average surface roughness of aluminum AA3003-O part is predicted. The MAE of 1.068% and the performance index of the model are coming as 0.95 which is a very close agreement to the measured surface roughness [13].

In this paper, the ANN model is developed for the prediction of Ra-value of the aluminum AA3003-O. This material has good formability characteristic and resistance to corrosion; therefore, it is required in every manufacturing industry.

## 2 SPIF Experiment

The Taguchi design of experiment (DOE) is adopted for conducting the experiments. The SPIF experiment has a simple fixture in which the workpiece material is clamped precisely at its edges. The forming tool that follows a CNC programmed path as per part design and deformed the clamped sheet plastically. The machine, forming tool, and material used in the SPIF are discussed below.

### *2.1 Machine*

The SPIF is conducted in a precise multifunctional CNC machine DT-110, Mikrotool Pvt. Ltd, Singapore (Fig. 1), placed in the department of production engineering,

**Fig. 1** Mickrotool DT-110 and typical fixture of SPIF

Birla Institute of Technology, Mesra, India. The machine has the capability to turn, mill, drill, electric discharge machining (EDM), wire electric discharge machining (WEDM), and electric discharge grinding (EDG) at micro-level.

## 2.2 *Forming Tool*

The customized tool is made of mild steel 40C6. The material is turned to a dimension of 50 mm long and 7 mm diameter. On one end of this turned workpiece, a hemispherical groove is been made to hold 06 mm diameter ball bearing. The balls are being inserted into the groove to make the hemispherical head of the tool. The standard 06 mm diameter bearing ball (BOHLER W300) was inserted in the groove. It is chosen because of sustainability during SPIF and high resistance to wear. This material usually used to manufacture the press die and cutting tools.

**Table 1** Ultimate tensile testing results

| Stress at ultimate (MPa) | Strain at Max. load (%) | Stress at Auto. break (MPa) | Maximum percent strain (%) |
|---|---|---|---|
| 161.48 | 2.66 | 130.6 | 5.46 |

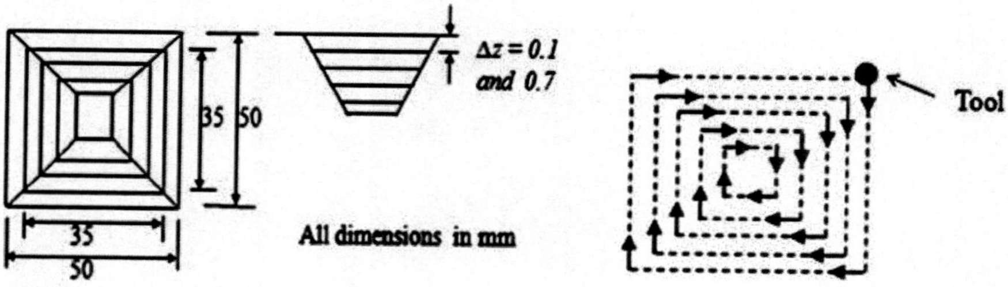


**Fig. 2** Proposed dimension of the square pyramid and tool traverse direction

### 2.3 *Workpiece Material*

The commercialized aluminum grade AA3003-O is taken for SPIF. The stresses and strains of AA3003-O are measured in an automated materials testing system 'Instron Series-9.' The test samples are prepared as per ASTM standard B-557-M. The results are given in Table 1.

During the literature review, It is noticed that the symmetric components which are having no corners are formed easily in SPIF such as conical frustum [14–16], whereas the shapes having any corners are suffering through-thickness at corners and blank get failed at corners, e.g., square pyramid, pentagonal shapes, etc. [17–19]. Since the thickness deviation and strain distribution at the corners of a square frustum during SPIF are different from conic [20–22], therefore research interest towards the forming of the square pyramidal frustum is taken into consideration. The square pyramid frusta with specified dimensions are shown in Fig. 2.

After an extensive review and pivot experiment, the significant input parameters, e.g., step downsize ($\Delta z$), feed rate ($f$), RPM ($R$), the thickness of the metal sheet ($T$), wall angle ($\theta$), and type of lubricant (LL) are considered for SPIF. The Taguchi design of experiment (DOE) is used for conducting the experiments. The orthogonal array L16 design which is adopted for the experiment and the numeric value of input variables in two levels are presented in Table 2.

## 3 Surface Roughness Measurement

The test samples are prepared by cutting 5 mm*5 mm section from the base of the pyramid frusta. The average surface roughness Ra is measured across the direction

**Table 2** Orthogonal array (L16) and numeric value of input variables

| Exp No | $\Delta z$ (mm) | $f$ (mm/min) | $R(\times 102)$ | $\Theta$ (°) | $T$ (mm) | $L$ (kg/m$^3$) |
|---|---|---|---|---|---|---|
| 1 | 0.1 | 20 | 5 | 15 | 0.2 | 15 |
| 2 | 0.1 | 20 | 5 | 45 | 0.4 | 49 |
| 3 | 0.1 | 20 | 20 | 15 | 0.4 | 49 |
| 4 | 0.1 | 20 | 20 | 45 | 0.2 | 15 |
| 5 | 0.1 | 100 | 5 | 15 | 0.4 | 49 |
| 6 | 0.1 | 100 | 5 | 45 | 0.2 | 15 |
| 7 | 0.1 | 100 | 20 | 15 | 0.2 | 15 |
| 8 | 0.1 | 100 | 20 | 45 | 0.4 | 49 |
| 9 | 0.7 | 20 | 5 | 15 | 0.4 | 49 |
| 10 | 0.7 | 20 | 5 | 45 | 0.2 | 15 |
| 11 | 0.7 | 20 | 20 | 15 | 0.2 | 15 |
| 12 | 0.7 | 20 | 20 | 45 | 0.4 | 49 |
| 13 | 0.7 | 100 | 5 | 15 | 0.2 | 15 |
| 14 | 0.7 | 100 | 5 | 45 | 0.4 | 49 |
| 15 | 0.7 | 100 | 20 | 15 | 0.4 | 49 |
| 16 | 0.7 | 100 | 20 | 45 | 0.2 | 15 |

of tool feed, since the change in layer formation by moving forming tool downwards formed a wavy profile in the finished part. The average surface roughness Ra is measured in nanoscale through atomic force measurement (AFM) technique on a sophisticated machine (NT-MDT) at 5X magnification (Fig. 3).

## 4 ANN Modeling for Surface Roughness

The variability in the output response is a common issue in the manufacturing process. These variations may be due to machine efficiency, personnel skill, or even due to poor quality of raw materials. There are some chance causes which are also responsible for the variation. Presently, the artificial intelligence (AI) is used for overcoming these complex situations in manufacturing processes. ANN is a module of AI which is used in a very simple way without having complete knowledge of relationships between inputs and outputs [23]. However, large data is required for its training and to validate the model. Also, the better ANN modeling needed an analysis of neural network (NN) configurations. Therefore, it is required to set the input parameters carefully during the investigation phase. During the training of data, over-fitting should be avoided. The data for validation is recommended for measuring the error function during the analysis as it freezes the NN at its minimum value (early stopping method).

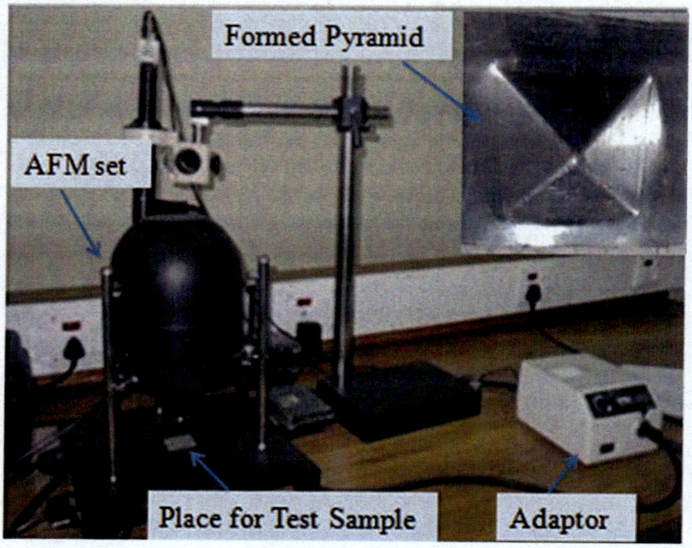


**Fig. 3** Atomic force measurement equipment used for measuring Ra-value and formed pyramid

The feed-forward backpropagation (FFBP) algorithm is widely utilized by researchers for solving the manufacturing processes over other ANN such as cascade-forward backpropagation (CFBP), Elman backpropagation (EBP), time-delay back-propagation (TDBP), perception, radial basis, and self-organizing map (SOM). The ANN model is used for the prediction of the maximum height of forming part with varying wall angle in SPIF. They developed two learning algorithms, e.g., error backpropagation (EBP) with momentum and Levenberg–Marquardt (LM) theory [24]. Further, the FFBP is used for predicting the surface roughness SPIF [25, 26]. In ANN modeling, Ra obtained from each experiment is taken as output and different experiment set of input parameters are considered as input for Ra prediction and mean square error (MSE) FFBP is used for modeling in which the network topology 6–10-1 is developed in MATLAB R2010a (Fig. 4).

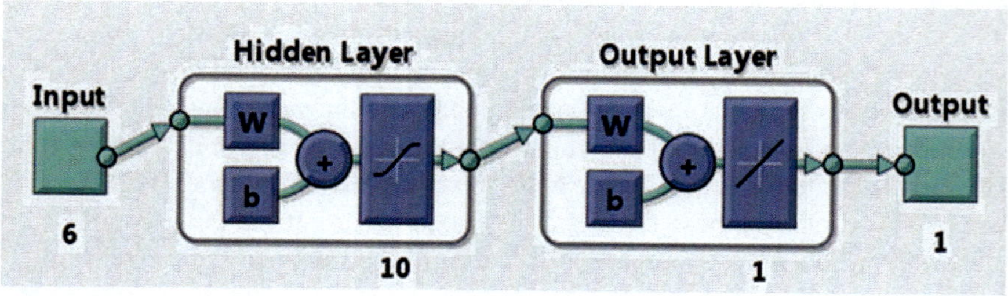


**Fig. 4** Network topology 6–6-1 developed in MATLAB R2010a

Sigmoid transfer functions (sigmoid) at the hidden layer and linear activation function (Purelin) are used at the output layer. Randomly, 55% of experimental data are used for training (Exp. No. 1,2,3,4,5,6,7,9,13,14,15), whereas 25% data are used for testing (Exp. No. 4,5,12,16) and the rest 20% are used for validation (Exp. No. 8,10,11) of the BP model without normalizing the input data. The network has started the training with one neuron, and it is varied to 15 neurons and compared the coefficient of regression (R-value). The ANN model adopted for the present study is summarized below.

Network type: Feed-forward backpropagation.
Training: Levenberg–Marquardt (LM) algorithm.
No. of layers: 3
Output: 1
No of neurons ($n$): 0–15.
Performance: Mean square error (MSE).
Training function: TRAINL M.
Hidden layer transfer function: Tran sigmoid.
Output layer transfer function: Pure linear.
Adaption of learning rate: LEARNGD M.

## 5 Result and Discussion

The modeling of the SPIF process through ANN is showing a good agreement to the measured Ra-values. It is observed that the best-fitted function is achieved with 10 neurons in the hidden layer and the minimum error function is found at epochs 05. At this iteration stage, the natures of curves for training, validation, and test data are the same. The performance plot (training, testing, and validation) is shown in Fig. 5. An acceptable testing regression coefficient of $0.916 \geq 0.900$ is observed as compared to training regression coefficient of 0.999 and validation regression coefficient of 0.994. The overall coefficient of regression ($0.989 = 98.9\%$) indicated the confidence level of the developed model. The present ANN model results the $R$-value and MAE of 0.9899 and −0.136, respectively (Table 3).

## 6 Conclusion

The soft computing tool, i.e., MATLAB, is a user-friendly modeling tool which may be used for the prediction of the output responses without knowing the processing environment, even though with entering new input variables. The feed-forward neural network along with backpropagation learning algorithm finds the best-fitted model for surface roughness with the limited experimental data set. The overall regression performance of ANN found as 98.99% (0.9899), and an acceptable error of −0.136 is showing the capability of the artificial intelligence tool with the limited data set.

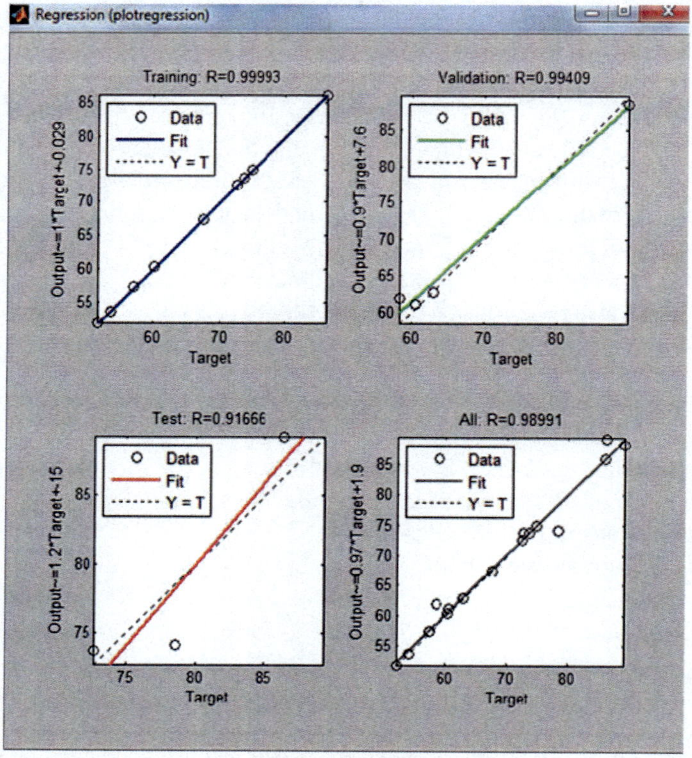


**Fig. 5** Plot of training, testing, validation, and overall performance

**Table 3** Comparing experimental and predicted Ra-value and MAE

| Exp. No. | Measured *Ra* (nm) | Predicted *Ra* (nm) | Error | Mean error |
|---|---|---|---|---|
| 1 | 57.244 | 57.354 | −0.110 | −0.136 |
| 2 | 62.951 | 62.858 | 0.092 | |
| 3 | 58.29 | 61.927 | −3.637 | |
| 4 | 60.477 | 61.161 | −0.684 | |
| 5 | 67.654 | 67.347 | 0.306 | |
| 6 | 60.259 | 60.367 | −0.108 | |
| 7 | 53.735 | 53.618 | 0.116 | |
| 8 | 51.844 | 51.741 | 0.102 | |
| 9 | 89.637 | 88.446 | 1.190 | |
| 10 | 72.573 | 72.636 | −0.063 | |
| 11 | 78.532 | 74.1567 | 4.375 | |

(continued)

**Table 3** (continued)

| Exp. No. | Measured *Ra* (nm) | Predicted *Ra* (nm) | Error | Mean error |
|---|---|---|---|---|
| 12 | 86.358 | 89.264 | −2.906 | |
| 13 | 72.751 | 73.698 | −0.947 | |
| 14 | 74.845 | 74.863 | −0.018 | |
| 15 | 86.102 | 86.034 | 0.067 | |
| 16 | 73.735 | 73.700 | 0.034 | |

Therefore, with the help of previous results, ANN can be used as a tool for modeling and prediction of output responses before conducting new experiments.

## References


1. A. Maruta, A New Trial on Incremental Forming of Sheet Metal Parts. Toyota Central R&D Labs. Inc 34(3) (1999)
2. Y.H. Kim, J.J. Park, Effect of process parameters on formability in incremental forming of sheet metal. Int. J. Mater. Process. Technol. **130–131**, 42–46 (2002)
3. E.S. Masuku, A.N. Bramley, A.R. Mileham, G.W. Owen, Incremental sheet metal forming: a die-less rapid prototyping process For sheet metal. In: Advances in Integrated Design and Manufacturing in Mechanical Engineering (2005), pp. 305–314
4. M. Oraon, V. Sharma, Sheet metal micro forming: future research potentials . Int. J. Prod. Ind. Eng. **01**, 31–35 (2010)
5. R. Aerens, P. Eyckens, A.B. Van, J.R. Duflo, Force prediction for single point incremental forming deduced from experimental and FEM observations. Int. J. Adv. Manuf. Technol. **46**, 969–982 (2010). https://doi.org/10.1007/s00170-009-2160-2
6. M. Oraon, V. Sharma, Effectiveness of tool profile in sheet metal incremental forming. In: 5th International Scientific and Expert Conference of the International TEAM Society (2013) pp. 55–59
7. J. Duflou, Y. Tunckol, A. Szekeres, P. Vanherck, Experimental study on force measurement for single point incremental forming. J. Mater. Process. Technol. **189**, 62–72 (2007)
8. A. Petek, K. Kuzman, J. Kopac, Deformations and forces analysis of single point incremental sheet metal forming. Arch. Mater. Sci. Eng. **35**(2), 107–116 (2009)
9. M. Oraon, V. Sharma, Predicting force in single point incremental forming by using artificial neural network. Int. J. Eng. Trans. A **31**(1), 88–95 (2018)
10. V. Oleksik, A. Pascu, R. Deac, R. Fleaca, O. Bologa, G. Racz, Experimental study on the surface quality of the medical implants obtained by single point incremental forming. Int. J. Mater. Form. **3**(1), 935–938 (2010)
11. M.C. Radu, I. Cristea, Processing metal sheets by spif and analysis of parts quality. Mater. Manuf. Process. **28**,. 287–293 (2013)
12. P.B. Uttarwar, S.K. Raini, D.S., Malwad, Optimization of process parameter on Surface Roughness (Ra) and Wall Thickness on SPIF using Taguchi method. Int. Res. J. Eng. Technol **2**(9), 781–784 (2015)
13. M. Oraon, V. Sharma, Prediction of surface roughness in single point incremental forming of AA3003-O alloy using artificial neural network. Int. J. Mater. Eng. Inn. **9**(1), 1–19 (2018)
14. G. Ambrogio, L. Filice, F. Micari, A force measuring based strategy for failure prevention in incremental forming. J. Mater. Process. Technol. 413–416 (2006)

15. M. Ham, J. Jeswiet, Single point incremental forming and the forming criteria for AA3003. Ann. CIRP**55**(1) (2006)
16. G. Hussain, L. Gao, N.U. Dar, An experimental study on some formability evaluation methods in negative incremental forming. J. Mater. Process. Technol. 45–53 (2007)
17. J.J. Park, Y.H. Kim, Fundamental studies on the incremental sheet metal forming technique. J. Mater. Process. Technol. 447–453 (2003)
18. M. Pohlak, J. Majak, R. Kuttner, Manufacturability and limitations in incremental sheet forming. Proc. Estonian Acad. Sci. Eng **13**(2), 129–139 (2007)
19. J. Duflou, Y. Tunckol, A. Szekeres, P. Vanherck, Experimental study on force measurement for single point incremental forming. J. Mater. Process. Technol. 62–72 (2007)
20. Q.M. Alattaby, T.F. Abaas, A.S. Bedan, The effect of tool path strategy on mechanical properties of brass(65–35) in single point incremental sheet metal forming (SPIF). J. Eng. **19**(5), 629–637 (2013)
21. S. Kurra, S.P. Regalla, Analysis of formability in single point incremental forming using finite element simulations. Procedia Mater. Sci. 430–435 (2014)
22. S. Ai, B. Lu, J. Chen, H. Long, H. Ou, Evaluation of deformation stability and fracture mechanism in incremental sheet forming. Int. J. Mech. Sci. **124–125**, 174–184 (2017)
23. V. Kecman, *Learning and Soft Computing* (MIT Press, New York, USA, 2001)
24. G. Ambrogio, L. Filice, F. Guerriero, R. Guido, D. Umbrello, Prediction of incremental sheet forming process performance by using a neural network approach. Int. J. Adv. Manuf. Technol (2010)
25. M. Vahdati, M. Sedighi, R. Mahdavinejad, Prediction of applied forces in incremental sheet metal forming (ISMF) process by means of artificial neural network (ANN), J. Automot. Appl. Mech. **2**(2) (2014)
26. R. Varthini, R. Gandhinathan, C. Pandivelan, A.K. Jeevanantham, Modelling and optimization of process parameters of the single point incremental forming of aluminium 5052 alloy sheet using genetic algorithm-back Propagation neural network. Int. J. Mech. Prod. Eng. **2**(5), 55–62 (2014)

# Prediction of Droplet Size Distribution For Viscoelastic Liquid Sheet

**Saurabh Sharma, Debayan Dasgupta, Sujit Nath, and Dipankar Bhanja**

**Abstract** A maximum entropy formulation (MEF) has been applied to predict the droplet size distribution for a viscoelastic planar liquid sheet. A modified approach has been introduced in the formulation to solve the set of nonlinear equations which result in less computational time and increase probability of convergence of Newton–Raphson method. The effect of gas–liquid density ratio ($\rho$) also has been studied over the droplets size distribution at different gas–liquid velocity ratio ($U$). Analysis shows that droplet size distribution becomes narrower and peak of the curve increases and shift towards the finer droplet size with increase in $\rho$ and $U$.

**Keywords** Atomization · Sprays · Droplet size distribution · MEF

## 1 Introduction

Sprays are formed after the disintegration of the bulk liquid sheet into the small droplets. A liquid spray consists of millions of droplets of different sizes. Typically, the size of the droplets in a spray varies from 0.1 to 1000 μm and depends on its final application. Droplets are present all around us in different forms such as fog, rain, and mist. Sprays are playing an important role in human life since hundreds of year in the various field such as health care (drug delivery and respirable sprays), agriculture (herbicides spray), fire exhausting systems, industrial use (painting, aerosols), automobiles (fuel injection system), and household use. The efficiency of such systems depends on the size of the droplets produced [1]. Thus, it is becoming a major interest to precise control over the size of the droplets. In any calculation, it is convenient to use the average or mean diameter of droplets besides individual droplets size. Thus, it is favorable to represent the spray in terms of average diameter like surface, volume, and Sauter mean diameter (SMD). SMD is the diameter for which volume to surface ratio is the same as that of the entire droplet sample [2].

S. Sharma (✉) · D. Dasgupta · S. Nath · D. Bhanja
Department of Mechanical Engineering, National Institute of Technology, Silchar, India
e-mail: sharmasaurabh2711@gmail.com



K. M. Pandey et al. (eds.), *Recent Advances in Mechanical Engineering*, Lecture Notes in Mechanical Engineering, https://doi.org/10.1007/978-981-15-7711-6_26

However, due to the continuous distribution of the droplets of varying sizes in a spray, only mean/average diameter does not provide sufficient information about the spray but also droplets size distribution or probability density function is required. It is important to measure the droplet size distribution for the design and operation of a spray generating system. Initially, an empirical approach to modeling drop size distribution uses the curve fitting to fit the experimentally obtained data. The problem with this empirical approach is that it becomes difficult to extrapolate this experimental data outside the experimental range. Thus, it becomes very tedious to use this approach for the modeling of droplet size distribution.

As an alternative to this classical approach, an analytical approach widely known as maximum entropy formulation (MEF) was introduced for modeling of droplet size distribution. MEF is based on the concept of information entropy given by Shannon (1948) [3]; later, it was introduced by Jaynes [4] as a tool for statistical analysis. Later, it found its application in various fields such as thermodynamics concept, biomedical, information technology, economic and atomization, and spray. MEF is applicable for the probability associated problems where sufficient information is available. MEF is commonly used approach for the prediction of droplets size and velocity distribution in a spray forming process. It was firstly introduced by Sellens and Brzustowski [5] and Li and Tankin [6] in the area of droplet formation. Droplets formation is a state-changing phenomena, thus according to the thermodynamics concept, in a state-changing process in a control volume its mass, momentum, and energy remains conserved and entropy maximization happens [2].

Accuracy of the maximum entropy formulation depends on the number of constraints considered and their definition. Here in this paper, normalization, mass, momentum, and energy constraint have been considered. In this method besides the satisfaction of constraint equations, the entropy of the system also is to be maximized. Thus, the most suitable distribution will be which satisfies the conservation equation as well as the entropy principle too. In this approach, Newton–Raphson method has been considered for the solution of the constraints equation. Previously, many issues have been reported by the several researchers regarding the convergence of the Newton–Raphson method [5, 6]. In this paper, a modified formulation has been introduced related to the solution of the constraint equations and result in improved convergence and less computational time required.

## 2 Mathematical Formulation

Maximum entropy formulation's (MEF) credibility depends on the definition of the constraints consider in the formulation. Here in this paper, normalization constraint as well as mass, momentum, and energy balance between the liquid sheet and surrounding medium have been considered. The energy balance constraint consists of sum of the surface and kinetic energy of the liquid sheet following the previous work of Li and Tankin [7], Li et al. [8, 9], Bodaghkhani et al. [10].

As we know that total sum of probability $P_{ij}$ will be equal to unity. Thus, normalization constraint in the spray defines as

$$\sum_{i=1}^{n}\sum_{j=1}^{k} P_{ij} = 1 \quad (1)$$

where $P_{ij}$ is the number based probability of the droplets, $n_{ij}$ is number of droplets having diameter,$d_i$ and velocity $v_j$, and $N$ = total number of droplets formed.

Another constraint in the work is that mass flow rate of the liquid sheet will be equal to the total mass of all the droplets formed per unit time. Here no vaporization or condensation has been considered at liquid and gas interface; thus, mass source term $s_m$ will be equal to the zero. Therefore, mass balance constraint is

$$\sum_{i=1}^{n}\sum_{j=1}^{k} P_{ij} D_i^3 = 1 + s_m \quad (2)$$

where $D_i = d_i / D_{30}$.

On injecting the liquid sheet into the gaseous medium, due to high relative velocity, some momentum transfer will happen between the two medium, which is due to the frictional force acting at the interface. On applying the conservation of momentum, the final expression obtained is as follows,

$$\sum_{i=1}^{n}\sum_{j=1}^{k} P_{ij} D_i^3 V_j = 1 \pm s_{mu} \quad (3)$$

where $s_{mu}$ is the non-dimensionalized momentum source term, which represents the amount of momentum transfer from bulk liquid to the gas or vice versa. Momentum source term $s_{mu}$ can be calculated as

$$s_{mu} = \frac{1}{2}\rho(U-1)^2 L C_f$$

Here $C_f$ is the drag coefficient and can be calculated as [11]

$C_D = \frac{1.328}{\sqrt{\text{Re}}}$ For Re > 1000

$C_D = \frac{1.328}{\sqrt{\text{Re}}} + \frac{2.3}{\text{Re}}$ For Re < 1000

Energy balance constraint equation (Eq. 4) can be obtained after equating the energy of the liquid sheet coming out at exit of the nozzle to total energy of the droplet formed resulting to the disintegration of bulk liquid sheet. The energy balance equation entertains the both gain and loss in energy due to drag force on the interface of liquid and gaseous medium.

$$\sum_{i=1}^{n}\sum_{j=1}^{k} P_{ij}(D_i^3 V_j^2 + BD_i^3)_i = 1 \pm s_e \tag{4}$$

where $B = 12\sigma/\rho_l u_l^2 d_{30}$ and $s_e$ is the non-dimensionalized energy source term, which represent the amount of energy transfer from bulk liquid to the gas or vice versa. Energy source terms can be calculated as

$$s_e = \rho(U - 1)^3 LC_f$$

The normalization, mass, momentum, and energy balance constraint equations can be represented in form as discussed below (Eqs. 5–8)

$$f = \sum_{i=1}^{n}\sum_{j=1}^{k} P_{ij} - 1 \tag{5}$$

$$g = \sum_{i=1}^{n}\sum_{j=1}^{k} P_{ij} D_i^3 - S_m \tag{6}$$

$$h = \sum_{i=1}^{n}\sum_{j=1}^{k} P_{ij} D_i^3 V_j - S_{mu} \tag{7}$$

$$m = \sum_{i=1}^{n}\sum_{j=1}^{k} P_{ij}\left(D_i^3 V_j^2 + BD_i^2\right) - S_e \tag{8}$$

where $S_m = 1 \pm s_m$,$S_{mu} = 1 \pm s_{mu} S_e = 1 \pm s_e$

As there is the infinite number of probability distribution possible but most unbiased distribution is which maximize the Shannon entropy [3], which can be expressed in the form,

$$S = \sum_{i}^{m}\sum_{j}^{n} P_{i,j} \ln P_{i,j}$$

After maximizing the entropy, the obtained expression for $P_{ij}$ is,

$$P_{ij} = \exp[-(1 + \lambda_1 + \lambda_2 D_i^3 + \lambda_3 D_i^3 V_j + \lambda_4 (D_i^3 V_j^2 + BD_i^2))] \tag{9}$$

$$\sum_{i=1}^{n}\sum_{j=1}^{k} P_{ij} = \sum_{i=1}^{n}\sum_{j=1}^{k} \exp[-(1 + \lambda_1 + \lambda_2 D_i^3 + \lambda_3 D_i^3 V_j + \lambda_4 (D_i^3 V_j^2 + BD_i^2))]$$

From Eqs. (1) to (9), the final expression for the $\lambda_1$ is

$$\lambda_1 = \ln\left[\sum_{i=1}^{n}\sum_{j=1}^{k} \exp[-(1 + \lambda_2 D_i^3 + \lambda_3 D_i^3 V_j + \lambda_4 (D_i^3 V_j^2 + B D_i^2))]\right] \tag{10}$$

On putting the value of $\lambda_1$ from Eqs. (10) to (6),

$$g = \sum_{i=1}^{n}\sum_{j=1}^{k} P_{ij} D_i^3 - S_m$$

$$= \sum_{i=1}^{n}\sum_{j=1}^{k} \exp[-(1 + \lambda_1 + \lambda_2 D_i^3 + \lambda_3 D_i^3 V_j + \lambda_4 (D_i^3 V_j^2 + B D_i^2))] D_i^3 - S_m = 0$$

$$= \sum_{i=1}^{n}\sum_{j=1}^{k} \exp[-(1 + \lambda_1 + \lambda_2 D_i^3 + \lambda_3 D_i^3 V_j + \lambda_4 (D_i^3 V_j^2 + B D_i^2))] D_i^3 - S_m = 0$$

$$= \sum_{i=1}^{n}\sum_{j=1}^{k} \exp[-(1 + \lambda_2 D_i^3 + \lambda_3 D_i^3 V_j + \lambda_4 (D_i^3 V_j^2 + B D_i^2))] D_i^3 - S_m \exp(\lambda_1) = 0$$

$$g = \sum_{i=1}^{n}\sum_{j=1}^{k} \exp[-(1 + \lambda_2 D_i^3 + \lambda_3 D_i^3 V_j + \lambda_4 (D_i^3 V_j^2 + B D_i^2))](D_i^3 - S_m) = 0$$

Let, $kk = \exp[-(1 + \lambda_2 D_i^3 + \lambda_3 D_i^3 V_j + \lambda_4 (D_i^3 V_j^2 + B D_i^2))]$

$$g = \sum_{i=1}^{n}\sum_{j=1}^{k} (D_i^3 - S_m) kk = 0 \tag{11}$$

Similarly Eqs. (7) and (8) can be modified to Eqs. (12) and (13), respectively,

$$h = \sum_{i=1}^{n}\sum_{j=1}^{k} (D_i^3 V_j - S_{mu}) kk = 0 \tag{12}$$

$$m = \sum_{i=1}^{n}\sum_{j=1}^{k} (D_i^3 V_j^2 + B D_i^2 - S_e) kk = 0 \tag{13}$$

Set of the above nonlinear Eqs. (11–13) is to be solved with the help of the Newton–Raphson method. After finding out the value of $\lambda_2$, $\lambda_3$ and $\lambda_4$, value of $\lambda_1$ can be find out by using normalization constraint as discussed in Eq. (10).

## 3 Result and Discussion

In this paper, we have studied the effect of liquid–air density for different gas–liquid velocities on the droplets size distribution. All these studies have been carried out at Re = 1000. Obtained results after the second-order stability analysis for a viscoelastic liquid sheet have been shown in Table 1. Considering the brevity of the paper, stability analysis has not been discussed here.

Break-up length is defined as the distance between the nozzle exit and point of break-up of the liquid sheet. It has been non-dimensionalized with the help of liquid sheet thickness which has been considered as a reference length parameter. All the obtained results (Table. 1) have been plotted (Fig. 1). Study has been carried out at different value of gas–liquid density ratio ($\rho$), viz. 0.0010, 0.0015, 0.0020, and 0.0030. It has been seen that at particular gas–liquid velocity ratio ($U$), peak of the droplet size distribution is shifting towards the finer droplets size with increasing $\rho$ as well as droplet size distribution is also getting narrower (Fig. 1a). Studies show that with higher density more uniform droplets will produce. This same kind of trend also has been noticed on different $U$, viz. 2.00, 2.25, 2.50, and 2.75 (Fig. 1a–d). Peak of the curve is also increasing with increment in the gas–liquid velocity ratio.

To see the effect of $U$, a comparison analysis also has been done for $\rho = 0.003$ (Fig. 2). Thus, it is found out that peak of the curve has being shifted towards the

**Table 1** Result of nonlinear stability analysis for Re = 1000, at different $\rho$ and $U$

| Gas–liquid velocity ratio ($U$) | Weber No. | Gas–liquid density ratio ($\rho$) | Non-dimensional breakup length | Non-dimensional breakup area of ligament after breakup |
|---|---|---|---|---|
| 2.00 | 200 | 0.0010 | 799.29 | 62.83186 |
| | | 0.0015 | 516.88 | 41.88790 |
| | | 0.0020 | 396.65 | 31.41593 |
| | | 0.0030 | 215.04 | 20.26835 |
| 2.25 | 200 | 0.0010 | 495.44 | 39.26991 |
| | | 0.0015 | 332.40 | 26.17995 |
| | | 0.0020 | 206.13 | 19.63496 |
| | | 0.0030 | 109.63 | 12.82283 |
| 2.50 | 200 | 0.0010 | 359.9 | 27.31820 |
| | | 0.0015 | 180.77 | 17.95196 |
| | | 0.0020 | 116.30 | 13.36850 |
| | | 0.0030 | 61.68 | 8.49080 |
| 2.75 | 200 | 0.0010 | 214.02 | 20.26834 |
| | | 0.0015 | 112.50 | 13.08997 |
| | | 0.0020 | 71.80 | 9.519980 |
| | | 0.0030 | 37.77 | 5.872140 |

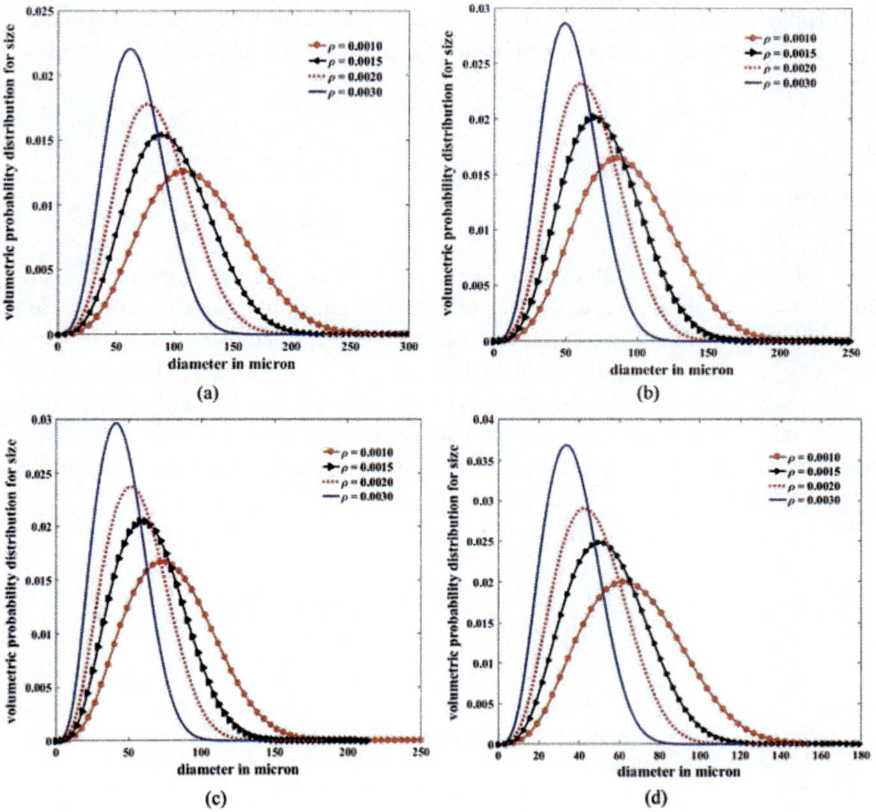


**Fig. 1** Effect of gas–liquid density ($\rho$) on volumetric probability size distribution for **a** $U = 2$ **b** $U = 2.25$ **c** $U = 2.5$, **d** $U = 2.75$ at $\mathrm{Re} = 1000$

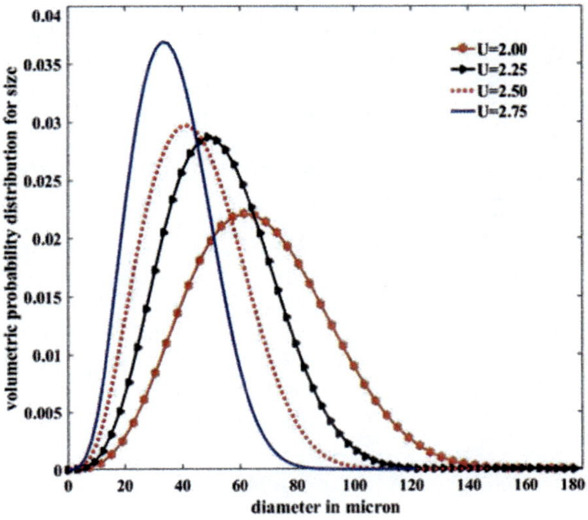


**Fig. 2** Effect of gas–liquid velocity ($U$) on volumetric probability size distribution at $\rho = 0.003$ and $\mathrm{Re} = 1000$

finer droplets with increasing $U$. Hence, it can be said that droplet size distribution became narrower and more uniform droplets would be produce with increase in $\rho$ and $U$.

## 4 Conclusions

A maximum entropy formulation has been applied to predict the droplet size distribution for a viscoelastic liquid sheet. The effect of gas–liquid density ratio ($\rho$) has been studied at different gas–liquid velocity ratio ($U$). It has been seen that distribution curve is getting narrower and peak of curve is increasing and shifting toward the smaller droplets size with increase in $\rho$. Thus, smaller droplets will produce with increase in $\rho$. This trend is consistent with all the $U$. Same effect also has been encountered with increasing $U$ at particular $\rho$. Thus, it can be concluded that smaller droplet would produce with increase in $\rho$ and $U$.

## References

1. C. Dumouchel, On the experimental investigation on primary atomization of liquid streams. Exp. Fluids **45**(3), 371–422 (2008). https://doi.org/10.1007/s00348-008-0526-0
2. C. Dumouchel, The maximum entropy formalism and the prediction of liquid spray drop-size distribution. Entropy **11**(4), 713–747 (2009). https://doi.org/10.3390/e11040713
3. C.E. Shannon, A mathematical theory of communications. Bell Labs Tech. J. **27**, 379–423 (1985)
4. E.T. Jaynes, Information Entropy and Statistical Mechanics. Phys. Rev. **108**(2), 171–190 (1957)
5. R.W. Sellens, T.A. Brzustowski, A prediction of the drop-size distribution in a spray from first principles. Atom. Spray Tech. **1**, 89–102 (1985)
6. X. Li, R.S. Tankin, Droplet Size Distribution: A Derivation of a Nukiyama-Tanasawa Type Distribution Function. Combust. Sci. Technol. **56**(1–3), 65–76 (1987). https://doi.org/10.1080/00102208708947081
7. X. Li, R.S. Tankin, Derivation of Droplet Size Distribution in Sprays by Using Information Theory. Combust. Sci. Technol. **60**(4–6), 345–357 (1988). https://doi.org/10.1080/00102208808923992
8. X. Li, R.S. Tankin, M. Renksizbulut, Calculated Characteristics of Droplet Size and Velocity Distributions in liquid sprays. Part. Part. Syst. Charact. **7**(1–4), 54–59 (1990). https://doi.org/10.1002/ppsc.19900070111
9. X. Li, L.P. Chin, R.S. Tankin, T. Jackson, J. Stutrud, G. Switzer, Comparison between experiments and predictions based on maximum entropy for sprays from a pressure atomizer. Combust. Flame **86**(1–2), 73–89 (1991). https://doi.org/10.1016/0010-2180(91)90057-I
10. A. Bodaghkhani, B. Colbourne, Y.S. Muzychka, Prediction of droplet size and velocity distribution for spray formation due to wave-body interactions. Ocean Eng. **155**, 106–114 (2018). https://doi.org/10.1016/j.oceaneng.2018.02.057
11. F.M. White, *Viscous Fluid Flow* (McGraw-Hill, New York, 1974)

# Linear Stability Analysis of Viscoelastic Liquid Sheet in Presence of Moving Gas Medium

**Debayan Dasgupta, Saurabh Sharma, Sujit Nath, and Dipankar Bhanja**

**Abstract** A linear analysis of sinuous instabilities in a two-dimensional planar viscoelastic sheet subjected to two inviscid gas streams of equal non-zero velocities is performed. The rheological model of the viscoelastic sheet is considered to be corotational Jeffrey's model. Perturbation technique is employed to obtain the linear governing equation and boundary conditions. Solution of the first-order dispersion equation yields the maximum growth rate and corresponding dominant wave number. Parametric investigation of the effects of elasticity number and time constant ratio is performed for different gas-to-liquid velocity ratios. Linear analysis predicts that elasticity number enhances the maximum growth rate. On the contrary, time constant ratio is observed to dampen the maximum growth rate. Hence, elasticity number and time constant ratio exhibit a destabilizing and stabilizing effect on the liquid sheet, respectively

**Keywords** Instability · Linear · Viscoelastic · Non-zero gas velocity

## 1 Introduction

Viscoelastic fluids exhibit fluid like properties at in steady shear over long time scales but behave like elastic solids in presence of deformation over short time scales. Viscoelastic fluids are often employed in applications concerning polymer processing operations like spray coating, spray drying, food processing, etc. Early investigation on viscoelastic fluid was performed by Liu et al. [1] for symmetric and antisymmetric disturbances in the presence of still gas medium. Brenn et al. [2] performed a comparatively analysis of 2D and 3D disturbances on viscoelastic liquid sheet and observed that growth of 2D disturbances was relatively higher than that of 3D disturbances. Combined influence of sinuous and varicose instabilities on viscoelastic gel propellant sheet was conducted by Yang et al. [3]. It was observed that

---

D. Dasgupta · S. Sharma · S. Nath (✉) · D. Bhanja
Department of Mechanical Engineering, National Institute of Technology, Silchar, Assam 788010, India
e-mail: sujitnath2008@gmail.com; sujitnath@mech.nits.ac.in



K. M. Pandey et al. (eds.), *Recent Advances in Mechanical Engineering*, Lecture Notes in Mechanical Engineering, https://doi.org/10.1007/978-981-15-7711-6_27

viscoelastic sheet was relatively more unstable than Newtonian liquid sheet. Atalik and Keunnings [4] reported dual effect of elasticity for liquid sheet with Oldroyd B model. Initial increase in elasticity number destabilized the sheet, but a further increase showed the existence of a restabilizing regime. Wang et al. [5] and Xie et al. [6] analyzed sinuous and varicose disturbances in viscoelastic liquid sheet, respectively.

All the literature discussed above have considered the gas velocities to be zero. However, applications dealing with viscoelastic fluid often employ twin fluid atomizer, where the kinetic energy of the fast moving gases is utilized to disintegrate the liquid sheet in small droplets. As a result, it is imperative to study the effect of viscoelasticity on sheet behavior in the presence on nonzero gas velocities. The current study performs linear stability analysis of temporal instabilities in a planar viscoelastic liquid sheet sandwiched between two inviscid gas streams of equal nonzero velocities. Only antisymmetric or sinuous disturbances have been considered for the present case.

## 2 Mathematical Formulation

A two-dimensional viscoelastic sheet having of thickness 2 h has been considered. Liquid density, surface tension, and viscosity are represented as $\rho_l, \sigma_l$ and $\mu_l$, respectively. Initially, the unperturbed liquid sheet has velocity only in the $x$-direction, represented by $u_l$. The surrounding inviscid gases have density $\rho_g$ and are flowing with velocity $u_g$. The sheet is subjected to an initial sinuous mode of disturbance. Both liquid and gas velocities are considered to be incompressible and the effect of gravity has been neglected. Non-dimensionalization of constitutive relation, governing equation, and boundary conditions is performed using the following scale: [length, time, density, velocity, stress] $= [h, \frac{h}{u_l}, \rho_l, u_l, \rho_l u_l^2]$. Reynold's number is expressed as $\mathrm{Re} = \frac{\rho_l u_l a}{\mu}$. The schematic of the configuration is presented in Fig. 1.

The convected Jeffrey's model for corotational model in its non-dimensional form is expressed as [5]

$$\tau + \lambda_1[\tilde{\tau} - W.\tau + \tau.W] = \frac{1}{\mathrm{Re}}[\gamma + \lambda_2[\tilde{\gamma} - W.\gamma + W.\gamma]] \quad (1)$$

Here, $\lambda_1$ and $\lambda_2$ represent stress relaxation time and deformation retardation time, respectively. $\tau$ is extra stress tensor and its material derive is expressed as

$$\tilde{\tau} = \frac{\partial \tau}{\partial t} + (v.\nabla)\tau \quad (2)$$

$\gamma$ is the strain tensor such that

$$\gamma = \nabla v + (\nabla v)^T \quad (3)$$

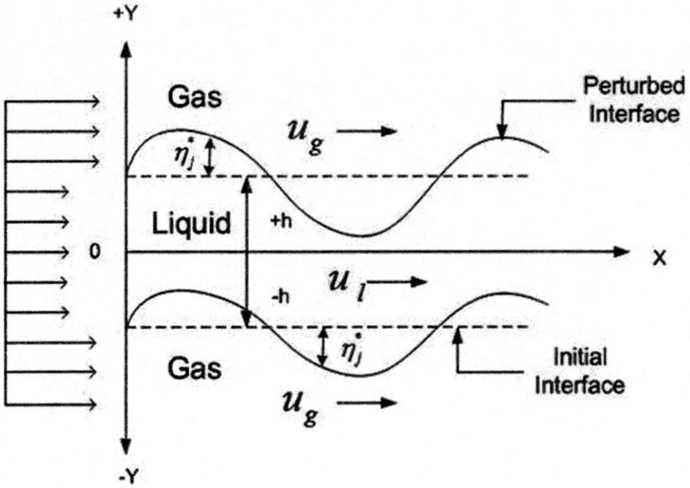


**Fig. 1** Schematic of planar viscoelastic liquid sheet surrounded by two moving gas streams with equal velocities subjected to sinuous mode of instability

$W$ represents the vorticity tensor such that

$$W = \frac{1}{2}\left[\nabla v - (\nabla v)^T\right];\ W_{xx} = W_{yy} = 0;\ W_{xy} = W_{yx} = u_y - v_x; \tag{4}$$

Two characteristic numbers viz. elasticity number ($El$) and time constant ratio ($\lambda$) are introduced to define the rheological property of the viscoelastic fluid. Corresponding expressions are gives an

$$El = \frac{\lambda_1}{Re};\ \lambda = \frac{\lambda_2}{\lambda_1}; \tag{5}$$

Gas phase is considered to be irrotational and is represented by velocity potential $\phi_{gj}$, where $j = 1$ and 2 represent upper and lower gas liquid interface, respectively. Gas velocity potential at undisturbed state is given as $\phi_{gj}\big|_{t=0} = Ux$, where $U$ is the non-dimensional gas velocity $\left(\frac{u_{gj}}{u_l}\right)$. Liquid and gas pressures in their non-dimensional form are represented as $P_l$ and $P_g$, respectively. Location of non-dimensional interfaces after initial perturbation is given as

$$y(x, t) = (-1)^{j+1} + \eta_j(x, t) \tag{6}$$

where $\eta_j(x, t)$ is non-dimensional surface deformation.

All the parameters used in governing equations and boundary condition are presented using power series of $\eta_0$, where $\eta_0$ is the initial perturbation and its value is considered as 0.1 [5]. The governing equations and boundary conditions are approximated around the unperturbed interface using Taylor series expansion. Finally, by comparing the power of $\eta_0$, the first-order governing equations and boundary conditions are obtained as

$$\tau_1 + \lambda_1(\tau_{1,t} + \tau_{1,x}) = \frac{1}{\mathrm{Re}}[\gamma_1 + \lambda_2(\gamma_{1,t} + \gamma_{1,x})] \tag{7}$$

$$u_{1,x} + v_{1,y} = 0 \tag{8}$$

$$u_{1,t} + u_{1,x} = -P_{l1,x} + \left(\frac{\partial}{\partial x}\tau_{1,xx} + \frac{\partial}{\partial y}\tau_{1,xy}\right) \tag{9}$$

$$v_{1,t} + v_{1,x} = -P_{1,y} + \left(\frac{\partial}{\partial x}\tau_{1,yx} + \frac{\partial}{\partial y}\tau_{1,yy}\right) \tag{10}$$

$$v_1 = \eta_{j1,t} + \eta_{j1,x} \tag{11}$$

$$\tau_{yx1} = 0 \tag{12}$$

$$\varphi_{gj1,xx} + \varphi_{gj1,yy} = 0 \tag{13}$$

$$\varphi_{gj1,y} - \eta_{j1,t} - U\eta_{j1,x} = 0 \tag{14}$$

$$\varphi_{gj1,y} = 0 \tag{15}$$

$$-P_{l1} + \tau_{yy1} + (-1)^j \frac{1}{We}\eta_{j1,xx} - \rho\varphi_{gj1,t} - \rho U\varphi_{gj1,x} = 0 \tag{16}$$

## 3 Solution Procedure

The first-order solutions are expressed as [5]

$$\left(u_1, v_1, P_{l1}, \varphi_{gj1}, \eta_{j1}\right) = \left[\hat{u}_1, \hat{v}_1, \hat{P}_{l1}, \hat{\tau}_1, \hat{\gamma}_1, \hat{\varphi}_{gj1}, \hat{\eta}_{j1}\right] \exp\left[i(k_1 x - \omega_1 t)\right] + c.c \tag{17}$$

Here, $\omega_1$ is linear complex frequency whose real ($\omega_{1r}$) and imaginary part ($\omega_{1i}$) signify angular frequency and growth rate, respectively. The symbol '^' represents the components that are function of '$y$' only and '$c.c$' is complex conjugate. Substituting

Eq. (17) in the linear constitutive relation (7) yields

$$\left(u_1, v_1, P_{l1}, \varphi_{gj1}, \eta_{j1}\right) = \left[\hat{u}_1, \hat{v}_1, \hat{P}_{l1}, \hat{\tau}_1, \hat{\gamma}_1, \hat{\varphi}_{gj1}, \hat{\eta}_{j1}\right] \exp\left[i(k_1 x - \omega_1 t)\right] + c.c \tag{18}$$

$$\tau_1 = \frac{1}{\mathrm{Re}_1}\gamma_1 \text{ where } \mathrm{Re}_1 = \frac{1 + \lambda_1 i(k_1 - \omega_1)}{1 + \lambda_2 i(k_1 - \omega_1)}\mathrm{Re} \tag{19}$$

Components of strain tensor obtained from Eq. (2) are

$$\gamma_{xx} = 2u_x; \quad \gamma_{yx} = \gamma_{xy} = u_y + v_x; \quad \gamma_{yy} = 2v_y \tag{20}$$

Using Eqs. (17) and (19) in Eqs. (7)–(16), solutions for liquid and gas phase are obtained as

$$\hat{u}_1 = iA_1 \sinh(k_1 y) + \frac{il_1}{k_1} B_1 \sinh(l_1 y) \tag{21}$$

$$\hat{v}_1 = A_1 \cosh(k_1 y) + B_1 \cosh(l_1 y) \tag{22}$$

$$\hat{p}_1 = \frac{(w_1 - k_1)}{k_1} i A_1 \sinh(k_1 y) \tag{23}$$

where $A_1 = \frac{(k_1^2 + l_1^2)\hat{\eta}_j}{\mathrm{Re}_1 \cosh(k_1)}$; $B_1 = \frac{-2k_1^2 \hat{\eta}_j}{\mathrm{Re}_1 \cosh(l_1)}$;

$$l_1^2 = k_1^2 + i\mathrm{Re}_1(k_1 - \omega_1) \tag{24}$$

$$\hat{\varphi}_{gj1} = (-1)^{j+1}\left[\frac{i\omega_1}{k_1} - iU\right]\hat{\eta}_j \exp\left[k_1 + (-1)^j k_1 y\right] \tag{25}$$

It can be noticed that when $\lambda_1 = \lambda_2 = 0$, first-order effective Reynolds number $\mathrm{Re}_1$ reduces to Re and the solutions of the liquid phase reduces to the solutions of viscous liquid [7]. The first-order dispersion equation is obtained by substituting Eqs. (22)–(25) in the dynamic boundary condition (Eq. 16) and is expressed as

$$-\rho\left[(\omega_1 - Uk_1)^2\right] + \frac{k_1^3}{We} + \frac{\left(l_1^2 + k_1^2\right)^2}{\mathrm{Re}_1^2}\tanh(k_1) - \left(\frac{2}{\mathrm{Re}_1}\right)^2 l_1 k_1^3 \tanh(l_1) = 0 \tag{25}$$

## 4 Results and Discussions

Solution of the first-order dispersion equation provides the first-order complex root $\omega_1$. The imaginary part of $\omega_{i1}$ yields the linear growth rate ($\beta$). This linear growth rate is plotted against wave number $k$ to obtain the maximum growth rate ($\beta_{max}$) and corresponding dominant wave number ($k_{dominant}$). Figures 1a and 2b show variation of maximum growth rate ($\beta_{max}$) and corresponding ($k_{dominant}$) with elasticity number (*El*), respectively, for different gas to liquid velocity ratios ($U$) keeping $We = 300$ and $\rho = 0.0014$.

Figure 2a shows that an increase in elasticity number (*El*) increases the maximum growth rate ($\beta_{max}$) However, the influence of *El* on $\beta_{max}$ is very minor at $U = 2$. But as $U$ increases, the effect of *El* on $\beta_{max}$ becomes more significant. Hence, it can be concluded that elasticity produces a destabilizing effect on linear growth rate. Similar effect of *El* can also be noticed on the most dominant wave number. An increase in *El* at higher values of $U$ shifts $k_{dominant}$ toward higher wave number.

Figure 3a and b shows variation of $\beta_{max}$ and $k_{dominant}$ with λ at different values of

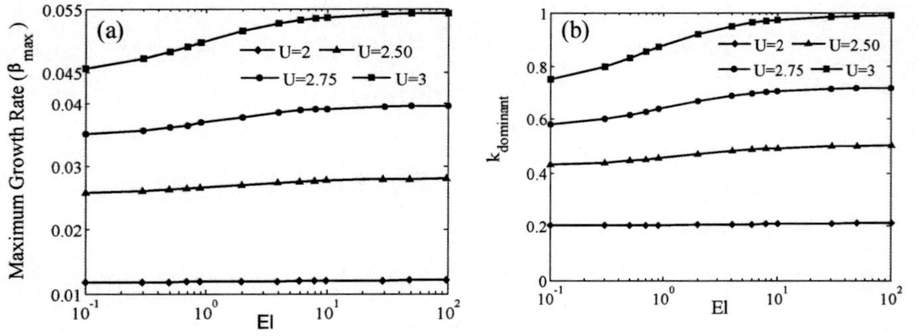


**Fig. 2** Variation of **a** maximum growth rate **b** dominant wave number with *El* at different $U$ keeping $\lambda = 0$, $We = 300$ and $\rho = 0.0014$

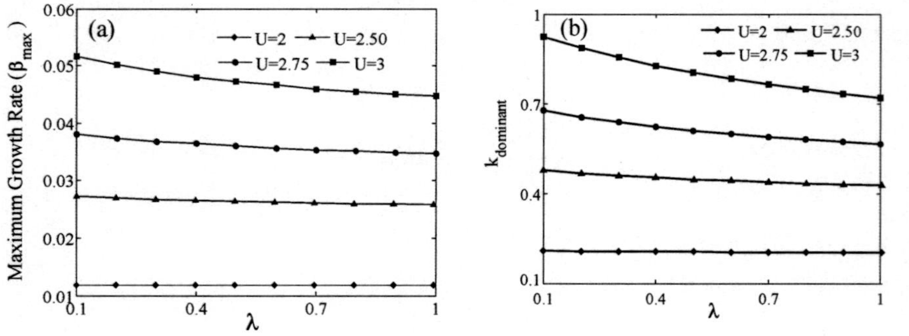


**Fig. 3** Variation of **a** maximum growth rate **b** dominant wave number with λ at different $U$ keeping $El = 10$, $We = 300$ and $\rho = 0.0014$

$U$. It can be observed that an increase in λ reduces both $\beta_{\max}$ and $k_{\text{dominant}}$. However, the effect of λ is very minor at $U = 2$ and increases with increase in $U$. Hence, it can be said that time constant ratio produces a stabilizing effect on linear growth rate.

## 5 Conclusion

A temporal linear analysis of sinuous disturbances in a two dimensional planar viscoelastic liquid sheet has been performed. The effect of elasticity number and time constant ratio on linear growth rate in presence of nonzero equal gas velocities has been discussed. Results show that elasticity number and time constant ratio exhibit a destabilizing and stabilizing effects on linear growth rate, respectively. However, their corresponding effects are very minor at low gas velocities and become increasingly significant with increase in gas velocity. Similar effect of elasticity number and time constant ratio can also be observed on the most dominant wave number. Elasticity number and time constant ratio shift the dominant wave number toward large and small wave number at relatively higher gas velocities. Their corresponding effects are very negligible at small gas velocities.

## References

### *Journal Articles*

1. Z. Liu, F. Durst, Linear analysis of the instability of two-dimensional non-Newtonian liquid sheets. J. Nonnewton. Fluid Mech. **78**, 133–166 (1998)
2. G. Brenn, Z. Liu, F. Durst, Three dimensional temporal instability of non-Newtonian liquid sheets . Sprays **11**, 49–84 (2001)
3. L.Yang, Y. Qu, Q. Fu, B. Gu, Linear stability analysis of a non-Newtonian liquid sheet J Propuls Power**26**
4. K. Atalık, R. Keunings, Non-linear temporal stability analysis of viscoelastic plane channel flows using a fully-spectral method. J. Nonnewton. Fluid Mech. **102**, 299–319 (2002)
5. C. Wang, L. Yang, L. Xie, P. Chen, Weakly nonlinear instability of planar viscoelastic sheets Phys. Fluid**27** (2015)
6. L. Xie, L. Yang, J. Wang Jie, L. Qin, Weakly nonlinear instability of viscoelastic planar sheets with initial varicose disturbances Aerosp. Sci. Technol. **79**, 373–382 (2018)
7. D. Dasgupta, S. Nath, D. Bhanja, Dual-mode nonlinear instability analysis of a confined planar liquid sheet sandwiched between two gas streams of unequal velocities and prediction of droplet size and velocity distribution using maximum entropy formulation Phys. Fluid**30** (2018)

# Development of Feature Extraction-Based Currency Recognition System Using Artificial Neural Network

**Deep Singh, Rahul Kumar, Rutupurna Choudhury, Ashutosh Padhan, and Yogesh Singh**

**Abstract** Development of paper currency detection system is one kind of smart framework which is of significant requirement in today's modern world. Currency recognition means denomination classification and counterfeit detection. There are many currencies all over the world. Every currency appears unique based on features such as variation in size, texture, colour etc. In the present study, a system is developed to reduce counterfeit detection time. Image processing and feature extraction technique in the presence of UV light are implemented for currency identification. Some of the features of real banknote are only visible in the presence of UV light. In case of Indian currencies, security thread looks green in presence of UV light. This study trains a feedforward backpropagation neural network with 20 samples of each denomination. The validation resulted 100% accuracy with reduced processing time for currency recognition and counterfeit detection.

**Keywords** Artificial neural network · Currency recognition · Feature extraction · Image processing

D. Singh · R. Kumar · R. Choudhury · A. Padhan · Y. Singh (✉)
Department of Mechanical Engineering, National Institute of Technology Silchar, Silchar 788010, Assam, India
e-mail: yogeshsingh15@gmail.com

D. Singh
e-mail: deepsingh.0318@gmail.com

R. Kumar
e-mail: rahul.me.nitdgp@gmail.com

R. Choudhury
e-mail: rutupurna.choudhury10@gmail.com

A. Padhan
e-mail: ashutoshpadhan.1994@gmail.com



K. M. Pandey et al. (eds.), *Recent Advances in Mechanical Engineering*, Lecture Notes in Mechanical Engineering, https://doi.org/10.1007/978-981-15-7711-6_28

## 1 Introduction

The Indian currency framework is predominant for quite a while. The Government of India has presented its first paper currency-issuing 10 rupees notes in 1861. As of November 1, 2016, the present circulating paper currencies were in denominations of 5, 10, 20, 50, 100, 500, and 1000 rupees. These denominations have been in use for a long time. However, to overcome the problem of money laundering, the government decided to ban a few currency denominations. In 2016, the denomination of 500 and 1000 was banned and introduced new paper currencies of denominations 10, 50, 100, 200, 500, and 2000.

The methods for recognizing banknote includes artificial neural networks (ANN), hidden Markov chains, and dynamic templates that mimic human brain behavior. Currency check frameworks are utilized for separating authentic currency from the fake. These frameworks have a wide scope of utilization, for example, automatic teller machines (ATMs), auto-vender machines, cash trade offices, lodgings, banks, and shops. The headways in imaging advancements for checking and printing have made the creation of fake banknotes a lot simpler and hard to identify. Besides, the complex procedures utilized by forgers in the arrangement of fake banknotes lead to provoking assignment in separating real and fake banknotes physically.

Many researchers have contributed to the domain of feature recognition technique so far and performed currency recognition. Feature extraction has been proven very helpful in differentiating the fake currency and the classification of currency denominations.

Lee and Kims [1] suggested a recognition algorithm for the identification of paper currency using the coordinate extraction technique of a particular point from the banknotes of the same color. Five neural networks were developed for identification purposes. The aim of the framework design was to decrease the recognition time. The test resulted in an outstanding recognition level under a small training duration [1].

Debnath et al. [2] introduced a paper currency identification system by means of ensemble neural network (ENN) through negative correlation learning (NCL). The currencies used were fresh, noisy, and old. The banknote image is converted into grayscale and compressed. Then, each pixel of the compressed image is transferred to the network as an input. This technique identifies the noisy currencies and delivers excellent outcomes. Results indicate the precision of identification range from 100 to 54% on the basis of the distortion level of the input images [2].

Sharma et al. [3] proposed a system on the basis of local binary patterns (LBP) to recognize Indian paper currency. The outcome demonstrates that the framework provides superior performance for low-noise images with 99% precision using a smart phone [3].

Sargano et al. [4] proposed a new smart framework for the identification of Pakistani currency which required lesser time as compared to other frameworks. For classification, three layers of feedforward backpropagation neural network (BPN) were used. The system was tested with 350 Pakistani banknotes. The outcome indicated that the precision of identification of the system was 100%. This technique is introduced without distortion to banknotes [4].

Vijay and Jain [5] suggested an image processing method to extract the denomination of banknotes. The region of interest (ROI) is obtained by the feature recognition method and the corresponding neural networks. In this technique, they obtained the images with a defined size by the easy flat scanner and then a few filters were applied to obtain the banknote denomination value [5].

Currencies show much more resistance to brightening because of the absence of specular reflection [6]; likewise, they have many subtleties that should be distinguished. Currency acknowledgment in a changeable domain is a complex problem since we have numerous unrestrained conditions that influence the quality of image. On the other hand, due to the lack of specular reflection, bills demonstrate much more tolerance to lighting. Currency possesses many details to be recognized. Currency recognition, however, has several troubles such as warping of shape due to folding or wrinkling [7].

### *1.1 Contribution in the Present Paper*

In this regard, the present paper proposed a currency recognition system with the help of image processing, feature extraction, and neural network. The dataset is the images of Indian currencies (Rs. 200, Rs. 500, and Rs. 2000). This system has high accuracy and low cost as compared to other currency recognition system. For counterfeit detection, image processing is done on the image captured in the presence of ultraviolet and white light against the currency. This makes the hidden attributes (visible under UV light) of the currency visible. For currency identification, the image fed into the algorithm undergoes various image processing and feature extraction. The extracted features are further compared using a trained neural network. The neural networks were trained by the dataset of original currency with denominations Rs. 200, Rs. 500, and Rs. 2000.

## 2 Image Processing

Image processing is a simulation process that evolves one or more images input into an image output. There are many types of image processing steps such as image histograms, monadic operations, dyadic operations, spatial operations, template matching, nonlinear operations, and shape changing [8]. The present study uses two types of operation, monadic and spatial, for the image processing in currency recognition.

A monadic operation can transform a color picture into a gray picture where each pixel yields a scalar value describing the mid-tones of the respective pixel input. It helps in reducing the image size in order to reduce the processing time.

Spatial operation (for edge detection) acquires a broad area in image processing such as linear spatial filtering, template matching, and nonlinear operations. In the present study, the identification of the edge of features present in the currency is detected using edge detection technique, linear spatial filtering. The two-dimensional edge detection technique operates with the help of three masks—Robert operator, Prewitt operator, and Sobel operator [8].

## 3 Artificial Neural Network

The artificial neural network (ANN) is a scheme that processes data and is considered as a generalization of the human brain's mathematical model. The neural network comprises an input layer, a concealed (hidden) layer, and an output layer. The number of neurons in the hidden layer is determined on the grounds of trial and error [9].

## 4 Methodology

To structure the total framework in an appropriate manner, a graphical user interface (GUI) is utilized in MATLAB. A GUI provides the user with a recognizable environment to work. The procedure followed for the recognition of original currency is depicted in Fig. 1. In this methodology, the execution is finished for denominations of Rs. 200, Rs. 500, and Rs. 2000. The validation is performed on these three denominations only.

### *4.1 Image Acquisition*

In order to obtain an image of the paper-based Indian currency, a model depicted in Fig. 2 is developed as per the required environment. The model depicted in Fig. 2 has been developed using the integration of various devices as specified in Table 1.

The camera is placed 14 cm above the screen or the currency to capture a focused image (2.2 focus aperture) with clear visibility of each currency features.

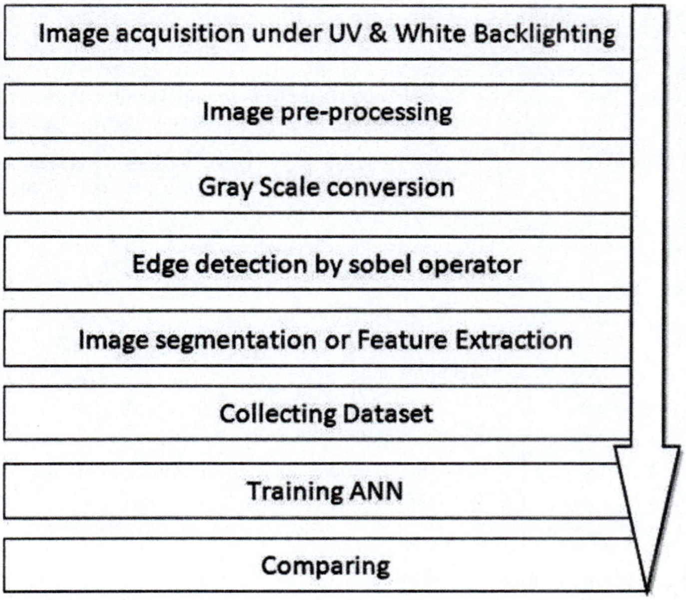


**Fig. 1** Steps followed in the present study for currency recognition

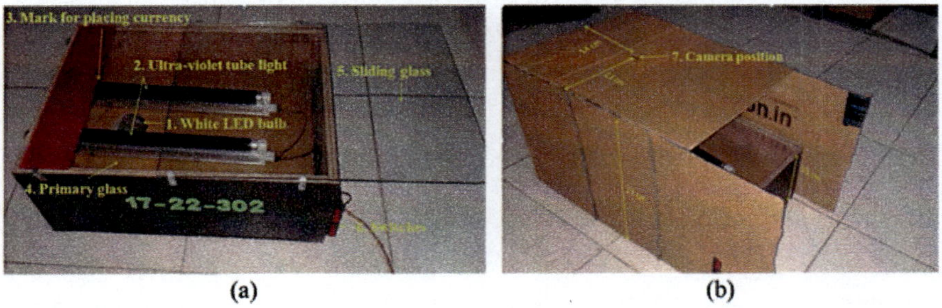


**Fig. 2** **a** Model setup to capture the image and **b** model setup fixed inside a box

## *4.2 Image Preprocessing*

The captured image is too large, and hence, the image is cropped and resized for further processing which leads to reduced processing time. During image preprocessing, three features are extracted from each denomination which are (a) denomination logo, (b) angular bleed lines, and (c) security thread. The features are cropped using the command "imcrop (a, [co-ordinates])".

**Table 1** Devices integrated to develop the model setup

| Device | Purpose |
|---|---|
| White LED bulb | A 5 W LED bulb used in the setup provides white backlighting behind the screen to make the currency attributes visible |
| Ultraviolet tubelight | Two numbers of 8 W UV neon fluorescent lights are fixed behind the screen which also makes the hidden features of the currency visible |
| Primary glass | The primary glass is used as a screen |
| Sliding glass | The sliding glass ensures that the paper currency is unwrinkled and straight |
| Switch | Two switches are used to provide alternate white and UV backlighting |
| 13 MP digital camera (mobile) | It is used to take the image of the paper currency under different lighting conditions |
| Cover | The model placed inside a cover ensures a dark environment for better results |

### 4.3 Grayscale Conversion

The acquired RGB image is converted to grayscale using the command "rgb2gray (a)" to reduce the processing time. For perfection in grayscale image, the threshold value is selected using the command "im2bw (a < (1–277))" such that the image is visible along with the hidden attributes. Further, the complement image is obtained which inverts the pixel value using the command "imcomplemnt (a)".

### 4.4 Edge Detection

During the edge detection process, the edges are recognized from the complement image of the paper currency by incorporating a Sobel operator, the first-order derivative in edge detection. It helps to extract thick and smooth edges from the processed image using the command "Edge (a, 'sobel')".

### 4.5 Image Segmentation or Feature Extraction

Edge-based segmentation is incorporated here to differentiate the region of interest from the background. These segmented images are usually extracted features of the currency referring to a zone with a few amusing geometric or topological properties. These extracted features are used for making the dataset. For each denomination, there are four dataset files referring to four different features. These dataset files are used to train the neural network.

### 4.6 *Training of Artificial Neural Network (ANN)*

In the present study, a multilayer feedforward neural network is formed which is also known as multilayer perceptron (MLP). The MLP consists of three layers namely: input layer, hidden layer, and output layer. The learning procedure of MLP can be summarized in the following three steps:

(i) The patterns of training data are forward propagated through the network using the input layer to generate network output.
(ii) On the basis of network's output, the calculated error is further minimized using the cost function.
(iii) The error is back propagated and its derivative with respect to each weight in the network is determined and the model is updated.

The extracted features of 200, 500, and 2000 rupees were used to build a neural network in order to classify the different currencies. The concerned features of the currency, in order to build the differentiating model, are the security thread and the denomination markings. This study implements a single hidden layer.

## 5 Results and Discussion

In the present study, MATLAB has been used to perform computation. The developed algorithm is tested on the Indian paper currencies of denominations Rs. 200, Rs. 500, and Rs. 2000. The model classifies and recognizes the Indian currencies accurately based on trained datasets.

### 5.1 *Image Acquisition*

The developed model has been implemented to capture two images (one under white backlighting and the other under UV backlighting) for each denomination. Figure 3 depicts the images of Rs. 200 currency note. Similar images were obtained for the other denominations Rs 500 and Rs. 2000.

(a) (b)

**Fig. 3** Rs. 200 currency under **a** white backlight and **b** UV backlight condition

## *5.2 Image Preprocessing*

Based on the region of interest (ROI), the features of the currency note as discussed in Table 2 are cropped based on different lighting conditions to reduce image size. Figure 4 depicts the cropped images of the discussed currency features for the denomination Rs 200. Similar images were obtained for the other denominations as well.

**Table 2** Features of the currency cropped under different lighting conditions

| Sl. No. | Region of interest | Lighting condition |
|---|---|---|
| 1 | Denomination logo | White backlight |
| 3 | Angular bleed lines | |
| 4 | Security thread | Ultraviolet light |

(a) (b) (c)

**Fig. 4** Cropped images of currency features **a** denomination logo, **b** angular bleed lines, and **c** security thread for the denomination Rs. 200

### 5.3 Conversion of an Image to Grayscale and Complement Images

The cropped image is in RGB color format with large file size which is converted to grayscale by setting a threshold value. Figure 5 and 6 depicts the converted threshold grayscale image and the complement image respectively, for the denomination Rs. 200. Similar images were obtained for the other denominations also.

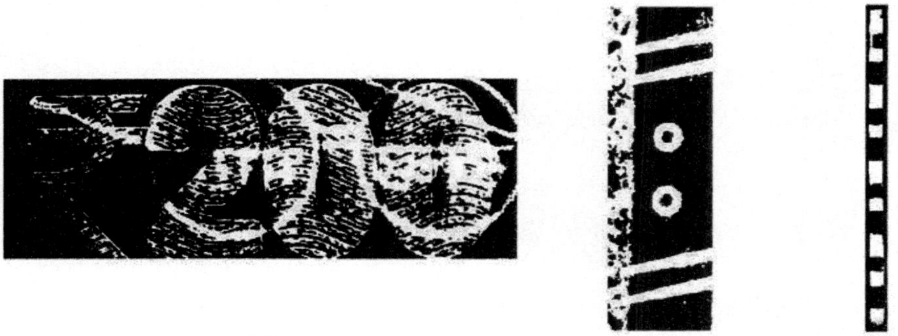

(a) (b) (c)

**Fig. 5** Grayscale images of currency features **a** denomination logo, **b** angular bleed lines, and **c** security thread for the denomination Rs. 200

(a) (b) (c)

**Fig. 6** Complement images of currency features **a** denomination logo, **b** angular bleed lines, and **c** security thread for the denomination Rs. 200

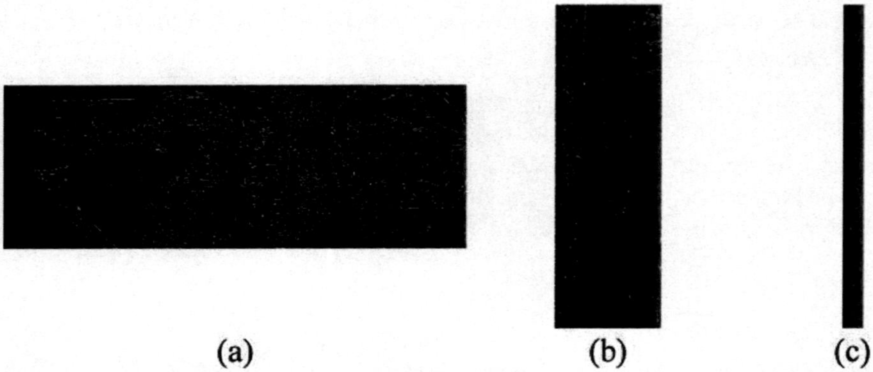


**Fig. 7** Sobel operated images of currency features **a** denomination logo, **b** angular bleed lines, and **c** security thread for the denomination Rs. 200

## 5.4 *Edge Detection Using Sobel Operator*

Edge detection for the currency features has been performed on the obtained complement images by the Sobel operator based on gradient method. The edge detection technique obtains the slope at the edges of the features, as depicted in Fig. 7, and is completely visible.

## 5.5 *Training ANN from Extracted Features*

In the previously discussed steps, the converted images of the currency features are obtained based on the image processing technique. These images serve as input to the ANN. For each feature, a set of matrices is present whose rows and columns depends on the number of pixels in the image.

In the present study, a three-layered feedforward back propagation neural network is implemented for training. ANN is applied in the system for two purposes: (i) to detect the counterfeit currency and (ii) recognition of currency denomination. Hence, the system requires two separate networks for the said purposes.

### 5.5.1 Network to Detect Counterfeit Currency

The UV light detects the counterfeit currency. Presence of UV light makes the hidden attributes ("RBI" and 'भारत' present on the security thread) complete visible as a green small text over the security thread in the original currency as depicted in Fig. 8. Whereas the counterfeit currency shows only a black or blue thread.

(a) (b)

**Fig. 8** Security thread of **a** original and **b** counterfeit currencies

### 5.5.2 Network to Recognize Currency Denomination

The extracted features of the angular bleed lines are used to recognize denomination value as it easily differentiates the different valued currencies. The number of line depends on the value of currency such as four angular bleed lines for Rs 200, five angular bleed lines for Rs 500, and seven angular bleed lines for Rs 2000. Its complement image makes the recognition easy as it results in a matrix with element values as 0 or 1.

In the network, we set three inputs for Rs. 200, Rs. 500, and Rs. 2000 to train based on one output of a particular denomination. Three-layer neurons are used as an input layer, output layer, and a hidden layer. Figure 9 depicts the regression result of the trained neural network. The regression clearly states that the target data fit perfectly and the obtained neural network model effectively predicts the paper currencies.

## 6 Conclusion

In the present study, a currency recognition model has been developed based on feature extraction and ANN. The developed model detects the counterfeit currency based on the hidden currency features visible under UV light. The ANN algorithm is trained using twenty images of each currency feature for the denominations Rs. 200, Rs. 500, and Rs. 2000. The algorithm when tested resulted in 100% accuracy and more efficient as compared to previous research.

The automated version of the model when implemented will give rise to a new era in the currency recognition system. It can further be implemented for the identification of objects present with holograms.

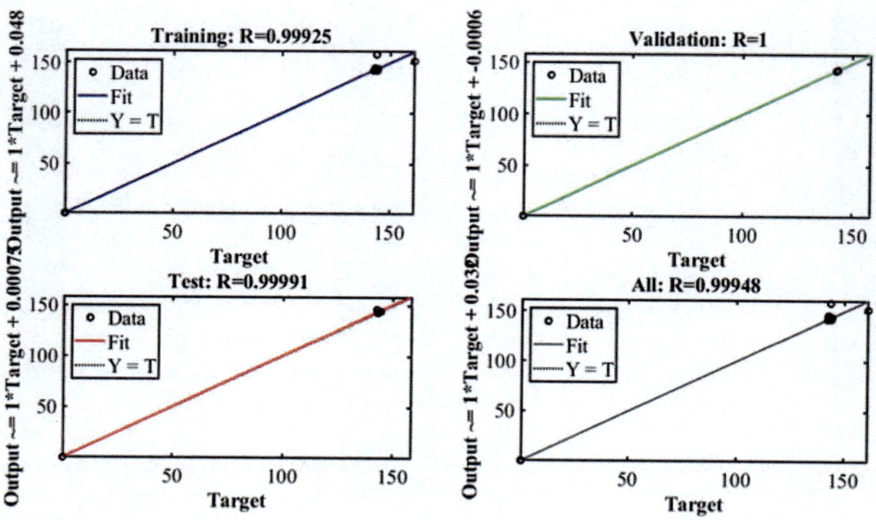


**Fig. 9** Data fitting through regression of neural network model

## References


1. J. Lee, H. Kim, New recognition algorithm for various kinds of Euro banknotes. in *Proceeding of the IEEE 29th Annual Conference of Industrial Electronics Society* (2003), pp. 2266–2270
2. K. Debnath, A. Ahmed, M. Shahjahan, A paper currency recognition system using negatively correlated neural network ensemble. J. Multimedia **5**(6), 560–567 (2010)
3. B. Sharma, A. Kaur, Vipan, Recognition of Indian Paper Currency based on LBP. Int. J. Comput. Appl. **59**(1), 24–27 (2012)
4. A. Sargano, M. Sarfraz, N. Haq, Robust features and paper currency recognition system, in *The 6th International Conference on Information Technology (ICIT 2013)* (2013)
5. R. Vijay, V. Jain, Indian currency denomination identification using image processing technique. Int. J. Comput. Sci. Inf. Technol. **4**(1), 126–128 (2013)
6. A. Khashman, B. Sekeroglu, K. Dimililer, Coin identification using neural networks. in*Proceedings of the 5th WSEAS International Conference on Signal Processing (SIP'06)* (Istanbul, Turkey, 2006) pp. 27–29.
7. I. Toytman, J. Thambidurai, Banknote Recognition on Android Platform (2011)
8. P. Corke, *Robotics*, 2nd edn. (Vision and Control, Springer International Publishing, AG, 2017)
9. S. Modi, B. Bawa, Automated coin recognition system using ANN. Int. J. Comput. Appl. **26**(4) (2011)

# Design and Fabrication of Android Application-Based Grass Cutter Robotic System

**Rutupurna Choudhury, Deep Singh, Anuj Kumar, and Yogesh Singh**

**Abstract** The study presents the design and fabrication of the Android application-based robotic grass cutter. A smartphone is used as remote, and it is able to control the movement of four omni wheels to perform grass cutting operation with the smooth movement of the cutter installed on the system. This apparatus composed of five DC motors, two motor drivers, an Arduino board, a Bluetooth module and a battery. The DC motor aligned vertically, compounded with a blade, cuts the rough grass as the robot moves in a particular direction. Remaining four DC motors are connected to the four wheels of the robotic grass cutter vehicle providing three planar motion, i.e., motion along $x$-axis, $y$-axis and rotation about $z$-axis. The direction of rotation of these motors can be controlled by the Arduino board. The battery supplies power to the motors. Bluetooth module is used as a transmitter as well as a receiver to follow up the input/output signals from the Android phone to the Arduino board controller. As the mobile remote sends input data to the Arduino, the motors are driven accordingly by the motor drivers. Thus, the remote is able to move the grass cutter in all the directions in the plane like forward, backward, diagonal motion and can also turn around. The mobile-operated grass cutter presented here is a very convenient robotic device which is simple in assembly and construction. It is employed to maintain and conserve the lawns in gardens, schools and colleges etc. Design is intended to be simple and effective thereby making grass cutting inexpensive.

**Keywords** Grass cutter · Omni wheels · Arduino board · Servo motors

---

R. Choudhury · D. Singh · A. Kumar · Y. Singh (✉)
Department of Mechanical Engineering, National Institute of Technology Silchar, Silchar, India
e-mail: yogeshsingh15@gmail.com

R. Choudhury
e-mail: rutupurna.choudhury10@gmail.com

D. Singh
e-mail: deepsingh.0318@gmail.com

A. Kumar
e-mail: anujpal096@gmail.com


K. M. Pandey et al. (eds.), *Recent Advances in Mechanical Engineering*, Lecture Notes
in Mechanical Engineering, https://doi.org/10.1007/978-981-15-7711-6_29

# 1 Introduction

Grass cutting machines are highly popularized today due to the wide application in the agriculture fields, sports complex, schools, public centers, industries, and hotels. In the past few years, several advanced manual grass cutting machines are developed. The past grass cutting technology consisted of physically operated hand devices like scissors, which required more human effort and time to accomplish the desired work. Also, in such methods, a lack of uniformity of the residual grass is often seen. The use of engine-powered cutting machines increases the air and noise pollution and such machines also require high maintenance. The traditional grass cutting mowers thus produces a variety of spiteful effects on humans as well as on the society. This paper presents the mobile-operated automated robotic grass cutting machine which is capable to resolve the above-said demerits. The prime advantage of this robotic device is that it can be operated by a smartphone. In the world of smartphone technology, the use of an Android-based grass cutter will always be more preferred over the conventionally used models. This is more flexible in use. Edwin Beard [1] first introduced the concept of grass cutting machine, which consists of a gear mechanism in association with a large roller and cutting cylinder for cutting the grass. This is the prime step toward the innovation of grass cutters. In recent years, various research were performed on the robotic grass cutters. Maity et al. [2] developed an Android application-based Bluetooth that controlled the automated vehicle. The whole vehicle is controlled by an Android application. This device composed of the Bluetooth module (HC-05) and Arduino Uno (ATMEGA 328P) for smooth control. A detailed review of the mobile-controlled robots has been examined using audio channels of the cell phone for the closed loop control systems and presented by Pahuja et al. [3]. Rahul et al. [4] investigated the Android-controlled Bluetooth robot in association with the 8051 controller. It was shown that robots and smartphones are the perfect matches for advanced robots. Eshita et al. [5] presented an Android application-controlled robot imparts by means of Bluetooth-to-Bluetooth module available on the robotic device. A smartphone is used as a remote to the robot. Embedded C language is used to achieve the objective. Jain et al. [6] presented the autonomous robotic grass cutter which can perform with minimum effort. The key advantage of this design requires no perimeter wires for the motion of the robot within the grass. Solar energy is used as a power source. Hariya et al. [7] developed a solar energy-based autonomous grass cutter. It is very simple in construction and ease in use. The battery is used to reserve the energy directly from sunlight. Atmega8 microcontroller operates the device by driving the motors connected to the wheels. Syeda Asra et al. [8] developed an automated renewable energy-based robotic device for the guidance and obstacle detection application. This device consists of an ultrasonic sensor for obstacle detection, DE MCU for networking, a cutter to cut the grass and DC motor to drive the robot. Rao et al. [9] developed a solar lawnmower monitoring mechanism which can be controlled by computer and Bluetooth device. The real-time monitoring system is achieved by the LabVIEW software. LabVIEW software is used for the current coverage of the grounds along with the magnitude of voltage developed in the photovoltaic cell. More et al. [10] developed an automatic and portable grass cutting machine. This is controlled by a microcontroller 8051 and work on solar energy. The speed and direction of the device are controlled by a Bluetooth module device which made the device fully autonomous. Kumar et al. [11] developed a grass cutting device with 12 V batteries to provide motion to the vehicle as well as the grass cutting motor. An attached solar panel charges the battery. Microcontroller 8051 is used to control the motion and an IR sensor is used for object detection. The key advantage of this device is that it can detect the object and change the direction of motion. This paper presents the Android application-based robotic grass cutter. A smartphone is used as remote, and it is able to control the movement of four omni wheels to perform grass cutting operation with smooth movement. This apparatus composed of five DC motors, two motor drivers, an Arduino board, a Bluetooth module, and a battery. A DC motor located vertically is compounded with a blade to cut the rough grass with the motion of the robot. Remaining four DC motors are connected to the four wheels of the grass cutting machine which rotates 360°. The direction of rotation of these motors can be controlled by the Arduino. The attached battery supplies power to the motors. The Bluetooth module is used as a transmitter as well as a receiver to follow up the input/output signals from the Android phone to the Arduino. As the mobile remote sends input data to the Arduino, the motors are driven accordingly by the motor drivers. The novelty of this work is that it can move in omnidirectional motion. The Bluetooth module controls the grass cutter system. The ultrasonic sensor is used to protect the cutting blades from the external obstacles. The linear mechanism is used to adjust the cutting operation. At last, this design is simple and reliable in structure. The remaining topics of the paper are described in the following manner. The objectives of the robotic device are explained in Sect. 2. Section 3 explains the methodology and terminology of the robotic setup. Section 4 comments on the outputs or the results obtained in the research work. At last in Sect. 5, the conclusion of the present work is presented.

## 2 Contribution of the Present Paper

The sole objective of this paper is to develop an Android-based autonomous robotic grass cutter which requires minimum human effort. It is designed to move in all possible directions. Omni wheels are used to obtain the multidirectional movement. In this device, the cutter can be controlled by the Bluetooth module device which becomes more flexible in cutting operation. In the literature, the grass cutters already in place do not have sensors to detect objects in front and avoid unwanted accidents and damage to the cutter model. This device used the ultrasonic sensor to detect nearby objects and avoid them. This protects the grass cutter body and blades from damage. The grass cutters which are in use and the models which have been proposed cannot trim the grass according to the desired height. This device use a linear actuator mechanism to achieve such a purpose of the adjustable cutting operation. This is the most important feature of the grass cutter. The device is simpler as compared to the existing literature.

## 3 Development of the Android-Based Grass Cutter

Grass cutter is a key machine for the maintenance of the yards. This machine can be designed on the basis of various parameters like an energy source, size, mode of control, and mode of operation, size, material and human effort. The proposed device can work with the help of the battery. The novelty of the proposed system is that it can be controlled by an Android application-based smartphone. Bluetooth module device is used to control the height of the grass as per the user requirement. This device is composed of five DC motors, two motor drivers, an Arduino controller board, a Bluetooth module and a battery. The Android application control is achieved with the help of Arduino Uno board. The grass cutter that has been developed involves the fabrication and assembly of all its components and subsequently installing commands into the microprocessor board, i.e., the Arduino Uno to run the robotic system. The primary components of the device are Arduino board, omni wheels, Bluetooth device, DC gear motors, motor drivers, lead screw for the linear actuator and a cutter blade.

### 3.1 *Arduino Uno (Atmega328)*

Arduino Uno is a prime device for the control scheme. It is a microcontroller board. It consists of 14 digital i/o pins out of which 6 can be employed as PWM outputs, a USB connection, 6 analog inputs, a 16 MHz ceramic resonator, an ICSP header, a power jack and a reset button.

The physical diagram of the Arduino Uno is shown in Fig. 1. It contains all the

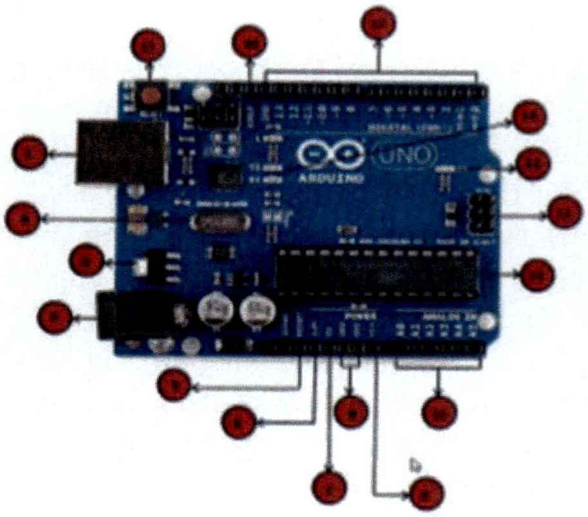

**Fig. 1** Schematic diagram of the Arduino Uno

**Table 1** Different pins of Android Uno microcontroller board

| Pin No. | Purpose |
|---|---|
| Pin-1 | Power button |
| Pin-2 | Power barrel jack |
| Pin-3 | Voltage regulator |
| Pin-4 | Crystal oscillator |
| Pin-5, 17 | Arduino reset pin |
| Pin-6, 7, 8, 9 | Pins (3.3, 5, GND, Vin) |
| Pin-10 | Analog pin |
| Pin-11 | Microcontroller |
| Pin-12 | ICSP pin |
| Pin-13 | Power LED indicator |
| Pin-14 | TX and RX LEDs |
| Pin-15 | Digital I/O |
| Pin-16 | AREF |

useful supports to the microcontroller. It can be connected to any computer with the help of USB cable and can be powered with the help of an AC-to-DC adapter. The detailed components of the Arduino are presented in Table 1.

## 3.2 *Omni Wheel*

Omnidirectional wheels are uniquely designed that can roll in two directions. There are wheels present on the circumference of the wheel which helps for the lateral motion. The prime advantage of the omni wheel is that it can convert the robotic device from non-holonomic to holonomic. A non-holonomic robot is a device that can move only forward, backward, and rotation motion. The lateral movement is the demerit of this robot. The proposed omni wheel has overcome this problem by allowing the lateral motion. The holonomic omnidirectional robot has the ability to travel in any arbitrary direction continuously without any alteration to the direction of the wheels. It can move backward, forward, lateral and can turn in the same direction (Fig. 2).

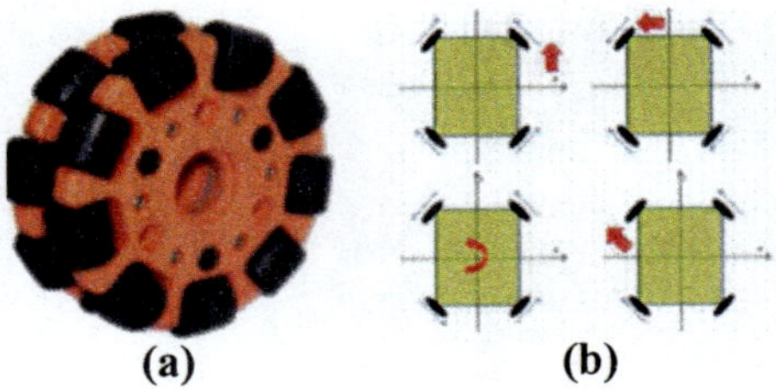


**Fig. 2** **a** Schematic diagram of the omni wheel and **b** working principle of omni wheel

(a) (b)

**Fig. 3** **a** HC-05 Bluetooth module and **b** pin diagram of HC-05 module

## 3.3 Bluetooth Module

The Bluetooth module, HC-05, controls the motion of the device. It can be connected to the device by a smartphone. HC-05 module is simple and reliable to use Bluetooth serial port protocol (SPP) module for wireless connection. The module is advanced and can be employed for master configuration, a great solution for wireless connections. The HC-05 serial Bluetooth device has V2.0 + EDR (Enhanced data rate), 3 MBPS modulation and 2.4 GHz radio baseband. The schematic diagram is illustrated in Fig. 3.

## 3.4 Geared Motor

A geared motor is a definite type of electrical device that is planned to generate high torque while maintaining low speed, or, low horsepower motor output. Here, DC motor (18,000 rpm) and operating voltage of 6–18 V are used. The schematic diagram is presented in Fig. 4a.

A DC motor driver 20 A of model RKI1341 is used in this device. This device is designed for a specific application where two motors need up to 20 amperes of current during starting and during the normal operating condition. It is designed by considering TTL/CMOS-based interface which helps to directly connect the IOs of an MCU. The schematic diagram is presented in Fig. 4b. It has a prime advantage that most high-power applications can ensure an immediate halt to the shaft of motors and also has safety circuitry to prevent any electrical deviations affecting the general operation of an MCU.

(a) (b)

**Fig. 4** **a** DC gear motor and **b** DC motor driver RKI-1341

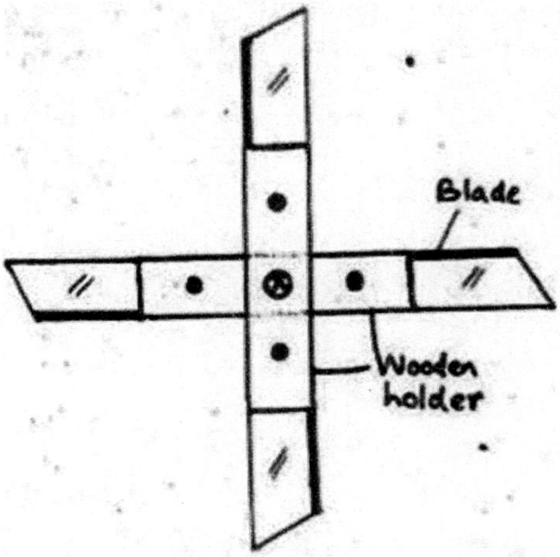


**Fig. 5** Geometric diagram of the cutting blade

### *3.5 Cutting Blade*

A steel sheet has been hammered followed by grinding operation to develop the cutter blade for the prototype model. Grinding is done to provide sharp edges on all sides of the cutter blade. This ensures an easy and efficient cutting operation of grass. The schematic diagram is presented in Fig. 5.

The experimental setup for the grass cutter robotic system is presented in Fig. 6. It is the assembly of all the components discussed in the previous section.

## 4 Results and Discussion

The present study proposes an Android application-based robotic grass cutter. This device is fabricated considering the ease availability of components. This is capable to overcome the previous problems appeared in the literature. The wheels have been assembled in a cross shape for the better movement of the grass cutter, which is clearly seen when operated in Fig. 6. The rotation of wheels is presented in Fig. 7 for motion in different directions. Movement of the wheels can be operated in the Android platform by the following chart.

The Android application runs the grass cutting robotic system according to the directional buttons present, as shown in Fig. 8. The Bluetooth module is connected to a smartphone using an open-source Android application from the Internet.

The Android application used consists of the following directional symbols which are represented by string functions, such as, for forward movement string function '*f*', backward string '*b*', left '*l*', and right '*r*'. The above string functions are used while programming to the Arduino board. The movement and direction of rotation of the motors are controlled in this manner.

**Fig. 6** Experimental setup of grass cutter robotic system

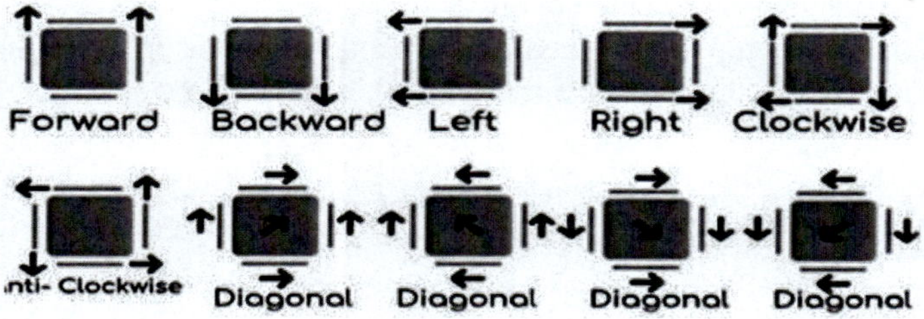


**Fig. 7** Different direction and motion of the system

**Fig. 8** Android application for grass cutting robotic system

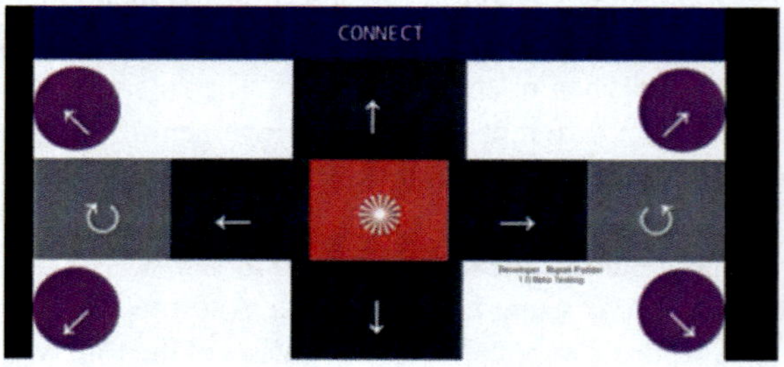

This enables the robot to be controlled from a comfortable distance of 1–2 m using a smartphone. The integration of the ultrasonic sensors, Bluetooth module and linear actuators makes the device autonomous to be controlled by smartphone. The various devices used for the development of the model along with individual prices have been listed in Table 2. The overall price of the developed model is INR 15,000. Adequate space has been provided on the base metal board to install the cutter for the primary operation. The motor of the cutter is controlled via the same Android application. The height adjustment mechanism is welded at the front of the vehicle which works on the principle of linear actuator, i.e., the rotary motion is converted to linear motion with the help of lead screw and nut as shown in Fig. 9. The lead screw

**Table 2** Devices incorporated for the development of grass cutting model

| Sl. No. | Devices | Quantity (Nos.) | Cost (INR) |
|---|---|---|---|
| 1 | Omni wheels | 4 | 10,000.00 |
| 2 | Arduino Uno | 1 | 850.00 |
| 3 | Bluetooth module | 1 | 250.00 |
| 4 | DC Gear motor | 5 | 1,500.00 |
| 5 | Motor driver board | 1 | 700.00 |
| 6 | Lead screw | 1 | 650.00 |
| 7 | Battery | 1 | 470.00 |
| 8 | Miscellaneous (including fabrication) | - | 580.00 |
| Total cost (INR) | | | 15,000.00 |

**Fig. 9** Height adjustable cutter setup

is attached with the motor through coupler which resulted in the height adjusting chamber. The net weight of the device was 5 kg.

## 5 Conclusions

The proposed Android application-based robot control framework makes a huge commitment to the field of the versatility of robot application, effectively incorporating and using a conventional Arduino board-based robot configuration to develop a model that would help the general population to cut grass/mow land from a comfortable location. While thinking about the quick progression of brilliant gadgets with different implanted sensors, it is decided to implement this technology to trim grass without the necessity of any kind of skill for this job. The use of such a grass cutter robot greatly decreases the input effort required for the intended operation. The maintenance and working costs are minimal and so is the pollution generated by this robot. Such a household gadget can thus be fabricated without much fuss. Moreover, the overall cost of such a device is much less compared to the other existing grass cutter robotic systems. Usually, the minimum market price of robotic lawnmower is INR 50,000 but the proposed system cost is INR 15,000.

The proposed grass cutter model can move in all the directions due to the presence of omnidirectional wheels. The presence of Bluetooth module enables it to be controlled by smartphone. The model avoids obstacles due to the presence of the ultrasonic sensor. Also, the presence of a linear actuator mechanism helps to cut the grass at any desired height.

## References

1. A. Shukitis, *History of Lawn Mowers* (2012)
2. A. Maity, A. Paul, P. Goswami, A. Bhattacharya, Android application based bluetooth controlled robotic car. Int. J. Intell. Inform. Syst. **6**, 62–66 (2017)
3. R. Pahuja, N. Kumar, Android mobile phone controlled bluetooth robot using 8051 microcontroller. Int. J. Sci. Eng. Res. (IJSER) **2**, 2347–3878 (2014)
4. R. Kumar, P. Ushapreethi, P.R. Kubade, H.B. Kulkarni, Android phone controlled bluetooth robot. Int. J. Sci. Eng. Res. (IJSER) **3**(4), 2395–2472 (2016)
5. R.Z. Eshita, T. Barua, A. Barua, A.M. Dip, Bluetooth based android controlled robot. American J. Eng. Res. (AJER) **5**, 195–199 (2016)
6. V. Jain, S. Patil, P. Bagane, Solar based wireless grass cutter. Int. J. Sci. Technol. Eng. **2** (2016)
7. A. Hariya, A. Kadachha, D. Dethaliya, Fully automated solar grass cutter. Int. J. Sci. Technol. Eng. **3** (2017).
8. N.S. Asra, Automated grass cutter robot based on IOT. Int. J. Trend Sci. Res. Develop. **2** (2018)
9. B.H. Rao, K. Varma, S. Jainwar, H. Tamrakar, Bluetooth cum PC controlled solar lawn mower with real time monitoring mechanism. Int. Res. J. Eng. Technol. **3** (2016)
10. A.D. More, S.N. More, V.V. Shetty, S.V. Patil, A portable and automatic weed cutter device, IOSR J. Electri. Electron. Eng. 14–17 (2017)

11. S. Kumar, A. Sharma, R. Sharma, S. Kesarwani, A review paper on grass cutter device using bluetooth. J. Electron. Electromag. Technol. **1**, 01–03 (2018)

# Review on Various Coating Techniques to Improve Boiling Heat Transfer

**Amatya Bharadwaj and Rahul Dev Misra**

**Abstract** Boiling has got prominence in the recent decades for its effectiveness in cooling of micro-electronic devices due to its superior heat extraction ability as compared to air or single-phase liquid cooling. Numerous works have been published regarding augmentation of boiling heat transfer by developing modified surfaces. Micro- and nano-surfaces have been developed for this purpose. These surfaces are engineered either by surface coating or by micro-machining. The present review attempts to elaborate the various coating techniques and methods that have been used to fabricate surfaces to improve pool and flow boiling heat transfer. The experimental studies have been primarily focused in this paper. The results obtained using the modified surfaces and the mechanisms responsible for them have been discussed.

**Keywords** Critical heat flux (CHF) · Heat transfer coefficient (HTC) · Boiling incipient superheat

## 1 Introduction

The miniaturization of electronic components has led to the advent of micro-sized chips. These micro-chips have to satisfy the requirement of very high heat flux dissipation of the order of $10^3$–$10^5$ W/cm$^2$ [1], which could not be adequately provided by air cooling or by single-phase liquid cooling. Due to this, two-phase cooling mechanisms such as boiling heat transfer has been explored. Two-phase cooling is also used in fusion reactors, rocket engines, hybrid electronic vehicles, and various defense sectors [2, 3]. Boiling has a superior heat transfer coefficient compared to the natural and single-phase forced convection as it utilizes the sensible heat as well as the latent heat of the coolant, to extract heat.

A. Bharadwaj (✉) · R. D. Misra
Department of Mechanical Engineering, National Institute of Technology, Silchar, India
e-mail: amatyabharadwaj@gmail.com

R. D. Misra
e-mail: rdmisra@gmail.com



K. M. Pandey et al. (eds.), *Recent Advances in Mechanical Engineering*, Lecture Notes in Mechanical Engineering, https://doi.org/10.1007/978-981-15-7711-6_30

The aim of electronic cooling is to extract the requisite heat from the device while simultaneously maintaining the device temperature within permissible limits. This has led to exploring avenues for surface modifications to improve boiling performance. Recent developments in surface modification techniques have facilitated the manufacture of micro/nanostructures (having sizes ranging from hundreds of nanometers to several micrometers) on boiling surfaces. This has expedited the enhancement of heat transfer coefficient and CHF. Numerous reviews have been published in this regard [4–7]. A variety of working fluids have been used, for boiling, ranging from water and nanofluids to refrigerants and dielectric fluids. A comprehensive review has been presented by Leong et al. [8] on pool boiling and flow boiling by using dielectric fluids on modified surfaces and also highlighted the fabrication techniques used.

Numerous mathematical correlations have been proposed by various researchers to predict boiling characteristics. Rohsenow [9] developed a correlation, shown in [Eq. (1)], which is widely used in pool boiling analysis, by considering bubble departure diameter as the characteristic length and bubble departure velocity as the characteristic velocity.

$$\frac{c_p(T_w - T_s)}{h_{fg}\,\mathrm{Pr}^n} = C_{sf}\left[\frac{q}{\mu_l h_{fg}}\left(\frac{\sigma}{g(\rho_l - \rho_g)}\right)^{1/2}\right]^n \tag{1}$$

Kandlikar [10] proposed a theoretical model [Eq. (2)] to show that CHF is arrived at when the evaporation momentum force causing the bubble to leave the interface exceeds the sum total of the surface tension and gravitational forces, keeping the bubble in place. The effect of contact angle and surface orientation was also taken into account while formulating the model.

$$q_c'' = h_{fg}\rho_g^{1/2}\left(\frac{1+\cos\beta}{16}\right)\left[\frac{2}{\pi} + \frac{\pi}{4}(1+\cos\beta)\cos\varnothing\right]^{1/2} \times \left[\sigma g(\rho_l - \rho_g)\right]^{1/4} \tag{2}$$

The factors causing boiling enhancement by surface modifications are improved wettability, porosity, higher surface area available for boiling, and higher nucleation site density. The present study emphasizes on various surface modification techniques that have been adopted for developing surfaces showing enhanced boiling characteristics. Both pool and flow boiling have been discussed in this review with regard to the surface modification techniques. The mechanisms responsible for the enhancement have been discussed from the prism of contact angle, surface morphology, roughness, and the surface area enhancement factor.

## 2 Pool Boiling

The process of pool boiling can be described through a typical boiling curve as shown in Fig. 1. The boiling phenomenon is sub-divided into four phases—free convection followed by nucleate boiling, transition boiling, and film boiling. The system, however, functions best in the nucleate boiling regime, because it facilitates maximum heat removal due to the dominance of bubble formation and detachment processes. However, the CHF phenomenon, caused by bubble coalescence, limits the heat transfer process. Beyond CHF, there is a rapid transition phase of temperature rise leading up to film boiling which could cause severe burnout of the device.

The surface modification modifies the surface properties such as roughness, wettability, porosity, and surface area such that CHF is enhanced and surface superheat is reduced.

There has been number of experimental researches to explain roughness effects on pool boiling [11–13]. These studies have observed increment in heat transfer and CHF with increase in surface roughness and have ascribed this trend to the increased active nucleation sites and capillary wicking. Experimental studies [14–16] have established that heat transfer is improved on hydrophilic surfaces having hydrophobic spots compared to pure hydrophilic surfaces.

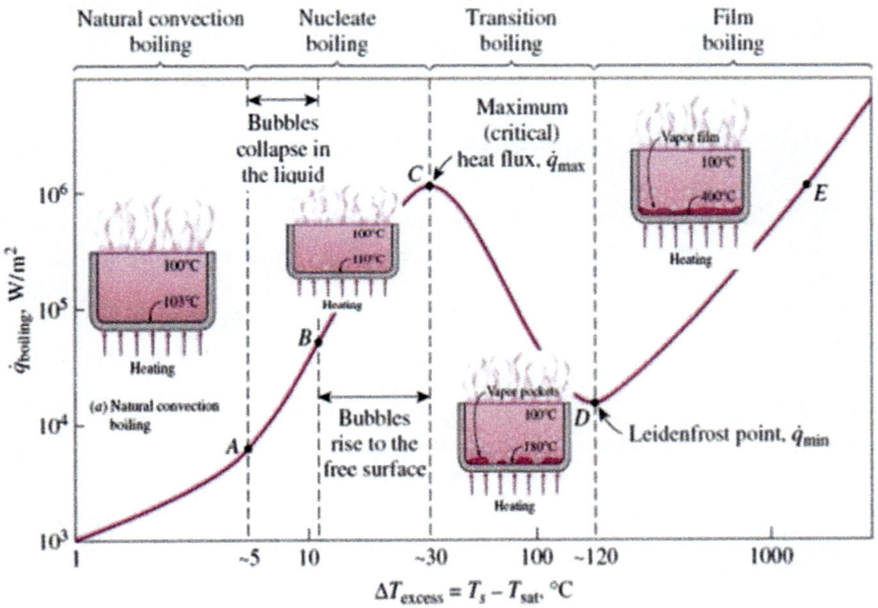


**Fig. 1** Typical boiling curve

## 3 Flow Boiling

Flow boiling involves two modes, namely forced convection-dominated boiling and nucleate boiling. In this case, the liquid mass flux and vapor quality are important factors in determining the rate of heat transfer. Though flow boiling delivers a high rate of heat removal, a key issue associated with it is the pressure drop that occurs in the direction of flow. At high heat fluxes ranges, the drop in pressure occurred across the channel is very high. This leads to high pressure in the upstream region of the channel which causes vapor backflow preventing any further liquid from entering the channel. This causes a condition of momentary dry out within the channel, thus leading to CHF. This continues until the inlet liquid pressure is able to overcome the drop in pressure across the channel. Kim and Mudawar [17] presented a consolidated database and developed universal correlations for predicting pressure drop in single- and multi-channelled flow for a wide variety of fluids and operating conditions.

Further, Kandlikar [18] suggested a general correlation to predict heat transfer coefficients for flow boiling in horizontal and vertical tubes. This study reviewed 5246 data points with ten fluids and also introduced a fluid-dependent parameter $F_{fl}$ that could be used to apply this correlation to other fluids as well. Kandlikar [19] also performed a scaling analysis of the effects of inertia, shear, evaporation, surface tension, and gravitational forces on heat transfer in micro-channels. It also brought to light the similarity in heat transfer mechanism between an evaporating bubble in nucleate boiling and vapor slug in flow boiling.

## 4 Surface Coating Techniques

Surface modification techniques are of two types—one is surface coating in which layers are coated on a substrate and the other involves machining of the boiling surface to generate intrinsic features on it. This paper discusses the surface coating techniques used to develop boiling surfaces and the associated results yielded.

For surface coating, powdered coatings of same or different composition are deposited on a substrate material through appropriate techniques. The boiling surfaces include copper, aluminum, silicon, steel, etc. Micro/nanoparticles of copper, aluminum, titanium, and zinc have been widely used as coating materials. Copper micro/nanoparticles have found a major use in electrochemical deposition and sintering processes because of its superior heat transfer properties. Other coating materials include oxides or nanocomposites of these metallic powders.

A number of deposition techniques have been adopted to modify boiling surfaces. The techniques discussed in this paper include:

- Electrochemical deposition
- Spray coating
- Chemical vapor deposition

- Particle sintering
- Dipping and dripping

## 4.1 Electrochemical Deposition

Electrochemical deposition is a widely used process in the field of surface modification. In this method, powdered form of metallic powder is deposited by mixing it with an electrolyte. This technique has been adopted by El-Genk and Ali [32] to deposit copper micro-porous layers and by Gupta and Misra [20] to form Cu-$Al_2O_3$ nano-composite coating on copper surface. Hydrophilic CuNW was fabricated on Si substrates by Yao et al. [21] using this method. Two-stage deposition process has been adopted by Patil et al. [22]and Gupta and Misra [23] to promote stability of the layer formed.

In flow boiling research, Gupta and Misra [24] formed four different surfaces by coating Cu-$TiO_2$ nano-composite on copper minichannels of 1.5 mm height through a two-step electrodeposition process. Deionized water was flown at different mass fluxes. This study obtained HTC increment with increase in heat flux in two-phase zone, but it remained unaffected in single-phase zone. CHF enhancement upto 92% & HTC enhancement upto 94% have been obtained. Morshed et al. [25] developed a surface having Cu-$Al_2O_3$ nano-composite coating on Cu microchannel. The surface grew more hydrophilic with contact angle reduction from 97° to 67°. They conducted flow boiling of sub-cooled water at various mass fluxes. They reported CHF enhancement in the range of ~35–55% and two-phase HTC improvement of ~30–120% due to improved surface wettability & capillarity. Higher wettability, porosity, roughness, and thickness seemed to enhance heat transfer.

## 4.2 Spray Coating

This technique involves accelerating the micro-particles in specialized apparatuses and impacting them on the substrate where they undergo plastic deformation and get adhered to it. Earlier work in this field has been done by Connor et al. [26] where they modified a simulated electronic chip by spraying alumina particles (0.3–3 $\mu$m) and painting the surface with varying thickness of diamond micro-structures. Pool boiling experiments performed on it using FC-72 saw reduction of incipient superheat by 33–55% for alumina and by 63–85% and painted diamond micro-structure were reported. Liu and Yang [27] developed a micro-porous surface on a Cu plate by spraying Al particles, mixed with a binder (OB-200) and a carrier (methyl ethyl ketone). They analyzed the effect of confined spaces in boiling phenomenon with methanol as the boiling fluid and obtained that confined spaces greatly reduce heat transfer performance. Reduced graphene oxides (rGO) flakes were sprayed supersonically over a copper substrate by An et al. [28], and pool boiling was conducted using FC-72.

This surface obtained had superior properties such as better wettability, roughness, repeatability, and permeability which enhance boiling.

### 4.3 Chemical Vapor Deposition (CVD)

Phan et al. [29] analyzed the impact of surface wettability on the nucleate boiling process through deposition of a variety of nanocoating materials on stainless steel (grade-301). They deposited Pt, $Fe_2O_3$, SiOx, SiOC, and Teflon by various chemical vapor deposition processes. The contact angles obtained with water varied from 22° to 112°. The inferences made from the tests are that, for a hydrophobic surface, the bubbles form at a lower superheat, but they do not detach from the surface and coalesce with each other, thus impeding the process of heat transfer. For a hydrophilic surface, the bubble departure diameter increases, and frequency of bubble departure decreases with increasing wettability. Dharmendra et al. [30] applied an intermediate diamond layer between copper substrate and CNTs to improve the adhesion. The process adopted for CNT deposition was hot filament chemical vapor deposition (HFCVD) with $CH_4$ and $H_2$ precursor gas. The CHF increased by 38% and HTC improved by 80% compared to bare Cu. This is suggested to be due to better fin effectiveness, as the axial thermal conductivity of the CNTs is better than the copper substrate, and also intermediate resistance was lower between substrate and coating due to better adhesiveness.

On the other hand for flow boiling, Khanikar et al. [31] fabricated rectangular microchannel of Cu and coated with multi-walled carbon nanotubes (MWCNTs) of 60 nm diameter by CVD process. Deionized water with varying mass fluxes was flown through the channel. They observed that CNT arrays provide more surface area for heat transfer and also serve as high conductivity fins. The percentage of CHF enhancements observed was more at low mass fluxes. High mass velocities, though showed higher CHF, tend to fold the CNTs toward the wall. They conducted another study [32] using similar setup, but the catalyst layer thicknesses were varied. They observed that as the liquid is converted to vapor at high heat fluxes, significant pressure drop occurs. A momentary increase in upstream pressure causes vapor backflow which leads to dryout in the channel and thus CHF occurs. Fish-scale patterns are formed due to bending of CNTs at high flow velocities. The CHF results were repeatable at low mass fluxes, but showed degradation on repetition at high mass fluxes, though enhancement happens initially. The use bending of the CNTs at high heat fluxes causes lack of stability in the boiling surface.

### 4.4 Particle Sintering

Particle sintering improves the porosity of the substrate which provides additional nucleation site cavities. Thiagarajan et al. [33] prepared microporous surfaces on copper by particle sintering to study saturated pool boiling and bubble dynamics with HFE-7100. Copper-rich micro-particles were fused on plain copper surface to prepare micro-porous surfaces of thickness 100, 360, and 700 μm. The boiling incipience temperature reduced and HTC improved by 50–270% and CHF improved by 33–60%. Weibel et al. [34] and Qu et al. [35] developed boiling surfaces by sintering copper powder of various sizes on copper substrate. They concluded that decrease in pore size and diameter increased the density of nucleation sites and thus improved boiling. But, below a certain pore size, the permeability decreased and thus heat transfer. Sarangi et al. [36] compared pool boiling of FC-72 between copper-free particle and sintered particle-coated surfaces and established that the better heat transfer result is shown by the sintered coatings which resulted in a 95% reduction in wall superheat compared to 32% reduction in case of free-particle coating.

Bai et al. [37] fabricated 15 micro-channels of 400 μm in width. These channels were coated with copper powders of 30, 55, and 90 μm in diameter by solid-state sintering to form a porous structure. Boiling was conducted with anhydrous ethanol. The experimentations led them to conclude that 55 μm diameter was the optimum particle size for maximum heat transfer. Porous coating was found to increase pressure drop but mitigates pressure drop fluctuations. Sun et al. [38] sintered Cu particles of size 20, 50, and 120 μm on minichannels of 0.29, 0.93, & 1.26 mm hydraulic diameter to form micro-porous coatings. With dielectric FC-72 as the working fluid, maximum heat flux enhancement was ~20%, and heat transfer enhancement was 7–10 times. Best performance was shown by the 50 μm particle size which is in compliance with results of previous literature showing the medium-sized particle being the optimum size for heat transfer. Deng et al. [39] fabricated 14 re-entrant porous micro-channels (RPM) having hydraulic diameter 786 μm. Spherical Cu powder of particle size 75–110 μm was sintered on a graphite mold, patterned to the Ω-shaped profile, and later demoulded. The coolant used was deionized water at different inlet temperatures and mass fluxes. In addition to the higher nucleation sites provided by the porous structure, the ONB came down to 0.5–2.1 °C, and a 2–5 times augmentation in two-phase heat transfer was recorded as opposed to reentrant copper microchannels. High and moderate sub-cooling caused rapid decline of two-phase heat transfer with increase in heat flux and vapor quality while it undergoes moderate decline with low sub-cooled liquid. Nucleate boiling was identified as the dominant heat transfer mode in the low heat flux regime and forced convection with thin-film evaporation being the dominant mode in moderate-to-high heat flux regime. This was also revealed in another analysis [40] by using anhydrous ethanol. Also, onset of nucleate boiling (ONB) initiates at large heat flux for large inlet sub-cooling, and large inlet sub-cooling causes low pressure drop and high two-phase instability.

## 4.5 Dipping and Dripping

Wu et al. [41] prepared a surface by depositing a droplet of $TiO_2$-ethanol solution on copper and then heating the surface, so that the ethanol evaporates and $TiO_2$

gets bonded to the surface to form a 1 μm thick layer. The $TiO_2$ coating made the surface super-hydrophilic with water and FC-72 showing contact angles of 9° and 0°, respectively. This increased the CHF of water and FC-72 by 50.4% and 38.2%, respectively and also improved the heat transfer coefficient up to 36.2% and 91.2%, respectively. Tang et al. [42, 43] implemented a hot-dip galvanizing and dealloying process to develop a nano-porous metallic surface. They reported maximum HTC improvement of 172.7% and a reduction of wall superheat by 63.3% by boiling deionized water. They also observed that nano-structured surfaces displayed better boiling performances at low heat flux which was corroborated by observation of smaller bubble diameter and higher departure frequency at low heat fluxes. Kim and Kim [44] proposed that the capillary wicking has a major role in enhancing CHF of nano-fluids. Boiling of nanofluids causes nanoparticles to be deposited on the surface. The CHF increased from 950 to 1500 kW/m$^2$ with contact angle coming down from 70° to 20°. But, it further increased to 2500 kW/m$^2$ with the contact angle remaining constant at 20°. This led to the investigation of capillary wicking effect and was observed that capillary wicking increased drastically with concentration variation from $10^{-3}$–$10^{-1}$%. The coatings developed by this process are relatively thin compared to the coatings developed by other processes.

## 5 Conclusion

This review brings into perspective the effects of various techniques used for surface coating. It also focusses on the effect of various parameters on boiling heat transfer, i.e., surface morphology, wettability effect, roughness, porosity, coolant/boiling fluid used, type of powder/coating. Micro/nano-pores improve the heat transfer performance due to increased density of nucleation sites. These pores trap air/vapor within them which serve as bubble nucleation sites. Change in micro-structures due to change in fabrication technique or parameters cause difference in heat transfer performance. Decrease in pore size increases available area for heat transfer, but further decrease cause reduction in permeability and thus reduce heat transfer. Therefore, an optimum pore size for heat transfer exists. Surface parameters such as coating thickness, surface roughness, and porosity increase the CHF and HTC and reduce boiling incipient temperature. Hydrophilic surfaces promote better wetting and increase CHF by preventing/delaying dry out. Liquid spreadabiliy and capillary wicking effects promotes rewetting of the surface after bubble nucleation. On the other hand, in hydrophobic surfaces, the bubbles do not detach and coalesce with each other and thus form a film which inhibits heat transfer.

The study also shows that coatings have been formed of metallic powders such as Al, Cu, Zn; metal oxides such as $Al_2O_3$, $TiO_2$, ZnO; and also nanotubes/nanowires. It has been observed that use of composite coatings (such as Cu-$Al_2O_3$, Cu-$TiO_2$) could improve heat transfer characteristics because of contributions of their individual heat transfer properties. This could prove to be a promising aspect for surface coating material.

The surface coating techniques used so far includes electrochemical deposition, chemical vapor deposition, spray coating, MEMS, dipping, particle sintering, etc. Nano-wires and micro-structures formed by electrochemical deposition are generally hydrophilic which is beneficial for heat transfer. However, the electrolyte used could alter the boiling performance. CNTs developed by CVD improve HTC due to better fin effectiveness and thermal conductivity. But, repeated tests at high mass fluxes cause bending of CNTs which cause poor stability. Moreover, they require sophisticated apparatus and controlled environment such as nitrogen or argon environment which is also the case with Spray coating, sputtering, etc. On the other hand, relatively simpler processes such as dipping generate surfaces having low coating thicknesses. In order to address these issues, an appropriate/alternative technique of surface preparation could be devised that can result in surfaces with improved mechanical properties and higher surface roughness. These could cause improved stability of the boiling surface as well as increased nucleation sites, respectively, thus leading toward improved boiling heat transfer rate.

## References


1. I. Mudawar, Assessment of high-heat-flux thermal management schemes. IEEE Trans. Components Packag. Technol. **24**(2), 122–141 (2001)
2. I. Mudawar, Two-phase microchannel heat sinks: theory, applications, and limitations. J. Electron. Packag. **133**(4), 041002 (2011)
3. I. Mudawar, Recent advances in high-flux, two-phase thermal management. J. Therm. Sci. Eng. Appl. **5**(2), 021012 (2013)
4. D.E. Kim, D.I. Yu, D.W. Jerng, M.H. Kim, H.S. Ahn, Review of boiling heat transfer enhancement on micro/nanostructured surfaces. Exp. Therm. Fluid Sci. **66**, 173–196 (2015)
5. Y.W. Lu, S.G. Kandlikar, Nanoscale surface modification techniques for pool boiling enhancementa—A critical review and future directions. Heat Transf. Eng. **32**(10), 827–842 (2011)
6. M. Shojaeian, A. Koşar, Pool boiling and flow boiling on micro- and nanostructured surfaces. Exp. Therm. Fluid Sci. **63**, 45–73 (2015)
7. C.M. Patil, S.G. Kandlikar, Review of the manufacturing techniques for porous surfaces used in enhanced pool boiling. Heat Transf. Eng. **35**(10), 887–902 (2014)
8. K.C. Leong, J.Y. Ho, K.K. Wong, A critical review of pool and flow boiling heat transfer of dielectric fluids on enhanced surfaces. Appl. Therm. Eng. **112**, 999–1019 (2017)
9. W.M. Rohsenow, A method of correlating heat transfer data for surface boiling of liquids. ASME **74**, 969–976 (1952)
10. S.G. Kandlikar, A theoretical model to predict pool boiling CHF incorporating effects of contact angle and orientation. J. Heat Transfer **123**(6), 1071 (2001)
11. M.G. Kang, Effect of surface roughness on pool boiling heat transfer. Int. J. Heat Mass Transf. **43**(22), 4073–4085 (2000)
12. J. Kim, S. Jun, R. Laksnarain, S.M. You, Effect of surface roughness on pool boiling heat transfer at a heated surface having moderate wettability. Int. J. Heat Mass Transf. **101**, 992–1002 (2016)
13. B.J. Jones, J.P. McHale, S.V. Garimella, The influence of surface roughness on nucleate pool boiling heat transfer. J. Heat Transfer **131**(12), 121009 (2009)
14. H. Jo, H.S. Ahn, S. Kang, M.H. Kim, A study of nucleate boiling heat transfer on hydrophilic, hydrophobic and heterogeneous wetting surfaces. Int. J. Heat Mass Transf. **54**(25–26), 5643–5652 (2011)

15. A.R. Betz, J. Xu, H. Qiu, D. Attinger, Do surfaces with mixed hydrophilic and hydrophobic areas enhance pool boiling? Appl. Phys. Lett. **97**(14), 1–4 (2010)
16. B.J. Suroto, M. Tashiro, S. Hirabayashi, S. Hidaka, M. Kohno, Y. Takata, "Effects of Hydrophobic-Spot Periphery and Subcooling on Nucleate Pool Boiling from a Mixed-Wettability Surface, J. Therm. Sci. Technol. **8**(1), 294–308 (2013)
17. S.M. Kim, I. Mudawar, Review of databases and predictive methods for pressure drop in adiabatic, condensing and boiling mini/micro-channel flows. Int. J. Heat Mass Transf. **77**, 74–97 (2014)
18. S.G. Kandlikar, A general correlation for saturated two-phase flow boiling heat transfer inside horizontal and vertical tubes. J. Heat Transfer **112**(1), 219 (2008)
19. S.G. Kandlikar, Scale effects on flow boiling heat transfer in microchannels: a fundamental perspective. Int. J. Therm. Sci. **49**(7), 1073–1085 (2010)
20. S.K. Gupta, R.D. Misra, Experimental study of pool boiling heat transfer on copper surfaces with Cu-$Al_2O_3$ nanocomposite coatings. Int. Commun. Heat Mass Transf. **97**, 47–55 (2018)
21. Z. Yao, Y.W. Lu, S.G. Kandlikar, Effects of nanowire height on pool boiling performance of water on silicon chips. Int. J. Therm. Sci. **50**(11), 2084–2090 (2011)
22. C.M. Patil, K.S.V. Santhanam, S.G. Kandlikar, Development of a two-step electrodeposition process for enhancing pool boiling. Int. J. Heat Mass Transf. **79**, 989–1001 (2014)
23. S.K. Gupta, R.D. Misra, An experimental investigation on pool boiling heat transfer enhancement using Cu-$Al_2O_3$ nano-composite coating. Exp. Heat Transf. **32**(2), 133–158 (2019)
24. S.K. Gupta, R.D. Misra, An experimental investigation on flow boiling heat transfer enhancement using Cu-$TiO_2$ nanocomposite coating on copper substrate. Exp. Therm. Fluid Sci. **98**(June), 406–419 (2018)
25. A.K.M.M. Morshed, T.C. Paul, J. Khan, Effect of Cu-$Al_2O_3$ nanocomposite coating on flow boiling performance of a microchannel. Appl. Therm. Eng. **51**(1–2), 1135–1143 (2013)
26. D.C. Price, A dielectric surface coating technique to enhance boiling heat transfer from high power microelectronics. IEEE Trans. Components Packag. Manuf. Technol. Part a **18**(3), 656–663 (1995)
27. C.F. Liu, C.Y. Yang, Effect of space distance for boiling heat transfer on micro porous coated surface in confined space. Exp. Therm. Fluid Sci. **50**, 163–171 (2013)
28. S. An et al., Supersonically sprayed reduced graphene oxide film to enhance critical heat flux in pool boiling. Int. J. Heat Mass Transf. **98**, 124–130 (2016)
29. H.T. Phan, N. Caney, P. Marty, S. Colasson, J. Gavillet, Surface wettability control by nanocoating: the effects on pool boiling heat transfer and nucleation mechanism. Int. J. Heat Mass Transf. **52**(23–24), 5459–5471 (2009)
30. M. Dharmendra, S. Suresh, C.S. Sujith Kumar, Q. Yang,"Pool boiling heat transfer enhancement using vertically aligned carbon nanotube coatings on a copper substrate. Appl. Therm. Eng. **99** 61–71 (2016)
31. T. Fisher, Flow boiling in a micro-channel coated with carbon nanotubes flow boiling in a micro-channel coated with carbon. Nanotubes **32**, 639–649 (2014)
32. V. Khanikar, I. Mudawar, T. Fisher, Effects of carbon nanotube coating on flow boiling in a micro-channel. Int. J. Heat Mass Transf. **52**(15–16), 3805–3817 (2009)
33. S.J. Thiagarajan, R. Yang, C. King, S. Narumanchi, Bubble dynamics and nucleate pool boiling heat transfer on microporous copper surfaces. Int. J. Heat Mass Transf. **89**, 1297–1315 (2015)
34. J.A. Weibel, S.V. Garimella, M.T. North, Characterization of evaporation and boiling from sintered powder wicks fed by capillary action. Int. J. Heat Mass Transf. **53**(19–20), 4204–4215 (2010)
35. Z.G. Qu, D.G. Li, J.Y. Huang, Z.G. Xu, X.L. Liu, W.Q. Tao, Experimental investigations of pool boiling heat transfer on horizontal plate sintered with metallic fiber felt. Int. J. Green Energy **9**(1), 22–38 (2012)
36. S. Sarangi, J.A. Weibel, S.V. Garimella, Effect of particle size on surface-coating enhancement of pool boiling heat transfer. Int. J. Heat Mass Transf. **81**, 103–113 (2015)
37. P. Bai, T. Tang, B. Tang, Enhanced flow boiling in parallel microchannels with metallic porous coating. Appl. Therm. Eng. **58**(1–2), 291–297 (2013)

38. Y. Sun, L. Zhang, H. Xu, X. Zhong, Flow boiling enhancement of FC-72 from microporous surfaces in minichannels. Exp. Therm. Fluid Sci. **35**(7), 1418–1426 (2011)
39. D. Deng, Y. Tang, D. Liang, H. He, S. Yang, Flow boiling characteristics in porous heat sink with reentrant microchannels. Int. J. Heat Mass Transf. **70**, 463–477 (2014)
40. D. Deng, W. Wan, H. Shao, Y. Tang, J. Feng, J. Zeng, Effects of operation parameters on flow boiling characteristics of heat sink cooling systems with reentrant porous microchannels. Energy Convers. Manag. **96**, 340–351 (2015)
41. W. Wu, H. Bostanci, L.C. Chow, Y. Hong, M. Su, J.P. Kizito, Nucleate boiling heat transfer enhancement for water and FC-72 on titanium oxide and silicon oxide surfaces. Int. J. Heat Mass Transf. **53**(9–10), 1773–1777 (2010)
42. Y. Tang, B. Tang, J. Qing, Q. Li, L. Lu, Nanoporous metallic surface: Facile fabrication and enhancement of boiling heat transfer. Appl. Surf. Sci. **258**(22), 8747–8751 (2012)
43. Y. Tang, B. Tang, Q. Li, J. Qing, L. Lu, K. Chen, Pool-boiling enhancement by novel metallic nanoporous surface. Exp. Therm. Fluid Sci. **44**, 194–198 (2013)
44. H.D. Kim, M.H. Kim, Effect of nanoparticle deposition on capillary wicking that influences the critical heat flux in nanofluids. Appl. Phys. Lett. **91**(1), 1–4 (2007)

# Effect of Particulate Type Reinforcements on Mechanical and Tribological Behavior of Aluminium Metal Matrix Composites: A Review

**V. S. S. Venkatesh and Ashish B. Deoghare**

**Abstract** The intrinsic properties of particulate types of reinforcements such as isotropy, good machinability, withstanding capacity of high tensile, compressive, shear stresses make it suitable as a reinforcement in Aluminium metal matrix composites. This paper mainly discussed the different types of reinforcements such as TiC, Boron Carbide, Coconut shell powder, Aloe Vera powder, SiC which affects the mechanical properties such as yield strength, ultimate strength, fracture toughness, and tribological properties such as wear resistance of composite. It has been found that the harder ceramic particulate reinforcement enhances the ultimate tensile strength and hardness of the composite under the phenomenon of pilling of dislocations at the grain boundaries. Reduction in wear resistance due to softening of matrix material at higher loads has also been noticed in the literature. This study demonstrates that the Aluminium metal matrix composites can be treated as the superior materials for the design of automobile components such as piston and cylinder assembly and crank shafts which require high wear resistance and specific strength.

**Keywords** Al MMC · Particulate reinforcements · Mechanical properties · Wear resistance

## 1 Introduction

Composite is a material made by combining two or more constituent materials having different physical or chemical properties, so that when combined better characteristics can be achieved from that individual materials. Aluminium metal matrix composites (AMC) are implicit materials for numerous applications in automobile industry like pistons, cylinder liners, crankshafts, etc., because of their higher strength to weight ratio and higher specific strength [1]. Reinforcements are generally added to

V. S. S. Venkatesh (✉) · A. B. Deoghare
Department of Mechanical Engineering, National Institute of Technology Silchar, Silchar, Assam, India
e-mail: vssvenkateshnits@gmail.com



K. M. Pandey et al. (eds.), *Recent Advances in Mechanical Engineering*, Lecture Notes in Mechanical Engineering, https://doi.org/10.1007/978-981-15-7711-6_31

improve the properties of base material like hardness, creep, fatigue, and wear resistance properties. The reinforcing materials are selected in such a way that they must be stable at given operating temperature and withstanding capacity of higher loads. There are different types of reinforcements available for manufacturing of composite like Fibers, Whiskers, Flake, particulates et al., [2]. Among all these reinforcements Particulate type reinforcements are widely used in MMC because of high density and less percentage of elongation. There are different techniques available for manufacturing of composite materials like stir casting, powder metallurgy, and semi-solid processing. The main criterion during the fabrication of composite is uniform distribution of reinforcements in the matrix material. The type of manufacturing method, size, and shape of reinforcement mainly influence the microstructure and strength of composite [3]. The hardness of composite specimen can be improved by the incorporation of reinforcements having hardness higher than the matrix material as discussed by Kumar et al. [4]. Similar results reported by Admile et al. [5], by the addition of 20 wt.% SiC reinforcement to Aluminium matrix increases the hardness to 45.06VHN when compared to the unreinforced Aluminium which has the hardness of 24VHN. This is attributed to the restriction of dislocation movement of SiC reinforcements which improves the hardness of composite. James et al. [6], reported that upto the incorporation of 2.5% of TiC reinforcements the hardness enhanced. But beyond 2.5% addition of TiC hardness decreases due to the formation of clusters between the Al and TiC particles. Chandra et al. [7], concluded that by the addition of $Al_2O_3$ particles into Al6061 matrix hardness of Al6061/$Al_2O_3$ composite improved from 61.15 VHN at 5% $Al_2O_3$ to 89.91 VHN at 20% due to the stoppage of dislocations at the grain boundaries of matrix and reinforcements. Table 1 represents the change in mechanical properties of composite with type of reinforcement and processing technique.

This review mainly focuses on the particulate type of reinforcements like TiC, SiC, $B_4C$, coconut shell powder, etc. on the mechanical and Tribological behavior of the Al MMC were discussed.

## 2 Types of Reinforcements

### 2.1 *TiC Reinforcement*

The superior properties of TiC such as High Young's Modulus (400 Gpa), Low friction coefficient (0.29–0.34) and high melting temperature (3067 °C) make it suitable as a reinforcement material for manufacturing of cutting tools, abrasion-resistant surface coating on metal parts, cylinder liners, and automobile bearings. The effect of TiC reinforcement in Al MMC is explained as follows. Thangarasu et al. [18], fabricated Al6082/$X$% TiC ($X = 0, 6, 12, 18, 24$) metal matrix composite using friction stir processing technique. It was observed that the wear rate of composite decreases from 0.00697 to 0.00303 mg/m with the addition of TiC reinforcement

**Table 1** Change in mechanical properties of composite with processing technique

| S. No. References | Particle size. reinforcement material Matrix type | Fabrication technique | Stirring speed/melting point | Effect |
|---|---|---|---|---|
| 1 Thangarasu et al. [8] | 2 μm TiC AA6082 | Friction stir processing | 1200 Rpm | Micro hardness, UTS (T) increased |
| 2 Albiter et al. [9] | 1.12 μm TiC Al2024 | Infiltration | 1200 °C | Micro hardness, UTS (T) increased |
| 3 Sabbaghian et al. [10] | 25 μm TiC copper-based composite | Friction Stir processing (Fsp) | 1000 Rpm | Micro hardness increased |
| 4 Sheibani et al. [11] | TiC AMC | Reactive slag in-situ/powder metallurgy | 800 °C | Mechanical properties improved |
| 5 Gopalakrishnan et al. [12] | TiC AA6061 | Stir casting mathod | 650 °C | Mechanical properties improved |
| 6 Kishore et al. [13] | TiC AA6061 | Flux assisted synthesis | 900 °C | Micro hardness increased |
| 7 Jafarian et al. [14] | TiC less than 100 nm Al1050 | Accumulative Roll bonding (ARB) | — | UTS(T) increased |
| 9 Jerome et al. [15] | TiC in situ Al/Tic | Resistant heating furnace containing a stirrer | 900 °C | Tensile strength, hardness improved |
| 10 Bauri et al. [16] | TiC50 μm in situ Al/sic | Friction stir processing (FSP) | 1000 Rpm | Tensile strength, hardness improved |
| 12 Kennedy et al. [17] | TiC 10 μm Al/Tic MMC | Flux-casting process, powder processing (PM), | 400 °C | Hardness improved |

from 0 to 24%. This is associated with higher hardness of TiC reinforcements which offers greater abrasion resistance to the composite material. It was further noticed that with the change in speed from 40 to 80 mm/min the hardness of composite can be improved from 112 to 135 HV. This was attributed to with increase in linear velocity the contact time between micro reinforcement particles and tool is less which generates less amount of heat causes the TiC reinforcements dispersed in lesser area resulting in higher hardness of the composite. Similar results reported by Krishnan et al. [19], studied the wear behavior stir casted Al6061/TiC metal matrix composite at different velocities ranging from 1.5 to 4.5 m/s. It was noticed that at the higher speeds due to the existence of continuous contact between the surfaces material softening takes place which improves the wear rate linearly with sliding speed and the maximum wear rate of $52.13 \times 10^{-6}$ mm$^3$/m occurs at 4.5 m/s. Sai Chaitanya Kishore et al. [20], analyzed the Al6061/TiC composite by chemical reaction of $K_2TiF_6$ (Potasium hexafluorotitanate) with graphite powder. Initially, 3 kg of Al is taken in a crucible, after melting this Al powder premixed $K_2TiF_6$ and C is added to the molten slurry so that TiC is formed by chemical reaction of $K_2TiF_6$ with Al. Vickers hardness test is performed by applying load of 300 g with 15 s dwell time. Results concluded that due to the existence of strong bonding between matrix and reinforcement material, hardness of composite is increased from 52.3 VHN at 0% reinforcement to 62.7 VHN at 4% reinforcement. Jeyasimman et al. [21] analyzed the mechanical behavior of Al 6061-X%TiC ($x = 0.5, 1.0, 1.2, 2.0$) nanocomposite fabricated through cold compaction technique. It was found that due to the presence of flake-like and irregular morphology of crystalline powders, the green compaction strength of the composite increased from 100 to 233 MPa for unreinforced Al alloy to 2% TiC reinforced composite. This subsequently improves the dispersed strength of the Al matrix. Kumar et al. [22], Studied the microstructural behavior of As Cast and Hot forged AA6061-TiC composite. The grain size of hot forged composite is found to be lesser when compared to As cast specimens, this is due to the material deformation during forging operation which originates the new α-Al grains and simultaneously the dislocation density of this α-Al grains increases. These dislocations arrest the driving force for additional growth of α-Al grains which significantly exerts pinning effect on α-Al and TiC grain boundaries causes finer microstructure and enhances the mechanical properties of composite.

## 2.2 Boron Carbide ($B_4C$) Reinforcement

Boron carbide is a robust material which processes higher Vickers hardness (38 GPa), fracture toughness (3.5 MPa·$m^{1/2}$) and Elastic Modulus (460 GPa). $B_4C$ is the third strongest metal next to diamond and cubic boron nitride. The effect of $B_4C$ reinforcement on the aluminium metal matrix composite is explained as follows. Shirvanimoghaddam et al. [23], studied the mechanical behavior of stir casted Al-356/$B_4C$ composite. The composite samples of 5, 10, 15% $B_4C$ are fabricated at 800 °C and 1000 °C. XRD analysis indicates that the phases of $AlB_2$ and $Al_3BC$ are formed at 1000 °C, these phases enhance the wettability between aluminium matrix and reinforcement phase. The existence of lesser density Boron Carbide reinforcements reduces the density of composite from 2.69 to 2.65 g/$cm^3$ with the addition of reinforcements from 5 to 15%. Tensile strength of composite at 800 °C is enhanced from 176 to 197 Mpa for 5–10% reinforcement and then decreased to 194 Mpa for 15% of reinforcement because of formation of slag in molten metal. But at 1000 °C the UTS of composite is proportional to volume of reinforcement percentage. Hardness of composite increases with $B_4C$ percentage at 8000 and 1000 °C. At higher temperature, the hardness value of 95 BHN for 15% $B_4C$ is found because of uniform distribution of reinforcements and good wettability between the particles. Similar results reported by Nic et al. [24], For Al2024/$B_4C$ composite by mechanical Alloying-hot extrusion technology. Initially, Aluminium and $B_4C$ powders are ball milled with Argon gas atmosphere and stearic acid is added as process control agent

(PCA). The effect of PCA on composite powder is examined and concluded that the obtained particle size was minimum when 2% PCA for 5 h of milling time and 28.7% of improvement of UTS is observed from pure Al 2024 at 10%reinforcement.

## 2.3 Coconut Shell Microparticle Reinforcement

Coconut shell particles are naturally available reinforcements in southern states of India which processes lower density and higher wear resistance property which is suitable for dispersion phase in AMCs for automobile brake disc application. Bello et al. [25], developed Aluminium/coconut shell micro particle composite (AL/CMP) by compo cast technique. Composite was prepared by adding 2, 4, 6, 8, 10% of CMP reinforcements. It was observed that the due to presence of lesser density coconut shell particles the density of composite declines from 2.7 to 2.2 g/cm$^3$ with addition of reinforcements from 0 to 10%. The impact energy of 19 J is obtained at 0% reinforcement and it decreases gradually upto 6% reinforcement and then declines drastically above 6% of CMP the minimum value of 7 J is achieved at 10% reinforcement. It is also noticed that because of existence of chemical reaction between coconut shell powder and Al matrix, new compounds like $TaO_2$, $CO_3F_7$, $Mg_2Al_3$, CoFe, are formed and these compounds occupied the interstitial spaces in Al matrix, which improves the 88% of ultimate tensile strength of composite than the unreinforced Al alloy. Sankararaju et al. [26], studied the wear behavior of stir casted Al 1100/X% ($X = 5$, 10, 15) Coconut shell powder metal matrix composite at different pressure levels. It was noticed that due to the presence of higher volume of softer material for pure Al-1100 and 5% reinforcement the wear rate is higher at low-pressure values of 2 N/mm$^2$ when compared to 10% CSP and 15% CSP. SEM analysis reveals that the presence smoother and finer particles CSP on the composite specimen which undergone abrasive wear at a pressure of 2 N/mm$^2$. With increasing the pressure value to 10 N/mm$^2$ rough surfaces are observed which causes Adhesive wear.

## 2.4 Aloe Vera Powder Reinforcement

Aloe vera powder is reinforced with Aluminium matrix because of its lesser density and eco-friendly material and abundantly available at lesser cost. Aloe vera powder contains considerable wettability with better mechanical and physical properties compared to Fly ash reinforcement. Hima Giresh et al. [27] developed AMC/10% Aloe Vera composite by stir casting method at a speed of 300 rpm. The presence of good bonding between Al matrix and Aloe vera reinforcements is noticed which improves the 55.62% of UTS when compared to Pure Aluminium. Hardness value of 33.8 BHN is achieved which is 43.7% more than unreinforced alloy. Because of spherical nature of aloe vera microparticles the energy absorption capability of Aloe vera composite is 1.80 J/mm$^2$ which is quiet higher than pure Aluminium specimen having 0.1 J/mm$^2$.

### 2.5 SiC Reinforcement

Silicon Carbide is a low-density ceramic semiconductor contains Silicon and carbon. In modern manufacturing industries SiC used in abrasive machining processes such as honing, grinding, water-jet machining. Jaya Prasad et al. [28], investigated the microstructural behavior of stir casted Al5083/SiC composite. Reinforcements of 3, 5, 7% are added to molten melt along with 1% of Mg in order to improve the wettability between the matrix and reinforcement particles. SEM micrographs reveal that pore-free and dense microstructure of composite samples and there is no evidence of formation of intermetallic phases between SiC and Al 5083 particles. Kalyankumar Singh et al. [29], studied wear behavior and frictional properties of Stir Casted Al 7075/8% SiC MMC at various sliding distances of 1979.2 and 2262 m by varying loads from 10 to 30 N. The coefficient of friction value is found to decrease about 30–40% by varying the loads from 10 to 30 N. This was attributed to material softening takes place at higher temperatures due to large contact between the mating surfaces causing deduction of coefficient of friction at higher loads [17]. It was also noticed that with increasing sliding distance from 1979.20 to 2262 m at 15 N load coeff: of friction value enhances from 0.6 to 0.7. This happened because at higher sliding distance the temperature-induced is more, which causes Al matrix undergoing material softening so that the hard ceramic SiC particles break and clogged between specimen and wheel surface, which results high coeff: of friction. Mohanavel et al. [30] fabricated Al6351/SiC composite by using stir casting method, the reinforcements are added from 0 to 20% in stages of 4%. It was found that Hardness of composite increases from 40 RHN to 64.7 RHN with unreinforced Aluminium to 20% SiC reinforcement. This is due to increase in dislocation density with SiC particles which arrests the resistance to deformation. In addition to this, the Yield strength of the composite material enhances from 109 to 194 MPa because of decrement in SiC particle grain size with increase in % of reinforcements according to Hall–patch equation. Ramgopalreddy et al. [31] fabricated Al6082/SiC/Fly ash hybrid composite by using stir casting method. The fabricated MMC contains 2.5, 5, 7.5% of sic and fly ash reinforcement. Tensile test results concluded that strength increases from 110 to 128 MPa with reinforcement of 0–7.5%, because of increase of bonding between matrix and reinforcements.

## 3 Conclusions

Several challenges must be considered in order to strengthen the usage of AMCs such as influence of reinforcements, processing technology, applied load on the mechanical and Tribological behavior of the MMC. The prior conclusions derived from the literature can be depicted as follows

- Hot forged TiC reinforced MMC exhibits better mechanical properties compared to As cast Al-TiC AMC for the same volume fraction of reinforcements and particle size.
- The optimum temperature for Al-$B_4C$ MMC fabricated through stir casting process is 1000 °C which attains higher hardness value of 95 BHN for 15% reinforcements.
- Presence of lower density coconut shell reinforcements in Al MMC the maximum impact energy of 19 J is achieved at 0% reinforcement. At lower pressure values of 2 N/mm$^2$ SEM analysis revealed the existence of abrasive wear and adhesive wear is undergone by the composite sample at a pressure of 10N/mm$^2$.
- The energy absorption capacity of Aloe vera AMC is improved to 1.80 J/mm$^2$ at 10% of reinforcement due to the presence of spherical shaped reinforcement particles.
- The wear rate of $B_4C$ reinforced AMC is higher than the SiC reinforced composite for all applied load and sliding distances.
- The wear resistance of composite is not only depending on type of reinforcement but also depends on the applied load and sliding distance of composite. For nanocomposite, the strength of composite is depending on milling time and the wettability between the matrix and reinforcement material.

## References


1. G. Manohar, K. Abhijit Dey, M. Pandey, S.R. Maity, Fabrication of metal matrix composites by powder metallurgy: a review. AIP Conf. Proc. 1952, 020041 (2018)
2. S. Dinesh Kumar, M. Ravichandran, M. Meignanamoorthy, Aluminium metal matrix composite with zirconium diboride reinforcement: a review, vol. 5, Issue 9, Part 3, pp. 19844–19847 (2018)
3. M. Lei, H. Ledbetter, Communications: elastic constants of SiCp/Al: measurements and modeling. Metall. Mater. Trans. **25A**, 2832–2835 (1994)
4. T. Satish Kumar, R. Subramanian, S. Shalini, Microstructure, mechanical properties and corrosion behaviour of Al–Si–Mg alloy matrix/zircon and alumina hybrid composite. J. Mater. Res. Technol. **4**, 333–347 (2015)
5. H. Admile, S.G. Kulkarni, S.A. Sonawane, Review on mechanical & wear behavior of aluminum-fly ash metal matrix composite. Int. J. Emer. Technol. Adv. Eng. **4**, 41–45 (2014)
6. S. Johnyjames, K. Venkatesan, P. Kuppan, R. Ramanujam, Hybrid aluminium metal matrix composite reinforced with SiC and TiB2. Proc. Eng. **97**, 1018–1026 (2014)
7. B. Chandrakandpal, J. Kumar, H. Singh, Production technologies of metal matrix composite: a review. IJRMET **4**, 52–57 (2014)
8. A. Tangarasu, N. Murugan, I. Dinaharan, S.J. Vijay, Influence of traverse speed on microstructure and mechanical properties of AA6082-TiC surface composite fabricated by friction stir processing. Proc. Mater. Sci. **5**, 2115–2121 (2014)
9. A. Albiter, C.A. León, R.A.L. Drew, E. Bedolla, Microstructure and heat-treatment response of Al-2024/TiC composites. Mater. Sci. Eng. A **289**(1–2), 109–115 (2000)

10. H.R. Akramifard, M. Shamanian, M. Sabbaghian, M. Esmailzadeh, Microstructure and mechanical properties of Cu/SiC metal matrix composite fabricated via friction stir processing. Mater. Design (1980–2015) **54**, 838–844 (2014)
11. S. Seibani, M. Fazel, Najafabadi, In situ fabrication of Al-TiC metal matrix composites by reactive slag process. Mater. Design **28**(8), 2373–2378 (2007)
12. S. Gopalakrishnan, N. Murugan, Production and wear characterisation of AA 6061 matrix titanium carbide particulate reinforced composite by enhanced stir casting method. Compos. B Eng. **43**(2), 302–308 (2012)
13. D. Sai Chaitanya Kishore, K. Prahlada Rao, A. Ramesh, Optimization of machining parameters for improving cutting force and surface roughness in turning of Al6061-SiC in-situ metal matrix composites by using Taguchi method. Mater. Today Proc. **2**(4–5), 3075–3083 (2015)
14. H. Jafarian, J. Habibi-Livar, S.H. Razavi, Microstructure evolution and mechanical properties in ultrafine grained Al/TiC composite fabricated by accumulative roll bonding. Compos. Part B Eng. **77**, 84–92 (2015)
15. S. Jerome, B. Ravisankar, P.K. Mahato, S. Natarajan, Synthesis and evaluation of mechanical and high temperature tribological properties of in-situ Al-TiC composites. Tribol. Int. **43**(11), 2029–2036 (2010)
16. Rangitkaury, D. Yadav, G. Suhas, Effect of friction stir processing (FSP) on microstructure and properties of Al-TiC in situ composite. Mater. Sci. Eng. A **528**(13–14), 4732–4739 (2011)
17. A.R. Kinnedy, D.P. Weston, M.I. Jones, Reaction in Al-TiC metal matrix composites. Mater. Sci. Eng. A **316**(1–2), 32–38 (2001)
18. A. Tangarasu, N. Murugan, I. Dinaharan, S.J. Vijay, Synthesis and characterization of titanium carbide particulate reinforced AA6082 aluminium alloy composites via friction stir processing. Arch. Civil Mech. Eng. **15**(2), 324–334 (2015)
19. K. Narasimhan, D.G. Bhat, Effect of chemical vapor deposition process parameters on the growth aspects of titanium carbide whiskers. Surf. Coat. Technol. **61**(1–3), 171–176 (1993)
20. D. SaiChaitanya Kishore, K. Prahlada Rao, A. Ramesh, Optimization of machining parameters for improving cutting force and surface roughness in turning of Al6061-TiC in-situ metal matrix composites by using Taguchi method. Mater. Today Proc. **2**(4–5), 3075–3083 (2015)
21. S. Sivasankaran, K. Sivaprasad, R. Narayanasamy, R.S. Kambali, An investigation of the synthesis, consolidation and mechanical behaviour of Al 6061 nanocomposites reinforced by TiC via mechanical alloying. Mater. Design **57**, 394–404 (2014)
22. G.S. Pradeep Kumar, P.G.K. Koppad, K. Ramaia, M. Alipour, Microstructure and mechanical behaviour of in situ fabricated AA6061-TiC metal matrix composites. Arch. Civil Mech. Eng. **17**(3), 535–544 (2017)
23. K. Shirvanimoghaddam, H.K. Abdizadeh, M. Karbalaei Akbari, M. Naebe, Boron carbide reinforced aluminium matrix composite: Physical, mechanical characterization and mathematical modelling. Mater. Sci. Eng. A **658**, 135–149 (2016)
24. C.-Z. Nie, Gu. Jia-Jun, J.-L. Liu, Di. Zhang, Production of boron carbide reinforced 2024 aluminum matrix composites by mechanical alloying. Mate. Trans. **48**(5), 990–995 (2007)
25. S.A. Bello, I.A. Raheem, N.K. Raji, Study of tensile properties, fractography and morphology of aluminium (1xxx)/coconut shell micro particle composites. J. King Saud Univ. Eng. Sci. **29**(3), 269–277 (2017)
26. R. Siva SankaraRaju, M.K. Panigrahi, R.I. Ganguly, G. SrinivasaRao, Tribological behaviour of al-1100-coconut shell ash (CSA) composite at elevated temperature. Tribol. Int. **129**, 55–66 (2019)
27. C. Hima Gireesh, K.G. Durga Prasad, K. Ramji, P.V. Vinay. Mechanical characterization of aluminium metal matrix composite reinforced with aloe vera powder. Mater. Today Proc. **5**, 3289 (2018)
28. V. Jaya Prasad, K. Narasimha Rao, N. Kishore Bab, Mechanical and tribological characterization of aluminum metal matrix composite reinforced with micro ceramic particles (TiB2/SiC). Mater. Today Proc. (In press) corrected proof, Available online 13 Aug 2019
29. K.K. Singh, S. Singh, A.K. Shrivastava, Comparison of wear and friction behaviour of aluminum matrix alloy (Al 7075) and silicon carbide based aluminum metal matrix composite under dry condition at different sliding distance. Mater. Today Proc. **4**(8), 8960–8970 (2017)

30. V. Mohanavel, K. Rajan, S. Suresh Kumar, S. Udishkumar, C. Jayasekar, Effect of silicon carbide reinforcement on mechanical and physical properties of aluminum matrix composites. Mater. Today Proc. 5, 2938–2944 (2018)
31. B. Ramgopal Reddy, C. Srinivas, Fabrication and characterization of silicon carbide and fly ash reinforced aluminium metal matrix hybrid composites. Mater. Today Proc. 5, 8374–8381 (2018)

# A Review on Solar Drying Applications Using Latent Heat as Energy Storage Media

**Supreme Das, Agnimitra Biswas, and Biplab Das**

**Abstract** Solar energy may be considered as the most feasible renewable energy source with a wide range of applications in today's world. A suitable energy storage medium, capable of storing and supplying this energy as per requirement provides for an effective thermal management of solar energy devices. Latent heat storage systems have gained significant attention from researchers and academicians due to its higher storage density and smaller temperature difference between storing and releasing heat as compared to sensible heat storage. This chapter attempts to summarize the previous work carried out in solar drying applications implementing phase change materials as latent heat energy storage medium. It is concluded that such materials are capable of reducing the heat losses associated with solar dryers and significantly increase its efficiency.

**Keywords** Solar dryer · Latent heat · PCM · Thermal performance · Efficiency

## 1 Introduction

The world energy demand is estimated to be doubled by 2050 and approximately tripled by the end of the century owing to a rapid increase in energy consumption and economic development of the nations around the world. This has resulted in an expeditious exhaustion of fossil fuel resources, such as oil and gas. The degradation of our environment from greenhouse gas emissions and in cessantsur charge in fuel prices has instigated the interest of researchers around the world towards renewable

S. Das (✉) · A. Biswas · B. Das
Department of Mechanical Engineering, National Institute of Technology Silchar, Silchar, Assam 788010, India
e-mail: supremedas.19031987@gmail.com

A. Biswas
e-mail: agnibis@yahoo.co.in

B. Das
e-mail: biplab.2kmech@gmail.com


K. M. Pandey et al. (eds.), *Recent Advances in Mechanical Engineering*, Lecture Notes in Mechanical Engineering, https://doi.org/10.1007/978-981-15-7711-6_32

energy prospects, such as solar, wind, geothermal, biomass and tidal energy sources [1].

Solar energy is the most abundant amongst these energy sources and because of its eco-friendly nature and equitable distribution; it may be considered the most propitious renewable energy source. The sun emits energy at a rate of $3.8 \times 10^{23}$ kW, of which approximately $1.8 \times 10^{14}$ kW is intercepted by earth. Around 60% of this amount is reached the earth's surface, the rest being absorbed by the atmosphere and reflected back into space.

A mere 1% efficient conversion of this energy could generate as much as 400 times the total energy production of the earth [2].

The utilization of solar energy to dispense hot air provides significant prospects in the drying of agricultural (fruits, vegetables, medicinal plants, coffee and tea products), textile and marine products, seasoning of timber, regenerating dehumidifying agents and space heating, especially in low-temperature regions [3]. Moreover, a solar air heater is cheaper and uses less material as compared to other solar collectors.

Solar dryers may be classified as direct, indirect, and mixed mode. Direct mode utilizes air heated by solar radiation directly in the dryer enclosure, whereas an indirect mode employs a separate drying chamber and a blower to duct the heated air from the heater to the chamber. The mixed mode engages both these techniques for heating the material [4].

## 2 Phase Change Materials

Solar energy is intermittent in nature. As such, in order to provide an even output and increase their reliability, solar energy systems implement a number of energy storage methods. Phase change materials (PCM) provide an effective thermal energy management solution by storing the excess energy during peak radiation hours to be used when solar radiation is in adequate. PCMs have high-energy storage density and have the ability of constant temperature heat addition and rejection during phase change. They use latent heat of fusion/vaporization for a desired temperature control. During the charging phase, a PCM is sensibly heated till saturation (i.e. melting) following which it absorbs its latent heat at a constant temperature till it is completely melted. It is then sensibly heated further in its molten state. While discharging, a PCM is capable of releasing its sensible heat and latent heat to the surrounding [5].

The suitability of a PCM is determined from its thermophysical properties like specific heat capacity, heat of fusion, thermal conductivity at its melting temperature. Moreover, a PCM should possess certain physical properties like high density, non-toxic, stable composition and chemically inert. The use of PCMs is still very limited owing to its higher initial cost and limited space availability. Irregular sunshine also limits the efficiency of PCMs in solar energy systems [6].

## 3 Thermal Performance of a Conventional Solar Air Heater

Figure 1 illustrates a typical solar air heater showing the heat and energy transfer phenomenon. The construction and design features of such solar air heater systems are described by Garg and Prakash [7]. Thermal performance of these systems can be analysed by considering the energy balance between the solar energy absorbed by the absorber plate and useful thermal energy output and can be demonstrated with the help of Hottel-Whillier-Bliss equation reported by Duffie and Beckman [8].

$$Q_u = A_c F_R[I(\tau\alpha)e - U_L(T_i - T_a)] \quad (1)$$

where $F_R$ is the collector heat removal factor and may be defined as a ratio of actual useful energy gain to the useful energy gain if the whole absorbing surface of the collector were at the fluid inlet temperature.

The rate of useful energy gains by air flowing through the duct may be computed as follows

$$Q_u = \dot{m} C_P(T_o - T_i) = h A_c(T_{p_m} - T_a)_m \quad (2)$$

The value of heat transfer coefficient can be represented in non-dimensional form by using the relationship of Nusselt number (Nu)

$$Nu = \frac{hL}{K} \quad (3)$$

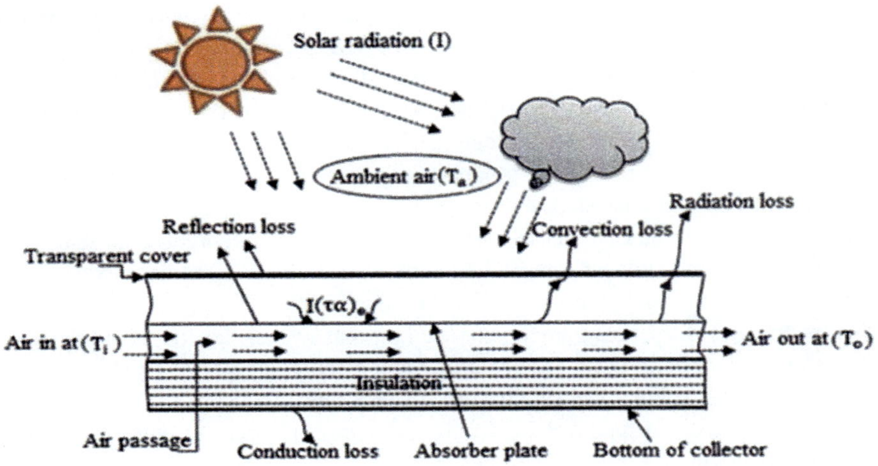


**Fig. 1** A conventional solar air heater

The thermal efficiency of a solar air heater can be expressed by the following equation

$$\eta_{th} = \frac{Qu/Ac}{I} = F_R[(\tau\alpha)_e - U_L - \frac{(T_i - T_a)}{I}] \tag{4}$$

From the above equation, it can be inferred that the collector efficiency decreases linearly with the parameter $\frac{(T_i - T_a)}{I}$ and can be approximated by a straight line, provided $U_L$ is constant. The collector parameters $F_R(\alpha\tau)$ and $F_R U_L$ can be determined from the intercept and slope of the graph respectively.

## 4 Solar Dryers with Latent Heat Storage Medium

Drying processes are normally used to decrease the moisture content of food and agricultural products, thereby increasing their shelf life and making their preservation easier. By regulating the humidity level and operating temperature (usually 40–60 °C), the nutritional properties of the food products can be appropriately controlled [9]. Solar dryers have a huge potential from technical and energy saving standpoint. Numerous types of solar dryers with varying degrees of technical performance have been designed worldwide. They may be broadly categorized as passive dryers (natural convection), active dryers (forced circulation) and hybrid dryers. Fudholi et al. provided a comprehensive review of these drying systems on the basis of their technical and economic parameters for a wide array of agricultural products including apple, banana, cashew nuts, cocoa, paddy, tobacco, turmeric, etc. The technical developments include compact collector design, better efficiency, integrated storage and long life [10]. Fudholi et al. also studied the energy and exergy efficiencies of different configurations of solar air flat plate collectors for drying applications [11].

PCMs have the potential to improve the thermal performance of solar dryers due to their latent heat storage capacity. Devahastin and Pitaksuriyarat [12] investigated the viability of using a latent heat storage (LHS) with paraffin wax for studying the drying kinetics of sweet potato energy and energy conservation during the process. Their proposed system consisted of a compressor, a temperature controller, a heater and a cylindrical, acrylic LHS vessel (0.10 m diameter, 0.20 m height) connected to the heater via a finned copper tube (tube diameter: $1.27 \times 10^{-2}$ m, no. of fins: 18, fin diameter: 0.08 m, fin thickness: $5 \times 10^{-3}$ m, fin spacing: 0.01 m). They analysed the temperature profiles, heat transfer characteristics as well as the effects of inlet air temperature and velocity during the charging and discharging period. It was calculated that the energy savings were approximately 40% and 34% using inlet air velocity of 1 $ms^{-1}$ and 2 $ms^{-1}$, respectively. Cakmak and Yildiz [13] experimentally investigated the drying kinetics of seeded grapes using a novel type of solar air dryer using swirl elements in the entrance and inner part of the drying chamber. An expanded surface solar air collector with 15-mm-drilled holes was used to achieve high heat transfer and turbulence effect. Another collector using calcium chloride hexahydrate

(PCM) in the lower portion was included in the design to ensure a continued drying process even after sunset. Experiments were conducted under natural conditions and by the dryer with/without swirl flow at three different air velocities. It was found that the drying was more uniform and faster for the system with PCM in swirl flow media. Moreover, the drying time shortens with increasing air velocity. Moisture ratio curves obtained from the experimental values were compared with six different moisture value correlations, and it was observed that Midilli model provided the most relevant results for each seeded grape drying status (Figs. 2 and 3).

Esakkimuthu et al. [14] developed an indirect solar dryer, which utilized a LHS unit with HS58, an inorganic salt-based PCM. A double-pass V-corrugated solar air collector of total area 6 $m^2$ was used. The absorber plate was constructed from 3 embossed-type aluminium sheets of 2 m × 1 m area, 1.1 mm thickness, V-corrugation height of 38 mm and an included angle of 60°. The air passage between the absorber and bottom surface was kept as 15 cm. Rock wool of thickness 25 mm and 50 mm was

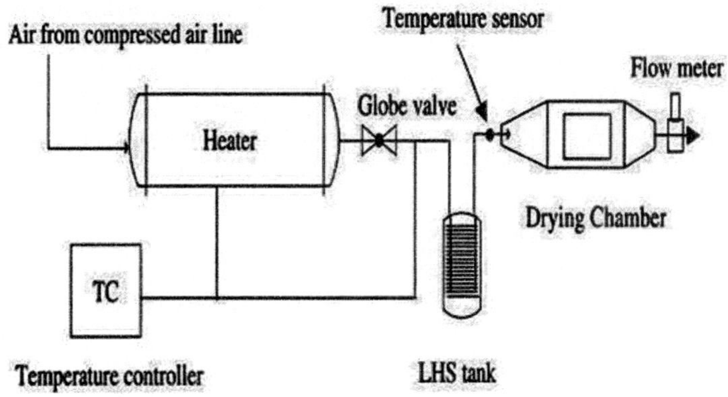


**Fig. 2** Schematic diagram of the experimental solar air dryer [12]

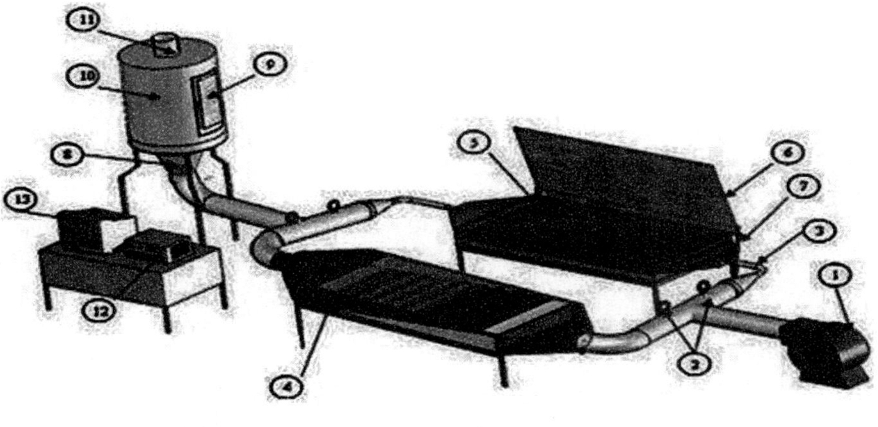

**Fig. 3** Illustration of the solar air dryer using swirl elements [13]

used for side and bottom insulation, respectively. The absorber surface was glazed with a 5-mm-toughened glass with 5 cm air gap. The PCM was packed in spherical capsules made of high-density polyethylene and kept inside the thermal energy storage tank (0.4 m diameter, 1.8 m length). The tank itself was made of galvanized iron sheets and insulated with glass wool. An aluminium cladding was provided above the insulation to prevent its deterioration from atmospheric conditions. Experiments were conducted to investigate the charging and discharging characteristics of the storage unit. It was found that a mass flow rate of 200 kg/h was able to provide a near uniform heat transfer rate during the charging and discharging processes. At higher mass flow rates, the average temperature of the collector reduced owing to a decrease in heat losses and increase in the value of heat transfer coefficient. This led to an increase in the collector efficiency. Lower mass flow rates led to a longer duration of heat supply with better utilization of the storage system capacity. Furthermore, a proper selection of PCM with suitable phase change temperature could prevent overheating during peak sunshine hours, thus avoiding the spoilage of food products. The performance of an indirect forced convection and desiccant integrated solar dryer for drying green peas and pineapple slices was assessed by Shanmugam and Natarajan [15] with and without a reflective mirror. The system was operated in sunshine hours as well as non-sunshine hours and consisted of a flat plate solar air collector, a drying chamber, a desiccant bed and a centrifugal blower. A south facing solar air collector of dimensions 1.2 m × 2.4 m and 6 mm transparent glass cover with a tilt angle of 30° was used. The bottom and sides were insulated with fine saw dust. The wooden drying compartment was 1 m high (1.2 m × 1.2 m X-sectional area) with double glazing of 50 mm air gap having the same inclination as the collector and consisted of 10 shelves for holding the products. A perforated tray capable of stacking 75 kg desiccant material (60% bentonite, 10% calcium chloride, 20% vermiculite and 10% cement, moulded into cylindrical shapes) was provided just below the glazing. A thick insulation board was positioned below the tray during sunshine hours and above the tray during non-sunshine hours. It was found that the inclusion of a reflective mirror increased the drying potential of the desiccant material by 20% with faster regeneration and reduced drying rate. The average dryer thermal efficiency was calculated to be 43–55% with a pickup efficiency of 20–60%. About 60% of the moisture was removed by the air heated by solar energy and the remainder by the desiccant (Figs. 4 and 5).

Shalaby and Bek [16] experimentally investigated a novel indirect solar air dryer using PCM for drying medicinal plants. The system consisted of two identical solar air heaters, a drying chamber, a blower and PCM storage units. The solar air heater was mounted on the top of a room surface at 30° tilt angle facing south. A copper flat plate of dimensions 0.83 m × 2 m and 1 mm thickness was used as an absorber plate. It was provided with two glass covers of 5 mm thickness having an air gap of 25 mm. The bottom and sides were insulated with 40 mm layer of foam. The drying chamber was fabricated using galvanized iron of length 1.2 m, width 0.6 m and height 1.7 m with similar insulation and fitted with retractable drying trays. Heated air was forced to pass through a series of vertical copper tubes (32 total) planted inside PCM filled plastic containers at the bottom of the chamber. Experiments were conducted for mass

**Fig. 4** Photograph of the indirect solar air dryer with Sunit [14]

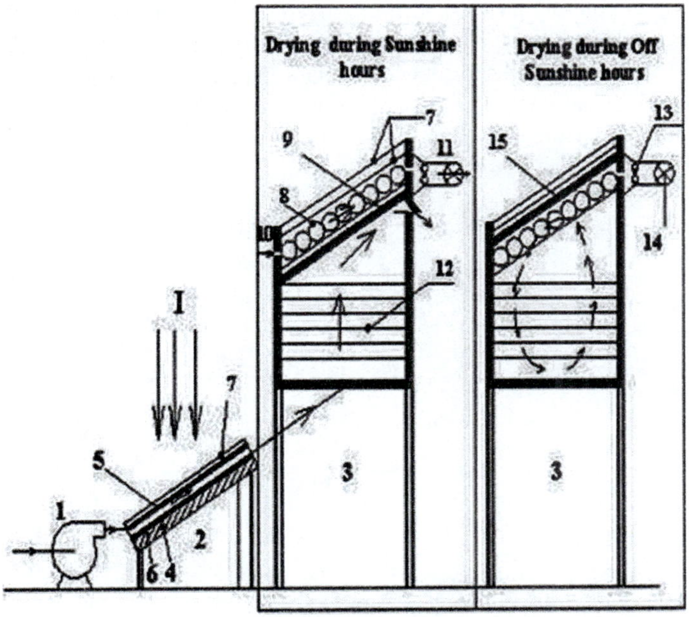


**Fig. 5** Schematic diagram of the desiccant integrated solar air dryer [15]

flow rates of air ranging from 0.0664 to 0.2182 kg/s both with and without PCM. The maximum drying temperature was obtained for a mass flow rate of 0.1204 kg/s with PCM and 0.0894 kg/s without PCM. Implementation of PCM provided drying air temperature higher than the ambient by 2.5–7.5 °C after sunset for at least five hours. Shringi et al. [17] analysed the thermodynamic processes involved in the solar drying of garlic cloves using PCM as energy storage. The system consisted of a drying chamber cum dehumidifier unit, heat storage and heat exchanger unit and evacuated-tube heat-pipe collector. The drying chamber was made from galvanized iron of dimensions 1.2 m × 0.6 m × 0.5 m with a layer of polystyrene insulation and fitted with seven trays, three of which were for desiccant material and the remaining for drying product. Charging and discharging processes of the PCM were done through copper coils of diameter 1.2 cm and 1.9 cm, respectively. Solar fluid (propylene glycol 60% and water 40% v/v) was used to transfer heat from the collector to the heat storage unit via a manifold. The heat from the storage unit is released to the PCM, and the fluid is circulated back to the collector, thus completing the charging cycle. A dehumidifier blower was used to force hot air coming out of the drying chamber to the heat storage and heat exchange run it and back to the bottom of the drying chamber to complete the cycle of discharging. Experimental results depicted a decrease in the moisture content of garlic cloves from 55 to 6.5% (wetbasis) over a period of 8 h. The drying data obtained were fitted to five different drying kinetics models, and it was observed that the Midilli model was the best description for the drying behaviour of garlic cloves. The energy and exergy efficiencies of the drying chamber with circulating air were also improved by 14.9% and 88.2%, respectively (Figs. 6 and 7).

Tewari [18] studied the performance of a natural convective solar crop dryer with phase change thermal energy storage. The dryer consisted of a flat plate solar air collector, an energy storage system (packed bed), natural draft system and a drying chamber. The matt black absorber plate had a surface area of 1.5 m$^2$ and a tilt angle

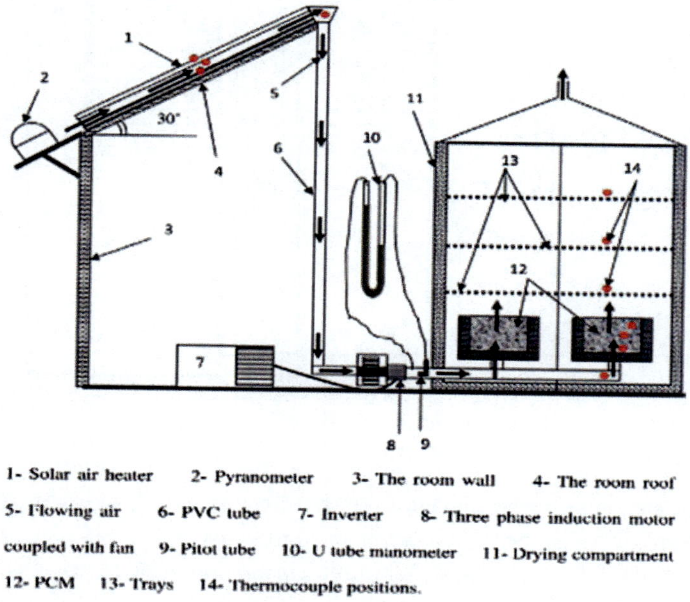


**Fig. 6** Sectional details of a PCM collector [16]

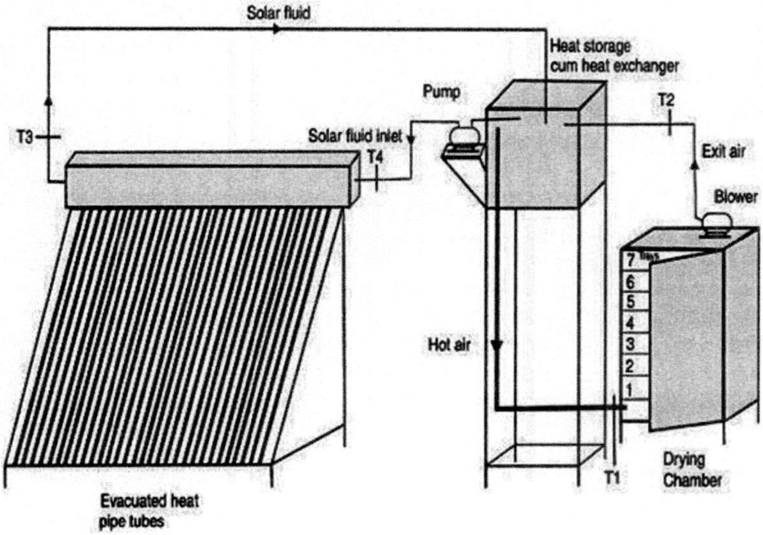


**Fig. 7** Illustration of a PCM integrated solar dryer using evacuated-tube heat-pipe collector [17]

of 25°. It was glazed with toughened glass with a spacing of 0.05 m to allow air flow inside the dryer. A total of 48 PCM (paraffin wax) filled cylindrical tubes of 0.05 m diameter, 0.75 m length and 1 kg capacity were tightly packed in zigzag orientation below the drying chamber. The drying chamber consisted of six rectangular drying trays of 10–12 kg leafy herbs. Another absorber plate with toughened glass glazing was placed above the drying chamber at an inclination of 23°. A PCM packed bed placed 0.05 m below the absorber plate provided vent space for enhanced convection and hot air movement. The dryer efficiency was further enhanced by a south facing mirror was placed adjacent to the drying chamber to reflect most of the incident radiation on the flat plate collector below. This dryer could effectively function up to 6 h after sunset with air temperature approximately 6 °C higher than ambient. The thermal efficiency of the dryer was calculated to be 28.2%. Economic analysis of the dryer performance suggested its financial viability with a payback period of 1.5 year. The prospect of a natural convection-type solar dryer with latent heat storage for drying of ginger was investigated by Sain et al. [19] in load and unload conditions. The dryer was constructed using a solar collector of area 2 $m^2$ and a drying chamber of 0.8 $m^2$. Paraffin wax of volume 0.006$m^3$ was used for the experiments. The moisture content of ginger was reduced from 74 to 3% (w.b) under full load conditions over a period of 24 h. The collector efficiency varied from 53–96% under no load and 40–55% under full load conditions. The estimated drying efficiency was 12.4% with an overall system efficiency of 22.7% (Fig. 8).

Dina et al. [20] studied the effectiveness of a continuous solar dryer integrated with desiccant thermal storage for drying cocoa beans. Two types of desiccants were tested, viz. (Na86[(AlO2)86 · (SiO2)106]0.264H2O) as an adsorbent type and CaCl2

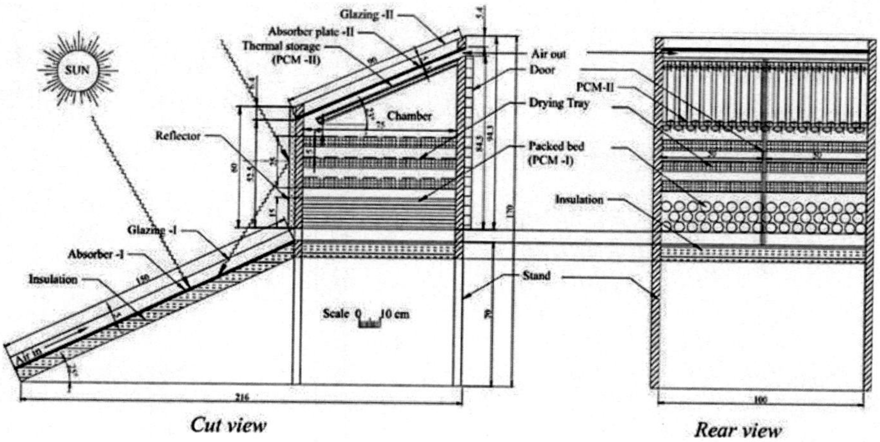


**Fig. 8** Cross-sectional view of a natural convective solar air dryer with latent heat storage [18]

as absorbent type. A north-facing flat plate-type solar air collector with a tilt angle of 60° of dimensions 2 m × 0.5 m × 0.1 m was used. Two transparent glass covers having 2 cm air gap were used to prevent top losses. The heated air was forced to a drying chamber of volume 50 $cm^3$ and integrated with perforated aluminium sheet drying tray of surface area 49$cm^2$. The thermal storage was placed in an open steel container of dimensions 30 cm × 30 cm × 5 cm just below the drying tray. It was found that there was a 25% reduction in drying time using adsorbent-type desiccant, whereas for the absorbent type, the drying time reduced by 45.45%, when compared to traditional intermittent direct sun drying. Moreover, there was a 68% and 78% improvement in specific energy consumption over direct sun drying for the adsorbent type and absorbent type, respectively. A hybrid solar air dryer for dehydration of mushrooms using paraffin wax as latent heat storage was proposed and designed by Reyes et al. [21]. The system comprised of a solar panel, a solar energy accumulator, an electrical heater, a centrifugal fan and a drying chamber. The 3 m ×1 m solar panel was composed of a glass sheet of 5 mm thickness and a black wavy zinc plate 30–50 mm below it to facilitate airflow between them. The incident solar radiation was further increased by introducing 40 parallel zinc fins of 3 m length and 3 cm height on the plate. The solar energy accumulator placed beside the solar panel contained 14 kg PCM distributed in 100 copper tubes with external aluminium fins for enhanced heat transfer to the drying air. Airflow from the solar panel and accumulator was regulated with the help of valves. The drying chamber consisted of a total of 10 perforated stainless steel trays. 70–80% of the expended air was recycled using a centrifugal fan and thus mixed with the incoming heated air before reaching the electrical heating system to maintain the drying air temperature at 60 °C. It was found that thin mushroom slices and lower amount of recycled air statistically favoured the mushroom drying process. The drying kinetics adequately conformed to the specifications of simplified constant diffusivity model and Page's model. The maximum thermal efficiency obtained was 67% with the accumulator

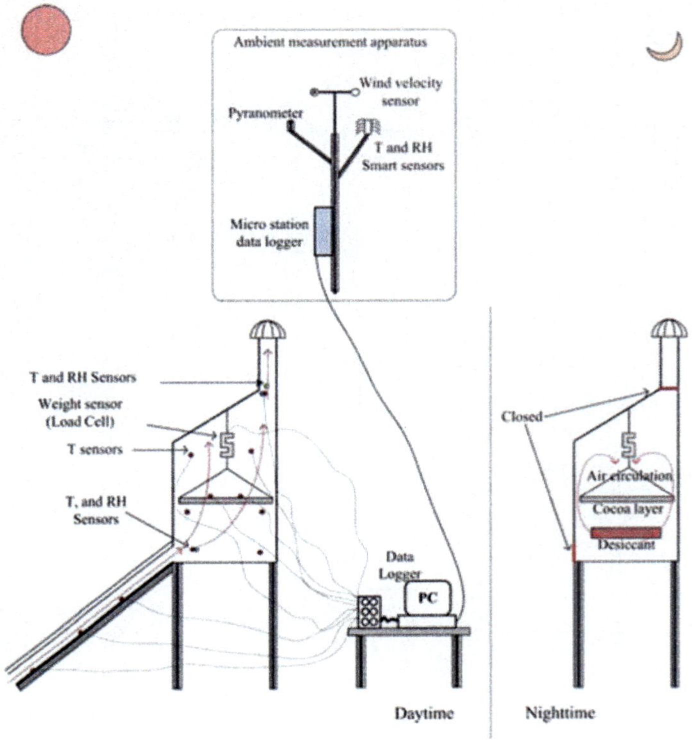


**Fig. 9** A continuous solar air dryer integrated with desiccant thermal storage [20]

accounting for 20% of the maximum energy fraction. This efficiency could be further augmented by increasing the mass of paraffin wax in the accumulator (Figs. 9 and 10).

## 5 Conclusions

This chapter reviews the various research activities carried out in solar drying applications implementing latent heat storage media for rational and effective energy management. Latent heat storage systems have the potential to improve system performance and minimize the discrepancies between energy supply and demand due to their higher value of energy storage density as compared to sensible heat storage devices. PCM-based energy storage devices have been found to be particularly advantageous in the drying of medicinal plants due to their capability of maintaining a constant temperature during the process. The thermal performance of a phase change material can be further improved by optimizing its thermal conductivity and heat transfer with the working fluid. It has been found that integration of carbon

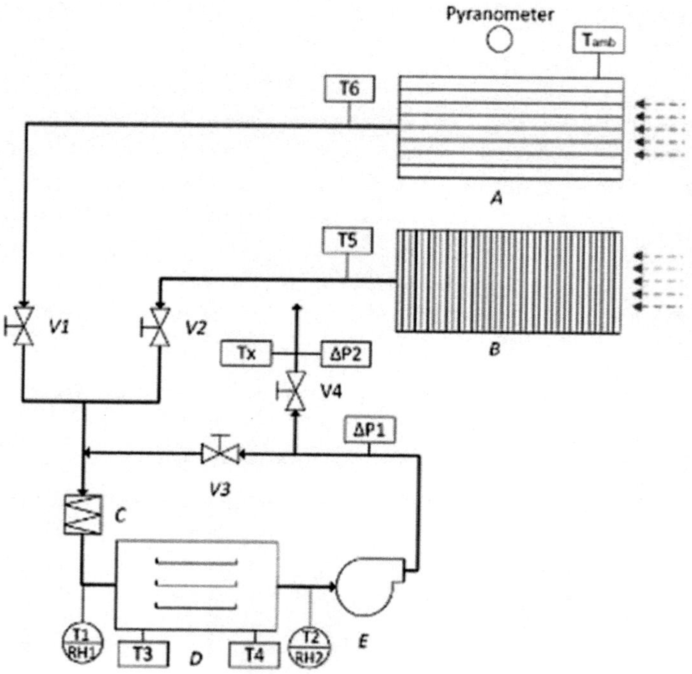


**Fig. 10** A PCM integrated hybrid solar air dryer [21]

fibres, graphite foam and expanded graphite enhances the thermal conductivity of paraffin wax significantly. Computational modelling software can be efficiently used to improve dryer designs using PCM. A number of research studies based on design parameters have also been conducted to improve the solar air heater connected to the drying chamber. There is significant scope for further research on the development of efficient, cost-effective PCMs for solar thermal energy storage applications.

## References

1. F. Kreith, D.Y. Gowami, *Handbook of Energy Efficiency and Renewable Energy* (New York, Taylor and Francis Group, London, 2007)
2. M. Thirugnanasambandam, S. Iniyan, R. Goic, A review of solar thermal technologies. Renew. Sustain. Energy Rev. **4**(1), 312–322 (2010)
3. R. Tchinda, A review of the mathematical models for predicting solar air heaters systems. Renew. Sustain. Energy Rev. **13**(8), 1734–1759 (2009)
4. M.A. Wazed, Y. Nukman, M.T. Islam, Design and fabrication of a cost effective solar air heater for Bangladesh. Appl. Energy **87**(10), 3030–3036 (2010)
5. M.C. Browne, B. Norton, S.J. McCormack, Heat retention of a photovoltaic/thermal collector with PCM. Sol. Energy **133**, 533–548 (2016)
6. L.M. Bal, S. Satya, S.N. Naik, V. Meda, Review of solar dryers with latent heat storage systems for agricultural products. Renew. Sustain. Energy Rev. **15**(1), 876–880 (2011)

7. H.P. Garg, J. Prakash, *Solar Energy Fundamentals and Applications* (Tata McGraw-Hill, New Delhi, 1997)
8. J.A. Duffie, W.A. Beckman, *Solar Engineering of Thermal Processes* (Wiley, New York, 1980)
9. K. Kant, A. Shukla, A. Sharma, A. Kumar, A. Jain, Thermal energy storage based solar drying systems: a review, in *Innovative Food Science and Emerging Technologies*, vol. 34 (2016), pp. 86–99
10. A. Fudholi, K. Sopian, M.H. Ruslan, M.A. Alghoul, M.Y. Sulaiman, Review of solar dryers for agricultural and marine products, 2010. Renew. Sustain. Energy Rev. **14**(1), 1–30 (2010)
11. Fudhol, K. Sopian, A review of solar air flat plate collector for drying application. Renew. Sustain. Energy Rev. **102**(1), 333–345 (2019)
12. S. Devahastin, S. Pitaksuriyarat, Use of latent heat storage to conserve energy during drying and its effect on drying kinetics of a food product. Appl. Therm. Eng. **26**(14–15), 1705–1713 (2006)
13. G. Cakmak, C. Yildiz, The drying kinetics of seeded grape in solar dryer with PCM-based solar integrated collector. Food Bioprod. Process. **89**(2), 103–108 (2011)
14. S. Esakkimuthu, A.H. Hassabou, C. Palaniappan, M. Spinnler, J. Blumenberg, R. Velraj, Experimental investigation on phase change material based thermal storage system for solar air heating applications. Sol. Energy **88**, 144–153 (2013)
15. V. Shanmugam, E. Natarajan, Experimental study of regenerative desiccant integrated solar dryer with and without reflective mirror. Appl. Therm. Eng. **27**(8–9), 1543–1551 (2007)
16. S.M. Shalaby, M.A. Bek, Experimental investigation of a novel indirect solar dryer implementing PCM as energy storage medium. Energy Convers. Manage. **83**, 1–8 (2014)
17. V. Shringi, S. Kothari, N.L. Panwar, Experimental investigation of drying of garlic clove in solar dryer using phase change material as energy storage. J. Therm. Anal. Calorim. **118**(1), 533–539 (2014)
18. D. Jain, P. Tewari, Performance of indirect through pass natural convective solar crop dryer with phase change thermal energy storage. Renew Energy **80**, 244–250 (2015)
19. P. Sain, V. Songara, R. Karir, N. Balan, Natural convection type solar dryer with latent heat storage, in *Proceedings of 2013 International Conference on Renewable Energy and Sustainable Energy* (2014), pp. 9–14.
20. S.F. Dina, H. Ambarita, F.H. Napitupulu, H. Kawai, Study on effectiveness of continuous solar dryer integrated with desiccant thermal storage for drying cocoa beans. Case Stud. Thermal Eng. **5**, 32–40 (2015)
21. A. Reyes, A. Mahn, F. Vásquez, Mushrooms dehydration in a hybrid-solar dryer using a phase change material. Energy Convers. Manage. **83**, 241–248 (2014)

# Finite Element Analysis of Stamping Process of Maraging Steel Built-Up Wing Panel of a Missile

**P. Sridhar Reddy, B. V. R. Reddy, S. R. Maity, and K. M. Pandey**

**Abstract** Wing manufacturing is one of the critical tasks in aerospace industry. Built-up-wing panel is manufactured by forming sheet blank into the desired shape using a stamping die. The design of a stamping die essentially begins with the geometric development of die-faces, which are forming interfaces of punch, die, and binder. Nevertheless, ensuing forming interface geometry is hardly possible right at the first time, and depending on the experience and skills of the methods engineer, several costly physical tryouts may be required to ensure a die-face design that deforms the blank into the required stamping part. Adding to this, the wing panel geometry is a complex 3D surface and uses maraging steel as the material which is an ultra-high strength steel, forming behavior of which is unknown to the industry. Therefore, it becomes very difficult to develop a stamping die which forms the desired part and also consumes relatively low process time and other valuable resources. Finite Element (FE) simulation of the process of sheet metal forming, on the other hand, shifts the costly press-shop try-outs to virtual environment and provides the essential information on material forming behavior, part formability, spring back deformation, forming process feasibility, etc. This information is useful in designing the forming interface of the die. In the present study, the plastic anisotropy parameters are determined based on the tensile tests and these values are used in the FE analysis. The study and analysis of the present design of die and the forming process are done and the results are presented. The FE analysis of the stamping process of maraging steel wing panel has been performed. Based on FE analysis, a new die-face design is suggested, and comparison of the present and the suggested forming processes are presented together with the conclusions.

**Keywords** FEA · Built-up-wing · Sheet metal forming · Spring back

P. Sridhar Reddy (✉)
Sukhoi Engine Division, Hindustan Aeronautics Limited, Koraput, Odisha 7636002, India
e-mail: reddymeta.iitr@gmail.com; sridharreddy2032@gmail.com

B. V. R. Reddy · S. R. Maity · K. M. Pandey
Department of Mechanical Engineering, National Institute of Technology, Silchar 788010, India
e-mail: kmpandey2001@yahoo.com


K. M. Pandey et al. (eds.), *Recent Advances in Mechanical Engineering*, Lecture Notes in Mechanical Engineering, https://doi.org/10.1007/978-981-15-7711-6_33

# 1 Introduction

Wing is one of the most critical components of a supersonic missile. The basic purpose of the wing is to generate the required lift to the missile. The ideal wing must be light in weight, having higher bending stiffness, and should have modal frequencies completely separated from the missile system frequency. Also, the wing design must be suitable for mass production because, ultimately, these wings are required to be produced in large numbers. The earlier wing, which was machined out of bar stock, had lower bending stiffness, heavyweight, and high deflection at the tip. These factors lead to the need for higher control forces to keep the missile on its designed path and caused instability in the missile system. Its modal frequencies were very close to the missile system frequency and this resulted in "resonance", which had detrimental effect on missile during flight. Also, the manufacturing process was uneconomical, involved long cycle time and high cost of production. Built-up-Wing is designed to overcome the above drawbacks and to reduce the weight of the wing. Built-up wing is an assembly of three components, namely, wing panel top, wing panel bottom and stiffener. Sheet metal forming process is chosen to manufacture all the components of built-up wing considering the cost of production, economy, and cycle time. All these components are made up of 1 mm thick Maraging Steel sheets. Maraging Steel is a low carbon ultra-high strength steel, principal alloying element being Nickel (15–25%). It possesses superior strength and toughness without losing malleability and offers good weldability and machinability. It is chosen because of two main reasons; one is its high tensile strength and stiffness and the other, its corrosion resistance. As the rocket motor is also made of maraging steel, welding defects can also be minimized [1]. Material model describes the material behavior. Earlier, simple models like Tresca and von Mises were used. These models no longer appear to have unlimited validity since anisotropy, kinematic hardening, and so on are described inadequately. From Fig. 1 the influence of appropriate material model on the accuracy of the simulation results, especially spring back, can be understood.

Firat et al. [3] shows that shell element formulation for sheet metal forming simulation with explicit- dynamic time integration schemes gives desired accuracy with optimum computer resources, whereas, spring back simulation is better performed using shell elements with implicit-static time integration scheme. Based on the aforementioned studies, the following guidelines are presented for FE analysis of stamping process of a maraging steel wing panel. (i) Four-node shell elements are

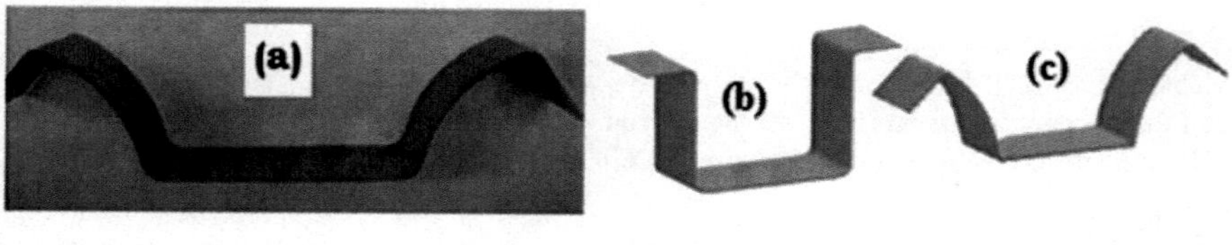


**Fig. 1** FE simulation of spring back using different material models **a** experimental, **b** isotropic model, **c** Y-U model [2]

used in FE mesh of the blank. (ii) Three and four node rigid shell elements are used in the geometric mesh of forming interfaces. (iii) Hill orthotropic material law with isotropic hardening rule is used. (iv) Five integration points are used in the through thickness direction to account for bending stresses, which gives better spring back prediction. (v) Explicit dynamic time integration scheme is employed in the forming simulation. (vi) Implicit static time integration scheme is used for spring back simulation. (vii) A constant coefficient of friction of 0.125isassumed.

## 2 Experimental Procedure

### *2.1 Selection of the Material*

Maraging Steel is a low carbon ultra-high strength steel with the principal alloying element being Nickel (15–25%). It possesses superior strength and toughness without losing malleability and offers good weldability and machinability. The most important criteria in selecting a material are related to the function of the part—qualities such as strength, density, and stiffness and corrosion resistance. For sheet metal, the ability to be shaped in a given forming process, often called its formability, should also be considered. To assess formability, the behavior of the sheet has to be described precisely and the properties are to be expressed in a mathematical form, which are derived from various tests.

### *2.2 Tensile Testing*

The aim of this test is to find the Young's Modulus ($E$), Ultimate Tensile Strength, Tensile Strain-Hardening Exponent ($n$), and Strength Coefficient ($K$) of a 0.8 mm thick Maraging steel sheet specimen. Tensile test specimen preparation and the tensile test were done according to ASTM E8/E8M-09 standard [4]. The data collected from the test is used to calculate strain hardening exponent ($n$) and strength coefficient (K) following the procedure given in ASTM E 646-07 standard [5]. Test is conducted on 100 kN INSTRON 5500R Universal Testing Machine. Least count of the 100 kN Load Cell used is 1 N. A 50 mm gauge length INSTRON make extensometer of least count 0.001 mm is used for strain measurement. Mitutoyo makes digital Vernier calipers with least count 0.001 mm was used for measuring specimen dimensions. Firstly, the thickness and width of the reduced section of the specimen were measured and recorded. Then, specimen was gripped in the testing machine in a manner to ensure axial alignment with cross-heads and extensometer is attached. The test was conducted at 23 °C with 50% humidity at a crosshead speed of 1 mm/min and the test continues until the specimen breaks. The procedure is repeated for the remaining samples too. Specimen represented in Fig. 2

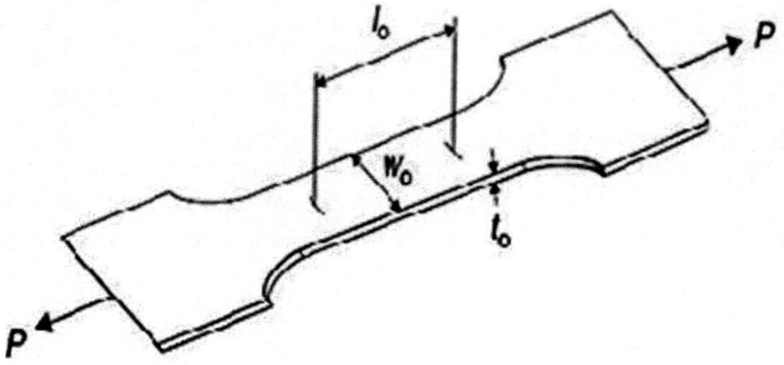


**Fig. 2** Typical tensile test specimen [2]. Where G—gauge length = 50 mm; T—thickness = 0.8 mm; R—radius of Fillet = 13.5 mm; W—width = 12.5 mm; A—length of reduced section = 60 mm; B, L—overall length = 200 mm; B—length of grip section = 50 mm; C—width of grip section = 20 mm

### 2.3 *Stamping Simulation Procedure of Maraging Steel Built-Up Wing Panel*

With the finite element analysis and design of stamping process of a maraging steel wing panel. The present stamping analysis consists of mainly two parts namely forming analysis and spring back analysis. Forming analysis involves forming simulation and the formability analysis of the part by observing the values of stamping criteria like, thinning, splitting, etc. Springback analysis involves study of the deformation caused in the blank after the forming loads are removed. The FE modeling of die-face and the blank are described. Also, forming process simulation and springback simulation are presented.

**Blank**: A flat, precut metal shape ready for subsequent press operation. This is a piece of sheet metal, produced in cutting dies or by any other means such as Wire EDM, etc., that is to be subjected to further press operations. A blank may have a specific shape developed to facilitate forming or to eliminate a trimming operation subsequent to forming.

**Blank Development**: Determination of the optimum size and shape of a blank for a specific part. Blank size estimation is done based on the volume constancy assumption and using the stamping criteria like allowable thinning and the amount of stretch of the part. As the present forming process of wing panel involves mainly bending; the thinning and stretching effects are negligible. Hence, 3D CAD model of blank shown in Fig. 5 is developed in the CAD software Pro/Engineer by unfolding the 3D CAD model of the wing panel shown in Figs. 3 and 4. The 3D CAD model of the blank is imported into HyperForm—an FE based sheet metal forming simulation software. The FE mesh of blank is shown in Fig. 6.

**Fig. 3** Front view and top view of 3D wing panel

**Fig. 4** Isometric view of 3D CAD model of wing panel

The die-face for a sheet metal forming die may be defined as the composition of a complete surface geometry that deforms a sheet metal blank plastically into a desired stamping shape by ensuring a rigid tooling construction. In the present case, die-face comprises forming interfaces of die, punch and binder. The design of forming process of the part stamping form is done following a pure geometric modeling approach. All forming interfaces are assumed rigid and hence their deformations are not considered. Undercut check is conducted and ensured that the die locking is avoided. Using the 3-D CAD model of the part, a set of surfaces for the lower die geometry are generated as shown in Fig. 6; this 3-D composite surface forms the forming interface geometry from which the other tooling elements are obtained. The lower die forming interface is imported into Hyper Form and a geometric mesh

**Fig. 5** 3D blank developed from the part geometry

**Fig. 6** Isometric view of lower die forming interface

is generated using three-node and four-node shell elements as shown in Fig. 7 The punch and binder interface geometric meshes are generated as shown in Fig. 8 and Fig. 9 respectively, by a simple duplication using their geometric counterpart, i.e., lower die geometry. At this stage, the die-face design is completed and in the present

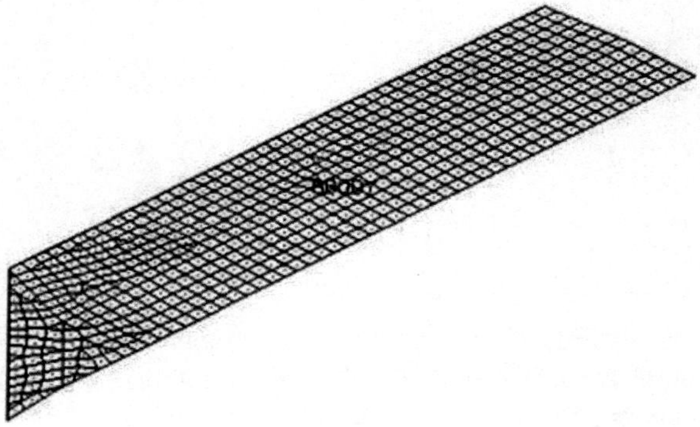

**Fig. 7** Rigid mesh of lower die forming interface

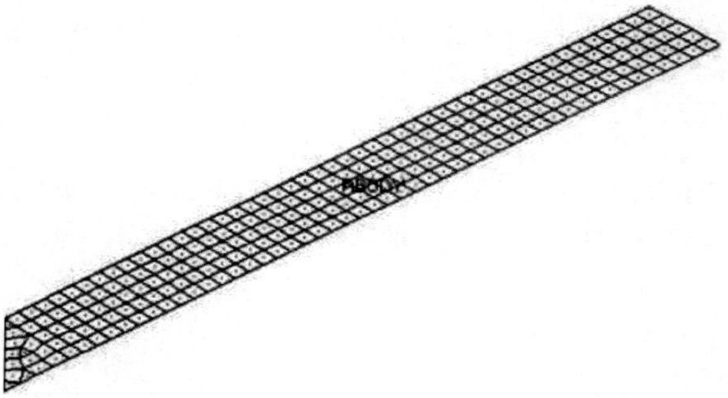

**Fig. 8** Rigid mesh of binder die-face

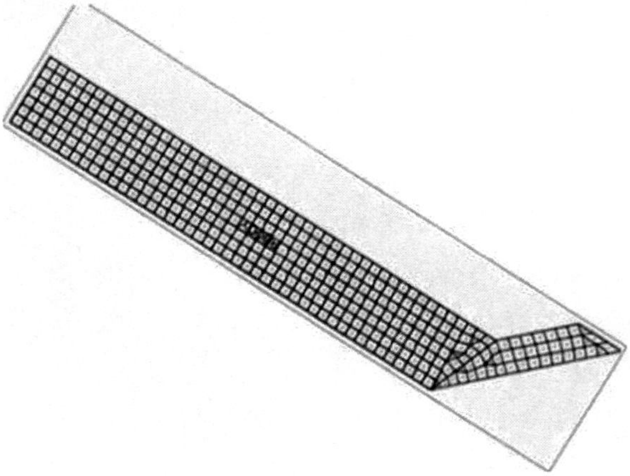

**Fig. 9** Rigid mesh of punch forming interface

**Table 1** Tensile test I result

| S. No. | Width (mm) | Thickness (mm) | Extensometer gauge length (mm) | Yield stress (MPa) | Tensile strength (MPa) | *E* (GPa) | *n* |
|---|---|---|---|---|---|---|---|
| 1 | 12.50 | 0.80 | 50 | 878 | 1013 | 200 | 0.13 |
| 2 | 12.50 | 0.82 | 50 | 851 | 1006 | 199.7 | 0.16 |
| 3 | 12.48 | 0.82 | 50 | 796 | 991 | 203 | 0.24 |

Averages of the above values are given: *Y*—Yield Strength—**841.61 MPa**, *T*—Tensile Strength—**1003.33**, *MPaE*—Young's Modulus—**200.9**, *GPan*—Strain-Hardening Exponent—**0.17** **K**—Strength Coefficient—**1822 MPa**

case the punch, binder, and the die surfaces consist of 1,231 three-node and four-node shell elements in total. The model can now be employed for the formability assessment and springback analyses.

## 3 Results and Discussions

### 3.1 Tensile behavior

Obtained tensile properties of maraging steel in Tables 1 and 2.

The above results are obtained from the simple tensile tests conducted on 0.8 mm thick M250 grade cold rolled annealed maraging steel sheets. From the above, it can be seen that maraging steel is an ultra-high strength and therefore, the parts made of it have high strength to weight ratio. As the *Y*/*E* ratio is high, springback related issues will arise during forming. Lankford parameters show that it is anisotropic and since $\mathbf{\Delta R}$ is of considerable value the orientation of the sheet with respect to die or the part to be formed is important. Since, $\mathbf{R_m}$ is greater than unity, the material offers high resistance to thinning. From the above discussions, it can be concluded that, in the stamping process of maraging steel wing panel, thinning effects are negligible but there will be considerable amount of springback distortions after forming.

### 3.2 Forming Analysis

The below-shown wing panel is made by the forming process in practice. This is a physical trial and error method. The form of the wing shown in Fig. 10 was achieved after two physical tryouts. In the figure, the top and bottom wing panels are placed on each other and considerable amount of gap can be seen between the corresponding tips and leading edges. Ideally, there should be no gap and corresponding edges should rest one on each other. It also shows that the gap is also not uniform and

**Table 2** Tensile test II results

| | Specimen 1 | Specimen 2 | Specimen 3 |
|---|---|---|---|
| Initial width, $w_0$ (mm) | 12.48 | 12.33 | 12.50 |
| Initial thickness, $t_0$ (mm) | 0.80 | 0.80 | 0.82 |
| Initial gauge length, $l_0$ (mm) | 50 | 50 | 50 |
| Max. load (kN) | 9.45 | 9.11 | 9.85 |
| Stress at max. load, $P$ (MPa) | 946 | 923 | 961 |
| Final width, $w_f$ (mm) | 12.36 | 12.22 | 12.38 |
| Final length, $l_f$ (mm) | 51.01 | 50.98 | 50.8 |
| Plastic strain ratio, $R$ | $R_0 = 1.03$ | $R_{45} = 0.858$ | $R_{90} = 1.524$ |
| Planar anisotropy, $\Delta R$ | $(R_0 + R_{90} - 2R_{45})/2 = 0.419$ | | |
| Normal plastic anisotropy, $R_m$ | $(R_0 + 2R_{45} + R_{90})/4 = 1.068$ | | |

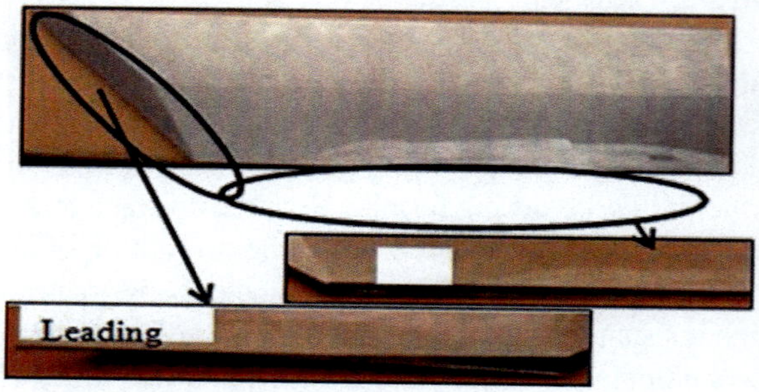


**Fig. 10** Views of leading edge and tip of wing panel

varying along leading edge and tip. From this it can be said that the wing surfaces are not planar but are distorted, which is attributed to the spring back effect (Fig. 11).

In Fig. 12, vertical axis represents the position of LE from Root and the horizontal axis represents various points on the LE starting from Tip and towards Root. Ideally, LE should be vertically5mm below the root, to obtain this 8.5 mm is provided on the die to compensate springback. But, LE obtained has lot of deviation.

In Fig. 13, vertical axis represents the vertical distance between the Root and Tip and the horizontal axis represents the various points on the wing which starts from rear end and towards LE. Ideally, the vertical distance between the Tip and Root should be 5 mm but the case is not so. The tip is not uniform and a deviation can be observed.

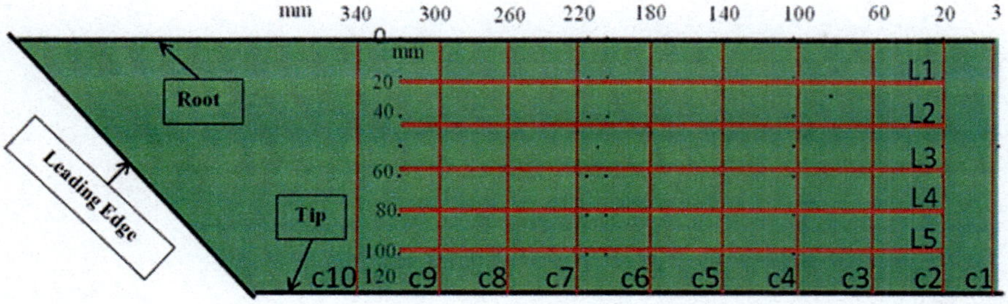


**Fig. 11** Description of wing

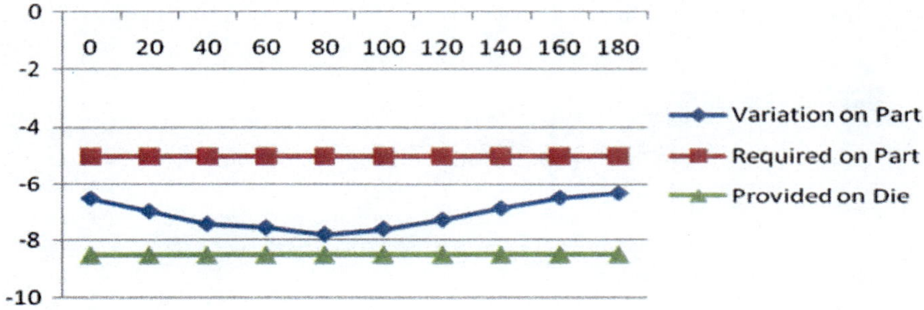


**Fig. 12** Leading edge profile of wing

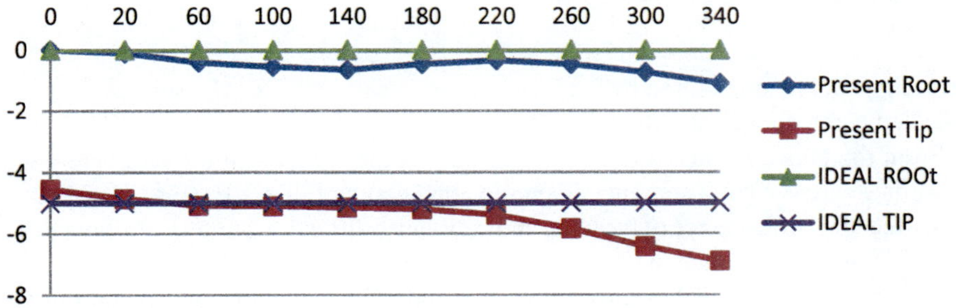


**Fig. 13** Root and tip variation in wing panel

In Fig. 14, vertical axis represents the vertical distance between the Root and Tip and the horizontal axis represents the various points on the wing which starts from rear end and towards LE. The surface profile matches with the Ideal at the rear of the wing and it is increasingly deviating towards LE.

In the Fig. 15, the horizontal axis represents the various points along cross-section of the wing starting from Root and towards Tip; vertical axis represents the position of the above points. It can be observed the variation in the cross-section of wing from rear end and towards LE. The above graphs are obtained by scanning the surface of

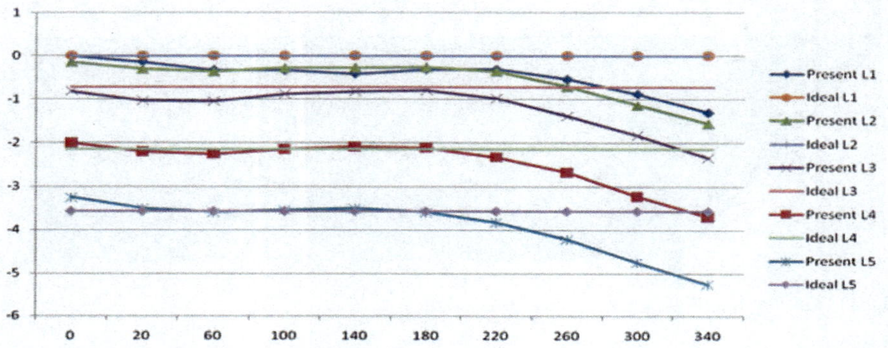


**Fig. 14** Longitudinal section variation of wing panel

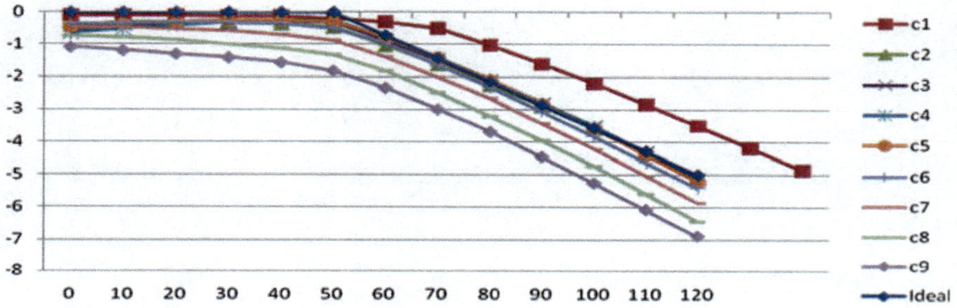


**Fig. 15** Cross-section variation of wing panel

the wing panel manufactured by the forming process in practice. The surface data is collected using coordinate Measuring Machine.

From the above results, it can be observed that the surfaces of the wing are not uniform. There is a large amount of deviation on Leading Edge (LE) (Fig. 13) which may affect the performance of the wing and may cause difficulty while assembling the wing. This variation is due to the presence of compound angle on the LE surface and calls for proper springback compensation. From the Tip Variation and Longitudinal plots, (Fig. 14 and Fig. 15 respectively), the trend in the graphs shows that the surface profile is satisfactory from rear side of the panel, i.e., 0 mm and up to 180 mm towards LE and thereafter there is considerable deviation between the required surface profile and the obtained surface profile. It is also observed that the springback deformation is more at the rear side and it decreases towards LE side. It shows that there is a non-uniform springback deformation. In the current forming practice uniform springback compensation is provided on the die, which leads to overcompensation around LE portion. Due to the complex 3D part geometry and the use of anisotropic ultra-high strength steel, non-uniform springback deformation is present. As the part design and the material are fixed and changes cannot be done, necessary changes are to be made to the forming process and the design of die considering the non-uniform springback. This forming process is suggested based on the FE analysis results. In this process, dies are provided with proper springback compensation. More than 10% thinning is not acceptable. Leading-edge and tip should lie on the same plane. The normal distance between the root plane and tip plane is 5 mm with a tolerance of ±1 mm.

From Fig. 16 it can be said that maximum thinning is less than 0.5% which is negligible (well below the 10% limit)

From Fig. 17, it can be observed that the portion of leading edge near root has deformed more compared to the portion of leading edge near tip. This is because of relatively greater spring back observed at that portion. As discussed previously, the spring back is not uniform and the compensation should be in accordance with the spring back.

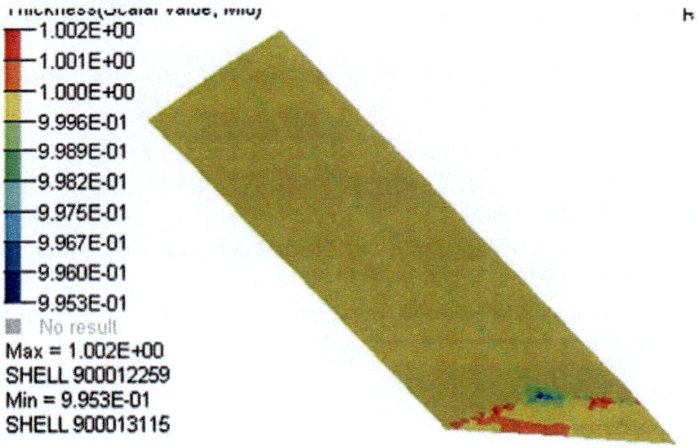


**Fig. 16** Thickness plot at the end of forming

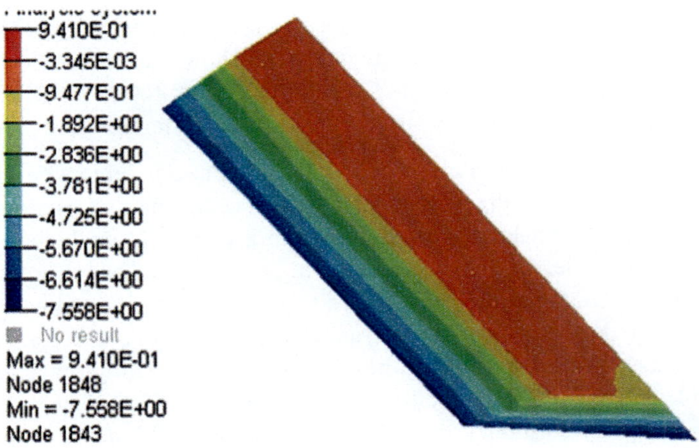


**Fig. 17** Z displacement plot at the end of forming

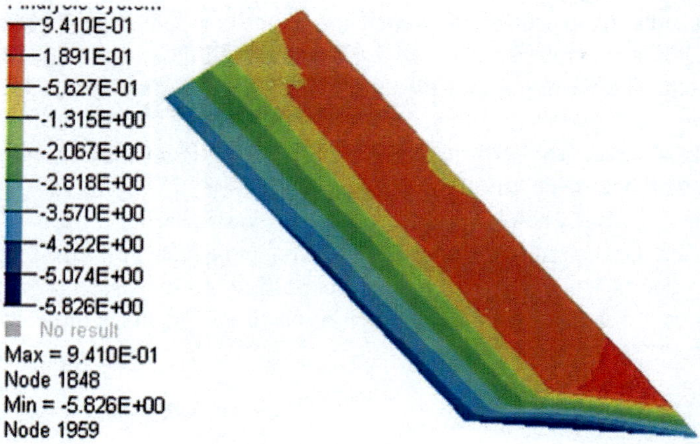


**Fig. 18** Z displacement plot after spring back

From Fig. 18 it can be seen that the leading edge and tip nearly lie on the same plane with acceptable deviation from the required. The portion at the centre of the leading edge seems to be deviating but the deviation is acceptable or can be sort out manually. Also, the third criterion that is the normal distance between the root plane and tip plane is also satisfied. From the above discussion, it can be concluded that the suggested forming process and die design are acceptable in the manufacturing of wing panel and the process can be controlled better compared to the previously discussed one.

## 4 Conclusions

The material is anisotropic and the part produced with it will have a higher strength to weight ratio because of its high yield strength. Since, flow stress to elastic modulus ratio is higher; springback deformation is predominant after forming. As Rm value is greater than 1, material exhibits resistance to thinning during forming process.

1. With the forming process in practice, the part produced exhibits non- uniform springback deformation after forming, but uniform compensation is provided on the die which over-compensates the wing at some portions. Hence, high distortion is observed on the leading edge due to the improper springback compensation. One of the important requirements is Leading edge and tip should be on same plane but large deviations are observed and this is because of the uniform spring-back compensation provided on the die. The surfaces of the wing panel produced are not planar and high amount of deviations are observed.
2. With the suggested forming process, negligible amount of thinning is observed, but springback is predominant. As the non-uniform compensation is provided on the die based on the FE analysis results, less distortion is observed on the leading

edge. Leading edge and tip are on same plane which meets one of the important criteria. The surfaces of the wing panel produced are comparatively better and very less amount of deviations are observed.

3. Hence, better wing would be obtained with suggested forming process. As the binder is used, the process can be controlled in a better manner. With the FE simulation of stamping process, better forming process design and die-face design estimate can be obtained before first try-out.

## References


1. K. Abhishek, K. Hemant, S. Saikumar, D.R. Yadav, Development of built up wing of a missile using sheet metal forming process, in *Proceedings of the 3rd Int. & 24th AIMTDR Conference* (2010), pp. 557–563
2. Z. Marciniak, J.L. Duncan, S.J. Hu, *Mechanics of Sheet Metal Forming*, 2nd edn. (Butterworth-Heinemann, 2002)
3. M. Firat, Computer aided analysis and design of sheet metal forming processes: part I—the finite element modeling concepts. Mater. Des. **28**, 1298–1303 (2007)
4. E8/E8M—09, Standard test methods for tension testing of metallic materials (ASTM, 2009)
5. E 646—07, Standard test method for tensile strain-hardening exponents (n-Values) of metallic sheet materials (ASTM, 2007)

# Effect of Non-uniform Heating on Forced Convective Flow Through Asymmetric Wavy Channel

**Sumit Kumar Mehta and Sukumar Pati**

**Abstract** We explore the heat transfer characteristics for forced convective flow of a Newtonian fluid through the wavy channel under the effect of non-uniform heat flux. The findings are presented for different values of Reynolds number (Re) and dimensionless wavelength of the non-uniform heating ($\gamma$) in the range of $100 \leq \text{Re} \leq 500$ and $0.25 \leq \gamma \leq 4$, respectively. The non-uniform heating decreases the hot spot intensity in the circulatory flow zone and moreover it induces additional maxima of Nusselt number as compared to the constant heating case. The average Nusselt number for sinusoidal heating case is much more than the constant heating case for the smaller undulation of the non-uniform heating.

**Keywords** Wavy channel · Nusselt number · Non-uniform heating · Forced convection

## 1 Introduction

The increase in power consumption in the form of thermal energy eagerly needs efficient heat exchanging devices. There are two methods of heat transfer enhancement technique: active and passive methods. The passive method requires geometrical modifications like waviness and corrugation, porous media, nanofluid; whereas, the active method needs an external force to perturb the flow other than pumping power. In the last few decades, researchers have studied the passive heat transfer enhancement techniques using waviness. Mehta and Pati [1] found that the formation of the recirculation zone in the sinusoidal wavy channel causes local heat transfer enhancement. Wang and Chen [2] numerically explored the forced convective flow through sinusoidal wavy channel for the laminar regime. They found that the wavy

S. K. Mehta (✉) · S. Pati
Department of Mechanical Engineering, National Institute of Technology, Silchar, Silchar, India
e-mail: sumit090391@gmail.com

S. Pati
e-mail: sukumarpati@gmail.com



K. M. Pandey et al. (eds.), *Recent Advances in Mechanical Engineering*, Lecture Notes in Mechanical Engineering, https://doi.org/10.1007/978-981-15-7711-6_34

channel always increases the average Nusselt number in the considered regime of Reynolds number. Pati et al. [3] numerically compared the thermo-hydraulic performance for two types of the wavy channel and found that serpentine channel has always higher performance factor. Mehta and Pati [4] numerically investigated the effect of triangular corrugation on thermo-fluidic transport characteristics for flow of air in laminar regime. It is found from their work that the higher value of the performance parameter depends on both the amplitude of triangular corrugation and the flow regime (Re) at a higher wavelength. Shubham et al. [5] analyzed the effects of rheology on the heat and fluidic transport characteristics for forced convective flow through sinusoidal corrugated channel. The effects of non-uniform heating on heat transfer characteristics have many practical applications, for example, electronic architecture creates discrete hot spot due to the heat generation by corresponding electronics components [6]. Mehta and Pati [7] investigated the effect of sinusoidal heating on the forced convective flow through raccoon wavy channel, with a change in phase lag and amplitude of sinusoidal heat flux. It is found from their work that the average Nusselt number is higher for smaller amplitude and zero phase lag. Alawadhi and Bourisliy [8] numerically analyzed the effect of discrete heat source for flow through symmetric wavy channel and found that the placing of discrete heat source at minimum cross-section gives the higher performance. Pati et al. [9] used different heat flux profile to identify the optimal heating strategy for minimization of peak temperature and irreversibility generation for forced convective flow through a circular pipe. From the reported work in the literature, it is found that the influence of sinusoidal heating on the flow transport characteristics for flow through asymmetric or serpentine wavy channel is not yet done although asymmetric wavy channel has higher performance factor for laminar flow regime [3]. Hence, the objective of the current work is to examine the effect of sinusoidal heating on the heat transfer characteristics for laminar forced convective flow through asymmetric serpentine wavy channel.

## 2 Theoretical Formulation

The two-dimensional, incompressible and steady flow of a Newtonian fluid through the asymmetric or serpentine wavy channel is considered for the current work as shown in Fig. 1. The undulation surface profiles of the top and bottom wall have

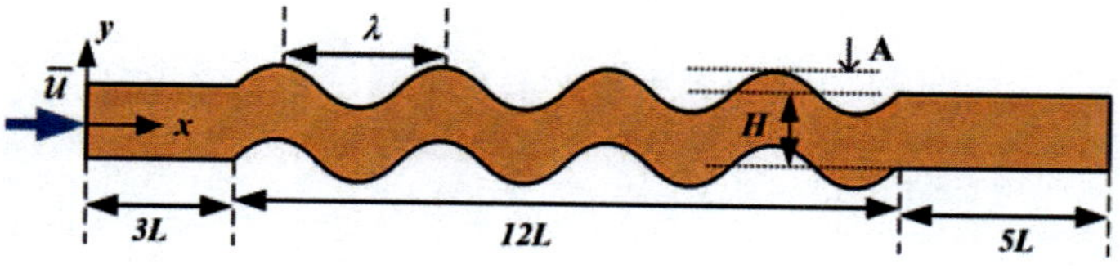


**Fig. 1** Physical domain

been represented mathematically as [2, 3]:

$$s_T(x) = L + 0.3A\sin\left[2\pi(x - 3L)/\lambda\right] \tag{1}$$

$$s_B(x) = -L - 0.3A\sin\left[\left(2\pi(x - 3L)/\lambda\right) + \pi\right] \tag{2}$$

where $L$, $A$, and $\lambda$ are the inlet half-height, amplitude, and wavelength of the sinusoidal wall, respectively. The normalized profile of the sinusoidal walls with the scale $L$ can be represented as:

$$S_T(X) = 1 + 0.3\sin\left[2\pi(X - 3)/\Lambda\right] \tag{3}$$

$$S_B(X) = -1 - 0.3\sin\left[\left(2\pi(X - 3)/\Lambda\right) + \pi\right] \tag{4}$$

For the present case, the value of dimensionless wavelength ($\Lambda$) and amplitude are taken as 2 and 0.3, respectively [2, 3]. The fluid at a temperature of $T_C$ enters the channel with uniform velocity ($\overline{u}$). The flat part of the wall is thermally insulated, whereas the wavy part imposed with the sinusoidal temperature distribution with upper and lower limit $T_H$ and $T_C$, respectively. The mathematical form of this sinusoidal wall temperature is given as [10]:

$$T_W = T_C + \left(0.5\Delta T\left[1 + \sin\left((2\pi[x - 3L])/\lambda_T\right)\right]\right) \tag{5}$$

Here, $\Delta T = (T_H - T_C)$ and $\lambda_T$ are the wavelengths of the sinusoidal wall temperature. The viscous dissipation and radiation effect are neglected. The temperature-independent thermo-physical properties are taken for the analysis. Under these assumptions, the governing transport equations in normalized form can be represented as follows [3, 4]:

Continuity equation:

$$\nabla \cdot \mathbf{U} = 0 \tag{6}$$

Momentum equation

$$(\mathbf{U} \cdot \nabla)\mathbf{U} = -\nabla P + (1/\mathrm{Re})\nabla^2\mathbf{U} \tag{7}$$

Energy equation

$$\mathbf{U} \cdot \nabla\theta = [1/(\mathrm{Re\ Pr})]\nabla \cdot (\nabla\theta) \tag{8}$$

Here, $\nabla = (\hat{i}\partial/\partial X + \hat{j}\partial/\partial Y)$ and normalized velocity vector $\mathbf{U} = (u\hat{i} + v\hat{j})/\overline{u} = U\hat{i} + V\hat{j}$. For the above governing transport equations, the normalizing scale for space is $L$, for velocity vector ($\mathbf{U}$) is $\overline{u}$, for pressure is $\rho\overline{u}^2$, and the normalized temperature is $\theta = (T - T_C)/(T_H - T_C)$. Where $P$ is the normalized pressure, $\mu$, $c_p$, $\rho$ and $k$ are dynamic viscosity, specific heat capacity, density, and thermal conductivity of the working fluid, respectively. The mathematical definition of Reynolds number and Prandtl numbers is:$\text{Re} = (\rho\bar{u}L)/\mu$ and $\text{Pr} = \mu c_p/k$, respectively. The imposed boundary conditions for above-mentioned governing transport equations are the following:

*At the inlet*:

$$U = 1, V = 0, \theta = 0 \tag{9}$$

*At the outlet*:

$$P_G = 0,\ \partial\theta/\partial X = 0, \tag{10}$$

*At the walls*:

$$U = V = 0 \tag{11a}$$

$$\theta = 0.5\left(1 + \sin\left[(2\pi(X - 3))/\gamma\right]\right)\text{for} 3 \le X \le 15 \tag{11b}$$

$$\partial\theta/\partial Y = 0 \text{ for } X < 3 \text{ or } X > 15 \tag{11c}$$

where $\gamma\left(= \lambda_T/L\right)$ is the dimensionless wavelength of the sinusoidal undulated wall temperature. The calculated temperature field is characterized by local Nusselt number as defined by [3, 7]:

$$Nu = -\partial\theta/\partial n \tag{12}$$

where $n$ is the outward normal to the undulated surface.

The total effect of heat transfer augmentation is written as in the form of mean Nusselt number and mathematically defined as [3, 7]:

$$\overline{Nu} = \int_{X=3}^{X=15} Nu\, dS \Big/ \int_{X=3}^{X=15} dS \tag{13}$$

## 3 Numerical Methodology and Model Validation

We implemented the finite element method numerical technique to solve the governing transport equation. The domain under consideration is divided into subdomains of unequal size, within each of which, the transport variables are approximated by using proper interpolation functions. As a result, the finite element equations are generated which are evaluated in a closed form. The iterative technique is employed to solve these equations. The details of this numerical procedure is presented in [4]. The grid independency has been done for the present study and the difference in calculating the average Nusselt number for the selected mesh is less than 1% as compared to the very fine mesh system. The selected mesh system is with the number of elements 85,998. The presented numerical model is validated with the work of Wang and Chen [2]. The same validation has been presented in [4].

## 4 Results and Discussion

The influence of sinusoidal heating on the thermo-fluidic characteristics for forced convective flow through an asymmetric wavy channel has been investigated numerically for laminar regime. The flow pattern, isotherms, and Nusselt number have been discussed for the suitable range of Reynolds number (Re) and dimensionless wavelength of the non-uniform heating ($\gamma$):$100 \leq \text{Re} \leq 500$ and $0.25 \leq \gamma \leq 4$[2–4].

Analysis of the flow pattern in the asymmetric wavy channel is made first to explain the related flow field. Figure 2 depicts the streamlines contour at Re = 100. The recirculation zone is generated in the concave surface (see from the core) due to the increase in momentum loss as advection strength decreases near the diverging-curved

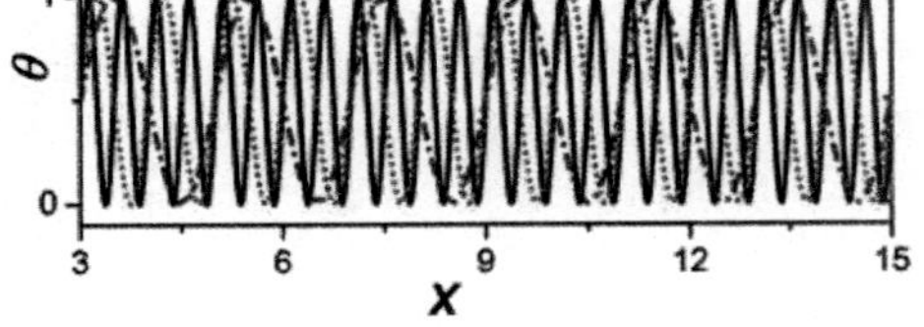


**Fig. 2** Streamline contours at Re = 100 and variation of dimensionless temperature at wavy wall with different dimensionless wavelength ($\gamma$)

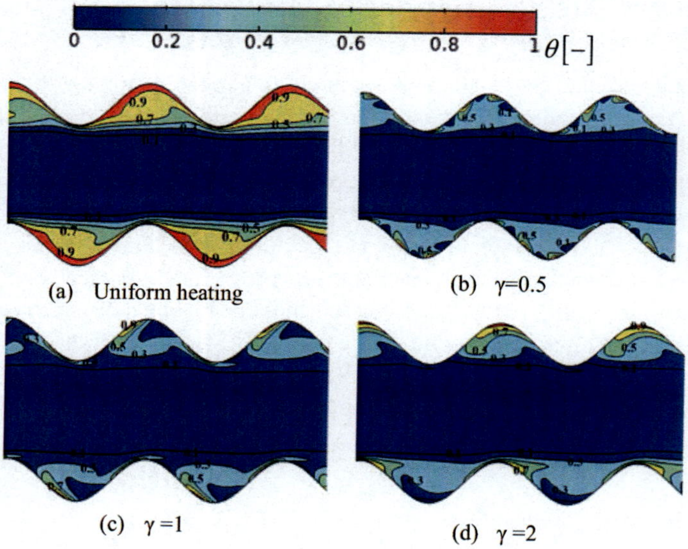


**Fig. 3** Normalized temperature and isotherm contours for **a** constant heating, **b** sinusoidal heating with $\gamma = 0.5$, **c** $\gamma = 1$ and **d** $\gamma = 2$ at Re = 100

wall [4, 11, 12]. The positioning of minima and maxima of non-uniform heating near the wall with the recirculation zone affects the temperature field. The corresponding positioning of non-uniform temperature distribution on wall is depicted in Fig. 2 for $\gamma = 0.5$, 1 and 2. It can be noted that $\gamma = 2$ represents the case where the wavelength of the undulation of temperature distribution and the wavy wall are the same. The cases of $\gamma < 2$ and $\gamma > 2$ represent the situation where the number of undulation of non-uniform heating is higher and smaller than the geometrical undulation, respectively. The corresponding effects on the isotherms contours are presented in Fig. 3.

For uniform heating, isotherms contour is presented in Fig. 3a. At the diverging part of the wall, higher intensity of isotherms are found due to the recirculation zone thus creating the hot spot zone due to the trapping of hot fluid by the circulatory flow. At the convex part (see from the core), higher velocity gradient results in smaller temperature intensity due to higher heat removal locally. The isotherms contour for non-uniform heating with $\gamma = 0.5$ is depicted in Fig. 3b. For this case, one geometrical undulation has four undulations of non-uniform heating. Hence, the decrease in heating intensity near the wall of the recirculation zone decreases the temperature intensity of the trapping hot fluid. For $\gamma = 1$, each converging and diverging parts contain maxima and minima of non-uniform heating. Hence, the similar arrangements of temperature field are obtained for top and bottom part as shown in Fig. 3c. For $\gamma = 2$, the maxima of non-uniform heating is placed at diverging and converging parts for the top and bottom walls, respectively. Hence, the larger region of higher temperature intensity is obtained for the top part.

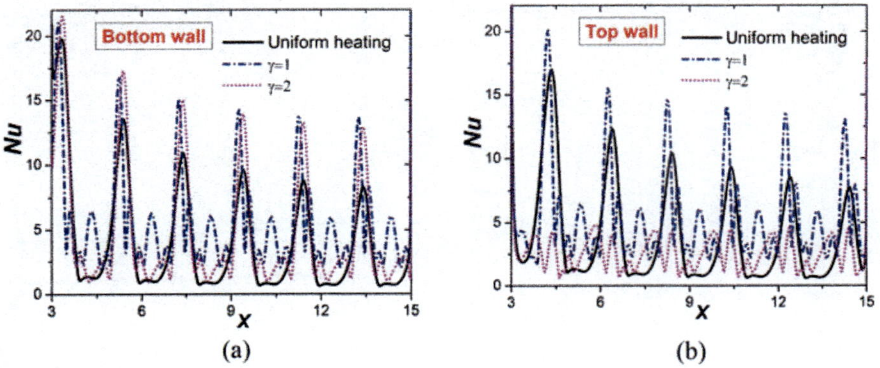


**Fig. 4** Variation of local Nusselt number at **a** bottom and **b** top wall for constant heating case and sinusoidal heating case with $\gamma = 1$ and 2, Re = 100

The variation of local heat transfer enhancement on the bottom and top wall at Re = 100, are illustrated in Fig. 4a and b, respectively, in terms of local Nusselt number ($Nu$). It can be observed from Fig. 4a that the maxima of $Nu$ at bottom wall convex part is higher for the sinusoidal heating case and these maxima are same for $\gamma = 1$ and 2 for every undulation. The maximum $Nu$ for sinusoidal heating case is found due to the combination of higher velocity gradient and relatively smaller temperature near the convex (see from the core) part.

For the uniform heating case, minima of $Nu$ is found at the primary flow detachment point where temperature intensity is high. On the other hand, for $\gamma = 1$, the additional $Nu$ maxima is found at the midpoint of concave (see from the core) part. It is because of the presence of the interface of maxima and minima of wall temperature, along with the significant temperature gradient created by the cold circulatory flow (See Fig. 3c). Whereas for $\gamma = 2$, the local minima of $Nu$ is found at the midpoint of concave undulation. It is because the cold wall temperature at this section creates a smaller temperature gradient (Refer Fig. 3d). The variation of $Nu$ at the top wall is presented in Fig. 4b. For this case, the maxima at the convex wall is higher for $\gamma = 1$ and lower for $\gamma = 2$ as compared to the constant heating case.

Whereas the decrease in maxima for $\gamma = 2$ can be explained by the fact of the presence of maximum wall temperature near the recirculation zone, as well as the presence of minimum temperature at the convex surface results in smaller temperature gradient and decreases the heat transfer rate.

The presentation of $\overline{Nu}$ with Re is depicted in Fig. 5a at different $\gamma$. The value of $\overline{Nu}$ increases with Re as advection strength increases. It can be noted that the value of $\overline{Nu}$ decreases with increase in $\gamma$ and the rate of decrement is higher for smaller value of $\gamma$ and approaches to asymptotic value at higher values of wavelength of sinusoidal heating. This observation can be clarified as the smaller values of $\gamma$ results in additional maxima of $Nu$ in the circulatory flow zone, as well as increase in maxima of the convex surface as compared to the constant heating case as discussed. These effects get smaller when $\gamma$ increases. The ratio of $\overline{Nu}$ for non-uniform and

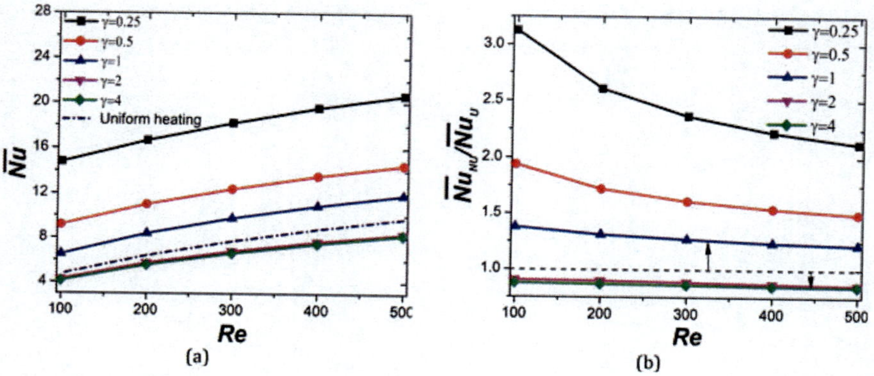


**Fig. 5** **a** Average Nusselt number versus Re at different normalized wavelength of sinusoidal wall temperature ($\gamma$), for constant heating, **b** average Nusselt number ratio for sinusoidal and constant heating case versus Re at different $\gamma$

uniform heating case is presented in Fig. 5b. This ratio decreases with Re as for both the cases $\overline{Nu}$ increases with Re linearly. It can be noted that $\overline{Nu}$ for non-uniform heating case with $\gamma = 0.25$ increases up to 2–3 times compared to uniform heating case in considered regime of Re.

## 5 Conclusions

In the presented work, the effect of sinusoidal heating on the heat transfer characteristics for laminar forced convective flow through asymmetric serpentine channel has been explored numerically. The key outcomes from the present work are as follows:

- The sinusoidal heating decreases the intensity of hot fluid trapping in the re-circulatory flow regions.
- The sinusoidal heating results in additional maxima of local Nusselt number in the re-circulatory flow region, where minima is found for the constant heating case.
- The cumulative Nusselt number decreases with increase in undulation of sinusoidal heating ($\gamma$) and the rate of decrement is higher for smaller value of $\gamma$ and approaches to asymptotic value at higher values of $\gamma$. Average Nusselt number for sinusoidal heating case with $\gamma = 0.25$ increases up to 2–3 times as compared to the constant heating case in the considered range of Re.

## References


1. S.K. Mehta, S. Pati, Thermo-hydraulic and entropy generation analysis for magnetohydrodynamic pressure driven flow of nanofluid through an asymmetric wavy channel. Int. J. Numer. Meth. Heat Fluid Flow. https://doi.org/10.1108/HFF-05-2020-0300
2. C.C. Wang, C.K. Chen, Forced convection in a wavy-wall channel. Int. J. Heat Mass Transf. **45**, p2587 (2002)
3. S. Pati, S.K. Mehta, A. Borah, Numerical investigation of thermo-hydraulic transport characteristics in wavy channels: comparison between raccoon and serpentine channels. Int. Commun. Heat Mass Transf. **88**, 171 (2017)
4. S.K. Mehta, S. Pati, Analysis of thermo-hydraulic performance and entropy generation characteristics for laminar flow through triangular corrugated channel. J. Therm. Anal. Calorim. **136**, 49 (2019)
5. A. Shubham, A. Saikia, A. Dalal, S. Pati, Thermo-hydraulic transport characteristics of non-Newtonian fluid flows through corrugated channels. Int. J. Thermal Sci. **129**, 201 (2018)
6. D. Lorenzini, C. Green, T.E. Sarvey, X. Zhang, Y. Hu, A.G. Fedorov, et al., Embedded single phase microfluidic thermal management for non-uniform heating and hotspots using microgaps with variable pin fin clustering. Int. J. Heat Mass Transf. **103**, 1359 (2016)
7. S.K. Mehta, S. Pati, Effect on non-uniform heating on heat transfer characteristics in wavy channel, 2018, in *Proceedings of the 5th International Conference on Computational Methods for Thermal Problems*, vol. 5 (IISc Bangalore, India), p. 498. ISSN 2305-6924
8. E. Alawadhi, R. Bourisliy, Forced convection in a wavy channel with discrete heat sources. Int. J. Numer. Meth. Heat Fluid Flow **22**, 215 (2012)
9. S. Pati, R. Roy, N. Deka, M.P. Boruah, M. Nath, R. Bhargav, P.R. Randive, P.P. Mukherjee, Optimal heating strategy for minimization of peak temperature and entropy generation for forced convective flow through a circular pipe. Int. J. Heat Mass Transf. **150**, 119318 (2020). https://doi.org/10.1016/j.ijheatmasstransfer.2020.119318
10. S. Bhardwaj, A. Dalal, S. Pati, Influence of wavy wall and non-uniform heating on natural convection heat transfer and entropy generation inside porous complex enclosure. Energy **79**, 467 (2015)
11. B. Mondal, S.K. Mehta, P.K. Patowari, S. Pati, Numerical study of mixing in wavy micromixers: comparison between raccoon and serpentine mixer. Chem. Eng. Process. **136**, 44 (2019)
12. S.K. Mehta, S. Pati, Numerical study of thermo-hydraulic characteristics for forced convective flow through wavy channel at different Prandtl number. J. Therm. Anal. Calorim. https://doi.org/10.1007/s10973-020-09412-5

# Effect of Sintering Temperatures on Mechanical Properties of AA7075/B4C/Graphite Hybrid Composite Fabricated by Powder Metallurgy Techniques

**Guttikonda Manohar, Saikat Ranjan Maity, and Krishna Murari Pandey**

**Abstract** Materials which are having high strength-to-weight ratio, high stiffness, hardness, and corrosion resistance should attain high priority in case of material selection for industrial applications, defense, and aerospace sectors. There are so many materials which are selected depending upon type of applications; they are super alloys, shape memory alloys, high entropy alloys, and composite materials. Among all, composite materials find major applications in automobile, military and aerospace, especially aluminum metal matrix composites are in demand because of its high strength-to-weight ratio and high corrosion resistance compared to other composite materials. In the present work, experimental investigation on aluminum (AA7075) hybrid composite with $B_4C$ (6%) and graphite (2%) as reinforcements by altering the sintering temperatures and their effect on mechanical properties of the composite material were studied. From the results, it was concluded that added graphite acts as binder material that helps in enhancement in strength of the composite material at high sintering temperatures and also by increasing the sintering temperatures mechanical properties of the composite material improves along with reducing porosity levels but beyond the critical sintering temperatures porosity levels increases that effects the composite material in negative way. From the results, it was concluded that added graphite acts as binder material that helps in enhancement in strength of the composite material at high sintering temperatures and also by increasing the sintering temperatures mechanical properties of the composite material improves along with reducing porosity levels but beyond the critical sintering temperatures porosity levels increases that effects the composite material in negative way.

**Keywords** Metal matrix composites · Powder metallurgy · Ball milling · Sintering

G. Manohar (✉) · S. R. Maity · K. M. Pandey
Department of Mechanical Engineering, National Institute of Technology Silchar, Silchar, Assam 788010, India
e-mail: manohar_rs@mech.nits.ac.in

S. R. Maity
e-mail: Saikat.jumtech@gmail.com

K. M. Pandey
e-mail: kmpandey2001@yahoo.com


K. M. Pandey et al. (eds.), *Recent Advances in Mechanical Engineering*, Lecture Notes in Mechanical Engineering, https://doi.org/10.1007/978-981-15-7711-6_35

# 1 Introduction

Composite materials show complex behavior because of its ability to use one or more materials added in to it that forms interfaces and able to share properties with matrix material while load transferring mechanisms, many numbers of reinforcements are there which are suitable to add in to ductile metal matrix materials [1]. Majorly added reinforcement materials 3 are hard ceramic particles having strength, hardness and high corrosion properties. Generally used reinforcements are $B_4C$, graphite, SiC, graphene, CNT, TiC, WC, ZrC, $Al_2O_3$, AlN, etc. [2–6]. Hard ceramic boron carbide is perfectly suitable to reinforce in to soft ductile aluminum matrix and forms good interface. There are many processing techniques to fabricate composite materials like stir casting, compo casting, powder metallurgy, etc., and among them, powder metallurgy attains high priority because of its simple processing techniques that make material defect less by avoiding unnecessary intermetallic compounds that formed at high temperature processing conditions [7]. In some cases, formed intermetallic compounds acts as precipitation hardeners that enhance the strength of the composite and in some cases, it acts as brittle metallic compounds that makes interface bonds weaker between matrix and reinforcements [4], hybrid composites having two or more reinforcements in a matrix material lead to complex chemical reactions at interface regions that forms unnecessary intermetallic compounds that may degrade the mechanical properties of the composite material [8].

Sintering mechanism plays a vital role in powder metallurgy process, diffusion and grain refinement occur that lead to formation of metallic bond and strengthen the material [9]. Sintering temperatures effect the properties of the material to a greater extent, higher temperatures result in high diffusion rates, and grain growth along with high interfacial energies of powder particles at high temperatures tend toward high reaction mechanisms between matrix and reinforcements [10] that form intermetallic compounds that makes interface chemistry more complex. From the literature work, it is observed that optimum sintering temperature and time are needed to enhance the mechanical properties of the material in a positive way [11], at higher temperatures of sintering chemical reactions between matrix and reinforcements are at higher rate lead to consumption of reinforcement materials that effects the composition levels of composite material that directly alters the properties of the material [12] and in another side, interface should be with unnecessary intermetallic compounds that effects the effective load transfer mechanisms, low sintering mechanisms form just a physical bond that was weak to withstand or transfer high loads from matrix material [13, 14].

Sintering temperatures affect the material properties and change the grain sizes, interface chemistry [15]. Present work focuses on effect of sintering temperature on mechanical properties of AA7075/$B_4C$/graphite hybrid composite was analyzed.

## 2 Materials and Methods

AA7075 powder was used as a matrix material with an average size of 50 microns, $B_4C$ powder with an average particle size (APS) of 50 microns and graphite powder with an APS of 60–70 μm were used as reinforcement particles. 6 wt.% of $B_4C$ powder and 2 wt.% of graphite powder were added to AA7075 powder to get maximum enhancement results. All matrix and reinforcement particles were loaded in a tungsten carbide ball milling vial with ball to powder ratio of 10:1, acetone as process control agent (PCA) and stain less steel balls as milling media, milled for 5 h to uniformly distribute the reinforcement particles in the matrix material and a regular intervals of 5 min was allowed to cool down the milling chamber to avoid cold welding effects [16] and addition of PCA if required. To make green compacts from composite powders, cold compaction process was used with the help of manual hydraulic pellet press. H13 die steel as die material and D2 steel as plunger material to withstand high compaction loads were used to compact composite powders in to green compacts and a total of twelve samples was prepared; among them, four samples for compression, four samples for tensile, and four samples for hardness and porosity tests.

Sintering was performed on green compacts in a muffle furnace at four different temperatures, i.e., 590, 610, 630, and 640 °C for 120 min and allowed to cool in the oven itself. Prepared composite samples are mechanically tested for tensile test machined accordingly as mentioned in the above Fig. 1 and compression strength according to ASTM E9 by using universal testing machine and hardness test by brinell hardness testing machine.

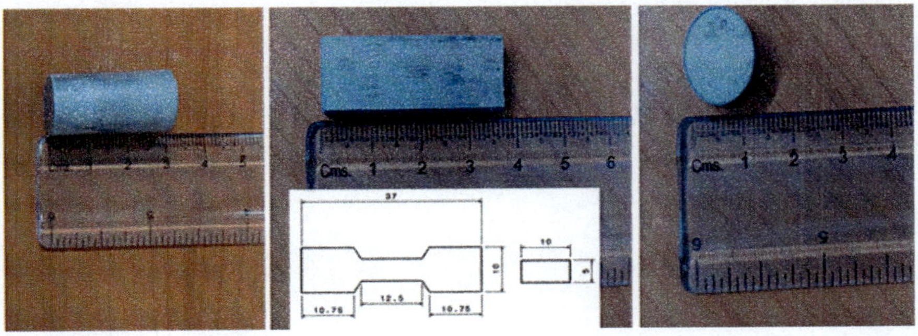


**Fig. 1** Green compacts from left to right for compression, tensile, and hardness tests

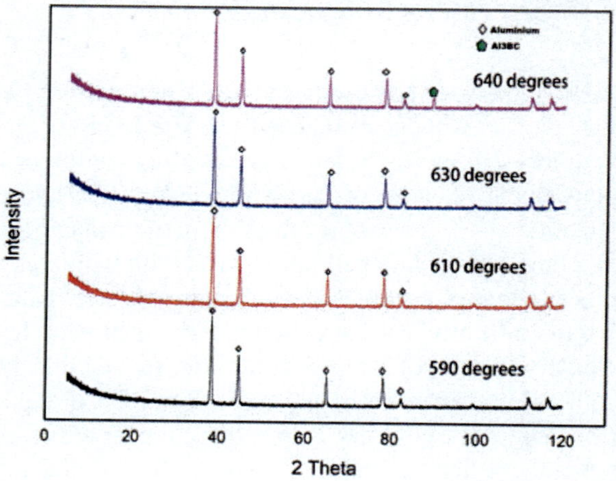


**Fig. 2** XRD analysis of AA7075/$B_4C$/graphite hybrid composite samples sintered at different temperatures

# 3 Results and Discussions

## 3.1 XRD Analysis

AA7075/$B_4C$/graphite samples were tested mechanically and also investigated by XRD to find out the intermetallic compounds formed at different sintering temperatures. Composites sintered at 640 °C shows additional phase $Al_3BC$ as shown in Fig. 2.

At high sintering temperatures, there is a high chance of formation of intermetallic compounds because of excess availability of Gibbs free energy that makes reaction rates faster between matrix and reinforcements [17]. From the mechanical test analysis, it was observed that composite sintered at 640 °C shows low mechanical strength compared to composite sintered at 630 °C, and this was due to the additional intermetallic phase formed at matrix reinforcement interface [18].

## 3.2 Porosity

Pores are generated because of air gaps between powder particles; while compaction process, it makes material weaker in a way that pores acts as a crack initiation sites and stress concentration sites while applying loads. Tendency of pores formation arises because of thermal expansion mismatch between matrix and reinforcement particles [19] while sintering process and also uneven particle geometries that makes particles adhesion improper leaving pores around the particles [20]; another factors that effects the porosity levels are compaction pressures, sintering temperatures, and time. In AA7075/$B_4C$/graphite composite porosity levels are decreased with sintering temperature but after 630 °C (Fig. 3), porosity levels rise and it may be due to grain

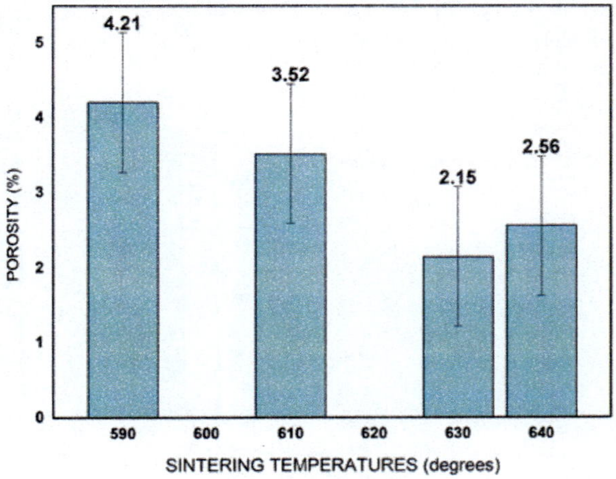


**Fig. 3** Porosity levels versus sintering temperatures of hybrid composite material

growth of matrix particles and having more air gaps between them and also thermal mismatch between matrix and reinforcement particles [21].

## 3.3 Mechanical Properties

Materials which are suitable for industrial application must have superior mechanical properties interms of tensile, compression, and hardness, and that is the reason, composites are most suitable materials for industrial applications. In AA7075/$B_4C$/graphite hybrid composites sintered at various temperatures show superior enhancement in mechanical properties with raising sintering temperatures but after critical temperatures, means after 630 °C, strength of the composite degraded as compared to composite sintered at 630 °C as shown in Fig. 4. This was explained by considering the results from XRD analysis that formation of brittle intermetallic compound ($Al_3BC$) after 630 °C makes load transferring mechanisms weaker.

One of the factor that effects the mechanical strength was compaction pressure, studies find that while initial compaction process particles start to slide and rearrange and after critical pressures particles start to interlock between particles and start to plastic deformation finally undergo cold welding and adhere to each other [22], in this process there is a high chance of pores formation that effects the mechanical properties of the composite material. From the above tests, it was concluded that 630 °C temperature was optimum temperature for AA7075/$B_4C$/graphite hybrid composite.

Hardness of the composite was increased with increasing sintering temperature (Fig. 5) because of hard ceramic reinforcement particles in it and another factor was grain pinning effect that makes matrix grains resist to deform that are adjacent to reinforcement particles and also interfaces between matrix and reinforcements act as barriers to dislocation motions [23], with out dislocations material was unable to

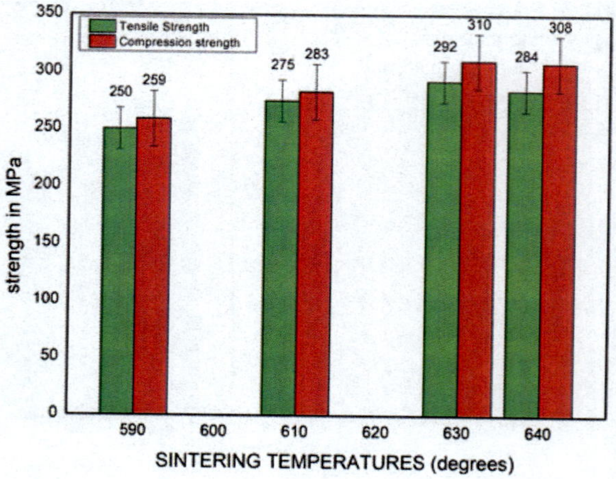


**Fig. 4** Strength of composite versus sintering temperatures

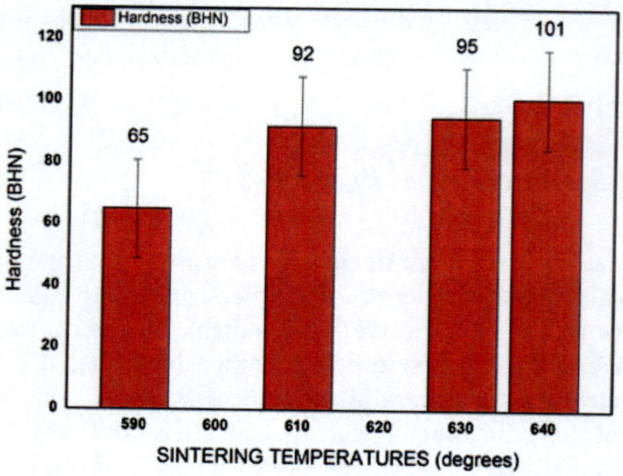


**Fig. 5** Hardness versus sintering temperatures of composites

deform makes material harder. These all mechanisms make composite material to attain high strength and hardness.

## 4 Conclusion

- $AA7075/B_4C$/graphite hybrid composite was fabricated by powder metallurgy techniques by varying sintering temperatures and their effect on mechanical properties was investigated.
- From the XRD analysis, it was observed that intermetallic compounds are formed at 640 °C sintered composite.
- Hardness of the composite material was raised with sintering temperatures along with added hard ceramic reinforcements help to increase it.

- Tensile and compression strength raise along with sintering temperature up to 630 °C, and after that, it starts to fall because of hard brittle intermetallic compounds formed at high temperatures.
- Porosity levels start to fall up to 630 °C, and after that, its value raised this corelates with strength of the composite material degradation.
- Finally concluded that sintering temperature effects the composite material in a positive way up to the critical point after that, it start to degrade the properties of the material.

## References


1. A.E. Nassar, E.E. Nassar, Properties of aluminium matrix nano composites prepared by powder metallurgy processing. J. King Saud Univ. Eng. Sci. **29**, 295–299 (2017)
2. H. Alihosseini, K. Dehghani, J. Kamail, Microstructure characterization, mechanical properties, compressibility and sintering behaviour of Al-$B_4C$ nano composite powders. Adv. Powder Technol. **28**, 2126–2134 (2017)
3. A. Javdani, A.H. Daei-Sorkhab, Microstructural and mechanical behaviour of blended powder semisolid formed Al7075/$B_4C$ composites under different conditions. Trans. Nonferrous Met. Soc. China **28**, 1298–1310 (2018)
4. R. Zheng, J. Chen, Y. Zhang, K. Ameyama, C. Ma, Fabrication and characterization of hybrid structured Al alloy matrix composites reinforced by high volume fraction of $B_4C$ particles. Mater. Sci. Eng. A **601**, 20–28 (2014)
5. X. Pang, Y. Xian, W. Wang, P. Zhang, Tensile properties and strengthening effects of 6061Al/12wt%$B_4C$ composites reinforced with nano-$Al_2O_3$ particles. J. Alloy. Compd. **768**, 476–484 (2018)
6. R. Harichandran, N. Selvakumar, Effect of Nano/micro $B_4C$ particles on the mechanical properties of aluminium metal matrix composites fabricated by ultrasonic cavitation-assisted solidification process. Arch. Civil Mech. Eng. **16**, 147–158 (2016)
7. R. Liu, Wu. Chuandong, J. Zhang, G. Luo, Q. Shen, L. Zhang, Microstructural and mechanical behaviours of the ultrafine grained AA7075/$B_4C$ composites synthesized via one-step consolidation. J. Alloy. Compd. **748**, 737–744 (2018)
8. A. Abdollahi, A. Alizadeh, H.R. Baharvandi, Dry sliding tribological behaviour and mechanical properties of Al2024-5 wt.% $B_4C$ nanocomposites produced by mechanical milling and hot extrusion. Mater. Design **55**, 471–481 (2014)
9. I. Topcu, H.O. Gulsoy, N. Kadioglu, A.N. Gulluoglu, Processing and mechanical properties of $B_4C$ reinforced Al matrix composites. J. Alloys Comp **482**, 516–521 (2009)
10. L. Zhang, Z. Wang, Q. Li, Wu. Junyan, G. Shi, F. Qi, X. Zhou, Microtopography and mechanical properties of vaccum hot pressing Al/$B_4C$ composites. Ceram. Int. **44**, 3048–3055 (2018)
11. R.G. Vogt, Z. Zhang, T.D. Topping, E.J. Lavernia, J.M. Schoenung, Cryomilled aluminium alloy and boron carbide nano-composite plate. J. Mater. Process. Technol. **209**, 5046–5053 (2009)
12. S. Karabulut, H. Karakoc, R. Citak, Influence of $B_4C$ particle reinforcement on mechanical and machining properties of Al6061/$B_4C$ composites. Compos. Part B **101**, 87–98 (2016)
13. G. Manohar, A. Dey, K.M. Pandey, S.R. Maity, Fabrication of metal matrix composites by powder metallurgy: A review, in *AIP Conference Proceedings 1952* (2018), p. 020041. https://doi.org/10.1063/1.5032003
14. J.C. Viala, J. Bouix, G. Gonzalez, C. Esnouf, Chemical reactivity of aluminium with boron carbide. J. Mater. Sci. **32**, 4559–4573 (1997)

15. K. Soorya Prakash, P.M. Gopal, D. Anburose, V. Kavimani, Mechanical, corrosion and wear characteristics of powder metallurgy processed Ti-6Al-4V/$B_4C$ metal matrix composites. Ain Shams Eng. J. **9**, 1489–1496 (2018)
16. E. Mohammad Sharifi, F. Karimzadeh, M.H. Enayati, Fabrication and evaluation of mechanical and tribological properties of boron carbide reinforced aluminium matrix nanocomposites. Mater. Design **32**, 3263–3271 (2011)
17. M.F. Ibrahim, H.R. Ammar, A.M. Samuel, M.S. Soliman, F.H. Samuel, On the impact toughness of Al-15% $B_4C$ metal matrix composites. Compos. B **79**, 83–94 (2015)
18. E. Ghasali, M. Alizadeh, T. Ebadzadeh, A. Hossein Pakseresht, A. Rahbari, Investigation on microstructural and mechanical properties of $B_4C$ aluminium matix composites prepared by microwave sintering. J. Mater. Res. Technol. **4**(4), 411–415 (2015)
19. S. Ozkaya, A. Canakci, Effect of the $B_4C$ content and the milling time on the synthesis, consolidation and mechanical properties of AlCuMg-$B_4C$ nanocomposites synthesized by mechanical milling. Powder Technol. **297**, 8–16 (2016)
20. A. Atrian, G.H. Majzoobi, Characterization of Al7075-$B_4C$ composite fabricated by powder compaction techniques under different densification rates, in *10th International Conference on Composite Science and Technology ICCST/10*
21. A. Alizadeh, E. Taheri-Nassaj, H.R. Baharvandi, Preparation and investigation of Al-4 wt% $B_4C$ Nano composite powders using mechanical milling. Bull. Mater. Sci. **34**(5), 1039–1048 (2011) © Indian Academy of Sciences
22. A.K. Bodukuri, K. Eswaraiah, K. Rajendar, V. Sampath, Fabrication of Al-SiC-$B_4C$ metal matrix composite by powder metallurgy technique and evaluating mechanical properties. Persp. Sci. **8**, 428–431 (2016)
23. M.R. Moradi, A. Moloodi, A. Habibolahzadeh, Fabrication of nano-composite Al-$B_4C$ foam via powder metallurgy-space holder technique. Proc. Mater. Sci. **11**, 553–559 (2015)

# Natural Convection from Two Cylinders in an Enclosure with Sinusoidal Bottom Wall: A Numerical Study

**Dhrijit Kumar Deka, Gopal Chandra Pal, Sukumar Pati, and Pitambar R. Randive**

**Abstract** Unsteady laminar natural convection within a square enclosure with wavy bottom wall embedded with a couple of circular cylinders placed in the vertically symmetric mid-plane is numerically investigated. Present study numerically quests for four different interspacings of the cylinders ranges $0.1 \leq S \leq 0.4$ and for Rayleigh number (Ra) in the range of $10^3$–$10^6$. The results are discussed based on distribution of isotherms, streamlines, and temporal distribution of surface-averaged Nusselt number ($Nu_s$) along with the time-averaged Nusselt number ($Nu_t$). The outcome of the investigation infers that heat transfer takes place solely due to conduction up to $Ra = 10^3$, whereas the heat transfer shifts the mode from conduction to convection with increase in Ra. Further, $Nu_t$ is found to increase with increase in $S$ at lower values of Ra. Moreover, for higher values of Ra, the variation of $Nu_t$ evaluated at the surface of the cylinders shows contrasting features with changing cylinder spacing.

**Keywords** Natural convection · Nusselt number · Rayleigh number · Enclosure · Sinusoidal bottom wall

## 1 Introduction

Owing to its versatile applications, the buoyancy-driven convection draws most of the scientific attention in the domain of heat transfer in the current research field. Such major applications of natural convections are found in nuclear reactors, cooling system of electronic gadgets, thermal safety processes, etc. Numerous investigations [1–6] have been performed on this regard to analyse the effect of various influencing

D. K. Deka (✉) · S. Pati · P. R. Randive
Department of Mechanical Engineering, National Institute of Technology Silchar, Silchar 788010, India
e-mail: dhrijitdeka@gmail.com

G. C. Pal
Department of Mechanical Engineering, Indian Institute of Technology Ropar, Rupnagar, Punjab, India



K. M. Pandey et al. (eds.), *Recent Advances in Mechanical Engineering*, Lecture Notes in Mechanical Engineering, https://doi.org/10.1007/978-981-15-7711-6_36

flow parameters, geometrical orientation, imposed boundary conditions, etc., on the flow and thermal features along with the characteristics of entropy generations.

Among the major earlier studies, Ghaddar [7] studied the naturally convected heat transfer in a rectangular cavity embedded with a uniformly heated cylinder. Later, Warringtone and Powe [8] experimentally investigated the buoyancy-driven heat transfer generated from cylinders placed within both spherical as well as cubical enclosure. From the geometrical aspects, Roychowdhury et al. [9] numerically investigated the influence of cylinder-to-enclosure ratio in the heat transfer characteristics. Bhowmick et al. [10] recently investigated similar study with non-uniform temperature distribution on the walls. Hussain and Hussein [11] performed similar study by giving isoflux boundary condition on the enclosed cylindrical element.

The study of natural convection heat transfer has been extended by incorporating more than one cylinder within the cavity and got serious attention from the researchers [12–15]. Park et al. [12] performed numerical investigation on the buoyancy-induced convection generated from two heated circular cylinders embedded in a cavity for different vertical locations. They observed that both Rayleigh number and the cylinders interspacing influence the state of the flow fields. Karimi et al. [13] investigated similar study for transient conditions. They reported that for lower values of Ra ($<10^4$), the cylinder spacing strongly alters the rate of heat transfer, whereas the cylinder spacing does not have much influence on heat transfer when $\mathrm{Ra} > 10^4$. Pal et al. [14] conducted similar study within a porous enclosure, however, the study was averted from sinusoidal bottom wall.

Most of the studies as delineated above have analysed buoyancy-driven convection from two cylinders embedded within a cavity or regular geometry. However, the influence of sinusoidal bottom wall on the flow and thermal features has not been reported so far, although wavy wall geometry holds great significance in some passive heat transfer applications viz., solar heating, heat exchanger, etc [16]. Hence, we analyse the unsteady buoyancy-driven convection generated from a couple of symmetrically oriented heated cylinders on the vertical mid-plane of a square cavity with sinusoidal bottom wall geometry.

## 2 Theoretical Formulation

The schematic diagram of the computational domain for the present study is displayed in Fig. 1. The square cavity of length $L$ with sinusoidal bottom wall is enclosing two circular cylinders with diameter $d$ placed on the vertical mid-plane of the enclosure. The separating distance between the cylinders is $s$. The working fluid is water. The surface of the cylinders is at temperature $T_h$, whereas the walls of the cavity are kept at $T_c$ ($<T_h$).

The assumptions made for the present study are:

i. The flow is laminar, 2D, and incompressible.
ii. The thermal radiation and viscous dissipation effects are not considered.

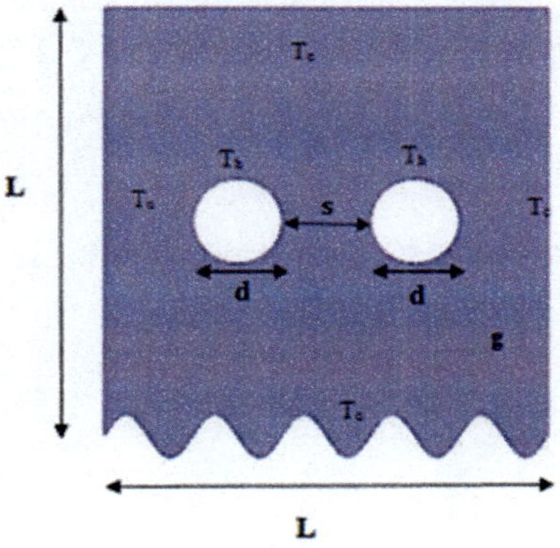


**Fig. 1** Computational domain

iii. The fluid properties are considered to be constant except the density for which the Boussinesq approximation is considered [17].

Based on the above assumptions, the conservation equations of mass, momentum, and thermal energy in the non-dimensional form can be written as [12],

$$\frac{\partial U^*}{\partial X^*}+\frac{\partial V^*}{\partial Y^*}=0 \tag{1}$$

$$\frac{\partial U^*}{\partial \tau}+U^*\frac{\partial U^*}{\partial X^*}+V^*\frac{\partial V^*}{\partial X^*}=-\frac{\partial P}{\partial X^*}+\Pr\left(\frac{\partial^2 U^*}{\partial X^{*2}}+\frac{\partial^2 V^*}{\partial X^{*2}}\right) \tag{2}$$

$$\frac{\partial V^*}{\partial \tau}+U^*\frac{\partial U^*}{\partial Y^*}+V^*\frac{\partial V^*}{\partial Y^*}=-\frac{\partial P}{\partial Y^*}+\Pr\left(\frac{\partial^2 V^*}{\partial X^{*2}}+\frac{\partial^2 V^*}{\partial Y^{*2}}\right)+\mathrm{Ra}\Pr\theta \tag{3}$$

$$\frac{\partial \theta}{\partial \tau}+U^*\frac{\partial \theta}{\partial X^*}+V^*\frac{\partial \theta}{\partial X^*}=\left(\frac{\partial^2 \theta}{\partial X^{*2}}+\frac{\partial^2 \theta}{\partial Y^{*2}}\right) \tag{4}$$

Different parameters are defined as,
$X^*=\frac{x}{L}$, $Y^*=\frac{y}{L}$, $\tau=\frac{t\alpha}{L^2}$, $U^*=\frac{uL}{\alpha}$, $V^*=\frac{vL}{\alpha}$, $\theta=\frac{T-T_h}{T_h-T_c}$, $\mathrm{Ra}=\frac{g\beta L^3(T_h-T_c)}{\nu\alpha}$ and $\Pr=\frac{\nu}{\alpha}$.

here, $u$ and $v$ are the velocity components along the direction of $x$ and $y$, respectively. $T$ is the temperature. The time is denoted with $t$, $\alpha$ is the thermal diffusivity, $\nu$ is the kinematic viscosity, $g$ is the acceleration due to gravity, $\beta$ is the volume expansion coefficient. Ra and Pr can be termed as Rayleigh number and Prandtl number, respectively.

The boundary conditions imposed on the computational domain are as follows:

$$\text{Cylinders's surface:}\theta=1;\ U^*=V^*=0 \tag{5}$$

$$\text{Enclosure's walls: }\theta=0;\ U^*=V^*=0 \tag{6}$$

The local Nusselt number can be evaluated as [12]:

$$\mathrm{Nu}_\phi = -\frac{\partial \theta}{\partial n}\bigg|_{\text{Cylinder surface}} \tag{7}$$

where $n$ is outward normal to the surface. The surface-averaged Nusselt number is calculated as [12]:

$$\mathrm{Nu}_s = \left(\frac{1}{2\pi}\right)\int_0^{2\pi} \mathrm{Nu}_\phi \mathrm{d}\varphi \tag{8}$$

## 3 Numerical Solution Methodology

Finite volume method [10] has been implemented to solve the governing transport equations. The grid used for the numerical simulations is of non-uniform type where the number of cells near the walls of the enclosure and cylinder surfaces are more as compared to the rest part of the domain. For discretizing the diffusion and advection terms, second-order upwind scheme is used, whereas for the temporal terms, discretization implicit scheme is employed. The temporal step-size is considered to be $10^{-2}$. For coupling the pressure field with velocity, SIMPLE algorithm is adopted.

Prior to the present investigation, it is important to ascertain the accuracy of the chosen grid system. Thus, grid independent test has been carried out for four different grid systems G1 (7095 elements), G2 (8385 elements), G3 (9709 elements), and G4 (16,289 elements) for $\mathrm{Ra} = 10^6$ and $S = 0.1$. Table 1 depicts the results of the grid independency test and confirms that the G3 is insensitive with further change elements. Thus, G3 grid system is considered for the entire subsequent investigation.

Figure 2 displays the comparison of the temporal variation of $\mathrm{Nu}_s$ with the computational results of Karimi et al. [13]. For the same, we consider two different cylinder interspacing distances ($S = 0.1$ and 0.4) and four Rayleigh numbers within the considered range. The results with the present numerical approach are found to be in close agreement with the benchmark results of Karimi et al. [13].

**Table 1** Values of $\mathrm{Nu}_t$ for different grid for $\mathrm{Ra} = 10^6$ and $S = 0.1$

| Grid system | $\mathrm{Nu}_t$ | % difference |
|---|---|---|
| G1 | 3.2 | 1.5 |
| G2 | 3.223 | 0.8 |
| G3 | 3.2367 | 0.4 |
| G4 | 3.2496 | |

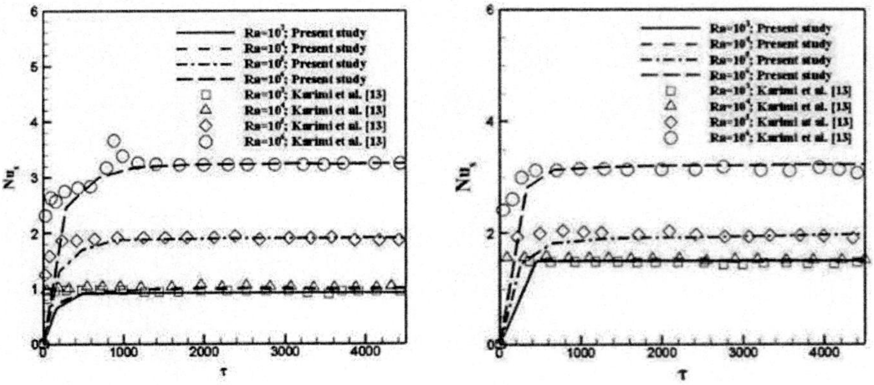


S = 0.1 S = 0.4

**Fig. 2** Comparison of $Nu_S$ with time for different Ra with Karimi et al. [13]

## 4 Results and Discussions

Figure 3 displays the streamline distributions for different Ra at varying cylinder spacing *S*. It is observed that the fluid within the close proximity of the cylinders surface lifts up due to the buoyancy effect and subsequently descends along the side walls due to density variation forming primary recirculation vortices. Symmetrical orientations of the streamlines are maintained along the vertical centre line of the enclosure at Ra = $10^3$ and $10^4$ as the conduction mode of heat transfer dominates. With increase in Ra (=$10^5$), the buoyancy force prevails and the mode of heat transfer shifts to convection due to which the vortices formed get intensified. The strength of the vortices goes on increasing as Ra reaches $10^6$ and a distorted streamline distribution is observed.

Figure 4 shows the isotherms distribution for different Ra at different locations of the cylinders. Figure 4a, b show almost uniform distribution isotherms as at Ra = $10^3$ and $10^4$ indicating the dominance of diffusion heat transfer. Highly dense isotherms are seen around the cylinders with increase in *S*. For the combination of Ra = $10^5$ and $S = 0.1$, a rising thermal plume can be identified. The strength of the convection becomes stronger and the thermal plumes lift up towards the upper cold wall more intensely at Ra = $10^6$.

The temporal variations of $Nu_s$ for different Ra and *S* are shown in Fig. 5. It can be observed that $Nu_s$ attains the steady state from a null value within a definite settling period which is different for different Ra. Moreover, with increase in Ra, the magnitude of $Nu_s$ also increases as buoyancy forces get enhanced. Further, the variation of $Nu_t$ at different *S* with varying Rayleigh number is presented in Fig. 6 which conveys that for lower values of Ra ($10^3$ and $10^4$), $Nu_t$ increases with increase in *S* as conduction heat transfer prevails. However, owing to the fact that the thickness of the thermal boundary is reduced with increase in *S*, thereby, the thermal gradient near the cylinders surface becomes steeper. However, $Nu_t$ decreases first with an

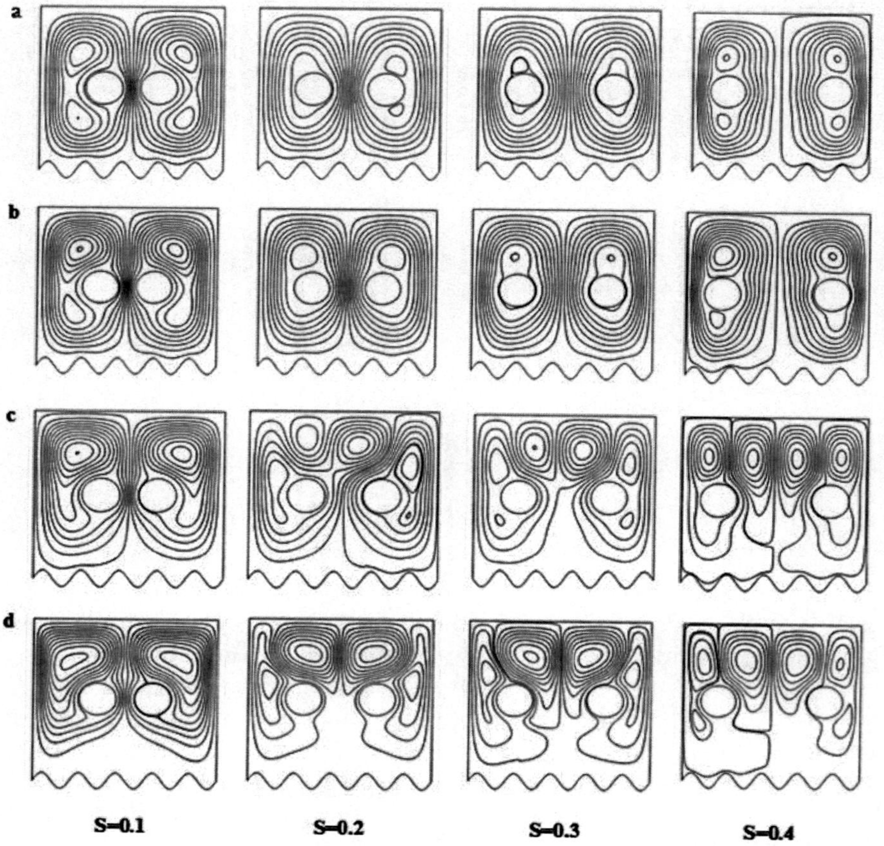


**Fig. 3** Streamlines orientations for different $S$ at $\tau = 2500$ **a** $\mathrm{Ra} = 10^3$, **b** $\mathrm{Ra} = 10^4$, **c** $\mathrm{Ra} = 10^5$, **d** $\mathrm{Ra} = 10^6$

increase in the cylinder interspacing and subsequently increases as the Rayleigh number goes higher and moreover the rate of increase in $\mathrm{Nu}_t$ with $S$ is lesser as compared to lower Ra. The buoyant plumes is the reason for it.

## 5 Conclusion

In this work, laminar natural convection within a square enclosure with wavy bottom wall embedded with a couple of circular cylinders placed in the vertically symmetric mid-plane is numerically investigated in an unsteady framework. The enclosure walls are kept at constant temperature which is lower than that of the temperature of the cylinder surfaces. The influence of Ra along with the cylinder interspacing is extensively investigated. The following are the conclusions from the present work:

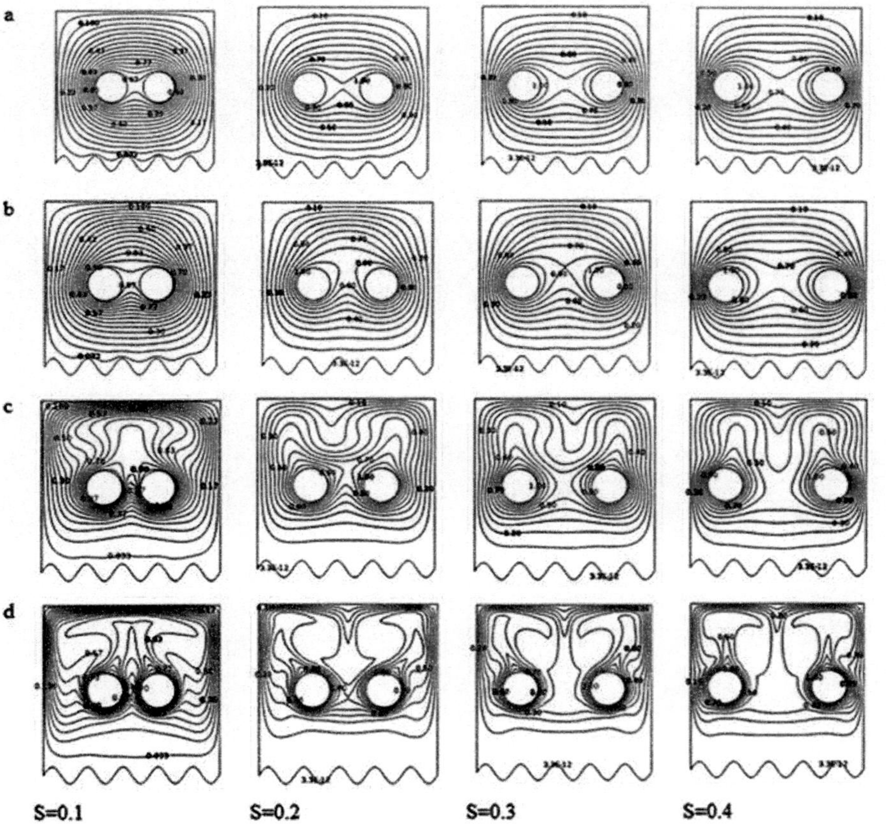


**Fig. 4** Isotherms orientations for different $S$ at $\tau = 2500$ **a** Ra $= 10^3$, **b** Ra $= 10^4$, **c** Ra $= 10^5$, **d** Ra $= 10^6$

- Heat transfer is due to conduction for lower values of Ra, whereas for higher values of Ra, heat transfer is prevailed by convection.
- For lower Ra, $\text{Nu}_t$ increases with the cylinder spacing. However, $\text{Nu}_t$ decreases first with an increase in the cylinder interspacing and subsequently increases as the Rayleigh number goes higher.
- At higher Ra, the rate of increase in $\text{Nu}_t$ with $S$ is lesser as compared to lower Ra.

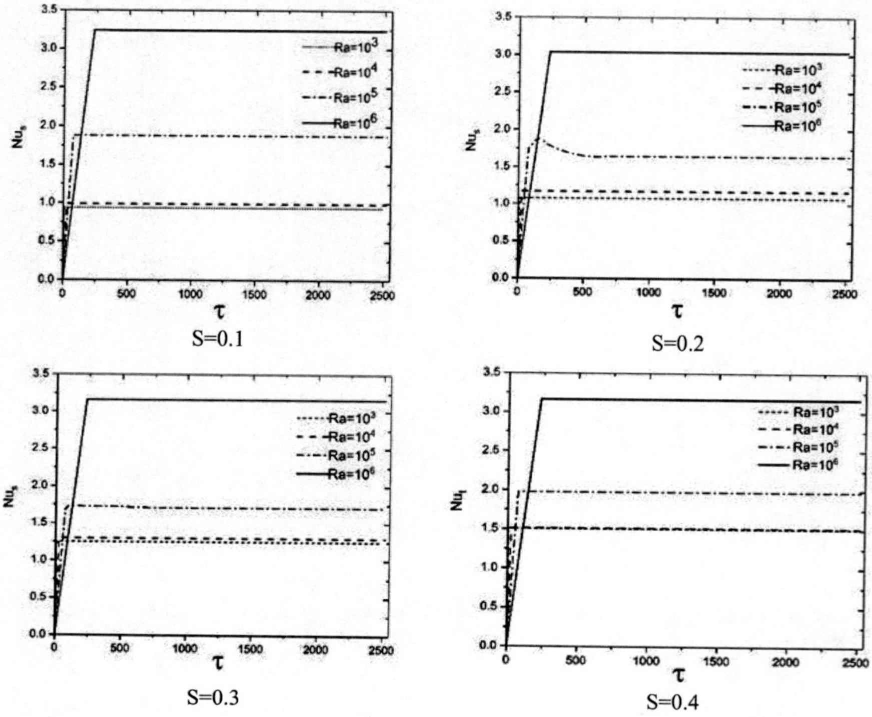


**Fig. 5** Temporal evolution of $Nu_s$ for different Ra

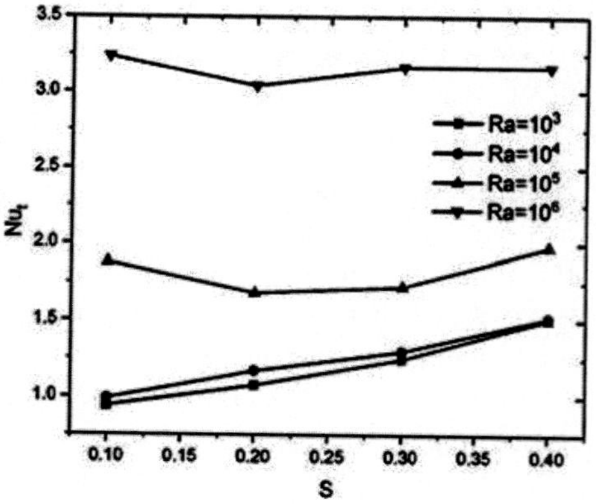


**Fig. 6** Variation of $Nu_t$ at different $S$ and Ra

## References

1. S. Bhardwaj, A. Dalal, S. Pati, Influence of wavy wall and non-uniform heating on natural convection heat transfer and entropy generation inside porous complex enclosure. Energy **79**, 467–481 (2015)

2. P. Meshram, S. Bhardwaj, A. Dalal, S. Pati, Effects of inclination angle on natural convection heat transfer and entropy generation in square porous enclosure. Numer. Heat Transf. Part A Appl. **70**, 1271–1296 (2016)
3. S. Dutta, A.K. Biswas, S. Pati, Analysis of natural convection in a rhombic enclosure with undulations of the top wall—a numerical study. Int. J. Ambient Energy (2019). https://doi.org/10.1080/01430750.2019.1630304
4. S. Dutta, A.K. Biswas, S. Pati, Numerical analysis of natural convection heat transfer and entropy generation in a porous quadrantal cavity. Int. J. Numer. Methods Heat Fluid Flow **29**, 4826–4849 (2019)
5. S. Dutta, N. Goswami, A.K. Biswas, S. Pati, Numerical investigation of magnetohydrodynamic natural convection heat transfer and entropy generation in a rhombic enclosure filled with Cu-water nanofluid. Int. J. Heat Mass Transf. **136**, 777–798 (2019)
6. S. Dutta, A.K. Biswas, S. Pati, Natural convection heat transfer and entropy generation inside porous quadrantal enclosure with non-isothermal heating at the bottom wall. Numer. Heat Transf. Part A Appl. **73**, 222–240 (2018)
7. N.K. Ghaddar, Natural convection heat transfer between a uniformly heated cylindrical element and its rectangular enclosure. Int. J. Heat Mass Transf. **35**(10), 2327–2334 (1992)
8. R.O. Warrington Jr., R.E. Powe, The transfer of heat by natural convection between bodies and their enclosures. Int. J. Heat Mass Transf. **28**(2), 319–330 (1985)
9. D.G. Roychowdhury, S.K. Das, T. Sundararajan, Numerical simulation of natural convective heat transfer and fluid flow around a heated cylinder inside an enclosure. Heat Mass Transf. **38**, 565–576 (2002)
10. D. Bhowmick, P.R. Randive, S. Pati, H. Agrawal, A. Kumar, P.K. Srivastava, Natural convection heat transfer and entropy generation from a heated cylinder of different geometry in an enclosure with non-uniform temperature distribution on the walls. J. Therm. Anal. Calorim. **141**, 839–857 (2020)
11. S.H. Hussain, A.K. Hussein, Numerical investigation of natural convection phenomena in a uniformly heated circular cylinder immersed in square enclosure filled with air at different vertical locations. Int. Commun. Heat Mass Transf. **37**(8), 1115–1126 (2010)
12. Y.G. Park, M.Y. Ha, H.S. Yoon, Study on natural convection in a cold square enclosure with a pair of hot horizontal cylinders positioned at different vertical locations. Int. J. Heat Mass Transf. **65**, 696–712 (2013)
13. F. Karimi, H.T. Xu, Z. Wang, M. Yang, Y. Zhang, Numerical simulation of unsteady natural convection from heated horizontal circular cylinders in a square enclosure. Numer. Heat Transf. Part A Appl. **65**(8), 715–731 (2014)
14. G.C. Pal, N. Goswami, S. Pati, Numerical investigation of unsteady natural convection heat transfer and entropy generation from a pair of cylinders in a porous enclosure. Numer. Heat Transf. Part A Appl. **74**(6), 1323–1341 (2018)
15. D. Bhowmick, S. Chakravarthy, P. Randive, S. Pati, Numerical investigation on the effect of magnetic field on natural convection heat transfer from a pair of embedded cylinders within a porous enclosure. J. Therm. Anal. Calorim. **141**, 2405–2427 (2020)
16. S. Dutta, S. Pati, A.K. Biswas, Thermal transport analysis for natural convection in a porous corrugated rhombic enclosure. Heat Transf. **49**, 3287–3313 (2020)
17. S. Dutta, N. Goswami, S. Pati, A.K. Biswas, Natural convection heat transfer and entropy generation in a porous rhombic enclosure: Influence of non-uniform heating. J. Therm. Anal. Calorim. https://doi.org/10.1007/s10973-020-09634-7

# Identification of Parameters for Ultrasonic Machining (USM) on Drilling of Graphene Oxide/Pineapple Leaf Filler-Reinforced Epoxy Hybrid Composite Using TODIM Method

**Angkan Bania and Saikat Ranjan Maity**

**Abstract** This paper aims at selecting the optimal parameters for drilling of graphene oxide/pineapple leaf filler-reinforced epoxy hybrid composite using ultrasonic machining (USM). The selection of optimal parameters has been done using the interactive multi-criteria decision-making (TODIM) method. The subsequent process parameters are kept constant: boron carbide abrasive, amplitude of vibration (3–5 μm), and frequency (20 kHz) and the other parameters abrasive grit size, abrasive flow rate, power rating, and slurry concentration are considered for evaluation. The corresponding response parameters viz material removal rate (MRR), tool wear rate (TWR), and overcut (OC) are measured for every experimental runs. Two different levels of abrasive grit size and four different levels of flow rate, power rating, and slurry concentration are selected for detailed experimentation. CRITIC method has been applied for assigning weights to the process parameters. By implementing the TODIM method, the highest rank of the parametric combinations of process parameters is found out which tends to maximize MRR and minimize TWR and OC.

**Keywords** Hybrid composite · USM · Optimization · MCDM

## 1 Introduction

USM is largely employed for machining materials having high hardness and materials which are brittle. In USM, the material is removed by using a shaped tool which vibrates with a very high frequency and low amplitude and circulating abrasive slurry between tool tip and workpiece. Compared to other non-traditional machining processes, USM does not affect the workpiece thermally, chemically, or electrically and also the microstructure of the material. These features of USM also make it more suitable for machining of advanced materials which use has been increasing day by

A. Bania · S. R. Maity (✉)
Department of Mechanical Engineering, National Institute of Technology Silchar, Silchar, Assam 788010, India
e-mail: saikat.jumtech@gmail.com



K. M. Pandey et al. (eds.), *Recent Advances in Mechanical Engineering*, Lecture Notes in Mechanical Engineering, https://doi.org/10.1007/978-981-15-7711-6_37

day. Through gravitational search algorithm (GSA) and fireworks algorithm (FWA), Goswami and Chakraborty [1] have optimized surface roughness and MRR of USM process and concluded that the optimization capability of FWA is superior to GSA and other population-based algorithm. Kataria et al. [2] have optimized machining features in USM for WC–Co composite applying Gray Relation Analysis method and is found that the power rating, abrasive grit size, and tool material have dominating effect on MRR and TWR.

This paper also reveals that the maximum MRR is obtained at a combination of coarse grit size and high power. Through Taguchi method, Singh et al. [3] have optimized the MRR in USM process for polycarbonate bulletproof glass and acrylic heat-resistant glass. This paper concludes that MRR is greatly influenced by abrasives types and HF acid concentrations. Sindhu et al. [4] have applied response surface methodology for optimizing the surface roughness and MRR of rotary ultrasonic machining on quartz glass. It is found that MRR increases whenever there is an increase in power, tool rotational speed, and tool feed rate while feed rate is the most influential one and as the feed rate increase,s surface roughness also increases but as the tool rotational speed and power increases surface roughness decreases. A number of multi-criteria decision-making (MCDM) tools [5–9] have also been utilized to address number of decision-making problems. Biswas et al. [10] have conducted experiments on zirconia composite by Taguchi method and optimized USM parameters by multi-objective optimization on basis of ratio analysis (MOORA). In the present study, an exertion has been made to focus the impact of the process parameters on response parameters and accordingly optimizing the input process parameters using TODIM method. This method is an adequate approach for solving multi-criteria decision-making (MCDM) problems.

## 2 Methodology

The fabrication of the hybrid composite is carried out by hand lay-up technique for filler loadings, viz 0.1% graphene oxide and 5% pineapple leaf filler of the weight of epoxy resin. The drilling experiments are carried out on table top USM of Sonic mill for the hybrid epoxy polymer composite. Tool heads which are available are at 0.5 and 1 in. and the tool is constructed using stainless steel as tool material. Stainless steel microtubes are carved to the desired weight and length by the limitation that for sonic turning of the machine employing wire electric discharge machine (WEDM). The microtube is then connected to the replaceable tool head by brazing to bear the ultrasonic vibration throughout machining which is then soldered on the top of the tool holder. Boron carbide ($B_4C$) is used as abrasive with water as a slurry medium which can be changed depending upon the workpiece medium and application. The abrasive grit sizes of three variations are available viz 220, 400, and 600.

## 3 Results and discussions

Gomes and Lima introduced the TODIM method, an interactive multi-attribute decision-making method, measures the dominance degree of each alternative over others. This method can capture the decision maker's psychological behavior and the ranking of the alternatives is based on the dominance degree. The algorithm of TODIM for set of alternatives $A = (A_1, \ldots, A_n)$ and set of criteria $C = (C_1, \ldots, C_m)$; are as follows:

Step 1: Normalize the decision-making matrix $X = (x_{ij})_{nxm}$ into $Z = (z_{ij})_{nxm}$.
Step 2: Calculate the relative weight of criterion $C_k$ to a reference criterion $C_r$ by

$$w_{kr} = \frac{w_k}{w_r} \tag{1}$$

Step 3: Calculate the dominance degree of alternative $A_i$ over alternative $A_j$ under criterion $C_k$ as follows

$$\Phi_k(A_i, A_j) = \begin{cases} \sqrt{\dfrac{w_{kr}(z_{ik} - z_{jk})}{\sum_{l=1}^{m} w_{lr}}}, & if\ z_{ik} \geq z_{jk} \\ -\dfrac{1}{\theta}\sqrt{\dfrac{\sum_{l=1}^{m} w_{lr}(z_{ik} - z_{jk})}{w_{kr}}}, & if\ z_{ik} < z_{jk} \end{cases} \tag{2}$$

where $\theta > 0$ and represents the effect of the losses. If $\theta > 1$, losses are reduced and if $\theta < 1$, losses are amplified.

Step 4: Calculate the overall dominance degree of alternative $A_i$ over alternative $A_j$

$$\Phi(A_i, A_j) = \sum_{k=1}^{m} \Phi_k(A_i, A_j) \tag{3}$$

Step 5: Calculate the overall performance of alternative $A_i$

$$\Phi(A_i) = \sum_{j=1}^{n} \Phi(A_i, A_j) \tag{4}$$

4. Step 6: Calculate the normalized overall performance of alternative $A_i$

$$\xi(A_i) = \frac{\Phi(A_i) - \min \Phi(A_i)}{\max \Phi(A_i) - \min \Phi(A_j)} \tag{5}$$

Step 7: Rank the alternatives based on the ascending values of $\xi(A_i)$. Higher value of $\xi(A_i)$ than $\xi(A_j)$ indicates alternative $A_i$ is better than $A_j$.

In this study, numbers of experiments are conducted using USM process for optimizing MRR, TWR, and overcut which is a multi-response optimization problem. The input process parameters boron carbide abrasive, amplitude of vibration 3–5 μm, and frequency of 20 kHz are kept constant while the process parameters grit size, slurry flow rate, power rating, and slurry concentration with different levels are used for evaluation.

Some of the experimental results of the response parameters are given in Table 1. Here, grit size, flow rate, and slurry concentration are beneficial criteria and power rating is a non-beneficial criterion. By using CRITIC method, weights have assigned to the process parameters as follows GS = 0.226, SFR = 0.101, PR = 0.126, SC = 0.152, MRR = 0.182, TWR = 0.103, and OC = 0.11. The decision matrix is first converted into normalized matrix as shown in Table 2 and then the relative weights of each criterion is calculated refer to (1). The dominance degree of each alternative over

**Table 1** Experimental output performance

| Experiment no | Grit size | Flow rate ($cm^3/s$) | Power rating (kw) | Slurry concentration | MRR | TWR | OC |
|---|---|---|---|---|---|---|---|
| 1 | 400 | 158 | 35 | 60 | 0.100093 | 44.58095 | 0.14885 |
| 2 | 600 | 125 | 35 | 50 | 0.091304 | 38.09535 | 0.14387 |
| 3 | 400 | 90 | 35 | 40 | 0.087222 | 47.51542 | 0.1433 |
| – | – | – | – | – | – | – | – |
| – | – | – | – | – | – | – | – |
| 14 | 600 | 125 | 20 | 50 | 0.05937 | 23.96115 | 0.11538 |
| 15 | 400 | 90 | 20 | 40 | 0.058418 | 23.21508 | 0.11337 |
| 16 | 600 | 65 | 20 | 30 | 0.061442 | 18.49008 | 0.10939 |

**Table 2** Normalized decision matrix

| S. No. | Grit size | Flow rate | Power rating | Slurry concentration | MRR | TWR | OC |
|---|---|---|---|---|---|---|---|
| 1 | 0 | 1 | 1 | 1 | 0.634409584 | 0.767768737 | 1 |
| 2 | 1 | 0.64516129 | 1 | 0.666666667 | 0.500616523 | 0.576918799 | 0.873796249 |
| 3 | 0 | 0.268817204 | 1 | 0.333333333 | 0.438477113 | 0.854120565 | 0.859351242 |
| – | – | – | – | – | – | – | – |
| – | – | – | – | – | – | – | – |
| 14 | 1 | 0.64516129 | 0 | 1 | 0.014492092 | 0.160995647 | 0.15179929 |
| 15 | 0 | 0.268817204 | 0 | 0.666666667 | 0 | 0.139041254 | 0.100861632 |
| 16 | 1 | 0 | 0 | 0.333333333 | 0.046033703 | 0 | 0 |

**Table 3** Ranking result obtained by TODIM method

| Alternatives | Overall performance values | Normalized overall performance | Ranking |
|---|---|---|---|
| – | – | – | – |
| 8 | −66.8272626 | 0.543037542 | 7 |
| – | – | – | – |
| 11 | −12.04897823 | 0.899280518 | 3 |
| – | – | – | – |
| 15 | 3.438317809 | 1 | 1 |
| 16 | 2.362944228 | 0.993006457 | 2 |

the other alternatives is firstly calculated and then the overall dominance degree of each alternative is calculated refer to (2) and (3). After that, the overall performance of each alternative is calculated and these performance values are then normalized refer to (4) and (5), respectively. Ranking is done based on the highest normalized values of the alternatives as shown in Table 3. Among all the alternatives, alternative 15 has the highest normalized value and hence has been ranked 1. The values of input process parameters related to alternative 15 are grit size of 400, flow rate of 125 $cm^3/s$, power rating of 20 kW, and slurry concentration of 40% and the corresponding response parameters values are MRR = 0.058418, TWR = 23.21508, and OC = 0.11337.

## 4 Conclusion

In this present study, drilling of graphene oxide/pineapple leaf filler-reinforced epoxy is executed with four process parameters as grit size, flow rate, power rating, and slurry concentration and the response parameters as MRR, TWR, and OC in USM process. TODIM method, which is an interactive MCDM method, is used for obtaining optimal parametric combination for USM process. It has been found that abrasive grit size of 400, abrasive flow rate of 125 $cm^3/s$, power rating of 20 kW, and slurry concentration of 40% hold the highest rank which results in optimizing the response parameters.

## References

1. D. Goswami, S. Chakraborty, Parametric optimization of ultrasonic machining process using gravitational search and fireworks algorithms. Ain Shams Eng. J. **6**(1), 315–331 (2015)
2. R. Kataria, J. Kumar, B.S. Pabla, Experimental investigation and optimization of machining characteristics in ultrasonic machining of WC–Co composite using GRA method. Mater. Manuf. Process. 31(5), 685–693 (2016)

3. K.J. Singh, I.S. Ahuja, J. Kapoor, Optimization of material removal rate in ultrasonic machining of polycarbonate bulletproof glass and acrylic heat-resistant glass by Taguchi method. Multidisc. Model. Mater. Struct. **13**(4), 612–627 (2017)
4. D. Sindhu, L. Thakur, P. Chandna, Parameter optimization of rotary ultrasonic machining on quartz glass using response surface methodology (RSM). *Silicon*, 1–15 (2019)
5. D. Zindani, S.R. Maity, S. Bhowmik, S. Chakraborty, A material selection approach using the TODIM (TOmada de Decisao Interativa Multicriterio) method and its analysis. Int. J. Mater. Res. **108**(5), 345–354 (2017)
6. D. Zindani, S.R. Maity, S. Bhowmik, Fuzzy-EDAS (evaluation based on distance from average solution) for material selection problems, in *Advances in Computational Methods in Manufacturing* (Springer, Singapore, 2019), pp.755–771
7. D. Zindani, S.R. Maity, S. Bhowmik, A decision-making approach for material selection of polymeric composite bumper beam, in *Composites and Advanced Materials for Industrial Applications* (IGI Global, 2018), pp.112–128
8. D. Zindani, K. Kumar, Material selection for turbine seal strips using PROMETHEE-GAIA method. Mater. Today Proc. **5**(9), 17533–17539 (2018)
9. D. Zindani, S.R. Maity, S. Bhowmik, Selection of material for wind turbine blade using PROMETHEE-GAIA method, in *AIP Conference Proceedings*, vol. 1998, no. 1 (AIP Publishing, 2018), p. 020008
10. J.H. Biswas, Jagadish, and Amitava Ray, Experimental investigation and optimisation of ultrasonic machining parameters on zirconia composite. Int. J. Mach. Mach. Mater. **21**(1–2), 115–137 (2019)

# Optimization of Planetary Gearbox Using NSGA-II

**Abhishek Parmar, P. Ramkumar, and K. Shankar**

**Abstract** In this study, optimization of planetary gearbox considering regular mechanical and critical tribological constraints such as wear and scuffing is done. The design variables considered are the number of teeth in the sun, planet, ring gear, module, face width, diameter of shafts, planet pin diameter, the hardness of gear material and kinematic viscosity of the oil. The two objective functions are the weight and total power loss of the planetary gearbox. The analysis is done for the unmodified gear tooth profile with various input power from 30 to 90 kW with an incremental of 30 kW at the input speed of 1000 rpm. The result of multi-objective optimization with tribological constraints is compared with the single objective optimization considering the weight of the gearbox as the objective with and without tribological constraints. The result obtained considering multi-objective optimization shows that there is a significant reduction in weight and power loss as compared to results obtained by considering single objective optimization with and without tribological constraints. Further, the risk of gear failure in wear is in moderate to high range when the tribological constraints are not considered.

**Keywords** Multi-objective optimization · NSGA-II · Planetary gearbox · Lubrication · Scuffing · Wear

## 1 Introduction

Planetary gearboxes owing to greater torque capability, greater power density and compact size for the same torque application compared with simple gearboxes, are

---

A. Parmar · P. Ramkumar (✉) · K. Shankar
Machine Design Section, Department of Mechanical Engineering, Indian Institute of Technology Madras, Chennai, India
e-mail: ramkumar@iitm.ac.in

A. Parmar
e-mail: abhishekparmar19957@gmail.com



K. M. Pandey et al. (eds.), *Recent Advances in Mechanical Engineering*, Lecture Notes in Mechanical Engineering, https://doi.org/10.1007/978-981-15-7711-6_38

commonly used for wind turbines, aircraft engines, etc. In modern energy transmission systems, lightweight and long life are the key requirements. Thus, multi-objective optimization problem is formulated considering total weight and total power loss as two conflicting objective functions with inclusion of wide range of design constraints (wear, scuffing, bearing life, etc.). Discrete version of Non-Dominated Sorting Genetic Algorithm-II (NSGA-II) [1] is used to solve mixed-integer type optimization problem for both multi-objective optimization and single objective optimization.

Filiz et al. [2] investigated the single objective optimization problem using Matlab® Optimtool optimization toolbox in order to minimize the kinetic energy of the planetary gearbox. Bending, pitting, and geometrical constraints were considered for the analysis. Reduction in volume was also observed but analysis was not done for the power loss. Milojevic et al. [3] considered center distance minimization and efficiency maximization as two objective functions which were first solved using Genetic Algorithm (GA), and later weighted method was used to convert the Multi-objective optimization problem into single objective optimization problem. However, the design variables were considered as the continuous type whereas, in practical situation, the design variables for planetary gearbox could be mixed-integer type. Further, tribological constraints such as scuffing and wear were not considered in the analysis. Tripathi et al. [4] used NSGA-II and sequential quadratic programming (SQP) techniques to minimize the surface fatigue life factor and volume of planetary gearbox. Based on the result NSGA-II was recommended for the analysis. On the other hand, the volume of gears was only analyzed in volume objective function and did not include the volume of shafts and bearings. Maruti et al. [5] worked on two-stage helical gearbox for minimization of volume and power loss using NSGA-II considering mechanical and tribological constraints. 50% reduction in power loss with slight increase in weight was observed. High probability of gears failing in wear was also noticed. But the analysis was done on two-stage gearbox rather than planetary gearbox and did not consider bearing constraint in analysis. Based on the above literature, it is necessary to include the weight of gears, bearing and shafts in the total weight objective function for better analysis of total planetary gear system. Hence, new weight objective function was formulated considering the aforementioned parameters. Further, power loss has also been considered in this problem to find the trade-off solution between weight and power loss. Ultimately, this study would give a better understanding of the whole planetary system along with tribological constraints.

## 2 Objective Functions

The two objective functions which are considered are as follows.

## 2.1 Minimization of Weight of the Planetary Gearbox

The first objective function is minimization of total weight $g_1(x)$ which include the weight of the gears $W_g$, planet gear bearing $W_{pb}$ and shaft $W_s$ given as.

$$\mathrm{Min} g_1(X) = W_g + W_s + W_{Pb} \tag{1}$$

where the weight of shafts and gear is given by Eq. (2),

$$W_g = \frac{\pi}{4} b\left[\left(d_x^2 - d_{i1}^2\right) + N_P\left(d_{ps}^2 - d_{pd}^2\right) + \left(d_{ro1}^2 - d_{ri1}^2\right)\right]\rho g \tag{2}$$

$$W_s = \frac{\pi}{4}[d_{i1}^2 l_{p1} + N_P d_{pd}^2 l_{pd} + d_{o1}^2 l_{o1}]\rho g \tag{3}$$

where

$d_x$ pitch diameter of sun gear.
$d_{i1}$ input shaft diameter.
$d_{ps}$ pitch circle diameter of planet gear.
$d_{pd}$ planet pin diameter.
$d_{ro1}$ outside diameter of ring gear.
$d_{ri1}$ inside diameter of ring.
$\rho$ density of gear material.
$l_{p1}$ input shaft length.
$l_{pd}$ planet pin length.
$l_{o1}$ output shaft length.
$N_p$ number of planet gears.

## 2.2 Minimization of Total Power Loss

The second objective function is minimization of power loss $g_2(X)$ which includes both non-load dependent [7] (contact oil seals loss, rolling bearing oil churning loss, and gear oil churning loss) $P_{\mathrm{NLD}}$ (kW) and load dependent [7] (rolling bearing friction and gear mesh friction) $P_{\mathrm{LD}}$ (kW) losses.

$$\mathrm{Min} g_2(X) = P_{LD} + P_{NLD} \tag{4}$$

## 3 Design Variable Vector

The design variable considered in the analysis are mixed-integer type which consists of discrete, integer and continuous type. Here, hardness of gear material $(H)$, face width of gear $(b)$ are continuous type, number of teeth in sun gear $(Z_s)$, planet gear $(Z_p)$, ring gear $(Z_r)$,$d_{i1}$, $d_{o1}$ are integer type and module $(m)$, viscosity $(\nu)$, and $d_{pd}$ are discrete type. The design vector is shown below.

$X = \{Z_s, Z_p, Z_r, m, d_{i1}, d_{o1}, d_{pd}, H, \nu, b\}$.

## 4 Constraints

### 4.1 Bending Constraint [6]

$$F_t K_o K_v K_s \frac{1}{bm} \frac{K_H K_B}{Y_J} - \frac{\sigma_{FP} Y_N}{S_F Y_\theta Y_Z} \leq 0 \tag{5a}$$

Here, $\sigma_{FP}$ is allowable bending strength which is function of hardness and is given as:

$$\sigma_{FP} = 0.533H + 88.3 \tag{5b}$$

### 4.2 Pitting Constraint [6]

$$Z_E \sqrt{F_t K_o K_v K_s \frac{K_H}{d_w b} \frac{Z_R}{Z_I}} - \frac{\sigma_{HP} Z_N Z_W}{S_H Y_\Theta Y_Z} \leq 0 \tag{6a}$$

Here, $\sigma_{HP}$ is allowable contact strength which is function of hardness, given as

$$\sigma_{HP} = 2.22H + 200 \tag{6b}$$

### 4.3 Factorizing Constraint [7]

$$\frac{Z_R}{N_P} = \text{aninteger} or \frac{Z_S}{N_P} = \text{aninteger} \tag{7}$$

### 4.4 Equally Spaced Planets Constraint [7]

$$\frac{Z_R + Z_S}{N_P} = \text{aninteger} \tag{8}$$

### 4.5 Bearing Life Constraint [7]

$$L_{10h} = \frac{10^6}{60n_{pc}}\left(\frac{C(N_p)^{7/9}}{P_{lb}}\right)^{10/3} \geq 50,000 \tag{9}$$

### 4.6 Face width constraint [5]

$$3\pi \mathrm{m} \leq b \leq 5\pi \mathrm{m} \tag{10}$$

### 4.7 Shaft Diameter Constraint [8]

$$\left[\frac{32\mathrm{S_{sf}}}{\pi}\sqrt{\left(\frac{T}{S_y}\right)^2 + \left(\frac{M_\mathrm{m}}{S_e}\right)^2}\right]^{1/3} - d_i \leq 0 \tag{11}$$

where, $T$ is maximum torque acting on shaft, $M_m$ is maximum bending moment on shaft, $S_y$ and $S_e$ is yield strength and endurance strength respectively.

### 4.8 Eletrohydrodynamic Lubrication Constraints for Bearing

$$\lambda = \frac{h_{\min}}{R_s} > 1 \tag{12a}$$

The minimum film thickness equation for the line contact [9] is given below,

$$h_{\min} = \frac{1.714\overline{U_r}^{0.694}(\alpha E')^{0.568}}{\overline{W}^{0.128}} \tag{12b}$$

### 4.9 Scuffing Constraint

The scuffing constraint [10] for sun-planet external gear mesh and planet-ring internal gear mesh is given as,

$$P_{se} = f(Z_s, Z_p, m, b, \nu) \tag{13a}$$

$$P_{si} = f(Z_p, Z_r, m, b, \nu) \tag{13b}$$

### 4.10 Wear Constraint

Wear constraint [10] for sun-planet external gear mesh and planet-ring internal gear mesh is given as,

$$P_{we} = f(Z_s, Z_p, m, b, \nu) \tag{14a}$$

$$P_{wi} = f\left(Z_p, Z_r, m, b, \nu\right) \tag{14b}$$

## 5 Results

The analysis is done for various input power at input speed of 1000 rpm considering the unmodified gear tooth profile. Table 2 shows the optimal design variables obtained after multi-objective optimization and single objective optimization with and without tribological constraints. Whereas, Table 3 shows the value of both the objective functions obtained after multi-objective optimization and the weight value obtain after single objective optimization. It is to be noted that the power loss value obtained in single objective optimization case is by substituting the design variables value in power loss function.

Table 4 shows percentage reduction or increase (two cases) in weight and power loss value as compared to single objective optimization with and without tribological constraints. It can be observed from Table 4 that there is reduction of both weight and power loss for multi-objective optimization as compared to single objective optimization with tribological constraint. The maximum percentage reduction of weight and power loss is 6.76% and 11.61% respectively as both are obtained at 90 kW input power. However, it can also be observed that on comparing with single objective optimization without tribological constraint, there is an increase in weight and power at 30 and 60 kW. The reason for increase in weight and power loss is

**Table 1** Design variable details

| Design variable | Lower bound | Upper bound |
|---|---|---|
| $Z_s$ | 20 | 50 |
| $Z_p$ | 30 | 60 |
| $Z_r$ | 100 | 180 |
| $m$ | 2, 2.5, 3, 3.5, 4, 4.5, 5, 5.5, 6, 8 | |
| $d_{i1}$ | 30 | 150 |
| $d_{o1}$ | 80 | 200 |
| $d_{pd}$ | 40, 45, 50, 55, 60, 65,70, 75, 80 | |
| $H$ | 200 | 400 |
| *v os ISO index* | 100, 150, 220, 320, 460 | |
| $b$ | 50 | 120 |

due to selection of higher grade oil index (due to presence of tribological constraints in multi-objective optimization) which in turn increases the power losses, as higher grade oil has higher viscosity. Further, reduction in weight and power loss is observed for 90 kW in this case.

Table 5 shows the wear risk values obtain after single objective optimization without tribological constraints. It can be observed that the probability of gear failure in wear is moderate to high for each input power at both sun-planet and planet-ring gear mesh. Although, from Table 4 it is observed that there is an increase in weight and power loss value for multi-objective optimization case but these gears are safe in wear. Thus, in order to prevent wear failure in both sun-planet and planet-ring gear mesh the inclusion of tribological constraint is a must.

Figure 1 shows the Pareto optimal curve (locus of trade-off solution) which end result is obtained after multi-objective optimization for three different input powers using NSGA-II.

## 6 Conclusion

The analysis was done for three input powers (30, 60, 90 kW) for unmodified gear tooth profile at the input speed of 1000 rpm. The results from multi-objective optimization (MOO) were compared against single objective optimization (SOO) with and without scuffing constraints. Significant reduction in weight and power loss was obtained in MOO when compared with SOO with tribological constraints. Further, reduction in weight and power loss value got increased with the increasing power. Maximum reduction in weight and power loss as 6.76% and 11.71% was obtained respectively for 90 kW power. Similarly, results of SOO without tribological constraints showed reduction in weight and power loss value for 90 kW power

**Table 2** Optimal design variables values obtain after optimization at 1000 rpm for unmodified gear profile for three cases

| Input power (kW) | $Z_s$ | $Z_p$ | $Z_r$ | $m$ (mm) | $d_{i1}$ (mm) | $d_{o1}$ (mm) | $d_{pd}$ (mm) | $H$ | $\nu$ (ISO grade oil) | $b$ (mm) |
|---|---|---|---|---|---|---|---|---|---|---|
| Single objective optimization with tribological constraint | | | | | | | | | | |
| 30 | 33 | 54 | 141 | 4.5 | 97 | 119 | 50 | 291 | 220 | 70.52 |
| 60 | 30 | 45 | 120 | 5.5 | 102 | 128 | 55 | 240 | 320 | 75.94 |
| 90 | 27 | 42 | 111 | 6 | 108 | 141 | 65 | 272 | 320 | 88.83 |
| Single objective optimization without tribological constraint | | | | | | | | | | |
| 30 | 30 | 48 | 126 | 4.5 | 89 | 115 | 45 | 261 | 150 | 70.27 |
| 60 | 39 | 42 | 123 | 5.5 | 111 | 131 | 55 | 294 | 100 | 72.49 |
| 90 | 36 | 45 | 126 | 5.5 | 117 | 155 | 65 | 312 | 150 | 77.95 |
| Multi-objective optimization with tribological constraint | | | | | | | | | | |
| 30 | 33 | 48 | 129 | 5 | 90 | 113 | 45 | 219 | 220 | 71.78 |
| 60 | 30 | 45 | 120 | 5 | 97 | 132 | 55 | 278 | 320 | 75.10 |
| 90 | 33 | 48 | 129 | 5.5 | 114 | 132 | 65 | 328 | 320 | 76.20 |

**Table 3** Objective functions value obtain after single and multi-objective optimization

| Input power, *P* (kW) | Values obtained after single-stage optimization considering tribological constraints | | Values obtained after single-stage optimization without considering tribological constraints | | Values obtain after multi-objective optimization considering tribological constraints | |
|---|---|---|---|---|---|---|
| | Weight (N) | Power loss (kW) | Weight (N) | Power loss (kW) | Weight (N) | Power loss (kW) |
| 30 | 1132.87 | 1.67 | 1063.38 | 1.57 | 1100.71 | 1.62 |
| 60 | 1213.40 | 2.20 | 1149.54 | 2.03 | 1165.03 | 2.09 |
| 90 | 1476.23 | 2.56 | 1455.19 | 2.41 | 1376.32 | 2.26 |

**Table 4** Multi-objective optimization functions values comparison with single objective optimization with and without tribological constraints

| Input power, *P* (kW) | With tribological constraint | | Without tribological constraint | |
|---|---|---|---|---|
| | Reduction in weight (%) | Reduction in power loss (%) | Reduction in weight (%) | Reduction in power loss (%) |
| 30 | 2.83 | 3.08 | 3.54 (increase) | 3.18 (increase) |
| 60 | 3.98 | 5 | 1.34 (increase) | 2.95 (increase) |
| 90 | 6.76 | 11.71 | 5.41 | 6.22 |

**Table 5** Wear risk obtains considering single objective optimization without tribological constraints

| Input power, *P* (kW) | Sun-planet gear mesh, $P_{we}$ (%) | Planet-ring gear mesh, $P_{wi}$ (%) |
|---|---|---|
| 30 | 26.72 | 41.77 |
| 60 | 22.32 | 49.61 |
| 90 | 22.71 | 40.82 |

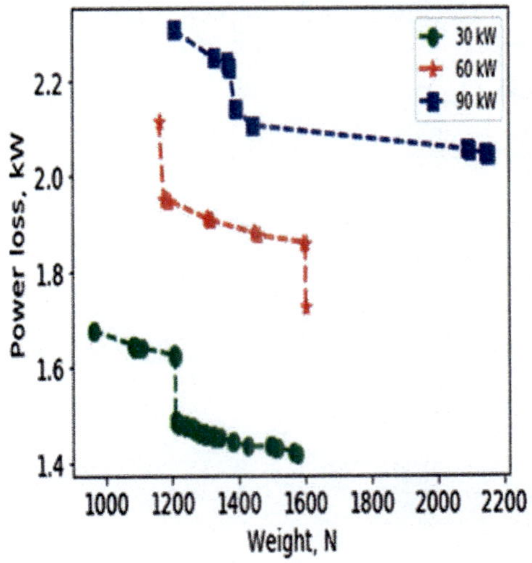


**Fig. 1** Pareto front obtain after multi-objective optimization for unmodified gear profile at three different input power

but there was increase in power loss and weight value for 30 and 60 kW power respectively. Later, it was found that gears failed in wear aspect for both sun-planet (external) and planet-ring (internal) gear mesh. Thus inclusion of wear constraint becomes essential for safer design of planetary gearbox.

## References

1. K. Deb, Multi-objective optimization using evolutionary algorithms, vol. 16 (Wiley, 2001)
2. I.H. Filiz, S. Olguner, E. Evyapan, A study on optimization of planetary gear trains. Acta Physica Polonica, A. **132**(3) (2017)
3. B. Rosic, S. Radenovic, L. Jankovic, M. Milojevic, Optimisation of planetary gear train using multiobjective genetic algorithm. J. Balkan Tribol. Assoc. **17**(3), 462–475 (2011)
4. V.K. Tripathi, H.M. Chauhan, Multi objective optimization of planetary gear train, in *Asia-Pacific Conference on Simulated Evolution and Learning* (Springer, 2010), pp. 578–582
5. M. Patil, P. Ramkumar, K. Shankar, Multi-objective optimization of the two-stage helical gearbox with tribological constraints. Mech. Mach. Theory **138**, 38–57 (2019)
6. Fundamental rating factors and calculation methods for involute spur and helical gear teeth, ANSI/AGMA 2001-D04
7. AGMA 6123-C16 Design manual for enclosed epicycle gear drives, ANSI/AGMA6123-C16 (2016)
8. R.L. Norton, Machine design: an integrated approach, 2nd edn. (HongKong, Pearson Education, Asia, 2001), pp. 366–383
9. R.W. Bruce, Handbook of lubrication and tribology, in *Theory and Design*, vol. II (CRC press, 2012)
10. American Gear Manufactures Association, "AGMA 925-A03 Effect of Lubrication on Gear Surface Distress" 90 (2003)

# Conjugate Heat Transfer Analysis for Flow Through Microduct Subjected to Non-uniform Heating

**Abhijit Borah and Sukumar Pati**

**Abstract** A numerical analysis is conducted to examine the effect of non-uniform heating for forced convective flow through a circular microduct having finite wall thicknesses. Two different arrangements of applied heat flux in the form of stepwise constant are studied for two different wall thickness, Reynolds number in the range $20 \leq Re \leq 200$ and for a given substrate material. The results in the form of temperature and Nusselt number are compared with that of the non-conjugate analysis and uniform heating strategy. The analysis suggests that the effect of wall thickness is significant for the analysis of forced convective heat transfer in microduct. Moreover, the arrangement of non-uniform heating should be such that lower heat-producing sources be at the entrance and followed by higher heat sources.

**Keywords** Conjugate heat transfer · Microduct · Non-uniform heating

## 1 Introduction

Modern Electronic circuitry generates high heat flux, which needs to be removed in order to ensure smooth functioning of the devices. This is one of the most important application areas for microscale cooling devices because it possesses a large working surface in a very small volume. Microscale cooling devices significantly improve the heat transfer coefficient [1] and therefore it is an obvious choice among thermal engineers. The thermo-hydraulic characteristics of such devices are discussed by many researchers to examine the effect of various heat flux profile [2], flow conditions [3] etc. The complex feature of microchannel is further complicated by having diverse heat sources in a compact space [4]. As a result, the cooling devices have to handle non-uniform heating along the length of the channel. Transport properties of

A. Borah (✉) · S. Pati
Mechanical Engineering Department, National Institute of Technology Silchar, Silchar, India
e-mail: abhijit1d2@gmail.com

S. Pati
e-mail: sukumarpati@gmail.com



K. M. Pandey et al. (eds.), *Recent Advances in Mechanical Engineering*, Lecture Notes in Mechanical Engineering, https://doi.org/10.1007/978-981-15-7711-6_39

microchannel cooling devices are analyzed by number of researchers to find the effect of increasing–decreasing heat flux [5], sinusoidal heat flux [6, 7] and step heat flux [8]. Both developing as well as fully developed flow through circular pipe subjected to step heat flux is examined by Hazmohammadi et al. [8]. The authors reported a contradiction between the results of hot spot minimization analysis and entropy analysis with respect to the arrangements of applied heat flux. Therefore, further studies are needed to clear the contradiction and implement it in modern thermal devices. In addition to challenges put forth by applications of microscale cooling device, heat transfer process in a microchannel itself possesses various complexity because of the existence of multiple forms of heat transfer in this process [9]. Here, the dimension of hydraulic diameter is comparable to that of the wall thickness and so the heat transfer in the wall cannot be neglected. The necessity of conjugate heat transfer in microchannel heat transfer is pointed out by Hong et al. [10].

From the existing literature, it seems that there is scope to study the effect non-uniform heating for flow through microduct with finite solid thickness. Accordingly, in this work, we analyze the effect of non-uniform heating in the form of stepwise constant heat flux on the forced convective heat transfer in a microduct having finite wall thickness.

## 2 Numerical Analysis

Numerical analysis of conjugate heat transfer for flow through a circular tube with finite wall thickness and subjected to step heat flux is considered. The length of the computational domain (Fig. 1) is 0.01 m and the inner diameter is considered according to the ratio, $L/d_i = 100$. In order to analyze the effect of wall thickness the outer diameter of the duct is varied according to the ratio, $d_o/d_i = 1$, 2, and 4. The heat flux in the first and the second half of the duct are represented by $q_1$ and $q_2$, respectively and their ratio $q_1/q_2$ is varied in order to create different configurations of applied heat flux. The flow is assumed to be laminar, steady, and incompressible. The viscous heating and thermal radiation are not considered. Assuming constant thermo-physical properties of the working fluid (water), the governing equations, i.e., continuity, momentum, and energy equations are mathematically expressed below.

Continuity equation is

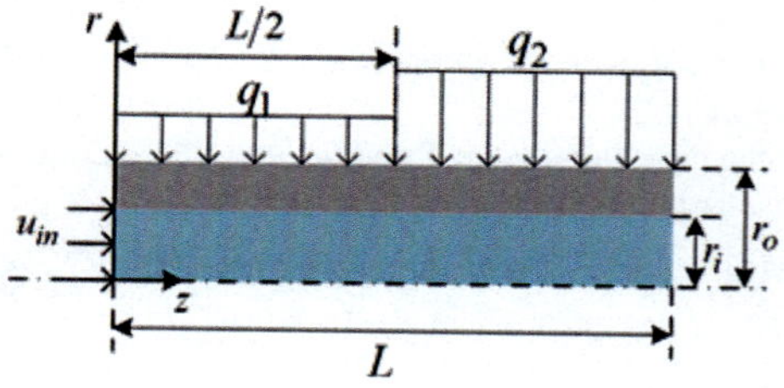


**Fig. 1** Computational domain of the circular microchannel

$$\frac{\partial v_r}{\partial r} + \frac{v_r}{r} + \frac{\partial v_z}{\partial z} = 0 \tag{1}$$

Momentum equation for the fluid in radial direction is

$$\rho_f\left(v_r\frac{\partial v_r}{\partial r} + v_z\frac{\partial v_r}{\partial z}\right) = -\frac{\partial P}{\partial r} + \mu_f\left\{\frac{\partial}{\partial r}\left(\frac{1}{r}\frac{\partial}{\partial r}(r v_r)\right) + \frac{\partial^2 v_r}{\partial z^2}\right\} \tag{2}$$

Momentum equation for the fluid in axial direction is

$$\rho_f\left(v_r\frac{\partial v_z}{\partial r} + v_z\frac{\partial v_z}{\partial z}\right) = -\frac{\partial P}{\partial z} + \mu_f\left\{\frac{1}{r}\frac{\partial}{\partial r}\left(r\frac{\partial v_z}{\partial r}\right) + \frac{\partial^2 v_z}{\partial z^2}\right\} \tag{3}$$

The energy equation for the fluid domain is

$$\rho_f C_p\left(v_r\frac{\partial T}{\partial r} + v_z\frac{\partial T}{\partial z}\right) = k_f\left\{\frac{1}{r}\frac{\partial}{\partial r}\left(r\frac{\partial T}{\partial r}\right) + \frac{\partial^2 T}{\partial r^2}\right\} \tag{4}$$

The energy equation for the solid domain is

$$\frac{1}{r}\frac{\partial}{\partial r}\left(r\frac{\partial T}{\partial r}\right) + \frac{\partial^2 T}{\partial r^2} + \frac{\partial^2 T}{\partial z^2} = 0 \tag{5}$$

At the liquid–solid interface no-slip [11] and no temperature jump boundary conditions are employed. The following boundary conditions are employed:

At the inlet:

$$v_z = u_{in}\,,\ v_r = 0,\ T = T_a \tag{6a}$$

At the outlet:

$$\frac{\partial v_z}{\partial z} = \frac{\partial v_r}{\partial z} = 0, \frac{\partial T}{\partial z} = 0 \tag{6b}$$

At the outer surface of the wall:

$$-k_s\frac{\partial T}{\partial y} = \begin{cases} q_1'' & 0 \le x \le L/2 \\ q_2'' & L/2 \le x \le L \end{cases} \tag{6c}$$

At the fluid–solid interface:

$$k_s \frac{\partial T}{\partial r}\bigg|_s = k_f \frac{\partial T}{\partial r}\bigg|_f \tag{6d}$$

$$T_s = T_f \tag{6e}$$

$$v_z = v_r = 0 \tag{6f}$$

The thermal characteristics of the conjugate heat transfer at the fluid–solid interface can be quantified by Nusselt number. The local Nusselt number (Nu) at the fluid–solid interface is

$$\mathrm{Nu}_x = \frac{q_i''\delta_f}{k_f(T_w - T_b)} \tag{7}$$

where,$q_i''$ and $T_w$ are heat flux and temperature at the fluid–solid interface, $T_b$ is the bulk fluid temperature which is calculated as

$$T_b = \int_{-\delta_f/2}^{\delta_f/2} uT\mathrm{d}y \bigg/ \int_{-\delta_f/2}^{\delta_f/2} u\mathrm{d}y \tag{8}$$

The expression for the local and average convective heat transfer coefficient, respectively, are

$$h = \frac{Nuk_f}{\delta_f} \tag{9}$$

$$\overline{h} = \frac{1}{L}\int_0^L h\,\mathrm{d}x \tag{10}$$

## 2.1 Numerical Procedure and Grid Independence

As the problem is symmetry with respect to the axis of the duct, half of the duct as shown in Fig. 1 is considered for the computation. The computational domain is divided into number of elements and the governing equations are solved by finite element method. The differential governing equations are transformed into integral equations assuming approximate values of the variables and then solve the obtained equations by iterative method. Here, the residual criteria is chosen as $10^{-6}$ for all the variables. The simulation is tested for any dependence on grid size by simulating the results in four different grid sizes. For the same, average Nusselt number is calculated for four different grid sizes using same input parameter and with reference to the outcome, grid sizes of 90,842, 137,524, and 141,251 elements are chosen for $d_o/d_i$ = 1, 2, and 4, respectively.

## 3 Results and Discussions

The numerical model used in this work is validated by remodeling the published work of Aydin et al. [5] and comparing local Nusselt number profile. The comparison is shown in Fig. 2 displaying strong similarity between both the results. Moving ahead, the forced convection in a circular microduct having thick walls and subjected to step heating is modeled. Heat flux is applied in the form of two steps formed by dividing the length of the duct into two equal parts. This study is carried out for the heat flux ratio ($q_1/q_2$) of 0.5, 1, and 2. All three ratios are analyzed for different $d_o/d_i$. The ratio of thermal conductivity of the fluid to that of the solid ($k_{sf}$) is taken as 646.0767. The results are presented in the form of transverse temperature, axial temperature, local Nusselt number, and average Nusselt number profiles.

Figure 3 displays the transverse temperature profile at two axial locations, showing the effect of $q_r$ for Re = 200, $k_{sf}$ = 646.0767 and $d_o/d_i = 2$. The axial locations are $x$ = 0.0045 m and 0.0055 m which are, respectively, before and after the location of change in applied heat flux. It can be clearly viewed from Fig. 3 that the fluid temperature near the fluid–solid interface is higher and gradually decreases towards the centreline at both cross-section. This is because of the conduction taking place within the substrate having high value of thermal conductivity with respect to water. It is seen in Fig. 3a that $q_r = 0.5$, i.e., higher heat flux is applied at the first section of the duct results in higher interface temperature when compared to that of $q_r$ =

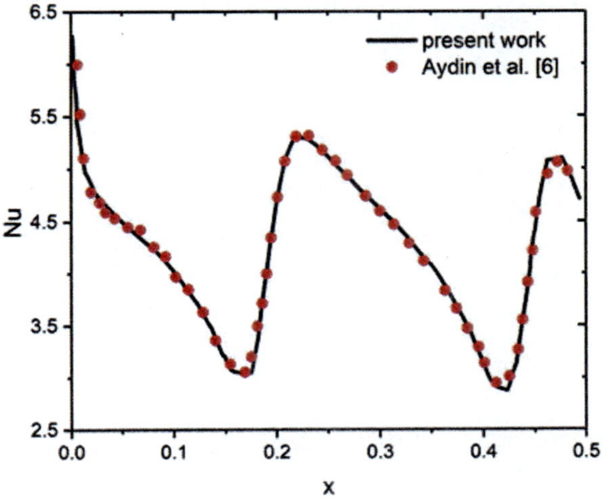


**Fig. 2** Comparison of variation of Nusselt number along axial direction with Aydin et al. [6] for $A = 1$, $k_s/k_f = 100$, $d_o/d_i = 2$ and Pe = 200

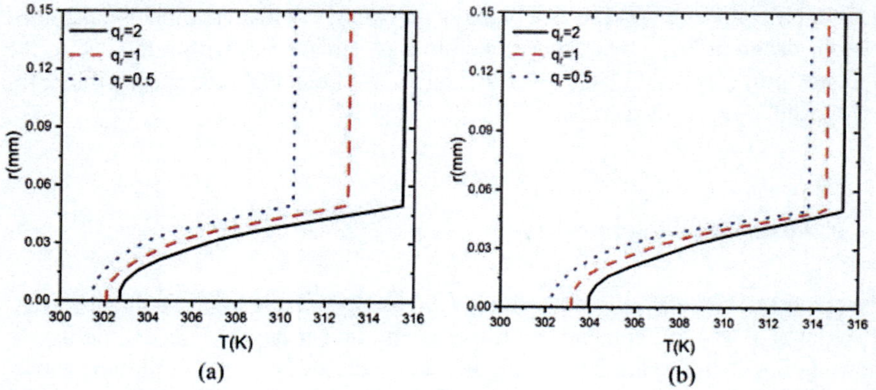


**Fig. 3** Transverse temperature profile at **a** $x = 0.0045$ m and **b** $x = 0.0055$ m showing the effect of $q_r$ for Re = 200, $k_{sf} = 646.0767$ and $d_o/d_i = 2$

2. The cross-section considered in Fig. 3b lies at the second half of the duct where magnitude of the applied heat flux for $q_r = 0.5$ and 2 become reverse when compared to that in the first half of the duct. In spite of this, interface temperature for $q_r = 0.5$ is higher than that for $q_r = 2$. This suggests that axial heat conduction within the solid exists because of a temperature gradient between the two halves of the channel.

This axial conduction has a significant effect on the axial temperature profile too. So, the axial temperature profile at the fluid–solid interface is plotted for both conjugate ($d_o/d_i = 2$) and non-conjugate ($d_o/d_i = 1$) analysis and presented in Fig. 4a and b, respectively. These results are generated at Re = 200 and $k_{sf} = 646.0767$. For the case of $q_r = 1$ (uniform heat flux) the temperature profile in both the figures are uniform whereas, the temperature profiles for $q_r = 0.5$ and 2 differ from one another. Sudden change in temperature at the point of change in applied heat flux is observed

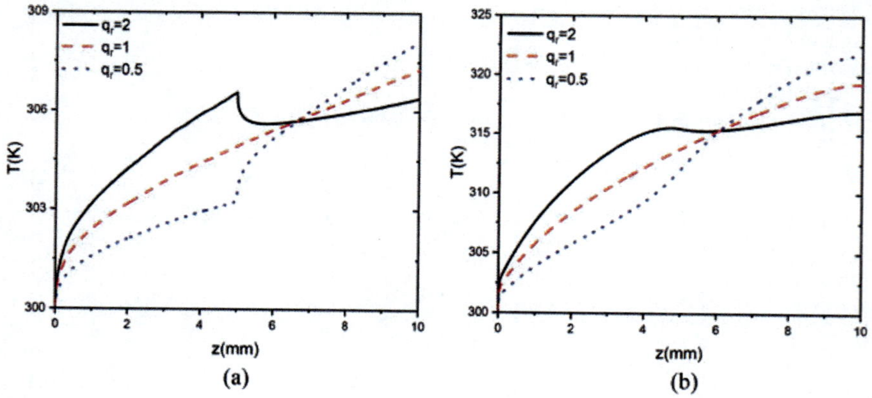


**Fig. 4** Temperature profile at the fluid–solid interface showing the effect of $q_r$ for **a** $d_o/d_i = 1$ and **b** $d_o/d_i = 2$ at Re = 200 and $k_{sf} = 646.0767$

for $d_o/d_i = 1$ (Fig. 4a) whereas, for $d_o/d_i = 2$ (Fig. 4b), a smooth transition is observed starting before the point of change in applied heat flux. This is because of the axial heat conduction within the substrate as discussed above. Another observation from Fig. 4 shows that at $q_r = 0.5$ and 2, temperature after the point of change in applied heat increases and decreases, respectively. This can be attributed to the respective increase and decrease of heat flux in the same axial region.

The local convective heat transfer coefficient is calculated by using Eq. 9 to understand the heat transfer characteristics at the fluid–solid interface. It is presented in Fig. 5 for $d_o/d_i = 2$ at Re = 200. It can be seen that $h$ attains the maximum value at the entrance and gradually decreases along the direction of flow. This is because of the decreasing temperature gradient along the direction of the flow. It can be seen from Fig. 5 that at the transition point of the applied heat flux, Nu abruptly increases for $q_r = 0.5$ whereas, the same decreases for $q_r = 2$ and continues with the usual decreasing trend along the length thereafter. High heat flux is applied after the transition point in case of $q_r = 0.5$ which leads to increase in temperature gradient thereby increasing heat transfer. It also needs to be mentioned that as the $h$ for $q_r = 2$ at the second half of the duct starts decreasing because of the cumulative effect of lesser applied heat flux and decrease of thermal convection with the increase in axial length.

In order to get a clear picture on the effect of step heating on the rate of heat transfer, average convective heat transfer coefficient ($\overline{h}$) on the fluid–solid interface is presented in Fig. 6 for $d_o/d_i = 2$ and 4. It clearly shows that there is a significant effect of $q_r$ on $\overline{h}$. Importantly, $\overline{h}$ becomes higher for $q_r = 0.5$ as compared to uniform heating for both the considered thickness and any value of Reynolds number, while for $q_r = 2$, $\overline{h}$ is less as compared to uniform heating. It implies that the rate of heat transfer is more for non-uniform asymmetric heating with lower value of step heat flux at the inlet and is recommended. The percentage increase in $\overline{h}$ when $q_r$ changes from 1 to 0.5 at Re = 50 and 200 are 9.64% and 9.03%, respectively, for $d_o/d_i = 2$.

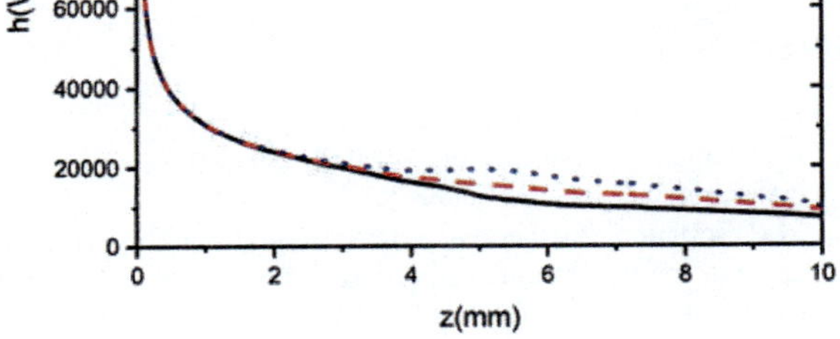


**Fig. 5** Local convective heat transfer coefficient showing the effect of $q_r$ for $d_o/d_i = 2$ at Re = 200 and $k_{sf} = 646.0767$

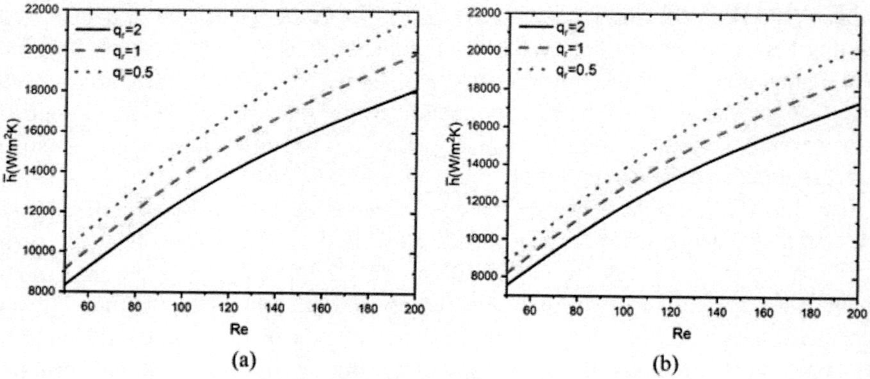


**Fig. 6** Average convective heat transfer coefficient within a range of Re = 50–200 showing the effect of $q_r$ for **a** $d_o/d_i = 2$ and **b** $d_o/d_i = 4$ at Re = 200 and $k_{sf} = 646.0767$

The percentage increase in $\overline{h}$ at Re = 50 and 200 for $d_o/d_i = 4$ are 7.62% and 7.77%, respectively. A smaller wall thickness should also be preferred in order to achieve higher rate of heat transfer.

# 4 Conclusions

Conjugate heat transfer analysis is performed for laminar flow through a circular microduct to investigate the effects of wall thickness and step heat flux on the rate of heat transfer. The results generated are interpreted to find the significance of considering finite wall thickness and arrangement of heat sources with respect to the direction of flow. The significant findings are concluded within the following points.

- The effect of wall thickness is significant for the analysis of forced convective heat transfer in microduct and cannot be neglected.
- The arrangement of non-uniform heating in the microduct should be in such a way that it is subjected to lower heat-producing sources near the entrance and followed by higher heat sources.

# References

1. M. Richter, P. Woias, D. Weiβ, Microchannels for applications in liquid dosing and flow-rate measurement. Sens. Actuators **A62**, 480 (1997)
2. M.P. Boruah, P.R. Randive, S. Pati, Effect of non-uniform asymmetric heating on the thermal and entropy generation characteristics for flow of Al2O3-water nanofluid in a micro-channel. Int. J. Numer. Methods Heat Fluid Flow. **29**, 981 (2019)

3. B. Mondal, S. Pati, P.K. Patowari, Influence of confluence angle between inlets on the mixing performance of micro-mixer with obstacles, Techno-Societal (2018), p. 275
4. D. Lorenzini, Y.K. Joshi, Computational fluid dynamics modeling of flow boiling in microchannels with nonuniform heat flux. J. Heat Transf. **140**, 1 (2018)
5. H. Yapici, B. Albayrak, Numerical solutions of conjugate heat transfer and thermal stresses in a circular pipe externally heated with non-uniform heat flux. Energy Convers. Manag. **45**, 927 (2004)
6. O. Aydin, M. Avci, T. Bali, M.E. Arıcı, Conjugate heat transfer in a duct with an axially varying heat flux. Int. J. Heat Mass Transf. **76**, 385 (2014)
7. A. Borah, M.P. Boruah, S. Pati, Conjugate heat transfer in a duct using nano fluid by two-phase Eulerian—Lagrangian method: effect of non-uniform heating. Powder Technol. **346**, 180 (2019)
8. M.R. Hajmohammadi, S. Poozesh, M. Rahmani, A. Campo, Heat transfer improvement due to the imposition of non-uniform wall heating for in-tube laminar forced convection. Appl. Therm. Eng. **61**, 268 (2013)
9. K.K. Ambatipudi, M.M. Rahman, Analysis of conjugate heat transfer in microchannel heat sinks. Numer. Heat Transf. Part A **37**, 711 (2000)
10. F.J. Hong, P. Cheng, H. Ge, G.T. Joo, Conjugate heat transfer in fractal-shaped microchannel network heat sink for integrated microelectronic cooling application. Int. J. Heat Mass Transf. **50**, 4986 (2007)
11. S. Pati, *A Textbook on Fluid Mechanics and Hydraulic Machines* (McGraw-Hill Education (India) Pvt. Ltd., New Delhi, 2012)

# Micro Electrical Discharge Milling of Titanium: Effects of Voltage and Tool Rotation Speed

**Siddhartha Kar, Pallab Sarmah, and Promod Kumar Patowari**

**Abstract** Micro electrical discharge milling (μED-milling) is gaining a tremendous reputation due to its capability of fabricating intricate shapes with a simple rotating tool. Its unique advantage of machining any electrically conductive material makes it applicable for processing high strength materials like titanium (Ti) and its alloys. The present study investigates μED-milling of Ti grade 2 alloy using tungsten carbide as a tool electrode. The effect of important process parameters like voltage and tool rotation speed (TRS) is examined by evaluating response measures such as material removal rate (MRR), tool wear rate (TWR) and electrode wear ratio (EWR). MRR and TWR increase with an increase in both voltage and TRS due to a rise in discharge energy and better flushing of eroded particles from the machining zone. Highest MRR of 268,232 $\mu m^3/s$ and lowest TWR of 27,845 $\mu m^3/s$ is achieved at highest (200 V and 2000 rpm) and lowest (110 V and 500 rpm) parametric combination, respectively, considered in this study. EWR decreases with an increase in voltage, whereas EWR has a negligible variation with an increase in TRS. The lowest EWR of 0.199 is achieved at a voltage of 200 V and TRS of 500 rpm.

**Keywords** EDM · Micromachining · Micro electrical discharge milling · Milling · Titanium

## 1 Introduction

Micro electrical discharge milling (μED-milling) is a variant of micro electrical discharge machining (μEDM) process possessing the exceptional capability to produce microchannels and three-dimensional (3D) cavities on hard and high strength materials. In this process, a rotating tool electrode is fed towards the workpiece material to actuate spark, which forms the desired shape by melting and evaporation

S. Kar (✉) · P. Sarmah · P. K. Patowari
Department of Mechanical Engineering, National Institute of Technology Silchar, Silchar, Assam 788010, India
e-mail: siddkar.nita@gmail.com



K. M. Pandey et al. (eds.), *Recent Advances in Mechanical Engineering*, Lecture Notes in Mechanical Engineering, https://doi.org/10.1007/978-981-15-7711-6_40

of the workpiece material [1]. This process has certain advantages over conventional milling that it can produce burr-free microchannels and 3D complex profile at micro-level with higher accuracy and precision, and high aspect ratio (AR). Also, μED-milling can achieve complex 3D profiles with the use of a cylindrical tool accompanied by rotational motion [2]. Thus, unlike sinking μEDM, it avoids the manufacturing of any complex shaped tools for the fabrication of intricate cavity. The process parameters of μED-milling such as voltage, capacitance, feed rate (FR), tool rotation speed (TRS), layer thickness (LT) decide the efficiency and quality of the machined features in terms of material removal rate (MRR), tool wear rate (TWR), electrode wear ratio (EWR), dimensional deviation, etc. A brief review of μED-milling on various materials is discussed herein.

Jafferson et al. [3] analyzed the influence of non-electrical parameters such as LT, FR and TRS on MRR and TWR in μED-milling of stainless steel using tungsten (W) as a tool. The study revealed that higher TRS and FR yields higher MRR and TWR. Elsewhere, LT influenced both MRR and TWR significantly, where the layer-by-layer approach of machining was observed to automatically dress the tool. Kuriachen and Mathew [4] investigated MRR and TWR in μED-milling of Ti–6Al–4V using tungsten carbide (WC) tool. They observed that capacitance, TRS and FR had a significant influence on MRR. Elsewhere, TWR increased with increment in voltage, capacitance, TRS and FR. But, the voltage had negligible influence on TWR beyond 115 V. From the surface topography, re-solidification of molten metal was observed on the machined surface due to repeated cooling and melting, which ultimately resulted in the emergence of recast layer. In another study, Kuriachen and Mathew [5] studied the influence of silicon carbide (SiC) powder dispersed on dielectric while μED-milling of Ti–6Al–4V. The analysis of variance revealed that high MRR and low TWR were attained simultaneously at powder concentration of 5 g/L, voltage of 115 V and capacitance of 0.1 mF. The surface of Ti–6Al–4V undergoing μED-milling exhibited a uniform distribution of modified layer consisting W and SiC. But, the surfaces were exposed to microholes and microcracks toward their ends. Lin et al. [6] examined the influence of spark gap (SG), peak current ($I_p$), pulse on-time ($T_{on}$) and pulse off-time ($T_{off}$) on electrode wear (EW), MRR and working gap (WG) in μED-milling of Inconel 718 using WC tool of 200 μm diameter. By Grey-Taguchi method, the experimental optimal condition for μED-milling was achieved at $I_p$ of 0.5 A, $T_{on}$ of 3 μs, $T_{off}$ of 3 μs, and SG of 60 V. Karthikeyan et al. [7] successfully performed μED-milling on EN 24 of 1.7 mm thickness by W tool of 500 μm diameter. MRR and TWR decreased due to poor flushing at a higher value of AR, wherein AR of 1.5 was utilized for mass removal of material in a single cut without undermining the MRR. A lower value of TRS and FR resulted in larger debris at the edges. In another study, Karthikeyan et al. [8] observed that rotation of tool electrode not only disturbed the plasma but also affected the quality (dimensional accuracy) of the channel. At a low value of energy (50 μJ), milling marks were found on the surface of the channel, but it disappeared at high value of energy, which was attributed to redeposition of molten materials. Unune and Mali [9] fabricated microchannels on Inconel 718 and studied the effect of low-frequency vibration of the workpiece in μED-milling. From the parametric analysis, capacitance emerged as the parameter affecting both MRR and frontal electrode wear (FEW) significantly by 76.62 and 77.61%, respectively. Elsewhere, the frequency of vibration also affected MRR and FEW by 13.47% and 4.42%, respectively, which indicated a favourable impact of vibration in μED-milling. From the microscopic analysis, it had been revealed that microchannel with uniform and smooth surface was machined at lower discharge energy and vibrational frequency. Vidya et al. [10] performed μED-milling on steel of nominal dimension 1000 μm using 200 μm WC electrode and studied the machining performance of different cavities like channel, cross channel, triangular, square and circular. They observed that cavities of square and triangular shape exhibited accurate and smooth surface as compared to the others. The microscopic study of the cavities revealed that rotation of tool electrode played a key role in the flow of the molten metal, the variation of globule formation, deposition of recast layer and final shape of the cavities.

From the literature, it can be revealed that process parameters play a significant role in the performance of μED-milling. Although a handful of research had been carried out in this domain, still, there are no concrete facts about the effect of process parameters on μED-milling. Moreover, the μED-milling process is highly dependent on the physical properties of a material such as thermal conductivity, thermal diffusivity, melting point and boiling point [11]. In this regard, very less work has been carried out on titanium (Ti) grade 2 alloy, which requires further investigation to establish its behaviour in μED-milling. Apart from Ti being the main constituent of Ti grade 2, it also contains iron, carbon, hydrogen, nitrogen and oxygen [12]. It is mainly used in airframe components, cryogenic vessels, heat exchangers, condenser tubing, pickling baskets, etc. In this study, μED-milling is performed on Ti grade 2 with WC tool electrode. Voltage and TRS are varied to examine their effects on μED-milling response measures such as MRR, TWR and EWR. The overall experimentation is intended to enhance the machining efficiency by maximizing MRR and minimizing TWR and EWR.

## 2 Materials and Methods

The μED-milling experiments are performed on a tabletop μEDM setup (Make: Sinergy nano systems, Mumbai, India; Model: Hyper-15). Experimental condition such as process parameters and other fixed parameters are depicted in Table 1. Voltage and TRS are varied as process parameters, whereas MRR, TWR and EWR are chosen as response measures. WC rod (diameter = 495$\mu$m) is selected as the tool electrode for μED-milling on Ti alloy. The workpiece and tool electrodes are submerged in hydrocarbon oil, which acts as a dielectric medium. Fresh dielectric is supplied to the machining zone by providing jet flushing, which also avoids deflection and bending of the thinner micro tool. Based on pilot experiments, it has been observed that μED-milling works well at highest capacitance ($10^4$ pF) available in the present μEDM setup. Thus, capacitance has been fixed at $10^4$ pF throughout the experiments. Elsewhere, significant variation of the process has been observed with a change in

**Table 1** Experiment condition

| Variable parameters | | |
|---|---|---|
| Parameter | Level | Value |
| TRS (rpm) | 4 | 500, 1000, 1500, 2000 |
| Voltage (V) | 4 | 110, 140, 170, 200 |

| Fixed parameters | |
|---|---|
| Parameter | Value |
| Tool | WC (diameter: 495 μm) |
| Workpiece | Ti grade 2 (thickness: 2200 μm) |
| Polarity | Workpiece (+); Tool (-) |
| Dielectric medium | Hydrocarbon oil |
| Capacitance | $10^4$ pF |
| Depth of cut | 250 μm |
| Feed rate | 5 μm/s |
| Flushing type | Jet flushing |

voltage and TRS. Thus, four levels of voltage (110–200 V) and TRS (500–2000 rpm) are selected for the present investigation. After experimentation, the workpiece and tool are visualized in an optical microscope (Make: Leica; Model: DM 2500 M) to measure the dimension of their eroded volume. Using these dimensional data, 3D images of the eroded parts of both workpiece and tool are created in Creo Parametric 3D Modelling software. Then, the volume of material eroded from the workpiece and tool electrode is evaluated from the 3D images. Thereafter, MRR, TWR and EWR are calculated by Eqs. (1), (2) and (3), respectively.

$$\mathrm{MRR} = \frac{\mathrm{Volumeofworkpieceremoved}}{\mathrm{Machiningtime}} \tag{1}$$

$$\mathrm{TWR} = \frac{\mathrm{Toolwearvolume}}{\mathrm{Machiningtime}} \tag{2}$$

$$\mathrm{EWR} = \frac{\mathrm{TWR}}{\mathrm{MRR}} \tag{3}$$

## 3 Results and Discussion

The full factorial design of process parameters based on which the experiments are performed and the result of response measures, are depicted in Table 2. The microscopic images of WC tool used for 16 experimental conditions are shown in Fig. 1. The worn-out tip of the tools is visible, which occurs mainly due to lateral wear

**Table 2** Experimental results of μED-milling on Ti grade 2

| Exp. no. | Process parameters | | Response measures | | |
|---|---|---|---|---|---|
| | Voltage (V) | TRS (rpm) | MRR ($\mu m^3/s$) | TWR ($\mu m^3/s$) | EWR |
| 1 | 110 | 500 | 130,464 | 27,845 | 0.213 |
| 2 | 110 | 1000 | 138,868 | 35,462 | 0.255 |
| 3 | 110 | 1500 | 137,665 | 39,358 | 0.286 |
| 4 | 110 | 2000 | 149,956 | 39,637 | 0.264 |
| 5 | 140 | 500 | 140,438 | 41,607 | 0.296 |
| 6 | 140 | 1000 | 151,396 | 47,485 | 0.314 |
| 7 | 140 | 1500 | 145,836 | 47,644 | 0.327 |
| 8 | 140 | 2000 | 171,266 | 48,507 | 0.283 |
| 9 | 170 | 500 | 193,415 | 43,458 | 0.225 |
| 10 | 170 | 1000 | 199,719 | 48,713 | 0.244 |
| 11 | 170 | 1500 | 203,850 | 49,547 | 0.243 |
| 12 | 170 | 2000 | 214,328 | 51,891 | 0.242 |
| 13 | 200 | 500 | 234,202 | 46,707 | 0.199 |
| 14 | 200 | 1000 | 236,624 | 51,702 | 0.218 |
| 15 | 200 | 1500 | 241,002 | 57,852 | 0.240 |
| 16 | 200 | 2000 | 268,232 | 62,316 | 0.232 |

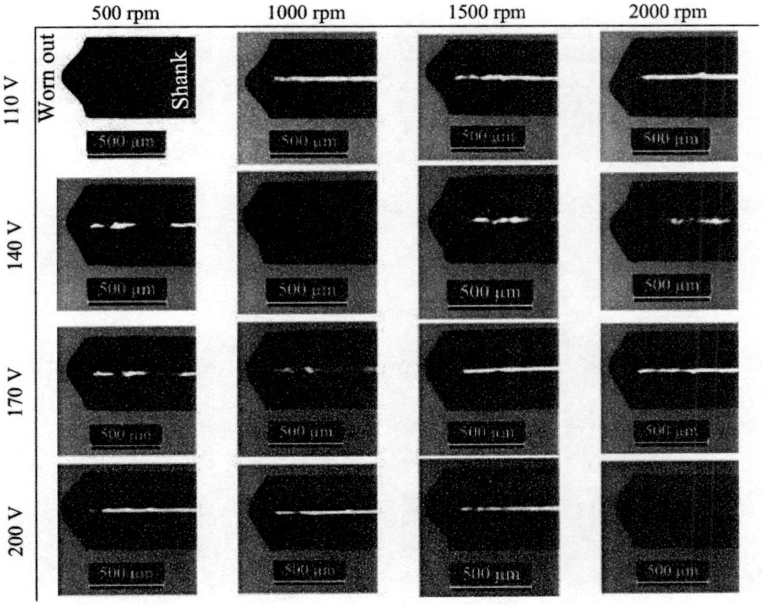


**Fig. 1** Microscopic images of WC tool after performing μED-milling

during μED-milling. Figures 2, 3 and 4 show the microscopic images at start side, end side and top view of the microslots machined by μED-milling. The parametric effect of response measures is also depicted and discussed in the following sub-sections.

Parametric effect of the process parameters on MRR and its mean effect plot is shown in Fig. 5a and b, respectively. MRR rises with increment in both voltage and TRS, as evident from Fig. 5a. Increment in voltage from 110 to 200 V upsurges the discharge energy of the process, which in turn increases the spark intensity and removes material at a faster rate [13]. With an increase in TRS from 500 to 2000 rpm, the centrifugal force and agitation effect exerted by the tool electrode increases in the machining zone. The increase in centrifugal force and agitation effect regulates the dielectric fluid and flushes the eroded particles more efficiently, which ultimately leads to an increase in the MRR [14]. The increment in MRR with the increase in both voltage and TRS is also reflected from the mean effect plot as shown in

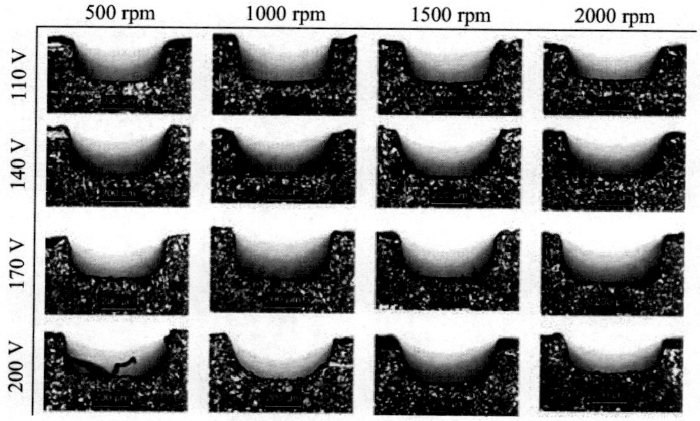


**Fig. 2** Start side of the microslots machined by μED-milling

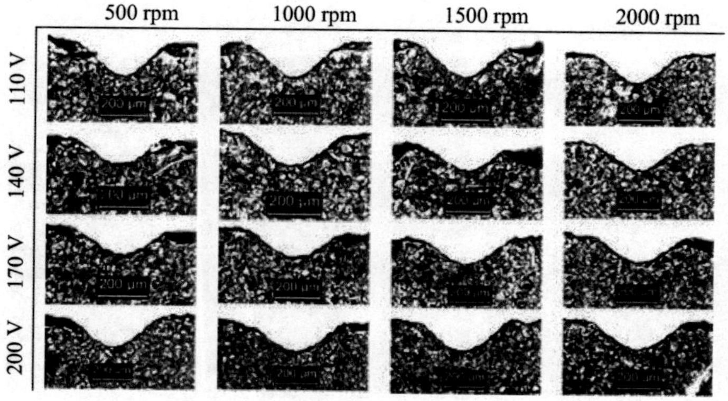


**Fig. 3** End side of the microslots machined by μED-milling

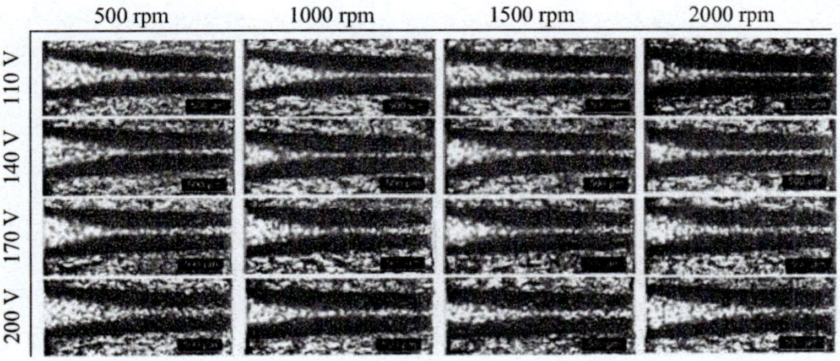


**Fig. 4** Top view of the microslots machined by μED-milling

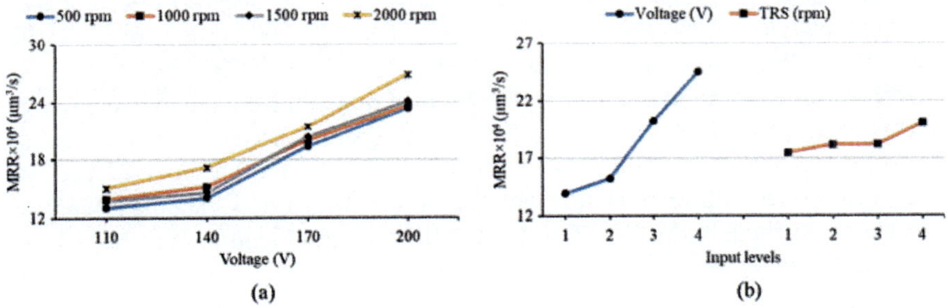


**Fig. 5** Variation of MRR with voltage and TRS **a** parametric effect and **b** mean effect

Fig. 5b. Highest MRR of 268,232 $\mu m^3/s$ is achieved at a voltage of 200 V and TRS of 2000 rpm.

In any electrical discharge machining process, tool wear is inevitable. The intrinsic mechanism of the process leads to wear from the tool electrode along with material removal from the workpiece. Parametric effect of voltage and TRS on TWR is depicted in Fig. 6a. Like MRR, an increment in voltage from 110 to 200 V leads to the increase in TWR due to rise in discharge energy. Elsewhere, an increase in TRS (500–2000 rpm) also leads to the increment in TWR, which can be credited to better

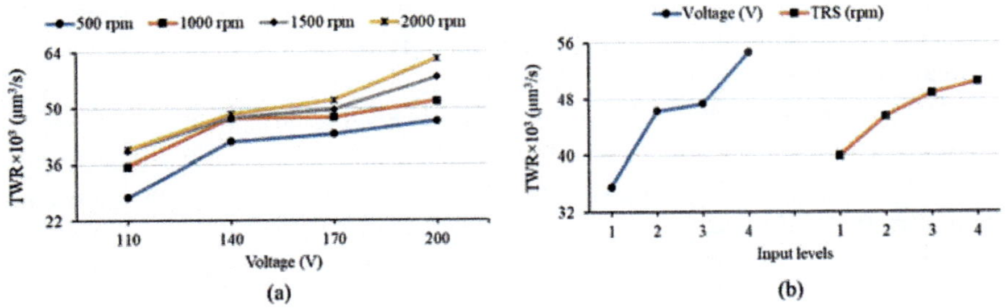


**Fig. 6** Variation of TWR with voltage and TRS **a** parametric effect and **b** mean effect

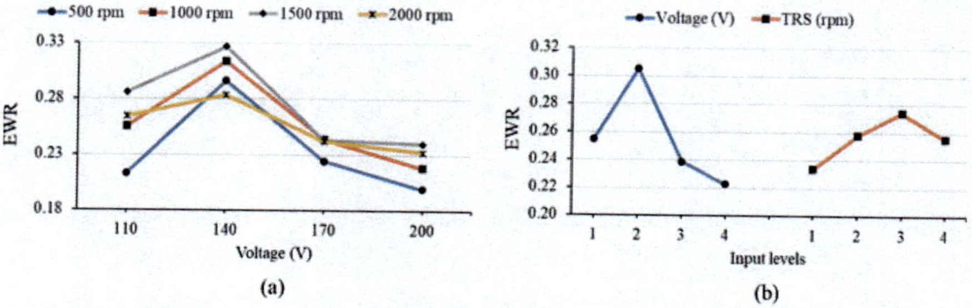


**Fig. 7** Variation of EWR with voltage and TRS **a** parametric effect and **b** mean effect

flushing of eroded particles due to higher centrifugal force and agitation effect. The mean effect plot of TWR (Fig. 6b) also shows a similar trend, where TWR increases with increment in both voltage (110–200 V) and TRS (500–2000 rpm). Lowest TWR of 27,845 $\mu m^3/s$ is achieved at a voltage of 110 V and TRS of 500 rpm.

EWR signifies the ratio of the eroded volume of material from the tool and workpiece as depicted in Eq. 3. Figure 7a and b show the parametric effect plot and mean effect plot of voltage and TRS on EWR, respectively. An infinitesimal increase in EWR is observed with increment in voltage from 110 to 140 V. The increment in EWR can be attributed to higher TWR, lower MRR or a combination of both at 140 V. With a further increment in voltage from 140 to 200 V, EWR decreases. This is due to very high MRR at upper levels of voltage, although TWR is higher as compared to that of low voltage [15]. Elsewhere, very minimum change in EWR has been found with an increase in TRS from 500 to 2000 rpm, which suggests the negligible effect of TRS on EWR. Lowest EWR of 0.199 is achieved at a voltage of 200 V and TRS of 500 rpm. TWR does not increase at a higher rate as compared to MRR in higher voltages. This gives an idea for the selection of process parameter having high MRR and lower TWR simultaneously.

## 4 Conclusions

Microslots were successfully machined on Ti grade 2 alloy using μED-milling. The effect of variation of process parameters (voltage and TRS) on response measures like MRR, TWR and EWR had been investigated. Both MRR and TWR increased with increment in voltage and TRS. The rise in discharge energy due to increment in voltage was accountable for higher material removal from both the workpiece and tool. Increment in TRS contributed to better flushing of eroded particles from the machining zone, which in turn enhanced the material removal phenomenon from both the electrodes. Barring an infinitesimal increment in EWR at 140 V, a decreasing trend of EWR was observed with an increase in voltage. Elsewhere, the variation of EWR with a change in TRS was very minimum.

Thus, a preliminary study of the fabrication of microslots in Ti grade 2 alloy using μED-milling was done successfully. Future studies may be concentrated on the fabrication of a linear microchannel possessing higher dimensional accuracy. Furthermore, the effect of variation of other important parameters for instance capacitance, depth of cut, FR, etc., may also be considered in evaluating the machining efficiency and dimensional accuracy of the microslots machined by μED-milling.

## References


1. S. Kar, P.K. Patowari, Electrode wear phenomenon and its compensation in micro electrical discharge milling: a review. Mater. Manuf. Process. **33**(14), 1491–1517 (2018)
2. S. Kar, P.K. Patowari, Effect of non-electrical parameters in fabrication of micro rod using BEDG. Mater. Manuf. Process. **34**(11), 1262–1273 (2019)
3. J.M. Jafferson, P. Hariharan, J. Ram Kumar, Effect of non-electrical parameters in μED milling: an experimental investigation. Int. J. Adv. Manuf. Technol. **85**(9–12), 2037–2047 (2016)
4. B. Kuriachen, J. Mathew, Experimental investigations into the effects of microelectric-discharge milling process parameters on processing Ti–6Al–4V. Mater. Manuf. Process. **30**(8), 983–990 (2015)
5. B. Kuriachen, J. Mathew, Effect of powder mixed dielectric on material removal and surface modification in microelectric discharge machining of Ti-6Al-4V. Mater. Manuf. Process. **31**(4), 439–446 (2016)
6. M. Lin, C. Tsao, C. Hsu, A. Chiou, P. Huang, Y. Lin, Optimization of micro milling electrical discharge machining of Inconel 718 by Grey-Taguchi method. Trans. Nonferrous Met. Soc. China **23**(3), 661–666 (2013)
7. G. Karthikeyan, J. Ramkumar, S. Dhamodaran, S. Aravindan, Micro electric discharge milling process performance: an experimental investigation. Int. J. Mach. Tools Manuf. **50**(8), 718–727 (2010)
8. G. Karthikeyan, A.K. Garg, J. Ramkumar, S. Dhamodaran, A microscopic investigation of machining behavior in μED-milling process. J. Manuf. Process. **14**(3), 297–306 (2012)
9. D.R. Unune, H.S. Mali, Experimental investigation on low-frequency vibration-assisted μ-ED milling of Inconel 718. Mater. Manuf. Process. **33**(9), 964–976 (2018)
10. S. Vidya, Vijay, S. Barman, A. Chebolu, Nagahanumaiah, A.K. Das, Effects of different cavity geometries on machining performance in micro-electrical discharge milling. J. Micro Nano-Manuf. **3**(1), 11007 (2015)
11. J.C. Pilligrin, P. Asokan, J. Jerald, G. Kanagaraj, Effects of electrode materials on performance measures of electrical discharge micro-machining. Mater. Manuf. Process. **33**(6), 606–615 (2018)
12. Titanium Grade 2. Available: https://www.matweb.com/search/datasheet.aspx?MatGUID=24293fd5831941ec9fa01dce994973c7. Accessed 09 Jan 2020
13. K. Mondol, M.S. Azad, A.B. Puri, Analysis of micro-electrical discharge drilling characteristics in a thin plate of Ti–6Al–4 V. Int. J. Adv. Manuf. Technol. **76**(1–4), 141–150 (2015)
14. S. Kar, P.K. Patowari, Experimental investigation of machinability and surface characteristics in microelectrical discharge milling of titanium, stainless steel and copper. Arab. J. Sci. Eng. **44**(9), 7843–7858 (2019)
15. M.P. Jahan, Y.S. Wong, M. Rahman, A study on the quality micro-hole machining of tungsten carbide by micro-EDM process using transistor and RC-type pulse generator. J. Mater. Process. Technol. **209**(4), 1706–1716 (2009)

# Bio-plastic from Yam—An Ecofriendly Prospective

**Susheel Kumar, S. K. Pattanayak, and Krishna Murari Pandey**

**Abstract** Nowadays, plastics has become an integral part of the modern world. Its utilization in different fields is increasing rapidly because of its excellent "long-life" property. But the conventional plastics produced from non-renewable resources like coal, petroleum and natural gas need decades to degrade in nature. Biodegradable plastics or bio-plastics are form of plastics which are derived from plant resources and are "naturally" degraded. The present work investigates the extraction of the optimum quantity of starch and develops a method for producing bio plastic from yam (Dioscoreaalata). It also aims to access mechanical and physical properties of the starch-based plastic film. Tensile strength and percentage elongation plays important role in plastics use, so to prepare a plastic film that has good strength and which can elongate without fracture is desirable. Also, biodegradability test is evaluated in this work which is necessary for environment-friendly plastics. Optimum quantity of starch was extracted from a given quantity of yam by different processes like blending, filtration, drying. Preparation of plastic from starch overall follows a polymerization reaction but is divided into processes like plasticization, gelatinization, neutralization, heating, cooling. Hydrochloric acid and plasticizer (glycerol and sorbitol) play a vital role in plastic film preparation. Sodium Bicarbonate ($NaHCO_3$) is used as preservative for the plastic after manufacturing.

**Keywords** Dioscoreaalata · Bio plastic · Biodegradable · Yam · Plasticizers

## 1 Introduction

Plastics are one of the most important things that we come across in our day to day lives. Plastics are most widely used for our daily needs products for example

S. Kumar (✉) · S. K. Pattanayak · K. M. Pandey
Department of Mechanical Engineering, National Institute of Technology Silchar, Silchar, Assam 788010, India
e-mail: susheelkumar549@gmail.com

K. M. Pandey
e-mail: kmpandey@mech.nits.ac.in



K. M. Pandey et al. (eds.), *Recent Advances in Mechanical Engineering*, Lecture Notes in Mechanical Engineering, https://doi.org/10.1007/978-981-15-7711-6_41

containers for lunch boxes to toys or any furniture, carry bags, water bottles, etc. There is such a widespread application of plastics that we just can't ignore it, so its proper disposal management is necessary. The huge application of plastics is found in food industry for packaging (e.g. soft drink bottles, containers, boxes its lids, shampoo bottles, etc.), but they are also used in electronics industries (e.g. appliances, furniture) and medical fields (e.g. diapers, trash bags, cups and utensils, syringes, stitches and various medical devices).

There are many such polymers, among which starch is the one that is of prominent interest. Starch is recovered from carbon dioxide and water by the process of photosynthesis occurring in green plants. Its main property is that it is totally biodegradable, minimal effort of producing and its sustainability; cheap, abundant, edible, etc. Starch is one of the best alternatives for creating feasible biodegradable materials. All these properties of starch have been very much accepting and are in further developing consideration since the ages of 1970s. Numerous endeavours to create starch-based polymers have been applied for monitoring the decreasing petrochemical assets, decreasing natural effect and seeking more applications. Plasticizer's are used to increase film flexibility and prolonged usefulness without using suitable additives the film produced will be brittle and fragile that's of no practical use. Several methods are used to inspect the starch film produced such as XRD, SEM, polarized light microscopy, etc. Based on studying its microstructure we can define various physical and chemical properties of the film produced.

The plastics made from petrochemicals are the conventional ones that are nonbiodegradable. But the problem is that there are only limited assets of oil resources also the increasing use of nonbiodegradable plastics has to raise alarming concerns of environment pollution, also using plastics for temporary use are not at all reasonable therefore materials which are degradable in nature have gained lot of attention of researchers since times of 1970s. Omotoso et al. [1] effectively arranged biodegradable films from tuber and root starches and on inspecting synthetic, useful and mechanical properties it's been presumed that Starch can be utilized as substitute to deliver plastics that are biodegradable. Pimpan et al. [2] studied the Preparation of Biodegradable Plastic from Modified Cassava Starch. What's more, reasoned that Cassava starch can be altered by utilizing malefic anhydride as modifier, sodium hydroxide as impetus and water as dissolvable solvent at 50 °C. Souza et al. [3] contemplated Cassava starch biodegradable films: Influence of glycerol and mud/clay nanoparticles content on tensile and boundary properties. They inferred that Films arranged with lower glycerol content introduced preferred tensile and a hindrance property over films with higher amounts of glycerol. Ezeoha et al. [4] took a shot at Production of Biodegradable Plastic Packaging Film from Cassava Starch. They came about Biodegradable plastic packaged films from cassava starch that could fill in as a decent substitute for the regular plastic packaging films.

### 1.1 *Yam (***Dioscoreaalata***)*

Yam is a taproot, from underground root which is wealthy in starch and one of the mainstream edible root vegetables in parts of Asia, Pacific Islands, West Africa, and Amazonian areas of South America. The regular names of genus (*Dioscorea*) in the species (*alata*) is yam, purple yam, tube, more noteworthy Yam is an expansive lasting herbaceous plant developing below earth surface. It develops in marshy, wet soil and warm damp atmospheres. The Yam develops to a size of turnip, has globular or oval shape with dark colour brownish skin. Its surface is set apart with round rings demonstrating the purposes of connection of layered leaves. Inside, its tissue is white to cream-yellow, yet may include diverse colours relying on cultivator types. A normal size Yam weighs around 2–4 kg. Its fresh finished substance has nutty flavour simply like in water chestnuts. The Yams, nonetheless, are not free from gluten. They convey top-notch phyto-nourishment profile involving dietary fiber, and cell reinforcements with moderate extents of minerals, and nutrients. Yam is one of the best source dietary fibers; 100 g of yam gives 4.1 g or 11% of day by day necessity of dietary fiber. Gaind et al. and Taki et al. [5] reported that Yam is also rich in gums (mucilages). When the Yam corns and tubers were extracted there was 10.7% crude Yam mucilages Purified gums (100 g) also are isolated from fresh Yams (1 kg). The drum drying properties and storage stability of tropical fruit-Yam mixtures were reported by Nip (1979). Also, they studied the use of crude Yam gums as binding and emulsifying agents.

### 1.2 *Starch*

Starches are fundamentally sugars, known as polysaccharides. For business use, starches are derived from different types of grains like rice, wheat, sorghum, corn and tubers like potato, custard and sweet potato. Universally mainstream types of starch are for the most part derived from corn and custard/cassava because of their simple accessibility. At the point when the starch is treated with high temp water the starch transforms into thick glue and when cooled with specific added substances, it structure turns into a gel. This gives it high consistency which shapes the premise of its numerous uses. Starch is a polymer of hexacarbon monosaccharide-D-glucose. It is very plentiful in corn seeds, potato tubers and the roots and stems of the plants.

Extensive research work has been done in the area of modern-day biodegradable plastics like making plastics from cassava plants, banana peel, potato skin, etc. Plastics produced from these sources differ in strength and elongation and biodegradability. Some of them showed good tensile strength but poor biodegradability. And some showed less elongation property. In this present work, we will try to produce plastic film from another starch-containing tuber, with the help of various combinations. Various tests are being performed to check for its tensile strength and elongation property along with its biodegradability. Anova analysis is done to show the best combination to be used for preparing plastic film which has good strength and elongation and biodegradable properties for useful operations.

## 2 Materials and Methods

### *2.1 Raw Materials*

Yam is purchased from local market. Starch powder was extracted from yam and is used for film formation to provide biodegradable films. Glycerol was used as plasticizer of the films. Distilled water was used as solvent. Dilute 0.1 M Hydrochloric acid (HCl) was used as hydrolyzer. Also, 0.1 M Sodium hydroxide solution was used as neutralizer. Sodium met bisulphate which is optional was used to increase the life of biodegradable films.

### *2.2 Film Preparation*

About 1000 g of Yam was manually washed with potable water; peeled, cut into cubed and then blended. The blended yam was mixed with water (2 times the volume of the blended yam). The mixture was squeezed out by wrapping it with a piece of cotton cloth and finally strained by using a filter paper or sieve. Wait for 5–6 h to allow the starch to settle at the bottom of the beaker which is contained in the strained water. The supernatant was decanted in Petri dish and allowed for sun drying for one day (24 h). The dried starch was then allowed to cool at room temperature. Then it is weighed. The starch was crushed into fine powder in a grinder. Then, weight is measured.

Starch powder (5 and 10 g) is taken in a beaker. 100 ml of distilled water is added to the starch powder and stirred with a glass rod. 5 ml of hydrochloric acid is added to the mixture and stirred. Glycerol (2, 3 and 4%) is added to the above mixture and again stirred. 5 ml of NaOH solution is added to neutralize the mixture. The beaker of mixture is heated at 80 °C for 15 min until it becomes gelatinized. As the mixture starts to boil, it is poured onto an acrylic glass plate or aluminium foil and evenly spread on the surface. Then the mixture was labelled and left it to dry out at room temperature for about 3–4 days.

### *2.3 Tensile Strength and Percentage Elongation*

The plastic samples were submitted to SAIC Tezpur University, and tests were done. Universal Testing Machine was utilized to test for the elasticity or tensile strength of the biodegradable film. The six samples with various proportions of glycerol and starch were cut into a similar shape according to ASTM D-882 standards. In the lab, the prolongation and strength was tested by clipping the two finished ends and extending it upwards until it breaks. The machine recorded and gathered the outcomes while the test was going on.

### *2.4 Biodegradability*

The samples were weighted and placed independently in a container with soil collected from college campus. Consistently every week reading of weight was taken by digging out the samples, washing it with distilled water to remove soil particles and oven dried them for 24 h. The weights of the samples were recorded for several weeks.

**Formulation of Biodegradability Test.**

The percent weight loss of the specimen as a function of time was determined using the Eq. 1:

$$\text{Weightloss, } W_1 = \frac{Wb - Wa}{Wb} \times 100 \tag{1}$$

Where:

$W_1$ Weight loss, %
$W_b$ Weight of the specimen before burying, g
$W_a$ Weight of the specimen after digging and conditioned, g.

## 3 Results and Discussion

### *3.1 Results of Tensile Strength and Elongation Test*

The film produced was tested for tensile strength and elongation. Each sample was cut into desired shape for tensile test.

Table 1 demonstrated the results obtained for tensile strength (MPa) and the elongation (%) tests performed on the biodegradable plastic film sample. From the result table, it very well may be seen that the most noteworthy maximum stress (MPa) of the film produced was from 10 g starch; 2% glycerol combination which is 3.55 MPa, and this could be the consequence of the glycerol that has been used in the film which was added to increase its mechanical strength. It can also be seen from the chart plotted that the biodegradable film created demonstrated higher tensile strength for 10 g starch than 5 g starch. The effect of glycerol changed the percentage stretching

**Table 1** Tensile strength (TS) and percentage elongation (%E) of biodegradable films

| Level of starch | Level of glycerol (%) | Thickness (mm) | Width (mm) | Max. stress (Mpa) | % elongation |
|---|---|---|---|---|---|
| 5 g starch | 2 | 0.17 | 12.50 | 2.52 | 16.50 |
| 5 g starch | 3 | 0.15 | 12.50 | 1.95 | 20.14 |
| 5 g starch | 4 | 0.17 | 12.50 | 1.50 | 21.50 |
| 10 g starch | 2 | 0.36 | 12.50 | 3.55 | 12.88 |
| 10 g starch | 3 | 0.25 | 12.50 | 2.85 | 15.44 |
| 10 g starch | 4 | 0.20 | 12.50 | 1.74 | 19.80 |

at break of the films. As indicated by Detduangchan et al. [6], revealed that the cause of increase in Tensile strength is because of more and more number of cross-linking responses, density in starch films that is caused by the cross-linking reaction between hydroxyl groups and cross-linking agents bringing about cross-linked starches.

It is demonstrated that the higher percentage elongation was obtained for 5 g starch than the 10 g starch. The combination 5 g starch; 4% glycerol showed maximum percentage elongation which is 21.50% and the lowest percent elongation was showed by 10 g starch; 2% glycerol combination which is 12.88%. Thus it can be concluded that the lower the starch content the higher the percent extension will be obtained. The presence of plasticizer makes the weak films progressively adaptable and more flexible, but also makes them less strong [7]. As indicated by Thakore et al. [8], the reduction of tensile strength happened because of shortcoming of between inter adhesion bond between the starch-polymer. The higher the amount of the starch and the lower the amount of glycerol increments the tensile strength and it was strong and rigid. The lower the amount of the starch and higher the amount of glycerol declines its tensile strength and it was fragile but more elongation percentage. The decrease in tensile strength by increasing glycerol is because of disruption of molecular cohesiveness of starch as well as lowering of intermolecular interaction in the film. The high tensile strength at the low glycerol percentage is the effect of strong hydrogen bonds between starch-starch intermolecular interactions over starch-glycerol attraction.

Obviously, adding higher amount of plasticizer increases the film elongation limit and diminished its tensile strength. Yam plastic film samples with higher level of glycerol content demonstrated lower tensile strength. This behaviour can be related with the atomic structure of glycerol, which has small chains, increasingly ready to go into the polymeric net. As the amount of glycerol is increased, the Tensile strength of films essentially decreases but the elongations at which the film breaks altogether are increased. Also, it is seen that with higher elongation property more often than not it requires a lower load to cause film breaking, this is because of the fact that its ductility is increased and the film becomes less brittle. The Plasticizer is commonly required for development of strong cohesive films. Glycerol and sorbitol are the most commonly used plasticizers for the making of biodegradable film. As a rule, using glycerol as plasticizer brought about better plastic films then using sorbitol because it causes structural modifications in the polymeric network and increases the molecular mobility, which facilitates the diffusivity of water. The conclusion of this analysis is almost similar and supports the discoveries, i.e. increasing the amount of plasticizer brought about increasingly adaptable flexible films, yet tensile strength was associatively decreased. Main effects plot for maximum stress and percentage elongation is shown in Figs. 1 and 2 respectively.

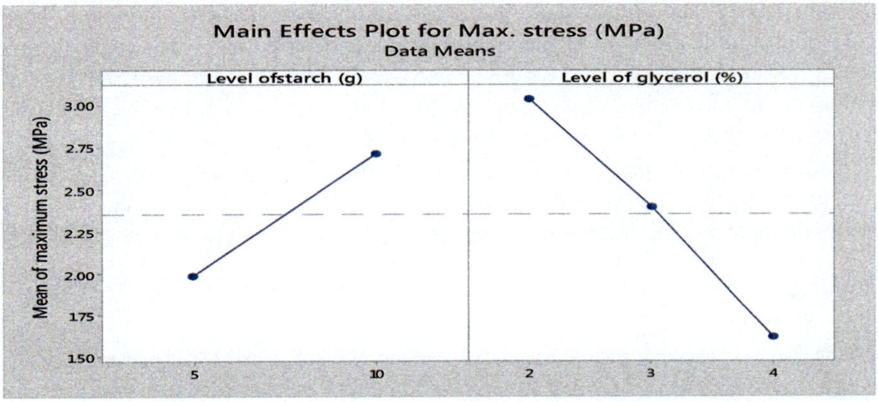


**Fig. 1** Effect of parameters at different levels on mean of maximum tensile strength

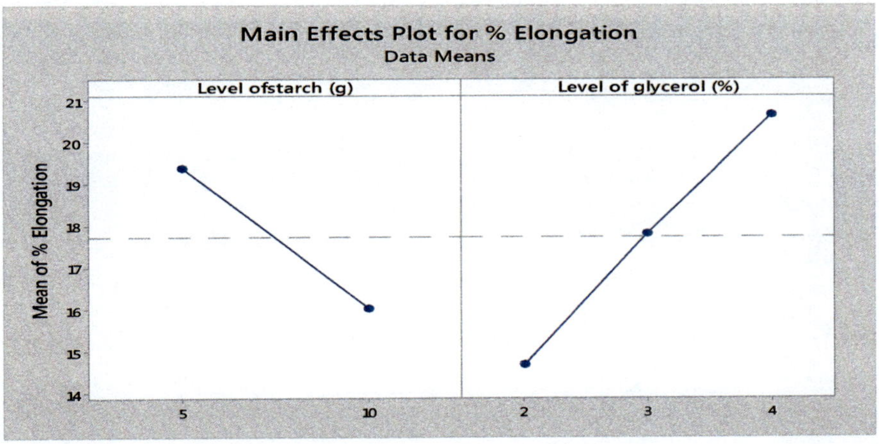


**Fig. 2** Effect of parameters at different levels on mean of % elongation

## 4 Biodegradability Test

Biodegradability of the film was found out. One sample made of various treatments of starch and glycerol was utilized to test the biodegradability of the yam plastic film. The highest tensile strength plastic showed loss in weight after testing and its average biodegradability for 3 weeks was found out to be 58.36%.

Table 2 demonstrated the results of biodegradability test. At the point when the film was covered and after that when it was dug out from the soil, the weight was light which implies the film decays and degrades the following one week. In the event, if lower amount of glycerol will be used, the biodegradability will be lesser in week as compared to the sample having higher level of glycerol. It likewise suggests that the film created is biodegradable because of the fact that it was degraded by soil with the help of microbes like *T. harzianum* etc. The average weight loss in 3 weeks is shown in the table for all the samples. The starch-based biodegradable plastics degrade quicker than conventional Plastics. Also Fig. 3 signifies that glycerol

**Table 2** Percent weight loss of film over 3 weeks

| Level of starch | Level of glycerol (%) | Thickness, mm | Width, mm | Biodegradability (%) |
|---|---|---|---|---|
| 5 g starch | 2 | 0.17 | 12.50 | 56.866 |
| 5 g starch | 32 | 0.15 | 12.50 | 67.860 |
| 5 g starch | 4 | 0.17 | 12.50 | 78.360 |
| 10 g starch | 2 | 0.36 | 12.50 | 58.360 |
| 10 g starch | 3 | 0.25 | 12.50 | 70.250 |
| 10 g starch | 4 | 0.20 | 12.50 | 83.360 |

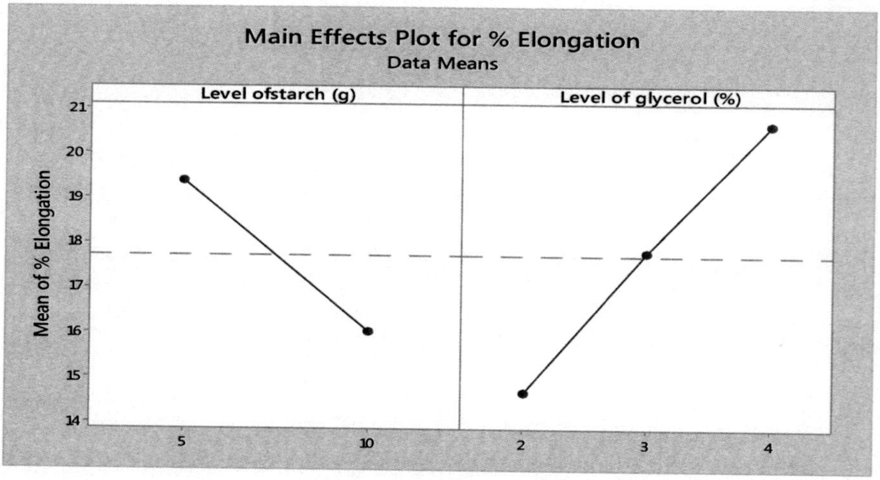


**Fig. 3** Effect of parameters at different levels on mean of % weight loss

percentage plays a significant role than the amount of starch in biodegradability of the samples.

## 5 Conclusion

Yam starch is a feasible component in the production of biodegradable films. The flexibility of the film is affected by the amount of the glycerol. Higher level of glycerol increased tensile strength and decreased elongation. Glycerol is a plasticizer that is compatible with starch. The level of starch at 5 g and level of glycerol at 2% was the best treatment which is less in water absorption and is same as in 5 g starch and 4%. Maximum tensile strength obtained was 3.55 MPa for 10 g starch and 2% level of glycerol with a percentage elongation of 12.88%. Although the composition showed less biodegradability 58.360% which is less than others, still it showed better strength and appreciable elongation. The highest biodegradability is nearly up to 83.360% which is from 10 g starch and 4% level of glycerol but it has low tensile strength of 1.74 MPa.

Some future research is recommended are as following; Test the barrier properties such as oxygen gas and water vapour permeability. Evaluate the film under a Scanning Electron Microscopy (SEM) to know the morphology of the film and the dispersion of starch and glycerol of the film. Evaluate the toxicity of the biodegradable films made from yam starch. Use blown film extruder to produce a stronger film for commercial purposes.

## References

1. M.A. Omotoso, G.O. Ayorinde, O.A. Akinsanoye, Preparation of biodegradable plastic films from tuber and root starches. IOSR J. Appl. Chem. 10–20 (2015)
2. V. Pimpan, K. Ratanarat, M. Pongchawanakul, Preliminary study on preparation of biodegradable plastic from modified cassava starch. J. Sci. Res. Chula. Univ. **26** (2001)
3. A.C. Souza et al., Cassava starch biodegradable films: influence of glycerol and nanoparticles content on tensile and barrier properties and glass transition temperature. LWT Food Sci. Technol. 110–117 (2011)
4. S.L. Ezeoha, J.N. Ezenwanne, Production of biodegradable plastic packaging film from cassava starch. IOSR J. Eng. (IOSRJEN) 14–20 (2013)
5. K.N. Gaind et al., Study of mucilages of corn and tuber of *Colocasiaesculentia*. Linn. Part 1. Emulsifying properties. Indian J. Pharm. 30–208 (1968)
6. N. Detduangchan et al., Enhancement of the properties of biodegradable rice starch films by using chemical crosslinking agents. Department of Material Product Technology, Faculty of Agro-Industry. Hat Yai, Songkhla, 90112, Thailand. Int. Food Res. J. **21**(3), 1225–1235 (2014)
7. W. Tongdeesoontorn, L.J. Mauer, S. Wongruong, P. Sriburi, P. Rachtanapun, Effect of carboxymethyl cellulose concentration on physical properties of biodegradable cassava starch-based films. Chem. Cent. J. **5**, 6 (2011)
8. I.M. Thakore et al., Morphology, thermochemical properties and biodegradability of low density polyethylene/starch blends. J. Appl. Polym. Sci. **74**, 2701–2802 (1999)

# Application of Box-Behnken Method for Multi-response Optimization of Turning Parameters for DAC-10 Hot Work Tool Steel

**Sunil Kumar, Saikat Ranjan Maity, and Lokeswar Patnaik**

**Abstract** Turning parameters for cutting DAC-10 tool steel was optimized using surface response methodology (RSM). Turning was performed with TiAlN coated single point tool bit on CNC lathe. Cutting speed, feed rate, and depth of cut were considered as the cutting parameters and relative effect of process parameters on surface roughness and tool wear rate was analyzed. Outcomes revealed that feed rate and cutting speed are the governing parameters for surface quality and cutting speed for tool wear rate respectively. Optimization method confirms reasonable zone for responses and gives optimal condition for turning with cutting speed 150 m/min, feed rate 0.1 mm/rev and depth of cut 0.4 mm.

**Keywords** DAC-10 tool steel · Cutting speed · Feed rate · Depth of cut · Surface roughness · Tool wear rate

## Abbreviations

| | |
|---|---|
| CS | Cutting speed |
| FR | Feed rate |
| DOC | Depth of cut |
| Ra | Surface roughness |
| TWR | Tool Wear rate |
| FW | Flank wear |
| CNCLM | CNC lathe machine |
| DOE | Design of experiment |

S. Kumar · S. R. Maity (✉) · L. Patnaik
Department of Mechanical Engineering, National Institute of Technology Silchar, Silchar 788010, Assam, India
e-mail: saikat.jumtech@gmail.com



K. M. Pandey et al. (eds.), *Recent Advances in Mechanical Engineering*, Lecture Notes in Mechanical Engineering, https://doi.org/10.1007/978-981-15-7711-6_42

# 1 Introduction

Turning is the oldest material removal process wherein a tool moves along the axis of the lathe to facilitate material removal in the form of chips. Surface characteristics of a product is the one of the essential attribute for determining the quality of the product. There are many applications which are directly affected by quality of surface such as tribological properties, aesthetic appearance, fatigue behavior of the product and corrosion resistance, etc.[1]. Right combination and level of turning parameters is important to get desirable surface finish with minimum tool wear rate. To improve the surface quality, CS needs to be increased. With rise in CS, TWR gets affected. Also, at lower CS, FW increases due to rubbing between tool flank and work material. Higher CS affects the crater wear which leads to tool breakage [2, 3]. Cutting fluid is widely used in machining process for cooling the tool and to improve surface roughness and machinability [4, 5]. Cutting speed can be increased upto 30% without having any effect on tool life [6].

Hard coating like TiC, TiN, and TiAlN minimize tool wear rate and improve surface roughness [7]. Thin coating is widely used on cutting and forming tools to improve the tool life and their performance. The PVD hard coating has a variety of applications. These coatings are used as protective layer for cutting tool, forming tool, gears and bearing [8, 9]. Tribological performance of TiAlN is superior to the TiN, AlCrN and TiCN [10, 11]. It is a third-generation coating among the ceramic hard coating [12]. Cutting speed is the major governing parameter for tool failure. Rise in temperature at higher cutting speed causes softening of outer layer which subsequently leads to delamination of the coating [13, 14].

L15 orthogonal array gives 15 sets of different combinations of cutting parameters for experiments. All 15 experiments were performed on CNCLM and the responses are measured. The experimental result was analyzed and optimal parameters are described.

# 2 Methodology

## *2.1 Material*

In the present study, DAC-10 tool steel was used as work material. It is a hot work tool steel used for manufacturing of die casting die elements for their excellent heat crack and wear resistance at high temperatures. The chemical composition of the tool steel is C-0.3%, Si-0.3%, Mn-0.6%, Cr-5.2%, Mo-2.7 and V-0.9% [15].

## 2.2 Box-Behnken Design of Experiment

Turning was conducted on a CNCLM, total 24 number of experiments were performed including pilot experiments. Number of experimental run was designed by Box-Behnken DOE for L15 orthogonal array. CS, FR, and DOC were considered as the cutting parameter and their levels are tabulated in Table 1. Investigations were performed according to the experimental run and values of the responses were collected and tabulated in Table 2.

**Table 1** Process parameters and their levels

| S. No. | Process parameters | Level 1 | Level 2 | Level 3 |
|---|---|---|---|---|
| 1 | CS (m/min) | 150 | 200 | 250 |
| 2 | FR (mm/rev) | 0.1 | 0.15 | 0.2 |
| 3 | DOC (mm) | 0.4 | 0.6 | 0.8 |

**Table 2** Experimental run and responses value

| Run order | CS (m/min) | FR (mm/rev) | DOC (mm) | Ra (μm) | TWR (g/min) |
|---|---|---|---|---|---|
| 1 | 150 | 0.2 | 0.6 | 2.4 | 0.0015 |
| 2 | 200 | 0.1 | 0.4 | 1.3 | 0.00175 |
| 3 | 200 | 0.15 | 0.6 | 1.9 | 0.00198 |
| 4 | 200 | 0.2 | 0.8 | 2.3 | 0.0022 |
| 5 | 200 | 0.15 | 0.6 | 1.8 | 0.00198 |
| 6 | 200 | 0.2 | 0.4 | 2.1 | 0.00211 |
| 7 | 150 | 0.15 | 0.4 | 1.7 | 0.00117 |
| 8 | 150 | 0.1 | 0.6 | 1.3 | 0.00139 |
| 9 | 200 | 0.1 | 0.8 | 1.7 | 0.00215 |
| 10 | 200 | 0.15 | 0.6 | 1.8 | 0.00212 |
| 11 | 250 | 0.1 | 0.6 | 1.1 | 0.00235 |
| 12 | 250 | 0.15 | 0.8 | 1.4 | 0.00252 |
| 13 | 250 | 0.15 | 0.4 | 1.2 | 0.00229 |
| 14 | 250 | 0.2 | 0.6 | 1.4 | 0.00244 |
| 15 | 150 | 0.15 | 0.8 | 1.9 | 0.00159 |

## 3 Results and Discussion

### *3.1 Relative Effect of Cutting Parameter on Response*

Quality of surface is an important outcome of turning operation which is affected by FR and CS. Increase in CS and decrease in FR gives preferable surface finish. Figure 1a shows that better surface finish is obtained at 250 m/min CS and 0.1 mm/rev FR. Paengchit et al. [16] performed turning on AISI4140 steel with $Al_2O_3$ + TiC cutting tool to obtain the minimum Ra at CS 220 m/min and FR 0.06 mm/rev. Ibrahim et al. [17] performed machining on D2 steel and suggested that surface quality depends on CS and FR. Arefi et al. [18] conducted similar experiment on lead alloy and obtained similar consequences of CS and FR on Ra. Oehaia et al. [19] conducted machining of C62D cold rolled steel and found that higher CS and lower FR give better Ra. DOC also has an impact on Ra; surface quality improves with lower DOC. Combined effect of DOC and CS is expressed in Fig. 1b, it was observed that surface quality improves at 250 m/min CS and 0.54 mm DOC. Chandra et al. [20] has conducted an investigation on the consequences of process parameters on Ra of alloy steel and observed that Ra improve at higher CS and lower DOC. Surface quality reduces with lower FR and DOC as shown in Fig. 1c.

Tool wear affects the cost of production. Combined effect of process parameters are shown in Fig. 2a–c. TWR is directly affected by CS, FR, and DOC. With increase in CS, FR, and DOC, TWR also increases. Zheng et al. [21] had examined the wear behavior of TCVD-TiCN-$Al_2O_3$ coated tool for machining of 40CrNi2SiMoV steel

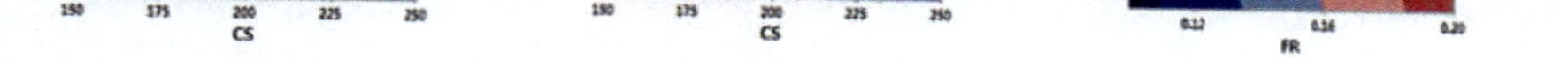


**Fig. 1** Relative effect of cutting parameter on Ra

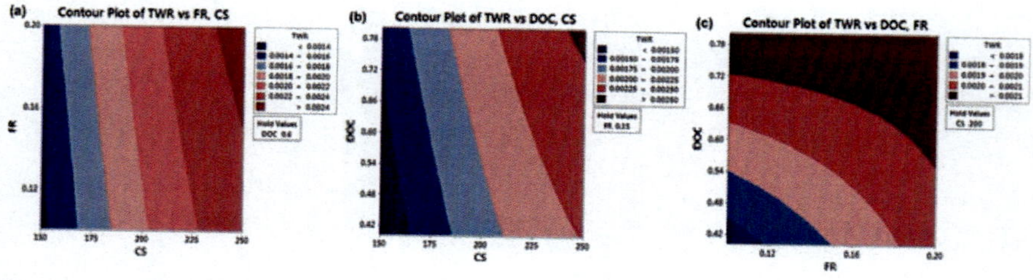


**Fig. 2** Relative effect of cutting parameter on tool wear rate

and found that increase in CS and FR results increase in TWR. Similar experiments were performed by Korade et al. [22] on H21 tool steel and Thamizhmanii et al. [23] on titanium, stainless steel, and inconel, they found similar consequences of turning parameters on tool wear rate. Zhou et al. [24] had performed experiments on titanium alloy with Co10Ti3-CAT coated tool and found that tool wear reduces by 24% with decrease of CS, FR, and DOC. Kuntoglu et al. [25] investigated machining of AISI 1050 carbon steel and found that feed rate directly affects quality of machined surface. Similar experiment was done by Thiyagu et al. [26] and suggested the similar consequences of process parameter on TWR.

## 3.2 TWR Analysis

Worn surface of single point cutting tool was investigated using optical microscopy (OM). The OM images of worn tool tip were arranged according to the experimental run order (see Table 2). Figure 3a, g, h, and o are showing tool wear for the combination of different process parameters for which CS (150 m/min) is constant. The least TWR of 0.0015 g/min was observed at CS 150 m/min, FR 0.2 mm/rev and DOC 0.6 mm (Fig. 3a). A least TWR value of 0.00175 g/min was observed at CS 200 m/min, FR 0.1 mm/rev, and DOC 0.4 mm as expressed in (Fig. 3b) among the other combination of parameters where CS 200 m/min was kept constant. Worn surface of tools for these combinations are presented in Fig. 3b–f, i, j. Maximum

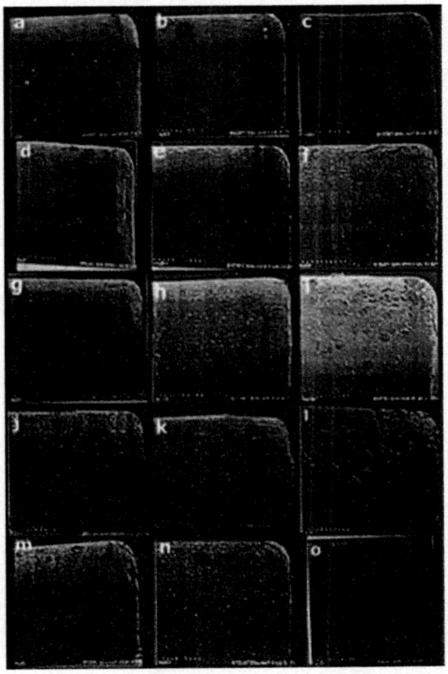


**Fig. 3** Tool wear analysis

TWR of 0.00252 g/min was noticed at CS 250 m/min, FR 0.15 mm/rev and DOC 0.8 mm (Fig. 3l). Worn surfaces of tools after turning at highest CS (250 m/min) are shown in Fig. 3k–m, n.

Least TWR of 0.0015 g/min was observed at CS 150 m/min, FR 0.2 mm/rev, and DOC 0.6 mm among all the experimental runs. It was also noticed that TWR is directly proportional to CS and FR.

In addition, adhesion and abrasion are the main wear mechanism of the tool surface. Adhesive wear is due to high temperature generated at tool work interface under high pressure during cutting. This resulted in adhesion of small chips or fragments on to the tool surface causing the coating to gets torn away by itself and adhere to the tool surface. On the other hand, presence of hard particles such as oxide compounds, built-up fragments along with nitrides are responsible for abrasive wear.

Further, flanking of coating was observed (see Fig. 3i, l and n) which is also responsible for tool wear. When the average flank wear reached to 300 μm of tool wear criterion then flanking becomes more apparent [27]. Similar observations were found in a study by Bhatt et al. [28].

## 3.3 *Examination of Data and Acceptability of the Model*

Residual plot for Ra and TWR is presented in Fig. 4a and b respectively. Errors are normally distributed as the residual fall in straight line. Acceptability of responses is tabulated in Table 3. Value of $R^2$, $R^2$ (adj) shows that models fits the data, which reinforces the expectation capacity of the model. $R^2$ (pred) values are well above 95%, which build them fit for forecasting the solution.

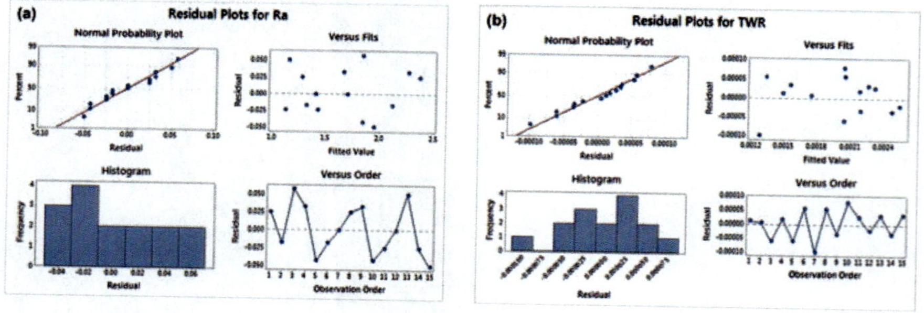


**Fig. 4** Residual plot of Ra and tool wear rate

**Table 3** Process parameters and their levels

| Response | Standard deviation | $R^2$ (%) | $R^2$ (adj) (%) | $R^2$ (pred) (%) |
|---|---|---|---|---|
| Ra | 0.0462910 | 99.23 | 98.65 | 97.24 |
| TWR | 0.00000636 | 98.41 | 97.53 | 95.53 |

**Table 4** ANOVA for responses

| Response | Source | DF | Seq SS | Contribution (%) | Adj SS | Adj MS | *F* | *P* |
|---|---|---|---|---|---|---|---|---|
| Ra | Model | 6 | 2.20019 | 99.23 | 2.20019 | 0.366698 | 171.13 | 0 |
| | Error | 8 | 0.01714 | 0.77 | 0.01774 | 0.002143 | | |
| | Total | 14 | 2.21733 | 100.00 | | | | |
| TWR | Model | 5 | 0.000002 | 98.41 | 0.000002 | 0.00 | 111.42 | 0 |
| | Error | 9 | 0.00 | 1.59 | 0.00 | 0.00 | | |
| | Total | 14 | 0.000002 | 100.00 | | | | |

## *3.4 Anova*

Analysis of variances for the responses is tabulated in Table 4. $P$ is <0.05 for responses and $F$-value is remarkable at 95% confidence limit. It established that the generated model is sufficient. Expected value and measured data are also acceptable.

## *3.5 Optimization of Parameter*

Optimized cutting parameter for the responses is shown in Fig. 5. TWR and Ra are minimum at the initial value of all three parameters which fulfils the condition of

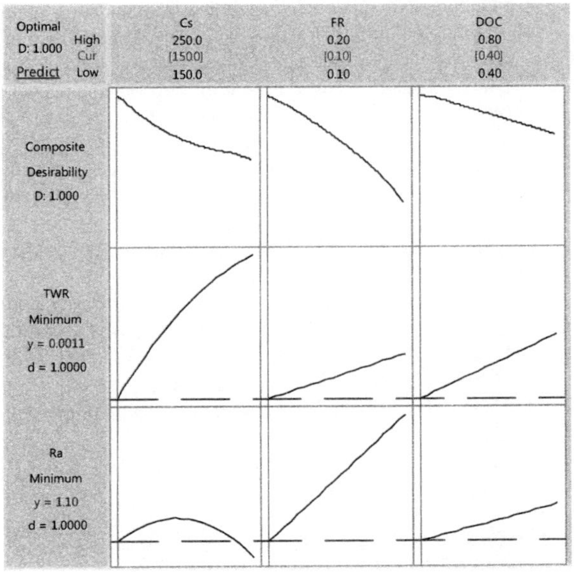


**Fig. 5** Optimization of parameter

optimization. The optimal values of cutting parameters are 150 m/min CS, 0.1 mm/rev FR and 0.4 mm DOC.

## 4 Conclusions

This study has furnished an approach of Box-Behnken design for experiment and optimize the turning process parameters for cutting DAC-10 tool steel. Following conclusions are drawn:

- The contour plots show that CS and FR are the most influencing parameter for surface roughness. Surface quality enhances with increase in CS and decrease in FR.
- CS is the most influential parameter for determining TWR. Increase in CS leads to increase in TWR whereas FR and DOC have least effect on TWR.
- The optimized parameters were CS 150 m/min, FR 0.1 mm/rev, and DOC 0.4 mm for smaller Ra and TWR for turning of DAC-10 tool steel with TiAlN coated tool.
- P value less than 0.05 suggests the authenticity of the model.

## References

1. M. Thomas, Y. Beauchamp, A.Y. Youssef, J. Masounave, Effect of tool vibrations on surface roughness during lathe dry turning process. Comput. Industr. Eng. **31**(3–4), 637–644 (1996)
2. A. Siddhpura, R. Paurobally, A review of flank wear prediction methods for tool condition monitoring in a turning process. Int. J. Adv. Manuf. Technol. **65**(1–4), 371–393 (2013)
3. L. Vela-Martínez, J.C. Jáuregui-Correa, E. Rubio-Cerda, G. Herrera-Ruiz, A. Lozano-Guzmán, Analysis of compliance between the cutting tool and the workpiece on the stability of a turning process. Int. J. Mach. Tools Manuf. **48**(9), 1054–1062 (2008)
4. S. Debnath, M.M. Reddy, Q.S. Yi, Influence of cutting fluid conditions and cutting parameters on surface roughness and tool wear in turning process using Taguchi method. Measurement **78**, 111–119 (2016)
5. T. Leppert, Effect of cooling and lubrication conditions on surface topography and turning process of C45 steel. Int. J. Mach. Tools Manuf. **51**(2), 120–126 (2011)
6. S.K. Khrais, Y.J. Lin, Wear mechanisms and tool performance of TiAlN PVD coated inserts during machining of AISI 4140 steel. Wear **262**(1–2), 64–69 (2007)
7. S.S. Gill, J. Singh, H. Singh, R. Singh, Investigation on wear behaviour of cryogenically treated TiAlN coated tungsten carbide inserts in turning. Int. J. Mach. Tools Manuf. **51**, 25–33 (2011)
8. Y.C. Chim, X.Z. Ding, X.T. Zeng, S. Zhang, Oxidation resistance of TiN, CrN, TiAlN and CrAlN coatings deposited by lateral rotating cathode arc. Thin Solid Films **517**(17), 4845–4849 (2009)
9. S.Y. Lee, S.D. Kim, Y.S. Hong, Application of the duplex TiN coatings to improve the tribological properties of electro hydrostatic actuator pump parts. Surf. Coat. Technol. **193**(1–3), 266–271 (2005)
10. J.D. Bressan, R. Hesse, E.M. Silva Jr., Abrasive wear behavior of high speed steel and hard metal coated with TiAlN and TiCN. Wear **250**(1–12), 561–568 (2001)

11. T. Leyendecker, O. Lemmer, S. Esser, J. Ebberink, The development of the PVD coating TiAlN as a commercial coating for cutting tools. Surf. Coat. Technol. **48**(2), 175–178 (1991)
12. P.C. Jindal, A.T. Santhanam, U. Schleinkofer, A.F. Shuster, Performance of PVD TiN, TiCN, and TiAlN coated cemented carbide tools in turning. Int. J. Refract Metal Hard Mater. **17**(1–3), 163–170 (1999)
13. D. Zhu, X. Zhang, H. Ding, Tool wear characteristics in machining of nickel-based superalloys. Int. J. Mach. Tools Manuf. **64**, 60–77 (2013)
14. S.Y. Luo, Y.S. Liao, Y.Y. Tsai, Wear characteristics in turning high hardness alloy steel by ceramic and CBN tools. J. Mater. Process. Technol. **88**(1–3), 114–121 (1999)
15. Hitachi metal, https://www.hitachi-metals.co.jp/e/products/auto/ml/pdf/dac_b.pdf
16. P. Paengchit, C. Saikaew, Feed rate affecting surface roughness and tool wear in dry hard turning of AISI 4140 steel automotive parts using TiN+ AlCrN coated inserts. IOP Conf. Ser. Mater. Sci. Eng. **307**(1), 012024 (2018)
17. M.R. Ibrahim, T. Sreedharan, F. Hadi, N. Aisyah, M.S. Mustapa, A.E. Ismail, M.F. Hassan, T. Arifin, A. Mubarak, The effect of cutting speed and feed rate on surface roughness and tool wear when machining D2 steel. Mater. Sci. Forum **909**, 80–85 (2017)
18. G.A. Arefi, R. Das, A.K. Sahoo, B.C. Routara, B.K. Nanda, A study on the effect of machining parameters in turning of lead alloy. Mater. Today Proc. **4**(8), 7562–7572 (2017)
19. N. Qehaja, K. Jakupi, A. Bunjaku, M. Bruçi, H. Osmani, Effect of machining parameters and machining time on surface roughness in dry turning process. Proc. Eng. **100**, 135–140 (2015)
20. P. Chandra, C.R. Prakash Rao, R. Kiran, V. Ravi Kumar, Influence of machining parameter on cutting force and surface roughness while turning alloy steel. Mater. Today Proc. **5**, 11794–11801 (2018)
21. G. Zheng, R. Xu, X. Cheng, G. Zhao, L. Li, J. Zhao, Effect of cutting parameters on wear behavior of coated tool and surface roughness in high-speed turning of 300M. Measurement **125**, 99–108 (2018)
22. D.N. Korade, K.V. Ramana, K.R. Jagtap, Study of effect of population density of carbides on surface roughness and wear rate of H21 tool steel. Mater. Today Proc. **19**, 228–232 (2019)
23. S. Thamizhmanii, C. Yuvaraj, J.S. Senthilkumar, I. Arun, Effect of feed rate on difficult to cut metals on surface roughness and tool wear using surface treated and untreated tools. Proc. Manuf. **30**, 216–223 (2019)
24. X. Zhou, K. Wang, C. Li, Q. Wang, Wu. Shen, J. Liu, Effect of ultrafine gradient cemented carbides substrate on the performance of coating tools for titanium alloy high speed cutting. Int. J. Refract Metal Hard Mater. **84**, 105024 (2019)
25. M. Kuntoğlu, H. Sağlam, Investigation of progressive tool wear for determining of optimized machining parameters in turning. Measurement **140**, 427–436 (2019)
26. M. Thiyagu, L. Karunamoorthy, N. Arun Kumar, Thermal and tool wear characterization of graphene oxide coated through magnetorheological fluids on cemented carbide tool inserts. Arch. Civil Mech. Eng. **19**(4), 1043–1055 (2019)
27. R.M. Arunachalam, M.A. Mannan, Performance of CBN cutting tools in facing of age hardened Inconel 718. Papers Presented at NAMRC **32**, 525–532 (2004)
28. A. Bhatt, H. Attia, R. Vargas, V. Thomson, Wear mechanisms of WC coated and uncoated tools in finish turning of Inconel 718. Tribol. Int. **43**(5–6), 1113–1121 (2010)

# Cost Analysis of an Off-Grid Solar/Wind/Battery Based Renewable Energy System for Variable Load

**Sujeet Singh, Krishna Murari Pandey, and K. K. Sharma**

**Abstract** Off-grid renewable energy systems have been fascinating to provide energy to different sectors in all the directions like sustainability, viability and environmental safe-conduct, particularly for the societies living in remote areas where expansion of grid is not relevant. Renewable energy system shows numerous combinations built on the basis of renewable sources that can be practiced together to provide power in the form of a dedicated off-grid system supported by battery-bank storage and diesel generator as backup systems. In this article, wind turbine-PV-battery storage-inverter was used as system components and these were simulated and optimized for the entire NIT Silchar campus in the state of Assam, India. The primary load demand of the entire campus is 11,378.94 kWh/day and peak load of 671.62 kW. A popular freeware HOMER modelling software has been used to analyze the stand-alone RES system. Solar energy and wind are used as prime sources to generate power and supply it straight to the load. If excess electricity is produced, it is used to charge the battery bank. The campus's load consists of power required for lighting, pumping of water, hotel electricity load, different department electricity load and various quarters load which are situated inside the campus. While analyzing this energy system, the simulation is done and results are optimized on the basis of power load, meteorological data sources. The economics of energy components and other parameters in which the net present cost (NPC) is to be minimized to select an economically viable energy system. However, other criteria, such as additional power generation, capacity shortage, COE, were also acknowledged to investigate the technological ability to choose an excellent system in techno-economic perspectives. The two approaches are used as a comparative criterion to select an energy system from the chosen options that give sufficient merit to one of the measuring tools (Net present cost and low cost of energy).

S. Singh · K. M. Pandey (✉) · K. K. Sharma
Department of Mechanical Engineering, National Institute of Technology Silchar, Silchar, Assam 788010, India
e-mail: kmpandey@mech.nits.ac.in

S. Singh
e-mail: sujeet138@gmail.com



K. M. Pandey et al. (eds.), *Recent Advances in Mechanical Engineering*, Lecture Notes in Mechanical Engineering, https://doi.org/10.1007/978-981-15-7711-6_43

**Keywords** COE · HOMER · NPC · Off-grid power system · RES

# 1 Introduction

Energy demand is growing at an increasing rate and can not be fully met by conventional energy systems because of inadequate supply. As we all know that fossil fuels are depleting, and energy needs are growing day by day, the production of renewable energy is attracting worldwide attention. Specially, for stand-alone applications, the hybrid solar photovoltaic system and the wind power generation system become a desirable solution. Combining solar and wind power sources together can deliver much better reliability, and their renewable energy system becomes more cost-effective because one system's deficiency can be complemented by another's power. The combination of solar and wind generation systems into a grid should further enhance the overall economy as well as reliability of renewable energy sources in delivering their energy. Similarly, integrating solar and wind power into an off-grid system can decrease the amount of power storage required for continuous power supply. There are several storage components and power generators for hybrid power systems. It is designed to meet the remote area's energy needs. Wind generators, PV generators and other electricity sources are added to meet local geographical conditions and other specifications. It is important to know the specific energy demand and available resources for that location in order to develop a hybrid system for a specific location. Energy planners, therefore, need to explore the potentially available resources for solar, wind and hydropower for a specific site location.

# 2 Literature Review

A detailed study was needed to gather intelligible information on the country's renewable energy potential, hybrid power systems, and rural electrification techniques using combined resources. Renewable energy opportunities for access to renewable energy potential and stand-alone hybrid systems have been implemented in various research efforts. The following different authors were conducted at different times, locations and countries for a number of hybrid systems; the simulation methodologies used are Genetic algorithm and HOMER.

Nfah et al. [1] describe the solar photovoltaic system project information and the wind generator hybrid electricity generation system to supply power to societies of 110 households, a public health centre and a primary school. Research has started by exploring the potential of interesting solar and wind sources in our area. In optimized simulation, the result revealed that the system was configured by PV/wind turbine/diesel generator/battery and converter. The total value of NPCs and COEs for this configuration is \$113,814 and \$0.312/kWh, respectively, an 84% renewable share, and 1,945 L of diesel consumed per year, corresponding to 635 h per year.

A study conducted by [2] to evaluate the potential for electrification in rural areas of villages in Bhutan. The research work was conducted across the country at four different locations. Furthermore, only for lighting and communication services was the load requirement taken. The article's main purpose was to optimize aggregates of hybrid energy. The PV/battery power system is Gasa and Lunana's cheapest technology, while the Getena area's cheapest hybrid diesel/battery system. It is better to implement the introduction of a wind and rechargeable battery system instead of Yangtse. According to the study [3] on the technical and economic aspects of rural community schemes in Algeria in the hybrid (wind and diesel generators). In order to reduce fuel consumption, a detailed research was conducted to include a wind generator to the current diesel power plants. The author concludes that the wind speed of 0.04–0.189 $/L is economically viable from the hybrid system below 6 m/s of existing diesel fuel. The hybrid system's feasibility is guaranteed at a wind velocity of 6.58 m/s, a maximum yearly capacity shortage of 0%, a minimum return of 0% and a fuel price of $0.151/L. Kasuakana et al. [4] explore the feasibility of a renewable energy system as a primary source of electricity for mobile phone stations in the DRC. The survey was carried out for three different non-network-connected areas: Fireplace, Mbuji-Mayi and Cabinda. Possible setting options are PV-WT-diesel generator, clean PV and clean wind energy systems; the techno-economic and environmental effects have also been studied. Bilal et al. [5] presented a multi-purpose PV-battery-wind hybrid system that fulfils two basic objectives to minimize LPSP and annual costs. The production of the produced hybrid system (kWh/year) is 63,035 and the annual cost system (EUR million) is 0,110 with 95 kW h/day load. Bilal et al. [6] approach MOGA which has been used to optimize and design a stand-alone hybrid PV- diesel hybrid system that reduces energy costs and $CO_2$ emissions in Senegal.

## 3 Cost Data and Size Specifications of Each Component

The basic measure of choosing the right energy system components in this current work is the cost of the components, as the primary objective of the paper is to find the optimal energy system configuration that will meet the demand with minimum COE and NPC. The cost of the components was estimated based on current market values.

### *3.1 Solar PV Size and Cost*

The following panel was selected after surveying various products taking cost into consideration. The reason why the stated company chooses the product is because of its low cost delivered until and unless until efficiency becomes a major concern here. A 1 kW solar panel was used with four solar module numbers having 250 W capacity.

**Table 1** Size and cost of PV panel

| PV size (kW) | Capital cost (₹) | O&M cost (₹/year) | PV life (year) |
|---|---|---|---|
| 1 | 80,000 | 800 | 25 |

**Table 2** Size and cost of wind turbine

| Capacity (kW) | Capital cost (₹) | Replacement cost (₹) | O&M cost (₹/year) |
|---|---|---|---|
| 3 | 450,000 | 315,000 | 9000 |

**Table 3** Size and cost of battery

| Capacity (kW) | Capital cost (₹) | Replacement cost (₹) | O&M cost (₹/year) |
|---|---|---|---|
| 1 | 8000 | 5600 | 800 |

In this analysis, the installation cost is considered to be 8% of the photovoltaic module cost, and the cost of operation and maintenance will be 1% per annum (Table 1).

### 3.2 Wind Turbine Size and Cost

The wind turbine taken for this paper has a power output of 3 kW. The wind turbine is manufactured by Vikktor & Co. KUG, Germany. The proposed wind turbine costs for operation and operation are about 2% of its initial capital expenditure (Table 2).

### 3.3 Cost and Size of Battery

Like other power system components, battery cost and number are input parameters that are introduced in the software. Pay attention to the definitions; the nominal capacity of the battery is the amount of power the battery emitted. The minimum battery charging status is the charging state under which the battery will never discharge to prevent any potential which can cause damage to the battery. The recommended minimum charge state is 30–50% (Table 3).

### 3.4 Power Converter Size and Cost

The flow of power between the DC and AC power system components must be maintained by a converter. The power rating of the inverter must be equal to or greater than the peak value of the load, but since both the renewable and non-renewable

**Table 4** Size and cost of power converter

| Capacity (kW) | Capital cost (₹) | Replacement cost (₹) | O&M cost (₹/year) | Life (year) |
|---|---|---|---|---|
| 10 | 75,000 | 52,500 | 750 | 15 |

loads will supply, it will be installed below the peak. In this case, there is no need for estimating operating and maintenance costs. The capital cost of the converter is taken as 75,000, the cost of replacement is around 52,500, the efficiency of the converter is about 92%, and the lifetime of the converter will last for 15 years (Table 4).

# 4 Methodology

The methodology used in this present analysis is primarily two-fold. In case of the first one, four different models of RES are analyzed while taking their fundamental costs, technological specifications into account. Wind and solar data are used to simulate the RES model mentioned above. The load characteristics of the RES system are matched with the hourly energy demand so that the load profile can be matched. Subsequently, net present cost (NPC), lifetime installation cost and operation and maintenance costs and energy costs (COE) are calculated to estimate the economic feasibility of RES.

## *4.1 Cost Function*

A key objective of this paper is to minimize several fixed and operating costs for chosen RES systems and to examine their cost performance. An optimal RES system is selected based on the minimum cost of energy production. An objective function is defined, which is to be minimized using HOMER software.

An objective function [7] has been generated, which is expressed as

$$\min\{C_{\mathrm{NPC}}(x)\} - \min\{C_{\mathrm{pv}} + C_{\mathrm{wt}} + C_{\mathrm{aux}}\} \tag{1}$$

where $x$ is a particular RES configuration vector. $C_{\mathrm{pv}}$ denotes the photovoltaic module total cost along with the accessories, $C_{\mathrm{wt}}$ represents the wind generator total cost along with its other accessories, and $C_{\mathrm{aux}}$ represents all the other required components cost that is not either sized by the HOMER or not enlisted in the component list.

$$C_{\mathrm{pv}}(x) = N_{\mathrm{pv,total}}(C_{\mathrm{pv}} + nM_{\mathrm{pv}} + I_{\mathrm{pv}}) + N_{\mathrm{inv}}(C_{\mathrm{inv}} + I_{\mathrm{inv}}) \tag{2}$$

where $N_{pv}$ represents the amount of the photovoltaic module in $x$ vector; $N_{inv}$ represents amount of the inverter module; $C_{inv}$ represents an inverter/converter cost. $M_{pv}$ is the cost of maintenance per year of the PV module, $I_{pv}$ and $I_{inv}$ are the respective component cost of installation and n is lifespan of component, taken as 25 years. $C_{wt}$ represents the wind generator system cost, and is expressed as

$$C_{wt} N_{inv,total}(C_{inv} + I_{inv}) + N_{wt,total}(C_{wt} + nM_{wt} + I_{wt}) \tag{3}$$

where $N_{inv}$ is the number of inverter model; $N_{wt}$ represents the number of the wind generator system; $C_{inv}$ denotes the cost of a single inverter; $C_{wt}$ is WT model cost; $M_{wt}$ is the maintenance cost of the wind generator; $I_{inv}$ and $I_{wt}$ are cost of installation of inverter and WT respectively and n is lifetime, which is taken as 25 yr. The final term, i.e., $C_{aux}$, represents all additional system-specific components costs

$$C_{aux} = N_{bat} C_{bat} \tag{4}$$

where $N_{bat}$ represents the number of battery and $C_{bat}$ represents capacity of battery.

## 4.2 Modelling Strategies and HOMER Simulation

By examining the meteorological inputs and load profiles of the selected location, unit sizes of system components are selected on the basis of the accessibility of standard renewable systems in the market with due attention to their economics. Subsequent to choosing the components required in the RES, input data are inserted for every single component, which essentially depicts the cost of components, technical features, and resource data. HOMER analyzes demand to the power that system delivers and estimates the flow of power to and from every single component of the system (Fig. 1).

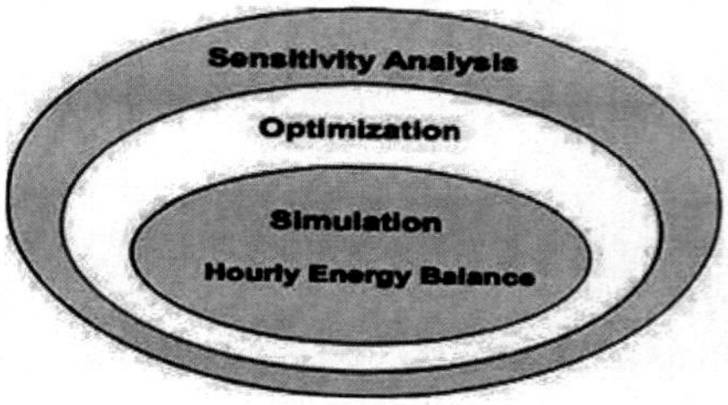


**Fig. 1** Interaction between simulation, optimization and sensitivity analysis

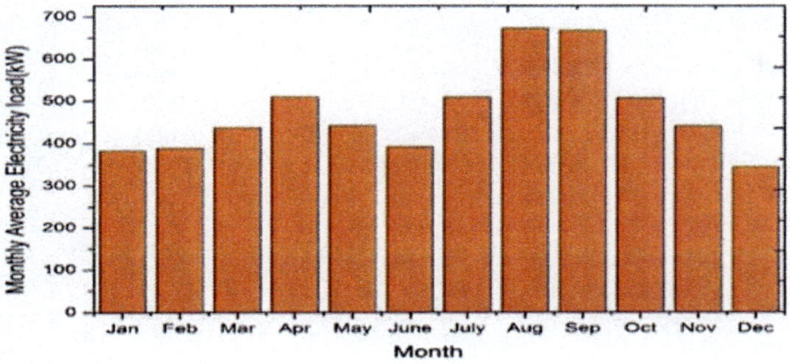


**Fig. 2** Monthly average electricity load

### *4.3 Electricity Load Input*

The electrical load is first entered into the modeling tool when selecting the component technology from the HOMER software library. The primary load input was entered daily (data for 24 h) and the software subsequently modelled the peak load (Fig. 2).

## 5 Results and Discussions

Though HOMER has simulated many configurations of power system components, however, it only shows the scenario of energy systems possible for detailed study. Complexity and computation time relate to the amount of parameters and the total number of implicit values included in the design. Two different scenarios have been suggested for further study and to develop the possibility of finding the most optimized system. Power schemes (scenarios) with low NPC, low COE, low capacity constraints, small standby power would be suggested as an optimal system.

(a) Solar Photovoltaic- Battery (scenario A)
(b) Solar Photovoltaic-wind generator-Battery (scenario B).

### *5.1 Optimization Analysis of the Selected Scenario*

On the basis of inputs provided, 42,670 simulation runs were executed using a 64-bit OS, 3.8 GHz processor and 16 GB RAM desktop computer. While considering operating reserve as a safety margin, the energy system was created in order to allow the energy system to feed reliable power and make it easier for further load expansion in the upcoming future. Optimization result of all the possible configurations of viable energy scheme established on total net present cost (NPC) of the system are shown is Table 5.

**Table 5** Overall optimization results using HOMER software

| PV size (kW) | Wind turbine quantity | Battery quantity | Converter size (kW) | Initial capital cost (crores) | Q& M cost (crores/year) | Total NPC (crore) | Cost of energy (₹/kWh) | Capacity shortage (%) |
|---|---|---|---|---|---|---|---|---|
| 3469 | 20 | 4458 | 684 | 40.14 | 0.54 | 50.04 | 8.56 | 9 |
| 3582 | 22 | 4512 | 632 | 40.84 | 0.56 | 50.13 | 8.61 | 7 |
| 3671 | 19 | 4496 | 652 | 41.14 | 0.57 | 50.34 | 8.72 | 9 |
| 3645 | 28 | 4317 | 721 | 42.31 | 0.58 | 51.32 | 8.86 | 8 |
| 3678 | 26 | 4389 | 726 | 42.98 | 0.61 | 53.32 | 8.91 | 6 |
| 3572 | 31 | 4485 | 690 | 42.16 | 0.55 | 52.61 | 8.89 | 9 |
| 3601 | 31 | 4369 | 668 | 42.21 | 0.56 | 52.85 | 8.94 | 5 |
| 3254 | 0 | 4256 | 676 | 37.90 | 0.52 | 45.21 | 7.56 | 0 |
| 3262 | 21 | 4360 | 712 | 38.23 | 0.53 | 46.32 | 7.72 | 0 |

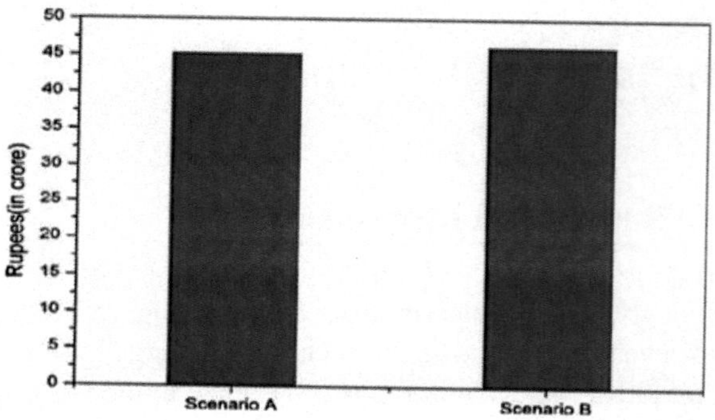


**Fig. 3** Comparison of scenarios based on NPC

## 5.2 Comparison of Scenarios

### 5.2.1 Based on Total Net Present Cost

The principal load of 11,378.94kWh/day for which the parameters are specified, highest power shortage of 10% is considered for which comparison of parameters are addressed for the selection of the techno-economic feasible power system. Referring to Table 5, the NPC in case of scenario B is slightly greater than the scenario A, which is 45.21 crores (Fig. 3).

### 5.2.2 Based on Cost of Energy

For detailed information of the COE for every energy system arrangements (scenarios) refer to Fig. 4 and Table 5, two scenarios (scenario A and scenario B)

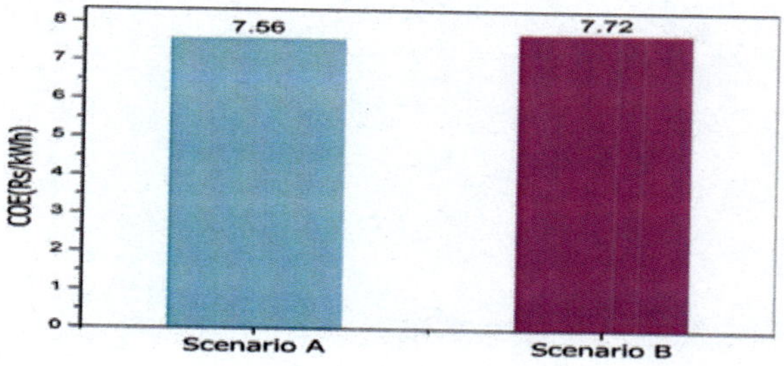


**Fig. 4** Comparison of scenarios based on COE

there is a slight difference in cost of electricity. Scenario A showed the lowest value of COE. For case A the COE is ₹7.56/kWh and for case B, it is ₹7.72/kWh.

## 6 Conclusion

An optimization process was performed during the analysis of the stand-alone RES system set-up based on the power demand, climate data inputs, and economy of the energy components in which the net present cost must be minimized in order to select an economically feasible energy system. The result of HOMER simulation showed the most cost-effective system sorted by NPC from top to bottom. Two optimized results are found, one having configuration solar/battery while others having solar/wind/battery. The cost of energy in the first configuration is ₹7.56/kWh and ₹7.72/kWh in the second configuration respectively.

## References

1. E.M. Nfah, J.M. Ngundam, M. Vandenbergh, J. Schmid, Simulation of off-grid generation options for remote villages in Cameroon. Renew. Energy **33**(5), 1064–1072 (2008)
2. G. Boneya, Design of a photovoltaic-wind hybrid power generation system for ethiopian remote area. Ph.D. thesis, Institute of Technology Department of Electrical and Computer Engineering, Addis Ababa University (2011)
3. G. Bekele, B. Palm, Feasibility study for a standalone solar wind-based hybrid energy system for application in Ethiopia. Appl. Energy **87**(2), 487–495 (2010)
4. K. Kusakana, H.J. Vermaak, Hybrid renewable power systems for mobile telephone base stations in developing countries. Renew. Energy **51**, 419–425 (2013)
5. B.O. Bilal, V. Sambou, P.A. Ndiaye, C.M.F. Kébé, M. Ndongo, Optimal design of a hybrid solar wind battery system using the minimization of the annualized cost system and the minimization of the loss of power supply probability (LPSP). Renew. Energy **35**(10), 2388–2390 (2010)
6. B.O. Bilal, D. Nourou, C.M.F. Kébé, V. Sambou, P.A. Ndiaye, M. Ndongo, Multiobjective optimization of hybrid PV/wind/diesel/battery systems for decentralized application by minimizing the levelized cost of energy and the $CO_2$emissions. Int. J. Phys. Sci. **10**(5):, 92–203 (2015)
7. S. Rehman, I.M. El-Amin, F. Ahmad, S.M. Shaahid, A.M. Al-Shehri, J.M. Bakhashwain, A. Shash, Feasibility study of hybrid retrofits to an isolated off-grid diesel power plant. Renew. Sustain. Energy Rev. **11**(4), 635–653 (2007)

# Design and Development of a Mini Sugarcane Harvester

M. A. Nagarjun, N. C. Mahendra Babu, and Manjunath M. Ullegaddi

**Abstract** Harvesting is a crucial activity of sugarcane production, which affects overall productivity. Timely harvest is very essential to achieve better quality and high yield, but shortage of labours during this time incurs major losses to farmers. Manual harvesting of sugarcane is a very labour-intensive and arduous activity, intervention of mechanical systems for harvesting frees harvest labours from drudgery and helps to improve productivity. Existing harvesting machinery is of huge size and is not suitable for Indian farming canopy and also not affordable to farmers with small landholdings. The proposed work was aimed at developing a low-cost mini sugarcane-harvesting machine, which has a very small footprint and a simple mechanism for cutting sugarcane at the base. Detailed design of the subsystems and components have been carried out to realize the final working prototype. The developed machine is capable of harvesting sugarcane with a single cut leaving partial to no edge damage on the cut surface. Different varieties of sugarcane of diameter up to 40 mm can be harvested with the developed harvester. The developed machine can also be employed for harvesting other similar crops with same cutter assembly or by using separate attachments. Cost and time of harvesting can be significantly reduced by employing the developed harvester, leading to increased productivity.

**Keywords** Sugarcane harvesting · Kinematics of base cutter · Stalk damage · Quality of cut

M. A. Nagarjun (✉) · N. C. Mahendra Babu · M. M. Ullegaddi
Department of Mechanical and Manufacturing Engineering, Ramaiah University of Applied Sciences, Bengaluru, India
e-mail: nagarjun.arjun1111@gmail.com

N. C. Mahendra Babu
e-mail: ncmbabu@gmail.com

M. M. Ullegaddi
e-mail: mmullegaddi.k@gmail.com



K. M. Pandey et al. (eds.), *Recent Advances in Mechanical Engineering*, Lecture Notes in Mechanical Engineering, https://doi.org/10.1007/978-981-15-7711-6_44

## 1 Introduction

Sugarcane is an important commercial crop in India. India is the second-largest global producer of sugarcane next to Brazil and the world's leading sugar consumer. India's total sugarcane cultivation area is 5.1 million ha and production is 352 million tons with an average productivity of 71 tons/ha (ICAR-SBI Report 2017). Uttar Pradesh, Maharashtra, Karnataka, Gujarat, Andhra Pradesh and Tamil Nadu are the major sugarcane growing states contributing about 81% of the total production in India. About fifty million farmers are engaged in sugarcane cultivation across the country. Sugar industry in India is the second-largest agroprocessing industry, which has a huge share in the growth of industrialization and socio-economic changes in rural parts of the country. Harvesting is one of the most labour intensive and arduous operation in sugarcane cultivation. Manual harvesting of sugarcane is most common in India. About 850–1000 man-hours per hectare is required for manual sugarcane harvesting with traditional tools (Yadav 2002). Sugarcane harvesting involves base cutting of sugarcane, stripping, detopping, bundle making and finally transporting the sugarcane to sugar mills. Manual sugarcane harvesting employing traditional tools is highly labour intensive and involves high cost of operation out of all sugarcane production operations. Timely harvesting is critical and delayed harvesting affects the quality of sugarcane, yield, juice quality and sugar recovery. Under these circumstances, the sugar industry is looking for replacing manual harvesting with alternate mechanical means at a reasonable cost.

Although huge sized imported/locally produced harvesting machines are available in the market, they are finding very limited acceptance due to large-sized footprint, huge investment and unsuitable method of harvesting. Performance evaluations (Yadav et al. 2002) carried out on the sugarcane chopper harvester have proven that these machines are very expensive. Due to their high costs and initial investment, presently available mechanical harvesters are out of reach for small and medium farmers. Apart from this, existing harvesters are not suitable for harvesting in small-sized fields and due to narrow spacing of rows which is practiced in India. In addition, billets from chopper harvesting cause deterioration of cane quickly and must be processed within 24 h post-harvest, which is not possible in Indian practice. In view of the above-listed findings, a partially mechanized, small to medium-sized harvesting machine would seem more suitable for Indian farming practices and canopy. A compact and cost-effective machine is preferable to address the issue of non-availability of labour and the present work is an attempt in this direction.

## 2 Problem Definition

Sugarcane being one of the most important cash crop in India, needs mechanization intervention to increase productivity. One of the important operations in sugarcane cultivation is harvesting which is presently done manually. Manual sugarcane harvesting is a very labour-intensive and arduous activity. Harvest labourers experience fatigue due to excessive stress on the joints and muscles. In other countries adaption of mechanical harvesting systems has freed harvest labourers from this drudgery. Typically to harvest one hectare of sugarcane, it requires 3.3–4.2 machine-hours by mechanical harvesting whereas 850–1000 man-hours by manual harvesting (Yadav et al. 2002). In the Indian context, mechanical harvesting of sugarcane is expensive due to large-sized machineries which are capital intensive and not affordable to small/medium farmers. Hence, development of a small-sized compact harvester that is affordable would help Indian farmers in reducing labour cost and drudgery involved in sugarcane harvesting.

## 3 Methodology

After identifying the need for a small and compact sugarcane harvester, a systematic approach was formulated to design and develop the proposed harvester. The aim was to design and develop a walk-behind type mini sugarcane harvester suitable for small/medium landholdings. To execute the systematic development, specific objectives were established. The main objectives are, to review literature on sugarcane harvesting practices, physical–mechanical properties of sugarcane and existing machineries, arrive at the specifications, develop different concepts for harvesting mechanism and lastly to synthesize the finalized concept. Further, detailed design of the synthesized concept was carried out to facilitate fabrication of components, sub-assemblies and the same were integrated to realize the proposed sugarcane harvester.

## 4 Design and Development

In order to develop the specification of the harvester, a detailed literature survey and field visit was carried out. The literature review covered the current practices and status of production, varieties of sugarcane produced, and existing machinery. The field visit and interaction with farmers helped to identify farming and planting practices and to establish the canopy of sugarcane (Fig. 1). The field visit also gave insight into farmers' requirements about harvesting. The data gathered was interpreted and converted to design specifications applying quality function deployment tool. These specifications are useful in detailed designing of components and subsystems. Data such as row to row distance, cane distribution length, in line cane spacing length, average stalk height, average number of shoots, average stalk diameter, etc. were gathered. These data were useful in generation of concept of cutter assembly and also in design of the cutter.

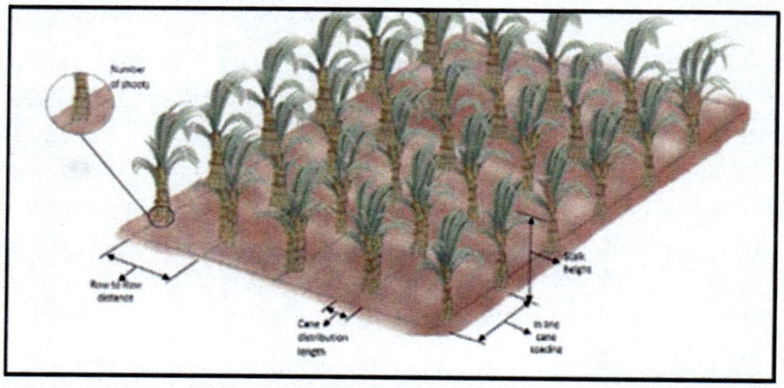

**Fig. 1** Typical sugarcane canopy

## *4.1 Sugarcane Harvester System*

Based on the developed specifications, typical conceptualized harvester system consists of three important subsystems. The subsystems are represented in block diagram as shown in Fig. 2, which consist of

- Transmission System
- Traction System
- Base cutter Assembly.

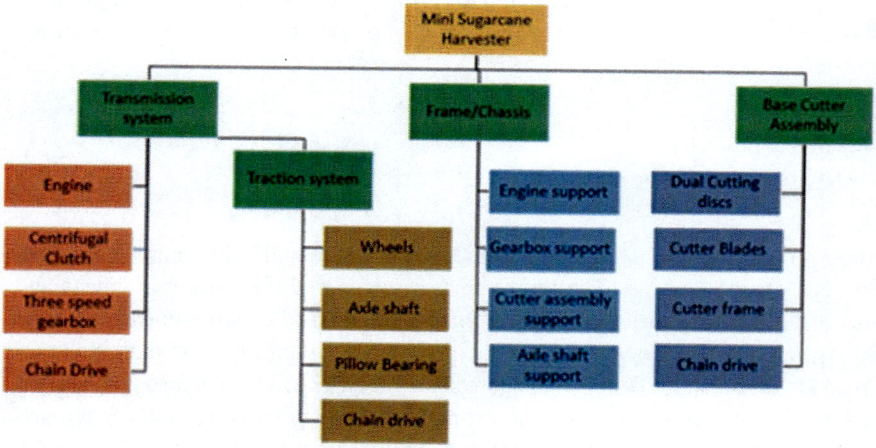


**Fig. 2** Block diagram of mini sugarcane harvester

**Fig. 3** CAD concept of the dual base cutter sugarcane harvester

## *4.2 Concept Generation*

Based on the study of existing machineries such as John Deere CH330 and Case IH Auto Soft sugarcane harvesters, it was evident that these machinery cut the sugarcane stalks into billets which is a drawback as it needs to be processed within a day after harvesting. This method of cutting into billets is not ideal for Indian scenario of sugar production as there will be delay in processing. Hence, a compact harvesting machine capable of cutting sugarcane at ground level which can be topped later and can be conveniently operated in a small field is required to cater to the needs of farmers who have small land holdings. Based on the derived specifications four different concepts were generated to suit the canopy requirements and finally the dual disc base cutter sugarcane harvester concept which met overall necessary requirements was selected.

**Final Concept**

The selected concept of dual base cutter sugarcane harvester shown in Fig. 3 facilitates maximum area coverage and high productivity. Single cut of sugarcane stalk can be achieved by conveniently optimizing the harvester speed. Various other kinematic parameters can also be optimized to achieve still higher performance of the cutter. This dual cutter system offers high inertia and impact forces on the stalk during cutting ensuring complete cut. Also, manufacturing and assembly of the cutter to the system is comparatively easy for the selected concept.

## *4.3 Synthesis of Base Cutter*

Design of the dual base cutter system is outlined in this section. A mathematical model developed by Kroes [1], which describes the kinematics of dual base cutter has been employed to determine the maximum permissible velocity ratio i.e. ratio of harvester forward speed to disc rotational speed. This model calculates the ratio to maintain total blade coverage of the sugarcane crop and to prevent contact between the uncut stalks and discs based on the base cutter and crop parameters.

Figure 4 shows the path traversed by consecutive blades with inner and outer blade radii. The model is applicable to double horizontal discs where the crop is centred between these two discs. The blades on both the discs are phased in order to prevent contacting each other. The maximum permissible velocity ratio for the harvester is defined as follows.

***Ra*** The maximum permissible velocity to ensure that all the cane row area, which does not pass through the gap between the discs is covered by the blade paths.

***Rc*** Maximum permissible velocity ratio to prevent contact between the base cutter discs and any stalk prior to completion of cut.

***Rs*** Maximum permissible velocity ratio to ensure that all the cane row area which passes through the gap between the discs is covered by the blade paths.

**Area coverage**—The maximum permissible velocity ratio necessary for the complete harvest of a row centred between two discs [1],

$$R_a = \frac{(r_0^2 - Y^2)^{\frac{1}{2}} - (r_1^2 - Y^2)^{\frac{1}{2}}}{\beta + \cos^{-1}\left(\frac{Y}{r_0}\right) - \cos\left(\frac{Y}{r_1}\right)} \tag{1}$$

Substituting the parameters from Table 1 to Eq. 1 we get,

$$\mathbf{R_a = 0.79}$$

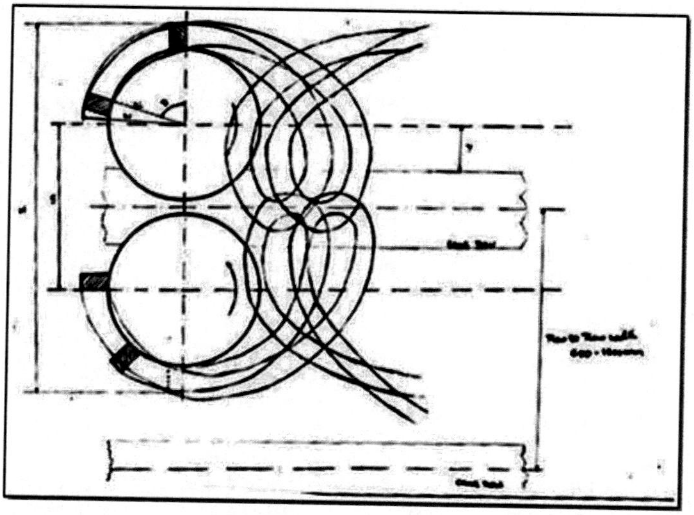

**Fig. 4** Dual rotary disc cutting system [1]

**Table 1** Base cutter parameters

| | |
|---|---|
| Inner blade radius ($r_i$) | 125 mm |
| Outer blade radius ($r_o$) | 175 mm |
| Distance from outside of the row to disc center(Y) | 100 mm |
| Disc incline angle ($\varphi$) | $15^0$ |
| Number of blades ($n$) | 5 |
| Cane diameter ($c$) | 50 mm |
| Disc separation distance ($S$) | 300 mm |
| Blade separation angle ($\beta$) | $72^0$ |
| Blade width ($h$) | 60 mm |
| Stalk base angle ($\zeta$) | $10^0$ |

**Disc contact**—Maximum permissible velocity ratio for the requirement that all the crop area be covered with no disc contact [1] is,

$$R_c = \frac{(r_0^2 - Y^2)^{\frac{1}{2}} - (r_1^2 - Y^2)^{\frac{1}{2}} - c*}{\beta + \cos^{-1}\left(\frac{Y}{r_0}\right) - \cos\left(\frac{Y}{r_1}\right)} \tag{2}$$

Substituting the parameters from Table 1 to Eq. 2 we get,

$$\mathbf{R_c = 0.25}$$

**Disc gap**—For the blade tip radius of one disc does not overlap the radius of the opposing disc [1] then,

$$R_s = \frac{(r_z^2 - Z^2)^{\frac{1}{2}}}{2\cos^{-1}\left(\frac{Z}{r_z}\right) + \beta} \tag{3}$$

Substituting the parameters from Table 1 to Eq. 3 we get,

$$\mathbf{R_s = 1.52}$$

The velocity ratio of the dual base cutter for the condition that all the area of the row is within the disc gaps ($R_a$), No disc contact for the incomplete cut stalk ($R_c$) and disc gap ($R_s$) [1] is given by,

$$R_{\min}\,(R_a,\ R_b,\ R_c) \tag{4}$$

From the above condition,

$$\mathbf{R_{min} = R_c}$$

## 4.4 Cutting Speed of the Disc Cutter

Critical cutting speed reported for sugarcane by Gupta and Oduori (1992) was in the range of 13.8–18.4 m/s for a straight cut. Mathanker [2] reported that the cutting speed below 10 m/s, the bending resistance of the partially cut stalk section was rigid to oppose the cutting arm impact. Based on the study to determine the cutting speed for sugarcane by Mathanker [2] it was noted that the critical speed was in the range of 12–15 m/s. Based on the above findings, a critical cutting speed of disc = 15 m/s for $30^0$ oblique angle is adapted.

By using equations $V = R * \omega$ and $\omega = 2\pi N/60$ we get the required speed of cutter as **N = 573 rpm** and maximum harvester forward speed of **3.06 km/h**.

## 4.5 Power Required for Cutting Sugarcane Stalks

Based on the mechanical testing carried by Kanchan et al. (2015) to assess the mechanical properties of sugarcane stalk, it is reported that the specific shearing energy for variety CO80632 by averaging the test results for 5 stalk samples having an average diameter of 22 mm is 18,010 J/m$^2$.

Sureshkumar et al. [3] have carried out tests to determine mechanical properties of sugarcane for variety CO80632 of average diameter 30 mm and reported that the specific cutting energy ranged from 27,000 to 37,000 J/m$^2$. For a tilt angle of $20^0$ and oblique angle of $30^0$ the specific cutting energy was found to be 23,000 J/m$^2$. Therefore, considering the specific cutting energy = 23,000 J/m$^2$, equivalent torque is 16.25 N m. Hence, the power required to operate each cutter is,

$$P = 1.3\,\text{hp}$$

To operate two cutters power required is 2.6 hp.

## 4.6 Synthesis of Traction System

As per the canopy requirements, the minimum ground clearance of cutting disc from ground is set at 200 mm. Farm tractor tires of specification 6.00″–12″ (Width of the tire in inches—Diameter of the rim in inches) 6 ply has an overall diameter of 635 mm. The overall height from ground to axle shaft centre with 200 mm cutter ground clearance is approximately 315 mm. This required height can be achieved by using the tires with the above specifications and the same has been selected.

The overall weight of the harvester system is estimated to be 100 kg considering all the subsystems. Considering this dynamic vertical load, the power required to pull the system is **P = 1.48 hp**. The overall power required for both traction and

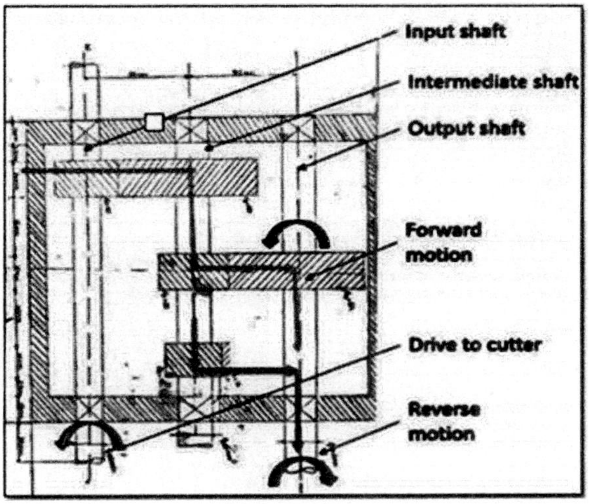


**Fig. 5** Layout of three-speed gearbox

base cutter assembly is **4.08 hp**. Hence, a standard available petrol engine of **5 hp** is selected and incorporated.

## 4.7 *Synthesis of Transmission System*

The function of the gearbox was to transfer the drive from engine to the wheels and the base cutter as per required reduction. There were no readily available standard gearboxes, which provides one forward, reverse and cutter speeds as a single system. To cater to this need a three-speed two-stage reduction compound gearbox has been designed and realized. Figure 5 illustrates the layout of the designed gearbox indicating the different speeds.

## 4.8 *Prototype of the Developed Mini Sugarcane Harvester*

See Fig. 6.

**Fig. 6** Prototype of mini sugarcane harvester

### *4.9 Testing of Mini Sugarcane Harvester*

Testing of the machine needs to be carried out to check whether the design is working as per the set requirements of the harvester. These test results indicate the effectiveness of working for the intended application and help in further refinements to increase the efficiency.

A test setup was prepared mimicking the canopy requirements, where a wooden plank of dimension 1200 mm × 300 mm × 75 mm was selected as a base for the sugarcane stalks. The plank was divided into three 100 mm columns, and holes of 32 mm in diameter were drilled along the length at four to five locations each separated by a distance of approximately 250 mm as shown in Fig. 7.

Sugarcane stalks with an average diameter of 38 mm were firmly inserted into the holes in upright position. After inserting the stalks, the test setup was placed in the

**Fig. 7** Wooden plank and sugarcane stalk used for testing

Fig. 8 Test setup placed in the field

field and locked into the ground with the help of hooks at two positions as shown in Fig. 8.

## 5 Results and Discussions

The developed harvesting machine has been tested for intended application and the performance has been found to be satisfactory. Mello [4] reported the classification of the damages caused to stalks after mechanized harvest indicating the upper and lower limits of damage as shown in Fig. 9. The most important factor in determining the quality of cut is the first partial cut and is dependent on both the harvester forward

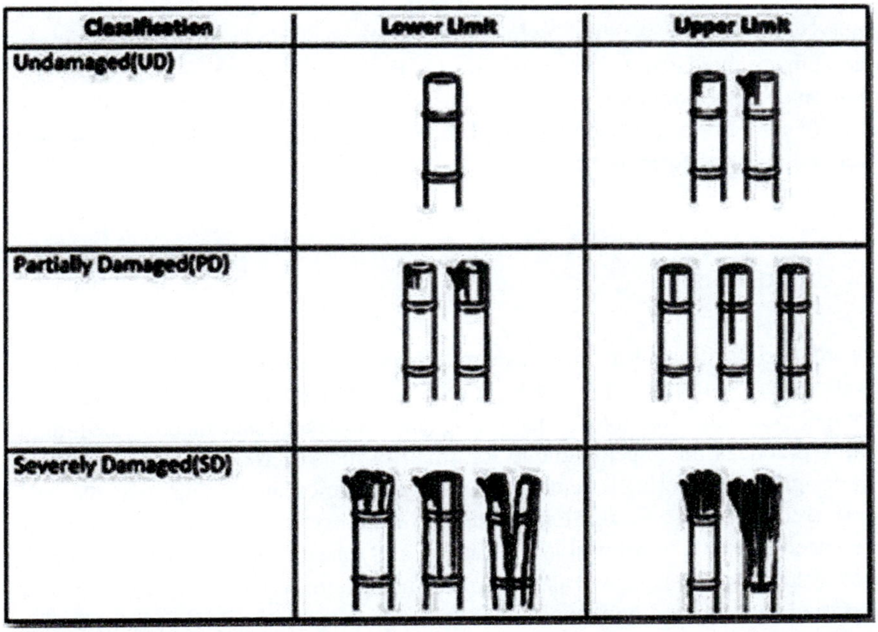


Fig. 9 Classification of stalk damages [4]

**Fig. 10** Results of quality of cut—test-1

**Fig. 11** Results of quality of cut—test-2

speed and rotational speed of the cutter. Test results for two trials are shown in Figs. 10 and 11. When the harvester speed is more, it causes initial partial cut to penetrate deep into the cane, which leads to shatter of the stalk indicated in the results of first test (Fig. 10). During the second test with decreased harvester speed, the stalk damage was very low as initial partial cut was less resulting in minor edge damage to no damage indicated as shown in Fig. 11. Factors like sugarcane varieties, soil conditions, stalk orientation and the maturity of crop play a significant role in the quality of cut. Further, blade oblique angle of $30^0$ incorporated for blades has resulted in better cutting performance.

## 6 Conclusions

Detailed specification of a mini sugarcane harvester suitable for Indian farming conditions has arrived through literature review and field study. The harvester with these specifications is suitable for harvesting one row of sugarcane meeting the Indian farming practices making it suitable for small farmers. Base cutter assembly has been designed considering kinematics, cutting mechanics and cutting coverage area to ensure complete harvesting along a single row. Transmission system consisting of three-speed gearbox, transmission for traction and transmission for cutter assembly has been successfully designed and realized. The developed harvesting machine has been tested for intended application and the performance was found to be satisfactory. Manual harvesting of sugarcane which is achieved in one day can be reduced to three hours by the use of developed harvesting machine leading to reduction of harvesting cost by 70% compared to manual harvesting.

## References

1. S. Kroes, H.D. Harris, A kinematic model of the dual basecutter of a sugarcane harvester. J. Agric. Eng. Res **62**, 163–172 (1995)
2. S.K. Mathanker, T.E. Grift, A.C. Hansen, Effect of blade oblique angle and cutting speed on cutting energy for energycane stems. Biosys. Eng. **133**, 64–70 (2015)
3. P.K. Sureshkumar, D.M. Jesudas, Physico-mechanical properties of sugar cane stalks related to mechanical harvesting. J. Tropic. Agric. **53**(1), 48–55 (2015)
4. R.D.C. Mello, H. Harris, Performance of basecutter of sugarcane harvester with angled and serrated blades. Revista Brasileira De Engenharia Agrícola E Ambiental **7**(2), 355–358 (2003)

# Simulation and Experimental Performance Studies of a Computerized Spark Ignition Engine

**Pritam Jyoti Saikia, Partha Pratim Dutta, and Paragmoni Kalita**

**Abstract** The present work is simulation and experimental performance analysis of a single cylinder spark ignition engine. The simulation model was prepared as a MATLAB code consisting of various input parameters like bore, stroke, compression ratio, spark angle, load, RPM, fuel–air equivalence ratio, inlet pressure, temperature, valve timings, etc. The model uses Wiebe's heat release model, Annand's heat transfer model and Blair's friction model in order to predict the properties with respect to crank angle. Using these data, various performance parameters like brake power, brake thermal efficiency, torque and brake specific fuel consumption were determined by varying the engine speeds, equivalence ratios and loads. The results obtained by simulation were verified by an experimental analysis on a four-stroke single cylinder Honda GX200 computerized Spark Ignition engine. It was observed that the trends in variation of the performance parameters were similar for both the simulation and experimental data. Brake power, torque and brake thermal efficiency were higher in the simulation model whereas brake specific fuel consumption was higher for experimental data. Maximum values of brake power obtained were 2.33 kW for simulation and 1.85 kW for experiment both at 2780 RPM for an engine load of 2.1 kg. Maximum values of torque obtained were 7.35 N m for simulation and 6.35 N m for experiment both at 2780 RPM at an engine load of 2.1 kg. Maximum values of brake thermal efficiency were 13.8% for simulation and 11.3% for experiment both at 2769 RPM for engine load of 1.9 kg. Maximum values of brake specific fuel consumption were 1.92 kg/kWh for simulation and 2.8 kg/kWh for experiment both at 2874 RPM for an engine load of 0.5 kg.

P. J. Saikia · P. P. Dutta (✉) · P. Kalita
Mechanical Engineering, Tezpur University, Tezpur, India
e-mail: ppdutta06@gmail.com

P. J. Saikia
e-mail: psaikia755@rediffmail.com

P. Kalita
e-mail: paragmoni@gmail.com


K. M. Pandey et al. (eds.), *Recent Advances in Mechanical Engineering*, Lecture Notes in Mechanical Engineering, https://doi.org/10.1007/978-981-15-7711-6_45

**Keywords** SI engine · Brake power · Brake thermal efficiency · Brake-specific fuel consumption

## Nomenclature

| | |
|---|---|
| $V$ | Cylinder volume ($m^3$) |
| $Q_{in}$ | Heat released inside the cylinder (kW) |
| $V_c$ | Clearance volume ($m^3$) |
| $m_f$ | Fuel flow rate (kg/s) |
| $B$ | Engine bore diameter (m) |
| $f$ | Residual gas fraction |
| $l_r$ | Connecting rod length (m) |
| $r$ | Compression ratio |
| $a$ | Crank radius (m) |
| $CV$ | Calorific Value (kJ/kg K) |
| $s$ | Distance of crank from piston pin (m) |
| $T_{\text{corr}}$ | Corrected temperature (K) |
| $x_b$ | Mass fraction burned |
| $T_{CE}$ | Temperature at the end of cycle (K) |
| $N$ | Engine speed (rpm) |
| $\theta$ | Crank angle (°) |
| $T_w$ | Cylinder wall temperature (K) |
| $\theta_s$ | Start of combustion (°) |
| $A_w$ | Heat transfer area ($m^2$) |
| $\theta_d$ | Duration of combustion (°) |
| $h_c$ | Convective heat transfer coefficient ($W/m^2$ K) |
| $\mu_{\text{gas}}$ | Dynamic viscosity of gas (N $s/m^2$) |
| $h_r$ | Radiative heat transfer coefficient ($W/m^2$ K) |
| $\rho_{\text{gas}}$ | Density of gas ($kg/m^3$) |
| $Nu$ | Nusselt number |
| $\varnothing$ | Fuel–air equivalence ratio |
| $Q_w$ | Heat loss from wall (kW) |
| $\gamma$ | Specific heat ratio of mixture |

# 1 Introduction

The internal combustion engines have been prime mover for transportation over the century. Engine performances, fuel economy have been significantly improving over time and it will continue to increase in the future. Lot of researches are going on worldwide experimentally to bring new fuel injection and combustion technologies for the gasoline and diesel engines to obtain more energy-efficient units with lesser exhaust emissions. The mathematical model based simulation using diverse environments like MATLAB makes the analysis easy, less time consuming and economical, for the different technologies. The results of these simulations may be applied for new technologies in engine modelling. This will confirm the validity of the model and can be modified to contain more complex engines that need a lot of time, money, and resources for their testing. The present work uses a simple 4-stroke, single cylinder computerized SI engine tested with the gasoline fuel.

Various researches are going on in the field of spark ignition engines. İlhak et al. [1] did an experimental analysis on an SI engine fuelled by gasoline, ethanol and acetylene by conducting experiments on a four-cylinder, four-stroke and water-cooled engine at partial loads. They found that using ethanol and acetylene, UHC and NO emissions were significantly reduced as compared to gasoline. It was also observed that acetylene had higher thermal efficiency and lower emissions than that of gasoline and ethanol in high EAR. Mourad et al. [2] did an investigation of SI engine performance and emissions fuelled with ethanol/butanol-gasoline blends. The experimental results showed a reduction in pollutant emissions from the engine by 13.7% for carbon monoxide and 25.2% for hydrocarbons and fuel consumption by 8.22%. However, there was a reduction in engine power output up to 11.1% for the fuel blends. Wittek et al. [3] performed an experimental investigation on a variable compression ratio (VCR) gasoline engine. It was observed that reduction of fuel consumption up to 4% was found at part-load with an optimum combustion phasing. Due to knock-limited spark timing, penalties in fuel consumption were present at higher loads. At even higher compression ratio, fuel intake would be further reduced at both higher and lower loads. Qian et al. [4] studied performance of a dual fuel spark ignition (SI) engine using ethanol/gasoline substitutes as a fuel. The results showed that for a direct injection of toluene reference fuels having research octane number of 90 with the rise of ethanol injection ratio, the flame development duration and rapid combustion duration sustained. Nitrogen oxide emissions and total hydrocarbon emissions gradually decreased. The indicated thermal efficiency of a dual fuel spark ignition mode with direct fuel injection having research octane number 75 combined with 35% ethanol ratio was identical to that of direct injection spark ignition mode fuelled having research octane number of 95 under respective knock limited spark timings. Efemwenkiekie et al. [5] performed a comparative analysis of a four-stroke spark ignition engine using local ethanol and gasoline blends. The results showed that better engine performance was obtained with locally-produced ethanol—gasoline blends. Maximum engine performance was obtained with E3 (3% ethanol and 97% gasoline) at full load throttle conditions. The study proved that locally formed ethanol can be used in gasoline alternative to highly purified ethanol as an additive. Sugiarto et al. [6] made comparison of the gasoline fuels with octane number variations 88, 92 and 98 on the performance of four-stroke single cylinder 150 cc SI engine. Results showed that the largest torque value on gasoline 88 fuel was 34.26 Nm at 2000 engine RPM. The largest power value was 7.89 kW by using gasoline 88 fuels or equivalent to 14.35 HP at 2500 RPM. The highest specific fuel consumption was 1075 g/kWh that occurred

**Table 1** Engine technical specifications

| | |
|---|---|
| Engine type | 4-stroke single cylinder, 25° inclined cylinder, horizontal shaft |
| Bore × Stroke | 68 × 54 mm |
| Displacement | 196 cc; |
| Compression ratio | 8.5: 1 |
| Net power | 4.1 kW (5.5 HP)/3600 rpm |
| Max. net torque | 12.4 N-m/1.26 kgfm/2500 rpm |
| Fuel tank capacity | 3.1 L |
| Fuel cons. at cont. rated power | 1.7 L/h—3600 rpm |
| Engine oil capacity | 0.6 L |
| Dimensions (L × W × H) | 321 × 376 × 346 mm |
| Orifice Coefficient of Discharge | 0.60 |
| Dynamometer Arm Length (mm) | 155 |

with the gasoline 92 fuels at an engine speed of 1000 RPM. Dutta et al. [7, 8] experimented with preheated biodiesel in a coil tube heat exchanger and producer gas in a slow speed diesel engine after modification.

This study aims at: (i) simulation modelling of the performance of a single cylinder, four-stroke spark ignition engine and checks the validity of the model, (ii) perform the experiment on an actual engine to determine brake power, torque, brake thermal efficiency and brake specific fuel consumption and (iii) comparison of the performance parameters obtained by both computation and experimental methods.

## 2 Methodology

The methodology used is that firstly, the engine modelling is done through a MATLAB code by using the engine specifications as given in Table 1. The model produces outputs like brake power, brake thermal efficiency, brake specific fuel consumption and torque. The results obtained by simulation were verified by performing an experiment on an actual engine. By comparing the results of both methods, the model verification was done.

## 2.1 Governing Equations

The principal governing equations to be used for modelling the engine are as follows.

Volume: Using the geometrical parameters of the engine, the cylinder volume at any crank angle is given by the following relation: [9]

$$V(\theta) = V_c + \frac{\pi B^2}{4}(l_r + a - s) \tag{1}$$

Heat release during combustion: The combustion analysis in the engine is done by analyzing the cylinder pressure in a single-zone thermodynamic model using Wiebe function for calculating heat release during combustion. The expression for mass fraction burned inside the cylinder at any crank angle is given by Eq. (2) [10]

$$x_b(\theta) = 1 - exp\left[-a\left(\frac{\theta - \theta_s}{\theta_d}\right)^n\right] \tag{2}$$

where $n$ is the form factor and $a$ is the efficiency factor. These parameters determine the shape and accuracy of the Wiebe function to predict the actual heat release during combustion.

Cylinder temperature: The rate of change of cylinder temperature w.r.t crank angle is given by: [10]

$$\frac{\mathrm{d}T}{\mathrm{d}\theta} = T(\gamma - 1)\left[\left(\frac{1}{PV}\right)\left(\frac{\mathrm{d}Q}{\mathrm{d}\theta}\right) - \frac{1}{V}\left(\frac{\mathrm{d}V}{\mathrm{d}\theta}\right)\right] \tag{3}$$

Heat transfer rate: Using Annand's method, Newton's law of cooling can be expressed as follows: [11]

$$\delta Q = (h_c + h_r)A_w(T - T_w)\mathrm{d}t \tag{4}$$

The convective heat transfer coefficient can be obtained as

$$h_c = \frac{k_{gas}Nu}{B} \tag{5}$$

The thermal conductivity of the cylinder gas can be modelled using a polynomial curve-fitting of experimental data as: [11]

$$k_{\text{gas}} = 6.1944 \times 10^{-3} + 7.3814 \times 10^{-5}T - 1.2491 \times 10^{-8}T^2 \tag{6}$$

For a four-stroke engine, Nusselt no can be expressed as: [11]

$$Nu = 0.49Re^{0.7} \tag{7}$$

The Reynolds number is expressed as:

$$Re = \frac{\rho_{\text{gas}}\ \bar{S_p}\ B}{\mu_{\text{gas}}} \quad (8)$$

The radiative heat transfer coefficient is given by: [11]

$$h_r = 4.25 \times 10^{-9}\left(\frac{T^4 - T_w^4}{T - T_w}\right) \quad (9)$$

Pressure rise: Ideal gas law for a closed system is used to obtain the pressure rise in the combustion chamber and this equation can be differentiated with respect to crank angle [10].

$$\frac{\mathrm{d}P}{\mathrm{d}\theta} = \frac{-\gamma P}{V}\left(\frac{\mathrm{d}V}{\mathrm{d}\theta}\right) + \left(\frac{\gamma - 1}{V}\right)Q_{\text{in}}\frac{\mathrm{d}x_b}{\mathrm{d}\theta} + (\gamma - 1)\frac{1}{V}\left(\frac{\mathrm{d}Q_w}{\mathrm{d}\theta}\right) \quad (10)$$

Engine friction: In order to calculate the losses due to friction, friction mean effective pressure needs to be calculated which is given by Blair as follows [11]:

$$f\text{mep} = 100,000 + 350(L)(\text{RPM}) \quad (11)$$

Residual gas equations: The thermodynamic relations for the residual exhaust gases are as follows [10]:

$$T(\theta) = T_{EVO}\left(\frac{P_{BDC}}{P_{EVO}}\right)^{\frac{\gamma-1}{\gamma}} \quad (12)$$

$$f = \frac{1}{r}\left(\frac{P_e}{P_i}\right)^{\frac{1}{\gamma}} \varnothing \quad (13)$$

$$T_{\text{corr}} = T_{CE} f + T_{BDC}(1 - f) \quad (14)$$

Torque: The torque produced by the engine in N m is given by: [9]

$$T = \text{Load(kg)} \times g \times \text{Dynamometerarmlength(m)} \quad (15)$$

Brake Power: The brake power produced by the engine in kW is given by: [12]

$$bp = \frac{2\pi NT}{60000} \quad (16)$$

Brake Thermal Efficiency: The brake thermal efficiency is given by: [12]

$$\eta_{bth} = \frac{bp}{m_f \times CV} \tag{17}$$

Brake Specific Fuel Consumption: The brake specific fuel consumption in kg/kWh is given by: [12]

$$bsfc = \frac{m_f}{bp} \times 3600 \tag{18}$$

## 2.2 *Uncertainty Analysis*

The uncertainty analysis helps to detect the errors in the estimated quantities from the measured ones. The uncertainty in the dependent variable is determined by Eq. (19) [13].

$$\Delta Y = \left[\left(\frac{\partial Y}{\partial X_1}\Delta X_1\right)^2 + \left(\frac{\partial Y}{\partial X_2}\Delta X_2\right)^2 + \ldots\left(\frac{\partial Y}{\partial X_n}\Delta X_n\right)^2\right]^{1/2} \tag{19}$$

where, $\Delta Y$ is the uncertainty in the estimated value and $\Delta X_1$, $\Delta X_2$ ... $\Delta X_n$ are the errors in the independent variables.

The total uncertainty in brake thermal efficiency of the gasoline engine is given by Eq. (20) [13].

$$\Delta\eta = \pm\left[\left(\frac{\partial\eta_{bth}}{\partial N}\Delta N\right)^2 + \left(\frac{\partial\eta_{bth}}{\partial T}\Delta T\right)^2 + \left(\frac{\partial\eta_{bth}}{\partial m_f}\Delta m_f\right)^2\right]^{1/2} \tag{20}$$

where, $\Delta\eta$ is the uncertainty in the brake thermal efficiency of the engine, and $\Delta N$, $\Delta T$, $\Delta m_f$ are the uncertainties in RPM, torque, and mass flow rate of gasoline into the engine, respectively. The average uncertainty in the brake thermal efficiency is computed as $\pm 1.5278\%$.

## 2.3 *Simulation Model*

The following flowchart shows the process flow of the simulation model. The model takes inlet conditions, fuel properties, and engine geometry and engine parameters as inputs. Using these values, the model calculates the various properties like pressure, temperature, volume, heat transfer, etc. Finally, the model produces output values of

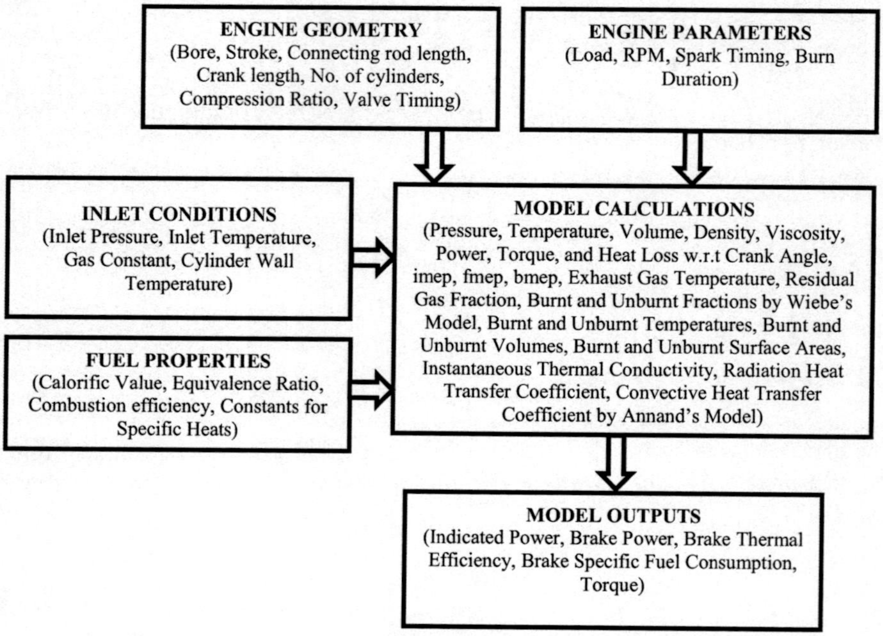


**Fig. 1** Flowchart of the engine model inputs and outputs

brake power, torque, brake thermal efficiency and brake specific fuel consumption (Fig. 1).

## 2.4 *Experimental Work*

The experiment was conducted in the Department of Mechanical Engineering, Tezpur University. Figure 2 shows the experimental setup of the computerized engine. The experiment was performed on a single cylinder, four-stroke spark ignition Honda GX200 196 cc engine. The engine specifications are given in Table 1.The in-cylinder and the fuel pressure are sensed by two piezo sensors. There are also provisions in the setup for measuring airflow, fuel flow and load. The engine setup is coupled with an eddy current dynamometer for controlling the engine torque through a computer. A software package 'ICEngineSoft' records all the required results on the computer. The engine was run for 10 cycles. The loads and speeds were varied steadily and the fuel supply was controlled manually. Cold water supply to the calorimeter was maintained continuously. The engine was loaded upto a maximum load of 2.1 kg and maximum speed of 2874 RPM.

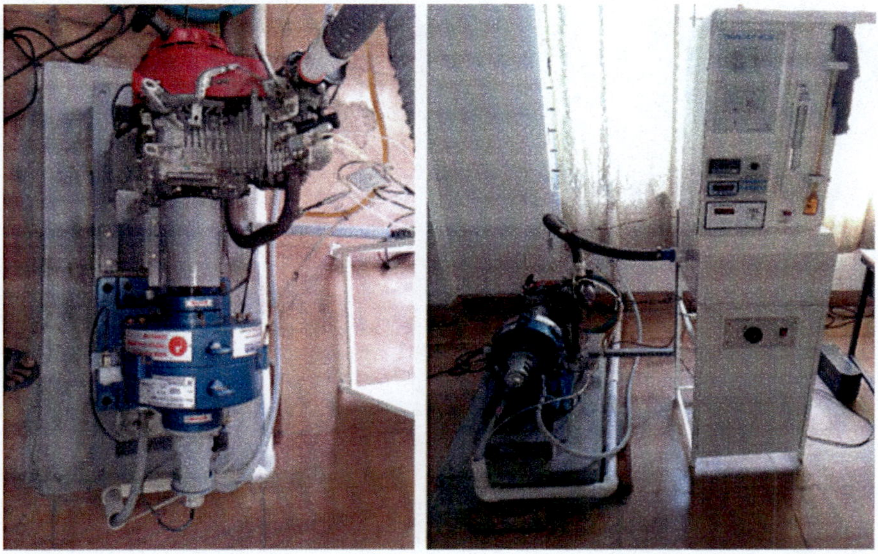

**Fig. 2** Computerized engine setup

# 3 Results and Discussions

The results obtained from computation and experimental work are as follows:

## *3.1 Mass Fraction Burned*

Figure 3 shows the simulation plot of mass fraction burned inside the cylinder at any crank angle. The curve is almost similar to theoretical curve. There is a sudden increase in the burned fraction at 150–250 degrees crank angle due to the combustion process. At the end of the combustion, all the gases completely burned, so percent mass fraction burned becomes 100% and remained constant for the remaining process.

## *3.2 Indicated Cylinder Pressure*

Figure 4 shows the simulation plot of indicated cylinder pressure at any crank angle. Maximum pressure value of approximately 4000 kPa is obtained at nearly 200° crank angle. There is a rapid increase in cylinder pressure near to 150°–250° crank angle due to the combustion process. At exhaust valve opening, the cylinder pressure drops to atmospheric pressure.

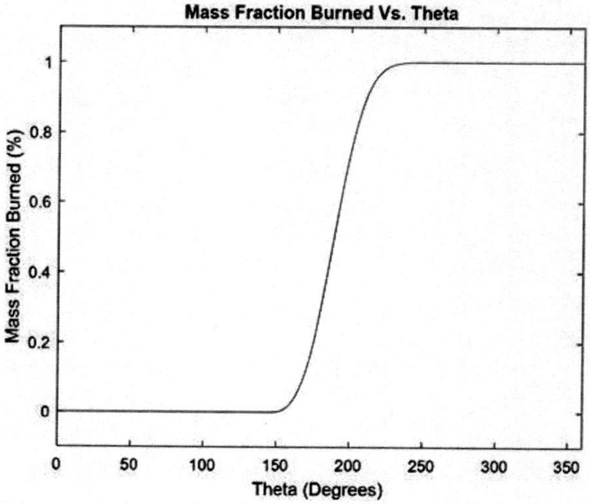


**Fig. 3** Mass fraction burned versus crank angle

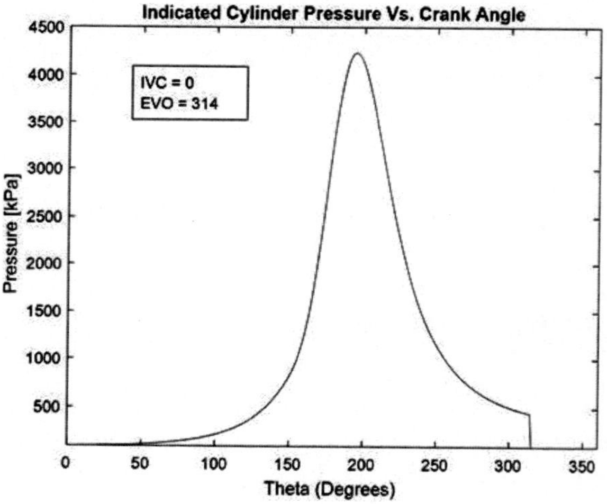


**Fig. 4** Cylinder pressure versus crank angle

## *3.3 Brake Power*

Figures 5shows the plot of brake power with engine speed. Maximum values are obtained at 2780 RPM for an engine load of 2.1 kg. Initially, there is a rapid increase in the brake power with increase in speed. It is because the engine load was increased in steps in the initial part of the experiment. But at higher speeds, the brake power decreases rapidly due to increase in frictional effects and decrease in engine load. Average error obtained between the predicted and experimental values is 25%.

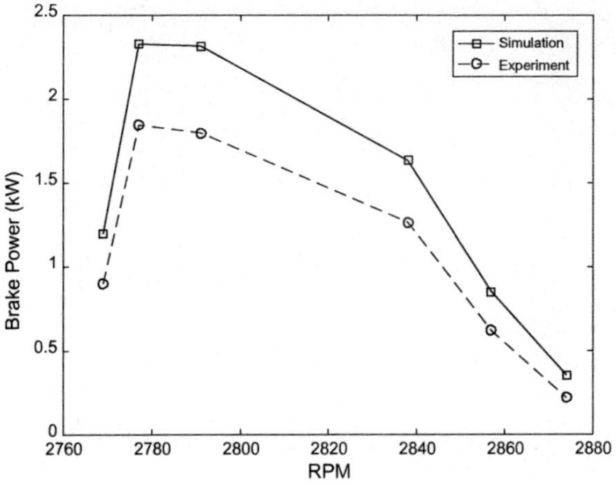


**Fig. 5** Brake power versus speed

## *3.4 Torque*

Figures 6 shows the plot of brake power with engine speed. Maximum values are obtained at 2780 RPM for maximum engine load of 2.1 kg. As torque is a function of brake power, so trend in the variation of torque is the same as that of brake power. At higher speed, torque decreases due to increase in friction and decrease in engine load. Average error obtained between the predicted and experimental values is 24%.

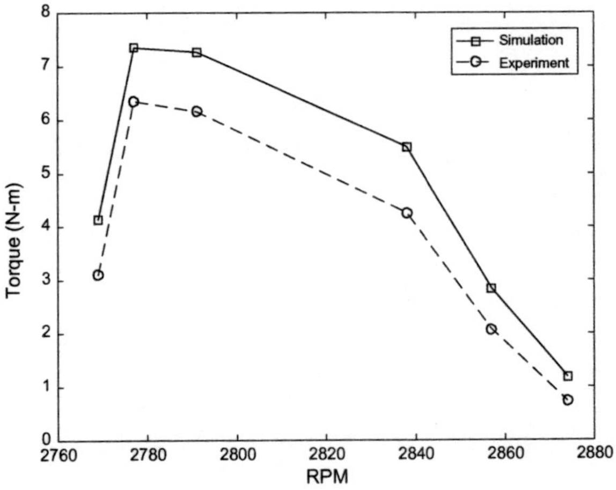


**Fig. 6** Torque versus speed

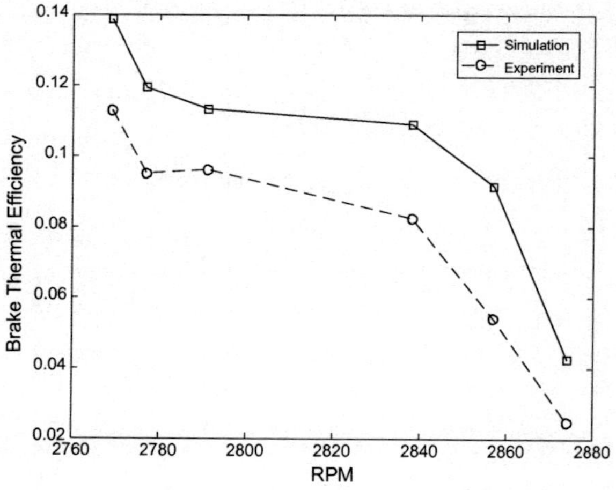


**Fig. 7** Brake thermal efficiency versus speed

## *3.5 Brake Thermal Efficiency*

Figure 7 shows the plot of Brake Thermal Efficiency with speed for simulation and experimental results. The shapes of both curves are similar to each other. Maximum values are obtained at a speed of 2769 RPM for an engine load of 1.9 kg. Initially, as the speed increases, brake thermal efficiency decreases, becomes somewhat constant in the middle, and decreases rapidly at higher speeds. Average error obtained between the predicted and experimental values is 26%. The decrease in thermal efficiency with increase in engine speed can be due to increase in fuel consumption, decrease in power output, frictional losses, incomplete combustion and heat transfer effects.

## *3.6 Brake Specific Fuel Consumption (BSFC)*

Figure 8 shows the plot of BSFC with speed for simulation and experimental results. The shapes of both the curves are similar to each other. Maximum values are obtained at a speed of 2874 RPM for engine load of 0.432 kg. At lower speeds, BSFC has lower values. It is due to high engine load used at lower engine speeds during the experiment. As the speed increases, BSFC increases rapidly. It is due to reduction in engine loads, incomplete combustion, and engine friction and heat transfer effects. Average error obtained between the predicted and experimental values is 26%.

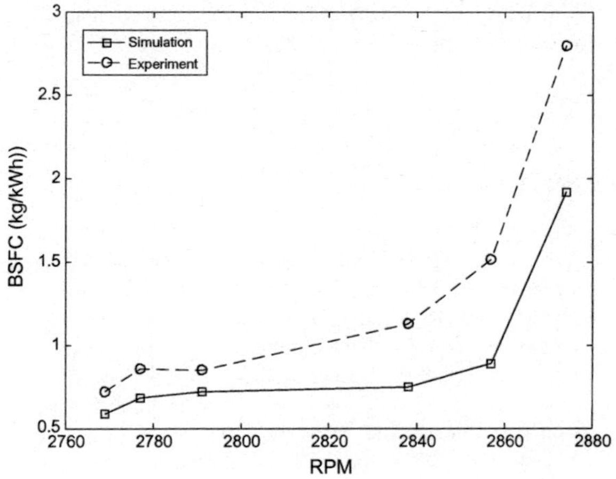


**Fig. 8** BSFC versus speed

# 4 Conclusions

In this study, experimental performance analysis on a four-stroke single cylinder computerized gasoline engine was carried out. Performance parameters assessed were engine torque, brake power, brake thermal efficiency and brake specific fuel consumption. Experiments were conducted under part-load conditions at different engine speeds. The following conclusions are made:

- The model correctly predicted the pressure trace and Wiebe's function curve and heat transfer with respect to the crank angle, which proved that the model can correctly simulate the performance of the engine.
- With an increase in speed of the engine, brake power and torque rised rapidly in initial stages, and decreased steadily at higher speeds.
- As the speed was increased, brake thermal efficiency decreased, and the rate of decrease of efficiency was higher at higher speeds due to change in engine load and fuel–air flow rates.
- Brake specific fuel consumption increased as the speed was increased. It increased rapidly at higher speeds due to increase in frictional effects at higher speeds.
- Maximum value of brake power obtained was 2.33 kW for simulation and 1.85 kW from experiment both at 2780 RPM for an engine load of 2.1 kg.
- Maximum values of torque obtained were 7.35 N m for simulation and 6.35 N m for experiment both at 2780 RPM for an engine load of 2.1 kg.
- Maximum values of brake thermal efficiency obtained were 13.8% for simulation and 11.3% for experiment both at 2769 RPM for an engine load of 1.9 kg.
- Maximum values of BSFC obtained were 1.92 kg/kWh for simulation and 2.8 kg/kWh for experiment both at 2874 RPM for an engine load of 0.5 kg.
- The relative errors obtained in simulation and experimental data varied from 23 to 28% for all the cases.

- The trend in the variation of the curves for computation and experimental data are quite similar to each other.

## References


1. M.I. İlhak, R. Doğan, S.O. Akansu, N. Kahraman, Experimental study on an SI engine fueled by gasoline, ethanol and acetylene at partial loads. Fuel **261**, 116–148 (2019)
2. M. Mourad, K. Mahmoud, Investigation into SI engine performance characteristics and emissions fuelled with ethanol/butanol-gasoline blends. Renew. Energy **143**, 762–771 (2019)
3. K. Wittek, F. Geiger, J. Andert, M. Martins, V. Cogo, T. Lanzanova, Experimental investigation of a variable compression ratio system applied to a gasoline passenger car engine. Energy Convers. Manage. **183**, 753–763 (2019)
4. Y. Qian, G. Liu, J. Guo, Y. Zhang, L. Zhu, X. Lu, Engine performance and octane on demand studies of a dual fuel spark ignition engine with ethanol/gasoline surrogates as fuel. Energy Convers. Manage. **183**, 296–306 (2019)
5. U.K. Efemwenkiekie, S.O. Oyedepo, U.D. Idiku, D.C. Uguru-Okorie, A. Kuhe, Comparative analysis of a four stroke spark ignition engine performance using local ethanol and gasoline blends, in *2nd International Conference on Sustainable Materials Processing and Manufacturing (SMPM 2019)* (Sun City, South Africa, 8–10 Mar. 2019, Procedia Manufacturing), pp.1079–1086
6. B. Sugiarto, C.S. Wibowo, A. Zikra, A. Budi, T. Mulya, M. Muchar, Comparison of the gasoline fuels with octane number variations 88, 92 and 98 on the performance of 4 strokes single cylinder 150CC spark-ignition engine, in *The 10th International Meeting of Advances in Thermofluids (IMAT 2018)*, vol. 2062, issue 1 (Grand Inna Hotel, Kuta, Bali, Indonesia, 16–17 Nov 2018, AIP Publishing)
7. P.P. Dutta, N. Kalita, S. Sagar, J. Kumar, C. Das, N. Goswami, J. Bora, M. Pathak, Design of a coil tube heat exchanger for preheating of biodiesel. Adv. Res. Electr. Electron. Eng **3**(5), 381–386 (2017)
8. P.P. Dutta, B.C. Jain, R. Das, S. Mahapatra, Modification and performance testing of a slow speed diesel engine for purely producer gas operation, in *Proceeding of National Conference on Renewable Energy Technology Issues and Prospects*, 14-09/2010 to 15/09/2010 at NERIST, Arunachal Pradesh, India (2010)
9. J.B. Heywood, *Internal Combustion Engine Fundamentals* (McGraw Hill Education Private Limited, India, 2011)
10. C.R. Ferguson, A.T. Kirkpatrick, *Internal Combustion Engines: Applied Thermosciences,* 3rd edn. (Wiley Ltd, 2016)
11. G.P. Blair, *Design and Simulation of Four Stroke Engines [R-186]* (Society of Automotive Engineers Inc, 1999)
12. V. Ganeshan, *Internal Combustion Engines*, 4th edn. (McGraw Hill Education Private Limited, India, 2012)
13. J.P. Holman, *Experimental Methods for Engineers* (The McGraw-Hill Companies, New Delhi, 2007)

# Machinability Study of Rubber using USM for Microdrilling

**Pallab Sarmah, Tapas Debnath, and Promod Kumar Patowari**

**Abstract** Precision machining of the difficult to machine materials is the most challenging task of the present manufacturing industry. Rubber is one of the most commonly used materials and can be cut easily by conventional cutting process in macrolevel. But to perform precision machining and microfeatures on rubber is quite a difficult task for its well-known properties like elasticity and resilience. On the contrary, ultrasonic machine is making use of machining of tough and brittle materials. In this paper, drilling operation has been done in microlevels on rubber using ultrasonic machining. During experimentation, the influence of process parameters on the performance measures material removal rate (MRR), and overcut (OC) has been studied. After successful drilling in rubber material, it has been observed that diameter and MRR varies from 398 to 535 μm and 0.0053 to 0.0175 $mm^3$/min, respectively, with a stainless steel tool of 550 μm in diameter.

**Keywords** Microholes · Microtool · Rubber · USM

## 1 Introduction

Among the different advanced machining processes, ultrasonic machining (USM) is one variant which is working to machine tough and brittle materials (both conductive and non-conductive). In USM, tool is oscillating normal to the work surface at ultrasonic frequency and material is removed by the abrasive grains which are impinged on the work surface. This process is preferable than the other machining process due to its minimal thermal effect on machined surface. Rubber is an elastic material which is used in different applications. Punching on rubber has disadvantages due to piercing effect. Moreover, rubber has high resilience, elasticity, and abrasion resistance.

P. Sarmah (✉) · T. Debnath · P. K. Patowari
Department of Mechanical Engineering, National Institute of Technology, Silchar, Assam, Silchar, India
e-mail: pallabsarmah92@gmail.com


K. M. Pandey et al. (eds.), *Recent Advances in Mechanical Engineering*, Lecture Notes in Mechanical Engineering, https://doi.org/10.1007/978-981-15-7711-6_46

Many researchers have observed the machining precision and accuracy of the holes produced by USM. For this, different parameters like type of abrasives, grit size, static load, and machining time have been considered. They reported that with fine abrasives accuracy of holes increased [1]. Machining efficiency and accuracy can determine from the produced microhole diameter and material removal rate (MRR). Material removal from workpiece takes place due to the striking of the tool and the impression of high velocity abrasives particles on work area [2]. Different varieties of tool material also stimulate the precision of the machining in USM. Stainless steel and high carbon steel can be used as tool material, which retain its shape while machining using USM [1, 3]. In ultrasonic machining material, removal of glass takes place due to the microbrittle fracture on the surface. The materials like glass, ferrite, and alumina of higher brittle in nature given high MRR [4–6]. Moreover, ultrasonic machining can be possible on different materials like WC–Co composite, titanium (Grade 2 and 5), and titanium diboride/silicon carbide. Micromachining of carbon fiber-reinforced polymer CFRP/Ti stacks using USM process resulted microholes on it without CFRP entrance elimination or Ti exit burr formation [7]. Without any change in properties of metallic glass, material can be drilled using USM process in room temperature. Metallic glass has good machinability nature for USM with higher MRR and very lower tool wear rate (TWR) [8]. Moreover, the array of multiple holes can also be performed at a time on ultrasonic machining process [9, 10].

From the literature, it has been observed that lots of works are found on micro-drilling of variability brittle material like ceramics, glass, and silicon using USM. But drilling operation has not been observed on rubber materials using USM. In this investigation, microholes have been drilled in rubber using ultrasonic machining. During experimentation, performance measures MRR and overcut (OC) has been studied with the different control parameters.

## 2 Materials and Methods

Initially rubber of thickness 1 mm is selected as workpiece material and stainless steel of 550 $\mu$m as tool material. Figure 1 presents the USM machine setup that makes use of machining in rubber material. Depending on different pilot experiments, feed rate (FR), slurry concentration, and slurry flow rate (SFR) are selected at different levels as process parameters for USM machining. Table 1 displays the selected process parameters and their corresponding levels for experimentation. In this work, sixteen experimental runs are conducted generated on Taguchi $L_{16}$ design of experiment (DOE) which is shown in Table 2.

The effect of the process parameters on MRR and OC has been observed while experimentation. After drilling operation, the diameter of the microholes is assessed by an optical microscope (Make: Leica; Model: DM 2500 M). Image of the drilled holes using conventional and ultrasonic method has been taken in SEM set up (Make: HITACHI, Model: TM 4000 plus). Thereafter, MRR is calculated by Eq. (1) considering the ratio of the total volume of work material removal to the machining time

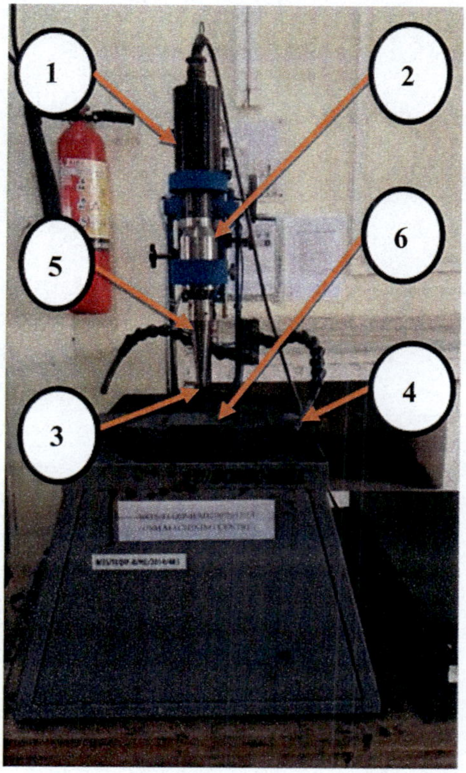


**Fig. 1** Ultrasonic machine set up for microdrilling (1. transducer, 2. acoustic head, 3. cutting tool, 4. abrasive flow nozzle, 5. horn, and 6. magnetic chuck)

**Table 1** Selected process parameters and their corresponding levels

| Parameters | Levels | Values |
|---|---|---|
| Feed rate (μm/s) | 4 | 40,50,60,70 |
| Slurry concentration (%) | 4 | 30,40,50,60 |
| Slurry flow rate ($cm^3$/s) | 2 | 50,100 |

and OC is evaluated by Eq. (2).

$$\text{MRR}\left(\text{mm}^3/\text{min}\right) = \frac{\text{Total volume of the drilled holes}}{\text{Machining time}} \tag{1}$$

$$\text{Overcut} = \frac{\text{Hole diameter} - \text{Tool diameter}}{2} \tag{2}$$

**Table 2** Experimental results of the Taguchi L16 orthogonal array

| Exp. no. | Process parameters | | | Response measures | |
|---|---|---|---|---|---|
| | FR (μm/s) | Slurry concentration (%) | SFR ($cm^3/s$) | OC (μm) | MRR ($mm^3/min$) |
| 1 | 40 | 30 | 50 | −27.5 | 0.0116 |
| 2 | 40 | 40 | 50 | −22.5 | 0.0120 |
| 3 | 40 | 50 | 100 | −13.5 | 0.0141 |
| 4 | 40 | 60 | 100 | −7.5 | 0.0175 |
| 5 | 50 | 30 | 50 | −41.0 | 0.0095 |
| 6 | 50 | 40 | 50 | −39.0 | 0.0110 |
| 7 | 50 | 50 | 100 | −34.5 | 0.0113 |
| 8 | 50 | 60 | 100 | −31.0 | 0.0116 |
| 9 | 60 | 30 | 100 | −52.5 | 0.0070 |
| 10 | 60 | 40 | 100 | −50.0 | 0.0070 |
| 11 | 60 | 50 | 50 | −42.5 | 0.0081 |
| 12 | 60 | 60 | 50 | −41.0 | 0.0093 |
| 13 | 70 | 30 | 100 | −76.0 | 0.0053 |
| 14 | 70 | 40 | 100 | −67.0 | 0.0054 |
| 15 | 70 | 50 | 50 | −62.5 | 0.0057 |
| 16 | 70 | 60 | 50 | −58.0 | 0.0066 |

## 3 Results and Discussion

Table 2 depicts the experimental results and the impact of process parameters on the response measures are explained in subsequent subsections.

The response measure OC is found more negative due to getting smaller diameter of the hole than the tool diameter. It is for the well-known piercing effect of the rubber material. As a result, MRR is observed very less which varies from 0.0053 to 0.0175 $mm^3/min$. The photographic image of the rubber sample after drilling operation is shown in Fig. 2. Figure 3 represents microscopic image of the machined microhole under an optical microscope.

### *3.1 Impact of Process Parameters on OC*

Figure 4 specifies influence of process parameters on OC of rubber material. Due to the general characteristics of rubber, the hole dimension is always lower than that of the tool diameter, indicating the negative OC. It is observed that diameter reduces with the increment in FR because of the faster machining rate results in less

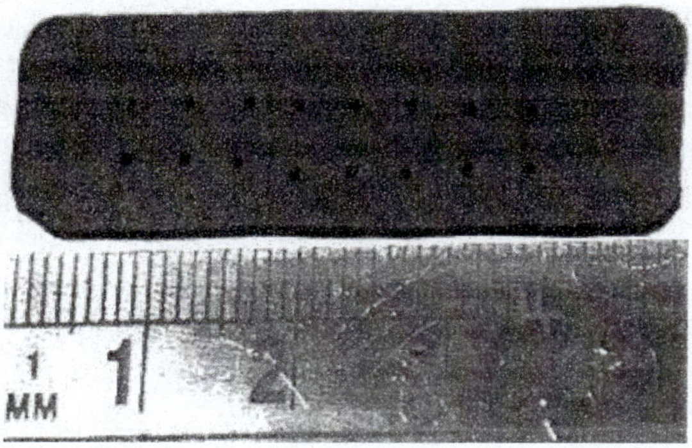


**Fig. 2** Drilled microholes on rubber samples

**Fig. 3** Micrographic image of the rubber sample

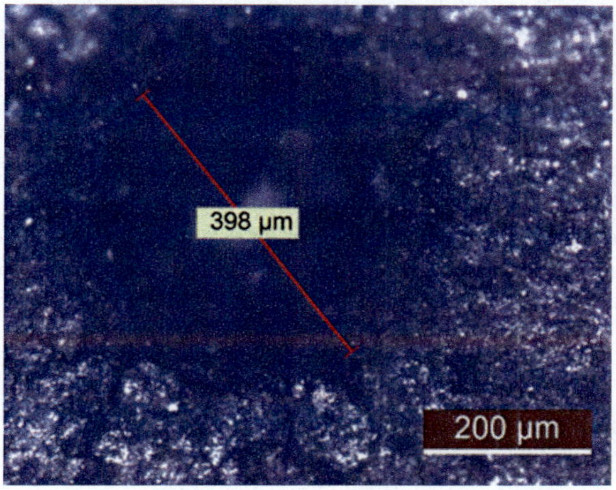


**Fig. 4** Influence of control parameters on OC

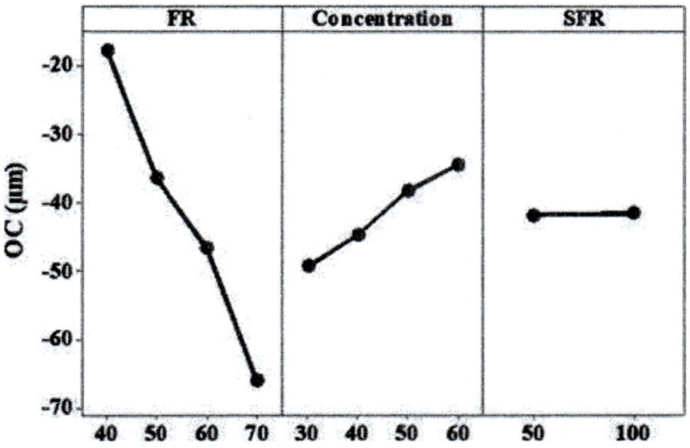

abrasion and prominent by the abrasive particles on the work surface. So, smaller hole dimension compared to tool diameter indicates the increase in negative OC. On the contrary, the negative OC reduces with increase in slurry concentration as MRR increases due to the its good impact on workpiece, helping to get near net shape. With an increase in slurry flow rate, no significant change has been observed on OC.

## 3.2 *Impact of Process Parameters on MRR*

Figure 5 demonstrates variation of MRR with the control parameters. It is perceived that MRR decreases with the increment of feed rate because of faster machining rate and results in slight abrasion and lower MRR. With increment in slurry concentration, MRR increases as abrasive slurry concentration has impact on workpiece which increases the MRR. On the contrary, MRR increases with an increment in slurry flow rate as the complete active particle increases in the slurry marks a good number of indentations.

As OC is a non-beneficial criterion, the dimension of hole closer to the tool dimension is desirable. On the other hand, the general characteristics of the rubber OC are negative, i.e., lower than that of the tool dimension. So, higher value of OC in Fig. 4 is better. Also, higher value of MRR is better for the microdrilling on rubber using USM. From Figs. 4 and 5, it is observed that the optimum value of OC and MRR is achieved at FR at lowest level, whereas slurry concentration and SFR at highest level belong to the experimental condition 4. Table 3 shows the optimal

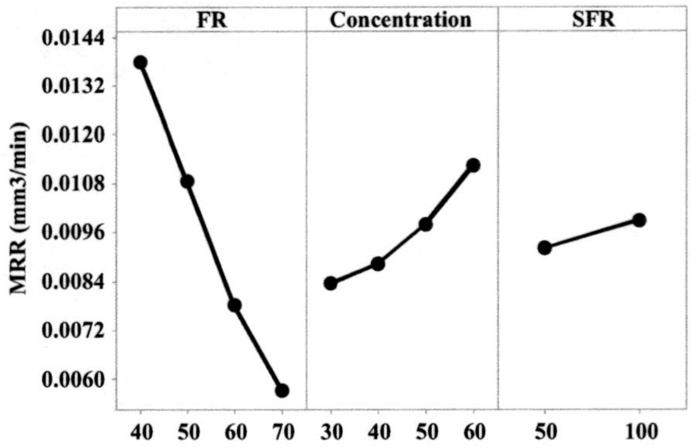


**Fig. 5** Influence of control parameters on MRR

**Table 3** Optimal condition and their results for OC and MRR

| Optimum condition of process parameters | | | Performance measures | |
|---|---|---|---|---|
| FR (μm/s) | Slurry concentration (%) | SFR ($cm^3$/s) | OC (μm) | MRR ($mm^3$/min) |
| 40 | 60 | 100 | −7.5 | 0.0175 |

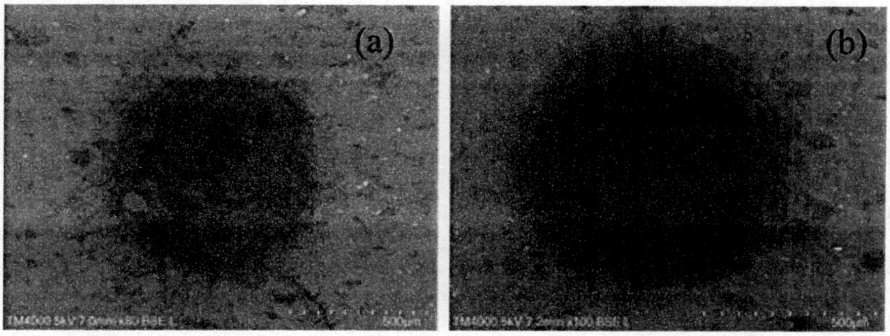


**Fig. 6** SEM images of the drilled hole using. **a** Conventional punching. **b** Ultrasonic drilling

combination of the control parameters and the obtained performance measures.

Figure 6 displays the SEM images of the drilled holes on rubber material using conventional and ultrasonic method. Drilling holes on rubber using ultrasonic machining results good as compared to the conventional drilling even though there are some burrs observed on the periphery of the exit of the hole.

## 4 Conclusions

The drilling operation on rubber has been successfully carried out using ultrasonic machining. After machining on rubber material, it is observed that hole diameter varies from 398 to 535 μm. The variation of response measures OC and MRR with respect to the change in the process parameters is also studied. Various conclusions drawn from this experimental work are as follows.

1. With the increment in feed rate, the tool advancement toward the workpiece at a faster rate in a lowest abrasion and striking on the work surface by the abrasive particles as a result MRR and OC decreases.
2. MRR and OC increase as per increase in slurry concentration due to which it gives good impact on the workpiece.
3. MRR and OC increase a little with increment in slurry flow rate as more impact of the abrasive particles is on the work material.

## References

1. V. Soundararajan, V. Radhakrishnan, An experimental investigation on the basic mechanisms involved in ultrasonic machining. Int. J. Mach. Tool Des. Res. **26**(3), 307–332 (1986)

2. R.S. Jadoun, P. Kumar, B.K. Mishra, R.C.S. Mehta, Optimization of MRR in ultrasonic drilling (USD) based on Taguchi's robust design methodology. Int. J. Mach. Mach. Mater. **1**(4), 445 (2006)
3. M. Komaraiah, M.A. Manan, P.N. Reddy, S. Victor, Investigation of surface roughness and accuracy in ultrasonic machining. Precis. Eng. **10**(2), 59–65 (1988)
4. S. Agarwal, On the mechanism and mechanics of material removal in ultrasonic machining International. J. Mach. Tools Manuf. **96**, 1–14 (2015)
5. M. Ramulu, Ultrasonic machining effects on the surface finish and strength of silicon carbide ceramics. Int. J. Manuf. Tech. Manag. **7**(2/3/4), 107–125 (2005)
6. A. Haashir, T. Debnath, and P.K. Patowari, A comparative assessment of micro drilling in boron carbide using ultrasonic machining. Mater. Manuf. Proces. **35**, 1–9 (2019)
7. S. James, Experimental study on micromachining of CFRP-Ti stacks using micro ultrasonic machining process. Int. J. Adv. Manuf. Technol. **95**, 1539–1547 (2017)
8. S. Kuriakose, P.K. Patowari, J. Bhatt, Machinability study of Zr-Cu-Ti metallic glass by micro hole drilling using micro-USM. J. Mater. Process. Technol. **240**, 42–51 (2017)
9. K.K. Patra, T. Debnath, and P. K. Patowari, Fabrication of an array of square micro-holes on glass using ultrasonic machining, in AIP Conference Proceedings, August 2018, vol. 1998, no. 1, p. 020013
10. T. Debnath, K.K. Patra, P.K. Patowari, Gang drilling of square micro-holes on glass using USM, in Advances in *Unconventional Machining and Composites* (2020), pp. 549–557

# Design, Fabrication and Experimentation of Aqua Silencer for Diesel Genset

**Neelutpal Ghosh, Sameer S. Gajghate, Sagnik Pal, and Swapan Bhaumik**

**Abstract** Air pollution is rising day by day and it is a serious threat to society and the environment. One of the main reasons for air pollution is exhaust gas emission from automobiles and industries. To reduce air pollution from the exhaust emission a device is introduced called Aqua silencer. With the help of this air is purified from the pollutants such as Carbon monoxide, Unburnt Hydrocarbons, Oxides of Nitrogen, etc. and it also reduces the damping noise. In this present paper, designing, fabrication, and testing of modified aqua silencer for the diesel Genset are discussed. Moreover, testing was conducted for three different conditions of silencers and the result shows that the exhaust emission of the aqua silencer with and without lime water reduces considerably.

**Keywords** Aqua silencer · Air pollution · Emission · Pollutants · Diesel engine

## 1 Introduction

At present, air pollution is a serious problem faced by the environment. The main sources of air pollution are Automobile, Power generators, Industrial and domestic fuel consumption, etc. and among all of these automobiles cause more air pollution in the atmosphere. Also, to admit the fact that these industries have been the major source of livelihood for a large number of common people throughout the world. The exhaust gases from the mentioned industries are polluting the environment rapidly,

N. Ghosh · S. S. Gajghate (✉) · S. Pal · S. Bhaumik
Mechanical Engineering Department, National Institute of Technology Agartala, Agartala, India
e-mail: mtech_sameer@yahoo.in

N. Ghosh
e-mail: neelutpalghosh@gmail.com

S. Pal
e-mail: sagnikpal181@gmail.com

S. Bhaumik
e-mail: drsbhaumik@gmail.com



K. M. Pandey et al. (eds.), *Recent Advances in Mechanical Engineering*, Lecture Notes in Mechanical Engineering, https://doi.org/10.1007/978-981-15-7711-6_47

which ultimately deplete the ozone layer of the atmosphere, which inturn become a cause of natural disasters in the upcoming years. In recent years, the world is facing the scarcity of fossil fuel and parallelly the excess utilization of fossil fuels in industries causes a tremendous amount of air pollution. But, nowadays the diesel engines are playing an important role in transport industries and as well as in other industries like agriculture, minning, etc. Considering, the available fuel resources and the present technological developments the diesel fuel is evidently indispensable. In general, the feeding of fuel is an index for finding out the economic asset of any country. In spite of this, no one can ignore the lethal effects of the exhaust gases, which vanishes the cleanliness and freshness of our environment. Automobiles contribute dangerous pollutants like Unburnt Hydrocarbons (UHCs), CO, Pb, $CO_2$, $SO_2$ and $NO_x$ etc. into the atmosphere. These gases are very harmful to the environment and for human health. So, serious steps must be taken in the direction to save the environment from degradation and conservation of energy from exhaust gases. The present work is an effort in this direction, which mainly deals with emission control, noise pollution & with the extra arrangement for the conservation of energy from the exhaust gases.

## 2 Literature Review

This section includes the International and National authors reported works related to present experimental investigation. Rawale et al. [1] have conducted the trails on aqueous ammonia in Silencer to remove the $CO_2$, $SO_2$ and $NO_x$ from exhaust gases of I. C. Engines. It was observed that ¼ part of carbon-dioxide is reduced from the exhaust gases compared to the ordinary silencer. Also, no such improvement in reduction was observed for other gases from the exhaust gases. Hatami et al. [2] used SST $k$–$\omega$ and RNG $k$–$\varepsilon$ model to recover the engines exhaust waste heat using the finned type heat exchangers numerically. Two HEXs is used in a SI and CI engine with $H_2O$ as a cold fluid, and with a mixture of 50% $H_2O$ and 50% ethylene glycol as cold fluid. Results show that the viscous models have the best heat recovery and the experimental results are better than the RSM results. Zhang et al. [3] have done the mathematical modelling on diffused pneumatic silencer to know the flow structure and gas flow through the interior and exterior of the silencer to understand the mechanism of the silencer's noise reduction. The porous media model and the Darcy–Forchheimer principle is applied to the model and found that the experimental result fits well with the numerical results. Mankhiar et al. [4] used a combination of titanium nano-tubes and charcoal in aqua silencer to overcome the drawback of charcoal and the reduction in noise pollution. Patel et al. [5] and Sharma et al. [6] worked on aqua silencer for two-stroke engine using water as a fluid and activated carbon pallets instead of charcoal for reduction noise pollution and emission control and observed the reduction in emissions. The researchers developed the aqua silencer [7–25] using adsorption technique with charcoal, lime water for automobile, diesel engine and portable twin filter aqua silencer found that the reduction in noise and emission in the I. C. engines. The authors have conducted review [26–31]on reported studies of aqua silencer and suggested that it should be utilized in mass quantity in the engines so as to minimize the noise and air pollution. Mohamed et al. [32] analyzed the back pressure fault in motorcycles using new wedge system and accordingly found the exhaust gas back pressure using ANSYS software to which the non-return valve to be designed and to be used. After the extensive study of reported works, it is clear that no discussion was included in finding the water quantity for moving vehicles, which is to be overcome! Also, few researchers are focused on with and without lime water, emission analysis for the various engine load and, the low power consumption from the exhaust gases and also from the activated carbon charcoal. The current research paper experimentally investigated the effect of lime, engine load based emission analysis with the low power generation system, i.e. Vapour adsorption with the use of high-pressure exhaust gases of the diesel genset engine.

## 3 Calculation of Dimensions for Designing

The Diesel engine was considered for further dimensions calculation for Aqua Silencer. The consider parameters and its dimensions are Bore ($D$)—0.073 m, Stroke ($L$)—0.0795 m, No. of Cylinder—3, Engine Power—67.07 bhp @ 6200 rpm, Max. RPM (N)—8000 rpm, Transmission Loss Noise Target—30 dB.

1. **To find Fundamental Frequency**:

$$\begin{aligned} \text{Cylinder Firing Rate(CFR)} &= 8000/120\ (\text{for } 4-\text{cycle engines}) = 66.67\ \text{Hz} \\ \text{Engine Firing Rate(EFR)} &= \text{No. of Cylinder} \times \text{CFR} \\ &= (3 \times 66.67)\ \text{Hz} \approx 200\ \text{Hz} \end{aligned}$$

2. **Muffler Volume Calculation**:

$$\begin{aligned} \text{Swept vol.}(V_s) &= \left(\pi \times D^2 \times L\right)/4 \\ &= \left(3.14 \times 0.073^2 \times 0.0795\right)/4 \\ &= 3.32569 \times 10^{-4}\ \text{m}^3 \\ &= 0.33261 \\ \text{Vol. of cylinder}(V) &= \text{No. of cylinder} \times V_s \\ &= 3 \times 0.3326 = 0.9978 \approx 11 \end{aligned}$$

3. **Silencer Volume ($V_m$)**:

$$\begin{aligned} &= \text{Factor}^* \times \text{vol. of the cylinder}(V) \\ &= 40 \times 11 \\ &= 40{,}000\ \text{cc} \end{aligned}$$

*Assumed factor = 40, i.e. for the volume of silencer the factor should be at least 30 to 40 times to the volume to be considered. Volume can be changed considering space constrain.

4. **Internal Configuration of Muffler and Concept Design**:

$$\begin{aligned} &\text{Vol. taken by Air filter and Perforated tube} \\ &= \left(\pi \times 0.06^2 \times 0.41\right)/4 \\ &= 1158.66\,\text{cc} \end{aligned}$$

where, dia. of air filter = 0.06 m, Length of air filter = 0.41 m.

$$\begin{aligned} \text{Vol. taken by inlet pipe} &= \left\{\pi \times 0.036^2 \times (0.15 + 0.044)\right\}/4 \\ &= 197.4\ \text{cc} \end{aligned}$$

where, dia. of inlet pipe = 0.036 m, Length of inlet air pipe of exhaust pipe = (0.15 + 4.4) m.

Now, Vol. taken by a charcoal layer of 0.03 m thickness, around the air filter of dia. 0.06 m

$$\begin{aligned} &= \left[\pi \times \left\{(0.06 + 0.03)^2 – 0.06^2\right\} \times 0.41\right]/4 \\ &= 1448.32\,\text{cc} \\ &\text{Total vol. needed for cylinder} \\ &= (40{,}000 + (1158.66 + 197.4) + 1448.32) \\ &\approx 42804\,\text{cc} \end{aligned}$$

Now, after studying various cylinder dimensions, we selected the length of the cylinder as twice of the dia. of the cylinder, $L = 2\text{D}$.

42,804 = ($\pi \times D^2 \times L$)/4 or

$$\begin{aligned} 42{,}804 &= \left(\pi \times D^2 \times 2D\right)/4 = 42{,}804 \text{ or} \\ D^3 &= 27{,}263.69\,\text{cc or} \\ D &= 30.09 \sim 0.3\,\text{m} \\ \text{Therefore, } L &= 2D = 2 \times 0.3 = 0.6\,\text{m} \end{aligned}$$

As per the calculated dimension for Aqua silencer and the available dimension for a tail pipe of considering model is 0.036 m and accordingly, the 2D model is designed as shown in Fig. 1. The details dimensions considered for fabrication are postulated in Table 1 along with the selection of material.

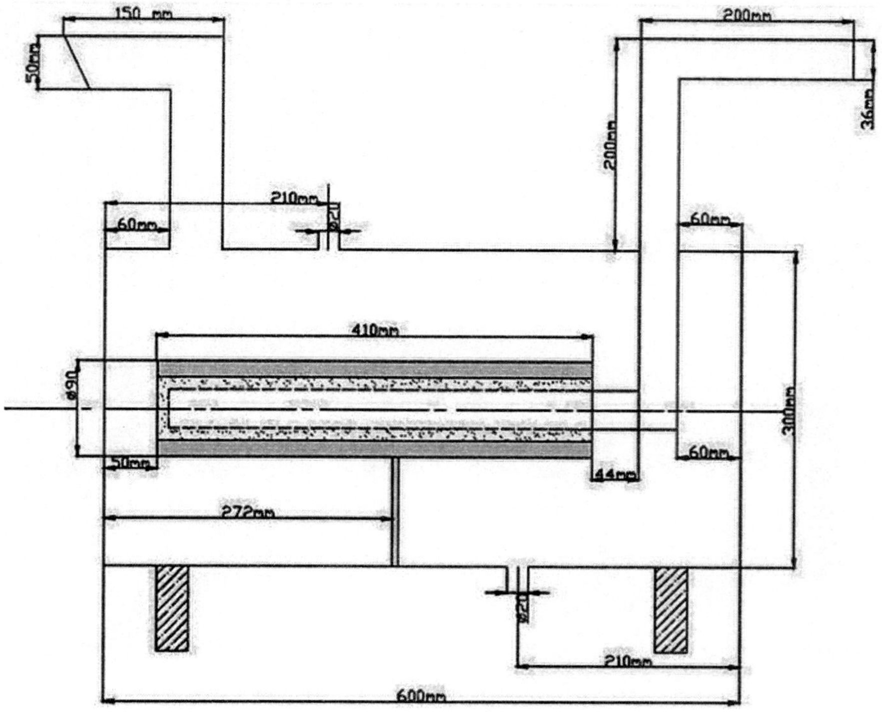


**Fig. 1** Free-hand sketch of aqua silencer setup showing different parts dimension

**Table 1** Dimensions used for designing the different parts for aqua silencer

| Different parts | Length, m | Dia., m | Thickness, m | Proposed material |
|---|---|---|---|---|
| Outer shell | 0.6 | 0.3 | 0.005 | Cold rolled stainless steel |
| Perforated tube | 0.4 | 0.036 | 0.004, 0.006 and 0.009 | Stainless steel pipe |
| Water inlet/Outlet port | – | 0.036 | – | Stainless steel pipe |
| Exhaust port | – | 0.050 | 0.005 | Stainless steel pipe |
| Air filter | Polypropylene (product name)—Purerite PS-05 (5 μm) | | | |

As per Fig. 1. Overall 2D drawing has been prepared, which gives a detailed study about the parts with their dimension. To fabricate the Aqua silencer different parts, the raw material has been purchased and fabrication work carried out at Central workshop, National Institute of Technology Agartala. Before the fabrication complete aqua silencer has been designed in Solid Works 2018 version, as shown in Fig. 2.

Figures 2 and 3 shows the outer shell, inlet and outlet exhaust section, water inlet and outlet port, the internal part of the perforated tube packed with activated carbon

**Table 2** Engine exhaust results for normal silencer

| Load (kg) | Time take to fuel consumption for 10 ml (sec) | Engine rpm | Vol % of CO | HC, ppm | Vol % of $CO_2$ | Vol. % of $O_2$ | NOx, ppm |
|---|---|---|---|---|---|---|---|
| 0 | 21 | 1482 | 0.20 | 30 | 2.70 | 17.59 | 180 |
| 2 | 20.54 | 1443 | 0.13 | 50 | 3.50 | 16.22 | 260 |
| 4 | 19.93 | 1430 | 0.09 | 38 | 5.01 | 15.85 | 340 |
| 6 | 19.10 | 1422 | 0.07 | 62 | 5.97 | 16.33 | 420 |

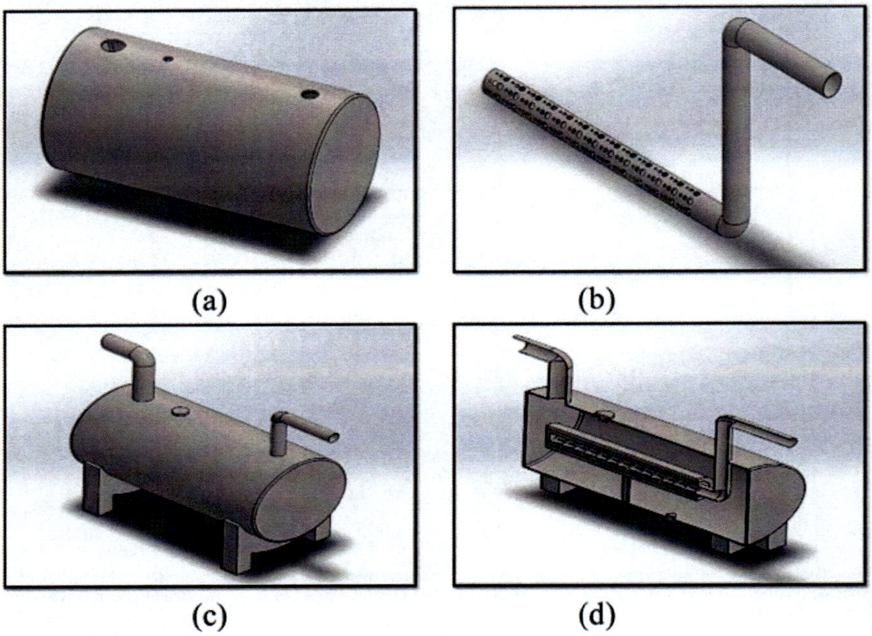


**Fig. 2** 3D design of **a** ouert shell, **b** perforated tube, **c** 3D view of aqua silencer, and **d** cut section of the aqua silencer

charcoal inside the tube and, also the polypropylene air filter in front of the perforated tube along with the stand arrangement.

## 4 Experimental Method

The experiments are conducted on a single-cylinder diesel engine test rig (Kirloskar Engine Ltd. make) as shown in Fig. 4 along with digital measuring parameters

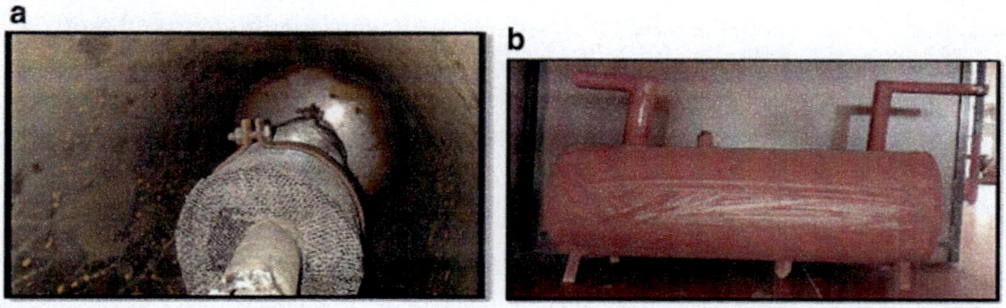

**Fig. 3** Fabricated parts **a** clamp holding the perforated tube, polypropylene air filter and the charcoal layer together, **b** the final product Aqua Silencer along with the supporting stands

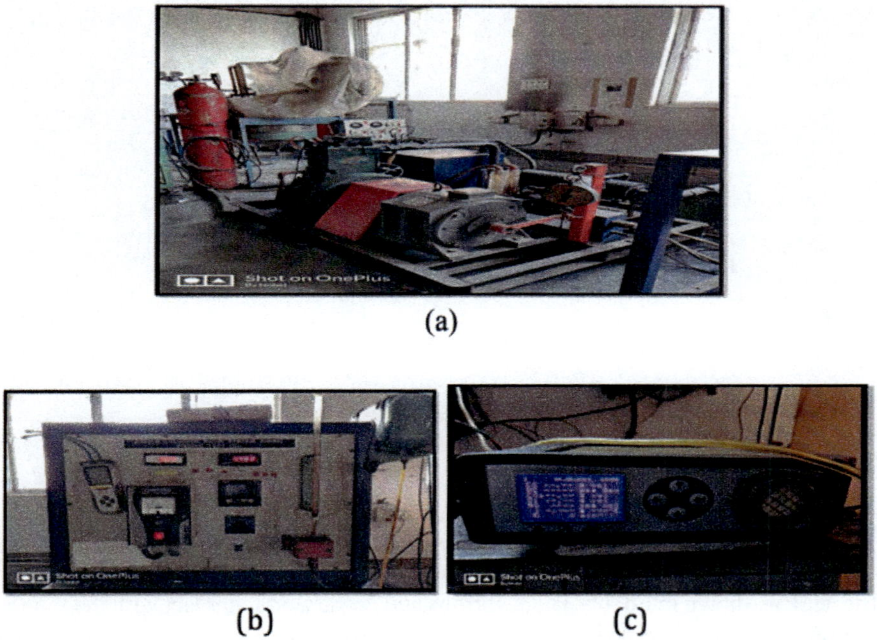


**Fig. 4** **a** Single-cylinder four-stroke diesel engine test rig, **b** digital display of test rig and **c** smoke analyser

displays (in Fig. 4b, and to assess the feasibility of the fabricated Aqua silencer. To measure the pollutants, a Smoke Analyser (AVL make) is attached in the test rig, as shown in Fig. 4c. As this is the stationary engine, the water quantity in the aqua silencer has been overcome.

The experimental procedure is followed as mentioned in reported studies [1, 4, 7] with three different conditions of silencers (a) for a conventional silencer and (b) aqua silencer without lime water, and (c) aqua silencer with lime water at 0–6 kg of engine load. Inside the tank, water is used to dissolve the unburned hydrocarbons

(UHCs). By this method, the UBHC, even if it is in glowing conditions, it is dissolved in water; thereby it is suppressing a spark, which could escape from the engine to the inflammable environment. Hence, the chemical reaction is taking parts when the exhaust gases are passing through the aqua silencer are as follows.

$$NO_2 + 2H_2O \rightarrow 2HNO_3(\text{Dilute}) + 2HNO_2 \quad (1)$$

$$Ca(OH)_2 + 2HNO_2 \rightarrow Ca(NO_3)_2 + 2H_2O \quad (2)$$

$$Ca(OH)_2 + 2HNO_2 \rightarrow Ca(NO_2)_2 + 2H_2O \quad (3)$$

$$Ca(OH)_2 + CO_2 \rightarrow CaCO_3 + H_2O \quad (4)$$

$$CaCO_3 + H_2O + CO_2 \rightarrow Ca\ (HCO_3)_2 \quad (5)$$

$$Ca(OH)_2 + SO_2 \rightarrow CaSO_3 + H_2O \quad (6)$$

$$CaCO_3 + H_2O + SO_2 \rightarrow Ca(NO_3)_2 + CO_2 + H_2O \quad (7)$$

$$4NO_2 + 2H_2O \rightarrow 2HNO_3 + 2HNO_2 \quad (8)$$

$$CaCO_3 + 2HNO_3 \rightarrow Ca(NO_3)_2 + CO_2 + H_2O \quad (9)$$

$$CaCO_3 + 2HNO_2 \rightarrow Ca(NO_2)_2 + CO_2 + H_2O \quad (10)$$

The $NO_2$ is reacting with dissolved in the water and gives the Nitrous & diluted Nitric acid, as shown in Eq. (1).Then, Nitrous and Nitric acid will react with the lime water present in the scrubber and gives the Calcium Nitrate, as shown in Eq. (2–3).Also, the lime water will react with $CO_2$ which is present in the exhaust gases and will precipitate the $CaCO_3$. Thus, when the $CaCO_3$ further exposed to $CO_2$, $Ca(HCO_3)$ will be precipitated, as shown in Eq. (4–5). Further, the $SO_2$ also react with the lime water, as shown in Eq. (6) will precipitate calcium carbonate and reacts further to gives the $CaSO_3$ and $CO_2$ as the byproduct as shown in Eq. (7). Because CO is chemically balanced, negligible vol. (0.2%) and stable, it won't readily react with water or any bi-products, which is resulted from the above reactions. Even though the lime water absorbs a part of the oxides of N and C, the time constraint for the reaction to take place allows a considerable percentage for emission. But, the stone container, which is provided with limestone ($CaCO_3$) encourages the chemical reaction further in the presence of hot water/steam, which evaporates from the scrubber tank due to high exhaust temperature. Because of the little percentage of

$SO_2$ presence, the liberation of $CO_2$ is very less. But the liberated $CO_2$ will again combine with $CaCO_3$ to form $Ca(HCO_3)$ as shown in Eq. (5).The presence of hot water possibly reacts with oxides of nitrogen, as shown in Eq. (8). The $Ca(NO_3)_2$ and $Ca(NO_3)_2$ are the bi-products and carbon-dioxide is liberated, as shown in Eq. (9–10). The liberated $CO_2$ again combines with $CaCO_3$ to form $Ca(HCO_3)$ (Eq. 5). In this way, the actual reaction takes place inside the aqua silencer to control the emission and noise pollution.

## 5 Result and Discussion

The exhaust gas passed through the exhaust pipe from the diesel engine, and the nozzle part of the Smoke Analyser is placed at the exhaust port of the engine, and the readings are noted down.

1. **Observation Without Aqua Silencer**

The single-cylinder four-stroke Diesel Engine shows the emission parameter values for the different loads of the engine, as shown in Table 2.

It is observed that as the load, increases the carbon monoxides increase along with the unburnt HC, $CO_2$ and $NO_x$.

2. **Observation with Aqua Silencer (Without Lime Water)**

The polypropylene air filter is used in the silencer because of its high absorption capacity. It absorbs some portions of the toxic gas present in the exhaust. During the experimentation, it is found that certain amount of HCs and oxide of nitrogen have been reduced. This happens only because of the air filter implanted inside the silencer, which absorbed a certain amount of the gas as depicted in Table 3.

The changes in the number of other gasses are negligible. Also, Fig. 5 is drawn as a bar diagram to analyze the emission of pollutants and noise control at different load for better understanding. These results are compared with the normal silencer results, which are mentioned in Fig. 6. It is the comparative study of both the silencer

**Table 3** Engine exhaust results for aqua silencer (without lime water)

| Load (kg) | Time take to fuel consumption for 10 ml (sec) | Engine rpm | Vol % of CO | HC, ppm | Vol. % of $CO_2$ | Vol. % of $O_2$ | NOx, ppm |
|---|---|---|---|---|---|---|---|
| 0 | 28.17 | 1490 | 0.11 | 21 | 2.60 | 17.03 | 172 |
| 2 | 24 | 1479 | 0.08 | 39 | 3.30 | 16.04 | 218 |
| 4 | 20.33 | 1442 | 0.05 | 31 | 4.20 | 15.02 | 314 |
| 6 | 19.40 | 1418 | 0.05 | 50 | 5.30 | 13.32 | 404 |

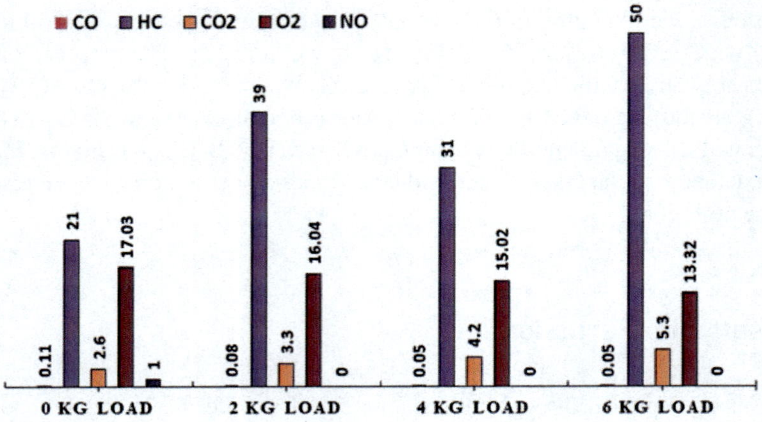


**Fig. 5** Emission results at different engine load for aqua silencer (without lime water)

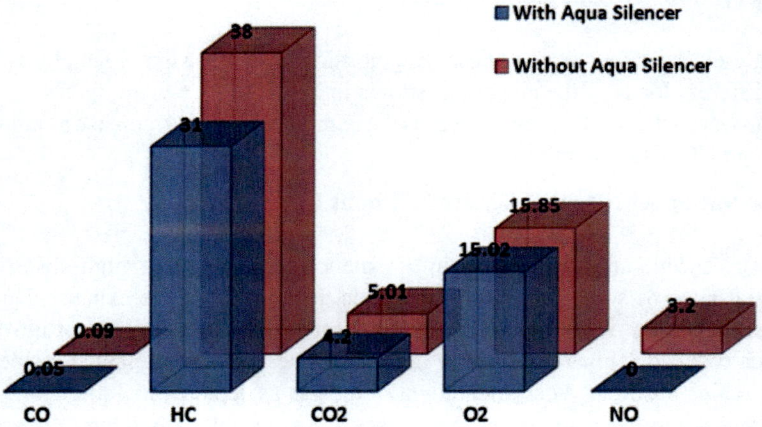


**Fig. 6** Comparison between normal engine exhaust results and engine using aqua silencer exhaust results for 4 kg load

at 4 kg of the load to the effective analysis of the performance of the normal and aqua (without lime water) silencer. And, it is found that the emission of pollutants and noise of the engine reduced due to use of an aqua silencer.

3. **Observation with Aqua Silencer (With Lime Water)**

In this test, lime water is filled inside the silencer. This test is found to be more efficient. Approximately, no pollution is observed as depicted in Table 4.

This is because, from the previous reading, it is clear that, the air filter used in Aqua Silencer absorbs to a great extent of the pollutants in the gas, thereby decreasing the number of pollutants.

**Table 4** Engine exhaust results for aqua silencer (with lime water)

| Load (kg) | Time take to fuel consumption for 10 ml(s) | Engine rpm | % of CO | HC, ppm | % of $CO_2$ | % of $O_2$ | NOx, ppm |
|---|---|---|---|---|---|---|---|
| 0 | 28.17 | 1490 | – | – | – | – | – |
| 2 | 24 | 1479 | – | – | – | – | – |
| 4 | 20.33 | 1442 | – | – | – | – | – |
| 6 | 18.40 | 1418 | – | – | – | – | – |

When lime water is used, then the remaining polluted gas reacts with lime water, thus again, the amount of pollutants in the gas decreases considerably. But due to the blockage in the silencer by bubble behaviour of lime water, unable to measure the exhaust gas concentration. So no pollution is observed.

## 6 Conclusion

It is concluded that the aqua silencer is useful for minimizing the emission of gases from the engine exhaust.

- The sound levels have been reduced by using lime water as a medium.
- It is found to be smokeless emissions and pollution-free by using Polypropylene Air Filter in water, and is also economical, considering the persistent use.
- Consumption of fuel is as same as the conventional system.
- Contamination in the water is negligible.
- No vibration is produced when the engine is running.

One can urge to a human being to use Aqua Silencer and can raise awareness about the increasing pollution that Aqua Silencer is one of the ultimate remedies to this problem.

## References


1. S.S. Rawale, S.S. Patil, A.A. Nandrekar, A.S. Kabule, Use of aqueous ammonia in silencer for removal of $CO_2$, $SO_2$ and NOx from exhaust gases of I.C. Engines. Int. J. Eng. Sci. Innov. Tech. **2**(5), 157–160 (2013)
2. M. Hatami, D.D. Ganji, M. Gorji-Bandpy, Numerical study of finned type heat exchangers for ICEs exhaust waste heat recovery. Case Stud. Ther. Eng. **4**, 53–64 (2014)
3. X.W. Zhang, Z.H. Yao, F. He, Numerical simulation of the flow field of a diffused pneumatic silencer. Appl. Math. Model. **33**, 3896–3905 (2009). https://doi.org/10.1016/j.apm.2009.01.006
4. A.B. Mankhiar, L.S. Sindhu, G. Sasikala, An advancement to reduce pollution effectively by using TI nanotubes in aqua silencer. Int. J. of Eng. Sci Res. Tech. **3**(3), 1741–1744 (2014)

5. K.I. Patel, S.R. Gajjar, Design and development of aqua silencer for two stroke petrol engine. Int. J. Innov. Res. Sci. Tech. **1**(1), 31–38 (2014)
6. P.K. Sharma, S.V. Kasar, S.A. Patil, N.A. Jadhav, M. Deore, Design and manufacturing of aqua silencer for two stroke petrol I. C. engine. Int. J. Adv. Eng. Res. Dev. **4**(3), 39–47 (2017)
7. R.R. Thakare, A.J. Khushwah, Design and fabrication of aqua silencer for automobiles. Int. J. Mech. Prod. Engg. Res. Dev. **5**(6), 139–148 (2015)
8. S. Raj, A.K. Aniyan, A. Aji, A. Raj, A. Mohan, T.R. Sharon, Fabrication and testing of portable twin filter aqua silencer. Int. J. Mech. Indus. Tech. **3**(2), 177–186 (2015)
9. M.A. Alen, M. Akshay, R.P. Sankar, M. Shafeeque, Fabrication and testing of aqua silencer. Int. Res. J. Eng. Tech. **2**(5), 1315–1320 (2015)
10. A.A. Kumar, N. Anoop, J.P.P. Aquib, E. Bijoy, T.V. Midhun, N.P.M. Shiyas, K.P.T. Ranjith, Design and development of aqua silencer. Int. J. Eng. Innov. Tech. **5**(11), 35–41 (2016)
11. R. Acharekar, D. Bhujbal, O. Bhingole, N. Sherkar, R.R. Argade, Design, development and experimentation of aqua-silencer for four stroke petrol engine. Int. Res. J. Eng. Tech. **3**(4), 62–66 (2016)
12. A.M. Gawande, P.K. Ingle, Aqua silencer. Int. Res. J. Eng. Tech. **3**(10), 153–155 (2016)
13. R.S. Padval, N.V. Patil, M.P. Pachare, Aqua silencer, in *Proceedings of International Conference on Emerging Trends in Engineering and Management Research*, 23rd March 2016. ISBN: 9788193207475
14. S. Rajesh, V. Akshay, M. Priyadharshini, A.K.V. Vijay, Silencer powered by milk of lime for emission control. Adv. Nat. Appl. Sci. **11**(7), 804–808 (2017)
15. S. Gite, A. Ingole, M.M. Kulkarni, Experimental investigation and testing of diesel engine and analysis of exhaust gases by using aqua silencer. Int. Eng. Res. J. **2**(8), 2496–2500 (2017)
16. D. Parmar, P.K. Dabhi, P. Soni, F.H. Pathan, Experimental investigation on an aqua silencer. Int. J. Adv. Res. Tech Eng. Sci. Tech. **4**(4), 867–872 (2017)
17. V.D. Chavan, T.V. Desai, P.R. Gidaye, T.V. Surve, Experimental investigation on single cylinder four stroke engine using aqua-silencer system, in *Proceedings of National Conference on Changing Tech and Rural Development*, pp. 169–173 (2017)
18. S.S.S. Kumar, S.R. Brinton, G.S. Kumar, S. Saravanamanoj, M. Babu, Design and fabrication of eco-friendly aqua silencer. Int. J. Sci. Tech. Eng. **3**(10), 409–415 (2017).
19. A. Deshpande, U. Kadam, S. Shah, R. Shukla, C. Srinidhi, Y.R. Ingole, Aqua silencer—a possible solution for reducing engine emission and IC Engine. Int. Res. J. Eng. Tech. **4**(1), 327–328 (2017)
20. Y.C. Vikas, A. Holkar, P.K. Mali, Design, fabrication and testing of aqua silencer with twin filters. Int. Res. J. Eng. Tech. **4**(1), 795–798 (2017)
21. A. Saraf, T. Khese, T. Shah, G. Gaikwad, S.D. Bhaisare, Design and analysis of aqua silencer. Int. Res. J. Eng. Tech. **4**(2), 1432–1436 (2017)
22. S.B. Said, S.P. Wagh, P.D. Gaikwad, S. Kondo, Aqua silencer. Int. Res. J. Eng. Tech. **3**(3), 1829–1832 (2017)
23. R. Sambasivam, P. Devendran, G. Prasanth, S.K. Thirunthaiyan, Design and performance evaluation of aqua silencer. Int. Res. J. Eng. Tech. **5**(1), 95–98 (2018)
24. O. Mali, P. Shinde, S. Patil, A. Masal, A. Ravi, P.V. Kadam, V.P. Patil, Research paper on aqua silencer. Int. Res. J. Eng. Tech. **5**(4), 743–745 (2018)
25. P.R. Sansare, A.R. Kasture, K.A. Umale, P.S. Ghodekar, Design and fabrication of aerodynamic aqua silencer. Int. Res. J. Eng. Tech. **6**(4), 5124–5131 (2019)
26. V. Kachare, V. Kannan, P. Taware, P. Ghayal, A review on design and development of aqua silencer, in *Proceedings of International Conference on Emanations in Modern Tech. and Engineering* (ICEMTE-2017), vol. 5(3), pp 59–64.
27. V. Gulhane, S. Hande, V.M. Chavan, Emission & noise control device aqua silencer: a review, in *Proceedings of 3rd International Conference on Ideas, Impact and Innovation in Mechanical Engineering* (ICIIIME 2017), vol. 5, no. 6, pp. 1008–1015
28. D. Parmar, P. Dabhi, P. Soni, F.H. Pathan, Reduction of air and noise pollution by aqua silencer—a review. Int. J. Adv. Res. Eng. Sci. Tech. **4**(2), 50–52 (2017)

29. J. Shah, V. Patel, Effect of aqua silencer & catalytic converter on exhaust emission: a review. Int. J. Innov. Res. Sci Tech. **3**(12), 233–238 (2017)
30. O. Mali, P. Shinde, S. Patil, A. Masal, A. Ravi, P.V. Kadam, N.V. Hargude, Review paper on design and development of aqua silencer. Int. Res. J. Eng. Tech. **4**(10), 166–168 (2017)
31. B.S. Sagar, P. Hullur, A. Sangappa, V. Kannolli, V. Rayakar, Review paper on aqua silencer. Int. J. Adv. Res. Innov. Ideas Educ. **5**(2), 2376–2382 (2019)
32. F.M.A. Mohamed, S. Sankar, C. Mathew, K.V. Sreejith, Design and development for exhaust back pressure reduction with noise control for motorcycles. Int. J. Innov. Res. Sci. Eng. Tech. 7(5), 5316–5323 (2018).

# Numerical Analysis OF PCM Within a Square Enclosure Having Different Wall Heating Conditions

**Pallab Bhattacharjee, Sujit Nath, and Dipankar Bhanja**

**Abstract** In the present study, the melting of phase change material (PCM) in a square enclosure with different wall heating conditions has been studied numerically. It has been found that at the early stage of melting, PCM melts faster in case of bottom wall heating. Melting for sidewall heating transcends the melting of bottom wall heating in a later stage. However, the increase of solid–liquid interface length leads to higher convective heat transfer for side wall heating in later part of melting. On the other hand, isothermal heating from the upper wall results in thermal stratification of PCM layers which leads to no convection. So, the melting time is quite large for this case.

**Keywords** PCM · Thermal energy reservoir · Melt fraction

## Nomenclature

| | |
|---|---|
| $c_p$ | Specific heat at constant pressure (KJ/kgK) |
| $g$ | Gravitational acceleration |
| $k$ | Thermal conductivity |
| $T$ | Temperature (K) |
| $H$ | Length of square arm (m) |
| $\alpha$ | Coefficient of thermal diffusivity ($m^2$ /s) |
| $\beta$ | Coefficient of thermal expansion (1/K) |
| $\mu$ | Dynamic viscosity (Pa-s) |
| $\rho$ | Density ($Kg/m^3$ ) |

P. Bhattacharjee (✉) · S. Nath · D. Bhanja
Mechanical Engineering, National Institute of Technology, Silchar, Assam, Silchar 788010, Assam, India
e-mail: bhattacharjeepallab27@gmail.com



K. M. Pandey et al. (eds.), *Recent Advances in Mechanical Engineering*, Lecture Notes in Mechanical Engineering, https://doi.org/10.1007/978-981-15-7711-6_48

## Subscripts

| | |
|---|---|
| l | liquid |
| m | melting |
| w | wall |
| s | solid |
| avg | average |

# 1 Introduction

High amount of energy can be stored during phase change process of phase change material with a negligible change in temperature. Thus, PCM has the potential to be used as energy reservoir in thermal systems, refrigeration, and air conditioning unit. But the challenge associated with PCM is very low thermal conductivity which results in slow charging and discharging. Researchers have used different techniques such as mixing high conductive nanoparticles with PCM, using different types of finned structures [2, 3], using metal matrix, incorporation of heat pipe, and using multiple PCM at a time [1].

Biwole et al. [2] studied the influence of fin size and distribution on solid–liquid phase change in a rectangular enclosure. They found that increasing the number of fin reduces the temperature of the plate and also accelerates melting process. Using thinner and longer fins also improves melting marginally. Ji et al. [3] explored the effect of fin placement at different inclination angles in a rectangular enclosure. Minimum melting time was obtained for 15° downward tilted fin. Upward tilted fin gives rise to uneven heating thus increasing the melting time. Kamkari et al. [1] have done both numerical and experimental study on inclined rectangular enclosures. It was found that for complete melting 0° and 45°, enclosures take 52 and 37% shorter time, respectively, in comparison with vertical enclosure. Asl et al. [4] had investigated the solid and porous fin in inclined rectangular enclosure. It has been predicted that porous fins are more effective with high Raleigh number and with highly conductive material. Ebadi et al. [5] did a numerical analysis and finally validated experimentally on melting of nano-PCM inside a cylindrical thermal energy reservoir (TES). In their work, the bio-based coconut oil PCM with dispersed copper oxide nanoparticles inside a cylindrical container is heated isothermally from lateral walls and from the top. Adding copper oxide nanoparticles improved the melting at later stage when dispersion of nanoparticle starts. Arena et al. [6] numerically analyzed a latent heat thermal energy storage system under partial load operating conditions. A substantial reduction in the duration of the TES (up to 50%) process against a small decrease in stored energy (up to 30%) has been noticed. Shokouhmand et al. [7] tried to apprehend the temperature field with melt front in a rectangular enclosure. They interpreted that the conduction is dominant at the beginning but convection is dominant in the later part of melting. The convection heat transfer in the later stage also reduces due to thermal stratification. Dhar et al. [8] analyzed the remelting phenomenon of eutectic aqueous ammonium chloride solution in a top cooled rectangular cavity during solidification process. It has been observed that remelting occurs during the progress of solidification however it depends on cavity height as well as imposed temperature gradient.

In the present work, the effect of heating at different walls of a rectangular enclosure is studied to find out the best heating strategy for getting higher melting rate. For this, a numerical model is developed with specific heat capacity formulation by COMSOL Multiphysics 5.3. Lauric acid is taken as PCM and kept in a square enclosure which is heated isothermally from different sides like left wall, top wall, and bottom wall, respectively.

## 2 Description of Numerical Model

A 2D numerical model is built with finite element-based software COMSOL Multiphysics 5.3. The model consists of a square enclosure filled with lauric acid. An implicit time stepping scheme along with a parallel sparse direct solver (PARDISO) is used to solve the governing equations. The absolute tolerance for scaled residues of the variables is $10^{-2}$. A total 1375 triangular and 232 quadratic mesh have been generated for the model. The results are independent with average element quality 0.815. The properties of lauric acid [2] are given in Table 1. The container is heated isothermally from different sides at 343 K wall temperature ($T_{\mathrm{w}}$). The square container has a dimension of 0.06 m × 0.06 m. There is a thin aluminum plate of 0.002 m width at heating surface.

*Governing equations*

For the present PCM model, the following equations [6] have been used to simulate the charging and discharging process.

*Continuity equation*

$$\frac{\partial \rho}{\partial t} + \nabla.(\rho \vec{u}) = 0 \tag{1}$$

**Table 1** Thermophysical properties of lauric acid

| Property | Value | Property | Value |
|---|---|---|---|
| $c_{p,s}$ | 2180 J/(kg K) | $\rho_{\mathrm{s}}$ | 940 kg/m$^3$ |
| $c_{p'l}$ | 2390 J/(kg K) | $\rho_{\mathrm{l}}$ | 885 kg/m$^3$ |
| $L$ | 187,200 J/(kg K) | $\mu$ | $8 \times 10^{-3}$ kg/(m s) |
| $k_{\mathrm{s}}$ | 0.16 W/(m k) | $\beta$ | $8 \times 10^{-4}$ K$^{-1}$ |
| $k_{\mathrm{l}}$ | 0.14 W/(m k) | $T_{\mathrm{m}}$ | 316.65 K |

*Momentum equation*

$$\frac{\partial \rho \vec{u}}{\partial t} + \rho(\vec{u}.\nabla)\vec{u} - \nabla.[\mu(\nabla \vec{u} + (\nabla \vec{u})^T)] + \nabla p = \vec{F} \tag{2}$$

*Energy equation*

$$\rho c_p \frac{\partial T}{\partial t} + \rho c_p \vec{u} \nabla T = \nabla.(k \nabla T) \tag{3}$$

where $\rho$ is density, $\vec{u}$ is velocity vector, $\mu$ is dynamic viscosity, $p$ is pressure, $T$ is temperature, $t$ is time, and $\vec{F}$ is the body force. For modeling the phase change process, Eq. (3) is modified using specific heat capacity formulation by Eq. (4) [6]. In the specific heat capacity formulation, the latent heat ($L$) term is added. The phase change process takes between $(T_m - \Delta T_m/2)$ to $(T_m + \Delta T_m/2)$ temperature. Here $\theta$ is defined as melt fraction, which is 0 for temperature less than $(T_m - \Delta T_m/2)$ and 1 for temperature more than $(T_m + \Delta T_m/2)$. The range of phase change temperature is taken $\Delta T_m = 2$ °C.

$$c_p = \frac{1}{\rho}[(1-\theta)\rho_{\text{phase1}} c_{\text{p, phase1}} + \theta \rho_{\text{phase2}} c_{\text{p, phase2}}] + L\frac{\delta \alpha}{\delta T} \tag{4}$$

The effective conductivity is calculated as [6]

$$k = (1-\theta)k_{\text{phase1}} + \theta k_{\text{phase2}} \tag{5}$$

The effective density is evaluated as

$$\rho = (1-\theta)\rho_{\text{phase1}} + \theta \rho_{\text{phase2}} \tag{6}$$

The mass fraction$\alpha$ is expressed as [6]:

$$\alpha = \begin{bmatrix} -\frac{1}{2}\left\{T < T_m - \frac{\Delta T_m}{2}\right\} \\ \frac{1}{2}\frac{\theta_{\rho\,\text{phase2}} - (1-\theta)\rho_{\text{phase1}}}{\rho}\left\{T_m - \frac{\Delta T_m}{2} < T < T_m + \frac{\Delta T_m}{2}\right\} \\ \frac{1}{2}\left\{T > T_m + \frac{\Delta T_m}{2}\right\} \end{bmatrix} \tag{7}$$

As the total PCM is treated as liquid always, the body force term of the Navier–Stokes equation is modified to calculate velocity of the PCM even if in solid state. One part of the body force as expressed in Eq. (8) is to cater buoyant force ($F_b$) which is responsible for natural convection, where Boussinesq approximation is used.

$$\vec{F}_{\mathrm{b}} = g\rho_1\beta(T - T_{\mathrm{m}}) \tag{8}$$

where $\rho_l$, $\beta$, $g$, and $T_{\mathrm{m}}$ are density of liquid PCM, thermal expansion coefficient, gravitational constant, and melting temperature, respectively. Along with this part of the body force, Carman–Kozeny equation [6] is also used to add one more volume force in the momentum equation as expressed in Eq. (9).

$$\vec{F}_{\mathrm{v}} = -A(T)\vec{u} \tag{9}$$

$$A(T) = A_{\mathrm{mush}}\frac{(1-\theta)^2}{(\theta^3+\varepsilon)} \tag{10}$$

where the constants $A_{\mathrm{mush}}$ and $\varepsilon$ are having the values of $10^5$and $10^{-3}$, respectively [6]. The term $\varepsilon$ is used to avoid division by 0 when $\theta$ is equal to zero. The value of $A(T)$ is very high in solid state of PCM, when melt fraction tends to zero [2]. But its value tends to 0 for liquid state of PCM. This volume force term actually makes the velocity approximately 0 in solid PCM and to a finite value in liquid region for solving the momentum equation. The viscosity as expressed in Eq. (11) is also modified for temperature less than $T_{\mathrm{m}} - \frac{\Delta T_{\mathrm{m}}}{2}$.

$$\mu(T) = \mu_{\mathrm{liquid}}(1 + A(T)) \tag{11}$$

Nusselt number gives a measure of approximate heat transfer from the heating surface. In the present study, the average Nusselt number for the side wall and upper/lower wall is expressed as Eqs. (12) and (13), respectively [9].

$$Nu_{\mathrm{avg}} = -\frac{1}{T_{\mathrm{h}} - T_{\mathrm{m}}}\int_0^H \left(\frac{\partial T}{\partial x}\right)_{x=x_{\mathrm{wall}}} \mathrm{d}y \tag{12}$$

$$Nu_{avg} = -\frac{1}{T_h - T_m}\int_0^H \left(\frac{\partial T}{\partial y}\right)_{y=y_{wall}} \mathrm{d}x \tag{13}$$

## 2.1 Boundary and Initial Condition

No-slip boundary condition at all solid walls, i.e., $u = v = 0$; isothermal boundary condition ($T = T_{\mathrm{w}}$) at the heated wall; rest of the walls are insulated, i.e., $\frac{\partial T}{\partial n} = 0$ where $n$ is the normal to the insulated wall; at $t = 0$, heated wall temperature $T_{\mathrm{w}} =$ 298 K.

## 2.2 Model Validation

The numerical model is validated with the experimental work [7] where melting of lauric acid inside a rectangular enclosure was considered. A comparative study is done on the variation of melt fraction with time for the left wall temperature of 333 K as shown in Fig. 2 which shows a good agreement. The maximum error in numerical results is less than 10%. Different assumptions such as melting takes place in 2 °C temperature range and constant thermophysical property are the reason of deviation of numerical results.

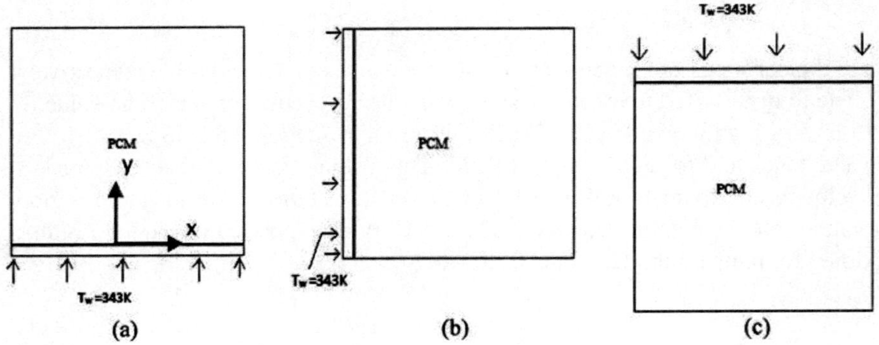


**Fig. 1** Schematic of PCM enclosure under different wall heating condition (**a**) bottom wall (*case* 1) (**b**) side wall (*case* 2) (**c**) upper wall (*case* 3)

**Fig. 2** Validation of numerical model

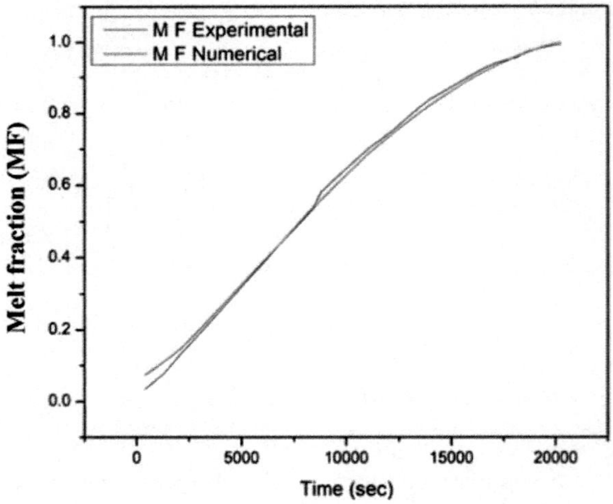

## 3 Results and Discussion

Figures 3 and 4 show the contour plots of melting front for all three cases at 1800s and 3600 s, respectively. It has been observed that for side wall heating the interface between liquid and solid PCM increases significantly at the later stage of melting. This increases the convective heat transfer significantly. But for the case of bottom wall, heating interface does not increase to that extent and the least rate of melting is observed for upper wall heating. The upper wall heating leads to thermal stratification as a result convective current is not generated and conduction is the only predominant mode of heat transfer for that case.

Figure 5 represents the melt fraction at different times for the considered all three cases. Heating from the side wall shows the highest melting rate among them, though at the beginning of melting process heating from bottom wall shows higher melting rate than that of heating from side wall. Because convective current encounters lesser restrictions at the beginning of convection process for bottom wall heating in comparison to side wall heating. The convective current is confined in the bottom portion of

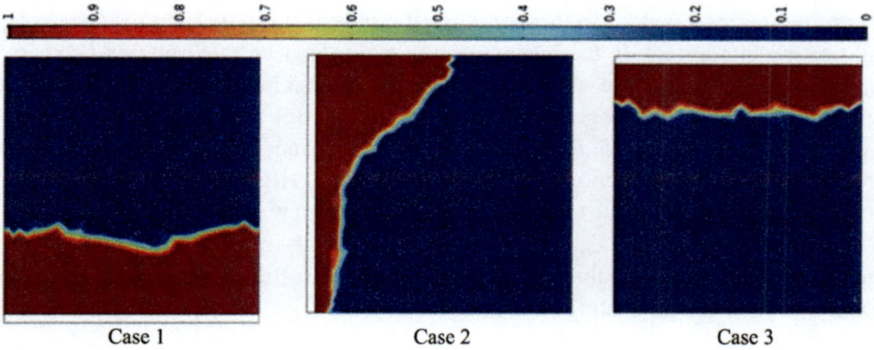


**Fig. 3** Liquid fraction contour plots for all the three cases at 1800s

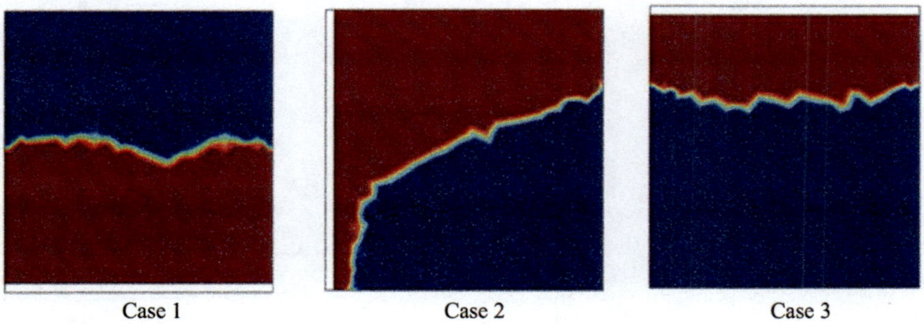


**Fig. 4** Liquid fraction contour plots for all the three cases at 3600 s

Fig. 5 Melt fraction variation with time for different heating conditions

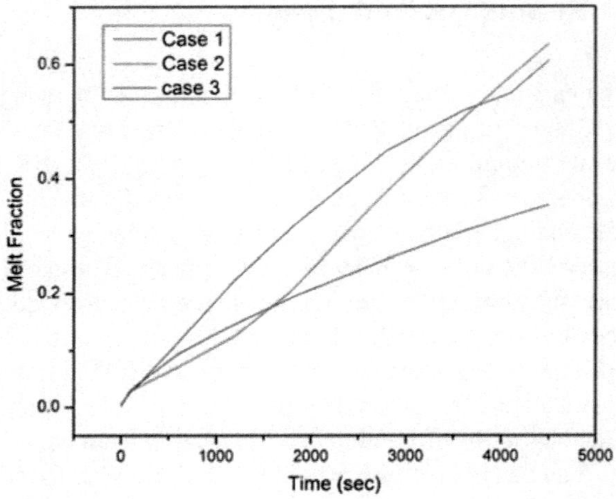


the container for bottom wall heating. On the contrary in case of side wall heating, convective current passes through the narrow channel produced by melting of PCM.

Figure 6 represents the variation of Nusselt number (*Nu*) with time for different wall heating conditions. In the heater surface, a thin concentrated melted layer is formed in the starting phase of melting which results in a higher near wall temperature gradient. So at the beginning, *Nu* is very high. But as the thickness of the layer increases with time *Nu* decreases [9]. As time advances, the melting process is regulated by both conduction and convection process. Higher *Nu* signifies higher convective heat transfer in the melting process. For bottom wall heating, *Nu* increases up to certain time and then decreases. On the other hand, for side wall heating, *Nu* attains a very low value and then increases as time passes. But consistently lower *Nu* is observed for top wall heating case.

Fig. 6 Nusselt number variation with time for different heating conditions

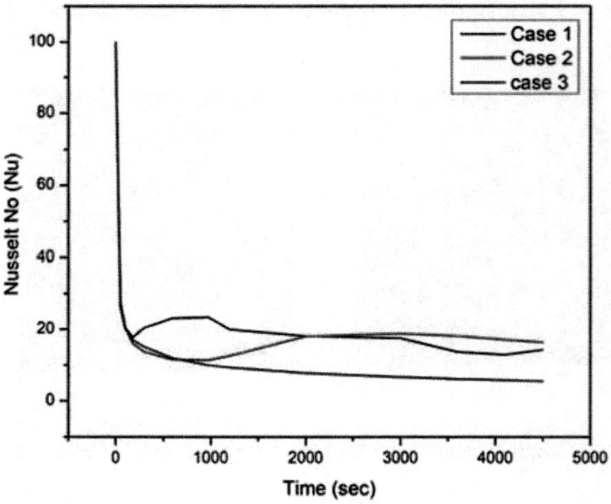

## 4 Conclusion

Two-dimensional analysis of PCM melting in different wall heating condition is analyzed numerically in the present study. Finite element method is used to solve the governing equations. Highest melting is observed in case of side wall heating. Though in beginning of melting process bottom wall heating gives highest melting rate, with the advancement of time, side wall melting overshoots than that of the bottom wall heating. On the other hand, top wall heating results in least melting advancement as only conduction is the predominant mode of heat transfer in this case.

## References

1. B. Kamkari, H.J. Amlashi, Numerical simulation and experimental verification of constrained melting of phase change material in inclined rectangular enclosures, International Communications of. Heat Mass Transf. **88**, 211–219 (2017)
2. P.H. Biwole, D. Groulx, F. Souayfane, T. Chiu, Influence of fin size and distribution on solid-liquid phase change in a rectangular enclosure. Int. J. Therm. Sci. **124**, 433-446 (2018)
3. C. Ji, Z. Qin, Z. Low, S. Dubey, F.H. Choo, F. Duan, Non-uniform heat transfer suppression to enhance PCM melting by angled fins. Appl. Therm. Eng. **129**, 269–279 (2018)
4. A.K. Asl, S. Hossainpour, M.M. Rashidi, M.A. Sheremet, Z. Yang, Comprehensive investigation of solid and porous fins influence on natural convection in an inclined rectangular enclosure. Int. J. Heat Mass Transf. **133**, 729–744 (2019)
5. S. Ebadi, S.H. Tasnim, A.A. Aliabadi, S. Mahmud, Melting of nano-PCM inside a cylindrical thermal energy storage system: numerical study with experimental verification. Energy Conserv. Manage. **166**, 241–259 (2018)
6. S. Arena, E. Casti, J. Gasia, L.F. Cabeza, G. Cau, Numerical analysis of a latent heat thermal energy storage system under partial load operating conditions. Renew. Energy **128**, 350–361 (2018)
7. H. Shokouhmand, B. Kamkari, Experimental investigation on melting heat transfer characteristics of lauric acid in a rectangular thermal storage unit. Exp. Therm. Fluid. Sci. **50**, 201–212 (2013)
8. M. Dhar, N. Barman, S. Mandal, H. Chattopadhyay, Remelting and interface dynamics during solidification of a eutectic solution in a top-cooled rectangular cavity: a numerical study. Int. J. Heat Mass Transf. **77**, 730–737 (2014)
9. M. Alomair, Y. Alomair, S. Tasnim, S. Mahmud, H. Abdullah, Analyses of Bio-based nano-PCM filled concentric cylindrical energy storage system in vertical orientation. J. Energy Storage **20**, 380–394 (2018)

# Numerical Modelling and Thermohydraulic Analysis of Circular Pipe Having Internal Vortex Generators

**Binay Bhushan Bora, Mohd Zeeshan, Sujit Nath, and Dipankar Bhanja**

**Abstract** In the present work, a numerical investigation has been performed to investigate the thermohydraulic performance of a circular pipe used in a heat exchanger with and without insertion of vortex generators. Vortex generators (VGs) which are used inside the pipe having dimensions of blockage ratio 0.3 and attack angle of 0°. The investigation has been carried out for Reynolds number (*Re*) ranges from 6000 to 24,000. Two cases of implementing vortex generators namely two VGs configuration and four VGs configuration have been investigated and compared with a smooth pipe which is of the same geometric parameters. Nusselt number (*Nu*) and friction factor (*f*) have been considered for representing the thermal and hydraulic characteristics, respectively. The overall performance of the two VGs and four VGs configuration has been calculated using the thermal performance enhancement factor (TPE). It has been found from the results that using VGs, heat transfer rate of smooth pipe can be improved while pressure drop also increased compared to the smooth pipe. With the increase in *Re*, *Nu* of all the cases increased with a significant decrement in friction factor (*f*). The overall performance in terms of TPE is maximum for four VG configuration at Reynolds number of 6000.

**Keywords** Circular pipe · Vortex generators · Thermal performance enhancement

## Nomenclature

| | |
|---|---|
| $A$ | Heat transfer area ($m^2$) |
| $B$ | Blockage ratio (H/D) |
| $c_p$ | Specific heat at constant pressure (kJ /kg K) |
| $D$ | Pipe inner diameter (m) |
| $f$ | Friction factor |
| $H$ | Height of the winglet (m) |

B. B. Bora (✉) · M. Zeeshan · S. Nath · D. Bhanja
Department of Mechanical Engineering, National Institute of Technology Silchar, Silchar, India
e-mail: binaybhushanbora1995@gmail.com



K. M. Pandey et al. (eds.), *Recent Advances in Mechanical Engineering*, Lecture Notes in Mechanical Engineering, https://doi.org/10.1007/978-981-15-7711-6_49

$h$ Heat transfer coefficient (W/m$^2$K)
$k$ Air thermal conductivity (W/m K)
$L$ Pipe length (m)
$Nu$ Nusselt number
$Pr$ Prandtl number
$\Delta p$ Pressure drop (Pa)
$Re$ Reynolds number
$U$ Mean axial velocity (m/s)

## Greek Symbols

$\beta$ Attack angle (°)
$\alpha$ Thermal diffusivity (m$^2$/s)
$\rho$ Fluid density (kg/m$^3$)
$\mu$ Dynamic viscosity (Pa s)
$\nu$ Kinematic viscosity (N/m$^2$)
$\Gamma_k$ Effective diffusivity of $k$
$\Gamma_\omega$ Effective diffusivity of $\omega$

## Subscripts

$k$ Turbulence kinetic energy
$\omega$ Specific dissipation rate
$o$ Smooth pipe

## 1 Introduction

Application of heat exchangers (HX) is very wide in various fields like power plants, chemical plants, refrigeration and air conditioning, automobiles, etc. Therefore, the improvement of HX in terms of heat transfer is very important to develop an HX of high performance and high efficiencies but with the low cost and lightweight. Vortex generator (VGs) as a passive heat transfer improvement technique has a big role in improving the heat transfer of an HX. The VGs can be of wings or winglets type for various shapes like rectangular, delta, curved, trapezoidal, etc. which are projected into the flow with an attack angle to the flow direction. There are two types of vortices, produce by VGs which are called longitudinal and transverse. Flow direction and vortex are parallel in the case of longitudinal vortices while they are perpendicular in the case of transverse vortices.

Xu et al. [1], numerically, investigated the wall friction with heat transfer of VGs inside a pipe for different configurations and determined the best configuration of VGs. Thermal enhancement factor (TEF) and flow behaviour of VG inserted circular pipe were determined for turbulent flow and TEF decreases with the increasing *Re* [2–4]. Promvonge et al. [5] did an experiment to find the thermal performance of vortex rings (VR) inside a heat exchanger (HX) pipe for various blockage ratio (B) and pitch ratio. Thermohydraulic performance increases with increasing *B* and decreasing pitch ratio. Yakut et al. [6] did an experimental investigation using conical ring turbulators inside HX pipe for turbulent flow and maximum heat transfer was found for the smallest pitch of conical rings. Performance of central slant rod inside a pipe was investigated, numerically by Liu et al. [7] for laminar flow and both Nu and f increases when compared to a plain pipe. A numerical investigation for heat transfer and pressure drop evaluation of ribbed channel inside a high-temperature HX was done by Ma et al. [8] and it was found that heat transfer is more for larger rib heights but inlet temperature has more impact on increasing heat transfer as compared to rib heights. Islam et al. [9] did an experiment for investigating the effect of rectangular plate fins of four different patterns which are named co-rotating, co-angular, zigzag and co-counter rotating in a duct with Re ranges from 15,700 to 104,500. Thermo-hydraulic performance of rectangular duct having different shapes ribs was experimentally investigated and results were concluded that transverse ribs give better heat transfer performance [10, 11]. Numerical study for heat transfer enhancement of an HX tube using different types of twisted tape was done by researchers and the best types of twisted tape with proper configuration were determined [12, 13].

Most of the researches for HX pipe has been done using various insets like helical rods, circular rings, twisted tape, conical rings, etc. In this work, the thermo-hydraulic performance of vortex generators in a circular pipe of HX has been studied numerically for two configurations of VGs. The two configurations are two VGs configuration where two delta winglet VGs are inserted in the pipe surface and four VGs configuration where four delta winglet VGs are inserted in the pipe surface. A comparison of two VGs and four VGs configuration has not been noticed in the literature review which includes in this work.

## 2 Description of the Model

### *2.1 Physical Model*

Two types of vortex generator (VG) arrangements are considered in this work which is named as two VGs configuration and four VGs configuration as shown in Fig. 1. The vortex generators have been mounted on the surface of a circular pipe having a diameter (D) and length (L) of 50 mm and 0.28 m respectively, from the inlet with a distance of 40 mm. Blockage ratio (B) of the vortex generator which can be defined

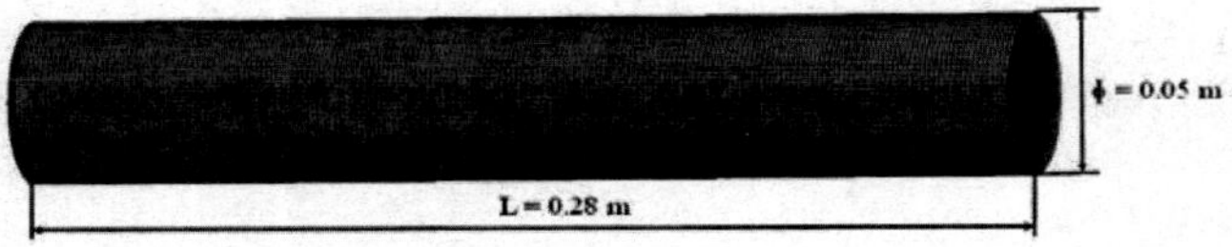


(a) Model framework and generated mesh for smooth tube

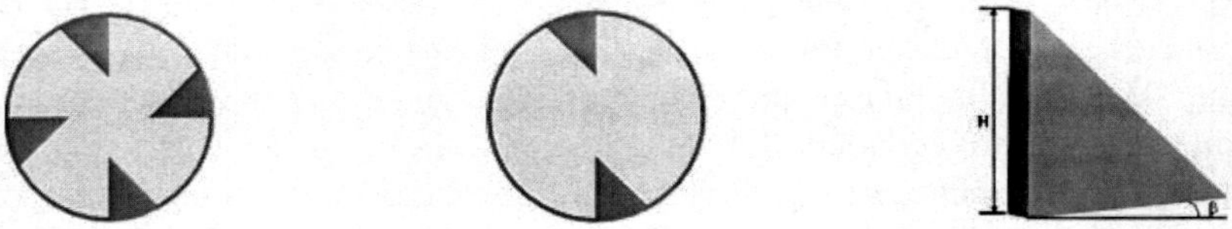


(b) Two VG pipe (front view) (c) Four VG pipe (front view) (f) VGs geometry

**Fig. 1** Geometry, mesh and placement of VGs inside the pipe

as the ratio between the height of VG (H) and pipe diameter (D) is considered as 0.3 and attack angle (β) of VGs is equal to 0° as shown in Fig. 1f.

## 2.2 *Mathematical Modelling*

The three dimensional, Steady, incompressible, and turbulent flow have been considered for developing the numerical model to describe the flow and thermal behaviour of the proposed problem. FLUENT (ANSYS) has been used for solving the model and a turbulence model, SST $k$–$\omega$ was imposed [14]. The governing equations which were used for solving the problem are given below:

Continuity equation:

$$\frac{\partial u_i}{\partial x_i} = 0 \tag{1}$$

Momentum equation:

$$\frac{\partial(\rho u_i u_j)}{\partial x_i} = \frac{\partial}{\partial x_i}(\mu \frac{\partial u_j}{\partial x_i} - \rho \overline{u_i' u_j'}) - \frac{\partial p}{\partial x_j} \tag{2}$$

Energy equation:

$$\frac{\partial(\rho u_i T)}{\partial x_i} = \frac{\partial}{\partial x_i}\left(\frac{\lambda}{c_p}\frac{\partial T}{\partial x_i}\right) \tag{3}$$

*k*-equation:

$$\frac{\partial(\rho k u_i)}{\partial x_i} = \frac{\partial}{\partial x_j}\left(\Gamma_k \frac{\partial k}{\partial x_j}\right) + G_k - Y_k \quad (4)$$

ω-equation:

$$\frac{\partial(\rho \omega u_i)}{\partial x_i} = \frac{\partial}{\partial x_j}\left(\Gamma_\omega \frac{\partial \omega}{\partial x_j}\right) + G_\omega - Y_\omega \quad (5)$$

where, $\Gamma_k = \mu + \mu_t/\sigma_k$; $\Gamma_\omega = \mu + \mu_t/\sigma_\omega$ and $\mu_t = \frac{\rho k}{\omega}\frac{1}{\max[\frac{1}{\alpha^*}, \frac{SF_2}{a_1\omega}]}$.

Also, $\sigma_k = \frac{1}{F_1/\sigma_{k,1} + (1-F_1)/\sigma_{k,2}}$ and $\sigma_\omega = \frac{1}{F_1/\sigma_{\omega,1} + (1-F_1)/\sigma_{\omega,2}}$.

In the above equations, $G$ and $Y$ indicates generation and dissipation terms respectively; $\Gamma$ and $\mu_t$ represents effective diffusivity and turbulence viscosity respectively; $S$ is the strain rate magnitude and $F_1$, $F_2$ represents blending functions. Also $\alpha^*$, $\alpha_1$, $\sigma_{k,1}$, $\sigma_{k,2}$, $\sigma_{\omega,1}$ and $\sigma_{\omega,2}$ are model constants. A detailed description of the governing equation can be found in Ref. [14].

## 2.3 Boundary Conditions (B.C.)

The boundary conditions used to solve the numerical problem are given below:

(a) Inlet: Velocity inlet (Fully developed flow) $Re$ = 6000 to 24,000, Inlet temperature = 300 K.
(b) Pipe surface: Solid Wall of constant heat flux ($q$ = 694 W/m$^2$) and at outlet, outflow B.C. has been applied.
(c) The intensity of turbulence which is equal to $0.16 \times \mathrm{Re}^{-1/8}$ was used for different Reynolds numbers [1].
(d) The working fluid is considered as an air with following properties: $Pr = 0.73$, $\rho = 1.16$ kg/m$^3$, $K = 0.02588$ W/m K, $\mu$ = 1.87e-05 Pa S, $C_p$ = 1007 J/kg K.

## 2.4 Parameter Definition

Different parameters that are required for determining heat transfer performance for VGs inside a circular pipe have been discussed in this section.

The expression for local and area average Nusselt number:

$$Nu_x = h_x D/k \quad (6)$$

and

$$Nu = \frac{1}{A}\int Nu_x \mathrm{d}A \tag{7}$$

The expression for Reynolds number:

$$Re = UD/\nu \tag{8}$$

The expression for f with pressure drop for internal flow:

$$f = 2\Delta p/(L/D)\rho U^2 \tag{9}$$

Thermal performance enhancement (TPE) [15] can be expressed as the ratio of heat transfer coefficient for VG inserted pipe and heat transfer coefficient of smooth pipe.

$$TPE = (Nu/Nu_o)/(f/f_o)^{1/3} \tag{10}$$

### 2.5 *Numerical Modelling*

The three-dimensional models of the interest have been developed in WORKBENCH DESIGN MODELLER. The mesh has been generated in WORKBENCH (ICEM CFD Meshing). The grid has been made finer by using body sizing operation. The prescribed governing equation has been solved by commercial software package FLUENT with the SIMPLE algorithm for velocity and pressure coupling. The quick scheme is used for solving the momentum equation while the upwind scheme of second-order is employed for solving turbulent kinetic energy and specific dissipation rate.

## 3 Results and Discussion

### 3.1 *Grid Independence Test and Validation for Numerical Results*

In order to make the numerical results free from the grid effect, the gid independence test has been done for a smooth pipe with $Re = 9000$ as shown in Fig. 2. It can be seen in Fig. 2 that $Nu$ does not vary more than 1% for the number of cells more than 426,000. As the difference is very insignificant and to reduce the computational efforts the simulation has been performed for a mesh with the number of cells equal to 426,000.

**Fig. 2** Grid independence test for $Re = 9000$

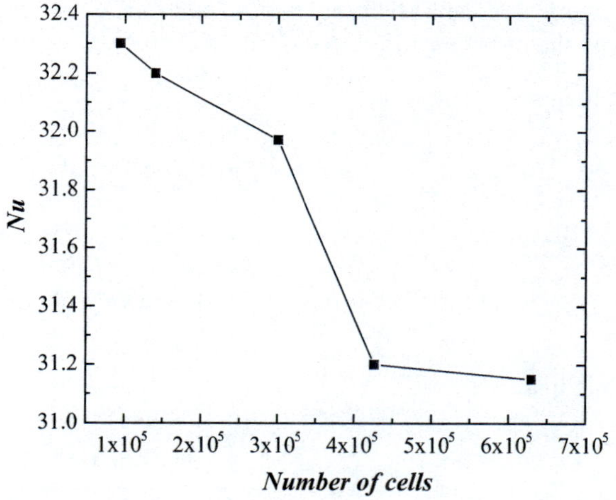


The current numerical results are compared with existing results in order to validate the numerical model. A comparison of *Nu* and *f* for the smooth pipe of the current model is presented in Fig. 3. The friction factor (*f*) is compared with the results of Xu et al. [1] and the empirical correlations which are given by Blasius and Petukhov [16]. Whereas, *Nu* is compared with the results of Xu et al. [1] and empirical correlations which are given by Dittus-Boelter and Gnielinski [16]. The average differences in the results of friction factor (*f*) are 6.23, 18, and 17% compared to Xu et al. [1], Petukhov and Blasius [16] correlation, respectively. Also, the average differences in the results of *Nu* are 6.13, 6.67, and 5.11% when compared to Xu et al. [1], Gnielinski and Dittus-Boelter [16] correlation respectively. Though there exists a difference between the present numerical results and the existing results due to the difference of idealization in numerical modelling. The fairly good agreement between the current results and existing results shows the reliability of the present

**Fig. 3** Validation for smooth pipe

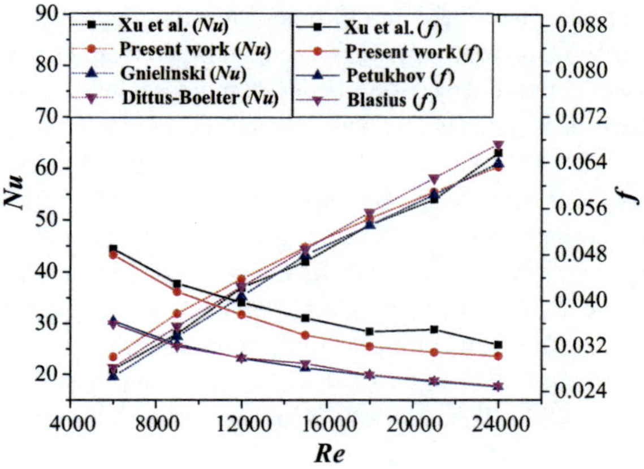

numerical model. All the equations which are given below are valid for smooth walls with the turbulent-fully developed flow.

Petukhov:

$$f = (0.79 \ln Re - 1.64)^{-2} \tag{11}$$

and Blasius:

$$f = 0.316 Re^{-0.25} \tag{12}$$

Gnielinski:

$$Nu = \frac{\left(\frac{f}{8}\right)(Re - 1000)Pr}{1 + 12.7(f/8)^{1/2}(Pr^{2/3} - 1)} \tag{13}$$

and Dittus-Boelter:

$$Nu = 0.023 Re^{0.8} Pr^{0.4} \tag{14}$$

### 3.2 *Effect of VGs on* Nu, F, *and TPE*

The implementation of VGs in the flow domain considerably increases the heat transfer rate by creating the swirls/vortices in the flow domain due to the pressure difference of upstream and downstream regions. Velocity streamlines have been plot in the cross-sectional plane for smooth pipe and VGs pipe at a distance of 30 and 50 mm from the inlet i.e. before and after VGs which are shown in Fig. 4.

It is clear from Fig. 4 that VGs are creating turbulence in the flow region on comparing with the smooth pipe. Figure 4c, e, and d, f represents before VGs and after VGs, respectively. It can be concluded from Fig. 4 that the velocity magnitude, as well as vortices created by VGs, increases with the increasing no. of VGs due to which heat transfer rate will also be increased. The effect of VGs on *Nu* for *Re* of 6000–24,000 is shown in Fig. 5. It has been seen that *Nu* for VGs pipe increases when compared with the smooth pipe. Four VGs configuration gives a higher increment in *Nu* compared to two VGs configuration with 11%. An increase in *Nu* is more with increasing *Re* because high *Re* tends to more turbulence. The average *Nu* increases for two VGs configuration is 10.8% while four VGs configuration increases 23% of average *Nu* when compared to the smooth pipe.

Effect of VGs on $f$ for VGs of $B = 0.3$ and $\beta = 0°$ are shown in Fig. 5 for $Re =$ 6000–24,000. The value of $f$ is higher for two VGs and four VGs configuration when compared to the smooth pipe. It can be said from the results that the friction factor

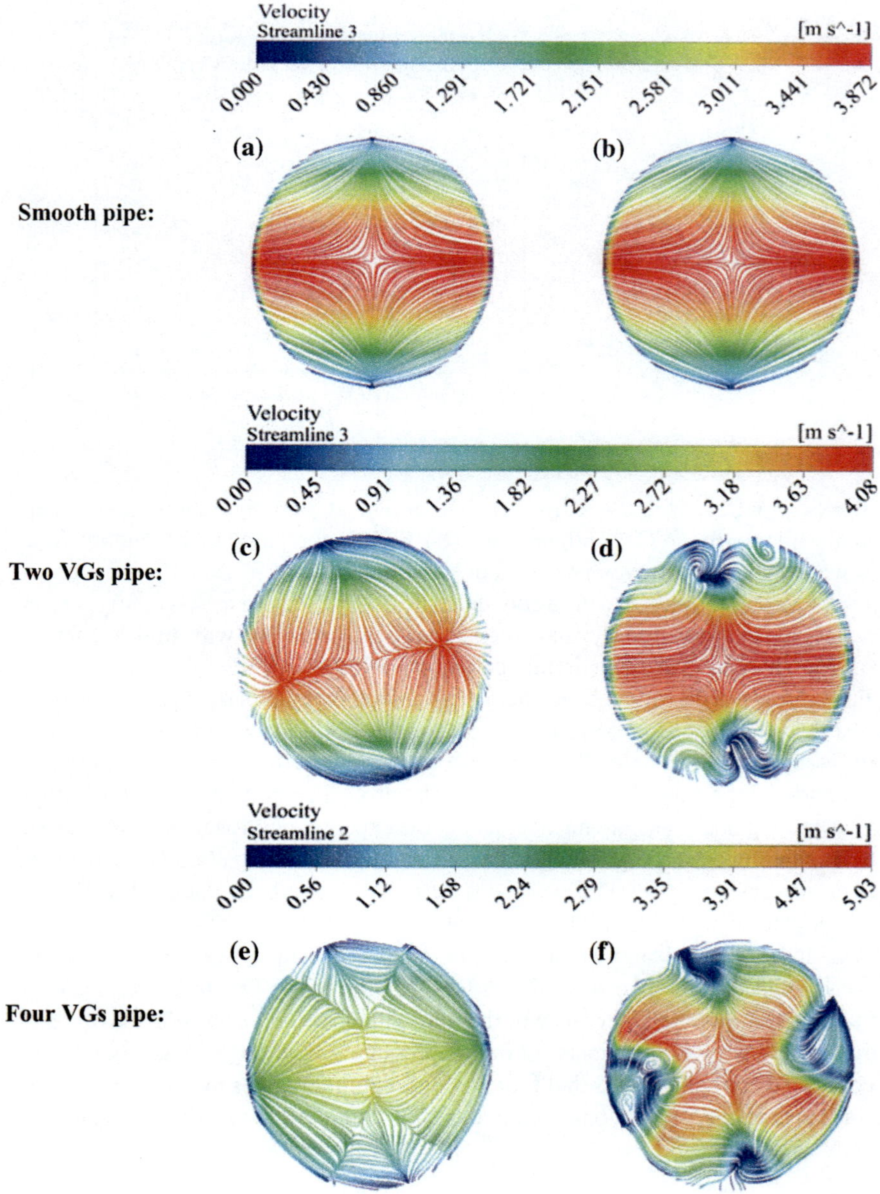


**Fig. 4** Representation of velocity streamlines at two cross-sectional planes where **a**, **b** for smooth pipe; **c**, **e** for before VGs and **d**, **f** for after VGs at $Re = 12{,}000$

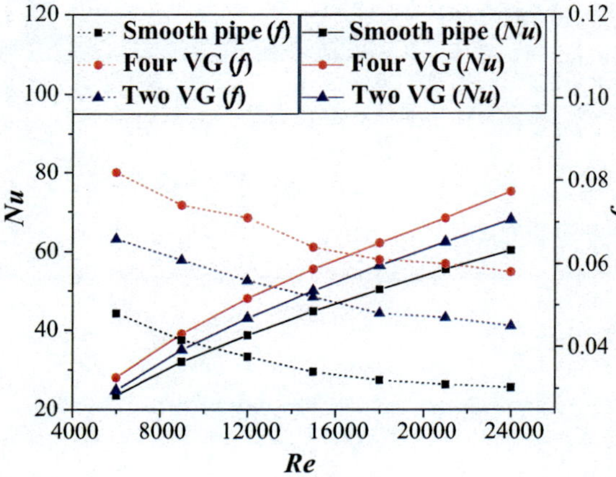


**Fig. 5** VGs effect on *Nu* and *f*

decreases with increasing *Re*. The average *f* for two VGs configuration increases up to 46% while for four VGs configuration is 84% when compared to the smooth pipe. However, the increase in average *f* for four VGs configuration is 25% when compared to two VGs configuration. It can be concluded from Fig. 5 that four VGs configuration provides better thermal performance characteristics (*Nu*) but with higher friction factor (*f*) which lowers the hydraulic performance.

Furthermore, a performance evaluation criterion, thermal performance enhancement (TPE) has been introduced to evaluate the performance of the considered configurations. TPE is defined as the ratio of thermal performance characteristics (*Nu*) divide by the flow characteristics (*f*). Therefore TPE provides detail information on gaining heat transfer enhancement at the expanse additional energy required to overcome the excess $\Delta p$ which is caused due to the use of VGs. The effects of VGs on increasing *Re* for TPE = $(Nu/Nu_o)/(f/f_o)^{1/3}$ have been depicted in Fig. 6. TPE is greater than 1 for both the cases which indicate the overall performance of the VGs inserted pipe is more than the smooth pipe. It can also be seen in Fig. 6 that the performance of the two VGs pipe is less than the four VGs configuration. *TPE* is maximum at $Re = 6000$ because of the minimum pressure drop. Whereas TPE is minimum at $Re = 24{,}000$ because of higher pressure drop though *Nu* is large. It can be concluded from the results that four VGs configuration gives the highest thermal performance enhancement compared to two VGs configuration for $B = 0.3$ and $\beta = 0°$ at $Re = 6000$.

## 4 Conclusion

A numerical investigation for enhancement of thermal–hydraulic performance of an HX pipe with the insertion of VGs was carried out. Two VGs and four VGs configurations were considered for the simulation and results were compared with

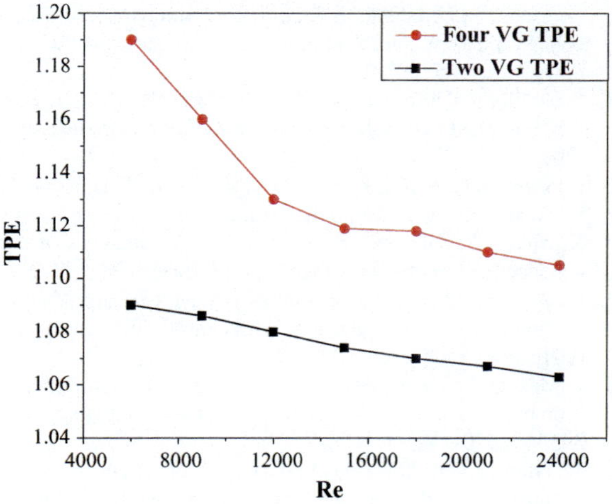


**Fig. 6** VGs effect on *TPE*

the smooth pipe. The thermal performance enhancement (TPE) has been considered for overall performance evaluation. Results can be encapsulated after evaluation of performance measuring parameters as follows:

- Heat transfer is increasing after implementing the VGs inside the pipe where four VG configuration gives better improvement compared to two VGs configuration.
- *Nu* increases for all the configuration of VGs with the increasing *Re* while *f* decreases with the increasing *Re*.
- Average *Nu* increases for four VGs configuration is 23% compared to the smooth pipe while for two VGs configuration increment is 10.8%. Average *f* increases for two VGs and four VGs configuration is 46% and 84% respectively when compared to the smooth pipe.
- TPE is maximum for both two VGs and four VGs configuration at $Re = 6000$ and minimum at $Re = 24{,}000$. The current study by introducing different attack angles, blockage ratio, and placements of VGs, further research can be done and heat exchanger of better performance may be designed.

**Acknowledgements** Authors thank TEQIP-III (N.I.T. Silchar, Assam) for their financial support. We also thankful to the CFD lab, M.E. Department (N.I.T. Silchar, Assam), for providing the computational facility in carrying out the present numerical investigation.

## References

1. Y. Xu, M.D. Islam, N. Kharoua, Numerical study of winglets vortex generator effects on thermal performance in a circular pipe. Int. J. Therm. Sci. **112**, 304–317 (2017)
2. C. Zhai, M.D. Islam, R. Simmons, I. Barsoum, Heat transfer augmentation in a circular tube with delta winglet vortex generator pairs. Int. J. Therm. Sci. **140**, 480–490 (2019)

3. G. Liang, M.D. Islam, N. Kharoua, R. Simmons, Numerical study of heat transfer and flow behavior in a circular tube fitted with varying arrays of winglet vortex generators. Int. J. Therm. Sci. **134**, 54–65 (2018)
4. S. Skullong, P. Promvonge, C. Thianpong, N. Jayranaiwachira, Thermal behaviors in a round tube equipped with quadruple perforated-delta-winglet pairs. Appl. Therm. Eng. **115**, 229–243 (2017)
5. P. Promvonge, N. Koolnapadol, M. Pimsarn, C. Thianpong, Thermal performance enhancement in a heat exchanger tube fitted with inclined vortex rings. Appl. Therm. Eng. **62**, 285–292 (2014)
6. K. Yakut, B. Sahin, Flow-induced vibration analysis of conical rings used for heat transfer enhancement in heat exchangers. Appl. Energy **78**, 273–288 (2004)
7. P. Liu, N. Zheng, F. Shan, Z. Liu, W. Liu, Numerical study on characteristics of heat transfer and friction factor in a circular tube with central slant rods. Int. J. Heat Mass. Transf. **99**, 268–282 (2016)
8. T. Maa, Q. Wang, M. Zeng, Y. Chen, Y. Liu, V. Nagarajan, Study on heat transfer and pressure drop performances of ribbed channel in the high temperature heat exchanger. Appl. Energy **99**, 393–401 (2012)
9. M. Islam, K. Oyakawa, M. Yaga, Heat transfer enhancement from a surface affixed with rectangular fins of different patterns and arrangement in duct flow. J. Enhanc. Heat Transf. 15 (2008)
10. S. Singh, S. Chander, J.S. Saini, Investigations on thermo-hydraulic performance due to flow-attack-angle in V-down rib with gap in a rectangular duct of solar air heater. Appl. Energy **97**, 907–912 (2012)
11. G. Tanda, Heat transfer in rectangular channels with transverse and V-shaped broken ribs. Int J Heat Mass Transf **47**, 229–243 (2004)
12. P. Li, Z. Liu, W. Liu, G. Chen, Numerical study on heat transfer enhancement characteristics of tube inserted with centrally hollow narrow twisted tapes. Int J Heat Mass Transf **88**, 481–491 (2015)
13. J. Guo, A. Fan, X. Zhang, W. Liu, A numerical study on heat transfer and friction factor characteristics of laminar flow in a circular tube fitted with centercleared twisted tape. Int. J. Therm. Sci. **50**, 1263–1270 (2011)
14. ANSYS Inc. Fluent user guide and fluent theory guide, version 14.1 (2011)
15. P. Promvonge, N. Koolnapadol, M. Pimsarn, C, Thianpong. Thermal performance enhancement in a heat exchanger tube fitted with inclined vortex rings. Appl. Therm. Eng. **62,** 285–292 (2014)
16. P.P. Incropera, D.P. Dewitt, T.L. Bergman, A.S. Lavine. *Principles of Heat and Mass Transfer* (Wiley, Hoboken, 2011)

# Effect of Non-uniform Heating on Electroosmotic Flow Through Microchannel

**T. Sujith, Sumit Kumar Mehta, and Sukumar Pati**

**Abstract** The effect of non-uniform heating on the heat transfer characteristics for electroosmotic flow through a microchannel has been investigated numerically. The temperature field and Nusselt number are studied by changing the normalized wavelength of non-uniform heat flux ($\gamma$) and thermal Peclet number (Pe) in the range of $1.5 \leq \gamma \leq 6$ and $1 \leq \mathrm{Pe} \leq 100$, respectively. It is found that the intensity of maximum temperature reduces for non-uniform heating as compared to the uniform heating. The maxima of local Nusselt number increases with a decrease in the wavelength of the non-uniform heat flux. The critical Peclet number ($\mathrm{Pe_c}$) is found such that average Nusselt number shows the monotonic and non-monotonic variation with $\gamma$.

**Keywords** Electroosmotic flow · Peclet number · Non-uniform heating · Nusselt number

## 1 Introduction

Electroosmotic flow (EOF) has several advantages over the pressure-driven flow (PDF) in the microchannel. The micro-electro-mechanical systems (MEMS) gain more importance in recent years and these types of devices are used in many biochemical and biomedical industries. For the micro-level flow, interfacial interaction is important as the surface to volume ratio is high. Therefore, the mechanical pump may fail for such systems. To overcome these problems, electroosmotic flow is the

T. Sujith (✉) · S. K. Mehta · S. Pati
Department of Mechanical Engineering, National Institute of Technology Silchar, Silchar 788010, Assam, India
e-mail: sujitht996@gmail.com

S. K. Mehta
e-mail: sumit090391@gmail.com

S. Pati
e-mail: sukumarpati@gmail.com



K. M. Pandey et al. (eds.), *Recent Advances in Mechanical Engineering*, Lecture Notes in Mechanical Engineering, https://doi.org/10.1007/978-981-15-7711-6_50

best solution option for such transport [1, 2]. The thermal transport characteristics for EOF through microchannel have got immense interest from research community in the past few years. Dey et al. [3] investigated the effect of steric effect on heat transfer characteristics for EOF through a microchannel and found that the steric factor alters the results drastically when the zeta potential is very high as compared to point charge assumption. For many applications of micro-level-flow in heat sinks like CPUs and ICs, the discrete heat generation occurs due to the micro-arrangement of electronics components [4–6]. Azari et al. [7] investigated the distributed wall heat flux on the heat transfer characteristic for combined EOF and PDF in micro-slit. It is found that the Nusselt number decreases with EDL thickness and PDF velocity for any wall heat flux distribution. Pati et al. [8] used different heat flux profiles to identify the optimal heating strategy for minimization of peak temperature and entropy generation for forced convective flow through a circular pipe. From the available work in literature, it is found that the effect of non-uniform heating for EOF needs to be explained in detail, as the effects of wavelength or undulation of hot spots are not investigated till now. Hence, the objective of this work is to investigate the effect of wavelength of non-uniform heating on the temperature field and heat transfer characteristics for EOF through a microchannel.

## 2 Theoretical Formulation

The EOF of a Newtonian fluid through a microchannel is considered. The non-uniform heat flux $q(x) = 0.5q_o[1+\sin(2\pi x/\lambda)]$ is imposed at the walls [5] as shown in Fig. 1. The applied electric field strength is $E$. It is assumed that the flow is steady, laminar, hydrodynamically fully developed, and incompressible and thermo-physical properties are temperature independent. The charge is considered as a point, as well as, zeta potential is assumed to be uniform at the wall for the entire length. The convective term is neglected in the momentum equation as the Reynolds number is very small. The viscous dissipation term is also neglected for the energy equation. Under these assumptions, for single valance ions (e.g. KCl, NaCl, etc.), transport equation can be written as [3, 7]:

$$\frac{\mathrm{d}^2\psi}{\mathrm{d}Y^2} = \kappa^2 \sinh(\psi) \tag{1}$$

$$0 = -\frac{\partial^2 U}{\partial Y^2} + \frac{\kappa^2}{\zeta} \sinh(\psi) \tag{2}$$

$$\mathrm{Pe}\, U\frac{\partial \theta}{\partial X} = \frac{\partial^2 \theta}{\partial X^2} + \frac{\partial^2 \theta}{\partial Y^2} + G \tag{3}$$

Equations (1) to (3) represent the Poisson-Boltzmann equation for EDL potential, momentum equation, and energy equation, respectively. Here, $\psi$, $U$ and $\theta$ are

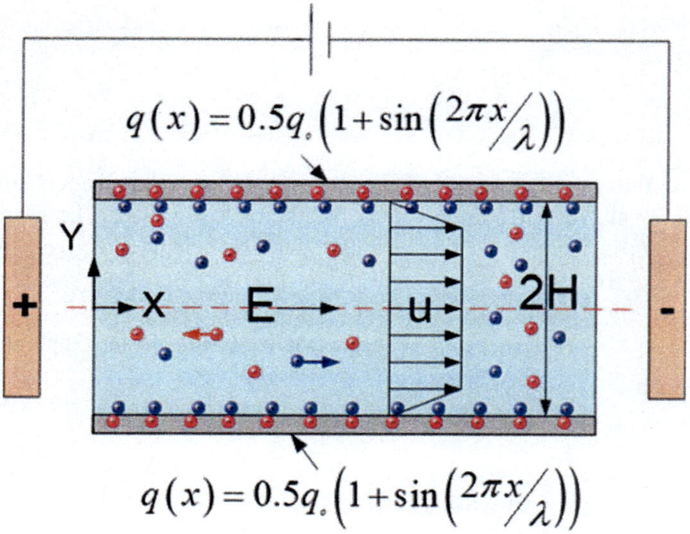


**Fig. 1** Physical domain

normalized EDL potential, flow velocity, and dimensionless temperature, respectively, $X$ and $Y$ are the dimensionless axial and transverse co-ordinate normalized with $H$. The other normalized parameters are: $\psi = \psi^*/\zeta_{\text{ref}}$, $U = u/u_{HS}$, $\theta = (T - T_{\text{in}})k/q_o H$, Peclet number is $\text{Pe} = u_{HS}H/(k/\rho c_p)$, Joule heating term is $G = HE^2/q_o\sigma$. Reference velocity is $u_{HS}(= -\varepsilon\varepsilon_o E\xi_{\text{ref}}/\mu)$, $\zeta_{\text{ref}}$ is the reference zeta potential, dimensionless Debye parameter is $\kappa = H\left(2n_o z^2 e^2/\varepsilon\varepsilon_o k_B T\right)^{0.5}$. Note that $\mu,\ \rho,\ c_p,\ k,\ \sigma,\ k_B,\ z, e$ and $\varepsilon_o$ are the dynamic viscosity, density, specific heat capacity, thermal conductivity, electric conductivity, Boltzmann constant, valency, the charge on single electron and permittivity of the free space, respectively.

To solve the above transport equations, the imposed boundary conditions are the following [7]:

At inlet

$$\frac{d\psi}{dX} = 0,\ \ \frac{\partial U}{\partial X} = 0,\ \theta = 0 \tag{4}$$

At wall ($Y = 1$):

$$\psi = \zeta,\ U = 0,\ \frac{\partial \theta}{\partial Y} = 0.5(1 + \sin(2\pi X/\gamma)) \tag{5}$$

At outlet:

$$\frac{d\psi}{dX} = 0,\ P_G = 0,\ \frac{\partial \theta}{\partial X} = 0 \tag{6}$$

At centerline ($Y = 0$):

$$\frac{d\psi}{dY} = 0, \ \frac{dU}{dY} = 0, \ \frac{\partial\theta}{\partial Y} = 0 \tag{7}$$

where $P_G$ is the gauge pressure. The heat transfer rate can be calculated in terms of the local Nusselt number ($Nu$). Mathematically, it can be represented as: $\mathrm{Nu} = hH/k$. Here, $h$ is the heat transfer coefficient and given by $h = q(x)/(T_\mathrm{W} - T_\mathrm{m})$ [4, 7], where $T_\mathrm{W}$ and $T_\mathrm{m}$ is the wall and bulk mean temperature, respectively. Hence, local Nusselt number can be expressed as: $\mathrm{Nu} = [0.5(1 + \sin(2\pi X/\gamma))]/(\theta_\mathrm{W} - \theta_\mathrm{m})$. The normalized bulk mean temperature can be represented as: $\theta_\mathrm{m} = \int_0^1 \theta U \mathrm{d}Y \Big/ \int_0^1 U \mathrm{d}Y$.

## 3 Numerical Method and Validation

The finite element method is used to solve the governing transport equations. The domain under consideration is divided into smaller subdomains of triangular geometry of non-uniform size. These subdomains are known as elements. The flow variables are approximated by using proper interpolation functions within each subdomains. Thus, the governing equations are converted into closed-form, which are then solved using iterative techniques. The relative convergence criterion for the residuals of all the transport variables is set to $10^{-6}$. The details of the numerical procedure can be seen in [9, 10]. The grid independency has been done for the present study and the maximum difference in calculating Nusselt number for the selected mesh is less than 0.1% as compared to very fine mesh. For the selected fine mesh, number of nodes and elements are 35,500 and 68,614, respectively. Before presenting the findings of the current investigation, we first ascertain the accuracy of the numerical scheme used to stimulate the results. For validation purpose, we consider electroosmotic flow through a square microchannel similar to the work of Hsieh et al. [11]. They have conducted an experimental study on the velocity distribution of EOF using micro-particle velocimetry. Figure 2 depicts the comparison of velocity distribution for the present study and by Hsieh et al. [11]. The comparison is shown for electroosmotic flow of TAE buffer for 1000 and 5000 V/m intensity of electric field, zeta potential is −42.24 mV and dielectric constant 77.232. The comparison shows a good match with the available results [11]. Hence, the present model is accepted for the numerical simulation.

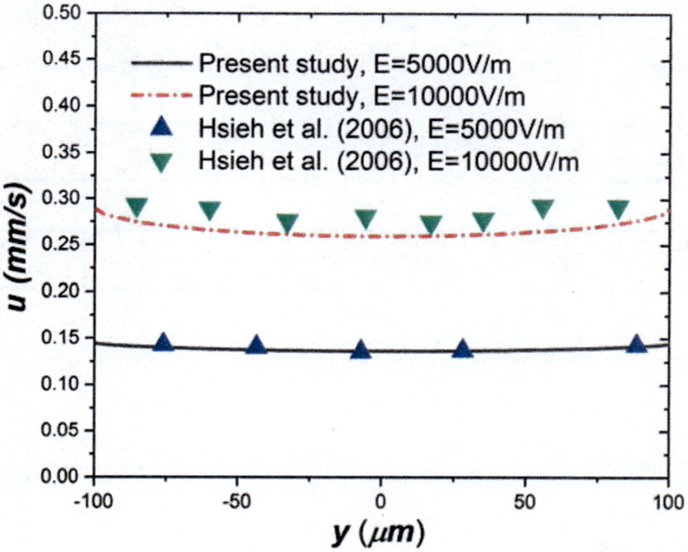


**Fig. 2** Validation for velocity profile along the transverse direction of the rectangular microchannel with the experimental results of Hsieh et al. [11] for TAE buffer at $E = 5000$ V/m and 1000 V/m, zeta potential is −42.24 mV and dielectric constant 77.232

## 4 Results and Discussions

The effect of non-uniform heating on the temperature field and heat transfer characteristics for EOF through a microchannel has been investigated by changing the normalized wavelength of non-uniform heat flux ($\gamma$) and thermal Peclet number (Pe) in the range $1.5 \le \gamma \le 6$ and $1 \le \text{Pe} \le 100$, respectively [2, 4, 7]. The values of other parameters are taken as: $\zeta = -2,\ G = 1,\ E = 10{,}000\text{V/m}$ and $\kappa = 20$ [4, 12].

To study the effect of non-uniform heating on the temperature field for EOF, isotherms contours are presented in Fig. 3 at different $\gamma(= 1.5$ and 6) and uniform heating case at $\text{Pe} = 10$. It can be seen that the uniform heating case causes higher temperature intensity as compared to the non-uniform heating cases. It can be also noted that in transverse direction temperature distribution is more uniform in total length for the non-uniform heating case for smaller wavelength $\gamma = 1.5$, and this uniformity is smaller at higher wavelength $\gamma = 6$ as compared to uniform heating case. It can be explained as the heating load in given length is distributed for smaller $\gamma$, whereas at higher $\gamma$ it is concentrated. Thus for smaller $\gamma$ gives more uniform transverse temperature distribution.

The variation of wall temperature ($\theta_W$) and bulk mean temperature ($\theta_m$) along the channel length is presented in Fig. 4 for non-uniform heating at $\gamma = 1.5$ and 6, and uniform heating case at $\text{Pe} = 100$. It can be seen that wall temperature for smaller $\gamma(= 1.5)$ is more uniform as the line of local maxima is inside the corresponding

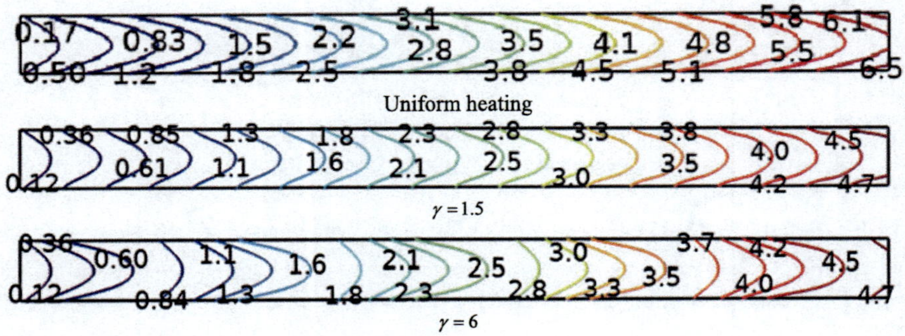


**Fig. 3** Isotherm contours for non-uniform case with $\gamma = 1.5$ and 6, and uniform heating at Pe = 10

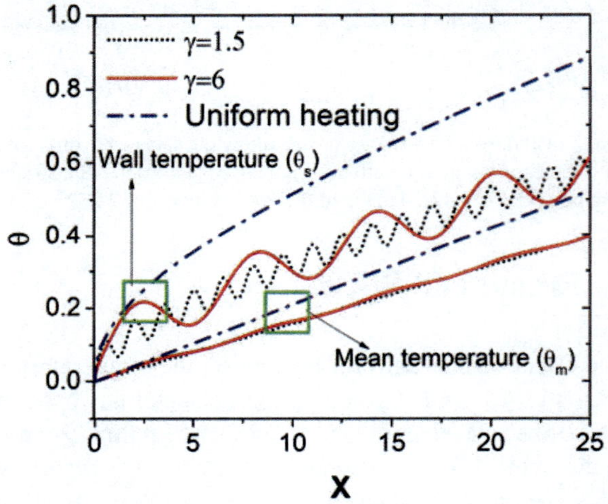


**Fig. 4** Variation of bulk mean and surface temperature for $\gamma = 1.5$, $\gamma = 6$ and uniform heating for Pe = 100

limit at $\gamma = 6$. It is because of the higher uniformity of temperature due to the higher undulation as discussed. On the other hand, the effect of $\gamma$ on $\theta_m$ is minimal. Although, the higher $\theta_m$ is found at the position of maximum heat flux undulation for $\gamma = 6$ due to the highly concentrated heat flux as compared to smaller $\gamma$ in given length. The uniform heating shows small difference in variation of wall temperature ($\theta_W$) and bulk mean temperature ($\theta_m$) near to inlet and gradually tends to a constant value in the downstream direction.

Variation of local heat transfer enhancement in terms of local Nusselt number (Nu) is presented in Fig. 5 for the uniform and non-uniform cases at different thermal Peclet number. Figure 5a depicts the variation of Nu at Pe = 1. It can be noted that the local minima and maxima of Nu are induced for the non-uniform heating case as reported in [13]. The minima of Nu for all $\gamma$ is zero due to the zero flux at that position, whereas maxima of Nu increases with decrease in $\gamma$. It can be noted that all maxima of Nu for non-uniform heating case is higher than the uniform heating case. The increase in Nu with the decrease in $\gamma$ can be explained by the fact of decrease in difference of wall and mean temperature with the decrease in $\gamma$ at that maxima heat flux position. (See Fig. 4). Whereas, for uniform heating case, the higher wall temperature creates higher difference in bulk mean and wall temperature, thereby reducing Nu. Similar effect has been observed for Pe = 10, only difference is that the similar undulation of Nu for uniform and non-uniform case, respectively has reached relatively far from the inlet as shown in Fig. 5b. For Pe = 100, the developed region of Nu increases far towards the downstream. It is because of the increase in Pe allows relatively higher advection strength, as compared to the conductive or diffusive strength. It can be noted that for the downstream region of Nu for Pe = 1 and 10, maxima of Nu increases with Pe for all $\gamma$. It is because of the decrease in the difference of wall and mean temperature with Pe. It can be also noted that with increase in Pe, the rate of decrease in maxima of Nu with $\gamma$ decreases. It can be explained by analying the rate of decrease in the difference of the bulk mean temperature and wall temperature at the corresponding location with increase in Pe. It is important to mention that the peak of Nu for higher wavelength ($\gamma = 6$) is more flat, whereas, increasing Pe corresponding peak becomes sharper. This can be explained by the difference between wall and bulk-mean temperature at lower and higher Pe. At lower Pe, for higher wavelength ($\gamma = 6$) ($\theta_w - \theta_m$) approaches near to zero at the maxima of heat flux position, whereas for higher values of Pe, the value of ($\theta_w - \theta_m$) is relatively higher at the maxima heat flux position. This variation affects the average Nusselt number variation with Pe.

The variation of average Nusselt number $\left(\overline{\text{Nu}}\right)$ with Peclet number is shown in Fig. 6 for uniform and non-uniform heating case for different $\gamma$. The value of $\overline{\text{Nu}}$ is always higher for the uniform heating case because of the minima of Nu for non-uniform heating approaches to zero, and $\overline{\text{Nu}}$ for both the cases increase with Pe. Form the figure it can be seen that the two regimes of Pe are found such that beyond the critical limit ($\text{Pe}_c$), [regime 2], $\overline{\text{Nu}}$ shows the monotonic trend with $\gamma$ and increases with the decrease in $\gamma$. For $\text{Pe} < \text{Pe}_c$ [regime 1], the non-monotonic behavior of $\overline{\text{Nu}}$ can be explained by the fact of occurrence of more flat peak of Nu for higher $\gamma (= 6)$ at smaller Pe as discussed above. On the other hand, the decrease in $\gamma$ results in smaller effect of flatness on the peak of Nu, as reduced wavelength itself increases its sharpness. Therefore, for $\text{Pe} < \text{Pe}_c$ $\overline{\text{Nu}}$ is higher for $\gamma = 1.5$ and smaller for $\gamma = 3$. For the present study the value of $\text{Pe}_c$ is 11.5.

## 5 Conclusions

The heat transfer for EOF through a microchannel has been investigated numerically under the effect of non-uniform heating. The temperature field and Nusselt number have been studied by varying the wavelength of non-uniform heat flux ($\gamma$) and thermal Peclet number ($Pe$). The important findings from the present work are summarized as follows:

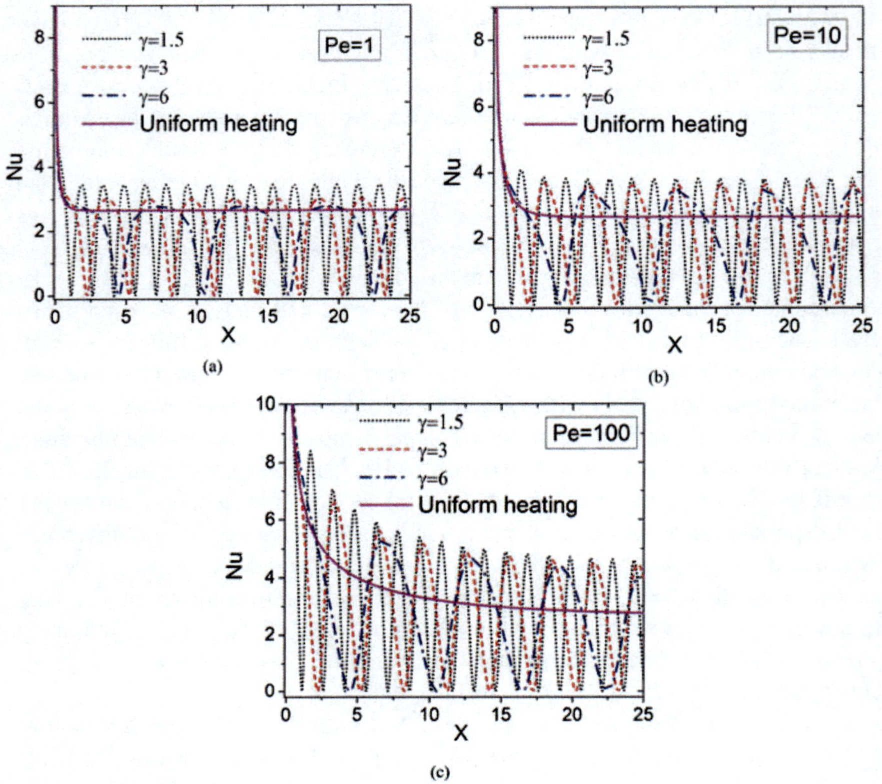


Fig. 5 Variation of local Nusselt number for uniform and non-uniform heating case with different normalized wavelength of heat flux at **a** Pe = 1, **b** Pe = 10 and **c** Pe = 100

**Fig. 6** Variation of average Nusselt number with Pe for uniform and non-uniform heating case with different normalized wavelength ($\gamma = 1.5, 3, 6$) of heat flux

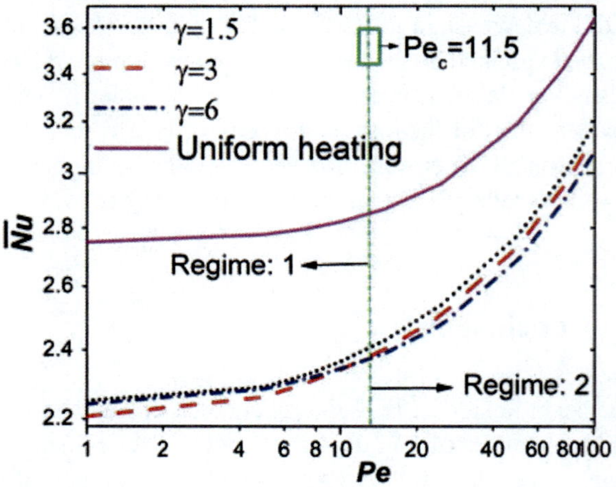

- The hot spot intensity in the domain decreases for the non-uniform heating as compared to the uniform heating case. The uniformity of temperature profile increases with a decrease in wavelength of the non-uniform heat flux.
- The maximum intensity of the wall temperature increases with an increase in wavelength of non-uniform heat flux.
- The maxima of local Nusselt number increases with a decrease in wavelength of non-uniform heat flux. The rate of decrement in local Nusselt number with the wavelength of non-uniform heat flux increases with an increase in thermal Peclet number.
- The average Nusselt number $\left(\overline{\mathrm{Nu}}\right)$ for uniform heating case is always higher as compared to the non-uniform heating case. The critical value of Peclet number ($\mathrm{Pe_c}$) is found such that for $\mathrm{Pe} < \mathrm{Pe_c}$, the value of $\overline{\mathrm{Nu}}$ shows the non-monotonic variation with non-uniform heat flux wavelength and opposite effect for $\mathrm{Pe} > \mathrm{Pe_c}$. For the present study the value of $\mathrm{Pe_c}$ is 11.5.

## References


1. X. Wang, J. Wu, Flow behavior of periodical electroosmosis in microchannel for biochips. J. Colloid Interf. Sci. **293**, 483 (2006)
2. R. Sarma, N. Deka, K. Sarma, P.K. Mondal, Electroosmotic flow of Phan-Thien Tanner fluids at high zeta potentials: an exact analytical solution. Phys. Fluids **30**, 062001 (2018)
3. R. Dey, T. Ghonge, S. Chakraborty, Steric-effect-induced alteration of thermal transport phenomenon for mixed electroosmotic and pressure driven flows through narrow confinements. Int. J. Heat Mass Transf. **56**, 251 (2013)
4. E.S. Cho, J.W. Choi, J.S. Yoon, M.S. Kim, Experimental study on microchannel heat sinks considering mass flow distribution with non-uniform heat flux conditions. Int. J. Heat Mass Transf. **53**, 2159 (2010)
5. M.P. Boruah, P. Randive, S. Pati, Effect of non-uniform asymmetric heating on the thermal and entropy generation characteristics for flow of $Al_2O_3$-water nanofluid in a micro-channel. Int. J. Numer. Methods Heat Fluid Flow **29**, 981 (2019)
6. S.K. Mehta, S. Pati, Effect on non-uniform heating on heat transfer characteristics in wavy channel, in *Proceedings of the 5 th International Conference on Computational Methods for Thermal Problems,, IISc Bangalore, India* (2018), vol. 5, p. 498. ISSN 2305-6924
7. M. Azari, A. Sadeghi, S. Chakraborty, Graetz problem for combined pressure-driven and electroosmotic flow in microchannels with distributed wall heat flux. Int. J. Heat Mass Transf **128**, 150 (2019)
8. S. Pati, R. Roy, N. Deka, M.P. Boruah, M. Nath, R. Bhargav, P.R. Randive, P.P. Mukherjee, Optimal heating strategy for minimization of peak temperature and entropy generation for forced convective flow through a circular pipe. Int. J. Heat Mass Transf. **150**, 119318 (2020)
9. S.K. Mehta, S. Pati, Analysis of thermo-hydraulic performance and entropy generation characteristics for laminar flow through triangular corrugated channel. J. Therm. Anal. Calorim. **136**, 49 (2019)
10. S.K. Mehta, S. Pati, Numerical study of thermo-hydraulic characteristics for forced convective flow through wavy channel at different Prandtl number. J. Therm. Anal. Calorim. **141**, 2429 (2020)
11. S.S. Hsieh, H.C. Lin, C.Y. Lin, Electroosmotic flow velocity measurements in a square microchannel. Colloid Polym. Sci. **284**, 1275 (2006)

12. J. Chakraborty, S. Pati, S. K. Som, S. Chakraborty, Consistent description of electrohydrodynamics in narrow fluidic confinements in presence of hydrophobic interactions, Phys. Rev E. **85**, 046305 (2012)
13. A. Borah, M. P. Boruah, S. Pati, Conjugate heat transfer in a duct using nanofluid by two-phase Eulerian-Lagrangian method: Effect of non-uniform heating, Powder Technol. **346**, 180 (2019)

# Temperature-Dependent Random Frequency of Functionally Graded Spherical shells—A PCE Approach

**Vaishali and S. Dey**

**Abstract** This paper presents the effect of temperature on random natural frequencies of spherical shells, composed of functionally graded materials (FGM) with zirconia (ceramic rich) and aluminium (metal rich). An eight noded isoperimetric quadratic element is considered for the finite element formulation. The power law is employed to construct the material modelling of Functionally Graded (FG) spherical shells. Monte Carlo Simulation (MCS) is carried out in conjunction to standard eigenvalue problems. The polynomial chaos expansion (PCE) model is constructed to reduce the computational iteration time and cost and validated it with the traditional MCS model. The statistical analyses are conducted to portray the first three random modes of frequencies. In the present analysis, the statistical results obtained are the first known results.

**Keywords** Functionally graded materials · Random natural frequency · Spherical shells · Monte Carlo simulation · Polynomial chaos expansion · Eight noded isoperimetric quadratic elements

## 1 Introduction

Functionally graded materials (FGM) are the advanced materials which have gained immense popularity because of its exclusive properties like heat and corrosion resistant in addition to high stiffness and strength. They are extensively used in aerospace, marine, civil construction and mechanical industries. These materials are the example of non-homogeneous materials which is graded at intervals, so throughout the thickness different properties can be seen. In the present study, considering top surface as ceramic rich surface while the bottom surface as metal-rich surface (see Fig. 1). Therefore at different sections, unique properties are present which doesn't resemble the properties of parent materials. By using power law, the gradations of material

Vaishali (✉) · S. Dey
Department of Mechanical Engineering, National Institute of Technology, Silchar, Assam, Silchar 788010, Assam, India
e-mail: vaishali765@gmail.com



K. M. Pandey et al. (eds.), *Recent Advances in Mechanical Engineering*, Lecture Notes in Mechanical Engineering, https://doi.org/10.1007/978-981-15-7711-6_51

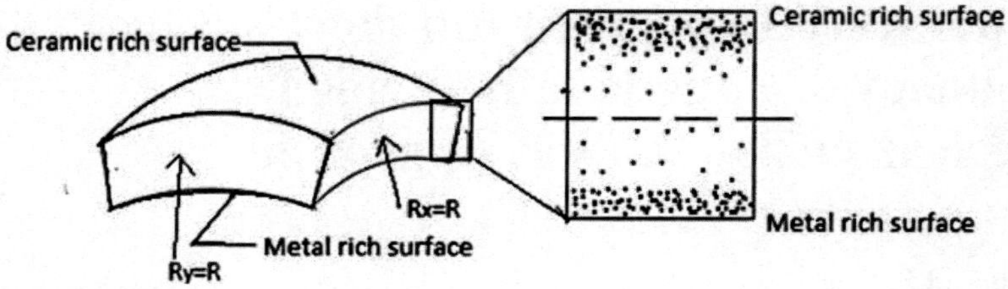


**Fig. 1** Functionally graded (FG) Spherical shell geometry

properties are varied across the thickness. In general, engineering structures are susceptible to vibration, so dynamic analysis plays an important role. Using a deterministic approach, various authors [1–4] have carried out a vibration analysis of FGM structures. Some researchers worked on the analytical solution [5, 6] of functionally graded plates. In various dimensions, many researchers worked on functionally graded (FG) structures [7–11]. But random natural frequency responses of FGM shells are yet not intrusively addressed. In the present study, it is aimed to assess the stochastic first three natural frequencies of functionally graded spherical shells considering the effect of variation of temperature. In this paper, various sections are presented after Sect. 1 as an introduction, such as Sect. 2: which illustrate the theoretical formulation of FGM spherical shells, Sect. 3: depicts the stochastic results obtained followed by discussion while Sect. 4: portrays the concluding remarks.

## 2 Theoretical Formulation

In FGM, the variability in material properties [12] vary with change in temperature which can be shown by given formulation,

$$Q = Q_0 + Q_{-1}T^{-1} + 1 + Q_1 T + Q_2 T^2 + Q_3 T^3 \tag{1}$$

where the temperature coefficients are represented by $Q_0$, $Q_{-1}$, $Q_1$, $Q_2$, $Q_3$ while T represents temperature (in Kelvin). The material properties variation across the depth can be expressed by various laws such as sigmoid law, exponential law and power law. In the present study, power law [13] is considered which can be expressed as,

$$R(\hat{w}) = R_m(\hat{w}) + [R_c(\hat{w}) - R_m(\hat{w})]\left[{}^{w}/_{t(\hat{w})} + 0.5\right]^P \tag{2}$$

where $R_m$ and $R_c$ represent properties of metal and ceramic respectively and $P$ indicates the power law exponent. Here, $\hat{w}$ represents the degree of randomness. The geometry of the spherical shell is shown in Fig. 1. Finally, the equation of motion for free vibration is obtained in the global form [7]

$$[M(\hat{w})]\{\ddot{\delta}_p\} + ([K(\hat{w})] + [K_\sigma(\hat{w})])\{\delta_p\} = 0 \quad (3)$$

where $\{\delta_p\}$ is the displacement vector, $[M(\hat{w})]$ is the mass matrix while $[K(\hat{w})]$ and $[K_\sigma(\hat{w})]$ are the random elastic stiffness matrix and the random geometric stiffness matrix, respectively. Considering eigenvalue problem [14], the random natural frequencies $(\omega(\hat{w}))$ can be obtained as,

$$[A(\hat{w})]\{\phi\} = \lambda(\hat{w})\{\phi\} \quad (4)$$

where $[A(\hat{w})] = ([K(\hat{w})] + [K_\sigma(\hat{w})]) - 1[M(\hat{w})]$.
and

$$\lambda(\hat{w}) = 1/[\omega(\hat{w})]^2 \quad (5)$$

The summary of the entire procedure is represented in the form of flow diagram as shown in Fig. 2. The uncertainty of each input parameter in the random variable approach is modeled by describing a probability density function (PDF), $f_x(a)$. In its least complex structure, the PCE of a stochastic response, $f\left(\vec{a}\left(\vec{\zeta}\right)\right)$ this depends on the randomness of the input variables, $\vec{a}\left(\vec{\zeta}\right)$ can be shown as

$$f\left(\vec{a}\left(\vec{\zeta}\right)\right) = \sum_{n=0}^{p} b_n \phi_n\left(\vec{\zeta}\right) \quad (6)$$

After extending the random response, the next step is to determine the expansion coefficients $b_n$. On obtaining the PCE coefficient, random response statistical moments can easily be obtained.

## 3 Results and Discussion

In the present study, Monte Carlo simulation is employed for random natural frequency analysis of functionally graded spherical shells having $R_x = R_y = R$. In addition, uncertain inputs of material properties like Young's Modulus ($E$), Shear Modulus ($G$), Poisson's ratio ($\upsilon$) and mass density ($\rho$) are considered as random input variables. In the present FE formulation, an eight noded isoperimetric quadratic element is considered. As a surrogate, polynomial chaos expansion (PCE) is employed. From the scatter plot (refer to Fig. 3) obtained for sample size 256, 512 and 1024, it is observed that sample size 256 shows maximum deviation (maximum scattering) and for sample size 1024, there is a least deviation (almost a diagonal line without any scattering) from mean deterministic value. Furthermore, the percentage

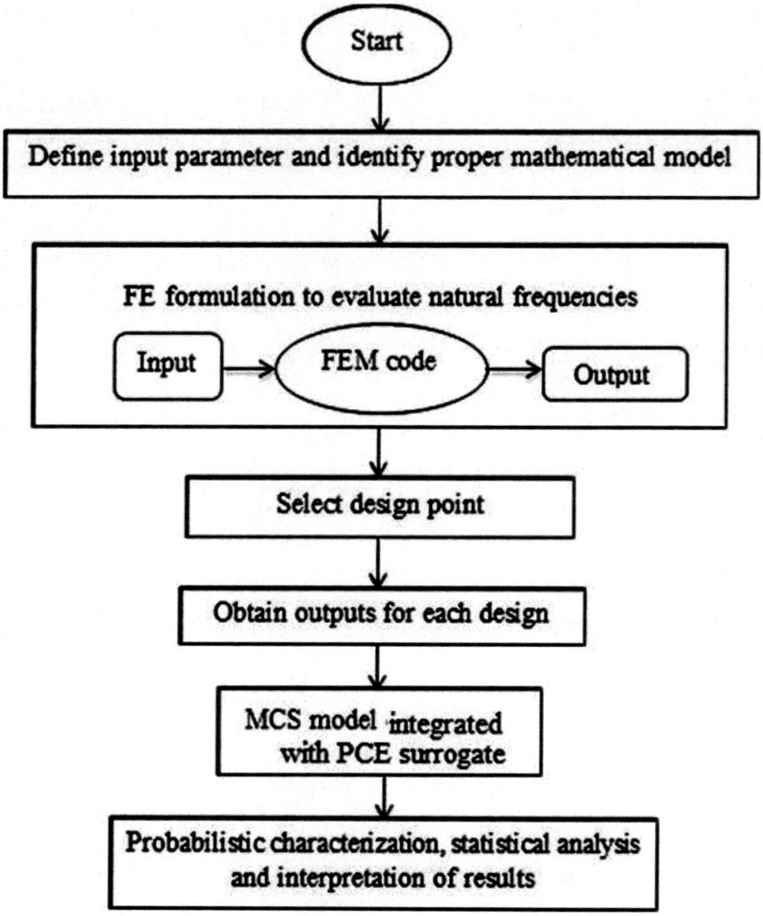


**Fig. 2** Flowchart for random natural frequency analysis

error plot (refer to Fig. 4) obtained for sample size 256, 512 and 1024. The error for sample size 256 is maximum (lies between ±0.2% for first mode, lies between ±0.5% for second mode and more than ±0.4% for third mode). For sample size 512, the errors lie between ±0.1% for first mode, lies between ±0.1% for second mode and more than ±0.2% for third mode. For sample size 1024 the error is observed to be minimum (less than ±0.1% for all the three modes). So for further study, sample size 1024 is considered. The first three random natural frequencies are determined by varying the temperature of functionally graded spherical shells i.e. 300, 600, 900 and 1200 K. The composition of the materials considered for the present analysis is namely, zirconium (ceramic) and aluminium (metal), whose properties are furnished in Table 1. The present study is validated with the previous research results obtained in past literature [11] as furnished in Table 2 where (for plate, $R_x = R_y = \infty$). The results show (refer to Fig. 5) that due to the increase of the temperature, the values of

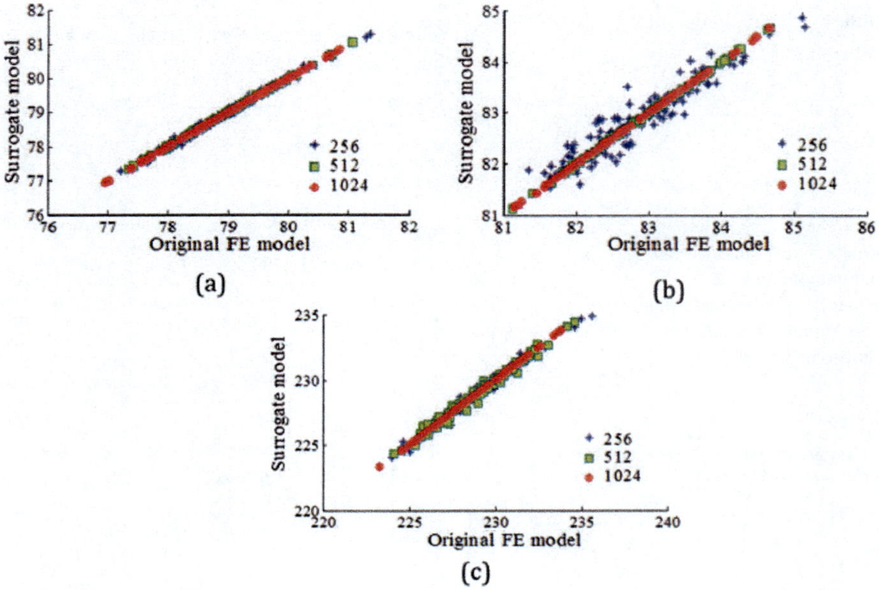


**Fig. 3** Scatter plot for the first three modes of natural frequencies (rad/s) considering PCE surrogate model for sample size of 256, 512 and 1024

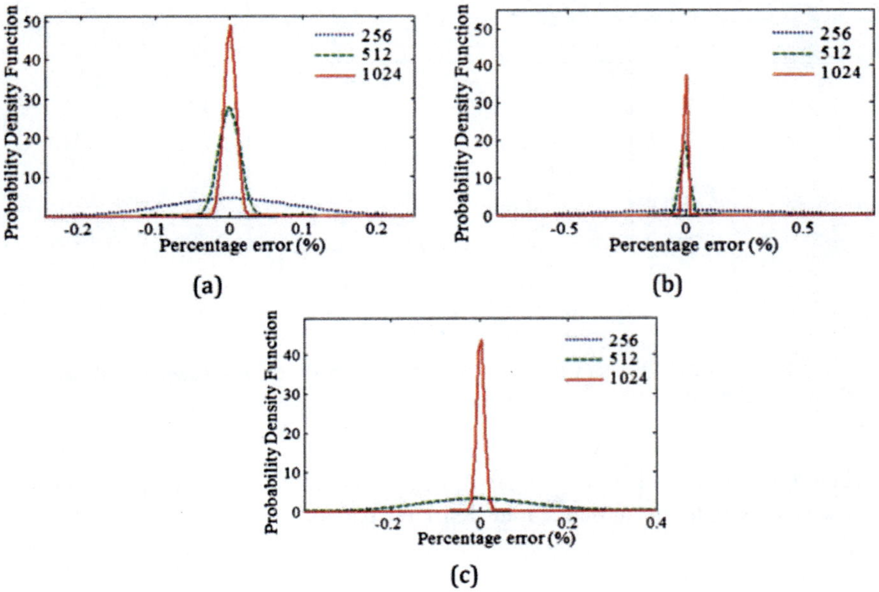


**Fig. 4** Percentage error obtained for the first three modes of natural frequencies (rad/s) considering PCE surrogate model for sample size of 256, 512 and 1024

**Table 1** Material properties [10]

| | E (Young's modulus) [GPa] | $\upsilon$ (Poission's ratio) | $\rho$ (Mass density) [kg/m$^3$] |
|---|---|---|---|
| Metal | 700 | 0.3 | 2707 |
| Ceramic | 168 | 0.3 | 5700 |

**Table 2** Non-dimensional fundamental natural frequency of FG square plate for symmetric boundary conditions

| *P* | *t/L* | Baferani et al. [11] | Present study |
|---|---|---|---|
| 1 | 0.1 | 0.0891 | 0.0883 |
| 2 | 0.1 | 0.0819 | 0.0797 |

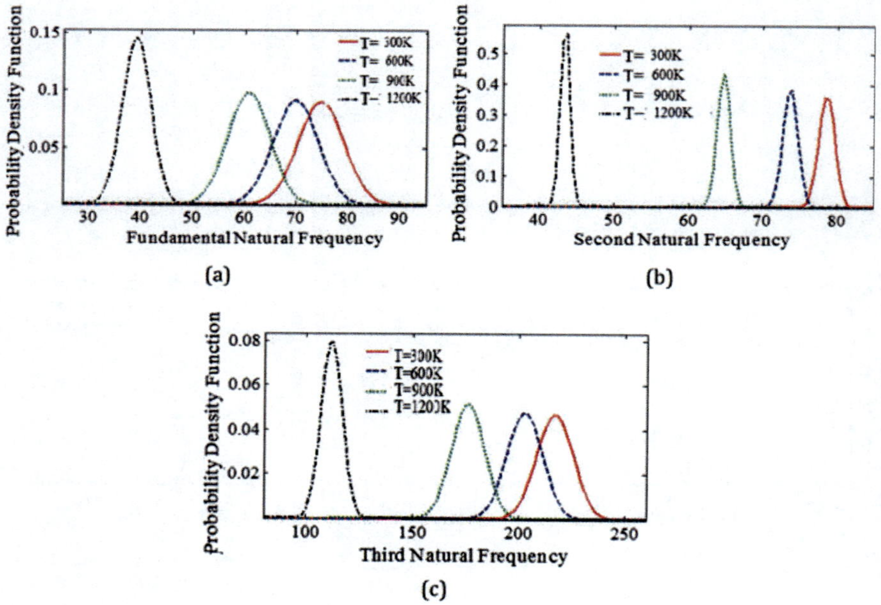


**Fig. 5** Stochastic natural frequency (rad/s) of FG spherical shells, (**a**) Fundamental, (**b**) Second and (**c**) Third modes considering temperature ($T$) = 300, 600, 900, and 1200 K

both deterministic, as well as the stochastic mean of the first three natural frequencies, decreases along with the subsequent decrease in sparsity.

## 4 Conclusions

Based on Polynomial Chaos Expansion (PCE) in combination with finite element formulation, the first three random natural frequencies of functionally graded (FG) spherical shell are studied. The PCE as surrogate model is employed to increase computational efficiency along with the reduction in computational cost. The novelty of present work includes the effect of temperature on random natural frequencies of FGM spherical shells. Due to unavoidable inherent randomness present in these structures due to variation in temperature, a band of deviation in the values of natural frequencies is observed compared to the deterministic mean value. Due to manufacturing inaccuracies, it is difficult to achieve exact deterministic value of material properties such as Young's Modulus, Shear modulus, Poisson's ratio and mass density. In FGM, the variability in material properties varies with change in temperature. The variations of each material property have an impact on influencing the output stochastic natural frequencies in different value of contribution. The initiation of variability of a material property causes inherent anisotropy throughout the entire geometry of FGM shell structures, resulting in different modes of sensitivities for different material properties. It causes stochastic variation in output quantity of interest (in this case, it is natural frequency). Based on these observations, the present work can be further extended to more complex geometries and structures.

**Acknowledgements** The first author acknowledges Ministry of Human Resource and Development (MHRD) and Technical Education Quality Improvement Programme (TEQIP-III), Govt. of India, for the financial support provided during research work.

## References

1. A.M.A. Neves, A.J.M. Ferreira, E. Carrera, M. Cinefra, C.M.C. Roque, R.M.N. Jorge, C.M.M. Soares, Free vibration analysis of functionally graded shells by a higher-order shear deformation theory and radial basis functions collocation, accounting for through-the-thickness deformations. Eur. J. Mech. Solids **37**, 24–34 (2013)
2. Y.Q. Wang, J.W. Zu, Nonlinear dynamics of functionally graded material plates under dynamic liquid load and with longitudinal speed. Int. J. Appl. Mech. **9**(04), 1750054 (2017)
3. S. Dey, S. Sarkar, A. Das, A. Karmakar, S. Adhikari, Effect of twist and rotation on vibration of functionally graded conical shells. Int. J. Mech. Mater. Des. **11**(4), 425–437 (2015)
4. A.K. Sharma, P.S. Chauhan, M. Narwaria, R. Pankaj, Vibration analysis of functionally graded plates with multiple circular cutouts. Int. J. Current Eng. Sci. Res. **2**(3), 15–21 (2016)
5. A. Apuzzo, R. Barretta, R. Luciano, Some analytical solutions of functionally graded Kirchhoff plates. Compos. B Eng. **68**, 266–269 (2015)
6. R. Barretta, R. Luciano, Analogies between Kirchhoff plates and functionally graded Saint-Venant beams under torsion. Continuum Mech. Thermodyn. **27**(3), 499–505 (2015)
7. P.K. Karsh, T. Mukhopadhyay, S. Dey, Stochastic dynamic analysis of twisted functionally graded plates. Compos. B Eng. **147**, 259–278 (2018)
8. S. Dey, S. Adhikari, A. Karmakar, Impact response of functionally graded conical shells. Latin Am. J. Solids Struct. **12**(1), 133–152 (2015)

9. P.K. Karsh, T. Mukhopadhyay, S. Dey, Stochastic investigation of natural frequency for functionally graded plates, in *IOP Conference Series: Materials Science and Engineering*, 2018, March, vol. 326, no. 1, p. 012003. IOP Publishing
10. F. Tornabene, E. Viola, Static analysis of functionally graded doubly-curved shells and panels of revolution. Meccanica **48**(4), 901–930 (2013)
11. A.H. Baferani, A.R. Saidi, H. Ehteshami, Accurate solution for free vibration analysis of functionally graded thick rectangular plates resting on elastic foundation. Compos. Struct. **93**(7), 1842–1853 (2011)
12. Y.S. Touloukian (ed.), *Thermophysical Properties of High Temperature Solid Materials*, vol. 1. (Macmillan, 1967)
13. C.T. Loy, K.Y. Lam, J.N. Reddy, Vibration of functionally graded cylindrical shells. Int. J. Mech. Sci. **41**(3), 309–324 (1999)
14. K.J. Bathe, E.L. Wilson, Solution methods for eigenvalue problems in structural mechanics. Int. J. Numer. Meth. Eng. **6**(2), 213–226 (1973)

# Microwave Processing of Polymer Matrix Composites: Review of the Understanding and Future Opportunities

**Tejas Pramod Naik, Ram Singh Rana, Inderdeep Singh, and Apurbba Kumar Sharma**

**Abstract** In the fast-changing world, the quest for the energy-efficient and environment-friendly processing method is getting attention while without compromising on the product properties, processing time and cost. Consequently, the processing of material with microwave energy is emerging as a novel method and it is used as an alternate processing route for various types of advanced materials like MMC, PMC and alloys for commercial applications in the field of joining, biomedical, powder metallurgy, coatings, claddings, etc. The microwave processing dominates over the conventional processing of materials due to its notable advantages like low processing time, less energy consumption, volumetric heating, negligible thermal gradient and better mechanical properties. The current study primarily focuses on microwave processing and its interaction phenomena with polymer, reinforced with different synthetic and natural fibers. It was observed that only electric field component of microwave is responsible for heating of PMC, due to its non-magnetic nature; the losses responsible for microwave heating like dipole and conduction further depend upon the type of reinforcements. It was found that the dipole loss mechanism predominates in natural fibers, while conduction loss is responsible in processing conductive fibers. A few initial results have been presented; future possibilities are indicated.

**Keywords** Microwave curing · Thermoset polymer · Thermoplastic polymer · Synthetic fibers · Natural fibers

---

T. P. Naik (✉) · R. S. Rana · I. Singh · A. K. Sharma
Department of Mechanical and Industrial Engineering, Indian Institute of Technology Roorkee, Roorkee 247667, India
e-mail: tejasnit@gmail.com

R. S. Rana
e-mail: ramsingh.mech17@gmail.com

I. Singh
e-mail: inderdeep.singh@me.iitr.ac.in



K. M. Pandey et al. (eds.), *Recent Advances in Mechanical Engineering*, Lecture Notes in Mechanical Engineering, https://doi.org/10.1007/978-981-15-7711-6_52

# 1 Introduction

## *1.1 Introduction to Microwaves*

The electromagnetic waves (EM) lying in a frequency range of 300 MHz–300 GHz are called microwaves. Only particular kinds of materials can be used while using this process. The properties of prime importance regarding this process are complex relative permittivity and loss tangent of the dielectric used. Higher loss tangent, dielectric loss leads to more heating and vice versa. Microwaves are EM waves having electric and magnetic field orthogonal to each other, having a wavelength range from 1 to 1000 mm. The interaction of the material with microwaves depends on the magnetic and dielectric property and property of the wave because both the electric field and the magnetic field components interact on their own differences with the material during irradiation [1, 2]. The PMCs are processed as well as joined by low-temperature microwave processing having a temperature between 500 and 1000 °C. Microwave welding is a type of electromagnetic welding. It is possible due to the interaction of the microwaves with the material leading to heating and the joining. Based on a different kind of material, many kinds of interaction take place leading to variable core mechanisms for heating to take place. For metal-based materials, this heating is influenced by a magnetic field but for nonmetals, this heating takes place due to the electric field usually. In metals, ceramics, etc., temperature gradients can be set up. Also, microstructure change can occur if proper attention is not given to the process parameters. Uniform bulk heating takes place in microwave for PMCs as compared to conventional heating methods. The uniform heating characteristic of microwave yields better polymeric products than the conventionally cured. For example, thin as well as thick PMCs were processed with better mechanical properties and better joining in PMC specimen by using variable microwave heating methods to concentrate more heat at the interface of the adhered [3–11]. The researchers have, in the past, used susceptor materials like charcoal powder, conducting particles, nanofiber coatings on composites to expose most of the microwaves to a specific area for most of the heat to generate there, called selective heating. Using this method, processing of various composites as well as the joining of different composites has taken place. In selective heating, only a selected volume of the target material is exposed to microwave radiation. This method of selective heating has led to better matrix fiber bonding [12–37]. Quite less work has been done on microwave joining and processing of natural fiber-based PMCs, though the results have proved improvement in the properties of the final specimen as well as better joining strength, making it a good alternative for joining of such materials [38–44].

## 1.2 Principle of Microwave Heating

Maxwell's equations have great importance in microwave heating phenomenon when microwave welding is taking place inside the microwave cavity. The interaction of the microwaves with the specimen is governed by Maxwell equations [45]. They take into consideration the effects of the magnetic field as well as the electric field harmonic variation on the specimen over some time and how its properties are varying because of it. The Maxwell equations are given below:

$$\nabla \cdot H = 0$$
$$\nabla \times E = -j\omega\varepsilon_0\varepsilon^* E$$
$$\nabla \cdot (\varepsilon E) = 0$$
$$\nabla \times E = j\mu\omega H$$

where,

$E$—time-harmonic electric field

$H$—time-harmonic magnetic field

$\varepsilon^*$—complex permittivity of material

$\varepsilon$—permittivity

$\varepsilon_0$—permittivity in air

$\mu$—magnetic permeability

$j$—imaginary unit

$\omega$—angular frequency of microwaves.

In the equation above, the complex permittivity $\varepsilon^* = \varepsilon' - j\varepsilon''$ is of prime importance which governs the heat generation due to microwaves [14, 17, 20, 22]. In this equation, $\varepsilon'$ represents the real segment of complex permittivity $\varepsilon^*$, which denotes quantity of electrical energy which could be stored inside the specimen and is termed as dielectric constant, and $\varepsilon''$ is imaginary segment termed as dielectric loss factor which signifies capacity of sample to give away the energy it received from microwaves. These Maxwell's equations can be solved with appropriate boundary conditions (BCs) for the electric and magnetic field distribution in the microwave applicator of a known dimension [46]. By knowing the EM field distribution, we can know the approximate position of the specimen inside the microwave for better heating by providing maximum coupling of EM waves with the material [46–49]. By increasing the size of a microwave cavity, the number of hotspots on the sample can be decreased leading to a higher volume of sample capacity [47–49]. But with an increase in the size of the specimen, the heating also becomes less uniform leading to the formation of many hotspots and the microwave heating will take place similar to the regular heating process [11, 46, 48]. The power absorbed and the penetration depth to which microwaves can go in a particular material depend on its properties of EM waves, material and its mass and thickness. Microwave penetration in a metallic material is nearly zero and termed as skin depth [50–52].

## 2 Microwave Interaction with Various Materials

Depending on the material, there is a variable variation of the electric and magnetic fields throughout the material thickness. In low-loss insulator materials like Teflon and generally most of the polymers, there is an almost negligible decrease in electric and magnetic fields throughout the thickness, so no need to consider skin depth because of negligible effect in case of polymers. But because of such high penetration depth which is generally because of very high dielectric loss also means that amount of energy absorbed per unit volume of the material is also meager, but as usually the polymers are low melting point materials so it is not much a problem in most of the cases. In the case of polymers having moderately high dielectric constant, the amount of energy absorbed per unit volume can be increased to balance the energy; the loss is taking place due to its moderately low dielectric loss to make it interact more with the microwaves. Most of conducting metals possess very high dielectric loss or also called no loss materials and negligible penetration depth and reflected the microwaves. In general, the advanced materials like PMCs are considered as mixed absorbers, because of having variable dielectric loss of fiber as well as a matrix.

Non-magnetic materials such as polymers, water and Al etc. are influenced only by the electric filed and not the magnetic field component of the EM waves, and heating takes place because of dipolar losses and conduction losses. Mainly, dipolar losses are predominant in dielectric insulators, but in the case of metals and conductors conduction losses take preference. In dielectric insulators like polymers, dipoles are produced in the material due to oscillating electric filed which during reorienting themselves generate heat and increase the temperature [45, 48, 49, 53–55].

In conduction loss, metals having free electrons move along the direction of applied electric field, and because of high conductivity, electric field decreases through the thickness leading to induced current and magnetic fields in a direction opposite to externally applied fields; this causes little motion because of these opposite fields leading to heating.

It is to be noted that in pure insulators, no microwave energy absorption is there because of deficient dielectric loss factor whereas as in dielectric insulators like polymers they have the property to polarize (due to displacement of electrons around nuclei, due to displacement of nuclei with respect to each other and due to aligning of dipole) in electric field leading to heating [49]. No matter how much a high dielectric loss factor (to absorb more power) still, the process can sometimes not be efficient enough because of the low penetration depth of the microwaves. This can happen in case if the size of the specimen is comparatively much larger than the microwave penetration depth inside it, which will lead only to surface heating. Penetration depth is also a function of the frequency of the microwaves, and with an increase in frequency, the microwave power absorption is found to increase but the penetration depth, in this case, will decrease [45, 47]. This dependency of loss factor and dielectric material size on power absorbed and depth of penetration is valid only up to moderate temperature up to nearly 500 °C, and when the temperature is increased further, the dielectric property of the material will start varying; e.g., SiC having 1.71 loss factor at room temperature jumps to 27.99 loss factor at 695 °C for a 2.45 GHz microwave [56].

The skin depth in the case of metallic materials can be increased by making them in powder form so that their penetration depth is comparable to their size, but in case of bulk metal rather than in powder form, the microwaves are reflected and can cause sparking too. Experimentally, the nano and submicron powders have been found to absorb better microwaves, improvement in the microstructures of the specimen has been found, and critical size of 100 $\mu$m has been found below which heating is uniform and above which heating becomes nonuniform [57].

In composite materials, microwave absorption is dependent on whether the fiber or the matrix phase is having a high dielectric loss, which further transfers the other one by conduction. It has been found that reinforcing with conductive particles has improved the microwave absorbing and other related properties of the composites [58, 59].

For microwave joining of PMCs, lot depends on the dielectric loss factor of fiber as well as a matrix. In case of synthetic fiber like carbon reinforced polymer matrix composites, fiber will act as the high loss factor constituent in the composite and will get heated selectively and transfers heat by conduction to the matrix leading to better bonding at their interfaces [33].

There is the temperature dependence of dielectric properties of fiber and polymer on temperature, which is varied during microwave joining. In case of thermosets, as they have cross-linked bonds, so during microwave heating, the viscosity increases which prevents the dipole motion in the presence of applied electric field, consequently no heating can take place by dipole loss, only conventional heating will take place and it will get burned [35–37], whereas in the case of thermoplastics, they have to be heated initially up to at least a critical temperature because of possessing a low degree of crystallinity and dielectric loss factors. The degree of crystallinity in the case of thermoplastics plays a vital role in the uniform heating of a PMC. It has been found that for more than a 45% degree of crystallinity, the specimen becomes nearly transparent to the incoming microwaves because of obstruction to the movement of dipoles on the application of the electric field [34].

In thermoplastic and thermoset PMC's way of absorption of energy is different. In the case of thermoplastic-based PMCs, mostly the energy is absorbed by fibers below the critical temperature when the PMC is having a high degree of crystallinity and this heat is later on transferred by conduction to the polymer part. But if the thermoplastic PMC possesses a low degree of crystallinity, then energy is absorbed by the fiber as well as the matrix too, and the more heat between the two is absorbed by the one having a high value of dielectric loss factor. This energy absorption by the thermoplastic-based PMCs is also dependent on the temperature and it increases as the temperature reaches above the critical temperature which leads to even heating of the PMC having high conductive fiber reinforcement. The curing of thermoset polymer-based PMCs is dominated with energy absorption by high dielectric loss factor constituent at room temperature; however, at elevated temperatures, energy absorption is dependent on the conductivity of reinforcement fibers due to the cross-linking of thermoset polymer matrix [44].

## 2.1 Microwave Curing Process

Initially, studies on microwave curing of thermoset as a polymeric material and synthetic fibers as reinforcement took place. Now, the research trend has been shifted toward the fabrication of natural fiber-reinforced thermoplastic composites primarily for biodegradable plastics like polylactic acid owing to their fully biodegradable nature of matrix as well as fibers. The uses of NFRPCs reduce the plastic uses, and it further contributes toward the reduction of global warming temperature. On the other hand, the composites reinforced with natural fibers fabricated by using the matrix like polypropylene and polyethylene can be successfully recycled as reported by the authors [72]. In one of the studies the short CFRP composite using epoxy as a matrix material was fabricated using three different ways: without curing, thermal curing and thermal curing followed by microwave irradiation. Mechanical properties of the fabricated composite were studied to find out the effect of microwave irradiation. Degree of polymerization was evaluated by infrared (IR) spectroscopy. The author reported that the microwave-irradiated composite samples exhibited ductile behavior, stronger physicochemical linkage at the interface between CF and epoxy resin [60]. The mechanical behavior of microwave-cured glass–epoxy composites was studied, and the results were compared with thermally cured composites. Microwave curing of a glass fiber reinforced with epoxy (LY556/HY951) was carried out in multi-mode, industrial microwave operating at a frequency of 2.45 GHz. The author stated that the mechanical properties of the specimen fabricated using both the curing routes showed close relationship; in fact, the microwave curing route was more time economic and energy saving [61]. As microwave processing is a less time-consuming as well as eco-friendly process, various researchers recently have started using microwave energy for the curing of natural fiber-based thermoplastics polymer composites. Natural FRPC were fabricated with microwave curing consisting of fibers—sisal and Grewia optiva—and thermoplastic polymers—PP and ethylene vinyl acetate. Microwave wattage and the exposure time were optimized and tensile, flexural, impact strength were evaluated. Tensile, flexural strength was enhanced because of better interfacial bonding, but impact strength reduced due to stress concentration regions [62]. In another study, the author developed polypropylene (PP)- and polyethylene (PE)-based composites using microwave curing setup as shown in Fig. 1 with 15% weight percentage of jute and kenaf fibers. The scanning electron microscope (SEM) fractography was used to study the mechanisms of failure, and mechanical characterization of the microwave-cured composites was carried out using various tests. Results

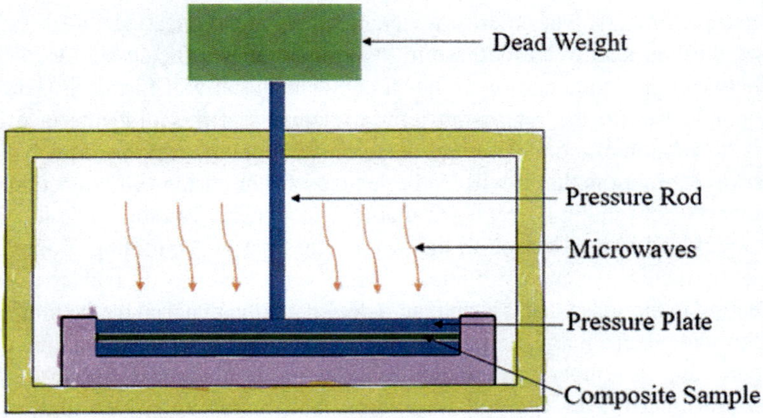


**Fig. 1** Schematic diagram of microwave curing setup [63, 64]

showed that the ultimate tensile strength was in the range of 44–50 MPa. Interlaminar shear strength of the PP-based composites was 62% higher than the PE-based composite impact energy was in the range of 18–24 kJ [63]. The influence of microwave power on mechanical properties of 20% by wt. coir-reinforced HDPE composite was also studied. Short fiber-reinforced HDPE/coir composite were fabricated using various power levels like 360, 540, 720, 900 W. Various tests very carried out like XRD, tensile and flexural, and it was reported that the mechanical properties of specimen cured at low power were high as compared to high power [64]. The effect of microwave treatment on the tensile properties of treated sugar palm with 6% NaOH-reinforced polyurethane composite was also investigated by group of researchers. The fibers were treated at three different temperatures like 70, 80 and 90°c; then, this treated fiber was mixed with the polymer in extruder and composite was fabricated using hot press machine. The author reported that the tensile strength of the microwave-treated composite is far excellent with respect to untreated and treated composite without microwave; this is because due to microwave treatment there is drop in excess moisture content. The highest tensile strength recorded was 18.42 MPa when treated at microwave temperature of 70 °C. On the other hand, tensile strain was found to be reduced dramatically in microwave-treated composite because of the enhancement in bonding between the fiber and matrix [65].

## 2.2 *Microwave Joining Process*

In microwave welding, any material in the form of a thin layer having functional electromagnetic absorbing capacity has to be placed at the bond line between the parts to be joined and simultaneously welding pressure has to be applied along with high-intensity microwave energy having frequency range depending upon the microwave parameters. The microwave energy will induce heating in the sample which would then lead to an increase in the temperature in the susceptor material placed between the joints and then this susceptor material will further transfer the heat by conduction to the remaining joining elements at the joint interface, thus leading to formation of molten polymer material at the interface. One of the benefits of microwave joining is that it will lead to usually faster curing times as compared to the other type of joining processes available. Due to faster heating rate microwave joining could lead to a reduction in the processing time of the joining process and consequently increase in energy efficiency. Microwave heating differs from conventional heating in the sense that it is delivered directly to the material by the molecular interaction with electromagnetic fields produced. Heating by conventional methods causes a buildup of temperature gradient only along the thickness direction of the specimen but microwaves can penetrate inside and supply energy throughout the sample resulting in heating the material uniformly, which is referred to as a volumetric heating process. So, microwave heating is quite helpful for fast and uniform heating of thicker materials. In conventional joining processes during heating, there is a need to select slow heating rates to prevent sharp thermal gradients which could lead to the development of induced stresses. The interaction of microwaves during joining of PMCs depends upon the dielectric properties of the matrix and fiber. Selective coupling of the microwaves depends upon the loss tangent of the material. The material possessing a high value of loss tangent will have more selective coupling with microwaves and would heat more than the other material. So, to generate more heat, at the joint interface, a material possessing high dielectric loss called susceptor, has to be used at the interface for selective heating to take place. Nowadays, a lot of work is going on the natural-based fiber composites and their joining. The possible setup which can be considered for joining the polymer composites through microwave energy is as shown in Fig. 2 [63, 64]. In one of the studies, green sisal fiber-reinforced polylactic acid composites were prepared with variable fiber content using compression molding and joined using a microwave energy process with charcoal as an accelerator. Time taken for bond formation was more for samples with

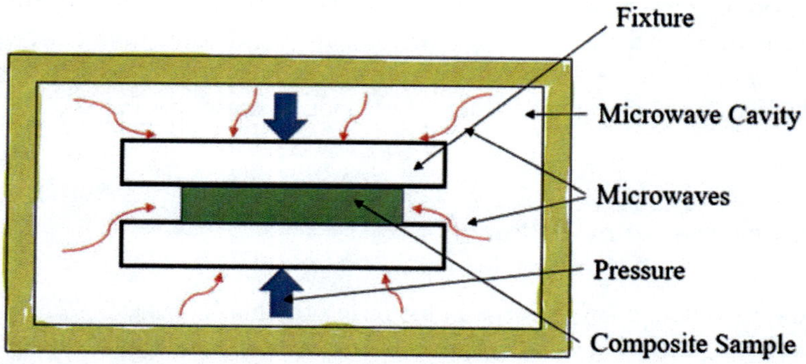


**Fig. 2** Schematic diagram of microwave joining setup [63, 64]

lower fiber content [66]. To promote green composites and to replace the petroleum-based resources, various studies have taken place. The various research works leading to the development and mechanical characterization of different green composites (PLA-based in detail) with different natural fibers and biodegradable polymers have been studied [67]. In another study, fully biodegradable natural fiber (Grewia optiva)-reinforced PLA composite has been developed and joining in lap configuration done using adhesive bonding and microwave joining and charcoal used as a susceptor and then the simulation is done. Tensile strength of neat PLA resin increased by 75% by the addition of Grewia optiva and better joint strength obtained using microwave than adhesive joining [68]. Grewia optiva (GO) and nettle fiber (NF) were also used in combination with PLA- and PP-based composite producing four combinations and joining done using adhesive as well as a microwave. PLA-based composite shows higher failure load (maybe because the failure strength of neat PLA resin was higher than PP). Joint strength was observed more in a microwave than adhesive joining. Power input, microwave exposure time, susceptor and location of the specimen in a microwave cavity are found to be having essential roles in governing the joining process [69]. Three different fibers (nettle, Grewia optiva and sisal) used with PLA polymer to develop composite using hot compression wear and frictional characteristics of developed composite observed under varying load, sliding speed and sliding distance revealed improvement in wear behavior and reduction in friction coefficient [70]. Processing of polymer matrix composites was done using variable frequency microwave (2–7 MHz) and compared with fixed frequency microwaves (915, 2.45 MHz) by joining nylon 66/GF (33%) in joint lap configuration using Araldite as an adhesive which is cured during the process. The bond strength of the test piece joined by VFM was found 1.5% higher than fixed frequency equipment. Also found that the higher the power of VFM equipment, the higher will be bond strength within some limits [71].

## 3 Conclusions

Capability of microwaves in material processing has been recognized a very long time back in 90's. In any case, due to limited understanding of the phenomenon, their utilization remained to a great extent limited to just a couple of materials. Over the most recent 65 years, the microwave processing of materials is getting well known because of its latent capacity points of interest over the conventional procedures. There is an increasing requirement in processing of the advanced materials using microwave process as this process is eco-friendly, less processing time involved. This paper has highlighted the basics of how the microwave energy is used for curing and joining of polymer-based composites. Various phenomena which lead to heating in polymer-based composite are also addressed. There are monstrous opportunities to utilize microwaves in material processing, developing inventive methodologies in design and its implementations will increase the scope of practical usage of microwave energy in various applications.

## References

1. D.E. Clark, W.H. Sutton, Microwave processing of materials. Annu. Rev. Mater. Sci. **26**, 299–331 (1996)
2. A.K. Sharma, P. Kumar, Advanced manufacturing processes, NPTEL e-learning course. Retrieved on date 28 Mar 2015
3. F.Y.C. Boey, W.L. Lee, Microwave radiation curing of a thermosetting composite. J. Mater. Sci. Lett. **9**(10), 1172–1173 (1990)
4. C.Y. Yue, H.C. Looi, Influence of thermal and microwave processing on the mechanical and interfacial properties of a glass/epoxy composite. Composites **26**(11), 767–773 (1995)
5. C. Nightingale, R.J. Day, Flexural and inter laminar shear strength properties of carbon fibre/epoxy composites cured thermally and with microwave radiation. Compos. A **33**(7), 1021–1030 (2002)
6. E. Marand, K.R. Baker, J.D. Graybeal, Comparison of reaction mechanisms of epoxy resins undergoing thermal and microwave cure from in situ measurements of microwave dielectric properties and infrared spectroscopy. Macromolecules **25**(8), 2243–2252 (1992)
7. R. Rao, S. Rao, B. Sridhara, Studies on tensile and inter laminar shear strength properties of thermally cured and microwave cured glass epoxy composites. J. Reinf. Plast. Compos. **25**(7), 783–795 (2006)
8. T. Varaporn, S. Kaew, Comparison of microwave and thermal cure of epoxy-anhydride resins: mechanical properties and dynamic characteristics. J. Appl. Polym. Sci. **97**(4), 1442–1461 (2005)
9. T. Varaporn, J. Dumrong, Comparison between microwave and thermal curing of glass fiber-epoxy composites: effect of microwave heating cycle on mechanical properties. J. Appl. Polym. Sci. **102**(2), 1059–1070 (2006)
10. E.T. Thostenson, T.W. Chou, Microwave and conventional curing of thick-section thermoset composite laminates: experiment and simulation. Polym. Compos. **22**(2), 197–212 (2001)
11. R.R. Mishra, A.K. Sharma, Microwave material interaction phenomena: heating mechanisms, challenges and opportunities in material processing. Compos. A Appl. Sci. Manuf. **81**, 78–97 (2016)
12. F.Y.C. Boey, T.H. Lee, Electromagnetic radiation curing of an epoxy/fibre glass reinforced composite. Int. J. Radiat. Appl. Instrum. Part C Radiat. Phys. Chem. **38**(4), 419–23 (1991)
13. P.S. Mooteri, B. Sridhara, S. Rao, Studies on mechanical behaviour of microwave and thermally cured glass fiber reinforced polymer composites. J. Reinf. Plast. Compos. **25**(5), 503–512 (2006)
14. M.S. Johnson, C.D. Rudd, D.J. Hill, Microwave assisted resin transfer moulding. Compos. A **29**(1–2), 71–86 (1998)
15. W.I. Lee, G.S. Springer, Microwave curing of composites. J. Compos. Mater. **189**(4), 387–409 (1984)
16. J. Wei, M. Hawley, J. Assmussen, Microwave power absorption model for composite processing in a tunable resonant cavity. J. Microwave Power Electromagn. Energy **28**(4), 234–242 (1993)
17. D.A. Papargyris, R.J. Day, A. Nesbitt, D. Bakavos, Comparison of the mechanical and physical properties of a carbon fibre epoxy composite manufactured by resin transfer moulding using conventional and microwave heating. Compos. Sci. Technol. **68**(7), 1854–1861 (2008)
18. C. Jordan, J. Galy, J.P. Pascault, Comparison of microwave and thermal cure of an epoxy/amine matrix. Polym. Eng. Sci. **35**(3), 233–239 (1995)
19. J. Mijovic, W.V. Corso, L. Nicolais, G. d'Ambrosio, In-situ real-time study of cross-linking kinetics in thermal and microwave fields. Polym. Adv. Technol. **9**(4), 231–243 (1998)
20. J. Mijovic, J. Wijaya, Comparative calorimetric study of epoxy cure by microwave vs. thermal energy. Macromolecules **23**(15), 3671–3674 (1990)
21. G. Saccone, E. Amendola, D. Acierno, Conventional and microwave curing process of epoxy systems: kinetic analysis and characterization. Microwave Opt Technol Lett **51**(11), 2777–2783 (2009)

22. S.L. Bai, V. Djafari, M. Andreani, D. Francois, A comparative study of the mechanical behaviour of an epoxy resin cured by microwaves with one cured thermally. Eur. Polym. J. **31**(9):875–884 (1995)
23. S.L. Bai, V. Djafari, Interfacial properties of microwave cured composites. Composites **26**(9), 645–651 (1995)
24. F.Y.C. Boey, I. Gosling, S.W. Lye, High-pressure microwave curing for epoxy– matrix/glass–fibre composite. J. Mater. Proces. Technol. **29**(1), 311–319 (1992)
25. S.W. Lye, F.Y.C. Boey, PC-based monitoring and control for microwave curing of polymer composites. Mater. Manuf. Process **9**(5), 851–868 (1992)
26. S.C. Joshi, S.K. Bhudolia, Microwave–thermal technique for energy and time efficient curing of carbon fiber reinforced polymer prepreg composites. J. Compos. Mater. **48**(24), 3035–3048 (2014)
27. B. Tanmay, Role of metallic, ceramic and composite plates on microwave processing of composite dielectric materials. Mater. Sci. Eng., a **457**(1), 261–274 (2007)
28. T. Chaowasakoo, N. Sombatsompop, Mechanical and morphological properties of fly ash/epoxy composites using conventional thermal and microwave curing methods. Compos. Sci. Technol. **67**(11), 2282–2291 (2007)
29. L.A. Ramajo, A.A. Cristóbal, P.M. Botta, J.M.P. López, M.M. Reboredo, M.S. Castro, Dielectric and magnetic response of $Fe_3O_4$/epoxy composites. Compos. A **40**(4), 388–393 (2009)
30. K.Y. Park, J.H. Han, S.B. Lee, J.W. Yi, Microwave absorbing hybrid composites containing Ni–Fe coated carbon nanofibers prepared by electroless plating. Compos. A **142**(5), 573–578 (2011)
31. V.K. Rangari, M.S. Bhuyan, S. Jeelani, Microwave curing of CNFs/EPON-862 nano composites and their thermal and mechanical properties. Compos. A **42**(7), 849–858 (2011)
32. C. Ageorges, L. Ye, M. Hou, Advances in fusion bonding techniques for joining thermoplastic matrix composites: a review. Compos. A **32**(6), 839–857 (2001)
33. L.T. Drzal, K.J. Hook, R.K. Agrawal, Enhanced chemical bonding at the fiber– matrix interphase in microwave processed composites, in *Microwave Processing of Materials II, Materials Research Society Proceedings*, ed. by W.B. Snyder Jr, W.H. Sutton, M.F. Iskander, D.L. Johnson(Materials Research Society, Pittsburgh, 1991), p. 449–454
34. M. Chen, E.J. Siochi, T.C. Ward, J.E. McGrath, Basic ideas of microwave processing of polymers. Polym. Eng. Sci. **33**(7), 1092–1109 (1993)
35. M. Martinelli, P.A. Rolla, E. Tombari, A method for dynamic dielectric measurements at microwave frequencies: applications to polymerization process studies. IEEE Trans. Instrum. Meas. **34**(3), 417–421 (1985)
36. D. Kranbuel, S. Delos, E. Yi, J. Mayer, T. Jarvie, W. Winfree, Dynamic dielectric analysis: nondestructive material evaluation and cure cycle monitoring. Polym. Eng. Sci. **26**(5), 338–345 (1986)
37. J. Mijovic, J.M. Kenney, A. Maffezzoli, F. Bellucci, L. Nicolais, The principles of dielectric measurements for in-situ monitoring of composite processing. Compos. Sci. Technol. **49**(3), 277–290 (1993)
38. I. Singh, P.K. Bajpai, D. Malik, A.K. Sharma, P. Kumar, Feasibility study on microwave joining of green composites. Akademeia **1**(1), 1–6 (2011)
39. I. Singh, P.K. Bajpai, D. Malik, J. Madaan, N. Bhatnagar, Microwave joining of natural fiber reinforced green composites. Adv. Mater. Res. **410**, 102–105 (2012)
40. P.K. Bajpai, I. Singh, J. Madaan, Joining of natural fiber reinforced composites using microwave energy: experimental and finite element study. Mater. Des. **35**, 596–602 (2012)
41. S. Ali, P.K. Bajpai, I. Singh, A.K. Sharma: Curing of natural fibre-reinforced thermoplastic composites using microwave energy. J. Reinf. Plast. Composit. (2014)
42. A. Mali, A.K. Sharma, I. Singh, Microwave curing of natural fiber and synthetic fiber reinforced polymer matrix composites. Imanager's J. Mater. Sci. **1**(1), 8–14 (2013)
43. N. Sgriccia, M.C. Hawley, Thermal, morphological, and electrical characterization of microwave processed natural fiber composites. Compos. Sci. Technol. **67**(9), 1986–1991 (2007)

44. N. Sgriccia, M.C. Hawley, M. Misra, Characterization of natural fiber surfaces and natural fiber composites. Compos. A **39**(10), 1632–1637 (2008)
45. H. Zhang, A.K. Datta, Microwave power absorption in single-and multiple item foods. Food Bioprod. Process. **81**(3), 257–265 (2003)
46. G.S.J. Sturm, G.D. Stefanidis, M.D. Verweij, T.D.T. Van Gerven, A.I. Stankiewicz, Design principles of microwave applicators for small-scale process equipment. Chem. Eng. Process. Process Intensification **49**(9), 912–22 (2010)
47. E.T. Thostenson, T.W. Chou, Microwave processing: fundamentals and applications. Compos. A **30**, 1055–1071 (1999)
48. C. Saltiel, A.K. Datta, Heat and mass transfer in microwave processing. Adv. Heat Transf. **33**(1), 1–94 (1999)
49. A.C. Metaxas, R.J. Meredith, *Industrial Microwave Heating, Series 4* (Institution of Engineering Technology, 1983)
50. M. Gupta, E.W.W. Leong, W.L. Wong, *Microwaves and Metals* (Wiley, Hoboken, 2007)
51. E. Kubel, Advancement in microwave heating technology. Industr. Heating **62**, 43–53 (2005)
52. M. Sparks, *Ferromagnetic-Relaxation Theory* (McGraw Hill, New York, 1964)
53. D.E. Clark, D.C. Folz, J.K. West, Processing materials with microwave energy. J. Mater. Sci. Eng. A **287**, 153–158 (2000)
54. D.M. Pozar, *Microwave Engineering*, 2nd edn. (Wiley, Toronto, 2001), pp. 1–49
55. A.J. Moulson, J.M. Herbert, *Electroceramics: Materials, Properties, Applications*, 2nd edn. (Wiley, Toronto, 2003)
56. W.H. Sutton, Microwave processing of ceramics—an overview, in *Microwave processing of materials III, materials research society proceedings*. ed. by R.L. Beatty, M.F. Iskander, W.H. Sutton (Mater. Res. Soc., Pittsburgh, 1992), p. 3
57. R.M. Anklekar, K. Bauer, D.K. Agrawal, R. Roy, Improved mechanical properties and microstructural development of microwave sintered copper and nickel steel. Powder Metall. **48**(1), 39–46 (2005)
58. A. Moulart, C. Marrett, J. Colton: *Polymeric Composites for Use in Electronic and Microwave Devices*. Polymer Engineering and Science
59. D. Pramila, D.S.A. Nair, T. Jabin, S.K. Kutty, Mechanical, thermal, and microwave properties of conducting composites of polypyrene/polypyrene-coated short nylon fibers with acrylonitrile butadiene rubber. J. Appl. Polym. Sci. **126**(6), 1965–1976 (2012)
60. K. Sato, Y. Hotta, Carbon fiber/epoxy composite materials cured thermally and with microwave irradiation. Compos. Interfaces **22**(1), 67–74 (2015)
61. P.S. Mooteri, B.K. Sridhara, S. Rao, M.R. Prakash, R.M.V.G.K. Rao, Studies on mechanical behavior of microwave and thermally cured glass fiber reinforced polymer composites. J. Reinf. Plast. Compos. **25**(5), 503–512 (2006)
62. S. Ali, P.K. Bajpai, I. Singh, A.K. Sharma, Curing of natural fiber-reinforced thermoplastic composites using microwave energy. J. Reinf. Plast. Compos. **33**(11), 993–999 (2014)
63. M.K. Singh, S. Zafar, Development and mechanical characterization of microwave-cured thermoplastic based natural fibre reinforced composites. J. Thermoplast. Compos. Mater. **32**(10), 1427–1442 (2019)
64. M.K. Singh, S. Zafar, Influence of microwave power on mechanical properties of microwave-cured polyethylene/coir composites. J. Nat. Fibers **00**(00), 1–16 (2018)
65. A.A. Mohammed, D. Bachtiar, M.R.M. Rejab, J.P. Siregar, Effect of microwave treatment on tensile properties of sugar palm fibre reinforced thermoplastic polyurethane composites. Def. Technol. **14**(4), 287–290 (2018)
66. I. Singh, P.K. Bajpai, D. Malik, A.K. Sharma, P. Kumar : Feasibility study on microwave joining of green composites. J. Akademeia (2011)
67. I. Singh, P.K. Bajpai, J. Madan: Development and characterization of PLA-based green composites. J. Thermoplast. Compos. Mater. **27**(1), 52–81 (2014)
68. I. Singh, P.K. Bajpai, D. Malik, J. Madan, N. Bhatnagar, Microwave joining of natural fiber reinforced green composites. Adv. Mater. Res. **410**, 102–105 (2012)

69. P.K. Bajpai, I. Singh, J. Madan, Joining of natural fiber reinforced composites using microwave energy: experimental and finite element study. Mater. Des. **35**, 596–602 (2012)
70. P.K. Bajpai, I. Singh, J. Madan, Tribological behaviour of natural fiber reinforced PLA composites. Int. J. Sci. Technol. Frict. Lubric. Wear **297**, 829–840 (2013)
71. H.S. Ku, E. Siores, Elias, J.A.R. Ball, F. Siu: Processing of polymer matrix composites using variable frequency microwave, in 13th International Conference on Composite Materials (ICCM-13) 2001, Beijing, China
72. M.K. Lila, A. Singhal, S.S. Banwait, I. Singh, A recyclability study of bagasse fiber reinforced polypropylene composites. Polym. Degrad. Stab. **152**, 272–279 (2018)

# A Comparative Study on the Hydrodynamic and Heat Transfer Behaviour of Conical Fluidized Bed with that of a Columnar Pressurized Circulating Fluidized Bed

**Hirakh Jyoti Das, Rituraj Saikia, and Pinakeswar Mahanta**

**Abstract** The velocity distribution in the axial direction of a conical fluidized bed is not uniform due to the expansion of cross-sectional area along the height. As such, the hydrodynamic and heat transfer characteristics of a tapered fluidized bed differ from that of the columnar fluidized bed. Various operating parameters have a significant effect on these characteristics. In this paper, the effect of bed inventory and airflow rate on the hydrodynamic and heat transfer characteristics in a columnar and conical are investigated and compared both experimentally and numerically. A Eulerian-Eulerian model is employed to investigate the hydrodynamics and heat transfer behaviour for both the types of bed. The Syamlal-O'Brien model is used as a drag model. The pressure drop across the bed is found to higher for the columnar bed. The heat transfer is also found to be better for the conical bed. Results obtained by experiments are seen to be excellent agreement with the numerical results.

**Keywords** CFB · CFD · Heat transfer · Hydrodynamic · Pressure drop

## Nomenclature

| | |
|---|---|
| $A_h$ | Surface area of heat transfer |
| $C_D$ | Drag coefficient |
| $C_{pg}$ | Specific heat capacity of gas |
| $D_{ij}$ | Rate of strain tensor for solid phase |

H. J. Das (✉) · R. Saikia · P. Mahanta
Department of Mechanical Engineering, Indian Institute of Technology Guwahati, Guwahati 781039, Assam, India
e-mail: hirakh.das@iitg.ac.in

R. Saikia
e-mail: r.saikia@iitg.ac.in

P. Mahanta
e-mail: pinak@iitg.ac.in



K. M. Pandey et al. (eds.), *Recent Advances in Mechanical Engineering*, Lecture Notes in Mechanical Engineering, https://doi.org/10.1007/978-981-15-7711-6_53

| | |
|---|---|
| $\frac{D\overrightarrow{v_g}}{Dt}$ | Substantial time derivative for gas velocity |
| $V$ | Voltage |
| $I$ | Current |
| $\frac{D\overrightarrow{v_s}}{Dt}$ | Substantial time derivative for solids velocity |
| $e_{ss}$ | Particle–particle restitution coefficient |
| $\overrightarrow{F_{dg}}$ | Drag forces for gas phase |
| $\overrightarrow{F_{d,s}}$ | Drag forces for solid phase |
| $\overrightarrow{v_s}$ | Local velocity of the solid phase |
| $v_s'^2$ | Mean square velocity of particles |
| $\overrightarrow{F_{vm,g}}$ | Virtual masses for gas phase |
| $\nabla P_g$ | Gas-phase momentum equation |
| $\nabla P_s$ | Solids phase momentum equation |
| $\overline{S_s}$ | Solid-phase source term |
| $T_s$ | Surface temperature (K) |
| $\overrightarrow{v_g}$ | Local velocity of the gas phase |
| $d/D$ | Non-dimensional radial distance |
| $\overrightarrow{F_{vm,s}}$ | Virtual masses for solid phase |
| $g$ | Acceleration due to gravity (m/s$^2$) |
| $g_{o,ss}$ | Solid radial distribution function |
| $h$ | Heat transfer coefficient |
| $h_g$ | Enthalpy of gas |
| $Q_{sg}$ | Heat transfer between gas and solids |
| $h_s$ | Enthalpy of solids |
| $I_{2D}$ | Second invariant of the deviatory stress tensor |
| $k_g$ | Thermal conductivity of gas (W/m-K) |
| $k_s$ | Thermal conductivity of solids (W/m-K) |
| $P_{fr}$ | Frictional pressure (Pa) |
| $P_g$ | Gas-phase pressure (Pa) |
| $P_s$ | Solid-phase pressure (Pa) |
| $\Delta P$ | Pressure drop (Pa) |
| $q$ | Heat transfer between wall to bed (J) |
| $T_b$ | Bed temperature (K) |
| $v$ | Velocity vector |
| $v_{r,s}$ | Relative velocity of solid particles |

## Greek Symbol

| | |
|---|---|
| $\rho_s$ | Density of solid |
| $\varepsilon_g$ | Voidage of gas |
| $\varepsilon_{s,cr}$ | Critical solid volume fraction |
| $\phi$ | Angle of internal friction |
| $\mu_g$ | Viscosity of gas |

| | |
|---|---|
| $\overline{\overline{\tau_g}}$ | Gas-phase stress tensor |
| $k_{\Theta_s}$ | Diffusion coefficient |
| $\Theta_s$ | Granular temperature |
| $\gamma_w$ | Collisional dissipation at the wall |
| $\rho_g$ | Density of gas |
| $\varepsilon_s$ | Volume fraction of the solid phase |
| $\varepsilon_{s,max}$ | Maximum solid volume fraction |
| $\beta_{gs}$ | Drag coefficient between the gas and solid phase |
| $\mu_s$ | Solid viscosity |
| $\overline{\overline{\tau_s}}$ | Solid-phase stress tensor |
| $k_{\Theta_s}\nabla\Theta_s$ | The transport of energy due to diffusion |
| $\gamma_{\Theta s}$ | Dissipation of energy due to collision |
| $\left(-P_s\bar{I}+\bar{\tau}_s\right):\nabla\bar{v}_s$ | Generation of energy by the solid stress tensor |

## 1 Introduction

Design and development of Circulating fluidized beds (CFBs) is a key research subject in present times. CFBs are widely used in the power generation, chemical, petroleum and energy industries. Various designs of CFBs have been studied by researchers and results based on these have been presented. Appropriate designs play a significant role in the performance of a CFB. The gas-solid flow in the riser column of a conventional CFB consists mainly of a dilute core and a dense annulus region. But this core-annular property is seen to differ in case of a converging fluidized bed [1]. Also, the core annulus structure is observed to change due to the exit bend angle [2]. The core annulus structure also changes with the introduction of secondary lateral air jets [3]. An abrupt exit in the shape of a 'T' has a substantial restriction on the dilution of flows in a CFB [4]. The outlet configuration also affects the flow pattern in the lower and middle region of the riser apart from the exit region [5]. Sau and Biswal [6] made some investigations both experimentally and numerically on the hydrodynamic behaviours such as pressure drop and bed fluctuation ratio in a conical bubbling fluidized bed. Meer et al. [7] experimentally found that the solid down flux is much higher at the corners in a square cross-sectional CFB riser. Several researchers also reported the various aspects of particle distribution and mixing. Minimization of gas and particles back mixing prevails if the bed is divided into sub-sections [8]. Chalermsinsuwan et al. [9] numerically studied the hydrodynamic behaviour in three different types of bed. They found that turbulence of mixing of particles is more in the tapered-out riser. However, the particle residence time in tapered-in riser is found to be more enhanced and also the temperature distribution more uniform in the system. Chen et al. [10] also studied the behaviour of the particles in a convergent channel and observed that the convergent angle of the bed inevitably influences the concentration of particles.

From the literature review, it can be concluded that many investigations have been done on the hydrodynamics and chemical reactions in conventional fluidized beds. Moreover, some researches have also been done the same in conical fluidized beds. Abdelmotalib et al. [11] experimentally observed the hydrodynamic and heat transfer characteristics in bubbling conical fluidized bed combustor of height 0.8 m with 30° cone angle and compared their results with a numerical model. Gan et al. 2014 [12] also investigated the effect of cone angle on the minimum fluidization velocity and full fluidization velocity in a bubbling conical fluidized bed only. However, none has reported a comparative study on the heat transfer and hydrodynamics of a columnar and conical beds. Hence, an attempt is made to compare these two types of fluidized beds in terms of heat transfer and hydrodynamic characteristics, both experimentally and numerically. The pressure drop along riser and wall to bed heat transfer characteristics in the upper splash region in a columnar and conical pressurized circulating fluidized bed riser have been studied.

## 2 Experimental Method

Two different types of CFBs, a columnar and a conical CFB are used in the present investigation. A CFB unit comprising of a riser of 2000 mm height, a downcomer, and a cyclone separator with dimensions as shown in Fig. 1. The elutriation of particles from the riser is recirculated through the cyclone separator and returned to the riser with the help of a returned leg. The conical CFB is taken to be of the same height i.e. 2000 mm. The diameter at the middle of the riser was 50 mm which is same as the columnar CFB. The inlet and Outlet are 40 and 60 mm respectively. The cone angle is 0.3°. Air is supplied through the inlet of the fluidized bed from the high discharge blower (Fig. 2).

A rotameter is attached at the inlet of the riser to measure the airflow rate and a regulating valve regulates the airflow rate. A distributor plate having opening area of 16.9% is attached with the inlet to ensure the uniform distribution of airflow. Pressure taps are located along the riser height at 600 mm below and 35, 50, 200, 230, 370, 500, 650, 800, 1100, 1400, and 1700 mm above the distributor plate to measure the pressure drop. Water filled U-tube manometer is used to measure the pressure drop. A heat transfer probe of inlet diameter and height 54 and 550 mm respectively are located at the upper splash region of the riser as shown in Fig. 1. K type thermocouples are attached at riser heights of 1400, 1500, 1600, 1700 and 1800 mm from the distributor plate to measure the surface and bed temperatures. A heater coil of 1000 W capacity is wound over the probe. An autotransformer supplies electric power to the heater coil and the supplied power is measured with a multifunction meter. Ceramic wool and mica sheet of 1 mm thickness are used to provide adequate insulation. Agilent 34972A LXI data acquisition system is used to record the temperatures. Local sand granular particle is used as a bed inventory in this present investigation. The properties of air and granular sand particles is shown in Table 1. Experiments are performed at a constant energy flux of 2000 W/m$^2$ with

Fig. 1 Schematic of the columnar CFB

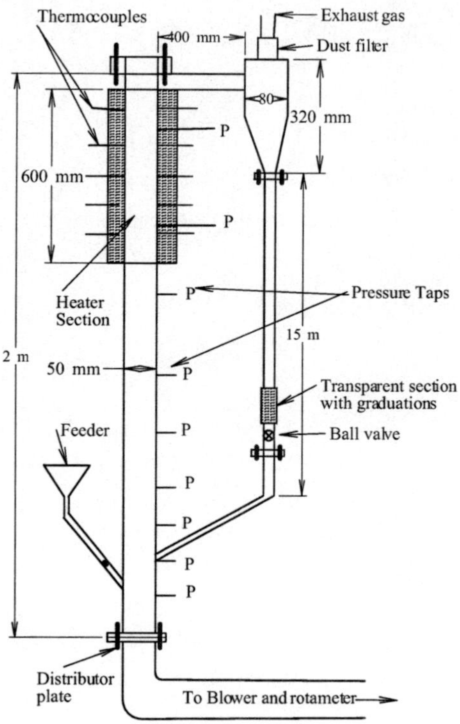


primary fluidizing airflow rates of 6.6 and 5.1 l/s. In all the experiments both axial and radial heat transfer coefficient is evaluated. The readings are noted after the steady-state is achieved. The wall-to-bed heat transfer coefficient is calculated from the following relationship (1).

$$h = \frac{q}{T_s - T_b} = \frac{V \times I}{A_h(T_s - T_b)} \tag{1}$$

# 3 Numerical Study

## 3.1 CFD Model

In this work, CFD simulations with FLUENT 14.5 are accomplished to investigate the hydrodynamics behaviour such as pressure drop, voidage of air as well as solids, suspension density along the riser height of conical fluidized bed. 2D simulations are implemented in the investigation of present riser of CFB. However, it is evident from the literature that 3D simulations are more accurate than the 2D simulation. A

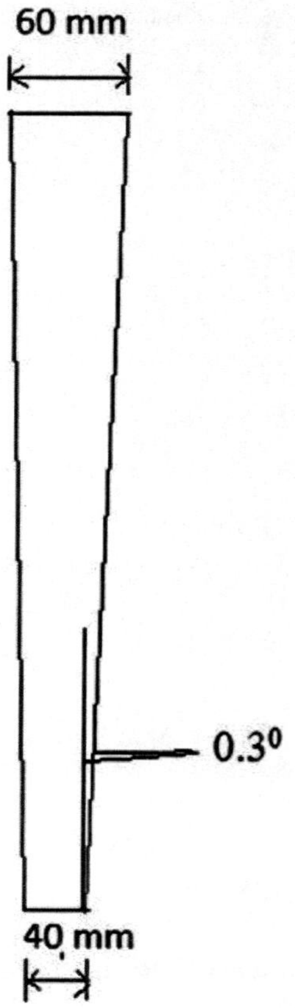


**Fig. 2** Schematic of the conical

**Table 1** Properties of air and sand particles

| Fluid and solid | Density (kg/m$^3$) | Thermal conductivity (W/m-K) | Specific heat (J/kg-K) |
|---|---|---|---|
| Air | 1.2 | 0.024 | 1005 |
| Sand | 2520 | 0.25 | 800 |

two-fluid approach Eulerian-Eulerian model is utilized to study the hydrodynamic behaviour in a conical fluidized bed. This model uses the kinetic theory of granular flow (KTGF) for the particle phase in which both phases are assumed to be continua. The gas phase is considered as the primary phase, whereas the particle phases are considered as the secondary or dispersed phases. The conservation equations are solved for both the phases. Details of the conservation equations are provided in Sect. 3.2. In the granular flow of two-phase simulation, a new era of concept is introduced as a volume fraction. The volume fraction of both phases cannot hold their region at the same time and these volume fractions are space and time-dependent (Table 2).

**Table 2** Simulation parameters

| Flow type | Turbulent, RNG method |
|---|---|
| Two-phase model | Eulerian-Eulerian |
| Time step used | 0.0002 s |
| Convergence criteria | 0.0001 |
| Solution controls | Pressure = 0.5<br>Momentum = 0.2<br>Volume and granular temperature = 0.2 |
| Maximum solid volume fraction | 0.63 |
| Pressure-velocity coupling | SIMPLE |
| Discretization scheme | Second-order upwind |
| Solid volume fraction | 0.59 |
| Mean particle diameter (μm) | 355–455 |

## *3.2 Governing Equation*

For gas,

$$\frac{\partial}{\partial t}\left(\varepsilon_{\mathrm{g}}\rho_{\mathrm{g}}\right)+\nabla.\left(\varepsilon_{\mathrm{g}}\rho_{\mathrm{g}}\vec{v}_{\mathrm{g}}\right)=0 \tag{2}$$

For solid

$$\frac{\partial}{\partial t}(\varepsilon_{\mathrm{s}}\rho_{\mathrm{s}})+\nabla.(\varepsilon_{\mathrm{s}}\rho_{\mathrm{s}}\vec{v}_{\mathrm{s}})=0 \tag{3}$$

Volume fraction equation:

$$\varepsilon_{\mathrm{g}}+\varepsilon_{\mathrm{s}}=1 \tag{4}$$

Momentum equation:
For gas

$$\frac{\partial}{\partial t}\left(\varepsilon_{\mathrm{g}}\rho_{\mathrm{g}}\vec{v}_{\mathrm{g}}\right)+\nabla.\left(\varepsilon_{\mathrm{g}}\rho_{\mathrm{g}}\vec{v}_{\mathrm{g}}\vec{v}_{\mathrm{g}}\right)=-\varepsilon_{\mathrm{g}}\nabla P_{\mathrm{g}}+\nabla\bar{\bar{\tau}}_{\mathrm{g}}+\varepsilon_{\mathrm{g}}\rho_{\mathrm{g}}\vec{g}+\vec{F}_{\mathrm{d.g}} \tag{5}$$

For solid

$$\frac{\partial}{\partial t}(\varepsilon_{\mathrm{s}}\rho_{\mathrm{s}}\vec{v}_{\mathrm{s}})+\nabla.(\varepsilon_{\mathrm{s}}\rho_{\mathrm{s}}\vec{v}_{\mathrm{s}}\vec{v}_{\mathrm{s}})=-\varepsilon_{\mathrm{s}}\nabla P_{\mathrm{g}}-\nabla P_{\mathrm{s}}+\nabla\bar{\bar{\tau}}_{\mathrm{s}}+\varepsilon_{\mathrm{s}}\rho_{\mathrm{s}}\vec{g}+\vec{F}_{\mathrm{d.s}}+\bar{S}_{\mathrm{s}} \tag{6}$$

Energy equation:
For gas,

$$\frac{\partial}{\partial t}\left(\varepsilon_g \rho_g h_g\right) + \nabla.\left(\varepsilon_g \rho_g \vec{v}_g h_g\right) = -\varepsilon_g \frac{\partial P_g}{\partial t} + \bar{\bar{\tau}}_g : \nabla \vec{v}_g + \nabla.\left(\varepsilon_g k_g \nabla T_g\right) + Q_{sg} \tag{7}$$

For solid,

$$\frac{\partial}{\partial t}(\varepsilon_s \rho_s h_s) + \nabla.(\varepsilon_s \rho_s \vec{v}_s h_s) = -\varepsilon_s \frac{\partial P_s}{\partial t} + \bar{\bar{\tau}}_s : \nabla \vec{v}_s + \nabla.(\varepsilon_s k_s \nabla T_s) - Q_{sg} \tag{8}$$

Constitutive equations for closure of the governing equations:

$$\vec{F}_{d.g} = \beta_{g.s}(\vec{v}_s - \vec{v}_g),\ \vec{F}_{d.s} = -\vec{F}_{d.g} \tag{9}$$

Syamlal-O'Brien drag model [13]:

$$\beta_{gs} = \frac{3}{4}\frac{\varepsilon_g \varepsilon_s \rho_g}{v_{r,s}^2 d_s} C_D\left(\frac{Re_s}{v_{r,s}}\right)\left|\vec{v}_s - \vec{v}_g\right| \tag{10}$$

The granular temperature is a function of collision frequency of particles and is defined by the following expression

$$\Theta_s = \frac{1}{3} v_s'^2 \tag{11}$$

Kinetic fluctuation energy:

$$\frac{3}{2}\left[\frac{\partial}{\partial t}(\rho_s \varepsilon_s \Theta_s) + \nabla.(\rho_s \varepsilon_s \vec{v}_s \Theta_s)\right] = \left(P_s \bar{I} + \bar{\tau}_s\right) : \nabla \vec{v}_s + \nabla.(k\Theta_s \nabla \Theta_s) - \gamma \Theta_s + \phi_{gs} \tag{12}$$

$$\phi_{gs} = -3 k_{gs} \Theta_s \tag{13}$$

Collisional dissipation of energy:

$$\gamma \Theta_s = \frac{12\left(1 - e_{ss}^2\right) g_{o,s}}{d_s \sqrt{\pi}} \rho_s \varepsilon_s^2 \Theta_s^{3/2} \tag{14}$$

Diffusion coefficient of energy:

$$k\Theta_s = \frac{150 \rho_s d_s \sqrt{\Theta_s \pi}}{384(1 + e_{ss}) g_{o,ss}}\left[1 + \frac{6}{5} g_{o,ss} \varepsilon_s (1 + e_{ss})\right]^2 + 2 \rho_s \varepsilon_s^2 d_s (1 + e_{ss}) g_{o,ss} \sqrt{\frac{\Theta_s}{\pi}} \tag{15}$$

Solid pressure:

$$P_s = \rho_s \varepsilon_s \Theta_s + 2 \rho_s (1 + e_{ss}) \varepsilon_s^2 g_{o,ss} \Theta_s \tag{16}$$

Radial distribution function [14]:

$$g_{o,ss} = \frac{1 + 2.5\varepsilon_s + 4.59\varepsilon_s^2 + 4.52\varepsilon_s^3}{\left[1 - \left(\frac{\varepsilon_s}{\varepsilon_{s,max}}\right)^3\right]^{0.678}} \tag{17}$$

Collision viscosity:

$$\mu_{s.col} = \frac{4}{5}\varepsilon_s d_s \rho_s (1 + e_{ss}) g_{o,ss} \sqrt{\frac{\Theta_s}{\pi}} \tag{18}$$

Solids bulk viscosity [15]:

$$k_s = \frac{4}{3}\varepsilon_s d_s \rho_s (1 + e_{ss}) g_{o,ss} \sqrt{\frac{\Theta_s}{\pi}} \tag{19}$$

The frictional viscosity as defined by Schaeffer [16] is

$$\mu_{s.fr} = \frac{P_{fr}\sin(\phi)}{\sqrt{I_{2D}}} \tag{20}$$

### *3.3 Solution Procedure*

The governing equations given are solved by commercial CFD Software FLUENT 14.5 and bed hydrodynamics across the riser are investigated in a conical fluidized bed by considering sand as a bed inventory. The inlet boundary condition is the velocity inlet where air is distributed uniformly and solid are patched up to certain height. However, both the walls are considered as stationary wall in which specific shear condition is considered for solid particles and no-slip condition for air phase. The 2D computational domain is discretized by three different elements sizes such as 6000, 9600 and 14,340 elements having rectangular cells.

## 4 Results and Discussions

### *4.1 Grid Independence Test*

The grid independence test is performed for three mesh sizes having 6000; 9600 and 14340 elements. The axial solid volume fraction obtained for these grids are found to be not significantly different. Hence 9600 elements are used for the numerical

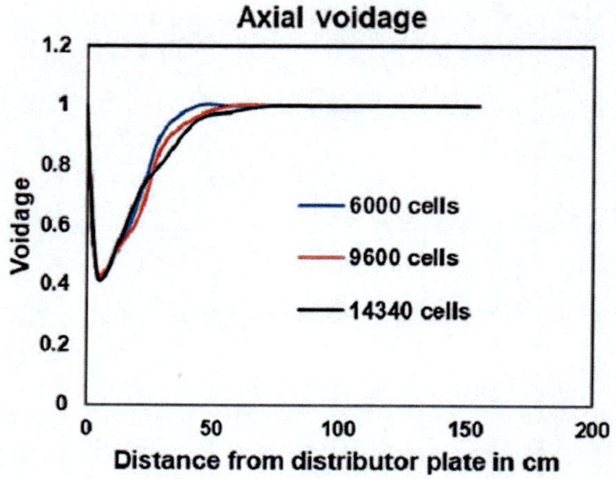


**Fig. 3** Axial voidage at different grid sizes

simulation. The experimental results obtained from both the different types of CFBs are shown below. Figure 4 shows a comparison of pressure drop in the axial direction for the columnar and the conical beds for different flow rates. Figure 5 shows the same for different bed inventories (Fig. 3).

From Fig. 4, it is observed that the pressure drop increases up to about 250 mm and then falls in the bottom region at 500 mm from the distributor plate, after that, it remains almost constant in the upper zone. This result is in agreement with earlier literature as observed by Kwauk et al. [17] and explained by Basu [18]. The pressure drop in the axial direction is also seen to decrease with an increase in airflow rate from 5.1 L/S to 5.6 L/S for both the beds. This is due to the decrease in the concentration of solid particles with an increase in air flow rate. Further, it is noticed that the pressure drop is comparatively more in the columnar bed than that in the conical bed. This is because of the drop in axial velocity of the fluidizing air in conical bed with the increasing cross-sectional area. The bed inventory plays a vital role in the

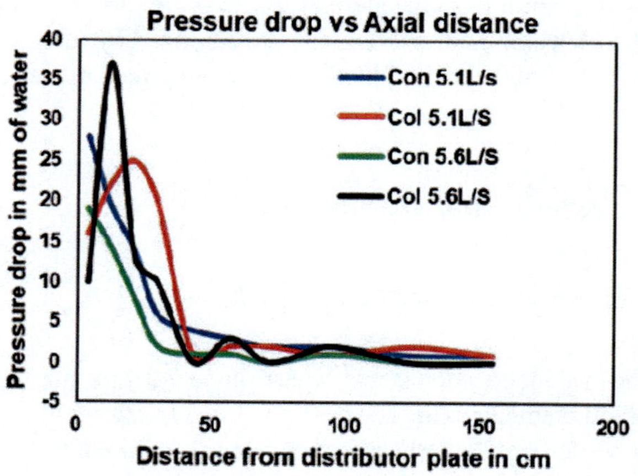


**Fig. 4** Comparison of pressure drop for various airflow rate with bed inventory of 400 g

pressure drop along the axial direction for both the CFBs as observed in Fig. 5. It is seen that the bed pressure is more when the bed inventory is increased from 400 to 450 g. It is because with the increase in bed inventory, the concentration of solid particles increases resulting in higher pressure. The comparison of pressure drop along the axial direction for the two types of CFBs shows that the pressure drop is less for conical bed than that for the columnar bed. The comparison of the axial heat transfer coefficient for the columnar and the conical beds for two flow rates and bed inventories are shown below in Figs. 6 and 7. The trend indicates that the heat transfer coefficient increases axially along the height for both the beds. This is because of the increased particle concentration near the riser exit due to more drag force acting on particles being lifted to the riser exit from the lower splash. The same trend was also observed by Kalita et al. [19]. The effect of convection may also be another reason for increased heat transfer in the upper region. Airflow rate also plays a significant

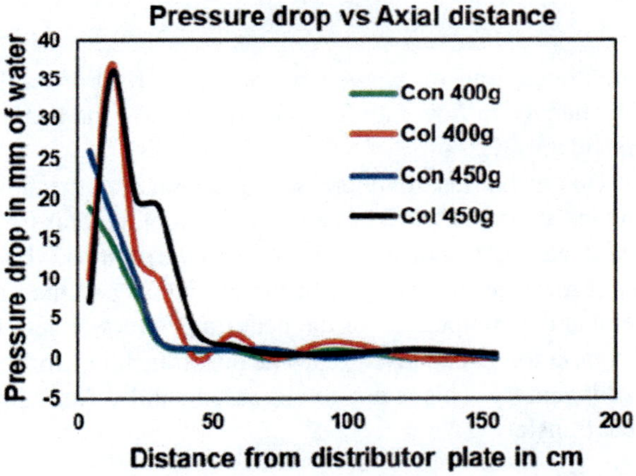


**Fig. 5** Comparison of pressure drop for various bed inventories with an airflow rate of 5.1 L/s

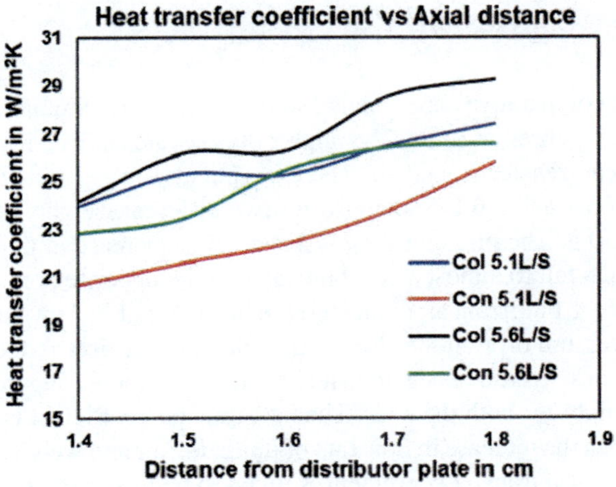


**Fig. 6** Comparison of heat transfer coefficient for various airflow rate with the bed inventory of 400 g

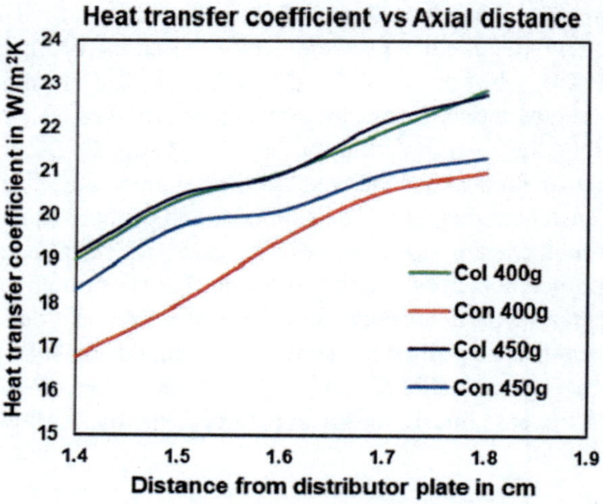


**Fig. 7** Comparison of heat transfer coefficient for various bed inventories with flow rate of 5.1 L/s

role in heat transfer phenomena along the riser. From Fig. 5, it is noticed that with the changes in flow rate from 5.1 to 5.6 L/S, the heat transfer coefficient increases considerably.

This is because of increased turbulence of the flow at a higher flow rate and enhanced circulation. The heat transfer coefficient is found to be higher for columnar bed in comparison to conical bed in all the regions. This is because for the same heat input the heat flux is more in the columnar bed than in the tapered bed. Figure 7 depicts the comparison of the heat transfer coefficient between the two CFBs with the variation in bed inventory. The heat transfer is observed to be increasing with the bed inventory. This is due to increased solid volume fraction which results in more heat transfer.

## 5 Conclusions

A comparative experimental study is made on a circular columnar and a conical CFB in this present study. The impact of flow rate and bed inventory on pressure drop and heat transfer is studied. The experiments are performed at two different flow rates of 5.1 and 5.6 L/S and also for two different weights of bed inventory viz 400 and 450 g. The pressure drop is observed to increase initially up to about 250 mm and then fall to almost a constant value in the upper region. An increase in the flow rate of the fluidizing air or an increase in weight of bed inventory is found to decrease the pressure drop in the risers. Also, the conical riser is found to show lesser pressure drops. In case of heat transfer, it is found an increasing trend along the axial direction height for both the beds. The heat transfer coefficient is also found to be increasing with the increase in flow rate of fluidizing air and weight of bed inventory. Moreover, the heat transfer coefficient is found to be less in the riser of the conical bed.

## References

1. L. Huilin, Z. Yunhua, S. Zhiheng, J. Ding, J. Jiying, Numerical simulations of gas-solid flow in tapered risers. Powder Technol. **169**(2), 89–98 (2006)
2. S.K. Gupta, F. Berruti, Evaluation of the gas-solid suspension density in CFB risers with exit effects. Powder Technol. **108**(1), 21–31 (2000)
3. A. Marzocchella, U. Arena, Hydrodynamics of a circulating fluidized bed operated with different secondary air injection devices. Powder Technol. **87**(3), 185–191 (1996)
4. Y. Cheng, F. Wei, G. Yang, Y. Jin, Inlet and outlet effects on flow patterns in gas-solid risers. Powder Technol. **98**(2), 151–156 (1998)
5. J.R. Grace, Influence of riser geometry on particle and fluid dynamics of circulating fluidized beds. Circ. Fluid. Bed Technol. 16–23 (1997)
6. D.C. Sau, K.C. Biswal, Computational fluid dynamics and experimental study of the hydrodynamics of a gas-solid tapered fluidized bed. Appl. Math. Model. **35**(5), 2265–2278 (2011)
7. E.H. van der Meer, R.B. Thorpe, J.F. Davidson, Flow patterns in the square cross-section riser of a circulating fluidised bed and the effect of riser exit design. Chem. Eng. Sci. **55**(19), 4079–4099 (2000)
8. S.R.A. Kersten, W. Prins, B. van der Drift, W.P.M. van Swaaij, Principles of a novel multistage circulating fluidized bed reactor for biomass gasification. Chem. Eng. Sci. **58**(3–6), 725–731 (2003)
9. B. Chalermsinsuwan, P. Piumsomboon, D. Gidaspow, Kinetic theory based computation of PSRI Riser: Part II—computation of mass transfer coefficient with chemical reaction. Chem. Eng. Sci. **64**(6), 1212–1222 (2009)
10. R.Y. Chen, H.C. Chiou, D. Sun, Deposition of particles in a convergent channel. Powder Technol. **87**(1), 83–86 (1996)
11. H.M. Abdelmotalib, D.G. Ko, I.-T. Im, A study on wall-to-bed heat transfer in a conical fluidized bed combustor. Appl. Therm. Eng. **99**, 928–937 (2016)
12. L. Gan, X. Lu, Q. Wang, Experimental and theoretical study on hydrodynamic characteristics of tapered fluidized beds. Adv. Powder Technol. **25**(3), 824–831 (2014)
13. M. Syamlal, T.J. O'Brien, Simulation of granular layer inversion in liquid fluidized beds. Int. J. Multiph. Flow **14**(4), 473–481 (1988)
14. D. Ma, G. Ahmadi, An equation of state for dense rigid sphere gases. J. Chem. Phys. **84**(6), 3449–3450 (1986)
15. C.K.K. Lun, S.B. Savage, D.J. Jeffrey, N. Chepurniy, Kinetic theories for granular flow: inelastic particles in couette flow and slightly inelastic particles in a general flowfield. J. Fluid Mech. **140**(1), 223–256 (1984)
16. D.G. Schaeffer, Instability in the evolution equations describing incompressible granular flow. J. Differ. Equ. **66**(1), 19–50 (1987)
17. M. Kwauk, W. Ningde, L. Youchu, C. Bingyu, S. Zhiyuan, *Fast Fluidization at ICM. Circulating Fluidized Bed Technology* (Elsevier, Amsterdam, 1986), pp. 33–62
18. P. Basu, *Combustion and Gasification in Fluidized Beds* (Taylor & Francis Group, LLC, Halifax, Nova Scotia, 2006)
19. P. Kalita, U.K. Saha, P. Mahanta, Parametric study on the hydrodynamics and heat transfer along the riser of a pressurized circulating fluidized bed unit. Exp. Therm. Fluid Sci. **44**, 620–630 (2013)

# Study of the Effect of Welding Current on Heat Transfer and Melt Pool Geometry on Mild Steel Specimen Through Finite Element Analysis

**Mohd Aslam and Chinmaya Kumar Sahoo**

**Abstract** Mild steel is one of the economical and widely used raw materials in engineering applications. The mild steel specimen could be easily fabricated using conventional arc welding, MMA welding, and gas welding process. TIG welding is one of the conventional welding techniques used in precision welding both for ferrous and nonferrous material. However, it is difficult and uneconomical to detect welding penetration and effect of the welding parameter through metallurgical analysis. Numerical analysis of welding process with the help of computational facility is economical with time saving. In this work, a finite elemental model has been developed to simulate the TIG welding process using bead on plate welding. The Gaussian heat model has been used to find the effect of welding current on weld pool geometry. The results showed that welding current has a maximum effect on weld pool's temperature and geometry. Welding using high current produced high temperature in the weld pool, which leads to high penetration depth in the material. Similarly, the results of the numerical study also showed that with increase in current, the temperature at the center of mild steel increases.

**Keywords** Gaussian heat source moving heat source · TIG welding · Numerical simulation

## Nomenclature

$V$ Voltage
$I$ Current
$n$ Overall efficiency

---

M. Aslam · C. K. Sahoo (✉)
Department of Mechanical Engineering, National Institute of Technology, Silchar, Assam, Silchar 788010, Assam, India
e-mail: chinmaya.cks@gmail.com

M. Aslam
e-mail: aslammech98@gmail.com



K. M. Pandey et al. (eds.), *Recent Advances in Mechanical Engineering*, Lecture Notes in Mechanical Engineering, https://doi.org/10.1007/978-981-15-7711-6_54

$e$ Effectiveness
$r$ Radial coordinate
$r_0$ Radius of electrode
$v$ Velocity of welding speed
$T$ Temperature
$T_0$ Ambient temperature
$Q$ Effective heat flux
$k$ Thermal conductivity
$\rho$ Density of material
$C_p$ Heat capacity
$\sigma$ Young's modulus
$t$ Time
Y Young's modulus
$\alpha$ Coefficient of thermal expansion

# 1 Introduction

Arc-based welding TIG and MIG are widely used in automobiles, heavy constructions, and shipbuilding sectors. The TIG-based welding is famous for its economical availability and ability to weld nonferrous material like Al, Mg, and Cu. TIG welding is widely used in industry for material processing applications. Arc welding heat source can be approximated by Gaussian, double ellipsoidal, and equivalent double Gaussian heat source [1]. However, it is difficult and time consuming to check the size of the weld pool after the experiment of the cladding process. The numerical analysis is one of the preferred methods for the analysis of welding process. The quality of the numerical analysis depends on the assumptions and close approximation of the welding process. The quality of welding depends on the arc welding parameter and external parameters like the welding area and preheating of the welding area. Yue et al. and Artinov et al. suggested that the actual laser/TIG heat source could be approximated mathematically by considering heat source as cylindrical, conical, and combination of both cylindrical and conical for numerical analysis [1, 2]. The process like laser treatment and TIG welding/cladding can be numerically and analytically solved by considering the moving heat source model [2]. Yue et al. [1] developed numerical model considering Gaussian heat source and moving heat load for simulation of TIG welding process. Authors have also checked the validity of the model experimentally. Chen et al. [3] developed model to simulate the welding process and predict the bead geometry, and the prediction of the model showed a good agreement with the experimental result. Edgar et al. [4] developed a numerical model, (finite pointset method) suitable for welding process using moving heat source and numerical simulation result are comparable with finite element method.

Ninpetch and Kowitwarangkul performed a numerical study by moving a laser heat source on steel plate, and the laser heat source was assumed to be Gaussian-type heat flux [5]. The experimental results established the utilization of Gaussian heat flux for the numerical analysis. Fu et al. used a double ellipsoidal heat source model to predict multi-pass welding and the productive model found to be suitable for real-time application [6]. Parkitny and Winczek developed analytical model considering Gaussian power density distribution and considering moving heat source [7]. The model was reported to be suitable for cladding, laser treatment, and welding applications. Similarly, Liu et al. also used Gaussian heat source and ellipsoid heat source for considering equivalent heat source for study of the heat effect of the arc [8]. Chen et al. developed a three-dimensional numerical model for laser-MIG hybrid welding [9]. The authors validated the mathematical model with the help of experimental results. The present study considers the Gaussian power density distribution for FEM analysis of TIG welding process considering transient thermal heat transfer model.

## 2 Analytical Design of Numerical Simulation

The numerical simulation performed at an ambient temperature of 293.15 K and numerical simulation based on the Gaussian moving heat source on the mild steel substrate material. The numerical simulation of the Gaussian moving heat source is carried out at 70, 80, and 90 A, respectively, and at a constant arc scan speed of 3 mm/s during the welding process. The radius of the heat source is considered as 3 mm. The substrate dimension of mild steel was 45 mm × 45 mm × 6 mm. Due to thermal conductivity behavior, the temperature rises rapidly at moving welding spot place and it take time for cooling. All numerical investigations are carried out at 14 V, the effectiveness is 0.5, the convective heat transfer rate is 80 w/m$^2$/K, and overall efficiency is assumed to be 0.48. Details of physical and thermal properties of mild steel are in Table 1, and the welding parameters used for the analysis are in Table 2.

**Table 1** Physical properties of mild steel

| Material | Thermal conductivity (W/m/K) | Density (kg/m$^3$) | Heat capacity ($C_p$) (J/kg/K) | Electrical conductivity (MS/m) | Relative permittivity (unitless) | Young's modulus (Pa) | Coefficient of thermal expansion (1/K) |
|---|---|---|---|---|---|---|---|
| Mild steel | 46 | 7850 | 4200 | 1.74 | 1 | 200 × $10^9$ | 17.3 × $10^{-6}$ |

**Table 2** Welding parameters used for

| Peak current $I$/(A) | Voltage $V$/(V) | Emissivity ($e$) | Efficiency( $\eta$) | The radius of the hot spot $r$ (mm) | The velocity of hot spot $v$ (mm/s) |
|---|---|---|---|---|---|
| 70 A, 80 A, 90 A | 14 | 0.5 | 0.48 | 3 | 3 |

# 3 Methodology

## *3.1 Finite Element Modeling*

A finite element model was developed to find the effect of welding current for moving heat source in TIG welding using COMSOL Multiphysics. For the numerical simulation, a mild steel plate of dimension 45 mm × 45 mm × 6 mm was selected. Bead on Plate welding was numerically simulated to observe the effect of the welding parameter. For the numerical simulation, the material properties and welding parameters used are in Tables 1 and 2, respectively. In the numerical analysis, Gaussian-welding parameters such as welding current, voltage, overall efficiency, the radius of the heat source, and velocity of heat source are assumed as a constant. A CAD model of the mild steel workpiece was generated in the COMSOL Multiphysics environment and meshed for numerical analysis. Finer meshing was used for the heat source moving area and relatively coarser meshing for the rest of the workpiece to reduce the computational time. Figure 1 shows the CAD model with finer meshing for heat source moving area and coarse meshing for the rest of the area. The models contain the 44411-domain element, 5824, and 421 for boundary elements and edge element, respectively.

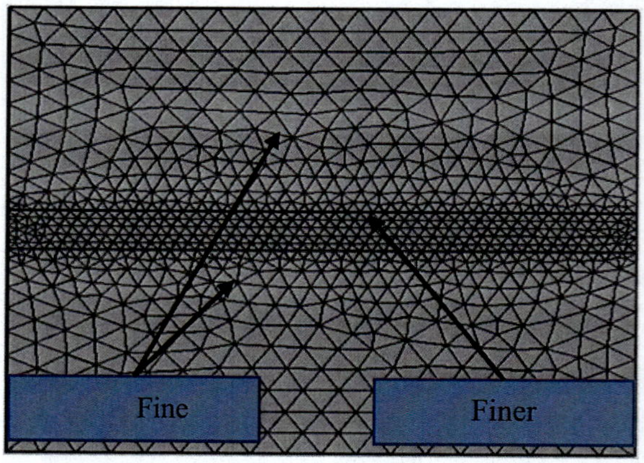


**Fig. 1** Top view of Meshed geometry of CAD model

## 3.2 *Mathematical Model of the Gauss Equation*

The effective heat input ($Q_0$) during the TIG welding process is represented by

$$Q_0 = V I \eta \tag{1}$$

where $V$, $I$ and $\eta$ are applied volage, current and welding efficiency respectively. The TIG/MIG heat source can be assumed as Gaussian in nature, and the effective heat input ($Q$) can be represented by Eq. (2) [1].

$$Q = \frac{3VIn}{\pi Rq^2} \times e^{\left(-3\frac{\left((x-vq\times t)^2+y^2\right)}{Rq^2}\right)} \tag{2}$$

where $Q$ is heat flux (Effective heat input/area of effective heat source). $t$ is the computation time in second, $v_q$ is the welding/scanning speed, $R_q$ is the radius of moving heat spot, where $Q$ is heat flux input, and $v_q$ is the welding/scanning speed.

# 4 Results and Discussion

Figure 2 represents the effect of the Gaussian heat source on the mild steel workpiece during the welding process at 70, 80, and 90 A current and 3 mm/s scan speed. To visualize the effect of the heat source, the cross-sectional view for 70, 80, and 90 A current and 3 mm/s scan speed was represented in Fig. 2a–c, respectively. The cross-sectional view has been represented when the TIG heat source reached at the center of the plate, i.e., after 7.5 s of welding initiation from the edge. The numerical simulation result showed that at 70 A current, the maximum temperature of the weld pool reached up to 1800 K, with an increase in current the maximum temperature of the weld pool increased to 1920 and 2200 K for 80 and 90 A, respectively. The melt pool temperature reduces with an increase in the depth of the weld pool. For the 70, 80, and 90 A current, the melt pool depth is observed to be 0.212, 0.489, and

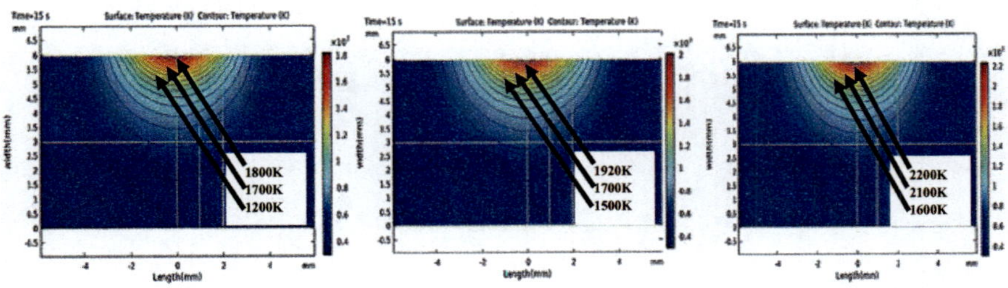


**Fig. 2** Cross-sectional image of the weld pool at **a** 70 A, **b** 80 A, **c** 90 A current and 3 mm/s welding speed at 7.5 s

0.788 mm, respectively. Similarly, the width of the weld pool and heat-affected zone has been governed by the welding current. During welding, the input heat is observed to be distributed uniformly, similar to the reverse of the Gaussian heat source. The maximum melting temperature on the workpiece was observed in the middle of the applied heat source, and the temperature decreases away from the center of the workpiece. Similar results were published earlier [1]. Figure 3a shows a top view of the heat profile built during numerical simulation of welding. The image represents the temperature distribution with the help of isothermal lines. For simplification of the analysis, the average temperature from the center zone (marked as '1') and point marked as '2' and '3' are 1 and 2 mm away from point '1', respectively. Figure 3a represents the cross-sectional view of Fig. 3b representing the temperature distribution in the thickness of the mild steel specimen. Figure 3a clearly shows that the temperature decreases away from the weld area. For a clear understanding of the heating and cooling phenomena, change of temperature in the middle section of the workpiece is represented with the help of Fig. 4a–c for 70, 80, and 90 A, respectively. The graphs (Fig. 5a–c) have been representing the cooling/heating rate

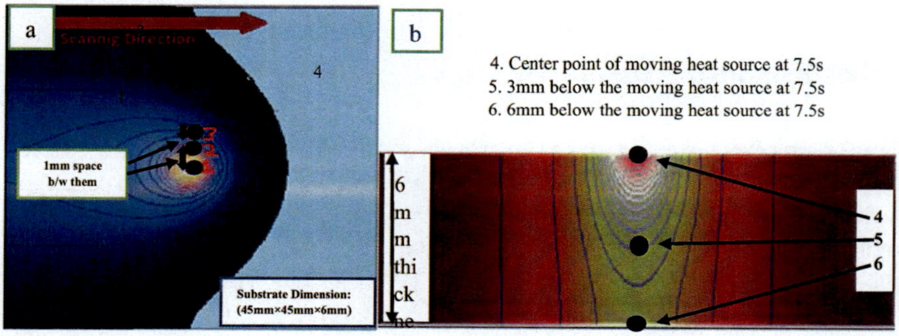


**Fig. 3** **a** Pictorial view at **a** top view. **b** cross-sectional view during welding representing isothermal lines showing temperature distortion during numerical simulation of TIG welding at welding current 70 A, and welding scan speed of 3 mm/s.

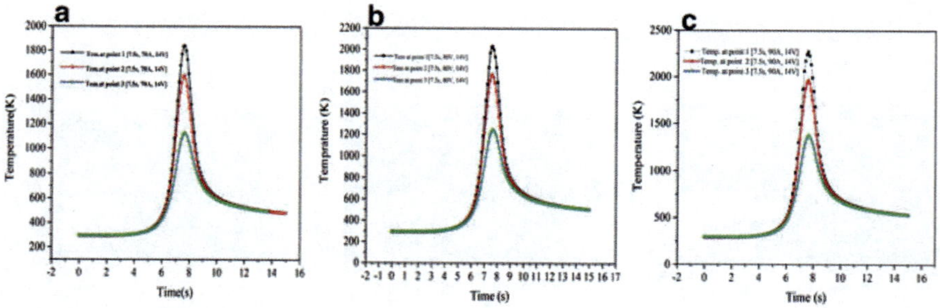


**Fig. 4** Thermal heating/cooling cycle at different observed point (marked as on Fig. 3a for welding at **a** 70 A current, **b** 80 A current, **c** 90 A current, and a scan speed of 3 mm/s

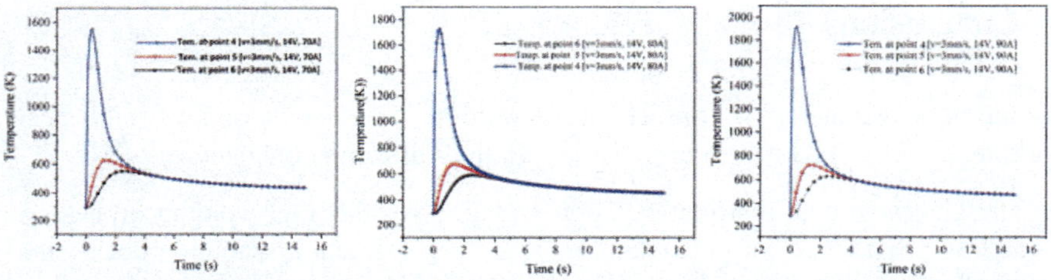


Thermal heating/cooling cycle at different observed point (marked as on Fig. 3. b) for welding at (a) 70A current (b) 80A current (c) 90A current and a scan speed of 3mm/s

**Fig. 5** Thermal cycle curve at different observation point 4, 5, 6 as shown in Fig. 3b. Thermal heating/cooling cycle at different observed point (marked as on Fig. 3b) for welding at **a** 70 A current, **b** 80 A current, **c** 90 A current, and a scan speed of 3 mm/s

as mentioned in Fig. 4a–c. Figure 4 shows the rise in temperature with an increase in the current. The numerical simulated result shows the maximum temperature in the weld pool which is about 1900, 2100, and 2200 K for sample processed using the welding current of 70, 80, and 90 A, respectively. Figure 5 clearly shows the sudden rise in temperature when the heat source reached the point '1', which is 22.5 mm distance from initiation of the welding. The weld pool temperature is quite high during the welding, and after welding the temperature of the heated surface takes considerable time for cooling. The heat is observed to penetrate three dimensionally inside the workpiece. However, the rise of temperature in the front section of the workpiece material is quite low as compared to post-heated portions. The heat input during the welding process is governed by parameters such as current, voltage, and scan speed. As the welding voltage and scan speed are constant for the study, so the heat input during the welding process only depends on the welding current, i.e., the heat input is directly proportional to the welding current. With the increase in welding current from 70 to 90 A, overall heat input during welding increases, which raises the temperature from 1800 to 2200 K. Figures 4a–c and 5a–c represent similar phenomena. The heat supplied during the welding raises the temperature of the mild steel specimen above its melting temperature.

However, heat conduction in the mild steel specimen distributes heat to a larger area. The heat is taken away from the weld pool by the adjacent material. Therefore, during welding, the temperature is quite high (Fig. 4a–c), and it decreases due to heat transfer through conduction, convection, and radiation. The cooling curve marked with '1', '2' '3' in Fig. 4a–c showed the same phenomena. After posting the welding period, the overall temperature of the workpiece is considerably high, and the temperature remains high for a considerable amount of time. Similar phenomenon observed is represented in Fig. 3a, b along the thickness direction of the weld pool. The melting point of mild steel generally used is between 1623 and 1803 K.

# 5 Conclusions

The numerical analysis of the effect of welding parameters on the mild steel workpiece for the TIG heat source has drawn the following conclusion.

1. The Gaussian heat distribution approximation for the TIG welding process is able to provide a clear understanding regarding the heat distribution during the welding process.
2. The welding current governs the heat input during the welding process. Welding using low TIG current showed low penetration of heat in the mild steel specimen, and with an increase in welding current the heat penetration in the mild steel workpiece increases. The melt pool increases from 0.212 to 0.788 mm along with depth and 1.377 to 3.270 mm along with width for increasing welding current from 70 to 90 A.
3. During the welding process, the high amount of the temperature is observed at the center of the melt pool and the temperature decreases away from the melt/weld pool.
4. After welding, the temperature of the welding decreases. However, the workpiece remains in high temperature for a considerable amount of time.

**Acknowledgements** The National Institute of Technology, Silchar, Assam India, of Mechanical Department, this work funded by MHRD through TEQIP-III.

**Conflict of Interest** The author declares that there is no conflict of interest.

# References

1. J. Yue, X. Dong, R. Guo, W. Liu, L. Li, Numerical simulation of equivalent heat source temperature field of asymmetrical fillet root welds. Int. J. Heat Mass Transf. **130**, 42–49 (2019)
2. A. Artinov, M. Bachmann, M. Rethmeier, Equivalent heat source approach in a 3D transient heat transfer simulation of full-penetration high power laser beam welding of thick metal plates. Int. J. Heat Mass Transf. **122**, 1003–1013 (2018)
3. J. Chen, C. Schwenk, C.S. Wu, M. Rethmeier, Predicting the influence of groove angle on heat transfer and fluid flow for new gas metal arc welding processes. Int. J. Heat Mass Transf. **55**, 102–111 (2012)
4. E.O Resendiz-Flores, F.R. Saucedo-Zendejo, Two-dimensional numerical simulation of heat transfer with moving heat source in welding using the Finite Pointset Method. Int. J. Heat Mass Transf. **90**, 239–245 (2015)
5. P. Ninpetch, P. Kowitwarangkul, A numerical study on the thermal transient model with moving laser heat source of AISI 304 stainless steel plate, in Materials Today Proceeding 2019, vol. 17, pp. 1761–1767
6. G. Fu, J. Gu, M.I. Lourenco, M. Duan, S.F. Estefen, Parameter determination of double-ellipsoidal heat source model and its application in the multi-pass welding process. Ships Offshore Struct. **10**, 204–217 (2014)
7. R. Parkitny, J. Winczek, Analytical solution of temporary temperature field in half-infinite body caused by moving tilted volumetric heat source. Int. J. Heat Mass Transf. **60**, 469–479 (2013)

8. K. Liu, F. Yang, S. Wang, B. Gao, Xu. Chong, The research on the heat source characteristics and the equivalent heat source of the arc in gaps. Int. J. Heat Mass Transf. **124**, 177–189 (2018)
9. X. Chen, Yu. Gang, X. He, S. Li, Z. Li, Numerical study of heat transfer and solute distribution in hybrid laser-MIG welding. Int. J. Therm. Sci. **149**, 106182 (2020)

# Study of Granular Food Material Drying in a Pilot-Scale Rotating Fluidized Bed with Static Geometry Dryer

**Pavitra Singh, Pankaj Kalita, Pinakeswar Mahanta, and Hirakh Jyoti Das**

**Abstract** A rotating fluidized bed with static geometry (RFB-SG) drying is a promising technique that is useful for various operations, such as agglomeration, food grain drying, particle coating, separation, and combustion. The advantage of this technique is that a large volume of hot air is circulated across the particles in a very small geometry, which results in higher heat and mass transfer. The higher heat and mass transfer through the RFB-SG dryer makes the drying process faster. Initially, the high-velocity air is injected into the vortex chamber through multi-air inlets, and then the solid particles are inserted into the vortex chamber. The high-velocity air injected into the reactor forces the solid particle to rotate in the form of a solid bed. The air entering into the vortex chamber carries away the moisture of food grains via a centrally located chimney outlet. In the present work, performance of scaled-up RFB-SG dryer has been evaluated considering parameters, such as temperature (55–65 °C), airflow rate (600–800 $m^3$/h), inventory (400–1000 g), and drying time. The RFB-SG dryer is found to be more efficient than the conventional fluidized bed (CFB) dryers as this dryer works on a higher airflow rate. Drying efficiency is improved by better utilization of the drying air at a temperature of 65 °C.

**Keywords** RFB-SG · Drying · Fluidization · Inventory

P. Singh (✉) · P. Kalita
Centre for Energy, Indian Institute of Technology Guwahati, Guwahati 781039, Assam, India
e-mail: pavitrasingh.mech@gmail.com

P. Kalita
e-mail: pankajk@iitg.ac.in

P. Mahanta
Department of Mechanical Engineering, National Institute of Technology Arunachal Pradesh, Yupia 791112, India
e-mail: pinak@iitg.ac.in

H. J. Das
Department of Mechanical Engineering, Indian Institute of Technology Guwahati, Guwahati 781039, Assam, India
e-mail: hirakh.das@iitg.ac.in



K. M. Pandey et al. (eds.), *Recent Advances in Mechanical Engineering*, Lecture Notes in Mechanical Engineering, https://doi.org/10.1007/978-981-15-7711-6_55

## Nomenclature

| | |
|---|---|
| $A$ | Cross-sectional area ($cm^2$) |
| $C_D$ | Drag coefficient |
| $D$ | Diameter of the vortex chamber (cm) |
| $d_p$ | Particle mean diameter (m) |
| $g$ | Acceleration due to gravity ($m/s^2$) |
| $H$ | Height of the vortex chamber (cm) |
| $I$ | Solid inventory (g) |
| $k$ | Thermal conductivity (W/m °C) |
| $n$ | Number of slots |
| $T_a$ | Temperature of fluidization air (°C) |
| $T_o$ | Ambient temperature (°C) |
| $v_a$ | Velocity of the fluidizing air (m/s) |
| $X_{initl}$ | Initial moisture content (kg water/kg dry paddy) |
| $X_{req}$ | Required moisture content (kg water/kg dry paddy) |

## Subscripts

| | |
|---|---|
| a | Air |
| i | Inlet |
| o | Ambient |
| p | Particle |

## Greek letters

| | |
|---|---|
| $\beta$ | Interphase momentum transfer coefficient ($Ns/m^4$) |
| $\rho_g$ | Density of dry gas ($kg/m^3$) |
| $\rho_s$ | Density of solid (($kg/m^3$) |
| $\lambda$ | Force ratio |
| $\varphi$ | Angle of initial fraction |
| $\theta$ | Injection angle (°) |

## 1 Introduction

More than half of the world population depends upon staple food source rice, and it is obtained from the post-harvesting process of paddy. An important process of drying is required to preserve nutrients of food grains and reduction of their losses.

Nygaard and Pellett [1] reported that the food grain losses in the post-harvesting process and the drying operation were found 10% and 1–5%, respectively. In order to reduce the food grain losses in improper drying operations, a promising technique of rotating fluidized bed with static geometry dryer is used in the present study. This drying technique is applicable in multiple operations, such as mixing, separation, agglomeration, particle coating, and rapid drying of granular materials. The most important feature of this dryer is its static reactor due to which the cyclic maintenance is almost negligible compared to conventional dryers. Various researches have been carried out on granular drying by using conventional and non-conventional methods. Drying is a process of moisture removal in which the moisture available in food grains is reduced to achieve its safe moisture (MC) level, such that the percentage of MC in paddy is maintained at 13–14% (WB) [2]. Mujumdar [3] stated that by reducing the percentage of moisture content in paddy, the possibilities of fungus, pests, germs, and the growth of biological activities were reduced. Conventional fluidized bed dryers are of two types, such as direct and indirect dryers. In the direct drying methodology, heat is directly exposed to the food grains surfaces, while the indirect drying method involves heat transfer through the heat exchanger wall. Sreekumar et al. [4] developed a solar energy-based forced convection dryer and has successfully conducted drying operations using paddy. Syahrul et al. [5] stated that paddy drying is successfully carried out at a commonly maintained temperature from 55 to 65 °C. They reported that air temperature beyond 65 °C affects the texture of the paddy and its nutritional value. The air is the most commonly used medium in the drying process, but heat can also be supplied by superheated steam, while the commercial application of steam is limited due to high cost and technology complexity [3, 6, 7]. To reduce energy consumption in drying operations, farmers use solar energy, biomass energy, or both types of energy sources [8]. However, conventional fluidized beds (CFBs) have salient features such as excellent heat and mass transfer characteristics [9]. Nevertheless, gas–solid slip velocities bound the process intensification and gas flow at high gas velocities. The width of the solid bed limits the specific gas flow rate per unit mass to its height, which is comparatively low in CFB and RFB, these limitations being due to the Earth's gravitational effect [10]. Quevedo [11] developed a rotating fluidized bed in dynamic geometry and successfully conducted an agglomeration process on it. De Wilde and de Broqueville [12] have given a proof of concept of a rotating fluidized bed with static geometry (RFB-SG) in which all the drawbacks of a rotating fluidized bed with dynamic geometry were eliminated. In this reactor, a rotating solid bed is formed by injecting gas–solids through several gas inlets. The solid particles start moving in the reactor, and the radial outward centrifugal force was experienced due to the effect of the tangent gas–solid drag force. The guidelines for the design of the reactor are given by [13, 14]. Pati et al. [15] have done comparative studies of paddy drying in a bubbling fluidized bed (BFB) and RFB-SG reactor and reported that the RFB-SG dryer was faster than the BFB dryer due to higher process intensity. Singh et al. [16] have developed the RFB-SG dryer for agricultural product drying and reported that both the drying capacity of the reactor and drying efficiency was improved. An experimental study of large cardamom drying has been conducted in traditional Bhatti of the Phu-Joram village farmers in Arunachal Pradesh. Also, they found that the annual cultivation of cardamom can be increased using a scientific approach and technical methods [17]. The thermal study of a natural convection dryer was conducted experimentally, and it was reported that the air heat rating increased and heat loss was reduced using sensible heat storage material in the rectangular chamber of the dryer [18].

In the present work, the performance of a scale-up RFB-SG drying chamber using a staple food paddy has been experimentally evaluated. Experiments were carried out on the developed reactor to investigate the effects of various parameters on drying time, such as air temperature, airflow rate, and inventory. The developed RFB-SG chamber dimensions such as the internal diameter of the chamber, the length of the chamber, the width of the slot, and the number of air inlets are 480, 65, 2.5, and 72 mm, respectively.

## 2 Experimental Setup and Procedure

### *2.1 Design of Air Inlet*

air inlet slots are designed to meet various airflow rate requirements considering various parameters of the reactor, such as chamber diameter, a width of the slot, and number of slots. Based on an essential factor λ (or force ration) given by Kochetov et al. [13] is employed in the present design. This factor λ (force ratio) varies from 0.025 to 0.038 and can be written as;

$$\text{Force ratio, } \lambda = \frac{\text{Centrifugal force}}{\text{Drag force}} = \frac{n.s}{\pi.D} \tag{1}$$

where *n*, *s*, and *D* are the number of gas slots, slot width, and vortex chamber diameter, respectively.

Figure 1 shows the schematic diagram of the scale-up RFB-SG drying chamber used in paddy drying operation. Various elements of this dryer, such as centrifugal blower, coupling, airflow control valve, air bypass valve, manometer, orifice plate, air heating chamber, rheostat, air distribution unit, air inlet, solid inlet, solid outlet, solid bed region, and freeboard region have been listed systematically. Initially, the air flowing through the centrifugal blower is heated in the air heating chamber, and this conditioned air is injected through the multi-air inlet into the vortex chamber in the tangential direction of the chamber periphery. The inlet air injected into the reactor creates an air vortex and then feeds food grains (paddy) into the reactor using a high-pressure air compressor. High-velocity air causes solid particles to rotate in the form of a rotating solid bed. The moisture available in the food grains is evaporated by hot air and is thrown out with air through a chimney outlet. At the frequency of 5 min, various samples were collected to observe moisture content in the food grain until the percentage of moisture content level reaches 13–14% (Table 1) [2, 14].

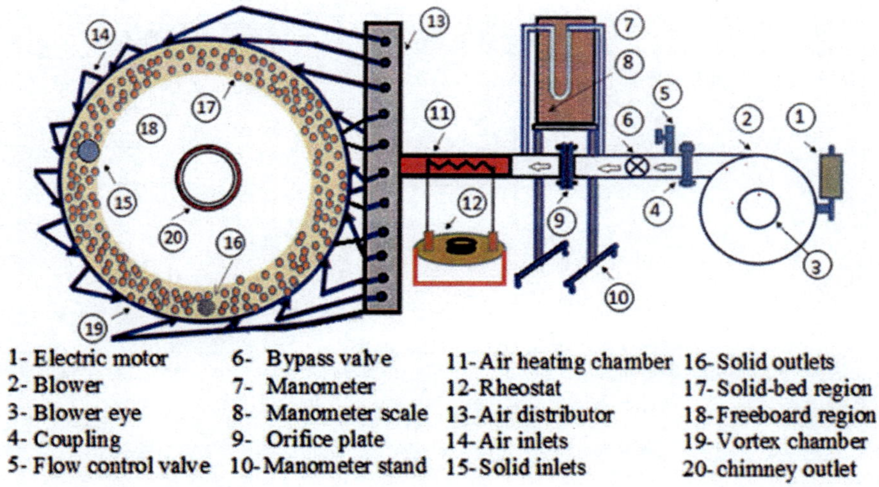


**Fig. 1** Schematic of the experimental setup of RFB-SG dryer

**Table 1** Input parameters for RFB-SG dryer and environmental conditions [1, 2]

| SL-RFBSG chamber | |
|---|---|
| Dimensions | Vortex chamber diameter ($D$): 480 mm, chamber height ($L$): 65 mm, air inlet port width ($s$): 2.5 mm, and chimney diameter: 120 mm |
| Characteristics | Particle diameter $d_p$= 0.0025 mm, initial value of MC ($X_{\text{intl}}$) = 0.33 kg water/kg dry paddy and required value of MC ($X_{\text{req}}$) = 0.13 kg water/kg dry paddy) |
| Operating conditions | |
| Inlet air temperature | 55, 60 and 65 °C |
| Inventory | 400, 600, 800, and 1000 g |
| Ambient temperature | 27 ± 5 °C |
| Airflow rate | 600, 700, and 800 $m^3$/h |

## 3 Results and Discussion

Experimentation has been conducted on a scale-up RFB-SG dryer to evaluate its performance, using locally available food grain (paddy). Drying characteristics of paddy in the scale-up RFB-SG chamber are presented and discussed in the following subsections.

### *3.1 Effect of Air Inlet Temperature on Drying Time*

Figure 2 shows the effect of temperature on the drying time for the inventory of

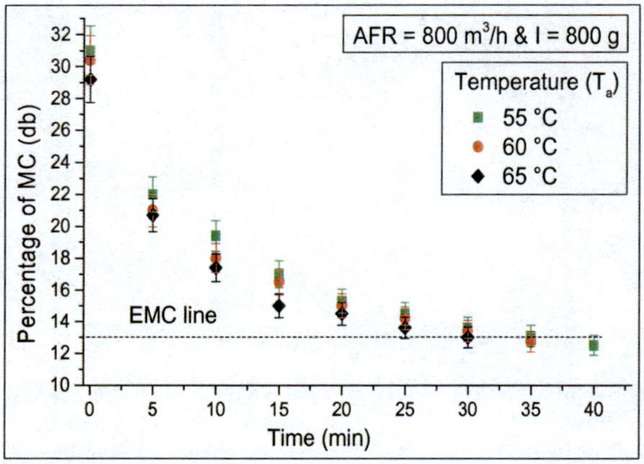


**Fig. 2** Variation of MC with time at different air inlet temperature for an inventory of 800 g

800 g and the airflow rate of 800 $m^3$/h. It is observed that by increasing the air inlet temperature from 55 to 65 °C, the drying time is reduced by 21.62%. Furthermore, an increase in temperature beyond 65 °C reduces the nutritional value of food grain (paddy) [5].

## 3.2 *Effect of Airflow Rate on Drying Time*

Figure 3 shows the effect of airflow rate on drying time for the inventory of 800 g paddy and inlet air temperatures of 65 °C. The drying time is found to be decreased by 21%, with an increase in the airflow rate from 600 to 800 $m^3$/h. Furthermore, the drying time reduced by increasing airflow rate.

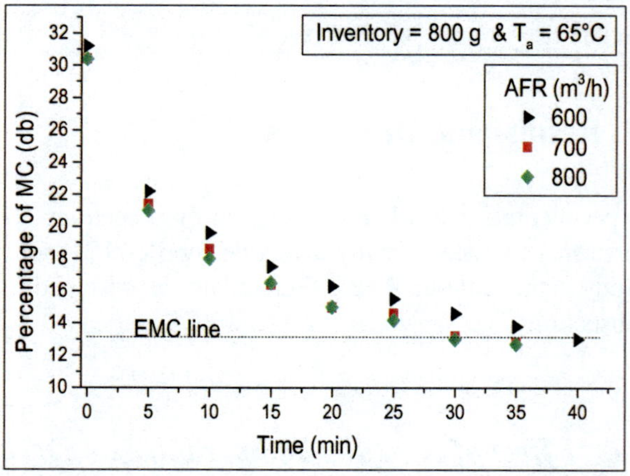


**Fig. 3** Effect of airflow rate at an air temperature of 65 °C and inventory of 800 g

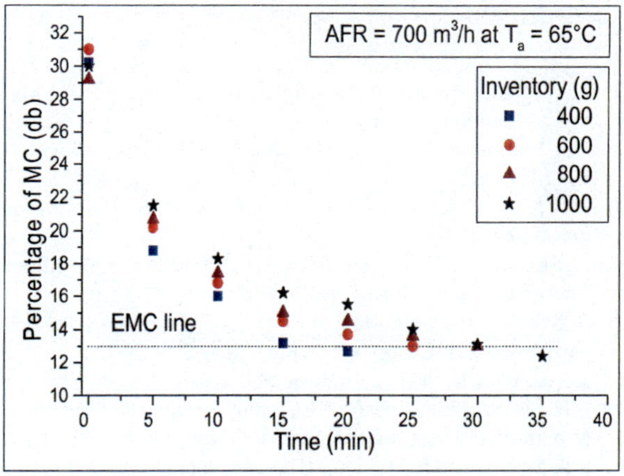


**Fig. 4** Effect of inventory on drying time at air temperature 65 °C and inventory of 800 g

### 3.3 *Effect of Inventory on Drying*

Figure 4 shows the effect of inventory on the drying time at the airflow rate of 700 $m^3$/h and air temperature of 65 °C. With an increase in inventory, the time required for drying is observed to be increased. Pati et al. [15] have reported the drying time of 21 min for 400 g of inventory at inlet air temperature 65 °C and the airflow rate of 700 $m^3$/h, whereas drying time for the same parameters in the current study is 17 min. Hence, drying time is reduced by 19% (4 min).

## 4 Conclusions

To enhance the drying capacity of the RFB-SG dryer, a pilot-scale setup has been developed and the effects of various parameters on drying time are analyzed throughout the drying process, as the findings of the research are summarized below:

- The drying time is assumed to be significantly reduced by increasing the air inlet temperature as well as the airflow rate, whereas paddy erosion is seen when the air temperature rises above 65 °C.
- Although drying time is increased when paddy inventory is improved, the rate of increase in paddy inventories is comparatively higher than drying time. Hence, the drying capacity and efficiency of paddy are improved.
- A challenging drying process of an unevenly shaped paddy has been conducted successfully.

## References


1. D. Nygaard, P.L. Pellett, *Dry Area Agriculture, Food Science, and Human Nutrition* (Pergamon Press, New York, 1986)
2. E.J. Wimberly, *Technical Handbook for the Paddy Rice Postharvest Industry in Developing Countries* (International Rice Research Institute, 1983)
3. A.S. Mujumdar, *Book Review: Handbook of Industrial Drying* (CRC Press, Boca Raton, 2007), pp. 1133–1134
4. A. Sreekumar, P.E. Manikantan, K.P. Vijayakumar, Performance of indirect solar cabinet dryer. Energy Convers. Manag. **49**(6), 1388–1395 (2008)
5. S. Syahrul, F. Hamdullahpur, I. Dincer, Energy analysis in fluidized-bed drying of large wet particles. Int. J. Energy Res. **26**(6), 507–525 (2002)
6. J.F. Smith, Microwave fluidized bed dryer, U.S. Patent 3,528,179, issued September 15, 1970
7. L.N. Kahyaoglu, S. Nesrin, G. Sumnu, Spouted bed and microwave-assisted spouted bed drying of parboiled wheat. Food Bioprod. Proces. **90**(2), 301–308 (2012)
8. S.S. Mohapatra, P. Mahanta, Thermodynamic evaluation of natural convection paddy dryer, in *2nd International Conference on the Developments in Renewable Energy Technology (ICDRET 2012)*, pp. 1–4. IEEE (2012)
9. H.J. Das, P. Mahanta, R. Saikia, A *Future Trend on Research Scope of Numerical Simulation on Conical Fluidized Bed* (IGI Global, 2020), pp. 401–437 (ch. no. 017)
10. A.D. Broqueville, J.D. Wilde, Numerical investigation of gas-solid heat transfer in rotating fluidized beds in a static geometry. Chem. Eng. Sci. **64**(6), 1232–1248 (2009)
11. J. Quevedo, R. Pfeffer, Y. Shen, R. Dave, H. Nakamura, S. Watano, Fluidization of nanoagglomerates in a rotating fluidized bed. AIChE J. **52**(7), 2401–2412 (2006)
12. J.D. Wilde, A.D. Broqueville, Rotating fluidized beds in a static geometry: Experimental proof of concept. AIChE J. **53**(4), 793–810 (2007)
13. L.M. Kochetov, B.S. Sazhin, E.A. Karlik, Experimental determination of the optimal ratios of structural dimensions in the whirl chamber for drying granular materials. Chem. Pet. Eng. **5**(2), 106–108 (1969)
14. A. Dutta, R.P. Ekatpure, J.H. Geraldin, A.D. Broqueville, G.B. Marin, Rotating fluidized bed with a static geometry: guidelines for design and operating conditions. Chem. Eng. Sci. **65**(5), 1678–1693 (2010)
15. J.R. Pati, S. Dutta, P. Eliaers, P. Mahanta, P.K. Chatterjee, J.D. Wilde, Experimental study of paddy drying in a vortex chamber. Drying Technol. **34**(9), 1073–1084 (2016)
16. P. Singh, P. Kalita, P. Mahanta, Study of agricultural product drying in a rotating fluidized bed with static geometry, in *Post Harvest Technology and Value Addition*, vol. I, Issue I (College of Horticulture & Forestry, Central Agricultural University, Pasighat, 2019), pp 1–8. ISBN-978-93-5396-087-2
17. P. Tamuly, P. Mahanta, A study on the drying of large/ black cardamom in conventional bhatti dryer and the related constrains faced in Arunachal Pradesh, in *Post Harvest Technology and Value Addition* vol. I, Issue I (College of Horticulture & Forestry, Central Agricultural University, Pasighat, 2019), pp. 33–38. ISBN-978-93-5396-087-2
18. D. Kumar, P. Mahanta, Thermodynamic analysis of a natural convection dryer, in *Post Harvest Technology and Value Addition*, vol. I, Issue I (College of Horticulture & Forestry, Central Agricultural University, Pasighat, 2019), pp 156–161. ISBN-978-93-5396-087-2

# Experimental Evaluation of Sand-Based Sensible Energy Storage System

**Prasant, Sujit Roy, Biplab Das, and Sumita Debbarma**

**Abstract** Sun is supplying ample amount of solar energy throughout the day. But due to the intermittent nature of this solar energy, one storage is required. Sensible energy storage (SES) stores the heat energy during shining hours and supply that heat in the absence of solar energy. In the present study, an experimental work is carried out to observe the usability of sand as the SES material. Polycarbonate sheet is used as the glazing material and also acts as a container to storage material. The sheet is kept at different tilt angle (20°, 25°, 30°, 35°, and 40°) to observe the effect of inclination on energy stored in SES. Maximum global solar radiation is received during 11.00 AM to 11.30 AM. The maximum average temperature and energy stored in storage found when tilt angle is 30° due to receiving of higher solar radiation in that inclination. The average temperature and stored energy is lowest when SES placed at 20° tilt angle. Further, the conversion efficiency of SES is calculated for all the tilt angles. It is found that SES with 30° tilt angle provides maximum conversion efficiency around 20%, whereas efficiency is lowest at tilt angle 20°.

**Keywords** Sand · Sensible energy storage · Conversion efficiency · Tilt angle · Solar radiation

## 1 Introduction

The global increasing energy demand and depleting fossil fuel sources are forcing world toward the use of renewable energy sources. Solar energy is a clean and abruptly

Prasant · S. Roy (✉) · B. Das · S. Debbarma
Department of Mechanical Engineering, National Institute of Technology Silchar, Silchar, Assam 788010, India
e-mail: sujitme12@gmail.com

B. Das
e-mail: biplab.2kmech@gmail.com

S. Debbarma
e-mail: sumita.mech09@gmail.com



K. M. Pandey et al. (eds.), *Recent Advances in Mechanical Engineering*, Lecture Notes in Mechanical Engineering, https://doi.org/10.1007/978-981-15-7711-6_56

available energy source. But due to mismatch between energy supply in day time and night time, there is a requirement of energy storage. Thermal energy storage (TES) serves the purpose as it can deliver the stored heat in the absence of solar energy. TES is mainly divided into sensible energy storage (SES), latent heat storage (LES), and thermochemical energy storage (TCES) [1]. Hadorn [2] calculated the volume required to store same amount of energy and found that SES required more volume followed by LES and TECS. Due to simple fabrication, cheap system, easy in availability, SES is still considered as the better option for energy storage although it has lower energy density [3]. A material should possess higher heat capacity (to keep less volume requirement per unit heat storage), suitable operating temperature range, higher thermal conductivity (to make the charging/discharging faster), and higher cycle stability (to perform for long duration) to be a suitable SES [4, 5]. SES can serve as the short term (daily charging/discharging) or long term (seasonal charging/discharging) storage based upon the volume used as storage. Fernandez et al. [6] presented a methodology to evaluate the SES material for short- and long-term storage keeping cost minimizing as the objective criteria. Water, rock, and soil have been utilizing as SES material from long time ago in different storage application [7]. Khare et al. [8] mentioned alumina, silicon carbide, high-temperature concrete, graphite, cast iron, and steel as the SES material for high-temperature (>500 °C) application. Rao et al. [9] used concrete as the SES material and found that it takes more time to get charged than the cast-iron-based SES due to low thermal conductivity of concrete. Priya et al. [10] tested thermo-physical properties of smallest grained sand of Cauvery river using thermal gravimetric analysis (TGA), Fourier transform infrared analysis (FTIR), and differential scanning calorimeter (DSC). They found 1.013% degradation of sample weight after first heating cycle of TGA, whereas no significant decrease in weight is observed in the following cycles, which makes the sand as a prominent material for SES. Mahfoudi et al. [11] performed 2D simulation using COMSOL multiphysics software to observe the energy storage capacity of sand. They found SES of 1.15 MJ heat storage capacity takes about 5 h to be fully charged. Hassan et al. [12] mentioned sand as a heat storage material due to its heat capacity and heat-conducting ability. They added variable water quantity to saturate the dry sand and observed that heat capacity reduced with increase in saturation level. Jradi et al. [13] designed a soil-based seasonal SES system of 900 $m^3$ to store the heat in the summer months and delivers the stored heat in the winter months. They found 22.2% energetic efficiency with the modified system incorporating SES. Schlipf et al. [14] theoretically analyzed packed-bed SES system using gravels and found that small grained materials of diameter less than 2 mm show better heat storage potential. Phueakphum and Fuenkajorn [15] modeled a SES with basaltic rock available in the Thailand to keep the house warmer during night. They kept the heat transfer tubes to exchange the heat from SES to house indoor and observed that there is an increment of 4–6 °C indoor temperature compared to surroundings. The system efficiency is found to be 35%, and SES can store energy equivalent to 203 kJ electric energy.

It is found in the literatures that various naturally available solid materials like rock, concrete, soil, and sand can store energy by increasing material sensible temperature. In the present study, an experimental work has been carried out using sand as the heat storage material for the climatic condition of Silchar, Assam (24.8333° N and 92.78° E) [16]. The SES is kept at different tilt angles (20°, 25°, 30°, 35°, and 40°) which are nearby latitude of the place. The effect of tilt angle on the SES temperature and energy storage is observed in the present study. The conversion efficiency of SES from the incident global radiation is also calculated to observe the efficiency of the system.

## 2 Experimental Setup

In the present study, sand is selected as the storage material due to its easy availability and low cost. The procured sand is sieved with a 1.18 mm sieve and poured into a 1 m × 1 m × 8 mm transparent polycarbonate double-wall sheet to keep the SES system in outdoor condition. Silicon glue is provided at the bottom end of the sheet to seal it from one end. The sheet acts as a glazing material to restrict the heated sand from emitting heat to surrounding and also acts as the container to sand. Polycarbonate sheet is light weight, durable, strong, and easy in handling, which makes the SES system portable. Insulation is provided at the back of the sheet to restrict the heat from bottom side. Nine resistance temperature detectors (RTDs) are placed at different location inside the storage to get the average temperature of the storage as shown in Fig. 1. RTDs are fixed at top, middle, bottom, and sides at certain fixed distances. Pyranometer is used to get the temporal solar radiation throughout the day. One RTD is placed at open atmosphere to get the atmospheric temperature. All the sensors are connected to a data logger, which stores the readings. The layout of experimental setup is shown in Fig. 2. SES system is placed on an arrangement which helps to keep the system at different tilt angle as shown in Fig. 3. Tilt angle is the angle between the polycarbonate sheet and horizontal surface. Data logger is connected to the electric power source through an extension board. The experimental setup with all the apparatus is shown in Fig. 4.

## 3 Results and Discussion

Experiments are carried out in the month of December, 2019. The SES system is kept at different tilt angle or inclination angle of 20°, 25°, 30°, 35°, and 40°, and temperature variation is measured using temperature sensors (RTD). In this section, the effect of tilt angle on the SES temperature is observed. The temperatures are recorded from 6 AM to 8 PM. The energy storage process comprises of charging and discharging cycles. During sunshine hours, the heat stored in the SES is known as charging. While, discharging occurs in the absence of solar radiation or during

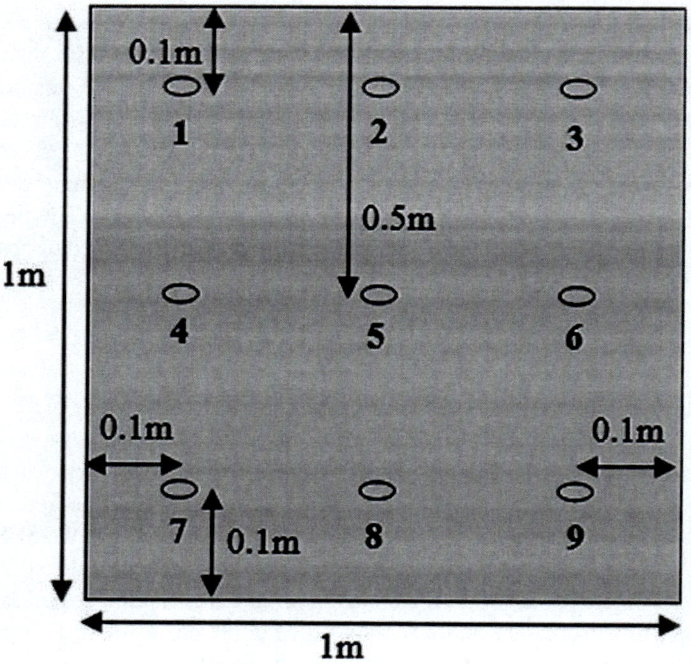


**Fig. 1** Location of RTDs in polycarbonate sheet

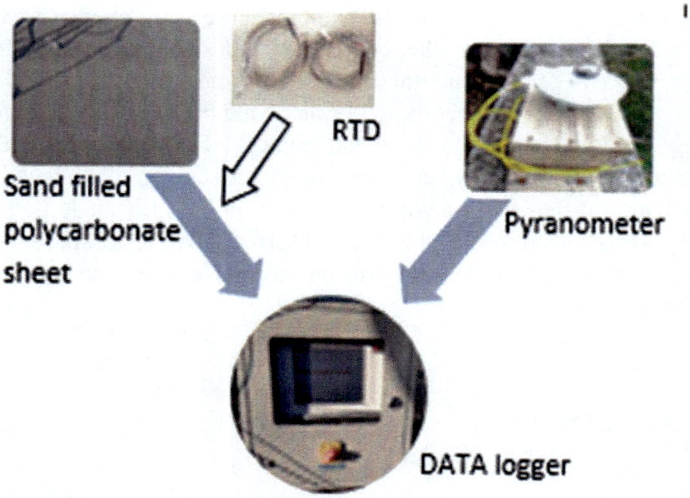


**Fig. 2** Layout of experimental setup

**Fig. 3** Arrangement for sheet inclination (side view)

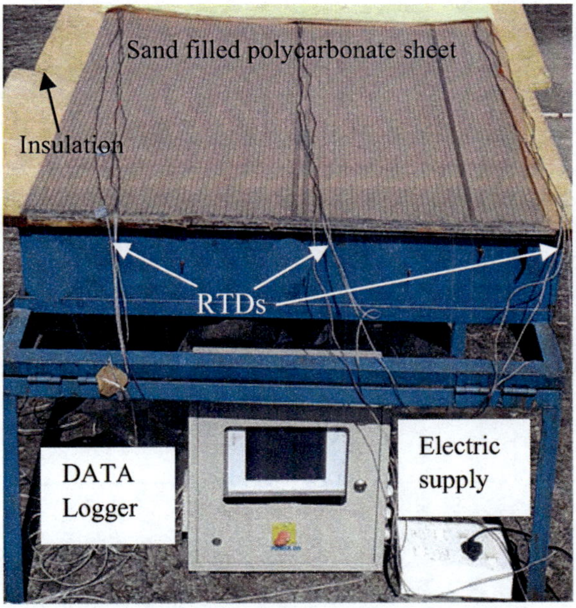


**Fig. 4** Experimental setup

low solar radiation. Figure 5 shows the average solar radiation throughout the day from 7th December to 9th December, 2019. It can be observed that maximum radiation is occurred between 11.00 and 11.30 AM for all the days and maximum of 1068 W/m$^2$ observed on 7th December, 2019. The effect of tilt angle on the sand temperature is shown in Fig. 6. The sand temperature is calculated taking average temperature of all the RTDs placed at different location in the sand. It can be observed

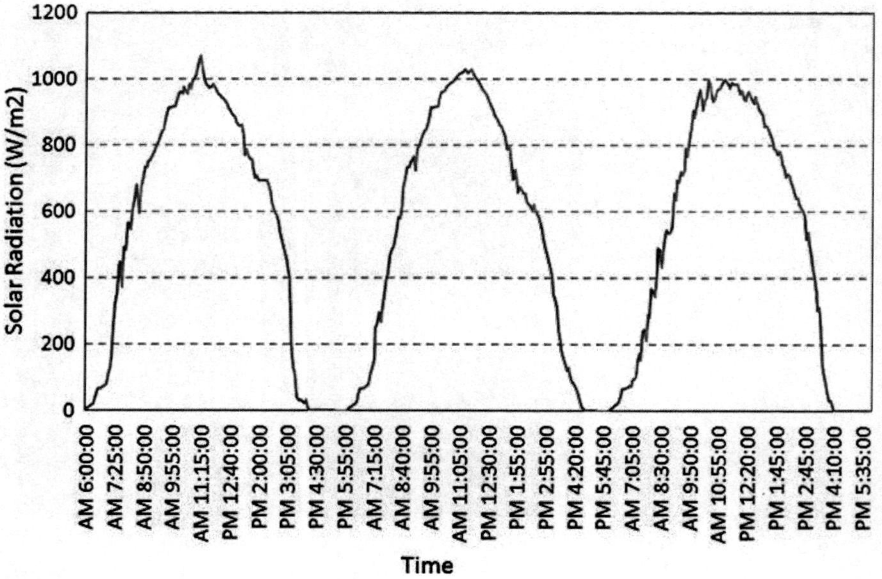


**Fig. 5** Daily solar radiation

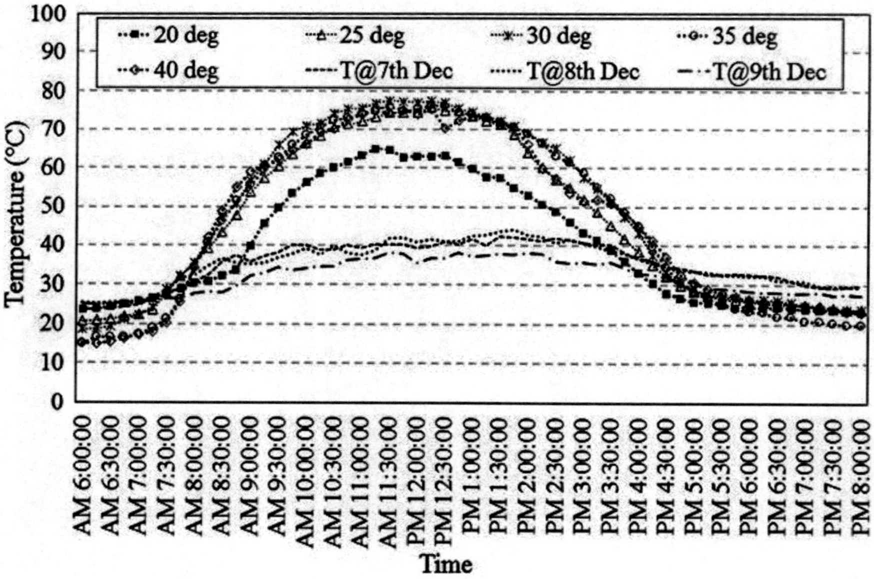


**Fig. 6** Atmospheric and sand average temperature

from the figure that the charging of SES occurred upto 12 PM during peak of solar radiation for all the cases (tilt angle 20°, 25°, 30°, 35°, and 40°). During charging, maximum average storage temperature (about 78 °C) is achieved when tilt angle is 30°. However, maximum average temperatures at tilt angle 40°, 35°, 25°, and 20° are 75.7 °C, 76.4 °C, 74 °C, and 65 °C, respectively. This is due to capturing more solar energy at tilt angle 30° followed by 35°, 40°, 25°, and 20°. Difference between average temperatures with 35° and 40° is observed to be very less. The atmospheric temperature on three consecutive days are also presented in Fig. 6. The atmospheric temperature on 7th and 8th December is almost same, while temperature on 9th December is less due to reduce solar radiation on that date compared to other two days. This may be also the reason along with the tilt angle for getting significant difference in maximum average temperature of SES. The average temperature starts decreasing as the solar radiation reduces and almost equal to atmospheric temperature after 5.00 PM.

## *3.1 Effect on Stored Energy and Conversion Efficiency*

The energy stored in the SES by increasing the material temperature without changing its phase. The heat storing is dependent on the solar radiation on a particular day. The energy stored in the SES can be calculated using Eq. (1) is dependent on the mass of the system, specific heat of the storage material, and temperature rise in the storage material [17]. The energy density or heat capacity ($\rho c_p$) of sand measured using thermal conductivity tester is equal to 1.2 MJ/m$^3$ K. Energy stored in the SES at different tilt angle is shown in Fig. 7. It can be observed that maximum energy stored when tilt angle is 30°. This is because of maximum average temperature obtained at this tilt angle. The maximum energy stored in the SES for 20°, 25°, 30°, 35°, and 40° are 341 kJ, 421 kJ, 434 kJ, 425 kJ, and 422 kJ respectively. With reduction of solar radiation and atmospheric temperature, SES starts discharging and get fully discharged around 7.00 PM. Conversion efficiency is the parameter which shows the fraction of total solar radiation utilized to increase the storage temperature. It is the ratio of energy accumulated in the SES to the energy received by the polycarbonate sheet [18, 19]. The transmissibility of polycarbonate sheet as the glazing material is about 10–20% which is less than the glass and is considered as 0.8 for the present efficiency calculation [20]. The conversion efficiency can be calculated using Eq. (2). Figure 8 shows the conversion efficiency of SES for different tilt angle. It can be observed from the figure that SES with 30° tilt angle provides maximum conversion efficiency around 20%, whereas efficiency with 20°, 25°, 35°, and 40° is 15.7%, 19.5%, 19.6%, and 19.5% respectively. The lower conversion efficiency of SES system may be due to heat losses and absence of black coating on the absorber surface.

$$Q_{\text{stored}}(\text{in joules}) = \rho V c_p \, \Delta T_{\text{storage}} \tag{1}$$

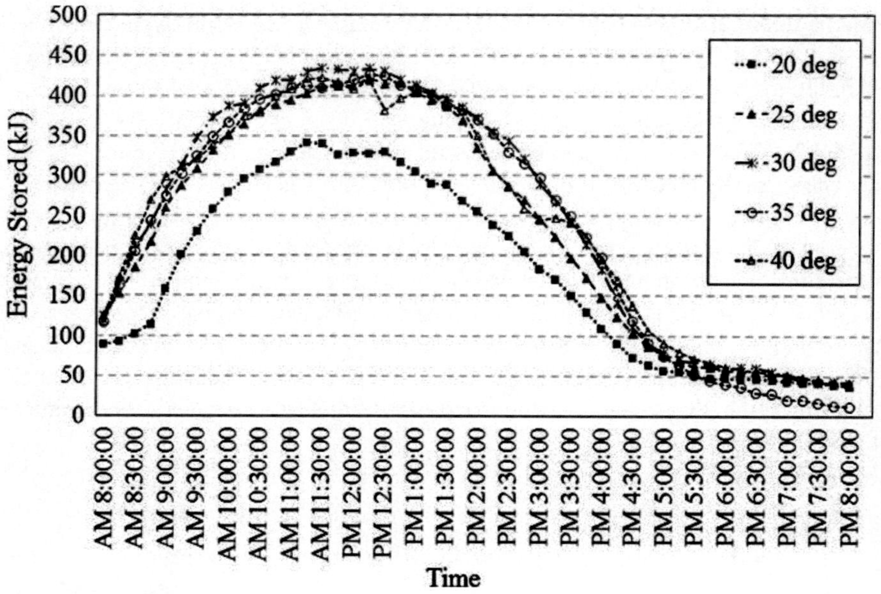


**Fig. 7** Energy stored in SES

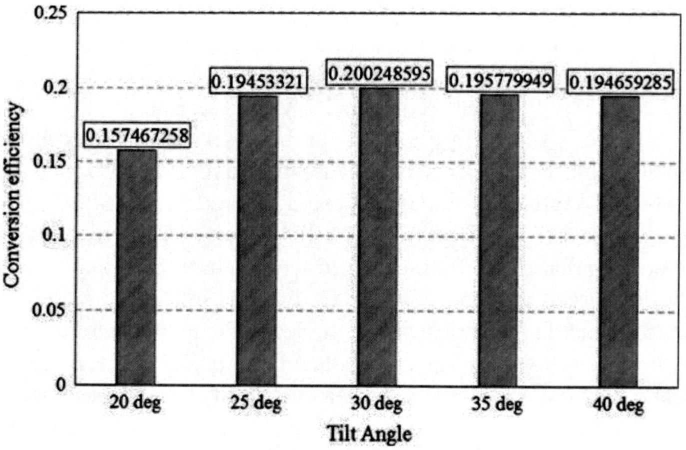


**Fig. 8** Conversion efficiency of SES

$$\eta_{\text{conversion}} = \frac{Q_{\text{stored}}(\text{in watts})}{Q_{\text{received}}} = \frac{\rho V c_p \, \Delta T_{\text{storage}}}{I\tau \, A_{\text{surface}}}\left({}^{1}/_{3600}\right) \quad (2)$$

where $\rho$ is density of sand (kg/m$^3$), $V$ is volume of sand inside the polycarbonate sheet (m$^3$), $c_p$ is specific heat of sand (J/kg K), $\Delta T_{\text{storage}}$ is difference between initial and final temperature of sand, $I$ is solar radiation (W/m$^2$), $\tau$ is the transmissivity of polycarbonate sheet, and $A_{surface}$ is the surface area exposed to solar radiation.

## 4 Conclusion

Experimental analysis is performed using sand as the energy storage material. Experiments are performed in the first week of December placing the SES system at five different tilt angle. Maximum solar radiation is received between 11.00 AM to 11.30 AM. It is found that maximum energy stored at tilt angle 30° is 434 kJ, whereas the maximum energy stored at tilt angle 40°, 35°, 25°, and 20° is 2.8%, 2.2%, 2.9%, and 21.4% lesser, respectively. Similarly, maximum solar energy conversion efficiency of 20% is achieved at tilt angle 30°. Therefore, SES is more efficient placing it at tilt angle 30°. Further, black coating may be applied on the top surface to increase the absorptivity of the polycarbonate sheet.

**Acknowledgements** The authors are thankful to TEQIP III, NIT Silchar, for providing financial assistance.

## References

1. I. Dincer, M.A. Rosen, *Thermal Energy Storage-Systems and Applications* (John Wiley & Son, US, New York, 2002)
2. J.C. Hadorn, Advanced storage concepts for active solar energy—IEA SHC Task 32 2003–2007, in *Eurosun 1st International Conference on Solar Heating, Cooling and Buildings* (Lisbon, Oct. 2008), pp. 1–8
3. M.E. Navarro, M. Martínez, A. Gil, I.A. Fernandez, L.F. Cabeza, R. Olives, X. Py, Selection and characterization of recycled materials for sensible thermal energy storage. Solar Energy Mater. Solar Cells **107**, 131–135 (2012)
4. A.S. Hariri, I.C. Ward, A review of thermal storage systems used in building applications. Build. Environ. **23**(1), 1–10 (1988)
5. I. Dincer, On thermal energy storage systems and applications in buildings. Energy Build. **34**(4), 377–388 (2002)
6. I.A. Fernandez, M. Martinez, M. Segarra, I. Martorell, L.F. Cabeza, Selection of materials with potential in sensible thermal energy storage. Solar Energy Mater. Solar Cells **94**(10), 1723–1729 (2010)
7. P. Pinel, C.A. Cynthia, I. Beausoleil-Morrison, A. Wills, A review of available methods for seasonal storage of solar thermal energy in residential applications. Renew. Sustain. Energy Rev. **15**(7), 3341–3359 (2011)
8. S. Khare, M. Dell'Amico, C. Knight, S. McGarry, Selection of materials for high temperature sensible energy storage. Sol. Energy Mater. Sol. Cells **115**, 114–122 (2013)
9. R.C.R. Chilaka, H. Niyas, P. Muthukumar, Performance tests on lab–scale sensible heat storage prototypes. Appl. Thermal Eng. **129**, 953–967 (2018)
10. L.R. Priya, S. Salim, B. Vaishnu, V.K.R.M. Chandrakala, Study on characterization of river sand as heat storage medium. Ind. J. Sci. Technol. **9**(30), 1–5 (2016)

11. N. Mahfoudi, M. Abdelhafid, G.E. Mohammed, Sand as a heat storage media for a solar application: simulation results. Appl. Mech. Mater. **621**, 214–220 (2014)
12. S. Hassan, R.S. Hernandez, K. Vonzabern, J. Pierce, K. Kantesaria, M. Rubin, Thermal inertia of sand at different levels of water saturation, in *44th Lunar and Planetary Science Conference*, Mar 2013, pp. 388–402
13. M. Jradi, C. Veje, B.N. Jorgensen, Performance analysis of a soil-based thermal energy storage system using solar-driven air-source heat pump for Danish buildings sector. Appl. Therm. Eng. **114**, 360–373 (2017)
14. D. Schlipf, P. Schicktanz, H. Maier, G. Schneider, Using sand and other small grained materials as heat storage medium in a packed bed HTTESS. Energy Procedia **69**, 1029–1038 (2015)
15. D. Phueakphum, F. Kittitep, A rock fills based solar thermal energy storage system for housing. Sci. Asia **36**(3), 237–243 (2010)
16. P. Jha, B. Das, R. Gupta, Energy and exergy analysis of photovoltaic thermal air collector under climatic condition of North Eastern India. Energy Procedia **158**, 1161–1167 (2019)
17. S. Roy, B.K. Debnath, Heat transfer enhancement of sensible energy storage for low temperature application, in *12th International Conference on Energy Sustainability and the ASME 2018 Nuclear Forum* (USA, 2018)
18. Z. Changji, X. Ding, Du. Xiaoming, Light transmissivity of polycarbonate multi-wall sheet used as greenhouse glazing material. Trans. Chinese Soc. Agric. Eng. **11**, 42 (2006)
19. P. Jha, B. Das, R. Gupta, An experimental study of a photovoltaic thermal air collector (PVTAC): a comparison of a flat and the wavy collector. Appl. Therm. Eng. **163**, 114344 (2019)
20. S. Debnath, B. Das, P.R. Randive, K.M. Pandey, Performance analysis of solar air collector in the climatic condition of North Eastern India. Energy **165**, 281–298 (2018)

# Purification of Biogas for Methane Enrichment Using Biomass Biochar and Biochar–Clay Composite

**Deep Bora, Lepakshi Barbora, and Pinakeswar Mahanta**

**Abstract** Biogas is mainly composed of methane ($CH_4$) and carbon dioxide ($CO_2$) with trace amounts of hydrogen sulfide ($H_2S$), of which $CO_2$ and $H_2S$ are impurities. Scrubbing of these two impurities are crucial for purification and upgradation of biogas, which would simultaneously also increase the calorific value of the treated biogas and address the issue of corrosion. Several studies have used expensive and environmentally harmful chemicals for the purification of biogas. This study reports a simple biogas purification system that utilizes biomass biochar and biochar–clay composites to remove $CO_2$ and $H_2S$ from biogas by the process of adsorption. The biomass biochar could enrich the methane content of raw biogas from 59.7 to 84.6%, which shows the potentiality and applicability of biomass biochar for the removal of $CO_2$ and $H_2S$ from biogas. This simultaneously enhanced the calorific value of the biogas and retarded the corrosiveness due to $H_2S$. The study also indicated that $CO_2$ adsorption by biomass biochar and biochar–clay composite is transient and has to be reloaded after saturation. Biochar and clay have the added advantage of being environment friendly and require no treatment for disposal. Observed results indicated that similar degree of enrichment, compared to commonly used chemical, could be achieved by application of biomass biochar and biochar–clay at a much lesser cost.

**Keywords** Biogas · Biochar–clay composite · Biomass biochar · $CH_4$ enrichment · $CO_2$ scrubbing

D. Bora · L. Barbora (✉)
Centre for Energy, Indian Institute of Technology Guwahati, Guwahati, Assam 781039, India
e-mail: lopab@iitg.ac.in

P. Mahanta
Department of Mechanical Engineering, Indian Institute of Technology Guwahati, Guwahati, Assam 781039, India

Department of Mechanical Engineering, National Institute of Technology Arunachal Pradesh, Yupia, Arunachal Pradesh 791112, India



K. M. Pandey et al. (eds.), *Recent Advances in Mechanical Engineering*, Lecture Notes in Mechanical Engineering, https://doi.org/10.1007/978-981-15-7711-6_57

# 1 Introduction

Rising concern about the environment, rapid depletion of fossil fuels and new environmental regulation has encouraged the search for new environmentally compatible materials obtained from natural resources [1, 2]. Thus, various forms of renewable energy such as solar, wind, tidal, wave, and biomass are getting attention worldwide during the last few decades. Harnessing energy from biomass is gaining popularity in developing countries due to the high availability of biomass and bio-waste. Biomass is an organic matter, which is available on a renewable basis and includes all plants and plant-derived materials. Lignocellulosic biomasses are abundantly available throughout the world and are recognized as a potential feed material for different applications like land filling, biogas, and biofuel productions [3, 4].

Biogas is a combustible mixture of gases produced from anaerobic digestion of organic matter by a community of microbial consortia. The digestion process takes place through various reactions and interactions among the methanogens and substrates fed into the digester [5, 6]. The biogas composition greatly depends on the type of feed materials, temperature, pressure, and pH inside the reaction digester [7]. The anaerobic digestion process generally comprises of four steps—hydrolysis, acidogenesis, acetogenesis, and methanogenesis [8]. Biogas is typically a mixture of 45–60% methane ($CH_4$), 40–55% carbon dioxide ($CO_2$), and a trace amount of other gases like nitrogen ($N_2$), hydrogen ($H_2$), and hydrogen sulfide ($H_2S$) [9, 10]. The methane in biogas is a high-valued source of energy, while the other gases are impurities that cause major impediments to the commercial use of biogas [11]. $CO_2$ has zero energy yield through combustion and greatly reduces the heating value due to its high concentration per volume of biogas. Thus, removal of carbon dioxide from biogas will enhance the fuel efficiency which could serve as a source of immense energy that can be used effectively for different purposes like cooking, lighting [12], vehicle fuel, and power generation [13]. A variety of processes are being used for removing carbon dioxide and hydrogen sulfide from natural gas and biogas in petrochemical and renewable industries. Several basic mechanisms are involved to achieve selective separation of gas constituents. These may include physical or chemical absorption, adsorption on a solid surface, membrane separation, and cryogenic separation [14].

Physical absorption of biogas is a simple and one of the cheapest methods to remove $CO_2$ and $H_2S$ simultaneously. It uses water as the scrubbing agent, but the only limitation is that it requires high pressure and a lot of water for this process which makes it impractical for areas with water scarcity. However, from chemical absorption process, it was seen that chemical absorbents are more efficient in low pressure, but the only limitation is that the regeneration of the solvents requires relatively high energy input and also chemicals are not eco-friendly. Membrane separation and cryogenic separation give high purity, but they are economically not viable for pilot plant as it consumes relatively more electricity per unit of gas production [14, 15]. Chemical adsorption on solid surface gives economy in production with comparatively high purity. It has good moisture removal capacities and simple in design.

Hence, in this context, this work proposes the use of biomass biochar and biochar–clay composite material as scrubbing agent for $CO_2$ removal targeting the rural households of the northeastern region of India. The scrubbing materials used here are anticipated to be beneficial over other conventional chemicals because the source material is abundantly available, has almost zero production cost, process of preparation is easy, and can be easily adopted in rural areas.

# 2 Materials and Methods

## *2.1 Materials*

### 2.1.1 Biomass Collection and Processing to Biochar

Solid waste banana peels were collected from canteens in the campus of IIT Guwahati. They were washed under tap water to make them free from any impurities and were kept in a hot air oven at 60 °C for 24 h for the removal of moisture. The dried materials were kept in a muffle furnace at 550 °C for 6 h to convert it to biochar.

### 2.1.2 Preparation of Biochar–Clay Composite

Clay was collected from a local pottery vendor which was combined with biomass in a 1:3 mass ratio to form a mixture. The mixture formed was kept in a muffle furnace at 550 °C for 6 h to form a powdered sample.

## *2.2 Characterization of Biomass Samples*

### 2.2.1 Proximate Analysis

Proximate analysis was used for the calculation of moisture, volatile matter, ash, and fixed carbon content of the biomass. The procedure used for the estimation of moisture and ash content was adopted from the National Renewable Energy Laboratory (*NREL*) protocol. American Society for Testing and Materials (*ASTM*) D 271-48 was followed for volatile matter determination. Fixed carbon was determined by difference from summation of moisture, volatile matter, and ash content.

### 2.2.2 Field Emission Scanning Electron Microscopy (FESEM)

The surface morphology of each adsorbent was determined by FESEM (*Make: Zeiss, Model: Sigma*) photography.

### 2.2.3 Elemental Analysis

Energy-dispersive X-ray spectroscopy (EDX) (*Make: Zeiss, Model: Sigma*) was used to determine the elemental composition of the selected adsorbents. The biomass samples were converted completely into ash by heating in a muffle furnace at 550 °C for 6 h and were kept in a desiccator to cool to room temperature. A spherical pellet of diameter 5 mm and height 1 mm was prepared with the help of a pelletizer for the analysis.

### 2.2.4 Gas Chromatography

Raw and biogas sample treated with adsorbents were collected in Tedlar® bags and immediately analyzed in a gas chromatograph (*Thermo Scientific Trace GC Ultra Gas Chromatograph, USA*) for the volume-percentage composition of biogas.

# 3 Results and Discussion

## *3.1 Proximate Analysis of Biomass Sample*

The dried biomass samples were used for proximate analysis. Proximate analysis results show that the biomass samples contains 8.52% moisture, 76.43% volatile matter, 10.25% ash, and 4.80% fixed carbon.

## *3.2 FESEM Images of Biomass Biochar and Biochar–Clay Composite*

The morphology of the biomass biochar and biochar–clay composite was determined by FESEM imaging. Figures 1 and 2 show the FESEM images at a magnification of 5.00 KX times. It can be observed from the micrograph that the porosity and the surface area of biomass biochar is comparatively higher than biochar–clay composite. The formation of the porous cylindrical structures is attributed to the formation and subsequent expulsion of volatile components [16]. However, in Fig. 2, the clay particles can be seen as small lumps. It is anticipated that the presence of porous

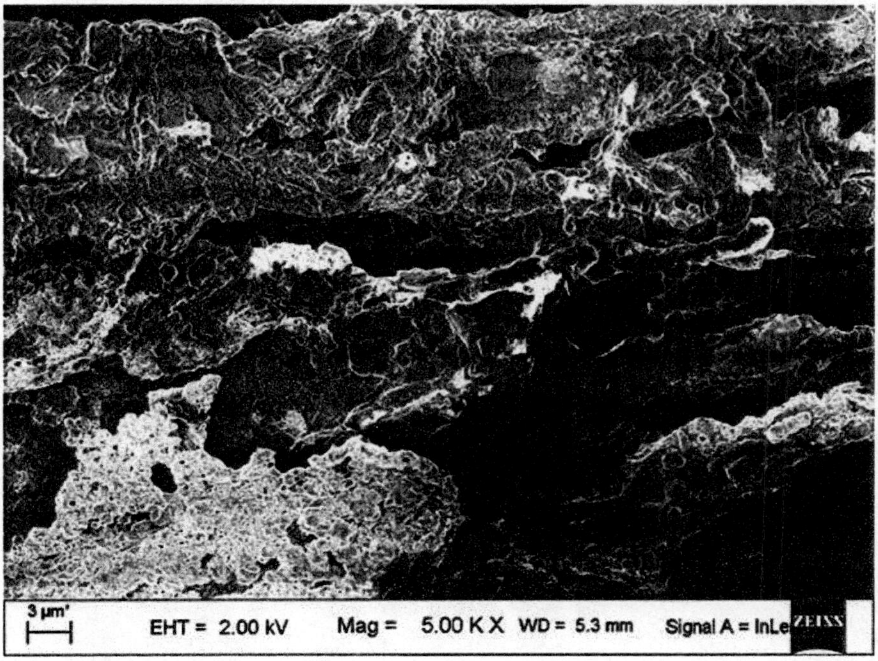


**Fig. 1** FESEM image of biomass biochar at 5 KX magnification

structures in biomass biochar and biochar–clay composite facilitates adsorption of $CO_2$ from biogas by chemical reactions within the pores as explained in Eqs. 1 and 2.

## 3.3 Elemental Analysis of Biomass Biochar and Biochar–Clay Composite (Wt%)

The elemental composition in biomass biochar and biochar–clay composite was determined with the help of energy-dispersive X-ray spectroscopy (EDX). The results are summarized in Table 1. The potassium content of biomass biochar (34.7 wt%) is found higher than biochar–clay composite (19.3 wt%). The potassium present in biomass biochar and biochar–clay composite is anticipated to react with the moisture present in biogas to form potassium hydroxide which in turn would help to adsorb $CO_2$ and hence enrich the methane content in biogas [17]. The chemical reactions are as follows:

$$H_2O_{(aq)} + CO_{2(g)} \leftrightarrow H_2CO_{3(aq)} \quad (1)$$

$$2KOH_{(aq)} + H_2CO_{3(aq)} \rightarrow K_2CO_{3(aq)} + 2H_2O_{(aq)} \quad (2)$$

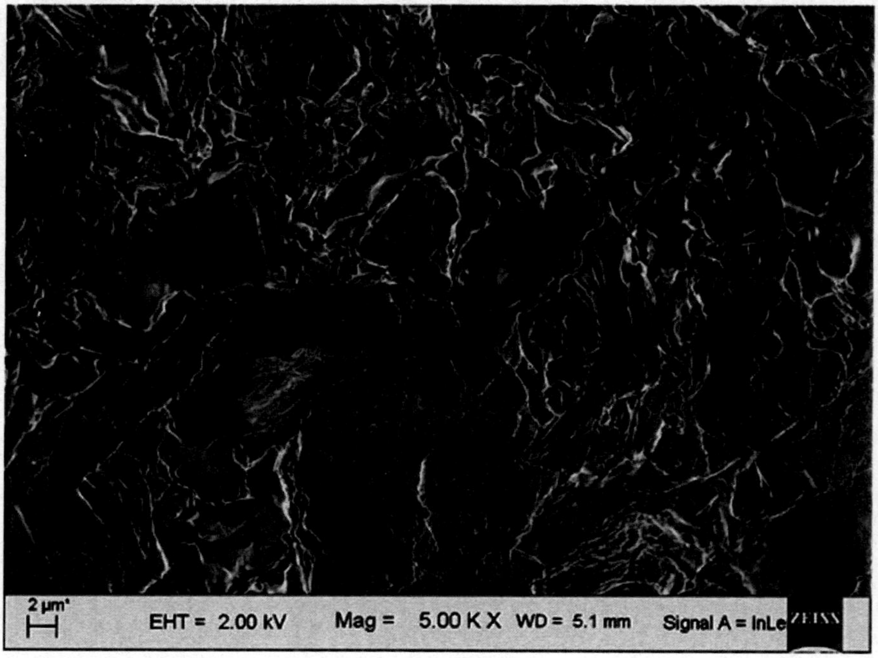


**Fig. 2** FESEM image of biochar–clay composite at 5 KX magnification

**Table 1** Elemental analysis of biomass biochar and biochar–clay composite using EDX

| Elements | Biomass biochar (wt%) | Biochar–clay composite (wt%) |
|---|---|---|
| K | 34.7 | 19.3 |
| Ca | 2.5 | 1.1 |
| Mg | 3.3 | 0.6 |
| Si | 1.3 | 0.6 |
| P | 3.1 | 0.9 |
| S | 0.6 | 0.2 |

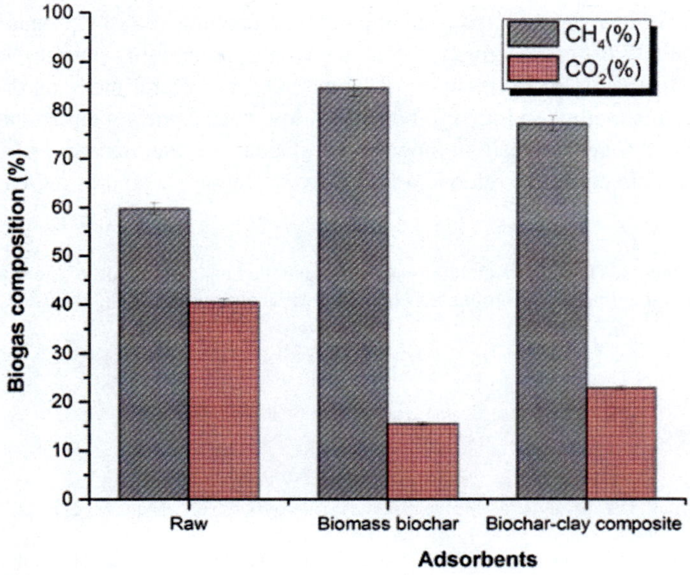


**Fig. 3** Composition of raw gas and purified gas

### *3.4 Purification of Biogas with Adsorbents*

The biogas was collected from a 3 $m^3$ Deenbandhu biogas plant located at AaoniatiSatra, North Guwahati, Assam. Cow dung was used as the only feed material in the biogas plant. The methane and carbon dioxide volume percentage of the raw biogas sample were found to be 59.7% and 40.3%, respectively. Biomass biochar and biochar–clay composite were used as adsorbents for the removal of $CO_2$and $H_2S$ from biogas. When compared with the raw biogas sample (having a methane content of 59.7%), the biogas samples treated with biomass biochar and biochar–clay composite showed 84.6% and 77.3% methane content, respectively (Fig. 3). Moreover, the $H_2S$ level, which was 286 ppm in the raw biogas, decreased to 32 and 46 ppm when treated with biomass biochar and biochar–clay composite, respectively.

## 4 Conclusions

This work reports the synthesis, characterization, and application of biomass biochar and biochar–clay composite as $CO_2$ and $H_2S$ scrubbers from biogas. The elemental characterization by EDX reveals that potassium is present in higher quantities as compared to the other elements. FESEM images supports the presence of pores in the structures of the adsorbents which facilitates adsorption of $CO_2$ from biogas. The gas chromatography results also show that the adsorbents, biomass biochar and biochar–clay composite, could enrich the methane content of a raw biogas sample by 24.9% and 17.6%, respectively, which shows the potentiality and applicability of natural adsorbents for purification of biogas. Out of several methods of biogas purification, this method is found to be simple, low cost, environment friendly, and suitable method for enrichment of biogas in rural areas. Further experiment needs to be done to enable complete removal of $CO_2$ from raw biogas and to increase their efficiency.

**Acknowledgements** The authors gratefully acknowledge the financial support of the Ministry of New and Renewable Energy, Government of India, for the above project (256/3/2017-BIOGAS).

## References

1. A. Demirbas, Use of algae as biofuel sources. Energy Convers. Manage. **5**(1), 2738–2749 (2010)
2. FM. Pelissari, AS.PauloJose´do, FC. Menegalli, Isolation and characterization of cellulose nanofibers from banana peels,*Cellulose*, 2014, 21, pp. 417–432.
3. A.J. Borah, M. Agarwal, M. Poudyal, A. Goyal, V.S. Moholkar, Mechanistic investigation in ultrasound induced enhancement of enzymatic hydrolysis of invasive biomass species. Biores. Technol. **213**, 342–349 (2016)
4. A.J. Borah, S. Singh, A. Goyal, V.S. Moholkar, An assessment of invasive weeds as multiple feedstocks for biofuels production. RSC Adv. **6**, 47151–47163 (2016)
5. P. Mahanta, A. Dewan, U.K. Saha, P. Kalita, B. Buragohain, Biogas digester: a discussion on factors affecting biogas production and field investigation of a novel duplex digester. J. Solar Soc. India SESI J. **15**(2), 1–12 (2005)
6. I. Diaz, S.I. Perez, E.M. Ferrero, M. Fdz-Polanco, Effect of oxygen dosing point and mixing on the microaerobic removal of hydrogen sulphide in sludge digesters. Biores. Technol. **102**(4), 3768–3775 (2011)
7. S. Rasi, A. Veijanen, J. Rintala, Trace compounds of biogas from different biogas production plants. Energy **32**, 1375–1380 (2007)
8. M. Gerardi, *The Microbiology of Anaerobic Digesters, Waste Water Microbiology Series* (John Wiley & Sons, Inc., United States of America, 2003), p. 7
9. G. Mann, M. Schlegel, R. Schumann, A. Sakalauskas, Biogas-conditioning with microalgae. Agron. Res. **7**(1), 33–38 (2009)
10. A.B. Karki, J.N. Shrestha, S. Bajgain, Biogas: as renewable source of energy in Nepal, theory and development. BSP-Nepal, 1–189 (2005)
11. N. Abatzoglou, S. Boivin, A review of biogas purification processes. Biofuels Bioprod. Biorefin. **3**(1), 42–71 (2008)
12. N.H.S. Ray, M.K. Mohanty, R.C. Mohanty, Biogas compression and storage system for cooking applications in rural Households. Int. J. Renew. Energy Res. **6**(2), 593–598 (2016)
13. Q. Sun, H. Li, J. Yan, L. Liu, Z. Yu, X. Yu, Selection of appropriate biogas upgrading technology-a review of biogas cleaning, upgrading and utilization. Renew. Sustain. Energy Rev. **51**, 521–532 (2015)
14. F. Osorio, J.C. Torres, Biogas purification from anaerobic digestion in a wastewater treatment plant for biofuel production. Renew. Energy **34**, 2164–2171 (2009)
15. M. Hagen, E. Polman, Adding gas from biomass to the gas grid, in *Final Report Submitted to Danish Gas Agency* (2001)

16. R. Yin, R. Liu, Y. Mei, W. Fei, X. Sun, Characterization of bio-oil and bio-char obtained from sweet sorghum bagasse fast pyrolysis with fractional condensers. Fuel **112**, 96–104 (2013)
17. M. Fernández-Delgado, Juárez, P. Mostbauer, A. Knapp, W. Müller, S. Tertsch, A. Bockreis, H. Insam, Biogas purification with biomass ash. Waste Manage. **71**, 224–232 (2018)

# Experimental Investigation of Drying Characteristics of Tea in a Conical Bubbling Fluidized Bed Dryer

**Plabon Tamuly, Hirakh Jyoti Das, and Pinakeswar Mahanta**

**Abstract** Tea has become an important crop in a many regions of the world. India is one of the best quality tea producers in the world. Despite of being an important cash crop, much research has not been made on tea, due to which modern drying technologies have not been employed in this regard. This work reflects the drying characteristics for tea in a bubbling conical bed dryer. Fresh Assam tea leaves have been processed into crush, tear, and curl (CTC) tea initially, followed by fermentation and drying in a conical bubbling fluidized bed dryer. The effect of various parameters, such as drying air temperature, superficial air velocity, inventory, and cone angle, on moisture removal rate from CTC tea in the conical bed dryer has been experimentally investigated and analyzed in this present work. Drying temperature and superficial air velocity are found to be the major parameter that affects the moisture removal rate from tea. The moisture removal rate from tea varies proportionally with the increase in drying temperature, but the quality deteriorates after a certain limit of temperature. Different cone angles of 0°, 5°, and 10° for the conical bubbling fluidized bed dryer are also inquired into for a static inventory bed height of 15 cm. It is found that moisture removal rate is highest for the case 10° cone angle of the dryer.

**Keywords** CTC tea · Drying · Inventory · Moisture removal · Superficial air velocity

## 1 Introduction

India is one of the best quality tea producers in the world. In general, tea is categorized into two types such as green tea and black tea, but on the basis of manufacturing

P. Tamuly (✉) · H. J. Das · P. Mahanta
Department of Mechanical Engineering, Indian Institute of Technology Guwahati, Guwahati 781039, India
e-mail: plabon@iitg.ac.in

P. Mahanta
Department of Mechanical Engineering, National Institute of Technology Arunachal Pradesh, Yupia 791112, India


K. M. Pandey et al. (eds.), *Recent Advances in Mechanical Engineering*, Lecture Notes in Mechanical Engineering, https://doi.org/10.1007/978-981-15-7711-6_58

process, tea can be processed into two main categories, as orthodox tea and CTC (crush, tear, and curl) tea [1]. Orthodox tea is the hand-processed tea or machine-rolled tea which is dried after the rolling and fermentation operation. On the other hand, CTC tea is processed with cutters which produce small tea particles which enhance the fermentation compared to orthodox tea [2]. Among the tea manufacturing processes, drying plays the most important role which defines the final quality of tea as well as the shelf life of tea [3]. Both the tea types are universally dried by convective drying employing hot air. For the drying of tea, some other experiments including radio frequency drying and vacuum drying also have been carried out, but hot air drying is the most practical method on commercial scale. Multi-stage fluid bed dryer has been identified as a very supreme dryer for tea [4]. All of the various methods of processing tea require drying as a major part of the process, where moisture content is reduced from initial 70–75% to the equivalent moisture content of 3–3.5% wet basis [5]. The duration of drying process and the temperature to which tea is exposed must be restricted to avoid damage to the quality [6]. Among the hot air drying methods, fluidized bed drying is a very promising aspect. Fluidized bed dryer has advantages of large capacity, low construction cost, easy operability, high thermal efficiency, and high moisture removal rate [7]. Fluidization refers to those fluid–solid systems in which the solid phase is subjected to behave more or less like a fluid by the streaming current of gas or liquid stream moving through the bed of solid particles. It is suited to any kind of wet solid as long as it is fluidized by hot gas. Fluidized bed drying is best suitable for heat-sensitive products [8]. It also has been found that fluidized and vibro-fluidized bed provide an effective drying for fresh leaves compared to any conventional hot air dryer [9]. In conventional drying of tea, it has been found that loss of the antioxidant properties occurs [10]. Hence, the drying aspect of tea in a fluidized bed dryer has become a research area of utmost importance.

In this research, experimental investigation of drying characteristics of tea in a conical bubbling fluidized bed dryer has been performed, and the effect of different operating parameters on the moisture removal rate has been analyzed.

## 2 Materials and Methods

### *2.1 Materials*

In this investigation, fluidized drying has been performed on a conical bed dryer for CTC tea. Fresh tea leaves were passed through crushers and cutters having 10 teeth per inch to obtain the CTC tea in the Sonapur Tea Co. Pvt. Ltd. in Guwahati. After this process, fermentation of tea was allowed to take place to achieve its antimicrobial activity [11]. The fermentation was perpetrated at a low atmospheric temperature of 18 °C since fermentation at low temperature improves the tea quality [12]. Finally, drying has been carried out in the conical bubbling fluidized bed dryer. During the

**Table 1** Dimensions of conical bubbling fluidized bed dryer

| S. No. | Cone angle | Inlet diameter (cm) | Outlet diameter (cm) | Height of riser (cm) |
|---|---|---|---|---|
| 1 | 0° | 15 | 15 | 120 |
| 2 | 5° | 15 | 25.48 | 120 |
| 3 | 10° | 15 | 36 | 120 |

drying process, effect of various parameters such as drying air temperature (T), superficial air velocity (U), inventory (I), and different cone angle (α) of the dryer was investigated, and further analysis has been executed.

## 2.2 *Experimental Method*

For the experimental investigation, three different conical risers of different cone angles are fabricated using MS sheet of thickness 2 mm. The bottom diameter of all these risers is taken as 15 cm. The area of cross section of the riser is increasing along the height of the riser to maintain the desired cone angle. However, the height of all three risers is taken as 120 cm. The three different riser cone angles (α) maintained here are 0°, 5°, and 10°. The dimensions of the setup are explained in Table 1. Experiments were carried out for three different superficial air velocities (U) such as 1, 1.5, 2 m/s for drying of tea. Air was introduced as a primary phase supplied from the high pressure and high discharge blower passes via pipes through the designed distributor plate with an opening area of 24.16%. Ten pressure tapings were inserted along the riser height keeping the same space as 5 cm between two successive pressure tapings in order to measure the pressure drop. Ten sets of k-type thermocouples were inserted in a row to measure the bed temperatures. These thermocouples were connected to the data acquisition system and computer. A schematic diagram of the experimental setup has been shown in Fig. 1. The initial moisture content of the tea is measured by a benchtop moisture analyzer, and it is found to be 70% wet basis. Equivalent moisture content of black tea varies from 3 to 3.5% wet basis [13]. Hence, in this investigation, the moisture percentage of fresh tea which contains up to 70% moisture wet basis is eventually brought to 3% wet basis. The percentage moisture content of the inventory has been recorded in a regular interval of 5 min throughout the drying. The effects of drying air temperature, superficial air velocity, inventory, and cone angle of the riser on the moisture removal rate are analyzed.

## 2.3 *Determination of Minimum Fluidization Velocity*

The minimum fluidization velocity is calculated by using the following Eq. 1 as given by [14]. The properties and parameters used in this experiment are mentioned

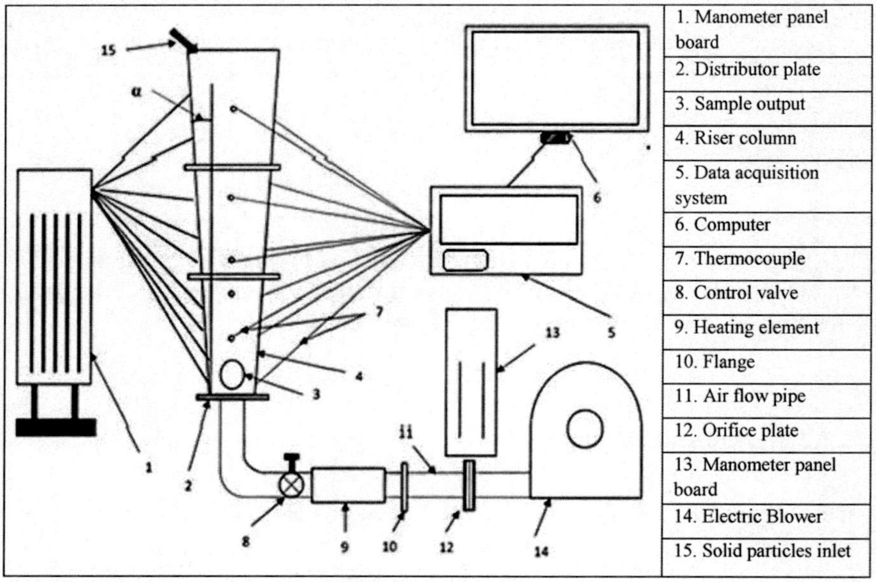


**Fig. 1** Schematic diagram of conical bubbling fluidized bed dryer

below as well (Table 2 and Table 3).

$$B\frac{r_0}{r_1}U_{\mathrm{mf}}^2 + \mathrm{AU}_{\mathrm{mf}} - (1 - \varepsilon_{\mathrm{mf}})(\rho_{\mathrm{s}} - \rho_{\mathrm{g}})g\frac{r_0^2 + r_0 r_1 + r_1^2}{3r_0^2} = 0 \quad (1)$$

**Table 2** Properties of air

| | |
|---|---|
| Density (kg/m$^3$) | 1.2 |
| Thermal conductivity (W/m k) | 0.025 |
| Specific heat (J/kg k) | 1005 |
| Viscosity (N s/m) | 1.86E−5 |

**Table 3** Parameters used in the experiment

| S. No. | Parameters | Value |
|---|---|---|
| 1 | Superficial air velocity (m/s) | 1, 1.5, 2 |
| 2 | Drying air temperature (°C) | 80, 85, 90 |
| 3 | Bed inventory (kg) | 0.75, 1, 1.5 |
| 4 | Initial moisture content of tea (% wet basis) | 70 |
| 5 | Equilibrium moisture content of tea (% wet basis) | 3 |
| 6 | Density of CTC tea (kg/m$^3$) | 370.3 |
| 7 | Cone angle of dryer (°) | 0, 5, 10 |

where

$$A = 150\frac{(1 - \varepsilon_{\mathrm{mf}})^2}{\varepsilon_{\mathrm{mf}}^3}\frac{\mu_{\mathrm{g}}}{(\phi_{\mathrm{s}} d_{\mathrm{p}})^2}$$

$$B = 1.75\frac{(1 - \varepsilon_{\mathrm{mf}})}{\varepsilon_{\mathrm{mf}}^3}\frac{\rho_{\mathrm{g}}}{\varphi_{\mathrm{s}} d_{\mathrm{p}}}$$

$r_0$ Inner radius of bed
$r_1$ Outer radius of bed
$\varepsilon_{\mathrm{mf}}$ Voidage at minimum fluidization = 0.45
$\rho_{\mathrm{s}}$ Density of solid
$\rho_{\mathrm{g}}$ Density of air
$\phi_{\mathrm{s}}$ Sphericity of particles = 0.86
$d_{\mathrm{p}}$ Particle diameter.

## 3 Results and Discussions

Drying characteristics for CTC tea particles at various operating conditions have been explained in this section. CTC unfermentaed tea collected as the bed material for this investigaton is shown in Fig. 2. It is observed that drying pattern follows a similar trend for all operating conditions. Tea is here dried with varying different parameters at once. The final dried product through this experiment is shown in Fig. 3. Firstly, the effect of different drying air temperature on the moisture removal rate is investigated. For this, three different inlet drying temperatures 80, 85, and 90 °C are taken, and the drying characteristics obtained for each drying temperatures are shown in Fig. 4. Inventory, cone angle, and superficial air velocity are set as 0.075 kg, 10°, and 1.5 m/s, respectively. It comes out as higher the drying temperature, shorter is the required drying time. Results (Fig. 4) have shown that drying time is the least for the

**Fig. 2** CTC tea

**Fig. 3** Final dried tea

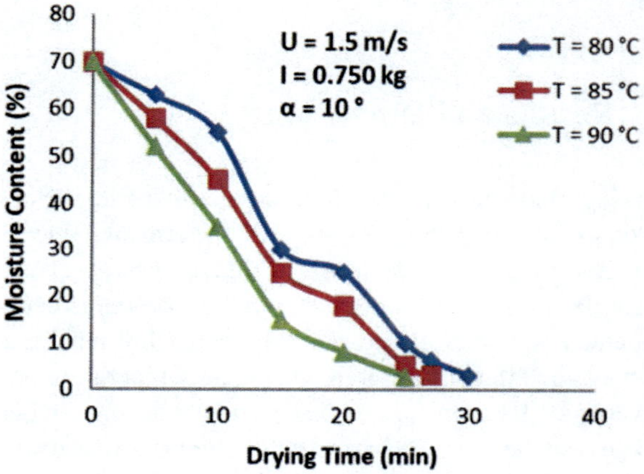


**Fig. 4** Moisture content versus drying time for different drying temperatures

highest inlet air temperature (in this case 90 °C). This can be explained by the fact that as the inlet air temperature is increased, the air carries more amount of heat which transfers to bed particles resulting in a superior moisture removal rate from the bed particles (here, tea). For the second case, effect of different superficial air velocities is looked into. Taking the inventory, cone angle, and drying temperature as 0.75 kg, 10°, and 80 °C, respectively, drying was performed at three different superficial inlet air velocity of 1 m/s, 1.5 m/s, and 2 m/s. For different inlet air velocities (1, 1.5 and 2 m/s), moisture removal rate is larger for the higher inlet air velocity (Fig. 5). This is because at high value of superficial air velocity, more heat is carried by the flowing air though the drying temperature provided is constant and vigorous mixing of air and tea particles occurs. This phenomenon leads to the high heat transfer between the hot air and the inventory of tea particles.

The effects of different bed inventory on drying characteristics are also investigated, and it is shown in Fig. 6. For a drying air temperature of 80° C, superficial inlet air velocity of 1.5 m/s, and a cone angle of 10°, drying is carried out for 0.075,

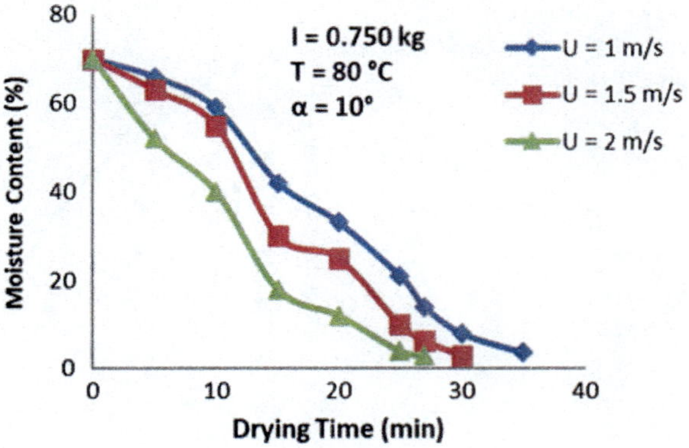


**Fig. 5** Moisture content versus drying time for different air velocities

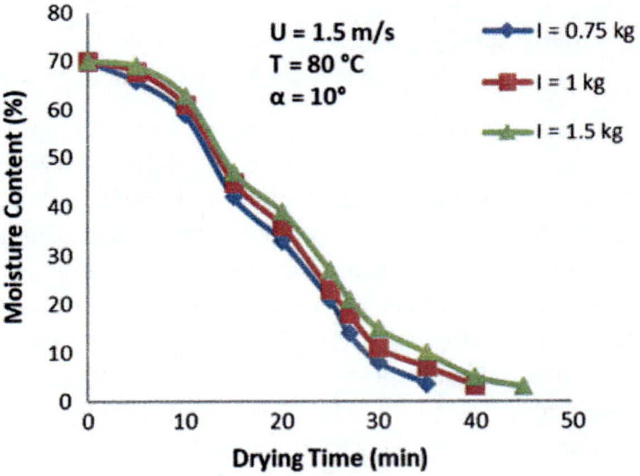


**Fig. 6** Moisture content versus drying time for different inventories

1, and 1.5 kg of inventory. It is found from the experimental data recorded that when the bed inventory is increased for a constant velocity and drying air temperature, the total drying time increases. The inventory of 1.5 kg here requires the highest drying time of 45 min to reduce its percentage moisture content from 70% wet basis to the final percentage moisture content of 3%. Finally, the investigations to reveal the effect of cone angle of riser on the drying characteristics of tea are performed. For this experiment, a 15 cm of static bed height (cone angles of 0°, 5°, and 10°) are taken into consideration. The drying inlet air temperature of 80 °C and superficial air velocity of 1.5 m/s are maintained. It can be seen from Fig. 7. that the moisture removal rate is becoming higher as we go from 0° to 10° cone angle. A minimum of 30 min total drying time is achieved while drying in a cone angle of 10°. This can be explained by the fact that with increasing cone angle there is more mass accumulated toward the bottom region of the drying riser, and hence, the mixing of bed particles and air occurs more precisely, as a result of which the rate of heat transfer increases between hot air and bed particles resulting in a higher moisture removal rate from bed particles.

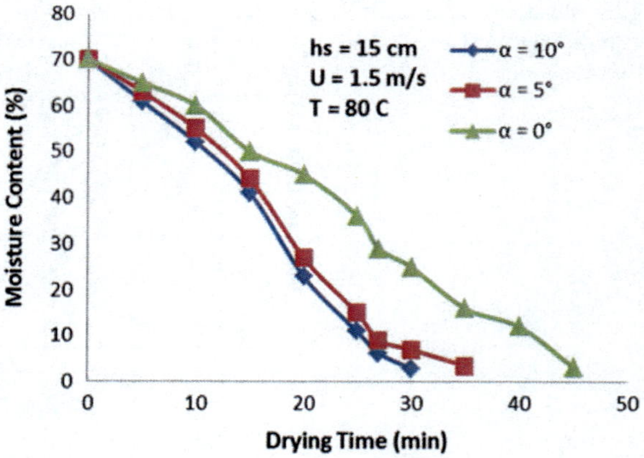


**Fig. 7** Moisture content versus drying time for different cone angles

## 4 Conclusion

In this present study, the experimental investigation of drying characteristics for CTC tea across a conical fluidized bed dryer has been performed. Effect of inventory, superficial velocity, cone angle, and drying air temperature on the performance of moisture removal rate has been investigated. Change in moisture removal rate has been noticed drastically with change in drying air temperature and superficial air velocity. It has been seen that for a specific cone angle of riser, inventory, and superficial inlet air velocity, higher the drying air temperature, shorter is the required drying time which implies a higher moisture removal rate. Also, as the superficial air velocity is increased, while keeping the other parameters of drying temperature, inventory, and riser cone angle constant, we obtain a shorter drying time. It is due to the reason that at a higher superficial air velocity, the amount of heat carried away by the air is more, and there occurs a robust mixing of air and tea particles, which lead to the high heat transfer between the two phases. The effect of inventory on the drying characteristic of tea particles obeys an inverse proportionality with the moisture removal rate for a specific drying temperature, cone angle, and superficial air velocity. The net drying time for a larger size of inventory under similar drying parameters is found to be larger. The cone angle of the drying riser is also found to be playing a vital role in the drying characteristics. A healthy heat transfer between air and bed particles and a well mixing of the hot air and tea are observed in case of the higher cone angle riser.

## References

1. B. Banerjee, *Botanical Classification of Tea* (Springer, Berlin, 1992), pp. 25–51
2. R. Ravichandran, R. Parthiban, Changes in enzyme activities (polyphenol oxidase and phenylalanine ammonia lyase) with type of tea leaf and during black tea manufacture and the effect of enzyme supplementation of dhool on black tea quality. Food Chem. **62**(3), 277–281 (1998)

3. A.M. Baruah, P.K. Mahanta, Fermentation characteristics of some assamica clones and process optimization of black tea manufacturing. J. Agric. Food Chem. **51**(22), 6578–6588 (2003)
4. S.J. Temple, A.J. Van Boxtel, PH-postharvest technology: a comparison of dryer types used for tea drying. J. Agric. Eng. Res. **77**(4), 401–407 (2000)
5. S.J. Temple, A.J. Van Boxtel, Equilibrium moisture content of tea. J. Agric. Eng. Res. **74**(1), 83–89 (1999)
6. S.J. Temple, C.M. Temple, A.J. Van Boxtel, M.N. Clifford, The effect of drying on black tea quality. J. Sci. Food Agric. **81**(8), 764–772 (2001)
7. S. Hovmand, Fluidized bed drying. Handbook Ind. Drying **1**, 195–248 (1995)
8. S.K. Chou, K.J. Chua, New hybrid drying technologies for heat sensitive foodstuffs. Trends Food Sci. Technol. **12**(10), 359–369 (2001)
9. RD. Lima, M. do Carmo Ferreira, Fluidized and vibrofluidized shallow beds of fresh leaves. *Particuology*, 2011, 9(2), pp.139–47.
10. C.C. Chou, L.L. Lin, K.T. Chung, Antimicrobial activity of tea as affected by the degree of fermentation and manufacturing season. Int. J. Food Microbiol. **48**(2), 125–130 (1999)
11. E.W. Chan, Y.Y. Lim, S.K. Wong, K.K. Lim, S.P. Tan, F.S. Lianto, M.Y. Yong, Effects of different drying methods on the antioxidant properties of leaves and tea of ginger species. Food Chem. **113**(1), 166–172 (2009)
12. P.O. Owuor, M. Obanda, Comparative responses in plain black tea quality parameters of different tea clones to fermentation temperature and duration. Food Chem. **72**(3), 319–327 (2001)
13. S.J. Temple, A.J. Van Boxtel, Thin layer drying of black tea. J. Agric. Eng. Res. **74**(2), 167–176 (1999)
14. S. Jing, Q. Hu, J. Wang, Y. Jin, Fluidization of coarse particles in gas solid conical beds. Chem. Eng. Process. **39**, 379–387 (2000)

# Design and Development of a *XY* Positioning Stage Using Shape Memory Alloy Spring Actuator

**Ashutosh Padhan and Yogesh Singh**

**Abstract** This paper demonstrates a working model of a 4PP (prismatic–prismatic) positioning stage where four sliders in the fixed base are associated with shape memory alloy (SMA) spring actuation individually. The end-effector of the robotic system developed has two degrees of freedom, i.e., translational in *X*- and *Y*-axis, respectively. The forward and backward movement of the developed platform is accomplished by coupling two sliders with the association of the SMA linear actuators. This model has a fixed part (the wooden base) and a movable part (end-effector). As the Nitinol SMA spring is lightweight in construction and has precise actuation, it has a wide range of applications over the other *XY* motion platform. Due to the deflection in the springs, the end-effector moves in both *X*- and *Y*-axis where the actuation of the SMA springs are studied with respect to the direction, and the workspace study is carried out. After the study, the comparison is done with the designed workspace. It is noted that this 4PP positioning stage has a precise actuation of springs due to the shape memory effect and provides a good workspace. Thus, it has a wide range of use in biomedical and optical applications where the motion stages are the primary requirement.

**Keywords** Shape memory effect · Nitinol · Positioning stage

## 1 Introduction

Smart materials such as piezoelectric, magnetostrictive, and shape memory alloy (SMA) are being widely used for the positioning stages. The Nitinol (Ni–Ti) spring which is most commonly used for its shape memory effect with negligible permanent deformations is used for the actuation of the *XY* positioning stage. As the SMA

A. Padhan (✉) · Y. Singh
Department of Mechanical Engineering, National Institute of Technology Silchar, Silchar, Assam 788010, India
e-mail: ashutoshpadhan.1994@gmail.com

Y. Singh
e-mail: yogesh@mech.nits.ac.in


K. M. Pandey et al. (eds.), *Recent Advances in Mechanical Engineering*, Lecture Notes in Mechanical Engineering, https://doi.org/10.1007/978-981-15-7711-6_59

spring actuates very precisely at an instant and once it gets voltage from the power supply, it has many advantages over previously used actuations like electric DC motor, hydraulic, etc. The SMA actuator has an advantage over the electrostatic actuator because the electrostatic actuators require a larger driving voltage which leads to the electrostatic breakdown [1]. Unlike the conventional motion stages, complaint mechanism stages provide a larger stroke and motion to end-effector [2]. Also, the stiffness and the resistivity of the SMA materials continuously increase with the temperature rise is an added benefit for the positioning stage because the rigidity of the stage is enhanced as discussed in Sect. 3. Smaller the dimension is, the shape memory alloy (SMA) generates a greater force and stroke lengths [3]. To increase performance and overcome limited strain, SMA springs are preferred over the wires [4]. With the rise in temperature, the force increases, residual displacement decreases, and the stiffness is greater at austenitic rather than martensitic state [3–5]. The displacement of the SMA spring increases with increase in temperature proportionally [6]. With the increase in the annealing temperature, the Young's modulus increases, and the pitch of the SMA springs a change in mechanical behavior because of the cold work done on the springs and also the stiffness is independent of temperature in the elastic region [7]. SMA can also be used to the self-optimizing support system theory where the parameters vary. The SMA can also be used in the field of vibration control for a rotating machine as it is feasible and effective [8]. The nonlinear geometric effects on the mechanical behavior of two different springs of different strengths with same modulus of shear has the higher austenitic stiffness than that of martensitic stiffness in the case of weaker spring and the reverse in the case of the stronger spring [9].

This paper deals with the fabrication of the 4PP motion stage using SMA springs as the actuator and calculating the workspace acquired by the end-effector. Further this paper presents the forward kinematics of the model in Sect. 2, and then, the development of the positioning stage is elaborated in Sect. 3. Section 4 discusses the results obtained and defines the workspace that is achieved. At last, the concluding remarks are made in Sect. 5.

## 2 Forward Kinematics of the Proposed Model

The direction of movement of 4PP motion stage is shown in Fig. 1. The kinematic model has a movable part (end-effector) and a fixed part (wooden base) connected with four prismatic joints which is the SMA spring. This motion stage has nine links, eight binary joints, and six sliding joints. According to the Gruebler's criterion, the motion stage has two degree of freedom as presented in Fig. 1. The $P$ in the figure represents the prismatic joint. The translational inputs $(a_1, a_2)$ are also the joint space displacements vector. The coordinates of the end-effector are represented as $O$ ($X$, $Y$), and the reference or origin is named as $D$ (0, 0).

Forward kinematics of the model.

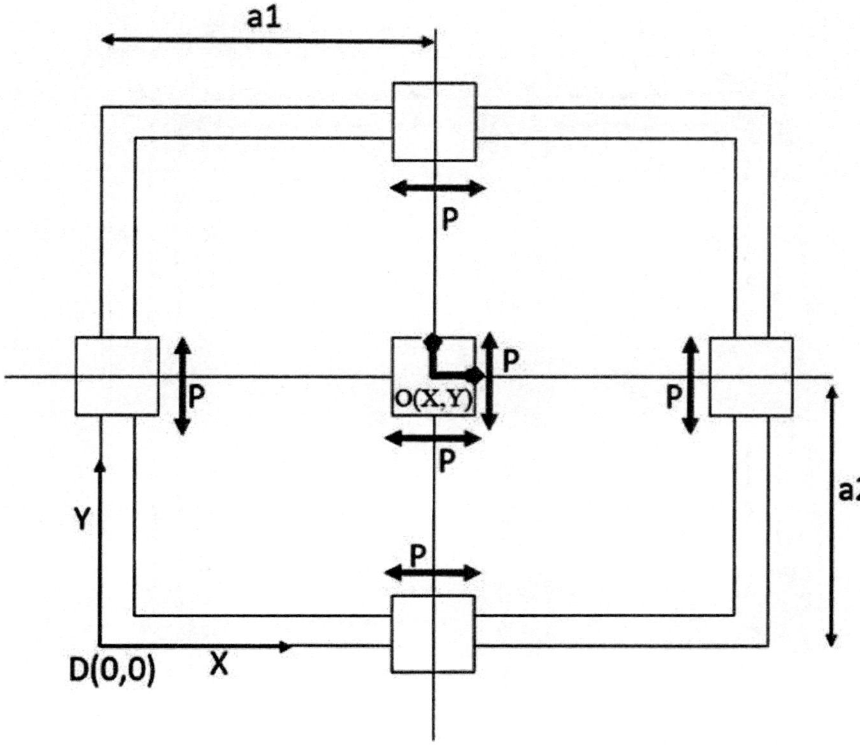


**Fig. 1** Block diagram of 4PP planar parallel manipulator

$$X = a_1. \tag{1}$$

$$Y = a_2. \tag{2}$$

## 3 Development of the 4PP Positioning Stage

This paper deals with the translational movement of the end-effector *E* which is caused due to the actuation of the SMA spring associated with the base. It has two degrees of freedom in *X*- and *Y*-direction as shown in Fig. 2. There are four sliders ($S1$, $S2$, $S3$, and $S4$) that are coupled to the end-effector by SMA springs as shown in Fig. 2. The 4PP SMA motion stage provides four independent motions both forward and backward in *X*- and *Y*-axis, respectively, as shown in Fig. 2. Also, there are ten guiding rods denoted as sliding rods (SR1, SR2, SR3, SR4, SR5, SR6, SR7, SR7, SR8, SR9, and SR10) where slider $S1$ slides on SR1 and SR2, slider $S2$ slides on SR3 and SR4, slider $S3$ slides on SR5 and SR6, and slider $S4$ slides on SR7 and SR8.

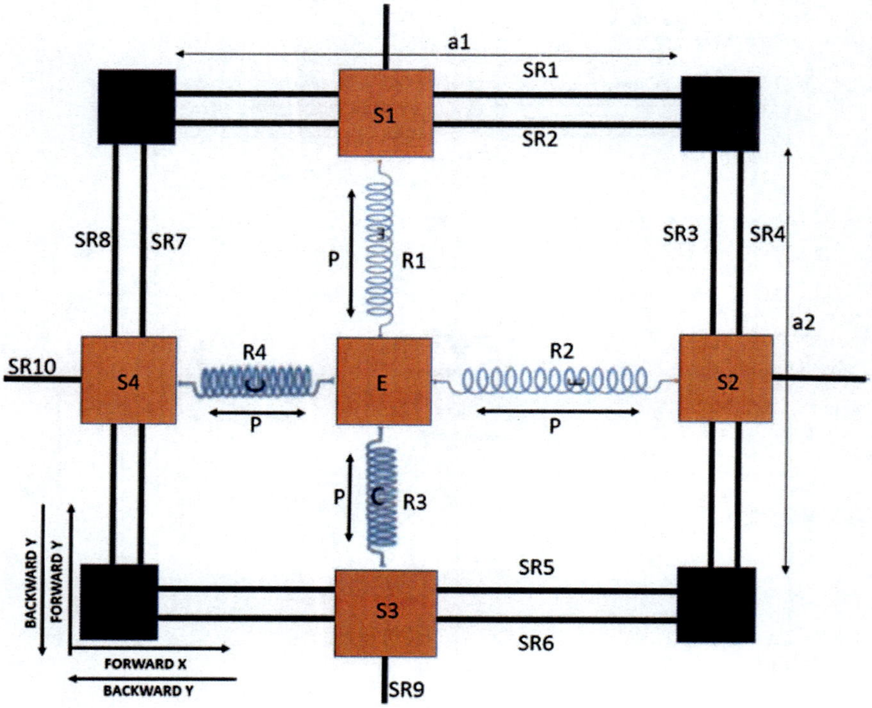


**Fig. 2** Schematic diagram of 4PP motion stage

The positioning stage carries two sliders SR9 and SR10. However, SR9 slides inside $S1$ and $S3$, and SR10 slides inside $S2$ and $S4$.

As there are two motions in $X$- and $Y$-direction, their joint space variables are denoted as $a_1$ and $a_2$.

As shown in Fig. 2, the initial position of the stage where springs $R3$ and $R4$ are in a compressed state and the other springs $R1$ and $R2$ are in the expanded form. So, in order to get a forward displacement in $X$-axis, spring $R2$ is actuated (SMA springs by joule heating effect) and $R4$ for the backward displacement in $Y$-axis.

Also, for $Y$-axis, spring $R1$ is actuated for forward, and $R3$ is actuated for backward displacement. Whenever the springs $R2$ and $R4$ are actuated, the joint space variable $\mathrm{a}_1$ is considered, and similarly, when springs $R1$ and $R3$ are actuated, the joint space variable $\mathrm{a}_2$ is considered.

The experimental setup is shown in Fig. 3. The setup has an available workspace of 200 × 200 mm$^2$, and the designed workspace is 35 × 35 mm$^2$. The setup has 200 mm as length and 200 as width. The SMA spring in its compressed state has 29 mm in length, and in fully expanded form it has 80 mm as its length. The property of the SMA spring is given in Table 1. For the measurement of the coordinates of the end-effector, a digital camera (CANON 1000D) is used having 3888 × 2592 pixels with a focal length of 50 mm at a shutter speed of 1/60 s.

**Fig. 3** Experimental setup of 4PP motion stager

**Table 1** Properties of SMA spring (Nickel–Titanium alloy)

| S. No. | Properties of SMA spring | Value |
|---|---|---|
| 1 | Density | 6.45 g/cm$^3$ |
| 2 | Young's modulus | 75–83 GPa |
| 3 | Poisson's ratio | 0.43 |
| 4 | Coefficient of thermal expansion (austenitic) | 11 × 10$^{-6}$/°C |
| 5 | Thermal conductivity | 0.18 W/cm K |

The external DC supply is provided for the expansion and contraction of the SMA spring which has an output of 0–30 V and current of 0–5A.

## 4 Result and Discussion

The shape memory effect of the SMA spring is seen when there is a current supplied to the springs. Initially, the Nitinol SMA spring was deformed. To regain the original heat, input is applied to the spring. The current input to the SMA spring develops a force in the reverse direction which contracts the spring and which further drives the positioning stage to give a forward and backward motion in *X*- and *Y*-direction. The coordinates of the end-effector are obtained due to the actuation of the SMA spring when the current is being input to the springs. Each SMA spring here acts as a prismatic joint in the 4PP motion stage as shown in Fig. 4. Also, at a time, only one spring is actuated because of which the other springs of the stage stays in inactive condition. In the fabricated model of 4PP motion stage which is actuated by SMA springs, different coordinates of the end-effector are noted by implementing different

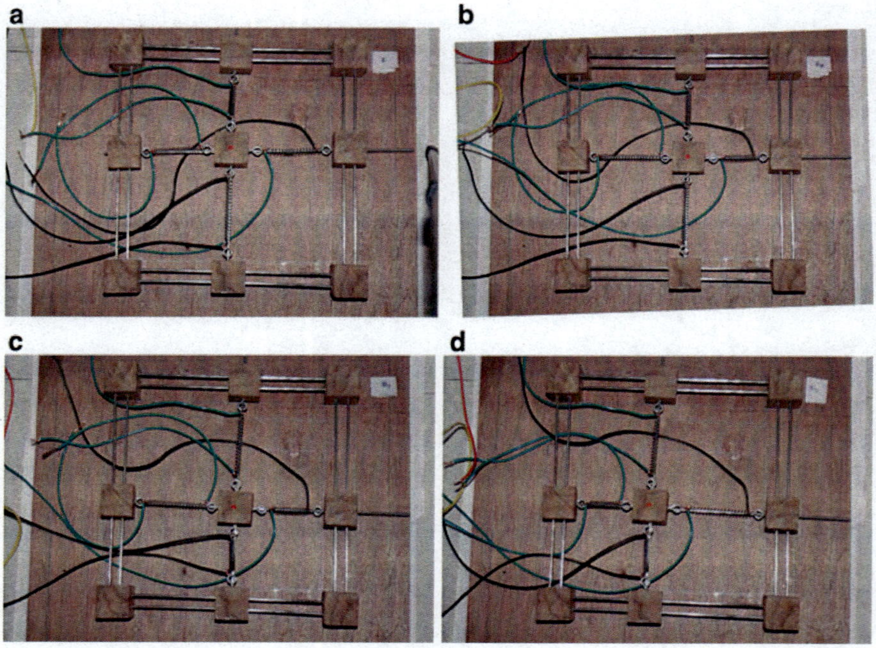


**Fig. 4** **a** Forward *Y* (*R*1), **b** forward *X* (*R*2), **c** backward *Y* (*R*3), **d** backward *Y* (*R*4)

experiments using power supply (current) to the springs (*R*1, *R*2, *R*3, and *R*4). As the SMA springs are coupled with the slider block, its contraction and elongation change the coordinate system of the end-effector. As when the current is supplied to the SMA springs, there is a change in coordinates. Also, the end-effector is connected to two SMA springs in a particular axis, and when one of the springs is contracted by the current input, the other spring gets expanded which leads to the forward and backward motion to the end-effector.

The movement of the sliders (*S*1, *S*3) in the forward and backward direction is the reason for the *X*-direction motion of the stage similarly sliders (*S*2, *S*4) for the *Y*-direction motion. SMA spring *R*2 is actuated to provide the end-effector motion in the forward *X*-direction as shown in Fig. 2, whereas for the backward *X*-direction the spring R4 is actuated. The slider *S*1 slides on the sliding rod (SR1, SR2), and slider S3 slides on sliding rod (SR5, SR6) as shown in Fig. 4. SMA spring *R*1 is actuated to displace the end-effector in the forward *Y*-direction as shown in Fig. 4, whereas for the backward *Y* direction the spring *R*3 is actuated. The slider S2 slides on the sliding rod (SR3, SR4), and slider S4 slides on sliding rod (SR7, SR8) as shown in Fig. 4.

The acquired coordinates of the end-effector in *X* and *Y* directions are tabulated in Table 2 by the actuation of the SMA springs of the 4PP motion stage. The digital camera installed on the top of the motion stage and used to track the coordinates of the end-effector is shown in Fig. 3. There are four different conditions where the

**Table 2** Coordinates of the end-effector with the actuation of spring

| Actuated spring | Reading 1 | | Reading 2 | |
|---|---|---|---|---|
| | $O_X$ (mm) | $O_Y$ (mm) | $O_X$ (mm) | $O_Y$ (mm) |
| Forward *Y* (*R*1) | 82.47 | 115.84 | 81.20 | 115.67 |
| Forward *X* (*R*2) | 113.56 | 115.84 | 113.90 | 113.3 |
| Backward *Y* (*R*3) | 112.9 | 81.8 | 113 | 82.47 |
| Backward *Y* (*R*4) | 82.67 | 81.8 | 82 | 81.67 |

coordinates of the end-effector are noted as given in Table 2. The first condition specifies that SMA spring *R*1 is actuated, and then, the sliders (S2, S4) and end-effector move in the forward *Y*-axis as shown in Fig. 4. For the second condition when the spring *R*2 is actuated, the initial position for this is the final coordinates achieved in the first condition. Similarly, *R*3 and *R*4 are also actuated, and the results obtained are given in Table 2.

It is also seen that every SMA actuator has achieved the maximum displacement required and returns to its original position in backward direction. The maximum and minimum displacement in *X*- and *Y*-direction is given in Table 2. All actuation of springs (*R*1, *R*2, *R*3, and *R*4) is shown in Fig. 4. All the forward and backward actuations are shown in Fig. 4.

There are three workspaces shown in Fig. 5. Red color shows the available workspace (200 × 200 mm$^2$) that is the maximum space that the end-effector can be moved, and the green color shows the designed workspace that is 35 × 35mm$^2$. Then, the experimental workspace is obtained which is shown in yellow color. Once the position of the end-effector is obtained, then the experimental workspace is plotted and compared with the designed workspace as shown in Fig. 5. It is observed that

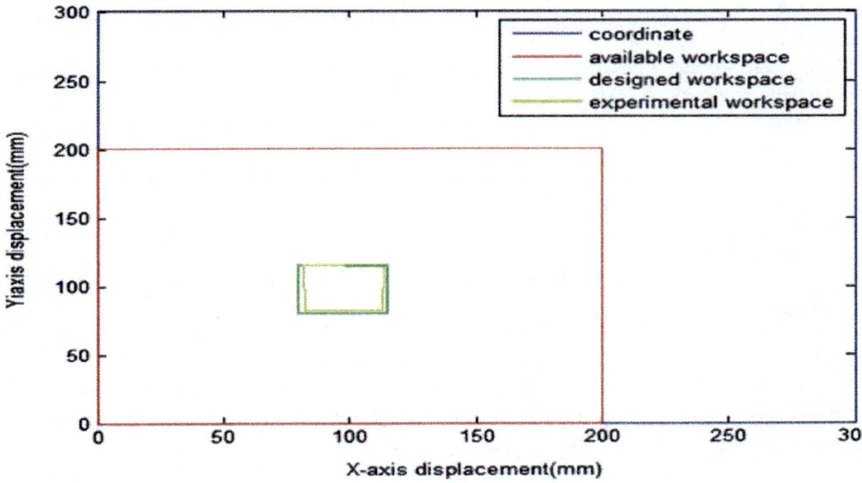


**Fig. 5** *X* versus *Y* displacement of the end-effector movement

the experimental results obtained are matched with the designed work space which comes out to be 34.5 × 34 $mm^2$. This obtained workspace is around 3% of the available workspace and 96% of the designed workspace of the positioning stage.

## 5 Conclusion

From all the discussion made, starting from the design and development of the 4PP motion stage actuated by SMA springs has been represented in this paper. Four set of readings have been investigated and tabulated. The average values were plotted *X* versus *Y* position, and it is found that this SMA-based actuators have good amount of workspace. This paper has achieved an overall workspace of 3.5% to that of the total feasible workspace. This investigation is completely done with the open loop principle where the constant current is applied individually to the SMA springs without any control to the magnitude of the current supplied to SMA spring. There is huge possibility in this research field where the closed loop control algorithm can be introduced to control the current supplied to the SMA springs. The further future research can be done on the use of SMA linear actuation technique in the different positioning device.

**Acknowledgements** This work was fully funded by National institute of technology, Silchar under the Technical Education Quality Improvement Program-3 of Government of India.

## References

1. Z.Wu, Q. Xu, Survey on recent designs of compliant micro-/nano-positioning stages. Actuators **7** (2018)
2. S. Wan, Y. Zhang, Xu. Qingsong, Design and development of a new large-stroke XY compliant micropositioning stage. J. Mech. Eng. Sci. **231**(17), 3263–3276 (2017)
3. R. Santhanam, Y. Krishna, M.S. Sivakumar, Behaviour of NiTi SMA helical springs under different temperatures and deflections. ISRN Mater. Sci. 320370 (2013)
4. R. Pillai, G. Murali, M. Gopal, Modeling and simulation of a shape memory alloy spring actuated flexible parallel manipulator. Int. Conf. Robot. Smart Manuf. **133**, 895–904 (2018)
5. J. Ma, H. Huang, J. Huang, Characteristics analysis and testing of SMA spring actuator. Adv. Mater. Sci. Eng. (2013)
6. B. Heidari, M. Kadkhodaei, M. Barati, F. Karimzadeh, Fabrication and modeling of shape memory alloy springs. Smart Mater. Struct. (2016)
7. Y.-Y. He, S. Oi, F.-L. Chu, H.-X. Li (2007) Vibration Control of a Rotor–Bearing System using Shape Memory Alloy: I Smart Materials and Structures Theory, pp. 114–121
8. M.A. Savi, P.M.C.L. Pacheco, M.S. Garcia, R.A.A. Aguiar, L.F.G. de Souza, R.B. da Hora, Nonlinear geometric influence on the mechanical behavior of shape memory alloy helical springs. Smart Mater. Struct. (2015)
9. S. Polit, J. Dong, Development of a high-bandwidth XY nano positioning stage for high-rate micro-nano manufacturing. Trans. Mech. (2011)

# Selective Etching of Aluminium 6068 Using Photochemical Machining

**Jaswant Kumar, Tapas Debnath, and Promod Kumar Patowari**

**Abstract** Photochemical machining (PCM) is one of the non-conventional machining processes that produces stress-free and burr-free components. It can be employed as micromachining manufacturing process to produce microsized components. This paper focuses on the selective etching of aluminium 6068 alloy using ferric chloride as an etchant. The primary aim of present work is to perform selective etching with greater accuracy. The input parameters/control parameters considered are etchant concentration, etching time and etching temperature. The performance measures have been chosen as material removal rate (MRR), edge deviation and undercut. The photochemical machining of aluminium has been carried out based on Taguchi L9 orthogonal array.

**Keywords** Aluminium · Chemicals · PCM · Selective etching

## 1 Introduction

Photochemical machining (PCM) is one of the advanced machining processes which is the combination of light and chemical energy and used to produce stress-free and burr-free components [1]. It is basically a controlled corrosion process in which material is removed by strong chemical solution known as "etchant" [2]. This process is preferable because it does not affect the physical and chemical properties of specimen. It is called by various names such as chemical etching, wet etching and photoetching.

J. Kumar (✉) · T. Debnath · P. K. Patowari
Department of Mechanical Engineering, National Institute of Technology Silchar, Silchar, Assam 788010, India
e-mail: jashigpj4895@gmail.com

T. Debnath
e-mail: nit.tapas11@gmail.com

P. K. Patowari
e-mail: ppatowari@yahoo.com



K. M. Pandey et al. (eds.), *Recent Advances in Mechanical Engineering*, Lecture Notes in Mechanical Engineering, https://doi.org/10.1007/978-981-15-7711-6_60

This process is widely used to machine thin and flat materials that produced geometrically complex and dimensionally very accurate components. Strength-to-weight ratio is very important to the automotive industries; therefore, if weight can be reduced from any area of part/product, keeping the same strength then the strength-to-weight ratio will be improved [3].

Wangikar et al. carried out the parametric optimization for photochemical machining of copper [4], brass and German silver [5] by using overall evaluation criteria (OEC). Agrawal and Kamble presented the optimization of PCM process parameters of SS-304 (microfluidic channel) [6], array of microhole on SS-304 [7] and Al/SiC composite [8] using grey relational analysis (GRA) method. Thorat and Sadaiah carried out experimental study on Cu–Co L605 alloy to find the effect of residual stress, grain size, grain orientation and hardness in photochemical machining [9]. Using grey relational analysis, multiobjective optimization of phosphor bronze was carried out by Agrawal et al. [10]. Misal and Sadaiah also used grey-based Taguchi method for optimization of the PCM process parameters while machining Inconel 601 [11]. Bhasme and Kadam carried out the experimental study and parametric optimization of process parameters for photochemical machining of SS316L [12]. Cakir investigated the regeneration process of cupric chloride [13]. Ekta et al. present the etching characteristics of aluminium while machining square cavity using photochemical machining [14].

From the literature review, it has been observed that no significant study has been reported on selective etching of aluminium alloy of grade 6068 using PCM. Therefore, the aim of this paper is to investigate the selective etching of aluminium alloy and study the performance measures for photochemical machining.

## 2 Materials and Methods

### *2.1 Material*

In this paper, aluminium alloy of grade 6068 is selected for the study. The major constituents of this alloy are aluminium, magnesium and silicon. It has good mechanical properties and exhibits good weld ability and also it has excellent electrical and thermal conductivity. The material removal from the work piece takes place by chemical reaction between etchant molecules with aluminium particles. The chemical reaction of aluminium with ferric chloride is given as.

$$3FeCl_3 + Al \rightarrow 3FeCl_2 + AlCl_3$$

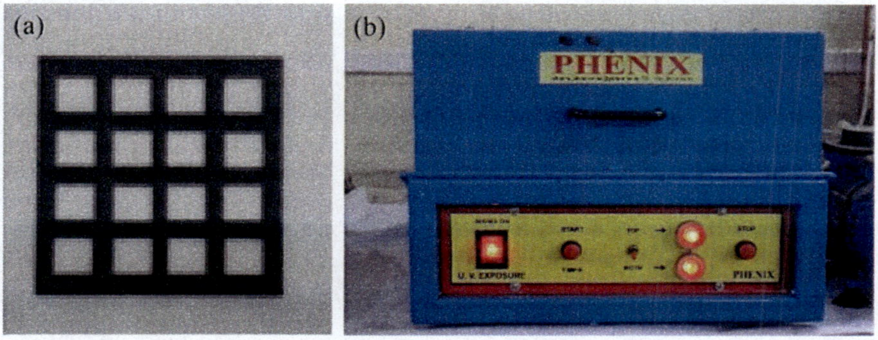


**Fig. 1** **a** Phototool; and **b** UV exposure unit

## 2.2 *Experimental Procedure*

Initially, the specimens have been prepared and cleaned with acetone to remove contaminants so that proper adhesion of photoresist is possible. The photoresist is applied on specimen by immersion process and allowed these to dry. After that, a phototool (designed by AutoCAD and printed out on a transparent paper) is used to create the impression on the specimen using UV exposure unit. A phototool and UV exposure unit are shown in Fig. 1a, b respectively.

After UV exposure, the specimen is immersed in a solvent-based developer which removes UV unexposed masks from the workpiece. Then after, the specimens have been washed in gently running water. The final step is machining, in which specimen gets chemically etched at required temperature which is controlled by heating plate as shown in Fig. 2. The steps followed in the chemical etching of aluminium are represented in Fig. 3.

## 2.3 *Selection of Process Parameters and Their Levels*

Based on the survey of literature and some pilot experiments, three control parameters with their levels have been chosen for the experimentation. Table 1 demonstrates the chosen control parameters with their levels.

The performance measures considered for the experimentation are discussed as below:

$$\text{MRR} = \frac{\text{Initial weight} - \text{Final weight}}{\text{Machining time}} \tag{1}$$

- Material removal rate (MRR) which indicates the rate at which material is removed from the specimen is calculated as

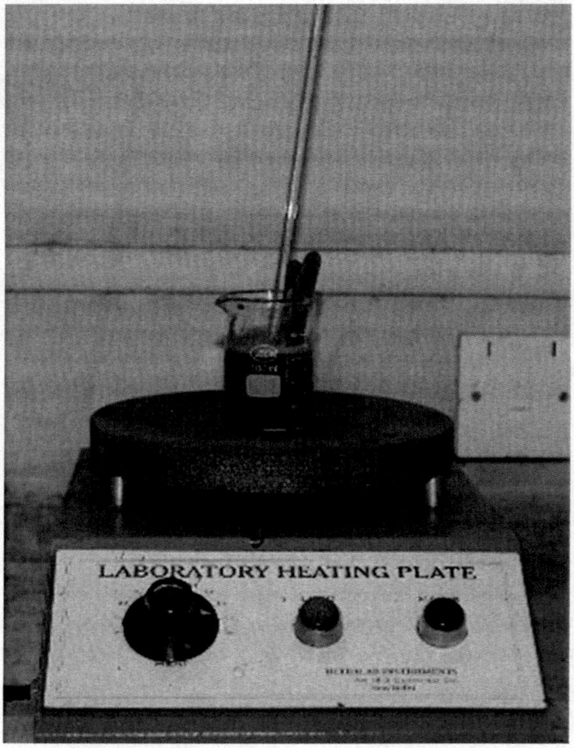


**Fig. 2** Heating plate with beaker containing etchant and thermometer

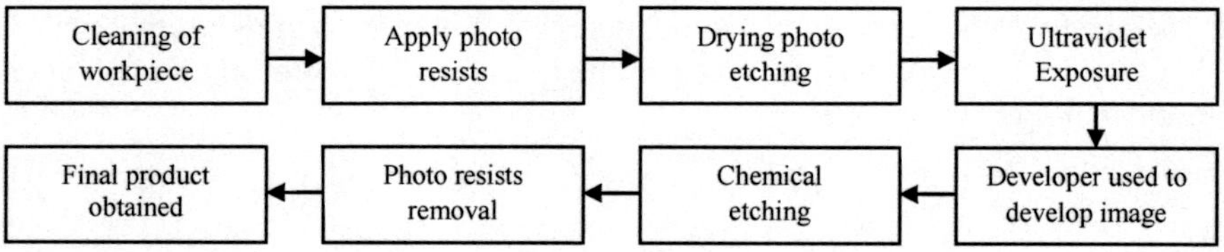


**Figure 3** Sequential steps followed in PCM

**Table 1** Control parameters with their levels

| Control parameters | Level 1 | Level 2 | Level 3 |
|---|---|---|---|
| Concentration (g/lit.) | 400 | 450 | 500 |
| Temperature (°C) | 35 | 40 | 45 |
| Time (min.) | 3 | 4 | 5 |

- Edge deviation (ED) is the non-conformity of the edge of the machined component. The edge deviation is measured at various points, and average value is considered for analysis.

**Table 2** Taguchi $L_9$ array with responses

| S. no. | Conc. (gm/lit) | Temp. (°C) | Time (min) | MRR (mg/min) | ED (μm) | UC (μm) |
|---|---|---|---|---|---|---|
| 1 | 400 | 35 | 3 | 0.167 | 13 | 10.0 |
| 2 | 400 | 40 | 4 | 0.275 | 24 | 11.0 |
| 3 | 400 | 45 | 5 | 0.300 | 28 | 12.8 |
| 4 | 450 | 35 | 4 | 0.175 | 16 | 8.4 |
| 5 | 450 | 40 | 5 | 0.280 | 22 | 20.7 |
| 6 | 450 | 45 | 3 | 0.425 | 36 | 18.8 |
| 7 | 500 | 35 | 5 | 0.240 | 19 | 10.5 |
| 8 | 500 | 40 | 3 | 0.310 | 32 | 13.0 |
| 9 | 500 | 45 | 4 | 0.500 | 33 | 28.4 |

- Undercut is the distance etched laterally under the photoresist layer. For analysis of undercut, the difference between the dimensions of machined area with that of the phototool is selected.

Both edge deviation and undercut are measured with the help of Leica Metallurgical Optical Microscope.

## 2.4 *Design of Experiments (DOE)*

Based on the chosen parameters, the design of experiment (DOE) has been prepared to conduct the experiment. Taguchi $L_9$ orthogonal array with response measures is shown in Table 2. DOE is a statistical technique which is used to study or analyse the effects of multiple response variables simultaneously. It refers to the layout that describes a combination of the factors included in the study.

## 2.5 *Experimentation*

Total nine experiments are carried out on aluminium to study the effects of input parameters, i.e. etchant concentration, etching temperature and machining time on performance measures, i.e. material removal rate (MRR), edge deviation and undercut. The edge deviation and undercut are measured by metallurgical optical microscope. MRR has been calculated by Eq. (1) as given above. Figure 4a shows the machined sample, Fig. 4b shows the microscopic image of complete machined area and Fig. 4c shows the measured edge deviation.

(a) Machined sample (b) Microscopic image of complete machined area (c) Measured Edge Deviation

**Fig. 4** Images of machined sample

# 3 Result and Discussion

The performance of PCM process has been analysed on the basis of microscopic images of machined samples for edge deviation and undercut, calculating the MRR by Eq. (1) as given above. The performance measures/responses are shown in Table 2.

## 3.1 *Effect of Control Parameters on MRR*

Figure 5 shows the variation of MRR with control parameters. MRR increases with increase in concentration as more molecules of ferric chlorides are present resulting in more interactions with workpiece. MRR also increases with increase in temperature because chemical energy of etchant molecules increases that leads to increase in the successful collisions. With the increase in time, MRR increases due to higher material

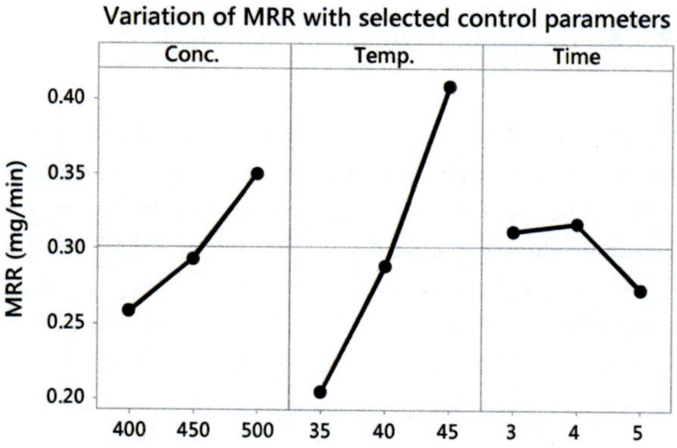


**Figure 5** Effect of input parameters on MRR

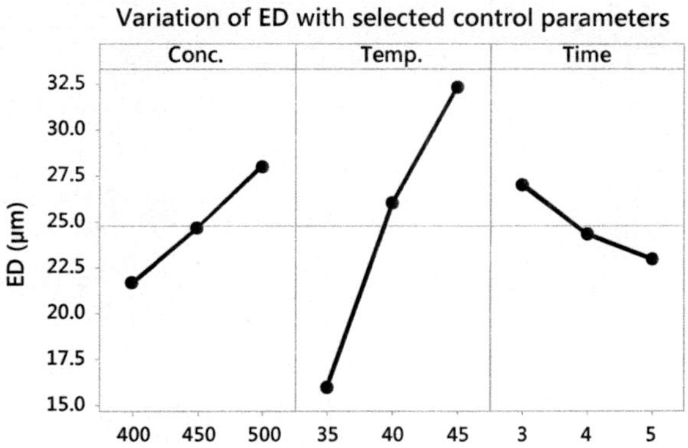


**Fig. 6** Effect of input parameters on ED

removal and after a certain period of time as the reaction scale becomes thicker, the removal of material slows down, hence MRR decreases with further increase in time.

### 3.2 *Effect of Control Parameters on Edge Deviation (ED)*

The variation of edge deviation with selected control parameters is shown in Fig. 6. Edge deviation increases with increase in concentration because of higher number of molecules of etchant that reacts with the workpiece particles. ED also increases with temperature because molecules at higher temperature are more reactive which leads to increasing in edge deviation. But the deviation of edge decreases with increase in etching time because at higher machining time the etchant gets sufficient time to flatten the sharp picks and ribs or making the edges dull that reducing the ED.

### 3.3 *Effect of Control Parameters on Undercut (UC)*

The effect of control parameters on undercut is shown in Fig. 7. From figure, it is clear that undercut increases with the increase in concentration and temperature because as explained in Sects. 3.1 and 3.2, molecules of etchant are more reactive at high temperature and with high concentration, more numbers of molecules react with the workpiece, resulting in increasing the undercut. The variation of undercut with time does not follow a proper path.

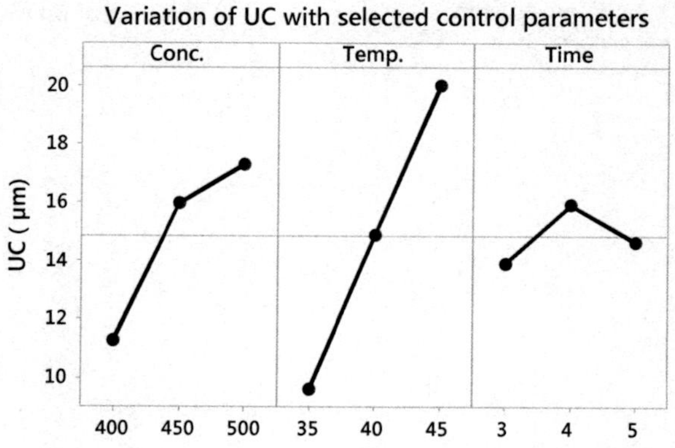


**Fig. 7** Effect of input parameters on UC

## 4 Conclusions

Square-type selective etching has been successfully machined on aluminium (grade 6068) using photochemical machining. The input parameters chosen are etchant concentration, machining time and etchant concentration whereas the performance measures are material removal rate, edge deviation and undercut. Based on the results and discussion, the following conclusions have been drawn:

- MRR increases with increase in concentration and temperature but with time, MRR initially increases and then decreases due to higher-scale formation with time.
- With the increase in concentration and temperature, edge deviation increases but it decreases with increase in etching time.
- Undercut increases with increase in temperature and concentration.
- Maximum material removal rate is observed at etchant concentration of 500 g/lit, etching temperature of 45 °C and etching time at 4 min.
- Minimum undercut and edge deviation are observed at time of 4 min, temperature of 35 °C and concentration of 450 g/lit and etching time of 3 min, temperature 35 °C and concentration 400 g/lit, respectively.
- Optimum edge deviation and undercut are observed at etching temperature of 35 °C.

## References

1. G.R. Rathod, S.U. Sapkal, R.M. Chanmanwar, Multi-objective optimization of photochemical machining by using GRA. Mater. Today Proc. **4**(10), 10830–10835 (2017)
2. O. Cakır, Chemical etching of aluminium. J. Mater. Process. Technol. **199**(1–3), 337–340 (2008)

3. P. Mumbareand, A.J. Gujar, Multi objective optimization of photochemical machining for ASME 316 steel using grey relational analysis. Int. J. Innov. Res. Sci. Eng. Technol. 5(7) , 12418–12425 (2016)
4. S.S. Wangikar, P.K. Patowari, R.D. Misra, Parametric optimization for photochemical machining of copper using overall evaluation criteria. Mater. Today Proc. **5**(2), 4736–4742 (2018)
5. S.S. Wangikar, P.K. Patowari, R.D. Misra, Effect of process parameters and optimization for photochemical machining of brass and german silver. Mater. Manuf. Processes **32**(15), 1747–1755 (2017)
6. D. Agrawal, D. Kamble, Optimization of photochemical machining process parameters for manufacturing microfluidic channel. Mater. Manuf. Processes **34**(1), 1–7 (2019)
7. D. Agrawal, D. Kamble, Effect and optimization of photochemical machining process parameters for manufacturing array of micro-hole. J. Br. Soc. Mech. Sci. Eng. **41**(4), 178 (2019)
8. D. Agrawal, D. Kamble, GRA and ANN integrated approach for photochemical machining of Al/Sic composite. Mater. Today Proc. **4**(8), 7177–7188 (2017)
9. S. Thorat, M. Sadaiah, The effect of residual stresses, grain size, grain orientation, and hardness on the surface quality of Co–Cr L605 alloy in photochemical machining. J. Alloy. Compd. **804**, 84–92 (2019)
10. D.P. Agrawal, K.V. Gurav, D.N. Kamble, Multi-objective optimization of photochemical machining process based on grey relational analysis method for spray etching. Appl. Mech. Mater. **612**, 77–82 (2014)
11. N.D. Misal, M. Sadaiah, Multi-objective optimization of photochemical machining of inconel 601 using grey relational analysis. Mater. Today Proc. **5**(2), 5591–5600 (2018)
12. A.B. Bhasme, M.S. Kadam, Parameter optimization by using grey relational analysis of photochemical machining. Int. Res. J. Eng. Technol. **3**(3), 992–997 (2016)
13. O. Cakir, Copper etching with cupric chloride and regeneration of waste etchant. J. Mater. Process. Technol. **175**(1–3), 63–68 (2006)
14. E. Tripathi, T. Debnath, P.K. Patowari, Etching characteristics of aluminium while machining square cavity using photochemical machining. AIP Conf. Proc. 1998(1) (2018). https://doi.org/10.1063/1.5049105

# Microstructure and Mechanical Characterisation of Friction Stir Welded Inconel 718 Alloy

**Sanjay Raj, Pardeep Pankaj, and Pankaj Biswas**

**Abstract** Joining with friction stir welding (FSW) of high-strength material is challenging work due to high tool wear/degradation and low tool life. Recently, developed tool WC–Co tool materials that withstand high temperatures and torsional stresses are used for welding of high-strength materials with FSW. In this article, friction stir welding (FSW) of Inconel plates is performed with WC–Co tool. The rotational speed of 300 rpm and a traverse speed of 90 mm/min were taken. In microstructural analysis, refinement of grain size of the base material (48 μm) to stir zone (18 μm) was observed, which results in enhancement of mechanical properties such as strength of the welded sample observed 77% of the base material strength, which shows partial welding of Inconel plate occurred and microhardness of the welded sample was 60% harder than base material hardness. FESEM of fractured tensile specimen informed the ductile fracture. Because of the high strength of Inconel alloy, after single weld, significant tool wear was observed. The EDS analysis was carried out at the stir zone of the weld, and it shows the presence of tungsten in the welded region, and also the presence of the Ni, Fe, Cr and Nb elements are found higher peaks in the EDS spectrum.

**Keywords** EDS analysis · Friction stir welding · Inconel 718 · Microstructure analysis · WC–co tool wear

S. Raj (✉) · P. Pankaj · P. Biswas
Department of Mechanical Engineering, Indian Institute of Technology Guwahati, 781039 Assam, India
e-mail: raj176103110@iitg.ac.in

P. Pankaj
e-mail: pankajpardeep22@gmail.com

P. Biswas
e-mail: pankaj.panu012@gmail.com



K. M. Pandey et al. (eds.), *Recent Advances in Mechanical Engineering*, Lecture Notes in Mechanical Engineering, https://doi.org/10.1007/978-981-15-7711-6_61

## 1 Introduction

Friction stir welding is an advance solid-state welding technique that utilises the thermomechanical effect of the rotating tool on the workpiece, without using consumable material to produce a weld. Nowadays, FSW of high melting point and high-strength material such as steel, titanium and nickel alloys is challenging research due to high tool wear/degradation and low tool life. Due to excellent properties, such as high strength, high toughness and resistance to thermal fatigue at an elevated temperature up to 700 $^{\circ}C$ with outstanding corrosion resistance, Inconel 718 is one of the most versatile and essential metals which has been used in aerospace, oil field, industrial turbine engine, cryogenic tankage and nuclear engineering. There are various joining methods that have been applied to join Ni-base superalloys such as the fusion welding in which gas tungsten arc welding, gas shielded metal arc welding, laser beam welding and electron beam welding are commonly used [1]. However, there are major problems related with fusion welding of Inconel 718 such as poor penetration, microfissuring in HAZ, the formation of intermetallic compound (known as) laves phase ((Ni, Fe, Cr)2 (Nb, Mo, Ti)) in the fusion zone and liquation cracking in the weld which affect the mechanical properties along with corrosion resistance [2]. Therefore, to overcome the above problem, friction stir welding (FSW), a solid-state joining process, is a feasible method, which generates lower heat as compared to fusion welding. Although due to high strength the use of FSW to weld Ni-based alloys has not been widely used, in this field, song et al. [3] successfully welded 2 mm thick Inconel 718 alloy sheet using WC–Co tool at a welding speed of 150 mm/min. They reported an 80% reduction of grain size, which results in the enhancement of mechanical properties. Hanke et al. [4] studied on process parameter of FSW for joining of 3.2 mm Inconel 625 alloy with pcBN tool and they reported wear mechanism of this tool. In addition, Ahmed et al. [5] also welded 4 mm thick Inconel alloy using a silicon nitride tool with different welding speed parameter and found the best parameter for enhancing mechanical properties. Sato et al. [6] used friction stir welding technique to join 4.8 mm thick Inconel 600 alloy plate with the use of a pcBN tool and they found defect-free joint with lower corrosion resistance in stir zone and HAZ. Ramkumar et al. [7] used friction stir welding to join dissimilar material Inconel 600 with Inconel 718 and studied the corrosion behaviour of the weld.

From the literature survey, it was observed that very few papers are written on solid-state welding of Inconel 718 alloy. Therefore, in this present work, friction stir welding was applied to the Ni-base Inconel 718 alloy, using a WC–Co tool. The microstructure and mechanical properties of the thin Inconel 718 alloy and welded sample were examined.

## 2 Experimental Procedures

In this study, Inconel 718 alloy was used as a base metal; its weight % of chemical composition is given in Table 1. The base material dimensions of 200 mm × 120 mm × 3 mm of each plate were prepared. The preparation of the surface was performed with light grinding using grid emery cloth to remove oxides and contaminants, followed by cleaning with ethanol. AW25 graded WC-11%Co cylindrical tool of diameter 25 mm with 2.6 mm long conical pin tool used. The dimension of the conical pin tool was 7 mm at the shoulder side and 5 mm at the free end side was used. The schematic diagram of the tool geometry and FSW of Inconel 718 is shown in Fig. 1. FSW was performed on a vertical milling machine engaged with 25 hp heavy-duty-powered servomotor. A tool plunge depth of 0.2 mm with constant backward tool tilt angle 2° has been taken for obtaining sound weld. The FSW was performed at a constant rotational speed 300 rpm with traverse speed 90 mm/min.

For finding the mechanical properties, Vickers microhardness test and tensile test were performed. The tensile samples were taken from the FSW welded Inconel plate as per the ASTM E8, using wire electric discharge machining (W-EDM) in the perpendicular direction to the weld line. The tensile test was performed on the universal tensile testing machine at a crosshead speed of 1 mm/min. For measuring the microhardness at 1.5 mm below from the upper layer of the welded joints, a Vickers indenter with a 300 kgf load for 20 s at an interval of 1 mm was applied on the transverse cross sections of the weld,

Optical microscopy (OM) and scanning electron microscopy (SEM) were used to analyse macrostructures and microstructures of the welds. The following steps prepared sample for metallographic investigation with OM, started with sample

**Table 1** Chemical composition (wt%) of Inconel 718

| Ni | Cr | Nb | Mo | Ti | S | Si | Al | Fe |
|---|---|---|---|---|---|---|---|---|
| 51.89 | 18.82 | 5.72 | 2.93 | 1.09 | 0.005 | 0.05 | 0.5 | Bal. |

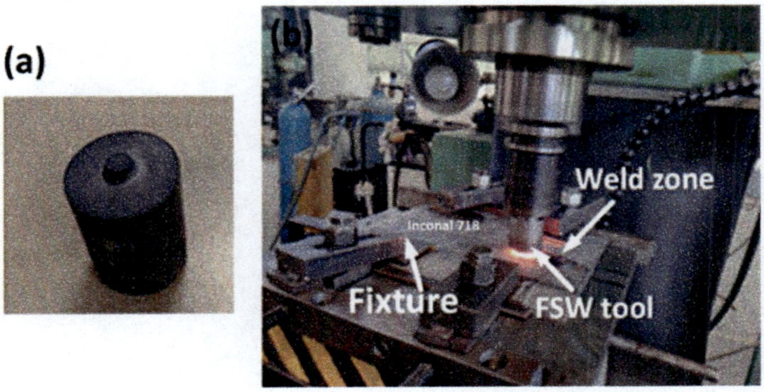


**Fig. 1** A pictorial view of the **a** WC–Co tool and **b** FSW of Inconel 718

sectioning, mounting, abrasive papers grinding, polishing and end with etching. The sample was etched with kalling's reagent (2 g $Cucl_2$ + 40 ml $CH_3OH$ + 35 ml HCL) for 40 s. For the study of the fracture morphology in the microstructure of welded sample, scanning electron microscopy (SEM) is used and the inclusion of external elements in the welded part was examined with energy-dispersive X-ray spectroscopy (EDS) analysis.

# 3 Results and Discussion

## *3.1 Microstructure Examination*

Figure 2 presents the external shape of the FSW welded Inconel 718. The material is welded at a traverse speed of 90 mm/min with a constant rotational speed of 300 rpm. The specimen welded at a low welding speed of 90 mm/min shows smooth weld without the presence of any defects on the surface. Furthermore, the welded specimen was observed that the partial penetration of tool, up to 2.01 mm from the top surface resulted in a sound weld, without causing any defects in the weld zone except for 0.84 mm from the bottom butt surface where insufficient welding happened as shown in Fig. 3. The stir zone was detected near the weld centre, and defect-free observation is found in the stir zone.

Optical micrograph of the base material (BM), stir zone (SZ) and thermo-mechanically affected zone (TMAZ) is shown in Fig. 3. The TMAZ shown in Fig. 3a has a slightly bigger grain structure than the SZ. FSW produces SZ near centre of the welded part shown in Fig. 3b, and this is generated due to frictional heat and severe plastic deformation of the BM. The microstructure of the SZ shows a fine equiaxed grain structure with an average grain size 18 μm. Since the FSW rotational tool produces stirring of the materials which refine the microstructure of the BM and produces the fine recrystallised grain structure [8], the BM has a coarse grain structure shown in Fig. 3c with an average grain size 48 μm. The friction stir welded

**Fig. 2** FSW welded Inconel 718

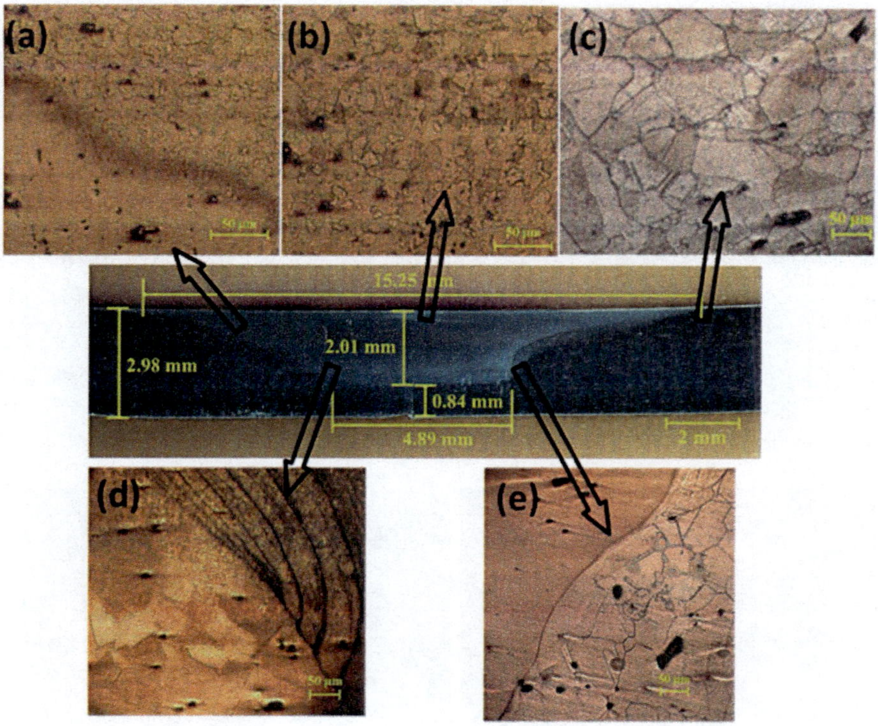


**Fig. 3** Microstructural view of welded cross section, **a** TMAZ, **b** SZ, **c** BM, **d** flow pattern of material, **e** interface pattern

zone experience recovery, recrystallisation and static grain growth in the course of the welding of the material. The region which is shown in Fig. 3d indicates the flow pattern of the material due to the high frictional heat input and stirring of the material at the tooltip. Interface structure observed in Fig. 3e indicates the distinguished pattern formation between TMAZ and SZ.

## *3.2 Mechanical Characterisation*

The welded sample of the tensile specimen and the fractured tensile specimen are shown in Fig. 4a, b, respectively. The result of the tensile test shows lower ultimate tensile strength (UTS), higher yield strength (YS) and lower elongation than the BM. Figure 5 shows a generalised fracture morphology of the welded tensile specimen, which occurs in the middle part of the welded region because of the lower tensile strength of the welded region than BM. Fracture morphology involves the ductile fracture. The presence of microdimples shows a characteristic ductile fracture. Figure 6 shows the Vickers hardness profile, which is obtained on the cross section at 1.5 mm

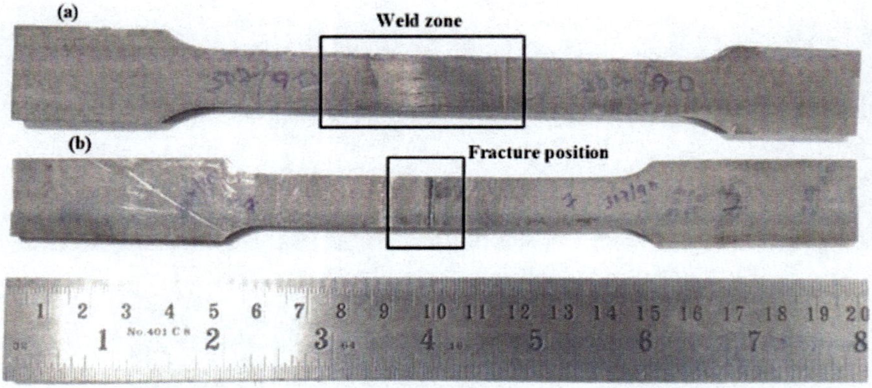


**Fig. 4** Top view of **a** welded specimen and **b** fractured specimen subjected to tensile tests

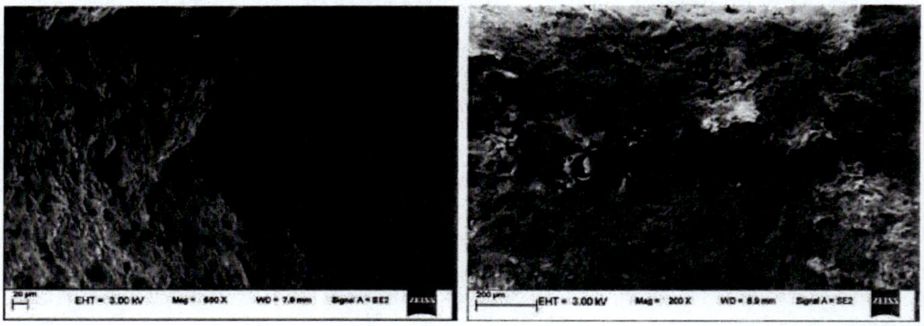

**Fig. 5** SEM photograph showing the ductile fracture at stir zone of the welded sample

**Fig. 6** Vickers hardness curves at 1.5 mm distance from the top surface of the welded sample

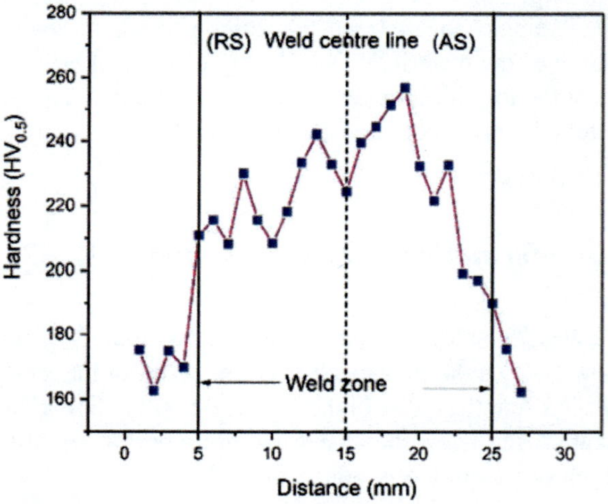

below from the top layer. The average hardness of BM and SZ obtained 160 HV and 256 HV, respectively. The differences in the microstructure graph within the deformed region of the FSW welded Inconel 718 resulted in an irregular distribution of the microhardness in the SZ as shown in Fig. 6. The maximum hardness in the SZ obtained 256 HV as a result of the finer grain size in the advancing side.

EDS analysis of the fractured tensile specimen shown in Fig. 7. This analysis revealed that the wear of the WC–Co tool. Besides the tungsten (W) inclusions grain also consists of the Ni, Fe, Nb and Cr peaks are found in the EDS spectrum.

Figure 8 presents the WC–Co tool wear views. After welding of 150 mm length Inconel 718 alloy plate, significant tool wear is observed. Due to the high strength of Inconel alloy, high frictional heat is generated between the tooltip and plate which results in 0.7 mm wear of tooltip

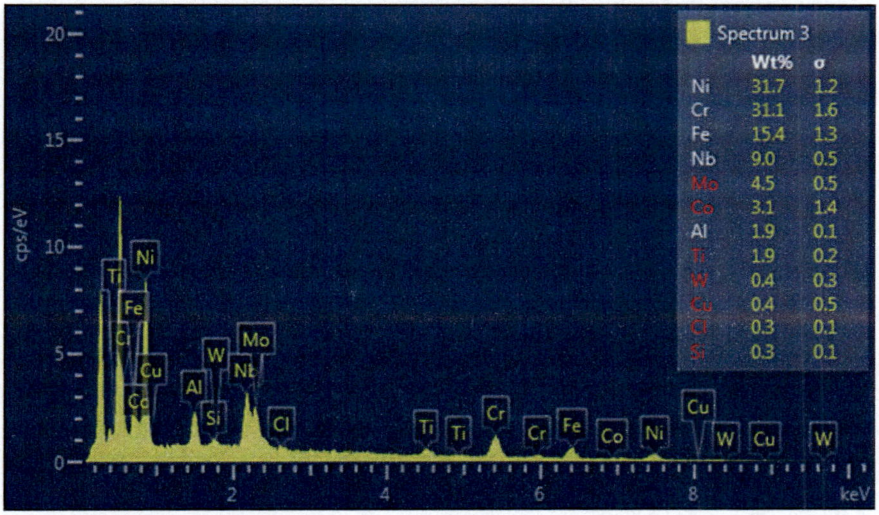


**Fig. 7** EDS analysis of the stir zone

**Fig. 8** FSW tooltip wear after welding of Inconel 718

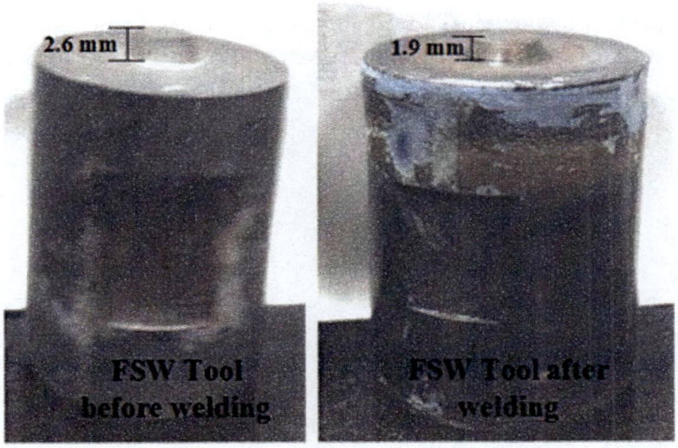

## 4 Conclusions

Joining of Inconel 718 alloy with FSW was successfully performed at a tool rotational and welding speed of 300 rpm and 90 mm/min, respectively, using WC–Co tool.

In the present study, microstructure and mechanical characteristics were examined. FSW made a smooth weld without any major defects in the welds except the bottom of the plate where insufficient penetration observed. The stir zone consists of 62% finer grain size than the base material. Tensile test failure occurred in the welded metal, and the fractured morphology shows ductile fracture. The hardness of the stir zone found maximum up to 256 HV which is 60% higher than the base material. 0.7 mm tool wear observed during a single pass of the tool in welding of Inconel alloy. Based on the results from the current studies, an additional heat source (hybrid FSW) recommended for joining of Inconel 718 alloy plates.

## References

1. J.T. Tharappel, J. Babu, Welding processes for Inconel 718-A brief review, in *IOP Conference Series: Materials Science and Engineering*, vol. 330, No. 1 (IOP Publishing, 2018), p. 012082.
2. B.G. Muralidharan, V. Shankar, T.P.S. Gill, *Weldability of Inconel 718-A Review*, No. IGC-175 (Indira Gandhi Centre for Atomic Research, 1996)
3. K.H. Song, K. Nakata, Microstructural and mechanical properties of friction-stir-welded and post-heat-treated Inconel 718 alloy. J. Alloy. Compd. **505**(1), 144–150 (2010)
4. S. Hanke, G.V.B. Lemos, L. Bergmann, D. Martinazzi, J.F. Dos Santos, T.R. Strohaecker, Degradation mechanisms of pcBN tool material during friction stir welding of Ni-base alloy 625. Wear **376**, 403–408 (2017)
5. M.M.Z. Ahmed, B.P. Wynne, J.P. Martin, Effect of friction stir welding speed on mechanical properties and microstructure of nickel based super alloy Inconel 718. Sci. Technol. Weld. Joining **18**(8), 680–687 (2013)
6. Y.S. Sato, P. Arkom, H. Kokawa, T.W. Nelson, R.J. Steel, Effect of microstructure on properties of friction stir welded Inconel Alloy 600. Mater. Sci. Eng. A **477**(1–2), 250–258 (2008)
7. A. Bansal, A.K. Sharma, S. Das, P. Kumar, On microstructure and strength properties of microwave welded Inconel 718/stainless steel (SS-316L). Proc. Inst. Mech. Eng. Part L J. Mater. Des. Appl. **230**(5), 939–948 (2016)
8. K.D. Ramkumar, W.S. Abraham, V. Viyash, N. Arivazhagan, A.M. Rabel, Investigations on the microstructure, tensile strength and high temperature corrosion behaviour of Inconel 625 and Inconel 718 dissimilar joints. J. Manuf. Processes **25**, 306–322 (2017)

# Effect of Dual-Height Plate Fins on Phase Change Material Cooling Technique: An Experimental Study

**Maibam Romio Singh, Asis Giri, and Pradip Lingfa**

**Abstract** The paper presents the study of phase change material for the cooling of electronic components. The phase change material (PCM) used in the study is eicosane. Thermal conductivity enhancers (TCEs), made of aluminum, are used in order to alter the low thermal conductive nature of PCM. The TCEs are divided into no fin heat sink, constant height plate heat sink, and dual-height plate heat sink. Three power inputs of 4, 6, and 8 W are used for the study. Thermal cooling capacity by all the three heat sinks setups is compared. The effect of parameters such as power input and volume of PCM is also discussed. Results indicate that the power input level and the volume of PCM are important factors that influence the thermal management of electronic components. The use of 5 dual-height plate fin heat sink elongates the charging period of the PCM filled setup, thereby maintaining the device temperature within a favorable limit for a longer duration, as compared to the no fin heat sink and 5 constant height plate fin heat sink.

**Keywords** Phase change material · Electronic cooling · Dual-height plate fins · Latent heat

## 1 Introduction

Recently, the study on phase change material (PCM) is booming in various areas where passive cooling is needed. Its application ranges from portable electronic devices to building insulation. Wang et al. [1] examine microencapsulated PCM for building material using aluminum honeycomb structures. The result is found very

M. R. Singh (✉) · A. Giri · P. Lingfa
Department Mechanical Engineering, NERIST, Nirjuli 791109, India
e-mail: mrsigh.mech@gmail.com

A. Giri
e-mail: measisgiri@rediffmail.com

P. Lingfa
e-mail: pradip.lingfa@gmail.com



K. M. Pandey et al. (eds.), *Recent Advances in Mechanical Engineering*, Lecture Notes in Mechanical Engineering, https://doi.org/10.1007/978-981-15-7711-6_62

promising, as it can restore the thermal condition of the wall even under different environmental conditions. Hawlader et al. [2] prepared encapsulated paraffin particles by complex coacervation as well as spray dying methods. Various parameters such as encapsulation efficiency, energy storage and release capacity were taken into the study. The results show that the energy storage or release capacity by either method counts about 145–240 J/g which indicates the encapsulated paraffin wax has a very good potential for thermal energy storage. Sanusi et al. [3] experimentally studied the performance of paraffin PCM embedded with graphite nanofibers (GNF) for energy storage and solidification. Results showed that the use of GNF-enhanced paraffin substantially reduced the solidification time for thermal containment units (TCUs) compared to the base material. As the sizes of electronic components are gradually decreasing, the study of the non-mechanical cooling method is becoming a prerequisite. Tan and Tso [4] experimentally studied the cooling of mobile electronic devices using a heat storage unit (HSU) filled with the PCM of n-eicosane inside the device. The high latent heat of n-eicosane in the HSU is found absorbing the heat dissipation from the chips and maintaining the chip temperature below the allowable service temperature of 50 °C for 2 h of transient operations of the PDA. It is found that different orientations of the HSU inside the PDA could affect significantly the temperature distribution. Kandasamy et al. [5] performed a numerical study for proper thermal management of the plastic quad flat package. It has been observed that the performance of thermal cooling is increased with the use of PCM. A good agreement is also acknowledged experimentally for the same problem. Arulmurugan et al. [6] investigated experimentally on thermal performance and effect of the PCM-based heat sink with different fins. The experimental setups are prepared with a graphical programming language (Lab VIEW). It has been found that the stability of temperature increases with the use of PCM and the increasing number of fins reduces the peak temperature of the heat sink. Srikanth et al. [7] investigated PCM-based composite pin fin matrix heat sink. n-eicosane and aluminum are used as PCM and heat sink material with a constant heat flux of 1.9 $kW/m^2$. The analysis is carried out using ANSYS FLUENT 14.0 software to determine the overall heat transfer coefficient. Tomizawa et al. [8] focus on the study of mobile phone cooling using PCM. Experimental and numerical analysis using finite element analysis concluded that the use of PCM sheets extends the saturation time. Use of a highly conductive copper sheet along with the PCM sheet further extends the saturation time of the devices. Lou et al. [9] examined PCM-based technique on Samsung SPH E2500 mobile phone. Baby and Balaji [10] extensively used a combination of artificial neural network and genetic algorithm to find a pin fin configuration for a maximum operational time of a device. Pakrouh et al. [11] highlighted that there is a complex relationship between the volume percentage of PCM and TCE. Taguchi method handled the optimization problem, and the results indicate a strong dependency on parameters such as fins height, fins thickness, fins number, and critical temperature. Hosseinizadeh et al. [12] also investigated on plate fin parameters for an efficient PCM heat sink. Kozak et al. [13] studied hybrid PCM—air heat sink. Effect of forced convection on a heat sink, supplied with constant heat flux, is discussed. Sensible heat accumulation is found increasing with the increase in power input. Saha et al.

[14] investigated for optimum distribution of fins heat sinks considering aluminum plate fins and pin fins as TCEs. Five different volume fractions of TCE are taken into consideration, and out of the five, TCE with a volumetric fraction of 8% is found performing better than the rest. Arshad et al. [15] discussed the effect of pin fin thickness with n-eicosane as PCM. Results show 2 mm thick pin fins performed better than 1 and 3 mm thick pin fins.

Many types of researches have been done on the cooling of electronic devices using PCM and fins. However, there is limited experimental finding on dual-height plate fins in the PCM cooling technique. Few numerical studies [16, 17] suggest that there is an improvement in heat transfer when variable height fins are used. Ji et al. [18] also numerically investigate double fin length, in a vertical position, for enhancement of PCM thermal storage system, and a promising result is obtained. However, there is still scarcity in experimental research of dual-height plate fins in the PCM cooling technique. With the miniaturization in the electronics industry, the weight of the heat sink plays a crucial role. The dual-height plate fins will reduce the weight of the heat sink, as well as accommodation of more thermal storage compounds, which can be made with proper design. Therefore, the purpose of the present study is to experimentally identify the effect of dual-height plate fins over the constant height plate fins for thermal management of portable electronic devices using eicosane as PCM material.

## 2 Experimental Setup

Figure 1 shows the schematic layout of the experimental setup. It consists of a DC power source for the heater, an insulated heat sink assembly, a data logger, and a computer to records the data collected through thermocouples. All the experiments are carried out in a closed and controlled temperature room.

### *2.1 Heat Sinks and Assembly*

The main components of the heat sink assembly consist of a heat sink filled with PCM, a silicon rubber heater of overall dimension of 53 × 63 mm$^2$, thermocouples, and insulating materials. The heat sink is insulated at all the sides and the bottom by rubber cork insulation. The insulation on the sides is further improved with the use of a thermocol insulator; this provides a one-directional heat flow from the heater to the PCM-filled heat sink. The radiation loss from the insulated sides is controlled by

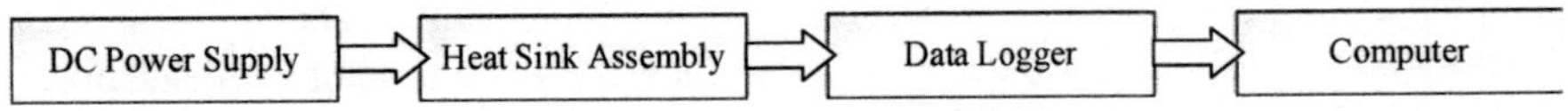


**Fig. 1** Schematic layout of the experimental setup

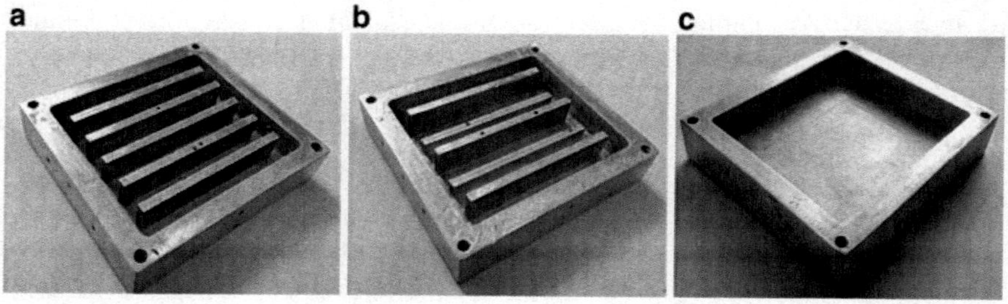


**Fig. 2** Photographic views of the heat sinks **a** 5 constant height plate-fin heat sink, **b** 5 dual-height plate-fin heat sink and **c** no fin heat sink

using a thin layer of aluminum foil in between the heat sink and the rubber cork. For a clear visualization of the melting phenomenon, a thin Plexiglas of 3 mm thickness is used to cover the PCM-filled heat sink. All the components are fastened together using a set of screws on a 20 mm thick acrylic plate.

Aluminum is considered as the heat sink material. It has a thermal conductivity of 202.4 W/m K and a density of 2719 kg/m$^3$. The heat sink considered has an overall dimension of 72 × 65 mm$^2$ base with a height of 15 mm. The experimental investigations have been done on three different sets of heat sink geometries. Two of the heat sinks have 5 numbers of plate fins, and the third one has no fin, which is for baseline comparison with the finned heat sinks. Again, one of the finned heat sinks has constant height plate fins and the other has dual-height arrangement of plate fins. For easier recognition, the heat sinks are named as no fin heat sink (or blank heat sink), 5 constant height plate-fin heat sink and 5 dual-height plate-fin heat sink. The cross-sectional dimensions of the plate fins are 3 × 52 mm$^2$. Figure 2 presents the photographic view of various heat sinks.

The PCM material use in the present investigation is eicosane from SIGMA ALDRICH, USA (CAS: 112-95-8), having a melting temperature ranges between 35 and 37 °C. This material has a density of 820 kg/m$^3$ (in the solid state) and 780 kg/m$^3$ (in liquid state) with a respective specific heat of 1.9 and 2.2 kJ/kg K. The thermal conductivity of the material in solid and liquid states is 0.39 and 0.16 W/m K, respectively. The material has a high latent heating capacity of 237.40 kJ/kg which makes it suitable for such a passive cooling method.

## 2.2 *Temperature Measurement*

The rise in temperature inside the PCM and the heat sink is mapped using calibrated K-type thermocouples. The thermocouples are attached using Araldite epoxy. The positioning of thermocouples in 5 dual-height plate-fin heat sink is shown in Fig. 3. The thermocouple H1 and H2 measures the temperature at the base of the heat sink. These temperatures represent the heat generated from the electronic device. In order

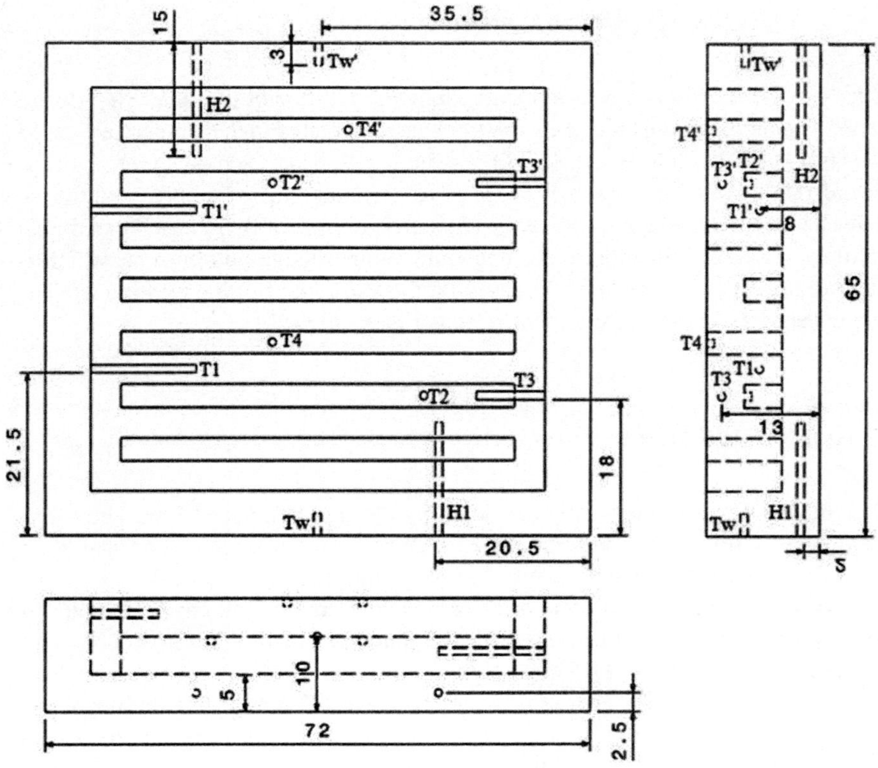


**Fig. 3** Position of thermocouples in the 5 dual-height plate fins

to place these thermocouples, a hole of 1 mm diameter and 15 mm deep is drilled at a height of 2.5 mm from the bottom surface of the heat sink. The thermocouples T1 and T1′ measure temperatures inside the PCM at a height of 8 mm from the base. The thermocouples T2 and T2′ measure the temperature at the top of 5 mm height plate fins. The thermocouples T3 and T3′ measure the temperature inside the PCM at 13 mm height from the base surface. The thermocouples T4 and T4′ measure the temperature at the top of 10 mm height plate fins. Apart from these, thermocouples Tw and Tw′ give the sidewall temperatures and another thermocouple detects the room temperature. A sixteen channel data logger (UniLogPro Plus) procured from Process Precision Instrument (PPI), India, served the purpose of recording the temperature from the thermocouples. Finally, a computer is used to display and store the data from the data logger.

### 2.3 Uncertainty in Measurement

Every precise or accurate measurement requires calibration of the measuring instruments. The thermocouples for the experiment are calibrated in the temperature range of 0–100 °C, according to the ASTM standard [19]. The discrepancy error in the thermocouple is found to be ± 1 °C. The DC regulated power supply used for the heater has a voltage range of 0–30 V and a current range of 0–5 A. A Fluke digital multimeter is used to calibrate the uncertainties in voltage and current, which is found to be ±0.1 V and ±0.01 A, based on the least count of the multimeter. The total uncertainty in the measurement of power is ±3.4% [20].

## 3 Results and Discussion

Three different power inputs are considered for the study; 4, 6, and 8 W. The corresponding heat fluxes are 1.19, 1.79, and 2.39 kW/m$^2$. Two different volumes of PCM have been considered for the analysis; these volumes are represented by $\phi = 0.5$ and $\phi = 1.0$. $\phi = 1$ is the volume of PCM which filled the TCE cavity completely (i.e., up to 10 mm height of fins) and $\phi = 0.5$ is the volume of PCM filled up to half height of the TCE cavity (i.e., 5 mm height of fins). For relatively low working load (upto 5 W power input), eicosane is a very effective PCM as seen from the above literature. To simplify the study, a well-defined temperature has been chosen as a set point temperature (SPT), and the time taken by each of the PCM-filled heat sink to reach this SPT is analyzed. Since eicosane has a latent heating phase between 35 and 37 °C, the upper limit temperature of phase transition, i.e., 37 °C, is chosen as one of the SPT (denoted in the figure by SPT37). This will show the nature of heat transfer during the melting phase. Apart from this temperature, another elevated temperature of 50 °C is also taken as SPT (represented as SPT50 in the figure) to see the trend of heat transfer after the latent heating phase.

### 3.1 Effect of Variable TCE

Figure 4 shows the time–temperature heating curve of eicosane in the constant and dual-height TCE. The average of the base temperatures H1 and H2 is used for the comparison. At the initial stage of the heating, there is a rapid rise in the time–temperature curve; this marked the sensible heating of the PCM material. With more addition of heat, the material starts a phase transition from solid to liquid. During this stage, the temperature inside the PCM becomes more or less stable even though heat is continuously supplied to the heat sink. The more the latent heating time, the better is the stability of the device temperature within a favorable limit. From the figures, it can be observed that the use of dual-height plate fins with PCM extends the latent

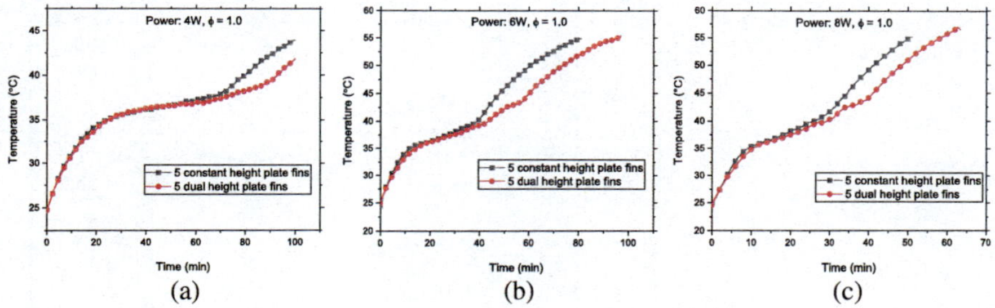


**Fig. 4** Heating inside the constant and variable TCE at **a** 4 W, **b** 6 W and **c** 8 W

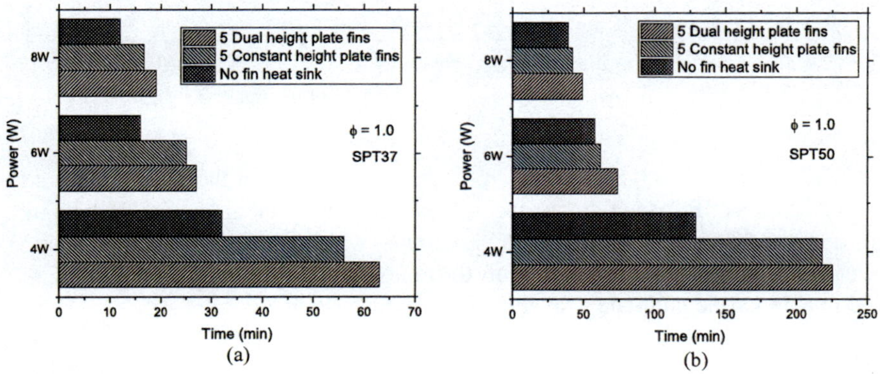


**Fig. 5** Charging time of PCM at different SPT **a** 37 °C and **b** 50 °C at different power input

heating time more than the constant height plate fins. Figure 5 compares the charging time required to reach SPT of 37 and 50 °C by the constant and dual-height plate fins filled with PCM. To reach an SPT of 37 °C at 4 W power input, the 5 dual-height plate-fin heat sink unit requires 63 min, while the 5 constant height heat sink unit takes 56 min to achieve the same temperature. This means the base of the heat sink unit for the dual-height heat sink will remain cooler for a longer duration compared to the constant height unit. Also, if the comparison is made with the no fin heat sink unit, elongation in operational time is much higher for the finned heat sink units.

## 3.2 *Effect of Volume of PCM*

The volume of PCM plays a major role in the latent heat thermal energy storage system. In this study, the effect of two different volumes of PCM is discussed and a comparison of heating time with the increase in the volume of PCM is made. When the PCM completely covers the fins, i.e., when $\phi = 1$, the heat carried away from the base by the plate fins is easily dissipated to the PCM. If the volume of PCM

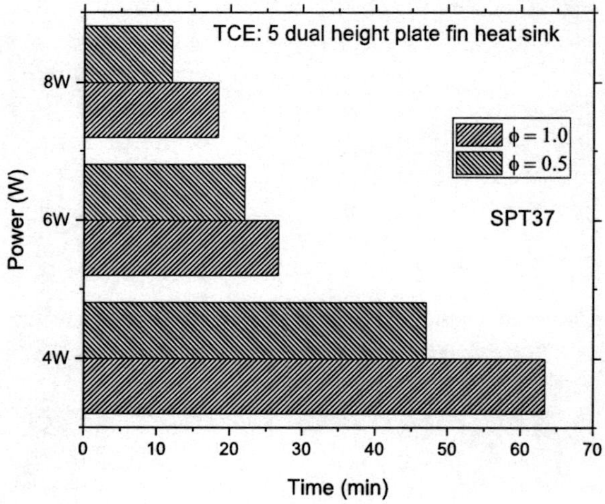


**Fig. 6** Variation in charging time due to volume of PCM

further reduced to $\phi = 0.5$, heat absorption by the PCM from the TCE reduces as the contact area of PCM and fins reduces. Consequently, the temperature of the base rises quickly, for the TCE with less PCM volume, which is not desirable in the case of electronic devices. Figure 6 highlights the time taken by different volumes of PCM filled in a 5 dual-height plate heat sink, to reach an SPT of 37 °C.

## 3.3 *Enhancement in Operational Time*

The enhancement achieved from both the finned units of PCM latent heat storage systems can be compare using enhancement ratio [10, 16]. By definition, it is the ratio of charging time of a finned PCM unit to no finned PCM unit. Figure 7 shows the enhancement ratios of the dual-height heat sink and constant height heat sink at various power inputs. In both cases, a higher enhancement ratio is achieved at lower power input. As the power level increase, the enhancement ratio also decreases. This suggests more effectiveness of the PCM cooling method with portable electronic devices with low power consumption. However, dual fin arrangement has higher enhancement in all the cases of the study.

## 3.4 *Heat Dissipation Inside the PCM*

As the heat flow from the base of the heat sink to the PCM, there is variation in temperature profile inside the PCM due to the difference in heat sink configuration. Figure 8 shows the special variation of the temperature inside the PCM, along with

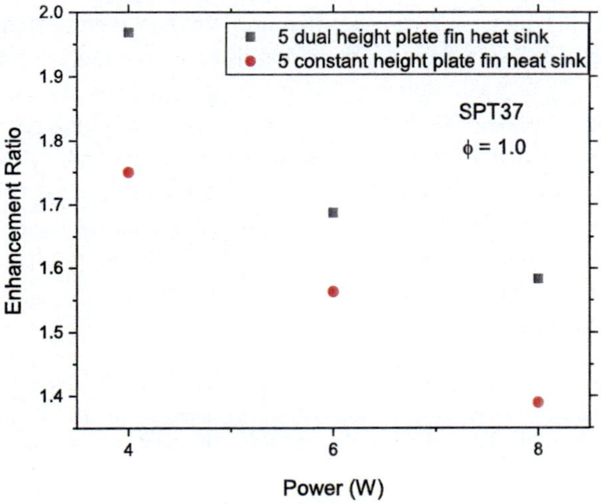


**Fig. 7** Enhancement ratio of the two finned heat sink unit to reach SPT of 37 °C

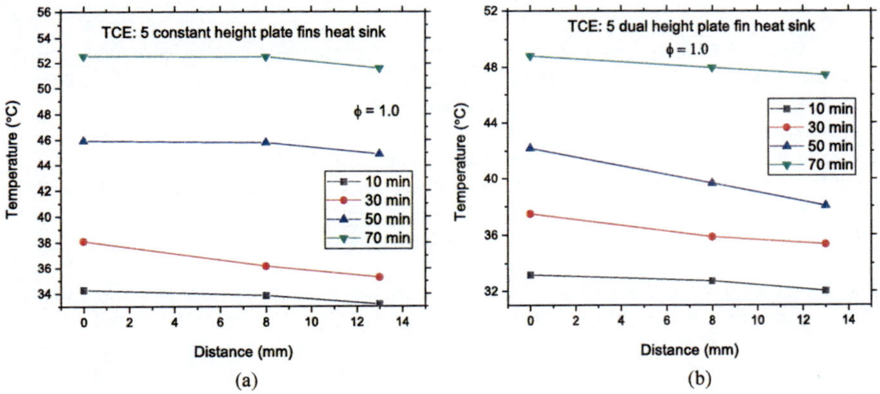


**Fig. 8** Variation of temperature at different locations inside the PCM with **a** constant height heat sink and **b** dual-height heat sink

the vertical height of the heat sink unit. The thermocouple positions correspond to the base, at 8 mm and at 13 mm height from the base are considered.

## 4 Conclusion

Effect of dual-height and constant height plate fin on thermal management using PCM is studied. Baseline comparison is also made with no fin heat sink filled with PCM. Results show the employment of plate fins enhanced the performance, and at the same time, the superiority of dual-height plate over the constant height plate fins is also witnessed. Studies are conducted at three power levels of 4, 6, and 8 W. The suitability of PCM techniques is found for portable devices with low power consumption. The volume of PCM is also found to strengthen the cooling effect. A maximum enhancement ratio of 1.97 is achieved at a power input of 4 W for the case with dual-height plate fins, whereas increasing power input is found reducing the enhancement effect. The dual-height fin arrangement also reduces the weight of the heat sink. So, the overall observation is that the implementation of dual-height plate fins in the PCM cooling technique gives better thermal management than the constant height plate fins.

**Acknowledgements** The authors acknowledged the financial support from MHRD. We are grateful to our institute as well as our Department of Mechanical engineering, NERIST, for all possible help. We also acknowledged SAIC, Tezpur University and TRTC, Guwahati, for their support in the work.

## References

1. S.M. Wang, P. Matiašovský, P. Mihálka, C.M. Lai, Experimental investigation of the daily thermal performance of a mPCM honeycomb wallboard. Energy Build. **159**, 419–425 (2018)
2. M.N.A. Hawlader, M.S. Uddin, M.M. Khin, Microencapsulated PCM thermal energy storage system. Appl. Energy **74**, 195–202 (2003)
3. O. Sanusi, R. Warzoha, A.S. Fleischer, Energy storage and solidification of paraffin phase change material embedded with graphite nanofibers. Int. J. Heat Mass Transf. **54**, 4429–4436 (2011)
4. F.L. Tan, C.P. Tso, Cooling of mobile electronic devices using phase change materials. Appl. Therm. Eng. **24**, 159–169 (2004)
5. R. Kandasamy, X.Q. Wang, A.S. Majumdar, Transient cooling of electronics using phase change material (PCM)—based heat sinks. Appl. Therm. Eng. **28**, 1047–1057 (2008)
6. L. Arulmurugan, M. Ilangkumaran, K.V. Prakash, Experimental investigation on thermal performance and effect of PCM based heat sink with different fins. IOSR J. Mech. Civil Eng. IOSR-JMCE, e-ISSN: 2278-1684, p-ISSN: 2320-334X, pp. 29–33 (2013)
7. R. Srikanth, N. Pavan, C. Balaji, Multi-objective geometric optimization of a PCM based matrix type composite heat sink. Appl. Energy **156**, 703–714 (2015)
8. Y. Tomizawa, K. Sasaki, A. Kuroda, R. Takeda, Experimental and numerical study on phase change material (PCM) for thermal management of mobile devices. Appl. Therm. Eng. **98**, 320–329 (2016)
9. Z. Luo, H. Cho, X. Luo, K. Cho, System thermal analysis for mobile phone. Appl. Therm. Eng. **28**, 1889–1895 (2008)
10. R. Baby, C. Balaji, Thermal optimization of PCM based pin fin heat sinks: an experimental study. Appl. Therm. Eng. **54**, 65–77 (2013)
11. R. Pakrouh, M.J. Hosseini, A.A. Ranjbar, R. Bahrampoury, A numerical method for PCM-based pin fin heat sinks optimization. Energy Convers. Manage. **103**, 542–552 (2015)
12. S.F. Hosseinizadeh, F.L. Tan, S.M. Moosania, Experimental and numerical studies on performance of PCM-based heat sink with different configurations of internal fins. Appl. Therm. Eng. **31**, 3827–3838 (2011)
13. Y. Kozak, B. Abramzon, G. Ziskind, Experimental and numerical investigation of a hybrid PCM-air heat sink. Appl. Therm. Eng. **59**, 142–152 (2013)

14. S.K. Saha, K. Srinivasan, P. Dutta, Studies on optimum distribution of fins in heat sinks filled with phase change materials. ASME J. Heat Transf. **130**, 034505/1–034505/4
15. A. Arshad, H.M. Ali, W.M. Yan, A.K. Hussein, M. Ahmadlouydarab, An experimental study of enhanced heat sinks for thermal management using n-eicosane as phase change material. Appl. Therm. Eng. **132**, 52–66 (2018)
16. A. Al-Sarkhi, Comparison between dual and constant height shrouded fin array subjected to forced convection heat transfer. Int. Commun. Heat Mass Transf. **32**, 548–556 (2005)
17. K.K. Pathak, A. Giri, P. Lingfa, A numerical study of natural convective heat transfer from a shrouded vertical dual height non-isothermal fin array. Appl. Therm. Eng. **130**, 1310–1318 (2018)
18. C. Ji, Z. Qin, S. Dubey, F.H. Choo, F. Duan, Simulation on PCM melting enhancement with double-fin length arrangements in a rectangular enclosure induced by natural convection. Int. J. Heat Mass Transf. **127**, 255–265 (2018)
19. B.N. Taylor, C.E. Kuyatt, *Guidelines for Evaluating and Expressing the Uncertainty of NIST Measurement Results* (US Department of Commerce Technology Administration, National Institute of Standards and Technology, Gaithersburg, MD, 1994)
20. S.J. Kline, F.A. McClintock, Describing uncertainties in single-sample experiments. Mech. Eng. **75**(1), 3–8 (1953)

# Dynamic Analysis of Parametrically Excited Coupled Beam-Based Piezoelectric Energy Harvester

**Ranit Roy, Anshul Garg, and Santosha Kumar Dwivedy**

**Abstract** In this present work, the dynamics of a piezoelectric energy harvester consisting of two vertical cantilever beams with tip mass and piezoelectric patch has been studied. The beams are coupled with a linear spring and the system is subjected to base excitation. Both of the beams are identical except length and tip mass. Tip masses are attached in such a way that the natural frequencies of the beams come closer to each other which helps both the beams to resonate with principal parametric resonance condition. It is assumed that the beams undergo large amplitude oscillations, hence geometric and inertial nonlinearities have been taken into account. Considering Euler–Bernoulli beam assumptions, the nonlinear electromechanical equations of motion have been obtained by using Lagrange's principle which is discretized to its temporal form by generalized Galerkin's method. Using 4th order Runge–Kutta method, these governing equations of motion have been solved. Method of multiple scales has also been used to obtain the approximate response and voltages in the beams. It has been observed that in the pre-buckling regime, with increase in load resistance, the output voltages and the output power of both the beams increase. While the load resistance has no effect on system frequency. It has also been observed that with increase in stiffness of coupled spring, output voltages and output power of both the beams increase. Also it can be observed that with increase in stiffness of the coupled spring, at lower frequency power can be generated.

**Keywords** Coupled beams · Nonlinear · Parametric excitation · Piezoelectric energy harvesting

---

R. Roy (✉) · A. Garg · S. K. Dwivedy
Department of Mechanical Engineering, Indian Institute of Technology Guwahati, Guwahati, India
e-mail: ranit174103128@iitg.ac.in

A. Garg
e-mail: anshul.garg@iitg.ac.in

S. K. Dwivedy
e-mail: dwivedy@iitg.ac.in



K. M. Pandey et al. (eds.), *Recent Advances in Mechanical Engineering*, Lecture Notes in Mechanical Engineering, https://doi.org/10.1007/978-981-15-7711-6_63

## Nomenclature

$A_j$ Cross-sectional area of each beam
$b_j$ Width of each beam
$b_p$ Width of piezo-patch
$C_a$ Damping coefficient
$C_p$ Capacitance of the piezo-patch
$E$ Young's modulus of the beam material
$h_j$ Thickness of each beam
$h_p$ Thickness of piezo-patch
$I_j$ Area moment of inertia of each beam
$I_{t_j}$ Mass moment of inertia of each tip mass
$K$ Spring stiffness
$L_1$ Length of first beam
$L_2$ Length of second beam
$L_p$ Piezo layer length
$M_{t_j}$ Tip mass of each beam
$R_L$ Load resistance
$\rho$ Density of beam material
$\omega$ Excitation frequency

# 1 Introduction

Vibrational energy harvesting has been generating greater attention since last few decades. As there are lots of ambient vibrations, researchers have proposed and developed various ways to harvest energy from these vibrations to run low-powered electronic devices, monitor structural health and operate wireless sensors. The purpose of this technology is to create remote sources of electrical power or to store the energy in some storage devices.

Masana and Daqaq [1] proposed a model of piezoelectric energy harvester with clamped–clamped beam condition where the beam is subjected to transverse excitation and static axial load. They have shown that the axial load gives the ability of tuning excitation frequency, increases the electrical damping, enhances effective nonlinearity of the system and amplifies the effect of external excitation. Daqaq et al. [2] proposed a lumped parameter electromechanical modeling of a parametrically excited energy harvester. They have investigated that there is a parametrically unstable region in which energy can be harvested when the coupling coefficient increases. They have also studied the effect of load resistance on harvested power. Abdelkefi et al. [3] proposed a distributed parameter model of an energy harvester. Depending on the nonlinearity and varying the system parameters, they have plotted softening and hardening effects. Rosa et al. [4] have studied piezoelectric energy harvesting for a cantilever beam with varying cross-sectional area and tip mass.

They have experimentally verified the model for tapered bimorph cantilever and found the optimal load resistance for maximum power output. Guo et al. [5] developed a piezo-magneto-elastic coupled cantilever-type harvester. They have studied the effect of spacing between two adjacent magnets on the static bifurcation characteristics of the system. They have found that interwell limit cycle motion of the beam around two centers corresponds to optimum power output; interwell chaotic motion and multi-periodic motion including intrawell oscillations are less effective. Friswell et al. [6] proposed an energy harvester consisting of a cantilever beam with tip mass. They have studied three cases: a linear system, a non-buckled beam with low natural frequency and a buckled beam. The system is highly nonlinear between double potential wells for post-buckled case. They have experimentally validated that the most practical configuration is the pre-buckled case, where the proposed system has a low natural frequency, high level of harvested power and increased bandwidth of operation over a linear system. Huang et al. [7] again designed and analyzed a harvester–absorber model consisting of two vertical cantilever beams coupled with a spring and set upon a spring mass system. They have mathematically derived the frequency–amplitude relationship.

While in most of the energy harvesters, a single cantilever beam is used, in the present work, two cantilever beams are coupled to form the piezoelectric energy harvester which is excited parametrically. Figure 1 shows the schematic diagram of the energy harvester. The beams are attached with tip mass and coupled by a weak spring. Nonlinear electromechanical equations of motion for the system have been derived using Lagrange's principle and discretized to its temporal form by using generalized Galerkin's method. Approximate solution is obtained using method of multiple scales (MMS). Effects of coupling spring and load resistance have been analyzed.

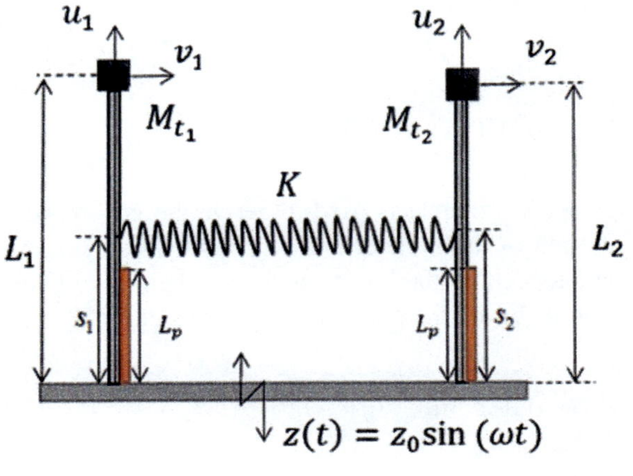


**Fig. 1** Schematic diagram of the parametrically excited piezoelectric energy harvester

## 2 Modeling

Here, as shown in Fig. 1, the piezoelectric-based energy harvester (PEH) consists of two vertical beams which are mounted to a base. Tip masses are attached to the beams individually. The beams are coupled with a linear spring of stiffness $K$. The spring is attached linearly with the beams at distance $s_1$ and $s_2$ from the base of the first and second beam, respectively.

Piezoelectric patches are attached at the base of both the beams. Euler–Bernoulli beam assumptions are considered to model the system mathematically. The transverse and axial displacements at the tip positions are $v_1, v_2$ and $u_1, u_2$. Subscripts 1 and 2 are used to represent the first and second beam, respectively. The displacement–curvature relation of the beams can be nonlinear due to large oscillation of the beams. This is represented by geometrical nonlinearity and inertial nonlinearity in the governing equation. Consider two arbitrary points $P_1$ at a distance $n_1$ and $P_2$ at a distance $n_2$ from the base of first and second beam, respectively. These points undergo rigid body translation due to the base excitation and further displacement due to the elastic deformation of the beams, which are given by transverse displacement ($v_{P_1}$ and $v_{P_2}$) and axial displacement ($u_{\mathrm{P}_1}$ and $u_{\mathrm{P}_2}$). Let $\phi_{\mathrm{P}_1}$ and $\phi_{\mathrm{P}_2}$ denote the rotation of the beams at a distance $n_1$ and $n_2$ from the base of each beam along the neutral axis, respectively, and hence, the rotation at both of the tip masses is $\phi_1 = \phi_{\mathrm{P}_1}(L_{\mathrm{t}1}, t)$ and $\phi_2 = \phi_{\mathrm{P}_2}(L_{\mathrm{t}2}, t)$ measured at the mass center. These are function of length of the beam $n_1$, $n_2$ and time $t$. Here, a single-mode approximation is considered. Displacement at any point of the beam can be represented as a function of tip mass displacement through beam deformation functions $\varphi_1(n_1)$ and $\varphi_2(n_2)$, as $v_{\mathrm{P}_1}(n_1, t) = v_{\mathrm{P}_1}(L_{\mathrm{t}_1}, t)\varphi_1(n_1) = v_1(t)\varphi_1(n_1)$ and $v_{\mathrm{P}_2}(n_2, t) = v_{\mathrm{P}_2}(L_{\mathrm{t}_2}, t)\varphi_2(n_2) = v_2(t)\varphi_2(n_2)$, where $\varphi_1(n_1)$ and $\varphi_2(n_2)$ are shape functions [6]. $\varphi_1(n_1) = \lambda_{t_1}(1 - \cos(\pi n_1/2L_1))$ and $\varphi_2(n_2) = \lambda_{t_2}(1 - \cos(\pi n_2/2L_2))$ where $\lambda_{t_1}$ and $\lambda_{t_2}$ are constants, such that $\varphi_1(L_{\mathrm{t}_1}) = \varphi_2(L_{\mathrm{t}_2}) = 1$.

### 2.1 Equations of Motion

Lagrange's principle is used to derive the governing electromechanical equations of motion of the system. It is to be noted that subscript $j$ and $k$ in the following equations denote the beam numbers, when $j = 1$ then $k = 2$ or vice-versa. For individual beam, the governing equations of motion are

$$\ddot{v}_j\left[\left(\rho A_j N_{j1} + M_{\mathrm{t}_j} + I_{\mathrm{t}_j} N_{j5}^2\right) + v_j^2\left(\rho A_j N_{j3} + M_{\mathrm{t}_j} N_{j4}^2 + I_{\mathrm{t}_j} N_{j5}^4\right) + v_j^4\left(\frac{1}{4} I_{\mathrm{t}_j} N_{j5}^6\right)\right]$$
$$+ \dot{v}_j^2\left[v_j\left(\rho A_j N_{j3} + M_{\mathrm{t}_j} N_{j4}^2 + I_{\mathrm{t}_j} N_{j5}^4\right) + v_j^3\left(\frac{1}{2} I_{\mathrm{t}_j} N_{j5}^6\right)\right] + \dot{v}_j(C_{\mathrm{a}})$$
$$+ v_j\left[EI_j N_{j6} - \rho g A_j N_{j9} - M_{\mathrm{t}_j} g N_{j4} + N_{j10}\right]$$

$$+ v_j^3[2EI_jN_{j7}] + v_j^5\left[\frac{3}{4}EI_jN_{j8}\right] - V_j[\theta_{j1} + \theta_{j2}v_j^2]$$
$$- v_k[N_{j11}] = -\ddot{z}[(\rho A_jN_{j2} + M_{t_j}N_{j4})v_j] \quad (1)$$

$$C_p\dot{V}_j + (V_j/R_L) + \theta_{j1}\dot{v}_j + \theta_{j2}v_j^2\dot{v}_j = 0 \quad (2)$$

where,

$$N_{j1} = \lambda_{t_j}^2((3\pi - 8)/2\pi)L_j, N_{j2} = \lambda_{t_j}^2(-0.25 + (\pi^2/16)),$$
$$N_{j3} = \lambda_{t_j}^2((\pi^2(2\pi^2 - 9)/384))(1/L_j), N_{j4} = \lambda_{t_j}^2(\pi^2/8L_j)$$
$$N_{j5} = \lambda_{t_j}(\pi/2L_{t_j}), N_{j6} = \lambda_{t_j}^2(\pi^4/32L_j^3)$$
$$N_{j7} = \lambda_{t_j}^4(\pi^6/512L_j^5), N_{j8} = \lambda_{t_j}^6(\pi^8/4096L_j^7)$$
$$N_{j9} = \lambda_{t_j}^2(-0.25 + (\pi^2/16)), N_{j10} = K[\lambda_{t_j}(1 - \cos(\pi s_j/2L_j))]^2$$
$$N_{j11} = K\lambda_{t_j}(1 - \cos(\pi s_j/2L_j))\lambda_{t_k}(1 - \cos(\pi s_k/2L_k))$$

$$\theta_{j1} = \gamma_c\lambda_{t_j}(\pi/2L_j)\sin(\pi L_p/2L_j),$$
$$\theta_{j2} = (1/2)\gamma_c\lambda_{t_j}^3(\pi^3/(32L_j^3))[3\sin(\pi L_p/2L_j) - \sin(3\pi L_p/2L_j)]$$

To obtain the relative contribution of the terms present in it, these equations of motion are now non-dimensionalized by taking the inverse of natural frequency ($\omega_n$) of the beam (as in present case natural frequencies of both the beams are optimized in such a way that they are almost equal) as time scale and the length parameter $l_g = 0.1$ as length scale. Considering the open-circuit voltage as $V_j$, the output voltage can also be non-dimensionalized. The constant parameters in the equations are scaled up in same order so that inertial, cubic, coupled and excitation nonlinear terms come in same order. After non-dimensionalizing, Eqs. (1) and (2) can be written in the following form, respectively.

$$\ddot{\bar{v}}_j + \varepsilon\alpha_{j0}\dot{\bar{v}}_j + \alpha_{j1}\bar{v}_j + \varepsilon\alpha_{j2}(\bar{v}_j^2\ddot{\bar{v}}_j + \bar{v}_j\dot{\bar{v}}_j^2)$$
$$+ \varepsilon\alpha_{j3}\bar{v}_j^3 - \varepsilon\alpha_{j4}\bar{V}_j - \varepsilon\alpha_{j5}\bar{v}_k = \varepsilon f_j[\sin(\Omega\tau)]\bar{v}_j \quad (3)$$

$$\dot{\overline{V}}_j + r_{j1}\overline{V}_j + \dot{\bar{v}}_j = 0 \quad (4)$$

where the non-dimensionalized transverse displacement is $\overline{v}_j = (v_j/l_g)$, the non-dimensionalized time is $\tau = \omega_n t$, the non-dimensionalized excitation frequency is $\Omega = (\omega/\omega_n)$, and the non-dimensionalized output voltage is $\overline{V}_j = [(C_pV_j)/(\theta_{j1}l_g)]$,

$$\alpha_{j0} = \left(C_a/\left(\varepsilon\left(\rho A_j N_{j1} + M_{t_j} + I_{t_j} N_{j5}^2\right)\omega_n \varepsilon\right)\right),$$
$$\alpha_{j1} = \left[\left(\mathrm{EI}_j N_{j6} - \rho g A_j N_{j9} - M_{t_j} g N_{j4} + N_{j10}\right)/\left(\left(\rho A_j N_{j1} + M_{t_j} + I_{t_j} N_{j5}^2\right)\omega_n^2\right)\right],$$
$$\alpha_{j2} = \left[l_g^2\left(\rho A_j N_{j1} + M_{t_j} N_{j4}^2 + I_{t_j} N_{j5}^4\right)/\left(\left(\rho A_j N_{j1} + M_{t_j} + I_{t_j} N_{j5}^2\right)\varepsilon\right)\right],$$
$$\alpha_{j3} = \left[\left(2\mathrm{EI}_j N_{j7}\right) l_g^2/\left(\rho A_j N_{j1} + M_{t_j} + I_{t_j} N_{j5}^2\right)\omega_n^2 \varepsilon\right],$$
$$\alpha_{j4} = \left[\theta_{j1}^2/C_p\left(\rho A_j N_{j1} + M_{t_j} + I_{t_j} N_{j5}^2\right)\omega_n^2 \varepsilon\right],$$
$$\alpha_{j5} = \left[N_{j0}/\left(\left(\rho A_j N_{j1} + M_{t_j} + I_{t_j} N_{j5}^2\right)\omega_n^2 \varepsilon\right)\right],$$
$$f_j = \left[z_0 \Omega^2\left(\rho A_j N_{j2} + M_{t_j} N_{j4}\right)/\left(\left(\rho A_j N_{j1} + M_{t_j} + I_{t_j} N_{j5}^2\right)\omega_n^2 \varepsilon\right)\right],$$
$$r_{j1} = \left(1/C_p R_L \omega_n\right)$$

where $\varepsilon$ is a book keeping parameter and assumed to be 0.1.

## 2.2 Approximate Solution

To get uniform first-order approximate solution of the non-dimensionalized equations of motion, method of multiple scales has been used. Considering multiple time scales, time dependence can be expressed by.

$T_n = \varepsilon^n \tau$. Time derivatives can be expressed as

$(\mathrm{d}/\mathrm{d}\tau) = D_0 + \varepsilon D_1 + O(\varepsilon^2)$ and $\left(\mathrm{d}^2/\mathrm{d}\tau^2\right) = D_0^2 + \varepsilon 2 D_0 D_1 + O(\varepsilon^2)$ where $D_n = (\partial/\partial T_n)$. Depending upon multiple scales, displacements and output voltages can be expressed as follows.

$$\overline{v}_j = \overline{v}_{j0}(T_0, T_1) + \varepsilon \overline{v}_{j1}(T_0, T_1) + O(\varepsilon^2) \tag{5}$$

$$\overline{V}_j = \overline{V}_{j0}(T_0, T_1) + \varepsilon \overline{V}_{j1}(T_0, T_1) + O(\varepsilon^2) \tag{6}$$

Substituting Eqs. (5) and (6) in Eqs. (3) and (4) and equating the coefficients of $\varepsilon^0$ and $\varepsilon^1$ to zero, one obtains the following equations.

Coefficients of $\varepsilon^0$:

$$D_0^2 \overline{v}_{j0} + \alpha_{j1} \overline{v}_{j0} = 0 \tag{7}$$

$$D_0 \overline{V}_{j0} + r_{j1} \overline{V}_{j0} + D_0 \overline{v}_{j0} = 0 \tag{8}$$

Coefficients of $\varepsilon^1$:

$$D_0^2 \bar{v}_{j1} + \alpha_{j1} \bar{v}_{j1} = \left\{ \begin{array}{l} -2 D_0 D_1 \bar{v}_{j0} - \alpha_{j0} D_0 \bar{v}_{j0} - \alpha_{j2}\left[\left(D_0^2 \bar{v}_{j0}\right)\bar{v}_{j0}^2 + \left(D_0 \bar{v}_{j0}\right)^2 \bar{v}_{j0}\right] \\ -\alpha_{j3} \bar{v}_{j0}^3 + \alpha_{j4} \bar{V}_{j0} + \alpha_{j5} \bar{v}_{k0} + f_j \sin(\Omega \tau) \bar{v}_{j0} \end{array} \right\} \tag{9}$$

$$D_1\overline{V}_{j0} + D_0\overline{V}_{j1} + r_{j1}\overline{V}_{j1} + D_0\overline{v}_{j1} + D_1\overline{v}_{j0} = 0 \tag{10}$$

Solution of Eqs. (7) and (8) can be given by

$$\overline{v}_{j0} = A_j(T_1)e^{i\omega_n T_0} + c.c \tag{11}$$

$$\overline{V}_{j0} = \left[-A_j(T_1)\left\{\left(i\omega_n r_{j1} + \omega_n^2\right)/\left(r_{j1}^2 + \omega_n^2\right)\right\}e^{i\omega_n T_0}\right] + c.c \tag{12}$$

Considering principal parametric resonance condition which arises when the excitation frequency is twice the natural frequency, then one may write $\Omega = 2\omega_n + \varepsilon\sigma$, where $\sigma$ is the detuning parameter. Now substituting Eqs. (11) and (12) in Eq. (9), to get a bounded solution, the coefficients of $e^{i\omega_n T_0}$, i.e., the secular terms must be equated to zero. This yields the following equation.

$$\begin{aligned} &2i\omega_n A_j' + \alpha_{j0} i\omega_n A_j - \left(2\alpha_{j2}\omega_n^2 - \frac{3\alpha_{j3}}{\omega_n^2}\right)A_j^2\bar{A}_j \\ &+ \alpha_{j4}\left(\frac{\omega_n^2 + i\omega_n r_{j1}}{\omega_n^2 + r_{j1}^2}\right)A_j - \alpha_{j5}A_k + \frac{f_j i e^{i\sigma T_1}}{2}\bar{A}_j = 0 \end{aligned} \tag{13}$$

Here, $\frac{\partial A_j}{\partial T_1} = A_j'$. Now, substituting $A_j(T_1) = \frac{a_j}{2}e^{i\beta_j}, \overline{A}_j(T_1) = \frac{a_j}{2}e^{-i\beta_j}, A_k(T_1) = \frac{a_k}{2}e^{i\beta_k}, \overline{A}_k(T_1) = \frac{a_k}{2}e^{-i\beta_k}$ in Eq. (13) and separating the real and imaginary part, one may obtain the following reduced equations.

$$\begin{aligned} a_j' &= -\alpha_{j0}\frac{a_j}{2} - \alpha_{j4}\left(\frac{r_{j1}}{r_{j1}^2 + \omega_n^2}\right)\frac{a_j}{2} \\ &+ \alpha_{j5}\frac{a_{\mathrm{k}}}{2\omega_n}\sin\left(\beta_{\mathrm{k}} - \beta_j\right) - \frac{f_j}{2\omega_n}\cos\left(\sigma T_1 - 2\beta_j\right)\frac{a_j}{2} \end{aligned} \tag{14}$$

$$\begin{aligned} \beta_j' &= -\alpha_{j2}\frac{a_j^2\omega_n}{4} + \alpha_{j3}\frac{3a_j^2}{8\omega_n} + \frac{\alpha_{j4}}{2}\left(\frac{\omega_n}{r_{j1}^2 + \omega_n^2}\right) \\ &- \alpha_{j5}\frac{a_{\mathrm{k}}}{2\omega_n a_j}\cos\left(\beta_{\mathrm{k}} - \beta_j\right) + \frac{f_j}{4\omega_n}\sin\left(\sigma T_1 - 2\beta_j\right) \end{aligned} \tag{15}$$

Taking, $\sigma T_1 - 2\beta_j = \gamma_j$ in Eqs. (14) and (15), and for steady state substituting $a_j' = \gamma_j' = 0$, one may obtain the following frequency–amplitude relation.

$$\sigma = \left(\frac{3}{4\omega_n}\alpha_{j3} - \frac{1}{2}\alpha_{j2}\omega_n\right)a_j^2 + \alpha_{j4}\frac{\omega_n}{\omega_n^2 + r_{j1}^2}$$

$$-\alpha_{j5}\frac{a_{\mathrm{k}}}{a_j\omega_n} \pm \sqrt{\left(\frac{f_j}{2\omega_n}\right)^2 - \left(\alpha_{j0} + \alpha_{j4}\frac{r_{j1}}{\omega_n^2 + r_{j1}^2}\right)^2} \quad (16)$$

Here, $\gamma_j = \gamma_{\mathrm{k}}$ is considered for simplification. Otherwise, one has to solve four equations numerically to obtain the response amplitude. Displacement–voltage relationship can be expressed as

$$a_j = \left[\left(C_{\mathrm{P}}\sqrt{r_{j1}^2 + \omega_n^2}\right)/\left(\theta_{j1}l_{\mathrm{g}}\omega_n^2\right)\right]\left|V_{j0}\right|. \quad (17)$$

## 3 Results and Discussion

In this paper, parametric resonance condition of the piezoelectric energy harvester has been analyzed. Here, time response and frequency responses of both the beams are plotted separately by varying different parameters like coupled spring stiffness and load resistance. Table 1 represents the material properties of the system, and Table 2 represents the geometric properties of the system.

It has been observed that without coupled spring, Euler buckling loads for the first and second beam are found to be 27 g and 30.2 g of tip mass, respectively. With coupled spring, the Euler buckling load increases and is found to be 29.8 g and 33.4 g of tip mass for the first and second beam, respectively.

**Table 1** Material properties of the harvester [6]

| Beam properties | | Piezoelectric (MFC) properties | |
|---|---|---|---|
| Density ($\rho$) | 7850 kg/m$^3$ | Capacitance ($C_{\mathrm{P}}$) | 51.4 nF |
| Young's modulus ($E$) | 210 GPa | $\gamma_C$ | $-4 \times 10^{-5}$ Nm/V |

**Table 2** Geometric properties of the harvester

| Beam and tip mass properties | | Piezoelectric (MFC) properties [6] | |
|---|---|---|---|
| Length of first beam | 0.2 m | Length of the patch | $28 \times 10^{-3}$ m |
| Length of second beam | 0.19 m | Width of patch | $14 \times 10^{-3}$ m |
| Width of each beam | $16 \times 10^{-3}$ m | Thickness of patch | $0.3 \times 10^{-3}$ m |
| Thickness of each beam | $0.254 \times 10^{-3}$ m | | |
| Inertia/tip mass | 0.000040087 m$^2$ | | |

### 3.1 Effect of Stiffness of the Coupled Spring

It is to be noted that the load resistance is fixed at 100 KΩ.

From time responses shown in Figs. 2 and 3, it can be observed that with increase in the coupled spring stiffness, the voltage increases significantly which is also shown in Table 3. From the frequency–voltage plots shown in Figs. 4 and 5, it can be observed that the stable trivial solution bifurcates with supercritical pitchfork bifurcation and the unstable trivial solution bifurcates with subcritical pitchfork bifurcation. It is to be noted that, here the solid lines show the stable solutions and dashed line shows the unstable solutions. From Figs. 4 and 5, it can also be observed that frequency-voltage

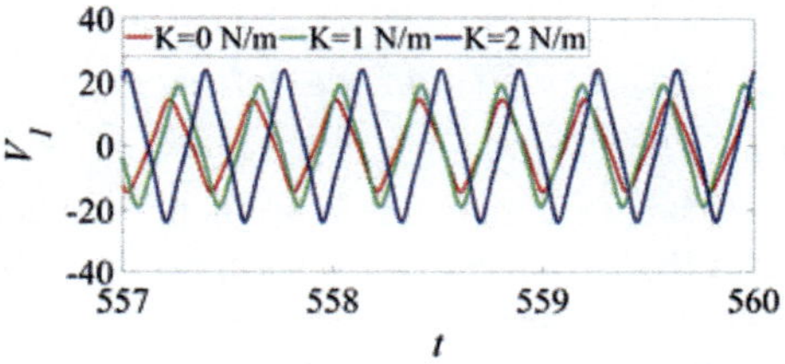


**Fig. 2** Time-voltage response of first beam for different values of spring stiffness

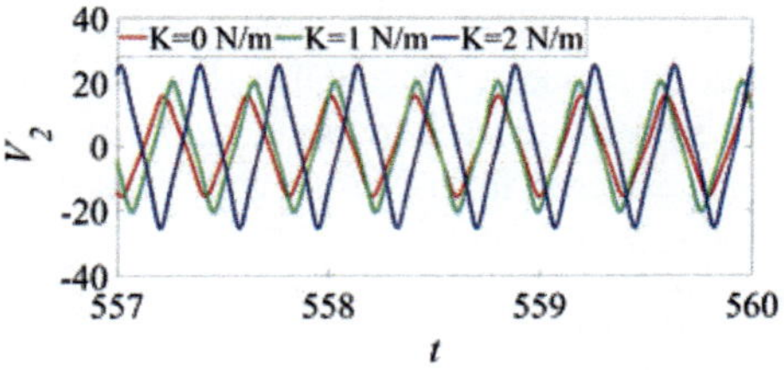


**Fig. 3** Time-voltage response of second beam for different value of spring stiffness

**Table 3** Output voltage and power with change in coupled spring stiffness

| $K$ (N/m) | $V_1$ (Volt) | $Power_1$ (mW) | $V_2$ (Volt) | $Power_2$ (mW) |
|---|---|---|---|---|
| 0 | 14.24 | 2.02 | 15.88 | 2.52 |
| 1 | 18.91 | 3.57 | 20.44 | 4.17 |
| 2 | 23.73 | 5.63 | 25.38 | 6.44 |

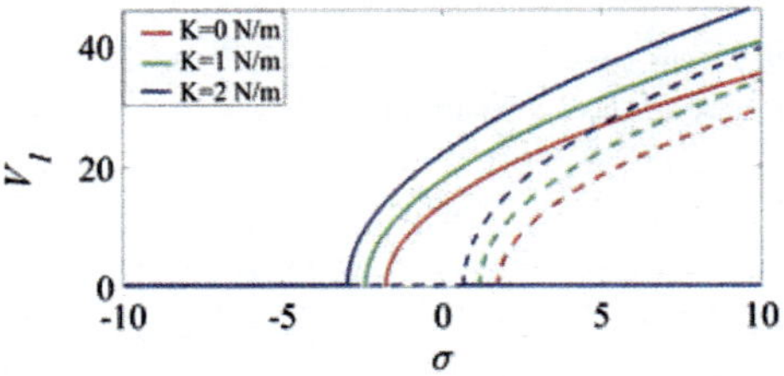


**Fig. 4** Frequency-voltage response of first beam for different values of spring stiffness

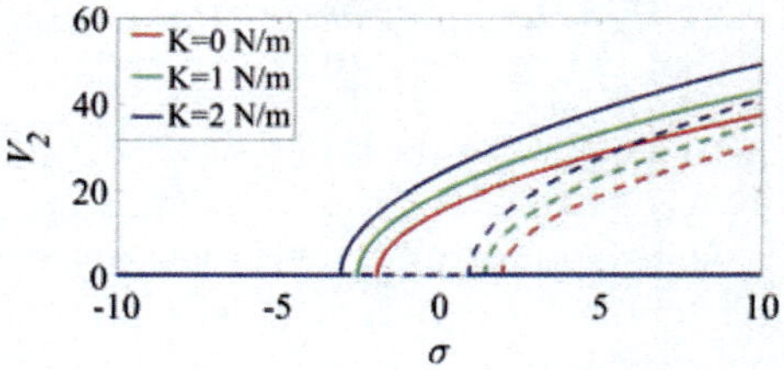


**Fig. 5** Frequency-voltage response of second beam for different values of spring stiffness

curves shift toward left with increase in the stiffness of the coupled spring. Hence, one may generate voltage at a lower frequency using a spring.

## 3.2 *Effect of Load Resistance*

It is to be noted that all the parameters are kept constant (except the load resistance) and stiffness of the coupled spring is taken as 2 N/m.

From time responses shown in Figs. 6 and 7, it can be observed that with increase in load resistance, voltage output increases. From Table 4, it can also be observed that the output power also increases. From Figs. 8 and 9, it can be observed that the load resistance has no effect on the supercritical and subcritical pitchfork bifurcation points. It is because the change in load resistance does not affect the natural frequency of the system, hence there is no frequency shifting of the curves. For the same reason, we can observe from Figs. 6 and 7 that the time period of the oscillation is same for different load resistances. The developed equations can further be used for many parametric studies.

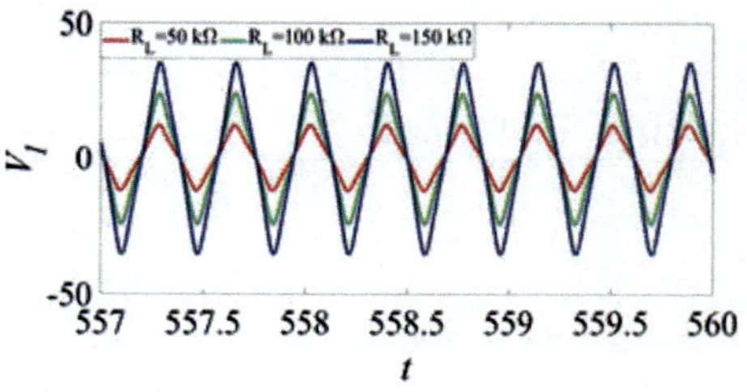


**Fig. 6** Time-voltage response of first beam for different values of load resistance

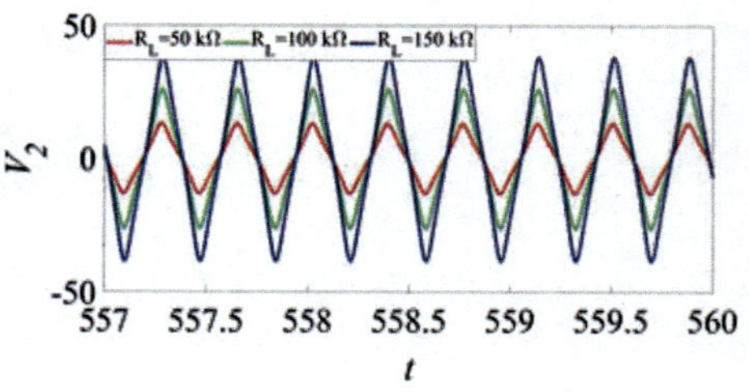


**Fig. 7** Time-voltage response of second beam for different values of load resistance

**Table 4** Output voltage and power with change in load resistance

| $R_L$ (kΩ) | $V_1$ (Volt) | $Power_1$ (mW) | $V_2$ (Volt) | $Power_2$ (mW) |
|---|---|---|---|---|
| 50 | 12.12 | 2.93 | 13.08 | 3.42 |
| 100 | 23.73 | 5.63 | 25.38 | 6.44 |
| 150 | 35.59 | 8.44 | 38.20 | 9.72 |

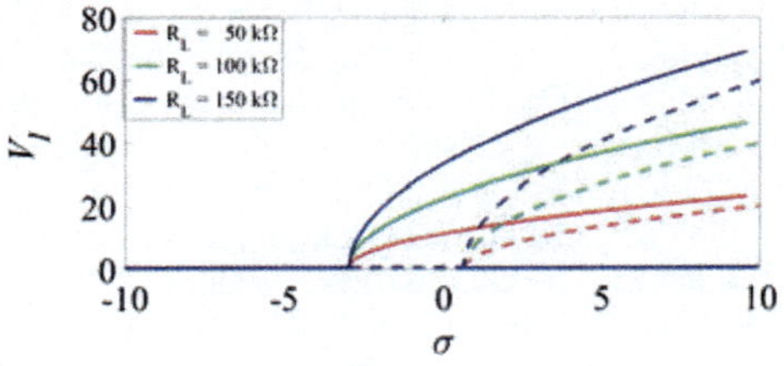


**Fig. 8** Frequency-voltage response of first beam for different values of load resistance

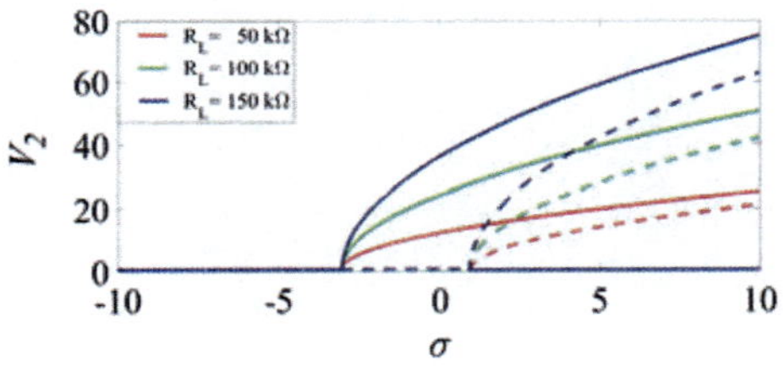


**Fig. 9** Frequency-voltage response of second beam for different values of load resistance

## 4 Conclusions

In the present work, a piezoelectric energy harvester consisting of two vertical cantilever beams coupled with a linear spring is considered, and it is shown that more output voltage and power can be generated by using this system than a conventional piezoelectric energy harvester. The system requires parametric excitation, which can be found naturally in bridges, suspended wires between two supports, tall buildings, etc. From the results, it can be observed that the coupled spring has significant impact on the output voltage and power. By using the coupled spring, one can generate voltage with a lower excitation frequency. Load resistance also has a wide impact on the system; as it can be observed from the result that with increase in load resistance output voltage and output power from the system increase. But the change in load resistance does not affect the natural frequency of the system. So, there is no possibility of tuning the harvester operating frequency. Hence, this study will help one to develop an efficient piezoelectric energy harvester.

## References

1. R. Masana, M.F. Daqaq, Electromechanical modelling and nonlinear analysis of axially loaded energy harvesters. J. Vib. Acoust. **133**(1), 011007–011010 (2010)
2. M.F. Daqaq, C. Stabler, Y. Qaroush, T. Seuaciuc-Osorio, Investigation of power harvesting via parametric excitations. J. Intell. Mater. Syst. Struct. **20**, 545–557 (2009)
3. A. Abdelkefi, A. Nayfeh, M.R. Hajj, Global nonlinear distributed-parameter model of parametrically excited piezoelectric energy harvesters. Nonlinear Dyn. **67**, 1147–1160 (2012)
4. M. Rosa, C. De Marqui Jr, Modelling and analysis of a piezoelectric energy harvester with varying cross-sectional area. Shock Vibr. 930503 (2014)
5. K. Guo, S. Cao, S. Wang, Numerical and experimental studies on nonlinear dynamics and performance of a bistable piezoelectric cantilever generator. Shock Vibr. 1–14 (2015)
6. M.I. Friswell, S.F. Ali, O. Bilgen, S. Adhikari, A.W. Lees, G. Litak, Non-linear piezoelectric vibration energy harvesting from a vertical cantilever beam with tip mass. J. Intell. Mater. Syst. Struct. **23**(13), 1505 (2012)
7. S.C. Huang, L.H. Nguyen, J.W. Liang, Y.M. Huang, Design and analysis of a collocated periodic vibration absorber-harvester. Int. J. Mech. Sci. **148** (2018)

# An Overview of Wire Electrical Discharge Machining (WEDM)

**Shatarupa Biswas, Yogesh Singh, and Manidipto Mukherjee**

**Abstract** WEDM is not a traditional machining process. The process is based on a thermo-electrical energy system. In this process, any mechanical connection between the electrode (wire) and the workpiece is not required for cutting any object. To establish a contact within the wire and the work object, different dielectric fluids are used. Generally, this approach refers only to conductive materials (such as silver, copper, iron, brass, bronze). Once the deionized water flows, metal ions are released from the workpiece, and electrons are released from the wire and formed a spark between the wire and workpiece. The range of the temperature produced due to this action lies in between 8000 °C and 12,000 °C. Due to this reason, the material is removed from the substrate. This article is about the investigation of the previous research work done on the different deionized water and its effect on the different output parameters acquired. From this study, researchers will be equipped with the summarized way to how to select the input parameters and enhance the output parameters using deionized water in the WEDM machining process.

**Keywords** WEDM · Deionized water · Dielectric fluid

## Nomenclatures

$A_{off}$ Arc off time

S. Biswas (✉) · Y. Singh
Department of Mechanical Engineering, National Institute of Technology Silchar, Silchar, Assam 788010, India
e-mail: supriticu@gmail.com

Y. Singh
e-mail: yogeshsingh15@gmail.com

M. Mukherjee
CAMM, CSIR-Central Mechanical Engineering Research Institute, Durgapur 713209, West Bengal, India
e-mail: m.mukherjee.ju@gmail.com


K. M. Pandey et al. (eds.), *Recent Advances in Mechanical Engineering*, Lecture Notes in Mechanical Engineering, https://doi.org/10.1007/978-981-15-7711-6_64

| | |
|---|---|
| $A_{on}$ | Arc on time |
| BPNN | Back propagation neural network |
| EDM | Electro discharge machining |
| KW | Kerf width |
| MRR | Material removing rate |
| $T_{off}$ | Pulse off time |
| $T_{on}$ | Pulse on time |
| RLT | Recast layer thickness |
| SV | Servo voltage |
| SH | Surface hardness |
| SR | Surface roughness |
| TWR | Tool wear rate |
| WEDM | Wire electrical discharge machining |
| WF | Wire feed |
| WP | Wire pressure |
| WT | Wire tension |

## 1 Introduction

The two Russian Scientists, B. R. Lazarenko and N. I. Lazarenko, invented the electro discharge machining (EDM) method in 1940 [1]. The Soviet Union scientists invented the first EDM after that in 1967, in which they used wire as the electrode [2]. EDM machining is a non-conventional process [3] which is used mainly for those (hard metals) which would be very difficult to machining with traditional processes (such as shaping, boring, spinning). EDM typically works with electrically conductive materials, while methods for using EDM in machine-insulating ceramics have also been proposed [4, 5]. Through pre-hardened steel, EDM can cut intricate contours or cavities without the need for heat treatment to soften and harden them. In this machining process, the metal alloy such as titanium, hastelloy, kovar and inconel can be used. EDM machining is dependent on input parameters such as current, pulse on time and pulse off time[6], and several types of dielectric fluids are typically used to boost the different types of material performance parameters, such as MRR, SR and TWR. Three types of EDM are commonly used, namely die sink EDM, wire EDM and micro EDM. WEDM [7, 8] is also known as spark EDM, which is usually used for cutting purposes. A thin wire is used as a tool in this kind of EDM. The wire mainly consists of brass or copper or zinc coated brass. The wire's diameter in this case lies in the range between 0.1 and 0.3 mm. And for machining purposes, the thickness of the plate is 300 mm in nature. This EDM is typically used when requires low residual stresses because there is no required high cutting forces to extract the material. WEDM process makes machining burr-free and provides electrical conductivity of the workpiece. WEDM machine generally has five axis

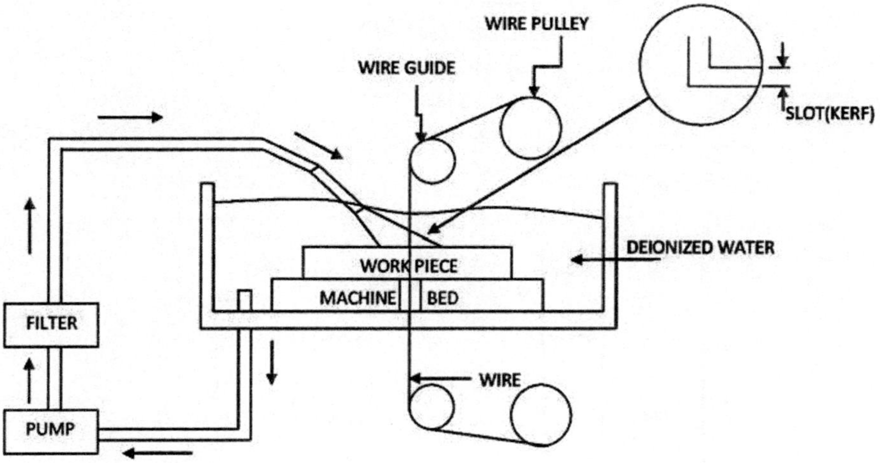


**Fig. 1** Block diagram of WEDM

movements such as *x* axis, *y* axis, *z* axis, *u* axis, *v* axis, respectively. The schematic representation of the method is shown in Fig. 1.

## 2 Important Parameters in Wire EDM

Wire EDM machining is generally depended on input and output parameters. Deferent types of wire EDM machining parameters are shown in Fig. 2. For wire EDM machining, some input parameters have to be chosen based on various types of materials, because the output parameters of the materials depend on the input parameters during wire EDM machining process.

### 2.1 Input Parameters

For the machining process, input parameters are used to control the output parameters. Types of input parameters are shown in Fig. 2.

- $\boldsymbol{T_{on}}$: The total duration of discharge time between the wire and material is known as $T_{on}$.
- $\boldsymbol{T_{off}}$: The total duration of non-discharge time between the wire and material is known as $T_{off}$.
- $\boldsymbol{A_{on}}$: The duration of electron flow time, i.e., the arc generating time is known as $A_{on}$.
- $\boldsymbol{A_{off}}$: The duration of non-flow time of electron is known as $A_{off}$.

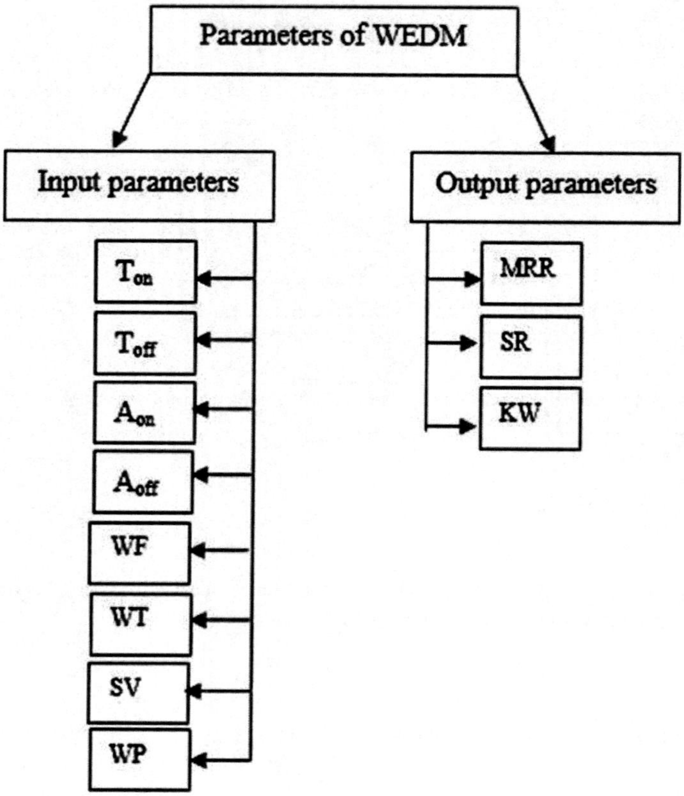


**Fig. 2** Parameters of wire EDM

- **WF**: It is expressed as a unit of wire length passing through a point per unit time.
- **WT**: The tensile force of wire in between upper nozzle and lower nozzle is known as wire tension.
- **SV**: The working voltage is known as servo voltage.
- **WP**: Water pressure is expressed as an amount of perpendicular force is applied on an object per unit area.

## 2.2 *Performance Parameters*

Various types of output parameters are shown below.

- **MRR**: It can be expressed as the volume of material eroded per unit time. It is a very important parameter for EDM machining. The working time with a job of EDM machining is depended on its MRR. MRR is closely related to the dielectric fluid which is used for machining purposes [9, 10]. MRR is calculated using this

Eq. (1) [11]

$$\mathrm{MRR} = \frac{\text{volume of the work material remove}}{\text{machining time}} = \frac{\mathrm{wb} - wa}{\rho * \mathrm{tm}} \quad (1)$$

where, wb denotes the work piece before machining weight and wa is weight after machining weight.$\rho$ is the density of work piece, and tm is machining time.

- **SR**: It is calculated by the deflection of the normal vector (direction of vector) to a real surface. If themeasured deflection is huge, then the surface is described as a rough, and if the measured deflection is little, then the surface is described as smooth [12, 13].
- **KW**: After machining process, the cutting width of material is known as kerf width.

## 3 Dielectric Fluid

The dielectric is used as a conductor between the electrode (wire) and the workpiece to promote a stable and regulated spark gap when ionizing, and it also serves as a flushing agent to clean and eliminate eroded debris from the region of the spark gap. For WEDM machining, basically the deionized water is used as a dielectric fluid. It is a substance that is high in flame, weak in viscosity, light in color and resistant to oxidation. It is easy to filter, non-corrosive and easily dissipates heat. It increases electrode(wire) life and enhances surface finish efficiency. It is very cost effective and easily available. Due to these reasons, deionized water becomes a very popular dielectric fluid in WEDM process.

### *3.1 Deionized Water*

The deionized water concentration of ions is zero. It can be created by taking normal water and exposing it to electrically charged resins that bind and absorb salts (such as calcium, magnesium, sodium) and extracting them from water. Thus, the materials are not affected by deionized water. It also acts as a semiconductor and flushing agent for cleaning and removing eroded debris from the work materials. Several researchers have performed their work on wire EDM using deionized water as a dielectric fluid [14, 15].

## 4 Electrode Used in Wire EDM

For cutting purpose in wire EDM, several types of electrode are used such as brass, zinc and copper. The performance of wire is depending on the following properties.

- Wire size
- Mechanical property (tensile strength, modulus of electricity)
- Electrical property (electrical conductivity)

## 5 Prior Art on WEDM

Various researches have been done on wire EDM using deionized water which is shown in Table 1.

## 6 Conclusions

From the previous research work, the following conclusions are drawn:

- Dielectric fluid is playing an important role for removing metals of the WEDM cutting process. And by using deionized water, the flushing efficiency is better.
- $T_{on}$ is an important parameter for finding KW among all other input parameters.
- For finding MRR, the important input parameters are SV and $T_{off}$.
- Deionized water gives a better SR for meso helical gear compared to the meso bevel gear.

**Table 1** Literature review on wire EDM

| Author and Year | Work material | Electrode (wire) | Finding |
|---|---|---|---|
| Kao et. al. (2007) [14] | AISI 6061 | Brass | It is found that flushing capability is better |
| Gupta and Jain (2013) [16] | Brass | Brass | It is observed that optimum ranges of the voltage and $T_{off}$ to minimize the pitch deviations |
| Gupta and Jain (2013) [17] | Copper | Brass | It is found that voltage, $T_{on}$, $T_{off}$and WF were to be highly significant parameters to obtaine the significance SR |
| Gupta and Jain (2014) [18] | Brass | Brass | It is obtained that the voltage, $T_{on}$, $T_{off}$, WF are critical variables influencing the normal and higher SR of the WEDM |
| Gupta. et al. (2014) [19] | Brass | Brass | It is observed that when the voltage is below 5 V and $T_{on}$ is below 0.6 μs, and it causes the slow cutting rate |
| Nayak and Mahapatra (2014) [20] | AISI 304 | Broncocut-W | In this study, it is found that the relation between input parameters and output parameters is developed by using BPNN for the taper cutting process |
| Gupta and Jain (2014) [7] | Brass | Brass | It is obtained that WEDM gear shows better surface texture, thus it had minimum RLT, burr-free (small pieces of job) tooth profile, good microstructure |

(continued)

**Table 1** (continued)

| Author and Year | Work material | Electrode (wire) | Finding |
|---|---|---|---|
| Shandilya et al. (2016) [21] | SiCp/6061 | Brass | It is observed that the SV and $T_{\text{off}}$ are the main factors of finding good amount of the MRR |
| Chaubey and Jain (2017) [22] | SS 304 | Brass | In this study, meso helical gear showed the betterSR (1.1 μm) compared to meso bevel gear (1.3 μm) |
| Chaubey and Jain (2017) [23] | SS 304 | Brass | It is found that WEDM gears (bevel and helical) showed better SR, infused low peak current, SV $T_{\text{on}}$ and high $T_{\text{off}}$, WF, WT, WA |
| Mahapatra and Sahoo (2017) [15] | Ti–6Al–4V | Brass | It is observed that small-scale auxiliary examination was done for the wires and the workpiece at the advanced settings of the reaction |
| Mahapatra and Sahoo (2018) [24] | Inconel 718 | Brass | It is observed that $T_{\text{on}}$is is most effective parameter to obtaine the higher KW and MRR |
| Mahapatra et. al. (2018) [8] | AISI 304 | Brass | It is discovered that $T_{\text{on}}$ is most significant parameter for this study.The equivalent stress at the middle of the wire is 397 MPa and low stress of 0.185 MPa |
| Majumder and Maity (2018) [25] | titanium alloy grade 6 | Brass, copper and zinc | The estimated error for the general regression neural network model was ±5%, but it is ±10% for the multiple regression analysis model |

## References


1. K.H. Ho, S.T. Newman, State of the art electrical discharge machining (EDM). Int. J. Mach. Tools Manuf. **43**(13), 1287–1300 (2003)
2. A. Pandey, S. Singh, Current research trends in variants of electrical discharge machining: a review. Int. J. Eng. Sci. Technol. **2**(6), 2172–2191 (2010)
3. G. Singh, D. Parkash Dhiman, Review: parametric optimization of EDM machine using Taghuchi and anova technique. Int. Res. J. Eng. Technol. 783–788 (2016)
4. N. Mohri, Y. Fukuzawa, T. Tani, N. Saito, K. Furutani, Assisting electrode method for machining insulating ceramics. CIRP Ann. Manuf. Technol. **45**(1), 201–204 (1996)
5. Y.H. Liu, X.P. Li, R.J. Ji, L.L. Yu, H.F. Zhang, Q.Y. Li, Effect of technological parameter on the process performance for electric discharge milling of insulating Al2O3 ceramic. J. Mater. Process. Technol. **208**(1–3), 245–250 (2008)
6. P.K.H.K. Kansal, S. Singh, Application of Taguchi method for optimisation of powder mixed electrical discharge machining. Int. J. Manuf. Technol. Manage. **7**(I), 329–341 (2005)
7. K. Gupta, N.K. Jain, Comparative study of wire-EDM and hobbing for manufacturing high-quality miniature gears. Mater. Manuf. Process. 37–41 (2014)
8. K.D. Mohapatra, V.B. Shaibu, S.K. Sahoo, Science direct modeling and analysis of wire EDM in a gear cutting process for a 2D model. Mater. Today Proc. **5**(2), 4793–4802 (2018)
9. H.N. Mehta, Modeling of electrical discharge machining process. Int. J. Eng. Res. Technol. **4**(06), 153–157 (2015)
10. B.B.B.C. Upadhyay, Rahul, S. Datta, S.S. Mahapatra, An experimental investigation on electro discharge machining of Inconel 601. Int. J. Ind. Syst. Eng. **29**(2), 223–251 (2018)
11. M. Niamat, S. Sarfraz, H. Aziz, M. Jahanzaib, E. Shehab, W. Ahmad, Effect of different dielectrics on material removal rate, electrode wear rate and microstructures in EDM. Procedia CIRP **60**, 2–7 (2017)
12. P.K. Patowari, P.K. Mishra, P. Saha, Surface modification of C40 steel using WC-Cu P/M green compact electrodes in EDM. Int. J. Manuf. Technol. Manage. **21**, 83–98 (2010)
13. S. Biswas, M.M. Prity Aniva Xess, Optimization of the EDM parameters on machining Ti-6Al-4V alloy with multiple quality characteristics. Appl. Mech. Mater. **619**, 89–93 (2014)
14. C.C. Kao, J. Tao, A.J. Shih, Near dry electrical discharge machining. Int. J. Mach. Tools Manuf. **47**, 2273–2281 (2007)
15. K.D. Mohapatra, S.K. Sahoo, Microstructural analysis of titanium alloy gear using WEDM process. World Sci. **1850112**, 1–20 (2017)
16. K. Gupta, N.K. Jain, Deviations in geometry of miniature gears fabricated by wire electrical discharge machining, in *International Mechanical Engineering Congress and Exposition*, 2013, no. ii, pp. 1–7
17. K. Gupta, N.K. Jain, Manufacturing of high quality miniature gears by wire electric discharge machining. DAAAM Int. Sci. B. 679–696 (2013)
18. K. Gupta, N.K. Jain, On surface integrity of miniature spur gears manufactured by wire electrical discharge machining. Int. J. Adv. Manuf. Technol. 1735–1745 (2014)
19. N.K.J. Kapil Gupta, S.K Chaube, Exploring Wire-EDM for manufacturing the high quality meso-gear. Int. Conf. Adv. Manuf. Mater. Eng. **5**, 1755–1760 (2014)
20. B.B. Nayak, S.S. Mahapatra, A hybrid approach for process optimization in taper cutting operation using wire electrical discharge machining. Appl. Mech. Mater. **619**, 83–88 (2014)
21. N.K.J. Pragya Shandilya, P.K. Jain, Modelling and process optimisation for wire electric discharge machining of metal matrix composites. Int. J. Mach. Mach. Mater. **18**(4), 377–391 (2016)
22. S.K. Chaubey, N.K. Jain, Investigations on microgeometry of meso bevel and meso helical gears manufactured by WEDM process. Int. J. Adv. Manuf. Technol. (2017)
23. S.K. Chaubey, N.K. Jain, Investigations on surface quality of WEDM-manufactured meso bevel and helical gears. Mater. Manuf. Process. 1–10 (2017)

24. K. Mohapatra, S. Sahoo, A multi objective optimization of gear cutting in WEDM of Inconel 718 using TOPSIS method. Decis. Sci. Lett. **7**, 157–170 (2018)
25. H.M.K.P. Maity, Predictive analysis on responses in WEDM of titanium grade 6 using general regression neural network (GRNN) and multiple regression analysis (MRA) (2018)

# Comparative Analysis of Combustion Noise, Performance and Emission of LTC Diesel Engine with Multiple Injections

**Sanjoy Biswas and Achintya Mukhopadhyay**

**Abstract** Fuel injection strategy has become the heart of modern diesel engine due to its better commend on combustion process which ensures improvement in performance (BSFC, torque) and combustion noise (CN) with simultaneous reduction of emissions. In this experimental research work, impact of quadruple (epMa), triple (pMa) and double (pM) injection strategies have been studied on a typical six cylinder low-temperature combustion (LTC) CRDI diesel engine. Experiments were conducted at five different speeds (low to high) and three engine load conditions (20%, 60% and 100% of maximum torque, respectively) with higher EGR fraction of 45% and fixed main injection timing (Crank angle) using conventional diesel (BS-IV) fuel. The comparative study shown that quadruple (epMa) injection strategy is superior to provide optimum (BSFC, overall emissions) results in comparison with triple and double injection strategies for all aspects. Smoke level is marginally higher at lower-speed range for quadruple injection scheduling, whereas NOx emission level is the lowest among the injection strategies. Quadruple injection is capable of reducing combustion noise around 2–3 dBA at low loads and speeds over other two injection strategies.

**Keywords** Multiple injection · Pilot · Main · After · Emission · EGR · Combustion noise (CN) · Brake-specific fuel consumption (BSFC) · Low-temperature combustion (LTC)

## 1 Introduction

Drastic reduction of NOx and PM emissions to meet the emission norms [1–4] without any compromisation of fuel economy (at par or better) is the key challenge for

---

S. Biswas (✉) · A. Mukhopadhyay
Department of Mechanical Engineering, Jadavpur University, Kolkata 700132, India
e-mail: s.biswas.me@gmail.com

A. Mukhopadhyay
e-mail: achintya.mukhopadhyay@jadavpuruniversity.in



K. M. Pandey et al. (eds.), *Recent Advances in Mechanical Engineering*, Lecture Notes in Mechanical Engineering, https://doi.org/10.1007/978-981-15-7711-6_65

diesel engine. Combustion noise is the inherent problem of any diesel engine which is annoying to buyer. Hence, all future trend development of OEMs/researchers has the goal of drastic reduction of exhaust emissions, decreasing of combustion noise and improvement of brake-specific fuel consumption. Fuel injection strategy is very important and can play the key role for simultaneous reduction of emissions [7–32] and combustion noise [5–6, 17–18 and 27–32] without any penalty on BSFC value as it has better control on combustion process of engine.

The impact of parameters of pilot injection on emissions level of compression ignition engine has been studied by Ishida et al. [7] and Carlucci et al. [8]. In one end, pilot injection helps to reduce the period of ignition delay of the main injection, hence decreases the burned-fuel fraction in the premixed combustion phase, which leads to the reduction in peak combustion temperature. Other end, bulk temperature rises in the course of the compression stroke, and consequently, the mean pressure value inside the combustion chamber because of the advanced ignition. Other researchers [10–12] also have enlightened minimum amount of NOx emissions locally with a small amount of pilot fuel injected following the aforesaid phenomena, and the pilot injection scheduling is not far advance in reference to the scheduling of main injection. At same period, particulate matter (PM) emission level displays the contradictory behavior. Also, Carlucci et al. [8] found increasing tendency of PM levels at lower speeds, while the contrary behavior witnessed at higher speeds of engine. Badami et al. [13] found in their work that increasing trend of the particulate emission level with increase of pilot injected fuel quantity at different operating conditions. Tow et al. [14] have researched the impact of double and triple injections strategies on fuel economy (BSFC).They found the increasing trend of BSFC by means of delay in injection timing at higher loads for all injection scheduling. In contrast, BSFC trends are inconsistent at lower loads. Generally, the triple injection provides a significant level of reduction of particulate emissions without any penalization of BSFC. In a comparative study, Yokota et al. [16] have changed the main injection timing in advance in reference to the pM injection strategy and using of early-main-after (eMa) injections scheduling simultaneously. End result indicates a declining trend in NOx emissions when shifting from pM to eM and to eMa, while particulate emission level and BSFC increase trends when shifting from pM to eM to eMa. Available scientific literature [26–32] on diesel engine combustion and emissions and combustion noise characteristics using multiple injections which comprises pilot, post or spilt injections along with main injection, exhibited opportunity to reduce CN along with pilot injection parameters and trade of between NOx and PM with post injection. Also, it is talked about the impact on BSFC performance. Rajkumar et al. [32] also developed a phenomenological model along with parametric study.

Lee et al. [22, 23] studied the impact of pilot and post injections on emissions level at different EGR condition (moderate and heavy) on typical diesel engine and concluded with combination to achieve optimum results.

In recent times, diesel engine combustion noise (radiation) has also drawn significant attention as it is associated with the passengers' and pedestrians' discomfort along with noise pollution [33, 34]. The three key sources of noise generation of any compression ignition (diesel) engine are, viz. a) gas flow, b) mechanical processes

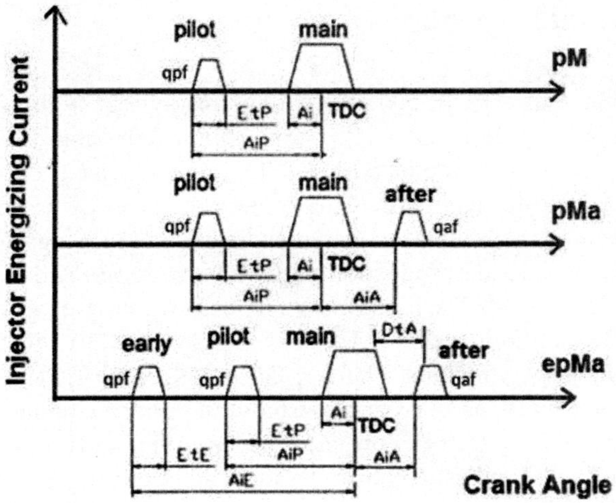


**Fig. 1** Injection strategies—i pM (pilot main), ii pMa (pilot-main-after) and iii epMa (early-pilot-main-after); TDC—top dead center; CA—crank angle; Ai—main injection advance w.r.t TDC; AiE—early injection advance w.r.t TDC, AiP—pilot injection advance w.r.t TDC; AiA—after injection w.r.t TDC; p—pilot injection or pilot injection 2; e—pilot injection 1 or early pilot injection; EtE—early injection duration; EtP—pilot injection duration; DtA—after injection Dwell, qpf—quantity of fuel at pilot or early pilot injection, qaf—quantity of fuel at after injection

and c) combustion [5, 18], respectively. By defining the correct rate of heat release (HRR) at steady-state or transient operating conditions, combustion chamber design and fuel injection parameters, e.g., timing, quantity and rate of fuel injected at pre and main injections play the major role in CI engine combustion noise [6]. According to Bharadwaj et al. [33], diesel engine combustion based and piston slapping noise should be the key focus area of engine acoustics as these contributed 80% of noise.

In this comprehensive work [Ref. Fig. 1], the potential of quadruple injections is assessed experimentally over two multi-injections on a typical six cylinder, heavy duty inline diesel engine (CRDI) at three different operating loads and five speed conditions using Taguchi's DOE approach. The results on emissions (NOx, particulate, smoke and THC levels) and BSFC performance and combustion noise (radiated) with fixed EGR% and main injection timing are compared at different steady-state engine working condition. Available scientific literatures are followed for, in detail, comparative analysis for the influence of aforementioned injection strategies on in-cylinder characteristics (combustion pressure, HRR (heat release rate) and temperature) of engine.

## 2 Experimental Procedure

The engine is with cooled EGR, and external cooling system is also attached with the engine. Details of experimental engine are mentioned in Table 1.

The typical engine is with a defined FMTC (fuel-mass-torque cycle) where broad outline of fuel demand for any particulate torque and speed is mapped by Bosch. Calibration, diagnostics and validation activities are monitored using INCA software. Allowable smoke limits for the production engine have been outlined for part loads as well as full-load application. The base BS-IV engine is with triple injections and has AiP 19.9 °CA BTDC, SOImain 6 to −3 °CA BTDC and AiA −12 °CA BTDC with qpf—1.5 mg/hub, qaf—2 mg/hub, Dtaf is 1350 ms. Keeping EGR % unchanged, key focus has been given on comparative study and understanding the effect of different multiple injection strategies on performance (BSFC and torque) and exhaust emissions trade-off and combustion noise. To adopt all three injection strategies (i.e. epMa, pMa, pM), delta optimization is done on base level calibration to reduce variable factors in experimentation and complexity. After treatment, arrangement remains same as in production engine during the experiments. Also, clutch, gearbox (transmission), external cooling/intake system and engine mounts adopted in this test bed from production models were the typical engine in use (Fig. 2; Table 2).

DOE method is used in this experimental research work for systematic approach of testing. The experimental tests are conducted as per the DOE matrix shown in Table 3 for performance and emissions. The data are captured under steady-state condition and average of 20 cycles for each set of data. Emission tests were done based on 13 modes ESC (steady-state cycle) with an ELR (dynamic load response) smoke test and more realistic operating condition ETC standard to check whether it is meeting the regulatory norms or not. Engine radiated (combustion Noise) noise level was measured taking reference of IS: 10399 [35] standard.

**Table 1** Engine specification

| Parameters | Specification |
|---|---|
| Engine type | BS-IV 6 cylinder inline |
| Displacement volume | 5.67L |
| Maximum speed | 2850 rpm |
| Minimum speed | 700 rpm |
| Compression ratio | 17.5:1 |
| Injection system | Bosch EDC-17 |
| Injection pressure | 120- 170 MPa |
| No. of holes at injector | 8 nos |
| Injector tip angle | 148° |
| Combustion bowl type | Shallow depth chamber |

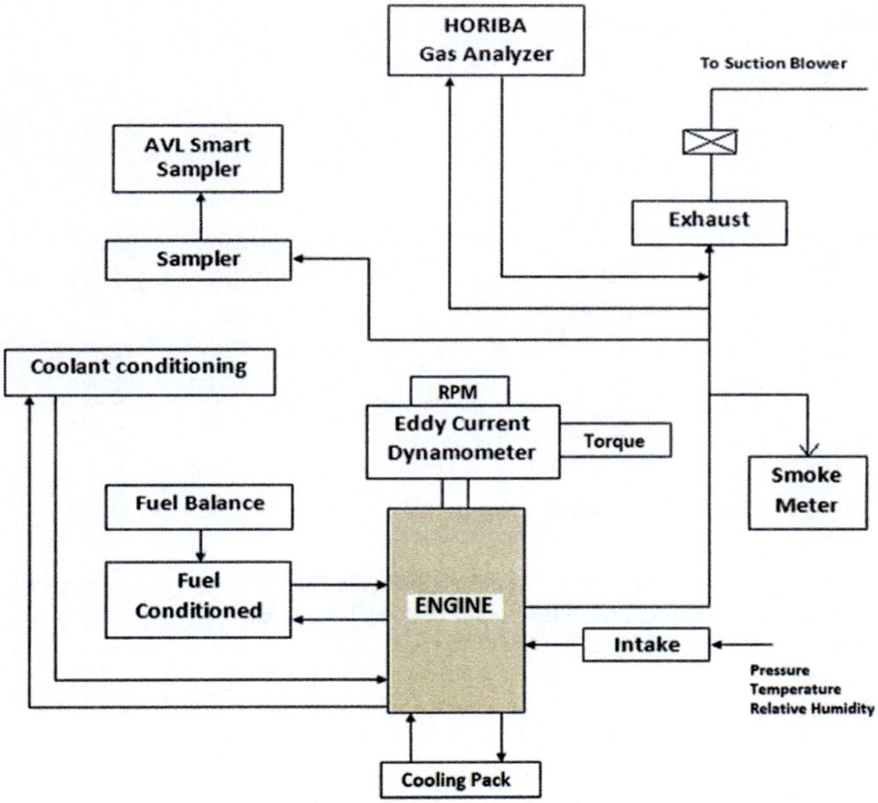


**Fig. 2** Schematic layout of experimental setup

**Table 2** DOE matrix inputs—factors, level and value

| Factors | Level | Value |
|---|---|---|
| Double injection | X | pM |
| Triple injection | Y | pMa |
| Quadruple injection | Z | epMa |
| Load (%) | L1, L2, L3 | 20, 60, 100 |
| Speed (rpm) | NI, N2, N3, N4, N5 | 1200, 1500, 1800, 2100, 2400 |
| Fixed factors | EGR | 45% |
| | Ai | 1 °CA BTDC |
| | AiP | 19 °CA BTDC |
| | AiE | 39 °CA BTDC |
| | AiA | −12 °CA BTDC |

**Table 3** DOE matrix for performance and emissions tests

| Injection strategy, load and speed combination | | Speed (rpm) | | | | |
|---|---|---|---|---|---|---|
| | | *N*1 | *N*2 | *N*3 | *N*4 | *N*5 |
| pM | *XL1* | *XL1N1* | *XL1N2* | *XL1N3* | *XL1N4* | *XL1N5* |
| | *XL2* | *XL2N1* | *XL2N2* | *XL2N3* | *XL2N4* | *XL2N5* |
| | *XL3* | *XL3N1* | *XL3N2* | *XL3N3* | *XL3N4* | *XL3N5* |
| pMa | *YL1* | *YL1S1* | *YL1N2* | *YL1N3* | *YL1N4* | *YL1N5* |
| | *YL2* | *YL2S1* | *YL2N2* | *YL2N3* | *YL2N4* | *YL2N5* |
| | *YL3* | *YL3S1* | *YL3N2* | *YL3N3* | *YL3N4* | *YL3N5* |
| epMa | *ZL1* | *ZL1N1* | *ZL1N2* | *ZL1N3* | *ZL1N4* | *ZL1N5* |
| | *ZL2* | *ZL2N1* | *ZL2N2* | *ZL2N3* | *ZL2N4* | *ZL2N5* |
| | *ZL3* | *ZL3N1* | *ZL3N2* | *ZL3N3* | *ZL3N4* | *ZL3N5* |

## 3 Results and Discussion

Here, all experimentally found test results are presented in tabular and graphical form for better understanding. Later stage interfered has been made.

### *3.1 Combustion Noise (CN) Performance*

CN level data are measured based on IS: 10399 [35] standard which is about noise measurement methodology at stationary vehicle. There is no other explicit experimental setups available for measurement of noise inside the combustion chamber. The noise data are captured for lower speeds (*N1* and *N2*). During noise trials, microphone was placed 0.5 m away from engine surface/envelop and positioned as per scheme shown in Fig. 3. Testing data were acquired only after stabilization of the exhaust out gas temperature at larger extend. Prior to start of typical experimental engine, ambient noise (dBA) level values are captured of inside of laboratory.

All tests data are represented in Table 4, and from this results, it observed around 2–3 dBA noise reduction with epMa found at low load condition.

**Fig. 3** Schematic of noise test setup

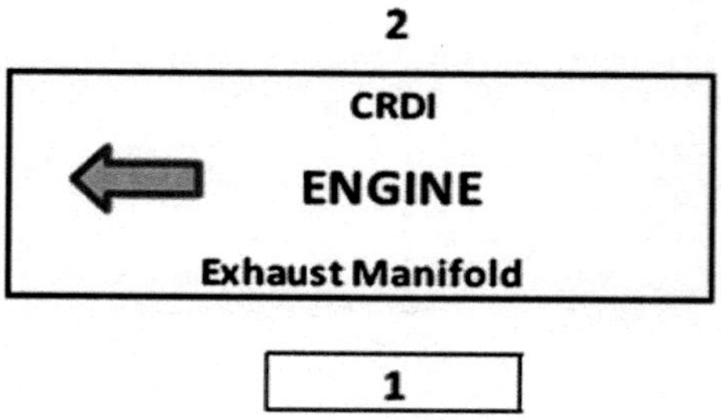

**Table 4** Nearby noise (CN) test results

| Injection strategy speed/location | | Load 20% | | Load 60% | | Load 100% | |
|---|---|---|---|---|---|---|---|
| | | 1200 | 1500 | 1200 | 1500 | 1200 | 1500 |
| pM | 1 | 89.8 | 91.2 | 93.1 | 93.5 | 95.3 | 96.3 |
| | 2 | 89.3 | 90.8 | 92.8 | 93.2 | 94.9 | 96.0 |
| pMa | 1 | 89.9 | 91.3 | 93.3 | 93.6 | 95.5 | 96.5 |
| | 2 | 89.5 | 90.9 | 93.0 | 93.3 | 95.1 | 96.1 |
| epMa | 1 | 87.5 | 89.4 | 90.9 | 91.4 | 93.1 | 94.2 |
| | 2 | 87.1 | 88.8 | 90.6 | 91.1 | 92.8 | 93.9 |

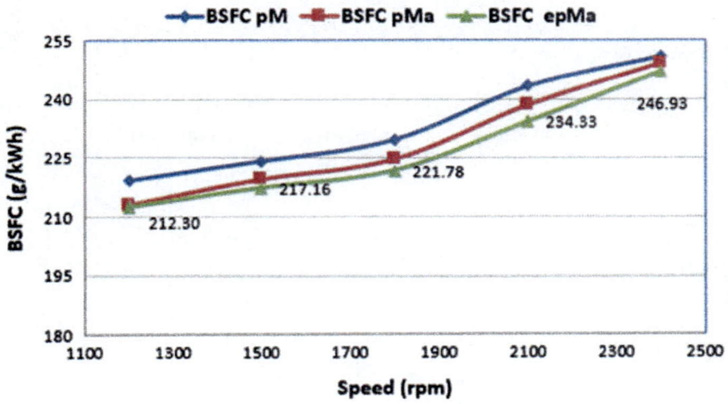


**Fig. 4** BSFC graph at 100% load

## 3.2 *BSFC Performance*

BSFC value is function of FER (Fuel flow Rate in kg/hr), Engine speed (rpm) and Torque (Nm). Hence, BSFC is calculated using this measured data during testing.

### 3.2.1 Fixed Load at Different Speeds

See Figs. 4, 5 and 6.

### 3.2.2 Average BSFC at Different Speeds

From the BSFC graphs, it has been observed that epMa provides better BSFC performance from medium to high-speed range for all load conditions with quite smoother curve. In the other hand, it exhibited relatively poor BSFC at low-speed range w.r.t

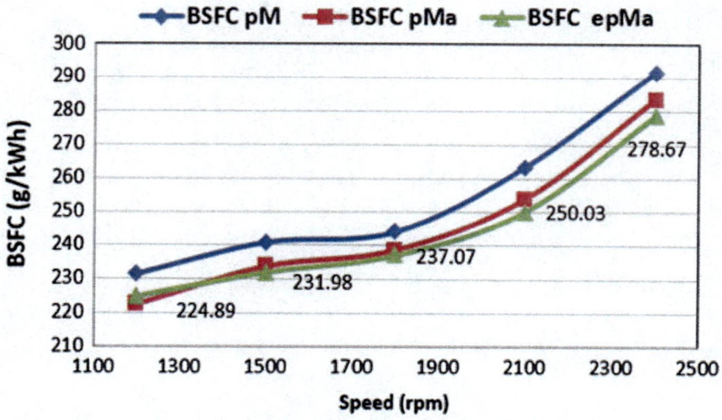


**Fig. 5** BSFC graph at 60% load

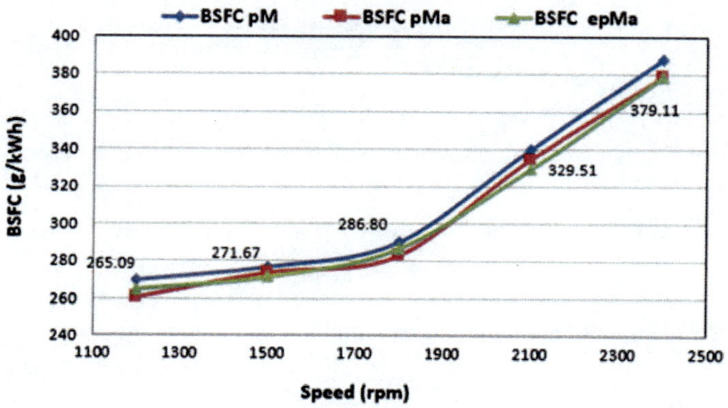


**Fig. 6** BSFC graph at 20% load

pMa injection strategy. Among these, double injection strategy (pM) shows worst BSFC performance (Fig. 7).

## 3.3 *Smoke (FSN) Performance—Fixed Load at Different Speeds*

The tests were conducted in reference to ELR. The smoke plots indicating that epMa produce (marginally) highest smoke level especially at low to medium speed ranges, whereas pMa gives intermediate results (Figs. 8, 9 and 10).

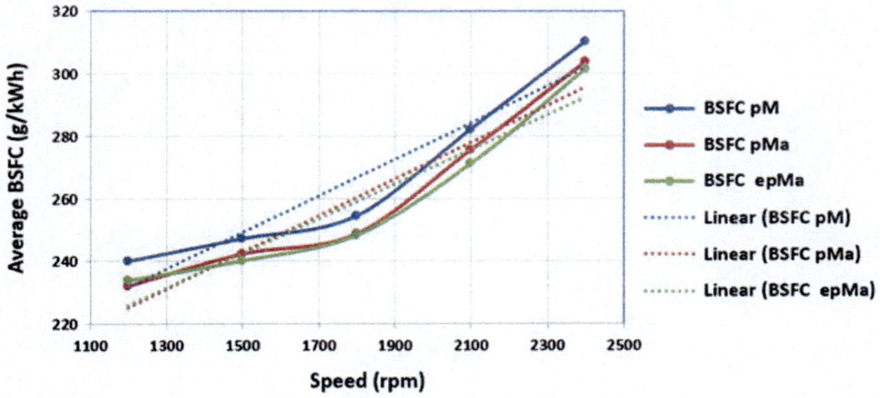


**Fig. 7** Average BSFC trend at different speeds

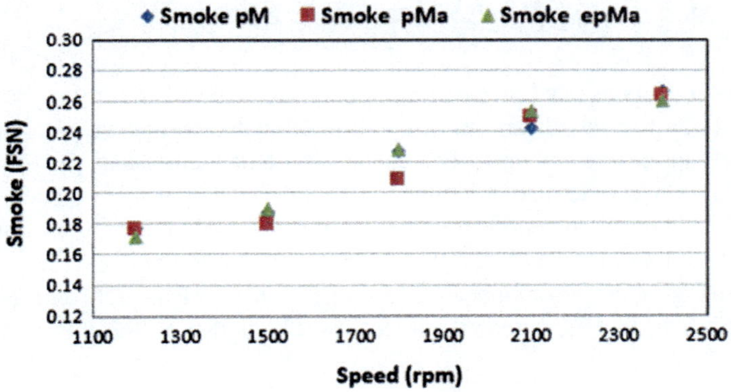


**Fig. 8** Smoke plot at 100% load

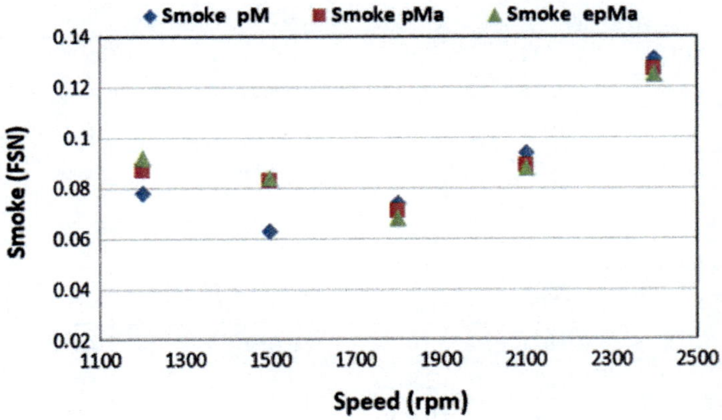


**Fig. 9** Smoke plot at 60% load

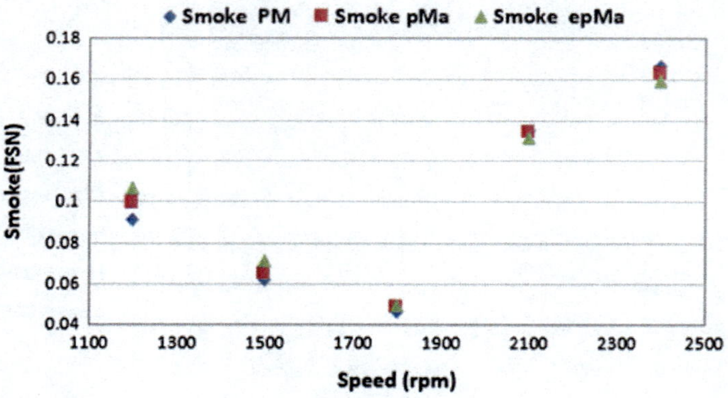


**Fig. 10** Smoke plot at 20% load

## 3.4 *Emission Test as Per ESC and ETC Standard*

Emission test results as per European Stationary Cycle (ESC) and European Transient Cycle (ETC) standards are represented in tabulated format below. These test standards are applicable for up to BS-V emission legalization and part of ARAI-MoRTH/CMVR/TAP-115/116 guideline.

Both pMa and epMa injection strategies meet the emission test norms among the injection scheduling but epMA results are optimum due to controlled heat release rate inside the combustion chamber. In case of pM, heat release rate and mean gas temperature are the highest among the injections which justifies the emission results (Tables 5 and 6).

**Table 5** ESC test results

| | pM | pMa | epMa | BS-IV limit | BS-V limit |
|---|---|---|---|---|---|
| PM | 0.015 | 0.014 | 0.017 | 0.020 | 0.020 |
| $NO_x$ | 3.623 | 3.243 | 2.812 | 3.500 | *2.000* |
| THC | 0.053 | 0.081 | 0.073 | 0.460 | 0.460 |
| CO | 0.088 | 0.077 | 0.011 | 1.500 | 1.500 |

**Table 6** ETC Test Results

| | pM | pMa | epMa | BS-IV limit | BS-V limit |
|---|---|---|---|---|---|
| PM | 0.020 | 0.021 | 0.024 | 0.030 | 0.030 |
| $NO_x$ | 3.641 | 3.075 | 2.918 | 3.500 | *2.000* |
| THC | 0.071 | 0.085 | 0.079 | 0.550 | 0.550 |
| CO | 0.089 | 0.078 | 0.012 | 4.000 | 4.000 |

## 4 Conclusion

From this experimental study, it can be concluded that double pilot with a post/after injection scheduling (epMa) quadruple injection strategy is better to provide optimum BSFC performance and emission level even at fixed main injection timing w.r.t double (pM) and triple (pMa) injection strategies. Also, this epMa strategy is the finest and capable of reducing combustion noise around 2–3 dBA at low loads and speeds over other two injection strategies. NOx emission level is also the lowest as a consequence of marginal higher smoke level observed, at low speeds and loads with quadruple injections. In totality, it is indicating that quadruple injection scheduling has better control on combustion process by means of reduction in the ignition delay, controlling the rate in cylinder pressure rise and heat release rate and combustion gas temperature at rapid or diffusion combustion phase. These are the key factors which control NOx, particulate matter (or Soot), THC, CO, combustion noise and BSFC. As a future scope of work, only quadruple injection strategy is to be considered for injection rate shaping and injection timing variation to achieve further improvement in emissions and performance level.

## References


1. J.B. Heywood, *Internal Combustion Engine Fundamentals* (McGraw-Hill, New York, 1998)
2. B.P. Pundir, *Engine Emissions-Pollution Formation and Advances in Control Technology* (Narosa Publishing House pvt Ltd., 2011)
3. F. Payri, J.M. Desantes, J. Benajes, Compression Ignition Engines: State-of-the-Art and Current Technologies, Future Trends and Developments, in *Handbook of Clean Energy Systems*, Online (John Wiley & Sons, Ltd., 2015)
4. India: Heavy-duty: Emissions; https://transportpolicy.net/index.php?title=India:Heavy-duty: Emissions
5. L. Zhang, *A Study of Pilot Injection in a DI Diesel Engine*. SAE Technical Paper 1999-01-3493, 1999. https://doi.org/10.4271/1999-01-3493
6. C.D. Rakopoulos, E.G. Giakoumis, *Diesel Engine Transient Operation* (Springer, London, 2009)
7. M. Ishida, Z. Chen, G. Luo, H. Ueki, The effect of pilot injection on combustion in a turbocharged D.I. Diesel engine, in SAE Technical Paper 941692, 1994. https://doi.org/10.4271/941692
8. P. Carlucci, A. Ficarella, D. Laforgia, Effects of pilot injection parameters on combustion for common rail diesel engines, in SAE Technical Paper 2003-01-0700, 2003. https://doi.org/10.4271/2003-01-0700
9. P.J. Tennison, R.D. Reitz, An experimental investigation of the effects of common-rail injection system parameters on emissions and performance in a high-speed direct-injection diesel engine. J. Eng. Gas Turbines Power **123**, 167–174 (2001)
10. M. Dürnholz, H. Endres, P. Frisse, Preinjection a measure to optimize the emission behavior of DI-diesel engine,SAE Technical Paper 940674, 1994. https://doi.org/10.4271/940674
11. T. Minami, K. Takeuchi, N. Shimazaki, Reduction of diesel engine NOx using pilot injection, in SAE Technical Paper 950611, 1995. https://doi.org/10.4271/950611
12. S. Chen, Simultaneous reduction of NOx and particulate emissions by using multiple injections in a small diesel engine, in SAE Technical Paper 2000-01-3084, 2000. https://doi.org/10.4271/2000-01-3084

13. M. Badami, F. Millo, D. D'Amato, Experimental investigation on soot and NOx formation in a DI common rail diesel engine with pilot injection, in SAE Technical Paper 2001-01-0657, 2001. https://doi.org/10.4271/2001-01-0657
14. T. Tow, D. Pierpont, R. Reitz, Reducing particulate and NOx emissions by using multiple injections in a heavy duty D.I. diesel engine, in SAE Technical Paper 940897, 1994. https://doi.org/10.4271/940897
15. D. Pierpont, D. Montgomery, R. Reitz, Reducing particulate and NOx using multiple injections and EGR in a D.I. diesel, in SAE Technical Paper 950217, 1995. https://doi.org/10.4271/950217
16. H. Yokota, Y. Kudo, H. Nakajima, T. Kakegawa, et al.,A new concept for low emission diesel combustion, in SAE Technical Paper 970891, 1997. https://doi.org/10.4271/970891
17. M. Badami, F. Mallamo, F. Millo, E.E. Rossi, Experimental investigation on the effect of multiple injection strategies on emissions, noise and brake specific fuel consumption of an automotive direct injection common rail diesel engine. Int. J. Engine Res JER. **4**(4), 0253 (2003)
18. S. Mendez, B. Thirouard, Using multiple injection strategies in diesel combustion: potential to improve emissions, noise and fuel economy trade-off in low CR engine. SAE Paper 2008-01-1329 (2008)
19. P. Carlucci, A. Ficarella, D. Laforgia, Effects on combustion and emissions of early and pilot fuel injections in diesel engines. Int. J. Engine Res. JER 02703 (2004 )
20. D. Nehmer, R. Reitz, Measurement of the effect of injection rate and split injections on diesel engine soot and NOx emissions, in SAE Technical Paper 940668, 1994. https://doi.org/10.4271/940668
21. J.W. Lee, H. Choi, K. Hong, S. Lee, S. Yu, S.M. Choi et al., Comparison of the effects of multiple injection strategy on the emissions between moderate and heavy EGR rate conditions: part 2-post injections. J. Mech. Sci. Technol. **27**(7), 2217–2223 (2013)
22. J.W. Lee, H. Choi, K. Hong, S. Lee, S. Yu, S.M. Choi et al., Comparison of the effects of multiple injection strategy on the emissions between moderate and heavy EGR rate conditions: part 1-pilot injections. J. Mech. Sci. Technol. **27**(4), 1135–1141 (2013)
23. H. Yun, R.D. Reitz, An experimental investigation of the effects of Post-Injection Strategies on Combustion and Emissions in the Low Temperature Diesel Combustion Regime. J. Eng. Gas Turbines Power **129**, 279–286 (2007)
24. J. Lee, J. Jinwoog Jeon, J. Park, C. Bae, Effect of multiple injection strategies on emission and combustion characteristics in a single cylinder direct-injection optical engine, in SAE Technical Paper 2009-01-1354
25. B. Mohan, W. Yang, S.K. Chou, Fuel injection strategies for performance improvement and emissions reduction in compression ignition engines—a review. Renew. Sustain. Energy Rev. **28**, 664–676 (2013)
26. S.J. Yoon, J. Park, B. Park, J. Park, S. Park, Effect of pilot injection on engine noise in a single cylinder compression ignition engine. Int. J. Autom. Technol. **16**(4), 571–579 (2015). https://doi.org/10.1007/s12239-015-0058-6
27. S. Busch, K. Zha, P.C. Miles, Investigations of closely coupled pilot and main injections as a means to reduce combustion noise in a small bore direct injection Diesel engine. Int. J of Engine Res. **16**(1) 13–22 (2015). https://doi.org/10.1177/1468087414560776
28. K.H. Suh, Study on the twin-pilot-injection strategies for the reduction in exhaust emissions in a low-compression-ratio engine. Proc. IMech. E Part D J. Autom. Eng. **228**(3), 335–343 (2014)
29. M.R. Ghaffarpour, A.R. Noorpoor, A numerical study of the use of pilot or split rate injection to reduce diesel engine noise. Proc. IMech. Eng. Part D J. Autom. Eng.**221**, 457–464 (2007). https://doi.org/10.1243/09544070JAUTO183
30. S. D'Ambrosio, A. Ferrari, Potentialities of boot injection combined with after shot for the optimization of pollutant emissions, fuel consumption and combustion noise in passenger car diesel engines, in SAE Int. J. Engines **10**(2) (2017). https://doi.org/10.4271/2017-01-9277
31. S. Biswas, M. Bakshi, G. Shankar, A. Mukhopadhyay, Experimental investigation on the effect of two different multiple injection strategies on emissions, combustion noise and performances

of an automotive CRDI engine, in SAE Technical Paper 2016-01-0871, 2016. https://doi.org/10.4271/2016-01-0871
32. S. Rajkumar, P. Mehta, S. Bakshi, Parametric investigation for NOx and soot emissions in multiple-injection CRDI engine using phenomenological model, in SAE Technical Paper 2011-01-1810, 2011. https://doi.org/10.4271/2011-01-1810
33. S. Bharadwaj, A. Gupta, S. Narayan, A review of various NVH sources of combustion engines. Int. J. Mech. Eng. Autom. **3**(6), 249–261 (2016)
34. S. Narayan, Analysis of noise emitted from diesel engines. Int. Conf. Vibr. Probl. ICOVP J. Phys. Conf. Ser. **662**, 012018 (2018). https://doi.org/10.1088/1742-6596/662/1/012018
35. IS: 10399, *Automotive Vehicles-Noise Emitted by Stationary Vehicles-Method of Measurement*, 1998 (2006-04), Bureau of Indian Standards

# Radial Basis Function-Based Probabilistic First-Ply Failure Analyses of Composite Spherical Shells

**Himanshu Prasad Raturi, Subrata Kushari, and Sudeep Dey**

**Abstract** This paper presents the first-ply failure analysis of laminated composite spherical shells by using the finite element method (FEM) in conjunction with Monte Carlo Simulation (MCS) approach. The material and geometric uncertainties caused by anisotropy and randomness inherent in the system configuration are predetermined in such structures and demand the stochastic analysis for realistic design. Finite element formulation is used to derive eight-noded iso-parametric quadratic elements. The input parameters include the ply orientation angle, assembly of ply, the number of layers, ply thickness and degree of orthotropy. The five failure criteria are considered, viz. maximum strain, maximum stress, Tsai-Hill, Tsai-Wu-Hahn and Tsai-Hill Hoffman theories. The variations in first-ply failure load are analysed. A deterministic study is carried out for the analyses of first-ply failure loads with respect to the mentioned failure criteria. The stochastic analysis is carried out firstly by the MCS approach. It is followed by implementation of the surrogate model. Radial basis function (RBF) is incorporated as the surrogate model. The current study predicts that the RBF surrogate model can be utilized to achieve similar computational efficiency with a significant reduction in computational time.

**Keywords** Monte Carlo simulation (MCS) · First-ply failure analysis · Radial basis function (RBF)

## 1 Introduction

The strength of a laminated composite spherical shell is predicted by first-ply failure analyses. The first-ply failure stress signifies the amount of stress the composite can endure until its fracture point. A curve-shaped laminated shell provides much

H. P. Raturi (✉) · S. Kushari · S. Dey
Department of Mechanical Engineering, National Institute of Technology Silchar, Silchar, Assam 788010, India
e-mail: himanshuraturi3@gmail.com

S. Kushari
e-mail: subrata734@gmail.com



K. M. Pandey et al. (eds.), *Recent Advances in Mechanical Engineering*, Lecture Notes in Mechanical Engineering, https://doi.org/10.1007/978-981-15-7711-6_66

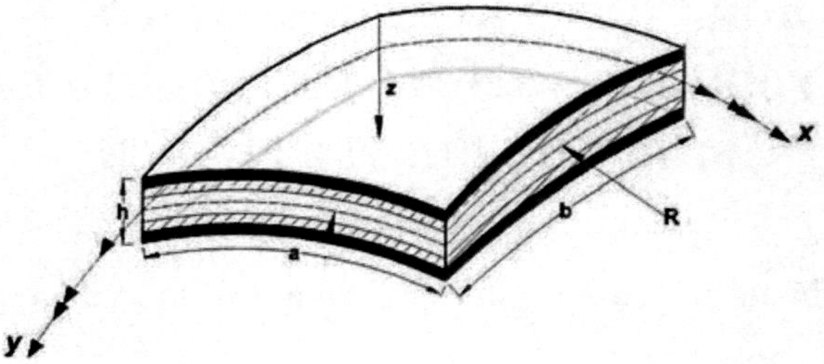


**Fig. 1** Isometric view of a spherical shell laminated composite

better agility and better aerodynamic structure as compared to flat ones. Some of its application can be found in rooftop structures, aerospace parts and automobiles. Although the deterministic regime provides a significant overview of the analyses performed, but in order to analyse the broader picture, probabilistic approach is necessary. Monte Carlo simulation is broadly utilized to perform probabilistic analyses. In order to achieve computational efficiency, the surrogate models are utilized to achieve computational efficiency. Based on Kirchhoff–Love shell theory, a nonlinear finite element analysis is developed [1] to predict the damage in a spherical shell. A similar study in conoidal shells by Bakshi and Chakravorty [2] for nonlinear and iso-parametric geometry presented various configuration of the ply orientation. They proposed FCFC boundary condition with cross-ply orientation maximizes the failure load. The World-Wide Failure Exercise (WWFE) by Soden et al. [3] provides a detailed failure analysis of composites. The advantages and drawbacks of various failure theories utilized worldwide are depicted broadly. A nonlinear finite element model is proposed by Kam et al. [4] based on von Karman-Mindlin plate theory and minimum total potential energy. They analysed the deflection and first-ply failure of the composite plate. Ghosh and Chakravorty [5] investigated the first-ply failure loads for industrial hypar shells for erratic aspect ratios and thicknesses under clamped boundary condition. It is suggested that shells with unit aspect ratio or square shells perform efficiently as compared rectangular shells. Stochastic analysis is investigated for sandwich shells [6] to analyse the failure in the shell by incorporating higher order zigzag theory. The model predicts the transverse shear stresses efficiently. The present study is focused on analysing the first-ply failure of a spherical shell in the stochastic regime by incorporating Monte Carlo simulation (MCS) and radial basis function (RBF) as the probabilistic tool (Fig. 1).

## 2 Mathematical Formulation

In order to determine the first-ply failure of the laminated composite, five failure criteria are taken into consideration as mentioned earlier. The finite element (FE) model is designed based on the failure criteria; followed by this, the surrogate model is implemented using RBF.

## 2.1 Failure Criteria for Laminated Composite

In the present work, a three-layered laminated spherical shell is considered to study the failure analysis. The orientation of the laminate is $[45°, -45°, 45°]$. The five failure criteria are utilized to investigate the first-ply failure load of the spherical shell and design the finite element model for the same.

***Maximum stress theory***

This theory involves two forms of stress (normal stress and shear stress) theories. It specifies that when a material exceeds its maximum stress enduring capacity in any of its axes it fails [7]. It can be expressed as,

$$(\sigma_1^c)_u < (\sigma_1) < (\sigma_1^T)_u \tag{1}$$

$$(\sigma_2^c)_u < (\sigma_2) < (\sigma_2^T)_u \tag{2}$$

$$(\tau_{12})_u < (\tau_{12}) < (\tau_{12})_u \tag{3}$$

where $\sigma_1$, $\sigma_2$ represents the normal stresses in *x*-axis and *y*-axis, respectively, while $\tau_{12}$ represents the shear stress. $\sigma_c$ and $\sigma^T$ represents the compressive stress and tensile stress, respectively, through the laminate. Here, the suffix '$u$' is used to signify the ultimate stress point.

***Maximum strain theory***

This theory is based on the maximum normal strain theory for isotropic materials as predicted by Sokolnikoff [8]. According to this theory, when the shear and principal strain exceeds the ultimate strain, the material tends to rupture or fail. The mathematical deduction for the same can be expressed as

$$(\varepsilon_1^c)_u < (\varepsilon_1) < (\varepsilon_1^T)_u \tag{4}$$

$$(\varepsilon_2^c)_u < (\varepsilon_2) < (\varepsilon_2^T)_u \tag{5}$$

$$\gamma_{12} < \Gamma_{12} \tag{6}$$

where $\varepsilon_1$ and $\varepsilon_2$ represent the normal strains in the *x*-axis and *y*-axis, while $\gamma_{12}$ represents the shear strain. $\varepsilon^c$ and $\varepsilon^T$ represent the compressive strain and tensile strain, respectively, through the laminate and $\Gamma_{12}$ represents the ultimate shear strain.

***Tsai-Hill (energy-based criterion) theory***

Tsai-Hill theory [9] for the failure of laminate is a combination of distortion energy (which is responsible for change of the shape) and dilation energy (which causes volumetric changes in the material). The failure in the material is depicted as follows.

$$f(\sigma_{ij}) = F(\sigma_2 - \sigma_3)^2 + G(\sigma_3 - \sigma_1)^2 + H(\sigma_1 - \sigma_2)^2 + 2L\sigma_4^2 + 2M\sigma_5^2 + 2N\sigma_6^2 = 1 \tag{7}$$

where *F, G, H, L, M* and *N* signify strength parameters of the material and $\sigma_4, \sigma_5, \sigma_6$ are the shear stress components.

***Tsai-Wu (Interaction tensor polynomial) theory***

The Tsai-Wu failure criterion is a special case of the general quadratic failure criteria developed by Gol'denblat and Kopnov. It is depicted as [10]

$$F_i\sigma_i + F_{ij}\sigma_i\sigma_j \geq 1 \tag{8}$$

where $F_i$ and $F_{ij}$ are the first-order and fourth-order strength tensors of the material. Here, $\sigma_i$ denotes the difference between compressive- and tensile-induced stress. The term $\sigma_i\sigma_j$ defines an ellipsoid along with the stress space.

***Tsai-Hill's Hoffman failure criteria***

The Tsai-Hill's Hoffman criterion is a special condition of Tsai-Hill failure criteria. In Hoffman's failure criteria, the difference between the strength of tension and compression is considered which is ignored in the case of Tsai-Hill failure criteria which is significant if brittle materials are considered. The modified criteria are established by adding the odd functions of the principal stress components ($\sigma_1$, $\sigma_2$ and $\sigma_3$) in the actual expression of Tsai-Hill criteria [11]. Thus,

$$C_1(\sigma_2 - \sigma_3)^2 + C_2(\sigma_3 - \sigma_1)^2 + C_3(\sigma_1 - \sigma_2)^2 + C_4\sigma_1 + C_5\sigma_2 + C_6\sigma_3 + C_7\sigma_4^2 + C_8\sigma_5^2 + C_9\sigma_6^2 = 1 \tag{9}$$

Here $C_1$ to $C_9$ denotes the material parameters.

## 2.2 Radial Basis Function-Based Surrogate Modelling

Radial basis function comprises of an input layer, output layer and a layer of RBF neurons in between them. The number of nodes in the output layer is equal to the types (category) of data filtered from a layer of RBF neurons. Each RBF neurons stores a prototype, and a new input variable is categorized based on Euclidean distance between input and prototype (computed by each neuron). The Euclidean distances of linear combination presented in the surrogate-based model are represented as Dey et al. [12]

$$\hat{Y}(x) = \sum_{p=1}^{M} w_p \varphi_p(X, x_p) \tag{10}$$

Weight determined by using the least-squares method is represented by $w_p$, the number of sampling points by $M$, while the pth basis function determined at the sampling point, $x_p$ is described by $\varphi_p\ (X,\ x_p)$. RBF model is represented by using a radial function, which is expressed as (Fig. 2),

$$F(x) = \frac{1}{\sqrt{1 + \frac{(x-c)^T(x-c)}{r^2}}} \quad \text{(For inverse multi – quadratic)} \tag{11}$$

$$F(x) = \exp\left(-\frac{(x-c)^T(x-c)}{r^2}\right) \quad \text{(For Gaussian)} \tag{12}$$

$$F(x) = \frac{1}{1 + \frac{(x-c)^T(x-c)}{r^2}} \quad \text{(For Cauchy)} \tag{13}$$

$$F(x) = \sqrt{1 + \frac{(x-c)^T(x-c)}{r^2}} \quad \text{(For multi – quadratic)} \tag{14}$$

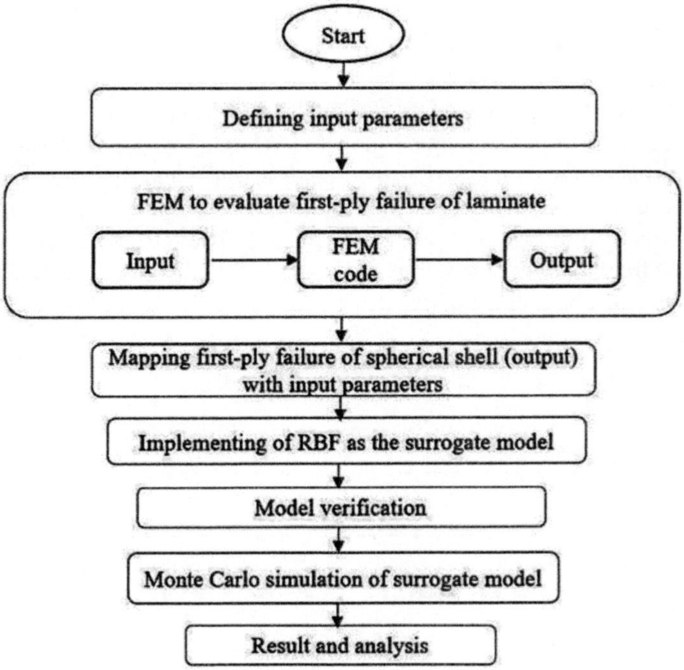


**Fig. 2** Flowchart of first-ply failure analyses of the laminated spherical shell

## 3 Result and Discussion

The present section involves the validation of the current model. A deterministic analysis is carried out at first for a three-layered laminated spherical shell with ply orientation [45°, −45°, 45°]. E glass–epoxy composite is considered as the present laminated composite spherical shell. The material properties of the composite for fabricating spherical shells and the geometric dimensions of the shells are as follows (Table 1).

Here, $R$ is the radius of curvature and $h$ implies the thickness of the spherical shell. The present model and finite element code used here can be accepted as a successful tool to explore the first-ply failure of composite spherical shells in future. The mesh plane area is considered of $(8 \times 8)$ configuration comprising of 64 elements and 225 number of nodes. The deterministic validation of the five different failure criteria from Ghosh and Chakravorty [14] is shown in Table 2.

Followed by the deterministic regime, stochastic analysis is carried out for first-ply failure analyses of the spherical shell. Monte Carlo simulation is performed considering 10,000 sample size. It is followed by incorporation of radial basis function (RBF) as the surrogate model in order to achieve computational efficiency. Four sample sizes are considered in the RBF surrogate model, viz. 32, 64, 128 and 256. The probability density function (pdf) is presented in Fig. 3.

## 4 Conclusion

The present study investigates the first-ply failure analysis of a laminated spherical shell. The deterministic finite element model is validated with previous literature. The probabilistic regime is analysed by incorporating MCS and RBF as the surrogate model. It can be predicted that the pdf plots of the RBF surrogate model with sample size 256 significantly merge with MCS from Fig. 3. The RBF surrogate model is computed in a fraction of time as compared to the MCS. It can be perceived that the RBF model can also be utilized in place of MCS.

**Table 1** Material properties *of* E glass–epoxy composite [13] with ply orientation [45°, −45°, 45°]

| $E_{11}$ (GPa) | $E_{22} = E_{33}$(GPa) | $G_{12}$ (GPa) | $G_{13}$ (GPa) | $G_{23}$ (GPa) | $\mu_{12} = \mu_{13}$ | $\mu_{23}$ | $\rho$ $(kg/m^3)$ | $R$ (mm) | $h$ (mm) |
|---|---|---|---|---|---|---|---|---|---|
| 40.0 | 8.27 | 4.13 | 4.13 | 3.03 | 0.26 | 0.25 | 1900 | 750 | 5 |

**Table 2** First-ply failure stresses (MPa) of an angle-ply spherical shell, validation of the present finite element model with experimental results [14] for in-plane loading of different failure theories

| Failure theory | Ghosh and Chakravorty [14] | Present FE model |
|---|---|---|
| Max. stress | 5.8519 | 5.4619 |
| Max. strain | 4.0396 | 4.8717 |
| Tsai-Hill | 5.5557 | 4.6336 |
| Tsai-Wu | 4.9793 | 4.6852 |
| Hoffman | 5.5212 | 4.4477 |

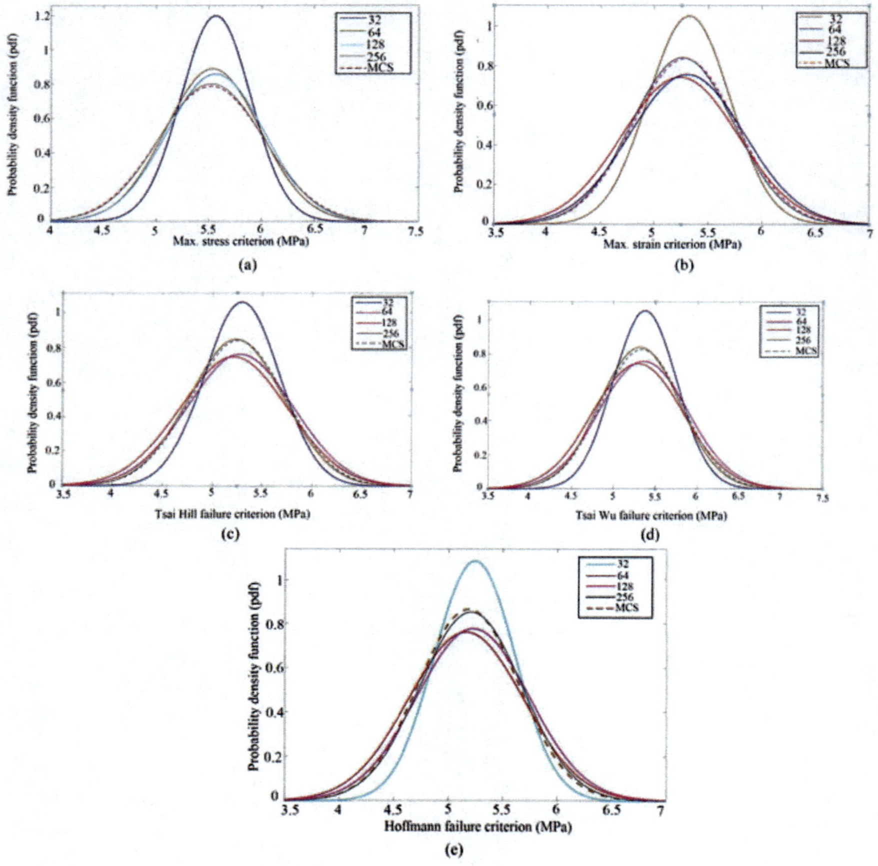


**Fig. 3** Probability density functions(PDF) of MCS and RBF model considering first-ply failure for **a** Maximum stress, **b** maximum strain, **c** Tsail-Hill, **d** Tsai-Wu and **e** Hoffman failure criteria

**Acknowledgements** The authors would like to acknowledge the Aeronautics Research and Development Board (AR&DB), Government of India (Project Sanction no.: ARDB/01/105885/M/I dtd. 26.06.2018) for the financial support for the present research work.

## References

1. S. Ganapathy, K.P. Rao, Failure analysis of laminated composite cylindrical/spherical shell panels subjected to low-velocity impact. Comput. Struct. **68**(6), 627–641 (1998)
2. K. Bakshi, D. Chakravorty, Geometrically nonlinear first ply failure loads of laminated composite conoidal shells. Procedia Eng. **173**, 1619–1626 (2017)
3. P.D. Soden, A.S. Kaddour, M.J. Hinton, Recommendations for designers and researchers resulting from the world-wide failure exercise, in *Failure Criteria in Fibre-Reinforced-Polymer Composites* (Elsevier, 2004), pp. 1223–1251
4. T.Y. Kam, H.F. Sher, T.N. Chao, R.R. Chang, Predictions of deflection and first-ply failure load of thin laminated composite plates via the finite element approach. Int. J. Solids Struct. **33**(3), 375–398 (1996)
5. A. Ghosh, D. Chakravorty, First ply failure analysis of laminated composite thin hypar shells using nonlinear finite element approach. Thin-Walled Struct. **131**, 736–745 (2018)
6. A. Kumar, A. Chakrabarti, P. Bhargava, R. Chowdhury, Probabilistic failure analysis of laminated sandwich shells based on higher order zigzag theory. J. Sandwich Struct. Mater. **17**(5), 546–561 (2015)
7. H.E. Tresca, *Sur l'ecoulement des corps solides soumis a de fortes pressions*. Imprimerie de Gauthier-Villars, successeur de Mallet-Bachelier, rue de Seine-Saint-Germain, 10, près l'Institut (1864)
8. I.S. Sokolnikoff, R.D. Specht, *Mathematical Theory of Elasticity*, vol. 83 (McGraw-Hill, New York, 1956)
9. R.V. Mises, Mechanik der festen Körper im plastisch-deformablen Zustand. Nachrichten von der Gesellschaft der Wissenschaften zu Göttingen, Mathematisch-Physikalische Klasse **1913**, 582–592 (1913)
10. S.W. Tsai, E.M. Wu, A general theory of strength for anisotropic materials. J. Compos. Mater. **5**(1), 58–80 (1971)
11. O. Hoffman, The brittle strength of orthotropic materials. J. Compos. Mater. **1**(2), 200–206 (1967)
12. S. Dey, T. Mukhopadhyay, S. Adhikari, Metamodel based high-fidelity stochastic analysis of composite laminates: a concise review with critical comparative assessment. Compos. Struct. **171**, 227–250 (2017)
13. L.P. Kollar, G.S. Springer. *Mechanics of Composite Structures* (Cambridge university press, 2003), pp. 464–467
14. A. Ghosh, D. Chakravorty, First-ply-failure performance of composite clamped spherical shells. Mech. Compos. Mater. **54**(2), 191–206 (2018)

# Experimental Investigation of Dimensional Deviation of Square Punches Cut by Wire EDM of A286 Superalloy

**Subhankar Saha, Saikat Ranjan Maity, and S. Dey**

**Abstract** A286 superalloy is a Fe–Ni-based superalloy is widely applicable in superchargers, gas turbines, jet engines, fasteners, after burner and turbine wheels because of its innate properties like high thermal resistance, high mechanical strength and substantial corrosion resistance. This study is particularly devoted to suggest optimal process parameter settings for this superalloy, where the dimensional deviation becomes minimum value after wire EDM cut. Measurement of dimensional deviation (DD) is important as it suggests the practitioners to set a proper wire offset during the CNC programming of wire tool path so that the dimension of the product after the Wire EDM cut and the required dimension of the product matches properly. Five important parameters such as pulse on time ($T_{\text{on}}$), pulse off time ($T_{\text{off}}$), peak current ($I_{\text{peak}}$), wire feed rate ($W_{\text{f}}$) and spark gap set voltage (SV) are controlled during the experiments, and the experimental layout is designed by L27 orthogonal array. Taguchi method in conjunction with ANOVA is adopted to obtain the significant control parameters. Parametric effect of the control factors on the dimensional deviation is explained. Furthermore, the optimal levels of the control factors are recommended based on higher signal-to-noise ratio values. Comparison between the experimental and the predicted values is evaluated to show reproducibility.

**Keywords** Wire EDM · A286 superalloy · Dimensional deviation (DD) · ANOVA · Signal-to-noise ratio

## 1 Introduction

The utilization of wire EDM technology has been remarkable in the industries for the last few decades. This technology is progressively replacing the conventional

---

S. Saha (✉) · S. R. Maity · S. Dey
Mechanical Engineering Department, National Institute of Technology Silchar, Silchar, Assam 788010, India
e-mail: sahamech90@gmail.com



K. M. Pandey et al. (eds.), *Recent Advances in Mechanical Engineering*, Lecture Notes in Mechanical Engineering, https://doi.org/10.1007/978-981-15-7711-6_67

machining processes with the advent of exotic materials in the high-technology industries which are capable of displaying appreciable mechanical properties at temperature even close enough to their melting points. Wire EDM offers a noncontact-based material removal of electrically conductive workpieces by electrical field-induced controlled sparks in between the continuously running wire which imitates here as tool and the workpiece, provided a channel of dielectric fluid exists in between the tool workpiece interface. Thus, minute amounts of work material which are melted and vaporized due to the energy transferred from the sparks to the workpiece material gets eroded and flushed away by the dielectric fluid from the surface leaving small craters on the machined surfaces. Therefore as stated above, the process does not make use of a cutting tool which makes it free from many undesirable anomalies usually encountered in practice during the conventional cutting of those materials such as excessive tool wear which is a serious issue and pose many challenges. Furthermore, conventional machining of exotic materials also leads to intensive burr formation on machined surfaces which are not acceptable from the standpoint of fatigue life of the material under cyclical loads and also require further post-processing, thereby incurring additional costs. In wire EDM, the wire follows a CNC programmed path that makes the process versatile for producing filigree geometries, however, due to improper process parameter settings and several forces which act on the wire electrode such as impact force induced by the sparks, damping force by the dielectric, electromagnetic force and uneven thermal stress, the process is susceptible to problems like dimensional deviation [1]. The interplay between these potential factors cause considerable dimensional deviation in the product geometry produced by wire EDM. Thus, the geometry sometimes may get oversized or undersized and sometimes fail to be within the tight tolerance limits which are a stringent requirement for critical components particularly in the context of turbo machinery industries. There are pile of research studies conducted over the last couple of decades where different modeling techniques to map the responses with the processing parameters were deployed and various optimization procedures had been adopted to optimize single and multiple process performances in wire EDM. Taguchi's robust design approach was performed for multi-objective optimization of three process performances such as material removal rate (MRR), surface roughness and wire wear ratio [2]. In another study, Taguchi grey relational analysis was employed for the optimization of multiple responses extracted by wire EDM of Inconel 718 [3]. Krishnan and Samuel also conducted optimization of MRR and surface roughness using non-dominated sorting genetic algorithm [4]. In a very recent study, optimal process parameter settings were suggested for four important process performances such as MRR, dimensional error, electrode wear and surface roughness in wire EDM of SS304 using Taguchi's method [5]. However, dearth of literatures exist which concentrates on the existing problems of geometric error and dimensional inaccuracy encountered in wire EDM. Therefore, this research is directed toward finding an optimal parametric setting for the dimensional deviation in wire EDM of A286 superalloy. In fact, there is no work which is reported on wire EDM of A286 superalloy [6, 7]. So the objective of this research as stated above is quite pertinent.

## 2 Experimental Methodology

The machining experiments were conducted on Ultra Cut F1 Model of Electronica Wire Cut EDM machine with 5-axis configuration (*X*, *Y*, *Z*, *U* and *V*) as shown in Fig. 1. Zinc-coated brass wire of 0.25 mm diameter was chosen as the tool electrode and the workpiece material which is considered for the study is of dimension (150 × 100 × 4 mm). Deionized water was employed as the dielectric fluid. Based on previous literatures and preliminary experimental trials, five input parameters with three levels were considered as control parameters during the experimentation. The dimensional deviation is measured using Mitutoyo digital micrometer with a least count of 0.001 mm.

The experimental setup and the dimensional deviation (DD) of a square punch are shown in Figs. 1 and 2, respectively. In Fig. 2, the half of the kerf width and DD

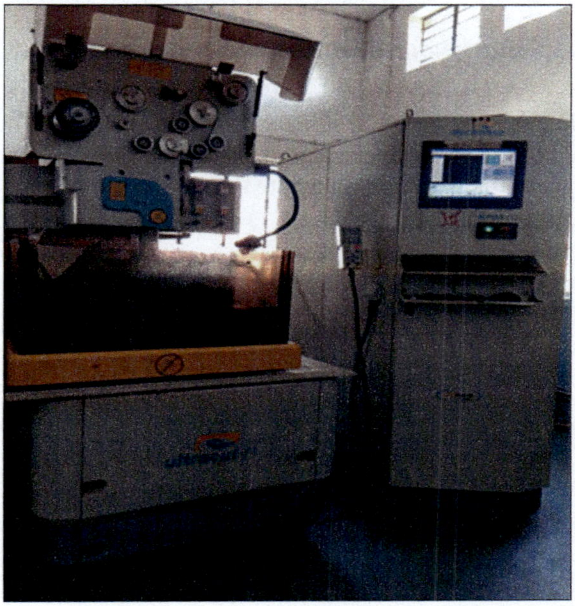

**Fig. 1** Experimental setup

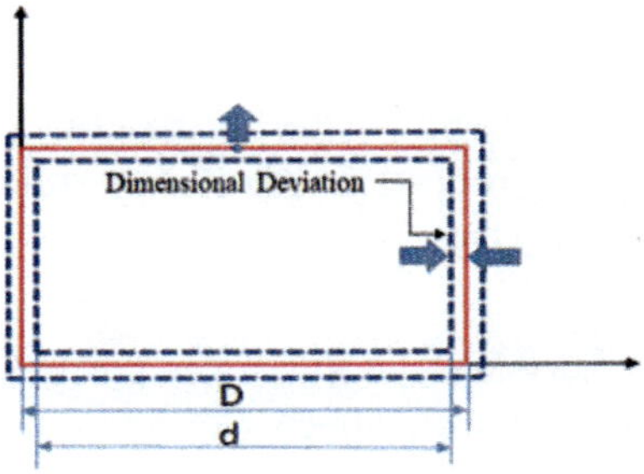


**Fig. 2** Dimensional deviation

can be calculated as below:

$$DD = 0.5 \times (D - d) \tag{1}$$

where, $D$ = actual size of the workpiece = 12 mm.
$d$ = size of the workpiece after machining.

## 3 Results and Discussion

The experimental values of DD and their associated signal-to-noise ratio values are presented in Table 1.

### 3.1 Statistical Analysis

The responses are transformed to signal-to-noise ratio values for the ANOVA to nullify the effect of uncontrollable factors. For smaller is better type of response, signal-to-noise ratio is computed as below [8]

$$S/N = -10 \times \log_{10}\left(\frac{\sum (y^2)}{n}\right) \tag{2}$$

where $y$ represents the measured data and $n$ is the number of replications.

The computed results for all the experimental settings are subjected to ANOVA to delineate the effects of the control parameters for 95% confidence interval. For a particular control parameter to be statistically significant for 95% confidence interval, the $F$-value (mean square between/mean squared error) should be higher and p-value must be less than 0.05. The ANOVA for dimensional deviation is displayed in Table 2.

From the Table 2, it can be inferred that $T_{\text{on}}$ is the most significant parameter followed by $I_{\text{peak}}$, $T_{\text{off}}$, SV and $W_{\text{f}}$. This is further supported by the response table in Table 3 for signal-to-noise ratios.

### 3.2 Effect of Control Parameters on Dimensional Deviation (DD)

To understand the effect of control parameters on dimensional deviation comprehensively, the main effects plot for means are shown in Fig. 3. It is a plot between the

**Table 1** Experimental results

| $T_{on}$ (μs) | $T_{off}$ (μs) | $I_{peak}$ (A) | $W_f$ (m/min) | SV (Volt) | Dimensional deviation (DD) in mm | Signal-to-noise ratio (dB) |
|---|---|---|---|---|---|---|
| 120 | 48 | 10 | 5 | 30 | 0.05 | 26.0206 |
| 120 | 48 | 11 | 7 | 35 | 0.07 | 23.0980 |
| 120 | 48 | 12 | 9 | 40 | 0.09 | 20.9151 |
| 120 | 52 | 10 | 7 | 35 | 0.03 | 30.4576 |
| 120 | 52 | 11 | 9 | 40 | 0.02 | 33.9794 |
| 120 | 52 | 12 | 5 | 30 | 0.06 | 24.4370 |
| 120 | 56 | 10 | 9 | 40 | 0.01 | 40.0000 |
| 120 | 56 | 11 | 5 | 30 | 0.02 | 33.9794 |
| 120 | 56 | 12 | 7 | 35 | 0.03 | 30.4576 |
| 125 | 48 | 10 | 7 | 40 | 0.06 | 24.4370 |
| 125 | 48 | 11 | 9 | 30 | 0.12 | 18.4164 |
| 125 | 48 | 12 | 5 | 35 | 0.14 | 17.0774 |
| 125 | 52 | 10 | 9 | 30 | 0.03 | 30.4576 |
| 125 | 52 | 11 | 5 | 35 | 0.04 | 27.9588 |
| 125 | 52 | 12 | 7 | 40 | 0.05 | 26.0206 |
| 125 | 56 | 10 | 5 | 35 | 0.06 | 24.4370 |
| 125 | 56 | 11 | 7 | 40 | 0.07 | 23.0980 |
| 125 | 56 | 12 | 9 | 30 | 0.14 | 17.0774 |
| 130 | 48 | 10 | 9 | 35 | 0.07 | 23.0980 |
| 130 | 48 | 11 | 5 | 40 | 0.08 | 21.9382 |
| 130 | 48 | 12 | 7 | 30 | 0.16 | 15.9176 |
| 130 | 52 | 10 | 5 | 40 | 0.06 | 24.4370 |
| 130 | 52 | 11 | 7 | 30 | 0.10 | 20.0000 |
| 130 | 52 | 12 | 9 | 35 | 0.12 | 18.4164 |
| 130 | 56 | 10 | 7 | 30 | 0.06 | 24.4370 |
| 130 | 56 | 11 | 9 | 35 | 0.08 | 21.9382 |
| 130 | 56 | 12 | 5 | 40 | 0.09 | 20.9151 |

different level response means shown in vertical axis and the corresponding levels in the horizontal axis for each control parameter. From the plots, it is shown that DD substantially increases with $T_{on}$. This is due to the fact that as the discharge energy increases with $T_{on}$, large-sized craters are formed on the surfaces which contribute to larger DD and also the other important cause contributing to larger DD is that various forces particularly the thrust forces resulting due to the explosion of sparks become prominent at higher pulse on time ($T_{on}$) which cause the wire to deflect from its programmed path. DD decreases significantly when pulse off time ($T_{off}$) increases from (48–52) but slightly increases at 56. This is attributed to decrease in the sparking

**Table 2** ANOVA

| Source | DF | Adj SS | Adj MS | $F$-value | $p$-value | Percentage contribution |
|---|---|---|---|---|---|---|
| $T_{on}$ | 2 | 314.628 | 157.314 | 14.02 | 0.000 | 36.34 |
| $T_{off}$ | 2 | 152.234 | 76.117 | 6.78 | 0.007 | 17.58 |
| $I_{peak}$ | 2 | 179.422 | 89.711 | 7.99 | 0.004 | 20.72 |
| $W_f$ | 2 | 2.259 | 1.129 | 0.10 | 0.905 | 0.26 |
| SV | 2 | 37.658 | 18.829 | 1.68 | 0.218 | 4.35 |
| Error | 16 | 179.558 | 11.222 | | | 20.74 |
| Total | 26 | 865.759 | | | | |

**Table 3** Response table

| Level | $T_{on}$ | $T_{off}$ | $I_{peak}$ | $W_f$ | SV |
|---|---|---|---|---|---|
| 1 | 29.26 | 21.21 | 27.53 | 24.58 | 23.42 |
| 2 | 23.22 | 26.24 | 24.93 | 24.21 | 24.10 |
| 3 | 21.23 | 26.26 | 21.25 | 24.92 | 26.19 |
| Delta | 8.03 | 5.05 | 6.28 | 0.71 | 2.78 |
| Rank | 1 | 3 | 2 | 5 | 4 |

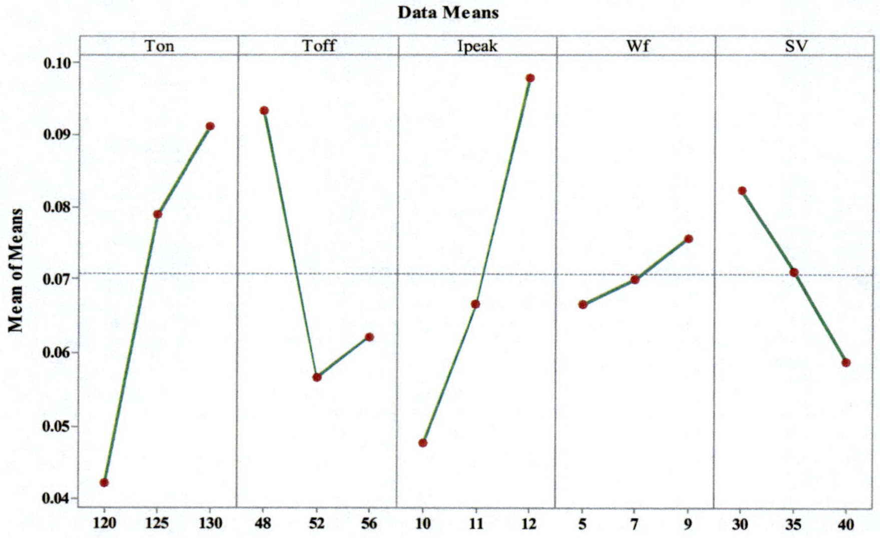


**Fig. 3** Main effects plot for means

frequency when pulse off time increases from (48–52) but at higher value of pulse off time (56), the wire lag effect becomes prominent because of excessive hydraulic forces during dielectric flushing which contributes to slight increase in DD. From the plot, it can be stated that DD increases significantly with increase in peak current ($I_{peak}$) which is directly related to the higher energy that is entrapped in the spark gap because of higher values of peak current ($I_{peak}$) which in turn contributes to large crater formation. Wire feed rate and servo voltage are found to have no significant effect on DD as per ANOVA.

### *3.3 Optimization of Dimensional Deviation (DD) Using Taguchi Technique*

The optimal levels for control parameters are suggested from main effects plot for SN ratios in Fig. 4. The levels are selected based on higher signal-to-noise ratio values and are $T_{on}$ (Level 1), $T_{off}$ (Level 3), $I_{peak}$ (Level 1), $W_f$ (Level 3) and SV (Level 3). This particular setting is found in the experimental design space with DD value of 0.01 and signal-to-noise ratio of 40. The predicted values by Taguchi method at similar settings are 0.0007407 and signal-to-noise ratio of 39.6842.

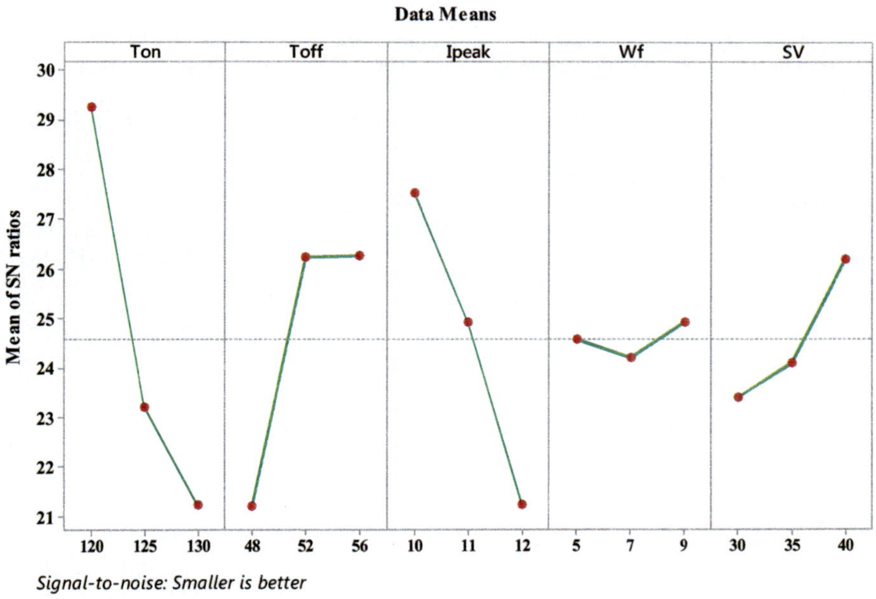


**Fig. 4** Main effects plot for SN ratios

# 4 Conclusions

It can be concluded from this research that the dimensional deviation increases with the pulse on time and peak current. Increase in pulse off time to a certain value is found to reduce the dimensional deviation, but extremely high values of pulse off time is not a good option as it may increase the dimensional deviation. Wire feed rate and servo voltage are found to have no effect on dimensional deviation. Moreover, at optimal setting, the dimensional deviation is found to be 0.01 which is exceptionally low.

# References

1. A.B. Puri, B. Bhattacharyya, Modelling and analysis of the wire-tool vibration in wire-cut EDM. J. Mater. Process. Technol. **141**(3), 295–301 (2003)
2. R. Ramakrishnan, L. Karunamoorthy, Multi response optimization of wire EDM operations using robust design of experiments. Int. J. Adv. Manuf. Technol. **29**(1–2), 105–112 (2006)
3. G. Rajyalakshmi, P.V. Ramaiah, Multiple process parameter optimization of wire electrical discharge machining on Inconel 825 using Taguchi grey relational analysis. Int. J. Adv. Manuf. Technol. **69**(5–8), 1249–1262 (2013)
4. S.A. Krishnan, G.L. Samuel, Multi-objective optimization of material removal rate and surface roughness in wire electrical discharge turning. Int. J. Adv. Manuf. Technol. **67**(9–12), 2021–2032 (2013)
5. G. Ugrasen, M.B. Singh, H.V. Ravindra, Optimization of process parameters for SS304 in wire electrical discharge machining using taguchi's technique. Mater. Today Proc. **5**(1), 2877–2883 (2018)
6. S.M. Muthu, M. Arivarasu, Investigations of hot corrosion resistance of HVOF coated Fe based superalloy A-286 in simulated gas turbine environment. Eng. Fail. Anal. **107**, 104224 (2020)
7. S.C. Liu, Y. Gao, Z.L. Lin, S.S. Guo, X.B. Zhang, X.J. Yin, Microstructure and properties after deformation and aging process of A286 superalloy. Rare Met. **38**(9), 864–870 (2019)
8. S.S. Mahapatra, A. Patnaik, Optimization of wire electrical discharge machining (WEDM) process parameters using Taguchi method. Int. J. Adv. Manuf. Technol. **34**(9–10), 911–925 (2007)

# Multi-optimization of μ-EDMed Arrayed Microrods Using Gray Relational Analysis

A. K. Singh, T. J. Singh, R. Pongen, and Promod Kumar Patowari

**Abstract** Fabrications of arrayed microrods using microelectrical discharge machining (EDM) are widely employed for drilling of multiple as well as arrays of microholes. It is commonly used in various applications such as perforated shadow mask, semiconductor device, and microheat exchanger. In the present work, gray relational analysis (GRA) has been proposed to optimize the multi-response performance characteristics (i.e., machining time and tool wear rate) of the process. GRA methodology is applied to optimize fabrication process of arrayed microrods to obtain the better dimensional accuracy with minimum tool wear and machining time using reverse micro-EDM (R-μEDM) process, a variant of micro-EDM process. In micro-EDM, tool wear and machining time are directly influence the dimensional accuracy of the microrods. The dimensional accuracy can be improved by reducing the tool wear and machining time during the fabrication process. The experimental investigation considers voltage, capacitance, and feed rate as input parameters. GRA optimization result shows that voltage of 150 V; feed rate of 25 μm/s; and capacitance of 10,000 pF are found as the optimum process parameters. The capacitance contributes 26.07% being highly influenced, followed by voltage (24.20%) and feed rate (5.17%). Voltage and capacitance have the statistical significance of 95% confidence level on overall performance toward the response parameters for getting the better dimensional accuracy with minimum time duration.

**Keywords** R-μEDM · Arrayed microrod · Dimensional accuracy · Process parameters · GRA

A. K. Singh (✉) · T. J. Singh · R. Pongen
Department of Mechanical Engineering, National Institute of Technology Nagaland, Dimapur, India
e-mail: amit.kumar965@yahoo.com

R. Pongen
e-mail: pongenrosang@gmail.com

P. K. Patowari
Department of Mechanical Engineering, National Institute of Technology Silchar, Silchar, Assam 788010, India
e-mail: ppatowari@gmail.com


K. M. Pandey et al. (eds.), *Recent Advances in Mechanical Engineering*, Lecture Notes in Mechanical Engineering, https://doi.org/10.1007/978-981-15-7711-6_68

# 1 Introduction

This electrical discharge machining (EDM) is a highly reliable machining process for machining of electrically conductive materials irrespective of their hardness [1]. The material removal mechanism takes place due to melting and thermal evaporation under the influence of high-intensity spark between the tool and the workpiece electrode. Microelectrical discharge machining (μEDM) is a variant of EDM, where the energy supplied in the interelectrode gap is very low compared to the conventional EDM process. The use of low energy enables machining of microfeatures with higher precision. The growing demand for microcomponents such as microprobes, microinjection, fuel nozzle, micromolds, etc., drove the evolution of micromachining techniques [2, 3]. μEDM is one of the micromachining technologies that have been progressing from leaps to bounds in the last decade. A μEDM machine tool has the capability to perform several machining operations such as grinding, milling, and drilling. The flexibility of using various techniques with a single machine tool enhanced the use of μEDM in micromachining.

R-μEDM is the technique which utilizes reverse polarity of μEDM, i.e., tool electrode is connected to the positive terminal and workpiece plate is connected to the negative terminal [4]. The whole process is conceptualized to fabricate tool of micron size. A predrilled hole on a plate acts as the tool electrode material, whereas the rod connected to the spindle acts as workpiece material. The workpiece is fed into the predrilled hole so that the rod resembles the shape of predrilled hole. This technique also facilitates fabrication of an array of microrods in a single pass, although this requires an array of predrilled holes on the tool plate [5]. Deposition of micropillars has also been carried out successfully following the technique of R-μEDM, where air has been used as the dielectric medium to diminish carbon adhesion to the tool electrode for maximizing tool wear [6, 7].

There are number of researchers who has used Taguchi and GRA methodology in order to optimize the process parameters [8, 9]. Taguchi has a limitation as it is only applicable for single-objective optimization problem. On the other hand, GRA is increasingly used by the researchers to optimize multi-response parameters. Eskandari et al. [10] have optimized surface roughness, volume of material and tool wear by using GRA, and found the optimum process parameters. Rajan et al. have successfully used GRA methodology to optimize the multi-response parameters. They identify the significant contributed parameters with the help of ANOVA [11].

In the present work, GRA has been proposed to optimize the multi-response parameters of the R-μEDM process, i.e., machining time and tool wear rate. Capacitance, voltage, and feed rate have been taken in different combination for the machining of arrayed microrods in order to establish a relationship with response parameters such as machining time and tool wear rate.

## 2 Materials and Methods

The margins' μEDM setup has been used for machining of arrayed microrods by R-μEDM process. μEDM machine tool is a miniature machine tool which is table-top-type and having RC-type pulse generator (Model: Hyper-15; Make: Synergy Nano Systems, India) shown in Fig. 1.

In the present work, total three control factors such as capacitance (pF), voltage (V), and feed rate (μm/sec) have been used, each of the factors having three levels as show in Table 1. A total 27 number of experiments have been conducted during the experimentation.

**Fig. 1** Micro-EDM experimental setup

**Table 1** Experimental conditions

| Process parameters | | |
|---|---|---|
| Parameter | Levels | Actual values |
| Feed rate (μm/sec) | 3 | 5, 15, 25 |
| Capacitance (pF) | 3 | 100, 1000, 10,000 |
| Voltage (V) | 3 | 100, 125, 150 |
| Workpiece | 1.8 mm diameter (brass bulk rod) | |
| Tool electrode | Electrolytic copper plate of 293 μm thickness, μEDM fabricated holes of 318 μm diameter | |
| Working liquid | Hydrocarbon oil | |
| Response parameters | Tool wear rate and machining time | |

## 2.1 Gray Relational Analysis (GRA)

GRA is used to solve multi-objective problem, which makes interrelationship between different response parameters. In this method, gray relational grade followed by rank is obtained in order to convert the multiple response problem to single response problem.

In the initial step, for normalizing the response parameters, three types of quality characteristics have been decided, i.e., larger-the-better, smaller-the-better and nominal-the-better [12] (Eqs. 1–3) based on whether the response parameters to be maximized or minimized.

$$z_i^*(m) = \frac{z_j(d) - \min z_j(d)}{\max z_j(d) - \min z_j(d)} \tag{1}$$

$$z_i^*(d) = \frac{\max z_j(d) - z_j(d)}{\max z_j(d) - \min z_j(d)} \tag{2}$$

$$z_i^*(d) = \frac{\left|z_j(d) - z_j(d)\right|}{\max z_j(d) - z_j(d)} \tag{3}$$

where $z_i^*(d)$ is the normalized value for $z_j(d)$, $z_j(d)$ is the $d$th dependent response of $j$th experiment.

In the second step, the normalized values of the response parameters have been used to calculate gray relational coefficient (GRC) using Eqs. (4) and (5).

$$\Delta z_j(d) = \left|z_R(d) - z_i^*(d)\right| \tag{4}$$

$$\xi_j(d) = \frac{\Delta \min + q \Delta \max}{\Delta z_j(d) + q \Delta \max} \tag{5}$$

where $\xi_j(d)$ is GRC and $q$ varies from 0 to 1, and it is called distinguishing coefficient. $q$ is usually consider as 0.5 [13]. $\Delta z_j(d)$ is deviation sequence, $z_R(d) = 1$, and comparability sequence, $z_i^*(d)$.

In this step, the response parameters have been transformed into single response factor, i.e., gray relational grade (GRG).

$$r_j = \sum\nolimits_{j=1}^{d} [u(d)\xi(d)] \tag{6}$$

where $C$ known as performance characteristics and $u(d)$ is the weight percentage.

## 3 Results and Discussions

In this section, GRA has been utilized for optimizing the multi-response parameters, for instance, machining time and tool wear rate. Process parameters and response parameters have been obtained after the experimentation on R-μEDM. One of the arrayed machined microrod is shown in Fig. 2.

### *3.1 Optimizing Response Parameters Using GRA*

The normalized values for machining time and tool wear rate have been made in Table 2. While normalizing, lower the better type has been considered in both the case (machining time and tool wear rate), as the objective is to minimize the machining time and tool wear rate, which is calculated using Eq. (2). The coefficient $q$ has been taken as 0.5. GRC, and GRG has been evaluated and shown in Tables 2 and 3, respectively. After the experimentation, it has been observed that, for 1st response, i.e., machining time, Exp. No. 27 is the optimum parametric condition, whereas for second response, i.e., tool wear rate, exp. No. 4 is the optimum parametric condition. The gray relational grade has been evaluated using Eq. 6, and it is shown in Table 3. While calculating GRG, 35% weightage has been given to tool wear rate and 65% weightage is given to machining time. The main objective is to minimize the tool wear rate as much as possible because tool wear rate is the main responsible for

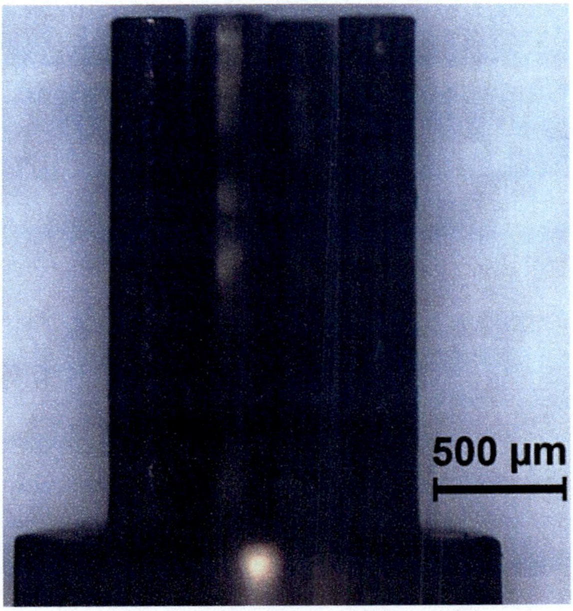


**Fig. 2** Pictorial view of EDMed arrayed microrod

**Table 2** Normalized and deviation sequence values

| | Experimental data | | Normalization | | Deviation sequence | |
|---|---|---|---|---|---|---|
| Exp. No. | Machining time (min.) | TWR ($10^{-5}$ μm$^3$/min) | Machining time (min.) | TWR ($10^{-5}$ μm$^3$/min) | Machining time (min.) | TWR ($10^{-5}$ μm$^3$/min) |
| 1 | 377.48 | 4.297 | 0.730 | 0.967 | 0.270 | 0.033 |
| 2 | 935.82 | 1.053 | 0.253 | 0.999 | 0.747 | 0.001 |
| 3 | 733.12 | 1.723 | 0.426 | 0.992 | 0.574 | 0.008 |
| 4 | 1232.70 | 0.900 | 0.000 | 1.000 | 1.000 | 0.000 |
| 5 | 695.03 | 2.355 | 0.459 | 0.986 | 0.541 | 0.014 |
| 6 | 492.57 | 2.577 | 0.632 | 0.984 | 0.368 | 0.016 |
| 7 | 650.78 | 2.319 | 0.497 | 0.986 | 0.503 | 0.014 |
| 8 | 243.03 | 5.072 | 0.845 | 0.960 | 0.155 | 0.040 |
| 9 | 196.08 | 7.586 | 0.885 | 0.936 | 0.115 | 0.064 |
| 10 | 995.18 | 2.129 | 0.203 | 0.988 | 0.797 | 0.012 |
| 11 | 828.07 | 2.519 | 0.345 | 0.984 | 0.655 | 0.016 |
| 12 | 772.93 | 3.159 | 0.392 | 0.978 | 0.608 | 0.022 |
| 13 | 587.75 | 3.546 | 0.550 | 0.975 | 0.450 | 0.025 |
| 14 | 466.22 | 5.703 | 0.654 | 0.954 | 0.346 | 0.046 |
| 15 | 461.92 | 5.558 | 0.658 | 0.955 | 0.342 | 0.045 |
| 16 | 446.47 | 4.847 | 0.671 | 0.962 | 0.329 | 0.038 |
| 17 | 305.05 | 8.250 | 0.792 | 0.929 | 0.208 | 0.071 |
| 18 | 289.83 | 5.887 | 0.805 | 0.952 | 0.195 | 0.048 |
| 19 | 163.08 | 35.410 | 0.913 | 0.669 | 0.087 | 0.331 |
| 20 | 158.75 | 37.820 | 0.917 | 0.646 | 0.083 | 0.354 |

(continued)

**Table 2** (continued)

| Exp. No. | Experimental data | | Normalization | | Deviation sequence | |
|---|---|---|---|---|---|---|
| | Machining time (min.) | TWR ($10^{-5}$ μm$^3$/min) | Machining time (min.) | TWR ($10^{-5}$ μm$^3$/min) | Machining time (min.) | TWR ($10^{-5}$ μm$^3$/min) |
| 21 | 162.75 | 35.993 | 0.913 | 0.663 | 0.087 | 0.337 |
| 22 | 111.48 | 51.128 | 0.957 | 0.518 | 0.043 | 0.482 |
| 23 | 106.73 | 50.607 | 0.961 | 0.523 | 0.039 | 0.477 |
| 24 | 104.25 | 53.863 | 0.963 | 0.492 | 0.037 | 0.508 |
| 25 | 68.98 | 92.464 | 0.993 | 0.122 | 0.007 | 0.878 |
| 26 | 63.20 | 105.139 | 0.998 | 0.000 | 0.002 | 1.000 |
| 27 | 60.93 | 98.599 | 1.000 | 0.063 | 0.000 | 0.937 |

**Table 3** Evaluate GRC and GRG

| Exp. No. | Gray coefficient | | GRG | Rank |
|---|---|---|---|---|
| | Machining time (min.) | TWR ($10^{-5}$ $\mu m^3$/min) | | |
| 1 | 0.649 | 0.939 | 0.751 | 14 |
| 2 | 0.401 | 0.997 | 0.610 | 25 |
| 3 | 0.466 | 0.984 | 0.647 | 22 |
| 4 | 0.333 | 1.000 | 0.567 | 27 |
| 5 | 0.480 | 0.973 | 0.653 | 21 |
| 6 | 0.576 | 0.969 | 0.713 | 16 |
| 7 | 0.498 | 0.973 | 0.665 | 20 |
| 8 | 0.763 | 0.926 | 0.820 | 2 |
| 9 | 0.813 | 0.886 | 0.838 | 1 |
| 10 | 0.385 | 0.977 | 0.592 | 26 |
| 11 | 0.433 | 0.970 | 0.621 | 24 |
| 12 | 0.451 | 0.958 | 0.629 | 23 |
| 13 | 0.527 | 0.952 | 0.675 | 19 |
| 14 | 0.591 | 0.916 | 0.705 | 18 |
| 15 | 0.594 | 0.918 | 0.707 | 17 |
| 16 | 0.603 | 0.930 | 0.717 | 15 |
| 17 | 0.706 | 0.876 | 0.766 | 9 |
| 18 | 0.719 | 0.913 | 0.787 | 3 |
| 19 | 0.852 | 0.602 | 0.764 | 11 |
| 20 | 0.857 | 0.585 | 0.762 | 13 |
| 21 | 0.852 | 0.598 | 0.763 | 12 |
| 22 | 0.921 | 0.509 | 0.777 | 6 |
| 23 | 0.927 | 0.512 | 0.782 | 4 |
| 24 | 0.931 | 0.496 | 0.779 | 5 |
| 25 | 0.986 | 0.363 | 0.768 | 8 |
| 26 | 0.996 | 0.333 | 0.764 | 10 |
| 27 | 1.000 | 0.348 | 0.772 | 7 |

dimensional inaccuracy while fabricating microrods using R-μEDM process, on the other hand, at lower parametric condition, machining time having very higher value. The GRG for machining time and tool wear rate has been shown in Fig. 3, and exp. No. 9 is the highest GRG.

The procedure for calculating the average gray relational grade (AGR) is shown below.

AGR is calculated by alliance the gray relational grade, factor-wise, and level-wise. For example, for factor A there are 3 level 1 GRG. The average GRG for this factor is shown below:

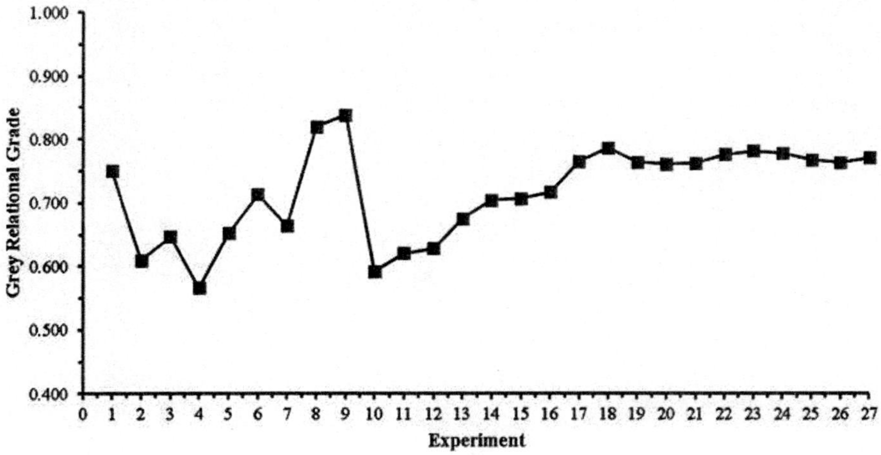


**Fig. 3** Grey relational grade for tool wear rate and machining time

$$\text{Avg. grey relational grade} = \frac{0.751 + 0.610 + 0.647 + 0.567 + 0.653 + 0.713 + 0.665 + 0.820}{8}$$

Similarly, the GRGs are evaluated for all input conditions and all levels shown in Table 4. Response graph based on average GRGs is shown in Fig. 4. It is detected from Fig. 4 that highest GRG for input condition *A* is level 3. Similarly, for input condition *B* and *C*, the highest GRGs are level 3 and level 3, respectively. The same can be presented in Table 4. The optimum process parameters, level 3; voltage, level 3; feed rate, level 3; capacitance, and the optimum conditions are A3, B3, C3 shown in Table 4. The values are 150 V voltage, 25 μm/s feed rate, and 10,000 pF capacitance. As per Table 4, input condition *A*, i.e., 0.0842 for voltage, has the maximum effect on multi-response characteristics. Followed by capacitance and feed rate having GRG values 0.0741 and 0.0400, respectively. Hence, the order can be listed as: condition *B* (voltage), condition *A* (capacitance), and condition *C* (feed rate), i.e., 0.0842 > 0.0741 > 0.0400. Table 4 also shows the optimum condition in which arrayed microrods have been fabricated with better dimensional accuracy and minimum time duration. 150 V voltage, 25 μm/s feed rate, and 10,000 pF capacitance is the best input parameter condition corresponding to multi-response

**Table 4** Response table of ARG

| Notation | Input parameters | *L*1 | *L*2 | *L*3 | Max–Min |
|---|---|---|---|---|---|
| *A* | Capacitance | 0.6959 | 0.6888 | 0.7700* | 0.0741 |
| *B* | Voltage | 0.6821 | 0.7064 | 0.7663* | 0.0842 |
| *C* | Feed rate | 0.6973 | 0.7202 | 0.7373* | 0.0400 |

*It indicates the value of highest grey relational grade (GRG)

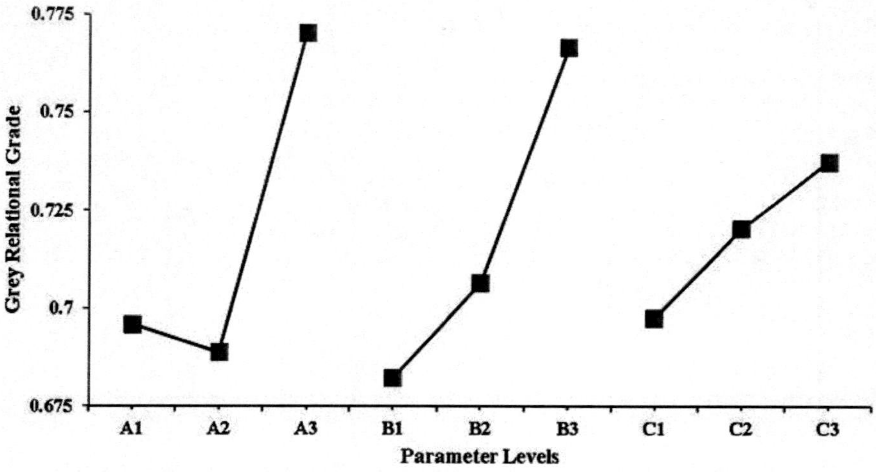


**Fig. 4** Response graph for ARG

performance characteristics. The result of this particular condition is available in the present experimental run. Hence, experiment has not been required to conduct.

## 3.2 ANOVA

ANOVA is basically used to point out the process parameters which are significant or which are not. 95% confidence level criteria have been used to identify the significant parameters, and it is depicted in Table 5. From the table, it is detected that capacitance (26.07%) is the utmost important parameter corresponding to multi-response performance characteristics followed by voltage (24.20%) and feed rate (5.17%).

**Table 5** ANOVA table of ARG

| Notation | Input parameters | DOF | Seq SS | Adj MS | *F* | % contribution | *P*-value |
|---|---|---|---|---|---|---|---|
| *A* | Capacitance | 2 | 0.036443 | 0.018221 | 5.85[a] | 26.07% | 0.010 |
| *B* | Voltage | 2 | 0.033837 | 0.016919 | 5.43[a] | 24.20% | 0.013 |
| *C* | Feed rate | 2 | 0.007230 | 0.003615 | 1.16 | 5.17% | 0.333 |
| *E* | Error | 20 | 0.062291 | 0.003115 | | | |
| Total | | 26 | 0.139802 | 0.04187 | | | |

[a]Significant at 95% confidence level

## 4 Conclusions

In this study, process parameters of the R-μEDM process have been optimized through GRA methodology for machining arrayed microrods with better dimensional accuracy and lower machining time. It is observed that the capacitance has most significant effect compared to other parameters on response parameters. Voltage and feed rate are the order of importance of the input parameters. Moreover, voltage contributes 24.20%, capacitance contributes 26.07%, and feed rate contributes 5.17%. 150 V voltage, 25 μm/s feed rate, and 10,000 pF capacitance are found to be the optimum condition.

## References


1. K.H. Ho, S.T. Newman, State of the art electrical discharge machining (EDM). Int. J. Mach. Tools Manuf. **43**, 1287–1300 (2003). https://doi.org/10.1016/S0890-6955(03)00162-7
2. E. Uhlmann, M. Roehner, Investigations on reduction of tool electrode wear in micro-EDM using novel electrode materials. CIRP J. Manuf. Sci. Technol. **1**, 92–96 (2008). https://doi.org/10.1016/j.cirpj.2008.09.011
3. S.H. Huang, F.Y. Huang, B.H. Yan, Fracture strength analysis of micro WC-shaft manufactured by micro-electro-discharge machining. Int. J. Adv. Manuf. Technol. **26**, 68–77 (2005). https://doi.org/10.1007/s00170-003-1974-6
4. A.K. Singh, P.K. Patowari, N.V. Deshpande, Experimental Analysis of Reverse Micro-EDM for Machining Microtool. Mater. Manuf. Processes **31**, 530–540 (2016). https://doi.org/10.1080/10426914.2015.1070426
5. A.K. Singh, P.K. Patowari, N.V. Deshpande, Effect of tool wear on microrods fabrication using reverse μEDM. Mater. Manuf. Processes **32**, 286–293 (2017). https://doi.org/10.1080/10426914.2016.1198015
6. Z. Peng, G. Chi, Z. Wang, Micro electrical discharge machining deposition in air for fabrication of micro spiral structures. Chin. J. Mech. Eng. English Edn. **23**, 154–160 (2010). https://doi.org/10.3901/CJME.2010.02.154
7. Z. Peng, Z. Wang, Y. Dong, H. Chen, Development of a reversible machining method for fabrication of microstructures by using micro-EDM. J. Mater. Process. Technol. **210**, 129–136 (2010). https://doi.org/10.1016/j.jmatprotec.2009.08.002
8. S. Shrestha, G. Manogharan, Optimization of Binder Jetting Using Taguchi Method. JOM. **69**, 491–497 (2017). https://doi.org/10.1007/s11837-016-2231-4
9. S. Swirad, R. Wdowik, Determining the effect of ball burnishing parameters on surface roughness using the Taguchi method. Procedia Manuf. **34**, 287–292 (2019). https://doi.org/10.1016/j.promfg.2019.06.152
10. B. Eskandari, B. Davoodi, H. Ghorbani, Multi-objective optimization of parameters in turning of N-155 iron-nickel-base superalloy using gray relational analysis. J. Br. Soc. Mech. Sci. Eng. **40**, 233 (2018). https://doi.org/10.1007/s40430-018-1156-y
11. B. Muthu Chozha Rajan, A. Senthil Kumar, T. Sornakumar, P. Senthamaraikannan, M.R. Sanjay, Multi response optimization of fabrication parameters of carbon fiber-reinforced aluminium laminates (CARAL): by Taguchi method and gray relational analysis. Polymer Compos. **40**, E1041–E1048 (2019). https://doi.org/10.1002/pc.24815
12. S. Liu, Y. Lin, Introduction to grey systems theory, in: *Understanding Complex Systems* (2010), pp. 1–399. https://doi.org/10.1007/978-3-642-16158-2_1

13. M. Sarikaya, A. Güllü, Multi-response optimization of minimum quantity lubrication parameters using Taguchi-based grey relational analysis in turning of difficult-to-cut alloy Haynes 25. J. Clean. Prod. **91**, 347–357 (2015). https://doi.org/10.1016/j.jclepro.2014.12.020; IEEE Trans. Instrument. Measur. **54**(6), 2462 (2005)

# Detection of Fault in a Bevel Gearbox Under Varying Speed Conditions

**Vikas Sharma, Anand Parey, Abhimanyu Pratap Singh, Atanu Paul, and Yogesh Singh**

**Abstract** A bevel gearbox is complicated in geometry and founds application where the shafts are perpendicular such as helicopter and automobiles. Unlike spur gears, three forces exert on bevel gears to transmit the motion. Due to complicated geometry of gears, nonlinear and nonstationary vibrations excited due to meshing exhibiting severe modulations in the gearbox vibration signals. Delay in detection of faults could be fatal; therefore, the condition monitoring and early fault detection techniques are of vital significance. In the present manuscript, a novel signal processing-based approach has been presented to detect the gear faults under varying speed conditions. The vibration signals acquired from the test bench were decomposed using variational mode decomposition (VMD) and the sensitive sub-band is selected upon estimating the instantaneous frequency (IF). The selected mode function (MF) upon decomposition is analyzed using the statistical condition indicators and its FFT is developed. The FFT of the selected mode exhibits the symptoms of the gear faults.

**Keywords** Bevel gearbox · Fault diagnosis · Variation mode decomposition · Instantaneous frequency

V. Sharma (✉)
Department of Mechanical Mechatronics Engineering, The LNM Institute of Information Technology, Jaipur, India
e-mail: vikas.sharma@lnmiit.ac.in

A. Parey
Discipline of Mechanical Engineering, Indian Institute of Technology Indore, Indore, India
e-mail: anandp@iiti.ac.in

A. P. Singh · A. Paul · Y. Singh
Department of Mechanical Engineering, National Institute of Technology Silchar, Silchar, Assam 788010, India
e-mail: abhimanyusingh02@outlook.com

A. Paul
e-mail: atanu3124@gmail.com

Y. Singh
e-mail: yogeshsingh15@gmail.com


K. M. Pandey et al. (eds.), *Recent Advances in Mechanical Engineering*, Lecture Notes in Mechanical Engineering, https://doi.org/10.1007/978-981-15-7711-6_69

## 1 Introduction

Bevel gearbox is one of the complicated mechanisms and any fault occurring in the gearbox leading to their failure, if missed to detect, could be fatal and catastrophic. Therefore, condition monitoring and early detection of gear faults are of primary significance. When a gear tooth meshes with its conjugate, the nonlinear and nonstationary vibration signals are generated [1]. The signals are very complicated in terms of their analysis as they exhibit severe amplitude modulation and frequency modulations generated due to meshing of gears [2].

Classical time-domain analysis and frequency domain analysis are incompetent to investigate the intrinsic features of the fault symptoms buried in strong noise. It is well known that time-domain analyses can only be used fundamentally for the analysis of time vs. amplitude change of the vibration signal; whereas, the frequency domain analysis gives an insight of frequency components present in the signal along with the amplitude change. However, frequency analysis based on discrete Fourier transform is restricted by the nonstationary signals [3, 4]. For nonstationary vibration signals, techniques such as wavelet transform (WT)-based decompositions [5, 6] empirical mode decomposition (EMD) [7] and variational mode decomposition (VMD) [8] have been widely used for condition monitoring of mechanical systems. WT uses filter bands to decompose signals; thus, it does not contains mono-frequency components with imperfect simultaneous time and frequency localization [9]. Issues such as mother-wavelet-selection criteria, border distortion, energy leakage also exist for WT [10]. Empirical mode decomposition is an improved signal processing technique capable of processing nonlinear and nonstationary vibration signals without any predefined basis. However, it has a serious problem of boundary distortion of IMFs, i.e., while performing statistical analysis of IMFs, boundary distortion may provide high values of CIs, even if a fault is absent [11]. Besides it lacks mathematical description and mode mixing, sensitivity to noise and sampling limits its performance [12]. Researchers such as Parey and Pachori [13], Sharma and Parey [14], Deng et al. [15] and Zhong and Shixiong [16] have tried to solve the problems of EMD. Methods such as ensemble empirical mode decomposition [17], synchrosqueezing [18], empirical wavelets [19] were partially successful in addressing the EMD problems, thereby detecting faults from vibration signal of the gearbox vibration signals by using additional noise. But, extra noise affects the performance relatively.

For nonstationary nonlinear vibration signals, variational mode decomposition (VMD) has been reported in the literature and moderately explored for vibration signals of rotary and reciprocating machines. To the best of our knowledge, despite of several studies reported, the researches on fault detection of bevel gears are very limited in the continuous growing literature. Therefore, this study attempts to develop a fault detection approach for bevel gears using VMD-based signal demodulation approach to identify and detect he fault symptoms hided by the overriding noise caused by system and surrounding. Further, upon decomposition of the vibration signal, to identify the most sensitive mode function (MF) is a very significant factor during investigation. Because, the selection of over-decomposed or under-decomposed MF may affect the entire detection process, as the relevant fault features may not be extracted. To identify the sensitive MF, instantaneous frequency (IF) of each MF will be evaluated. It is assumed that IF fluctuates around the gear mesh frequency (GMF) or its harmonics because the information related to the gear faults is expressed by either the sidebands about meshing frequency or its harmonics. Therefore, an IF-based selection of most sensitive MF is also proposed in this manuscript, which further will be analyzed by the statistical condition indicators and its FFT.

The paper is structured as follows: the main steps of the procedure of VMD and evaluation of IF are discussed in Sect. 2. The CIs used in this study have been also mentioned in subsequent part of Sect. 2, respectively. Section 3 illustrates the description of the test facility. The experimental findings, followed by the discussions, are presented in Sect. 4. Conclusions of the present study have been addressed in Sect. 5.

## 2 Theoretical Background

### 2.1 Variational Mode Decomposition

The VMD decomposes a real valued signal $r(t)$ into $K$ narrow-band components $s_k(t)$ by computing the center frequencies $\omega_k$ where $k = 1, 2, \ldots, K$ formulating constrained optimization problem. Before formulation of the optimization problem, the Hilbert transform is applied to the components $s_k(t)$ in order to compute the unilateral frequency spectrum. After that, modulation property is used to shift the frequency spectrum of these components based on the estimated center frequencies $\omega_k$. The bandwidth was estimated through the $H$ Gaussian smoothness of the demodulated signal [8]. Now, the constrained optimization problem is formulated as [8]:

$$\min_{\{s_k\},\{\omega_k\}} \left\{ \sum_k \left\| \partial_t \left[ \left( \delta(t) + \frac{j}{\pi t} \right) * s_k(t) \right] \mathrm{e}^{-j\omega_k t} \right\|_2^2 \right\} \tag{1}$$

such that $\sum_k s_k(t) = r(t)$ and $*$ denotes the convolution, $\|\cdot\|_2^2$ represents the squared $L^2$ norm, $\delta$ is the Dirac distribution. The Lagrangian multiplier ($\lambda$) has been applied to convert the constrained optimization problem (Eq. 2) into an unconstrained optimization problem. The unconstrained optimization problem can be expressed as [8]

$$L(\{s_k\}, \{\omega_k\},) := \alpha \sum_k \left\| \partial t \left[ \left( \delta(t) + \frac{j}{\pi t} \right) * s_k(t) \right] e^{-j\omega_k t} \right\|_2^2$$

$$+\left\| r(t) - \sum_k s_k(t) \right\|_2^2 + \left\langle \lambda(t), r(t) - \sum_k s_k(t) \right\rangle \tag{2}$$

In Eq. (3), parameter $\alpha$ is the penalty factor. The estimated narrow-band component and corresponding center frequency during $n+1$ iteration can be computed as follows [9]:

$$\hat{s}_k^{n+1}(\omega) = \frac{\hat{r}(\omega) - \sum_{i \neq k} \hat{s}_i(\omega) + \frac{\hat{\lambda}(\omega)}{2}}{1 + 2\alpha(\omega - \omega_k)^2} \tag{3}$$

$$\omega_k^{n+1} = \frac{\int_0^\infty \omega \left|\hat{s}_k(\omega)\right|^2 d\omega}{\int_0^\infty \left|\hat{s}_k(\omega)\right|^2 d\omega} \tag{4}$$

In Eqs. (4) and (5), $\hat{r}(\omega)$, $\hat{s}_k(\omega)$, $\hat{s}_k^{n+1}(\omega)$ and $\hat{\lambda}(\omega)$ represent the Fourier transform of $r(t)$, $s_k(t)$, $s_k^{n+1}(t)$ and $\lambda(t)$, respectively. The update in $\lambda$ can be expressed by following expression:

$$\hat{\lambda}^{n+1}(\omega) = \hat{\lambda}^n(\omega) + \tau \left[ r(\omega) - \sum_k \hat{s}_k^{n+1}(\omega) \right] \tag{5}$$

where $n$ is the number of iterations. More details about the methods can be found in Ref. [8].

In VMD method, the tolerance of convergence (tol) parameter is useful for controlling the relative errors corresponding to the estimated narrow-band components. Interestingly, the expression of $\hat{s}_k^{n+1}(\omega)$ in Eq. (4) contains wiener filter structure for denoising. The $\hat{s}_k^{n+1}[n]$ can be computed from the real part of the inverse Fourier transform applied on $\hat{s}_k^{n+1}(\omega)$. The advantages of VMD technique are as follows:

(1) The relative error is largely independent of the harmonic's frequency and is controlled by tolerance level.
(2) The VMD achieves good tones separation.

## 2.2 *Instantaneous Frequency*

It describes the concept of a time varying parameter defining the location of the signal's spectral peak with respect to time. Theoretically, for a signal under analysis, it is analogous to the frequency of a sinusoidal wave which locally fits the signal. Moreover, it can be expressed only for mono-component signals, having only one frequency or a narrow range of frequencies varying as a function of time. For a frequency modulated vibration signal $x(t)$, with instantaneous frequency $f_i(t)$ can be expressed as:

$$x(t) = a\cos\left[\int_0^t 2\pi f_i(t)\mathrm{d}t + \theta\right] \tag{6}$$

The whole argument of cosine function can be termed as $\emptyset(t)$; therefore, the IF will be calculated as:

$$f_i(t) = \frac{1}{2\pi}\frac{\mathrm{d}}{\mathrm{d}t}\emptyset(t) \tag{7}$$

IF provides a measure of the frequency domain signal energy concentration as a function of time. This measure equals the IF of the signal; this property explains the importance of the IF in signal recognition, tracking, estimation and modeling.

## 2.3 Statistical Condition Indicators

The brief descriptions of the CIs used in this study are mentioned as follows:

(1) Root mean square (RMS)—It is defined as the square root of the mean of the sum of the squares of the signal samples. It reflects the vibration amplitude and energy of the signal in time domain [2]

$$\mathrm{RMS} = \sqrt{\frac{1}{N}\sum_{i=1}^{N}(x_i)^2} \tag{8}$$

(2) Kurtosis—It is used to reflect the presence of impulses in the signal. It is the fourth-order normalized moment of a given signal and is given by [2]

$$k = \frac{N\sum_{i=1}^{N}(x_i - \overline{x})^4}{\left(\sum_{i=1}^{N}(x_i - \overline{x})^2\right)^2} \tag{9}$$

where $x$ is the original sample $i$ is the number of samples. A signal consisting exclusively of Gaussian distributed data will have a kurtosis of approximately 3. The increase in the kurtosis value highlights peakedness present in the distribution of the signal.

# 3 Experimental Investigation

## *3.1 Test Setup*

To detect the fault in the gearbox as well as to evaluate the performance of signal processing-based fault detection approach, the experimental analysis has been carried out. The vibration signals were recorded from an arrangement made on machinery fault simulator (MFS) as shown in Fig. 1. To make the test-setup funtionally a motor-drive-brake kind of, the test-bench used a 2.237 kW, 3 phase, 0–3000 rpm AC drive with motor controller for variable speed operation along with power transmitting belt drive, and a bevel gearbox. Further, the test rig is equipped with optical tachometers: to count the speed for the main shaft as well as the gearbox input speed. The bevel gearbox was driven by the belt drive from the main shaft. A magnetic spring-loaded brake was used to load the gear wheel with a maximum torque of 1.129 Nm was applied at the full-load condition. An accelerometer was mounted on the top of the bevel gearbox to acquire vibration signals (see Fig. 1). The severity of the bevel gear faults has been illustrated in Fig. 2; the following severity of faults has been considered to examine the proposed approach in this study. The details of gears are listed in Table 1. The faults were seeded into the pinion tooth by using wire EDM. For lubrication, SAE 40 oil was used.

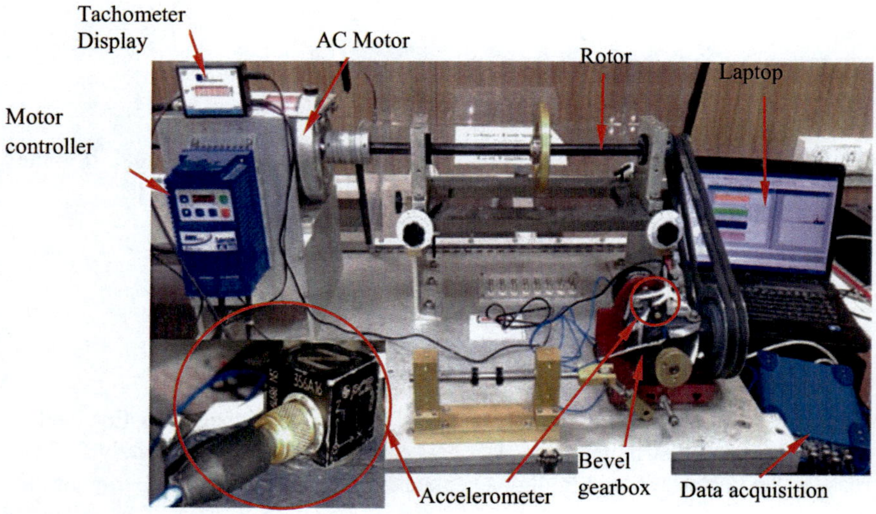


**Fig. 1** Experimental test setup with zoomed view of accelerometer

**Fig. 2** Different gear health: **a** healthy gear, **b** chipped tooth gear, **c** missing tooth gear

**Table 1** Technical specification of the gearbox

| S. No. | Parameter | Value | S. No. | Parameter | Value |
|---|---|---|---|---|---|
| 1 | Gear ratio | 1.5:1 | 6 | Number of teeth in gear | 27 |
| 2 | Pitch angle (gear) | 56° 19′ | 7 | Module | 2 mm |
| 3 | Pitch angle (pinion) | 33° 41′ | 8 | Pitch diameter (gear) | 42.8625 mm |
| 4 | Pressure angle | 20° | 9 | Pitch diameter (pinion) | 28.575 mm |
| 5 | Number of teeth in pinion | 18 | 10 | Material (gear and pinion) | Forged steel |

### *3.2 Test Conditions and Data Acquisition Settings*

The prime object of the proposed work deals with the detection of bevel gear faults using VMD and IF-based signal processing approach, thereby identifying the fault symptoms present in the vibration signals with respect to fault severity. Vibration data was acquired for all he aforementioned bevel gear health conditions. The vibration signals were acquired at a rate of 6.4 kHz and 5 datasets were acquired for each arrangement. The speed of the motor was fluctuated from 950 to 1150 rpm sinusoidally; but using belt drive having a ratio of 2.55, it reduced the speed of bevel gear to varying in a range of 372–450 rpm. This makes the angular frequency at input of bevel gear to fluctuate from 6.20 Hz to 7.516 Hz accordingly. The speed fluctuation was provided to the motor by motor controller via motor control setup available in the laptop used for data acquisition.

## 4 Result and Discussion

The vibration signals were acquired from the bevel gearbox using a triaxial accelerometer, i.e., all $X$, $Y$ and $Z$ direction signals were recorded corresponding to radial and axial vibrations. Out of all the three signals, vibration signals of $Y$ direction (radial) were found dominating and have been used for further investigations. The time-domain vibration signals for all the faults and their FFTs are shown in Fig. 3. The different peaks were noted in the FFT of the signals corresponding to pinion frequency, gear mesh frequency and its harmonics. However, sidebands in the FFT around the characteristic peaks were difficult to notice under the effect of varying speed conditions. Moreover, it was difficult to differentiate the time-domain

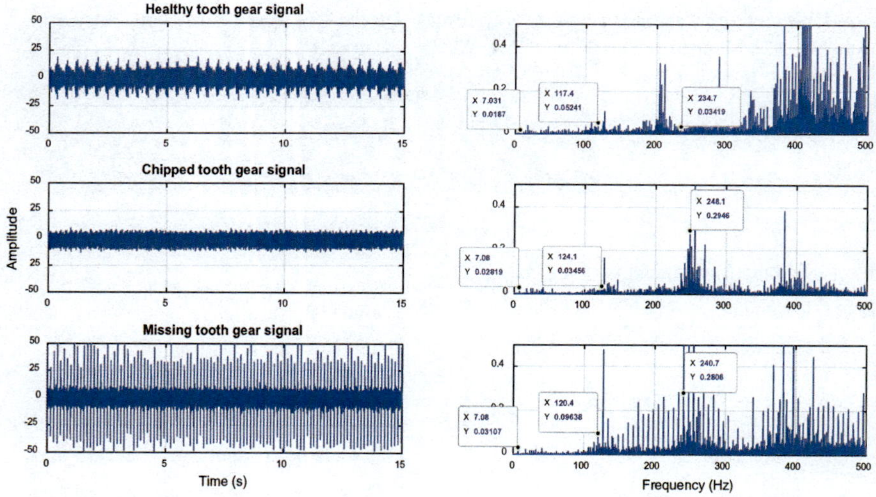


**Fig. 3** Acquired vibration signals and its FFT

vibration signals for healthy gear and chipped tooth gear. It is noticed form the time-domain vibration signals that the effect of modulation is very extensive as caused by both the varying speed and gear tooth profiles. Therefore, it is required to demodulate the gear vibration signals using adaptive decomposition. VMD is capable to decompose the signal adaptively and hence used to decompose the signals into six MFs as shown in Figs. 4, 5 and 6.

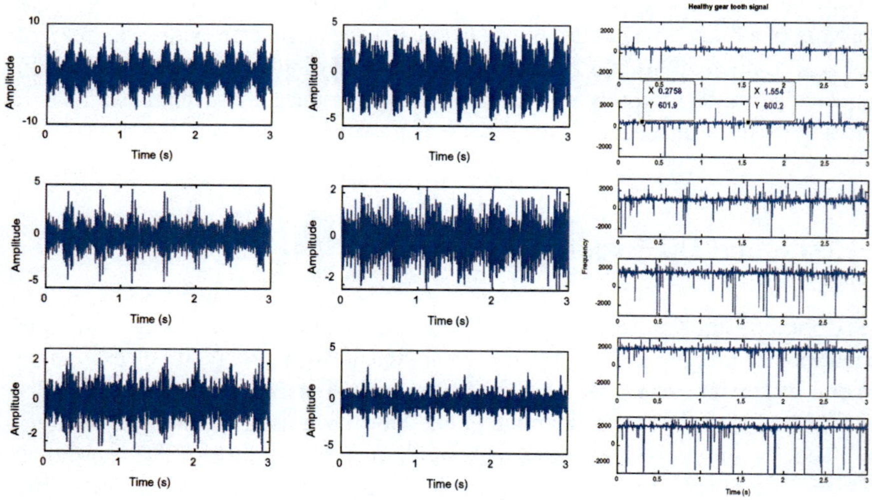


**Fig. 4** MF extracted by VMD for healthy tooth gear signal and its IF plots

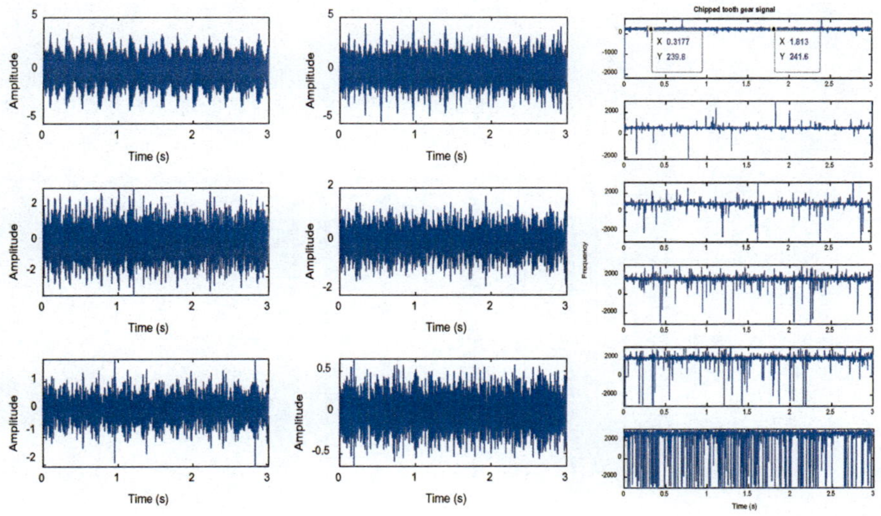


**Fig. 5** MF extracted by VMD for chipped tooth gear signal and its IF plots

The decomposed MFs exhibited the transients present in the gearbox vibration signals. For example, MF2 and MF3 for healthy gear, MF1 and MF2 for chipped tooth gear and MF1 and MF3 for missing tooth gear, the remaining other were observed to affected by amplitude modulation and noise. To identify the most sensitive MFs among all, the IF of all the MFs for all the gears was calculated as shown in Figs. 4, 5 and 6. The IFs were observed to be increased for the respective MFs. For healthy gear vibration signals, the IF of the second MF was noted as 600 Hz approximately, which corresponds to the fifth harmonic of the GMF (120 Hz $\approx$ 7.03 Hz $\times$17). Similarly, the IFs for chipped tooth and missing tooth were noted for first MF as 240 Hz approximately, equivalent to second harmonic. The calculation of IFs for the MFs and the transients exhibited by the same MFs were in agreement to each other. Thus, it can be said that the proposed signal processing-based approach works well for the fault detection of the bevel gears under varying speed conditions.

To further validate the results of the signal processing technique, the envelop spectrum was developed of the selected sensitive MF for all the aforementioned gear health conditions. The enveloped spectrum for healthy gear, chipped tooth gear and missing tooth gear was plotted subsequently as shown in Fig. 7. It is clear from the figure that the sidebands are clearly visible and can be easily distinguished. The envelop spectrum highlights the gear rotation frequency, pinion rotating frequency, GMF and their combinations. It can also be noted that with increasing fault severity, the number of sidebands is increasing, validating the symptoms of present of fault. Further, from Table 2, the performance of the RMS and kurtosis can also be noted. It was found that with proposed signal processing approach, the response of RMS and kurtosis was increasing with respect to fault severity. Thus, for a limited speed

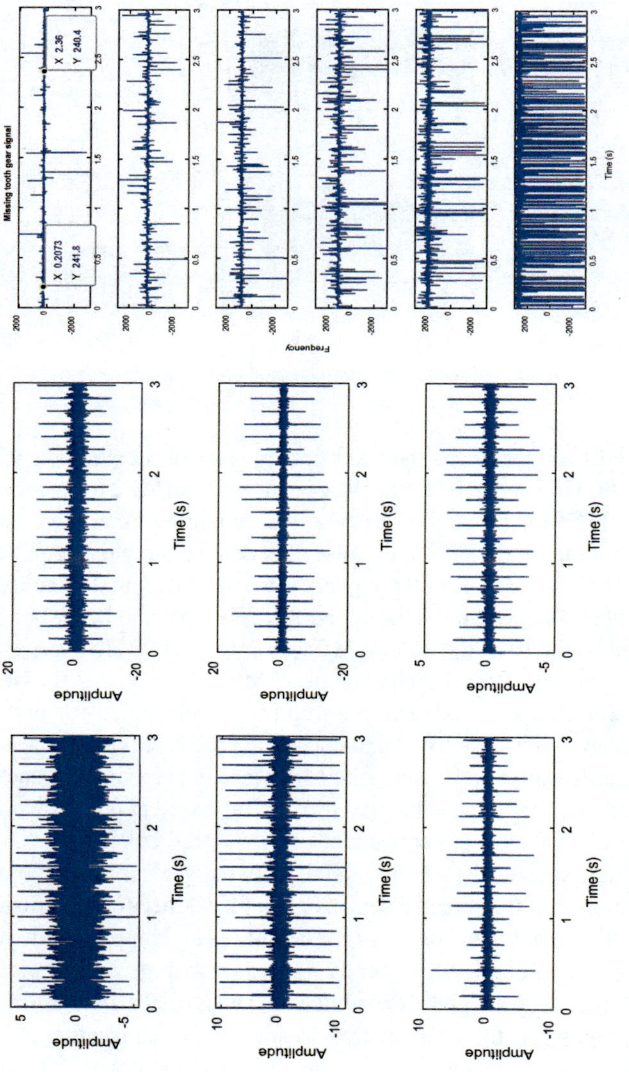


**Fig. 6** MF extracted by VMD for missing tooth gear signal and its IF plots

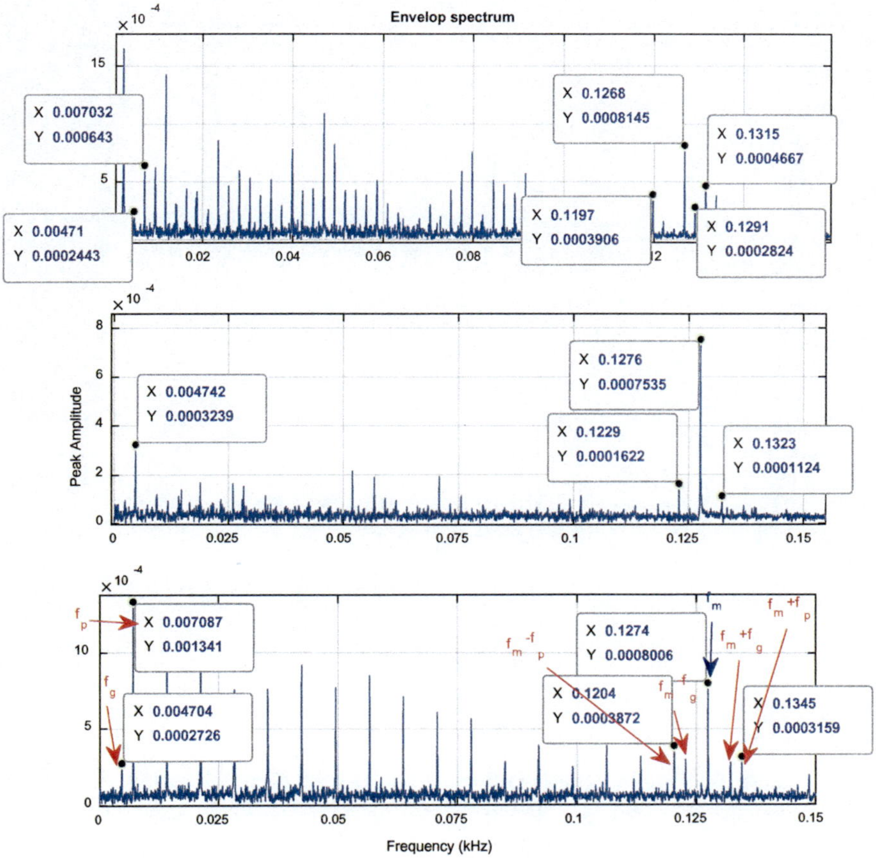


**Fig. 7** Envelop spectrum developed for healthy gear tooth (top), chipped tooth gear (middle) and missing tooth gear (bottom) vibration signals

variation, the proposed VMD and IF-based signal processing approach can be used for fault detection of bevel gears.

## 5 Conclusion

This manuscript reports a novel signal processing-based approach for fault detection of bevel gearbox. VMD decomposes the recorded vibration signal acquired from the test bench with different fault severity under limited variation of speed into different six mono-component MFs. VMD exhibited efficient separation of the different modes that corresponds to the effect of impulses due to fault severity. The IF was calculated for all the extracted MFs subsequently to identify the most sensitive MF. The selected MF was then examined using FFT and statistical condition indicators viz. RMS

**Table 2** Performance of statistical condition indicators

| Signals | Raw vibration signal | | Remark w.r.t. ref. value (increasing↑/decreasing↓) | Mode Functions | | Remark w.r.t. ref. value (increasing↑/decreasing↓) |
|---|---|---|---|---|---|---|
| | RMS | Kurtosis | | RMS | Kurtosis | |
| Healthy tooth gear | 3.4489 | 4.4543 | | 1.2363 | 2.8847 | |
| Chipped tooth gear | 2.4615 | 4.3145 | ↓ | 1.4860 | 4.9904 | ↑ |
| Missing tooth gear | 3.9896 | 40.9035 | ↑ | 1.8231 | 11.6293 | ↑ |

and kurtosis. Both RMS and kurtosis when used on raw vibration signals fail to characterize the level of fault present in signal. The FFT of the selected MF and statistical condition indicators validated the symptoms of the gear faults.

## References


1. X. Liang, M.J. Zuo, Z. Feng, Dynamic modeling of gearbox faults: a review. Mech. Syst. Sig. Process. **98**, 852–876 (2018)
2. V. Sharma, A. Parey, A review of gear fault diagnosis using various condition indicators. Procedia Eng. **144**, 253–263 (2016)
3. V. Sharma, A. Parey, Gear crack detection using modified TSA and proposed fault indicators for fluctuating speed conditions. Measurement **90**, 560–575 (2016)
4. V. Sharma, A. Parey, Gearbox fault diagnosis using RMS based probability density function and entropy measures for fluctuating speed conditions. Struct. Health Monit. **16**(6), 682–695 (2017)
5. M.S. Kan, A.C. Tan, J. Mathew, A review on prognostic techniques for non-stationary and non-linear rotating systems. Mech. Syst. Sig. Process. **62**, 1–20 (2015)
6. L. Cohen, *Time-Frequency Analysis*, vol. 778 (Prentice Hall, 1995)
7. N.E. Huang, Z. Shen, S.R. Long, M.C. Wu, H.H. Shih, Q. Zheng, H.H. Liu, The empirical mode decomposition and the Hilbert spectrum for nonlinear and nonstationary time series analysis, Proc. Roy. Soc. London A Math. Phys. Eng. Sci. **454**, 903–995 (1971) (1998)
8. K. Dragomiretskiy, D. Zosso, Variational mode decomposition. IEEE Trans. Signal Process. **62**(3), 531–544 (2014)
9. V. Sharma, A. Parey, Extraction of weak fault transients using variational mode decomposition for fault diagnosis of gearbox under varying speed. Eng. Fail. Anal. **107**, 104204 (2020)
10. X. Chen, Z. Feng, Iterative generalized time–frequency reassignment for planetary gearbox fault diagnosis under non-stationary conditions. Mech. Syst. Sig. Process. **80**, 429–444 (2016)
11. C. Junsheng, Y. Dejie, Y. Yu, Research on the intrinsic mode function (IMF) criterion in EMD method. Mech. Syst. Sig. Process. **20**(4), 817–824 (2006)
12. X. An, H. Zeng, C. Li, Envelope demodulation based on variational mode decomposition for gear fault diagnosis. Proc. Instit. Mech. Eng. Part E: J. Process Mech. Eng. **231**(4), 864–870 (2017)
13. A. Parey, R.B. Pachori, Variable cosine windowing of intrinsic mode functions: application to gear fault diagnosis. Measurement **45**(3), 415–426 (2012)
14. V. Sharma, A. Parey, Frequency domain averaging based experimental evaluation of gear fault without tachometer for fluctuating speed conditions. Mech. Syst. Sig. Process. **85**, 278–295 (2017)
15. Y. Deng, W. Wang, C. Qian, Z. Wang, D. Dai, Boundary-processing-technique in EMD method and Hilbert transform. Chin. Sci. Bull. **46**(11), 954–960 (2001)
16. C. Zhong, Z. Shixiong, Analysis on end effects of EMD method. J. Data Acquisition Process. **1**, 025 (2003)
17. T. Wang, M. Zhang, Q. Yu, H. Zhang, Comparing the applications of EMD and EEMD on time–frequency analysis of seismic signal. J. Appl. Geophys. **83**, 29–34 (2012)
18. Q. Jiang, B.W. Suter, Instantaneous frequency estimation based on synchrosqueezing wavelet transform. Signal Process. **138**, 167–181 (2017)
19. R.U. Maheswari, R. Umamaheswari, Application of Wavelet synchrosqueezing transform for wind turbine gearbox fault diagnosis, in *2016 IEEE International Conference on Computational Intelligence and Computing Research (ICCIC)* (IEEE, Dec. 2016), pp. 1–4

# CFD Analysis of the Performance of an H-Darrieus Wind Turbine Having Cavity Blades

**Y. Kumar, A. R. Sengupta, Agnimitra Biswas, H. M. S. M. Mazarbhuiya, and Rajat Gupta**

**Abstract** Straight-blade vertical axis wind turbines (VAWT) have the ability to exhibit self-starting features and improved aerodynamic performance in the built environment for significant wind speed conditions. But in low wind speeds, such turbines realize several constraints for improved performance in terms of various parameters like blade design, blade shape, solidity, tip speed ratio (TSR), thickness-to-chord ratio and others. Therefore, the motivation behind the present work is the need for performance improvement of H-Darrieus VAWT in the built environment, which has characteristically low wind speed. Cavity shapes on the airfoil surface might cause local flow acceleration leading to suppression of boundary layer separation, which might enhance VAWT's aerodynamic performance. In this paper, an attempt is made to investigate the effect of circular cavity shape on VAWT's aerodynamic performance. Cavities have been formed on NACA 0021 airfoil based H-Darrieus VAWT. A CFD study is carried out to understand the inherent flow physics of the turbine. Results showed that there is a significant improvement in starting characteristic of the turbine at wind speeds 5 and 6 m/s having cavity on pressure side. The optimal tip speed ratio for the H-Darrieus turbine has been obtained as 1.3 for which the considered NACA 0021 blade turbine shows maximum power coefficient of 0.16.

**Keywords** H-Darrieus wind turbine · VAWT · NACA · Cavity blade · CFD analysis

Y. Kumar · A. Biswas · H. M. S. M. Mazarbhuiya
Department of Mechanical Engineering, National Institute of Technology Silchar, Silchar, Assam 788010, India

A. R. Sengupta (✉)
Department of Mechanical Engineering, JIS College of Engineering, Kalyani, West Bengal, India
e-mail: analsengupta88@gmail.com

R. Gupta
National Institute of Technology Mizoram, Aizawl, Mizoram 796012, India



K. M. Pandey et al. (eds.), *Recent Advances in Mechanical Engineering*, Lecture Notes in Mechanical Engineering, https://doi.org/10.1007/978-981-15-7711-6_70

# 1 Introduction

Vertical axis wind turbines (VAWTs) have esthetical appeal as well as have some important advantages compared to horizontal axis wind turbine for, e.g. less noisy, less prone to structural failure, omni-directional, low starting torque, less costly and others. Its deployment is much easier in the built and remote places to harness wind energy in a sustainable way. But VAWT's performance efficiency is low in the built environment that has low wind speed. Entrepreneurs and scientists have attempted to improve its performance by different manners for, e.g. by optimizing the rotor aspect ratio and rotor solidity, introducing fixed or active blade pitching, increasing the number of blades, using augmenters, modifying the airfoil profile and others. Below are some of the literatures for VAWT's performance improvement from which the research gap can be identified.

Sabaeifard et al. [1] performed CFD simulations to observe the performance of NACA 0018 and DUW200 airfoil blade H-Darrieus VAWT rotor under different rotor solidities and operating tip speed ratios, only to report that the turbine's performance is increased for the solidity variation from 0.3 to 0.5. Sagharichi et al. [2] performed CFD simulations of VAWT by taking solidities between 0.2 and 0.8 with fixed and variable blade pitch conditions and having two, three and four blades. It was showed that high solidity and variable pitch rotor is preferred for initial self-staring torque requirement. Lee and Lim [3] reported that at low tip speed ratio, high solidity (0.8) rotor shows the highest torque and power coefficient. While at high tip speed ratio 2.4 or higher, the rotor with low solidity (0.4) shows the highest torque and power coefficient ($C$p). Subramanian et al. [4] studied numerically the effects of solidity and four different airfoil profiles, namely NACA 0012, NACA 0015, NACA 0030 and AIR 001. It was observed that NACA 0030, which is a thicker airfoil, performs better at low TSR and thinner airfoil NACA 0012 has a wider operating TSR range. At low TSR, higher solidity turbine performs better. Rezaeiha et al. [5] performed high fidelity CFD simulations to study the effect of solidity and number of blades under a wide range of TSR of 1.5–5.5. They found that due to low mean wind speed in urban areas, a low solidity turbine with moderate to high TSR is optimal. Castelli et al. [6] studied numerically the effect of number of blades of a straight-bladed NACA 0025 H-Darrieus-type rotor. They concluded that increasing the number of blades at low TSR and decreasing the same at high TSR lead to higher performance. The optimum number of blades was found to be three. Mabrouk and Hami [7] performed simulations to study the effect of number of blades on the global performance of H-Darrieus rotor. They also concluded that three bladed rotor generates more aerodynamic torque.

In addition to the improvement drives adopted by various researchers as discussed above, VAWT's performance can also be improved by varying the airfoil profile and its shape, which is a passive way of increasing its performance. Mohamed et al. [8] studied computationally 25 number of airfoil-based Darrieus rotors for several values of tip speed ratio to obtain some of the best profiles for its improved performance, for example, LS(1)-0413 airfoil for improved $C$p, NACA63-415 airfoil for wider range of operating TSR. Mohamed [9] again studied VAWT's performance with low rotor solidity (0.1) by considering symmetric and non-symmetric blade profiles. S-1046 airfoil-based rotor led to a maximum gain in power which was 26.83% among all the airfoils. It was also found that symmetrical airfoil has wider operating range than non-symmetric airfoil. Sengupta et al. [10] investigated experimentally the starting capability and dynamic performance by considering three types of blades viz. unsymmetrical S815 and EN0005 blades and symmetrical NACA 0018 blade. They concluded that unsymmetrical blade profile with high solidity improves starting characteristics of VAWT and also its dynamic performance. Sengupta et al. [11] again conducted CFD analysis of H-Darrieus VAWT and compared two different unsymmetrical airfoil blade profiles to obtain design information of blade camber and blade curvature at different low wind speeds 4–8 m/s. They found that S815 blade with more rounded curvature and thicker blade profile has a better influence on the turbine performance.

From the above literature, it can be inferred that VAWT's performance can be improved by working on its airfoil blade profile and its shape. In the present work, NACA 0021 airfoil profile has been considered for the blades of VAWT. In addition to this, a cavity shape is considered on inside of the airfoil blade surfaces. The performances of H-Darrieus VAWT with such airfoil profile blade is investigated in details numerically using commercial CFD software. From this present analysis, important performance insights are obtained and discussed here.

## 2 Model Geometry and Computational Modelling

The present 2D CFD analysis of the NACA 0021 blade H-Darrieus turbine is performed in ANSYS 14.0 package. After generating the turbine model, three control circles of diameter 160 mm are drawn outside each blade by considering centre as the centre of blades. After generating control circles, a large circle of diameter 800 mm named as the hub is generated. A rectangular domain of dimension 40,000 and 30,000 mm is generated to enclose the turbine and hub. The diagram of the required geometry is shown in Fig. 1. A circular cavity shape is formed on the airfoil surface whose centre is at a distance of 1/4th chord length from the leading edge. The diameter of the cavity is 0.08 times chord length and its top edge is at 0.04 times chord distance below the centreline of the blade as shown in Fig. 1c.

In the mesh generation process, all triangular method is used to mesh the whole body. The model is meshed at the relevance centre as 100 and proximity and curvature with curvature angle 180° with a minimum size of the cell as 0.01 mm. Edge sizing is given to blades, control circles and hub to get required mesh size. More and smaller elements are formed near the edge of blades to capture small changes in any properties. The unsteady flow is simulated using a sliding mesh model. The mesh around the airfoils, inside the hub and inside control circles is shown in Fig. 2. Here, transient-type analysis is selected. Realizable $k - \varepsilon$ turbulence model is chosen due to its fast CFD simulations and is recommended for rotating bodies. In $k - \varepsilon$ turbulence

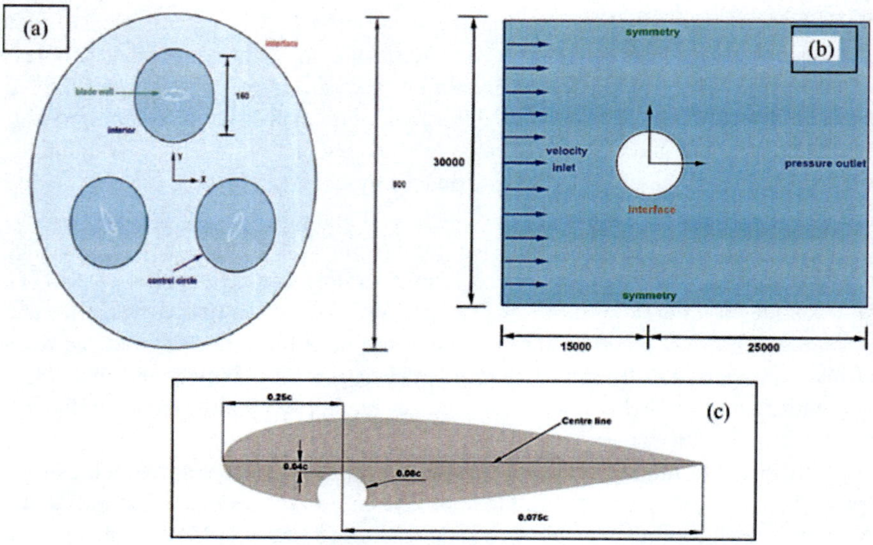


**Fig. 1** Dimensions (all dimensions are in mm) **a** hub and control circle, **b** rectangular domain, **c** size and position of the cavity

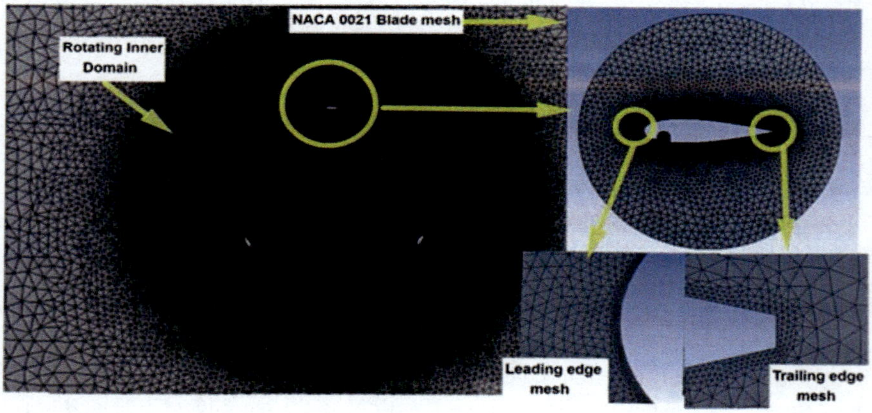


**Fig. 2** Mesh structure in the hub and near the airfoil section

model, $k$ is turbulent kinetic energy contained in fluctuations and ε is the turbulent dissipation rate at which turbulent kinetic energy is dissipated. Studies show that the $k-\varepsilon$ realizable model gives the best results compared to all other $k-\varepsilon$ models. This realizable model satisfies some constraints on normal stresses and consistent with the physics of turbulent flows [9, 12, 13]. Due to the complex blade designs of the rotors, unstructured triangular meshing has been done which is capable of showing good results for CFD simulation [10, 11].

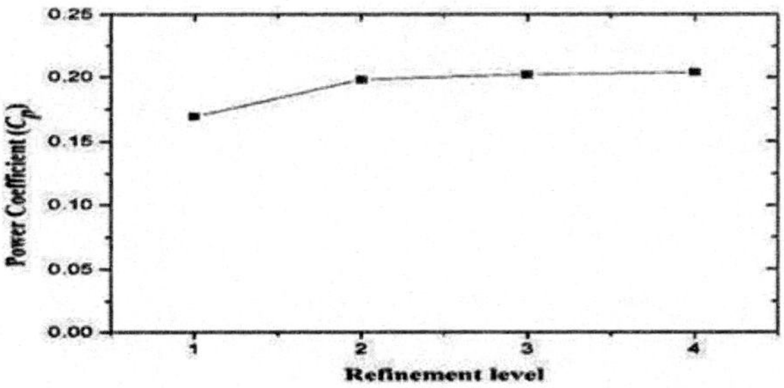


**Fig. 3** Mesh-independent analysis ($C_p$ vs. refinement level)

To optimize the analysis by controlling the variation and to achieve a more precise solution, mesh independence analysis has been done on the model. In the present work, after a long iteration by gradually decreasing the mesh size, i.e. by increasing the number of elements of mesh, the power coefficient values have been found nearly constant for the different number of elements. Thus, grid independence limit (GIL) mesh is achieved by computing several refinement levels.

Figure 3 shows graphically the mesh-independent analysis for the geometry. From the graph, it has been seen that the power coefficient is increasing gradually; finally, a constant value of power coefficient is observed at a particular element number.

## 3 Simulation Results and Discussion

Figure 4 shows the pressure contour distributions of the NACA 0021 blade H-Darrieus rotor with inside cavity at various azimuthal positions. As can be seen at 0° azimuthal position in Fig. 4a, a high pressure zone is formed only at the tip of the leading edge while very low pressure zone created on the suction side near the cavity, which is why the aerodynamic performance drops. As the advancing blade rotates and reaches 60° azimuthal position, strong pressure zone is created throughout the pressure side of the blade and also creates a large difference of pressure across the blade resulting into higher moment coefficient. This is also the reason why lift values are increased after this azimuthal position, and before 60°, lift is low due to lower pressure on suction side. However, in Fig. 4a, c, it is seen that, from 90° and 120° positions, performance is deteriorated from the aspect of moment as large recirculation zones are created near the advancing blade trailing edge. Again, lift is also found to drop after these azimuthal positions. Again at 180° position in Fig. 4b, moment coefficient again rises to a higher value, which is the result of strong pressure distribution right at the cavity on the blade suction side. In Fig. 4b, it can be noticed that, in second half of the cycle for 300° position, the pressure distribution on the leading edge of the returning blade increases, thereby increasing the moment coefficient value at this position.

The velocity streamlines along with velocity distributions are shown in Fig. 5 for the H-Darrieus rotor. These contours also corroborate with the findings of pressure

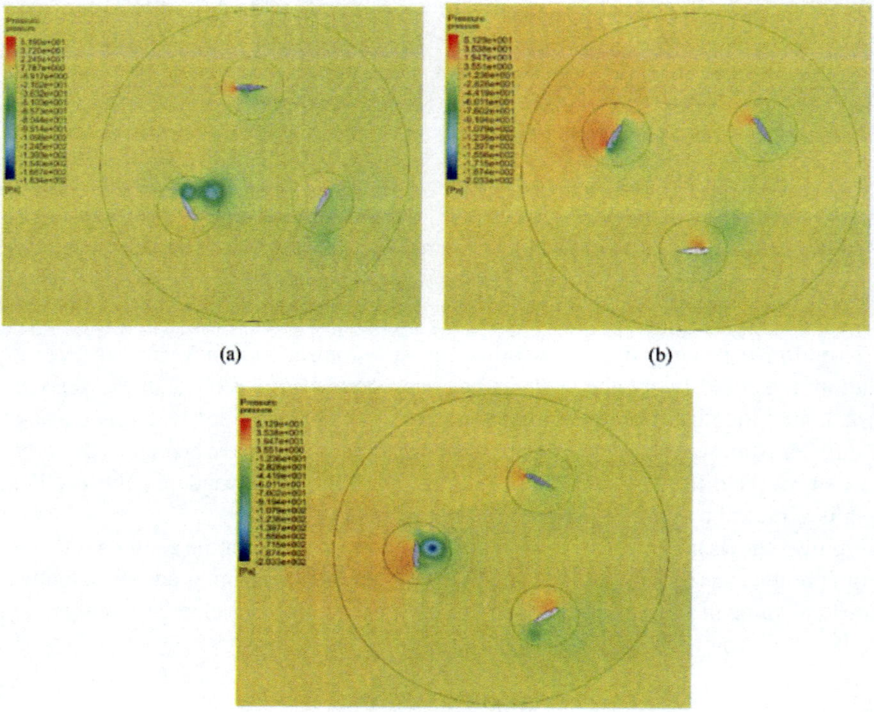

(a) (b)

(c)

**Fig. 4** Pressure contour plots at TSR = 1.3 **a** at azimuthal angle 0°, 120° and 240°, **b** at azimuthal angle 60°, 180° and 300° and **c** at azimuthal angle 90°, 210° and 330°

contour plots. In Fig. 5a, there is not very significant velocity difference between the suction and pressure side of the blade at 0° azimuthal position. But as the blade advances in Fig. 5b, at 60° azimuthal position, there is higher velocity on the advancing blade leading edge resulting into higher moment coefficient value. Due to this also, lift is found to increase after this angular position. However at 90° azimuthal position in Fig. 5c, this augmentation of velocity begins to subside as separation begins on the blade's suction side thereby pulling down the aerodynamic performance of the turbine.

CFD simulations are performed for H-Darrieus VAWT at various tip speed ratios 0.8, 1, 1.3, 1.5, 1.6 and 1.7 for 5 and 6 m/s wind speeds. Power coefficients are calculated for corresponding TSRs from CFD results and $C_{\mathrm{p}}$ versus TSR curves are plotted from the results obtained from computational simulations as shown in Fig. 6. It can be seen that the maximum power coefficient is obtained as 0.16 at TSR 1.3. The airfoil with cavity has a large gap on the leading edge, so the air was trapped within a larger surface. This resulted in further reduced flow speed and thereby increased pressure which enhanced the lift force employed to the blade, as well as improvement in the turbine's performance. Further, this NACA 0021 blade rotor shows higher $C_{\mathrm{p}}$

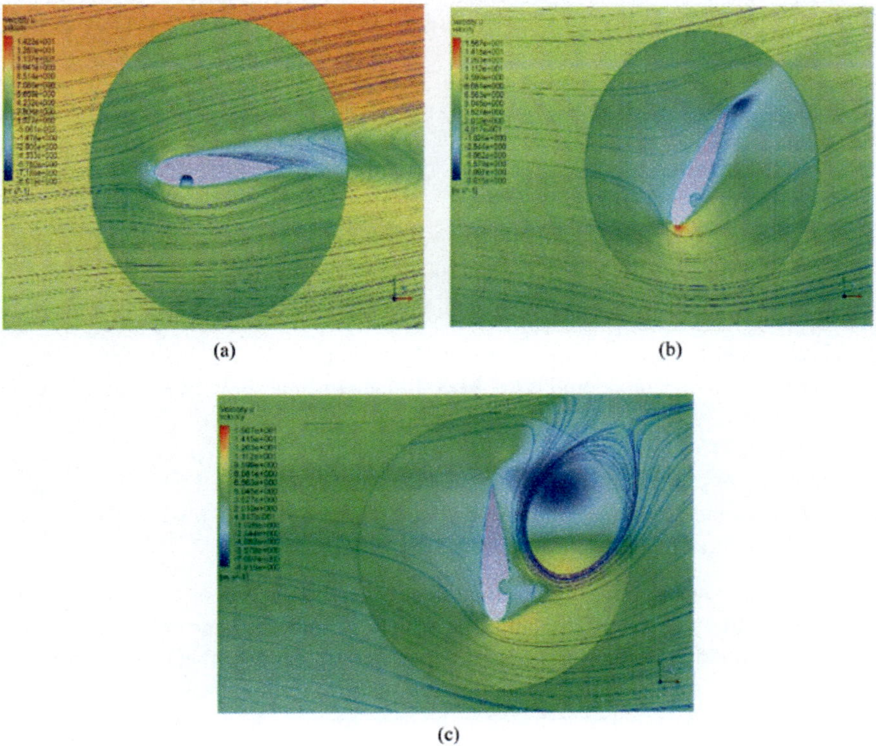

Fig. 5 Velocity contour plots near blades with streamlines at TSR = 1.3 for **a** at azimuthal angle 0°, **b** at azimuthal angle 60° and **c** at azimuthal angle 90°

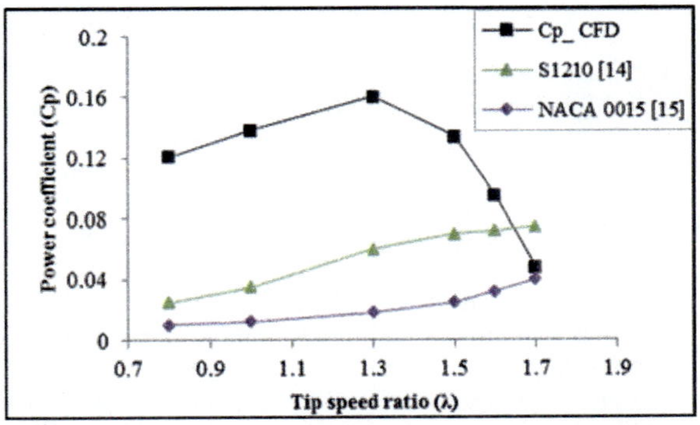


**Fig. 6** Validation of CFD results of NACA 0021 cavity blade with other blade shapes

value compared to some other published results of S1210 [14] and NACA0015 [15] blade rotors for the considered TSR range. The reason for such high difference in $C_p$ can be described as follows: in the circular cavity due to the absence of corners, the vortices get trapped and gradually vanish. The section with cavity increases the performance of the turbine in wide range of tip velocities and improves the torque of the turbine.

## 4 Conclusions

The performances of H-Darrieus VAWT with NACA 0021 airfoil profile blade along with a circular cavity formed on its surface are investigated numerically using commercial CFD software. From the study, the following conclusions are drawn

(a) The optimal tip speed ratio for the H-Darrieus turbines has been obtained as 1.3 for which the considered NACA 0021 blade turbine shows maximum power coefficient.
(b) NACA 0021 blade H-Darrieus VAWT has overall higher power coefficient compared to other published results of S1210 and NACA 0015 for the considered TSR range. The maximum power coefficient of 0.16 is obtained at a TSR of 1.3 for wind velocity 6 m/s.

The present investigation leads to the effect of cavity airfoils in low wind speed conditions to study their aerodynamic performance and self-starting features. In future, three-dimensional analysis can be performed to compare their performances considering struts, tip loss aspects and others with the current two-dimensional analysis. Again such cavity-based airfoils can be used for co-axial H-Savonius rotor to investigate its overall performances.

## References

1. P. Sabaeifard, H. Razzaghi, A. Forouzandeh, Determination of vertical axis wind turbines optimal configuration through CFD simulations. Int. Conf. Future Environ. Energy **28**, 109–113 (2012)
2. A. Sagharichi, M. Zamani, Am. Ghasemi, Effect of solidity on the performance of variable-pitch vertical axis wind turbine. Energy **161**, 753–775 (2018)
3. Y.-T. Lee, H.-C. Lim, Numerical study of the aerodynamic performance of a 500 W Darrieus-type vertical-axis wind turbine. Renew. Energy **83**, 407–415 (2015)
4. S. Abhishek Subramanian, A. Yogesh, H. Sivanandan, A. Giri, M. Vasudevan, V. Mugundhan, R.K. Velamati, Effect of airfoil and solidity on performance of small scale vertical axis wind turbine using three dimensional CFD model. Energy **133**, 179–190 (2017)
5. A. Rezaeiha, H. Montazeri, B. Blocken, Towards optimal aerodynamic design of vertical axis wind turbines: impact of solidity and number of blades. Energy **165**, 1129–1148 (2018)
6. M.R. Castelli, S.D. Betta, E. Benini, Effect of Blade Number on a Straight-Bladed Vertical-Axis Darreius Wind Turbine. Int. J. Aerosp. Mech. Eng. **6**(1), 68–74 (2012)
7. I.B. Mabrouk, A.E. Hami, Effect of number of blades on the dynamic behavior of a darrieus turbine geared transmission system. Mech. Syst. Signal Process. **121**, 562–578 (2019)
8. M.H. Mohamed, A.M. Ali, A. Hafiz, CFD analysis for H-rotor Darrieus turbine as a low speed wind energy converter. Eng. Sci. Technol. Int. J. **18**(1), 1–13 (2015)

9. H. Mohamed, Mohamed, Performance investigation of H-rotor Darrieus turbine with new airfoil shapes. Energy **47**(1), 522–530 (2012)
10. A.R. Sengupta, A. Biswas, R. Gupta, Studies of some high solidity symmetrical and unsymmetrical blade H-Darrieus rotors with respect to starting characteristics, dynamic performances and flow physics in low wind streams. Renew. Energy **93**, 536–547 (2016)
11. A.R. Sengupta, A. Biswas, R. Gupta, Comparison of low wind speed aerodynamics of unsymmetrical blade H-Darrieus rotors-blade camber and curvature signatures for performance improvement. Renew. Energy **139**, 1412–1427 (2019)
12. I. Hashem, M.H. Mohamed, Aerodynamic performance enhancements of H-rotor Darrieus wind turbine. Energy **142**, 531–545 (2018)
13. T.-H. Shih, W.W. Liou, A. Shabbir, Z. Yang, J. Zhu, A new k-$\epsilon$ eddy viscosity model for high Reynolds number turbulent flows. Comput. Fluids **24**(3), 227–238 (1995)
14. P.M. Kumar, K. Sivalingam, T.-C. Lim, S. Ramakrishna, H Wei, Strategies for enhancing the low wind speed performance of H-Darrieus wind turbine—Part 1. Clean Technol. **1**, 185–204 (2019)
15. A. Rossetti, G. Pavesi, Comparison of different numerical approaches to the study of the H-Darrieus turbines start-up. Renew. Energy **50**, 7–19 (2013)

# The Effect of Negative Hardening Coefficients on Yield Surface Evolution

**Praveen Kumar and Sivasambu Mahesh**

**Abstract** The evolution of the yield surface, as predicted by a polycrystal plasticity model of a face-centered cubic material, is studied. Grains in the model polycrystal are endowed with a classical hardening law, which accounts for interaction among the slip system through a hardening matrix. In the literature, the elements of the hardening matrix are assumed non-negative. In the present work, the effect of negative elements in the hardening matrix on the evolution of the yield function, particularly, during monotonic tensile and shear deformation, is systematically studied. In particular, it is shown that certain parametric values simulate a substantial kinematic hardening, similar to experimental observations. The greatest kinematic hardening is obtained when the latent hardening ratio of the reverse slip systems is taken to be $-1.2$.

**Keywords** Crystal-plasticity · Evolution · Hardening matrix/model · Isotropic · Kinematic hardening · Monotonic strain paths · Yield surface

## 1 Introduction

The yield surface of metals and alloys evolves during plastic deformation. Often the shape of the yield surface becomes quite complex even for relatively simple materials subjected to monotonic loading paths. For example, the initial yield surface of annealed commercially pure Al 1100 obeys the von Mises criterion to good approximation and plots as a circle centered at the origin. Figure 1a, b shows the yield surface measured by Khan et al. [1] of this material subjected to uniaxial tension and torsion, respectively. It is clear that the yield surfaces lose their von Mises character, acquire a sharp corner in the loading direction, and become blunt in the opposite direction. It is also clear that substantial kinematic hardening occurs so that the yield

P. Kumar (✉) · S. Mahesh
Department of Aerospace Engineering, Indian Institute of Technology, Madras, India
e-mail: praveenvirgo301@gmail.com

S. Mahesh
e-mail: smahesh@iitm.ac.in



K. M. Pandey et al. (eds.), *Recent Advances in Mechanical Engineering*, Lecture Notes in Mechanical Engineering, https://doi.org/10.1007/978-981-15-7711-6_71

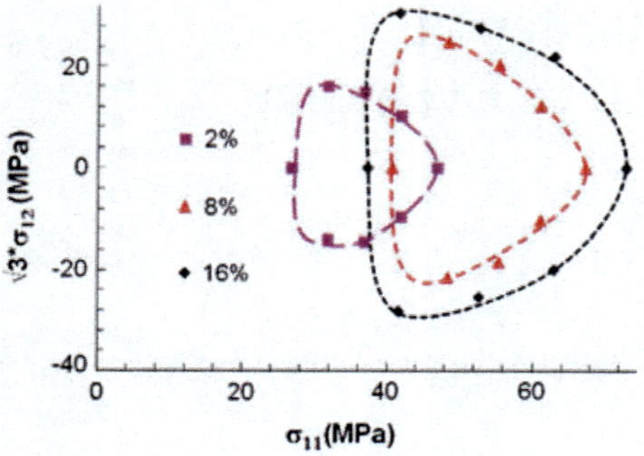


(a) Tensileloading

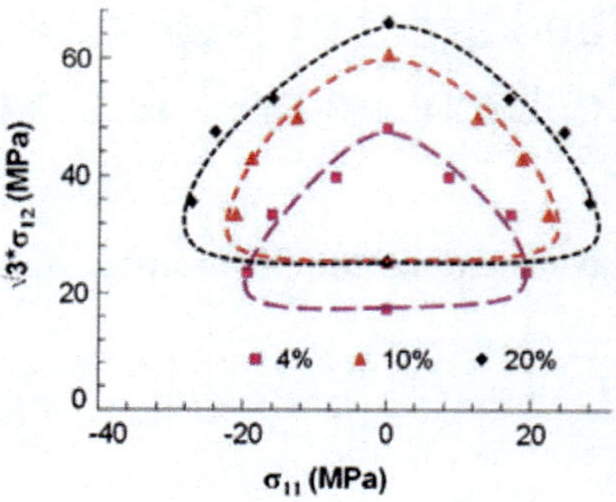


(b) Torsionalloading

**Fig. 1** Measured yield surfaces of a commercially pure Al alloy Al 1100 under **a** uniaxial tension, and **b** uniaxial shear. Reprinted from Khan et al. with permission from Elsevier Ltd.

surface after some plastic deformation does not include the origin, which represents complete unloading. These features are common in many metals and alloys and has been termed directional distortional hardening by Feigenbaum and Dafalias [2].

In order to successfully model metal forming processes numerically, it is important to accurately describe the evolution of the yield surface under arbitrary loading [3]. Various phenomenological models have been proposed recently for this purpose [2, 4, 5]. These models are, however, specific to certain materials and loading paths. Many of the more recent models also apply only to two-dimensional metal sheets.

A more general methodology to capture the evolving yield surface in both two-dimensional sheets, and three-dimensional solids is based on polycrystal plasticity [6]. In this method, the material is regarded as a polycrystalline collection of grains, each of which deforms plastically so that they collectively accommodate the imposed plastic deformation. The macroscopic yield surface of the model polycrystal can then be calculated from the state of hardening of all the grains making up the polycrystal. The particular polycrystal plasticity model adopted presently, due to Mahesh [7], enforces traction and velocity continuity between adjacent grains and sub-aggregates of grains.

The present work examines the effect of single crystal hardening on the yield surface predicted by the polycrystal plasticity model. Hardening of the grains is formulated in Sect. 2. Predictions of the yield surfaces under tension and torsion using the polycrystal plasticity approach are then described in Sect. 3. The significance of the simulation results in the context of the available experimental data is then discussed in Sect. 4.

## 2 Formulation

The Al 1100 grains of present interest have a face-centered cubic (fcc) structure and deform by $\{111\}\langle 110\rangle$ slip. Each slip system $s$ is identified by its slip direction $\mathbf{b}_s$, and slip plane normal $\boldsymbol{n}_s$. The Schmid tensor of a slip system is defined as $\boldsymbol{m}_s = (\mathbf{b}_s \otimes \mathbf{n}_s + \mathbf{n}_s \otimes \mathbf{b}_s)/2$. Every slip system is also associated with a critical resolved shear stress, $\boldsymbol{\tau}_s$. The slip rate of slip system $s$ is denoted $\dot{\gamma}_s$ and obeys Schmid's law,

$$\dot{\gamma}_s \begin{cases} = 0, \text{ if } \sigma : \boldsymbol{m}_s < \boldsymbol{\tau}_s, \\ \geq 0, \text{ if } \sigma : \boldsymbol{m}_s = \boldsymbol{\tau}_s, \end{cases} \tag{1}$$

where $\boldsymbol{\sigma}$ denotes the stress tensor in the grain, and: denotes the scalar product of two tensors. The accumulated slip in a slip system is denoted $\Gamma_s = \int \dot{\gamma}_s \mathrm{d}t$, and the total accumulated slip in the grain is denoted $\Gamma = \sum_s \Gamma_s$.

The critical resolved shear stress,$\tau_s$ evolves with plastic strain. This evolution is termed as slip system hardening. It has been modeled phenomenologically using the extended Voce law, proposed by Tome et al. [8]:

$$\dot{\tau}_s = \frac{\mathrm{d}\tau}{\mathrm{d}\Gamma} \sum_{s'=1}^{S} H_{ss'} \dot{\gamma}_{s'} \tag{2}$$

Here,$\boldsymbol{H}_{ss'}$, termed the hardening matrix [9], captures the interaction between the various slip systems, and $\tau(\Gamma)$ is taken to follow the extended Voce law of [8],

$$\tau(\Gamma) = \tau_0 + (\tau_1 + \theta_1 \Gamma) \left[ 1 - \exp\left( \frac{-\Gamma \theta_0}{\tau_1} \right) \right] \tag{3}$$

where $\tau_0, \tau_1, \theta_0,$ and $\theta_1$ are taken to be material hardening constants.

Two slip systems,$s_1$ and $s_2$,, are said to be reverse of each other if $\mathbf{n}_{s_1}.\mathbf{n}_{s_2} = \pm 1$, but $(\mathbf{n}_{s_1}.\mathbf{n}_{s_2})(\mathbf{b}_{s_1}.\mathbf{b}_{s_2}) = -1$. The reverse of a slip system $s$ is denoted by $r(s)$, so that

$$s_1 = r(s_2) \text{ and } s_2 = r(s_1)$$

Counting reverse slip systems as distinct, there are 24 $\{111\}\langle 110\rangle$ slip systems in the fcc lattice structure. In the next section, the consequence of simple choices for the elements of the hardening matrix, $H_{ss'}$, on the evolution of the yield surface in uniaxial tension and torsion, is examined.

## 3 Results

Simulations are performed on a synthetic fcc polycrystal comprised of 256 randomly oriented grains, each with volume fraction 1/256. A binary tree is constructed from these grains by pairing them at random and initializing a grain boundary between neighboring grains, oriented randomly over the unit sphere. As only qualitative insights into the evolution of the yield surface with plastic deformation are sought, it is assumed that $\tau_0 = 1$, $\theta_0 = 10$, and $\theta_1 = 0$ in Eq. (3). $\tau_1$, and $H_{ss'}$ represent the only remaining free parameters.

The evolution of the yield surface under tensile and torsion deformation with $\tau_1 = 1$, and

$$H_{ss'} = 1, \quad \text{for all}\, s, s' \tag{5}$$

Is shown in Fig. 2a, b, respectively. It is seen that the yield surface expands isotropically. This is entirely unlike the evolution found experimentally as shown in Fig. 1.

Next, consider the case where in $\tau_1 = 1$, as before, but

$$H_{ss'} = \begin{cases} -3, & \text{if}\, s = r(s'), \\ 1, & \text{otherwise} \end{cases} \tag{6}$$

Equation (6) differs from Eq. (5) only in the definition of softening in the reverse slip system corresponding to every activated slip system. Physically, this corresponds to the accumulation of polar dislocations on the activated slip system, which readily glide backward upon unloading.

The effect of assuming the negative hardening coefficient on the reverse slip systems is shown in Fig. 3a, b for tension and torsion, respectively. The predicted yield surfaces show both isotropic and kinematic hardening components. The predicted

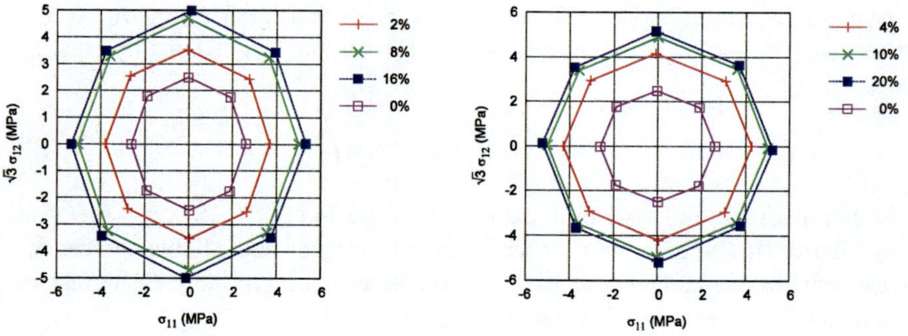


(a) Tension (b) Torsion

**Fig. 2** Evolution of the yield surface after **a** tensile and **b** torsion deformation predicted assuming the hardening matrix given by Eq. (5)

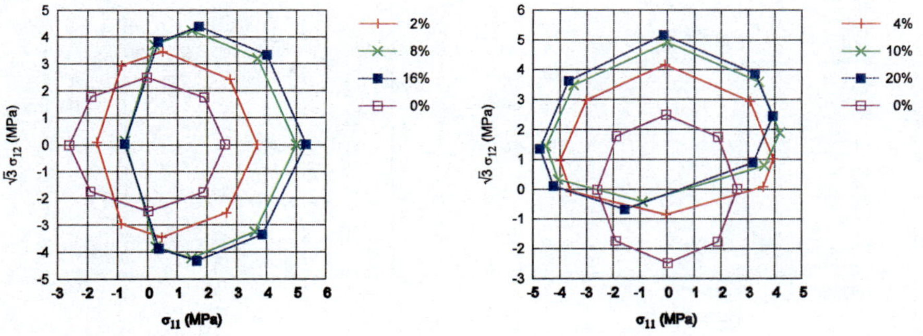


(a) Tension (b) Torsion

**Fig. 3** Evolution of the yield surface after **a** tensile and **b** torsion deformation predicted assuming the hardening matrix given by Eq. (6)

yield surfaces show a pronounced sharpening in the direction of loading and blunting in the opposite direction. They thus demonstrate directional distortional hardening [2]. The predicted yield surfaces develop an asymmetry that is especially pronounced under torsion, as shown in Fig. 3b. This is due to the development of texture [10] in the computer polycrystal.

Qualitatively, Fig. 3 resembles the experimental Fig. 1. However, it is seen that the predicted kinematic hardening component is much weaker than the experimental one. The experimental yield surfaces in Fig. 1 have a strong kinematic evolution component so that the yield surface does not include the origin for pre-strains as low as 2%.

Consider next the case where in $\tau_1 = 1$, as before, but

$$H_{ss'} = \begin{cases} -3, & \text{if } s = r(s'), \\ 0, & \text{otherwise.} \end{cases} \tag{7}$$

This is a case of self-hardening, wherein active slip systems only harden themselves and soften their reverse slip systems. The yield surfaces corresponding to this case are shown in Fig. 4. The kinematic component of the yield surface evolution here is similar to that in Fig. 3, but the isotropic component is not. In fact, the volume enclosed by the yield surface shrinks with deformation. This is not typical of Al 1100, but of precipitate hardened Al 6061 [11].

It is noteworthy that the predicted yield surfaces in both Figs. 3 and 4 include the origin, while the experimental Fig. 1 does not. As an example of a predicted yield surface that does not include the origin, Fig. 5 corresponding to $\tau_1 = 10$,

$$H_{ss'} = \begin{cases} -1.2, & \text{if } s = r(s'), \\ 0, & \text{otherwise.} \end{cases} \tag{8}$$

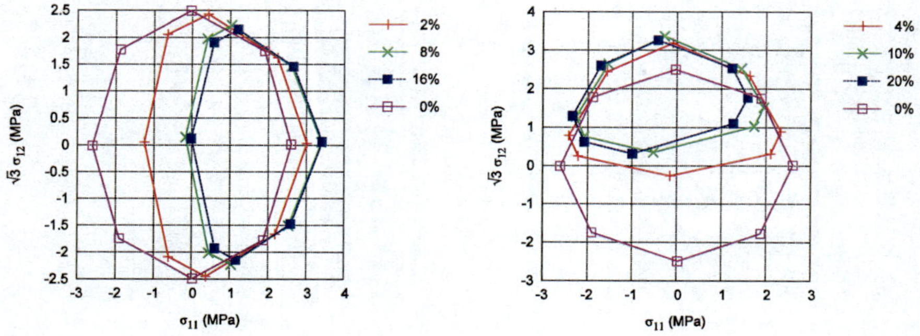


(a) Tension (b) Torsion

**Fig. 4** Evolution of the yield surface after **a** tensile and **b** torsion deformation predicted assuming the hardening matrix given by Eq. (7)

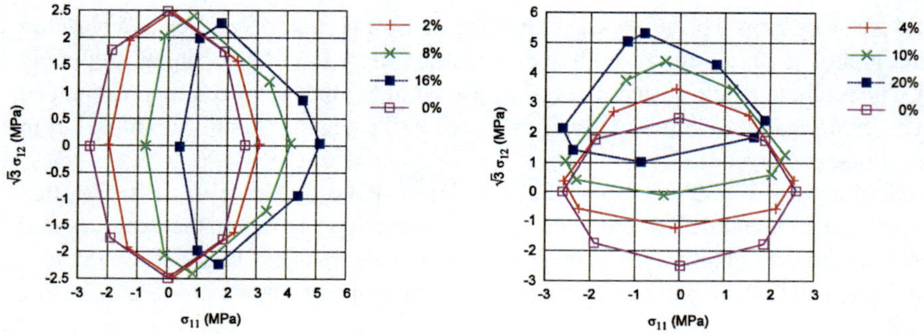


(a) Tension (b) Torsion

**Fig. 5** Evolution of the yield surface after **a** tensile and **b** torsion deformation predicted assuming the hardening matrix given by Eq. (8)

is presented. It is seen that the predicted yield surfaces for 16% tensile pre-strain and 20% shear pre-strain do not include the origin.

## 4 Discussion and Conclusion

The results presented above qualitatively confirm the ability of the polycrystal plasticity model with negative hardening coefficients to capture the directional distortional hardening experimentally observed by numerous authors in Al 1100. It has been shown that the greatest kinematic hardening is obtained when the latent hardening ratio of the reverse slip systems is taken to be $-1.2$. Even the greatest predicted kinematic hardening, however, is unable to match the kinematic hardening experimentally observed in Fig. 1. This suggests that the phenomenological hardening model, presented in Sect. 2, is inadequate for the present purpose. It is proposed to generalize the hardening formulation in the future work by introducing separate hardening matrices to represent the hardening due to polar and non-polar dislocation interactions.

## References


1. A.S. Khan, A. Pandey, T. Stoughton, Evolution of subsequent yield surfaces and elastic constants with finite plastic deformation. Part II: A very high work hardening aluminum alloy (annealed 1100 Al). Int. J. Plast. **54**(9), 2462–2468 (2010)
2. H.P. Feigenbaum, Y.F. Dafalias, Directional distortional hardening in metal plasticity within thermodynamics. Int. J. Solids Struct. **44**(22–23), 7526–7542 (2007)
3. P.M. Dixit, U.S. Dixit, *Modeling of Metal Forming and Machining Processes: By Finite Element and Soft Computing Methods* (Springer Science & Business Media, 2008)
4. M Zyczkowski, *Combined Loadings in the Theory of Plasticity* (Springer Science & Business Media, 1981)
5. O. Cazacu, B. Revil-Baudard, N. Chandola, *Plasticity-Damage Couplings: From Single Crystal to Polycrystal Line Materials* (Springer, Berlin, 2019)
6. R. Asaro, V. Lubarda, *Mechanics of Solids and Materials* (Cambridge University Press, 2006)
7. S. Mahesh, A binary-tree based model for rate-independent polycrystals. Int. J. Plast. **26**(1), 42–64 (2010)
8. C. Tome, G.R. Canova, U.F. Kocks, N. Christodoulou, J.J. Jonas, The relation between macroscopic and microscopic strain hardening in fcc polycrystals. Actametallurgica **32**(10), 1637–1653 (1984)
9. P. Franciosi, A. Zaoui, Multislip in fcc crystals a theoretical approach compared with experimental data. Acta Metall. **30**(8), 1627–1637 (1982)
10. U.F. Kocks, C.N. Tomé, H.R. Wenk, *Texture and Anisotropy: Preferred Orientations in Polycrystals and their on Materials Properties* (Cambridge university press, 1998)
11. A.S. Khan, R. Kazmi, A. Pandey, T. Stoughton, Evolution of subsequent yield surfaces and elastic constants with finite plastic deformation. Part-I: A very low work hardening aluminum alloy (Al6061-T6511). Int. J. Plast. **25**(9), 1611–1625 (2009)

# Influence of Camber Ratio and Thickness Ratio on the Airfoil Performance

**Sujit Roy, Biplab Das, and Agnimitra Biswas**

**Abstract** The aerodynamic performance of airfoil is certainly measured by the lift to drag ratio of that profile. In this paper, different NACA 4-digit airfoils are considered to perform the comparative analysis. The effects of camber ratio and thickness ratio on the aerodynamic coefficients are discussed. QBlade is used to calculate the lift coefficient and lift to drag ratio of each profile. Low Reynolds number (Re) of 3.6 $\times$ $10^5$ is used for the present analysis. The angle of attack is varied from 0 to 20° to observe its influence on the aerodynamic performances. It has been seen that the lift coefficient increases with the increase in camber ratio of airfoil which implies more gain in lift force. Simultaneously, drag force also increases with increase in camber ratio. Therefore, lift to drag ratio has been calculated at different angle of attack for various cambered airfoil. Similarly, influence of thickness ratio is also studied, and it is observed that airfoil with lower thickness ratio produces higher lift to drag ratio at lower angle of attack ($\alpha \leq 8°$). Whereas, airfoil with higher thickness ratio produces maximum lift to drag ratio at higher angle of attack.

**Keywords** Airfoil · Lift coefficient · Lift to drag ratio · Angle of attack · Camber ratio · Thickness ratio

## 1 Introduction

The scarcity of fossil fuels and increasing global warming direct the world towards the use of alternative energy sources to meet the day-to-day energy requirements. Wind, solar, geothermal, tidal, etc., are the various available renewable energy sources which are clean and freely available in nature. Due to the uneven temperature distribution throughout the earth surface wind flows from one end to another end. Wind has

S. Roy (✉) · B. Das · A. Biswas
Mechanical Engineering, National Institute of Technology Silchar, Silchar, Assam 788010, India
e-mail: sujitme12@gmail.com

B. Das
e-mail: biplab.2kmech@gmail.com



K. M. Pandey et al. (eds.), *Recent Advances in Mechanical Engineering*, Lecture Notes in Mechanical Engineering, https://doi.org/10.1007/978-981-15-7711-6_72

tremendous amount of kinetic energy in it and wind turbine can extract that energy to convert it to useful electric or mechanical energy. The horizontal wind turbine (HWT) blade contains airfoil cross section along the blade span. The energy extraction of turbine depends upon the aerodynamic forces generated due to the airfoil profiles. Due to pressure difference between upper and lower side of the airfoil, lift force is generated perpendicular to the air flow. This force is responsible for the rotation of the HWT. Another force generated in the direction of wind flow, known as drag force. This force tries to restrict the HWT rotation. Therefore, this is obvious that selected airfoil for HWT blade should have higher lift and lower drag. Kumar et al. [1] studied computationally NACA 23,024 airfoil using $k - \varepsilon$ turbulence model. They studied the lift and drag generation, pressure distribution in the upper and lower side of the airfoil for angle of attack ($\alpha$) ranges 0–20°. They found maximum lift coefficient of 1.58 at $\alpha = 18°$, which is over predicted compare to wind tunnel experimental results of maximum lift coefficient 1.3 at $\alpha = 14°$. Hossain et al. [2] studied NACA 6409 and NACA 4412 computationally to compare the aerodynamic performances at angle of attack 0° and 5°. They found that NACA 4412 produces higher negative pressure at suction side of airfoil. NACA 6409 shows 88.4% and 87.7% lesser lift to drag ratio than NACA 4412 at 0° and 5°, respectively. Eleni et al. [3] applied three different turbulence models ($k - \omega$ SST, realizable $k - \omega$ and S-A) to study NACA 0012 computationally at $-12° \leq \alpha \leq 20°$ and Reynolds number (Re) = $3 \times 10^6$ and concluded that $k - \omega$ SST turbulence model shows better agreement with the experimental result compared to other two turbulence model. Further, they also observed that turbulence models not able to predict the results at higher angle of attack. Kevadiya and Vaidya [4] performed computational analysis of NACA 4412 airfoil using S-A turbulence model at $0° \leq \alpha \leq 12°$ and Re = $1 \times 10^5$ and found that lift to drag ratio increases till $\alpha = 8°$ and then decreases. Gulzar et al. [5] used S-A turbulence model to perform the computational study of NACA 7420 airfoil at $\alpha = 4° - 6°$. Jun et al. [6] performed computational analysis of three airfoils (NACA 63,421, NACA 0021 and NREL S809) with same thickness ratio = 0.21 at Re = 5.5 $\times 10^5$ and found that NACA 63,421 shows better aerodynamic performance than the other two airfoils. Ren-nian et al. [7] computationally investigated the aerodynamic performance of S827, S902, S903 airfoils using RNG $k - \varepsilon$ turbulence model. Jha et al. [8] investigated NACA 0012 computationally and experimentally at $\alpha = 0° - 20°$. They observed the effect of Reynolds number on the aerodynamic performances of airfoil and found that lift generation is more at lower Reynolds number ($2.21 \times 10^5$). Further they observed that lift to drag ratio reduces with increasing Reynolds number. Jun et al. [9] used S-A turbulence model to predict the performance of S414 airfoil computationally and concluded that this turbulence model can predict the airfoil performance effectively. Yao et al. [10] compared four turbulence models to simulate NACA 0018 airfoil at $\alpha = -8° - 13°$ and Re = $5.5 \times 10^5$. They observed that five equation-based Reynolds stress turbulence model provides better results compared to other three models. Alaskari et al. [11] used QBlade software to compute the aerodynamics performances of SG6043 airfoil at different angle of attack. They observed that maximum lift to drag ratio is obtained at $\alpha = 2°$. Mahmuddin et al. [12] extrapolate the lift coefficient and drag coefficient using Viterna and Montgomerie methods available in the QBlade software. Different turbulence models were used by Villalpando et al. [13] to predict the aerodynamic characteristics of NACA 63-415 airfoil in the angle of attack $0° - 28°$. They concluded that SST $k - \omega$ is best suitable turbulence model to simulate the flow around an airfoil. RNG $k - \varepsilon$ turbulence model is used by Yan et al. [14] to predict the aerodynamics forces around FFA-W3- 211 airfoil. They simulated under the condition Reynolds number (Re) $= 1.5 \times 10^5$ and angle of attack ranges $-6°$–$60°$ and found maximum lift coefficient at $\alpha = 18°$. Fupeng et al. [15] designed a new airfoil with thick leading edge and providing more camber and compared with conventional thin leading edge airfoil. They simulate the airfoil under Re $= 4.8 \times 10^5$ and found that lift coefficient for modified airfoil is higher in modified airfoil for all the angle of attack from $-10°$–$50°$. However, the drag coefficient is also increased in the modified airfoil but it produces higher lift to drag ratio at $\alpha > 20°$. Yang et al. [16] used $k - \omega$ turbulence model to simulate the flow over S809 airfoil in the angle of attack ranges 0°–45°. They found maximum error of 37% and 7.8% in lift coefficient and drag coefficient, respectively, compare to experimental results. Matyushenko et al. [17] tried different turbulence model to simulate the various airfoils computationally. They found that the turbulence models matched well with the experimental results at initial angle of attack ($\alpha \leq 10°$) but none of the models is able to predict the maximum lift coefficient accurately. It can be seen from the available literatures that various analytical and computational studies are performed to predict the aerodynamic coefficients of airfoils. It is also noticed that shape of airfoil affects its performance. Therefore, it is required to perform airfoil study to know the best suitable airfoil for wind turbine blade design. In the present study, the effect of camber ratio and thickness ratio on the lift coefficient ($C_l$) and lift to drag ratio of airfoils are predicted using QBlade software.

## 2 Airfoil's Nomenclature and Performance Parameters

The nomenclature of an airfoil is shown in Fig. 1 [18]. The edge of the airfoil facing the fluid flow is known as leading edge, and other edge is known as trailing edge as shown in figure. The straight line connecting leading and trailing edge of the airfoil is known as chord line; whereas, the line bisecting the airfoil is known as mean camber line. The distance of maximum thickness and maximum camber from the leading edge of airfoil is denoted by $x_t$ and $x_f$, respectively. For the symmetric airfoil, the mean camber line coincides with chord line. The upper surface of the airfoil is known

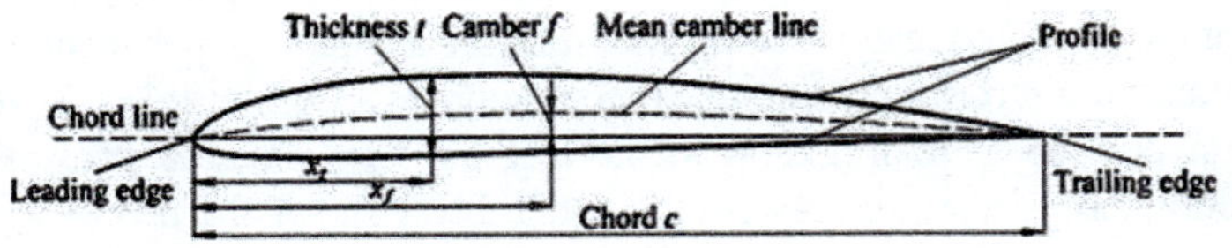


**Fig. 1** Airfoil nomenclature [18]

as suction side, and bottom surface is known as pressure side. The distance between pressure side and suction side is known as thickness of the airfoil. The angle between the fluid flow direction and chord line is known as angle of attack ($\alpha$). In the present study, NACA 4-digit airfoil is used for the comparative analysis. In 'NACA 1234' airfoil, first digit '1' denotes the maximum camber in percentage of chord, digit '2' denotes the position of maximum camber in tenths of chord, and digits '34' denotes the maximum thickness in percentage of chord. The performance of airfoil is often described by the ratio of lift force to the drag force. Lift force ($L$) and drag force ($D$) can be calculated using Eqs. (1) and (2) [19]. The generation of lift force is due to the pressure difference between the suction and pressure side. As the angle of attack increases, the pressure difference between two sides increases results in increase in lift force. After a certain angle of attack flow, separation occurred in the suction side of the airfoil leads to decrease in pressure difference between the sides and lift coefficient starts decreasing. This phenomenon is known as stalling of airfoil, and respective angle of attack is known as stall angle. The blade of HWT should operate at angle of attack where lift to drag ratio is maximum. Therefore, it is required to find an optimum range of angle of attack where lift force is maximum and drag force in minimum.

$$L = \frac{1}{2}\rho u_{\alpha}^{2} A C_{\mathrm{l}} \tag{1}$$

$$D = \frac{1}{2}\rho u_{\alpha}^{2} A C_{\mathrm{d}} \tag{2}$$

where, '$\rho$' is the fluid density, '$u_{\alpha}$' is the free stream velocity, '$A$' is the area of airfoil, '$C_l$' is the lift coefficient, and '$C_{\mathrm{d}}$' is the drag coefficient.

## 3 Results and Discussion

In the present study, QBlade is used to predict the lift coefficient and lift to drag ratio of different NACA 4-digit airfoils. The effect of geometrical shape of airfoil on the aerodynamics coefficient is discussed. QBlade works on the principle of blade element momentum theory (BEM) is an important method to obtain the result with good accuracy, and it has user-friendly interface [11]. It uses XFOIL to predict the pressure in the suction and pressure side of the airfoil. To validate the simulation tool, NACA 4415 is analysed under the condition of Re $= 3.6 \times 10^5$ and $0 \leq \alpha \leq 20°$. It is found that the predicted results are very close to the available experimental results [20]. The maximum error in predicting lift coefficient is 11.9% at $\alpha = 0°$, whereas, the maximum error in lift to drag ratio is 14.6% at $\alpha = 15°$. The more error is due to the less accuracy in predicting drag coefficient. Table 1 shows the percentage error between the experimental results and QBlade predicted results. It is also observed that negative error is occurred in lift/drag value at higher angle of attack ($\alpha = 15°$ and

**Table 1** Validation of QBlade

| Angle of attack | Lift coefficient (QBlade) | Lift coefficient [20] | % Error | Lift/drag (Qblade) | Lift/drag [20] | % Error |
|---|---|---|---|---|---|---|
| 0° | 0.47 | 0.42 | 11.9 | 50.63 | 44.6 | 13.5 |
| 5° | 1.03 | 1.01 | 1.9 | 86.36 | 80.3 | 7.5 |
| 10° | 1.43 | 1.34 | 6.7 | 77.28 | 72.8 | 6.2 |
| 15° | 1.45 | 1.42 | 2.1 | 22.54 | 26.4 | -14.6 |
| 20° | 1.38 | 1.33 | 3.8 | 10.03 | 11.7 | -14.3 |

20°). It indicates the over prediction of lift to drag ratio. But most importantly, it can be observed that this analysing tool can predict aerodynamics coefficients accurately at higher angle of attack or under the stall condition; whereas, other computational tools predict erroneous results in the stall region. In this section, the effects of camber ratio and thickness ratio on the lift coefficient and lift to drag ratio of different NACA airfoils are discussed.

### *3.1 Effect of Camber Ratio*

Camber ratio (CR) is defined by the ratio of maximum camber to the chord length of the airfoil. In the present study, NACA 0015 (CR = 0.00), NACA 1415 (CR = 0.01), NACA 2415 (CR = 0.02), NACA 3415 (CR = 0.04), NACA 4415 (CR = 0.04) airfoils with different camber ratios are considered keeping the maximum thickness same. The effect of camber ratio on lift coefficient is shown in Fig. 2.

It can be observed that all the axisymmetric airfoils generate positive lift at zero angle of attack due to presence of camber; whereas, symmetry airfoil or airfoil without camber generates zero lift at $\alpha = 0°$. The lift coefficient increases with increase in camber ratio and angle of attack. Maximum lift coefficient of 1.5 is observed with the highest camber ratio (CR = 0.04). It can also be observed that percentage reduction of lift coefficient in the post stall zone is higher in airfoil without camber; whereas, cambered airfoil experiences smooth stall occurrence. Further, the effect of camber ratio on lift to drag ratio can be seen in Fig. 3. The lift to drag ratio is maximum for the airfoil with the highest camber ratio at all the angle of attack. Maximum lift to drag ratio is 91.7 at $\alpha = 8°$ is found for airfoil with camber ratio 0.04. It can also be observed that the operating range of angle of attack (range of increasing lift to drag ratio) also increases with increase in camber ratio.

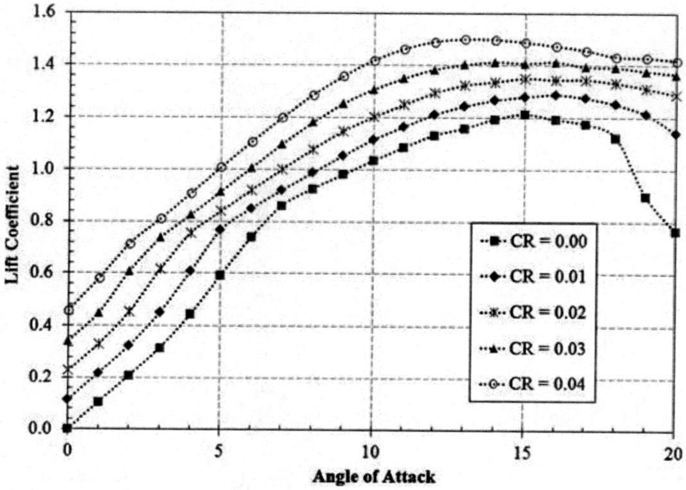


**Fig. 2** Effect of camber ratio on lift coefficient

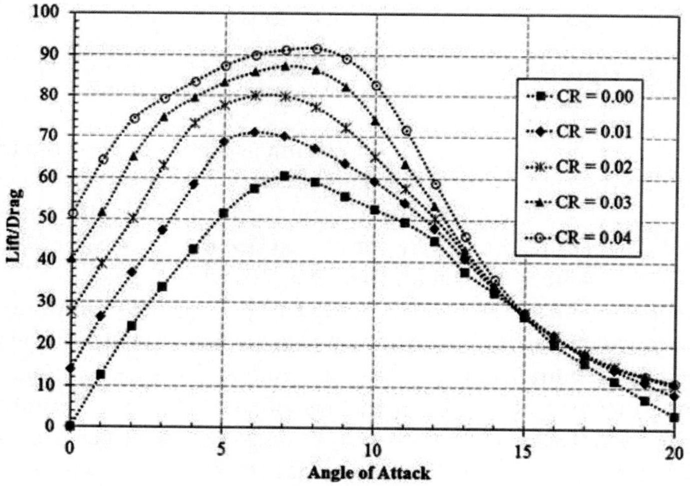


**Fig. 3** Effect of camber ratio on lift to drag ratio

## 3.2 *Effect of Thickness Ratio*

Thickness ratio (TR) is defined by the ratio of maximum thickness to the chord length of the airfoil. In the present study, NACA 4411 (TR = 0.11), NACA 4412 (TR = 0.12), NACA 4413 (TR = 0.13), NACA 4414 (TR = 0.14) and NACA 4415 (TR = 0.15) are studied maintaining same camber and location of maximum camber of airfoil to observe the effect of variable thickness ratio on the lift and drag generation. It can be seen from Fig. 4 that the lift coefficient is almost same in all the airfoils at

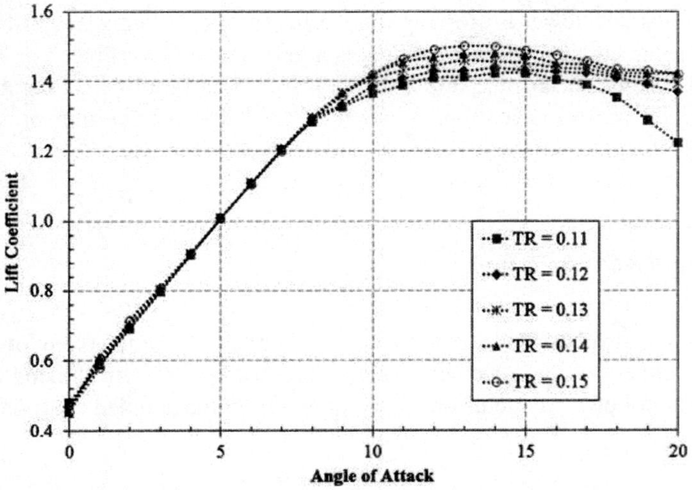


**Fig. 4** Effect of thickness ratio on lift coefficient

lower angle of attack ($\alpha = 8°$). But at higher angle of attack ($\alpha > 8°$), the lift coefficient is more in the airfoil with maximum thickness ratio. It indicates that at initial angle of attack, airfoil generates lift forces irrespective of thickness ratio; whereas, thicker airfoil generates more lift force at higher angle of attack. Maximum lift coefficient of 1.5 at $\alpha = 14°$ is found with airfoil of TR = 0.15. The airfoils show positive lift generation at 0° angle of attack is due to the selection of axisymmetric airfoil for the analysis. But generation of drag force also changes with the airfoil shape. Figure 5 shows the variation of lift to drag ratio with angle of attack. It can be observed that

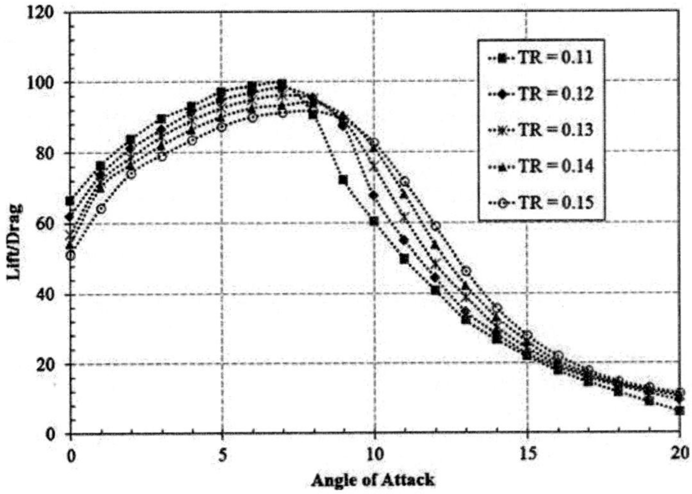


**Fig. 5** Effect of thickness ratio on lift to drag ratio

thicker airfoils have lower lift to drag ratio at initial angle of attack ($\alpha \leq 8°$). This is due to the percentage increment in drag is more compared to percentage increase in lift with the thicker airfoil. At $\alpha > 8°$, the lift to drag ratio is maximum for the airfoil with TR $= 0.15$. Therefore, it is recommended to use the thinner airfoil at lower angle of attack and thicker airfoil at higher angle of attack.

## 4 Conclusions

In the present study, NACA 4-digit airfoils are used to observe the effect of camber ratio and thickness ratio on the aerodynamic performance of airfoil using QBlade software. The following conclusion can be drawn from the detailed analysis.

- Airfoil with maximum camber produces the highest lift and lift to drag ratio at all the operating angle of attack. Cambered airfoil produces positive lift at 0° angle of attack; whereas, non-cambered airfoil produces zero lift. Maximum lift and lift to drag ratio are found to be 1.5 and 91.7, respectively, with airfoil of CR $= 0.04$.
- Thickness ratio has less significance in the lift generation till $\alpha = 8°$; whereas, lift coefficient increases with increase in thickness ratio at $\alpha > 8°$. Thinner airfoil produces maximum lift to drag ratio at $\alpha \leq 8°$; whereas, thicker airfoil provides maximum lift to drag ratio at $\alpha > 8°$.

**Acknowledgements** The authors are thankful to TEQIP III, NIT Silchar, for providing financial assistance.

## References

1. A.B.S. Kumar, S. Manjunath, R. Ganganna, Computational investigation of flow separation over NACA 23024 airfoil at 6 million free stream Reynolds Number using k-epsilon turbulence model. Mater. Today Proc. **5**(5), 12632–12640 (2018)
2. S. Hossain, M.F. Raiyan, M.N.U. Akanda, N.H. Jony, A comparative flow analysis of NACA 6409 and NACA 4412 aerofoil. Int. J. Res. Eng. Technol. **3**(10), 342–350 (2014)
3. D.C. Eleni, T.I. Athanasios, M.P. Dionissios, Evaluation of the turbulence models for the simulation of the flow over a National Advisory Committee for Aeronautics (NACA) 0012 airfoil. J. Mech. Eng. Res. **4**(3), 100–111 (2012)
4. M. Kevadiya, H.A. Vaidya, 2D analysis of NACA 4412 airfoil. Int. J. Innov. Res. Sci. Eng. Technol. **2**(5), 1686–1691 (2013)
5. O. Gulzar, S. Gulzar, S. Bhatele, N. Soni, Impact of variation in angle of attack on NACA 7420 airfoil in transonic compressible flow using Spalart- Allmaras turbulence model. Int. J. Res. Mech. Eng. Technol. **4**(2), 35–39 (2014)
6. J. Wang, J. Sheng, L. Shi, F. Zou, Analysis of aerodynamic performance of wind turbine airfoil under the same relative thickness. Water Resour. Power **3** (2011)
7. R.-N. Li, S.-A. Zhang, R. Yang, D. Li, Effect of aerofoil camber on airfoil aerodynamic performance. Fluid Mach. **5** (2009)

8. S. Jha, U. Gautam, S. Narayanan, L.A. Kumaraswami Dhas, Effect of reynolds number on the aerodynamic performance of NACA0012 aerofoil, IOP Conf. Ser. Mater. Sci. Eng. **377**(1), 012129 (2018)
9. Li. Jun, S. Wenlong, Numerical simulation on wind turbine airfoil aerodynamics performance. Int. J. Control Autom. **9**(12), 275–286 (2016)
10. Ji. Yao, W. Yuan, J. Xie, H. Zhou, M. Peng, Y. Sun, Numerical simulation of aerodynamic performance for two dimensional wind turbine airfoils. Procedia Eng. **31**, 80–86 (2012)
11. M. Alaskari, O. Abdullah, M.H. Majeed, Analysis of wind turbine using QBlade software. IOP Conf. Ser. Mater. Sci. Eng. **518**(3), 032020 (2019)
12. F. Mahmuddin, S. Klara, H. Sitepu, S. Hariyanto, Airfoil lift and drag extrapolation with viterna and montgomerie methods. Energy Procedia **105**, 811–816 (2017)
13. F. Villalpando, M. Reggio, A. Ilinca, Assessment of turbulence models for flow simulation around a wind turbine airfoil. Modell. Simul. Eng. **6** (2011)
14. C. Yan, Ye. Zhiquan, Li. Deyuan, He. Fupeng, Numerical simulation of large angle-of-attack separated flows over airfoils of HAWT rotors. Wind Eng. **30**(1), 35–42 (2006)
15. H. Fupeng, Li. Yuhong, C. Zuoyi, Suggestions for improving wind turbines blade characteristics. Wind Eng. **25**(2), 105–113 (2001)
16. S.L. Yang, Y.L. Chang, O. Arici, Navier-Stokes computations of the NREL airfoil using a k-ω turbulent model at high angles of attack. J. Solar Energy Eng. Trans. Actions ASME **117**(4), 304–310 (1995)
17. S. Zhang, H. Li, A.A. Abbasi, Design methodology using characteristic parameters control for low Reynolds number airfoils. Aerosp. Sci. Technol. **86**, 143–152 (2019)
18. A.A. Matyushenko, E.V. Kotov, A.V. Garbaruk, Calculations of flow around airfoils using two-dimensional RANS: an analysis of the reduction in accuracy. St. Petersburg Polytech. Univ. J. Phys. Math. **3**(1), 15–21 (2017)
19. J. Twidell, T. Weir, *Renewable Energy Resources* (Routledge, London, 2015)
20. Airfoil Tool: NACA 4415. [Online] Available from https://airfoiltools.com/airfoil/details?airfoil=naca4415-il#polars. Accessed 29 Dec 2019

# Development of Modified Cyclic Plasticity Model to Simulate Cyclic Behaviour for SA333C–Mn Steel Under LCF Loading Conditions

**Vibhanshu Pandey, Partha Pratim Dey, Niloy Khutia, and Suneel K. Gupta**

**Abstract** In this work, a modified cyclic plasticity model for FEM analysis of SA333 C–Mn steel has been developed. A memory stress based isotropic formulation is integrated into the modified Ohno-Wang model to revamp its nonmasing and hardening/softening characteristics. Accurate estimation of stress–strain hysteresis loops is a prerequisite for performing fatigue life analysis. The inbuilt classical cyclic plasticity model available in commercial FE software cannot precisely describe the material behavior. The proposed model has been implemented by using user-defined material (UMAT) subroutine with FORTRAN code on the ABAQUS platform. To evaluate the material constants for the proposed model, a set of experiments has been carried out under uniaxial loading condition. The same material constants are used for predicting simulation response for other loading conditions. To verify the proposed model, the simulation results are compared with experimental results which reveal good agreement under uniaxial loading condition

**Keywords** Cyclic plasticity · Memory stress · Non-masing behavior · UMAT

## 1 Introduction

SA333 Grade 6, C–Mn steel is primarily used in piping materials of pressurized heavy water reactors (PHWR) of Indian nuclear power plants [1]. Resistance to low cycle fatigue (LCF) is an essential aspect for the successful design of piping systems of PHWR. Due to variations in operating conditions or seismic actions or start-up and power failure, cyclic plastic loading can be activated on piping materials.

V. Pandey (✉) · P. P. Dey
Department of Mechanical Engineering, IIEST Shibpur, Howrah, India
e-mail: vibhanshupandey30@gmail.com

N. Khutia
Aerospace Engineering and Applied Mechanics Department, IIEST Shibpur, Howrah, India

S. K. Gupta
Reactor Safety Division, BARC, Mumbai, India



K. M. Pandey et al. (eds.), *Recent Advances in Mechanical Engineering*, Lecture Notes in Mechanical Engineering, https://doi.org/10.1007/978-981-15-7711-6_73

In the past decades, cyclic constitutive models have been extensively developed for various metallic materials and successfully used in industrial applications. Armstrong and Frederick [2] modified Prager's linear kinematic hardening formulation by adding a recall term to explain the dynamic recovery effect. Chaboche et al. [3] decomposed total back stress into several parts, each of which obeying the AF rule to gave a better description of transient hardening evolution and ratcheting effect. To predict less ratcheting as compared to Armstrong and Frederick model [2], Ohno and Wang [4, 5] inserted critical state in the formulation of dynamic recovery part of back stress. Further many researchers developed nonlinear kinematic hardening rules mainly by making changes in dynamic recovery part of back stress to widely discuss the problem in cyclic plasticity modeling studies [6, 7].

Sivaprasad et al. [8] performed various experiments on carbon-manganese steel using solid and hollow specimens to study the cyclic deformation behavior under strain-controlled loading. Both hollow and solid specimens were used for pure axial experiments and only hollow specimen were used for shear experiments. Higher axial stress was observed in the case of hollow specimens as compared to solid specimens under the same equivalent strain loading conditions. Xu et al. [10] proposed a model based on the concept of the combined isotropic kinematic hardening model using internal variables to incorporate the cyclic hardening and softening behavior with unified set of material parameters. Zhou et al. [9] modified the nonlinear kinematic hardening rule by introducing a kinematic hardening coefficient in the back stress to incorporate the hardening/softening effect for the 316L steel at room temperature. The main objective of this work is to develop a modified cyclic plasticity model to take care of nonmasing and hardening/softening behavior of SA333 C–Mn steel under uniaxial loading conditions using user-defined material subroutine (UMAT) program on ABAQUS software.

## 2 Constitutive Equations of Plasticity Model

Generally, cyclic plasticity models are considered as isotropic and rate-independent. It is assumed in this study that strain ($\underline{\varepsilon}$) is small and additively decomposed into elastic part ($\underline{\varepsilon^e}$) and plastic part ($\underline{\varepsilon^p}$).The formulation of plasticity models requires an initial yield criterion, plastic flow rule, and hardening rule. The initial yield condition is used to determine the state of yielding. After the yield condition is fulfilled, the plastic flow rule determines the magnitude and direction of the plastic strain increment tensor. Hardening rule modifies the initial yield condition in the plastic flow direction based on the loading magnitude and direction.

## 2.1 Yield Criterion and Plastic Flow Rule

Von Mises yield criterion is generally used for metallic material. Therefore, the Von Mises yield criterion is also used here. The evolution equation of plastic strain increment is governed by von Mises yield function $f$. The yield criterion is defined as follows:

$$f = \sqrt{\frac{3}{2}(\underline{S} - \underline{a}):(\underline{S} - \underline{a})} - Y \tag{2.1}$$

where $f$ is the yield function, $a$ is deviatoric back stress tensor, $S$ is deviatoric stress tensor, $Y$ is the current yield stress.Associative flow rule is expressed by

$$\underline{\dot{\varepsilon}}^p = \dot{p}\left(\frac{\partial f}{\partial \sigma}\right) \tag{2.2}$$

$\underline{\dot{\varepsilon}}^p$ the plastic strain increment vector and its direction is along the normal to the plastic potential surface. $\dot{p}$ is the positive factor for proportionality.

## 2.2 Kinematic Hardening Rule

The kinematic hardening rule shown in Eq. (2.3) is used for back stress increment [6]. In this model, Ohno Wang kinematic hardening model is modified by adding a new memory stress function in dynamic recovery term of back stress evolution rule to model non-masing behavior of the material [6].

$$\mathrm{d}\underline{a}^{(i)} = \frac{2}{3}C^{(i)}\mathrm{d}\underline{\varepsilon}^p - \gamma^{(i)}\exp^{(-b^{(i)}s)}\left\langle \underline{n}: \frac{\underline{a}^{(i)}}{f(\underline{a}^{(i)})}\right\rangle\left(\frac{f(\underline{a}^{(i)})}{r^{(i)}}\right)^{m^{(i)}}(\mathrm{d}p)\underline{a}^{(i)} \tag{2.3}$$

The factor $\zeta^{(i)}(s) = \exp^{(-b^{(i)}s)}$ is used as an exponential decaying function to integrate the fading memory properties of the material with plastic strain accumulation and $C$ and $\gamma$ are material constants. $b$ is material constant to account cyclic hardening/softening behavior.

## 2.3 Isotropic Hardening Rule

In the present model, the evolution of the yield surface is represented by the variable as a function of fading memory stress. In this formulation of isotropic hardening, variable $R$ is divided into two parts: (a) Primary Isotropic variable $R_1$ [6] (b) Secondary

Isotropic variable $R_2$ [10]. Primary isotropic variable and Secondary isotropic variable are used to predict hardening and softening behavior of the material based on loading conditions. Fading memory stress is added in the formulation of both primary and secondary isotropic variable.

$$R = \sum_{k=1}^{n} R_k \tag{2.4}$$

$$R_1 = x_1 Y_0 \exp^{x_2 s} \tag{2.5}$$

$$R_2 = \frac{Q_T^s Q_T^0 r[(x_3 s - 1)]p}{Q_T^s + \left[Q_T^0(rp - 1)\right]} x_4 \tag{2.6}$$

where, $s$ is memory stress, $x_1$,$x_2$,$x_3$ and $x_4$ are constants, p is plastic strain $Q_T^o$ and $Q_T^s$ are the initial and saturated value of the secondary hardening or softening. $r$ represents the rate of secondary hardening or softening. Derivation of these material parameters and constants will be discussed in Table 1. Final expression of yield stress is shown in Eq. 2.7.

**Table 1** Material parameters and their procedure

| S. No. | Parameter | Value | Procedure |
|---|---|---|---|
| 1 | Ohno-Wang kinematic hardening coefficients | $C^{1-12}$ = 91,610, 20,370, 23,635, 28,961, 23,343, 9861, 4567, 2810, 2156, 2341, 1283, 2575 $\gamma^{1-12}$ = 2925, 2780, 2015, 1454, 788, 422, 318, 256, 203, 171, 96, 69 | Determined with procedure as described by Bari and Hassan [7] |
| 2 | Cyclic hardening/softening coefficients | $b^{1-12}$ = 0.006024, 0.002321, 0.002071, 0.002559, 0.001732, 0.002531, 0.001765, 0.001684, 0.001643, 0.000328, 0.001154, 0.000112 | Calculated from stabilized uniaxial LCF of strain range 1.6% |
| 3 | Coefficients for modeling of non-masing behavior: | $x1 = 0.011$, $x2 = 0.011$, $x3 = 0.0146$, $x4 = 19$ QTO = 1.5, QST = 28, $r = 10$ | Obtained from calculation of yield stress at various loading amplitudes |

$$\sigma_y = Y_0 + R_1 + R_2 = Y_0 + x_1 Y_0 \exp^{x_2 s} + \frac{Q_T^s Q_T^0 r[(x_3 s - 1)]p}{Q_T^s + \left[Q_T^0(rp - 1)\right]} x_4 \tag{2.7}$$

## 3 Finite Element Simulation

Finite Element modeling of solid cylindrical specimen (gauge diameter = 7 mm and gauge length = 13 mm) and hollow tubular specimen (OD = 25.4 mm, wall thickness = 1.7 mm and gauge length = 30 mm) are shown in Fig. 1a, b respectively. Only the gauge portion (3D tube) of the specimen is modeled for finite element analysis. The fixed boundary condition is specified at the bottom of the specimen. A reference point (RP) is created on the axis of the geometry and it is set to be kinematically coupled (restricting all degrees of freedom with RP and loading surface) with the top surface of the specimen. Cyclically varying rotation is applied on that reference point. The 3D solid brick element of ABAQUS 14.0 (C3D8 element) is used for meshing both 3D specimens.

### *3.1 Material Parameters for the Proposed Model*

Material parameters for the proposed model are calculated from the initial and stable hysteresis loops obtained from the Uniaxial LCF test of ±1.6% strain amplitude.

The hysteresis loading curve is divided into twelve segments. Therefore, corresponding 12 values of $C$, $\gamma$ and $b$ are determined by the procedure as described by Bari and Hassan [7].

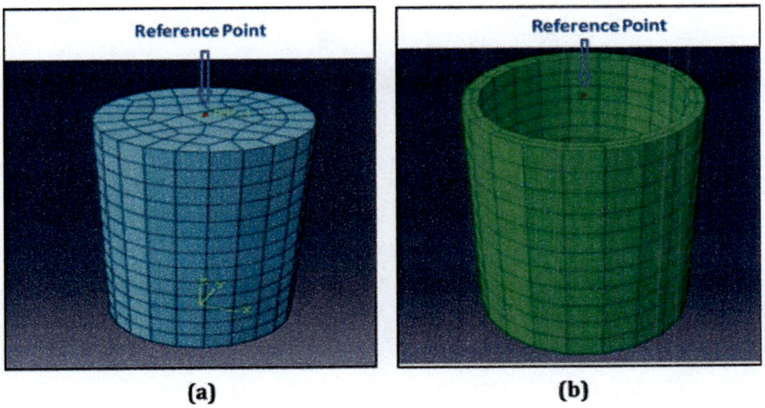


**Fig. 1** (**a**) and (**b**) Finite element model for (**a**) solid specimen, (**b**) hollow specimen

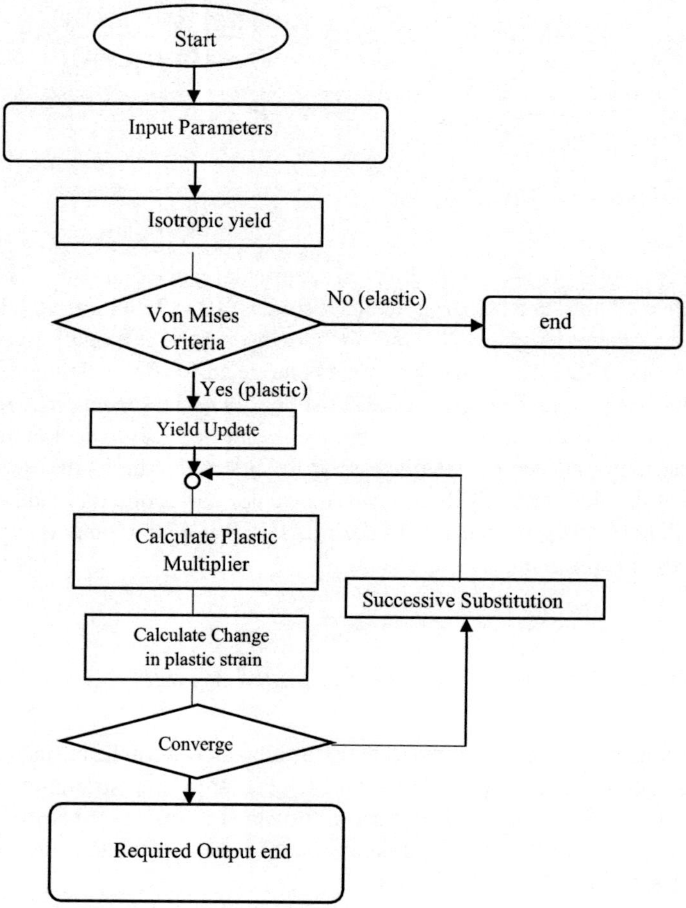


**Fig. 2** Flow chart of Implemented FE Model

### 3.2 *Flow Chart of FE Model*

The flow chart of FE model implemented in ABAQUS 6.14 using UMAT subroutine is given in Fig. 2.

## 4 Result and Discussion

Figure 3 presents the axial hysteresis loop of the saturated cycle under strain-controlled uniaxial tension– compression loading condition for strain amplitude of 1.6 and 1.0% for SA333 C–Mn steel. The predicated response from the simulation shows quite good correlation with the experiment response.

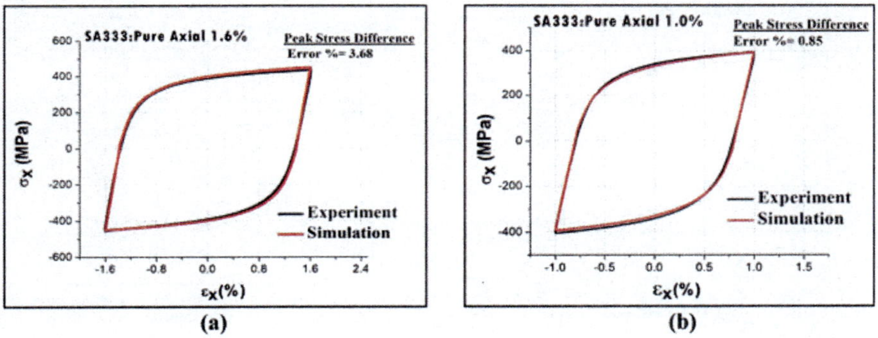


**Fig. 3** Axial stress versus axial strain for strain amplitude (**a**) 1.6% and (**b**) 1.0%

Simulation results for pure shear and pure axial with the hollow specimen are presented in Fig. 4a, b. Figure 4a shows shear stress versus shear strain response under shear strain-controlled for strain amplitude of 0.69%. Both the peak stress and nature of the simulated hysteresis loop matches well with the experiment hysteresis loop. Figure 4b shows hysteresis loop under pure axial with hollow specimen. The predicted hysteresis loop matches well with the experiment loop with a slight deviation in peak stress.

It is being observed that SA333 shows softening behavior at lower strain range and hardening behavior at higher strain range in case of uniaxial tension–compression for solid specimen [8]. Softening behavior at lower strain range indicates an adjustment in dislocation dynamics due to disparity in strain amplitude. Figure 5 illustrates that proposed model can simulate hardening/softening behavior at various strain amplitudes.

The FE model with the same set of kinematic hardening parameters calibrated from uniaxial solid specimen gives good matching in case of pure shear and pure axial with a tubular specimen. Additional hardening is observed in the case of a hollow axial [8] case may be the reason for little deviation in peak stress (refer to Fig. 4b).

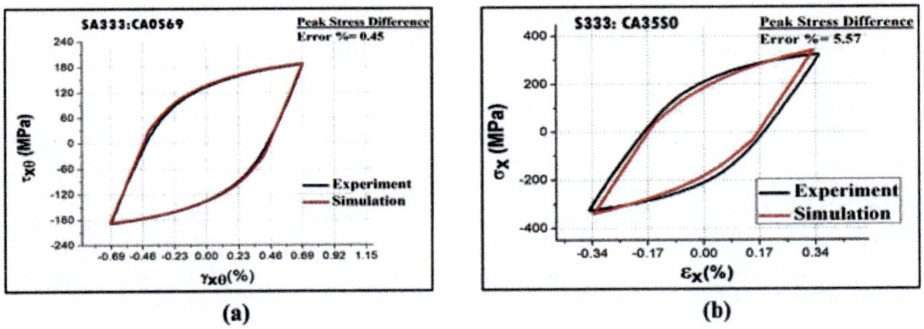


**Fig. 4** Experiment and simulation results (**a**) pure shear and (**b**) pure axial with tubular specimen

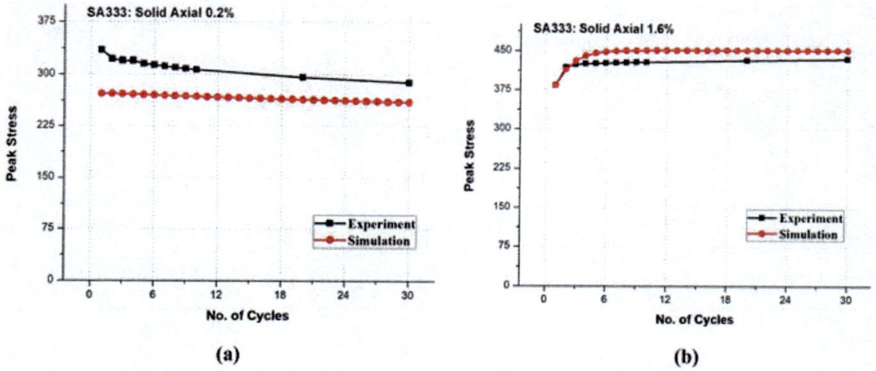


**Fig. 5** Peak stress versus no. of cycles for uniaxial solid specimen strain amplitude of (**a**) 0.5%, (**b**) 1.6%

## 5 Conclusion

- The cyclic hardening/softening features integrated into the model through isotropic hardening rule augmented the FE model strength in simulating various strain range.
- The kinematic hardening parameter calibrated from uniaxial solid specimen works well for a hollow specimen under uniaxial tension–compression and shear loading.
- The saturated hysteresis loop of the modified model matches well with saturated experiment hysteresis loop under various uniaxial loading conditions.

**Acknowledgements** The authors wish to acknowledge sincere thanks to the authorities of Bhabha Atomic Research Centre, Mumbai for financial assistance through a collaborative research project having sanction no 36(2)/14/06/2015/36009.

## References

1. P.K. Singh, V.R. Ranganath, S. Tarafder, P. Pradad, V. Bhasin, K.K. Vaze, Effect of cyclic loading on elastic-plastic fracture resistance of PHT system piping material of PHWR. Int. J. Pressure Vessels. Pip. **80**, 745–752 (2003)
2. P.J. Armstrong, C.O. Frederick, A mathematical representation of the multiaxial Bauschinger effect, in *CEGB Report RD/B/N731* (Berkeley Nuclear Laboratories, Berkeley, U.K, 1966)
3. J.L. Chaboche, K. Dang-Van, G. Cordier, Modelization of strain memory effect on the cyclic hardening of 316 stainless steel, in *International Transactions of the 5th International Conference on Structural Mechanics in Reactor Technology*, Berlin, No. Div. L in 11/3, 1979
4. N. Ohno, J.-D. Wang, Kinematic hardening rules with critical state of dynamic recovery, Part I: formulation and basic features for ratchetting behavior. Part II: application to experiments of ratchetting behavior. Int. J. Plast.**9**(3), 375–403 (1993)
5. N. Ohno, J.-D. Wang, Kinematic hardening rules for simulation of ratcheting behaviour. Euro. J. Mech. A Solids **13**(4), 519–531 (1994)

6. N. Khutia, P.P. Dey, P. Surajit, S. Tarafder, Development of Non Masing Characteristic Model for LCF and Ratcheting Fatigue Simulation of SA333 C-Mn Steel. Mech. Mater. **65**, 88–102 (2013)
7. S. Bari, T. Hassan, Anatomy of coupled constitutive models for ratcheting simulations. Int. J. Plast. **16**(3–4), 381–409 (2000)
8. S. Sivaprasad, H.N. Bar, S.K. Gupta, P. Arora, V. Bhasin, S. Tarafder, A comparative assessment of cyclic deformation behavior in SA333 Gr.6 steel using solid, hollow specimens under axial and shear strain paths. Int. J. Fatigue **61**, 76–86 (2014)
9. J. Zhou, Z. Sun, P. Kanoute, D. Retraint, Experimental analysis and constitutive modelling of cyclic behaviour of 316L steels including hardening/softening and strain range memory effect in LCF regime. Int. J. Plast. **107**, 54–78 (2018)
10. L. Xu, X. Nie, J. Fan, M. Tao, R. Ding, Cyclic hardening and softening behaviour of the low yield point steel BLY 160: Experimental response and constitutive modelling. Int. J. Plast. **78**, 44–63 (2016)

# High Speed Impact Studies of Kevlar Fabric with and without STF

M. Chinnapandi, Ajay Katiyar, Tandra Nandi, and R. Velmurugan

**Abstract** Kevlar fabrics have good energy absorption capacity and are widely used in defence, sports, and in impact resistance applications. To further improve these impact properties, materials such as shear thickening fluid (STF) could be used as they are believed to increase the composite's impact resisting property due to an increase in frictional interaction. The present work involves the simulation of impact studies of the Kevlar fabrics with and without STF. The modelling is done through LS-Dyna which is a multipurpose software and analysis is carried out for different number of Kevlar fabrics of size 150 mm × 150 mm with edges being fixed. Ballistic limits, residual velocities and energy absorption of these fabrics are obtained for different velocities ranging from 50 to 150 m/s. These results are compared with the experimental values obtained through a single-stage air gun an in house setup developed by us. The numerical values show good agreement with the experimental values and effect of STF on energy absorption is also studied and it is seen that there is improvement in energy absorption and ballistic limit which is observed both in numerical studies as well as in experiments.

**Keywords** High speed impact · Kevlar with and without shear thickening fluid · Finite element method · Impact behaviour · Ballistic limit · Residual velocity

## 1 Introduction

Materials used in automobile, aerospace, sports and defence sector are subjected to impact load of various kinds. Therefore, it is important to make sure that such materials are resistant towards such loads. Necessary testing needs to be performed

M. Chinnapandi · R. Velmurugan (✉)
Department of Aerospace Engineering, Indian Institute of Technology Madras, Chennai, Tamil Nadu 600036, India
e-mail: ramanv@iitm.ac.in

A. Katiyar · T. Nandi
Defense Materials and Stores Research and Development Establishment, Kanpur, Uttar Pradesh 208013, India



K. M. Pandey et al. (eds.), *Recent Advances in Mechanical Engineering*, Lecture Notes in Mechanical Engineering, https://doi.org/10.1007/978-981-15-7711-6_74

on material before the particular product or component gets qualified. In order to check for the ballistic impact resistance of materials, various test of high, low and medium velocity ballistic impact tests are to be performed where the material would be subjected to respective velocities by projectiles of various nose shapes. Kevlar fabrics are the most commonly used armour material for protection against bullets fired from hand guns because of their high impact resistance, high strength and low weight. These properties make them ideal materials to be used in bullet-proof vests when compared to other materials. An active area of research concerning the impact resistance of high strength fibres involves the study of the effect of shear thickening fluids (STF) on their ballistic performance [1, 3].

Kevlar being studied can be used in defence land systems, military aircraft, submarine hulls, body armours for soldiers as these structures are required to be resistant to impact load. The projectile may be bullets or particles due to explosions. These projectiles may damage the impinging structures due to high-velocity impact. Ballistic impact testing thus plays a very important role in this field. The study on the use of Kevlar fabric as an armour material is prevalent for quite some time. The impact behaviour of Kevlar and its polymer composites have been studied extensively. Ramadhan et al. [2] have studied Kevlar-polyester composites using a helium gas gun by using a cylindrical steel projectile. The energy absorption was found to increase with an increase in the number of layers of Kevlar at higher velocities. A numerical simulation has also been performed to validate the results.

Many techniques have been attempted in order to reduce the weight of the impact resistant structures, especially body armours. One such method studied in literature [2] is by adding a secondary phase to the fabrics. The use of shear thickening fluids for this purpose seems to be convenient as it enables the structure to act like a solid when under stress and as a liquid in normal conditions.

Brescianci et al. [5] numerically modelled woven fabric made of Kevlar 29 using two approaches, mesoscopic and macroscopic, and compared the result with experimentally obtained data. Numerical modelling was done using LS-Dyna. Solid element was used to model the fabric using macroscopic approach while woven fabric was modelled using texgen software for the mesoscopic approach. Similar studies were performed by Ha-minh et al. [8] using macroscopic and mesoscopic models and a comparison was drawn upon considering the variation in projectile velocity as well as the damage that occurs in the composite. It is also reported that shell element can be used to model composite plate by using both the approaches. Thickness of shell elements should be same as that of the fabric. Mass density of modelled plate has been considered in such a way that mass of the plate modelled equals the mass of fabric. Due to unaccountability of porosity and undulations, density of modelled plate comes out to be less than of the actual density. The analysis has been performed by modelling the plate as orthotropic with its properties similar in warp and weft directions. A comparison had been made on the two approaches based on residual velocity and damage in fabric. The study shows that both could predict few phases during impact, such as failure of only the yarns at the impact location while large area of fabric remain unaffected, and also narrowing of fabric edges was evident in both approaches. Interaction between the yarns during impact was not evident in the macroscopic approach which is a limitation when compared to the mesoscopic approach. Residual velocity that was obtained in both cases was plotted with time and comparison was made. For low velocity impact, deviation of macroscopic residual velocity from mesoscopic residual velocity was significant. Thus, macroscopic approach is good approximation to model a woven fabric and it also yields reasonable results.

In the present study, different numbers of Kevlar layers with and without STF are studied using LS-DYNA software to determine the number of layers needed to design a safe bullet-proof vest. The characterization of the Kevlar fabric was made using energy absorption and ballistic limit studies and the results were compared with experiment results which show good agreement.

## 2 Numerical Studies

A projectile modelled using mild steel material properties was used to compare the response of the woven fibre mats with and without STF. Impact response of the materials was compared for the same thickness and same mesh of fabrics with and without STF. Fibre mats with STF were modelled with a geometry similar to that of those without STF. The effect of STF was included in the model as a change in frictional parameters between the yarns and projectile to target plate. Analysing the residual velocity of the projectile after penetration and the internal energy rise in target plate are the main objectives of this study. The modelling parameters used for Kevlar fibre with and without STF are given in Table 1.

The frictional behaviour in the numerical model was modified to reflect the frictional behaviour of STF impregnated fabrics.

LS-Dyna was used for modelling and solving the problem of ballistic impact loading for different materials. Advantage of LS-Dyna includes explicit dynamic solver which is very useful to solve problem involving crash, penetration and large deformations.

Many contact types are available in LS-Dyna of which AUTOMATIC_SINGLE_SURFACE contact was used for the target plates and SURFACE_TO_SURFACE_INTERFERENCE was used as a contact between projectile and target plates. This contact is close to realistic case of composites where one of the failure mechanisms involved is delamination. Material model to define composite target plate in LS-Dyna was MAT_054/055 ENHANCED_COMPOSITE_DAMAGE. This model requires Young's modulus and Poison ratio, along with longitudinal and transverse failure stress in tension and compression and shear failure stress.

All the values were taken from the data available in [4]. Shell elements were used for modelling the fabrics since MAT 054/055_ENHANCED_COMPOSITE_DAMAGE holds well especially for shell element and also ensures low computational time compared to solid element.

**Table 1** Modelling parameters for with STF and without STF Kevlar fibre cases [4].

| Type | Density kg/m$^3$ | Young's modulus E1 and E2 (GPa) | Poison ratio (μ) | Shear modulus G12 GPa | Shear modulus G13 and G23 (GPa) | Layer thickness (mm) | Boundary condition | Static coefficient of friction ($F_s$) | Dynamic coefficient of friction ($F_d$) |
|---|---|---|---|---|---|---|---|---|---|
| Kevlar | 733 | 20 | 0.25 | 0.77 | 5.40 | 0.2 | $Y$, $R_y$ Bound all four sides | 0.4 (*y*-to-*y*) 0.26 (pro-to-fab) | 0.2 (*y*-to-*y*) 0.18 pro-to-fab |
| Kevlar STF | 825 | 18.5 | 0.25 | 0.77 | 5.43 | 0.2 | $Y$, $Ry$ Bound all four sides | 1.93 (*y*-to-*y*) 5.5 (pro-to-fab) | 1.62 (*y*-to-*y*) 0.9 (pro-to-fab) |

*y*-to-*y* = Yarn to yarn, pro-to-fab = projectile to fabric, $R_y$ = Rotational constraint at local *y*-axis

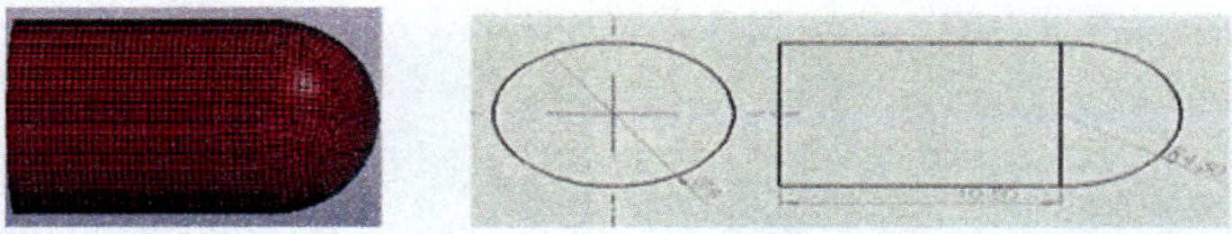

**Fig. 1** Projectile

The projectiles are assumed to be made of steel and defined as a rigid body in the high-velocity impact simulations. This is because the projectile does not deform significantly during penetration through the fabric. The density of the bullet is taken as 7850 kg $m^{-3}$, Young's modulus, 207 GPa, and Poisson's ratio, 0.3 [6]. The projectile was assigned the properties of mild steel using the MAT_020 RIGID model. Cylindrical projectile with a hemispherical nose was modelled using solid elements. The shape and dimensions of the results are shown in Fig. 1.

## 3 Results and Discussion

The experiments performed give us the information of the ballistic limit and the energy absorbed by the sample on impact. The LS-Dyna simulations are then used to validate these results. A progression of images showing the impact event simulation is shown in Fig. 2. Here, a–b show the initial mesh, c–f show the stress during impact, and finally g–h show the projectile penetration. The results of the various tests performed on Kevlar fabrics are given in this section.

### *3.1 Calculation of the Output Parameter*

To calculate the impact energy absorption for each specimen the following equation is used, which is derived assuming that the impact energy absorption of the specimen is the same as the energy loss of the impact projectile before and after impact.

$$E_{ab} = 1/2\, m_p (v_i^2 - v_r^2) \quad (1)$$

where

$E_{ab}$ Energy absorbed by the fabric (J).
$m_{\mathrm{p}}$ Mass of the projectile (kg)
$vi$ Initial velocity of the projectile (m/s).
$vr$ Residual velocity of the projectile (m/s).

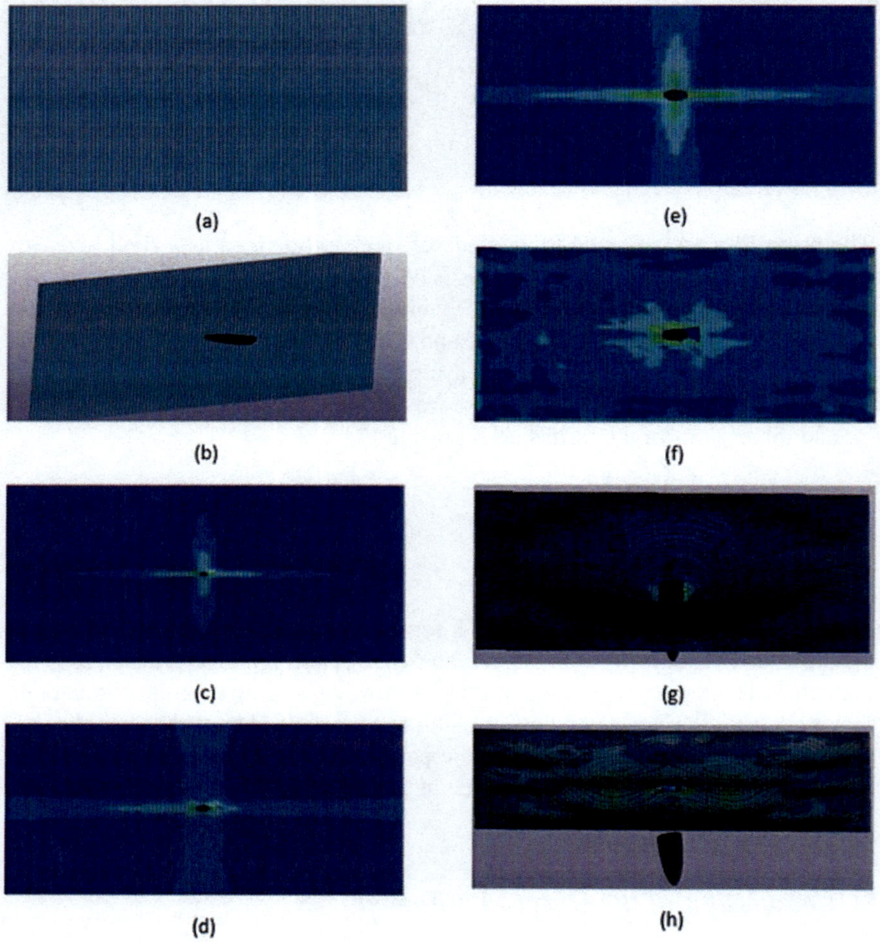


**Fig. 2** Impact penetration history

The impact parameters of 4, 6 and 8 layers of Kevlar fibre mats without and with STF are given in Tables 2 and 3. It is found that the addition of STF results in an improvement in the ballistic limit of the samples. The numerical results are comparable to the experiment results of [7]. The plots of the final velocities and energy absorption with the initial velocities are shown in Figs. 3 and 4.

The curves showing the change in initial velocity with final velocity as the bullet undergoes penetration for layers without STF are shown in Figs. 3. Similarly with STF is shown in Fig. 4.

The expression for residual velocity is given by

**Table 2** Initial velocity, residual velocity and energy absorption of 4, 6 and 8 layers of Kevlar fabrics without STF

| In/vel (m/s) | 4 Layers | | | | In/vel (m/s) | 6 Layers | | | | In/vel (m/s) | 8 Layers | | | |
|---|---|---|---|---|---|---|---|---|---|---|---|---|---|---|
| | Fi/vel (m/s) | | Energy (J) | | | Fi/vel (m/s) | | Energy (J) | | | Fi/vel (m/s) | | Energy (J) | |
| | Ex | L-D | Ex | L-D | | Ex | L-D | Ex | L-D | | Ex | L-D | Ex | L-D |
| 57 | – | – | – | 12.3 | 76 | – | – | – | 21.9 | 90 | – | – | – | 30.78 |
| 58.14 | – | 3.97 | 11.8 | 12.7 | 86.97 | – | 28.1 | 26.5 | 25.7 | 94.17 | – | 15.4 | 31.1 | 32.79 |
| 89.92 | 74.05 | 62.1 | 14.4 | 16.05 | 102.74 | 32.72 | 61.01 | 32.9 | 25.96 | 110.85 | 36.9 | 45.3 | 37.9 | 38.88 |
| 101.73 | 89.92 | 79.7 | 14.6 | 15.13 | 108.71 | 50.02 | 66.91 | 32.9 | 27.89 | 113.91 | 49.0 | 58.27 | 36.64 | 36.4 |
| 128.87 | 110.7 | 107 | 15.3 | 19.5 | 131.53 | 89.9 | 99.7 | 32.5 | 27.9 | 143.36 | 94.4 | 93.6 | 42.2 | 33.31 |
| 142.57 | 119.9 | 120.4 | 20.96 | 22.12 | 141.91 | 102.1 | 112.6 | 34.33 | 28.3 | 151.04 | 103.1 | 107.9 | 44.86 | 42.39 |
| 152.4 | 131.6 | 132.1 | 20.81 | 21.89 | 155.76 | 117.8 | 122.6 | 36.6 | 34.93 | | | | | |

In = Initial, vel = velocity, Fi = Final, Ex = Experimental, L-D = LS-DYNA

**Table 3** Initial velocity, residual velocity and energy absorption of 4, 6 and 8 layers Kevlar fabrics with STF

| In/vel (m/s) | 4 Layers | | | | In/vel (m/s) | 6 Layers | | | | In/vel (m/s) | 8 Layers | | | |
|---|---|---|---|---|---|---|---|---|---|---|---|---|---|---|
| | Fi/vel (m/s) | | Energy (J) | | | Fi/vel (m/s) | | Energy (J) | | | Fi/vel (m/s) | | Energy(J) | |
| | Ex | L-D | Ex | L-D | | Ex | L-D | Ex | L-D | | Ex | L-D | Ex | L-D |
| 70.15 | – | – | 17.29 | 18.69 | 94.17 | – | – | 31.16 | 33.13 | 109.22 | – | – | 41.92 | 44.73 |
| 87.34 | 30.74 | 42.52 | 23.61 | 22.1 | 100.28 | 12.45 | – | 34.75 | 38.21 | 113.48 | 12.69 | 15.22 | 44.55 | 47.6 |
| 98.79 | 34.48 | 50.87 | 30.2 | 27.25 | 110.22 | 23.06 | 26.5 | 40.68 | 43.49 | 120.01 | 20.5 | 23.15 | 49.14 | 51.99 |
| 110.77 | 43.46 | 56.87 | 36.68 | 34.33 | 114.48 | 37.69 | 35.52 | 41.3 | 44.41 | 128.01 | 46.08 | 51.21 | 50.94 | 51.61 |
| 137.45 | 88.11 | 97.74 | 39.32 | 35.48 | 128.53 | 50.73 | 45.92 | 49.29 | 54.76 | 156.95 | 81.06 | 78.41 | 62.97 | 70.24 |
| 148.37 | 109.65 | 110.01 | 35.3 | 37.66 | 143.37 | 67.87 | 95.79 | 56.34 | 43.24 | | | | | |

In = Initial, vel = velocity, Fi = Final, Ex = Experimental, L-D = LS-DYNA

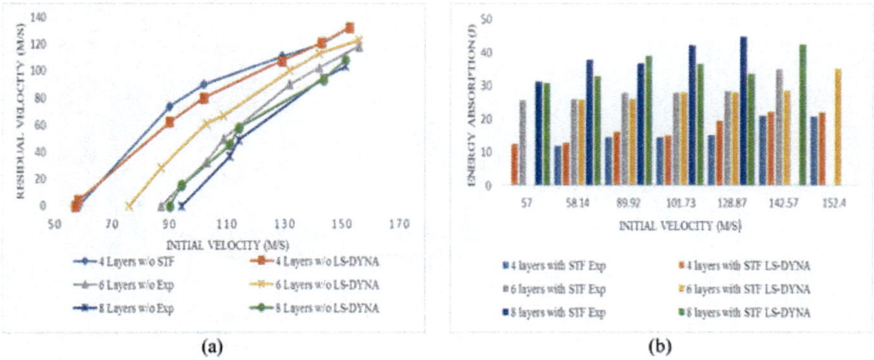


**Fig. 3** **a** Comparison of initial velocity with residual velocity and **b** energy absorption of 4, 6, 8 layers of Kevlar fabrics without STF

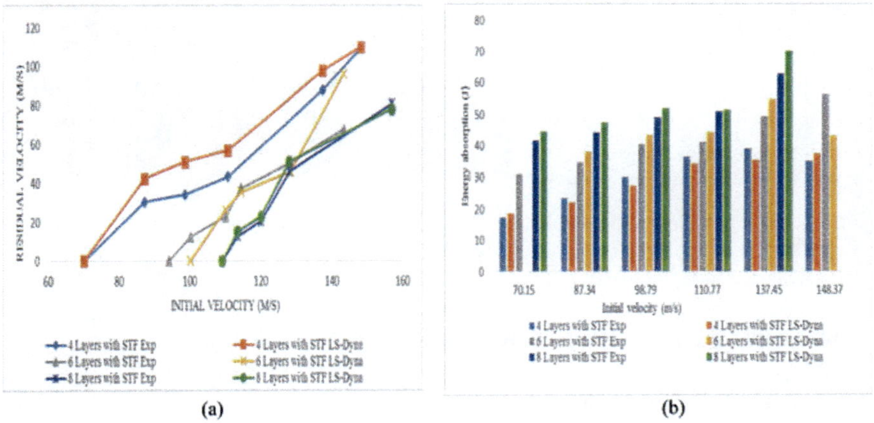


**Fig. 4** **a** Comparison of initial velocity with residual velocity and **b** energy absorption Kevlar with STF 4, 6, 8 layers

$$v_{\mathrm{R}} = \sqrt{\left(v_i^2 - \frac{2E_{\mathrm{C}}.V}{m}\right)} \tag{2}$$

And the expression for ballistic limit is given by

$$v_{\mathrm{B}} = \sqrt{\left(\frac{2E_{\mathrm{C}}.V}{m}\right)} \tag{3}$$

where

$v_i$ Initial velocity of projectile.
$v_{\mathrm{B}}$ Ballistic limit of the fabric.

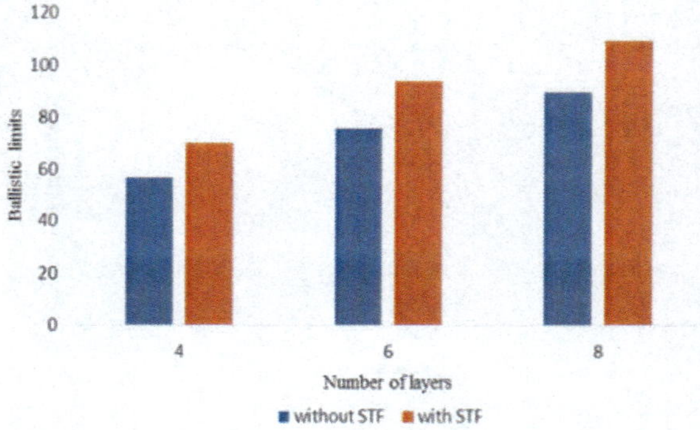


**Fig. 5** Variation ballistic limits for different layers of Kevlar fabric with and without STF

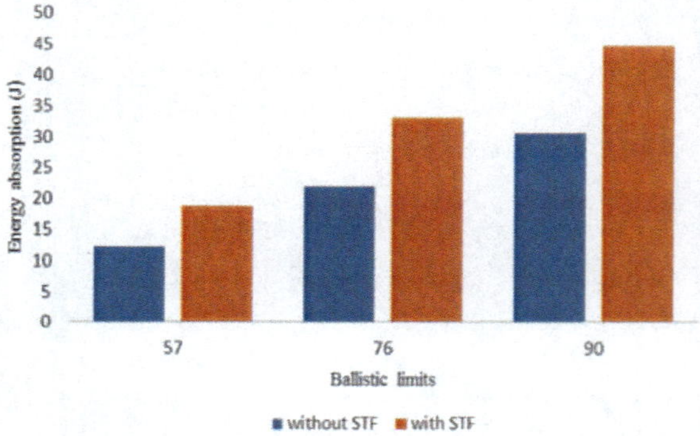


**Fig. 6** Energy absorption of the 4, 6, 8 layers Kevlar of fabric with and without STF

$E_c$ Area under the stress strain curve obtained from a tensile test.
$V$ 4 d.t,
$d$ Diameter of the projectile.
$t$ Thickness of the sample.
$R_D$ Radius of the damaged area in the primary yarns calculated from the sample.
$m$ Mass of projectile.

and the results are given in Tables 2 and 3 and in Figs. 3, 4.

The layers with STF show considerable improvement in energy absorption compared to the layers without STF. For example, in case of 4 layers, the energy absorption increases from 21 to 23%, which is a substantial increment and such increasing pattern is also seen with 6 and 8 layers Kevlar fabric with STF (Refer Figs. 5 and 6).

## 4 Conclusions

The behaviour of Kevlar fibre subjected to ballistic impact by a projectile has been analysed numerically. Samples with varying number of fibre layers are modelled and subjected to impact loading for different velocities. To improve the performance of these, fibre layers are impregnated with shear thickening fluid and were modelled. The impact resistance of Kevlar fabric has increased by adding STF. An improvement of 52%, 57%, 35% was observed for Kevlar fabric with STF to without STF, of 4, 6, 8 layers, respectively. Numerical results show good agreement with experimental results of Kevlar fabric with and without STF.

## References

1. P.A. Mossakovsky, A.M. Bragov, ME Kolotnikov, F.K. Antonov, Investigation of shear thickening fluid dynamic properties and its influence on the impact resistance of multilayered fabric composite barrier, in Proceedings of the 11th international LS-DYNA users conference, pp. 33–43 (2010)
2. A.A. Ramadhan, A.A. Talib, A.M. Rafie, R. Zahari, The Influence of impact on Composite armour system Kevlar-29/polyester-Al2O3. IOP Conf. Ser. Mater. Sci. Eng. **36**(1), 012028 (2012)
3. B.W. Lee, C.G. Kim, Computational analysis of shear thickening fluid impregnated fabrics subjected to ballistic impacts. Adv. Compos. Mater **21**(2), 177–192 (2012)
4. Y. Park, Y. Kim, A.H. Baluch, C.G. Kim, Numerical simulation and empirical comparison of the high velocity impact of STF impregnated Kevlar fabric using friction effects. Compos. Struct. **125**, 520–529 (2015)
5. L.M. Bresciani, A. Manes, A. Ruggiero, G. Iannitti, M. Giglio, Experimental tests and numerical modelling of ballistic impacts against Kevlar 29 plain-woven fabrics with an epoxy matrix: Macro-homogeneous and Meso-heterogeneous approaches. Compos. B Eng. **88**, 114–130 (2016)
6. A. Khodadadi, G.H. Liaghat, A.R. Sabet, H. Hadavinia, A. Aboutorabi, O. Razmkhah, M. Akbari, M. Tahmasebi, Experimental and numerical analysis of penetration into Kevlar fabric impregnated with shear thickening fluid. J. Thermoplast. Compos. Mater. **31**(3), 392–407 (2018)
7. A. Nashikkar, R. Boomurugan, R. Velmurugan, A. Katiyar, T. Nandi, Impact performance of kevlar fabrics with shear thickening fluid. J. Aerosp. Sci. Technol. **71**(3), 289–300. https://www.aerojournalindia.com/2019%20contents/Contents_August_2019.pdf
8. C. Ha-Minh, A. Imad, T. Kanit, F. Boussu, Numerical analysis of a ballistic impact on textile fabric. Int. J. Mech. Sci. **69**, 32–39 (2013)

# Comparative Study of Nano and Micro Fillers in EPDM/Silicone Rubber for Outdoor Insulator Application

**S. Bhavya, Unnam Mahesh, R. Velmurugan, and R. Sarathi**

**Abstract** Polymer outdoor insulators are widely replacing the conventional ceramic and porcelain insulators due to their compelling properties. This work, studies and compares the mechanical performance of an elastomer matrix, nano and micro filler reinforced, thermoset composite, EPDM Silicone rubber filled with Garamite nano clay and Micro Silica/Alumina Trihydrate, for their applicability as an outdoor insulator. The base composition used for the study is a 50–50 Ethylene Propylene Diene Monomer (EPDM) and Silicone rubber blend. The compatibilizer used in this study is Vinyl Trimethoxy Silane. The strength and weight are important parameters which are looked at, while choosing the material for the outdoor insulator application. Organic nanofillers due to their compact packing have been observed to improve the strength of rubber composites to a great extent even when added in small proportions. This paper discusses the variation in strength of the composite with varying concentrations of Garamite (organic nanoclay). Also, to understand its effect on composite, the concentrations of micro fillers such as Alumina Trihydrate and Fumed Micro Silica which yield comparable strength to that of Garamite are studied. The study reveals that there is not much variation in strength between ATH and Micro Silica fillers, while a distinct difference in strength is observed in the case of the composite with Garamite.

**Keywords** Alumina trihydrate · Ethylene propylene diene monomer · Garamite · Micro silica · Silicone rubber

S. Bhavya · U. Mahesh · R. Velmurugan (✉)
Department of Aerospace Engineering, Indian Institute of Technology Madras, Chennai, Tamil Nadu 600036, India
e-mail: ramanv@iitm.ac.in

S. Bhavya
e-mail: bs156@snu.edu.in

U. Mahesh
e-mail: unnam.mahesh2@gmail.com

R. Sarathi
Department of Electrical Engineering, Indian Institute of Technology, Madras, Chennai, India
e-mail: rsarathi@iitm.ac.in



K. M. Pandey et al. (eds.), *Recent Advances in Mechanical Engineering*, Lecture Notes in Mechanical Engineering, https://doi.org/10.1007/978-981-15-7711-6_75

## 1 Introduction

Owing to the better properties, such as lightweight, high mechanical strength, durability, good hydrophobicity, when exposed to stresses due to electric field, fluctuating temperature, moisture and pollution, elastomeric composites are rapidly replacing their ceramic counterparts. Since their introduction in the mid-twentieth century, they have gained attention of many researchers as they offer high scope for improvement in their properties [1, 2].

Addition of fillers to enhance the mechanical and electrical properties is a common choice among researchers. The size of the filler particles, alter the nature of the composite to a great extent. Nanofillers offer scope for further reduction in weight when added at optimum levels, while maintaining the required strength. These fillers apart from reducing the weight also enhance other properties. Optimizing the design and formulating a suitable material may considerably reduce the weight of the insulator.

The effect of addition of fillers on mechanical properties and their effect on the weight are studied. The fillers are investigated based on their size and effect. Micro Fillers: Fumed Micro Silica and Alumina Trihydrate (ATH) and Nano Filler: Garamite, are compared for their mechanical performance. The mechanical performance is studied by evaluating the tensile strength and hardness according to ASTM standards. In addition to the mechanical properties, the viscoelastic properties of the nanocomposites with Garamite are also studied.

The test results indicate an improvement in mechanical properties of nano filler at a particular composition in comparison to the Micro Fillers added in much greater concentrations. These results indicate the effectiveness of the organic nano fillers which can be appreciated from the data. This study helps in identifying a suitable concentration of the filler, which may help in choosing the optimized nano filler concentration in future studies.

Extensive research has been performed on silicone rubber insulators due its superior hydrophobicity and temperature resistance. But SiO has less alkyl group and larger bond length than that of C=C, which results in weak molecular forces and in poor mechanical properties. To overcome this shortcoming of silicone, EPDM is added to the elastomeric composite. EPDM, apart from enhanced mechanical properties, has good weathering properties such as oxidation and ozone resistance [3].

The phenomenon of surface tracking is of major concern in designing outdoor insulators as it deteriorates the dielectric properties of the material, making it a poor insulator.

Alumina Trihydrate (2 3 3 2), a micro filler, allows endothermic reactions which enable the release of water vapour on its surface, thereby reducing the level of thermal degradation thus limiting the formation of carbon tracks. This property makes ATH a filler of interest [4].

Silicone rubber composites are hydrophobic due to their low surface tension. This is the primary mechanism against the growth of leakage current which ultimately causes dry band arcing. Dry band arcing plays a crucial role in material degradation. Fillers act as secondary barriers against erosion and tracking. Studies show better erosion and tracking performance of fumed silica [5].

# 2 Material Specification

The raw material used for the preparation of the composites are given in Table 1.

## 2.1 Material Preparation

1. Garamite filled EPDM and Silicone rubber nano composite with variations in Garamite content. These composites are referred as ESG (EPDM Silicone Garamite)-Numeral (indicating the phr of Garamite).
2. Alumina Trihydrate (ATH) filled EPDM and silicone rubber composite (EPDM Silicone ATH).
3. Fumed micro silica filled EPDM and silicone rubber composite (EPDM Silicone Micro silica).

Tables 2, 3 and 4 show the compositions of the various rubber composites studied in this work.

**Table 1** Specifications of the raw materials used

| S. No. | Raw Material |
|---|---|
| 1 | Ethylene propylene diene monomer rubber (EPDM) S6090WF |
| 2 | Silicone rubber (SiR) XHG-2061 |
| 3 | VinylTrimethoxy silane (VMTO) |
| 4 | Dicumyl peroxide |
| 5 | Nanoclay, Nanomer I.44p, Nanomer® CLAy |
| 6 | Alumina trihydrate |

**Table 2** Composition of EPDM–Silicone rubber composite with Garamite nano filler (ESG)

| S. No. | Material | ESG-0 | ESG-1 | ESG-3 | ESG-5 | ESG-7 |
|---|---|---|---|---|---|---|
| 1 | EPDM | 50 | 50 | 50 | 50 | 50 |
| 2 | Silicone | 50 | 50 | 50 | 50 | 50 |
| 3 | Silane (VMTO) | 3 | 3 | 3 | 3 | 3 |
| 4 | Garamite | 0 | 1 | 3 | 5 | 7 |
| 5 | DCP | 6.25 | 6.25 | 6.25 | 6.25 | 6.25 |

**Table 3** Composition of EPDM–Silicone rubber composite with Alumina Trihydrate filler (ESA)

| Sample name | EPDM | Silicone | Vinyl trimethoxy silane (VMTO) (phr) | Alumina trihydrate (phr) | Dicumyl peroxide (phr) |
|---|---|---|---|---|---|
| Sample composition | 50 | 50 | 3 | 30 | 6.25 |

**Table 4** Composition of EPDM–Silicone rubber composite with micro silica filler (ESM)

| Sample name | EPDM | Silicone | Silane (VMTO) (phr) | Micro silica (phr) | Dicumyl peroxide (phr) |
|---|---|---|---|---|---|
| Sample composition | 50 | 50 | 3 | 30 | 6.25 |

### 2.1.1 Preparation of EPDM Silicone Nano Composites:

A Two Roll Mill is used to blend EPDM and silicone rubbers. EPDM due to its high rigidity is first made pliable by compressing it in the two roll mill at room temperature for 20 min. The EPDM which is white in colour is made into thin translucent sheet by milling. Following this, silicone rubber which is easy to blend is milled for less than two minutes and made into thin sheet. The two rubbers, taken in the ratio 1:1 are then mixed together until a uniform elastomeric blend is obtained. By blending it for half an hour, any irregularities can be avoided. Garamite nanoclay of varying compositions are added to the rubber mixture and blended until it is completely mixed for 15–20 min. Liquid silane (VTMO) is added to compatibilise the rubbers, following which Dicumyl Peroxide (DCP) is added. DCP acts as the vulcanizing agent.

### 2.1.2 Preparation of EPDM Silicone ATH/Micro Silica Composites:

The Two Roll Mill is heated to 45 °C to loosen the rubbers as to take in the fillers added. EPDM is milled first and made into a very thin sheet. Silicone rubber is then milled for a few minutes following which the two rubbers are blended for 12–15 min. Alumina trihydrate (30 phr)/ fumed micro silica (25 phr) in powder form is added slowly and blended (20–30 min) till it is uniformly dispersed in the rubber mixture. Silane is added to the mixture and blended to compatibilise the rubbers. Finally dicumyl peroxide is added to vulcanize the rubber mixture.

### 2.1.3 Vulcanization and Molding

Based on the literature survey the blended rubbers were vulcanized at 160 °C for 8 min. Flat plates of 3 mm are made using a mold of 300 mm × 300 mm in a hydraulic compression mold at a pressure of 100 bar for tensile testing and DMA Analysis. For surface charge accumulation analysis, sheets of 1 mm thickness are molded. The vulcanized rubber sheets are clamped to wooden boards and cut using water jet machines according to the ASTM Standards.

## *2.2 Characterization of Silicone EPDM Rubber Blends*

Mechanical and viscoelastic properties of the rubber composites prepared are evaluated for their suitability as Outdoor Insulators.

### 2.2.1 Mechanical Properties

Though not a structural member, the insulator should be able to resist the loads which may act during the processes of handling, packaging and while at service. The properties of strength and stiffness majorly constitute the mechanical properties apart from the hardness. This study particularly focuses on the enhancement of the strength with addition of fillers to the rubber blend.

Tensile Strength

An important characteristic of a material used for insulator purposes is the tensile strength. The samples are cut in dumb bell shape according to ASTM D-412. Universal Testing Machine UTM Instron is used at a cross head speed of 500 mm/min at room temperature. Each composition is tested with five samples and the average is considered. The average thickness is calculated by taking the average of the thicknesses at three different cross sections within the gauge length. The values are recorded till complete failure of the samples is observed.

Hardness

To withstand vandalism the material should have desirable hardness. It should resist the force at surface level. A Barcol hardness tester is used to measure the hardness of the rubber composite according to ASTM Standard D-2583. According to the standard, the sample should have a minimum thickness of 1/16th inch. The samples tested have an average thickness of 4 mm. The indenter of the instrument is pressed against the sample kept on a hard, flat horizontal surface. Readings are taken at five different points on the sample and the average calculated is taken as the final value.

#### 2.2.2 Thermal Properties

**Dynamic Mechanical Analyzer**

The insulators are exposed to varied stress conditions both electrical and thermal. Thus it is important to understand the behaviour of the material and their composites under cyclic loads. Dynamic Mechanical Analyzer (DMA) is used for this purpose [6]. The EPDM Silicone rubber composites with Garamite fillers were tested in DMA over a temperature range of −55 °C to 60 °C at 1 Hz frequency at a heating rate of 3 °C/min, with DMA 242 D NETZSCH. The samples were cut to a size of 40 mm × 10 mm × 4 mm and tested using bending fixture. As the material to be tested is thick and can take bending loads the flexural clamp setup gives a close approximation of the tensile properties [7].

## 3 Results and Discussions

### *3.1 Evaluation of Tensile Strength*

The tensile strength values obtained for different compositions of EPDM and silicone rubber are as given in Tables 5, 6 and 7.

The tensile strength values obtained for compositions of EPDM silicone and Garamite (0, 1, 3, 5, 7 phr) show that the tensile strength values increase up to 3 phr and decrease on further increasing the concentration of nano clay. The decrease in strength with increase in nano clay concentration might be due to improper dispersion of the fillers in the rubbers. Table 5 shows the tensile strength as a function of nano clay composition.

Table 6 gives the tensile strength of an EPDM silicone rubber composite added with fumed micro silica. It is observed that the average tensile strength of the

**Table 5** Variation in tensile strength of EPDM silicone rubber blends with Garamite

| Sample name | ESG-0 | ESG-1 | ESG-3 | ESG-5 | ESG-7 | Silicone |
|---|---|---|---|---|---|---|
| Sample composition | 0 phr | 1 phr | 3 phr | 5 phr | 7 phr | Silicone |
| Strength (MPa) | 1.6 ± 0.21 | 1.62 ± 0.13 | 4.6 ± 0.22 | 1.89 ± 0,29 | 0.84 ± 0.25 | 2.85 ± 0.19 |

**Table 6** Tensile strength of EPDM silicone rubber blends with fumed micro silica

| Sample No. | Thickness (mm) | Width (mm) | Strength (MPa) |
|---|---|---|---|
| 1 | 4 | 6.4 | 3.92 |
| 2 | 3.45 | 6.2 | 3.92 |
| 3 | 3.85 | 6.2 | 3.97 |

**Table 7** Variation in tensile strength of EPDM silicone rubber blends with alumina trihydrate

| Sample No. | Thickness (mm) | Width (mm) | Strength (MPa) |
|---|---|---|---|
| 1 | 3.3 | 6.35 | 4.26 |
| 2 | 3.6 | 6.1 | 3.64 |
| 3 | 3.65 | 6 | 3.78 |

**Table 8** Hardness of EPDM silicone rubber blends with Garamite nano filler

| Sample name | ESG-0 | ESG-1 | ESG-3 | ESG-5 | ESG-7 |
|---|---|---|---|---|---|
| Sample composition | 0 phr | 1 phr | 3 phr | 5 phr | 7 phr |
| Hardness | 28 | 15 | 5 | 10 | 15 |

composite ESM with 25% of fumed micro silica has equivalent or lesser strength when compared to the sample containing 3 phr Garamite.

Table 7 gives the Tensile strength of EPDM Silicone rubber composite added with 30 phr Alumina Trihydrate. It is observed that the tensile strength of ESA with 30 phr ATH is comparable to the strength of the nano composite with 3 phr Garamite.

Thus it is clear from the results, that the particle size of the filler plays an important role in shaping the property of the composite.

## 3.2 Evaluation of Hardness

The tensile strength values obtained for different compositions of EPDM and Silicone rubber are as given in Tables 8.

## 3.3 Evaluation of Thermal Properties

**Dynamic Mechanical Analysis: Storage Modulus (E')**

The energy stored per cycle against heat is denoted by the term Storage Modulus [6]. A higher storage modulus value is desired, as it indicates a better load bearing ability of the material while thermal scanning. Figure 1 shows the effect of temperature on storage modulus for the rubber composites with Garamite nanoclay. The graph shows a rapid decay in storage modulus as the temperature increases from −55 °C to −40 °C for nano composites, while the decrease is gradual for neat silicone rubber. On further increasing the temperature, it is observed that the decrease is slow and after 0 °C a plateau region is formed. It is evident that from −55 °C to −40 °C the storage modulus of the rubber composite containing 7 phr has higher storage modulus and 3 phr has the least. After −40 °C the storage modulus of 5 phr rubber composite increases.

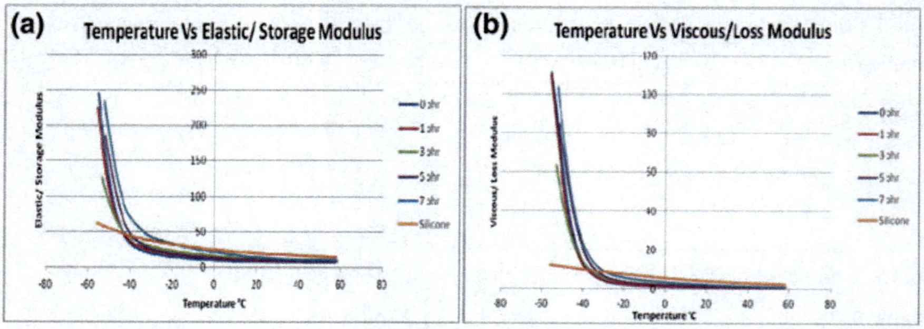


**Fig. 1** **a** Variation of elastic/storage modulus with temperature for Garamite added rubber composites; **b** variation of viscous/loss modulus with temperature

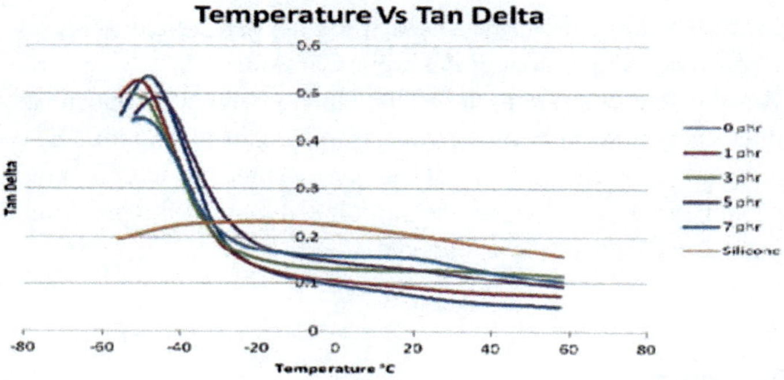


**Fig. 2** Variation of tan delta with temperature

**Loss Tangent (Tan $\delta$)**

The ratio of loss modulus to storage modulus gives a constant called Loss Tangent. Figure 2 shows the effect of temperature on loss tangent for the rubber composites with Garamite nanoclay. It is observed from the graph that the nano composite with 3 phr has the least glass transition temperature thus indicating a wider operating temperature range. A rapid decrease in tan delta value is observed from −45 °C to −20 °C. After −20 °C the tan delta values form a plateau.

## 4 Conclusion

The study results confirm that the addition of nano fillers to EPDM Silicone rubber blend has a great effect on the strength of the composite as a function of their concentration.

It is evident from the study, that the strength achieved with the addition of 25% micro silica and 30% alumina trihydrate can be achieved with 3% of Garamite nano filler. This could be attributed to the size factor. The nano composites have a higher surface area compared to the micro composites. This increased surface area enhances the immersion of nano fillers in the rubber matrix.

The lower the glass transition temperature of the material, the higher is the operational range of the insulator. From the results, it is observed that the composite with 3% nano filler has the least glass transition temperature of −53 °C.

**Acknowledgements** The work is supported by CPRI, Bangalore.

## References

1. F. James, History and Bibiliography of polymeric Insulators for Outdoor Applications. IEEE Trans. Power Delivery **8**(1), 376–385 (1993)
2. J.S.T. Looms, Insulators for High Voltages Peter Peregrinus Ltd. Copyright (1988), p. 19
3. Silicone Rubber. https://en.wikipedia.org/wiki/Silicone_rubber (2019)
4. A. Madi, Y. He, L. Jiang, B. Yan, Surface tracking on polymeric insulators used in electrical transmission lines. Indonesian J. Electr. Eng. Comput. Sci. **3**(3), 639–645 (2016)
5. H. Meyer, S.H. Jayaram, The role of inorganic fillers in silicone rubber for outdoor insulation aluminatri-hydrate or silica. Feature **20**(4)
6. A.S. Rana, M. Kiran Vamshi, K. Naresh, R. Velmurugan, R. Sarathi, Mechanical, thermal, electrical and crystallographic behaviour of EPDM rubber/clay Nanocomposites for out-door insulation applications. Adv. Mater. Process. Technol. (2019)
7. Characterization of Polymers using Dynamic Mechanical Analysis—By Debra Dunson Ph.d EAG Laboratories, Eurofins Materials Science. https://www.eag.com/wp-content/uploads/2017/09/M-022717-Characterization-of-Polymers-using-Dynamic-Mechanical-Analysis.pdf

# Machinability Study for Slot Cutting on Glass Using Ultrasonic Machining Process

**Anand Mohan Singh, Ranjan Majhi, and Promod Kumar Patowari**

**Abstract** With the advent of new technology in the manufacturing sector, the demand for micro products has been increased considerably. Having unique mechanical properties, glass has widespread application in the fabrication of micro products, but machining of glass at the micro-level, without any structural change is a challenging task. This work mainly emphasizes on slot-cutting/slotting of glass using ultrasonic machining. A stainless-steel tool of dimensions (12 mm × 8 mm × 0.6 mm) has been prepared and attached with a tool holder by the gas welding process and is used for slot cutting on a glass slide. The work also focuses on the effect of feed rate, power rating, and grit size on overcut, and edge deviation of the slot cut glass slide.

**Keywords** Ultrasonic machining · Slot cutting · Power rating · Feed rate · Grit size

## 1 Introduction

To meet the increasing demand for micro components in various industries like communication, biotechnology, optics, and aeronautics, etc., several micro-manufacturing processes have been developed. These processes fabricate micro products with features in the range of a few hundred microns.

Ultrasonic machining is one of such processes in which removal of material takes place due to continuous striking of abrasive particles, having high hardness. Themajor advantage of USM is that it does not impart any residual stress and surface damage to

A. M. Singh · R. Majhi (✉) · P. K. Patowari
Department of Mechanical Engineering, National Institute of Technology Silchar, Silchar, Assam 788010, India
e-mail: ranjanmajhi36@gmail.com

A. M. Singh
e-mail: singh.anandmohan16@gmail.com

P. K. Patowari
e-mail: ppatowari@yahoo.com



K. M. Pandey et al. (eds.), *Recent Advances in Mechanical Engineering*, Lecture Notes in Mechanical Engineering, https://doi.org/10.1007/978-981-15-7711-6_76

the workpiece as there is no direct contact between tool and workpiece. Any brittle conductive/non-conductive work material can be machined using a tough shaped tool. By opting suitable process parameters in USM high surface finish along with good dimensional accuracy can be produced. Thus, ultrasonic machining (USM) is utilized for machining hard and brittle material into complex shapes with good accuracy and high efficiency [1–3].

The accuracy of ultrasonically generated machined features is affected by slurry concentration, grit size, and feed rate, etc. The effect of some dominant parameters such as grit size, feed rate, and power rating has been considered for conducting experiments [4]. Finer abrasives resulted in reduced oversize and increased accuracy of the holes of the machined workpiece [5]. Taking Tungsten as tool material and working on glass material authors found that high power rating and high abrasive concentration resulted in more tool wear as abrasives hit the tool surface with impact. Further, they observed that very fine and very coarse both abrasives gave inaccurate surfaces [6]. Tool wear is non-uniform and depends on abrasive motion.Tool wear suffers three types of wear given as- longitudinal wear, i.e. reduction in length of tool which is found directly proportional to abrasive concentration.The second type is lateral wear i.e. taper formation on tool surface which is caused by abrasion and impact action of abrasives.Edge rounding is the third type of wear which is a reduction in the diameter of the tool face and is mainly caused due to repeated interaction of tool with abrasives [7]. It has been concluded that the accuracy of the machined surface primarily depends upon abrasive grit size. Finer abrasives not only reduced overcut but also enhanced the accuracy of holes of the machined surface [8].

During all experiments, a mixture of water along with boron carbide abrasive particle has been used in the fixed slurry concentration of 50%. After machining, the effects of input parameters have been analyzed on response variables. Many researchers performed experiments on different work materials like glass, ferrite, and alumina by taking different tool materials such as titanium and stainless steel. The effect of different abrasive sizes and hardness of the tool has been analyzed for surface roughness. The decrease in surface roughness observed with a decrease in grit size and harder tool material [9]. Due to improper slurry flow through cutting gap, machining on the workpiece of thickness greater than 12.5 mm is less effective [10, 11]. While performing drilling operation, it was observed that the hole of diameter as small as 76 μm can be drilled using USM [12]. Moreover, since USM is a machining process devoid of either thermal or electrical process, the internal properties of work material remain unaltered [4].

Apart from traditional glass, experimentation has also been carried out on $Zr_{60}Cu_{30}Ti_{10}$ metallic glass by drilling micro holes using micro-USM and it has been reported that slurry concentration of 50–60% gave good results; further edge deviation was found to be lesser for slurry containing higher sized abrasives [1]. From literature, it is concluded that several works have been reported on drilling of glass using USM. Minimal work has been performed as slot cutting operations on the glass. In this paper, the slot has been cut on glass by preparing tool of suitable dimension. For conducting experiments, the design of experiments has been done using

Taguchi L16 orthogonal array by taking different levels for selected input parameters which are tabulated under Table 2.

# 2 Methodology

## 2.1 Fabrication of the Tool

USM is a non-conventional machine with major parts as mentioned in Fig. 1a. Its main components are transducer, coupler, and horn. The transducer converts electrical energy into ultrasonic mechanical vibrations; coupler transmits the vibration further to horn and horn helps to amplify the vibrations. A threaded bolt acts as a tool holder that is attached to the horn. The tool of the desired shape is attached to this tool holder either by welding or soldering.

A rectangular stainless-steel sheet (12 mm × 8 mm × 0.6 mm) has been chosen as a tool to cut the slot on a glass slide of thickness 1.3 mm. The steel sheet has been gas welded on top of a tool holder to provide sufficient strength to the joint. The welded tool is attached to the horn to perform slot cutting on the glass slide. Figure 1b shows the tool before performing the slot cutting operation. Initially, the glass slide is fixed on the heat plate using glue. The plate on which the glass slide

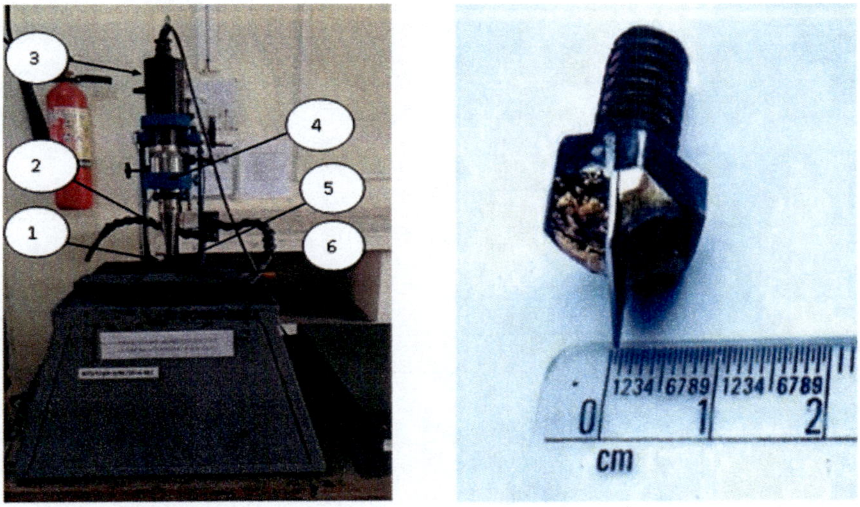


a) Showing various parts of USM

(b)The prepared tool attached with tool holder

1. Cutting tool 2. Horn 3. Transducer 4. Coupler 5. Electromagneticbase 6. Abrasive Flow Nozzle

**Fig. 1** **a** Showing various parts of USM, **b** the prepared tool attached with tool holder. 1—Cutting tool, 2—horn, 3—transducer, 4—coupler, 5—electro magnetic base, 6—abrasive flow nozzle

Table 1 Parameters selected and their ranges

| Parameters | Level | Range |
|---|---|---|
| Feed rate (μm/s) | 4 | 20, 30, 40, 50 |
| Grit size (mesh no.) | 2 | 400, 600 |
| Power rating (%) | 4 | 20, 25, 30, 35 |

has been attached is fixed on an electromagnetic base, applying a magnetic field. After performing experiments glass slides have been removed from the heating plate with the application of heat. Then acetone has been used for cleaning workpieces followed by observation using a microscope.

The input parameters which have been varied are grit size (Mesh no.), feed rate (μm/s) and power rating (watt). The power rating indicated as the percentage of input power fed to machine, i.e. if power rating is chosen as 20%, that means 20% of total input power is given to the machine. The response variables measured are overcut and edge deviation. Selected input variables along with their ranges and levels have been shown in Table 1. To conduct the experiments Taguchi L16 orthogonal array has been chosen. The experimental sequence has been followed as mentioned in Table 2 and detail output parameters are measured and shown in Table 2.

**Table 2** Experimental conditions and average output responses

| Exp. No. | Feed rate (μm/s) | Power rating (%) | Grit size (mesh no.) | Overcut (μm) | Edge deviation (μm) |
|---|---|---|---|---|---|
| 1 | 20 | 20 | 400 | 32 | 73.0 |
| 2 | 20 | 25 | 400 | 33 | 72.0 |
| 3 | 20 | 30 | 600 | 36 | 68.5 |
| 4 | 20 | 35 | 600 | 37 | 64.0 |
| 5 | 30 | 20 | 400 | 37 | 57.5 |
| 6 | 30 | 25 | 400 | 39 | 55.5 |
| 7 | 30 | 30 | 600 | 42 | 54.0 |
| 8 | 30 | 35 | 600 | 43 | 41.0 |
| 9 | 40 | 20 | 600 | 44 | 35.5 |
| 10 | 40 | 25 | 600 | 44 | 33.0 |
| 11 | 40 | 30 | 400 | 44 | 40.0 |
| 12 | 40 | 35 | 400 | 47 | 37.0 |
| 13 | 50 | 20 | 600 | 48 | 32.0 |
| 14 | 50 | 25 | 600 | 52 | 29.5 |
| 15 | 50 | 30 | 400 | 54 | 32.5 |
| 16 | 50 | 35 | 400 | 55 | 32.0 |

## 2.2 *Selection of Process Parameter*

Slot cutting of glass has been done using USM with the selected parameters which have been shown in Table 1.

# 3 Result and Discussion

Slot cutting on the glass slide has been successfully performed, Fig. 2a shows the glass slide machined using abrasive particles having grit size 400and Fig. 2b depicts the machined glass surface using grit size 600. The effect of feed rate, power rating and grit size on the overcut and edge deviation has been calculated for response variables, although the qualities of some slots are poor. Figure 3a shows the microscopic image of edge deviation along with the indicated dimension of one of the slots cut on the glass slide, while Fig. 3b depicts the inner surface of the slot.

For the grit size 400, at a feed rate of 20 μm/s and 20% of input power feed the minimum overcut is found to be 32 μm, while for same grit size minimum edge deviation of 32 μm has been observed at a feed rate of 50 μm/s and 35% of input power feed. For grit size 600, minimum overcut occurs as 36 μm at input values of 20 μm/s feed rate and 35% of feed power rate. While minimum edge deviation for grit size 600 is observed to be 29.5 μm at 50 μm/s feed rate and 25% of input power feed.

Figure 4 depicts the variation of edge deviation for various parameters such as feed rate, power rating and grit size. Edge deviation tends to increase with an increase

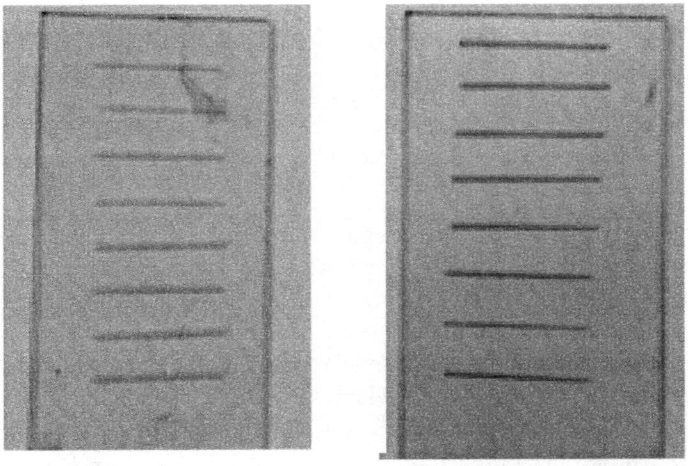

(a) With grit size 400 (b) With grit size 600

**Fig. 2** Workpiece after machining on USM using different grit size

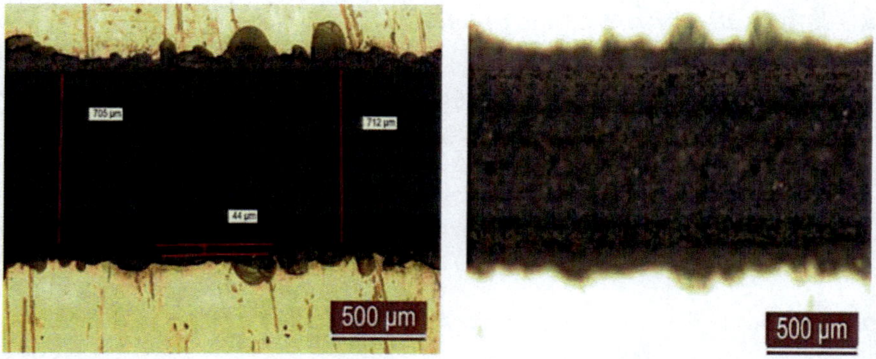


(a) Optical image of machined glass slide showing edge deviation (b) Optical image of slot machined on the glass slide.

**Fig. 3** Microscopic images of machined slots with dimensions

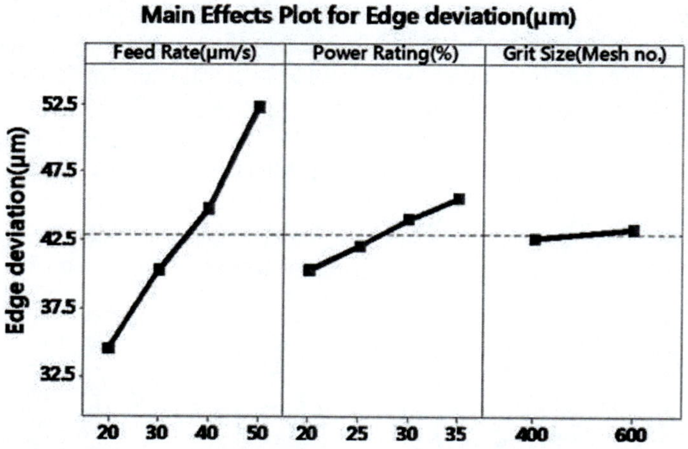


**Fig. 4** Variation of edge deviation with control parameters

in feed rate as tool penetration increases, the number of abrasive particles in contact with tool decreases and thus creating rough edges. With an increase in power rating edge deviation is observed to increase because of enhanced randomness in vibration. There is no significant variation found in edge deviation with grit sizes.

Figure 5 shows the variation of overcut for different control parameters. The overcut decreases with an increase in feed rate because of the high rate of tool penetration promoting less interaction of abrasives with the work material. A minimum overcut of 32 µm has been observed against feed rate of 50µ m/s. On increasing power rating, overcut is reduced because of the increased rate of vibration of the tool. With a decrease in grit size, the overcut has been found to decrease from 50–32 µm.

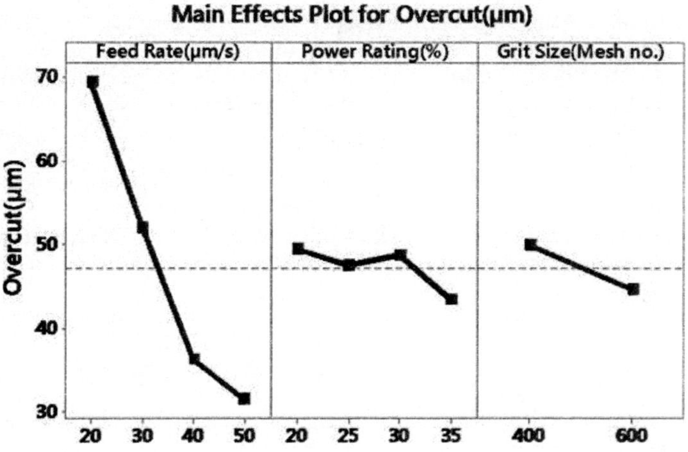


**Fig. 5** Variation of overcut with control parameters

## 4 Conclusions

Initially, the shaped tool has been prepared with the help of gas welding and then the tool has been employed to cut slots on a glass slide using micro-USM, the following conclusions can be drawn

- Overcut tends to decrease with an increase in all three control parameters, i.e. feed rate, power rating, and mesh number.
- Mean overcut of 50 μm is observed for 400 grit size and mean overcut corresponding to grit size 600 is 40 μm which is obvious because of the smaller grain size of abrasive particles with mesh number 600.
- Edge deviation increases with an increase in feed rate and power rating while almost no variation has been found in edge deviation with respect to grit size.
- Corresponding to the grit size of 400, a minimum edge deviation of 32 μm is observed at 50 μm/s feed rate and 35% of power rating. For grit size of 600 edge deviation is lower compared to the grit size of 400 and is found to be 29.5 μm at feed rate of 20 μm/s and 35% power rating.

## References

1. T. Debnath, A. Haashir, and P.K. Patowari, Parametric study of micro-hole drilling in glass using ultrasonic machining, in *International Conference on Precision, Meso, Micro and Nano Engineering* (2017)
2. A. Haashir, T. Debnath, P.K. Patowari, A comparative assessment of micro-drilling in boron carbide using ultrasonic machining. Mater. Manuf. Processes 1–9 (2019)

3. K.P. Rajurkar, Z.Y. Wang, A. Kuppattan, Micro removal of ceramic material ($Al_2O_3$) in the precision ultrasonic machining. Precis. Eng. **23**(2), 73–78 (1999)
4. S. Kuriakose, P.K. Patowari, J. Bhatt, Machinability study of Zr-Cu-Ti metallic glass by micro-holedrillingusingmicro-USM. J. Mater. Process. Technol. **240**, 42–51 (2017)
5. M. Adithanand, V.C. Venkatesh, Production accuracy of holes in ultrasonic drilling. Wear **40**(3), 309–318 (1976)
6. M.S. Cheema, P.K. Singh, O. Tyagi, A. Dvivedi, A.K. Sharma, Toolwear and form accuracy in ultrasonically machined microchannels. Measurement 81, 85–94 (2016)
7. T. Debnath, K.K. Patra, P.K. Patowari, Gang Drilling of Square Micro-Holes on Glass Using USM, in *Advances in Unconventional Machining and Composite*. (Springer, Singapore, 2020), pp. 549–557
8. M. Komaraiah, M.A. Manan, P.N. Reddy, S. Victor, Investigation of surface roughness and accuracy in ultrasonic machining. Precis. Eng. **10**(2), 59–65 (1988)
9. M.M. Barash, D. Watanapongse, On the effect of ambient pressure on the rate of removal in ultrasonic machining. Int. J. Mech. Sci. **12**(9), 775–779 (1970)
10. D. Goetze, Effectof vibration amplitude, frequency, and composition of the abrasive slurry on therate of ultrasonic machining in ketos tool steel. J. Acoust. Soc. Am. **28**(6), 1033–1037 (1956)
11. T.B. Thoe, D.K. Aspinwall, M.L.H. Wise, The effect of operating parameters when ultrasonic contour machining. rollsroyceplc-report-pnr (1995)
12. M. Moreland, Ultrasonic advantages revealed in the hole story. Adv. Ceramics **87** (1987)

# Mechanical Design of a Modular Underwater Rov for Surveillance and Cleaning Purpose

**Abhimanyu Pratap Singh, Atanu Paul, Yogesh Singh, and Koena Mukherjee**

**Abstract** The surveillance and cleaning of the cooling pipes in nuclear power plants is a tedious task, as these pipes are dimensionally huge and submerged under the sea. In this paper, we present the design process of a modular underwater remotely operated vehicle (ROV) for surveillance and cleaning purposes of the cooling pipelines at Nuclear Power Plant. We divided the design into five sections: structure, propulsion system, electronic system, illumination system and modular attachments. The iterative design process includes computer-aided design (CAD), finite elements analysis (FEA) and, computational fluid dynamics (CFD). We have tried to make the design modular while keeping the basic structure of ROV the same. This will save the costs of using different ROVs for different tasks while making the control easier as each of these modules will have their power supply and electronics and will only receive instructions from the parent ROV. Also, there are many ROVs available in the market that are being used for the surveillance purpose so here our objective is to use the materials and systems with minimal cost to make the design cost-effective for the purpose required.

**Keywords** AUV · Mechanical design · Modular design · Propulsion system · ROV · Underwater robotics

A. P. Singh (✉) · A. Paul · Y. Singh
Department of Mechanical Engineering, National Institute of Technology Silchar, Silchar, Assam, India
e-mail: abhimanyusingh02@outlook.com

Y. Singh
e-mail: yogeshsingh15@gmail.com

A. P. Singh · A. Paul · K. Mukherjee
Department of Electronics and Instrumentation Engineering, National Institute of Technology Silchar, Silchar, Assam, India
e-mail: koena@ei.nits.ac.in



K. M. Pandey et al. (eds.), *Recent Advances in Mechanical Engineering*, Lecture Notes in Mechanical Engineering, https://doi.org/10.1007/978-981-15-7711-6_77

## 1 Introduction

Over the last few decades, underwater robotic vehicles (URVs) have seen improvement and tremendous growth in use. Many are used for underwater inspection of subsea cables, oil and gas installations, structures, pipelines and also in seafloor observation. They are essential at depths where the use of human divers is impractical. Due to durability and capability, URV application is more suitable to be used in these conditions. Traditionally, URVs can be broadly classified into remotely operated vehicle (ROV) and autonomous underwater vehicle (AUV), depending on their designed tasks and modes of operations. AUVs are robots that have no remote control mechanism. They are programmed to perform certain tasks independently of direct human control. ROVs are remotely controlled vehicles.

Over the years, with advancements in computer processing power and manufacturing capabilities, the design and functions of ROV have improved and became complicated for the better. But all these design changes only dealt with more lightweight and aerodynamic ROV, we still use different robots for different applications. Though the concept of modularity is not new but the idea is still in its infancy when it comes to the ROV and the drone industry. There are few ROV which is modular like MARTA developed by the University of Florence [1] and MARIN by Maritime research institute Netherlands. Although these ROVs are modular, they are still not fully modular and adding different components needs a change in design and power requirements. Also, the added components are not independent of ROV.

Now about the Nuclear Power Industry, towards which our project is mainly aimed, worldwide nuclear power plants use seawater as a feasible means for the cooling purpose of nuclear reactors. These plants use giant suction pipes to suck water from the sea. Also, these pipes are made of concrete and can grow cracks within because of prolonged use, there may also be the growth of biological organisms and scales in the pipes. So inspection and cleaning of these pipes are of major concern for the proper functioning of such a cooling system [2]. These suction pipes can range from **a length of 500 to 800** m and of **a diameter of about 3–4** m and **at a depth of 40–50** m. Pressure at such depth ranges from **4 to 5** kgf/cm$^2$ which is not suitable for human systems [2]. So it is hazardous for divers to undertake cleaning and inspection operations in such a lengthy pipeline where dangerous and poising underwater organisms may exist. Here we feel the need for a remotely operated underwater vehicle **(ROV)** which can do the work of cleaning and inspection.

The proposed project aims to develop an ROV for surveillance and cleaning of the pipelines while being modular so that its different attachments can be attached to the existing ROV. This eliminates the cost of purchasing new ROV and training employees for proper control of ROV. If the ROV can be effectively used in the Nuclear power plants meeting all the safety requirements then the system can also be used in different industries and environments which are not as dangerous as a nuclear power plant. Moreover, if the module implements the defined tasks successfully, the idea can be implemented for other tasks also.

## 2 Design Constraints

The ROV has to operate at the shallow depth seawater, into the pipes used to supply the water required for reactor cooling. The constraints that are imposed on the ROV during its operation are addressed below.

### *2.1 Temperature, Salinity and, Density*

Seawater density is a function of salinity, pressure, and temperature. The temperature of seawater varies with depth but for a depth range of 0–200 m the operational temperature is equal to the surface's temperature [3], which is around 28 °C in the Indian Ocean. With an increase in hydrostatic pressure with depth, the density is likely to increase from 1021 kg $m^3$ at surface to 1070 kg/$m^3$ at 1000 m depth [4]. The seawater density at the surface is equal to 1024 kg/$m^3$ at the operational temperature. The Vehicle is supposed to work at a depth of around 50 m, which leads to an operational density of seawater is likely to be ~1024 kg/$m^3$.

### *2.2 Electromagnetic Radiation Propagation*

As depth increases, the penetration of light and radiation transmission capability of seawater decreases considerably. According to [4], the signal strength is inversely proportional to depth. For a depth of 100 m, there is minimal or better to say no visible light and radio wave propagation in freshwater. This phenomenon is even more severe in the case of seawater. This indicates that wireless communication systems cannot be used underwater [5] and the need for an illumination system and a tether cable for information transfer.

### *2.3 Working Depth*

The ROV should be designed to work at a depth of 50 m where the hydrostatic pressure on the ROV is around 0.6 MPa. Such high pressure is sufficient to collapse the hull of the ROV. Therefore the materials used in the system should be chosen accordingly so that they can sustain such a high amount of pressure. Moreover, internal ribbing is done inside the hull to provide better strength characteristics.

### 2.4 Design Depth

The ROV is designed for a depth greater than the working depth to assign a security margin to the design [3]. The usual working depth hydrostatic pressure varies between 0.5 and 0.6 MPa but to avoid failure of the system due to any fluctuations, the hydrostatic pressure for design is taken as 1 MPa. This gives a design depth of 90 m to the ROV. Moreover, this higher design depth gives us the ability to use the ROV at depths greater than 50 m.

### 2.5 Weight and Dimensions of the ROV

Our main objective is to design the ROV in such a way so that it can accommodate all the subsystems in an efficient way. The dimensions and weight should be as minimal as possible resulting in minimum power consumption during the movement. Moreover, a balance between the dimensions and weight is desirable for positive/neutral buoyancy of the system. Taking each aspect into consideration, it was decided that the ROV should be able to fit inside a cubic box of dimension 0.7 m.

## 3 Hydrostatic and Hydrodynamic Considerations

There are different hydrostatic and hydrodynamic components of forces that are acting on the ROV that must be taken into consideration for its design.

### 3.1 Static Equilibrium

Static equilibrium refers to the equilibrium of the system when it is at rest. At rest, the forces acting are its own weight and the buoyant force. It is always desired that the buoyant force be equal to the weight or greater, so that that the system is neutrally buoyant or slightly positively buoyant. Slightly positive buoyant system helps in easy retrieval of the vehicle to the surface in case of any failure making the system non-operational [3, 4].

### 3.2 Stable Equilibrium

When the system is given a slight angular displacement from its equilibrium position it should be able to attain its equilibrium position again. This condition is achieved when the center of gravity is positioned lower than the center of buoyancy. Efforts are made to keep the center of gravity low by concentrating most of the mass at the bottom. More the distance between the center of buoyancy and center of gravity (Metacentric height) more will be the stability [6, 7, 5].

### 3.3 Drag and Lift Force

Due to the relative motion between the System and the water, the system experiences drag and lift force. The magnitude and direction of these forces depend the shape of the body. The drag on the ROV consists of skin drag, which is related to surface area of ROV, and form drag, which is due to the shape of ROV. Streamlining of the body helps in reducing the form drag force and to produce the desired lift force which will subsequently help in yielding an energy-efficient design [8, 5]. The ROV can be designed to generate positive lift giving a positive buoyant design or can be designed to generate a negative lift which helps to keep the design positive buoyant in case of heavy systems.

### 3.4 Thrust

Thrust is provided by the propulsion system attached to the ROV for overcoming the drag and inertia force to keep the vehicle moving. The designed system needs to be 4-DOF, i.e. it should be able to move forward/backward (surge), ascend/descend (heave), move sideways (sway) and, rotate about the vertical z-axis to head in any direction. These movements are controlled using the propulsion system. According to the operating conditions, the maximum speed of the ROV is set at 0.5 m/s (Fig. 1; Table 1).

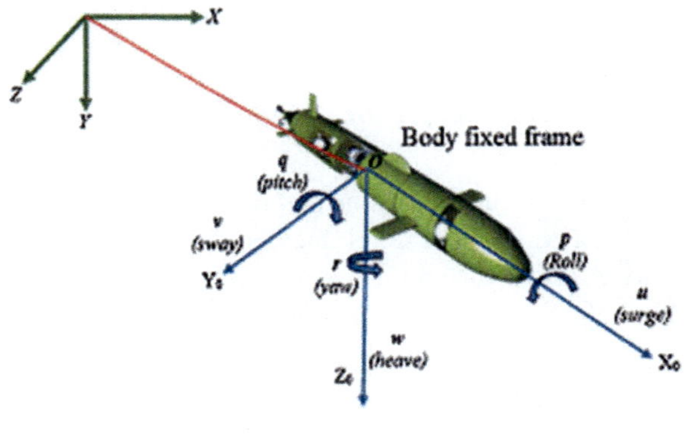


**Fig. 1** Schematic showing six different types of possible motion of an ROV

**Table 1** Design specifications

| S. No. | Design parameter | Specifications |
|---|---|---|
| 1 | Purpose | Surveillance and cleaning |
| 2 | Environment of operation | Seawater |
| 3 | Working depth | 40–50 m |
| 4 | Design depth | 90 m |
| 5 | Density of water | 1024 kg/m$^3$ |
| 6 | Speed | 0.5 m/s |
| 7 | Mobility | 4-DOF (surge, heave, sway, yaw) |
| 8 | Buoyancy | Positive |
| 9 | Communication system | Fiber optic cable |
| 10 | Power supply | External power supply with an on-board 12.8 V Li-ion battery (for electronic systems, propulsion and, manipulation in case of failure of external supply system) |
| 11 | Navigation system | DVL, gyroscope, SONAR |
| 12 | On-board sensors | CTD, pH, BOD sensor, leak detector |
| 13 | Controller unit | ATmega328p |
| 14 | Auxiliary systems | Camera (with manipulation system) and manipulator (for cleaning purposes) |

# 4 Design

The basic design is made in order to overcome the constraints. The basic design can be divided into five subsections- Structure, Propulsion system, power and electronic system, illumination system and camera, Modular attachments. Here we present two different designs that are suitable for the purpose. Each of their basic design details are discussed as follows.

Although both the designs are intended for the same operation, different design approaches are used for both designs.

## *4.1 Material Selection*

A comparative study was done before deciding the material for construction of ROV [9]. After taking into account factors like material price, strength, availability and, density eventually aluminum is decided as the material for construction due to its easy availability, low cost and high strength to density ratio (Table 2).

**Table 2** Comparative chart of available material for prototyping

| | Aluminum | HDPE plastic | ABS plastic |
|---|---|---|---|
| Cost (rupees/kg) | 180 | 183 | 282 |
| Density (g $cm^{-3}$) | 2.707 | 0.958 | 1.060 |
| Yield strength (MPa) | 276 | 25.9 | 43.3 |

## *4.2 Structure*

Most ROVs in existence so far are of two types: conventional rectangular box type hull which is intended for less speed and more working capability with more tools onboard and the other aerodynamic type which is intended for higher speed and surveillance purpose. We took both the type of designs into consideration and tried to come up with the design that fulfills both the purposes. Two of such designs are presented here. The thickness of material for each is decided after finite element analysis (Fig. 2).

In **design-1** a pressure hull of rectangular shape is designed to house the battery and electronic systems of the ROV. The shape of ROV is symmetrical form front and back, giving the operator the flexibility of operating the ROV from either side. The ellipsoidal shape provides a theoretical drag coefficient of 0.35. Aluminum brackets are provided on both sides for protection from collisions. The design is complete in itself and can be used for surveillance in this configuration. Different attachments can be added to the ROV in the lower portion. The design is kept as compact as possible and all the electronics and power systems are located inside the hull, making the ROV agile and efficient.

In **design-2** the pressure hull has a shape of an aero foil which will subsequently help in reducing the drag force and produce some lift force during the vehicle's movement [8]. The lift force generated will help the ROV to stay in its neutral equilibrium state during movement. The propulsion system and the auxiliary systems are mounted beneath the pressure hull in a separate frame. The frame offers a wide scope to attach the modular attachment as per the need.

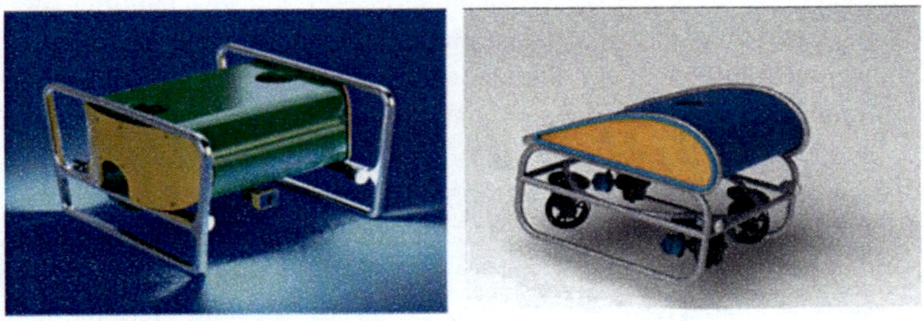

**Design 1** **Design 2**

**Fig. 2** Basic designs of vehicle

**Table 3** FE analysis results of hull

| Property/result | Design 1 | Design 2 |
|---|---|---|
| $E$ | 68 GPa | |
| $\mu$ | 0.33 | |
| Yield strength $S_y$ | 279 MPa | |
| Pressure @ depth = 90 m | 1 Mpa | |
| Thickness | 7 mm | 5 mm |
| Maximum von mises stress | 226.5 MPa | 237.2 Mpa |
| Maximum displacement | 0.9383 mm | 2.737 mm |
| Minimum factor of safety | 1.214 | 1.159 |

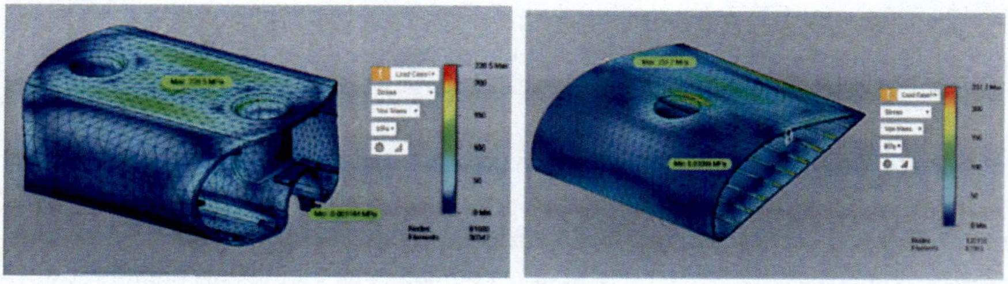

**Design 1** **Design 2**

**Fig. 3** FE analysis of hull for both designs

The finite element analysis of pressure hull for both designs is done and the findings are summarized in Table 3 (Fig. 3).

## *4.3 Propulsion System*

Blue Robotics T200 thrusters having maximum thrust of 51.5 N are used for propulsion. The thrusters are more expensive than other market offerings but offer superior power efficiency and thrust characteristics. Better efficiency enables us to utilize the available power more effectively for higher operation times. Also, high thrust outputs over a long range of rpm values help in making the ROV agile while giving easy control to the operator.

**Design-1** includes 5 thrusters in an orthogonal arrangement. Two thrusters are used for forward motion, one thruster is used for sideways motion and, two thrusters are used for upward and downward motion. Though this arrangement has some disadvantages as compared to vector thrust but the arrangement provides more thrust as compared to vector thrust when using same no. of thrusters. Also, the control scheme of direct arrangement is easy compared to vector thrust as controlling the thrusters in vector arrangement to operate at maximum efficiency while balancing the thrust at desired locations would be difficult task.

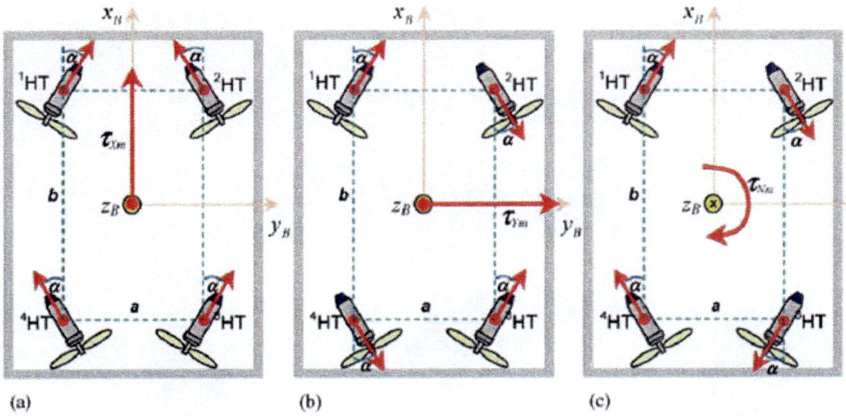


**Fig. 4** Motion using vector thruster arrangement

**Design-2** consists of a total of five thrusters to give the system 4-DOF movements.4 thrusters in Vector thrust arrangement is employed to provide the required surge, sway and yaw movement. One thruster is used to provide heave motion to the vehicle. The advantage of this type of propulsion system is that the surge, sway and heave motion can be obtained even using only two horizontal thrusters and thrust from individual thruster is added providing more thrust per similar-sized thruster as compared to direct thrust [5]. The horizontal thrusters are in the same horizontal plane and the vertical thruster passes through the center of mass thus restricting the roll and pitch motion of the vehicle (Figs. 4 and 5).

### *4.4 Thrust Requirements*

For a vehicle moving at a constant speed the thrust force required is equal to the Resistance due to drag [8, 6, 10]. Drag force is a function of the shape of the vehicle. Thus, thrust power is related to the vehicle shape.

Drag force can be calculated using the formula [6, 7], $\mathbf{F}_\mathrm{D} = 1/2 \times \rho \times \upsilon^2 \times C_\mathrm{d} \times A$.

Where $\rho$ is density of seawater, $\upsilon$ is velocity of the motion of ROV, $\boldsymbol{C}_\mathrm{d}$ is the coefficient of drag and, $A$ is the area in consideration of the ROV. The calculated drag forces for various velocities are plotted in graph given in figure for both the designs (Fig. 6).

The power consumed by the electronics (Hotel Load, $H$) on the ROV is approx. = 60 W.

The power provided by the thrusters = $\mathbf{F}_\mathrm{D} \times \mathbf{v}$ [10]

Therefore [8, 10],

$$\text{Total Power} = \left(H + \frac{1}{2} \times \rho \times \upsilon^3 \times C_\mathrm{d} \times A\right) \tag{1}$$

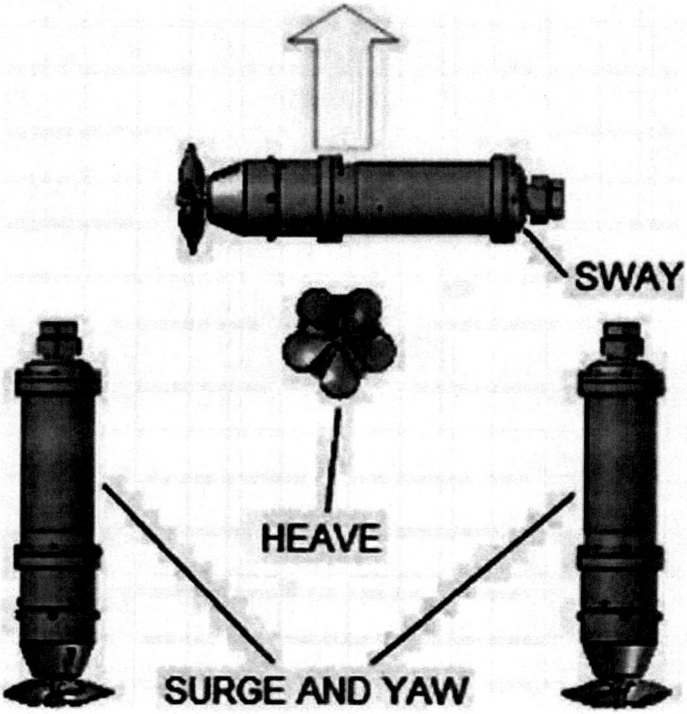


**Fig. 5** Motion using direct thrust arrangement

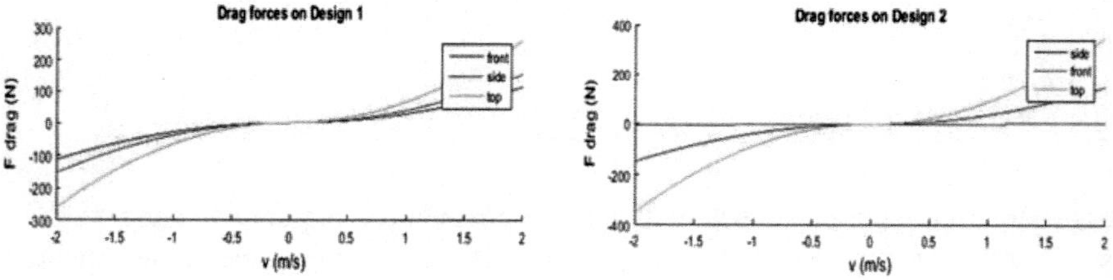


**Fig. 6** Drag forces on the ROV (without modular attachments) for different directions of motion

Now, Range $= v \times t$
Therefore,

$$\text{Range} = \frac{\text{Total Energy}}{\text{Total Power}} \times v \tag{2}$$

Differentiating (2) w.r.t. $v$ and equating it to zero yields $v = \sqrt[3]{\frac{H}{\rho \times C_d \times A}}$.

Therefore, the ROV must move at this $v$ to maximize range on full charge assuming 100% efficiency.

**Table 4** Range and velocity calculations

| | Design 1 | Design 2 |
|---|---|---|
| Optimal velocity | 1.43 m/s | 2.25 m/s |
| Range @ optimal velocity | 8.64 km/full charge | 18.93 km/full charge |
| Range @ 0.5 m/s | 6.79 km/full charge | 7.174 km/full charge |
| Drag force @ 0.5 m/s | 7.09 N | 0.42 N |

Energy of Li-ion battery (used in ROV) = 240 Wh.
The results of above calculations for both designs are summarized in Table 4.

## *4.5 Electronic System*

The electronic system is housed within the pressure hull which is made strong enough to withstand the hydrostatic pressure. The main electronic components are ATmega328p electronic control board, on-board Li-ion battery and electronic speed controllers (ESC). The sensors include CTD, temperature, pH, BOD sensor and leak detector whose ends are projected out of the hull to sense the environment. Proper waterproofing of the hull is done to protect the electronic components (Table 5).

**Table 5** Power consumption of various sensors on-board

| S. No. | Sensor/equipment | Power consumption (W) |
|---|---|---|
| 1 | CTD | 5 |
| 2 | BOD sensor | 0.5 |
| 3 | pH sensor | 0.5 |
| 4 | Doppler velocity log (DVL) | 3 |
| 5 | Leak detector | 3 |
| 6 | SONAR | 15 |
| 7 | Gyroscope | 0.5 |
| 8 | Controller board (Atmega328p) | 2 |
| 9 | Camera (with manipulator module) | 25 |

### 4.6 Illumination System

Four OrcaTorch D550 waterproof torches are used for illumination purposes. Two torches are used in front and two in the back providing a bright environment underwater. The torches are waterproof up to a depth of 150 m and have a brightness of 1000 lumens. All the torches are attached to the outer protection frame of both the ROVs.

### 4.7 Modular Attachments

Both the ROVs discussed above are made modular enough to eliminate the need for another vehicle to execute a different kind of task. Different modular attachments like spot cleaning attachment, manipulator system, and sample collection module can be attached to the system. Each of these modules will have their own power supply and electronics and will only receive instructions from the parent ROV.

### 4.8 Camera Module

The camera used is waterproof up to 100 m having a view of 170°. Therefore, there is no need to keep the camera inside the hull. A custom-designed 3D printed camera attachment is used with the ability to tilt 180 degrees using a waterproof servo motor. Two of such attachments are used- one at front and one at the back. Moreover, the camera module is designed such that it can be attached anywhere on the ROV according to task and visibility required. Using the two attachments a 360° view of the environment of ROV can be obtained leading to better control.

### 4.9 Design Integration

Design integration is done in such a way to ensure static and dynamic equilibrium and positive buoyant conditions.

First step is to measure the buoyant force, buoyancy center, weight, center of gravity, and comparing them to check its stability. Taking the origin at the front leftmost corner of the vehicle the parameters are measured (Figs. 7 and 8; Table 6).

It is seen that the buoyant force is greater than the weight of the vehicle by a good margin making the ROV positive buoyant. Dead masses are added to both the designs to decrease the buoyant force and making the ROV only slightly positive buoyant. Moreover, no modular attachments are considered in estimating the weight, modular

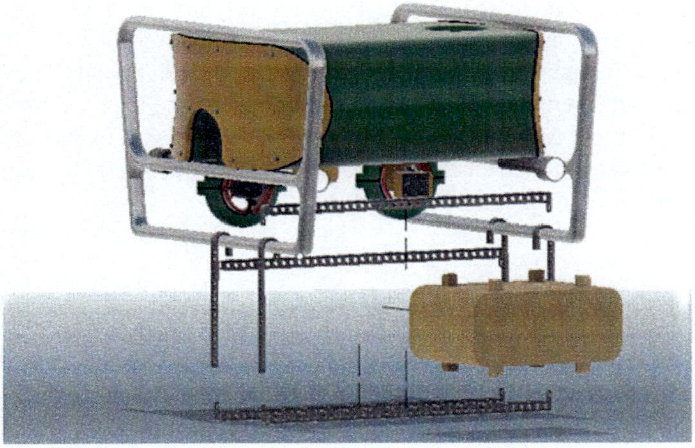

Fig. 7 Modular attachment

Fig. 8 Camera module

Table 6 Buoyancy and initial stability conditions

| | Design 1 | Design 2 | | Design 1 | Design 2 |
|---|---|---|---|---|---|
| Buoyant force | ~547 N | ~ 331.5 N | Weight | ~362 N | ~245 N |
| Buoyancy center $x$ | −0.28 m | −0.22 m | Mass center $x$ | −0.25 m | −0.24 m |
| Buoyancy center $y$ | 0.28 m | 0.27 m | Mass center $y$ | 0.28 m | 0.27 m |
| Buoyancy center $z$ | 0.25 m | 0.48 m | Mass center $z$ | 0.16 m | 0.37 m |

system attachment will increase the weight with no such difference in buoyant force. Thus, reducing the requirement of dead masses.

The Vehicle will be stable when the line of action of weight and buoyant force are both in the same vertical axis [5]. In our design there is a distance between the lines of action of these two forces. So we have used dead masses to make these two forces inline. Moreover, the center of mass is beneath the center of buoyancy by a considerable distance making the ROV stable.

## 5 Conclusion and Future Work

The design of Underwater ROV is done by first deciding the design constraints. Then static and hydrodynamic considerations were taken into account and finally two designs were proposed with the accompanying electronics. A modular approach is taken in the design process leaving space for further addition of electronics in the main hull as well as different attachments that can be used with both designs. These attachments can be generic or can be designed as per the user requirements. All the subsystems are designed using an iterative process wherein the parts are first designed and then changes were made according to the results obtained from the FE analysis.

Drag forces are calculated for motion in different directions and optimum velocity is calculated. Finally, weight, buoyant forces, center of mass and, center of buoyancy for both designs are obtained. Modifications were made according to the above numbers for stability.

The designs presented are preliminary and further work needs to be done for proper validation. Laboratory experiments are needed for validation of drag forces, thrust power and optimal velocity numbers. Therefore, final designs can differ from the ones presented here with improved dynamics and better range.

## References

1. B. Allotta, S. Baines et al., Design of a modular autonomous underwater vehicle for archaeological investigations, in *OCEANS 2015* (Genoa Italy, 18–21 May 2015), pp. 1–5
2. A.W.H. Turnpenny, J. Coughlan et al., *Evidence: Cooling Water Options for the New Generation of Nuclear Power Stations in the UK* (Environment Agency, Bristol, 2010), pp. 15–25
3. J.A. Ramirez, R.E. Vasquez et al., Mechanical/naval design of an underwater remotely operated vehicle (ROV) for surveillance and inspection of port facilities, in *IMECE 2007* (Washington, USA, 11–15 Nov 2007), pp. 1–10
4. F.A. Azis, M.S.M. Aras et al., Problem identification for underwater remotely operated vehicle (ROV): a case study, in *International Symposium on Robotics and Intelligent Sensors 2012 (IRIS 2012)* (Sarawak, Malaysia, 2012), pp. 554–560
5. R. Capocci et al., Inspection-class remotely operated vehicles—a review. J. Mar. Sci. Eng. **5**(13), 4–6, 12, 17–19 (2017)

6. Y.A. Cengel, J.M. Cimbala, *Fluid Mechanics—Fundamentals and Applications*, 13th Edn., (McGraw Hill, Chennai, 2019), pp. 619–625
7. S.K. Som, G. Biswas, *Introduction to Fluid Mechanics and Fluid Machines*, 2nd edn. (McGraw Hill, New Delhi, 2008), pp. 40–47
8. J.G. Bellingham, Y. Zhang et al., Efficient propulsion for the tethys long range autonomous underwater vehicle, in *Proceedings of IEEE AUVs 2010* (Monterey CA, Sept 2010), pp. 1–6
9. T. Graczyk, Methodology of remotely operated vehicle design. Trans. Built Environ. **42**, 573–577 (1999)
10. MIT, ATLANTIS II: Superman AUV, 2005. [Online] Available from https://web.mit.edu/12.000/www/m2005/a2/finalwebsite/equipment/robotics/superman.shtml. Accessed 20 Dec 2019
11. G. Meadows, L. Meadows, The marine environment, in Ed. by T. Lamb, *Ship Design and Construction* (Society of Naval Architects and Marine Engineers, Jersey City, NJ 2003), pp. 1–16

# Large Eddy Simulation of Turbulent Slot Jet Impingement on Heated Flat Plate

**Ghulam Rabbani and Dushyant Singh**

**Abstract** The present large eddy simulation (LES) study investigates heat transfer characteristics of a turbulent slot jet impingement on a smooth and flat target plate at a constant wall heat flux condition. The operating parameters are slot width ($S$) = 2.5 mm, nozzle to plate spacing ($H/D_h$) = 4, 8, 12 and Reynolds number (Re) = 4000, 8000, 12,000. The results show that Nusselt number (Nu) increases with an increase in Reynolds number. The results also show at a fixed nozzle to plate spacing, on increasing the Reynolds number, Nusselt number (Nu) will also increase. The results also show that at a fixed Reynolds number, on increasing the nozzle to plate spacing, Nusselt number will decrease. The Nusselt number will be highest at stagnation point and then it gets decreased along wall jet regions.

**Keywords** LES · Jet impingement · Wall jet · Stagnation point

## Nomenclature

| | |
|---|---|
| S | Slot jet width, mm |
| $D_h$ | Twice of slot jet width, m |
| $H$ | Distance between the slot jet exit and the targeted plate, m |
| Re | Reynolds number based on slot jet width and average jet exit velocity, $\frac{V2S}{\upsilon_k}$ |
| $h$ | Convective heat transfer coefficient, $W/m^2K$ |
| $k_f$ | Thermal conductivity of air, w/Mk |
| $B$ | Characteristics length, m |
| $Nu_{avg}$ | Nusselt number, $hB/k_f$ |
| $\Delta Y^+$ | Non-dimensional wall coordinate, $y\, u_\tau/\upsilon_k$ |

G. Rabbani (✉) · D. Singh
Department of Mechanical Engineering, National Institute of Technology Manipur, Imphal, India
e-mail: gulam95070@gmail.com

D. Singh
e-mail: dushyant7raghu@gmail.com



K. M. Pandey et al. (eds.), *Recent Advances in Mechanical Engineering*, Lecture Notes in Mechanical Engineering, https://doi.org/10.1007/978-981-15-7711-6_78

| | |
|---|---|
| $\Delta z^+$ | Size of the grid in *z*-direction in wall coordinates, *z* |
| $\Delta$ | Filter width or grid size |
| $\in$ | Dissipation rate, $m^2/s^3$ |
| $\eta$ | Kolmogorov length scale, m |
| $\upsilon_k$ | Kinematic viscosity, $m^2/s$ |
| $\upsilon$ | Dynamic viscosity, pa s |
| $\mu_{SGS}$ | Subgrid-scale eddy viscosity, $m^2/s$ |
| $C_w$ | WALE constant |
| $c_p$ | Specific heat of air, J/K |
| $i, j$ | Index of coordinate direction |
| $\Delta x^+$ | Size of the grid in *x*-direction in wall coordinates, *x* |

## 1 Introduction

Large eddy simulation (LES) has been a leading tool for representation of flow and heat transfer characteristics of jet impingement [1, 2] and solving many complex problem. The accuracy of LES results are claimed to be good [3, 4]. Different researcher [5–10] had worked on corresponding Reynold no. basically from low to high. These investigations make us to think strongly about replacement of RANS-based turbulence model with LES. Jet impingement physics is very complex consisting of free jet, wall jet, and stagnation region [11]. Difficulty arises in accuracy prediction by Reynolds–averaged Navier—stokes (RANS) equation [12] that reported the dependency of (RANS) with nozzle to plate spacing. This model does not capture the physics for high nozzle to plate spacing but agree well for low nozzle to plate spacing. Very few three-dimensional numerical investigation of slot jet impingement, with different slot width, on a flat plate with constant heat flux were conducted, thus the investigation of slot jet impingement is needed. The main objective of this paper is to validate the predicted constant heat flux results with experimental data [22] and study of flow field and thermal characteristics by large eddy simulation (LES) using the WALE model.

## 2 Governing Equation and Subgrid-Scale Modeling

The governing Eqs. (1), (2), (3), i.e., continuity, momentum, and energy are obtained after filtering operation [13].

$$\frac{\partial \overline{\rho}}{\partial t} + \frac{\partial \widetilde{(\overline{\rho} Uj)}}{\partial x_i} = 0 \tag{1}$$

$$\frac{\partial(\overline{\rho}\widetilde{Uj})}{\partial t}+\frac{\partial\overline{(\rho\widetilde{Ui}\widetilde{Uj})}}{\partial x_i}=\frac{\partial\overline{p}}{\partial x_i}+\frac{\partial\mu\left(\frac{\partial\widetilde{Uj}}{\partial x_j}+\frac{\partial\widetilde{Uj}}{\partial x_i}\right)-\frac{2}{3}\mu\frac{\partial\widetilde{Uk}}{\partial x_k}\delta_{ij}}{\partial x_i}+\frac{\partial\overline{\tau U_i U_J}}{\partial x_I} \tag{2}$$

$$\overline{\rho}c_p\frac{\partial\breve{T}}{\partial t}+\overline{\rho}c_p\tilde{U}_i\frac{\partial\tilde{T}}{\partial x_i}=\frac{\partial\left(\mu c_p/\Pr\right)\frac{\partial\tilde{T}}{\partial x_i}}{\partial x_i}+\frac{\partial\overline{\tau u_{jT}}}{\partial x_i}. \tag{3}$$

## 2.1 Modeling

The above governing equation for LES is obtained by performing filtering operations:

$$\widetilde{\phi(x_i,t)}=\int\limits_{\Delta}\phi(x_{i'},t)G(|x_i-x_{i'}|)\mathrm{d}x_{i'} \tag{4}$$

where $\widetilde{\phi\left(x_{i,}t\right)}$ is an arbitrarily filtered component, $x$ denotes the space coordinate, $x'$ is dummy space coordinate for each grid cell, and t is time coordinate. $\Delta$ is the filter width size. The filter width of the unresolved (subgrid) length is taken as [14].

$$\Delta=(\Delta x\Delta y\Delta z)^{\frac{1}{3}} \tag{5}$$

Favre averaging is used to account the density variation with temperature in flow field.

$$\widetilde{\phi(x_{i'},t)}=\frac{\overline{\rho\phi(x_{i'},t)}}{\rho} \tag{6}$$

The instantaneous variable can be written as follows:

$$\phi(x_{i'},t)=\widetilde{\phi(x_i,t)}+\phi'(x_{i'},t) \tag{7}$$

The unknown terms in the momentum and energy Eqs. (2) and (3), respectively, [13] are subgrid stress tensor $\overline{\tau_{uiuj}}$ and subgrid-scale heat flux $\overline{\tau_{ujT}}$. The unknown terms are obtained using SGS models.

The SGS tensor with the eddy–viscosity hypothesis is written as

$$\overline{\tau_{uiuj}}-\frac{1}{3}\overline{\tau}_{kk}\delta_{ij}=-2\mu_{\text{SGS}}=\left(\tilde{S}_{ij}-\frac{2}{3}\tilde{S}_{kk}\delta_{ij}\right)$$

where $\tilde{S}_{ij}=\frac{1}{2}\left(\frac{\partial\tilde{u}_i}{\partial x_j}+\frac{\partial\tilde{u}_j}{\partial x_i}\right),\delta_{ij}$ is the filtered strain rate tensor. $\overline{\tau}_{kk}$ can be neglected for incompressible flows suggested by Erlebacher et al. [15]. $\mu_{\text{SGS}}$ Subgrid stress eddy viscosity is computed using wall adapting eddy viscosity (WALE), and subgrid-scale model is proposed by Nicoud and Ducros [16]. An advantage of WALE model is that it retains a zero turbulent viscosity for laminar shear flows. In contrast, the Smagorinsky—lily model produces nonzero turbulent viscosity at the wall. Therefore, the WALE model is selected in the present study.

$$\mu_{\text{sgs}} = \overline{\rho} L_s^2 \frac{(\zeta_{ij}\zeta_{ij})^{3/2}}{(\widetilde{S_{ij}}\widetilde{S_{ij}})^{\frac{5}{2}} + (\zeta_{ij}\zeta_{ij})^{\frac{5}{4}}}$$

where $L_s$ is an SGS length scale given by

$$L_s = \min\left(Kr, C_w V^{\frac{1}{3}}\right)$$

where $K$ is von Karman's constant, $\boldsymbol{r}$ is the normal distance from the wall to the cell center, and $\mathbf{V}$ is the volume of the computational cell.

$\zeta_{ij}$ is the traceless symmetric part of the square of the filtered velocity gradient tensor which can be computed as

$$\xi_{ij} = \frac{1}{2}\left(\frac{\partial \tilde{U}_i}{\partial x_k} * \frac{\partial \tilde{U}_k}{\partial x_j} + \frac{\partial \tilde{U}_j}{\partial x_k} * \frac{\partial \tilde{U}_k}{\partial x_i}\right) - \frac{1}{3}\left(\frac{\partial \tilde{U}_i}{\partial x_k} * \frac{\partial \tilde{U}_k}{\partial x_k}\right)\delta_{ij}$$

Another unknown term SGS heat flux, $\overline{\tau_{ujT}}$ is computed on the basis of simple gradient diffusion hypothesis (SGDH) as

$$\overline{\tau_{ujT}} = \frac{\mu_{\text{SGS}}}{\text{Pr}_t} - \frac{\partial \tilde{T}}{\partial x_j}.$$

The value of following constant in the WALE model is:

$C_w = 0.325$, $K = 0.4187$, $\text{Pr}_t = 0.9$ [23]. The Fluent 18.1 is used as a solver for the present LES configuration. The SIMPLE algorithm was used in support for pressure velocity coupling of the equation, and the value of constant Cw which was computed using WALE model was selected as 0.325 to achieve a good quality.

Rate of extraction of turbulent kinetic energy from mean flow per unit mass ($\pi$)

$$\pi = \frac{u^3}{l}$$

$u$ and $l$ is size of largest eddy in terms of velocity and length scale.

Rate of dissipation of turbulent kinetic energy ($\in$)

$$€ = \upsilon_k \delta_{ij} \delta_{ij}$$

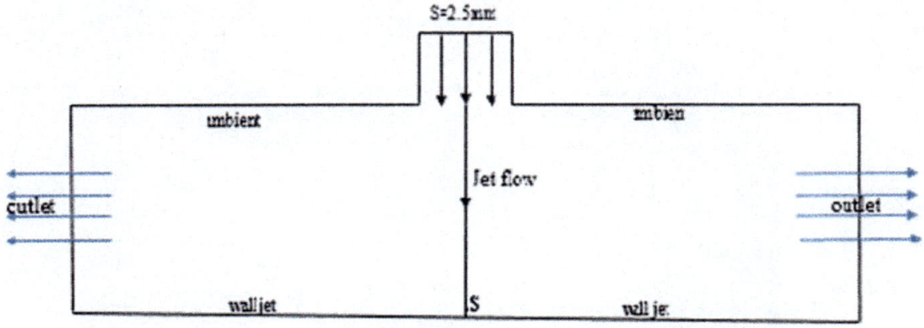


**Fig. 1** Schematic representation of the physical situation described

## 3 Problem

A numerical investigation using LES is carried out to investigate heat transfer characteristics of a turbulent slot jet impingement on a smooth and flat target plate at a constant heat flux condition. The operating parameters are slot width $(S) = 2.5$ mm, nozzle to plate spacing $(H/D_h) = 4, 8, 12$, and Reynolds Number $= 4000, 8000, 12{,}000$. Air is taken as an incompressible working fluid with density 1.225 kg/m$^3$, viscosity 1.78 * 10^(−5), and thermal conductivity 0.0242 w/m$^2$k.

The case presented in Fig. 1 describes the physical situation at three different nozzles to plate spacing and $S$ is the stagnation point in the impingement region.

## 4 Computational Setup

A schematic of present computational domain is shown in Fig. 2. The width of the computational domain in spanwise direction is taken as 5S, and this condition is appropriate for periodic boundary conditions, because fluctuation of velocity decorrelates at half of its width. The size of the domain in streamwise direction was taken as 60S which is sufficient for zero boundary condition [17]. A uniform velocity boundary condition at the jet inlet is taken based on Reynolds number 4000, 8000, 12,000, respectively. Turbulence technique namely no perturbation method is used at the jet inlet. No perturbation method shows good agreement with experimental results shown in [18]. The no-slip condition was specified for the impingement walls and nozzle wall. The temperature at the inlet was taken as ambient value and at impingement wall constant heat flux 3000 W/m$^2$ was taken. The symmetry conditions are imposed on the plane over the slot jet axis. The symmetry boundary condition is shown by dark sky in Fig. 2, i.e., which is throughout the left part of velocity inlet.

A bounded second-order implicit scheme was chosen for the time mandatory of the geometry equation. A non-dimensional time step $\Delta t\ V_0/B$ of 10^(−4) was taken which maintains the value of courant number below 1. The data presented here was

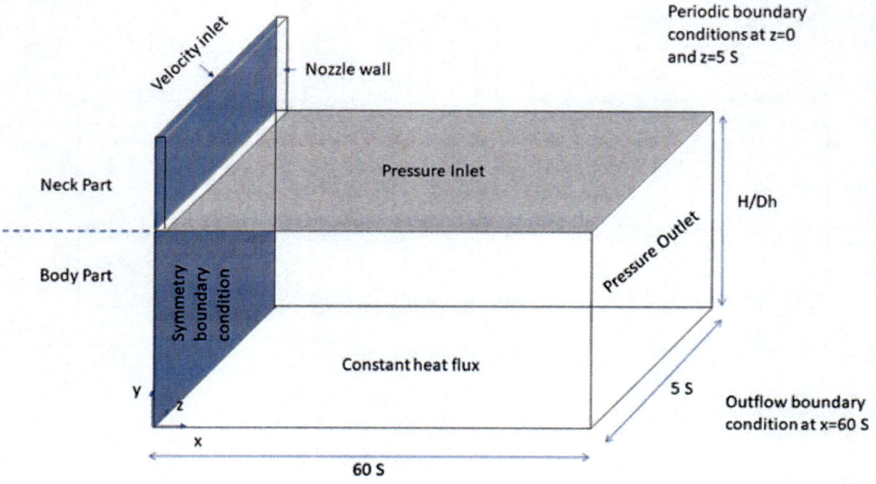


**Fig. 2** Computational domain for present LES study

time averaged for about non-dimensional units and also spatially averaged in the $z$ direction.

## 5 Grid Specification for Les

The optimum grid is considered as an important issues for LES. In the present study, final grid was designed based on the fine grid resolution [19–21]. The non-dimensional distance from the wall, i.e.,$\Delta Y^+$ was kept below 1 at the wall. $\Delta x^+$ and $\Delta z^+$ are limited to 100 and 80 near the impingement region. Table 1 presents fine grid computation for presenting number of nodes in $x$, $y$, $z$ directions. The computational domain is divided into two parts, i.e., neck part and body part shown in Fig. 2.

The final grid was done on the basis of criteria given by Pope [20]. Pope [20] stated that size of the grid should be limited to 12 times the Kolmogorov length scale ($\eta$). The $\eta$ was calculated by using the relation $\frac{\upsilon_k^3}{€}^{1/4}$, where € is the turbulent dissipation rate which is directly calculated using the SST-$k\omega$ turbulence model on

**Table 1** Grid specification presenting number of nodes

| Grid specification | $H/D_h$ | Number of nodes in $x$, $y$, $z$ directions | | Maximum $\Delta Y^+$ |
|---|---|---|---|---|
| | | Neck part | Body part | |
| Fine grid | 4 | 13 × 38 × 27 | 314 × 118 × 27 | 1 |
| Fine grid | 8 | 13 × 38 × 27 | 316 × 225 × 27 | 1 |
| Fine grid | 12 | 13 × 38 × 27 | 317 × 354 × 27 | 1 |

the same grid as that of the LES computation. Further, we have assessed the index of quality criterion for LES (LESI $Q_V$) given by celik et al. [21] as

$$\text{LESI } Q_V = \frac{1}{1 + 0.05\left(\left(\frac{\mu+\mu_{SGS}}{\mu}\right)^{0.53}\right)}$$

where $\mu$ and $\mu_{SGS}$ represent the dynamic viscosity of air and subgrid-scale viscosity of the flow, respectively. Celik et al. [21] mentions that LESI $Q_V$ should be greater than 0.8 for a wall-resolved LES. The value of the LESI $Q_V$ for the present grid falls within the range of 0.8–0.9.

## 6 Validation

The experimental data given by Asthforth [22] is validated by Dutta et al. by no perturbation method. Dutta et al. [18] taken full domain for the computational. We have validated the same experimental results shown by Ashforth [22] by taking half domain with symmetry boundary condition at the axis of slot jet, and rest of the boundary condition is same as taken in [22] by using LES with no perturbation method. Present validation results show that LES captures the experimental data at stagnation point very well and along the wall jet it shows the maximum error of 19% with respect to experimental data [22] (Fig. 3).

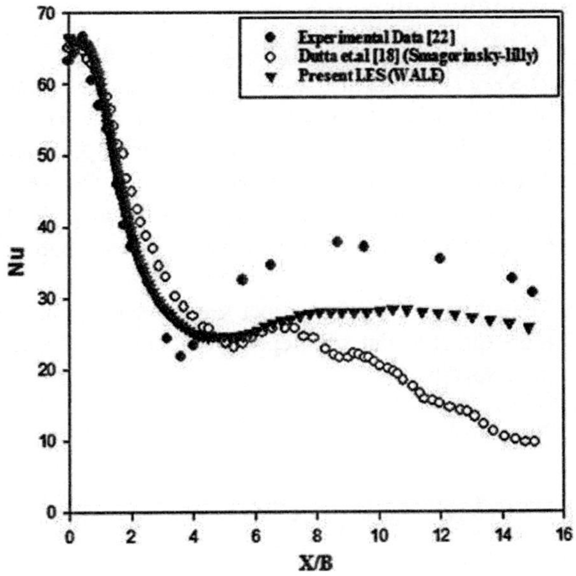


**Fig. 3** Comparison of surface Nusselt number for the two LES model with experimental data [22]

## 7 Results and Discussion

Zuckerman [11] stated that velocity profile remains unchanged within the potential core jet region and beyond that velocity jet profile, i.e., Gaussian curve becomes widens. It is understood that center line velocity of jet get decreased as it comes closer to impingement region. It is also understood from the below results that for high Reynold number, the center line velocity has higher effect in impingement region in comparison to lower Reynolds number due to which heat transfer characteristics, i.e., local Nusselt number is high for high Reynold number and decreases with decreasing Reynold number in the impingement region. And with decreasing Reynolds number, the boundary layer in the wall jet region gets thicker and thicker due to which heat transfer, i.e., local Nusselt number in the wall jet region becomes lower and lower with decreasing Reynold number as shown in Fig. 4. The below results in Fig. 5 show that at constant Reynolds number, with increasing nozzle to plate spacing, the local Nusselt number decreases. The reason behind this is that for high nozzle to plate spacing, the effect of center line velocity of jet will be very low in comparison to lower nozzle to plate spacing in the impingement region due to which boundary layer will be very thick in wall jet region which results in lower heat transfer. As the nozzle to plate spacing decreases, the effect of center line velocity will become high in the impingement region and due to which boundary layer will become thinner and thinner in the wall jet region. This results in high heat transfer, i.e., local Nusselt number becomes higher and higher in the wall jet region and impingement region with decreasing nozzle to plate spacing.

## 8 Conclusions

1. The effect of increase in Nusselt number was seen on increasing the Reynolds number at a constant nozzle to plate spacing. With increase in Reynolds number, Nusselt number increases at the stagnation point and along the wall jet region.
2. The effect of increase in Nusselt number was seen on decreasing the nozzle to plate spacing at a constant Reynolds number. With decrease in nozzle to plate spacing, Nusselt number increases at the stagnation point and along the wall jet region.
3. The effect in Nusselt number seen in stagnation point and in wall jet region is more effective due to effect of Reynolds number rather than effect of nozzle to plate spacing.

**Fig. 4** shows the effect of Re in **a**, **b**, **c**

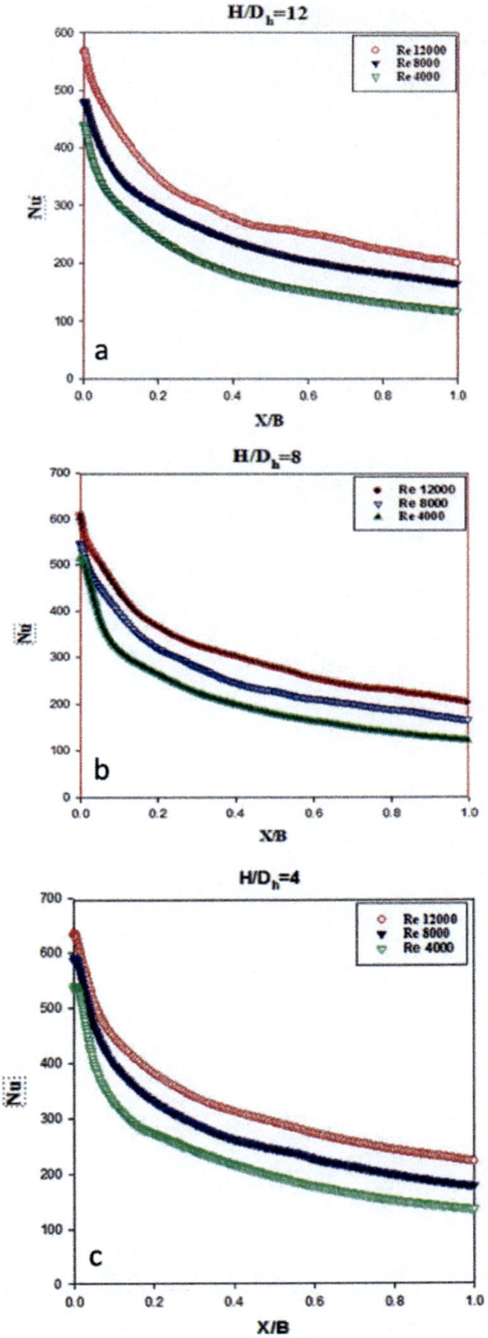

**Fig. 5** shows the effect of $H/D_h$ in **d**, **e**, **f**

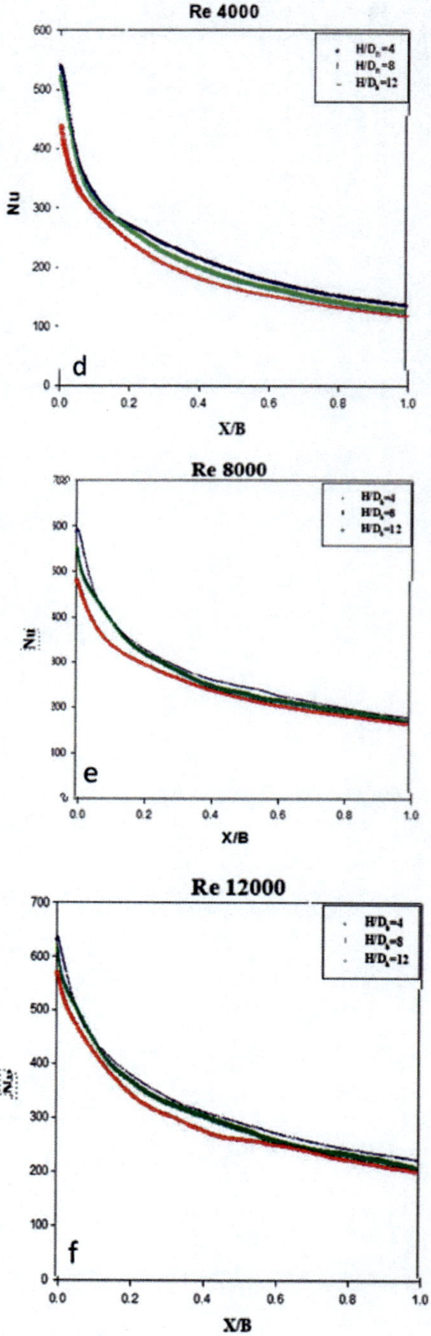

## References

1. M. Darksler, Numerical Simulations of Turbulence and Heat Transfer of Multiple Jets Impinging on the Heated Surface Ph.D. thesis , University of Ljubljana, Solvenia, 2015
2. T. Hallqvist, Large Eddy Simulation of Impinging Jets with Heat Transfer Ph.D. thesis, Royal Institute of Technology ,Sweden, 2006
3. M. Darksler, B. Niceno, B .Koncar, L. Cizelj, Large eddy simulation of multiple impinging jets in hexagon configuration—mean flow characteristics. Int. J. Heat Fluid Flow **46**, 147–157 (2014). https://doi.org/10.1016/j.ijheatfluidflow.2014.01.005
4. M. Darksler, B. Koncar, L. Cizelj, On the accuracy of large eddy simulation of multiple impinging jets. Int. J. Heat Fluid Flow (2018). https://doi.org/10.1016/j.ijheatmasstransfer.2018.12.125
5. G. Lodato, L. Vervisch, P. Domingo, A compressible wall-adapting similarity mixed model for large-eddy simulation of impinging round jet. Phys. Fluids **21**, 1–21 (2009)
6. S. Rhea, M. Bini, M. Fairweather, W.P. Jones, RANS Modeling and LES of a single-phase, impinging plane jet. Comput. Chem. Eng. **33**(8), 1344–1353 (2009)
7. M. Yu, L. Chen, H. Jin, J. Fan, Large eddy simulation of coherent structure of impinging jet. J. Therm. Sci. **14**(2), 150–155 (2005)
8. M. Hadziabdic, K. Hanjalic, Vortical structures and heat transfer in a round impinging jet. J. Fluid Mech. **596**, 221–260 (2008)
9. F. Beaubert, S. Viazzo, Large eddy simulations of plane turbulent impinging jets at moderate reynolds numbers. Int. J. Heat Fluid Flow **24**, 512–519 (2003)
10. T. Cziesla, G. Biswas, H. Chattopadhyay, N.K. Mitra, Large-eddy simulation of flow and heat transfer in an impinging slot jet. Int. J. Heat Fluid Flow **22**, 500–508 (2001)
11. N. Zuckerman, N. Lior, Jet impingement heat transfer: physics, correlations, and numerical modeling. Adv. Heat Transf. **39**, 565–631 (2006)
12. R. Dutta, A. Dewan, B. Srinivasan, comparision of various integration to wall (ITW) RANS models for predicting turbulent slot jet impingement heat transfer. Int. J. Heat Mass Transf. **65**, 750–764 (2013)
13. O.M. Knio, H.B. Najm, P.S. Wyckoff, A semi-implicit numerical scheme for reacting flow: II Stiff, operator-split formulation. J. Comput. Phys. **154**, 428–467 (1999)
14. U. Schumann, Subgrid scale model for finite difference simulations of turbulent flows in plane channels and annuli. J. Comput. Phys. **18**, 376–404 (1975)
15. G. Erlebacher, M.Y. Hussaini, C.G. Speziale, T.A. Zang, Toward the large-eddy simulation of compressible turbulent flows. J. Fluid Mech. **238**, 155–185 (1982)
16. F. Nicoud, F. Ducros, Sub grid-scale stress modeling based on the square of the velocity gradient tensor, Flow. Turbul. Combust. **62**, 183–200 (1999)
17. J.E. Jaramillo, F.X. Trias, C.D. Perez-Segarra, A. Oliva, DNS and RANS modeling of a turbulent plane impinging jet. Int. J. Heat Mass Transf. **55**, 789–801 (2012)
18. R. Dutta, A. Dewan, B. Srinivasan, Large eddy simulation of turbulent slot jet impingement heat transfer at small nozzle-to-plate spacing. Heat Transf. Eng. https://doi.org/10.1080/01457632.2015.1119592
19. U. Piomelli, J. Chasnov, *Large Eddy Simulations: Theory and Applications, Transition and Turbulence Modelling* (Kluwer Academic Publishers, Dordrecht, Netherlands, 1996), pp. 269–336
20. S.B. Pope, Ten questions concerning the large eddy simulation of turbulent flows, in *Large Eddy Simulation and Subgrid-Scale Modeling for Turbulent Mixing and Reactive Flows* (California Institute of Technology, Pasadena, CA, USA, 2003)
21. I. Celik, Z.N. Cehreli, I. Yavuz, Index of resolution quality for large eddy simulations. J. Fluids Eng. **127**, 949–958 (2005)
22. S. Ashforth-Frost, K. Jambunathan, C.F. Whitney, Velocity and turbulence characteristics of a semi-confined orthogonally impinging slot jet. Exp. Thermal Fluid Sci. **14**, 60–67 (1997)

23. S. Pachupate, B Premchandran, Experimental investigation and large eddy simulations of turbulent slot jet impingement cooling of a circular cylinder with and without a quadrilateral confinement. Appl. Thermal Eng. **144**, 854–876 (2018)

# Numerical Solution of Foreign-Gas Film Cooling in Supersonic Flow

**Hitesh Sharma, Dushyant Singh, and Ashutosh Kumar Singh**

**Abstract** The present numerical study is carried out to understand the film cooling characteristics of supersonic flow over a flat plate. The numerical investigation is carried out for the main stream Mach no. $M_\alpha$ of 2.67, secondary stream (coolant stream) Mach no. $M_C$ 0.05 and injection angle of 90°. The domain size is ($x/s$ = 277, $y/s$ = 22.8, $z/s$ = 11.4) where '$s$' is the slot width. The governing equations such as continuity, momentum and energy are solved using Ansys Fluent 18.1. Air is considered as working fluid for both main as well as secondary stream. The numerical results obtained using the LES modelling are validated with available Direct Numerical Simulation (DNS) results. The comparison of different LES Sub-grid scale models has been shown for providing standards for this type of problem.

**Keywords** Film cooling · Supersonic flow · Large-eddy simulation

## Nomenclature

| | |
|---|---|
| $T$ | Temperature |
| $s$ | Slot width |
| $\rho$ | Density |
| $x, y, z$ | Cartesian coordinates |
| $P$ | Pressure |
| $u, v, w$ | Velocity components |
| $U$ | Velocity |

H. Sharma (✉) · D. Singh · A. K. Singh
Department of Mechanical Engineering, NIT Manipur, Imphal, India
e-mail: sharmahitesh030@gmail.com

D. Singh
e-mail: Dushyant7raghu@gmail.com

A. K. Singh
e-mail: ashutosh317singh@gmail.com



K. M. Pandey et al. (eds.), *Recent Advances in Mechanical Engineering*, Lecture Notes
in Mechanical Engineering, https://doi.org/10.1007/978-981-15-7711-6_79

| | |
|---|---|
| $\mu$ | Dynamic viscosity |
| $\tau$ | DShear stress |
| $M_{\alpha}$ | Main stream Mach number |
| $M_{c}$ | Coolant stream Mach number |
| $T_{\alpha}$ | Freestream Temperature |
| $U_{\alpha}$ | Freestream Velocity |
| $T_{<Z>}$ | Non-dimensional temperature |
| $T_{<Z>}$ | $T/T_{\alpha}$ |
| $U_{<Z>}$ | Non-dimensional velocity |
| $U_{<Z>}$ | $U/U_{\alpha}$ |

# 1 Introduction

Rocket nozzle walls are prior subjected to extremely elevated temperatures of exhaust gases of combustion. High temperatures result in to high thermal loading including abrupt heating rate, thermal creep of material, sudden thermal expansion, etc. Due to lack of insufficiently capable materials the requirement of cooling becomes essential. Many techniques has been studied and implemented for this purpose one promising technique known as 'film cooling' has been investigated for several years due to its advantages over others. The technique is to inject a parallel stream of coolant on the surface, which causes formation of a film of coolant over the surface. The film of coolant isolates the wall or surface from the hot exhaust gases of combustion and thus protects it from high temperatures of exhaust stream. Goldstein et al. [1], Parthasarathy and Zakkay [2], and Richards and Strolley [3] conducted experimental studies on sonic flows and showed that film cooling is more effective in sonic and supersonic flows as compared to subsonic flows. Apart from that several numerical studies like that one of O'Connor and Haji-Sheikh [4] has supported the idea. Keller and Kloker [5], Performed Direct Numerical Simulations (DNS) to investigate supersonic film cooling in a mainstream flow of Mach no. of 3.3 and temperature of 564 K. Coolant Mach no was 0.05. Air is used as cooling gas injected in the wall normal direction. They showed several results of velocity and temperature profile.

## *1.1 Problem Statement*

In the present study Large Eddy Simulations has been performed to evaluate the heat transfer and flow characteristics in the supersonic film cooling. The computational domain addressed in the present problem is a three-dimensional domain, as shown in Fig. 1 The origin point lies at the centre of the injection slot in flat plate level. In streamwise direction the extension is $-230 < x/s < 47$, it has a width of $z/s = 3$ and height of $y/s = 22.8$. Where '$s$' being the slot width. Injection is provided through a

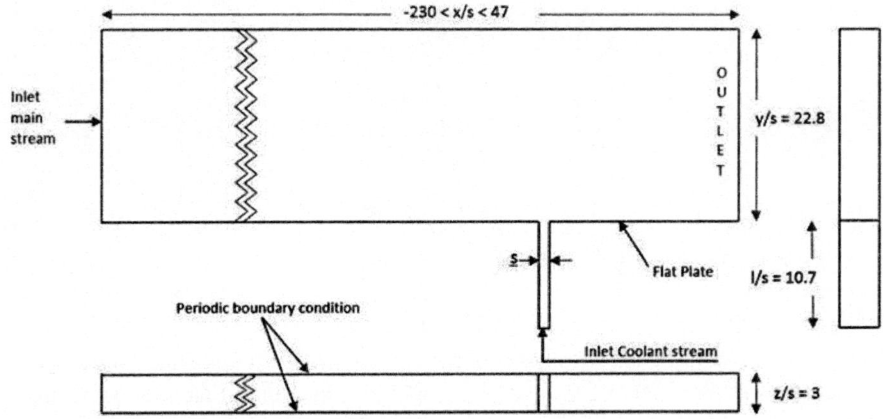


**Fig. 1** Geometry and boundary condition details of computational domain

slot extended below the flat plate by a depth of $l/s = 10.7$. The study is carried out for mainstream. Mainstream Mach number is 2.67 and coolant stream Mach number is 0.05. The mainstream static temperature is kept at 564 K and inlet temperature of cooling stream is 297 K. Temperature and velocity profiles are obtained at four observation stations situated at the distance of $x/s = 1$, 2, 4 and 6 from the slot. To understand the heat transfer and thermal effects in the study the film cooling effectiveness is observed at the centre of the flat plate after the slot for the length of 47 s.

## 2 Governing Equations

For a variable $\varphi$ if $\bar{\varphi}$ represents Grid filtered quantity and $\tilde{\varphi}$ represents Favre averaged quantity. The filtered continuity, momentum and energy equations are given as follows.

$$\frac{\partial\left(\bar{\rho}\right)}{\partial t} + \frac{\partial\left(\bar{\rho}\,\tilde{U}_j\right)}{\partial x_j} = 0 \tag{1}$$

$$\frac{\partial(\bar{\rho}\,\tilde{U}_j)}{\partial t} + \frac{\partial}{\partial x_j}(\bar{\rho}\,\tilde{U}_j\tilde{U}_i + \bar{P}\,\delta_{ij} - \tilde{\tau}_{ij} - \tau_{U_iU_j}) = 0 \tag{2}$$

$$\frac{\partial(\bar{\rho}\tilde{h})}{\partial t} + \frac{\partial}{\partial x_j}(\bar{\rho}\,\tilde{U}_j\,\tilde{h} - \bar{\rho}\,a\frac{\partial(\tilde{h})}{\partial x_j} - \tau_{U_ih}) = 0 \tag{3}$$

$$\bar{\rho} = \frac{P_0}{R\tilde{T}} \tag{4}$$

$\tilde{\tau}_{ij}$ = Shear stress tensor which for Newtonian fluids is given as

$$\tilde{\tau}_{ij} = \mu \tilde{s}_{ij} - \frac{2}{3}\mu\delta_{ij}\tilde{s}_{kk}$$

where $\tilde{s}_{ij}$ = Rate of stress tensor = $\left[\frac{\partial \tilde{U}_i}{\partial x_j} + \frac{\partial \tilde{U}_j}{\partial x_i}\right]$

and $\tau_{U_iU_j}$ is subgrid-scale stress and $\tau_{U_ih}$ is subgrid enthalpy flux. These can be considered as the effect of subgrid-scale and are modelled in simulations using different subgrid models. For example,

1. Smagorinsky–Lilly Model

$$\tau_{U_iU_j} = -2C\,\bar{\rho}\,\Delta^2\,\tilde{S}\,(\tilde{s}_{ij} - \frac{1}{3}\delta_{ij}\tilde{s}_{kk})$$

Here $\Delta = \sqrt[3]{\Delta x \Delta y \Delta z}$

2. WALE (Wall Adapted Local Eddy Viscosity) Model

$$\mu_t = \rho {L_s}^2 \frac{(S_{ij}^d S_{ij}^d)^{3/2}}{(\bar{S}_{ij}\bar{S}_{ij})^{5/2} + (S_{ij}^d S_{ij}^d)^{3/4}}$$

$\mu_t$ Here is known as the Eddy viscosity.

3. WMLES (Wall modelled LES) Model

In WMLES eddy viscosity is defined using hybrid length scale as

$$\vartheta_t = \min\left[(kd_w)^2, (C_{\mathrm{Smag}}\Delta)^2\right].S.\left[1 - \mathrm{e}^{\left[-\left(\frac{y^+}{25}\right)^3\right]}\right]$$

where $d_{\mathrm{w}}$ is the wall distance, $S$ is the strain rate, $k = 0.41$ and $C_{\mathrm{smag}} = 0.2$ are constants, and $y^+$ is the normal to the wall inner scaling.

## 3 Numerical Schemes

In the present numerical investigation, large eddy simulations are performed and the filtered continuity, momentum and energy equations are solved using finite volume method based CFD software, Ansys fluent 18.1. For discretization, implicit formulation with spatial discretization as least square cell based in gradient and second-order upwind scheme in flow is used. While for transient formulation second-order implicit scheme is preferred. Convergence of solution is considered to be achieved when residuals of velocity, continuity, turbulence and energy fall below $10^{-7}$. The time step size is maintained to ensure a Courant-Friedrichs-Lewy (CFL) number, CFL = 1.

## 3.1 Grid Resolution

LES requires sufficiently fine mesh in order to resolve the smaller size eddies. As the turbulent length scale is highly-mesh sensitive for LES based computation, a careful determination of computational grid becomes inevitable. For current problem, the grid size is maintained with the purpose of resolving up to 80% of turbulent kinetic energy. To maintain the requirement for LES study a grid which has 2.1 million nodes is considered for present study. In the entire region with number of nodes in $X$, $Y$ and $Z$ as 700, 150 and 20, respectively. In the domain the cell size is maintained as $\Delta X/L_{\text{ref}} \leq 0.002$, $\Delta Y/L_{\text{ref}} \leq 0.003$ and $\Delta Z/L_{\text{ref}} \leq 0.00054$. In the region, after the slot, the mesh on the plate is maintained such as $\Delta X_{\max}/L_{\text{ref}} = 0.0008$, $\Delta x^+ \leq 14.8$, $\Delta y^+ \leq 1$ and $\Delta z^+ = 90$.

## 3.2 Initial and Boundary Conditions

The mainstream inlet is provided with velocity inlet. Coolant stream inlet is provided with fully developed turbulent velocity profile. The test section plate is considered as no-slip adiabatic wall. Outlet is provided with pressure outlet at zero gauge pressure. Side domain surfaces in z direction are given translational periodic boundary condition. Upper domain surface is given pressure far field condition to treat it as freestream. Values of input parameters are given in Table 1.

**Table 1** Boundary conditions

| Input parameter | Values |
|---|---|
| Free stream Mach number | 2.67 |
| Free stream temperature | 564 K |
| Coolant stream Mach number | 0.05 |
| Coolant stream temperature | 297 K |
| Periodicity length | 3 s |

## 4 Result and Discussion

For comparison of present LES results of the fluid flow and heat transfer to DNS results. The DNS results presented by Keller and Kloker [1] were considered for validations. They obtained the non-dimensional temperature and velocity profile at various position of downstream locations at, $X/S = 1, 2, 4$ and 6 for $M_\alpha = 2.67$ and $M_c = 0.05$. Figure 2 shows the non-dimensional temperature profiles and Fig. 3 shows the non-dimensional velocity profile. The free stream temperature and velocity are used for non-dimensionalization.

The DNS results are plotted with various LES sub-grid scale models viz. WALE, Smagorinsky-Lilly and WMLES-s omega. It can be observed that WMLES-s Omega subgrid model predicts the flow and heat transfer very close to DNS results. Also, it is interesting to notice that the temperature profile exceeds the value of 1, which is quiet unusual. This is due to the viscous dissipation that the temperature rises near

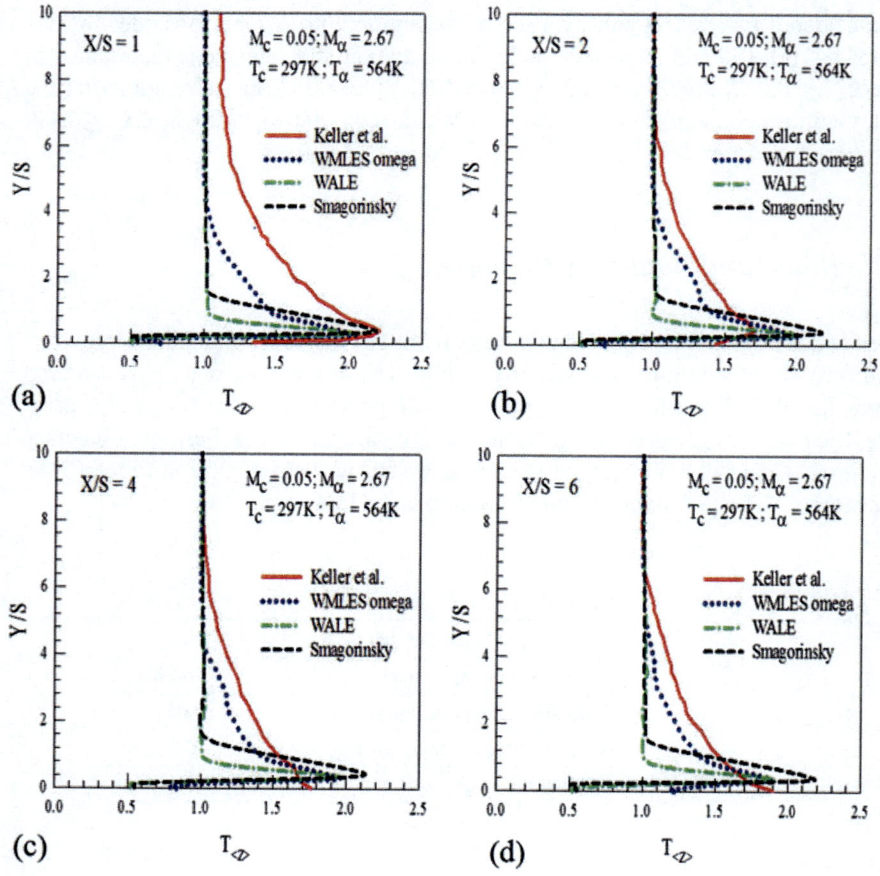


**Fig. 2** Temperature profile

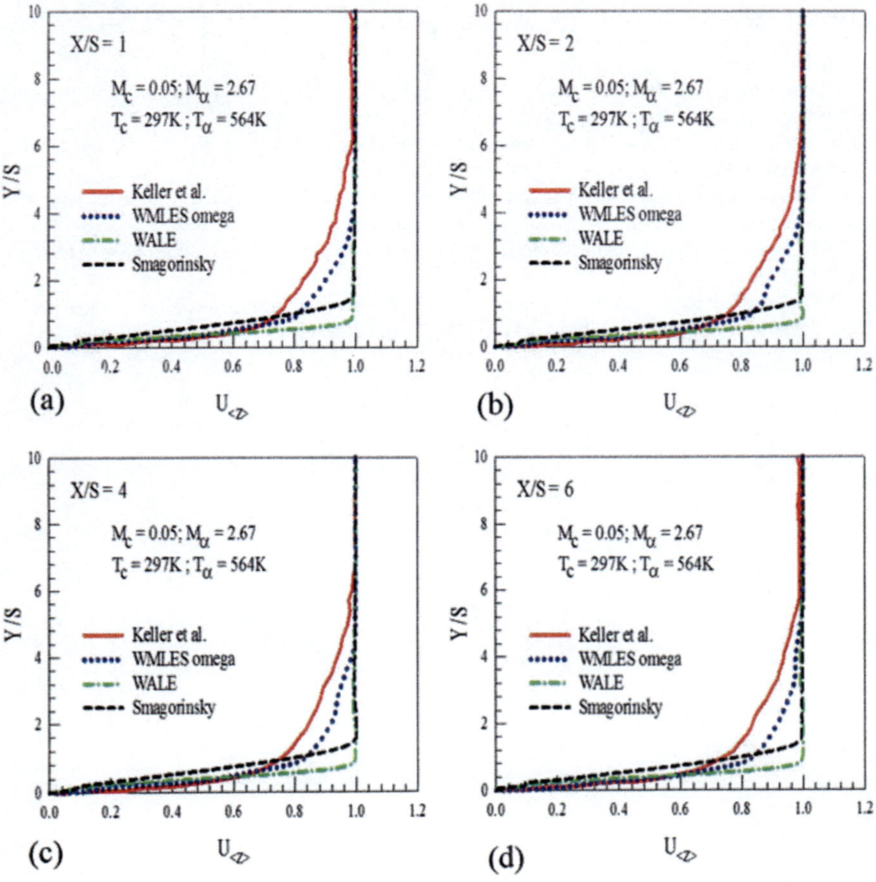


**Fig. 3** Velocity profile

the wall. Viscous dissipation while is negligible in subsonic flows leaves its remark in supersonic flow as seen clearly here.

## 5 Conclusion

Amongst all subgrid-scale models WMLES-s Omega shows the most promising results based on temperature and velocity profile. WALE model is not as good in these situations as it gave worst predictions of results. LES being highly grid-dependent solution scheme does not capture the flow physics as perfectly compared to DNS especially in near-wall region. Nevertheless, the phenomena of viscous dissipation is captured quiet profoundly.

## References:

1. R.J. Goldstein, E.R.G. Eckert, A. Haji-Sheikh, Film cooling with air and helium injection through a rearward-facing slot into a supersonic air flow. AIAA J. **4**(6) (1966)
2. S. Acharya, D.H. Leedom, An experimental investigation of turbulent slot injection at Mach. AIAA J. **8**(7). https://doi.org/10.2514/3.5889
3. B.E. Richardssand, J.L. Stollery, Laminar film cooling experiments in hypersonic. J. Aircraft **16**(3) (1979)
4. J.P. O'Connor, **A** Haji-Sheikh, Numerical study of film cooling in supersonic flow. AIAA J. **30**(10) (1992)
5. M.A. Keller, M.J. Kloker, Direct numerical simulation of foreign-gas film cooling in supersonic boundary-layer flow. AIAA J. https://doi.org/10.2514/1.J055115

# Deposition of Tungsten and Copper Particle on CFRP Composite

**Rashed Mustafa Mazarbhuiya and Maneswar Rahang**

**Abstract** This article describes the deposition of material onto the surfaces of the carbon fiber reinforced polymers (CFRP) material. For this purpose, powder metallurgical (P/M) green compact tool of tungsten-copper (W–Cu) particle is used in Electro Discharge Machining (EDM) process. The metal particles of P/M compact tool are transferred to the CFRP work surface under the reverse polarity. The material transfer rate (MTR) and tool wear rate (TWR) are observed by varying the process parameters (compact load, current, pulse on-time) in the suitable machining range. Grey relational analysis (GRA) is performed to determine the multi-objective optimal setting. Analysis of variance (ANOVA) and effect of factors in grey relational grade are conducted to determine the contribution of each factor to the multi optimal level. The experimental result shows that the multi optimum setting is obtained at compact load of 5 tons, current of 6A and pulse on-time of 1000 μs. Current is the most dominant factor whereas, pulse on-time is least dominant factor in multi optimal setting.

**Keywords** Deposition · Electro discharge machining (EDM) · Grey relational analysis (GRA)

## Abbreviations

| | |
|---|---|
| CFRP | Carbon fiber reinforced polymers |
| *P/M* | Powder metallurgical |
| EDM | Electro Discharge Machining |
| MTR | Material transfer rate |
| TWR | Tool wear rate |
| GRA | Grey relational analysis |

R. M. Mazarbhuiya · M. Rahang (✉)
Department of Mechanical Engineering, National Institute of Technology Meghalaya, Shillong 793003, India
e-mail: maneswar_rahang@nitm.ac.in



K. M. Pandey et al. (eds.), *Recent Advances in Mechanical Engineering*, Lecture Notes in Mechanical Engineering, https://doi.org/10.1007/978-981-15-7711-6_80

| | |
|---|---|
| ANOVA | Analysis of variance |
| CL | Compact load |
| $I$ | Current |
| $T_{on}$ | Pulse on-time |

## 1 Introduction

The carbon fiber reinforced polymers (CFRP) has wide applications such as, aerospace engineering, industrial engineering for reinforcement and sleeves in turbo molecular pumps or drive shafts, civil and automotive engineering applications, sports equipment, robot arms, etc. The CFRP is lightweight composite having improved corrosion resistance, improved fatigue, considerably greater rigidity, low coefficient of thermal expansion, sharply enhanced thermal and electrical conductivity at lower density [1, 2]. CFRP has other applications also such as making fishing rods, golf clubs, tripods etc.[3]. The inhomogeneous and anisotropic property of CFRP consisting of soft polymer (thermoplastics or epoxy) matrix and hard carbon fibers (diameters of few micrometers) make difficult to machine using conventional methods. It results in serious tool wear due to the high strength, delaminating, splintering, burrs of machined surface and shorting the life of the tool used [4, 5]. So it is a great challenge to meet the product cost-effective and high quality as well. However, the unconventional processes like, water-jet machining, Electro Discharge Machining (EDM), etc. makes it easier to achieve the desired applications [6]. The EDM is an efficient machining process that has capability to machine difficult to cut materials by other processes. Mechanism of removal of material is a process of continuous series of controllable spark energy occurred in between the workpiece and the tool in the submerged dielectric condition. Nowadays, die sinker EDM can be applied for surface modification purpose under reverse polarity conditions. Literature reports about the surface modification using tungsten-copper green tool [7]. Deposition of hBN + Cu layer using EDM process is also reported [8]. Using tungsten-copper P/M tool the electro discharge deposition is conducted by several researchers [9–11]. The deposition of tool constituents on the surface of the CFRP workpiece is a new approach in the domain of electro deposition process. The objective of this experimental work is to deposit the P/M green compact constituent materials on the surface of the CFRP workpiece.

## 2 Materials and Method

To conduct this experimentation, CFRP is selected as the workpiece material. A mixture 75%W–25%Cu is used to form green compact tool electrode. The combination of weight ratio is selected as per the literature to achieve sufficient amount of material transfer [12]. The CFRP workpiece fabricated by hand layup process. The reinforcing material is cut into appropriate size and positioned over the surface of the mold. Liquid epoxy (Araldite LY 556) and the hardener (HY 951) (1:0.9% by weight) mixture are dispersed uniformly in the entire mat of fiber. Then the second fiber layer is placed over the first fiber mat. Using hand roller excess resin and trapped gas bubbles present in between the fiber mats is removed. The process is repeated till a thickness of 4 mm is obtained. Over the top layup stack pressure is applied and allowed to hard up to 24hrs. The experimentation is carried out in die sinker EDM (model: S25 Sparkonix India Pvt. Ltd.) in immersed condition of EDM oil (DEF 92). Three input parameters at three different levels are considered as shown in the Table 1. Material transfer rate (MTR) and tool wear rate (TWR) are selected as the output parameters in this study. $L_9$ orthogonal array of Taguchi model is selected as design of experiment. Based on the pilot experiments the machining range of the input parameters is selected. All the experiments are conducted based on the selected orthogonal array. Using the Grey Relational Analysis (GRA) and analysis of variance (ANOVA) the multi-optimum setting and the influence of input parameters on output response are investigated.

The output parameters are selected for the experimentation desiring the maximum deposition of tool constituent owing to maximum wear of the tool. The $L_9$ orthogonal array selected for experimental run along with output responses are shown in Table 2. All the experimental runs are conducted at reverse polarity by keeping the voltage in between 40 and 45 V at constant 5 kgf/cm$^2$.

**Table 1** Selected input parameters and their levels

| Input parameters | Abbreviations | Level 1 | Level 2 | Level 3 |
|---|---|---|---|---|
| Compact load (tons) | CL | 5 | 10 | 15 |
| Current (A) | I | 4 | 5 | 6 |
| Pulse on-time (μs) | Ton | 400 | 600 | 1000 |

**Table 2** $L_9$ orthogonal array along with output responses

| S. No. | CL (tons) | $I$ (A) | $T_{on}$ (μs) | MTR (mg/min) | TWR (mg/min) |
|---|---|---|---|---|---|
| 1 | 5 | 4 | 400 | 27.3 | 36.1 |
| 2 | 5 | 5 | 600 | 27.2 | 73.0 |
| 3 | 5 | 6 | 1000 | 131.7 | 283.5 |
| 4 | 10 | 4 | 600 | 2.5 | 16.6 |
| 5 | 10 | 5 | 1000 | 9.6 | 42.0 |
| 6 | 10 | 6 | 400 | 6.1 | 122.0 |
| 7 | 15 | 4 | 1000 | 7.8 | 40.4 |
| 8 | 15 | 5 | 400 | 6.6 | 50.9 |
| 9 | 15 | 6 | 600 | 10.1 | 104.8 |

## 3 Results and Discussion

The CFRP workpiece samples after the material deposition are shown in Fig. 1. Figure 2 shows the pictorial view of the tool surface after EDM operation. The GRA method is applied to perform the multi-objective optimization to obtain the optimal set of input parameters for higher wear rate of tool and the consequent maximum transfer of material.

For this analysis initially S/N ratios of both the output parameters are calculated from the original output data. The Eq. 1 of "larger the better" is used for computing S/N ratio for MTR and TWR of the output responses. Table 3 shows the calculation process for computing grey relational grade. Figure 3 shows the microscopic view of the workpiece after deposition process. On the basis of these S/N ratio values subsequent analysis of normalization is carried out in the range in between zero to one. The pre-processing of the data depends on the characteristics of the responses [13]. For larger the better response i.e. MTR and TWR, the S/N ratio values are normalized using Eq. 2.

$$\text{S/N ratios(larger the better)} = -10\log\left[\frac{1}{n}\sum_{i=1}^{n}\frac{1}{Y_i^2}\right] \quad (1)$$

$$\text{Normalized S/N ratios (larger the better)} = Z_i = \frac{Y_i - Y_{i_{\min}}}{Y_{i_{\max}} - Y_{i_{\min}}} \quad (2)$$

$$\Delta_{0i}(k) = |x_0 * (k) - x_i * (k)| \quad (3)$$

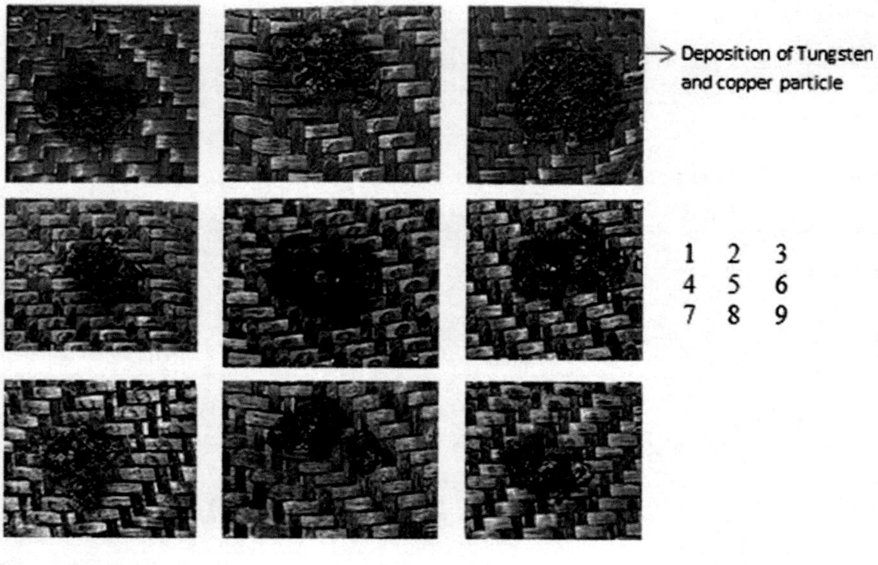


**Fig. 1** CFRP workpiece specimens after EDM deposition

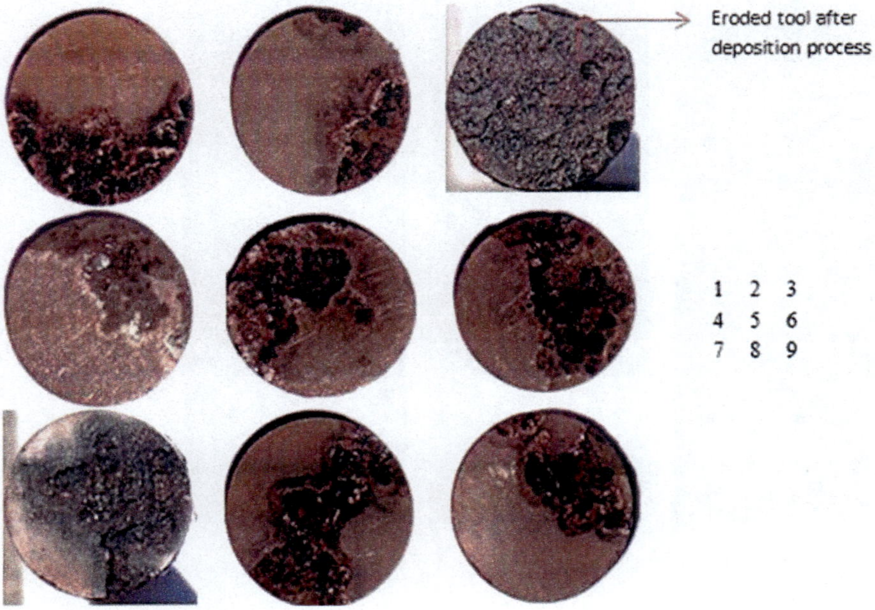


**Fig. 2** Tool specimen after deposition process

where, $x_0 * (k) = 1$ reference sequence and $\Delta_{0i}$ $(k)$ deviation sequence

$$\text{Grey Relation Coefficient}\ \xi_i = \frac{\Delta_{\min} - \lambda.\Delta_{\max}}{\Delta_i - \lambda.\Delta_{\max}} \tag{4}$$

After that deviation sequences for each output parameter is calculated by obtaining difference between deviation sequence and the reference sequence using Eq. 3. GRC is obtained using Eq. 4 from the normalize S/N ratio values. GRC is relating to the individual performance characteristics of the output response. Finally, GRG values are calculated for the evaluation of the multi-performance characteristic to know the optimal setting. GRG is obtained by averaging the corresponding GRC values of each experimental run using Eq. 5. The best value among all of the GRG values gives the optimal setting.

$$\text{Grey Relation Grade}\ \gamma_{i=}\frac{1}{3}\sum_{i=1}^{3}\xi i \tag{5}$$

From the above Table 3, it can be seen that Experiment no. 3 has the most value of GRG which gives the optimal setting. The level at compact load at 5 ton current at 6 A and pulse on-time at 1000 μs is optimal solution. As the motive of the work is to deposit maximum deposition of the tool material at improving tool wear rate, this setting has satisfied both the output response as shown in Table 3. At compact load

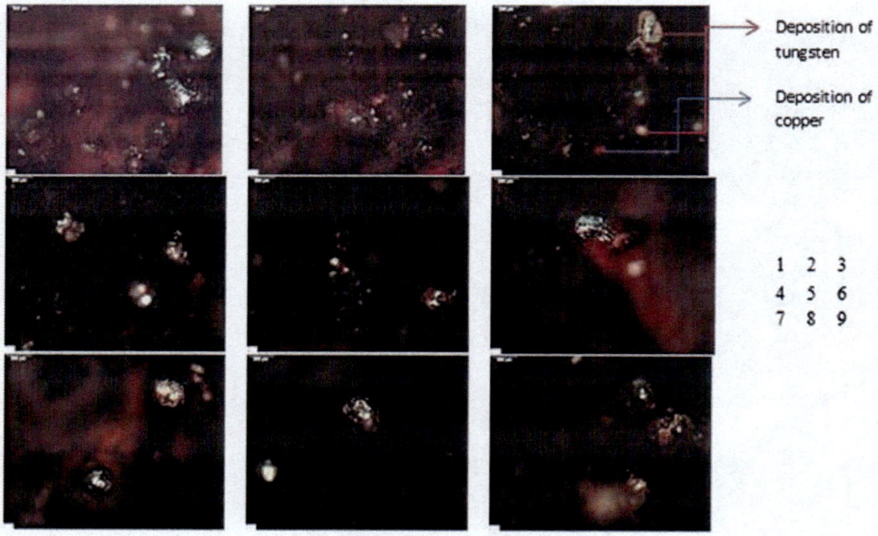


**Fig. 3** Microscopic view of the deposited specimen

**Table 3** $L_9$ orthogonal array along with output responses

| S/N ratios | | Normalize S/N ratios | | Deviation sequences | | GRC | | GRG | Rank |
|---|---|---|---|---|---|---|---|---|---|
| MTR | TWR | MTR | TWR | MTR | TWR | MTR | TWR | | |
| 28.72 | 31.15 | 0.603 | 0.273 | 0.396 | 0.726 | 0.557 | 0.407 | 0.482 | 5 |
| 28.69 | 37.26 | 0.602 | 0.521 | 0.397 | 0.478 | 0.556 | 0.511 | 0.534 | 2 |
| 42.39 | 49.05 | 1.000 | 1.000 | 0.000 | 0.000 | 1.000 | 1.000 | 1.000 | 1 |
| 7.95 | 24.40 | 0.000 | 0.000 | 1.000 | 1.000 | 0.333 | 0.333 | 0.333 | 9 |
| 19.64 | 32.46 | 0.339 | 0.327 | 0.660 | 0.672 | 0.430 | 0.426 | 0.428 | 6 |
| 15.70 | 41.72 | 0.225 | 0.702 | 0.774 | 0.297 | 0.392 | 0.627 | 0.509 | 4 |
| 17.84 | 32.12 | 0.287 | 0.313 | 0.712 | 0.686 | 0.412 | 0.421 | 0.416 | 8 |
| 16.39 | 34.13 | 0.244 | 0.394 | 0.755 | 0.605 | 0.398 | 0.452 | 0.425 | 7 |
| 20.08 | 40.40 | 0.352 | 0.649 | 0.647 | 0.350 | 0.435 | 0.587 | 0.511 | 3 |

of 5 tons, the tool particles are loosely bonded, so during spark energy generation at higher current and higher pulse on-time, the deposition is also more at this level. At high energy current i.e. at 6 A, the ernery to the spark generation is more results in more erosion of the tool and so more amount of material is deposited at this high level of current. At pulse on-time of 1000 μs, the amount of time for spark generation is more compare to other lower levels of pulse on-time, so the transfer of material to the workpiece is more, due to this the deposition at this level is more.

The mean of the GRG for each individual level of input parameters are calculated to determine the effect of each factor in GRG as shown in Table 4. From this Table 4

**Table 4** Effects of the input parameters in the Grey Relational Grade

| Input parameters | Level 1 | Level 2 | Level 3 | Delta | Rank |
|---|---|---|---|---|---|
| Compact load | **0.672** | 0.423 | 0.451 | 0.248 | **2** |
| Current | 0.410 | 0.462 | **0.673** | 0.262 | **1** |
| $T_{on}$ | 0.472 | 0.459 | **0.615** | 0.155 | **3** |

**Table 5** ANOVA for grey relational grade

| Source | Sum of square | DOF | Mean square | $F$-ratio | % Contribution | Rank |
|---|---|---|---|---|---|---|
| CL | 0.111 | 2 | 0.055 | 4.927 | 37.728 | **2** |
| $I$ | 0.116 | 2 | 0.058 | 5.155 | **39.471** | **1** |
| $T_{on}$ | 0.044 | 2 | 0.022 | 1.977 | 15.143 | **3** |

it is also shown that the compact load at lower level and current and pulse on-time at higher level gives the optimal setting. The rank of the input parameters is obtained by the difference of the smallest and biggest values of the individual level's responses from the rank of the effect of the factors, it also gives the contribution of the factors at the optimal setting.

Lastly the ANOVA for the GRG is computed (Table 5). From the ANOVA analysis the most contributing parameter is obtained. From the ranking of the input parameter it is showed that current is the most dominating parameter on the output responses. The result is also validated from the Table 4. As the optimal level exists in experiment number 3 of the $L_9$ orthogonal array, no confirmation experiment is required.

## 4 Conclusions

In this experimentation, the deposition of the tool constituent materials is successfully conducted on the CFRP workpiece using die sinker EDM at reverse polarity. The different combination of the levels of the input parameters is selected and the multi-optimal settings of the selected output responses are determined. From the above computation and research the following points can be concluded,

- At compact load of 5 tons, discharge current of 6 A and pulse on-time of 1000 μs the multi-optimum parameter setting is obtained.
- Increase in MTR results in increases in current and pulse on-time settings.
- Transfer of tool constituent is more at lower compact load then higher compact load settings.
- TWR decreases with the decreases in current and pulse on-time.
- It is observed that the same order of ranking of control parameters is obtained from the effects of the factors in the GRG and the ANOVA for GRG. The results show that current is the most dominating factor over multi-response optimization

of the selected output response. The second dominating factor is determined as compact load. The pulse on-time is the least significant over the output results.

**Acknowledgements** The authors would like to acknowledge their gratitude to SERB-DST for extending the facilities of tools and equipment for conducting this experimental work.

## References

1. J. Xie, Lu. Zhongyu, Y. Guo, Y. Huang, Durability of CFRP sheets and epoxy resin exposed to natural hygrothermal or cyclic wet-dry environment. Polym. Compos. **40**(2), 553–567 (2019)
2. M.Ö. Bora, E. Akman, O. Çoban, B.G. Oztoprak, A. Demir, The effect of $co_2$ laser-induced microhole formations on adhesive bonding strength of CFRP/CFRP joints. Polymer Compos. (2018)
3. S. Habib, A. Okada, S. Ichii, Effect of cutting direction on machining of carbon fibre reinforced plastic by electrical discharge machining process. Int. J. Mach. Mach. Mater. **13**(4), 414–427 (2013)
4. S.R. Karnik, V.N. Gaitonde, J. Campos Rubio, A. Esteves Correia, A.M. Abrão, J. Paulo Davim, Delamination analysis in high speed drilling of carbon fiber reinforced plastics (CFRP) using artificial neural network model. Mater. Des. **29**(9), 1768–1776 (2008)
5. H. Hocheng, C.C. Tsao, The path towards delamination-free drilling of composite materials. J. Mater. Process. Technol. **167**(2–3), 251–264 (2005)
6. G. Caprino, V. Tagliaferri, Maximum cutting speed in laser cutting of fiber reinforced plastics. Int. J. Mach. Tools Manuf **28**(4), 389–398 (1988)
7. M. Rahang, P.K. Patowari, Parametric optimization for selective surface modification in EDM using Taguchi analysis. Mater. Manuf. Processes **31**(4), 422–431 (2016)
8. R. Tyagi, K. Pandey, A.K. Das, A. Mandal, Deposition of hBN+ Cu layer through electrical discharge process using green compact electrode. Mater. Manuf. Processes **34**(9), 1035–1048 (2019)
9. M. Rahang, P.K. Patowari, Pattern generation by selective area deposition of material in EDM. Mater. Manuf. Processes, 1–8 (2019)
10. P.K. Patowari, P. Saha, P.K. Mishra, Taguchi analysis of surface modification technique using W-Cu powder metallurgy sintered tools in EDM and characterization of the deposited layer. Int. J. Adv. Manuf. Technol. **54**(5–8), 593–604 (2011)
11. M. Rahang, P.K. Patowari, Application of masking technique in EDM for generation of rectangular shaped pattern. Int. J. Precis. Technol. **5**(2), 140–156 (2015)
12. M. Eswara Krishna, P.K. Patowari, Parametric optimisation of electric discharge coating process with powder metallurgy tools using Taguchi analysis. Surf. Eng. **29**(9), 703–711 (2013)
13. K. Jangra, A. Jain, S. Grover, Optimization of multiple machining characteristics in wire electrical discharge machining of punching die using grey relational analysis. J. Sci. Ind. Res. **69**(08), 606–612 (2010)

# Performance Improvement of Turbine Blade Using Flow Control Techniques: A Review

**Hussain Mahamed Sahed Mostafa Mazarbhuiya, Agnimitra Biswas, and Kaushal Kumar Sharma**

**Abstract** Utilization of wind energy has become an emerging concept for clean energy generation due to its vast availability, low cost and its contribution to carbon dioxide reduction. Wind turbines are the prominent converter which are used in both onshore and offshore wind energy generation projects. The performance of the wind turbines manly depends on blade aerodynamic performance. The aerodynamic performance of blade can be enhanced by using flaps, vortex generator, airfoil modification on leading or trailing edge, slots in airfoils, etc. The present paper mainly focuses on different passive flow control techniques, especially Gurney flap and vortex generator are used for enhancing blade aerodynamic performance.

**Keywords** Flow control techniques · Gurney flap · Vortex generator

## 1 Introduction

The most important key element of a wind turbine is its blade which is of either symmetric or asymmetric. The aerodynamic performance depends on blade design and modification influence of the turbine performance. The wind energy projects are mainly focused on rural/urban and coastal regions. The large-scale wind turbines especially for coastal regions must be structurally and mechanically sound along with good aerodynamic performance. Different types of wind turbines have been designed and manufactured to enhance the performance and wind energy utilization. The lift-to-drag (l/d) ratio of wind turbine blade plays an important rule, and hence, importance has also been given to enhance this ratio. Flow separation occurs in the boundary layer due to viscous friction, and inverse pressure gradient [1] leads to dynamic stall after certain angle of attack (AOA) which is undesirable. Stall occurs

H. M. S. M. Mazarbhuiya (✉) · A. Biswas · K. K. Sharma
Department of Mechanical Engineering, National Institute of Technology, Silchar, Assam 788010, India
e-mail: mdhussain0309@gmail.com

K. K. Sharma
e-mail: kksharma1313@rediffmail.com


K. M. Pandey et al. (eds.), *Recent Advances in Mechanical Engineering*, Lecture Notes in Mechanical Engineering, https://doi.org/10.1007/978-981-15-7711-6_81

due to break in increase in lift coefficient after certain AOA. Separation vortex is periodically generated due to unstable flow separation in the boundary layer. This increases the fatigue load of turbine blade and degrades the turbine performance. Aerodynamic performance of the turbine can improve by controlling the flow separation and delaying the dynamic stall. Flow control technologies are mainly used to suppress or delay the separation by flow transition or increasing turbulence intensity. To keep flow attached to the airfoil, the near wall flow should have enough momentum to overcome the adverse pressure gradient and the stream-wise viscous dissipation. Different flow control technologies such as partial flexibility on airfoil suction side [2], leading edge rod [3], roughness element on the blade surface [4], microtabs at different blade locations [5]vortex generator, Gurney flap, etc., have been adopted to increase the blade's aerodynamic performance. Two types of flow control technologies are used: passive flow control and active flow control. The passive flow control is simple and is used extensively due to less mechanism. In the present paper, only two types of flow control techniques, such as Gurney flaps and vortex generators, are reviewed extensively. Some important research gap is found which are discussed in further sections.

## 2 Flow Control Techniques

Gurney flap is a passive flow control device consisting of small tabs which are extensively used due to its simplification to the trailing edge of the airfoil to increase aerodynamic performance. The length of the flap varies with blade's aerodynamics aspect. It generally enhances the lift coefficient of airfoil. Liebeck [6] recommended the optimal height of Gurney flap within the range of 0.01c–0.02c. Pastrikakis et al. [7] investigated the effect of Gurney flap with four different lengths (0.02c, 0.01c, 0.005c and 0.003c) and found 0.02c most effective. He et al. [8] investigated the performance of a thick airfoil considering six different sizes of flaps in the range of 0.0025c–0.03c and obtained an 2.7% increase in lift-to-drag ratio and 12.9% lift coefficient at 3° angle of attack for flap with 0.005c length. The use of Gurney flap can enhance both pre-stall lift and lift-to-drag ratio in a certain range of angle of attack. An increase in turbine's efficiency by 15% and energy extraction by 22% from wind under flap height 0.03c and reduced frequency 1.0 on flapping airfoil by using on both side at trailing edge [9]. The same height of flap gives an increase in power coefficient by 21.3% at 2.4 tip speed ratio when they applied at inner side of airfoils [10]. They have also studied the effect of Gurney flap on outer and both sides of the airfoil. But the maximum power coefficient is obtained where flap is applied at inner side of airfoil. A significant increase in lift is responsible for increase in power coefficient, although there is very small amount of drag. But increase in lift coefficient is considerably higher which leads to high lift-to-drag ratio, hence improving the performance [11]. Li et al.[12] obtained an increase in lift coefficient by 40% with 0.05c flap height. Zhang et al. [13] investigated the performance of horizontal axis wind turbine by applying two different sizes of Gurney flap and found that thrust increases with increase in Gurney flap height. But the flap with less height produces batter power compared to higher one due to completely submerge in the boundary layer. The above studies effloresce that Gurney flap might be in different height that depends on local boundary layer thickness. Daniel and Traub [14] concluded that if the wing aspect ratio is reduced then, a higher height of Gurney flap is required to maximize the aerodynamic efficiency. Meena et al. [15] concluded that high lift can be achieved at higher range of angle of attack using Gurney flap. Zhu et al. [16] extensively investigated the effect of Gurney flap to the inboard, outboard and both sides of the trail edge with normal and dimple-shaped airfoils. They found that all kinds of Gurney flap can enhance the turbine performance in a certain range of tip speed ratio (TSR). Moreover, the outboard–dimple Gurney flap can increase the power coefficient of turbine by 17.92% at TSR 3.1. So far the concept of Gurney flaps' concerned studies are limited to only to determine the optimum length of flap with using the flap to the end of airfoil trailing edge. Most of the investigations are focusing only inboard, outboard and both sides of the edge shown in Fig. 1a. A very few research is carried out for the flap position, including dimple on the airfoil shown in Fig. 1b. There is a scope for investigation with variation of dimple size, dimples of different patterns such as triangular, rectangular for optimum position on airfoil along with Gurney flaps.

The passive vortex generator (VG) is used extensively on airfoil to delay the onset of stall and improve the aerodynamic performance. It consists of pairs of vanes protruding from airfoil surface with certain height within boundary layer thickness. Passive VG is tested in wind tunnel and found to improve the post-stall behavior of pitching airfoil [17]. The VG is generally placed on the surface of the leading edge. Vahl et al. [18] recommended an optimal VG position which is 15%c–20%c and achieved a comparatively low parasitic drag. The chord position of airfoil for installing VG plays a vital rule in the post-stall behavior. Maximizing the vane height can yield strong stream-wise vortices, hence accelerating the momentum transfer to the wall surface and thereby increasing the lift. On the other hand, drag penalty is increased at low AOA with maximum vane height [19]. An increase in VG length

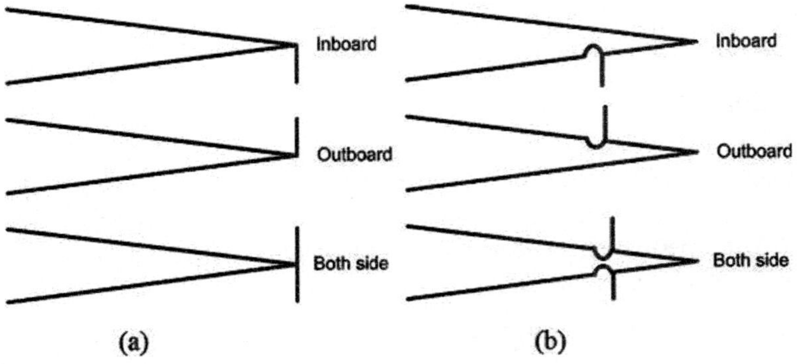


**Fig. 1** Geometries of Gurney flap and dimple Gurney flap

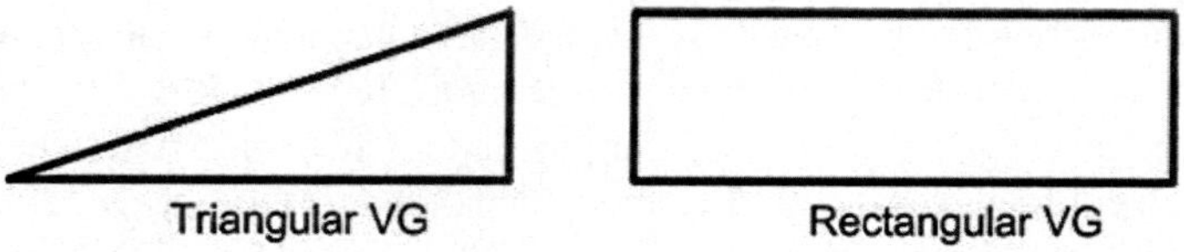


**Fig. 2** Vortex generators used for flow control

deteriorates both lift and drag [20]. Hence, the selection of vane height is a crucial part and a great deal between lift increase and drag penalty [18]. The installation location of VG should not be far away from the point of separation [21]. The lift coefficient become less at smaller AOA when VGs are installed below 20% chord position [22]. Wang et al. [23] reported that the stall angle of attack is increased from 14° to 18° after installing VG at 10% of chord position of airfoil. At certain AOA, the VG can reduce the boundary layer thickness of airfoil, thereby reducing drag coefficient. With the use of double VG arrangements, lift coefficients are further improved at higher AOA and the boundary layer thickness is reduced further. Hence, the stall AOA is also increased for S809 airfoil. Hence, double VG arrangements are recommended due to their batter performance in flow separation control [23]. Different VGs such as triangular and rectangular are installed, and their effect on aerodynamic performance is studied by [24]. They recommended the triangular-shaped VG due to their capability to control boundary layer separation. The optimum distance between the VGs is 3 mm, and the angle is 12° located at 50% of chord. Micro-VGs are found very efficient due to controlling the flow with less parasitic drag. Too far downstream VG position can lead to early sudden stall. The rectangular VG is found to delay the onset of dynamic stall and increase the maximum lift coefficient by 40% for NERL S809 airfoil in the range of angle of attack $(\alpha) = 13°–18°$ [25]. From the above literature review, some important glimpse is found. Most of the investigations are carried out using triangular and rectangular vortex generators shown in Fig. 2. The chord-wise position and size of VG are crucial for its installation. Future research may be carried out for design of VG for efficient wind turbine blade. There is also a gap of study for the combined effect of the above flow control techniques.

## 3 Conclusions

For enhancing the aerodynamic performance, it is required to improve the post-stall behavior. Different flow control techniques are used to delay the stall and can improve the aerodynamic performance. Some important research gap has come out for future investigation which are as follows:

(a) A very few research is carried out for the flap position, including dimple on the airfoil. There are scopes for investigation with variation of dimple size, dimples of different pattern such as triangular, rectangular for optimum position on airfoil along with gurney flaps.

(b) Most of the investigations are carried out using triangular and rectangular vortex generators. Research may be carried out for design of VG for efficient wind turbine blade. There is also a gap of study for the combined effect of the above flow control techniques.

## References


1. C.E. Willert, C. Cuvier, J.M. Foucaut, J. Klinner, M. Stanislas, J.P. Laval et al., Experimental evidence of near-wall reverse flow events in a zero pressure gradient turbulent boundary layer. Exp. Therm. Fluid Sci. **91**, 320–328 (2018). https://doi.org/10.1016/j.expthermflusci.2017.10.033
2. H.H. Açıkel, G.M. Serdar, Control of laminar separation bubble over wind turbine airfoil using partial flexibility on suction surface. Energy **165**, 176–190 (2018). https://doi.org/10.1016/j.energy.2018.09.040
3. J. Zhong, J. Li, P. Guo, Y. Wang, Dynamic stall control on a vertical axis wind turbine aerofoil using leading-edge rod. Energy **174**, 246–260 (2019). https://doi.org/10.1016/j.energy.2019.02.176
4. M. Serdar Genç, K. Koca, H.H. Açikel, Investigation of pre-stall flow control on wind turbine blade airfoil using roughness element. Energy **176**, 320–34 (2019). https://doi.org/10.1016/j.energy.2019.03.179
5. A. Ebrahimi, M. Movahhedi, Wind turbine power improvement utilizing passive flow control with microtab. Energy **150**, 575–582 (2018). https://doi.org/10.1016/j.energy.2018.02.144
6. R.H. Liebeck, Design of subsonic airfoils for high lift. J Aircr **15**, 547–561 (1978). https://doi.org/10.2514/3.58406
7. V.A. Pastrikakis, R. Steijl, G.N. Barakos, J. Małecki, Computational aeroelastic analysis of a hovering W3 sokol blade with gurney flap. J. Fluids Struct. **53**, 96–111 (2015). https://doi.org/10.1016/j.jfluidstructs.2014.06.014
8. X. He, J. Wang, M. Yang, D. Ma, C. Yan, P. Liu, Numerical simulation of Gurney flap on SFYT15thick airfoil. Theor. Appl. Mech. Lett. **6**, 286–292 (2016). https://doi.org/10.1016/j.taml.2016.09.002
9. Y.H. Xie, W. Jiang, K. Lu, D. Zhang, Numerical investigation into energy extraction of flapping airfoil with gurney flaps. Energy **109**, 694–702 (2016). https://doi.org/10.1016/j.energy.2016.05.039
10. A. Bianchini, F. Balduzzi, D. Di Rosa, G. Ferrara, On the use of gurney flaps for the aerodynamic performance augmentation of Darrieus wind turbines. Energy Convers. Manage. **184**, 402–415 (2019). https://doi.org/10.1016/j.enconman.2019.01.068
11. M.S. Chandrasekhara, Optimum Gurney flap height determination for "lost-lift" recovery in compressible dynamic stall control. Aerosp. Sci. Technol. **14**, 551–556 (2010). https://doi.org/10.1016/j.ast.2010.04.010
12. Y.C. Li, J.J. Wang, J. Hua, Experimental investigations on the effects of divergent trailing edge and Gurney flaps on a supercritical airfoil. Aerosp. Sci. Technol. **11**, 91–99 (2007). https://doi.org/10.1016/j.ast.2006.01.006
13. Y. Zhang, V. Ramdoss, Z. Saleem, X. Wang, G. Schepers, C. Ferreira, Effects of root gurney flaps on the aerodynamic performance of a horizontal axis wind turbine. Energy **187**, 115955 (2019). https://doi.org/10.1016/j.energy.2019.115955
14. L. Daniel, L.W. Traub, Effect of aspect ratio on gurney-flap performance. J. Aircr. **50**, 1217–1225 (2013). https://doi.org/10.2514/1.C032140
15. M. Gopalakrishnan Meena, K. Taira, K. Asai, Airfoil-wake modification with gurney flap at low reynolds number. AIAA J. **56**, 1348–1359 (2017). https://doi.org/10.2514/1.J056260

16. H. Zhu, W. Hao, C. Li, Q. Ding, Numerical study of effect of solidity on vertical axis wind turbine with Gurney flap. J. Wind Eng. Ind. Aerodyn. **186**, 17–31 (2019). https://doi.org/10.1016/j.jweia.2018.12.016
17. A. Choudhry, M. Arjomandi, R. Kelso, Methods to control dynamic stall for wind turbine applications. Renew. Energy **86**, 26–37 (2016). https://doi.org/10.1016/j.renene.2015.07.097
18. H. Mueller-Vahl, G. Pechlivanoglou, C.N. Nayeri, C.O. Paschereit, Vortex generators for wind turbine blades: a combined wind tunnel and wind turbine parametric study, pp. 899–914 (2012). https://doi.org/10.1115/GT2012-69197
19. P. Martínez-Filgueira, U. Fernandez-Gamiz, E. Zulueta, I. Errasti, B. Fernandez-Gauna, Parametric study of low-profile vortex generators. Int. J. Hydrogen Energy **42**, 17700–17712 (2017). https://doi.org/10.1016/j.ijhydene.2017.03.102
20. L. Gao, H. Zhang, Y. Liu, S. Han, Effects of vortex generators on a blunt trailing-edge airfoil for wind turbines. Renew. Energy **76**, 303–311 (2015). https://doi.org/10.1016/j.renene.2014.11.043
21. J.C. Lin, Review of research on low-profile vortex generators to control boundary-layer separation **38** (2002). https://doi.org/10.1016/S0376-0421(02)00010-6
22. L. Zhang, X. Li, K. Yang, D. Xue, Effects of vortex generators on aerodynamic performance of thick wind turbine airfoils. J. Wind Eng. Ind. Aerodyn. **156**, 84–92 (2016). https://doi.org/10.1016/j.jweia.2016.07.013
23. H. Wang, B. Zhang, Q. Qiu, X. Xu, Flow control on the NREL S809 wind turbine airfoil using vortex generators. Energy **118**, 1210–1221 (2017). https://doi.org/10.1016/j.energy.2016.11.003
24. O.M. Fouatih, M. Medale, O. Imine, B. Imine, Design optimization of the aerodynamic passive flow control on NACA 4415 airfoil using vortex generators. Eur. J. Mech. B/Fluids **56**, 82–96 (2016). https://doi.org/10.1016/j.euromechflu.2015.11.006
25. C. Zhu, J. Chen, J. Wu, T. Wang, Dynamic stall control of the wind turbine airfoil via single-row and double-row passive vortex generators. Energy **189**, 116272 (2019). https://doi.org/10.1016/j.energy.2019.116272

# Parametric Analysis for Machining of Stainless Steel AISI (SS-430) Using Photo Chemical Machining

**Gaijinliu Gangmei, Jaswant Kumar, Tapas Debnath, and Promod Kumar Patowari**

**Abstract** Parametric study of Photo Chemical Machining (PCM) is necessary to obtain excellent etching quality. It can be used to reduce weight of a component, hence it can improve strength to weight ratio of a component. In the current study, the effect of control variables such as etching time, etching temperature, and etchant concentration on material removal rate (MRR), surface roughness and undercut in PCM of Stainless Steel AISI (SS-430) has been investigated. Taguchi L-16 orthogonal array has been designed to conduct the experimental runs. Moreover, the analysis of variance (ANOVA) technique is used to evaluate the significance of control parameters and also percentage contribution of input parameters. From ANOVA table, it can be observed that temperature is highly influencing parameter for MRR and undercut; whereas time is for surface roughness.

**Keywords** ANOVA · PCM · SS-430 · Taguchi

## 1 Introduction

For manufacturing of geometrically complex machine components which are difficult by conventional machining processes are machined by non-traditional machining processes and photochemical machining is one of them [1]. Corrosion is assumed as destructive for material but it can be a process to machine particular region of a specimen by a controlled process named as "etching". PCM also uses the same principle for material removal by using a chemical reagent called "etchant". PCM is

G. Gangmei · J. Kumar · T. Debnath (✉) · P. K. Patowari
Department of Mechanical Engineering, National Institute of Technology Silchar, Silchar, Assam 788010, India
e-mail: nit.tapas11@gmail.com

G. Gangmei
e-mail: gangmei.jinjin@gmail.com

P. K. Patowari
e-mail: ppatowari@yahoo.com



K. M. Pandey et al. (eds.), *Recent Advances in Mechanical Engineering*, Lecture Notes in Mechanical Engineering, https://doi.org/10.1007/978-981-15-7711-6_82

mostly used for machining flat materials producing geometrically complex as well as dimensionally very accurate parts without residual/mechanical stress or deformation [2].

A proper selection of PCM control variables is important to improve etching quality and this can be done by obtaining the optimum process parameters setting which satisfies the requirements of outstanding quality of etching. Agrawal et al. investigated the effect of various selected control variables in PCM of phosphor bronze and using grey relational analysis (GRA) method their optimal conditions were calculated [3]. Wangikar et al. studied the effect of control parameters and their optimization for photochemical machining (PCM) of brass, german silver [4], and copper were carried out [5]. Rathod et al. also optimized the input variables of PCM used for the machining of copper material by using Taguchi method analysis technique [6]. Agrawal et al. developed aluminium based silicon carbide particulate AMCs and optimized the control parameters for photochemical machining of composites [7]. Misal et al. used grey-based Taguchi method for optimization of the PCM control parameters while etching of Inconel 601 [8]. Using weighted GRA, the optimization of PCM control parameters of SS-304 was evaluated by Agrawal and Kamble [9]. Optimization of control parameter of PCM for ASME 316 steel has been carried out by Mumbare and Gujar [10] and using Grey Relational Analysis method, the optimization of control parameters for PCM of SS316L was evaluated by Bhasme and Kadam [11].

From the literature review, it has been found that no significant study has been reported on parametric analysis for PCM of Stainless Steel AISI (SS-430).In this work, the machinability of SS-430 has been conducted using PCM by generating a square-shaped cavity, where ferric chloride ($FeCl_3$) is used as an etchant.

## 2 Materials and Methods

In the present study, stainless steel AISI (SS-430) is selected for the experimentation. It has good resistance to corrosion, heat, and to stress corrosion cracking. It is also widely used in vehicle exhausts, element supports and fasteners, scientific apparatus, etc.

Taguchi L-16 orthogonal array has been designed which is used to conduct the experimental runs and to evaluate optimum response variables. DOE is the statistical method used to study the effects of several control parameters on performance characteristics simultaneously. It lays out the combinations of factors integrated into the experimental study [2]. It plans, collects and analyzes the data, which reflects the optimum process parameters [6]. Table 1 illustrates the input parameters with their levels. After selecting the factors and their levels the next step is to form an orthogonal array. Table 2 shows Taguchi L-16 orthogonal array with responses.

In this paper, the control variables that have been selected are etching time, temperature and etchant concentration; and the response measures are discussed below:

**Table 1** Input parameters with their levels

| Input parameters | Level 1 | Level 2 | Level 3 | Level 4 |
|---|---|---|---|---|
| Time (min) | 10 | 15 | 20 | 25 |
| Temperature (°C) | 35 | 45 | 55 | 65 |
| Concentration (g/lit) | 475 | 550 | 625 | 700 |

**Table 2** Taguchi L-16 array with responses

| Exp. No | Time (min) | Temperature (°C) | Concentration (g/lit) | MRR (mg/min) | SR (μm) | UC (μm) |
|---|---|---|---|---|---|---|
| 1 | 10 | 35 | 475 | 0.070 | 0.47 | 44.78 |
| 2 | 10 | 45 | 550 | 0.113 | 0.85 | 187.62 |
| 3 | 10 | 55 | 625 | 0.310 | 0.93 | 272.47 |
| 4 | 10 | 65 | 700 | 0.270 | 1.17 | 303.21 |
| 5 | 15 | 35 | 550 | 0.073 | 1.16 | 93.93 |
| 6 | 15 | 45 | 475 | 0.180 | 0.94 | 214.81 |
| 7 | 15 | 55 | 700 | 0.226 | 1.22 | 293.93 |
| 8 | 15 | 65 | 625 | 0.330 | 1.25 | 330.50 |
| 9 | 20 | 35 | 625 | 0.079 | 1.39 | 164.94 |
| 10 | 20 | 45 | 700 | 0.290 | 1.70 | 241.28 |
| 11 | 20 | 55 | 475 | 0.117 | 1.77 | 246.03 |
| 12 | 20 | 65 | 550 | 0.384 | 1.83 | 315.55 |
| 13 | 25 | 35 | 700 | 0.086 | 2.29 | 185.55 |
| 14 | 25 | 45 | 625 | 0.210 | 2.43 | 218.66 |
| 15 | 25 | 55 | 550 | 0.223 | 2.73 | 266.96 |
| 16 | 25 | 65 | 475 | 0.344 | 2.73 | 331.91 |

- The MRR is a rate at which material is removed from the specimen. It is determined by the Eq. (1) as given below.

$$\text{MRR} = \frac{\text{Initial weight} - \text{Final weight}}{\text{Etching time}} \quad (1)$$

- The measurements of undercut of each specimen have been carried out with metallurgical optical microscope after photochemical machining. Figure 1 shows the typical image of undercut which is the distance etched laterally under the photo resist layer. The difference between the dimensions of the machined area with that of the photo-tool is considered as the undercut for observed samples.
- Surface roughness is the measure of irregularities on the surface texture. The surface roughness of each machined workpiece is measured using a surface profilometer.

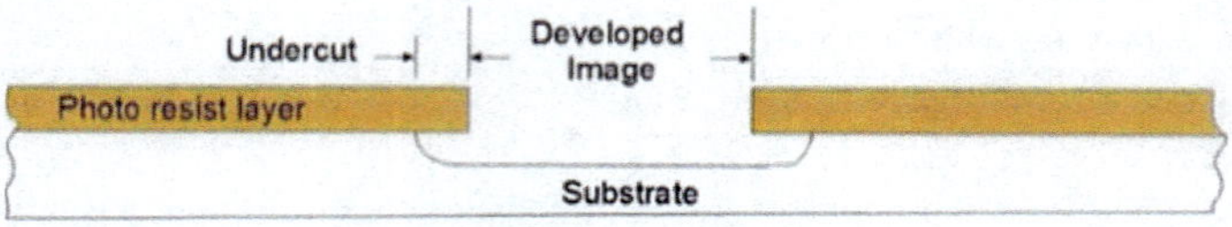


**Fig. 1** Schematic representation of undercut

## 2.1 Experimental Procedure

In photochemical machining, the prepared workpiece is first cleaned with acetone to make the samples free from contaminants which will provide maximum surface for adhesion of photo resist on the workpiece.

The photo resist is applied on the specimen using immersion process and is allowed to dry before proceeding to the next step. After that, a photo-tool is used to create the impression. The area to be etched is protected by a photo tool (square-shaped for the current study) as shown in Fig. 2a and it is exposed to UV light in a vacuum for about 2 min (using UV exposure unit) as shown in Fig. 2b.

After UV exposure, the specimen is immersed in a solvent-based developer which will remove unexposed areas of the photo resist (negative working photo resist) and then the specimen is washed in gentle running water. Next step is machining, i.e. etching in which metal gets chemically dissolved by etchant and the required temperature of etchant is attained by using a heating plate as shown in Fig. 3. After

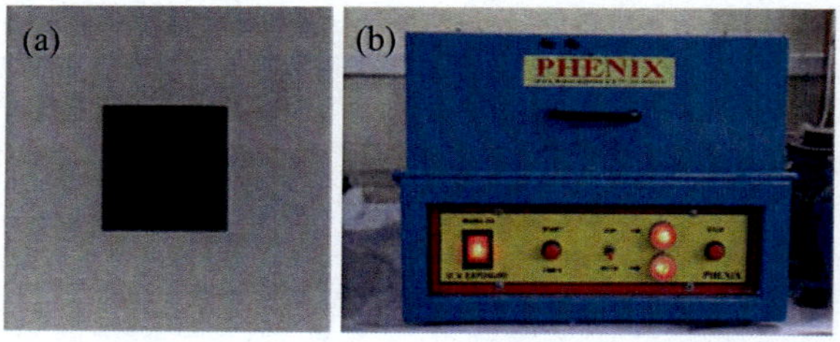


**Fig. 2** **a** Square photo tool; and **b** UV exposure unit

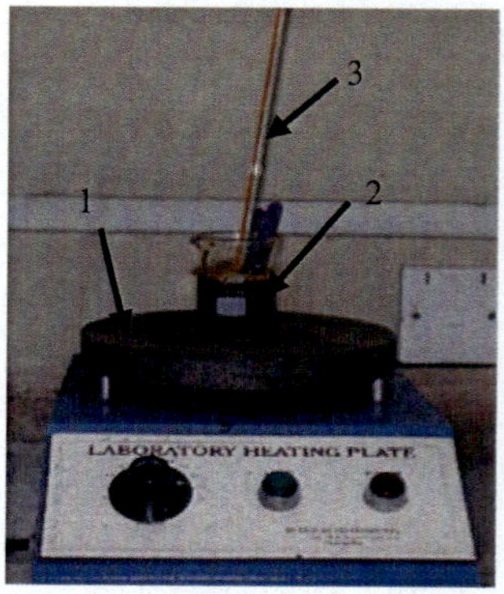


**Fig. 3** Machining set-up: i Heating plate; ii beaker containing etchant and; iii thermometer

etching, the specimens are cleaned again by washing in gentle running water followed by acetone.

## 3 Experimental Results and Discussion

The effect of control variables on material removal rate (MRR), surface roughness (SR) and undercut (UC) has been studied. Figure 4 shows the machined sample and microscopic view of the machined area. Figure 5 shows the detailed measurement of an edge of the machined sample. Based on the measurement and the data, the performance has been calculated for each experiment and tabulated in Table 2.

### *3.1 Effect of Control Parameters on MRR*

Figure 6 shows that MRR increases with increase in etching time because at higher time more reaction takes place between molecules of etchant and workpiece indicating higher removal of material. Similarly, with temperature, MRR increases as etchant is more active at higher temperature. From Fig. 6 it can be observed that MRR first increases and then decreases with concentration, this is due to the fact that ferric chloride solution becomes more viscous with time which reduces flow ability and the movement of the formed product. Moreover, the thicker reaction layer resists further reaction to take place.

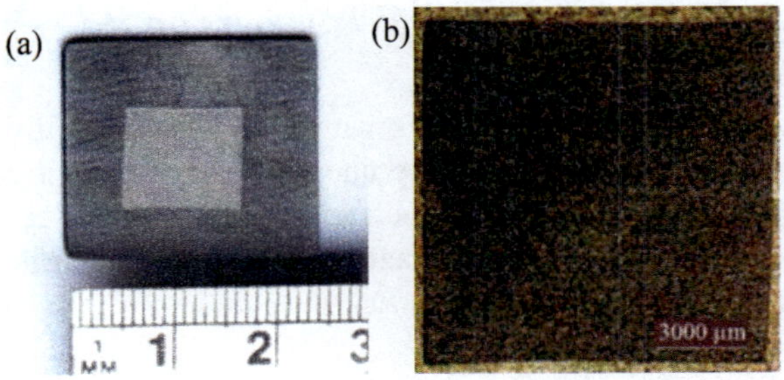


**Fig. 4** **a** Machined sample; **b** microscopic view of the machined area

**Fig. 5** Measurement of one side of the machined area

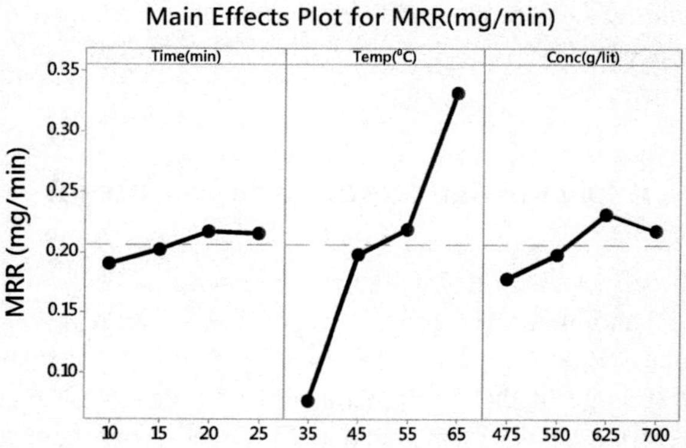


**Fig. 6** Variation of MRR with input parameters

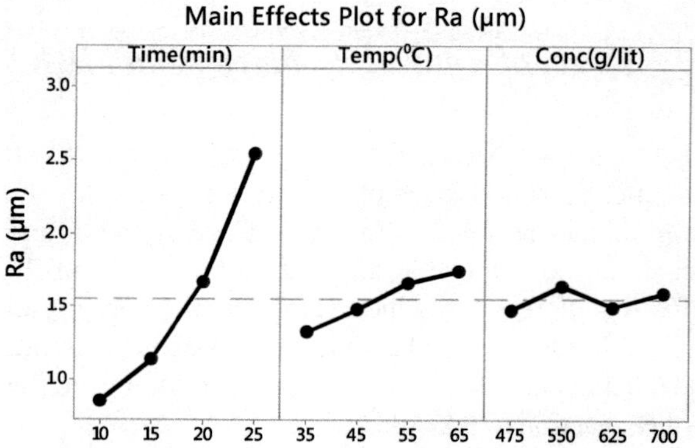


**Fig. 7** Variation of surface roughness with input variables

## 3.2 *Effect of Control Parameters on Ra*

Surface roughness increases with time because collisions for a longer period of time make the surface more uneven. From Fig. 7 it is also seen that the surface roughness increases with increase in temperature due to the higher energy of the etchant molecules. The average roughness does not follow a specific trend due to the change in etchant concentration.

## 3.3 *Effect of Control Parameters on Undercut (UC)*

Figure 8 shows that undercut increases with the increase in all control parameters, i.e. etching time, temperature and concentration. As the time increases more amount of material removes from the workpiece and with the increase in temperature and etchant concentration, due to the higher energy of the molecules the reaction is more

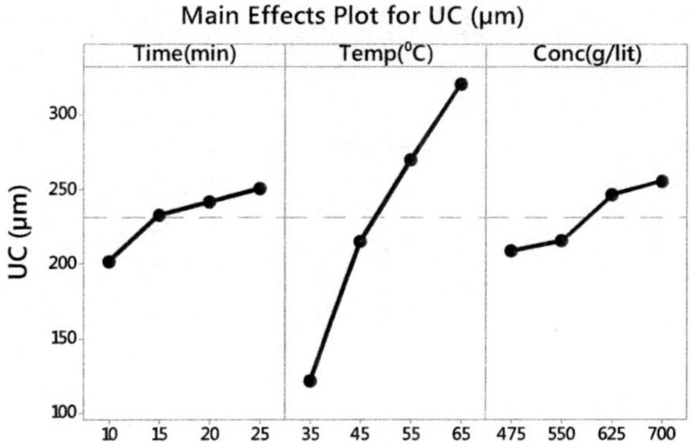


**Fig. 8** Variation of undercut with control parameters

prominent, resulting in higher undercut. Minimum undercut is observed at time of 10 min, etching temperature of 35 °C and concentration of 475 g/lit.

### *3.4 ANOVA Analysis*

ANOVA is used to investigate the effect of control variables on the performance measures. The percentage contribution of each parameter on the output parameters can also be evaluated using ANOVA [12]. Table 3 represents the analysis of ANOVA. From the ANOVA, it is found that temperature (75.83%) is most significant component for MRR, followed by concentration (3.92%) and time (1.09%). For surface roughness, Time (92.02%) is the highest influencing parameter followed by temperature (5.85%); whereas, for undercut, temperature (84.50%) is the maximum contributing input parameter, followed by concentration (6.11%) and time (5.30%). The correlation coefficient ($R^2$) and adjusted correlation coefficient [$R^2$(adj)] for MRR are (80.84% and 52.09%), for surface roughness (97.87% and 96.44%) and for undercut (95.91% and 87.78%), respectively. The higher $R^2$ value signifies a better accuracy and general ability of the model and Adjusted $R^2$ measures the percentage of variability and model quality. Percentage of error in MRR (19.16%), Ra (2.13%) and UC (4.09%) are lower than the significance of the variable parameters considered in the investigation, representing less effect of the interaction of the process parameters.

## 4 Conclusions

A square cavity was successfully machined on Stainless Steel (SS 430) using photochemical machining. The input parameters that have been chosen were etching time,

**Table 3** Analysis of variance for responses

| Source | DF | Seq SS | Adj MS | *F*-value | *P*-value | Contribution (%) |
|---|---|---|---|---|---|---|
| a. *Material removal rate (MRR)* | | | | | | |
| Time | 3 | 0.001891 | 0.000630 | 0.110 | 0.949 | 1.09 |
| Temp | 3 | 0.130979 | 0.043660 | 7.91 | 0.017 | 75.83 |
| Conc | 3 | 0.006760 | 0.002253 | 0.41 | 0.753 | 3.92 |
| Error | 6 | 0.033102 | 0.005517 | | | 19.16 |
| Total | 15 | 0.172732 | | | | 100 |
| $R^2 = 80.84\%$, $R^2$ (adj) $= 52.09\%$ | | | | | | |
| b. *Surface roughness (Ra)* | | | | | | |
| Time | 3 | 6.6162 | 2.20541 | 129.37 | 0.000 | 92.02 |
| Temp | 3 | 0.4201 | 0.14004 | 8.21 | 0.006 | 5.85 |
| Error | 9 | 0.1534 | 0.01705 | | | 2.13 |
| Total | 15 | 7.1898 | | | | 100 |
| $R^2 = 97.87\%$, $R^2$ (adj) $= 96.44\%$ | | | | | | |
| c. *Undercut (UC)* | | | | | | |
| Time | 3 | 5407 | 1802.4 | 2.60 | 0.148 | 5.30 |
| Temp | 3 | 86,125 | 28,708.4 | 41.35 | 0.000 | 84.50 |
| Conc | 3 | 6228 | 2076.2 | 2.99 | 0.118 | 6.11 |
| Error | 6 | 4166 | 694.3 | | | 4.09 |
| Total | 15 | 101,926 | | | | 100 |
| $R^2 = 95.91\%$, $R^2$ (adj) $= 89.78\%$ | | | | | | |

etching temperature, and etchant concentration. The response variables were material removal rate, surface roughness and undercut. From the results and discussion, the following conclusions have been observed:

- MRR increases with increase in temperature and time but with concentration, it initially increases up to a 625 g/lit and then after decreases.
- Undercut also increases with increase in all the control parameters, i.e. etchant concentration, machining time and etching temperature.
- Surface roughness does not follow a proper path with concentration but with time and temperature, it increases.
- Maximum material removal rate is observed at etching time of 20 min, temperature of 65 °C and concentration 550 g/lit.
- Minimum surface roughness and undercut are observed at etching time of 10 min, temperature 35 °C and of concentration 475 g/lit.
- From ANOVA table, temperature is most influencing parameter for MRR and undercut, time for surface roughness.

## References


1. O. Cakir, Photochemical machining of engineering materials, in *12th International Research/Expert Conference* (Trends in the Development of Machinery and Associated Technology TMT, Istanbul, Turkey, 26–30 Aug 2008)
2. E. Tripathi, T. Debnath, P.K. Patowari, Etching characteristics of aluminium while machining square cavity using photochemical machining. AIP Conf. Proc. **1998**(1) (2018). https://doi.org/10.1063/1.5049105
3. D.P. Agrawal, K.V. Gurav, D.N. Kamble, Multi-objective optimization of photochemical machining process based on grey relational analysis method for spray etching. Appl. Mech. Mater. **612**, 77–82 (2014)
4. S.S. Wangikar, P.K. Patowari, R.D. Misra, Effect of process parameters and optimization for photochemical machining of brass and german silver. Mater. Manuf. Processes **32**(15), 1747–1755 (2017)
5. S.S. Wangikar, P.K. Patowari, R.D. Misra, Parametric optimization for photochemical machining of copper using overall evaluation criteria. Mater. Today: Proc. **5**(2), 4736–4742 (2018)
6. G.R. Rathod, S.U. Sapkal, R.M. Chanmanwar, Multi-objective optimization of photochemical machining by using GRA. Mater. Today Proc. **4**(10), 10830–10835 (2017)
7. D. Agrawal, D. Kamble, GRA and ANN integrated approach for photochemical machining of Al/Sic composite. Mater. Today Proc. **4**(8), 7177–7188 (2017)
8. N.D. Misal, M. Sadaiah, Multi-objective optimization of photochemical machining of inconel 601 using grey relational analysis. Mater. Today Proc. **5**(2), 5591–5600 (2018)
9. D. Agrawal, D. Kamble, Optimization of photochemical machining process parameters for manufacturing microfluidic channel. Mater. Manuf. Processes **34**(1), 1–7 (2019)
10. P. Mumbare, A.J. Gujar, Multi objective optimization of photochemical machining for ASME 316 steel using grey relational analysis. Int. J. Innov. Res. Sci. Eng. Technol. **5**(7), 12418–12425 (2016)
11. A.B. Bhasme, M.S. Kadam, Parameter optimization by using grey relational analysis of photochemical machining. Int. Res. J. Eng. Technol. **3**(3), 992–997 (2016)
12. T. Debnath, P.K. Patowari, Fabrication of an array of micro-fins using Wire-EDM and its parametric analysis. Mater. Manuf. Processes **34**(5), 580–589 (2019)

# Fabrication and Characterization of Ramie Fiber Based Hybrid Composites

**Karanjit Kapila, Sutanu Samanta, and Sushen Kirtania**

**Abstract** The aim of the present work is to fabricate and strength analysis of pure ramie fabric/epoxy composite and its hybrids ramie-glass/epoxy and ramie-basalt/epoxy composites. Hand layup technique has been used for fabrication of all the composites. Tensile and three-point bending tests have been performed to obtain the tensile strength and flexural properties of the composites. Morphological studies have also been performed on the flexural specimen to investigate the defects in the composite. The present study reveals that hybridization of natural fiber (ramie) with synthetic fiber (glass) and mineral fiber (basalt) exhibits higher mechanical properties than pure ramie composite. It has also been observed that ramie-glass hybrid composite has higher tensile strength than ramie-basalt hybrid composite. Scanning electron microscope (SEM) has been used to find out the defects in the flexural specimen. It has been observed from the SEM images that less number of defect present in hybrid composites than pure ramie fabric composite. Therefore, it could be concluded that the inclusion of advanced fabric at the external layers of the natural fiber composite would enhance the strength of natural composites.

**Keywords** Natural fiber · Fabrication · Hybrid composites · Scanning electron microscope · Strength analysis

K. Kapila (✉) · S. Kirtania
Department of Mechanical Engineering, Tezpur University, Tezpur, India
e-mail: karanjitkapila@yahoo.com

S. Kirtania
e-mail: sushen.kirtania@gmail.com

S. Samanta
Department of Mechanical Engineering, North Eastern Regional Institute of Science and Technology, Nirjuli, India
e-mail: suta_sama@yahoo.co.in



K. M. Pandey et al. (eds.), *Recent Advances in Mechanical Engineering*, Lecture Notes in Mechanical Engineering, https://doi.org/10.1007/978-981-15-7711-6_83

# 1 Introduction

With the rapid development of science and technology, materials play an important role in the national economy and defense of a country. The three pillars of modern science and technology are materials science, energy technology and information science for the development of new material. There is an increase in use of natural fiber (NF) in automotive, aircraft, packaging industries due to its properties like light weight, low density, corrosion resistance, easy fabrication, good mechanical properties, and biodegradable [1]. Ramie is one of the oldest vegetable fibers and it can be used as reinforcement in wide variety of thermoplastic and thermosetting resin because of its excellent specific strength and modulus. In recent years, the use of natural fiber in hybrid composites has increased tremendously because hybridization of natural fiber with synthetic fiber which reduces the cost of production, increases economic growth of industries and implementing eco-friendly products. Simonassi et al. [2] studied the mechanical properties of Ramie/polyester composites and reported that with the addition of 30 volume % of ramie fibers in polyester the flexural strength of composite increase three times than the neat polyester. Ali et al. [3] fabricated the Kevlar-29 and ramie fiber reinforced polyester hybrid composite and performed impact and tensile test and concluded that the industrial woven ramie has good linkage with polyester resin which enhances the properties of ramie therefore it could be used for bullet-proof armor. Chu et al. [4] performed an experiment on basalt/ramie polypropylene (PP) hybrid composite and reported the optimum mixing ratio, optimum pressing temperature, optimum pressing time and optimal pressure of the hybrid composite. Gokulkumar et al. [5] reported that the Ramie fiber hybrid composite has higher values of tensile and bending strength as compared to the other natural fiber hybrid composites. The objectives of the present work are (i) fabrication and strength analysis of Ramie fabric based epoxy composite and its hybrids ramie-basalt epoxy and ramie-glass epoxy composites (ii) morphological study of the fractured specimens to observe the defects on the fractured surface, therefore it could be concluded that the hybridization of natural fiber with advanced fiber enhances the mechanical properties of natural fiber composite thus in future the natural fibers could be used in various industrial applications.

# 2 Materials and Methods

## 2.1 *Materials*

In the present work, epoxy is used as matrix material, natural fiber is used as reinforcement and the hardener is mixed with epoxy to increase the interfacial strength between fiber and matrix. The name of the matrix material for the present research work is Bisphenol-A-Diglycidyl-Ether. It belongs to the 'epoxide' family. Its commercial name is LB011 Epoxy resin Lapox B-11. Properties of cured resin and hardener,

**Table 1** Properties of Epoxy Resin

| Property | Tensile strength ($N/mm^2$) | Compressive strength ($N/mm^2$) | Flexural strength ($N/mm^2$) | Impact strength ($N/mm^2$) | Modulus of elasticity ($N/mm^2$) | Coefficient of linear expansion ($N/mm^2$) |
|---|---|---|---|---|---|---|
| Value | 50–60 | 110–120 | 130–150 | 17–20 | 4400–4600 | 64–68 |

**Table 2** Specification of hardener

| Specification | Visual appearance | Refractive index at 25 °C | Water content |
|---|---|---|---|
| Quantification | Pale yellow liquid | 14,940–15,000 | 1% max |

**Table 3** Properties of ramie fiber [6]

| Properties | Density ($g/cm^3$) | Young's modulus (GPa) | Tensile strength (MPa) | Elongation at break (%) |
|---|---|---|---|---|
| Ramie fiber | 1.5–1.56 | 60–128 | 400–1000 | 1.2–3.8 |

**Table 4** Properties of E-glass and basalt fibers [7]

| Properties | Density ($g/cm^3$) | Elastic modulus (GPa) | Tensile strength (GPa) | Elongation to fracture (%) | Specific modulus ($GPa/g/cm^3$) | Specific tensile strength ($GPa/g/cm^3$) |
|---|---|---|---|---|---|---|
| E-glass | 2.56 | 76 | 1.4–2.5 | 1.8–3.2 | 30 | 0.5–1 |
| Basalt | 2.8 | 89 | 2.9 | 3.15 | 31.78 | 1 |

which was procured from Haripa India Ltd are listed in Tables 1 and 2. The E-glass fiber and basalt fiber are used to fabricate the hybrid composites. E-glass fiber is the most common reinforcement for polymeric composites. Major advantages of E-glass fiber are low cost, high tensile strength, chemical resistance, and excellent insulating properties. Basalt fiber is also known as mineral fibers and its source is volcanic magma and flood volcanoes. Basalt is the most common rock found in the earth crust where the fibers are made from basalt rock by melting the rock at 1300–1700 °C and spinning it to make the fiber. Some of the properties of ramie fiber E-glass and basalt fibers are listed in Tables 3 and 4.

## 2.2 Fabrication

Hand layup method is used to fabricate ramie epoxy composite and its hybrids. The epoxy to hardener ratio is taken as 10:1 by weight and it is stirred for nearly 10 min. A smooth ceramic tile is laid and a releasing film is being placed on the tile and then

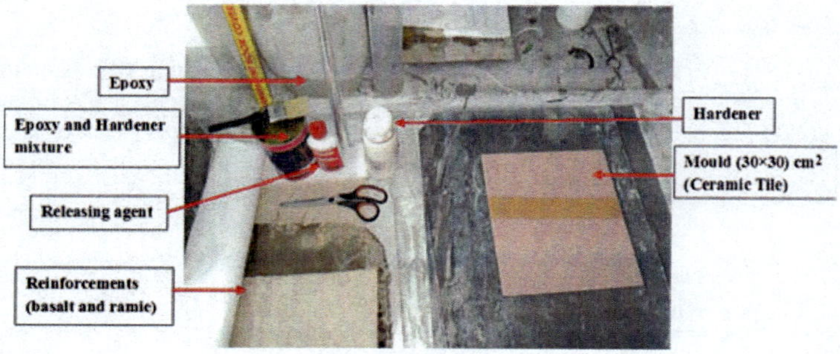


**Fig. 1** Hand layup process for the preparation of composite

film is coated with silicon liquid polish with brush. The prepared epoxy-hardener mixture is being applied slowly using brush on the mold releasing film. It is ensured that resin mixture is evenly distributed on the mold release film. Figure 1 shows the materials used to fabricate the composite by open mould technique.

After uniform distribution of resin on the releasing film, first layer of ramie fabric is placed and resin is applied on it. Rolling is performed with the help of a roller to remove the entrapped air that creates bubbles. In this way, six layers of unidirectional ramie mat fabric is placed in succession. Another mould releasing sheet is placed on the top surface coated with silicon liquid and final rolling is done on the top fabric. Curing is done at room temperature under a load of 40–45 kg to avoid any distortion during shrinkage for the preparation of ramie epoxy composite (RR). For fabrication of ramie basalt epoxy hybrid composite (RB), the external layer of ramie fabric is replaced by single basalt fabric on both sides. Similarly, ramie glass epoxy hybrid composite (RG) has also been fabricated by replacing external layers with glass fabric, thus RB and RG are called a sandwich composite. The type, number, orientation, and stacking sequence of the reinforcements are listed in Table 5.

Figure 2a–c show the fabricated pure ramie/epoxy, hybrid ramie-glass/epoxy and hybrid ramie-basalt/epoxy composites, respectively.

**Table 5** Designation and detailed compositions of the composites

| S. No. | Matrix | Reinforcements and number of fabric used for fabrication | | Fiber orientation | Stacking order | Designation | Type of composite |
|---|---|---|---|---|---|---|---|
| 1 | Epoxy | Ramie-6 | Nill | 0/90 | RRRRRR | RR | Monolithic |
| 2 | Epoxy | Ramie-4 | Basalt-2 | 0/90 | BRRRRB | RB | Sandwich |
| 3 | Epoxy | Ramie-4 | Glass-2 | 0/90 | GRRRRG | RG | Sandwich |

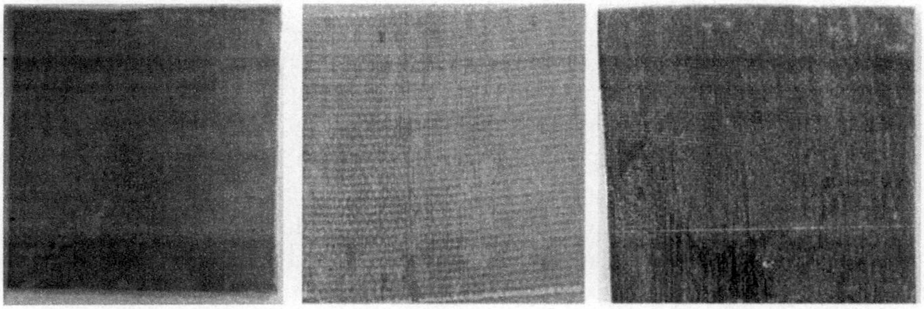

(a) (b) (c)

**Figure 2** **a** Pure ramie/epoxy (RR), **b** hybrid ramie-glass/epoxy (RG) and **c** hybrid ramie-basalt/epoxy (RB) composites

# 3 Experimental Procedure

Instron 8801 J4051 series of servo hydraulic testing system has been used to perform both tensile and flexural tests. Three samples of each type of composites have been tested according to their ASTM standards.

## *3.1 Tensile Test*

For tensile test, the specimens are cut from the composite plate as per ASTM D3039 standard. Figure 3b shows the pictorial views of the machine during tensile test

(a) (b) (c)

**Fig. 3** **a** Specimens before tensile test, **b** machine setup and specimen during tensile test and **c** specimens after tensile test operation. A sharp-toothed hand hacksaw has been used to cut the specimen for the required dimension. The length and breadth of the specimen are 250 mm and 25 mm respectively and thickness varies from composite to composite. The gauge length for the specimen is set to 140 mm and held properly in the grips of the machine. The maximum length of the gripper is 54 mm. The cross-head speed of the machine is 1.5 mm/min. Load is being applied until failure takes place and failure occurs in between the gauge lengths before 10 min, as mentioned in the standard. For each type of composite laminate, three samples have been tested and the images of specimen before and after tensile test are shown in Fig. 3a, c, respectively.

### 3.2 Flexural Test

According to ASTM D790—10 standards, three-point bending test is carried out and the machine setup for the test is shown in Fig. 4b. For flexural test the length and breadth of the specimen are 125 mm and 12.7 mm, respectively and thickness varies from composite to composite. The setup consists of three rollers whose diameter is 25 mm and maximum roller center limit is 240 mm. Load is applied continuously upon the specimen to deflect till rupture occurs due to deflection or maximum strain. The test is carried out at a constant head movement 1 mm/min. Figure 4a, c shows the pictorial view of the specimens before and after test.

(a) (b) (c)

**Figure 4** **a** Specimens before flexural test, **b** machine setup and specimen during flexural test and **c** specimens after flexural test

**Table 6** Tensile properties of fabric reinforced epoxy composite

| Composites | Tensile strength | | Tensile modulus | |
|---|---|---|---|---|
| | MPa | Percentage increase | MPa | Percentage increase |
| RR | 34.34 | Nill | 2890.63 | Nill |
| RB | 53.23 | 55.00 | 4131.09 | 42.91 |
| RG | 60.43 | 75.97 | 4800.35 | 66.06 |

# 4 Results and Discussion

Three types of fabric-reinforced epoxy composites have been fabricated. Tensile tests and three-point bending tests have been performed. Mechanical properties of the hybrid composites RB and RG have been compared with pure RR composite.

## 4.1 *Tensile Test*

Table 6 shows the tensile modulus, tensile strength of all the composites; and percentage increase of hybrid composites compared to pure ramie/epoxy composite. It can be observed from Table 6 that the tensile strength of the RB and RG hybrid composites are increased by 55% and 75.97%, respectively compared to the pure RR composite. It can also be seen that the tensile modulus of RB and RG hybrid composites are increased by 42.91% and 66.06%, respectively as compared to pure RR composite. Therefore, it could be concluded that hybridization of basalt and glass fiber shows positive effects by increasing the strength and modulus of ramie fiber composite. It could also be concluded that RG hybrid composite has greater tensile strength and modulus than RB composite. The reason for this is due to higher stiffness of glass fiber than basalt fiber.

The tensile strength and modulus of both the RG and RB hybrid composites are greater than the pure ramie/epoxy composite because placing advanced fiber at outer layers increases the load-carrying capacity of the hybrids. Similar observation has also been reported by Ahmed et al. [8] on jute/glass reinforced polyester composite by varying stacking sequence and concluded that inclusion of glass fiber increases the mechanical properties of composite by placing it on outer layers.

## 4.2 *Flexural Test*

Experimental values of flexural strength and flexural modulus of RR and its hybrid composites RB and RG are listed in Table 7. It can be seen from Table 7 that the flexural modulus of RB and RG hybrid composites are increased by 34.42% and 47.77%, respectively compared to pure RR composite. It could be observed from

**Table 7** Flexural properties of fabric reinforced epoxy composite

| Composites | Flexural strength | | Flexural modulus | |
|---|---|---|---|---|
| | MPa | Percentage increase | MPa | Percentage increase |
| RR | 69.89 | Nill | 4729.27 | Nill |
| RB | 88.98 | 27.31 | 6357.47 | 34.42 |
| RG | 104.53 | 49.56 | 6988.87 | 47.77 |

Table 7 that sandwiching of RR composite by glass fiber shows 49.56% increase in flexural strength. Similarly, sandwiching of RR composite by basalt fiber shows 27.31% increase in tensile strength. The flexural properties of ramie glass and ramie basalt hybrids have shown a rise in both modulus and strength values than pure ramie epoxy composite. Similar observations on flexural properties of ramie-glass epoxy hybrid composite have been reported by Giridharan [9]. Amico et al. [10] also studied the glass/sisal fiber reinforced hybrid composite and reported that the surface layers play an important role for hybrid composite.

## *4.3 SEM Observation*

The morphological study of the composites has been performed using SEM to observe the defects on fracture surface of the composite specimen after three-point bending test. The name of the SEM setup is JEOL, model no JSM-6390 which is a high-performance, low cost, SEM with a high resolution of 3.0 nm. Initially, the samples were taken out from the fractured surface and coating of the samples was done by platinum layer before placing in the equipment. Coating on samples are required in the field of electron microscopy to enable or improve the imaging of samples with 20 kV accelerating voltage. Figure 5 shows the SEM micrographs of fractured specimen of RR composite.

At initial stage of loading, fibers elongate and reach to its max value of elongation before failure starts. It can be observed from Fig. 5 that the tear of fibers, fiber matrix debonding, fibers pull out takes place at loading condition for RR composite. Similar observation has also been reported by Simonassi et al. [2].

The micrographs of RB and RG hybrid composites are shown in Figs. 6a, b, respectively. It can be observed from Figs. 5 and 6 that hybrid composite have less number of defects than RR composite. Therefore, it could be concluded that hybridization enhanced the flexural properties. Stress concentration on the surface layer of composites in flexural test causes initiation of damage such as fiber elongation, matrix debonding and fiber pull out as seen in Figs. 5, 6. Filho et al. [11] has also been reported similar types of defects of fractured composite as described in Fig. 6b.

The direction of matrix flow and the elongation of fibers take place in the direction of applied load as observed in Fig. 6a for RB hybrid. Debonding of matrices and

Fig. 5 Fractured surface SEM image of RR composite

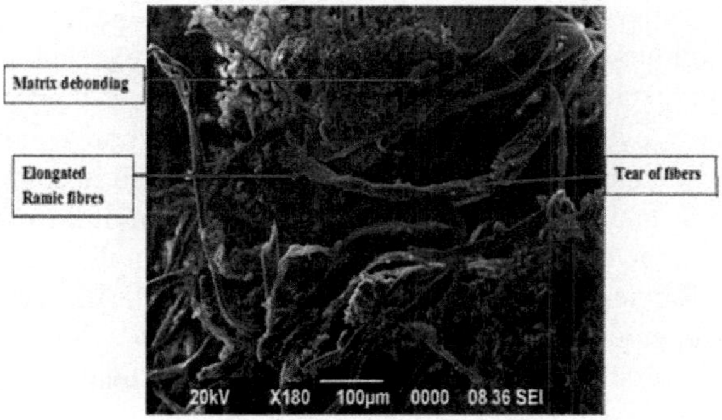

Fig. 6 Fractured surface SEM image of **a** RB and **b** RG composites

pull out of external fibers are also observed from Fig. 6a. Less flexural defects such as fibers pull out and matrix debonding are seen in Fig. 6b of RG compared to RB because of good fiber matrix adhesion, large load bearing capacity of glass fiber and proper stress transfer between fibers.

## 5 Conclusions

Fabrication and testing of pure ramie fabric/epoxy composite and its hybrid ramie-glass/epoxy and ramie-basalt/epoxy composites have been performed. Strength and Young's modulus of all the composites have been evaluated. The present experimental result has been compared with the earlier publication and the present results show a good agreement. Some of the important conclusions drawn from the present study are given below

- Tensile strength and modulus of the pure ramie/epoxy composite could be increased by the inclusion of basalt and glass fiber.

- Ramie-glass hybrid composite exhibits more tensile strength and modulus than ramie-basalt hybrid composite.
- Flexural strength and modulus of pure ramie/epoxy composite could be enhanced by hybridization and the ramie-glass hybrid composite exhibit more flexural strength and modulus than ramie-basalt hybrid composite.
- Scanning electron microscope observation of flexural specimens show that pure ramie/epoxy composite exhibit more flexural defects than hybrid composites.
- Ramie-glass hybrid composite shows less flexural defects than ramie-basalt hybrid composites.
- Overall, it could be concluded that the load-carrying capacity of the natural composite could be increased by adding the advanced fiber fabric at the outer layers of the natural composites.

## References


1. K. Begum, M.A. Islam, Natural fiber as a substitute to synthetic fiber in polymer composites: a review. Res. J. Eng. Sci. **2**(6), 46 (2013)
2. N.T. Simonassi, A.C. Pereira, S.N. Monteiro, F.M. Margem, R.J.S. Rodriguez, J.F. de Deus, C.M.F. Vieira, J. Drelich, Reinforcement of polyester with renewable ramie fibers. Mater. Res. **20**(2), 51 (2017)
3. A. Ali, Z.R. Shaker, A. Khalina, S.M. Sapuan, Development of anti-ballistic board from ramie fibers. Polymer Plast. Technol. Eng. **50**(6), 622–634 (2011)
4. F. Chu, H. Liu, Z. Chen, Q. Song, Manufacturing basalt ramie fibers hybrid reinforced PP thermoplastic panel. Adv. Mater. Res. **146–147**, 378 (2011)
5. S. Gokulkumar, P.R. Thyla, L. Prabhu, S. Sathish, N. Karth, A comparative study on epoxy based composites filled with pineapple/areca/ramie hybridized with industrial tea leaf wastes/GFRP. Mater. Today Proc. (2019)https://doi.org/10.1016/j.matpr.2019.09.221
6. K.M. Chandrashekar, C.G. Venkate, N.G.S. Uduppa, Evalution of mechanical properties of Hemp-Ramie fibers reinforced with epoxy hybrid composites, International Journal of Research in Engineering and Technology, 2016, 05 (13), p. 21–26.
7. V. Fiore, G.D. Bella, A. Valenza, Glass basalt epoxy hybrid composites for marine applications. Mater. Des. **32**(4), 2091–2099 (2011)
8. K.S. Ahmed, S. Vijayaranganb, Tensile, flexural and interlaminar shear properties of woven jute and jute-glass fabric reinforced polyester composites. J. Mater. Process. Technol. **207**(1–3), 330–335 (2008)
9. R. Giridharan, A preparation and property evaluation of glass/ramie fibers reinforced epoxy hybrid composites. Compos. Part B **167**, 342–345 (2019)
10. S.C. Amico, C.C. Angrizani, M.L. Drummond, Influence of the stacking sequence on the mechanical properties of glass/sisal hybrid composites. J. Reinf. Plast. Compos. **29**(2), 179–189 (2008)
11. G.C. de Oliveira Filho, R.C. de Sousa Mota, A.C. Rangel de Conceicao, M.A. Leao, O.O. de Araujo Filho, Effects of hybridization on the mechanical properties of composites reinforced by piassava fibers tissue. Compos. Part B, 73–79 (2019)

# Multiphase Numerical Modeling of PCM Integrated Solar Collector

Bharat Singh Negi, Satyender Singh, and Sushant Negi

**Abstract** In the present work, phase change material (paraffin wax) integrated double glazed rectangular finned solar air heater is investigated numerically using implicit discretization scheme. A MATLAB code is developed, capable of providing the solution for unsteady governing energy equations for each element of solar air heater and air flow. Numerical results are validated with published experimental results of a flat absorber plate collector design with PCM (Moradi et al. in Exp Thermal Fluid Sci 89:41–49, 2017 [1]). The idea of integration of PCM and fins with solar air heater is to provide thermal energy backup during off-sunshine hours and increase heat discharging rate of PCM, respectively. The average incident global radiation is the function of time and maximum incident radiation is taken as 960 W/m$^2$. Moreover, a correlation is developed to predict the ambient temperature as reported in Moradi et al. (Exp Thermal Fluid Sci 89:41–49, 2017 [1]) as a function of time. The results showed that the thermal backup of PCM lasted for about 12 h after sunset almost for all selected numbers of fins. The maximum outlet air temperature for 5, 15 and 25 fins is obtained as 42 °C, 47.16 °C and 51.25 °C, respectively, at mass flow rate of 0.0128 kg/s. Whereas, the maximum instant thermal efficiency for 5, 15, and 25 fins is obtained as 35.867%, 49.375%, and 58.198%, respectively.

**Keywords** Thermal storage · Fins · PCM · Heat transfer

---

B. S. Negi (✉) · S. Singh
Department of Mechanical Engineering, Dr. B. R. Ambedkar National Institute of Technology Jalandhar, Punjab, India
e-mail: negiindia@gmail.com

S. Negi
Department of Mechanical Engineering, National Institute of Technology Silchar, Silchar, Assam 788010, India


K. M. Pandey et al. (eds.), *Recent Advances in Mechanical Engineering*, Lecture Notes in Mechanical Engineering, https://doi.org/10.1007/978-981-15-7711-6_84

# 1 Introduction

Since time immemorial, energy has found amazingly a major role to play in lives of beings. Nature has always given ample resources to work with but with recent explosion of population, conventional energy sources are endangered and are expected to vanish soon. Therefore in order to nurture the needs, research is being focused on renewable energy sources. Solar energy being the most promising one has found its place in researchers list. Therefore, many solar air heater designs are created that utilizes the solar energy to heat the air [1–15]. One major limitation of solar energy is its availability only in the day time that limits its uses in thermal applications. Therefore, many researchers have identified latent heat storage as a solution and integrated PCM unit with solar air heaters [1–15]. Use of PCM in solar air heater as a part of it maintains the thermal performance during off sunshine hours during the day and also provides backup during the night time. In the direction, various designs are available such as single glass cover [1, 3–7] and flat [1, 3–7] and v-Corrugated absorber plate [1]. Thermal efficiency of about 30–70% is achieved with PCM integrated solar air heaters [1–15]. Moreover, effective thermal backup of 3–7 h after sunshine is obtained with PCM in solar air heater [1–15]. Moradi et al. investigated solar air heater with PCM and obtained that average off-sunshine temperature difference was 4.5 ° C using 23.5 kg of paraffin wax and the maximum efficiency was about 37% at mass flow rate of 65 kg/h.

It is observed in the literature that mostly experimental investigations [1–15] are performed and few are dedicated for numerical investigations [7, 10, 14, 15]. It is noticed in the literature that PCM is generally kept underneath the absorber plate, solar radiations incident on the absorber plate and increases the temperature of the absorber plate. Heat of the absorber is utilized to heat air and charging of PCM. While, PCM, receives the heat in the conduction mode of heat transfer. Moreover, paraffin wax as a PCM material is mainly used to store heat in solar air heaters. It is observed in case of PCM integrated solar collectors that heat storage and release is slow that results in delivery of heat to absorber plate at a slow rate. However, slow delivery rate can prolong the thermal backup for the more time, but the corresponding values after 3–5 h are almost ineffective as per the application requirement of solar air heaters. Therefore, in the present work, we have utilized the rectangular fins on the top surface of absorber plate and PCM is kept underneath the absorber plate. The idea is to increase the heat release rate of PCM and simultaneously provide the thermal backup. Fins on the absorber surface increases the heat transfer to air, extract heat at a higher rate from the PCM during off-sun shine and nocturnal hours. That increases instant heat release to fins and further to air. Although, the use of the fins on the absorber plate decreases the storage time of the PCM, but can provide the hot air at an effective temperature that can be used to support solar air heater applications.

## 2 Numerical Investigation

Figure 1 show a schematic diagram of proposed model of the single-pass PCM integrated solar air heater on which the study was carried out. Solar air heater consists of four elements, i.e., upper glass cover, lower glass cover, absorber plate, and PCM. Details of the physical and geometrical properties of solar air heater elements are presented in Tables 1 and 2, respectively.The air is flowing between finned absorber plate and lower glass cover. However, one-dimensional transient energy equations are developed for all four elements as well as air flow channel and fixed air between two glass covers. The following assumptions are made: (i) No temperature variation in transverse direction due to negligible thickness (ii) there is no heat transfer in the direction of air flow and energy is transferred only by mass (iii) losses from top, bottom and edges of collector are negligible (iv) sky is considered as black body for long wavelength radiation (v) air as an incompressible fluid. Correlations to calculate

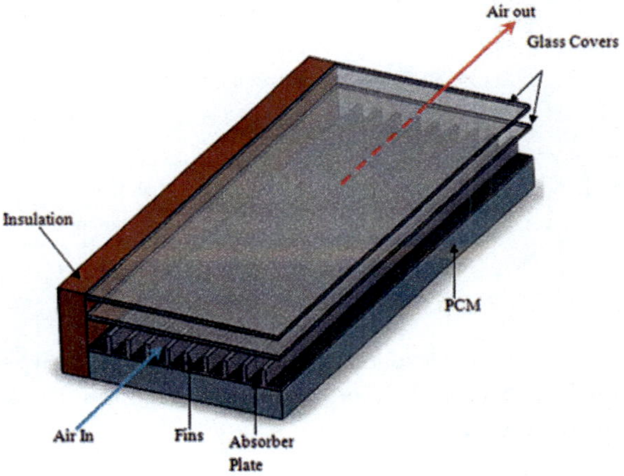


**Fig.1** Schematic of the PCM integrated solar air heater

**Table 1** Physical property of solar air heater elements [14]

| Input parameters | Value |
|---|---|
| Absorptivity of glass covers | 0.05 |
| Transmissivity of glass covers | 0.9 |
| Absorptivity of absorber plate | 0.95 |
| Emissivity of absorber plate | 0.8 |
| Emissivity of glass covers | 0.9 |
| Specific heat of glass | 840 kJ/kg K |
| Specific heat of absorber (galvanized iron) | 502.48 kJ/kg K |
| Density of absorber | 8030 kg/m$^3$ |
| Density of glass covers | 2600 kg/m$^3$ |
| Thermal conductivity of fins (galvanized iron) | 14.9 W/m k |

**Table 2** Geometric properties of solar air heater component [14]

| Geometric property | Value |
|---|---|
| Length of the collector | 1.1 m |
| Width of the collector | 0.6 m |
| Depth of the channel with air flow | 0.05 m |
| Depth of channel between glass covers | 0.03 m |
| Depth of PCM under the absorber plate | 0.02 m |
| Thickness of absorber plate | 1.25 mm |
| Thickness of glass covers | 6 mm |
| Height of fins | 2.5 cm |
| Thickness of fins | 3 mm |

**Table 3** Thermo-physical properties of Paraffin wax [14]

| Property | Value |
|---|---|
| Melting Temperature range | 52–56 °C |
| Congealing temperature range | 43–37 °C |
| Heat storage capacity in temp range between 35 and 50 °C | 174 kJ kg$^{-1}$ (±7.5%) |
| Specific heats in both solid and liquid states | 2.5 kJ kg$^{-1}$ K$^{-1}$ |
| Density in solid state at 15 °C | 890 kgm$^{-3}$ |
| Density in liquid state at 80 °C | 850 kgm$^{-3}$ |
| Heat conductivity | 0.2 Wm$^{-1}$ k$^{-1}$ |

heat transfer coefficients are used from Ref.[7, 10, 15–18]. Properties of PCM are presented in Table 3. Energy equations are formulated as given below for solar air heater elements:

**Upper Glass Cover**

$$\rho_{gu} V_{gu} c_{gu} \frac{dT_{gu}}{dt} = \left[ h_{c_{amb-gu}} \left( T_{amb} - T_{gu} \right) + h_{r_{gl-gu}} \left( T_{gl} - T_{gu} \right) + h_{c_{ag-gu}} \left( T_{ag} - T_{gu} \right) + \alpha G \right] W \Delta Z. \tag{1}$$

**Lower Glass Cover**

$$\rho_{gl} V_{gl} c_{gl} \frac{dT_{gl}}{dt} = \begin{bmatrix} h_{c_{ag-gl}} \left( T_{ag} - T_{gl} \right) + h_{r_{ab-gl}} \left( T_{ab} - T_{gl} \right) \\ + h_{r_{gu-gl}} \left( T_{gu} - T_{gl} \right) + h_{c_{air-gl}} \left( T_{air} - T_{gl} \right) + \tau \alpha G \end{bmatrix} W \Delta Z. \tag{2}$$

**Air gap Between Two Glasses**

$$\rho_{ag} V_{ag} c_{ag} \frac{dT_{ag}}{dt} = \left[h_{c_{gu-ag}}\left(T_{gu} - T_{ag}\right) + h_{c_{gl-ag}}\left(T_{gl} - T_{ag}\right)\right] W \Delta Z. \quad (3)$$

**Air Channel Air**

$$\begin{aligned}\rho_{air_L} V_{air_L} c_{air_L} \frac{dT_{air_L}}{dt} &= h_{c_{gl-air_L}}\left(T_{gl} - T_{air_L}\right) W \Delta Z \\ &+ h_{c_{ab-air_L}} A\emptyset\left(T_{ab} - T_{air_L}\right) - \dot{m} c_p\left(T_{air_L,t} - T_{air_L,t-1}\right). \quad (4)\end{aligned}$$

**Absorber Plate**

$$\begin{aligned}\rho_{ab} V_{ab} c_{ab} \frac{dT_{ab}}{dt} &= h_{c_{ab-air_L}} A\emptyset\left(T_{ab} - T_{air_L}\right) \\ &+ \left[h_{r_{ab-gl}}\left(T_{ab} - T_{gl}\right) + \left(\tau^2 \alpha\right) G\right] W \Delta Z + Q_{conduction}. \quad (5)\end{aligned}$$

**PCM**

$$\rho c_p \frac{\partial T}{\partial t} = \frac{\partial}{\partial x}\left(k \frac{\partial T}{\partial x}\right) + Q. \quad (6)$$

and

$$Q = \frac{\partial H}{\rho \partial T} = c_p - L \frac{\partial f_s}{\partial t} \frac{\partial t}{\partial T}. \quad (7)$$

where, $Q$ internal source of the latent heat of fusion.

## 2.1 *Iterative Solution Procedure*

Transient energy balanced equations are converted to linear algebraic form after discretization using implicit differencing scheme. Temperature for elements and fluid is obtained using matrix inversion method. To initialize the code, all temperatures are assumed equal to ambient temperature. A correlation is developed to predict the ambient temperature as that reported in Ref. [14] as a function of time. Mass flow rate as an input boundary condition is used for air flow channel. While absorber plate temperature is used as a boundary condition for PCM. Therefore, a function is created in the code to only solve the algebraic equation for PCM using absorber plate temperature. A convergence criterion of $10^{-3}$ is used for elements, outlet air, air between the gap, and PCM. Moreover, grid independence test is conducted. It was obtained that relative percentage error with increase in numbers of 50 grids is about

0.77%, therefore, 50 numbers of grids are used to predict the results in the present work. However, Courant-Friedrichs-Lewy (CFL) criteria shown below are used to provide the stability to the solution.

$$\Delta t = \frac{\Delta z}{v_{\text{air}}}. \tag{8}$$

## 2.2 Validation

Figures 2 and 3 show the results of outlet air temperature used for validation of the present numerical simulation with the reported experimental and numerical results of Moradi et al. [14], respectively. As it can be noticed in the figures that present numerical results are good agreement with experimental (Fig. 2) and numerical (Fig. 3) results of Ref. [14]. However, temperature of the outlet air is increasing with increase in solar flux and decreasing after 12:00 PM. This is due to the fact that solar radiations are incident on the absorber plate and increases its temperature. The air is flowing over the absorber plate extract most of the heat and raise its temperature along the length of the collector. While, the absorber plate temperature is a direct function of

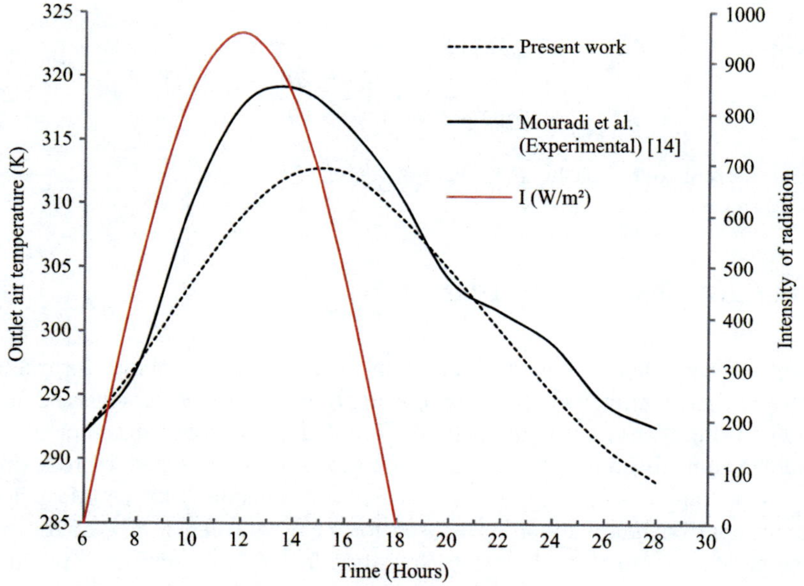


**Fig. 2** Results of validation of present numerical investigation with published experimental results of Mouradi et al. [14] for flat absorber plate solar air heater design with PCM, when mass flow rate is 0.0128 kg/s

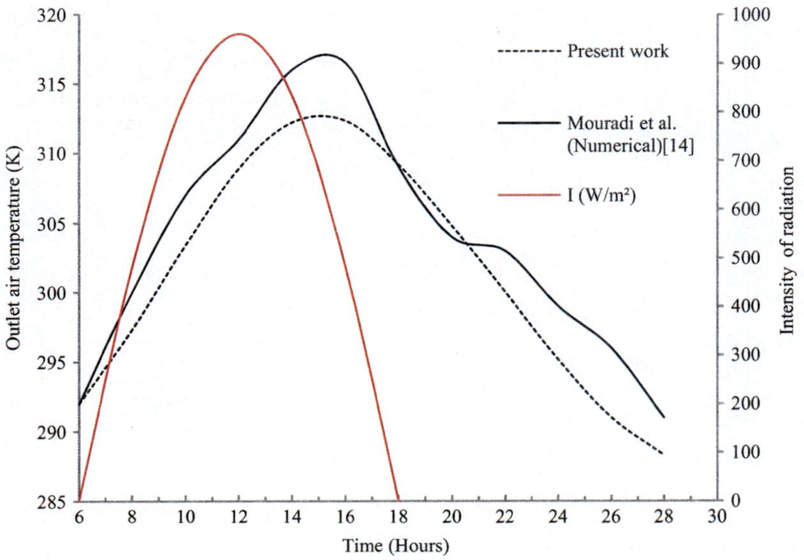


**Fig. 3** Results of validation of present numerical investigation with published numerical results of Mouradi et al. [14] for flat absorber plate solar air heater design with PCM, when mass flow rate is 0.0128 kg/s

incident solar radiation. However, about 20 °C rise in the outlet air temperature is obtained about 12:00 PM, which is an appreciable rise in the air temperature with absorber length of 1.1 m.

## 3 Results and Discussion

In this section, results and discussion of the outcome of this investigation for finned absorber plate solar air heater with PCM are presented. Investigation is carried out on the basis of geometrical and flow parameters (mass flow rate of 0.0128 kg/s) used for the validation with Ref [14]. and same values are extended as an input to investigate the finned solar air heater with PCM. However, comparison of the results obtained for finned design is presented with flat absorber plate (without fins) solar air heater design. Figure 4 shows the results of lower glass cover temperature as a function of numbers of fins, solar radiations, and time of the day (hours). It can be seen that maximum temperature of the glass cover is obtained about 3 PM and for flat absorber plate (zero fin) and followed by 5–25 numbers of fins. This is due to increasing thermal losses to surrounding due to emission of radiation more from the flat absorber plate, while radiation is utilized to heat the fins in finned absorber plate. Temperature of the glass cover can be considered as an indication of heat loss to surrounding.

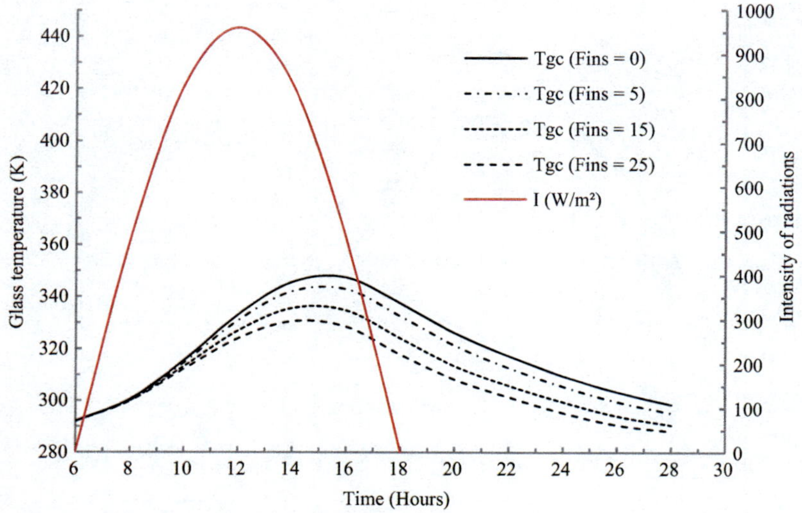


**Fig. 4** Results of temperature variation of lower glass cover as a function of numbers of fins, solar radiation, and time of the day, when mass flow rate is 0.0128 kg/s

Figure 5 presents the results of absorber plate temperature, it is obtained that maximum temperature of the absorber plate temperature is obtained for flat absorber plate solar air heater design compared to finned one. It can be noted that more heat transfer occurs between fins and air, which is maximum for 25 numbers of fins, as they have attained minimum temperature in Fig. 5. Figure 6 revealed the temperature of PCM. It can be noticed that maximum temperature of the PCM is obtained about 6 PM, that indicates changing time and after that it starts decreases till 28 h, which represents discharging time. The maximum temperature of PCM is obtained as 380 K for flat absorber plate and while it is obtained minimum for 25 numbers of fins. Figure 7 presents the results of temperature gain; it is obtained that the maximum temperature gain of about 20 K is obtained for 25 numbers of fins around 2 PM, while it is obtained minimum for flat absorber design (without fins). Figure 8 depicts the results of heat storage and release of PCM as a function of numbers of fins, solar radiations, and time of the day. It can be noticed that maximum charging takes place up to 16 h and after that due to decrease in the temperature of absorber plate due to air flow over it and decreasing solar intensity, temperature gradient forms and PCM starts discharging about 16 h (4 PM), when the solar intensity is about 250 W/m$^2$. However, discharging lasts for upto 28 h (6 AM of next day) for all the numbers of fins. While, less heat is left with 25 numbers of fins during discharging. However, it can be viewed in other way like during the day time if sun shine hours occur then 25 numbers of fins can be provided instant thermal energy backup by releasing heat from the PCM at a fast rate. Figure 9 presents the results of thermal efficiency, which is obtained higher for 25 numbers of fins, when mass flow rate of 0.0128 kg/s. Thermal efficiency of about 58% is obtained with 25 numbers of fins while it is about 27%

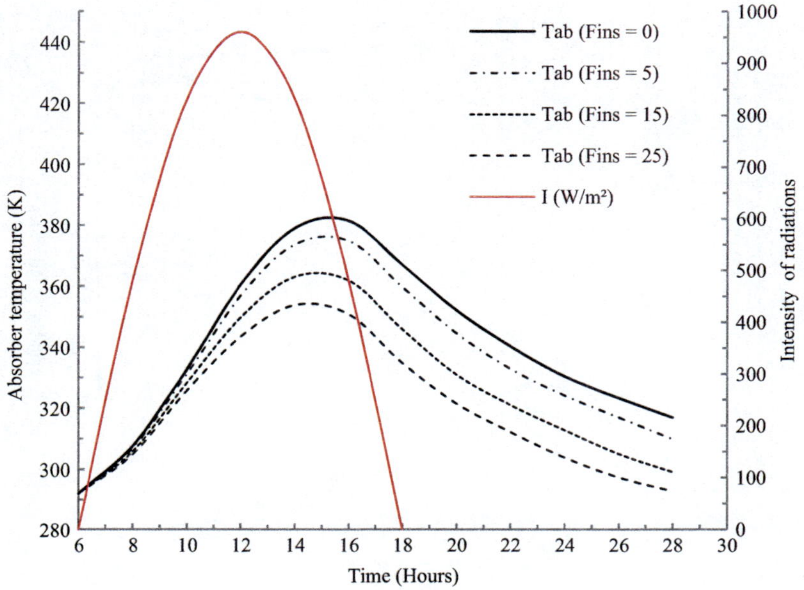


**Fig. 5** Results of temperature variation of absorber plate as a function of numbers of fins, solar radiation, and time of the day, when mass flow rate is 0.0128 kg/s

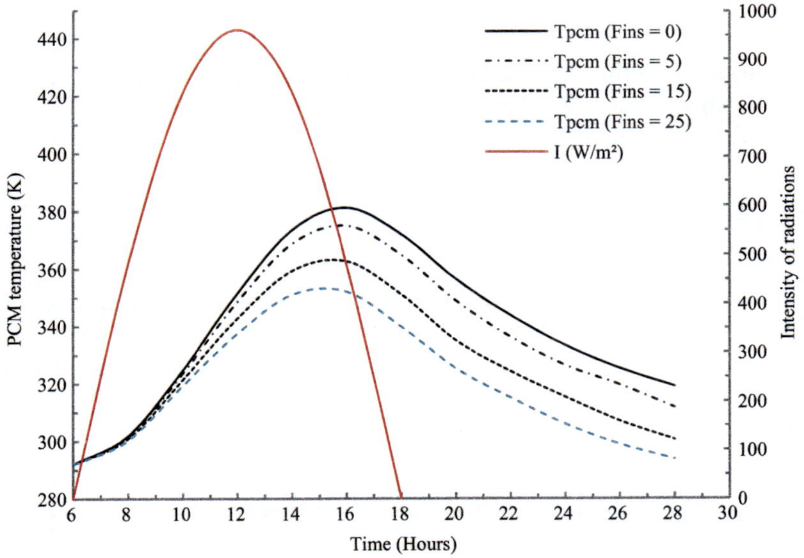


**Fig. 6** Results of temperature variation of PCMas a function of numbers of fins, solar radiation, and time of the day, when mass flow rate is 0.0128 kg/s

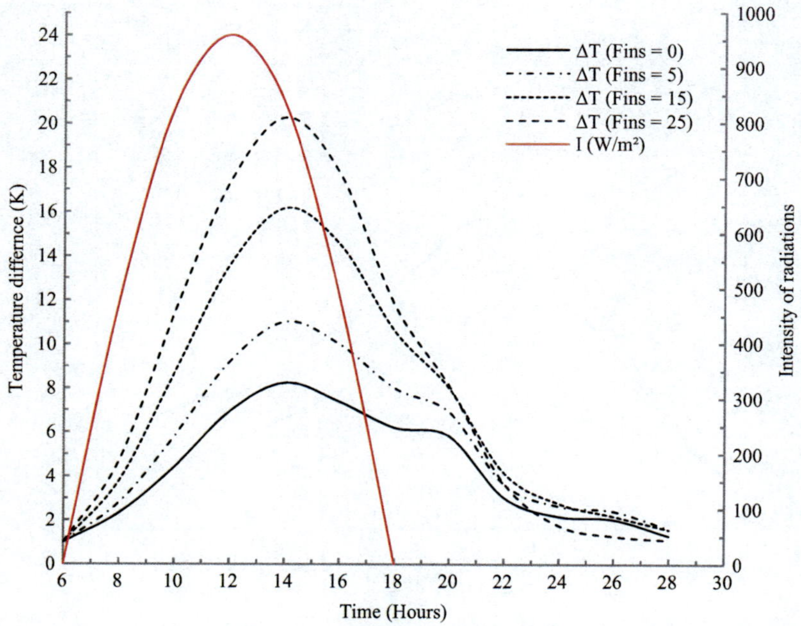


**Fig. 7** Results of temperature variation of temperature gain of air as a function of numbers of fins, solar radiation, and time of the day, when mass flow rate is 0.0128 kg/s

for solar air heater design without fins. This is due to the reason that fins increase heat transfer area and increases heat transfer to air.

## 4 Conclusions

In the present work, numerical investigation is carried out for a PCM integrated solar air heater design with and without fins. Comparison of two designs is performed and presented. A MATLAB code is developed capable of predicting the thermal performance of PCM integrated solar air heater with and without fins. Following are the conclusions of the present investigation:

- Implementation of finite-difference technique on the one-dimensional unsteady energy equation can best present the results of such heat and fluid flow problems.
- Use of fins increases the heat transfer to air and increases the thermal efficiency by 56% compared to without fin design of solar air heater.
- Use of fins decreases the heat stored at a fast rate during discharging, but essential during the off sun shine hours to provide instant thermal energy backup.

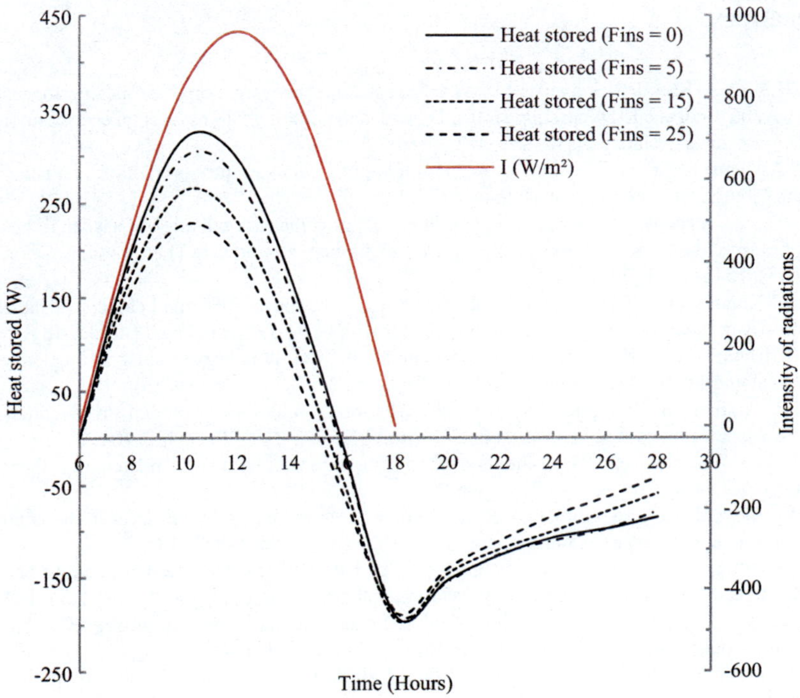


**Fig. 8** Results of heat store and release of PCM as a function of numbers of fins, solar radiation, and time of the day, when mass flow rate is 0.0128 kg/s

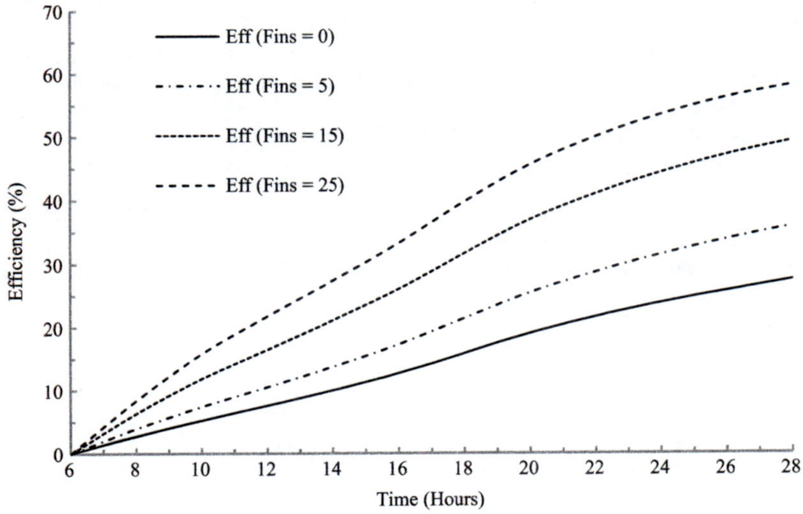


**Fig. 9** Results of thermal efficiency (Eff) of solar air heater designs (with and without fins) as a function of numbers of fins, solar radiation, and time of the day, when mass flow rate is 0.0128 kg/s

## References

1. A. Kabeel, A. Khalil, S. Shalaby, M. Zayed, Experimental investigation of thermal performance of flat and v-corrugated plate solar air heaters with and without PCM as thermal energy storage. Energy Convers. Manage. **113**, 264–272 (2016)
2. M. Alkilani, K. Sopian, S. Mat, M. Alghoul, Output air temperature prediction in a solar air heater integrated with phase change material. Euro. J. Sci. Res. **27**(3), 334–341 (2009)
3. V. Tyagi, A. Pandey, S. Kaushik, S. Tyagi, Thermal performance evaluation of a solar air heater with and without thermal energy storage: an experimental study. J. Therm. Anal. Calorim. **107**(3), 1345–1352 (2011)
4. A. El Khadraoui, S. Bouadila, S. Kooli, A. Farhat, A. Guizani, Thermal behavior of indirect solar dryer: Nocturnal usage of solar air collector with PCM. J. Clean. Prod. **148**, 37–48 (2017)
5. S. Krishnananth, K.K. Murugavel, Experimental study on double pass solar air heater with thermal energy storage. J. King Saud Univ. Eng. Sci. **25**(2), 135–140 (2013)
6. M.M. Alkilani, S. Kamaruzzaman, M. Sohif, Fabrication and experimental investigation of PCM capsules integrated in solar air heater. Am. J. Environ Sci. **7**(6), 542–546 (2011)
7. H.E. Fath, Transient analysis of thermosyphon solar air heater with built-in latent heat thermal energy storage system. Renew. Energy **6**(2), 119–124 (1995)
8. S. Shalaby, M. Bek, Experimental investigation of a novel indirect solar dryer implementing PCM as energy storage medium. Energy Convers. Manage. **83**, 1–8 (2014)
9. S. Bouadila, S. Kooli, M. Lazaar, S. Skouri, A. Farhat, Performance of a new solar air heater with packed-bed latent storage energy for nocturnal use. Appl. Energy **110**, 267–275 (2013)
10. P. Charvát, L. Klimeš, M. Ostrý, Numerical and experimental investigation of a PCM-based thermal storage unit for solar air systems. Energy Build. **68**, 488–497 (2014)
11. A. Mahmud, K. Sopian, M. Alghoul, M. Sohif, Using a paraffin wax-aluminum compound as a thermal storage material in a solar air heater. ARPN J. Eng. Appl. Sci. **4**(10), 74–77 (2009)
12. S. Esakkimuthu, A.H. Hassabou, C. Palaniappan, M. Spinnler, J. Blumenberg, R. Velraj, Experimental investigation on phase change material based thermal storage system for solar air heating applications. Sol. Energy **88**, 144–153 (2013)
13. S. Enibe, Thermal analysis of a natural circulation solar air heater with phase change material energy storage. Renew. Energy **28**(14), 2269–2299 (2003)
14. R. Moradi, A. Kianifar, S. Wongwises, Optimization of a solar air heater with phase change materials: experimental and numerical study. Exp. Thermal Fluid Sci. **89**, 41–49 (2017)
15. P. Theunissen, J. Buchlin, Numerical optimization of a solar air heating system based on encapsulated PCM storage. Sol. Energy **31**(3), 271–277 (1983)
16. J.A. Duffie, W.A. Beckman, *Solar Engineering of Thermal Processes* (John Wiley & Sons, 2013)
17. S. Singh, P. Dhiman, A numerical evaluation of thermal performance of double flow packed bed solar air heaters. Int. J. Renew. Energy Technol. **4**(3), 242–264 (2013)
18. S. Singh, P. Dhiman, Thermal and thermohydraulic efficiency of recyclic-type double-pass solar air heaters with fins and baffles. Heat Transfer Eng. **37**(15), 1302–1317 (2016)

# Computer-Aided Analysis of Solidification Time and its Effect on Hardness for Aluminium Copper Alloy

**Sasmita Tripathy and Goutam Sutradhar**

**Abstract** Aluminium alloys are known for their very good strength to weight ratio. Aluminium alloy finds its applications depending upon the alloying agent/agents added. Among other alloying materials Copper, Silicon, Magnesium are some of the alloying agents which are largely used with Aluminium, as they improve the strength of the alloy without affecting the density when compared to pure Aluminium. Addition of Silicon to pure Aluminium improves the strength of the resulting alloy at room temperature. But at higher temperatures (above 150 °C) strength of Aluminium–Silicon (Al–Si) alloy is very poor. It is a preferred alloying agent when it comes to casting as the Al–Si alloy shows good casting ability compared to other (alloy without having silicon). But in the present study Copper is used with Aluminium as the major alloying material and silicon is either absent or present only as impurity. Copper addition to the Aluminium improves the alloy strength at room as well as at high temperatures, but at the cost of quality of the cast component. To get good quality, defect-free castings and to improve yield 'casting simulation software' can be used which can predict the casting defects virtually. So in the present study effect of adding Copper in the absence of silicon is studied where casting simulation software helped to predict the solidification time and quality of final cast component. Prediction of solidification time for different compositions of alloy gives an idea about the hardness of the alloy. Simulation results from two types of sand castings (spiral shape, step) are compared for their solidification time. The variation in solidification time with variation of copper weight percentage which is 4, 8, 12% for this experiment is studied. It was found that for a fixed composition the solidification results from both types of sand casting (spiral shape and five-step component) are of similar nature. The shop floor casting of the five step component is done and Rockwall hardness for all three compositions tested. Prediction of solidification time and shrinkage defects results obtained through virtual casting process helps in improving

S. Tripathy (✉)
Mechanical Engineering, Jadavpur University, Kolkata, India
e-mail: sasmita_tr@yahoo.co.in

G. Sutradhar
Mechanical Engineering, NIT Manipur, Imphal, India
e-mail: cast_1963@rediffmail.com


K. M. Pandey et al. (eds.), *Recent Advances in Mechanical Engineering*, Lecture Notes in Mechanical Engineering, https://doi.org/10.1007/978-981-15-7711-6_85

process parameters which eventually can help in improving the yield of the component production. Hardness results obtained for different compositions goes well with the type of predicted solidification nature (time) for different shapes of casting for this study.

**Keywords** Aluminium-copper alloy · Hardness · Casting simulation software

## 1 Introduction

Aluminium alloy is the most used alloy in automotive industry. Among many other reasons of its use in automotive industry light weight, high strength to weight ratio, ease in choice of manufacturing process are few. Developments in aluminium alloy and optimization of casting techniques have lead to improved material properties and functional integration which enable aluminium castings to satisfy the new market requirements and have allowed to replace, in many cases, engine components made with heavy cast iron alloys. Aluminium when combined with Copper gives very good strength at room as well as at high operating temperatures. The strength of Al–Cu alloy is due to the tetragonal structure of the $\theta'$ phase. Precipitate-hardened Aluminium alloy A201 possesses the highest mechanical strength between room temperature and 200 °C among the casting aluminium alloy. It is reported that mechanical behaviour of AlCu5 alloys at high temperature despite the reported worse casting properties than conventional Ala–Si alloys showed better results [1–3].

Using Aluminium alloy for casting is subjected to limitations in using alloying elements to get defect-free casting and high yield. For example, using silicon along with copper add to the fluidity of the alloy composition which is necessary for casting process. But at the same time silicon, limits its use at high temperatures.

So combination of both alloy compositions and their optimized manufacturing techniques is the area of interest for researchers nowadays.

It is shown in studies that increase in solidification time decreases the hardness [4]. Higher freezing range attribute to lower hardness through cooling rate plays a vital role to determine the microstructure which eventually affects mechanical behaviour of any alloy [5].

In the present work with the help of simulation software hardness of aluminium alloy for different wt% of copper (4, 8 and 12%) is studied. Here depending upon the solidification time of alloy for spiral mould of different composition is compared. Also for all the three compositions 'five-step component' (Fig. 1) developed by sand casting. The computer-aided solidification time analysis showed that increasing the wt% of copper affects its freezing range which ultimately affects the hardness.

Use of simulation software in manufacturing in recent days gaining popularity as this can be used to predict rejection thereby improving yield. Simulation analysis before actual shop floor casting saves time, material, money, etc. as all the trials are done virtually. Study shows that replacing the trial and error with simulation software which involves a virtual process can be able to utilize our resources efficiently [6].

(a) (b)

**Fig. 1** Green sand mould and casting of the 5-step component

**Table 1** Alloy composition

| Alloy (wt%) | AL1 (%) | AL2 (%) | AL3 (%) |
|---|---|---|---|
| Cu | 4 | 8 | 12 |
| Al | 96 | 92 | 88 |

But as in casting of any component, lots of complex physical and chemical changes occur during solidification; validation of the simulation results is a vital part which too needs lots of research and study.

## 2 Experimental Procedure

### *2.1 Alloy Preparation*

Master alloy is developed having melting 50% of commercially pure (99.9%) aluminium and 50% of electrolytic copper. Then from the master alloy Alloy1, Alloy2 and Alloy3 has been developed having 4, 8 and 12 wt% of copper respectively by adding pure aluminium (Table 1).

### *2.2 Solidification Simulation*

For simulation of different shapes and compositions "Z cast simulation software" is used. Spiral shape simulation for all the three compositions, i.e. AL1, AL2, AL3 performed. Also Simulation for the above compositions done for the 'five-step component using "Z Cast simulation software" and the solidification time is noted. For the five-step cast component, the freezing range was calculated from the cooling curves.

The solidification simulation for the step component shows that there is shrinkage defects at the base of the riser. The shrinkage volume increases with the increase in wt% of copper. Shrinkage dimensions are shown in Figs. 2b, 3b, and 4b, respectively, for (AL1, AL2, AL3 alloy). Solidification cooling graph shows substantial variation in the cooling curve. For AL1, solidification begins at a temperature of 680 °C and it completes solidification zone at a temperature of 531 °C. For AL2, solidification begins at a temperature of 630 °C and it completes solidification zone at a temperature of 544 °C. For AL3, solidification begins at a temperature of 619 °C and it completes solidification zone at a temperature of 543 °C. As copper wt% increases the volume of the shrinkage keeps on increasing with defect location moving to other side Figs. 2, 3, 4 shows the dimensions of the shrinkage defects and solidification time for AL1, AL2, AL3 respectively. The results obtained from simulation verified with experiments. The results of shrinkage were very much similar to the predicted results.

Simulation solidification for spiral shape is shown in Fig. 5a–c respectively for alloy AL1, AL2, AL3.

All the results from simulation are tabulated in Table 2.

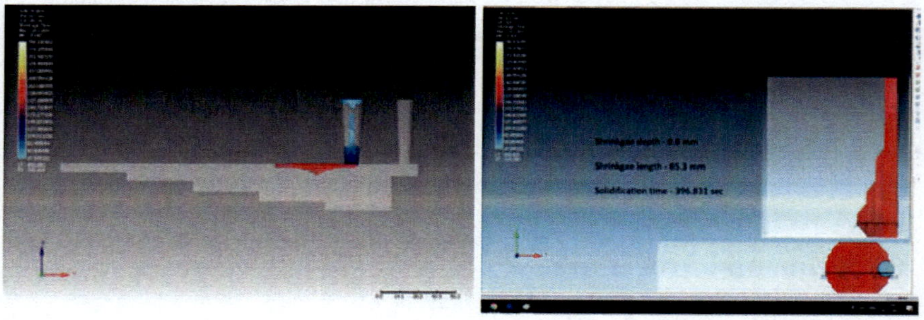

(a) (b)

**Fig. 2** Solidification simulation showing shrinkage and solidification time for AL1

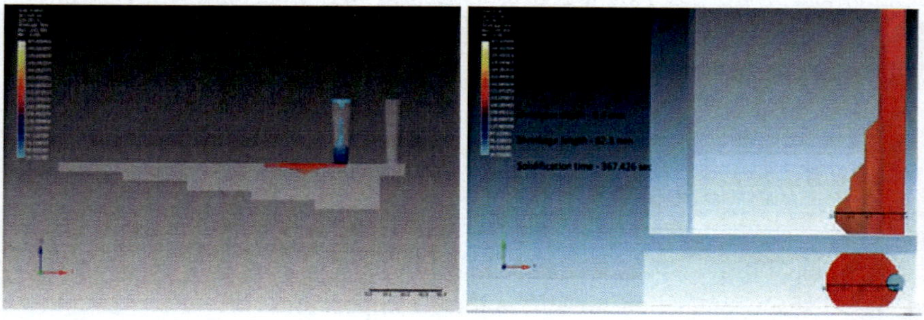

(a) (b)

**Fig. 3** Solidification simulation showing shrinkage and solidification time for AL2

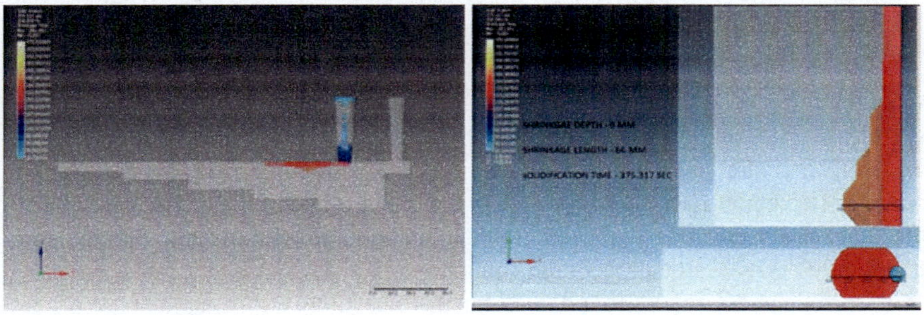

(a) (b)

**Fig. 4** Solidification simulation showing shrinkage and time for AL3

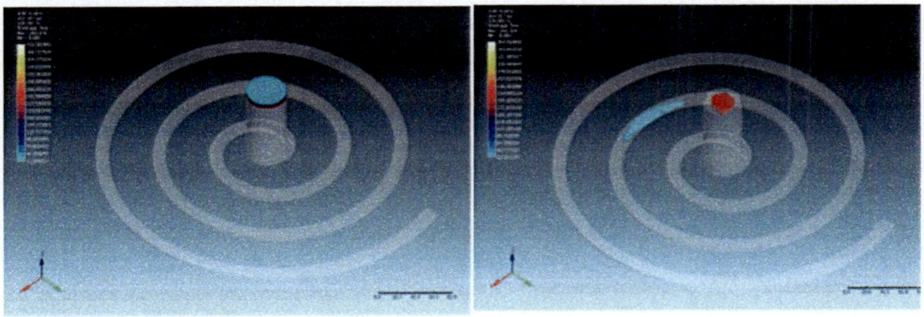

(a) (b)

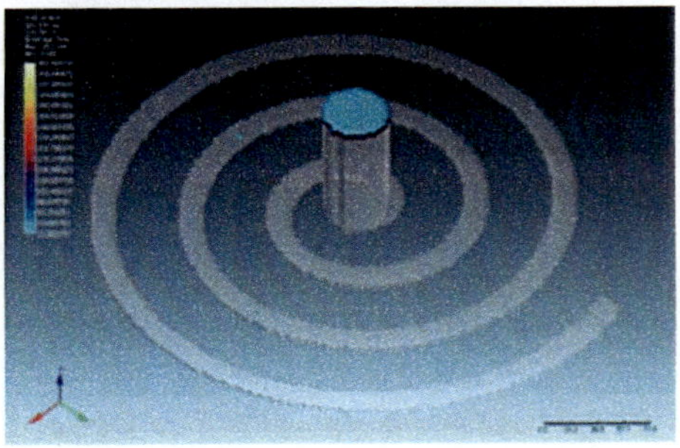

(c)

**Fig. 5** Spiral solidification simulation for solidification time

**Table 2** Simulation results of AL1, AL2, AL3 alloys

| Alloy | Step solidification time (in sec) | Spiral solidification time (in sec) | Freezing range for step solidification (in °C) |
|---|---|---|---|
| AL1 | 396 | 413 | 149 |
| AL2 | 367 | 382 | 86 |
| AL3 | 375 | 364 | 76 |

### 2.3 Shop Floor Casting of Five-Step Component

AL1 or Aluminium having 4 wt% of copper was melted in the furnace around temperature of 750 °C. The pouring temperature was measured with the k-type thermocouple. It was maintained around 720 °C. Green sand mould is prepared using the wooden 5-step pattern. Riser was located at the thickest part of the component. After sprue, gate and riser properly placed molten metal poured with the help of a ladle in to the mould cavity. Same procedure is maintained for AL2 and AL3. After solidification, shrinkage cavity was found in all the three castings Fig. 1b at the base of the riser.

### 2.4 Tests for Hardness

Samples of dimensions 20 × 20 × 10 mm (Fig. 6) were prepared from the five-step component developed from green sand casting. Test results for Rockwall hardness test for AL1, AL2, AL3 are shown in Table 3.

**Fig. 6** Hardness test of specimen AL1, AL2, AL3

**Table 3** Hardness for AL1, AL2, AL3 alloy

| Alloy | Hardness (HRB) |
|---|---|
| AL1 | 10 |
| AL2 | 25 |
| AL3 | 30 |

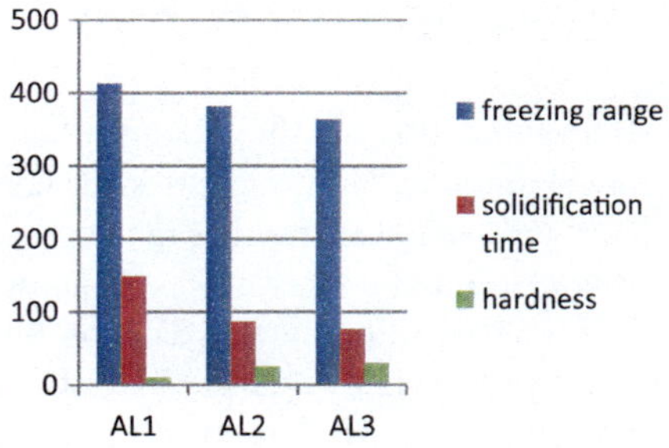


**Fig. 7** Comparison of hardness and simulation results

# 3 Results and Discussions

## *3.1 Simulation Results*

Simulation results of spiral test for all the compositions show that with increase in copper percentage the total solidification time decreases. The results of spiral shape give an idea about the variation of solidification time with composition Fig. 5a–c.

From the five-step component simulation results, the total solidification time for alloy AL3 is slightly more than AL2. But when the freezing range was compared it is seen that with increase in copper percentage the freezing range decreases. The graphical comparison of hardness and freezing range and total solidification time of AL1, AL2, AL3 (Fig. 7).

## *3.2 Experimental Results*

The hardness for different compositions of alloy shows that with increase in copper percentage in the alloy Rockwall hardness number(HRB)also increases. These values go well with the literature and also the predicted solidification time (Fig. 8). This increase in hardness is a result of lower freezing range and higher cooling rate for the given composition of the alloy

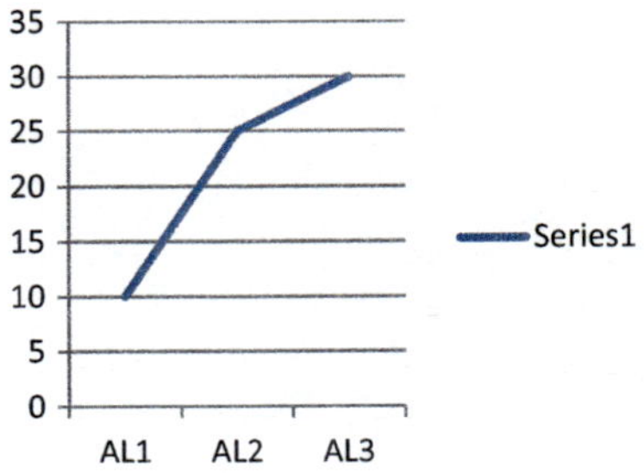


**Fig. 8** Hardness versus alloy composition

## 4 Conclusion

- Both simulation and shopfloor results show that with increasing copper percentage in Aluminium alloy increases its hardness. This increase in hardness can be attributed to lower freezing range of alloy with increase in copper.
- But addition of copper makes the alloy susceptible to casting defects. As we saw it from simulation as well as shop floor casting. Copper percentage affects the fluidity of the alloy due to which shrinkage defects arises.
- So to get defect-free casting and also maintaining good mechanical properties simulation software plays an important role. As it will reduce the actual shop floor trial to get defect-free casting and increasing yield.
- Further study and analysis of microstructre will give an idea about of the properties of the mentioned alloys.

## Referencess

1. R. Molina, P. Amalberto, Mechanical characterization of aluminium alloys for high temperature applications, Part 1: Al-Si-Cu alloys Teksid Aluminum M. RossoPolitecnico di Torino, vol. 29, no. 1 (2011)
2. J.E. Hatch, *Aluminium: Properties and Physical Metallaurgy* (ASM International, Metals Park, OH, 1984), pp. 322–388
3. R. Molina, P. Amalberto, Mechanical characterisation of aluminium alloy for high temperature applications Part 2: Al-Cu, Al-Mg all, Teksid Aluminum M. RossoPolitecnico di Torino, vol. 29, no. 2 (2011)
4. S. Ilangovan, Effects of solidification time on mechanical properties and wear behaviour of sand cast aluminium alloy. IJRET **3**(2), 74 (2014)
5. H. Akhyar, A. Farhan, Cooling rate, hardness and microstructure of aluminium cast alloys. www. Researchgate.net, May 2018, 2019
6. K. Shailendrakumar, K. Srinivasulu Readdy, Casting simulation of automotive wheel rim using aluminium alloy material. Int. J. Manuf. Mech. Eng. Int. Res. Publ. House **1**(1), 41–46 (2015)

# Attribute of SiC Powder Additive Mixed EDM on Machining Performance and Surface Integrity Aspects of Inconel 625

**Ankan Das, Bhavani Tharra, V. V. N. Siva Rao Sammeta, and John Deb Barma**

**Abstract** Superior properties such as stability at higher temperatures and corrosion resistance make nickel-based super alloys to use in aviation, marine, and processing industries. However, properties like chemical attraction and low thermal conductivity keep these super alloys under difficult to machine materials. In such cases, EDM is the one of best possible solutions to machine electrically conductive materials. It is difficult to achieve smooth surfaces and more material removal rate (MRR) using normal EDM due to instability in the process. To achieve better MRR and surface roughness (SR), silicon carbide (SiC) powders has been mixed in the dielectric fluid to study the effects of current, powder concentration, pulse on time, and pulse off time. Results indicated that 49.09% increment in MRR, 29.22% increment in surface finish, and better microstructure has been achieved by powder-mixed electro discharge machining when compared to the EDM with simple dielectric.

**Keywords** Additive-mixed EDM · Electro discharge machining (EDM) · Material removal Rate (MRR) · SiC powder · Surface roughness and microstructure

## 1 Introduction

Inconel 625 is nickel-based super alloy which posses properties like corrosion resistance and oxidation resistance at high temperatures which makes it widely used material in chemical processing, aviation, and marine industries. However, chemical

A. Das · B. Tharra · V. V. N. Siva Rao Sammeta · J. D. Barma (✉)
Mechanical Engineering, National Institute of Technology, Agartala, India
e-mail: johnneristian@gmail.com

A. Das
e-mail: dasankan93@gmail.com

B. Tharra
e-mail: bhavani2977.tharra@gmail.com

V. V. N. Siva Rao Sammeta
e-mail: sivaraoiit21@gmail.com



K. M. Pandey et al. (eds.), *Recent Advances in Mechanical Engineering*, Lecture Notes in Mechanical Engineering, https://doi.org/10.1007/978-981-15-7711-6_86

attraction and low thermal conductivity of Inconel 625 are restricting its application by making it difficult to machine by conventional processes. EDM place vital role in machining difficult to machine materials. Fig. 1 shows the mechanism of EDM process in which spark produces in between the work and tool due to the dielectric breakdown, the produced spark erode the work material, and continuing the process can be ended up with final required shape which is replica of the tool shape. Instability of the EDM process results in poor mechanical efficiency and SR, which limits the usage of EDM in industries [1]. Debris accumulation in the spark zone might be the reason for the reduction in the efficiency of the EDM process. To extend its usage, different techniques are being used by the researchers such as tool rotation [2] and powder-mixed EDM [1].

Improvement in flushing conditions was achieved by adopting the tool rotation, which ultimately results in increasing the spark efficiency [2]. Figure 2 shows the mechanism of powder-mixed EDM in which the mixed powder can form a chain-like

**Fig. 1** EDM mechanism

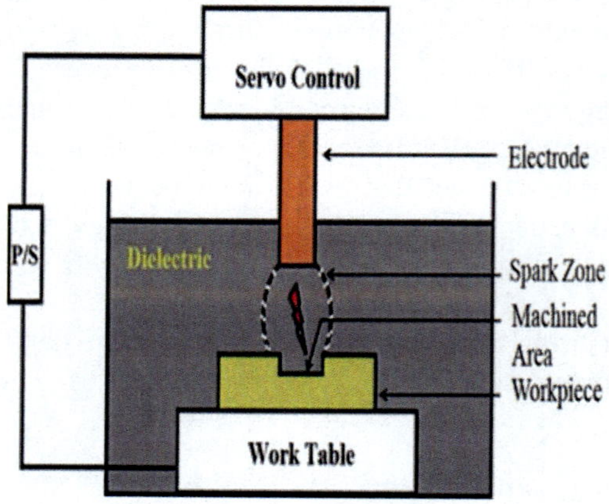


**Fig. 2** Powder-mixed EDM mechanism

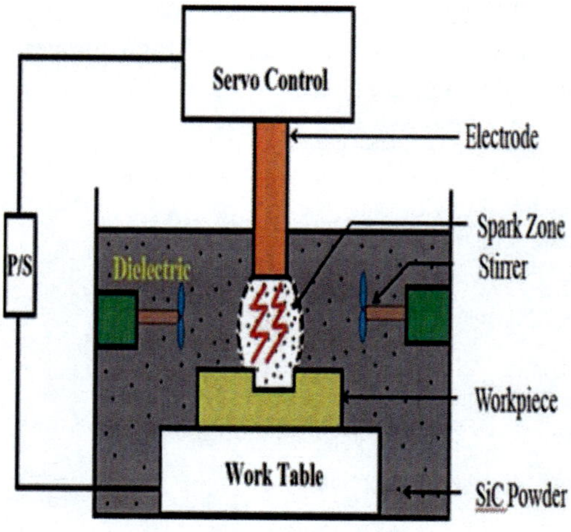

structures in between the tool and work piece due to which the multiple discharges can be possible. Thereby, low SR and high MRR can be achieved. Talla et.al [3]. found improved surface finish by adding 6 g/l graphite powder in the dielectric. Distributing of the discharge energy was observed over the powder particles and led to more number of shallower craters while machining Inconel 625. Anoop Kumar Singh et.al [1] observed smoother surface and improvement in micro-hardness while machining the Co605 super alloy by adding graphite powder to the dielectric. It was further said that suface finish improved to 1.99 from 2.23 μm using PMEDM. Pecas et.al [4] achieved smoother surface by adding 2 g/l silicon powder to the dielectric and concluded that adding powder decreases the unuseful discharges which can observe in case of normal EDM. Abdul Rani et.al [5] observed improvement in surface finish by adding nano-aluminum powders to the dielectric while machining titanium alloy and concluded that distribution of spark occurs due to the better suspension of powder particles in the spark zone because of its less density. Banh et.al [6] conducted research on die steel by mixing titanium powders in the dielectric and obtained improved productivity and quality of the surface.

In the present investigation, the effect of semiconductive SiC powder mixed EDM on Inconel 625 has been studied. Therefore, PMEDM is carried out to study the effect of SiC powder on EDM performance. Comparison has been made for both EDM and PMEDM in terms of MRR and SR by a varying current, pulse on time, pulse off time, and powder concentration. Microstructures of the machined samples have been studied using FESEM.

## 2 Experimental Procedure

Investigation has been done to study MRR, SR, and microstructure of Inconel 625 through additive-mixed electro discharge machining, and comparison has been made among the performance measures for both EDM with simple dielectric and additive-mixed EDM. All the experiments were conducted on die-sinking EDM (25A pulse generator Manual + ZNC, Sparoknix India Ltd.). Inconel 625 was selected as the workpiece material with 20X20X3 mm dimensions. Kerosene as the dielectric fluid and the electrolytic copper solid rod of diameter 12.5 mm and length of 60 mm as a tool were chosen to perform the experiments.

Semi-conductive ceramic silicon carbide powder with the average particle size of ~10 μm was used as an additive to the dielectric medium. Current, pulse on time, and pulse off time were selected as the process parameters after the pilot run. Negative tool polarity showed better results for the above-mentioned tool–work combination as per the trail run. The ranges of selected process parameters were fixed as per the result of pilot experiments and are listed in Table.1. The experiments were performed in two different stages in which one set of experiments were carried by the conventional mode of EDM, i.e., by using simple kerosene as the dielectric, as given in Table.2, and the other set was carried by SiC-mixed kerosene as dielectric. The first stage of experiments was done for getting the convincing set of process parameters by

**Table 1** Process parameters range

| Process parameters | Range |
|---|---|
| Polarity | Negative |
| Peak current | 10–16 amp |
| $T_{on}$ | 4–7 position |
| $T_{off}$ | 4–7 position |
| Powder concentration | 1–5.5 gm/liter |

**Table 2** Parametric design

| S. No. | Peak current (amp) | $T_{on}$ (μs) | $T_{off}$ (μs) |
|---|---|---|---|
| 1 | 10 | 100 | 30 |
| 2 | 12 | 100 | 30 |
| 3 | 14 | 100 | 30 |
| 4 | 16 | 100 | 30 |
| 5 | 14 | 50 | 30 |
| 6 | 14 | 100 | 30 |
| 7 | 14 | 150 | 30 |
| 8 | 14 | 200 | 30 |
| 9 | 14 | 100 | 20 |
| 10 | 14 | 100 | 30 |
| 11 | 14 | 100 | 40 |
| 12 | 14 | 100 | 75 |

varying one parameter at a time, keeping rest of the parameters constant to choose the effective level SiC powder concentration in the dielectric while varying it from 1 to 5.5 g/l. In powder-mixed EDM, the same set of experiments was carried out in the later stage by keeping the effective level of powder concentration in the dielectric.

$$\frac{Wb - Wa}{\rho * t} \tag{1}$$

where *Wb* and *Wa* are weights of the samples before and after the machining, respectively, $\rho$ is the density of the workpiece material, and t is the machining time.

The weights of each workpiece were measured before and after the experiments by using electronic balance (Shinko Denshi, Japan; Model: DJ 300S of 0.001 g precision), and time taken by each experimental run was being recorded using a digital stopwatch. Therefore, MRR was calculated using Eq. (1). Machined sample was cleaned with 99% pure alcohol to wash out the accumulated free debris over the machined surface, Taylor Hoboson 2D profilometer was used to find out the surface roughness values, and average surface roughness values of 5 that runs over 8 mm sample length were taken into account. The microstructures of the machined samples in both EDM and PMEDM were studied by using FESEM (Gemini, ZESISS Sigma 300) microscopic images at a magnification of 3.73Kx.

# 3 Results and Discussion

## *3.1 Estimation of Effective Powder Concentration*

After completion of conventional EDM experiments, effective current, pulse on time, and pulse off time were fixed for further PMEDM experiments to find the effective powder concentration in kerosene.

### 3.1.1 Effect of Powder Concentration on MRR

It is evident from Fig. 3 that MRR is increasing proportionally with the increasing of the powder concentration level in the dielectric. The increase in MRR maybe because of the reduction in instability of the process [5] (i.e., short-circuit discharges, open-circuit discharges as in case of EDM), and decrease in the dielectric strength leads to the early occurrence of spark. Beyond 3.5 g/l powder concentration, the length between tool and workpiece may be increased because of the bridging effect due to which spark might not be generated, and it leads to decrease in MRR.

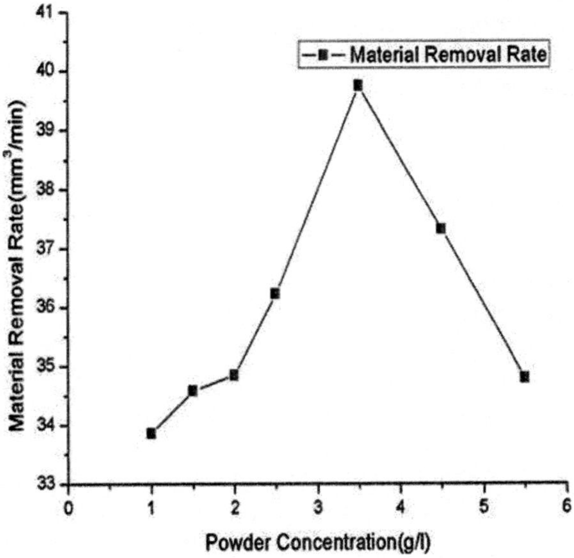


**Fig. 3** Effect of powder on MRR

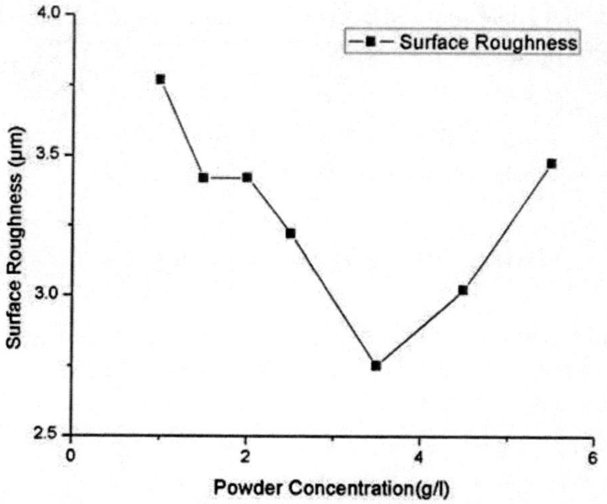


**Fig.4** Effect of powder on SR

### 3.1.2 Effect of Powder Concentration on SR

The effect of suspended powder particles in the kerosene on surface roughness can be seen in Fig. 4, where SR decreases with an increase in powder concentration up to 3.5 g/l. This is because the discharge density distributes into multiple discharges over the impregnated SiC particles which leads to decrease in the current density and ultimately forms shallow, wide crater on the workpiece. SR decreases with the further increase in powder concentration that leads to short-circuit pulses which results in deeper craters.

Therefore, 3.5 g/l was selected as an effective level of SiC powder concentration in the dielectric to conduct the PMEDM experiments with the parameters as incorporated in Table.2.

## *3.2 Analysis of Material Removal Rate*

A comparison has been done for MRR in both EDM and PMEDM processes, and effects of pulse on time, current, pulse off time have been assessed through graphical representation of results. It is evident from Fig. 5a, both EDM and PMEDM yield almost similar kind of results for MRR. As current increases, the discharge density increases, resulting in the melting of more amount of material from the work piece, thereby more MRR. In spark zone, at higher currents, debris were more in volume, resulting in un-stability of the process which leads to lesser MRR in case of conventional EDM. The aforementioned phenomenon was observed after 14A, but in the case of PMEDM as the gap between the work and tool increases debris can be evacuated, and thereby, increase in MRR can be observed even at higher current values.

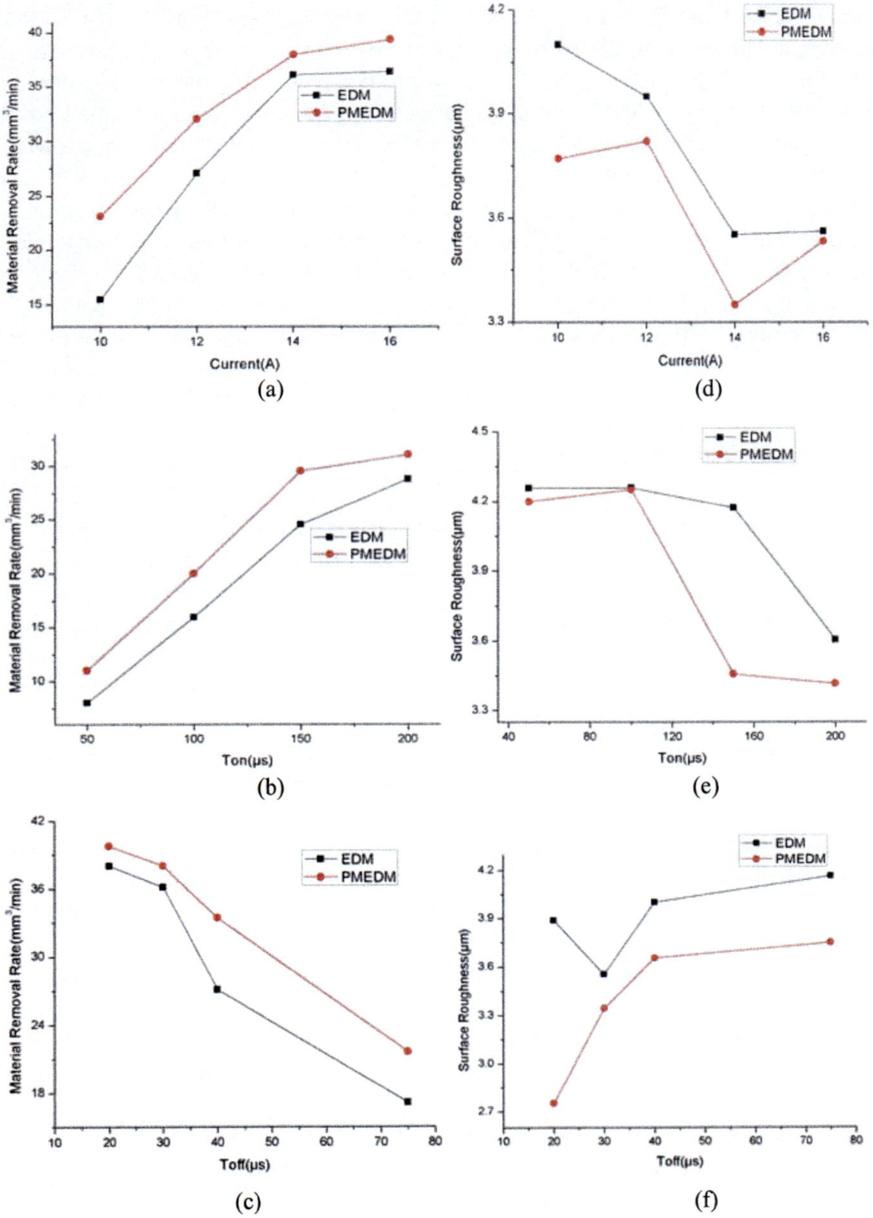


**Fig. 5** **a–c** Comparison of MRR for EDM and PMEDM w.r.t current, $T_{on}$, and $T_{off}$. **d–f** Comparison of SR for EDM and PMEDM w.r.t current, $T_{on}$, and $T_{off}$, respectively

In both EDM and PMEDM, as pulse on time increases, an increase in MRR was observed, as shown in Fig. 5b. Increase in pulse on time (i.e., holding the pulse energy for longer period) leads to more amount of material to erode from the parent material which results in higher MRR, but at higher pulse on time, the plasma channel gets enlarged and leads to loss of energy to the dielectric medium that results as a lower the MRR slope.

As pulse off-time increases, MRR decreases in both EDM and PMEDM, as shown in Fig. 5c. The reason may be exposure time of spark energy reduces, as pulse of time increases results in lesser MRR. As per Fig. 3, PMEDM yields more MRR with the variation of all parameters compared to EDM, which maybe because of the reduction of the instability of the process.

### *3.3 Analysis of Surface Roughness*

Surface roughness values for both conventional EDM and AEDM were compared, and the effects of current, pulse on time, and pulse off time have been assessed through graphical representation of the results.

SR follows a similar trend in both EDM and PMEDM, as shown in Fig. 5d. It was observed that SR increases upto a certain value of current and then decreases as current increases. As mentioned above, as the current increase, spark energy increases, which results in a deeper crater ended up with the rougher surface. Higher current yields more amount of debris, which is difficult to evacuate from the spark zone, might absorb some amount of spark energy, and ultimately results in decrease spark density, thereby lesser surface roughness was observed.

It was observed from Fig. 5e that SR decreases as pulse on time increases. This might be attributed to an increase in pulse on time enlarge the plasma channel, which results in decrease in discharge density creates shallow craters yields lesser surface roughness.

As pulse off time increases, increased SR was observed in both EDM and PMEDM, as shown in Fig. 5f. Spark zone may be free of debris as those were evacuated easily with an increase in pulse off time which maintains the stability in the process (i.e., more useful pulses) and creates deeper craters that result in high surface roughness.

In all the above cases, lesser surface roughness was observed in the case of PMEDM than EDM. This is maybe because of energy distribution over the SiC particle that produces the shallow crater compared to the conventional EDM process.

### *3.4 Improvement in MRR and SR in PMEDM*

The percentage of increase in MRR and reduction in SR has been calculated and graphically shown in Fig. 6 in which sets 1–4 represents the effect of current, sets

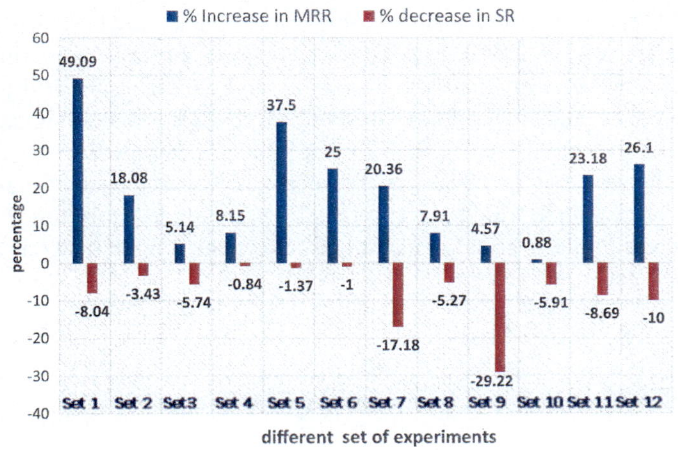


**Fig. 6** Increment and decrement of MRR & SR in PMEDM process

5–8 represents the effect of pulse on time, and sets 9–12 represents the effect off pulse off time. It was observed that in PMEDM, MRR was increased by about 49.09% at 10A of current, and SR was decreased by about 29.22% to that of EDM. This can firmly justify the improvement of the PMEDM process over EDM with a simple dielectric medium.

### *3.5 Microstructure Analysis of Machined Surface*

FESEM microscopic images of the machined surfaces at 3.7Kx magnification are shown in Fig. 7a, b, which represents the clear view of achieved machined surfaces

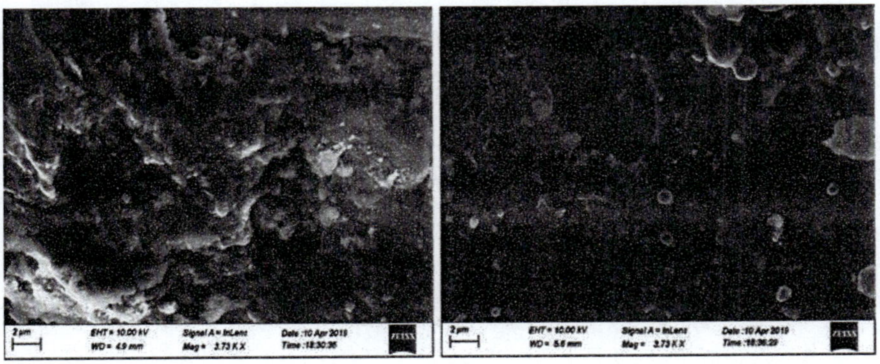

(a) (b)

**Fig. 7** Microstructures of **a** EDM and **b** PMEDM at peak current 10A, 100 μs $T_{on}$, and 30 μs $T_{off}$

both for EDM and PMEDM conducted at the same parametric condition. Uneven machined surface can be observed in Fig. 7a, which indicates the re-solidification of the molten material attribute to less flowability and the higher thermal gradient between the molten material and base material results in instant solidification of molten material. It can be seen in Fig. 7b that molten material more evenly gets solidified on the base material. This may be because SiC particle supplies the heat input to the work material, which was absorbed by SiC particles at the time of spark results in decreasing the thermal gradient and allows the molten material to flow evenly before it gets solidified.

## 4 Conclusions

In this work, current, powder concentration, pulse off time, and pulse on time have been considered as the process parameters, and SR, MRR, and study of microstructures are taken as output parameters.

The following are the conclusions of the study:

(a) As compared to the EDM process, in PMEDM process, MRR increases for each set of experiments, which clearly shows that the powder additive enhances the conductivity level in the inter-electrode gap, which results in multiple sparks generation with single discharge. As the peak current, pulse on time, MRR also increases and decreases when pulse off time increases.
(b) MRR and SR in PMEDM are largely dependent on the concentration level of powder particles in the dielectric fluid. In the present research work, the concentration level has been kept as 3.5 g/l, which has been maintained throughout.
(c) The surface roughness of machined samples achieved in PMEDM is also less as compared to EDM. Due to enlargement of the width of the plasma channel and increase in the discharge gap, it resulted in the formation of flat and large shallow craters on the machined surface.
(d) As compared to normal EDM, the maximum percentage increment of MRR achieved for PMEDM using SiC is about 49.09%, and maximum surface roughness has been found to improve by 29.22%.
(e) PMEDM results in more even surface than EDM because better flowability of molten material attributed to a less thermal gradient.

## References

1. A.K. Singh, S. Kumar, V. P. Singh, Effect of the addition of conductive powderin dielectric on the surface properties of superalloy Super Co 605 by EDM process. Int. J. Adv. Manuf. Technol. **77**, 99–106 (2015)

2. J.S. Soni, G. Chakraverti, Machining characteristics of titanium with rotary electro-discharge machining, Wear 51–58 (1994)
3. G. Talla, S. Gangopadhyay, C.K. Biswas, influence of graphite powder mixed EDM on the surface integrity characteristics of inconel 625. J. Part. Sci. Technol. 219–226 (2016)
4. P. Pecas, E. Henriques, Electrical discharge machining using simple and powder-mixed dielectric: the effect of the electrode area in the surface roughness and topography. J. Mater. Process. Technol. **20** 250–258 (2008)
5. A.M Abdul-Rani, A.M Nanimina, T.L. Ginta, M.A Razak, Machine surface quality in nano aluminum mixed electrical discharge machining. Procedia Manuf. **7**, 510–517 (2016)
6. T.-L. Banh, H.-P. Nguyen, C. Ngo, D.-T. Nguyen, Characteristics optimization of powder mixed electric discharge machining using titanium powder for die steel materials. J Process Mech. Eng. 1–18 (2017)

# Enhancement of Thermal Performance of Microchannels Using Different Channel Wall Geometries: A Review

Dipak Debbarma, K. M. Pandey, and Abhishek Paul

**Abstract** The current cooling demand and the size of the microchannel sink make it very much challenging for the removal of heat in modern electronics compact components. The heat sinks with simple smooth microchannels are not adequate to remove the required heat nowadays. Uses of rib, cavity, dimple, protrusions on channel wall are the techniques to improve the Nusselt number (Nu), thermal resistance which signify better performance of the heat sink. A significant number of works on single-layered microchannel heat sink are carried out and after the proposal of new concept of double-layered heat sink for the purpose of cooling of the electronic devices, works on this type of sink are also started due to its high prospect. Silicon and water as the substrate material and coolant, respectively, are the choices of most of the researchers due to the merits they have. The present work is a review of the available literatures and research papers relevant to microchannels having different geometry structures with the objective to find and analyze the effects of the channel geometry on the enhancement of performance of microchannels.

**Keywords** Rib · Cavity · Dimple · Protrusion · Thermal resistance · Nusselt number · Double-layered heat sink

## 1 Introduction

Cooling is the mandatory criteria for thermal design of microelectronics-mechanical system (MEMS) components and other electronic devices. The heat generated in these systems such as computer processors, sensors, actuators should be removed in

D. Debbarma (✉) · K. M. Pandey · A. Paul
Mechanical Engineering Department, National Institute of Technology, Silchar, India
e-mail: dipakdb.2003@gmail.com

K. M. Pandey
e-mail: kmpandey2001@yahoo.com

A. Paul
e-mail: v1.abhishek@gmail.com



K. M. Pandey et al. (eds.), *Recent Advances in Mechanical Engineering*, Lecture Notes in Mechanical Engineering, https://doi.org/10.1007/978-981-15-7711-6_87

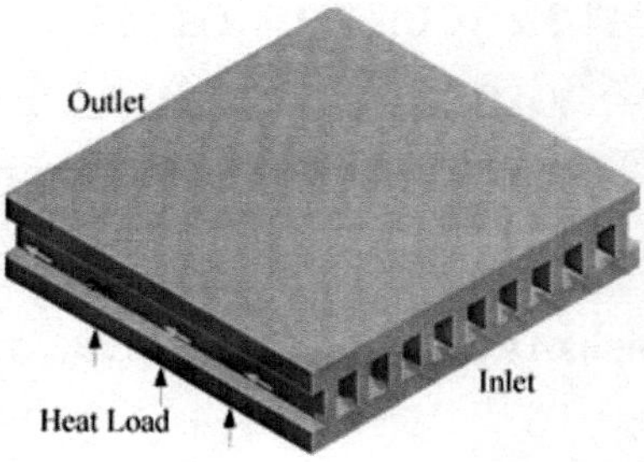


**Fig. 1** Microchannel heat sink (Courtesy: www.asme.org)

the necessary rate. The microchannel heat sink (MCHS) is the means to remove the heat generated in such devices. The MCHS contains many parallel microchannels which are having hydraulic diameter of less than 0.2 mm and fluid passes through these channels to take away heat from the heat generation region. However, with the advancement in the technology, the modern compact electronics products need comparatively high heat removal rate. Hence, several mechanisms and techniques in MCHS are predicted by many researchers to increase the heat transfer rate. Tuckerman and Pease [1], in 1981, first made the concept of microchannel and investigated to find the cooling of IC products. They fabricated various water-cooled silicon heat sinks of 10 mm × 10 mm area with the numbers of small channels having different channel depths and widths, into the heat sink. The Fig. 1 depicts such a typical heat sink.

They observed that the maximum heat transfer rate could reach 800 W/cm$^2$ in their experiments.

## 2 Studies on Single-Layered Microchannel Heat Sink

### *2.1 Studies of Microchannels with Different Grooves, Cavity*

Xia et al. [2] investigated the microchannel with cavities taking water as cooling fluid for the range of Re between 132 and 931. They studied the effects of parameters of cavities on heat transfer. Their finding was that the fluid flow became weak for the smaller number of the cavities; but the fluid flow became strong with the increase of the number of the cavities. Xia et al. [3] numerically studied the microchannel with triangular cavity and found the optimum geometric parameters. They obtained the best configuration which exhibited the heat transfer enhancement factor ($\eta$) of 1.6 with the corresponding value, $Nu/Nu_o = 1.8$. Chai et al. [4] performed numerical investigations on microchannel heat sink with the cavities of circular shape. The impacts of flow rate, heat flux on pressure drop, and heat transfer were studied. They could observe that the heat transfer increases significantly when the flow rate is increased. Chai et al. [5] in their experimental investigations followed by numerical studies, evaluated the pressure drop, thermal characteristics in microchannel for

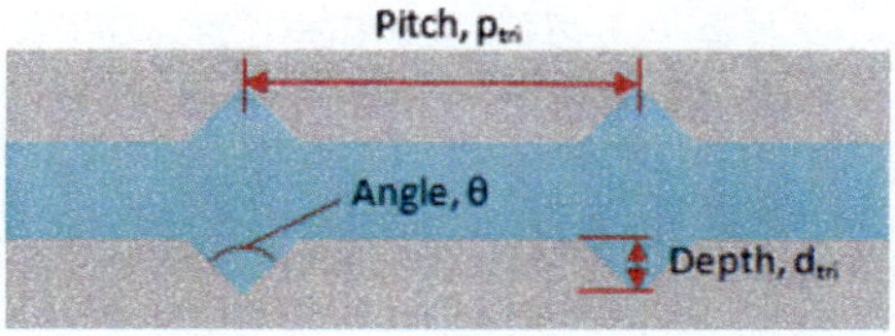


**Fig. 2** Channel with triangular cavity studied by Kuppusamy et al. [6]

different cross sections of channel. The channel with different cavities and the channel without cavity were compared.

Kuppusamy et al. [6]conducted study on microchannel having grooves of triangle-shaped. The increases of angle (ϴ) as well as depth ($d$) have significant impact in the performance of the channel, because with the increase of angle (ϴ) and depth ($d$), fluid–solid interface area and vortices are increased (Fig. 2).

Ahmed and Ahmed [7] did numerical investigation on the microchannel sinks with grooves on sidewall. The effects of groove on the triangular, trapezoidal, and rectangular shape were studied. They observed good heat transfer enhancement for some channels due to the formation of symmetrical eddies. In the work of Ma et al. [8] on microchannel having zigzag grooves, it was found that good performance of channel had been obtained in terms of heat transfer for the zigzag which were not very long and not much short. Debbarma et al. [9] performed numerical study on channel with cavity. The circular cavity on sidewall of channel has impressive impact on performance due to the improvement in fluid flow mechanism along the channel. The highest value of the heat transfer enhancement factor ($\eta$) as obtained in the study was 1.28 at Re 610.

## 2.2 *Studies of Microchannels with Different Ribs*

There are not much research works on only rib, since it alone does not show good results. Chai et al. [10] studied the thermal, hydraulic characteristics in channel with circular ribs. In their earlier investigations, the effects of circular-shaped ribs on thermal characteristics only were studied. The geometric parameters include the width $W_r$, height $H_r$, and spacing $S_r$ of ribs. The highest number (50) ribs in 10 mm channel provides maximum heat transfer rate. Chai et al. [11] conducted numerical study for the geometry of circular rib in the next series of the investigation of Chai et al. [10].In the study of aligned fan rib and offset fan rib, it was found that the ratio $f/f_o$ had always been smaller for latter for all Reynolds numbers.

## 2.3 Studies of Microchannels with Combination of Rib and Cavity

The combined application of cavity and ribs is very much promising technique. Xia et al. [12] did numerical investigation on microheat sink with cavities and ribs with different rib parameters. It was the extension work of Xia et al. [2], with the addition of the internal ribs. They found that the cavity and rib mounted together had impressive performance. At the entry of rib, the flow velocity of water increases which further decreases pressure in rib. This results in creation of flow disturbance. Zhai et al. [13] performed comparative study on microchannels with cavities and different ribs such as circular, trapezoidal, triangular, and rectangular ribs. Their findings showed that the heat sink was best configured with circular ribs. Zhai et al. [14] in their another work proposed six microchannels with ribs, cavities of different shapes and compared their heat transfer performance. They observed that all the microhannels of different models as studied had the best performance at the particular range of Re. Wang et al. [15] investigated the microchannel experimentally as well as numerically. The effects of ribs and different shapes of cavity were studied. It was observed that the rib as well as cavity, helped in better heat transfer rate even if there was a pressure drop. Li et al. [16] conducted the studies on microchannel with different cavities and ribs. They compared the results of simple microchannel, channels with cavity only, rib only and channel with both cavity and rib. They observed that laminar stagnation zones formed at cavity for cavity channel reduced heat transfer. The temperature distribution as well velocity distribution was uniform for channel with both rib and cavity.

## 2.4 Studies of Microchannels with Dimple and Protrusion

Few investigators worked on dimpled and protruded wall of microchannel. Wei et al. [17] performed numerical simulation of rectangular microchannel with the row of dimples at its bottom surface. They have considered different dimples at various positions in stream-wise direction of the channel and studied the phenomena of velocity vectors and the heat transfer. There was a good heat transfer rate at the end side of the dimple.

**Fig. 3** Channel with staggered and aligned dimple, protrusion studied by Xie et al. [18]

Xie et al. [18] investigated by numerical technique, the influences of the dimple as well as the protrusions on performance of microchannel. Their investigations were on the channel models with dimpled and smooth surface, dimpled and protruded surface which were further classified as staggered and aligned. They observed that the upstream portion of dimple was associated with low heat transfer. The heat sink they conducted study on, is shown in Fig. 3. Li et al. [19] studied the effects of pitch, diameter on heat transfer enhancement using $Al_2O_3$–water nanofluids as a working fluid and considering staggered and aligned pattern of microchannel. Dimpled and protruded model shows better heat transfer than other arrangements. Li et al. [20] in their investigation, used microchannel with both dimpled and protruded wall. They obtained the maximum $f/f_0$ for a region for which separation flow was increased. It showed that the thermal performance (TP), the significant parameter, improved with the larger value of depth. Xu et al. [21] performed numerical study for different position of dimples at the channel bottom wall. The aspect ratio, dimple spacing, and dimple depth were varied. At the increase of aspect ratio, higher performance was observed because the average velocity increases with increasing aspect ratio.

## 3 Studies on Double-Layered Microchannel Heat Sink

A new concept of double-layered microchannel heat sink for the purpose of cooling of the electronic devices is first proposed by Vafai et al. [22]. They demonstrated that this novel heat sink design was advantageous over the conventional single-layered heat sink. In their work, the performance along with the temperature distribution for these microchannels was predicted. As per their articulation, the unique feature of the double-layered sink design is that heat transfer also takes place from the heated coolant to the substrate at the outlet. They noted that maximum temperature difference was 5 °C for the double-layered sink in the stream-wise direction; whereas for the one-layered microchannel system, it was about 15 °C for the same parameters. The changes in pressure drop and thermal resistance were studied by varying the channel aspect ratio, width ratio, substrate materials, and pumping power in the work of Hung et al. [23]. It was found that copper exhibited better cooling performance in comparison with silicon, steel, and aluminum as substrate materials. Hung et al. [24] in their another work used the nanofluids such as aluminum oxide, titanium oxide, and copper oxide as coolants and analyzed the effects on the thermal resistance of the channel varying the particle volume fraction. The aspect ratio of the channel considered for the studies was in the range of 2–13. The effect of the aspect ratio was such that the thermal resistance had first decreased and then had started increasing on the increase of the aspect ratio.

Shen et al. [25] took novel structures having staggered flow alternation. They increased the flow alternation from 1 to 5 number and investigated parallel and counterflow in their numerical studies. The single-staggered channel with counter stream exhibits good overall convective thermal performance as per the result. Wu et al. [26] in their numerical work studied the thermal performance of double-layer channel with different width ratio and aspect ratio. Wei et al. [27] did experimental investigations on double-layer heat sink along with numerical simulations. The effects of different flow rate were studied. It showed that low thermal resistance was obtained for higher flow rate. The single- and two-layered heat sinks were compared by Levac et al. [28] where pumping power, thermal resistance had been evaluated on different flow arrangement and Reynolds numbers. One of the observations was that the two-layered sink with counterflow had the best results. Ahmed et al. [29] performed experiments on rectangular and triangular heat sink of aluminum substrate using nanofluids. Their results revealed that triangular channel had performed significant performances such as better temperature distribution which showed good agreement with the work of Wu et al. [26]. Some of the heat sinks tested in Ref. [29] are illustrated in Fig. 4.

Six heat sinks with different channel heights were investigated by Zhai et al. [30]. Two channels consist of triangular ribs and triangular cavities. For the case of microchannel with triangular ribs and triangular cavities, the heat transfer performance was highest in terms of the entropy generation rate and the temperature uniformity. Zhai et al. [31] investigated double-layered microchannel sink with cavities and ribs mounted on the wall. They found that channel with opposite cavities and aligned ribs gave the highest Nusselt number as the result of the formation of transverse vortices in the cavities areas for all the channels. The staggered flow is the center of the study of Zhai et al. [32]. Two types of staggered flow were proposed by them and compared with counterflow. These staggered flow provide better performance to the heat sink in terms of suitable temperature distribution.

Zhai et al. [33] investigated heat sink by field synergy principle and transport efficiency of thermal energy. The models of microchannel heat sink were similar to models taken in the work of by Zhai et al. [31]. Their work suggested the agreeable

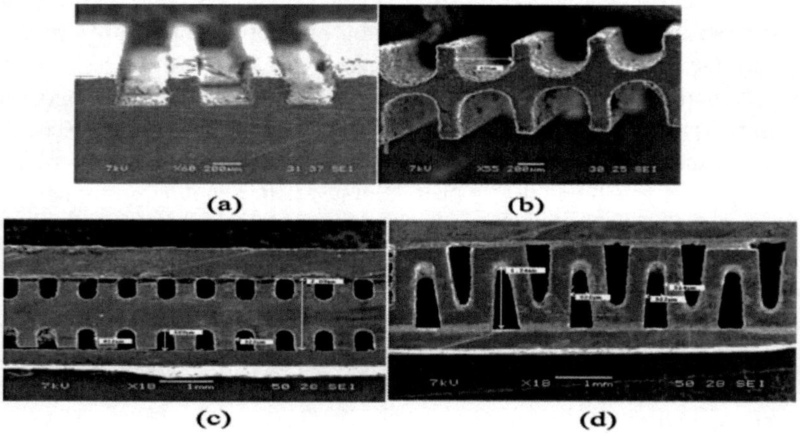


**Fig. 4** Ahmed et al. [29]: **a** rectangular microchannel, **b** rounded-end microchannel, **c** rectangular double-layered microchannel heat sink, and **d** zigzag triangular double-layered microchannel heat sink

results with that of other works as the new channel structure contribute in reduction of fluid temperature gradient.

Wong et al. [34] conducted numerical study to evaluate the thermal performance of microchannel with various channel aspect ratios for parallel flow and counterflow. They found that the high channel aspect ratio helped in reducing constriction to allow more flow. As a result of it, optimum heat removal was observed for the aspect ratio of 2.49 for counterflow configuration. In another investigation, Wong et al. [35] considered tapered and converging channel to estimate the thermal and hydraulic performance. The result suggested that although the effect of the taper channel had better thermal performance in comparison with simple straight channel, the higher pumping power was required for it.

Leng et al. [36] employed optimization method to predict the best performance of heat sinks. And for this purpose, they chose channel number, channel height, channel width, and coolant inlet velocity as search variables. They estimated that maximum bottom wall temperature change was only 3.23 K for the optimal channel which had number of channel = 54, channel width = 0.088 mm, channel height = 0.141 mm. Leng et al. [37] proposed a new double-layer sink with truncated top channel. They described that temperature of the coolant at top channel was always higher than that of the bottom coolant in the inlet region of bottom channels. Their simulation results showed that the overall thermal resistance was reduced by 37.5% for the truncated top channels. Leng et al. [38] applied the optimization technique on the heat sink similar to the sink of Leng et al. [37]. To minimize the overall heat resistance, search variables such as channel height, channel width, channel number, and truncation length of top channel were considered. The optimal channel was the channel with channel width = 0.116 mm, channel height = 0.525 mm, and dimensionless truncation length = 0.28. An improved heat sink with converging channels was presented by Osanloo et al. [39] where the temperature distribution, thermal resistance, and pumping power were obtained for different flow rates and convergence angles.

## 4 Conclusion

The width parameter, depth parameter and spacing of grooves, cavities, dimples, and protrusions have substantial impact on the flow characteristics of channel. Thus, the impact of these parameters regulates the thermal performance. The combination of cavity and rib in microchannel showed better results with significant heat transfer enhancement factor. The various research works on double-layered heat sink conclude that these new designs of heat sink are advantageous over the conventional single-layered heat sink. Furthermore, the performances such as the temperature distribution, thermal resistance for these microchannels can be enhanced by applying different channel wall geometry, flow arrangement, etc. Finally, the growing interest in such microchannel sinks indicates that research in this area is still developing [40].

## References

1. D.B. Tuckerman, R.F.W. Pease, High-performance heat sinking for VLSI. IEEE Electron. Device Lett. **2**(5), 126–129 (1981)
2. G.D.Xia, L. Chai, M. Zhou, H. Wang, Effects of structural parameters on fluid flow and heat transfer in a microchannel with aligned fan-shaped reentrant cavitie. Int. J. Thermal Sci. **50**, 411–419 (2011)
3. G.D. Xia, L. Chai, H. Wang, M. Zhou, Z. Cui, Optimum thermal design of microchannel heat sink with triangular reentrant cavities. Appl. Thermal Eng. **31**,1208–1219 (2011)
4. L. Chai, G.D. Xia, M. Zhou, J. Li, Numerical simulation of fluid flow and heat transfer in a microchannel heat sink with offset fan-shaped reentrant cavities in sidewall. Int. Commun. Heat Mass Transfer **38**, 577–584 (2011)
5. L. Chai, G.D. Xia, L. Wang, M. Zhou, Z. Cui, Heat transfer enhancement in microchannel heat sinks with periodic expansion–constriction cross-sections. Int. J. Heat Mass Transf. **62**, 741–751 (2013)
6. N.R. Kuppusamy, H.A. Mohammed, C.W. Lim, Thermal and hydraulic characteristics of nanofluid in a triangular grooved microchannel heat sink (TGMCHS). Appl. Math. Comput. **246**, 168–183 (2014)
7. H.E. Ahmed, M.I. Ahmed, Optimum thermal design of triangular, trapezoidal and rectangular grooved microchannel heat sinks. Int. Commun. Heat Mass Trans (2015)
8. D.D. Ma, G.D. Xia, Y.F. Li, Y.T. Jia, J. Wang, Effects of structural parameters on fluid flow and heat transfer characteristics in microchannel with offset zigzag grooves in sidewall. Int. J. Heat Mass Transf. **98**, 17–28 (2016)
9. D. Debbarma, S. Debbarma, S. Das, B.K. Deb, B. Datta, Numerical investigation on the effects of cavity on fluid flow and heat transfer characteristics in rectangular microchannel heat sink (MCHS). Int. J. Sci. Res. Revi. **07**, 650–662 (2019)
10. L. Chai, G.D. Xia, H.S. Wang,Parametric study on thermal and hydraulic characteristics of laminar flow in microchannel heat sink with fan-shaped ribs on sidewalls—half 1: Heat Transfer. Int. J. Heat Mass Trans (2016)
11. L. Chai, G.D. Xia, H.S. Wang, Parametric study on thermal and hydraulic characteristics of laminar flow in microchannel heat sink with fan-shaped ribs on sidewalls—half 2: Pressure drop. Int. J. Heat Mass Trans (2016).
12. G.D. Xia, Y. Zhai, Z. Cui,Numerical investigation of thermal enhancement in a micro heat sink with fan-shaped reentrant cavities and internal ribs. Appl. Thermal Eng. **58**, 52–60 (2013)
13. Y.L. Zhai, G.D. Xia, X.F. Liu, Y.F., Li, Heat transfer in the microchannels with fan-shaped reentrant cavities and different ribs based on field synergy principle and entropy generation analysis. Int. J. Heat Mass Trans. **68**, 224–233.
14. Y.L. Zhai, G.D. Xia, X.F. Liu, Y.F. Li, Exergy analysis and performance evaluation of flow and heat transfer in different micro heat sinks with complex structure. Int. J. Heat Mass Transf. **84**, 293–303 (2015)
15. G. Wang, D. Niu, F. Xie, Y. Wang, X. Zhao, G. Ding, Experimental and numerical investigation of a microchannel heat sink (MCHS) with micro-scale ribs and grooves for chip cooling Appl. Thermal Eng. Accept (2015)
16. Y.F. Li, G.D. Xia, D.D. Ma, J.T. Jia, J. Wang,Characteristics of laminar flow and heat transfer in microchannel heat sink with triangular cavities and rectangular ribs. Int. J. Heat Mass Trans. **98**, 17–28 (2016)
17. X.J. Wei, Y.K. Joshi, P.M. Ligrani,Numerical simulation of laminar flow and heat transfer inside a microchannel with one dimpled surface. J. Electron. Pack. (ASME) **129**, 63–70 (2016)
18. Y. Xie, J. Lan, D. Zhang,Flow and heat transfer in microchannels with dimples and protrusions. J. Heat Trans. (ASME). **134**, 1–9 (2012)
19. P. Li, D. Zhang, Y. Xie, Heat transfer and flow analysis of $Al_2O_3$–water nanofluids in microchannel with dimple and protrusion. Int. J. Heat Mass Transf. **73**, 456–467 (2014)

20. P. Li, Y. Xie, D. Zhang, Laminar flow and forced convective heat transfer of shear-thinningpower-law fluids in dimpled and protruded microchannels. Int. J. Heat Mass Transf. **99**, 372–382 (2016)
21. M. Xu, H. Lu, L. Gong, J.C. Chai, X. Duan, Parametric numerical study of the flow and heat transfer in microchannel with dimples. Int. Commun. Heat Mass Trans. **76**, 348–357 (2016)
22. K. Vafai, L. Zhu, Analysis of two layered microchannel heat sink concept in electronic cooling. Int. J. Heat Mass Transf. **42**, 2287–2297 (1999)
23. T.C. Hung, W.M. Yan, W.P. Li, Analysis of heat transfer characteristics of double-layered microchannel heat sink. Int. J. Heat Mass Transf. **55**, 3090–3099 (2012)
24. T.C. Hung, W.M. Yan, Enhancement of thermal performance in double-layered microchannel heat sink with nanofluids. Int. J. Heat Mass Trans. **55**, 3225–3238 (2012)
25. H. Shen, G. Xie, C.C. Wang, Heat transfer and thermodynamic analysis by introducing multiple alternation structures into double-layer microchannel heat sinks. Int. J. Thermal Sci. **145** (2019)
26. J.M. Wu, J.Y. Zhao, K.J. Tseng,Parametric study on the performance of double layered microchannels heat sink Energy Convers. Manage. **80**, 550–560 (2014)
27. X. Wei, Y. Joshi, M.K. Patterson, Experimental and numerical study of a stacked microchannel heat sink for liquid cooling of microelectronic devices. J. Heat Transf. ASME **129**, 1432–1444 (2007)
28. M.L.J. Levac, H.M. Soliman, S.J. Ormiston, Three-dimensional analysis of fluid flow and heat transfer in single- and two-layered micro-channel heat sinks. Heat Mass Transf. **47**, 1375–1383 (2011)
29. H.E. Ahmed, M.I. Ahmed, I. M.E. Seder, B.H. Salman, Experimental investigation for sequential triangular double-layered microchannel heat sink with nanofluids. Int. Commun. Heat Mass Transf (2016).
30. Y.L. Zhai, G.D., Xia, X.F., Liu, J. Wang,Characteristics of entropy generation and heat transfer in double-layered micro heat sinks with complex structure. Energy Convers. Manage. **103**, 477–486 (2015)
31. Y. Zhai, Z. Li, H. Wang, J. Xu, Analysis of field synergy principle and the relationship betweens econdary flow and heat transfer in double-layered microchannels with cavities and ribs. Int. J. Heat Mass Transf. **101**, 190–197 (2016)
32. Zhai, G.D. Xia, Z. Li, H. Wang, A novel flow arrangement of staggered flow in double-layered microchannel heat sinks for microelectronic cooling. Int. Commun. Heat Mass Transf. **79**, 98–104 (2016)
33. Y. Zhai, Z. Li, H. Wang, J. Xu, Thermodynamic analysis of the effect of channel geometry on heat transfer in double-layered microchannel heat sinks. Energy Convers. Manage. **143**, 431–439 (2017)
34. K.C. Wong, F.N.A. Muezzin, Heat transfer of a parallel flow two-layered microchannel heat sink. Int. Commun. Heat and Mass Transf. (2013)
35. K.C. Wong, M.L. Ang, Thermal hydraulic performance of a double-layer microchannel heat sink with channel contraction. Int. Commun. Heat Mass Transf (2016)
36. C. Leng, X.D. Wang, T.H. Wang, W.M. Yan, Optimization of thermal resistance and bottom wall temperature uniformity for double-layered microchannel heat sink. Energy Convers. Manage. **93**, 141–150 (2015)
37. C. Leng, X.D. Wang, T.H. Wang, An improved design of double-layered microchannel heat sink with truncated top channels. Appl. Thermal Eng. **79**, 54–62 (2015)
38. C. Leng, X.D. Wang, T.H. Wang, W.M. Yan,Multi-parameter optimization of flow and heat transfer for a novel double-layered microchannel heat sink. Int. J. Heat. Mass Transf. **84**, 359–369 (2015)
39. B. Osanloo, A.M. Ahmar, A. Solati, M. Baghani,Performance enhancement of the double-layered micro-channel heatsink by use of tapered channels. Appl. Thermal Eng. **102**, 1345–1354 (2016)
40. A.M. Adham, N.M. Ghazali, R. Ahmad, Thermal and hydrodynamic analysis of microchannel heat sinks: a review. Renew. Sustain. Energy Rev. **21**, 614–622 (2013)

# Experimental Investigations of Beeswax Based Composite Phase Change Material

**Durgesh Kumar Mishra, Sumit Bhowmik, and Krishna Murari Pandey**

**Abstract** Phase change material (PCM) can be utilized for thermal energy storage, where extra or waste heat is available. Extra heat generation is dangerous for various systems like electronics equipment, Solar panel, Li-ion battery, etc. that can reduce the efficiency or damage the systems. So, with the incorporation of PCM one can manage that extra heat. PCM can also store the available solar energy which may be utilized for solar water heating systems, solar stills, solar air heater, etc. In the current study, beeswax is used as PCM and graphite used as property enhancer. The fabrications of composite PCM have been done with melt-mixing method. Scanning electronic microscope (SEM) and X-ray diffractometer (XRD) were used to obtain morphology and chemical compatibility of the composite. Differential scanning calorimeter (DSC) and thermal conductivity meter were indicated the melting temperature, melting latent heat, and thermal conductivity of the sample. SEM and XRD results revealed that the fabricated composite was only the physical combination of BW and graphite there is no extra chemical form during the combination. The DSC data indicated that melting temperature values were decreased from 61 to 58.5 °C at the same time latent heat also decreased from 160 J/g to 141 J/g. Thermal conductivity result showed that the value of conductivity increased from 0.25 to 0.76 W/m–K. All the results favored that the developed composite can be utilized for water heating units.

**Keywords** Phase change material · Beeswax · Graphite powder · Latent heat · Thermal conductivity

---

D. K. Mishra · S. Bhowmik · K. M. Pandey (✉)
Department of Mechanical Engineering, National Institute of Technology Silchar, Silchar 788010, Assam, India
e-mail: kmpandey2001@yahoo.com

D. K. Mishra
e-mail: mishradurgesh34@gmail.com

S. Bhowmik
e-mail: bhowmiksumit04@yahoo.co.in



K. M. Pandey et al. (eds.), *Recent Advances in Mechanical Engineering*, Lecture Notes in Mechanical Engineering, https://doi.org/10.1007/978-981-15-7711-6_88

# 1 Introduction

Phase change material (PCM) is playing a very crucial role in the field of thermal energy storage and management [1]. In various engineering applications such as solar panel, electronic devices, Li-ion battery, etc., the temperature rise is a problematic issue that reduces efficiency of the system or can damage the system. So, by incorporating PCM with these devices the temperature of these systems can be controlled [2–5]. The working principle of PCM is shown in Fig. 1. where it can be easily understood that PCM works in a cyclic manner. First of all, it absorbs heat and become liquefied and after that if temperature of PCM environment decreases then it releases heat and become solidified [6].

PCM is classified basically in three categories organic, inorganic, and eutectic. Organic is further classified as paraffin and non-paraffin, inorganic is categorized as metal, alloys, salt, etc. and eutectic is the mixture of organic and inorganic [7–9]. Among all of them, organic PCM has excessive attention because of suitable phase transition temperature, high heat storage capacity, congruent melting, no super cooling, low vapor pressure, no volume alteration at the time of phase change, non-toxicity, excellent thermal and chemical stability [10]. Paraffin which is a kind of organic PCM is limited resource because it is a byproduct of petroleum product. So, it is required to find another non-paraffin organic PCM for energy storage in the place of paraffin. In this regards, beeswax is coming into picture for the replacement of paraffin. But the serious drawback of organic PCM is less thermal conductivity [11–15]. In this study, the focus is on the enhancement of thermal conductivity of beeswax (BW) based PCM. For enhancement of thermal conductivity, graphite

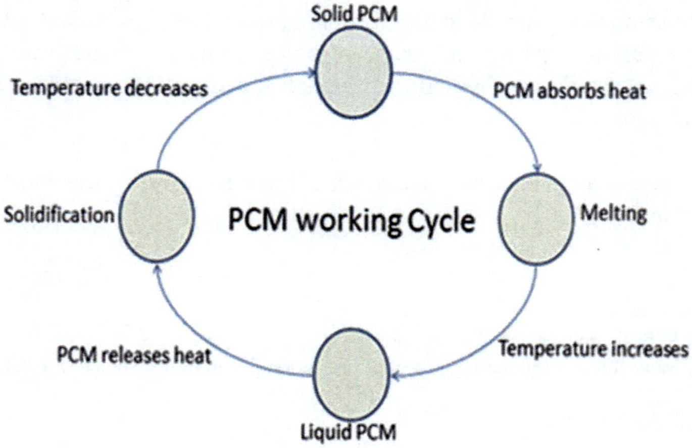


**Fig. 1** Working cycle of PCM

**Fig. 2** Physical appearance of **a** BW and **b** Graphite

has been incorporated with PCM which has excellent thermal conductivity. Scanning electron microscope (SEM), X-ray diffractometer (XRD), differential scanning calorimetry (DSC), and thermal conductivity meter were used to characterize physical, chemical, and thermal compatibility of the sample.

## 2 Materials and Methods

### *2.1 Materials*

BW ($C_{15}H_{31}COOC_{30}H_{61}$, melting point temperature 61 °C) was obtained from Hi-Tech Natural Products (India) Limited, Dilshad Garden, Delhi, which is used for energy storage media. Graphite is purchased from Intelligent Materials Pvt. Ltd. (Nanoshel), Punjab, which is used for properties enhancer. All the materials used without further purification that is shown in Fig. 2.

### *2.2 Fabrication Method*

The composite was fabricated through melt mixing method that is shown in Fig. 3. Equipment used at the time of fabrication were furnace, magnetic stirrer, and aluminum mold. First the BW is required to melt in the furnace at 100 °C. After the melting of BW, graphite powder was mixed with the help of magnetic stirrer at the temperature of 100 °C and 450 rpm for four hours. The main reason for mixing was proper dispersion of graphite powder in liquid BW. Finally, the mixed liquid was poured into the aluminum mold for the generation of samples. Aluminum mold

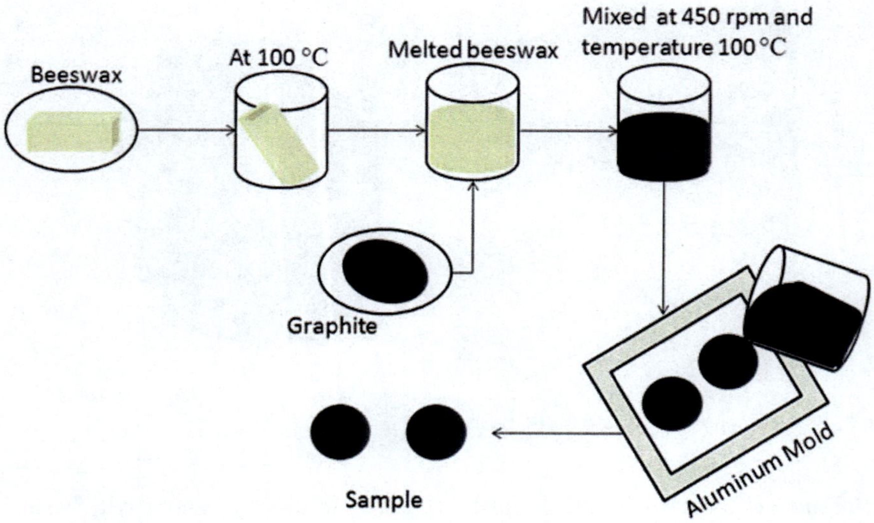


**Fig. 3** Fabrication of composite PCM

**Table 1** Sample label and composition

| S. No. | Sample label | Compositions in % |
|---|---|---|
| 1 | BW | 100% BW + 0% graphite |
| 2 | PCM-1 | 92.5% BW + 7.5% graphite |
| 3 | PCM-2 | 85% BW + 15% graphite |

was kept at room temperature only. The composition of the materials and the sample label is given in Table 1.

## *2.3 Characterization*

Surface morphology was characterized by SEM, JSM-6010LA, JEOL (IIT Kanpur). Chemical behavior of the material was examined by XRD (Xpert Pro, PANalytical XRD instrument, NIT Silchar). Thermal properties, containing melting point temperature and melting latent heat were measured through DSC (Perkin Elmer instrument, heating rate 10 °C/min, temperature range 15–150 °C, S.N. Bose institute Kolkata). Thermal conductivity of the specimen was observed by third-generation C-Therm thermal conductivity measuring device.

**Fig. 4** SEM image composite PCM-2

# 3 Results and Discussions

## 3.1 Morphology Characterization

SEM analysis was used to analyze the morphology of BW/graphite composite with 15% graphite is presented in Fig. 4. This figure exhibited the two different regions that is presenting two separate materials; BW and graphite powder. It can be observed from figure that the graphite powder is distributed into the BW, which utilized as matrix material. The proper dispersion of graphite into BW enhances the properties of BW. It can easily be identified that the graphite powder distributed into the BW as a darker area and uneven shape. The white area shows the BW as PCM.

## 3.2 Chemical Compatibility

Figure 5a and b shows the XRD graph of BW and BW/graphite powder composite. In figure (a), the peak is at 21.8° and 23.9° that belongs to BW and in figure (b) the peaks are at 21.5°, 23.6°, and 26.2° which belongs to composite. The peak of composite at 26.2° is due to graphite and at 21.5° and 23.6° due to BW. So, the XRD pattern revealed that no extra peak formation other than BW and graphite, this shows that there was only physical interaction within the material, no chemical reaction occurred. The peak intensity also demonstrates the better crystallinity of the composite.

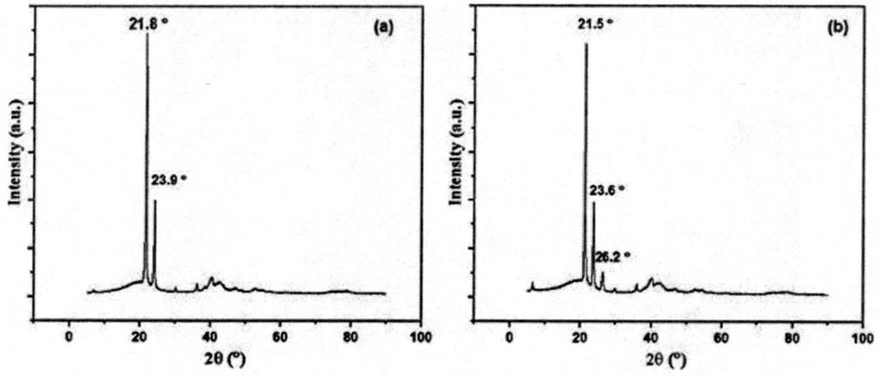


**Fig. 5** XRD Image of **a** beeswax **b** composite PCM

## *3.3 Thermal Properties of the Sample*

Thermal properties of the samples were estimated with the help of DSC. The DSC graph, which is in between heat flow and temperature for BW, PCM-1, and PCM-2 is shown in Fig. 6. It is revealed from the figure that there are two peaks in every graph. First one belongs to solid–solid phase transition and the second one belongs to solid–liquid phase transition. The solid–liquid phase transition peak value gives the melting temperature. Melting point temperature and latent heat of all the samples were obtained from DSC. The melting temperature of BW, PCM-1, and PCM-2 obtained was 61 °C, 59 °C and 58.2 °C, respectively. Latent heat value of BW, PCM-1, and PCM-2 was obtained 160 J/g, 152 J/g, and 141 J/g respectively which is exhibited in Table 2. Results showed that melting temperature and latent heat both showed reduction in the value with the incorporation of graphite powder.

## *3.4 Thermal Conductivity*

Thermal conductivity of the fabricated samples was obtained through thermal conductivity meter at room temperature and it was found that the incorporation of graphite enhanced the value of thermal conductivity. Since the available organic PCM contains less value of thermal conductivity and less thermal conductivity affects the working cycle of PCM. Therefore it is required to add properties enhancing material with PCM to enhance the working cycle efficiency. Each fabricated samples were measured three times and averaged value of obtained data is exhibited in Fig. 7. The thermal conductivity values of BW, PCM-1, and PCM-2 were obtained as 0.25 W/mK, 0.62 W/mK, and 0.72 W/mK respectively which is exhibited in Table 2.

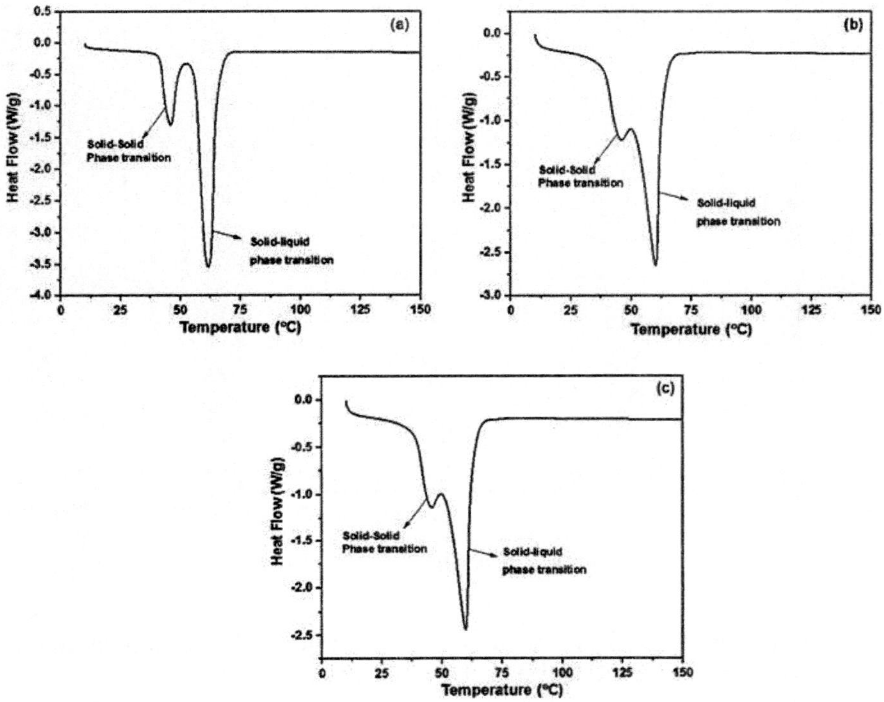


**Fig. 6** DSC curve of **a** Beeswax **b** PCM-1 (C) PCM-2

**Table 2** Thermal properties and thermal conductivity

| S. No. | Sample label | Melting temperature (°C) | Melting latent heat (J/g) | Thermal conductivity (W/mK) |
|---|---|---|---|---|
| 1 | BW | 61 | 160 | 0.25 |
| 2 | PCM-1 | 59 | 152 | 0.62 |
| 3 | PCM-2 | 58.2 | 141 | 0.71 |

# 4 Conclusions

BW is a kind of organic PCM with a good amount of heat storage capacity. However, BW shows less thermal conductivity, which hampers rate of heat storage. In the present study, graphite powder was used as properties enhancing material. For enhancing the properties of PCM, graphite mixed with PCM through melt-mixing fabrication method. After the characterization of all samples the following conclusions were obtained:

1. SEM and XRD results indicated that the composite of BW and graphite powder was physically and chemically compatible.

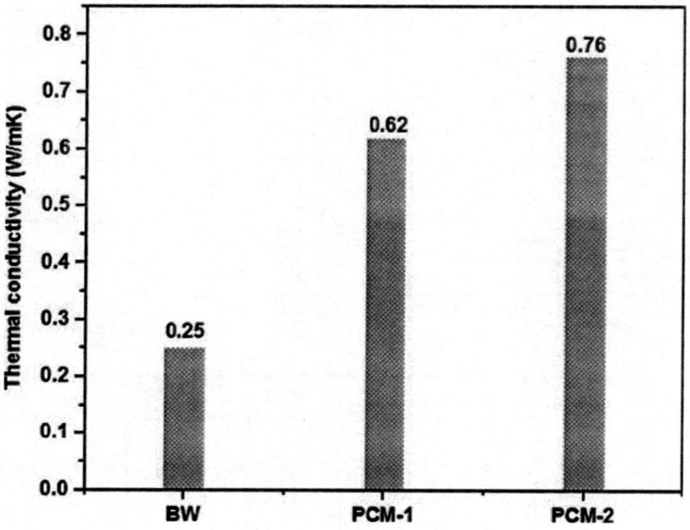


**Fig. 7** Thermal conductivity of fabricated composite

2. DSC test revealed that the phase transition temperature and latent heat of PCM decreased with the increasing value of graphite powder.
3. Thermal conductivity value of the fabricated composite affected with graphite powder addition and it increased with increment of graphite mass fraction.
4. The application of the developed composite significantly in the water heating units.

## References

1. K. Pielichowska, K. Pielichowski, Phase change materials for thermal energy storage. Prog. Mater Sci. **65**, 67–123 (2014)
2. N.I. Ibrahim, F.A. Al-Sulaiman, F.N. Ani, Solar absorption systems with integrated absorption energy storage–a review. Renew. Sustain. Energy Rev. **82**, 1602–1610 (2017)
3. M. Pan, Y. Zhong, Experimental and numerical investigation of a thermal management system for a Li-ion battery pack using cutting copper fiber sintered skeleton/paraffin composite phase change materials. Int. J. Heat Mass Transf. **126**, 531–543 (2018)
4. J. Wu, Y. Feng, C. Liu, H. Li, Heat transfer characteristics of an expanded graphite/paraffin PCM-heat exchanger used in an instantaneous heat pump water heater. Appl. Therm. Eng. **142**, 644–655 (2018)
5. X. Wang, Y. Xie, R. Day, H. Wu, Z. Hu, J. Zhu, D. Wen, Performance analysis of a novel thermal management system with composite phase change material for a lithium-ion battery pack. Energy **156**, 154–168 (2018)
6. M. Li, Z. Wu, A review of intercalation composite phase change material: preparation, structure and properties. Renew. Sustain. Energy Rev. **16**(4), 2094–2101 (2012)
7. N. Sarier, E. Onder, Organic phase change materials and their textile applications: an overview. Thermochim. Acta **540**, 7–60 (2012)

8. A.I. Fernández, C. Barreneche, M. Belusko, M. Segarra, F. Bruno, L.F. Cabeza, Considerations for the use of metal alloys as phase change materials for high temperature applications. Solar Energy Material Solar Cells **171**, 275–281 (2017)
9. S.Y. Kee, Y. Munusamy, K.S. Ong, Review of solar water heaters incorporating solid-liquid organic phase change materials as thermal Storage. Appl. Therm. Eng. **131**, 455–471 (2017)
10. S. Kahwaji, M.B. Johnson, A.C. Kheirabadi, D. Groulx, M.A. White, A comprehensive study of properties of paraffin phase change materials for solar thermal energy storage and thermal management applications. Energy **162**, 1169–1182 (2018)
11. A. Dinker, M. Agarwal, G.D. Agarwal, Preparation, characterization, and performance study of beeswax/expanded graphite composite as thermal storage material. Experimental Heat Transfer **30**(2), 139–150 (2017)
12. M. Amin, N. Putra, E.A. Kosasih, E. Prawiro, R.A. Luanto, T.M. Mahlia, Thermal properties of beeswax/graphene phase change material as energy storage for building applications. Appl. Therm. Eng. **112**, 273–280 (2017)
13. F. Cheng, R. Wen, X. Zhang, Z. Huang, Y. Huang, M. Fang, Y.G. Liu, X. Wu, X. Min, Synthesis and characterization of beeswax-tetradecanol-carbon fiber/expanded perlite form-stable composite phase change material for solar energy storage. Compos. A Appl. Sci. Manuf. **107**, 180–188 (2018)
14. X. Zhang, C. Zhu, G. Fang, Preparation and thermal properties of n-eicosane/nano-SiO2/expanded graphite composite phase-change material for thermal energy storage. Mater. Chem. Phys. **240**, 122178 (2020)
15. Y. Yu, J. Xu, G. Wang, R. Zhang, X. Peng, Preparation of paraffin/$SiO_2$ aerogel stable-stabilized phase change composites for high-humidity environment. J. Mater. Sci. **55**(4), 1511–1524 (2020)

# Investigation on Thermo-hydraulic Performance of Channel with Various Shapes of Rib Roughness: A Review

**M. K. Sahu, Kumari Ambe Verma, and K. M. Pandey**

**Abstract** To enhance the heat transfer performance, the advanced design of the equipment is essential as it can save energy, material, and cost. Among many techniques for the enhancement of the heat transfer rate, disturbance formation in the form of the roughness at the heated surface is suitable. This idea generally leads to generate turbulence. Ribs can highly be recommended for the placement on the heated surface to increase the heat transfer rate. The roughness in the form of ribs can be used in numerous applications such as solar air heater, gas turbine blade cooling, and nuclear reactors cooling, etc. Rough surface in the form of ribs create the disturbance in the laminar sub-layer. It leads to the increase in turbulence near the heated walls and this turned in to the enhancement of the heat transfer rate. However, rib roughness on the heated surface increases the flow resistance. So, this is the challenge to design a novel shape and configuration of ribs. The ribs should also be designed in the direction of enhancement of the heat transfer rate with the least pressure drop penalty. So, the objective of the article is to comprehend and explore the traditions behind the numerous design of ribs and their effect on the heat transfer rate. In recent years, a combination of more than one method to enhance the heat transfer are used simultaneously to achieve required heat transfer enhancement. The quest to enhance the heat transfer with a lesser drop in pressure by modification of heated surface is still a topic of research. So, there can be chances to enhance the heat transfer rate by using a novel design of ribs roughness. Lower heat transfer has been seen between the wall and the air because of the lower air heat transfer coefficient. Therefore, investigation on the heat transfer enhancement technique is still a research area when working fluid is air. It is also noticed that researchers have

M. K. Sahu · K. A. Verma · K. M. Pandey (✉)
Department of Mechanical Engineering, National Institute of Technology Silchar, Silchar, Assam, India
e-mail: kmpandey2001@yahoo.com

M. K. Sahu
e-mail: manojsahu123@gmail.com

K. A. Verma
e-mail: ambe.verma@gmail.com



K. M. Pandey et al. (eds.), *Recent Advances in Mechanical Engineering*, Lecture Notes in Mechanical Engineering, https://doi.org/10.1007/978-981-15-7711-6_89

mainly used single shaped ribs, but combination of different shapes of ribs is not seen in the literature.

**Keywords** Rib roughness · Heat transfer · Turbulators · Surface modifications

## 1 Introduction

The process of heat transfer is the most basic process associated with many devices. It is accessed that about 80% of the use of energy is associated with one or other kinds of heat transfer processes [1]. Therefore, it is important to investigate different techniques to boost the heat transfer rate and efficiency of heat transfer equipment. In recent years, the world is primarily focusing on enhancing the performance of the device. Improving the performance of a heat transfer device is important as it can save energy, material, and cost. Numerous turbulators are utilized to enhance the thermal performance of channels such as ribs, dimple, vortex generator, groove, etc. This turbulator rise the turbulence near the wall and this will turn to enhance the heat transfer rate. Rib in the form of roughness is used in different industrial usage such as air heater (Solar), an internal surface of a gas turbine blade, gas-cooled reactors, etc. Rib turbulators are placed on the wall to augment the heat transfer rate of equipment by enhancing turbulence near the region of the wall. The air (or the gas phase in general) has very low heat transfer coefficient. Surface modifications such as rib roughness create the disturbance in the viscous sub-layer region. In spite of several investigations on heat transfer performance; turbulence and turbulent boundary layer in a roughened surface and the in-depth physics of these types of flow problems are still a subject of research. General measurement methods have shown to be incorrect for the roughened surface primarily due to high turbulence close to the roughened surface. Therefore, computational investigations are generally used nowadays for the rib roughened surface. For the last few decades, rough surfaces have been made on the heated surface to augment the convection heat transfer by the increase of turbulence. The disadvantage of the roughness is that it leads to drag formation. Therefore, considerable effort has been done for the optimization of roughness configuration.

## 2 Enhancement Methods

Heat transfer enhancement is of main significance in the design of heat exchangers. Several enhancement methods have been invented in the last few decades and these are applied to different industrial applications. Different heat transfer augmentation methods are employed to increase the thermal performance of heat exchanger. These methods are mainly divided into three categories as given below [2].

### 2.1 Active Methods

In active methods, external power is required for enhancing the heat transfer rate. Active methods are complex from the application and design point of view.

### 2.2 Passive Methods

In passive methods, no external power is required for augmenting the heat transfer rate. Generally, the heat transfer surface is modified or some inserts are placed in the flow channel.

### 2.3 Compound Methods

In this method, two or more of these (active and passive) techniques are applied simultaneously to produce a heat transfer enhancement that is greater than the individual techniques used separately.

The objective of the article is to comprehend and explore the traditions behind the numerous design of ribs and their effect on the heat transfer rate. As the design of the ribs is still a major consideration to improve the performance. The major influence is commonly seen in thermal performance and fluid friction behaviour of the channel. Several investigations are reported in the below literature on the thermal performance of the channel by using several shapes of ribs.

## 3 Literature Review

### 3.1 V-Shaped Ribs

Jia et al. [4] explored numerically thermal and fluid flow features of square channel having V-shaped ribs with Re range of 4000–32,000. They suggested for turbine blade cooling to use V-shaped ribs pointing downstream to get maximum heat transfer. Ruck and Arbeiter [7] carried out numerical investigation for thermal behaviour of channel roughened using various dimensions of V-shaped and transverse ribs. They have taken helium gas as working fluid and Reynolds number considered is $1.05 \times 10^5$. Thakur et al. [18] performed both computational as well as experimental studies for thermal behaviour of channel mounted with hyperbolic ribs with different arrangements as inclined, V and W shaped. The inclination angle of ribs as 30°, 45°, 60°, and 90°. They found that V-shaped rib with 60° inclination at rib height 1 mm and pitch 10 mm at Re of 6000 has highest thermo-hydraulic performance because of enhanced mixing and secondary flow generation. Zheng et al. [29] studied numerically thermal and flow friction behaviour of channel mounted with V-shaped slit ribs. They investigated the influence of different shapes of slit such as one rectangular slit and four shapes of V-shaped slit on thermal performance. They noticed that broken slit ribbed channel has improved heat transfer performance due to higher turbulence intensity. Generally, hot spot and recirculation flow occur behind the solid ribs. However, in slit ribs, generation of secondary flow impinges directly to the recirculation zone behind the ribs and in turn decrease the recirculation flow, enhances the Nusselt number. Further, thermal performance index of broken slit is higher due to less pressure drop and increase in Nusselt number (Nu). Jin et al. [32] analyzed numerically the effect of multiple V-shaped ribs with staggered arrangement on thermal and flow behaviour of channel. It is observed that staggered arranged multiple V-shape ribs have higher heat transfer performance than equivalent inline arrangement.

### 3.2 Transverse Ribs

Singh et al. [6] analyzed computationally the thermal and fluid friction behaviour of channel roughened using transverse ribs of uneven cross-section, i.e. saw tooth and unchanging cross-section, i.e. trapezoidal, circular and square. They found Nusselt number (Nu) of uneven cross-section saw-tooth is higher than that of unchanging cross-section trapezoidal, circular and square ribs for Re No. beyond 7000 because of reduced low heat transfer area. Similarly, friction factor ($f$) of uneven cross-section saw-tooth is less than that of unchanging cross-section circular, square, and trapezoidal ribs. Chaube et al. [10] analyzed computationally the thermal and flow behaviour of channel roughened using different shapes of ribs. They observed that 2D analysis results are near to experimental results in comparison to 3D analysis because the influence of secondary flow in ribs 90° angle to flow is negligible. They suggested 2D analysis can be employed for transverse rib roughened channel with less consumption of memory and computational time. It is noticed that SST k-w turbulence model predicts satisfactorily with experimental data of heat transfer and flow performance. They noticed that the best heat transfer performance is for chamfered ribs and thermal performance factor is highest for rectangular rib of dimension $3 \times 5$ mm. Yadav and Bhagoria [14] implemented computational analysis for thermo-hydraulic behaviour of solar channel roughened with transverse square rib and found that performance is mainly influenced by the relative roughness height. They estimated the highest thermal performance factor value is 1.8. Yadav and Bhagoria [15] accomplished CFD investigation for thermal behaviour of solar channel roughened with transverse square-sectioned rib and found that performance is mainly influenced by the relative roughness height. The highest value of thermal performance factor is found to be 1.88 at e/D of 0.042 and p/e of 10.71. Singh et al. [31] performed both experimental and computational studies for thermal–hydraulic performance of rib roughened solar channel. They studied two different types of rib roughness such as square shape wave pattern and transverse ribs (multiple broken). Nusselt number for both rib shapes enhances with rise in Reynolds number. This situation is because of decrease in reattachment length and in turn decreases in low heat transfer zone behind the rib. Heat transfer performance of transverse ribs (multiple broken) is higher than square shape ribs with wave pattern because of flow through gaps and increase in local heat transfer surface. Further, thermal enhancement factor of transverse ribs (multiple broken) is higher than square shape ribs with wave pattern.

### *3.3 Circular, Conical and Curved Ribs*

Alam and Kim [8] investigated computationally the thermal and hydraulic behaviour of channel with conical protrusion rib on the heated surface. They observed higher heat transfer enhancement for conical protrusion rib as compared to spherical rib. They estimated 69.8% maximum thermal efficiency and maximum efficiency enhancement factor of 1.346 at relative roughness pitch 10. Yang et al. [3] performed computational investigation for thermal and flow friction behaviour of channel roughened using angled ribs. The author studied thermo-hydraulic performance by analyzing the ratio of rib pitch, rib height and rib angle. They observed the effect of rib height is more than other parameters on average Nusselt number. Experimental and numerical investigation has been done by Chung et al. [5] to investigate the thermal and fluid flow behaviour of channel roughened using semi-circular ribs. They have taken two turbulence models such as SST k-ω and $v^2$-f for actual flow physics near the wall. They observed that turbulence kinetic energy is directly associated with recirculation flow diffusion and a result of $v^2$-f turbulence model is conformity with experimental results. The thermo-hydraulic performance value of semi-circular ribs with pitch ratio of 13 has highest (1.35) from different pitch ratio considered. Xie et al. [11] conducted computational investigation for thermal and fluid flow behaviour of channel roughened employing three types of ribs. They found that curved ribs at depth to width ratio of 0.4 has maximum thermo-hydraulic performance from different ribs pattern considered. Singh et al. [28] performed both experimental and numerical analysis for thermal and fluid flow behaviour of channel roughened with criss-cross rib roughness placed in inline and staggered manner. It is observed that both inline and staggered criss-cross configuration has same level of enhancement of heat transfer from 2.7 to 3.1. The friction factor of staggered criss-cross rib configuration is higher than that of inline pattern. Further, thermal enhancement factor of both inline and staggered configuration are similar and value is from 1.2 to 1.5 at Reynolds number from 30,000 to 60,000.

### *3.4 Wavy Ribs*

Wang et al. [9] analyzed numerically thermal and flow friction characteristics of channel roughened using wavy ribs. The parameters of wavy ribs considered are height of rib, radius of rib round and rib angle. They found that rib height is the main parameter which decide the enhancement of heat transfer in ribbed surface. They also observed that wavy ribbed surface has 10–35% higher heat transfer performance than 45° V-shaped ribs is because of area of wavy rib surface is double that of 45° V-shaped ribs. Wang et al. [13] explored computationally the thermal behaviour of channel roughened using wavy ribs. They examined the influence of various geometric factors of ribs on thermal behaviour at Re range of 10,000–40,000. They observed that radius, angle and height of the ribs are the key factors that influence the thermal behaviour of channel whereas rib thickness is not so important rib parameter that influences the thermal behaviour of channel. Promthaisong and Eiamsa-ard [25] explored numerically the influence of wavy triangular ribs on thermal performance and flow behaviour of channel. It is noticed that wavy triangular ribs augment the Nusselt number due to generation of vortex flow in wavy triangular ribs that disrupt the thermal boundary layer. Nusselt number augments with increase in Re and blockage ratio, and reduction in pitch ratio, which increases the vortex flow, and thus boundary layer thickness decreases. They obtained highest thermal performance factor of 2.62. Singh and Singh [21] performed CFD investigation to find the thermo-hydraulic behaviour of solar channel roughed using non-uniform cross-section square-shape wave-structure rib. They noticed that Nu increases with increase in Re because of decrease of reattachment length with increase in Re. Further, it is noticed that highest increase in Nu and f with respect to smooth channel is 2.14 and 3.55 times, respectively. It is noticed that Nu and thermal enhancement factor of square wave ribbed channel (non-uniform cross-section) is higher than that of square wave ribbed channel (uniform cross-section). The highest thermal enhancement factor value is 1.43 for square wave ribbed channel (non-uniform cross-section).

## 3.5 *Longitudinal Ribs*

Betti et al. [24] performed RANS based numerical analysis for heat transfer performance of thrust chamber of rocket engine by placing longitudinal ribs on wall. They observed that Nusselt number is directly related to surface area. It is also noticed that with increase in rib height, efficiency decreases because of thermal stratification in the region between two ribs. Zhang et al. [33] performed computationally the thermal performance of channel placed with longitudinal intersecting ribs. The influence of number of longitudinal ribs on heat transfer rate is studied. They noticed that section of high heat transfer is just downstream of point of intersection of longitudinal ribs and angled ribs due to the creation of secondary flow. As the number of longitudinal intersecting ribs increases, heat transfer rate in increased due to more regions is covered with secondary flow. However, pressure drop is also increased because of increase in number of longitudinal intersecting ribs. Thus, it is indicated that heat transfer enhancement is maximum for two longitudinal intersecting ribs.

## 3.6 Different Ribs Shape

Du et al. [12] analyzed computationally the thermal–hydraulic behaviour of tube roughened using sinusoidal ribs with water as working fluid at Re from 400 to 1800. They observed that sinusoidal design based ribs enhance mixing and disturb the thermal boundary layer resulting in enhancement of thermal performance. They observed the maximum overall thermal performance of 3.64 with the use of sinusoidal ribs. Kumar et al. [19] performed computational analysis on triangular duct at different geometrical parameters of ribs. They found the highest value of Nu and f is 2.88 and 3.52, respectively, at e/D of 0.043. Kamali and Binesh [26] performed simulation to study the influence of various shapes of ribs on thermal and flow behaviour of square channel. Various shapes of rib considered are triangular, square, and trapezoidal. They noticed that heat transfer distribution between ribs is strongly influenced by rib shapes. Further, it is observed that trapezoidal rib offers the highest Nusselt number and pressure drop from the considered rib shapes. Shukla and Dewan [27] performed computational analysis for flow and thermal behaviour of square channel roughened using different ribs. Different ribs considered are transverse ribs and V-shaped ribs (60° angle broken thin and thick ribs). In transverse ribs, the recirculation is formed immediate after the ribs, which worsen the heat transfer performance. It is noticed that 60° V-shaped ribs (broken) has higher Nu and overall performance than continuous transverse ribs. Further, it is noticed that thin ribs have higher performance than the thick one. Xie et al. [34] analyzed computationally the thermal behaviour of rib-roughened channel with different types of deflector. The deflector is used to enhance the flow behaviour in the wake region and guide the fluid between ribs. It is observed that the deflector trips the boundary layer and enhance mixing. Further, it is found that cylindrical deflector has the highest TEF at higher Reynolds number (Re) from the considered shapes of deflectors. Lan et al. [35] performed numerical studies to find the roughness pattern in channel that has highest thermal performance. The roughness geometries considered are rib-dimple and rib-protrusion. It is noticed that rib-dimple channel offer negligible augment in heat transfer rate. Further, it is noticed that rib-protrusion channel has considerable heat transfer augmentation with small increase in friction factor.

## 3.7 Triangular, Elliptical and Rectangular Ribs

Yadav and Bhagoria [16] conducted computational investigation on performance of channel roughened using equilateral triangular ribs. They estimated the highest thermal performance factor as 2.11 for e/D of 0.042 and p/e of 7.14 at Re of 15,000. Gawande et al. [17] analyzed both experimentally and computationally the performance of channel roughened using L-shaped (reverse) roughness of rib on the heated surface. They estimated the highest TEF of 1.90 for e/D of 0.042 and p/e of 7.14 at Re of 15,000. Kumar et al. [20] performed two-dimensional CFD investigation for thermal performance and flow friction of channel having elliptical rib roughened on the heated surface at different geometrical parameters of ribs. They observed that vortex generation and mixing enhance the heat transfer performance. It is noticed that highest heat transfer and friction factor estimated at relative roughness width of 0.5, e/D of 0.045, and p/e of 6. Zheng et al. [30] performed computationally thermal behaviour of channel roughened using converging and diverging slit ribs of five different geometrical shapes such as rectangular slit and trapezoidal slit of different dimensions. It is observed that trapezoidal slit ribs of smallest angle have greater heat transfer performance due to advanced turbulence intensity. Further, thermal performance index of trapezoidal slit ribbed channel having smallest angle has highest value. Moon et al. [22] performed computational analysis for thermal and friction factor behaviour of channel roughened with sixteen shapes of rib-roughness. It is noticed that the new boot-shaped ribbed channel has the highest Nusselt number with average value of pressure drop. They observed that size of recirculating zone behind the rib roughness determines the heat transfer performance of channel. Yang and Chen [23] performed two-dimensional computational analysis and optimization method for selection of optimized geometric design of rib that has the highest thermal performance. They used response surface methodology and genetic algorithm for optimization of thermal performance. It is found that for optimized configuration of ribbed channel, thermal performance factor for inline and staggered arrangement are 1.1–1.5 and 2.68 respectively.

## 4 Summary

The following inferences are drawn from the literature review based on various research works carried out in the area of rib-roughened channel by different researchers:

- In recent years, a combination of more than one method to enhance the heat transfer are used simultaneously to achieve required heat transfer enhancement.
- The quest to enhance the heat transfer with a lesser drop in pressure by modification of heated surface is still a topic of research. So, there can be chances to enhance the heat transfer rate by using a novel design of ribs roughness.
- Lower heat transfer has been seen between the wall and the air because of the lower air heat transfer coefficient. Therefore, investigation on the heat transfer enhancement technique is still a research area when working fluid is air.
- It is also noticed that researchers have mainly used single shaped ribs, but combination of different shapes of ribs is not seen in any of the articles.

Roughness in the form of rib on the surface is one of the techniques of heat transfer augmentation. Generally, in a smooth surface flow; the boundary layer develops and grows along with the flow of fluid. Due to boundary layer growth, fluid particles near the heat transfer surface are not mixed with core bulk fluid particles and the heat transfer rate is low. Therefore, the combination of different shapes of rib roughness can be made on a heat transfer surface that can utilize to create the disturbance in the boundary layer. Also, reattachment of the free shear layer occurs due to rib roughness which enhance the heat transfer rate.

**Acknowledgements** Authors are grateful to the TEQIP III and Department of Mechanical Engineering, National Institute of Technology Silchar, Assam, India for the facilities.

## References

1. Z.-X. Li, Z.-Y. Guo, Optimization principles for heat convection. in *Advances in Transport Phenomena*, (Springer Berlin, Heidelberg 2011), 1–91
2. R.L. Webb, *Principles of Enhanced Heat Transfer* (Wiley New York, 1994)
3. Y.-T. Yang, H.-W. Tang, C.-J. Wong, Numerical simulation and optimization of turbulent fluids in a three-dimensional angled ribbed channel. Numer. Heat Trans. Part a Appl. **70**(5), 532–545 (2016)
4. R. Jia, S. Bengt, M. Faghri,Computational analysis of heat transfer enhancement in square ducts with v-shaped ribs: Turbine blade cooling. J. Heat Trans. **127**(4), 425–433 (2005)
5. H.S. Chung, G.H. Lee, M.J. Nine, K. Bae, H.M. Jeong, Study on the thermal and flow characteristics on the periodically arranged semi-circular ribs in a rectangular channel. Experimen. Heat Trans. **27**(1), 56–71 (2014)
6. S. Singh, B. Singh, V.S. Hans, R.S. Gill.CFD (computational fluid dynamics) investigation on Nusselt number and friction factor of solar air heater duct roughened with non-uniform cross-section transverse rib. Energy **84,** 509–517 (2015)
7. S. Ruck, F. Arbeiter, Detached eddy simulation of turbulent flow and heat transfer in cooling channels roughened by variously shaped ribs on one wall. Int. J. Heat Mass Transf. **118**, 388–401 (2018)
8. T. Alam, M.-H. Kim, Heat transfer enhancement in solar air heater duct with conical protrusion roughness ribs. Appl. Therm. Eng. **126**, 458–469 (2017)
9. L. Wang, S. Wang, W. Liu, F. Wen, X. Zhou, Z. Wang, Numerical predictions on heat transfer and flow characteristics in a straight channel with different geometric parameters wavy ribs. Appl. Therm. Eng. **140**, 245–265 (2018)
10. C. Alok, P.K. Sahoo, S.C. Solanki, Analysis of heat transfer augmentation and flow characteristics due to rib roughness over absorber plate of a solar air heater. Renew. Energy **31**(3), 317–331 (2006)
11. G. Xie, X. Liu, H. Yan, J. Qin, Turbulent flow characteristics and heat transfer enhancement in a square channel with various crescent ribs on one wall. Int. J. Heat Mass Transf. **115**, 283–295 (2017)
12. J. Du, Y. Hong, S.-M. Huang, W.-B. Ye, S. Wang, Laminar thermal and fluid flow characteristics in tubes with sinusoidal ribs. Int. J. Heat Mass Transf. **120**, 635–651 (2018)
13. L. Wang, S. Wang, F. Wen, X. Zhou, Z. Wang, Effects of continuous wavy ribs on heat transfer and cooling air flow in a square single-pass channel of turbine blade. Int. J. Heat Mass Transf. **121**, 514–533 (2018)
14. Y. Anil Singh, J.L. Bhagoria, Modeling and simulation of turbulent flows through a solar air heater having square-sectioned transverse rib roughness on the absorber plate. Sci. World J (2013)
15. Y. Anil Singh, J.L. Bhagoria, A numerical investigation of square sectioned transverse rib roughened solar air heater. Int. J. Thermal Sci. **79**, 111–131 (2014)
16. Y. Anil Singh, J.L. Bhagoria. A CFD based thermo-hydraulic performance analysis of an artificially roughened solar air heater having equilateral triangular sectioned rib roughness on the absorber plate. Int. J. Heat Mass Trans. **70**, 1016–1039 (2014)

17. G.B. Vipin, A.S. Dhoble, D.B. Zodpe, C. Sunil Experimental and CFD investigation of convection heat transfer in solar air heater with reverse L-shaped ribs. Solar Energy **131**, 275–295 (2016)
18. T. Deep Singh, M.K. Khan, and Manabendra Pathak. "Solar air heater with hyperbolic ribs: 3D simulation with experimental validation. Renewable Energy **113**, 357–368 (2017)
19. R. Kumar, V. Goel, A. Kumar, Investigation of heat transfer augmentation and friction factor in triangular duct solar air heater due to forward facing chamfered rectangular ribs: a CFD based analysis. Renewable Energy **115**, 824–835 (2018)
20. R. Kumar, A. Kumar, A parametric study of the 2D model of solar air heater with elliptical rib roughness using CFD. J. Mech. Sci. Technol. **31**(2), 959–964 (2017)
21. I. Singh, S. Singh, CFD analysis of solar air heater duct having square wave profiled transverse ribs as roughness elements. Sol. Energy **162**, 442–453 (2018)
22. M.-A. Moon, M.-J. Park, K.-Y. Kim, Evaluation of heat transfer performances of various rib shapes. Int. J. Heat Mass Transf. **71**, 275–284 (2014)
23. Y.-T. Yang, P.-J. Chen, Numerical optimization of turbulent flow and heat transfer characteristics in a ribbed channel. Heat Transfer Eng. **36**(3), 290–302 (2015)
24. B. Betti, F. Nasuti, M. Emanuele,Numerical evaluation of heat transfer enhancement in rocket thrust chambers by wall ribs. Numeric. Heat Trans. Part A Appl. **66**(5), 488–508 (2014)
25. P. Promthaisong, S. Eiamsa-ard, Fully developed periodic and thermal performance evaluation of a solar air heater channel with wavy-triangular ribs placed on an absorber plate. Int. J. Therm. Sci. **140**, 413–428 (2019)
26. R. Kamali, A.R. Binesh,The importance of rib shape effects on the local heat transfer and flow friction characteristics of square ducts with ribbed internal surfaces. Int. Commun. Heat and Mass Trans. **35**(8), 1032–1040 (2008)
27. A.K. Shukla, A. Dewan, Computational study on effects of rib height and thickness on heat transfer enhancement in a rib roughened square channel. Sādhanā **41**(6), 667–678 (2016)
28. P. Singh, Y. Ji, S.V. Ekkad, Experimental and numerical investigation of heat and fluid flow in a square duct featuring criss-cross rib patterns. Appl. Therm. Eng. **128**, 415–425 (2018)
29. D. Zheng, X. Wang, Qi. Yuan, Numerical investigation on the flow and heat transfer characteristics in a rectangular channel with V-shaped slit ribs. Infrared Phys. Technol. **101**, 56–67 (2019)
30. D. Zheng, X. Wang, Qi. Yuan, The flow and heat transfer characteristics in a rectangular channel with convergent and divergent slit ribs. Int. J. Heat Mass Transf. **141**, 464–475 (2019)
31. I. Singh, S. Vardhan, S. Singh, A. Singh, Experimental and CFD analysis of solar air heater duct roughened with multiple broken transverse ribs: A comparative study. Sol. Energy **188**, 519–532 (2019)
32. D. Jin, J. Zuo, S. Quan, Xu. Shiming, H. Gao, Thermohydraulic performance of solar air heater with staggered multiple V-shaped ribs on the absorber plate. Energy **127**, 68–77 (2017)
33. C. Zhang, Z. Wang, J. Kang, Flow and heat transfer in a high-aspect-ratio rib-roughed cooling channel with longitudinal intersecting ribs. J. Appl. Mech. Tech. Phys. **59**(4), 679–686 (2018)
34. G. Xie, S. Zheng, B. Sundén, Heat transfer and flow characteristics in Rib-/deflector-roughened cooling channels with various configuration parameters. Numerical Heat Transfer, Part a: Applications **67**(2), 140–169 (2015)
35. J. Lan, Y. Xie, D. Zhang, Heat transfer enhancement in a rectangular channel with the combination of ribs, dimples and protrusions. in ASME, 2011 Turbo Expo: Turbine Technical Conference and Exposition American Society of Mechanical Engineers Digital Collection (2011), 1447–1455

# Interfacial Instabilities in Rotating Hele-Shaw Cell: A Review

**Akhileshwar Singh, Krishna Murari Pandey, and Yogesh Singh**

**Abstract** We have reported the interfacial instabilities or viscous fingering instabilities in rotating radial Hele-Shaw cell (HSC). The cell is rotated with constant or time-dependent angular velocity about an axis perpendicular to the plane of the flow. Inside the cell, a low viscous fluid displaces a high viscous fluid. Due to this unstable displacement inside the cell fingers like pattern appears at the fluid–fluid interface. It is called viscous fingering (VF). Interfacial instabilities are usually undesirable in engineering applications (dendritic growth decreases the life of rechargeable lithium batteries, VF decreases sweep area due to this reason efficiency of enhanced oil recovery decreases) but it is also beneficial for some practical applications (improving the $CO_2$ mixing in saline aquifers for carbon sequestration, enhancing the mixing efficiency in microfluidics devices). Therefore active control (suppress or promote) of the interfacial instabilities is an important aspect. Many researchers have been done work on active control of interfacial instabilities in radial rotating HSC. A few of them we have taken in this paper. Researchers have been found fluid–fluid interfaces destabilized in radial rotation HSC mainly by centrifugal forces (density differences) and viscosity contrast. The interfacial instability in the radial rotating Hele-Shaw cell depends on rotational velocity ($\grave{\Omega}$), gap between plates ($b$), wettability of cell, viscosity, density and surface tension. We have reported many research works related to active control of interfacial instability by the above parameters in radial rotating radial HSC.

**Keywords** Interfacial instability · Rotating hele-shaw cell · Viscosity contrast · Centrifugal force · Wettability

A. Singh · K. M. Pandey (✉) · Y. Singh
Mechanical Engineering, National Institute of Technology Silchar, Silchar, Assam 788010, India
e-mail: kmpandey2001@yahoo.com

A. Singh
e-mail: akhileshwarsinghs@gmail.com

Y. Singh
e-mail: yogeshsingh15@gmail.com



K. M. Pandey et al. (eds.), *Recent Advances in Mechanical Engineering*, Lecture Notes in Mechanical Engineering, https://doi.org/10.1007/978-981-15-7711-6_90

## 1 Introduction

Interfacial instabilities or viscous fingering instabilities are a natural phenomenon. Interfacial instabilities are applied following industrial applications: geologic carbon dioxide sequestration [1], enhance oil recovery (EOR) from underground water [2], water infiltration into the soil, chromatographic separation, mixing of fluids at low Reynolds numbers in microfluidic devices, dendritic formation in metals solidification and electrodeposition [3], bacterial colonies pattern formation [4]. Interfacial instabilities have merit and demerit. The demerit of interfacial instability is that it is decreased the life of rechargeable lithium batteries due to dendritic formation and decreased the efficiency of enhanced oil recovery due to fingers or channels like pattern formation. The merit of interfacial instability is that it is increased the $CO_2$ mixing in saline aquifers for carbon sequestration and enhanced the mixing efficiency in microfluidics devices. Therefore active control (suppress or promote) of the interfacial instabilities is fascinating in industrial applications.

The Saffman-Taylor interfacial instability [5] or viscous fingering instability develops when a high viscous fluid is driven by a low viscous fluid in the Hele-Shaw cell (HSC). The fluid–fluid interface in the cell can be destabilized with the help of pressure force (viscosity contrast) or body force (density contrast). In the stationary horizontal Hele-Shaw cell, gravity effect (body force) is neglected at the interface. Therefore fluid–fluid interface can be destabilized by viscosity contrast only. Most of the research works have been done on rectilinear HSC and radial HSC. In the rectilinear HSC, the low viscous fluid is injected against the high viscous fluid along the direction of flow (parallel to length of plate) and the radial HSC (Fig. 1.) low viscous fluid is injected against high viscous fluid along the perpendicular of flow direction (perpendicular to plate) through a central hole which is located on the upper plate or lower plate. In both cells, the major cause of interfacial instability is viscosity contrast. When a low viscous fluid displaces a high viscous fluid in the cell then unstable displacement occurs, therefore, the fingers-like pattern appears on

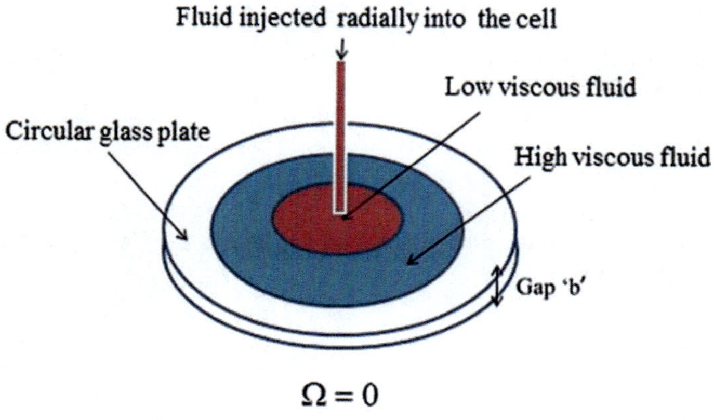


**Fig.1** Schematic diagram of stationary ($\dot{\Omega} = 0$) radial Hele-Shaw cell

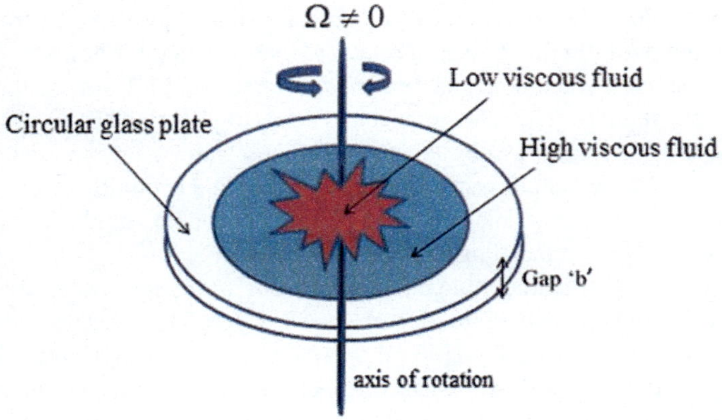


**Fig. 2** Schematic diagram of a rotating ($\grave{\Omega} \neq 0$) radial Hele-Shaw cell with fingering pattern

the fluid–fluid interface this phenomenon is called viscous fingering. Schwartz [6] has done experimental work of radial rotational HSC first time in 1989. After that many experimental, numerical and computation works have been done on interfacial instability of miscible [8] or immiscible fluids in radial rotation HSC. When a radial HSC is rotated about an axis perpendicular and symmetrical to glass plates of the cell is called radial rotating HSC (Fig. 2). In radial rotating HSC, the interfacial instabilities are driven by centrifugal forces (density differences) and pressure forces (viscosity contrast) but in stationary radial HSC only by pressure forces (viscosity contrast). Interfacial instability in radial rotating HSC depends on rotational velocity ($\grave{\Omega}$), gap between plates ($b$), wettability of cell [19, 20] viscosity, density and surface tension. In this paper, we have reported many research works related to active control (suppress or promote) of interfacial instability by the above parameters in radial rotating radial HSC.

## 2 A Few Novel Researches on Interfacacil Instabilities

Paiva et al. 2019 [7] performed a CFD analysis of fingering pattern formation in circular rotating HSC. ANSYS Fluent software is used in this investigation. The finite volume method has emerged for the discretization of non- linear equations. An inner phase of the cell is considered low viscous fluid and the outer phase of the cell is considered high viscous fluid. When the cell was rotating with uniform velocity the interface was destabilized by the combined effect of viscosity contrast and density contrast. Density contrast appears due to centrifugal force and this parameter plays an important role at high rotational velocity. They found, fingering pattern formation depends on viscosity contrast and dimensionless surface tension. If the angular velocity increases then Coriolis force becomes strong and this force tries to bend the fingers in the opposite direction of cell rotation.

Anjos et al. 2017 [8] investigated analytically, linear and weakly nonlinear dynamics of the fluid–fluid interface in rotating Hele-Shaw cell. They were considered immiscible fluids and time-dependent rotation instead of uniform angular speed. In most of the studies, researches vanishing effect of inertial but in this study, they have taken. This study was based on generalized Darcy's law and the second-order mode-coupling equation was accounted for the time evolution of the interfacial perturbation amplitudes. They concluded, time-dependent angular velocity and inertia also affect the interfacial instability.

Chen et al. 2011 [9] studied numerically, patterns formation on the fluid–fluid interface in rotating HSC in the presence of Coriolis force. They were accounted for the Boussinesq Hele-Shaw-Cahn–Hilliard approach for the understanding of topological singularities at a diffuse interface (pinch-off and coalescence of fingers). They have examined the effect of viscosity contrast (negative, zero, positive) on the emergence of such singularities. In the absence of Coriolis force, emergent fingering patterns calculated correctly. If Coriolis force is considered in the numerical simulation then these effects can be seen: stabilized radial growth, the existence of pattern phase drift, and finger bending.

Miranda et al. 2010 [10] examined the effect of Coriolis force on time evolution and development of cusp singularities in rotational HSC. The interfacial surface tension is considered zero. In this investigation, one of the fluids has negligible density and viscosity. The conformal mapping approach has emerged for the evolution of fluid boundaries. They found, the effect of Coriolis force is opposing the effect of centrifugal force and Coriolis force is responsible for time for cusp formation.

Gadelha et al. 2008 [11] examined the effect of Coriolis force in immiscible viscous fingering patterns formation in rotational HSC. They were accounted for vortex sheet formalism and two dimensional modified Darcy model. In this analytical study linear dispersion was considered to reveal the effects of the Coriolis force. Due to phase drift fingers travel opposite direction of the actual rotation direction of the Hele-Shaw cell. They found that radial growths of fingers are slow and fingers bend. The value of these deviations depends on the Reynolds number (Re) and the viscosity.

Abidate et al. 2007 [12] studied the impact of circular rigid boundaries and Coriolis forces on the viscous fingering instability in a rotating annular Hele-Shaw cell. This analytical study was based on linear stability and Newtonian immiscible fluids were accounted. They revealed how interfacial instability affected by Coriolis force and curvature parameters. They found that the Coriolis force slows down the maximum growth of interfacial instability and maximum instability growth rate increases by viscosity contrast.

Lacalle et al. 2006 [13] investigated the effect of wettability on the interfacial dynamic in rotating HSC. They were performed two series of experiments in a rotating HSC. In the first series, the cell was accounted dry and a circular blob of silicone oil was invaded through a hole at the center of the cell. The silicon blob was surrounded by air. When the cell was rotating without further invasion then the circular interface of oil was destabilized. In the second series, glass plates were pre-wetted. They found that fingering patterns develop in dry cell and pre-wetted are unlike in nature and if the wetting fluid is injected in the dry cell then interfacial pressure drop scales belong to $\mathbf{Ca^{2/3}}$ (capillary number).

Chen et al. 2006 [14] purposed a study on interfacial instability of miscible fluids in a rotating HSC. In this numerical simulation viscosity contrasts ($A$) changes. In this paper, they have shown the effect of a wide range of viscosity contrast ($-1 < A > 1$) and Korteweg stresses on the shape of the fingers pattern. From the results, miscible fluid finger patterns similar to immiscible fluids finger patterns when the value of Korteweg stress is large. Korteweg stress behaves like effective surface tension for miscible fluid systems. They concluded, immiscible patterns could be generated with the proper value of Korteweg stress and Péclet number.

Miranda et al. 2005 [15] examined numerically the effect of viscosity contrast ($A$) on viscous fingers pattern formation in rotating HSC. They have taken two immiscible fluids for fingering formation. This study has been done by numerical vortex sheet formalism. They have done a series of simulations by varying magnitude as well as a sign of viscosity contrast. In one set of simulation, they assumed inner fluid less viscous than the outer fluid it means that viscosity contrast negative. They concluded, viscosity contrast strongly affects the shape and size (width and length) of fingers and position of pinch-off singularities are chanced in case of less viscous inner fluid.

Gadêlha 2004 et al. [16] purposed an analytical study on the growth of finger competition in rotating HSC. This study was based on the weakly nonlinear mode-coupling approach. They were considered the effect of normal stresses at the interfacial pressure jump. In this latter, they reported a combined effect of viscosity contrast, surface tension and gab between plates on the development of finger competition. In this analytical study, they had taken the positive and negative value of viscosity contrast and found that viscosity and density drive the interfacial instability in case of positive viscosity contrast. They also found that finger competition more influence by viscosity contrast.

Chen et al. 2002 [17] purposed numerical simulation of interfacial stability of miscible fluids in the rotating HSC. The stationary fluid droplet (no injection) is presented in the center of the cell and the cell is rotated about the vertical axis of the cell plate with contact speed. Due to density difference centrifugal force acts on the fluids interface. The interfacial stabilities can be controlled by these parameters- rotating speed, viscosity contrast, diffusion rate, injection rate. The interfacial instability in the rotating cell is driven by centrifugal force and viscous contrast. When viscosity and density of fluid droplets are more than centrifugal force destabilized interface and viscous try to stabilize the front. More strong viscous fingerings develop at higher rotating velocity and lower viscosity contrast. The injecting strength tends to stabilize the interface of high viscous fluid.

Carrillo et al. 2000 [18] examined a linear stability analysis and an experimental study of interfacial instabilities in rotating annulus HSC. They had considered the following parameters for study—rotational velocity, gap between plates ($b$), wettability, viscosity, density, and interfacial tension. They were accounted for vaseline oil and silicon oil in the experiments. The experiments were performed both dry and wet conditions of the cell. The density contrast (density differences) drives the leading interface in the presence of centrifugal force and viscosity contrast drives the trailing interface in the interfacial instability analysis. It is observed that from experimental results fluids interface in case of pre-wet cell condition is stable and others hand dry cell fluids interface is unstable.

Carrillo et al. 1999 [19] performed an experiment on the annulus type of rotating HSC. They have examined the effect of wettability on unstable radial displacement and how the wetting conditions modify the flow of fluids in the rotating cell. The experiments were performed in two cases. In the first case of an experiment, the inner surfaces of the cell were pre-wetted with oil. It was prevented oil drop formation. In the second case of the experiment inner part of the cell was dry. The wetting case ensures that the contact angle will be zero. Therefore, the effect of capillary forces on the radial velocity due to the curvature of the interfaces in the vertical direction can be neglecting and the effect of surface tension can be also neglecting in evaluation. It is seen that whenever displaced fluid flows in a pre-wet cell then the mass of fluid in the annulus is constant but in case of dry cell mass loss at the inner interface due to capillary action at the outer interface.

Carrillo et al. 1996 [20] performed an experiment on rotating HSC. They were accounted immiscible fluids (air-oil). The cell was rotating with constant angular velocity about the vertical axis and fluid was injected at the vertical axis. When the cell is rotated then centrifugal force acts on the fluid–fluid interface. This force is created interfacial instability. The centrifugal force is depending on the radius of gyration therefore the value of this force is not uniform on the entire fluid–fluid interface. They observed that interfacial instability is driven by centrifugal force and control and density difference in rotating HSC. They were formulated linear dispersion formula and calculated with the maximum growth and experimental uncertainty.

## 3 Summary

We found the following conclusions:-

The interfacial instability in the radial rotating Hele-Shaw cell is driven by centrifugal force (density difference) and viscous contrast.

The interfacial instability in the radial rotating Hele-Shaw cell depends on rotational velocity, gap between plates ($b$), wettability, viscosity, density, and surface tension.

It is found that the fluid–fluid interface becomes stable in pre-wet cell conditions and unstable in dry cell conditions.

More strong viscous fingerings develop at higher rotating velocity and lower viscosity contras.

It is found that finger competition influence more by viscosity contrast.

The Coriolis force opposes the effect of centrifugal force and Coriolis force is responsible for time for cusp formation.

The Coriolis force slows down the maximum growth of interfacial instability and maximum instability growth rate increases by viscosity contrast.

The viscosity contrast strongly affects the shape and size (width and length) of fingers and position of pinch-off singularities are chanced in case of less viscous inner fluid.
The leading interface becomes unstable due to density contrast and trailing interface unstable by viscosity contrast.

## References

1. M.R. Soltanian et al., Dissolution trapping of carbon dioxide in heterogeneous aquifers. Environ. Sci. Technol. **51**, 7732–7741 (2017)
2. S.B. Gorell, G. Homsy, A theory of the optimal policy of oil recovery by secondary displacement processes. SIAM J. Appl. Math. **43**, 79–98 (1983)
3. R. Brady, R. BallvFractal, growth of copper electrodeposits. Nature **309**, 225 (1984)
4. M. Matsushita, H. Fujikawa, Diffusion-limited growth in bacterial colony formation. Phys. A **168**, 498–506 (1990)
5. P.G. Saffman, G.I. Taylor, The penetration of a fluid into a porous medium or Hele-Shaw cell containing a more viscous liquid. in *Proceedings of the Royal Society of London. Series A. Mathematical and Physical Sciences* (1958), 245(1242), pp.312–329
6. L.W. Schwartz, Instability and fingering in a rotating Hele-Shaw cell or porous medium. Phys. Fluids a **1**(2), 167–169 (1989)
7. A.S.S. Paiva, S.H.A. Lira, R.F.S. Andrade, Non-linear effects in a closed rotating radial Hele-Shaw cell. AIP Adv. **9**(2), 025121 (2019)
8. P.H. Anjos, V.M. Alvarez, E.O. Dias, J.A. Miranda, Rotating Hele-Shaw cell with a time-dependent angular velocity. Phys. Rev. Fluids **2**(12), 124003 (2017)
9. C.Y. Chen, Y.S. Huang, J.A. Miranda, Diffuse-interface approach to rotating Hele-Shaw flows. Phys. Rev. E **84**(4), 046302 (2011)
10. J.A. Miranda, H. Gadêlha, A.T. Dorsey, Coriolis effects on rotating Hele-Shaw flows: a conformal-mapping approach. Phys. Rev. E **82**(6), 066306 (2010)
11. E. Alvarez-Lacalle, H. Gadêlha, J.A. Miranda, Coriolis effects on fingering patterns under rotation. Phys. Rev. E **78**(2), 026305 (2008)
12. A. Abidate, S. Aniss, O. Caballina, M. Souhar, Effects of circular rigid boundaries and Coriolis forces on the interfacial instability in a rotating annular Hele-Shaw cell. Phys. Rev. E **75**(4), 046307 (2007)
13. E. Álvarez-Lacalle, J. Ortín, J. Casademunt, Relevance of dynamic wetting in viscous fingering patterns. Phys. Rev. E **74**(2), 025302 (2006)
14. C.Y. Chen, C.H. Chen, J.A. Miranda, Numerical study of pattern formation in miscible rotating Hele-Shaw flows. Phys. Rev. E **73**(4), 046306 (2006)
15. J.A. Miranda, E. Alvarez-Lacalle, Viscosity contrast effects on fingering formation in rotating Hele-Shaw flows. Phys. Rev. E **72**(2), 026306 (2005)
16. H. Gadêlha, J.A. Miranda, Finger competition dynamics in rotating Hele-Shaw cells. Phys. Rev. E **70**(6), 066308 (2004)
17. C.Y. Chen, S.W. Wang, Interfacial instabilities of miscible fluids in a rotating Hele-Shaw cell. Fluid Dyn. Res. **30**(5), 315–330 (2002)
18. L. Carrillo, J. Soriano, J. Ortín, Interfacial instabilities of a fluid annulus in a rotating Hele-Shaw cell Phys. Fluids **12**(7), 1685–1698 (2000)
19. L. Carrillo, J. Soriano, J. Ortın, Radial displacement of a fluid annulus in a rotating Hele-Shaw cell. Phys. Fluids **11**(4), 778–785 (1999)
20. L. Carrillo, F.X. Magdaleno, J. Casademunt, J. Ortín, Experiments in a rotating Hele-Shaw cell. Phys. Rev. E **54**(6), 6260 (1996)

# Performance Analysis of a Scramjet Combustor with Cavity for Mach Numbers 3.0, 3.25 and 3.50 with Hydrogen as a Fuel

**Namrata Bordoloi, K. M. Pandey, and K. K. Sharma**

**Abstract** The hypersonic industry has been developing with the invention of the scramjet engine. The researchers are making persistent attempts to understand the theory behind the complex flows generated due to supersonic combustion. In the current paper, a numerical study has been accomplished using Ansys 14-FLUENT code for studying the characteristics of the flame in a cavity-based supersonic combustor for a Mach number such as 3.0, 3.25, 3.50 with hydrogen is used a fuel. The simulations are carried out by considering SST K-omega turbulent model and compressible Reynolds Averaged Navier Stokes (RANS) equations. The present study is validated with an already available experimental study. The results are in good accord with the results of the experimental study. The results of the simulation are found by varying Mach numbers. The results of the study indicate that there is a movement in the oblique shock wave in the downstream of the hydrogen inlet with the increase in Mach number.

**Keywords** Combustion · Cavity flame holder · Hydrogen · K-omega model · Mach number · Oblique shock wave

## 1 Introduction

The evolution of the coherent hypersonic technology for the military applications has driven the interest for the development of air-breathing engines which is known as the Supersonic combustion ramjets or Scramjet [1] in the early 1960s. For perceiving

N. Bordoloi · K. M. Pandey (✉) · K. K. Sharma
Department of Mechanical Engineering, National Institute of Technology Silchar, Silchar, Assam 788010, India
e-mail: kmpandey2001@yahoo.com

N. Bordoloi
e-mail: namratabor@hotmail.com

K. K. Sharma
e-mail: kksharma1313@rediffmail.com



K. M. Pandey et al. (eds.), *Recent Advances in Mechanical Engineering*, Lecture Notes in Mechanical Engineering, https://doi.org/10.1007/978-981-15-7711-6_91

appropriate fuel for these engines, studies on different fuels such as hydrogen, ethylene, methane, etc. are found in the literature. However, amongst all the fuels, hydrogen is found to be most appropriate as it shows better ignition properties. A number of investigation on cavity-based combustion with hydrogen as a fuel has already been conducted [2–5]. Immense research is still going on in order to find a better candidate for the efficient scramjet combustor. The selection of fuel is an important factor in the combustion process. For a sustainable combustion process and efficient combustion process, deeper perforation of fuel into the hypersonic air stream is required. Micka et al. [6, 7] studied experimentally the combustor characteristics of a dual-mode scramjet combustor. The results indicated that at the forefront of the cavity the flame was produced by the heat released by the shear layer of the cavity. The flame then moved towards the downstream of the fuel injection. Lee et al. [8] studied different cavity-based flame holders in the scramjet combustor. The results indicated that the introduction of fuel nozzles without cavity yields better results than the configuration with the cavity. Masumoto et al. [9] studied the different modes of combustion experimentally. The results showed four different modes of combustion such as non-ignition, weak ignition, supersonic combustion, and dual-mode scramjet combustion. Pandey et al. [10] studied double cavity scramjet with hydrogen injection computationally. The results indicated that when the hydrogen ignition pressure was increased, they observed larger vortex structures bear the cavities which help in the enhancement of mixing of fuel/air. In another investigation, the length to depth effect on the combustion performance was studied [11, 12]. The results of the study indicated that for the optimum length to depth ratio the combustor performance was enhanced and stable combustion was achieved.

From the discussion above, it has been understood that the introduction of the flame holder in the combustor geometry enhances the combustion there by improving the combustor performance. In most literature, the use of hydrogen as a fuel has been observed due to its good ignition properties. Moreover, in the literature the use of the cavity based scramjet combustor with hydrogen as fuel is one of the most common combinations. But this combination has been addressed only for a range of Mach number till 2.50 and for high Mach number. Least literature is available that addresses this combination for a medium range of Mach numbers. Therefore, the present work aims to study the performance of the cavity based scramjet on using hydrogen as a fuel for Mach numbers 3.0, 3.25, and 3.50. Additionally, the considered simulation model is validated by comparing the results with already available experimental literature.

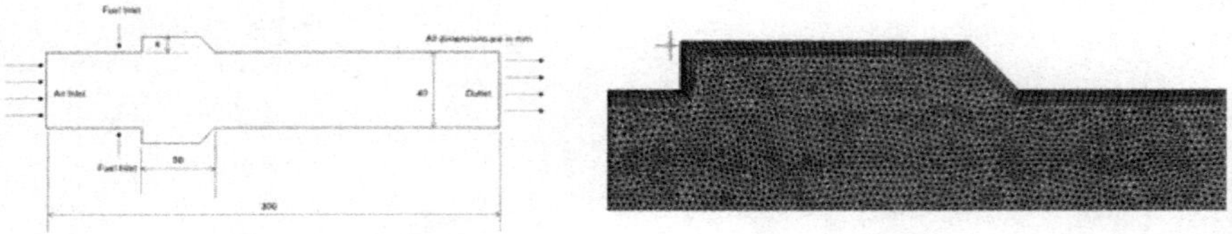

(a) (b)

**Fig. 1** **a** Illustrative diagram of the geometry[13, 14]; **b** unstructured triangular meshes

## 2 Formulation of the Problem

### *2.1 Modeling the Geometry*

A two-dimensional scramjet combustor model is developed which is similar to the experimental setup used by Yang et al.[13, 14]. The combustor consists of two parallel cavities place 40 mm away from the isolator section. Hydrogen fuel is injected through a small cavity of diameter 2 mm, which is located at 10 mm before the cavities. In this present study, the geometry is symmetric along the centerline. The detailed geometric dimension is given in Fig. 1a. As the considered geometric is symmetric in nature, therefore, only one portion of the geometry is modeled. In this paper, all computational simulations are analyzed in fluent Ansys-14 [15]code and the meshing is achieved using ICEM-CFD. The computational domain considered for the simulation is 300 mm. The simulations were initiated with a base grid of 90,000 elements. For clear shock formation near the walls, inflation layer is introduced into the meshing.[16] For the shear layer position, the refinement of mesh in the regions of incoming air and hydrogen jet was introduced. Finally, unstructured triangular mesh with 409,307 elements is adapted as shown in Fig. 1b.

### *2.2 Grid Independence Study*

In this study, for grid independence study variation of wall pressure along the wall for three different meshes such as coarse mesh, moderate mesh, and refined mesh as shown in Fig. 2. The numerical analysis for the considered model was first carried out on a base grid of 90,000 elements. The mesh is further enhanced to a grid of 248,873 elements for clear formation of the shock waves. Finally, the refined mesh grid consists of 409,307 elements along with inflation layer near the walls. It can be seen from Fig. 2, that there was no substantial change in the wall pressure variation along the length for different mesh grids. But more precise wall pressure is observed for the refined grid. Although in the numerical simulations stochastic, systematic and accumulated errors were observed but the present work does not consider any errors as these errors arise due to lack of accuracy in boundary conditions. Hence, the

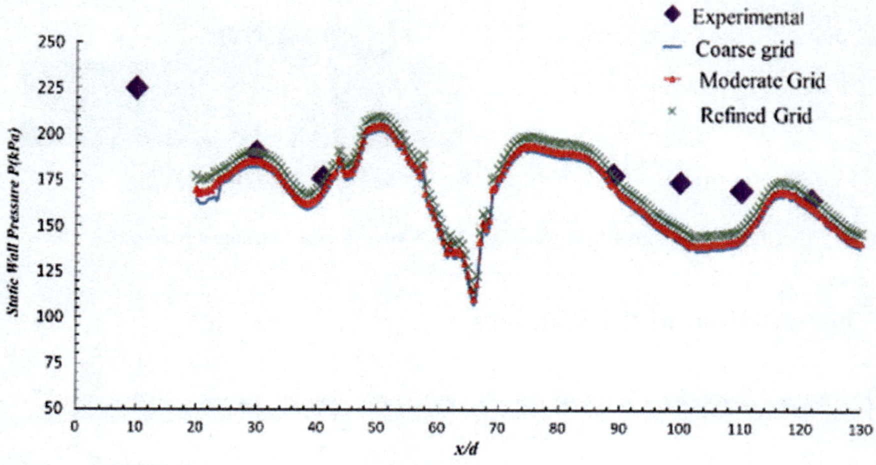


**Fig. 2** Grid independence test: static wall pressure variation along the wall

refined grid shows a balance between the precision and computational time consumed by the considered model.

# 3 Numerical Modelling and Boundary Condition

## 3.1 Governing Equations

For numerically solved problems, the selection of appropriate governing equations plays a prime role. The governing equations greatly affect the internal flow physics of the flow. The flow inside a scramjet combustor is complex. Therefore, it is necessary to use the compressible and turbulent flow equation. The Reynolds averaged Navier–Stokes equations (RANS) [17–20] are considered to describe the governing equations. The RANS equations are widely used to study the behavior of the scramjet such as the formation of boundary layer and shock waves. The following are the governing equation considered.

Mass conservation (Continuity equation)

$$\frac{\partial \rho}{\partial t} + \frac{\partial}{\partial x_i}(\rho u_i) = 0 \quad \text{where } i = 1, 2, 3 \tag{1}$$

Momentum Equation

$$\frac{\partial}{\partial t}(\rho u_i) + \frac{\partial}{\partial x_i}(\rho u_i u_j) + \frac{\partial P}{\partial x_i} = \frac{\partial}{\partial x_i}(\tau_{ij}) \text{ where } i, j = 1, 2, 3 \tag{2}$$

Energy Equation

$$\frac{\partial}{\partial t}(\rho H) + \frac{\partial}{\partial x_i}(\rho H u_i) = -\frac{\partial}{\partial x_i}\left(\tau_{ij} u_i\right) + \frac{\partial q_i}{\partial x_i} \text{ where } i, j = 1, 2, 3 \quad (3)$$

Species transport equation is given as

$$\frac{\partial}{\partial t}(\rho Y_i) + \nabla \cdot (\rho Y_i \vec{u}) + \nabla \cdot \left(\vec{J_i}\right) = \omega_i \quad (4)$$

## 3.2 Combustion Modeling

In the study, species transport along with eddy dissipation/finite rate reaction combustion model is used. The most commonly used kinetic model is the single-step kinetics model [21] which is simple to understand and an accurate model for species like hydrogen and ethylene. The reaction for hydrogen involves one step with three species. The reaction is given below

$$2H_2 + O_2 \xrightarrow{\text{yields}} 2H_2O \quad (5)$$

## 3.3 Boundary Condition

In the present study, three types of boundaries are defined. The flow simulations are computed for the entire computational domain. Dirichlet and Neumann boundary condition is applied for the inflow and outflow boundaries respective. At the fixed walls, no-slip condition with zero heat flux is considered. The symmetric boundary condition is applied at the symmetry. For the stability in the flow, the Courant Friedrich's Lewy number is considered as 0.5 [22].The SST k- omega model is used to address the turbulence. The turbulent kinetic energy ($k$) for air is 10($m^2/s^2$) and hydrogen is 2400 ($m^2/s^2$). The detailed boundary condition is given in Table 1.

**Table 1** Boundary condition considered

| Sl. No | Properties | Parameters | Air | Hydrogen |
|---|---|---|---|---|
| 1 | Total pressure | $P_0$ (MPa) | 1.6 | 0.63 |
| 2 | Mach number | M | 3.0, 3.25,3.50 | 1 |
| 3 | Total temperature | $T_0$ (K) | 1486 | 300 |
| 4 | Turbulent kinetic energy | k ($m^2/s^2$) | 10 | 2400 |

## 4 Validation

The adaptation of the parallel cavities scramjet combustor model is established by identical results with the experimental results conducted by Yang et al. [13, 14] The simulated contours of density and temperature of the model are compared with the schlieren and flame luminosity images obtained from the experiment, respectively, as shown in Fig. 3a and b. From the images extracted from the experiment and the density and temperature contours were in good agreement. Figure 4 shows the distribution of

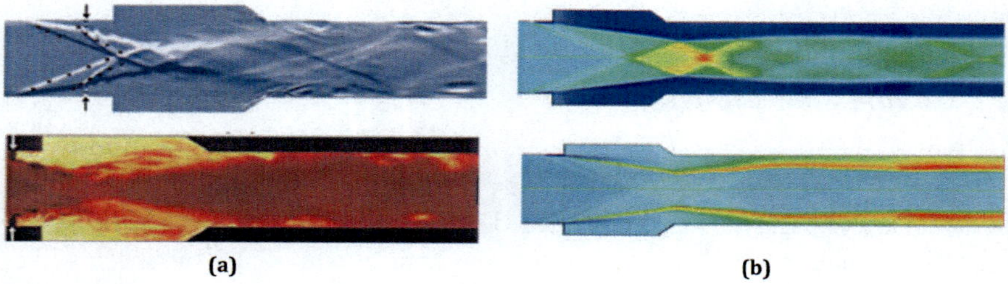


**Fig. 3** **a** Schlieren and Flame luminosity images obtained from the experiment [13]; **b** Simulated density and static temperature contours for M = 2.52

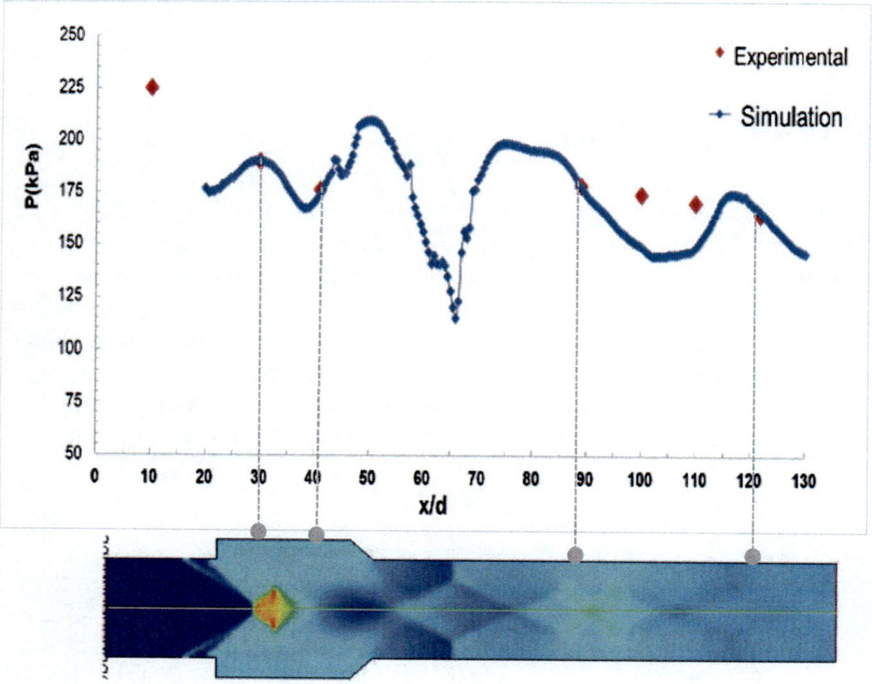


**Fig. 4** Wall static pressure distribution from the simulated results and experimental results

wall pressure of the simulation and experiment. The graph indicates the experimental results and computational results which show satisfactory remarks at four locations only. In the actual experiment, the static wall pressure measurement was done at six locations but for the computation analysis only four locations were selected for the pressure estimation. However, due to computational error, a slight deviation was observed in the prediction of shock waves. In the upper wall of the cavity, the wall pressure displayed by the first tap is about 180 kPa and in the remaining location is about 170 kPa. Also, due to continuous combustion, the maximum pressure is exhibited at the recirculation area formed.

# 5 Results and Discussion

On complete validation of the considered model, computational simulations of the considered problem have been carried out in ANSYS Fluent 14 code to understand the Mach number effect on the performance of the combustor. To under this effect, various contours such as static pressure, static temperature, density, etc. are studied. The pressure and density contours as represented in Fig. 5 shows the formation of shock waves. The temperature contours generally depict the performance of the combustor. The density and the pressure contours show the formation of the oblique

**Fig. 5** Pressure contours for Mach number = 3.0, 3.25, 3.50

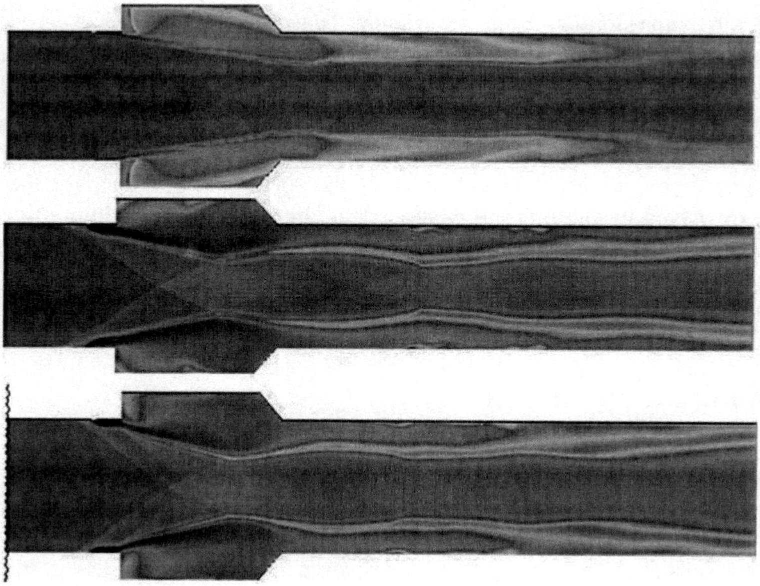

**Fig. 6** Temperature contours for Mach number = 3.0, 3.25, 3.50

shock waves, once the injection took place. These shock waves undergo multiple reflections at the combustor walls. Due to the presence of the parallel cavities, an additional shock wave and recirculation region is developed. Recirculation region is a region of high pressure and temperature and plays a major role in efficient combustion. In this present study, three Mach number such as 3.0, 3.25, 3.50 are examined. The temperature contours for M = 3.0, 3.25, 3.50 are shown in Fig. 6. From Fig. 6, it can be estimated that due to change in static temperature and velocity the heat addition is changing at the combustor. Due to this, there is a rise in temperature. The presence of parallel cavities results in the formation of recirculation region and recompression shock, which interacts with the fuel–air mixture. This causes the process of combustion to ignite. Figure 7 shows that for the considered Mach numbers the combustion process is observed to be shifting downstream while for the $M = 2.52$ the combustion process occurs at the combustor region, along with the formation of strong bow shock waves which is likely to be lacking for Mach 3 and above. Thus it can be observed that as the Mach number increases from 3 to 3.50 the strength of the shock waves are weakening.

Figure 7 represents the Mach number contour all the considered Mach number such as 3.0, 3.25, 3.50. Due to the presence of the parallel cavities on the combustor, the fuel was kept hypersonic for a longer interval of time. The shock waves formed after fuel injection expands and recirculation region is formed. The recirculation regions are formed due to the vortices present. These vortices are formed due to the $H_2$ injection from both the wall of the combustor. At this point boundary layer is also developed. Large vortices were formed due to the boundary layer interactions

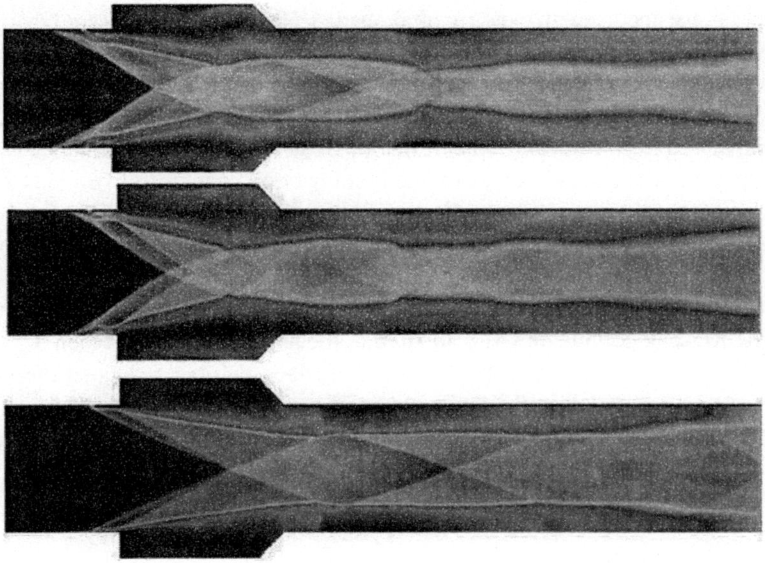

**Fig. 7** Mach number contours for incoming Mach number = 3.0, 3.25, 3.50

with the shock waves. The strength of the shock waves becomes weak as the process of combustion moves downstream. The Mach contours also show the recirculation regions which is formed away from the cavity.

Whereas, for Mach 2.52, the recirculation region is being observed in the middle part of the cavity causing the combustion process to occur at the combustor only. For Mach number 3 and above, the formation of recirculation zone becomes unsubstantial. For Mach numbers less than 3 results in stronger combustion than Mach numbers more than 3. Thus for implementing Mach 3 and above in this considered model to achieve efficient combustion different flame holders along with parallel cavity has to be incorporated.

## 6 Conclusions

The study attempts to numerically investigate the inlet Mach number effect on the performance of the considered scramjet combustor. The computational simulations are accomplished using software ANSYS Fluent 14 code. The schlieren and flame luminosity images of the experiment were used to validate the considered model. Following conclusions are drawn from the simulated results:

- The computational model is validated by comparing the schlieren and flame luminosity images of the experiment. It has been observed that the simulated results were in close accord with experimental results. The wall static pressure distribution of both the simulated as well as experimental results was studied. The graph shows satisfactory results at four different locations along the wall.

- The pressure contours for inlet Mach number = 3, 3.25, 3.50 are studied. The pressure contours clearly show the formation of the shock waves. The shock waves formed are oblique shock waves on the first injection of fuel.
- The temperature contours are studied for three different inlet Mach numbers. The contours show the formation of additional shock waves known as the recompression shock waves and recirculation region. But for inlet Mach number 3 and above the recirculation region appeared to be shifted from the middle of the cavities thus leading to weak combustion as compared to the Mach number less than 3.
- The Mach number profiles for different inlet Mach number indicated the weak shock waves formation in the downstream of the parallel cavities. Additionally, the profile indicates that the heat addition in the parallel is mainly concentrated near the wall of the combustor.

## 7 Scope of Future Work

- In this study, the scramjet combustor considered is modeled in 2D which can be extended to 3D model.
- This study is a computational analysis of the considered model. This model can be carried experimentally and compared the results with the computational results.
- For better precision in the turbulence, the Large eddy simulation with 3 D modeling can be considered.
- A detailed study on the interaction of the boundary layer and shock waves can be considered.

**Acknowledgements** The authors acknowledge the support furnished by Computational Laboratory of Mechanical Engineering Department of NIT, Silchar, and TEQIP-III for completing this work.

## References

1. E.T. Curran, Scramjet engines: the first forty years. J. Propuls. Power **17**(6), 1138–1148 (2001)
2. Y. Yang, Z. Wang, M. Sun, H. Wang, Numerical simulation on ignition transients of hydrogen flame in a supersonic combustor with dual-cavity. Int. J. Hydrogen Energy **41**(1), 690–703 (2016)
3. H. Wang, Z. Wang, M. Sun, N. Qin, Large-Eddy/Reynolds-averaged Navier-Stokes simulation of combustion oscillations in a cavity-based supersonic combustor. Int. J. Hydrogen Energy **38**(14), 5918–5927 (2013)
4. Y. Tian et al., Investigation of combustion and flame stabilization modes in a hydrogen fueled scramjet combustor. Int. J. Hydrogen Energy **41**(42), 19218–19230 (2016)

5. Z. Cai, Z. Wang, M. Sun, X.S. Bai, Effect of combustor geometry and fuel injection scheme on the combustion process in a supersonic flow. Acta Astronaut. **129**, 44–51 (2016)
6. D.J. Micka, J.F. Driscoll, Combustion stabilization, structure, and spreading in a laboratory dual-mode scramjet combustor. **3406341**, 227 (2010)
7. D.J. Micka, J.F. Driscoll, Combustion characteristics of a dual-mode scramjet combustor with cavity flameholder. Proc. Combust. Inst. **32**(II, 2), 2397–2404 (2009)
8. K. Lee, S. Kang, Y. Lee, B. Cha, B. Choi, Effects of fuel injectors and cavity configurations on supersonic combustion. J. Propuls. Power **29**(5), 1052–1063 (2013)
9. R. Masumoto, S. Tomioka, K. Kudo, A. Murakami, K. Kato, H. Yamasaki, Experimental study on combustion modes in a supersonic combustor. J. Propuls. Power **27**(2), 346–355 (2011)
10. K.M.M. Pandey, G. Choubey, F. Ahmed, D.H. Laskar, P. Ramnani, Effect of variation of hydrogen injection pressure and inlet air temperature on the flow-field of a typical double cavity scramjet combustor. Int. J. Hydrogen Energy **42**(32), 20824–20834 (2017)
11. N.K. Mahto, G. Choubey, L. Suneetha, K.M. Pandey, Effect of variation of length-to-depth ratio and Mach number on the performance of a typical double cavity scramjet combustor. Acta Astronaut. **128**, 540–550 (2016)
12. G. Choubey, K.M. Pandey, Effect of different strut + wall injection techniques on the performance of two-strut scramjet combustor. Int. J. Hydrogen Energy **42**(18), 13259–13275 (2017)
13. Y. Yang, Z. Wang, M. Sun, H. Wang, L. Li, Numerical and experimental study on flame structure characteristics in a supersonic combustor with dual-cavity. Acta Astronaut. **117**, 376–389 (2015)
14. H. Wang, Z. Wang, M. Sun, N. Qin, Large eddy simulation of a hydrogen-fueled scramjet combustor with dual cavity. Acta Astronaut. **108**, 119–128 (2015)
15. P. Kohnke, ANSYS Theory Reference-Release 5.6 (1999), p 1286
16. G. Choubey, K.M. Pandey, Effect of variation of angle of attack on the performance of two-strut scramjet combustor. Int. J. Hydrogen Energy **41**(26), 11455–11470 (2016)
17. W. Huang et al., Numerical investigation on the shock wave transition in a three-dimensional scramjet isolator. Acta Astronaut. **68**(11–12), 1669–1675 (2011)
18. B.E. Launder, D.B. Spalding, The numerical computation of turbulent flows Van Driest's constant Curte t number defined by ( 3 . 1 - 1 ) Coefficients in approximated turbulent transport equations Specific heat at constant pressure Diffusion coefficient for quantity ( p Rate of diffu) Science (80-. ), **3**, 269–289 (1974)
19. W.P. Jones, B.E. Launder, The effects of maternal diets, varying in fat content, on proximal hepatic and skeletal muscle insulin signalling in neonatal wistar rat offspring. Int. J. Heat Mass Transf. **15**, 301–314 (1972)
20. W. Huang et al., A parametric study on the aerodynamic characteristics of a hypersonic waverider vehicle. Acta Astronaut. **69**(3–4), 135–140 (2011)
21. C.L. Jiyuan Tu, G.H. Yeoh, Computational Fluid Dynamics: A Practical Approach
22. E. Erdem, K. Kontis, Numerical and experimental investigation of transverse injection flows. Shock Waves **20**(2), 103–118 (2010)

# Study of Fuel Injection Systems in Scramjet Engine—A Review

**Kumari Ambe Verma, K. M. Pandey, and K. K. Sharma**

**Abstract** The scramjet engine is categorized in air-breathing vehicles, and this engine utilizes the high-speed forward motion of the vehicle to compress the surrounding air for getting forward thrust. Air-breathing engines complete the cycle by using air from the atmosphere. In the scramjet engines, the atmospheric free stream air enters the combustor at supersonic speed. And the process of mixing and combustion both take place at the same speed. Due to the higher speed of incoming air, the residence time is less inside the combustor. This phenomenon leads to poor mixing of fuel and air. As the scramjet engine does not have any moving part, so there is very less way to improve the performance. One of the optimum ways is fuel injection strategies to enhance combustion performance. To incorporate the behavior of the fuel entrance inside the combustor with the performance of the engine, a detailed literature review has been done in three separate sections by their preferences, i.e., parallel, transverse, and combined fuel injection system. A detailed literature review has been performed to understand and explain the impact of the different fuel injection system at supersonic speed. The parallel fuel injection system at supersonic speed gives higher combustion performance when the additional vortices present. These stream wise vortices can be generated with the help of different shapes of a fuel injector. In the transverse fuel injection system, the total pressure recovery is found less nonetheless stable flame can be achieved by using cavities at the walls. The recirculation zone can easily be created near to this zone. As parallel and transverse fuel injection systems give the best performance among all in their own way however fewer research has been performed in the direction of combined fuel injection strategy at supersonic speed. Flame holding capacity with additional vortices formation can only be achieved by using a combined system.

K. A. Verma · K. M. Pandey (✉) · K. K. Sharma
Department of Mechanical Engineering, National Institute of Technology Silchar, Silchar, Assam, India
e-mail: kmpandey2001@yahoo.com

K. A. Verma
e-mail: ambe.verma@gmail.com

K. K. Sharma
e-mail: kksharma1313@rediffmail.com


K. M. Pandey et al. (eds.), *Recent Advances in Mechanical Engineering*, Lecture Notes in Mechanical Engineering, https://doi.org/10.1007/978-981-15-7711-6_92

**Keywords** Fuel injection strategies · Supersonic speed · Air-breathing engine

## 1 Introduction

The name supersonic combustion ramjet acronymed as a scramjet. High-speed regimes are essentially required to perform the scramjet engine. The scramjet engine cannot function independently. The system is usually associated with rocket, similar to its old predecessor, i.e., ramjet. The scramjet engine is categorized in an air-breathing vehicle, and this engine utilizes the high-speed forward motion of the vehicle to compress the surrounding air for getting forward thrust. Due to the lack of moving parts, compression can only be done by internal geometry. Fewer obstacles are beneficial for less pressure loss near the wall boundary. There is no need for any rotating machinery to increase the pressure inside the combustor, Change in combustor area is enough to complete the requisites. Thereby higher maximum cycle temperature can be achieved. Compressed air passes through the isolator before entering the combustor. Constant area duct isolator often used to maintain the homogeneity of the flow and also protect the operating condition of the engine. Free stream air while passing through the combustor allowed to mix and react with the fuel-injected from a certain location. The heat release inside the combustor causes local pressure rise during the chemical reaction of fuel and air, which is converted into a thrust. Simplicity of the combustor only operates with the help of shock waves at thousands of kelvin temperatures and 100 times greater pressure. The scramjet engine operates at the speeds of rockets. Its main difference from a rocket is that it collects air from the atmosphere to burn its fuel rather than carrying its oxidizer on board.

## 2 History

In the field of aviation, aircraft are a gift by the privileged scientists and researchers to accomplish the human's desire to fly. The Wright brothers are the ones who first introduced the aviation flight in 1903. Support of air is needed, to fly a machine like an aircraft. Since static or dynamic lifts are governing factors to lift the airfoil aircraft and thrust are used in jet engines to counter the gravitational force. In the present era, the high speed of the vehicle is a challenging issue to accomplish. This can be accomplished by air-breathing engines that are suitable for higher speed. To fly at hypersonic speed, supersonic ramjet engines are the only popular aircraft invented by France scientist Rene Lorin in 1913. The operational ramjet engine equipped airplane was first developed by France scientist Rene Lorin in 1995 [1].

## 3 Fuel Injection System

To develop the competent combustion chamber at a desired level of speed, a suitable means of fuel injection is required and also a key parameter for the scramjet combustor. The mixing of the fuel into the air is not fully achieved at a desirable level so far. So based on the performance criteria of the incoming air, the fuel system needs to be adjusted to get better mixing with higher combustion performance. The geometry of the combustor can be any type such as Rectangular, circular, elliptical, and combination of above, etc. Combustor geometry can be act as a secondary parameter compared to the fuel injection system. The geometrical prospect of fuel injectors has already been discussed by the present author.[2]. According to the fuel entrance into the compressed air high-pressure zone can be categorized into three parts, i.e., parallel, transverse, and combine fuel injection system. The schematic of types of the fuel injection system is shown in Fig. 1.

### 3.1 Parallel Fuel Injection System

The direction of the incoming air is the same as the direction of fuel entrance is known as a parallel fuel injection system. This injection system can be achieved by various means such as different types of strut, with the help of splitter plate or also known as co-flow system. Purely diffuse mechanism is utilized in the parallel system. Many researchers have focused their own work in these prospects. Aravind et al. [3] numerically investigated the mixing performance with the help of modified strut. Parallel fuel injection scheme was chosen for fuel injection. Three dimensional Reynolds Averaged Navier–Stokes equation with realizable k-ε turbulence model was selected. As the high-level turbulence in the shear layer is the key notion for improved mixing of selected fuel and incoming free stream air. Among the shear layer, vortices as disturbance often play a major role. Based on this idea author introduced innovative strut to enhance mixing of incoming air with hydrogen. It was identified that complete mixing of fuel–air with the help of modified strut took shorter

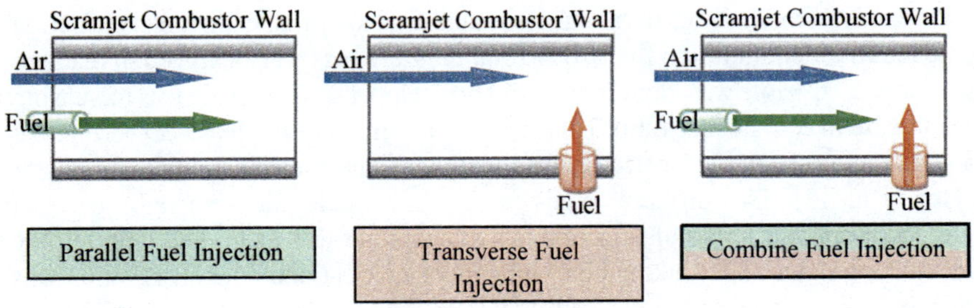


**Fig. 1** Schematic of types of fuel injection system

distance compared with a planar strut. 15 mm length reduction was remarked to reach above 95% of mixing efficiency. This observation can directly help in the reduction of combustor length and overall weight of the engine. The complete reaction of hydroxyl and all other radicals are desirable within the combustor span length. Wider reaction zone was recognized with modified strut and also shorter span of reaction zone was observed clearly by the temperature, mass fraction of hydroxyl radical contour plots. Vortices are commonly helpful for mixing, this can be generated with the help of the baroclinic effects. Density gradients and unregimented pressure lines are the cause of vortices generation thus it can be considered as a crucial technique for turbulence generation by means of baroclinic effect. Cao et al. [4] excavated the behavior of flow field in terms of mixing, combustion modes, and heat release from their keen experimental observation and numerical based solver. LES model was used for numerical simulation and OH chemiluminescence imaging and shadowgraph visualization were taken to understand the jet flame behavior. Baroclinic effects were observed near to jet flame propagation area while interacting with generated shocks, gave a major impact on mixing. Vortices are commonly helpful to transfer mass from shear layer to the jet spacing and vice-versa. This dynamics can be utilized to get better mixing performance. Numerical investigation has been performed to understand the effect of large scale vortices on mixing with the help of two different strut configuration by Soni and De [5]. Variable lip height of Straight and tapered strut was observed through Large eddy simulation based model. Author analyzed that the velocity gradient and strut lip height are more responsible to create large vortices, mixing layer growth and mixing efficiency. Compared to taper one, straight strut provided larger vortices structure. Ma et al. [6] was used mixed fuel (Hydrogen and Hydrocarbons mixture) to observe combustion characteristics. The comparison was done by two fuel types such as Fuel-A (Pure hydrogen) with Fuel-B (Mixed fuel). It was clearly observed by the mixing efficiency graph that mixed fuel took a shorter length to reach up to 100% compared to pure hydrogen however overall combustion efficiency was reduced. Nonetheless, disturbance of flame profile in flow field was highly influenced by vortices and shock waves. The noticeable comment was given by the same author that combustion intensity became weaker due to the addition of supplementary fuel (Methane and Ethylene). Numerical investigation has been performed to understand the effect of ethylene as a fuel during supersonic scramjet combustion by Dharavath et al. [7]. Reacting and non-reacting both flow field was analyzed. Also, leading edge of the strut was analyzed to evaluate the fuel distribution and penetration length of the fuel. 70% Total pressure loss was observed in reacting case however 10% loss was identified in cold flow. Upper wall pressure was increasing with increase in equivalence ratio. This is because of more heat release. Experimental investigation has been performed on strut wall kerosene-fueled scramjet combustor by Hua et al. [8]. Two different types of injection mode was selected, i.e., strut mode and strut/wall mode of injection. This technique was examined without any flame holding devices. By taking constant equivalence ratio, all the observation was performed. Both injection modes were taken as at the same position in axial position. Based on the various fuel feeding ration, combustion performance with wall temperature gradient was analyzed and strut/wall injection mode has been found for better combustion performance. Kerosene injection was also helpful for cooling the wall that's why lesser wall temperature gradient was identified. Supersonic combustion by means of parallel injection have taken hold for improved combustion performance however until now understanding behind the detailed characteristic of fuel-air mixing has not been explored completely. By giving light over this section, number of researchers have enlightened their works and these have concluded, Huang et al. [9] explored the dynamic characteristics behind the supersonic combustion through wedge-shaped strut fuel injector. Central focus was done overpressure oscillation and flame dynamics i.e. flame shifting inside the combustor. Pressure oscillation is often happened because of two reasons such as flame instability and heat release. Behavior of flame was identified unsteady due to shifting of flame location. Further comparison of flame has carried out and observed that flame behavior is directly related to transition zone created between shocks generated inside the combustor and flame propagation. To understand the flame stabilization inside the combustor, Wu et al. [10] selected 2D model of DLR based combustor. By detailed assessment of temperature and species concentration, author confirmed three different flame stabilization stages, i.e., induction, transition and intense stage. Three different oxidation mechanism were also carried out with changed stages of reductions, and from this analysis, it was found that chain reaction is majorly responsible for flame stabilization. Shan et al. [11] added new linearized correction model to analyze the compressibility effects overtemperature plot near the strut zone nonetheless combustion characteristics at supersonic level can predict better due to capturing the progress of chemical reaction. However, the effect of variable flow condition over the same model has not been resolved yet notify by the same author. Huang et al. [12] investigated combustion characteristics of liquid kerosene-based combine cycle combustor. Compared with RANS model, unsteady characteristics was well apprehended by LES-based modeling nonetheless it was also helpful to capture the behavior of turbulent flame and mixing performance. Mixing zone was divided into three zone, i.e. shear layer of the primary rocket jets, second was core region and third one was just after the strut. Premixed combustion was identified in the first mixing zone and there was no reaction identified in the second zone however non-premixed combustion or simply can say that autoignition is often done in the third zone of mixing. Among all three zones, first and last mixing zone are majorly responsible for flame stabilization.

## *3.2 Transverse Fuel Injection System*

As the name suggests that the fuel entrance direction should normal to the incoming air or simply it can summarized that both should be in perpendicular in nature except if the angled fuel injection is not used. Penetration mechanism is consumed during transverse injection system. This injection system is nowadays more dominating in scramjet research combined with cavity. Because of the flame holding capacity inside the cavity is most desirable situation for mixing. Recirculation zone often form inside the cavity is also helpful for better mixing. Flame Stabilization can be achieved with the help of flame holding devices such as wall injection, ramp, cavity, and strut. Low-speed recirculation zone is often present after the blunt end of the strut in combustion mode, which provides good agreement to sustain stable combustion. A similar trend has also been identified in the cavity. Increased Domkohlar number is realized in the low-speed recirculation zone. Increased Domkohler number can execute the flame stabilization. It can positively be accomplished by various flame holding devices. Experimental observation was performed to understand the flame stabilization characteristics of cavity based combustor by Yuan et al. [13]. Variation of temperature with equivalence ration was chosen to carry out further investigations. Flame stabilization was identified at four flame location i.e. inside the cavity, cavity shear layer, jet-wake and oscillation between jet-wake and cavity. Based on the observation performed, flame stabilization was a function was temperature and global equivalence ratio. Boundaries of these modes was moved at higher equivalence ratio. Nonetheless, correlation between combustion modes and flame stabilization modes got with the help of Mach number plot. Major difference was observed that cavity shear layer flame stabilization was helpful for scramjet mode and jet-wake at ramjet mode. Li et al. [14] selected dual mode scramjet combustor to understand the mode transition. Variation of fuel flow rate has been opted to understand the relationship between fuel equivalence ratio and combustion mode transition. Three consecutive mode pure scramjet, dual mode scramjet and dual mode ramjet were arrived at the isolator entrance due to change in equivalence ratio. Rapid conversion of pure scramjet mode to ramjet mode was also identified. Optimum fuel injector position has been investigated with the influence of micro air jets by Fallah [15]. Ethylene has selected as a fuel. Fuel location was more appropriate to enhance the mixing of the fuel in the cavity at the middle of the vertical wall. Nonetheless air micro jets was quite helpful for the uniform mixing of fuel with air inside the cavity and reaches up to flame holding situation. Double cavity based flame holder of dual mode scramjet combustor has been observed experimentally by Li et al. [16]. Two different fuel i.e. kerosene and hydrogen are used to complete the combustion. Hydrogen was injected from the inside of cavity and kerosene was injected ahead of the first cavity. Results reveal that the flame holding capacity is entirely dependent on the entrance pressure condition. Jeyakumar et al. [17] experimentally observed the behavior of aft angle in axisymmetric cavity wall. Comparison was performed through rectangular based cavity. Stable flow field with reduce cavity drag were identified by reducing aft wall angle. This is also helpful for flame holding. Nonetheless, pressure loss reduction can also be achieved. Experimental observation has been performed to understand the dynamics of an optical axisymmetric scramjet preferred for flame holding and combustion through Liu et al. [18]. 45 The heat addition from the cavity-stabilized flame mainly reduces the flow Mach number in the near-wall flow region. In the core flow region, the flow Mach number is decreased by the jet and cavity induced shockwaves and minimally affected by heat addition degree inclination. Experimental investigation has been performed by Wang et al. [19]. The effect of incoming vitiate species has been explained on supersonic combustion. Results reveal that the presence of vitiate species often reduces the pressure of the combustor by 3 to 30% at same equivalence ratio. Also increased molar fraction of these species is affect the flame spreading angle and illogical oscillation can see in the combustor. Ethanol as a fuel has been opted to observe the effect of combustion performance in cavity-based scramjet experimental setup by Nakaya et al. [20]. During the supersonic combustion, two different types of combustion mode, intensive and transient were identified. Same penetration height has been identified by comparing liquid and gaseous fuel. Numerical observation has been performed to understand the effect of fuel injection location after the cavity by Roos et al. [21]. The comparison was performed by taking no cavity model. The different shock interaction behavior was seen due to addition of cavity and also shock structure around the cavity was found to decrease the strength of the bow shock, reducing total pressure loss in the flow field close to the injector. Numerical investigation has chosen to understand the effect of flame flashback phenomena of ethylene based supersonic combustion by Zhao et al. [22]. It was found that the flame flashing is more dominated by temperature.

### 3.3 Combined Fuel Injection System

By analyzing the parallel and transverse fuel injection strategy, many authors used combination of both to get the associated effect of diffusion and penetration mechanism, Yan et al. [23] selected two separate models, one with single strut and another strut-cavity to analyze the nonlinear characteristics of dual mode combustor engine. Two-dimensional RANS based numerical simulation with RNG ($k$-$\varepsilon$) turbulent model has been selected to understand the hysteresis occurrences between mode transition. Nonlinear pressure graph has observed during transition. Stable flow field has been seen in strut-cavity based mode, due to this conversion of scram to ram jet mode is quite difficult. Hydrogen fuel based strut by introducing cavity-based combustor has selected and variable jet-to-crossflow pressure was chosen to analyze combustion performance by Lei et al. [24]. Reduced performance has been found during increased wall injection pressure, so author concluded that half of the wall injection pressure compared with strut injection pressure gives best performance. For flow stabilization point of view, cavity could help because it prevents pre-combustion shock, however, mixing and combustion performance is often reduced. Experimental observation has been performed for strut cavity based supersonic combustor to understand the induced pressure oscillation. Kerosene and pilot hydrogen was carefully chosen for fuel by Sathiyamoorthy et al. [25]. Oscillation between two tandem struts was seen during cold flow experiment. Pressure oscillation was affected by aspect ratio of cavity. This arrangement is more appropriate for mixing improvement. However flame stabilization can be achieved by using new passive control device called ventilated rear wall. Experimental observation was performed on strut-cavity based dual mode combustor by Zhang et al. [26] to understand the combustion modes through flow characteristics. In dual mode both ramjet and scramjet modes are available however rigorous observation showed three different mode such as scramjet mode, weak ramjet and strong ramjet mode. Weak boundary layer effect was observed during scramjet mode. By changing the fuel injection position, backpressure was improved in weak ramjet mode that was helpful to thicken the boundary layer. Formation of shock waves in the isolator before the wedge shape strut was observed in weak ramjet mode. Thermal throat formation was analyzed in strong ramjet mode. Experimental observation has been performed to get better flame holder and ignition of kerosene with the help of cone-strut, cavity and both by Zhang and Song [27]. Both ignition and flame holding feature could not obtained by cone type strut. By using larger cone strut, isolator entrance was affected by back pressure flow however this defect was not identified in smaller cone strut. Slight range of equivalence ratio in cavity based combustor was found good agreement towards ignition and flame holding. The range of equivalence ratio can be enlarged by using both cone-strut and cavity. Rui et al. [28] experimentally observed the rocket-based combine cycle propulsion system to see the behavior of combustion and its mode transition. Four combustion modes were identified clearly, i.e. weak combustion mode, rocket-scram mode, rocket ram mode with the help of wall pressure, Mach number. Both experimental and numerical observation have been performed by strut cavity arrangement at variable fuel equivalence ratio by Zhang et al. [26]. By exhaustive analysis of supersonic combustion, three different modes were seen such as scramjet mode, weak ramjet mode and string ramjet mode. Results reveal that the wall fuel injecting before strut would change the starting point of pressure rising. Experimental observation has been performed to explore the behavior of cone shaped strut with cavity arrangement by Zhang and Song [29]. Results showed that the ignition and flame holding cannot obtained buy solely strut arrangement. Back pressure disturbance can also observed by smaller to larger shapes of the cone-strut. Experimental investigation has been selected to get the best flame holder by comparing cavity to strut-cavity arrangement with non-reacting flow by Zhao et al. [30]. Axisymmetric flow field characteristics were observed due to strut-cavity based arrangement.

# 4 Summary and Future Scope

A detailed literature review has been performed to understand and explain the impact of the different fuel injection system at supersonic speed. Some of the following conclusive remarks have been summarized below:

- The parallel fuel injection system at supersonic speed gives higher combustion performance when the additional vortices present. These streamwise vortices can be generated with the help of different shapes of a fuel injector.
- In the transverse fuel injection system, the total pressure recovery is found less nonetheless stable flame can be achieved by using cavities at the walls. The recirculation zone can easily be created near to this zone.
- As parallel and transverse fuel injection systems give the best performance among all in their own way however fewer research has been performed in the direction of combined fuel injection strategy at supersonic speed. Flame holding capacity

with additional vortices formation can only be achieved by using a combined system.

So, flame is more stable in transverse fuel injection due to the presence of a cavity at the walls however the stream-wise vortices are more suitable for mixing. This can be generated with the help of different shapes of fuel injectors. This can only be possible in the Parallel fuel injection system. Based on these occurrences, the Combine fuel injection system has been found appropriate among all.

## References

1. R.P. Hallion, The Hypersonic Revolution, Volume II: From Max Valier to Project Prime, Aeronautical System Center, Air Force Material Command, Wright Patterson Air Force Base, ASC-TR-95–5010 (1995)
2. K.A. Verma, K.M. Pandey, K.K. Sharma, Computational investigation on design of scramjet combustor–A review. Int. J. Recent Technol. Eng. **7**(6), 544–548 (2019)
3. S. Aravind, R. Kumar, Supersonic combustion of hydrogen using an improved strut injection scheme. Int. J. Hydrogen Energy **44**(12), 6257–6270 (2019)
4. D. Cao, G. He, F. Qin, D. Michaels, Local supersonic and subsonic combustion mode transition in a supersonic jet flame. Proc. Combust. Inst. **37**(3), 3723–3731 (2019)
5. R.K. Soni, A. De, Investigation of mixing characteristics in strut injectors using modal decomposition. Phys. Fluids **30**(1), 016108 (2018)
6. S. Ma, F. Zhong, X. Zhang, Numerical study on supersonic combustion of hydrogen and its mixture with Ethylene and methane with strut injection. Int. J. Hydrogen Energy **43**(15), 7591–7599 (2018)
7. M. Dharavath, P. Manna, D. Chakraborty, Numerical exploration of mixing and combustion in ethylene fueled scramjet combustor. Acta Astronaut. **117**, 305–318 (2015)
8. J. Hu, J. Chang, W. Bao, Q. Yang, J. Wen, Experimental study of a flush wall scramjet combustor equipped with strut/wall fuel injection. Acta Astronaut. **104**(1), 84–90 (2014)
9. Z.W. Huang, G.Q. He, S. Wang, F. Qin, X.G. Wei, L. Shi, Simulations of combustion oscillation and flame dynamics in a strut-based supersonic combustor. Int. J. Hydrogen Energy **42**(12), 8278–8287 (2017)
10. K. Wu, P. Zhang, W. Yao, X. Fan, Numerical investigation on flame stabilization in DLR hydrogen supersonic combustor with strut injection. Combust. Sci. Technol. **189**(12), 2154–2179 (2017)
11. F. Shan, L. Hou, Z. Chen, J. Chen, L. Wang, Linearized correction to a flamelet-based model for hydrogen-fueled supersonic combustion. Int. J. Hydrogen Energy **42**(16), 11937–11944 (2017)
12. Z.W. Huang, G.Q. He, F. Qin, D.G. Cao, X.G. Wei, L. Shi, Large eddy simulation of combustion characteristics in a kerosene fueled rocket-based combined-cycle engine combustor. Acta Astronaut. **127**, 326–334 (2016)
13. Y. Yuan, T. Zhang, W. Yao, X. Fan, P. Zhang, Characterization of flame stabilization modes in an ethylene-fueled supersonic combustor using time-resolved CH* chemiluminescence. Proc. Combust. Inst. **36**(2), 2919–2925 (2017)
14. J. Li, K. Wang, G. Jiao, J. Luo, W. Song, Experimental study on mode transition in a dual-mode scramjet combustor. Combus. Sci. Technol. 1–19 (2019)
15. K. Fallah, M.B. Gerdroodbary, A. Ghaderi, J. Alinejad, The influence of micro air jets on mixing augmentation of fuel in cavity flameholder at supersonic flow. Aerosp. Sci. Technol. **76**, 187–193 (2018)

16. J. Li, G. Jiao, D. Shi, W. Song, Experimental study on combustion characteristics of kerosene-fueled scramjet combustor. Combust. Sci. Technol. **190**(10), 1772–1785 (2018)
17. S. Jeyakumar, S.M. Assis, K. Jayaraman, Effect of axisymmetric aft wall angle cavity in supersonic flow field. Int. J. Turbo Jet-Eng. **35**(1), 29–34 (2018)
18. Q. Liu, D. Baccarella, W. Landsberg, A. Veeraragavan, T. Lee, Cavity flameholding in an optical axisymmetric scramjet in Mach 4.5 flows. Proc. Combus. Instit. **37**(3), 3733–3740 (2019)
19. Y. Wang, W. Song, D. Shi, Q. Fu, Y. Wang, Experimental study of vitiation effects on combustion characteristics of a supersonic combustor. Combust. Sci. Technol. **189**(9), 1500–1509 (2017)
20. S. Nakaya, Y. Hikichi, Y. Nakazawa, K. Sakaki, M. Choi, M. Tsue, S. Tomioka, Ignition and supersonic combustion behavior of liquid ethanol in a scramjet model combustor with cavity flame holder. Proc. Combust. Inst. **35**(2), 2091–2099 (2015)
21. T. Roos, A. Pudsey, M. Bricalli, H. Ogawa, Cavity enhanced jet interactions in a scramjet combustor. Acta Astronaut. **157**, 162–179 (2019)
22. G. Zhao, M. Sun, J. Wu, X. Cui, H. Wang, Investigation of flame flashback phenomenon in a supersonic crossflow with ethylene injection upstream of cavity flameholder. Aerosp. Sci. Technol. **87**, 190–206 (2019)
23. L. Yan, L. Liao, W. Huang, L.Q. Li, Nonlinear process in the mode transition in typical strut-based and cavity-strut based scramjet combustors. Acta Astronaut. **145**, 250–262 (2018)
24. L. Liao, L. Yan, W. Huang, L.Q. Li, Mode transition process in a typical strut-based scramjet combustor based on a parametric study. J. Zhejiang Univ.-Sci. a **19**(6), 431–451 (2018)
25. K. Sathiyamoorthy, T.H. Danish, J. Srinivas, P. Manjunath, Experimental investigation of supersonic combustion in a strut-cavity based combustor. Acta Astronaut. **148**, 285–293 (2018)
26. C. Zhang, J. Chang, Y. Zhang, Y. Wang, W. Bao, Flow field characteristics analysis and combustion modes classification for a strut/cavity dual-mode combustor. Acta Astronaut. **137**, 44–51 (2017)
27. D. Zhang, W. Song, Experimental study of cone-struts and cavity flameholders in a kerosene-fueled round scramjet combustor. Acta Astronaut. **139**, 24–33 (2017)
28. R. Xue, G. He, X. Wei, C. Hu, X. Tang, C. Weng, Experimental study on combustion modes of a liquid kerosene fueled RBCC combustor. Fuel **197**, 433–444 (2017)
29. S. Zhang, D.W. Song, Experimental study of cone-struts and cavity flameholders in a kerosene-fueled round scramjet combustor. Acta Astronaut. **139**, 24–33 (2017)
30. Y. Zhao, J. Liang, Y. Zhao, Non-reacting flow visualization of supersonic combustor based on cavity and cavity–strut flameholder. Acta Astronaut. **121**, 282–291 (2016)

# A Quantitative and Qualitative Review of Sustainable Manufacturing

**Tejendra Singh and Jinesh Kumar Jain**

**Abstract** This article is written with the purpose to gather literature review on sustainable manufacturing. Many frameworks were analyzed and formulated. Various papers on this topic were searched and contents were explored analytically. Their association consists of correlation, differences, overlapping area, and integration of various forms and parts of sustainable manufacturing. Total of 60 papers of identical research were studied and reviewed for the research methodologies, their contributions, and relevant concepts. The article shows the foremost gaps in the field of research for sustainable manufacturing through different aspects. This article provides a quantitative analysis to offer an investigation of various concepts of sustainable manufacturing. In conclusion, here in this article, we emphasize over exclusive analysis of sustainable manufacturing through the recognization of various factors when many literature reviews are there with only the objective of pondering over sustainable manufacturing concepts. One more distinctive quality of this article is that total of 60 research papers have been studied before reviewing. The time frame of 18 years (2001–2018) is considered for this review.

**Keywords** Assessment · Descriptive analysis · Implementation · Environment · Manufacturing · Review · Sustainable manufacturing · Sustainability

## 1 Introduction

Various practices of manufacturing in which the environment remains clean and does not get harmed during any process are defined as sustainable manufacturing. Here main stress is laid on the use of various techniques that are harmless in case of end-users or the stakeholders of the community and less or not polluting at all in the

T. Singh (✉) · J. K. Jain
Department of Mechanical Engineering, MNIT Jaipur, Jaipur 302017, India
e-mail: rme9043@mnit.ac.in

J. K. Jain
e-mail: jineshjain.mech@mnit.ac.in



K. M. Pandey et al. (eds.), *Recent Advances in Mechanical Engineering*, Lecture Notes in Mechanical Engineering, https://doi.org/10.1007/978-981-15-7711-6_93

case of environment. Sustainable manufacturing combines up keeping, zero waste management, recycling, environmental protection, and pollution control. Sustainable manufacturing lays stress to a large extent on designing and delivering outputs that try to minimal the downbeat factors on the community by their generation, utilization, and discarding. In the present time, it is always better to develop the artifact for the economic also environmental possibilities for the organizations. Even the extension has compelled industries to advance their zero waste routine. The objective of the article is to find the current standing on sustainable manufacturing deals and advise for further research. This review is achieved after responding to the following six research queries:

(1) What is the position of various issues over sustainable manufacturing within overall time interval?
(2) What are the positions of these issues across countries?
(3) What are the various researched concepts in this field?
(4) What types of research tools are used for various research?
(5) Which industries already adopted this research?
6) What are the future scope and research gaps?

First three Questions will give out the answers about the growth of research in this domain worldwide. Another question will answer the various tools used by researchers and personnels in this field.

The composition of the paper is as under,

(A) Explaining the set of queries that will be answered by this article.
(B) Purposes the literature review in the area of sustainability.
(C) Methodology includes articles selection, classification, and analysis.
(D) Framework
(E) Future scope

## 2 Literature Review

The reasons of importance for the literature reviews, they are as under:

1. Providing base to the classification of an investigative topic, queries, and suggestion.
2. Study the literature in which the study will be helpful.
3. Improvising the terms into the literature.
4. Fabricating and knowing the hypothetical terms and concepts.
5. Forming a list of sources that have been used.
6. Referring to various methods that will help in studying and interpreting the results.

Also, all these points do not need to clarify by each review of the existing literature. The present study uses past research to concentrate on different ideas of sustainable manufacturing to provide and generate future scope for further research. In my view up to a certain extent a little work on concepts of sustainable manufacturing has been done so far, which has a quite large body of extent literature. Thus, it is needed to focus on concepts of sustainable manufacturing, and necessity of this paper is justified. Below is the table of the literature review of the selected papers (Table 1).

## 3 Method of Reviewing

This part of the article shows the broad plan for review of sustainable manufacturing concept in subsequent flow:

Duration of 2001–2018 has been taken into account. Papers were collected from esteemed journals. Catalog wise search is done for many articles, these were studied and analyzed and sorted in accordance with the relevance of our aim. finding, analyzing and finalization of the papers are based on the criterion that only those papers that have touch with concepts of sustainable manufacturing were selected. This results in the selection of 60 articles for final stage evaluation.

***Article Classification***

The Following are the contents on the basis of which research papers were selected.

1. Number of publication per year: Frequency of publishing the papers annually.
2. Journal Wise: classification of articles journal wise.
3. Nation wise: Detailed study nation wise.
4. Concept wise: Number of articles according to the relation of concept.
5. Tool used: Basis of the different types of tools used.

## 4 Quantitative Analysis

### *4.1 Paper Published Per Year*

It consists of the annual frequency of the selected papers including all the besieged articles. It is noticeable from the chart that in the year 2014, 2015, and 2016 maximum research papers was published (Fig. 1).

### *4.2 Journal Wise Classification*

This section shows the journal's name from where papers were published we opted to review. The figure depicts the quantity of papers year wise in accordance to journals. The maximum publication was in Elsevier that is 24, then 14 in Emerald, 10 in Taylor

**Table 1** Literature review

| Author | Year | Approach | Practical implementations |
|---|---|---|---|
| E Westkämper, L Alting, G Arndt | 2001 | Drivers and Barriers Assessment | Various tools used for design, database are needed by designers to incorporate environmental aspect in their propose work |
| C. O'Brien | 2002 | Any assessment only for Sustainable Manufacturing | Sustainability helps companies in generating additional revenues by tumbling the inputs use by them. Also, the process generates extra revenues from superior products or enables industries to create novel businesses |
| H. Kaebernick, S. Kara, M. Sun | 2003 | Any Assessment only for Sustainable Manufacturing | Enhanced models, metrics for sustainability estimation and optimized techniques at the product, method, and structure level is needed for manufacturing |
| Masaaki Kotabe Janet Y. Murray | 2004 | Drivers and Barriers Assessment | Need for customized product and services is a well-accepted business paradigm which tells the consumer that everything should or could be personalized including the learning process that will help the process of sustainable development |
| James Scott Baldwin<br>Peter M. Allen<br>Keith Ridgway | 2005 | Life cycle Approach | Notion of ideality in form of the ideal agenda for the exposed problem was important. While ideality is a dependent situation, it is vital to examine a case closer to actuality, developing the IFR which accounts the studied situation and is associated with its distinctiveness, compensation, and boundaries |

(continued)

**Table 1** (continued)

| Author | Year | Approach | Practical implementations |
|---|---|---|---|
| Vishesh Kumar Karl R. Haapala Julio L. Rivera | 2005 | Additive Manufacturing | For sustainable development, supportive policies and practices, balanced and integrated economic and environmental societal objectives are required |
| A. J. Thomas<br>B. Grabot | 2006 | Any Assessment only for Sustainable Manufacturing | Basic Studies on life cycle products, closed-loop manufacturing processes and integrated production systems applying to the complete supply chain operations is needed for implementing viable sustainable manufacturing |
| Samir K. Srivastava, | 2007 | Comparison of two techniques | Remanufacturing cannot be applied in every sector. It is useful only where there is a demand for remanufactured products. |
| Dr Rainer Seidel, Mehdi Shahbazpour | 2007 | Drivers and Barriers Assessment | Cost is an important factor during the implementation of an improvement in the product or process. Limitation of input resources implies that the capital accessible for the execution of sustainability at the plant is not important |
| Corinne Reich-Weiser, Athulan Vijayaraghava David A Dornfeld | 2008 | Performance analysis | Metrics plays an important role in engineering design and layout. Its help in the design process thus achieving environmental sustainability via design. Efficient and embattled metrics helps designers to concentrate on precise areas of attention throughout the design course |
| Sudarsan Rachuri, Ram D Sriram, Prabir Sarkar | 2009 | Drivers and Barriers Assessment | Strict rules and techniques are needed to help the rules to restrict harmful materials and responsibility for the end of life managing of completed products |

(continued)

**Table 1** (continued)

| Author | Year | Approach | Practical implementations |
|---|---|---|---|
| S.X. Zeng, X. H. Meng, H. T. Yin, C. M. Tamb, L. Sun | 2010 | Drivers and Barriers Assessment | Cleaner production activities can bring significant benefits to firms in the long run. But in a survey, it is found that they implemented the cleaner production only for short-run |
| M. Abramovici, A. Lindner | 2011 | Interrelation in two different concepts | Various new methods like product use information method, Feedback design assistant were implanted and commissioned to help the designer for the generation of new and better products |
| Marc A. Rosen and Hossam A. Kishawy | 2012 | Drivers and Barriers Assessment | Manufacturers adopting sustainability, focus and establish a sustainability culture within companies for successful implementation of enhanced design and manufacturing sustainability |
| Nurul Fadly Habidin, Suzaituladwini Hashim | 2012 | Any Assessment only for Sustainable Manufacturing | In this paper, a theoretical representation is proposed to inspect the association amid SMP implementation and SP measures relationship |
| Sergio Aguado a, Roberto Alvarez b, Rosario Domingo c, | 2013 | Drivers and Barriers Assessment | When environmental innovation is applied then social responsibility and sustainability can be improved. Also the growth of multipurpose and experienced workers, customer devotion, reduction in space needed, current stock reduction, reduction in expenses of non-observance and efficient management extra reduces the misuse |

(continued)

**Table 1** (continued)

| Author | Year | Approach | Practical implementations |
|---|---|---|---|
| Mélanie Despeisse a, Michael R. Oates b, Peter D. Ball | 2013 | Drivers and Barriers Assessment | Possibility of finding sustainable manufacturing enhancement opportunities in an orderly way using designing of manufacturing system throughout the disciplines in the examination are power, material, water, chemicals |
| Duygu Keskin, Jan Carel Diehl, Nelliene Molenaar | 2013 | waste management | Establishing new horizons from sustainability point of view is subjective to the firm's capability in accepting and talking the differences amid customer benefits and clearly defining the sustainability goals |
| Tom Bucherta, Alexander Kaluza, Friedrich A. Halstenberg, Kai Lindowa, Haygazun Haykab, Rainer Starka, b | 2014 | Drivers and Barriers Assessment | In spite of the recognized barriers, the joint essentials of the framework have the prospective to facilitate designpersonnel to choose and merge methods for sustainable design and thus for sustainable decision making |
| Ju Yeon Lee, Y. Tina Lee | 2014 | Drivers and Barriers Assessment | Research inventory provides an essential idea for the measurement by proposing about the factors we need to consider and how to access these factors for the betterment |
| Hao Zhang, Karl R. Haapala | 2015 | Drivers and Barriers Assessment | Existing weighting method (APH) and multi -criteria decision-making method are applicable to detect the probable parallel methods |
| D. Garg, S. Luthra, A. Haleem | 2014 | Validation through case study | DEMATEL (Decision-making trial and evaluation laboratory) approach been used to find and differentiate the drivers, those could be used into the case |

(continued)

**Table 1** (continued)

| Author | Year | Approach | Practical implementations |
|---|---|---|---|
| Danni Chang, C. K. M. Lee, Chun-Hsien Chen | 2014 | Drivers and Barriers Assessment | Various models and theories suggest that the growth of LCA must be pacing with the development of any product and to check the impact on the environment by the same |
| S. J. F. Roberts, P. D. Ball | 2014 | Drivers and Barriers Assessment | While developing a conceptual model describing the distinctiveness in the categorization and when applying it to sustainable development it has been noticed that much of the literature has been devoted to connecting performance development with practices |
| Norani Nordin, Hasbullah Ashari, and Mohamad Farizal Rajemi | 2014 | Comparison of two techniques | When the initiative of sustainable development is successfully implemented then the reduction of costs and employee satisfaction is achieved instead of a burden on companies |
| Vaidyanathan Jayaraman, Rakesh Singh & Ajay Anandnarayan | 2012 | Drivers and Barriers Assessment | Deeper research from all firms of all sectors is needed to overcome the problem of generating hazardous byproducts as waste |
| Danfang Chen, Steffen Heyer, Suphunnika Ibbotson, Konstantinos Salonitis | 2015 | Drivers and Barriers Assessment | DDM has a capable potential, particularly in 3D printing. DDM has a chance to merge the benefits of manufacture Paradigms into personalized high-value products with the lot size of one |

(continued)

**Table 1** (continued)

| Author | Year | Approach | Practical implementations |
|---|---|---|---|
| Michael D. Eastwood, Karl R. Haapala | 2015 | Drivers and Barriers Assessment | Unit process based methodology allows design and manufacturing engineers to examine trade-offs between available product design alternatives and to consider the sustainability performance of similar design and process resolution for finding the best sustainable product design alternative that can come expanding the presented methodology to complete a full cradle to grave assessment |
| Mohamed-Amine Abidi | 2016 | Drivers and Barriers Assessment | A new way is developed to simulate the industrial functions. This new path works on the VR tech and algorithm based on behavior and provides industries the scope to visualize the functioning of the factories |
| Gamage J. R. a, De Silva A. K. M. | 2015 | Validation through case study | Two approaches, first the screening approach and the second in-depth approach is used to help the LCI database problem |
| Kapil Gupta, R. F. Laubscher, J. Paulo Davim, N. K. Jain | 2016 | Drivers and Barriers Assessment | MQL reports improvement in output, rate, tool life, and waste management policies that all jointly perk up the manufacturing of gears |
| Deogratias Kibira, Charles McLean | 2015 | Validation through case study | Negative impact of industrialization on surroundings resulted in growing attention in sustainable manufacturing which can be achieved through simulation with the help of neutral and standard model |

(continued)

**Table 1** (continued)

| Author | Year | Approach | Practical implementations |
|---|---|---|---|
| Cauê Gonçalves Mançanaresa, Eduardo de Senzi Zancula, Paulo Augusto Cauchick Miguelb | 2015 | Drivers and Barriers Assessment | Various data from the journal impact factor report was considered and it was found that AM potentially impacts the dynamic process sustainability in different behavior |
| Neeraj Bhanota, P. Venkateswara Raoa, S. G. Deshmukha | 2015 | Drivers and Barriers Assessment | Government promotion and regulation, FDI is higher in researcher group than in industry professionals whereas the impact of improving quality and education and training system has been observed more in case of industry professionals |
| Dr. S. Nallusamy, G. B. Dinagaraj | 2015 | Performance analysis | Improving green lean manufacturing is a continuous process. Providing incentives to the small-scale industries and measures taken by the government for clean technologies are the steps to advance an energetic green wealth |
| Neeraj Bhanot, P. Venkateswara Rao, S. G. Deshmukh | 2016 | Drivers and Barriers Assessment | Foreign direct investment in industries has great potential in implementing sustainability. Improving the current education and training is needed to create the awareness of sustainable concepts and implementing them in the industry to boost sustainable performance |
| Simon Ford, Mélanie Despeisse | 2016 | Drivers and Barriers Assessment | Additive manufacturing concepts develop options for further sustainable production and utilization studies can offer deep insights into the possessions of additive manufacturing on sustainability |

(continued)

**Table 1** (continued)

| Author | Year | Approach | Practical implementations |
|---|---|---|---|
| Ian C. Garretson a, Mahesh Mani b, Swee Leong b, Kevin W. Lyons b, Karl R. Haapala | 2016 | Waste management | Organized discussion with the researchers and industry practitioners should be held in small groups to promote the concept of sustainable development |
| Erwin Rauch, Patrick Dallasega, Dominik T. Matt | 2016 | Interrelation in two different concepts | DMS plays an important role in sustainable manufacturing. More research is needed in this context |
| Miguel A. Salido, Joan Escamilla, Federico Barber, Adriana Giret | 2016 | Drivers and Barriers Assessment | Match up technique determines the instant at which the most optimal result could be obtained |
| Amir-mohammad, Yousef Nazzal, Hossam Kishawy | 2016 | Waste management | Energy required for various uses during material removal processes is discussed here |
| Varinder Kumar Mittal, Rahul Sindhwani, P. K. Kapur | 2016 | Drivers and Barriers Assessment | Analysis of different barriers between government policymakers and industrial sectors are discussed so that such policies should be framed in such a way that the industry can progress in harmony with competitiveness and environment |
| Shubham, Parikshit Charana and L. S. Murtyb | 2016 | Life cycle Approach | New regulation prevailing the sustainable manufacture techniques focuses on the undeviating result of manufacturing on the surroundings |
| Ming-Lang. Tseng, Kim Hua Tan, Yong Geng, Kannan Govindan | 2016 | Waste management | Studies have been conducted to emphasize the job of management to thrust through ecological, financial and communal concerns (bottom line, TBL) approaching through administration and management |

(continued)

**Table 1** (continued)

| Author | Year | Approach | Practical implementations |
|---|---|---|---|
| Kannan Govindana, Kiran Garg, Seema Gupta, P. C. Jha | 2016 | Implementation of concepts | Idea of sustainability is important to fight against the slow dilapidation of natural system and a rising variation in the set of living for the earth's inhabitants |
| K. Madan Shankar, DevikaKannan, P. Udhaya Kumar | 2017 | Implementation of concepts | A list of guided instruction can make the manufacturing process sustainable successfully. Here, in this paper, research aims in the context of India is achieved by a definite work frame. It also shows how to merge the concept of 6R in an organization to make manufacturing sustainable |
| Salwa Hanim, Hanim Abdul-Rashid, Novita Sakundarini, Raja Ariffin, Raja Ghazilla, Ramayah Thurasamy | 2017 | Implementation of concepts | Some insight into elements that can make manufacturing sustainable was proposed to organizations. The framework proposed will be helping the manufacturers to select the most appropriate practice. Also, help to decide the organization to improve the running practices. The study was done in the Malaysia context |
| Junfeng Ma, James D. Harstvedt, Daniel Dunaway, Linkan Bian, Raed Jaradat | 2017 | Drivers and Barriers Assessment | Firm is using additive manufacturing in their plants it reduces the cost to a large extent when done in mass and hence reducing the AMP price. Described process help to generate an optimized tool to measure and support the regulation for sustainability |
| Anastasiia Moldavska, Torgeir Welo | 2017 | Drivers and Barriers Assessment | In the definition of SM concept, there is a mix of performance-related features and sustainability-oriented instruments |

(continued)

**Table 1** (continued)

| Author | Year | Approach | Practical implementations |
|---|---|---|---|
| Biranchi Panda, Suvash Chandra Paul, Lim Jian Hui, Yi Wei Daniel Tay, Ming Jen Tan | 2017 | Drivers and Barriers Assessment | 3D printing can prove a breakthrough for faster and sustainable built environment for printing of geopolymer. Directional properties can help while designing a part for concrete printing application |
| Sunpreet Singh, Seeram Ramakrishna, Munish Kumar Gupta | 2017 | Drivers and Barriers Assessment | Sustainable Practices helps in controlling the alarming waste issues. Various methods such as established design for manufacturing, DOE, Simulation software can help in achieving this |
| A. Rajeev, Rupesh K. Pati, Sidhartha S. Padhi, Kannan Govindan | 2017 | Model development | An incorporated structure that captures the factors calculated on different TBL pillars and all of which have led to the development and progress of SCCM is presented in this study |
| Michael P. Brundage, William Z. Bernstein, Steven Hoffenson, Qing Chang, Hidetaka Nishi, Timothy Kliks, K. C. Morris | 2018 | Life cycle Approach | Finding opportunities where information from downstream stages can help designers to conduct environmental assessment of a product |
| Nikolay Dentchev, Romana Rauter, Lára Jóhannsdóttir, Yulia Snihur, Michele Rosano, Rupert Baumgartner, Timo Nyberg, Xingfu Tang, Bart van Hoof, Jan Jonker | 2018 | Model development | Need of creating further synergies with research stream and communities for multidisciplinary research with the objective to surpass the current understanding of SBMs |
| Mijoh A. Gbededo, Kapila Liyanage, Jose Arturo Garza-Reyes | 2018 | Drivers and Barriers Assessment | Now it is possible for the designers to mix aims that will help developing sustainability in development of a product by simulation-based sustainability impact analysis |

(continued)

**Table 1** (continued)

| Author | Year | Approach | Practical implementations |
|---|---|---|---|
| Aihua Huang, Fazleena Badurdeen | 2018 | Drivers and Barriers Assessment | Technique anticipated can be very helpful in fabrication line sustainability indicator and Plant sustainability index if the estimation is conducted by the similar persons and a reliable advance is followed in handing over subjective rating and weighs |
| Yoram Korena, Xi Gua, Fazleena | 2018 | Model development | Manufacturing processes and system performance are greatly influenced by RMS characteristics as they facilitate optimal and flexible adding capacity control to meet vigorously altering customer necessities |
| Gustav Sandin, Greg M. Peters | 2018 | Model development | Studies and research help in designing future reuse and recycling systems for optimal environmental performance. More studies and research is needed to make most of the recycled materials |
| Anuj Singla, Inderpreet Singh Ahuja, Amanpreet Singh Sethi | 2018 | Model development | Technology push plays an important practice to meet various challenges of competent markets worldwide |
| Mingzhou Jin, Renzhong Tang, Yangjian Ji, Fei Liu, Liang Gao, Donald Huisingh | 2018 | Implementation of concepts | New optimization modelling and algorithm can improve conventional manufacturing processes but more research is needed in this area |

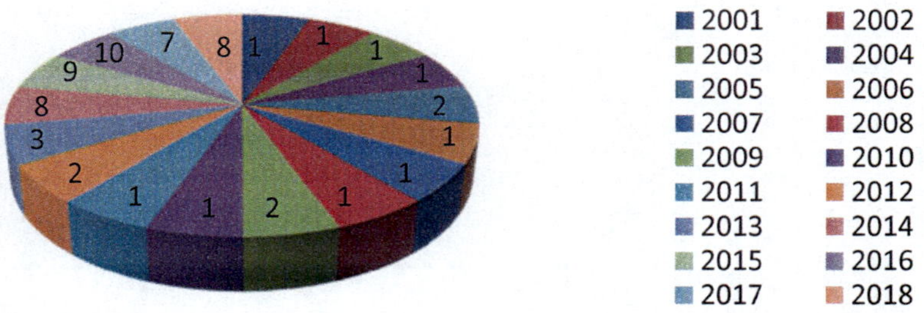


**Fig. 1** Year wise publications

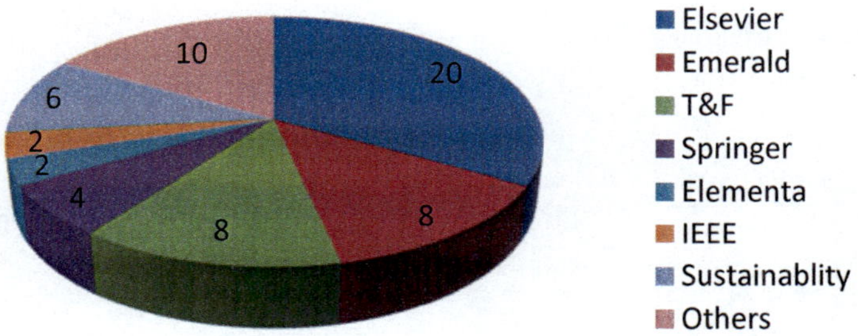


**Fig. 2** Journal wise publications

&Francis, 19 in others. These articles were published within the given time frame (Fig. 2).

## 4.3 *Nation Wise Classification*

This section shows various countries where work is completed on this topic. Figure 3 shows the papers from various countries in percentage form. Here, the papers in the respective domain are highest from India 35%, 7% from the US, 6% from the UK, and others from different but more than one country. Some country names were not shown in the papers.

## 4.4 *Classification Based on the Concepts*

Here numbers of articles are shown in the form of percentage for a particular tool/ technique used. About 20% of papers used literature review, 14% used to survey, 15% goes with fuzzy logics, 12% with MICMAC analysis, 8% incorporated discussion technique, and 10% goes with SPSS and others with various techniques (Fig. 4).

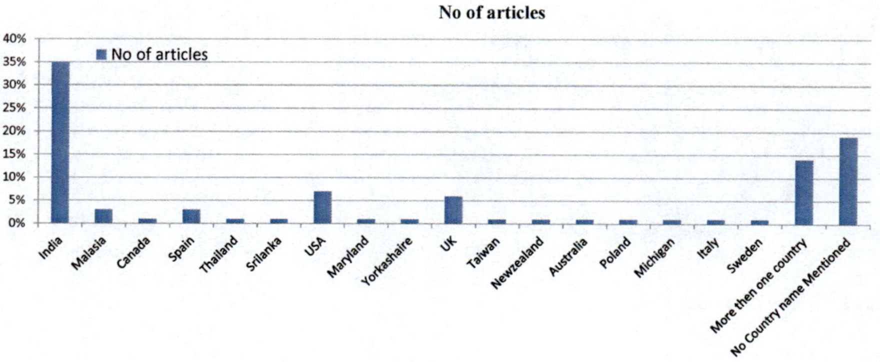


**Fig. 3** Nation wise publications

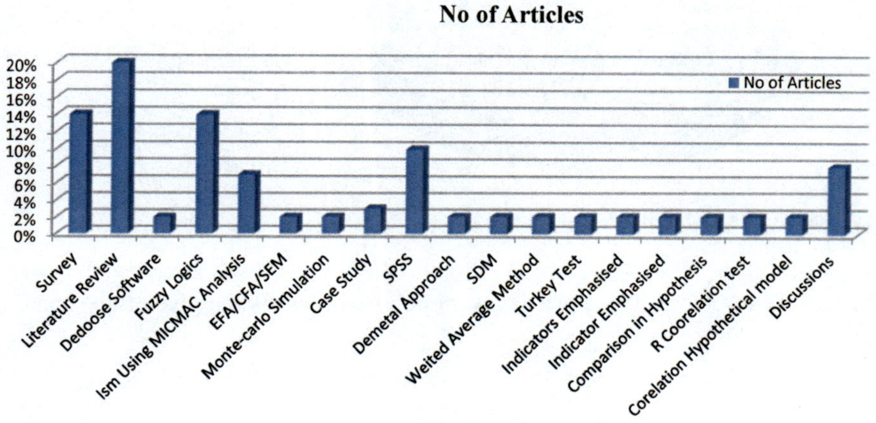


**Fig. 4** Concept wise publications

### *4.5 Classification Based on Research Tools and Techniques Used*

This section shows related concepts with no of articles published in the journals. 15 articles from sustainability, 12 in Green supply chain management, 15 in Manufacturing, 8 in Green Manufacturing, 4 in product design, and others (Fig. 5).

## 5 Qualitative Thematic Analysis

Here, thematic analysis is done based on the concepts used in articles. The (Table 2).

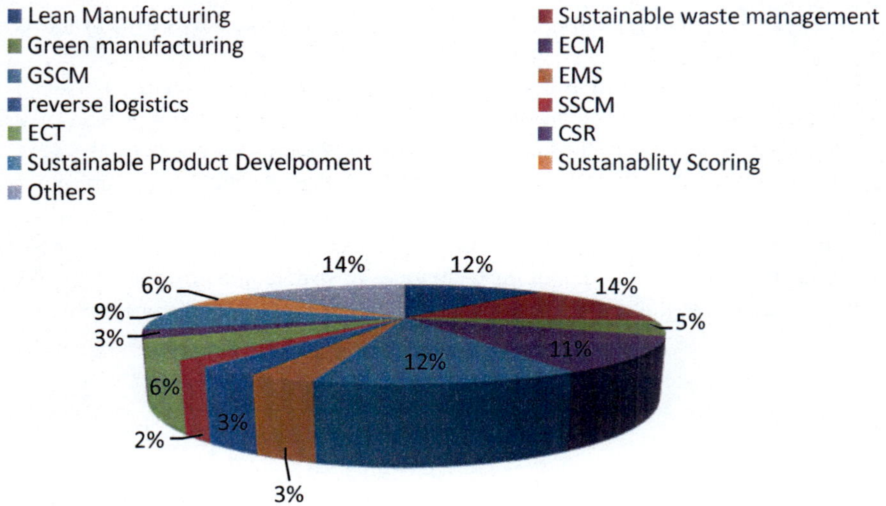


**Fig. 5** Publications on the basis of techniques

**Table 2** Thematic analysis

| Concepts | Articles | Nos. |
|---|---|---|
| Drivers and barriers assessment | 1, 4, 10, 11, 12, 14, 16, 17, 19, 20, 21, 23, 24, 26, 27, 28, 29, 31, 33, 34, 36, 37, 40, 42, 48, 49, 50, 51, 55, 56 | 30 |
| Assessment only for sustainable manufacturing | 2, 7, 3, 15 | 4 |
| Comparison of two techniques | 8, 25 | 2 |
| Performance analysis | 9, 35 | 2 |
| Validation through case study | 22, 30, 32 | 3 |
| Waste management | 18, 38, 41, 44 | 4 |
| Life cycle approach | 5,43,53 | 3 |
| Additive manufacturing | 6 | 1 |
| Interrelation in two different concepts | 13, 39 | 2 |
| Implementation of concepts | 45, 46, 47, 60 | 4 |
| Model development | 52, 54, 57, 58, 59 | 5 |

# 6 Conclusion

This paper proposes many new scopes for further work in sustainability manufacturing concepts. Through quantitative and qualitative analysis researchers find that more than six major publishers have been reviewed from different countries over a time span of 18 years. Where various concepts were founded and reviewed and concluded. Conclusions attained are given below.

Sustainable manufacturing is a rarely used term, related terms were used instead. Maximum amount of work is done on "Life cycle approach". Combination of survey and case study is not used anywhere in the papers, hence this could be the next big step to adopt sustainability manufacturing, as it has practical and conceptual knowledge simultaneously. There are very few papers available which are based on tools and technique without doing a survey, emphasis could be given to that. There is no industry-wise classification of the techniques of sustainable manufacturing. So researchers may take this research, further keeping the above conclusions in mind.

## References

1. E. Westkämper et al., Life cycle management and assessment: Approaches and visions towards sustainable manufacturing (2001). https://doi.org/10.1243/0954405011518557
2. C. O'Brien., Global manufacturing and the sustainable economy. (2002). https://doi.org/10.1080/00207540210157169
3. H. Kaebernik et al., Sustainable product development and manufacturing by considering environmental requirements. Robot. Comput.-Integrat. Manuf. (2003). https://doi.org/10.1016/s0736-5845(03)000565
4. M. Kotabe et.al., Global sourcing strategy and sustainable competitive advantage (2003). https://doi.org/10.1016/j.indmarman.2003.08.004
5. J.S. Baldwinet.al. Modelling manufacturing evolution: thoughts on sustainable industrial development. J Cleaner Prod. (2005).https://doi.org/10.1016/j.jclepro.2004.04.009
6. VisheshKumaret. al. (2005), " Infusing sustainability principles into manufacturing/mechanical engineering curricula", Journal of manufacturing systems, doi.org/https://doi.org/10.1016/s0278-6125(06) 80011-7
7. A.J. Thomas et al., Key technologies and strategies for creating sustainable manufacturing organisations. in *2nd I*PROMS Virtual International Conference* (2006). https://doi.org/10.1016/b978-008045157-2/50107-3
8. S.K. Srivastava, Green supply-chain management: a state-of-the-art literature review. Int. J. Manag. Rev (2007). https://doi.org/10.1111/j.1468-2370.2007.00202.x
9. C. Reich-Weiser et al., Metrics For Sustainable Manufacturing. in *Proceedings of the 2008 International Manufacturing Science and Engineering Conference, 5th Annual IEEE Conference on Automation Science and Engineering* (2008)
10. S. Rachuri et al. *Metrics, Standards and Industry Best Practices for Sustainable Manufacturing Systems*. IEEE (2009). https://doi.org/10.1109/coase.2009.5234090
11. R. Seidel et al., *Establishing Sustainable Manufacturing Practices in SMEs* (2009)
12. S.X. Zeng et al., Impact of cleaner production on business performance (2010). https://doi.org/10.1016/j.jclepro.20.https://doi.org/10.02.019
13. M. Abramovici et al., Providing product use knowledge for the design of improved product generations. CIRP Ann.-Manuf.g Technol (2011). 10.1016/j.cirp.2011.03.103
14. M.A. Rosen et al., Sustainable Manufacturing and Design: Concepts, Practices and Needs. Sustain (2012). https://doi.org/10.3390/su4020154
15. N.F. Habidin et al., Development of sustainable manufacturing practices and sustainable performance in malaysian automotive industry. J. Econ. Sustain. Develop (2012)
16. S. Aguado etal., Model of efficient and sustainable improvements in a lean production system through processes of environmental innovation. J. Cleaner Prod (2013). https://doi.org/10.1016/j.jclepro.2012.11.048

17. M. Despeisse et al., Sustainable manufacturing tactics and cross-functional factory modeling. J. Clean. Prod. (2013). https://doi.org/10.1016/j.jclepro.2012.11.008
18. D. Keskin et al., Innovation process of new ventures driven by sustainability. J. Cleaner Prod (2012). https://doi.org/10.1016/j.jclepro.2012.05.012
19. T. Buchert et al., Enabling product development engineers to select and combine methods for sustainable design. in *21st CIRP Conference on Life Cycle Engineering* (2014). https://doi.org/10.1016/j.procir.2014.06.025
20. J.Y. Lee et al., A framework for a research inventory of sustainability assessment in manufacturing. J. Cleaner Prod (2014). https://doi.org/10.1016/j.jclepro.2014.05.004
21. H. Zhang et al., Integrating sustainable manufacturing assessment into decision making for a production work cell. J. Cleaner Prod (2014). https://doi.org/10.1016/j.jclepro.2014.01.038
22. D. Garg et al., An evaluation of drivers in implementing sustainable manufacturing in India: using DEMATEL approach. Int. J. Mech. Mech. Eng. **8**(12) (2014)
23. D. Chang et al., Review of life cycle assessment towards sustainable product development. J. Clean. Prod. (2014). https://doi.org/10.1016/j.jclepro.2014.07.050
24. S.J.F. Roberts et al., Developing a library of sustainable manufacturing practices. in *21st CIRP Conference on Life Cycle Engineering* (2014). https://doi.org/10.1016/j.procir.2014.06.054
25. N. Nordin et al., A case study of sustainable manufacturing practices. J. Adv. Manag. Sci. **2**(1) (2014)
26. V. Jayaraman et al., Impact of sustainable manufacturing practices on consumer perception and revenue growth: an emerging economy perspective. Int. J. Prod. Res (2014). https://doi.org/10.1080/00207543.2011.571939
27. D. Chen et al. Direct digital manufacturing: definition, evolution, and sustainability Implications. J. Cleaner Prod (2015). https://doi.org/10.1016/j.jclepro.2015.05.009
28. M.D. Eastwood et al., A unit process model based methodology to assist product sustainability assessment during design for manufacturing. J. Clean. Prod. (2015). https://doi.org/10.1016/j.jclepro.2015.08.105
29. M.-A. Abidi et al. Contribution of virtual reality for lines production's simulation in a lean manufacturing environment (2015). https://doi.org/10.13140/2.1.2911.2642
30. G. J.R et al. Assessment of research needs for sustainability of unconventional machining processes. in *12th Global Conference on Sustainable Manufacturing* (2015) https://doi.org/10.1016/j.procir.2014.07.096
31. K. Gupta et al., Recent developments in sustainable manufacturing of gears: a review. J. Clean. Prod. (2016). https://doi.org/10.1016/j.jclepro.2015.09.133
32. D. Kibira et al., Modeling and simulation for sustainable manufacturing (2008) researchgate.net/publication/228343299
33. C.G. Mançanares et al., Sustainable manufacturing strategies: a literature review on additive manufacturing approach (2017). r esearchgate.net/publication/282523541
34. N. Bhanot etal., Enablers and barriers of sustainable manufacturing: results from a survey of researchers and industry professionals. in *22nd CIRP conference on Life Cycle Engineering* (2015). https://doi.org/10.1016/j.procir.2015.01.036
35. S. Nallusamy et al., Sustainable green lean manufacturing practices in small scale industries—A case study. Int. J. Appl. Eng. Res (2015)
36. P. NeerajBhanot et al., An integrated approach for analysing the enablers and barriers of sustainable Manufacturing. J. Clean. Prod. (2017). https://doi.org/10.1016/j.jclepro.2016.11.123
37. Simon Ford et al., Additive manufacturing and sustainability: an exploratory study of the advantages and Challenges. J. Clean. Prod. (2016). https://doi.org/10.1016/j.jclepro.2016.04.150
38. I.C. Garretson et al Terminology to support manufacturing process characterization and assessment for sustainable production. J. Cleaner Prod. (2016). https://doi.org/10.1016/j.jclepro.2016.08.103
39. E. Rauch et al., Sustainable production in emerging markets through distributed manufacturing systems (DMS). J. Clean. Prod. (2016). https://doi.org/10.1016/j.jclepro.2016.06.106

40. M.A. Salido et al., Rescheduling in job-shop problems for sustainable manufacturing systems. J. Cleaner Prod (2017). https://doi.org/10.1016/j.jclepro.2016.11.002
41. Amir-Mohammad et al., *Sustainable Manufacturing and its Application in Machining Processes*. IEEE (2016)
42. V.K. Mittal et al., Two-way assessment of barriers to lean–green manufacturing system: insights from India. Int. J. Syst. Assurance Eng. Manag (2016)
43. Shubham et al., Organizational adoption of sustainable manufacturing practices in India: integrating institutional theory and corporate environmental responsibility. Int. J. Sustain. Develop.World Ecol (2018) https://doi.org/10.1080/13504509.2016.1258373
44. M. Lang et al., Sustainable consumption and production in emerging markets. Int. J. Prod. Econ (2016). https://doi.org/10.1016/j.ijpe.2016.09.016
45. K Govindan et al., Effect of product recovery and sustainability enhancing indicator son the location selection of manufacturing facility. Ecological Indicators (2016). https://doi.org/10.1016/j.ecolind.2016.01.035
46. K. Madan Shankar et al., Analyzing sustainable manufacturing practices—A case study in Indian context. J. Clean. Prod. (2017). https://doi.org/10.1016/j.jclepro.2017.05.097
47. S. Hanim et al., The impact of sustainable manufacturing practices on sustainability performance: empirical evidence from Malaysia. Int. J. Operat. Prod. Manag. **37**(2) (2017)
48. J. Ma, J.D. Harstvedt et al., An exploratory investigation of additively manufactured product life cycle sustainability assessment. J. Cleaner Prod (2018). https://doi.org/10.1016/j.jclepro.2018.04.249
49. A. Moldavska et al., The concept of sustainable manufacturing and its definitions: a content-analysis based literature review. J. Cleaner Prod (2017). https://doi.org/10.1016/j.jclepro.2017.08.006
50. B. Panda et al., Additive manufacturing of geopolymer for sustainable built environment. J. Clean. Prod. (2017). https://doi.org/10.1016/j.jclepro.2017.08.165
51. S. Singh et al., Towards zero waste manufacturing: a multidisciplinary review. J. Clean. Prod. (2017). https://doi.org/10.1016/j.jclepro.2017.09.108
52. A. Rajeev et al., Evolution of sustainability in supply chain management: a literature review. J. Cleaner Prod (2017). https://doi.org/10.1016/j.jclepro.2017.05.026
53. M.P. Brundage et al., Analyzing environmental sustainability methods for use earlier in the product lifecycle. J. Clean. Prod. (2018). https://doi.org/10.1016/j.jclepro.2018.03.187
54. N. Dentchev et al., Embracing the variety of sustainable business models: a prolific field of research anda future research agenda. J. Cleaner Prod (2018). https://doi.org/10.1016/j.jclepro.2018.05.156
55. M.A. Gbededo et al., Towards a life cycle sustainability analysis: a systematic review of approaches to sustainable manufacturing. J. Clean. Prod. (2018). https://doi.org/10.1016/j.jclepro.2018.02.310
56. A. Huang et al., Metrics-based approach to evaluate sustainable manufacturing performance at the production line and plant levels. J. Cleaner Prod (2018). https://doi.org/10.1016/j.jclepro.2018.04.234
57. Y. Koren et al., Sustainable living factories for next generation manufacturing. in *15th Global Conference on Sustainable Manufacturing* (2018). https://doi.org/10.1016/j.promfg.2018.02.091
58. G. Sandin et al., Environmental impact of textile reuse and recycling—a review. J. Cleaner Prod (2018). https://doi.org/10.1016/j.jclepro.2018.02.266
59. A. Singla et al., An examination of effectiveness of technology push strategies for achieving sustainable development in manufacturing industries. J. Sci. Technol. Policy Manag (2019). https://doi.org/10.1108/jstpm-10-2017-0048
60. M. Jin et al., Impact of advanced manufacturing on sustainability: an overview of the special volume on advanced manufacturing for sustainability and low fossil carbon emissions. J. Cleaner Prod (2017)https://doi.org/10.1016/j.jclepro.2017.05.101

# Impact Dynamics of a Viscoelastic Ferrofluid Droplet Under the Influence of Magnetic Field

**Gaurav Kumar, Sudip Shyam, and Pranab Kumar Mondal**

**Abstract** In the present investigation, we explore the morphological evolution of a ferrofluid drop impacting on a solid substrate in the presence of a vertical magnetic field. The morphological evolution of the droplet is quantified by the change in its height, diameter, and contact line velocity. Two types of liquid drop are used in the present investigation viz. a water-based ferrofluid drop and a glycerol-based ferrofluid drop. The glycerol-based ferrofluid droplet acts as a viscoelastic drop. A comparative analysis of the impact dynamics is carried out between two different liquid drops in the presence/absence of a magnetic field. It is shown that the effect of fluid elasticity brings about a controllability on the impact dynamics in the presence of a magnetic field. The experimental investigations revealed that the viscoelastic ferrofluid drop in the absence of a magnetic field encounters a negligible recoiling effect. However, in the presence of a vertical magnetic field, the drop experiences significant recoiling of the contact line. Also, it is shown that the maximum deformation encountered by the drops is directly related to the strength of the applied magnetic field.

**Keywords** Ferrofluid · Magnetic field · Viscoelastic

## 1 Introduction

Ferrofluid is a colloidal suspension of Ferro/ferrimagnetic particles in a non-magnetic carrier fluid [1]. This typical fluid can be easily manipulated in the presence of a magnetic field. Owing to this fact, ferrofluid finds application in many engineering applications [2–4]. Many researchers have explored the implications of the thermophysical properties of the ferrofluid under the influence of the magnetic field in real-time engineering applications [5–8].

Droplet impact dynamics is an active area of research because of its wide range of scientific and industrial applications, such as cleaning of surfaces, inkjet printing,

G. Kumar · S. Shyam · P. K. Mondal (✉)
Microfluidics and Microscale Transport Processes Laboratory, Department of Mechanical Engineering, Indian Institute of Technology Guwahati, Guwahati, Assam 781039, India
e-mail: mail2pranab@gmail.com; pranabm@iitg.ac.in; mail2pranab@gmail.com



K. M. Pandey et al. (eds.), *Recent Advances in Mechanical Engineering*, Lecture Notes in Mechanical Engineering, https://doi.org/10.1007/978-981-15-7711-6_94

spray coating, pesticide delivery, processing and cutting of materials and forensic assay [9]. Droplet impact on liquid may result in floating, bouncing, coalesce with the reservoir, or splash. While droplet impact on a solid substrate may result in three stages bouncing, spreading, and splashing. Many researchers have investigated the droplet impact phenomenon mainly by exploiting the morphological evolution of the droplet height, diameter, and spreading velocity [10, 11]. Pertaining to this droplet impact phenomena, ferrofluid can be a suitable fluid due to its ability to stay under control in the presence of a magnetic field. Even though Ahmed et al. [12] have explored the ferrofluid droplet impacting phenomena in the presence of a magnetic field, the effect of a viscoelastic ferrofluid can be a suitable proposition in which more controllability in the droplet spreading dynamics can be achieved. To the best of the authors' knowledge, this aspect has remained unexplored in the literature to date. Therefore, in this investigation, we conduct a comparative analysis between a water-based ferrofluid droplet and a viscoelastic ferrofluid droplet. The evolution of the droplet morphology in absence/presence of magnetic field is investigated in detail.

## 2 Materials and Methods

In Fig. 1, we show the schematic of the fabricated experimental setup. Droplet of 10 μl volume is generated with the help of a droplet dispenser (Make: Apex instruments) and is allowed to freely fall on the treated substrate, solely by virtue of gravity. An initial distance of 1 cm is maintained between the droplet and the substrate for all the experiments. High-speed camera coupled with a macro lens (Make: Nikkor, Nikon) of focal length 105 mm (Make: Phantom) is used for the acquisition of the droplet impacting phenomena. Images.

are acquired at a frame rate of 1000 fps at a resolution of 1200 × 800 pixels$^2$. Illumination of the droplet is carried out with the help of a LED-backlight along with a diffuser. The substrates are coated with PDMS (Polydimethylsiloxane). PDMS solution is prepared by mixing it with a cross-linker in the ratio 10:1. The prepared solution is then kept in a desiccator for about 30 min. Following which a glass slide is mounted on spin coater (Make: Apex instruments) and coated with the prepared PDMS solutions at 3400 rpm for 50 s to prepare the final substrate. The coated substrate is then kept on a hot air oven at a temperature of about 80 °C for about 2 h. The magnetic field is provided with the help of a cylindrical shaped Ni–Cu permanent magnet (diameter 5 mm and height 3 mm). The magnet is placed below the substrate to impart the magnetic field.

For the preparation of the ferrofluid solutions, iron oxide (Fe3O4) nanoparticles were chemically synthesized by the co-precipitation method from an aqueous mixture of $Fe^{+3}/Fe^{+2}$ (2:1) solutions. The synthesized nanoparticles were then coated with surfactant (lauric acid) to prevent the agglomeration of nanoparticles. Nanoparticles were then dispersed in DI water with a volume fraction of 2.6% to synthesize the final ferrofluid solutions. To explore the effect of viscoelasticity, the Glycerol solution is

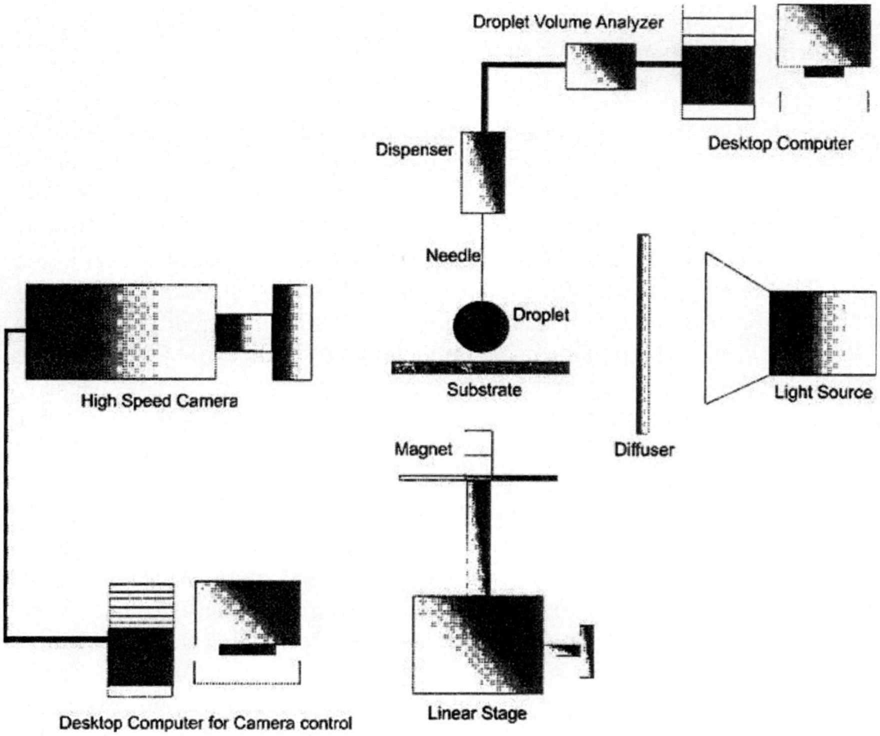


**Fig. 1** Schematic of experimental setup

mixed with the prepared ferrofluid sample in the ratio 3:7. This solution is further sonicated for 2 h in an ultrasonicator for a homogenous mixing. To calculate the change in diameter, height, velocity, and contact angle of the droplet, an in-house code is developed in Matlab®. The temperature and the humidity inside the laboratory are maintained at 25 $\pm$ 0.5 °C and 67 $\pm$1%respectively.

## 3 Result and Discussions

In this section, we explore the role of the magnetic field on the height, diameter, contact line velocity, and the contact angle of the droplet. The study is divided into two parts. In the first part, we discuss the role of the magnetic field on a water-based ferrofluid. In the second part, we explore the implications of the magnetic field on an impacting viscoelastic ferrofluid drop.

### 3.1 *Effect of Magnetic Field on a Water-Based ferrofluid Droplet*

In Fig. 2, we show the temporal evolution of the droplet morphology, as it impacts the PDMS substrate for different magnetic field strength. Two different scenarios are presented in Fig. 2, with and without a magnetic field. In the absence of magnetic, the droplet on striking the substrate spreads almost instantaneously. This instantaneous spreading is because of the rapid increase of pressure at the point of impact in the droplet. The droplet spreads until the maximum deformation is reached, following which the droplet retracts. Eventually, the droplet reaches its equilibrium contact angle, after which no deformation in its morphology takes place. All the stages of the droplet impacting dynamics remain the same, even on the application of the magnetic field. However, if we conduct a comparative analysis in Fig. 2, we can observe that the droplet spreads more in the presence of a magnetic field as compared to the absence of a magnetic field case.

In Fig. 3a, we show the variation of the diameter of the droplet. We can easily observe that the maximum diameter reached is directly related to the strength of the magnetic field. However, the time of reaching the maximum diameter (i.e., the stage of maximum deformation) is almost the same for all the investigating cases. In Fig. 3b, we show the change in the height of the ferrofluid droplet. The stage-A in Fig. 3b denotes the state in which the droplet is outside the influence of the magnetic field and exhibits its spherical shape. In Stage-B, the droplet undergoes the maximum deformation; thus, it encounters minimum height. In Stage-C, the droplet experiences the recoiling of the contact line, because of which the height increases. This spreading and retraction of the droplet contact line continue until the kinetic energy of the droplet is nullified.

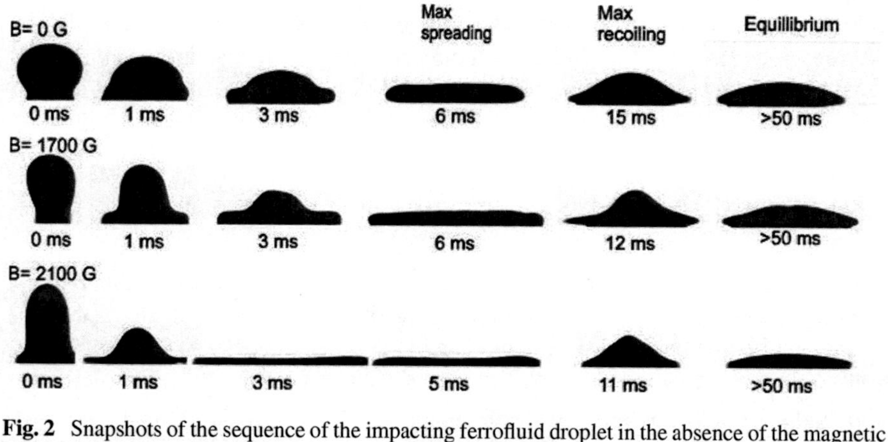


**Fig. 2** Snapshots of the sequence of the impacting ferrofluid droplet in the absence of the magnetic field and under the influence of the vertical magnetic field

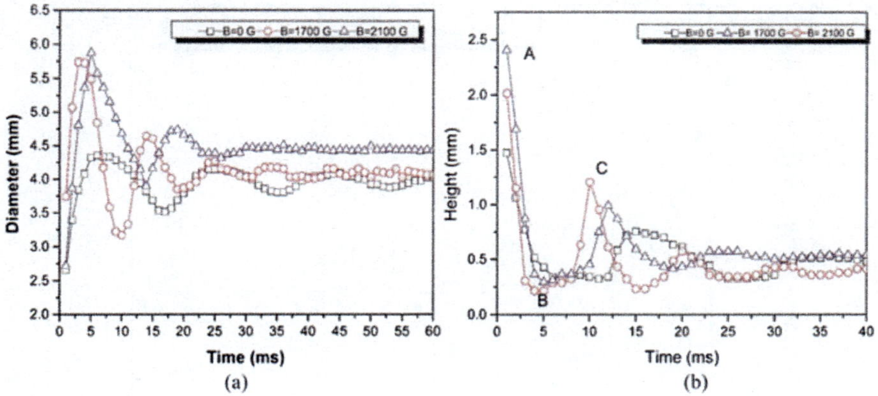


**Fig. 3** Plot depicts the temporal evolution of the droplet (a) Diameter and (b) Height, for the various cases under consideration

## *3.2 Effect of Magnetic Field on Viscoelastic Ferrofluid Droplet*

In Fig. 4, we show the morphological evolution of the viscoelastic ferrofluid drop in the presence and absence of a magnetic field. The droplet impact phenomena are almost the same for the viscoelastic and water base ferrofluid i.e., the droplet on striking the substrate undergoes a maximum deformation, following which it experiences a recoiling of the contact line. However, careful observation of Figs. 2

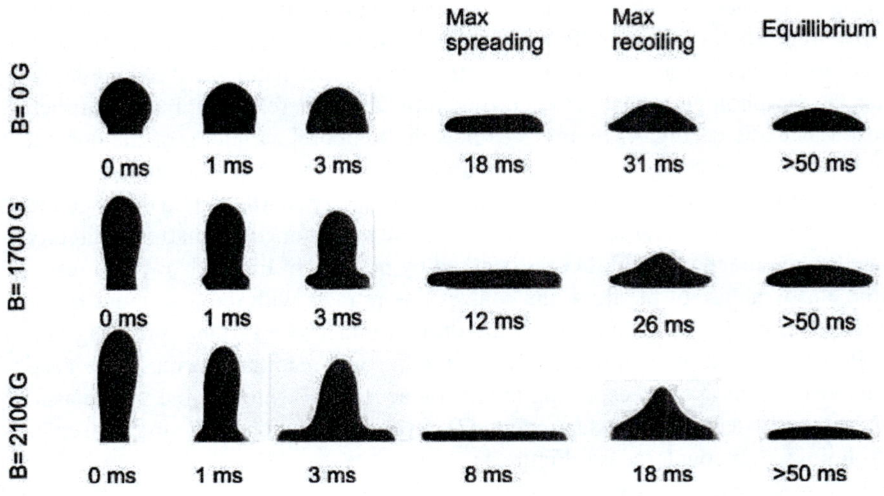


**Fig. 4** Snapshots of the sequence of the impacting viscoelastic ferrofluid droplet in the absence of the magnetic field and under the effect of the vertical magnetic field

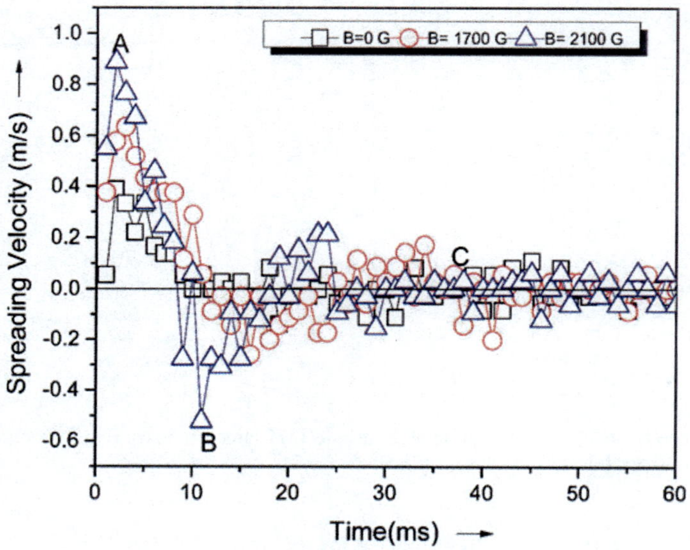


**Fig. 5** Plot depicts the temporal evolution of the viscoelastic ferrofluid droplet contact line velocity, for the various cases under consideration

and 4 will show that the viscoelastic drop takes more time to reach at the determining stages, i.e. the maximum spreading and maximum recoiling stage. Also, the maximum diameter experienced by the droplet in the absence of the magnetic field is less than the maximum diameter experienced by its Newtonian counterpart (i.e., water-based ferrofluid).

However, when the magnetic field is applied, significant deformation is encountered in the droplet diameter. This typical behavior of the fluid is mainly because of its viscous and elastic nature. This viscous and elastic nature of the fluid dominates the morphological changes occurring in the drop, thereby culminating in such a result as depicted Fig. 4. Figure 5 depicts the temporal evolution of the contact line velocity of the viscoelastic ferrofluid drop. Initially, the droplet on impacting the solid substrate encounters an instantaneous spreading. Primarily due to this fact, the contact line velocity increases for all the cases. The maximum deformation is directly related to the magnetic field flux, as denoted by point A in Fig. 5. Now, as a result of the elastic nature of the fluid, the contact line retracts with the maximum recoil velocity (Point B). Following which the droplet undergoes further spreading and recoiling. However, the amplitude of the spreading and recoiling decreases because of the viscous dissipation occurring at the contact line. Eventually, the viscoelastic drop reaches its equilibrium stage (point C) beyond which no change in the droplet morphology is encountered (cf. Figure 5).

## 4 Conclusion

In the present investigation, we have explored the evolution of the droplet morphology as it impacts a treated solid substrate, in the presence of a magnetic field. Two typical fluids are chosen for the present study viz., a water-based ferrofluid and a viscoelastic ferrofluid. The droplet impacting on the solid substrate undergoes two characteristic morphological changes: a maximum deformation stage and maximum recoiling stage, before reaching at its equilibrium state. These stages are also observed when a vertical magnetic field perturbs the impacting droplet domain. The maximum deformation experienced by the droplet is directly related to the strength of the magnetic field, for both the water and viscoelastic ferrofluid. The amplitude of the oscillation of the morphological changes dampens out once the strength of the magnetic field is increased. In the absence of a magnetic field, the viscoelastic ferrofluid droplet exhibits negligible recoiling. However, in the presence of a magnetic field, significant recoiling was seen.

## References


1. S. Odenbach, *Ferrofluids* (Springer, Berlin, 2002)
2. R.E. Rosensweig, *Ferrohydrodynamics* (Cambridge University Press, Cambridge, 1985)
3. S. Shyam, B. Mehta, P.K.P.K. Mondal, S. Wongwises, Investigation into the thermo-hydrodynamics of ferrofluid flow under the influence of constant and alternating magnetic field by InfraRed Thermography, Int J Heat Mass Transf. **135**, 1233–1247 (2019). https://doi.org/10.1016/j.ijheatmasstransfer.2019.02.050.
4. D. Singh, S. Shyam, B. Mehta, Heat transfer characteristics of ferrofluid flow between parallel plates under the influence of static and transient magnetic field, in Proceedings of the 24th National and 2nd International ISHMT-ASTFE Heat and Mass Transfer Conference (IHMTC-2017), BITS Pilani, Hyderabad, India , December 27–30, 2017. **DOI:** https://doi.org/10.1615/IHMTC-2017.80
5. S. Shyam, M. Asfer, B. Mehta, P.K. Mondal, Z.A. Almutairi, Magnetic field driven actuation of sessile ferrofluid droplets in the presence of a time dependent magnetic field. Colloids Surfaces A: Physicochemical Eng. Aspects. **586**, 124116 (2020) https://doi.org/10.1016/j.colsurfa.2019.124116
6. D. Singh, S. Shyam, B. Mehta, M. Asfer, A.S. Alshqirate, Exploring Heat Transfer Characteristics of Ferrofluid in the Presence of Magnetic Field for Cooling of Solar Photovoltaic Systems. J. Thermal Sci. Eng. Appl. **11** (2019). https://doi.org/10.1115/1.4044188
7. S. Shyam, B. Mehta, P.K. Mondal, Thermohydrodynamics of ferrofluidic flow with periodic pulsation under the effect of static and alternating magnetic field-: a numerical study. in Proceedings of the 16th International Heat Transfer Conference, IHTC- 16, Beijing, China, August 10–15, 2018. DOI: https://doi.org/10.1615/IHTC16.ctm.023439
8. Y.Y. Gawade, S. Shyam, B. Mehta, Pressure Characteristics of Single Isolated Ferrofluidic Slug under the Influence of External Magnetic Field, 7th International and 45th National Conference on Fluid Mechanics and Fluid Power, IIT Bombay, Mumbai, 10 - 12, December 2018.
9. M. Rein, Phenomena of liquid drop impact on solid and liquid surfaces. *Fluid Dyn. Res.* **12**(2), 61 (1993)
10. F. Boyer, E. Sandoval-Nava, J.H. Snoeijer, J.F. Dijksman, D. Lohse, Drop impact of shear thickening liquids. Phys. Rev. Fluids **1**(1), 1–9 (2016)

11. Y. Liu, L. Moevius, X. Xu, T. Qian, J.M. Yeomans, Z. Wang, Pancake bouncing on superhydrophobic surfaces. Nat. Phys. **10**(7), 515–519 (2014)
12. A. Ahmed, B.A. Fleck, P.R. Waghmare, Maximum spreading of a ferrofluid droplet under the effect of magnetic field. Phys. Fluids **30**, 7 (2018)
13. S. Shyam, A. Yadav, Y. Gawade, B. Mehta, P.K. Mondal, M. Asfer, Dynamics of a single isolated ferrofluid plug inside a micro-capillary in the presence of externally applied magnetic field. Experiments Fluids **61**, 210 (2020). https://doi.org/10.1007/s00348-020-03043-0
14. S. Shyam, P.K. Mondal, B. Mehta, Field driven evaporation kinetics of a sessile ferrofluid droplet on a soft substrate, Soft Matter. **16**6619–6632 (2020) https://doi.org/10.1039/D0SM00345J

# Reacting Flow Solver for Martian Atmosphere Conditions

**P. Vicky Kumar, Anil Kumar Birru, and Vinayak Narayan Kulkarni**

**Abstract** In the advancement of computational code, demonstration depicts essential elements. An attempt has been made to develop a finite volume in viscid non-equilibrium flow solver especially the flow of carbon-dioxide to study Martian atmospheric condition. The present study utilizes Venkat Krishnan limiter to provide second-order accuracy. The solver is incorporated by higher-order reacting convective or in viscid fluxes, AUSM scheme. The code is inspected by flowing carbon dioxide over sphere of diameter 25.4 mm and shock stand-off distance is measured at two different velocities, i.e., 4.220 km/s and 2.845 km/s. Similarly, for a ramp at angle 10 and 20 degrees the results obtained in terms of pressure ratio, temperature ratio, and wave angle by the solver are validated with an analytical approach. For all the cases studied, in house-solver exhibit satisfying agreement. Additionally, its capability can be enhanced by incorporating various flux evaluation schemes.

**Keywords** Martian atmosphere · AUSM scheme · Shock stand-off distance and wave angle

## 1 Introduction

Relative to the Earth, the Martian atmosphere is thin because it mainly consists of 95.70% $CO_2$, 1.6% Ar, and 2.7% $N_2$ is quite different from Earth's. The density is only 1.0% and the temperature is lower than Earth's atmosphere. The atmosphere

P. Vicky Kumar · A. K. Birru (✉)
Mechanical Engineering, NIT Manipur, Imphal, India
e-mail: birruresearch@gmail.com

P. Vicky Kumar
e-mail: papulappa@gmail.com

V. N. Kulkarni
Mechanical Engineering, IIT Guwahati, Guwahati, India
e-mail: vinayak@iitg.ernet.in



K. M. Pandey et al. (eds.), *Recent Advances in Mechanical Engineering*, Lecture Notes in Mechanical Engineering, https://doi.org/10.1007/978-981-15-7711-6_95

and climates on Mars vary severely and quasi-randomly with Mars geographic position and seasons, which makes the atmosphere parameters having evident scatter characteristics. As a result, these uncertainties of atmosphere parameter should not be neglected in entry vehicles aerodynamics computations in virtue of its interdependency with the trajectory design. A lot of probes supported from Mars exploration projects by the US and SU during the cold war period were launched to Mars. Unfortunately, most of them were failed until the "Viking" explorer belonging to NASA successfully entered Martian atmosphere in 1976. Then, NASA's other explorer, the "Pathfinder", successfully landed Mars once again in 1996. However, contemporary Mars probes from Russia and Japan all failed for different reasons. With the development of aerospace technology and understanding the environments of Mars atmosphere, the success probability of Mar exploration mission significantly grows in the twenty-first century, such as the well-known projects Phoenix and Mars Science Laboratory, Mars exploration attracts more and more countries concern. From the view of vehicle entry, there are still a lot of challenges, especially in accurately predicting aerodynamics characteristics concerning the Martian atmosphere. Candler et al. [1] conducted an experiment in an expansion tube facility at three different angles of attack: 0, 11, and 16°. In a hypervelocity, carbon dioxide flow was made to measure heat flux across blunt bodies and visualizes bow shock shapes. Whiteet al. [2] simulated Monte Carlo solver, known as dsmc Foam, is thoroughly examined for its ability to solve low and high-speed non-reacting gas flows in simple and complicated geometries. Two test cases were considered, i.e., flow over sharp and truncated flat plates, the Mars Pathfinder probe, a micro-channel with heated internal steps, and a simple micro-channel. Liao et al. [5] measure shock stand-off distances over hypersonic spheres in $CO_2$ have been conducted in the hypervelocity ballistic range of HAI, CARDC. It is thought from the calculated results that the flow over spheres of the present test is mainly nonequilibrium. Vital parameters like skin friction coefficient, mass fraction of various species, boundary layer thickness, entropy layer thickness, etc. are either impossible or very difficult and costly to obtain by performing experiments. Hence, it is advisable to device a methodology for known freestream conditions and geometry by employing a finite volume inviscid non-equilibrium flow solver for investigation and to obtain more detailed flow field information.

## 2 Numerical Methodology

### 2.1 Governing Equations

Computational fluid dynamic (CFD) is a methodology to solve the governing mathematical models for fluid flow using a suitable numerical technique. It is an efficient approach to simulate the variety of fluid problems in an economical way as compared to costly experimental procedures. Therefore, it has been an integral part of designing supersonic/hypersonic aircraft like reentry vehicles, missiles, etc. Simulation for these compressible flows involves the solution of Euler's equations for low enthalpy conditions. However, consideration of reacting gas flow is important for the precise prediction of high enthalpy flowfield. Therefore, species continuity equations also need to be solved along with the Euler's equations. The source term of these added equations accounts for the species production rate which is insignificant in the low enthalpy non-reacting gas flow. The coupled Euler's equation and species continuity equations for 2D axisymmetry laminar Inviscid compressible flows in vector form are presented as follows in Eq. 1:

$$\frac{\partial U}{\partial t} + \frac{\partial F}{\partial x} + \frac{\partial G}{\partial y} + S + \propto S_I = 0 \tag{1}$$

where,

$$\mathrm{U} = \begin{pmatrix} \rho \\ \rho u \\ \rho v \\ \rho E \\ C_1 \\ . \\ . \\ . \\ . \\ . \\ C_1 \end{pmatrix}, F = \begin{pmatrix} \rho \\ \rho u^2 + p \\ \rho u v \\ \rho E + p \\ u C_1 \\ . \\ . \\ . \\ . \\ . \\ u C_{N-1} \end{pmatrix}, E = \begin{pmatrix} 0 \\ \tau_{xx} \\ \tau_{xy} \\ u\tau_{xx} + v\tau_{xy} - q_x - \sum_{i=1}^{N} h_i C_i \overline{u_i} \\ -C_1 \overline{u_1} \\ . \\ . \\ . \\ . \\ . \\ . \\ -C_{N-1}\bar{u}N - 1 \end{pmatrix}$$

$$\mathrm{G} = \begin{pmatrix} \rho v \\ \rho u v \\ \rho v^2 \\ (\rho E + p)v \\ v C_1 \\ . \\ . \\ . \\ . \\ . \\ . \\ v C_{N-1} \end{pmatrix} \quad S_I = \frac{1}{y} \begin{pmatrix} \rho v \\ \rho u v \\ \rho v^2 \\ (\rho E + p)v \\ v C_1 \\ . \\ . \\ . \\ . \\ . \\ . \\ v C_{N-1} \end{pmatrix} \quad \mathrm{s} = - \begin{pmatrix} 0 \\ 0 \\ 0 \\ 0 \\ S_1 \\ S_2 \\ . \\ . \\ . \\ . \\ . \\ S_{N-1} \end{pmatrix}$$

With total energy $E$, is expressed as $E = e + \frac{1}{2}\rho\left(u^2 + v^2\right)$, where used for internal energy '$e$'. This internal energy of the mixture is able to calculate as,

$e = \sum_{i=1}^{N} e_1 \frac{C_i}{\rho MW_i}$. '$F$' and '$G$' denotes flux in vector form in $x$ and $y$ direction along with '$U$' as conservative vector or solution, respectively. '$S$' denotes reaction source term and '$S_I$' called axisymmetric source term. When '$\alpha$' is assigned as 1 and 0, it recognizes axis-symmetric and 2-D simulation. Additionally, $\rho$, $p$, and $T$ represent density, pressure, and temperature. Whereas $u$ and $v$ are velocities in $x$ and $y$ direction, respectively. The molar internal energy of the species is calculated as $e_i = h_{fi}^0 + \int_{T_R}^{T} C_{pi} dT - R_u T$ where $MW_i, C_i, C_{pi}$ *and* $h_{fi}^0$ expressed as the molecular weight, mass concentration, specific heat at constant pressure, and heat of formation, respectively. Apart from this, $R_u$ called universal gas constant, $T_R$ as character reference temperature and '$N$' denotes the number of species. In the present case, utilizes Venkat Krishnan limiter to provide second-order accuracy. The solver is incorporated by higher-order reacting convective or inviscid fluxes, AUSM scheme.

## 3 Validation

A present finite-rate chemistry model includes eight species $N_2$, NO, $O_2$, $CO_2$, C, O, CO and N and ten elementary chemical reactions. The rate coefficient and chemical reactions used in non-equilibrium conditions are mentioned in Table 1.97% $CO_2$ and 3% $N_2$ is considered for non-reacted freestream flow. The following reactions accord with rate constants have been studied by Maciel and Pimenta for the chemical non-equilibrium around Martian atmospheric conditions.

Numerical tests have been conducted to validate the accuracy of the present formulation. The test cases include a sphere of diameter 25.4 mm and ramp angles at 10 and 15° are implemented to validate the developed solver.

### *3.1 Flow Through Ramp*

In the computational solution, the geometry 2D ramp of angle 10° and 20° for the following Mach 4 and 5. Further freestream conditions includes a pressure of 199.45 Pa and a temperature of 131.70 K with mesh size of 280 × 150. The Mach contour attained is shown in Figs. 1, 2, 3, and 4. With the purpose of authenticating the solver, pressure, and temperature ratios along with shock wave angle ($\beta$) are compared with the analytic predictions used for low enthalpy test conditions. The comparisons have been discussed in detail and tabulated. By studying Table 2, it can be observed that the developed solver provides satisfactory results at low enthalpy conditions.